P9-AGK-472

FOUNDATIONS OF ASTRONOMY

ABOUT THE AUTHOR

Mike Seeds is Professor of Astronomy at Franklin & Marshall College, as well as director of the college's Joseph R. Grundy Observatory. An active researcher on photometry of short period variable stars, he includes among his other research interests archeoastronomy, photometry of giant stars in old open clusters, comets, and carbon stars. In addition to writing textbooks, Seeds frequently contributes to journals and enjoys writing popular magazine articles about his latest endeavor—the creation of educational software for toddlers! Seeds is also the author of *Horizons, Third Edition,* for Wadsworth, as well as *Astronomy: Selected Readings* (Benjamin/Cummings, 1980); and (with Joseph R. Holzinger) *Laboratory Exercises in Astronomy* (Macmillan, 1976).

FOUNDATIONS OF ASTRONOMY

1990 EDITION

MICHAEL A. SEEDS

Joseph R. Grundy Observatory
Franklin and Marshall College

Wadsworth Publishing Company
Belmont, California
A Division of Wadsworth, Inc.

Astronomy Editor: Anne Scanlan-Rohrer
Developmental Editor: Mary Arbogast
Editorial Assistant: Leslie Width
Print Buyer: Martha Branch
Art Editor: Marta Kongsle
Designer: Julia Scannell
Copy Editor: Betty Duncan-Todd
Technical Illustrators: Susan Breitbard, Cyndie Clark-Hugel, Innographics, Salinda Tyson
Unit Opening Photos: Atlas des Nordlichen gestirnten Himmels für den Anfang des Jahres 1855, p. 10
Chapter Opening Illustrations: Steve Campbell
Compositor: Graphic Typesetting Service, Los Angeles
Cover Photo: Images of Neptune. Neptune, Neptune's rings, and Neptune's moon Triton as recorded by the Voyager II spacecraft in August 1989. (Courtesy NASA/JPL)

Printed in the United States of America

1 2 3 4 5 6 7 8 9 10—94 93 92 91 90

Library of Congress Cataloging-in-Publication Data

Seeds, Michael A.
Foundations of astronomy / Michael A. Seeds—1990 ed.
p. cm.
Includes bibliographical references.
ISBN 0-534-12768-1
1. Astronomy. I. Title.
QB43.2.S43 1990
520—dc20 89-27118
CIP

ISBN 0-534-12768-1

To Emery and Helen Seeds

BRIEF CONTENTS

CONTENTS

BOXES

PREFACE

Imagine astronomy without new discoveries, new theories, new instruments, or new images of distant objects. Imagine an astronomy course that never changed year after year. Astronomy is exciting because it has the potential to astonish us with new insights into the secrets of the universe, but we pay a price for that excitement—we must struggle to keep up with a swiftly changing field.

This new edition of *Foundations of Astronomy* includes the newest discoveries and theories as well as the more gradual changes in general opinion among astronomers concerning such things as black holes, active galaxies, and planetary origins. Yet the book is unchanged in its overall approach. It attempts to present astronomy as a unified system of understanding that relates the student's personal existence to the basic processes in the universe.

MINIMUM TANGLE

The body of astronomical knowledge resembles a marionette with tangled strings. Each fact, observation, or principle is connected by inferences, assumptions, and theories to other facts, observations, and principles. The teacher's task is to untangle this marionette and make clear the logic of the connections. Just as a marionette has inherent in its design an untangled state, the body of astronomical knowledge has a state that makes the logic of the inferences, assumptions, and theories clearest. In this book, I have attempted to present astronomy in that state of minimum tangle.

By considering stars, galaxies, cosmology, and the solar system in that order, we simplify many of the evidential arguments and intuitive concepts so important to understanding astronomy. For example, by discussing star formation before the solar system, students recognize the solar nebula for what it is, a natural by-product of star formation. More important, by studying stars, galaxies, and cosmology first, students return to the solar system with a cosmic perspective on nature that places them and their planet into the universe, rather than constructing a universe around them.

Facts can be meaningful only when they are synthesized into a consistent description of nature. Thus, this book views the universe as a small set of natural processes that are responsible for a wide variety of phenomena and that explain a diverse assortment of objects. For example, galaxies, star clusters, individual stars, and planets are all expressions of the same process—gravitational contraction. Only the scale is different. This emphasizing of processes presents astronomy not as a collection of unrelated facts but as a unified body of knowledge.

In addition, this text presents astronomy as a case study in science. It distinguishes between observation and theory, between evidence and conclusion. We want our students to understand how science creates, tests, improves, or discards models of natural phenomena and thus more closely approximates a complete understanding of natural processes.

CHANGES IN THE NEW EDITION

Besides including new discoveries, we have carefully refined the presentation of each chapter, in response to comments from a great many instructors and students:

- A new Chapter 1, The Scale of the Cosmos, introduces students to the basic celestial objects and their relative scales. Through its strong graphical content, the chapter also presents the metric system, the astronomical unit, and the light-year.
- Material in the early chapters has been consolidated to better unify the discussions of the celestial sphere, seasons, and lunar phenomena. Coordinates and timekeeping have been retained in Appendix A. Optical telescopes, radio and space

telescopes are now discussed in a single chapter, to emphasize the parallels in their design.

- The material on relativity (once a separate chapter) has been merged into various chapters where it better supports the discussion of astronomical phenomena.
- At the suggestion of many instructors, a new chapter has been added to discuss neutron stars and black holes in greater detail.
- The chapter on peculiar galaxies has been extensively revised and updated to include new observations and the growing synthesis of those observations into a unified understanding of active galactic nuclei.
- The unit on the solar system now emphasizes the solar nebula theory as a unifying theme. Thus the discussion of each solar system object or phenomenon relates the descriptive facts to the solar nebula theory.
- While the solar system chapters follow a comparative plan, analyzing basic planetary processes by comparing planets with each other, the organization has been changed so that each planet is presented separately. *Data Files*—containing numerical, photographic, and graphical information describing the sun, moon, planets, and Comet Halley—have also been added. This arrangement presents the unique features of each body while still allowing a comparative analysis.
- Observational Activities in Chapters 2, 3, 10, 12, 22, 24, and Appendix A encourage students to make simple observations and guide them in analyzing the implications of what they see. The related Box 8-1 on observing the sun and the supplement, "Observing the Sky," have been retained, with updated tables.
- Students and instructors have responded with enthusiasm to the use of two novel types of illustrations: Film-strip figures, which illustrate temporal processes, and photographs paired with "key" drawings that act as guides to the content of the photographs. (These paired figures have proven especially useful in elucidating complex figures, and they avoid obscuring images with obtrusive labels.)
- At the urging of a number of reviewers, this edition has more end-of-chapter problems spanning a wider range of difficulty.

CONTINUING FEATURES

Many useful features have been retained and updated. "Perspectives," which appear at the end of several chapters, introduce new and interesting ideas that allow students to review and apply the principles covered in the chapter. Some Perspectives discuss the development of a theory, the synthesis of hypotheses from data, the testing of theories by observation, and the meaning of statistical evidence.

Study aids for each chapter include a chapter summary, a list of new terms, questions, problems, and recommended readings. The first time the new terms appear in the text they are **bold face**. These same terms are defined in the glossary. The questions are nonquantitative and could lead to essay answers or be used to stimulate class discussion. The problems are quantitative or involve mathematical reasoning. (Answers to even numbered problems appear at the end of the book, and to odd-numbered problems in the *Instructor's Manual*.) The recommended reading, which is intended for the student, ranges from *National Geographic* to *Science*. Instructors may wish to guide students in selecting appropriate reading material.

ACKNOWLEDGMENTS

My thanks to the many students and teachers who have responded so enthusiastically to *Foundations of Astronomy*. Their comments and suggestions have been very helpful in completing this new edition. I would especially like to thank the numerous reviewers whose careful analysis and thoughtful suggestions have been invaluable in refining *Foundations of Astronomy* as a teaching tool.

The people listed in the illustration credits were very kind in providing photographs and diagrams. Special recognition goes to the following, who were always ready to help locate unusual images from their institutions: Patricia Bridges, US Geological Survey; Jeff Butler, Lancaster County Planning Commission; Linda Carroll, U.S. Geological Survey; Helen S. Horstman, Lowell Observatory; David Malin, Anglo-Australian Observatory; Patricia A. Ross, National Space Science Data Center; Janet Sandland, Royal Observatory Edinburgh; Patricia Shand, Lick Observatory; Jeff L. Stoner, National Optical Astronomy Observatory; Margaret B. Weems, National Radio Astronomy Observatory; Donald E. Wilbur, Pennsylvania Department of Transportation; Denise R. Whitehead, National Optical Astronomy Observatory.

My appreciation also goes to the following institutions for their assistance in providing figures: The Anglo-Australian Observatory, AstroMedia Corporation, *The Astrophysical Journal,* Ames Research Center, Bell Laboratories, Brookhaven National Laboratories, Celestron International, U.S. Geological Survey, High Altitude Observatory, Jet Propulsion Laboratories, Johnson Space Center, Lick Observatory, Lowell Observatory, Lunar and Planetary Laboratory, Martin Marietta Aero-Space Corporation, Mount Wilson and Las Campanas Observatories, National Aeronautics and Space Administration, National Optical Astronomy Observatories, National Radio Astronomy Observatory, Palomar Observatory, Pennsylvania Department of Transportation, Royal Observatory, Edinburgh, and Yerkes Observatory. Many Voyager, Pioneer, and Mariner photographs were provided by the National Space Science Data Center.

Textbook publishing is a team sport, and the Wadsworth team deserves a trophy. Special thanks go to the editorial and production staff including Jack Carey, Gary Mcdonald, Julia Scannell, James Chadwick, Marty Kongsle, and Karen Hunt. I would especially like to thank Mary Arbogast for her detailed analysis and thoughtful guidance in the creation of this new edition.

Finally, I would like to thank my wife, Janet, and my daughter, Kathryn, for their patience and understanding. Books are made of time, and there is never enough.

Michael A. Seeds
Lancaster, Pennsylvania

A NOTE FROM THE PUBLISHER

The excitement of astronomy lies in new understanding. We see the nitrogen volcanos on Triton and suddenly understand better how planets work in the rigid cold of the outer solar system. To communicate that excitement to students, we must show them how astronomy progresses.

Since 1987 when the previous edition of *Foundations of Astronomy* was written, astronomers have made impressive advances. One team seems to have found an astonishing fast pulsar in the remains of Supernova 1987A, Voyager 2 revealed complex circulation in Neptune's atmosphere, complete rings with imbedded arcs, and a strange, icy geology on Triton. Quasars have been found with red shifts greater than 4, and striking infrared images reveal stars in the act of forming. These and other discoveries are changing astronomy.

In this 1990 edition of *Foundations of Astronomy,* the author has included updates on Neptune (Chapter 23), new telescopes (Chapter 6), Supernova 1987A (Chapter 13), gravity wave detectors (Chapter 14), quasars (Chapter 17), extra solar planets (Chapter 19), SETI (Chapter 25), and more. The author has added the most recent photographs and diagrams, and has included the latest information about space telescopes, probes to Venus, Jupiter, and Mars, the greenhouse effect, and so on. Following the suggestions of professors who have been using the book, he has also improved the clarity of many discussions. Also, thanks go to the Wadsworth editorial and production team for this edition—Ann Scanlan-Rohrer, Martha Branch, and Mary Douglas.

We view this revision as a necessary update, rather than as a major change (which could require substantial reorganization of an instructor's outline). Therefore, we have been careful to keep the original sequence, page numbering, chapter length, and book length intact, and we have integrated the new material smoothly into the book's original structure. We believe these changes will assist instructors in teaching a current, accurate, and effective course in astronomy. And we hope that students will continue to be fascinated and excited by learning about the universe with the help of *Foundations of Astronomy.*

TO THE READER

You will approach retirement around the year 2035, and your children about 2060. Your grandchildren will not be retiring until almost 2100. You and your family will live through a century of exploration unlike any in the history of this planet. You will see explorers return to the moon, and your children could be the first colonists in lunar cities. Your grandchildren may reach Mars, mine the asteroid belt, explore the icy satellites of Jupiter and Saturn, or leave the solar system bound for the stars. A century ago the airplane had not been invented. Whatever humanity is like a century in the future, we can guess that it will be deeply involved in the exploration of the solar system. Astronomy, the study of the universe beyond the clouds, helps us understand what we will find when we leave earth.

Living in the next century might be enough justification for taking an astronomy course, but there are other reasons. The coming years will see tremendous advances in science and technology, advances that will confuse anyone not familiar with how science progresses from data to hypothesis to theory to natural law. Should your state permit nuclear waste disposal sites? Should you support construction of orbiting solar power stations? Should you give your children massive doses of vitamin C to combat colds? To resolve such technical issues, you need to apply some of the methods of science. Thus, as you study astronomy in the pages that follow, look at it as an example of scientific reasoning. Distinguish between data and theory, and notice how hypotheses are tested over and over.

Yet another reason for taking an astronomy course is to satisfy your natural curiosity. Having heard about black holes, the expanding universe, or the rings of Saturn, you may want to know more about them. Satisfying your own curiosity is the most noble reason for studying anything.

Curiosity might lead you to consider astronomy as a career, but you should know that the field is very small and jobs are hard to find. Instead, you might consider astronomy as a hobby—an activity for personal satisfaction and enrichment. The magazines listed here will keep you up to date with the rapid advances in the field and give you some ideas for further projects, such as telescope building and astronomical photography:

Astronomy, 411 East Mason St., P.O. Box 92788, Milwaukee, WI 53202

The Griffith Observer, 2800 East Observatory Road, Los Angeles, CA 90027

Mercury, Astronomical Society of the Pacific, 1290 24th Ave., San Francisco, CA 94122

The Planetary Report, The Planetary Society, 65 N. Catalina Ave., Pasadena, CA 91106

Sky and Telescope, Sky Publishing Corporation, 49 Bay State Rd., Cambridge, MA 02238

All these reasons for taking astronomy are reasonable, but the most important reason is astronomy's cultural value. The one reason you should study astronomy, the reason your school goes to the expense of teaching the course for you, is that astronomy tells you about your place in nature. It shows you our tiny planet spinning in space amid a vast cosmos of stars and galaxies. It takes you from the first moment of creation to the end of the universe. You will see our planet form, life develop, and our sun die. This knowledge has no monetary value, but it is priceless if you are to appreciate your existence as a human being.

Astronomy will change you. It will not just expand your horizons, it will do away with them. You will see humanity as part of a complex and beautiful universe. If by the end of this course you do not think of yourself and society differently, if you don't feel excited, challenged, and a bit frightened, then you haven't been paying attention.

M.A.S.

MANUSCRIPT REVIEWERS

Wallace Arthur, *Fairleigh Dickinson University*
Timothy Barker, *Wheaton College*
Henry E. Bass, *University of Mississippi*
William P. Bidelman, *Case Western Reserve University*
Michel Breger, *University of Texas at Austin*
Michael F. Capobianco, *St. John's University*
Neil F. Comins, *University of Maine at Orono*
Peter S. Conti, *University of Colorado*
John J. Cowan, *University of Oklahoma*
Russell J. Dubisch, *Siena College*
Robert J. Dukes, Jr., *The College of Charleston*
T. Stephen Eastmond, *Rancho Santiago College*
L. X. Finegold, *Drexel University*
Jack K. Fletcher, *Eastern Kentucky University*
Marjorie Harrison, *Sam Houston State University*
Thomas G. Harrison, *North Texas State University*
Paul W. Hodge, *University of Washington*
Thomas O. Krause, *Towson State University*
Nathan Krumm, *University of Cincinnati*
Leonard Muldawer, *Temple University*
John Peslak, Jr., *Hardin-Simmons University*
C. W. Price, *Millersville University*
Robert Quigley, *Western Washington University*
W. L. Sanders, *New Mexico State University*
Robert F. Sears, *Austin Peay State University*
Michael L. Stewart, *San Antonio College*
Yervant Terzian, *Cornell University*
Leslie J. Tomley, *San Jose State University*
Raymond E. White, *University of Arizona*
Anne G. Young, *Rochester Institute of Technology*

CHAPTER 1

THE SCALE OF THE COSMOS

The longest journey begins with a single step.

Confucius

We are going on a voyage out to the end of the universe. Marco Polo journeyed east, Columbus west, but our voyage will take us away from our home on earth, out past the moon, sun, and planets, past the stars we see in the sky and past billions more that we cannot see without the aid of the largest telescopes. We will journey through great whirlpools of stars to the most distant galaxies visible from earth—and then we will continue on, carried only by experience and imagination—looking for the structure of the universe itself.

Besides journeying through space, we will also travel in time. We will explore the past: see the sun and planets form, search for the formation of the first stars and the origin of the universe. We will also explore forward in time to watch the sun die and the earth wither. Our imagination will become a scientific time machine to search for the ultimate end of the universe.

Though we may find an end to the universe, a time when it will cease to exist, we will not discover an edge. It is possible that our universe is infinite and extends in all directions without limit. Such vastness dwarfs our human dimensions, but not our intelligence or imagination.

Astronomy—this imagined voyage—is more than the study of stars and planets. It is the study of the universe in which we exist. Our personal lives are confined to a small corner of a small planet circling a small sun drifting through the universe, but astronomy can take us out of ourselves and thus help us understand what we are.

Our study of astronomy introduces us to sizes, distances, and times far beyond our common experience. The comparisons in this chapter are designed to help us grasp their meaning.

Michael A. Seeds

How big is a star? The answer—roughly 1 million miles in diameter—is meaningless. Such a large number tells us nothing. How can we humans, only 5–6 feet tall, hope to understand the vastness of the universe? The secret lies in the single word *scale*.

To understand the universe, we must understand the relative scale of planets, stars, galaxies, and the universe as a whole. Only when we can relate our own body size to the astronomical universe around us, can we begin to understand nature on the grandest scale.

To illustrate the scale of astronomical bodies, to fit ourselves into the universe, we will journey from a campus scene to the limits of the cosmos in 12 steps. Each step will widen our view by a factor of 100. That is, each successive picture in this chapter will show a region of the universe that is 100 times wider than the preceding picture.

This snowy scene shows a region about 52 feet in diameter. It is occupied by two human beings, a sidewalk, a tree, and a few bushes—all objects whose size we can understand. Only 12 steps separate this scene from the universe as a whole.

Pennsylvania Department of Transportation, Bureau of Design

Field of view enlarged 100 times from previous image

We now increase our field of view by a factor of 100, which is an area 1 mile across. The area of the preceding photograph is shown by the small white square (arrow). Individual people, trees, and sidewalks vanish, but now we can see a college campus and the surrounding streets and houses.

These dimensions are familiar. We have been in houses, crossed streets, and walked or run a mile. We have personal experience with such dimensions, and we can relate them to the scale of our bodies.

Although we have begun our adventure using feet and miles, we should use the metric system of units because it makes arithmetic simpler (Appendix B). One mile is 5280 feet and each foot 12 inches, so 1 mile is 5280 × 12 = 63,360 inches. If we use the metric system, this calculation is easier. One kilometer (km) is 1000 meters (m), and each meter is 100 centimeters (cm). Thus, 1 km is 1000 × 100 = 100,000 cm.

One mile equals 1.609 km, so our photograph is about 1.6 km in diameter. (See Appendix B for other conversion factors.) Only 11 more steps of 100 separate us from the largest dimensions in the universe. The next photograph will span 160 km.

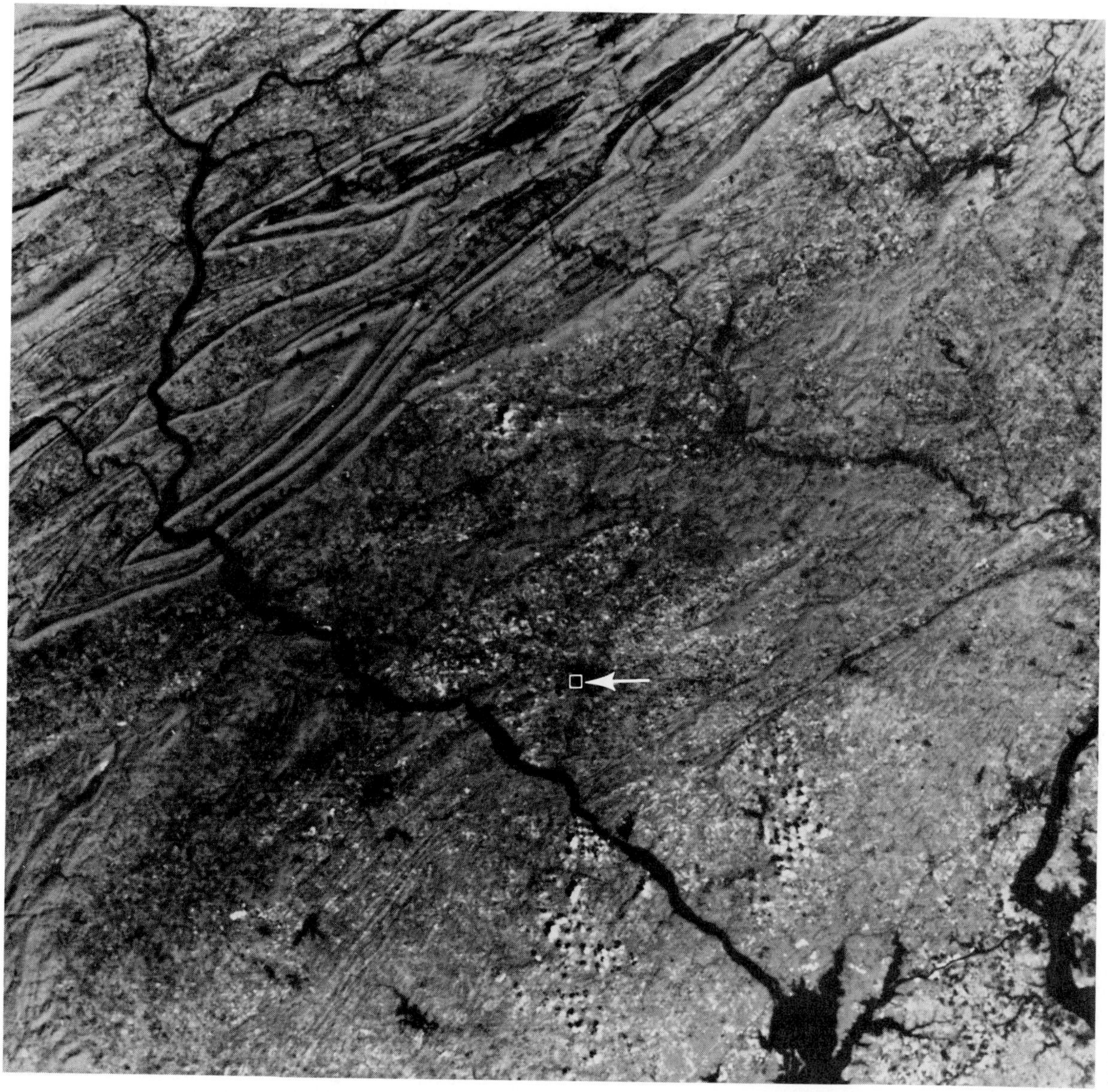

NASA

Field of view enlarged 100 times from previous image

Our field of view now spans 160 km, about 100 miles. The college campus is invisible, and we see only a few signs of human activity. Cities are visible as dark blotches, and farm lands are visible as tiny rectangular shapes. The suburbs of Philadelphia are visible at the lower right.

At this scale we see the natural features of the earth's surface. The Allegheny Mountains of southern Pennsylvania cross the photograph in the upper left, and the Susquehanna River flows southeast into the Chesapeake Bay. A few puffs of clouds dot the area.

These features remind us that we live on the surface of an evolving planet. Forces in the earth's crust pushed the mountain ranges up into parallel folds, like a rug wrinkled on a polished floor. The clouds remind us that the earth's atmosphere is rich in water, which falls as rain and erodes the mountains, washing material down the rivers and into the sea. The mountains and valleys that we know are only temporary features; they are constantly changing.

As we explore the universe we will see that it, like the earth's surface, is evolving. Change is the norm in our universe. We will see stars evolving and dying. And, we will discuss the possible end of the universe.

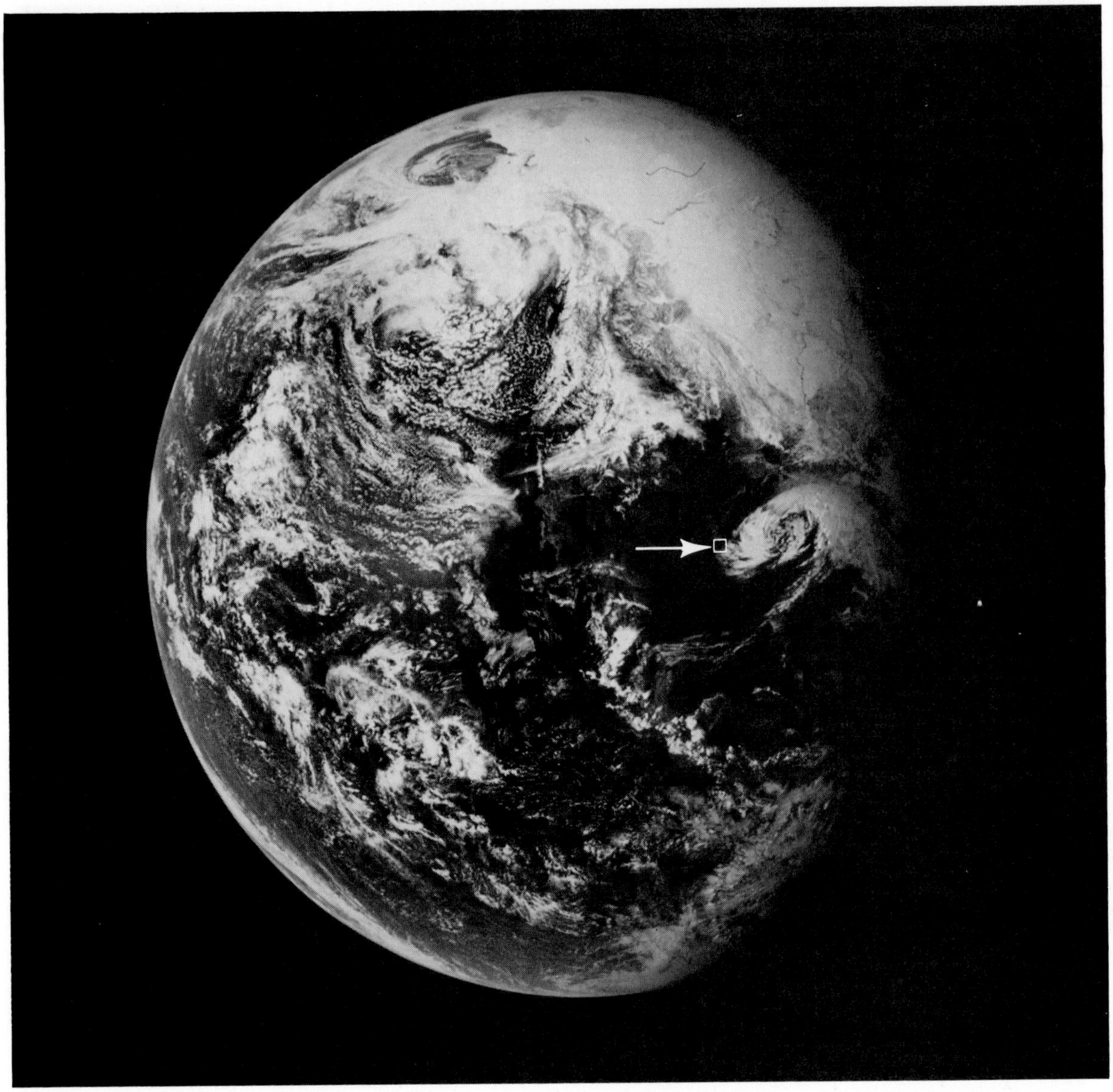

NASA

Field of view enlarged 100 times from previous image

The next step in our journey shows our entire planet. It is 12,756 km in diameter and rotates on its axis once a day.

This image shows most of the daylight side of the planet, with the sunset line at the extreme right. The rotation of the earth carries us eastward across the daylight side, and, as we cross the sunset line into darkness, we say the sun has set. Thus, the rotation of our planet causes the cycle of day and night.

We know that the earth's interior is made of iron and nickel and that its crust is mostly silicate rocks. Only a thin layer of water makes up the oceans, and the atmosphere is only a few miles deep. On the scale of this photograph, the depth of the atmosphere on which our lives depend is less than the thickness of a piece of thread.

The water and air on our planet has made life possible, but we know of no other planet where such conditions exist. Only eight other planets orbit the sun in our solar system, and none have liquid water on their surfaces. In addition, we can see no other planets orbiting other stars. Such planets probably exist, but they are too distant to detect. So far as we know, we are the only life in the universe.

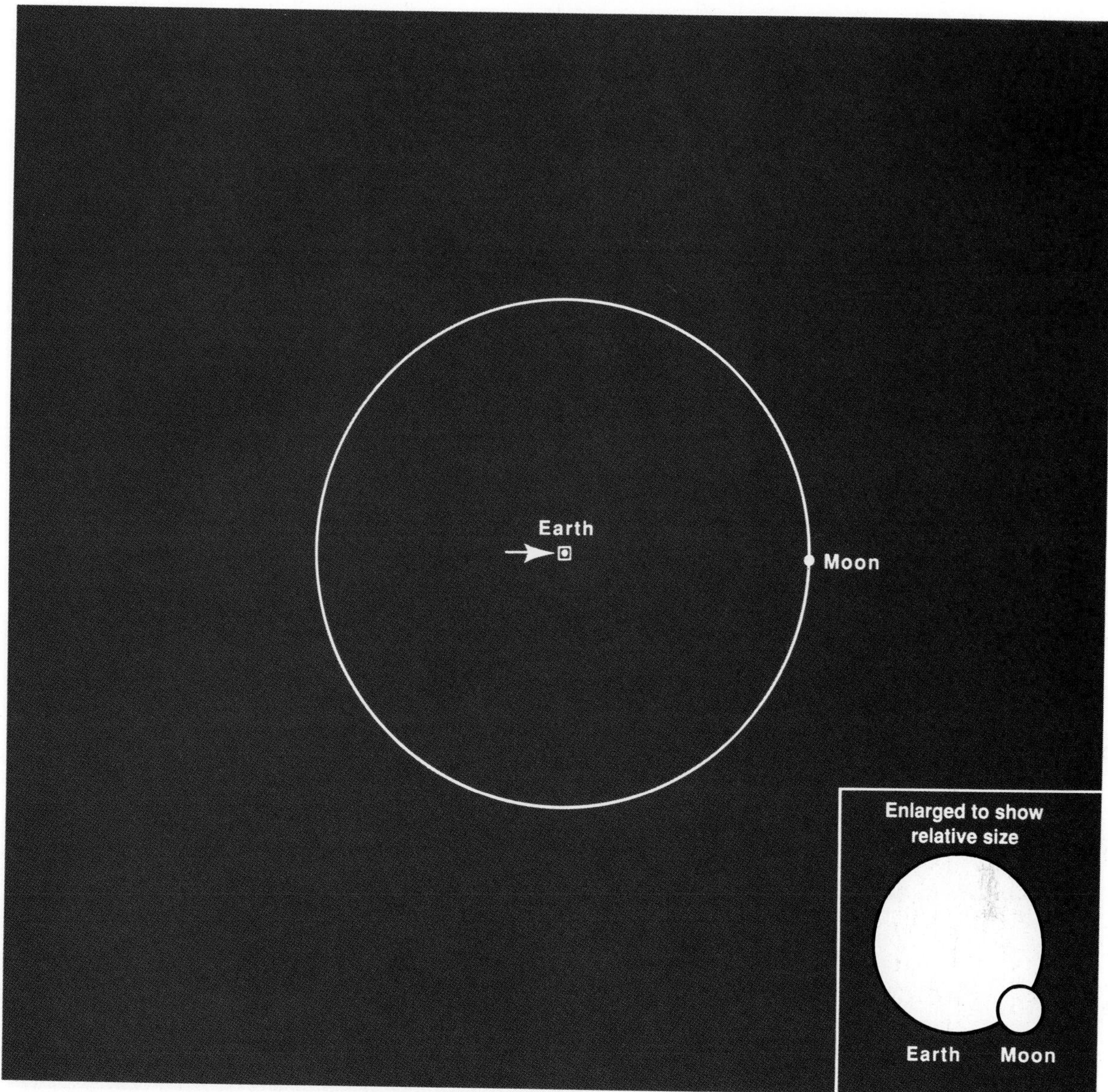

Field of view enlarged 100 times from previous image

Again we enlarge our field of view by a factor of 100, and we see a region of the universe 1,600,000 km wide. Earth is the small white dot in the center, and the moon, only one-fourth the diameter of Earth, is an even smaller dot along its orbit 380,000 km from Earth.

These numbers are so large that it is inconvenient to write them out. Astronomy is the science of big numbers, and we will use numbers much larger than these to discuss the depths of the universe. Rather than write these numbers out, it is convenient to write them in **scientific notation**. It is nothing more than a simple way to write big numbers without writing a great many zeros.

In scientific notation we would write 380,000 as 3.8×10^5. The 5 tells us to move the decimal point five places to the right. The 3.8 then becomes 380,000. In the same way, we would write 1,600,000 as 1.6×10^6. Notice that we could also write this as 16×10^5 or 0.16×10^7. It is the same quantity.

We can also use scientific notation to write very small numbers. If you are not familiar with scientific notation, consult Appendix B. The universe is too big to discuss without using scientific notation.

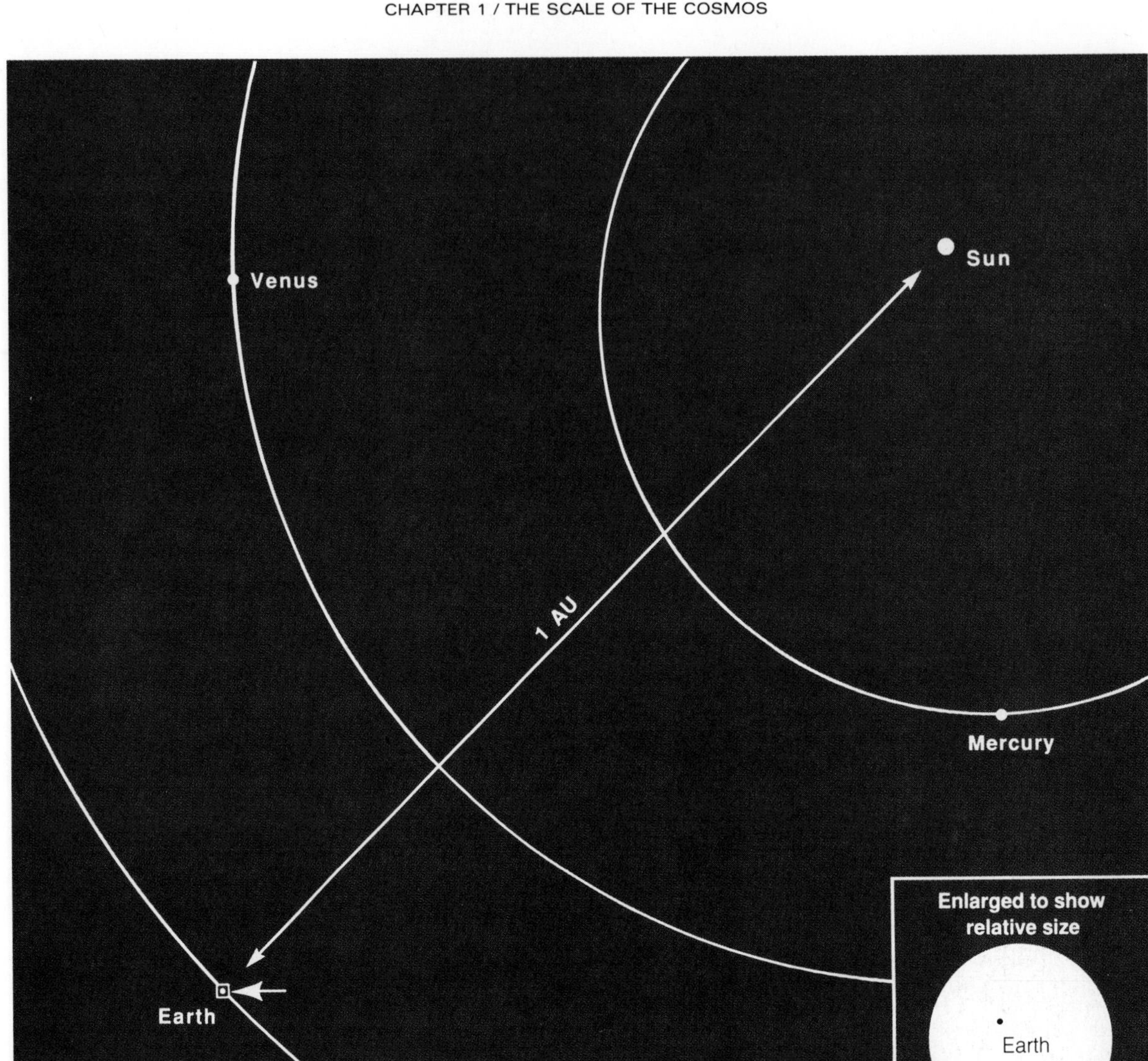

Field of view enlarged 100 times from previous image

When we once again enlarge our field of view by a factor of 100, Earth, the moon, and the moon's orbit disappear. They all lie in the small white box at lower left. But now we see the sun and two other planets.

Venus is about the size of Earth, and Mercury is a bit larger than the moon. On this diagram, they are both too small to show as anything but tiny dots. The sun is 107 times larger in diameter than Earth (inset), but it too is nothing more than a dot on this diagram.

This figure spans 1.6×10^8 km. One way to deal with such large distances is to define new units. Astronomers use the average distance from Earth to the sun as a unit of distance called the **astronomical unit** (AU). Using this unit we can say that the average distance from Venus to the sun is about 0.7 AU. The average distance from Mercury to the sun is about 0.39 AU.

The orbits of the planets are not perfect circles, and this is particularly apparent for Mercury. Its orbit carries it as close to the sun as 0.307 AU and as far away as 0.467 AU. Earth's orbit is more circular, and its distance from the sun varies by only 1.7 percent.

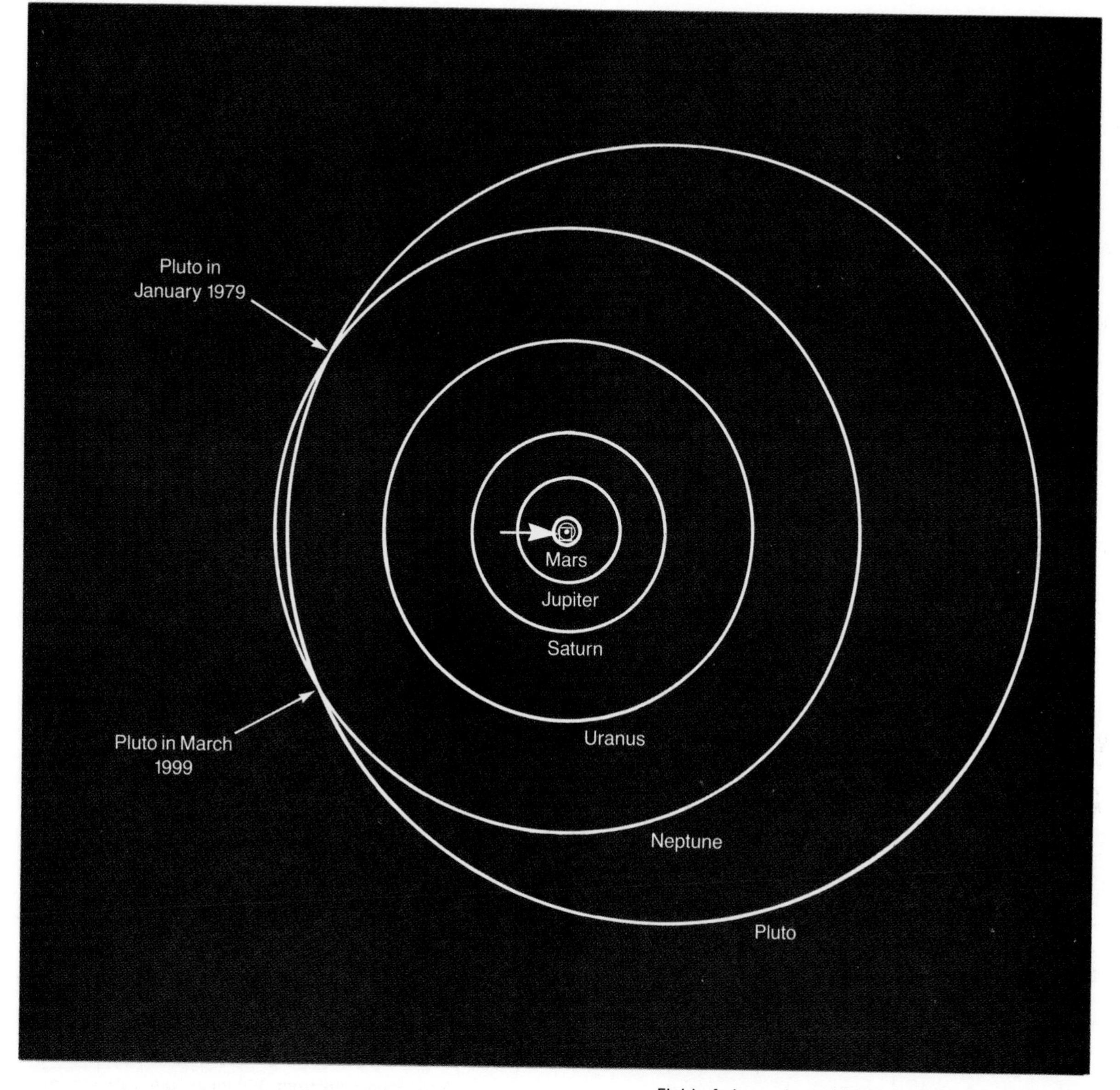

Field of view enlarged 100 times from previous image

When we began our journey, our field of view was only 52 feet (about 16 m) in width. After only six steps of enlarging our field of view by a factor of 100 at each step, we now see the entire solar system. Our field of view is 1 trillion (10^{12}) times wider than in our first view.

The details of the preceding figure are now lost in the tiny square at the center of this diagram. The sun, Mercury, Venus, and Earth lie so close together that we cannot separate them at this scale. Mars, the next outward planet, lies only 1.5 AU from the sun.

In contrast, the outer planets Jupiter, Saturn, Uranus, Neptune, and Pluto are so far from the sun that they are easy to place in this diagram. These are cold worlds far from the sun's warmth. Light from the sun takes over 4 hours to reach Neptune, which is slightly over 30 AU distant. In contrast, sunlight reaches Earth in only 8 minutes.

Notice that Pluto's orbit is so elliptical that it can come closer to the sun than Neptune. In fact, Neptune is now farther from the sun than Pluto and will remain the most distant planet in our solar system till nearly the end of this century.

Field of view enlarged 100 times from previous image

When we again enlarge our field of view by a factor of 100, our solar system vanishes. The sun is visible as a point of light, but all the planets and their orbits are now crowded into the small square at the center. None of the sun's family of planets are visible. They are too small and reflect too little light to be visible so near the brilliance of the sun.

Nor are any stars visible except for the sun. The sun is a fairly average star, and it seems to be located in a fairly average neighborhood in the universe. Although there are many billions of stars like the sun, none are close enough to be visible in this diagram of only 11,000 AU in diameter. The stars are typically separated by distances about ten times larger than this diagram. We will see stars in our next field of view, but, except for the sun at the center, this diagram is empty.

It is difficult to imagine the isolation of the stars. If the sun were represented by a golf ball in New York City, then the nearest star would be another golf ball in Chicago. Except for the widely scattered stars and a few atoms of gas drifting between the stars, the universe is nearly empty.

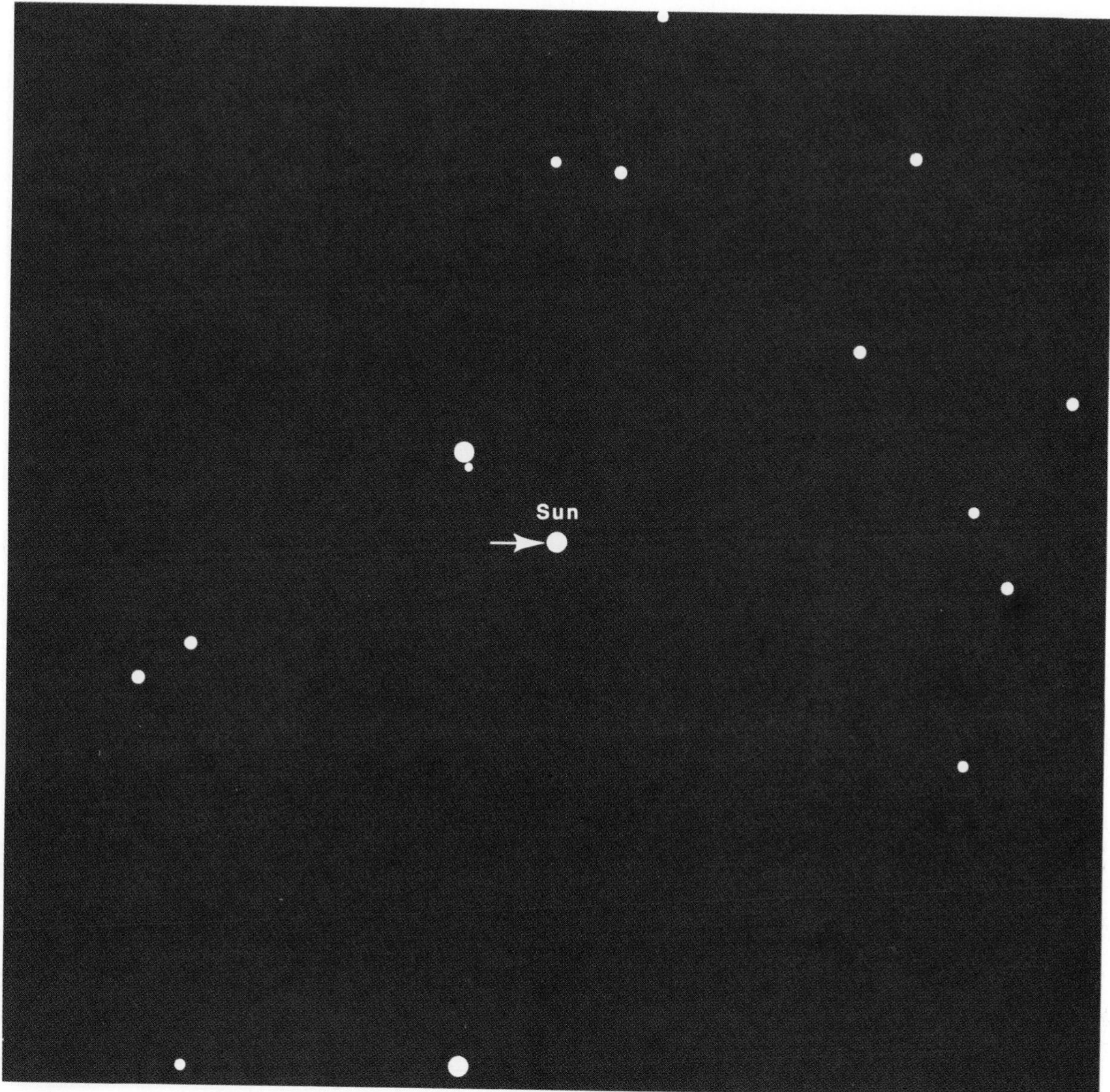

Field of view enlarged 100 times from previous image

Our field of view has now expanded out to a diameter of a bit over 1 million AU. The sun is located at the center, and we see a few of the nearest stars.

These stars are so distant it is not reasonable to quote their distances in astronomical units. We must define a new unit of distance, the light-year. One **light-year** (ly) is the distance that light travels in 1 year, roughly 10^{13} km or 63,000 AU. The nearest star to the sun is Proxima Centauri at a distance of 4.2 ly. Light from Proxima Centauri takes 4.2 years to reach Earth. The diameter of our field of view is now 17 ly.

Although these stars have diameters similar to the sun's, they are so far away we cannot see them as anything but points of light. Even looking through the largest telescopes on Earth, we still see only points of light. In this diagram, the diameters of the dots represent the brightness of the stars and not their actual diameters. This is the custom in astronomical diagrams, and it is also how star images are recorded on photographs. Bright stars make larger spots on the photographic plate than faint stars, although they may not be larger stars.

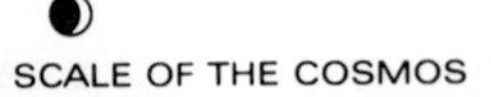

Lick Observatory

Field of view enlarged 100 times from previous image. This box ■ represents relative size of previous frame.

As we expand our field of view by another factor of 100, we find that the sun and its neighboring stars vanish into the background of thousands of stars. The field of view is now 1700 ly in diameter.

Of course, no one has ever journeyed thousands of light years from the sun to photograph the solar neighborhood, so we use a representative photo of the sky. Here the diameters of stars' images are related to the brightness of the stars and not their diameters. The sun is a relatively faint star, so we can no longer locate it on such a photo.

We notice a tendency for stars to occur in clusters. A loose cluster of stars lies in the left half of the photograph. We will discover that many stars are born in clusters and that both old and young clusters exist in the sky. Star clusters are forming right now.

What we do *not* see is critically important. We do not see the thin gas that fills the spaces between the stars. Although those clouds of gas are thinner than the best vacuum on earth, it is those clouds that give birth to new stars. Our sun formed from such a cloud about 5 billion years ago. We will see more star formation when we expand our field of view by another factor of 100.

Mt. Wilson and Las Campanas Observatories, Carnegie Institution of Washington

Field of view enlarged 100 times from previous image

If we expand our field of view by another factor of 100, we can see our own Milky Way galaxy. Of course, no one can journey far enough into space to photograph our galaxy, so we substitute a photo of a similar galaxy with an arrow pointing to a representative location for the sun.

Our sun and the neighboring stars of the previous figure would be lost among the 100 billion stars of the galaxy. Most of the stars are smaller and fainter than our sun, but some are larger and more luminous. Most are cooler, but a few are much hotter. Why some stars are larger, more luminous, or hotter than others is one of the mysteries of the universe that we will explore.

Typical of our galaxy are the graceful spiral arms marked by clusters of bright stars and clouds of gas. We will discover that stars are born in great clouds of gas and dust passing through these spiral arms.

Our galaxy is at least 100,000 ly in diameter, and until about 70 years ago, astronomers thought it was the entire universe—an island universe of stars in an otherwise empty vastness. Now we know that our galaxy is not unique. Indeed, ours is only one of billions of galaxies scattered throughout the universe.

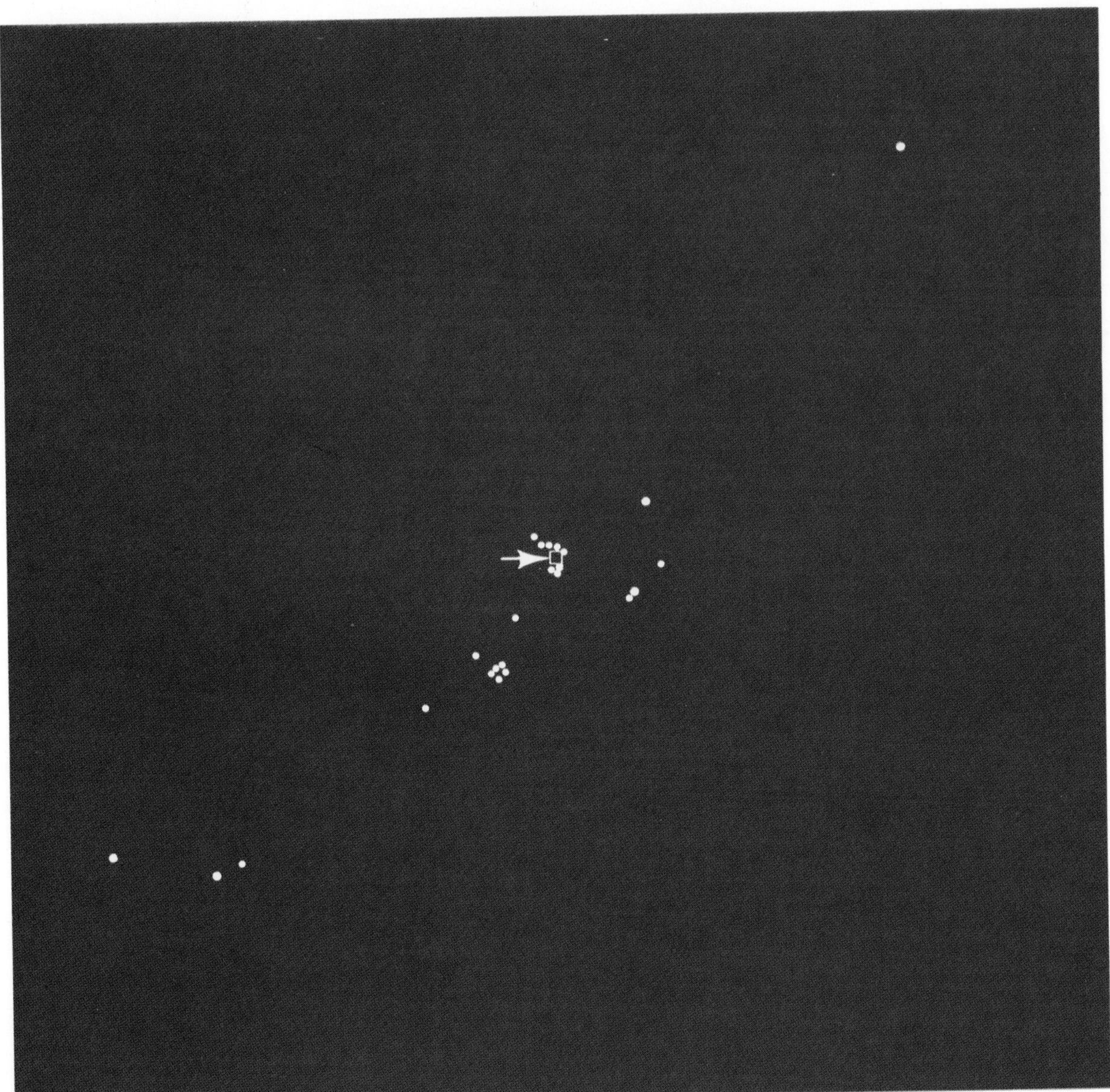

Field of view enlarged 100 times from previous image

As we expand our field of view by another factor of 100, our galaxy becomes a tiny luminous speck surrounded by other specks. This image includes a region 17 million ly in diameter. Each of the specks of light is a galaxy.

We see that our galaxy (arrow) is one of a small cluster of galaxies. This Local Group consists of roughly two dozen galaxies scattered throughout a region about 6 million ly in diameter.

Among the galaxies we see here, a few are as large as our own galaxy, but most are smaller. A few have the beautiful spiral features we see in our galaxy, but most do not. Among more distant galaxies we see a few galaxies twisted into peculiar shapes or wracked by violent eruptions. Although astronomers understand why stars differ from one another, no one is sure what makes one galaxy different from another. We will find some clues to the mystery when we compare the clusters of galaxies.

One theory holds that the centers of some galaxies contain supermassive black holes, which are capable of swallowing stars whole. Whatever the truth, the evolution of galaxies must occasionally be marked by events of titanic violence.

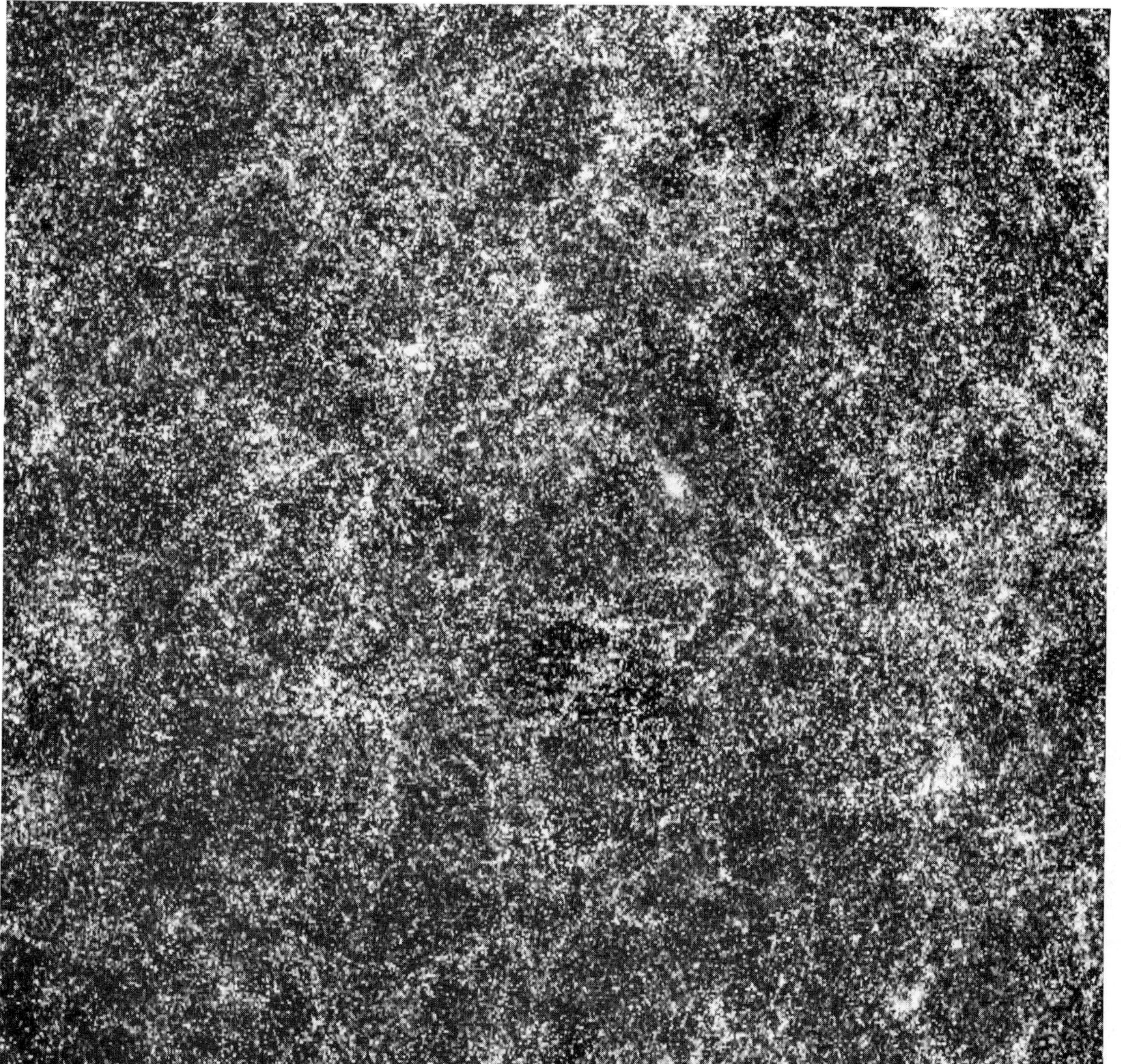

Detail from galaxy map from M. Seldner, B. L. Siebers, E. J. Groth, P. J. E. Peebles, Astronomical Journal 82 (1977).

Field of view enlarged 100 times from previous image. This box ■ represents relative size of previous frame.

Were we to again expand our field of view, we would see that our Local Group of galaxies is part of a large supercluster, a cluster of clusters. Other galaxies are not scattered at random throughout the universe but lie in clusters within larger superclusters.

To represent the universe at this scale, we use a diagram in which each dot represents the location of a single galaxy. At this scale we see superclusters linked to form long filaments outlining voids that seem nearly empty of galaxies. These appear to be the largest structures in the universe. Were we to expand our field of view yet another time, we would probably see a uniform sea of filaments and voids. When we puzzle over the origin of these structures we are at the frontier of human knowledge.

Our problem in studying astronomy is to keep a proper sense of scale. Remember that each of the billions of galaxies contains billions of stars. Most of those stars probably have families of planets like our solar system, and on some of those billions of planets liquid water oceans and a protective atmosphere may have spawned life. It is possible that some other planets in the universe are inhabited by intelligent creatures who share our curiosity and our wonder at the scale of the cosmos.

SUMMARY

Our goal in this chapter has been to preview the scale of astronomical objects. To do so, we journeyed outward from a familiar campus scene by expanding our field of view by factors of 100. Only 12 such steps took us to the largest structures in the universe.

The numbers in astronomy are so large it is not convenient to express them in the usual way. Instead, we use the metric system to simplify our calculations and scientific notation to write big numbers easier. The metric system and scientific notation are discussed in Appendix B.

We live on the rotating planet Earth, which orbits a rather average star we call the sun. We defined a unit of distance, the astronomical unit, to be the average distance from Earth to the sun. Of the eight other planets in our solar system, Mercury is closest to the sun, and Neptune is currently the most distant at about 30 AU.

The sun, like most stars, is very far from its neighboring stars, and this leads us to define another unit of distance, the light-year. A light-year is the distance light travels in 1 year. The nearest star to the sun is Proxima Centauri at a distance of 4.2 ly.

As we enlarged our field of view we discovered that the sun is only one of 100 billion stars in our galaxy and that our galaxy is only one of billions of galaxies in the universe. Galaxies appear to be grouped together in clusters, superclusters, and filaments, the largest structures known.

As we explored we noted that the universe is evolving. The earth's surface is evolving, and so are stars. Stars form from the gas in space, grow old, and eventually die. We do not yet understand how galaxies form or evolve.

Among the billions of stars in each of the billions of galaxies, many probably have planets, but even the nearest stars to the sun are too distant for us to see any planets they might have. We suppose that some of these planets are like the earth, and we wonder if a few are inhabited by intelligent beings like ourselves.

NEW TERMS

scientific notation
light-year (ly)
astronomical unit (AU)

QUESTIONS

1. Why are astronomical units and light-years more convenient for measuring astronomical distances than miles or kilometers?
2. In what ways is our universe evolving?
3. Why do all stars, except for the sun, look like points of light as seen from Earth?
4. Why are we unable to see planets beyond the nine in our solar system?
5. Which is the outermost planet in our solar system? Why does this change?
6. In photographs, some stars look larger than others. What does this tell us about the stars?
7. How long does it take light to cross the diameter of our galaxy? Of the Local Group of galaxies?
8. What are the largest known structures in the universe?
9. How many planets inhabited by intelligent life do you think the universe contains? Explain your answer.

PROBLEMS

1. How many inches are there in 100 yards? How many centimeters are then in 100 m?
2. If 1 mile equals 1.609 km, and the moon is 2160 miles in diameter, what is its diameter in kilometers?
3. The earth rotates once a day and has a radius of 6378 km. With what speed is the equator moving eastward in km/sec? In miles/hour?
4. If sunlight takes 8 minutes to reach the earth, how long does moonlight take?
5. If the earth were transported to the center of the sun, would the moon's orbit lie inside or outside the surface of the sun?
6. How many suns would it take, laid edge to edge, to reach the nearest star?
7. How many kilometers are there in a light-minute? (HINT: The speed of light is 3×10^5 km/sec.)
8. How many galaxies like our own, laid edge to edge, would it take to reach the nearest large galaxy (which is 2×10^6 ly away)?

RECOMMENDED READING

CALDER, NIGEL *Timescale.* New York: Viking Press, 1982.

DICKINSON, TERRENCE *The Universe and Beyond.* Ontario: Camden House, 1986.

MORRISON, PHILIP, and PHYLIS MORRISON *Powers of Ten.* New York: W. H. Freeman, 1982.

SAGAN, CARL *Cosmos.* New York: Random House, 1980.

WEISKOPF, VICTOR *Knowledge and Wonder: The Natural World as Man Knows It.* Cambridge, Mass.: MIT Press, 1979.

FILM

Powers of Ten. Santa Monica, Calif.: Pyramid Films, 1968.

UNIT 1

THE SKY

CHAPTER 2

THE EARTH AND SKY

The Southern Cross I saw every night abeam. The sun every morning came up astern; every evening it went down ahead. I wished for no other compass to guide me, for these were true.

Captain Joshua Slocum
SAILING ALONE AROUND THE WORLD

The sky is the rest of the universe as seen from our planet. When we look at the stars, we look through a layer of air only a few miles thick. Above that, space is nearly empty. The nearest star to Earth is the sun, and the next nearest star beyond the sun is 4 ly away, so distant light takes 4 years to reach us. The average bright star in our sky is about 300 ly away. Beyond the stars of our galaxy lie other galaxies billions of light-years away.

In the previous chapter, we took a quick journey through the universe to preview its scale. Now we are ready to repeat that journey more slowly. Our first step, in this chapter, is to try to understand why the sky looks the way it does.

Throughout this chapter, remember that we live on a moving planet. The earth spins on its axis once a day and revolves around the sun once a year. Because we feel stationary on the solid earth, the sky seems to spin around us in complicated ways. Our task is to try to understand the appearance and motions of the objects we see in the sky, but that task is complicated by the earth's motion. We are like observers on a roller coaster trying to sketch a map of the fairground from the careering car.

In our struggle to understand what we see, the most powerful tool we have is a concept that is thousands of years old. We can, like ancient astronomers, imagine that the sky is a great hollow globe surrounding the earth. Thus we begin our study of modern astronomy, by looking at the sky the way astronomers did three millennia ago.

This chapter will not answer all our questions about the universe. We are merely setting the scene for the challenging explorations to follow.

FIGURE 2–1 The constellations Orion and Taurus represent figures from mythology. (Adapted from Duncan Bradford, *Wonders of the Heavens,* Boston: John B. Russell, 1837.)

2.1 THE STARS

CONSTELLATIONS

Gazing at the night sky on a clear, moonless evening, we can see thousands of stars scattered in random groups. Some of these groups have been named and are called **constellations** (Figure 2-1). (Appendix C contains a listing of the constellations, and the Supplement "Observing the Sky" at the end of this book will help you locate the constellations using the star charts provided.)

Although we identify these patterns by name, we must remember that the stars in a constellation are usually not physically associated with each other. Some may be many times farther away than others (Figure 2-2). The only thing they have in common is that they lie in approximately the same direction as seen from earth.

Half of today's 88 constellations were named in ancient times. The oldest, including Taurus and Leo, seem to have originated in Mesopotamia more than 5000 years ago, and others were added by Babylonian, Egyptian, and Greek astronomers. In some cases, these star patterns had mystical or religious significance, but some appear to have been designed as practical navigation aids for sailors. Still others honored great heroes such as Orion and Hercules.

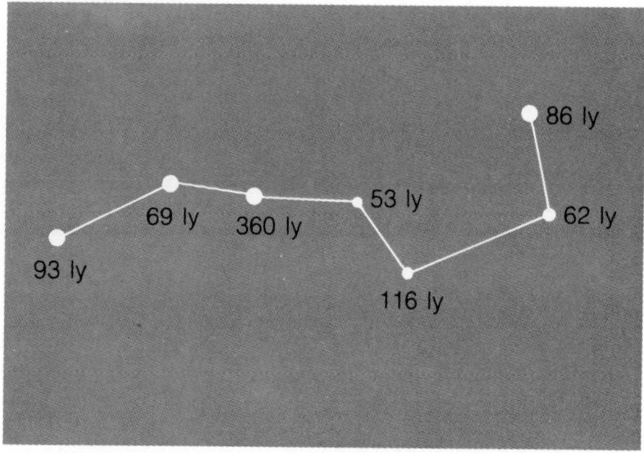

FIGURE 2–2 The stars we see as the Big Dipper, the brighter stars of the constellation Ursa Major, the Great Bear, are not all at the same distance. The distances to these stars are shown in light-years. The size of the dots represents the apparent brightness of the stars.

Different cultures drew and named their constellations differently. The constellation we know as Orion was

FIGURE 2–3 (a) The ancient constellations Andromeda and Pegasus share the star Alpheratz at the upper left corner of the great square of Pegasus. (Adapted from Duncan Bradford, *Wonders of the Heavens,* Boston: John B. Russell, 1837.) (b) The modern constellation boundaries assign Alpheratz to Andromeda.

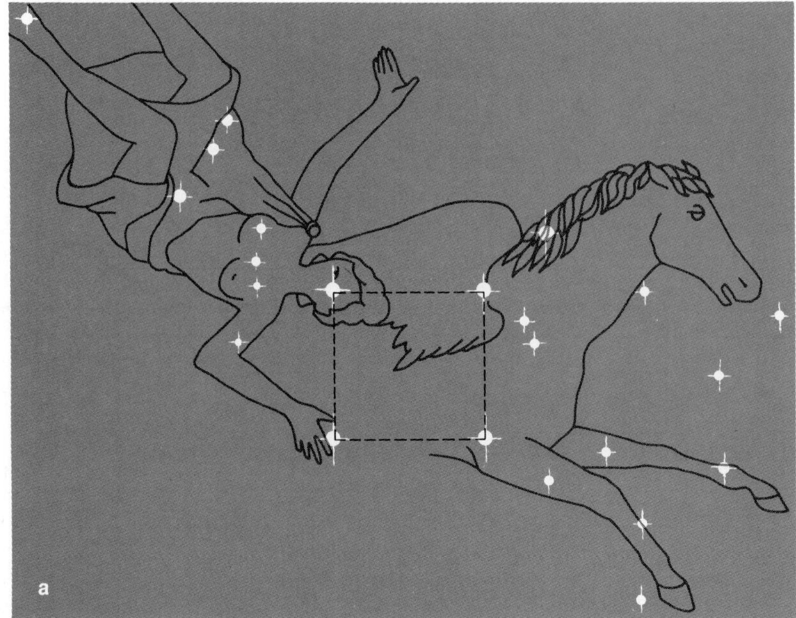

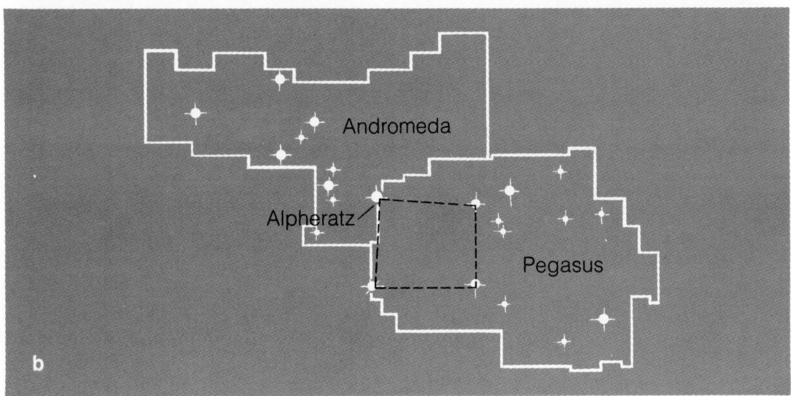

known as Al Jabbar, the giant, to the Syrians, as the White Tiger to the Chinese, and as Praja-pati in the form of a stag in ancient India. The Pawnee Indians knew the constellation Scorpius as two groupings. The long tail of the scorpion was The Snake, and the two bright stars at the tip of the scorpion's tail were The Two Swimming Ducks.

The ancient constellations were thought of as loose groupings of stars. A star could even belong to more than one constellation, as in the case of Alpheratz in the constellations Andromeda and Pegasus (Figure 2-3a). As a convenience, modern astronomers have given each constellation definite boundaries as defined by the International Astronomical Union in 1928. Thus, a constellation represents not just a group of stars but a region of the sky, and any star within the region is a part of the constellation. This convention places Alpheratz in the constellation of Andromeda (Figure 2-3b).

Not all regions of the sky were parts of ancient constellations. Some regions contain no bright stars, and the southern skies are never visible from the latitude of Greece. Consequently, the ancient astronomers never made these regions parts of constellations. Modern astronomers have invented 44 additional constellations to fill these spaces. Some of these modern constellations are Telescopium (the telescope), Microscopium (the microscope), and Antlia (the air pump).

Modern astronomers still use the names of constellations as a handy way to refer to regions of the sky.

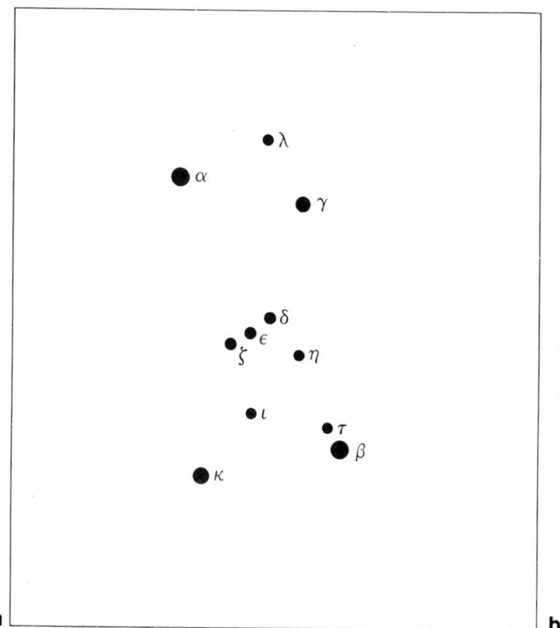

FIGURE 2–4 (a) The brighter stars in each constellation are assigned Greek letters in approximate order of brightness. In Orion, κ is brighter than its Greek letter suggests. (b) A long-exposure photograph reveals the many faint stars that lie within Orion's constellation boundaries. These are members of the constellation, but they do not have Greek letter designations. (Mount Wilson and Las Campanas Observatories, Carnegie Institution of Washington.)

THE NAMES OF THE STARS

The brightest stars were named thousands of years ago, and these names along with the names of the constellations found their way into the first catalogs of stars. We take our constellation names from Greek versions translated into Latin—the language of science from the fall of Rome to the nineteenth century—but most star names come from ancient Arabic. Such names as Sirius (the Scorched One), Capella (the Little She Goat), and Aldebaran (The Follower of the Pleiades) are beautiful additions to the mythology of the sky.

Giving the stars individual names is not very helpful because we see thousands of stars, and these names do not help us locate the star in the sky. In which constellation is Antares? Another way to identify stars is to assign Greek letters to the bright stars in a constellation in approximate order of brightness. (See Appendix C for a listing of the Greek alphabet.) Thus, the brightest star is usually designated α (alpha), the second brightest β (beta), and so on (Figure 2-4). To identify a star in this way, we give the Greek letter followed by the genitive form of the constellation name, such as α Scorpii (Antares). Now we know that Antares is in the constellation Scorpius and that it is probably the brightest star in that constellation.

This method of identifying a star's brightness is only approximate. In some constellations, the Greek letters were not assigned in exact order of brightness (Figure 2-4).To be precise, we must have an accurate way of referring to the brightness of stars, and for that we must consult one of the first great astronomers.

THE BRIGHTNESS OF STARS

Hipparchus, a Greek astronomer (160–127 BC), divided the stars into six classes. The brightest were first-class stars, and those slightly fainter were second-class stars. Continuing down to the faintest stars he could see, the sixth-class stars, he recorded his classifications in a great star catalogue that became a classic reference in ancient

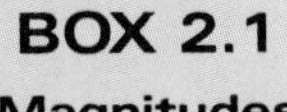

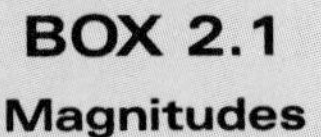

BOX 2.1
Magnitudes

Bright stars look bright because our eyes receive more energy per second from them than from faint stars. Thus, we can think of a star's brightness as equaling the light energy from the star that hits 1 square centimeter (cm^2) in 1 second. If we represent the brightness of two stars A and B as I_A and I_B, then the ratio of their brightnesses is I_A/I_B. This brightness ratio tells us how many times brighter star A is than star B.

Modern astronomers have defined the magnitude scale so that two stars whose magnitudes differ by 5 magnitudes have a brightness ratio of exactly 100. That is, if star A is 5 magnitudes brighter than star B, then star A must be 100 times brighter than star B. Two stars that differ by 1 magnitude have a brightness ratio of 2.512—that is, one star is about 2.5 times brighter than the other. Two stars that differ by 2 magnitudes have a brightness ratio of 2.512 × 2.512 or about 6.3, and so on (Table 2-1). Note that two stars that differ by 5 magnitudes have a brightness ratio of $(2.512)^5$ or 100.

A table of brightness ratios makes magnitude calculations simple. For example, consider two stars C and D. If star C is third magnitude and star D is ninth magnitude, how many times brighter is star C? It is 6 magnitudes brighter, and from the table we find that the brightness ratio is 250. Thus, our eyes will receive 250 times more light energy per second from star C than from star D.

This is a simple example because the magnitude difference is a whole number, 6. If it were not a whole number, we would have to estimate the proper brightness ratio from the table, or calculate the brightness ratio directly. The brightness ratio I_A/I_B is equal to 2.512 raised to the power of the magnitude difference $m_B - m_A$.

$$\frac{I_A}{I_B} = (2.512)^{(m_B - m_A)}$$

For example, if the magnitude difference is 6.32 magnitudes, the brightness ratio is

$$\frac{I_A}{I_B} = (2.512)^{6.32}$$

A pocket calculator shows that the brightness ratio is 337.

Table 2.1 Magnitude and brightness.

Magnitude Difference	Brightness Ratio
0	1
1	2.5
2	6.3
3	16
4	40
5	100
6	250
7	630
8	1,600
9	4,000
10	10,000
.	.
.	.
.	.
15	1,000,000
20	100,000,000
25	10,000,000,000
.	.
.	.
.	.

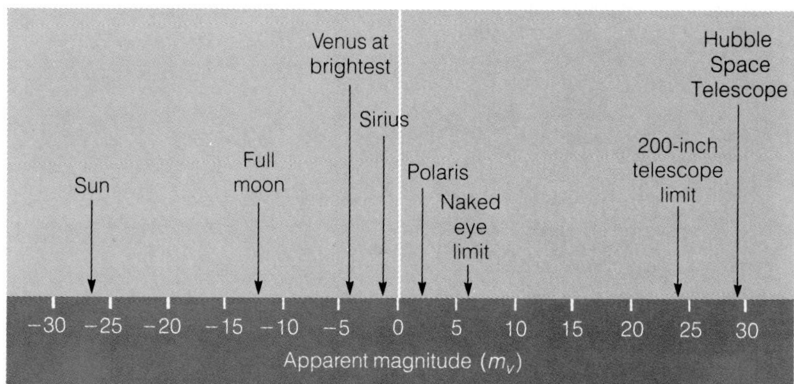

FIGURE 2–5 The scale of apparent visual magnitudes extends into negative numbers to represent the brighter objects.

astronomy. His method, slightly modified, is still in use today as the **magnitude scale**, the astronomer's brightness scale.

Despite its value, the magnitude scale can seem confusing. First, the fainter the star, the larger the magnitude number. For instance, sixth-magnitude stars are fainter than first-magnitude stars. This may seem backward at first, but think of it as Hipparchus did. The brightest stars are first-class stars, and the fainter stars are second- and third-class, and so on.

Another source of confusion is that the magnitude system is not linear—that is, a change of two magnitudes is not twice as big as a change of one magnitude. We must use logarithms to describe the magnitude system because Hipparchus designed it to represent the way stars look to our eyes, and our eyes work logarithmically. If they did not, we might be able to see subtle differences in brightness by sunlight, but would be totally blind in the shade. Box 2-1 discusses the mathematics of the magnitude scale.

Modern astronomers have made a major improvement in Hipparchus' magnitude system by measuring stellar brightness with sensitive instruments. For example, instead of merely saying that θ (theta) Leonis is a third-magnitude star, they can say specifically that its magnitude is 3.34.

If we measured the brightness of all of the stars in Hipparchus' first brightness class, some would be brighter than 1.0. For instance, Vega (α Lyrae) is so bright that its magnitude is almost zero at 0.04. A few stars are so bright the magnitude scale must extend past zero into negative numbers (Figure 2-5). On this scale, Sirius, the brightest star in the night sky, has a magnitude of −1.42.

The magnitude scale is open-ended. We can extend it far into negative numbers to include the moon at about −12.5 and the sun at about −26.5. We can also extend it to objects fainter than the faintest stars we can see with the naked eye. The 5-m (200-inch) telescope can detect stars as faint as twenty-fourth magnitude.

These are known as **apparent visual magnitudes** (m_v). Such magnitudes refer to how bright the stars look and does not compensate for their distance from earth. A star that is a million times more luminous than the sun might appear very faint if it were very far away, and a star that is much less luminous than the sun might look bright if it were nearby. Look at the brightness and distance of the stars in Figure 2-2. In Chapter 10 we will develop a magnitude scale that takes distance into account and tells us how bright the stars really are. Apparent visual magnitude only tells us how bright they appear.

So far we have discussed the sky as if it were static and unchanging. Now that we are familiar with constellations, star names, and magnitudes, we can look at the sky as a whole and note its motion.

2.2 THE CELESTIAL SPHERE

A MODEL OF THE SKY

Many ancient astronomers, including Hipparchus, thought of the sky as a great, hollow, crystalline sphere surrounding Earth. The stars, they imagined, were attached to the inside of the sphere like thumbtacks stuck in the ceiling. The sphere rotated once a day, carrying the sun, moon, planets, and stars from east to west.

We know now that the sky is not a great, hollow, crystalline sphere. The stars are scattered throughout space at different distances, and it isn't the sky that rotates once a day—Earth turns on its axis. Although we know that the crystalline sphere is not real, it is convenient as a model of the sky and is used daily by modern astrono-

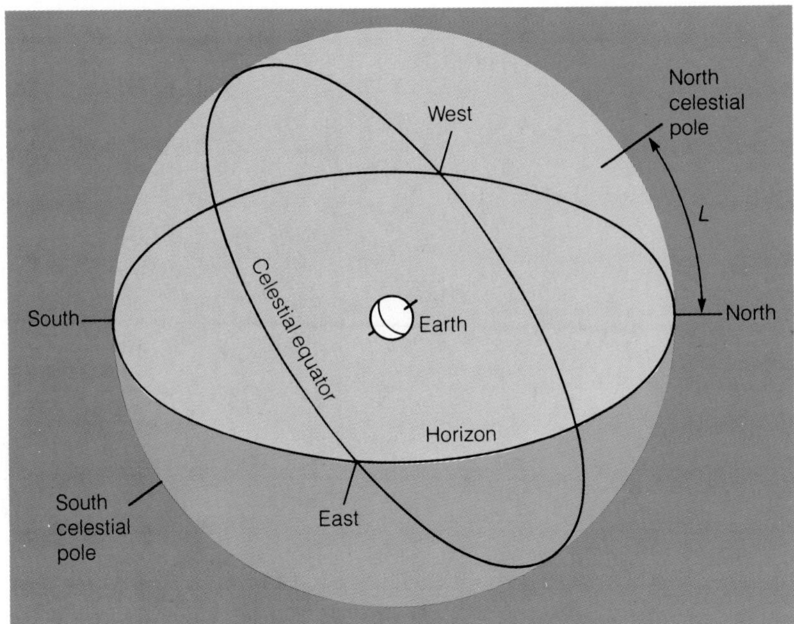

FIGURE 2–6 The modern celestial sphere models the appearance of the sky. The poles mark the pivots, and the equator divides the celestial sphere in half. Those objects below our horizon are invisible. The angle L is equal to the observer's latitude.

mers when they think about the locations and motions of celestial bodies.

Our model of the sky is called the **celestial sphere**, an imaginary sphere of a very large radius surrounding the earth and to which the stars, planets, sun, and moon seem to be attached (Figure 2-6). This sphere must have a large radius so that no part of the earth is significantly closer to a given star than any other part. Then it does not matter where on the earth we go: The sky always looks like a great sphere centered on our location.

If we watch the late afternoon sky for a few hours, we can notice movement. As the rotation of the earth carries us eastward, the sun appears to move westward and eventually sets. As darkness falls, we can see the stars, and in an hour or so it becomes obvious that the eastward rotation of the earth is making the sky appear to rotate westward (Figure 2-7a). As some constellations set in the west, others rise in the east.

This daily rotation of the sky is called its **diurnal motion.** The word *diurnal* means daily. The diurnal motion of the sky is only an illusion produced by the daily rotation of the earth on its axis.

REFERENCE MARKS ON THE SKY

The pivots about which the sky seems to rotate are called the celestial poles. The **north celestial pole** is the point on the sky directly above the earth's North Pole, and the **south celestial pole** is the point directly above the earth's South Pole. Stars located near the celestial poles seem to describe small circles about the poles as the earth turns. A time exposure photograph of the sky shows curved streaks made by the stars as the sky rotates (Figure 2-7b).

The location of the north celestial pole depends on the latitude of the observer. For example, if we stood in the ice and snow at the earth's North Pole, the north celestial pole would be directly overhead. If we stood on the earth's equator, the north celestial pole would lie on our northern horizon. At intermediate latitudes, such as those of the United States, the north celestial pole lies about halfway between overhead and the northern horizon, as in Figure 2-6. To be precise, the angular distance from the horizon to the north celestial pole equals the latitude of the observer. This relationship makes it simple for navigators in the earth's northern hemisphere to find their latitude by measuring the angle between the northern horizon and the north celestial pole.

Currently the star Polaris happens to lie very near the north celestial pole, and thus hardly moves as the sky rotates. (The Pawnee Indian name for this star means "Star-That-Does-Not-Walk-Around.") Figure 2-7b shows how Polaris, the brightest image, hardly moves as the sky rotates about its axis. At any time of the night, in any season of the year, from anywhere in the earth's northern hemisphere, Polaris always stands above the northern horizon

FIGURE 2–7 (a) The 4-day-old crescent moon and Venus setting. Exposures were made every 8 minutes, illustrating how the eastward rotation of the earth causes celestial objects to move westward across the sky. Note also the eastward motion of the moon relative to Venus. (b) A time exposure taken with a camera pointed at the north celestial pole shows star trails that demonstrate the apparent rotation of the sky. (a, William P. Sterne, Jr.; b, Lick Observatory photograph.)

and is consequently known as the North Star. In the next section, we will see that other stars have occupied this location in the past.

Because it lies below our horizon, the south celestial pole is never visible from the United States, and the constellations near the south celestial pole never rise. The constellations near the north celestial pole never set as seen from our latitude. These constellations are known as **circumpolar constellations.** Ursa Major, for example, containing the Big Dipper, is a north circumpolar constellation. In fact, we can always find the North Star easily by first locating the Big Dipper and then following the pointer stars to Polaris as shown in Figure 2-8.

Another important reference mark on the celestial sphere is the **celestial equator**, an imaginary line around the sky directly above the earth's equator (Figure 2-6). The celestial equator divides the celestial sphere into two equal halves, the northern and southern celestial hemispheres.

These reference marks on the sky will be useful trail markers as we explore. They will help us understand the motion of the sun and moon, the cycle of the seasons,

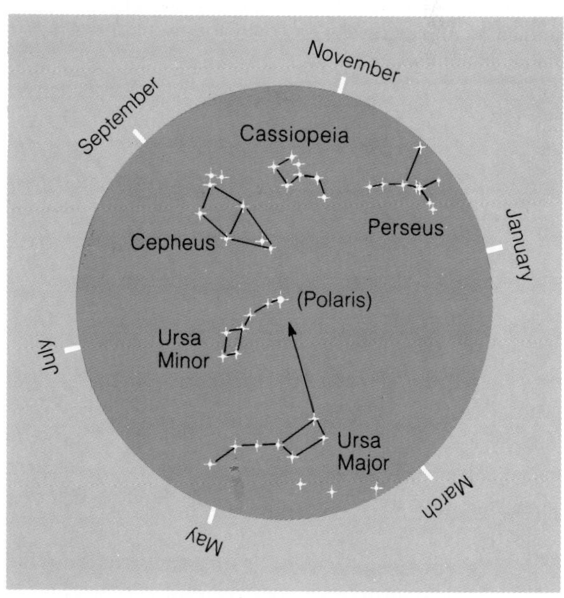

FIGURE 2–8 The northern constellations. To use the chart, face north soon after sunset and hold the chart in front of you with the current date at the top.

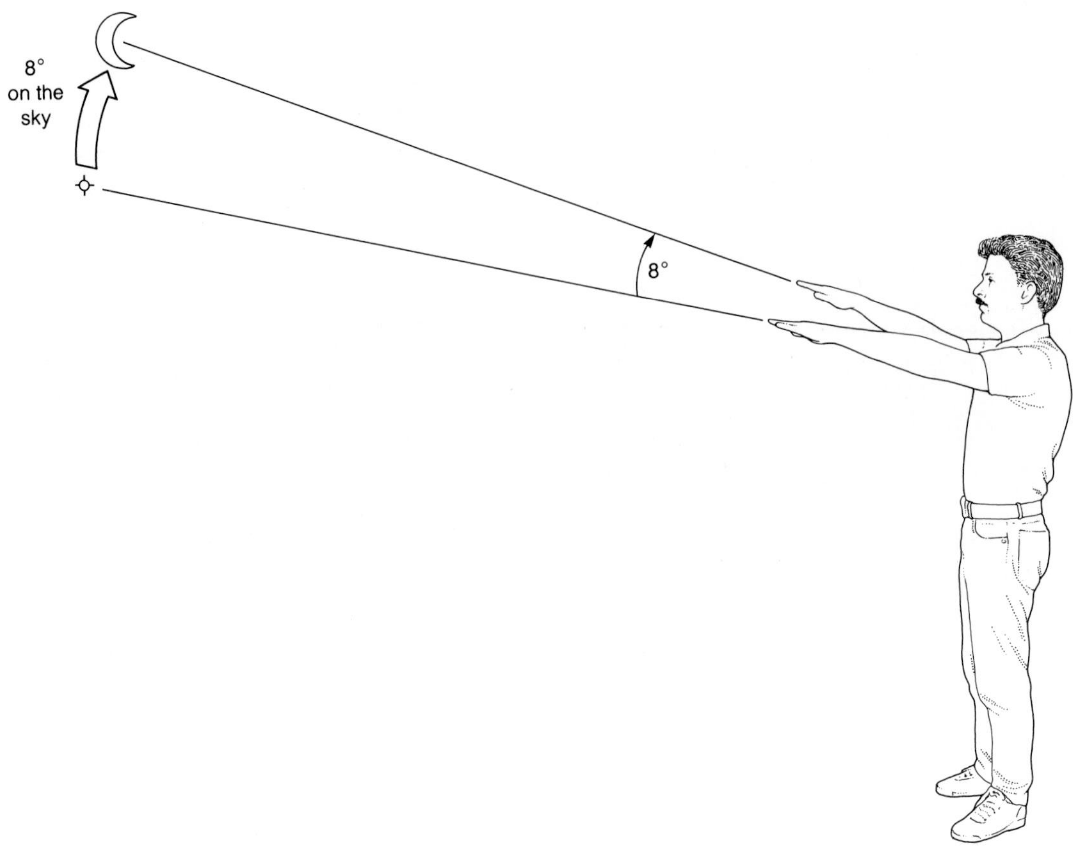

FIGURE 2–9 The angular separation between two objects is the angle our arms would make if we pointed at the two objects. Astronomers often refer to such angles as angles *on* the sky as if the two objects were painted on the inside of the celestial sphere.

and so on. But to use these reference marks, we must be able to think of angles on the sky.

ANGLES ON THE SKY

Astronomers often use angles to describe distance across the sky. They might say, for instance, that the moon is 8° north of a certain star, meaning that if we point one arm at the moon and the other arm at the star, the angle between our arms is 8° (Figure 2-9).

When astronomers speak of such angles they use the phrase "angles on the sky" as if the sky were a great plaster ceiling and the moon and star were spots painted on the plaster. We know the star is hundreds of light-years away, and we know the moon is much closer, so the true distance between them is immense. But if we imagine them painted on the celestial sphere, then we can think of their angular separation as an angle painted on the celestial sphere (Figure 2-9). Thus, we can discuss the angular distance between two objects even when we don't know their true distance from us or from each other.

We measure angles in degrees, minutes of arc, and seconds of arc. There are 360° in a circle and 90° in a right angle. Each degree is divided into 60 **minutes of arc** (sometimes abbreviated 60′). If you view a 25¢ piece from the length of a football field, it has an angular distance of about 1 minute of arc. Each minute of arc is divided into 60 **seconds of arc** (sometimes abbreviated 60″). If you view a dime edgeways from the length of a football field, the dime is about 7 seconds of arc thick.

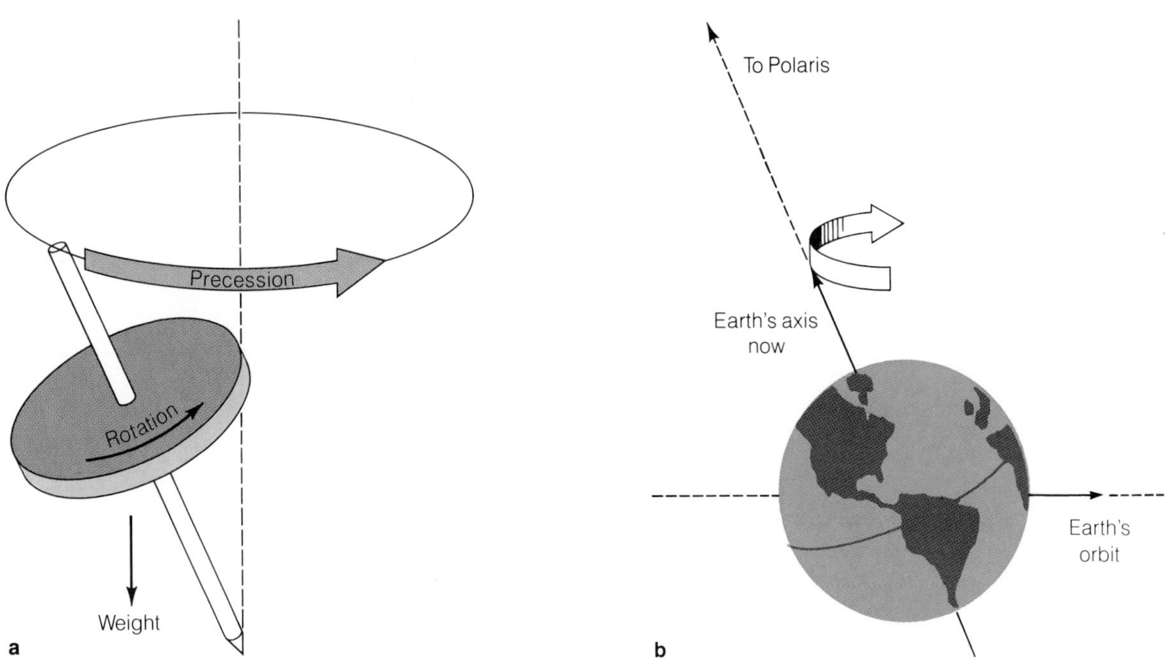

FIGURE 2–10 (a) The weight of a spinning gyroscope tends to make it fall over, and as a result it precesses in a conical motion about a vertical line. (b) The sun's gravity tends to twist earth's axis upright in its orbit, and as a result it precesses. In 13,000 years its axis of rotation will point toward Vega.

We can establish some angles on the sky that will be helpful in estimating angles. The sun and the moon are each about 0.5° in diameter. The pointer stars of the Big Dipper are about 5° apart, and the bowl of the Big Dipper is about 30° from the north celestial pole.

Angles on the sky combined with the reference marks we discussed in the preceding section make possible a coordinate system on the sky (Appendix A). Just as latitude and longitude specify the locations of points on the earth, the system of celestial coordinates can specify the locations of points on the sky. But we must be careful. The reference marks on the sky, the anchors for these coordinates, are defined by the earth's rotation, and the earth is wobbling like a toy gyroscope.

PRECESSION

If you have ever played with a gyroscope, you have seen how the spinning mass resists any change in the direction of its axis of rotation. The more massive the gyroscope and the more rapidly it spins, the more difficult it is to change the direction of its axis of rotation. But you may recall that the axis of even the most rapidly spinning gyroscopes does not remain absolutely fixed. A spinning gyroscope wobbles in a conical motion called **precession** (Figure 2-10a).

The gyroscope precesses because of the interaction of its rotation and its weight. Earth's gravity pulling on the gyroscope (its weight) tends to make the gyroscope tip over. The axis of rotation, however, tends to remain fixed. The result is that the axis moves in a direction perpendicular to the plane established by its axis of rotation and the direction of its weight.

The earth behaves like a giant gyroscope. Its large mass and rapid rotation keep its axis of rotation pointing near the star Polaris. But the earth is not a perfect sphere. It has a slight bulge around its middle, and the sun, pulling on that bulge, twists the earth's axis of rotation, tending to set it upright in its orbit. As a result, the earth's axis precesses in a conical motion, taking about 26,000 years for one cycle (Figure 2-10b).

Because the celestial poles and the equator are defined by the earth's rotation, precession changes these reference marks. We see no change at all from night to night or year to year, but precise measurements reveal the motion of the poles and the equator.

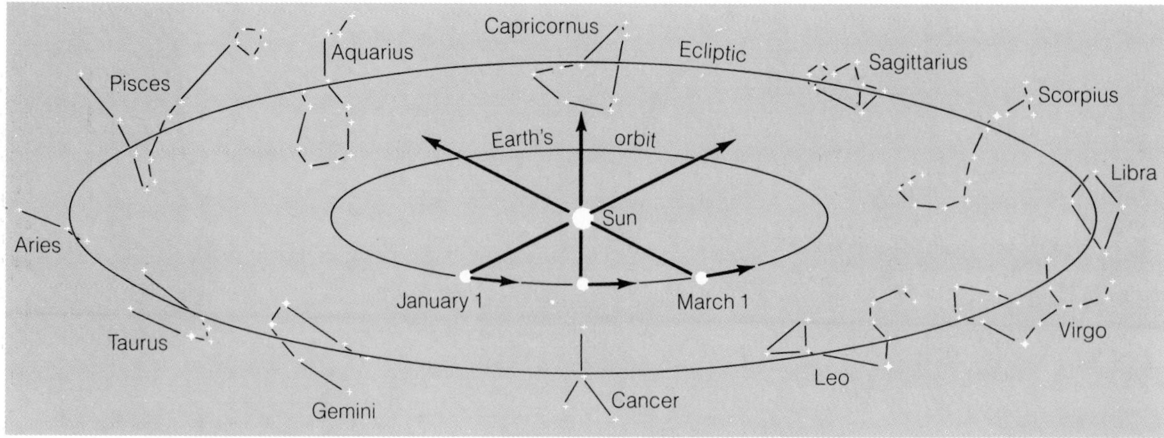

FIGURE 2–11 As the earth moves around its orbit, we see the sun in front of different constellations. In January, the sun is in front of Capricornus, but by March it has moved along the ecliptic into Aquarius. Compare this schematic diagram with the star chart shown in Figure S-3 in the supplement at the end of this book.

Over centuries, precession has dramatic effects. For example, it makes the celestial poles move across the sky. Egyptian records show that 4800 years ago, the north celestial pole was near the star Thuban (α Draconis). The pole is now approaching Polaris and will be closest to it about AD 2100. In about 13,000 years the pole will have moved away from Polaris and will be within 5° of Vega (α Lyrae).

These slow changes in the sky may seem to have little to do with our lives, but at the end of this chapter we will see how precession may have been one of the causes of the ice ages. Before we can think about ice ages, however, we must consider the orbital motion of the earth around the sun and the resulting apparent motion of the sun.

2.3 THE MOTION OF THE SUN

Everything in the sky is moving. The sun, moon, planets, and even the stars move along their various orbits. Because the stars are so distant, their motion is not obvious to us even over decades, but the sun, moon, and planets are closer and move noticeably against the background of more distant stars. In this section we will discuss the motion of the sun.

THE ECLIPTIC

The daily rising and setting of the sun, its diurnal motion, is caused by the eastward rotation* of the earth. But the sun has a second motion, a slow motion against the background of stars that is caused by the motion of the earth along its orbit.

To see how the motion of the earth could cause an apparent motion of the sun, imagine that we are riding on a raft, drifting smoothly around an island. As we begin, we would see the island against a background of more distant islands, but, as we drift, we would see the island from different directions against different backgrounds. If we did not know that our raft was moving, we would imagine that the island was moving around us. This is precisely what happens when we observe the sun from the moving earth. The earth moves so smoothly along its orbit that we feel motionless, and it appears that the sun moves around the sky.

In January we see the sun in the direction of the constellation Sagittarius (Figure 2-11). We can't see the stars of the constellation, of course, because the sun is too bright, but we can observe that the sun is located in that part of the sky merely by noting the time of sunset

*Astronomers distinguish between the words *revolve* and *rotate*. The Earth revolves around the sun, but rotates on its axis.

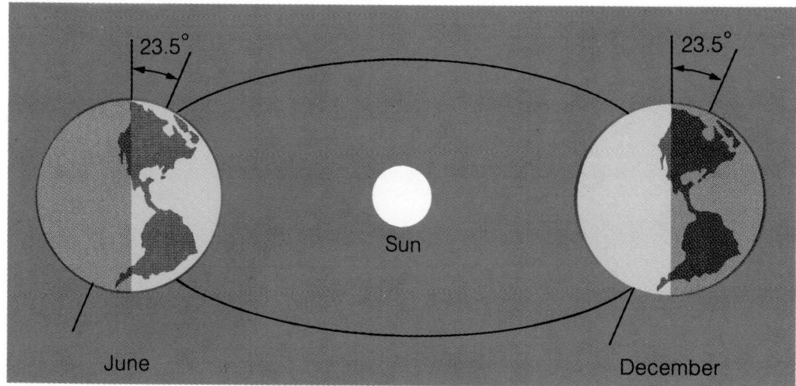

FIGURE 2–12 Spinning like a top, the earth holds its axis fixed as it orbits the sun. The earth's northern hemisphere is tipped toward the sun in June and away in December.

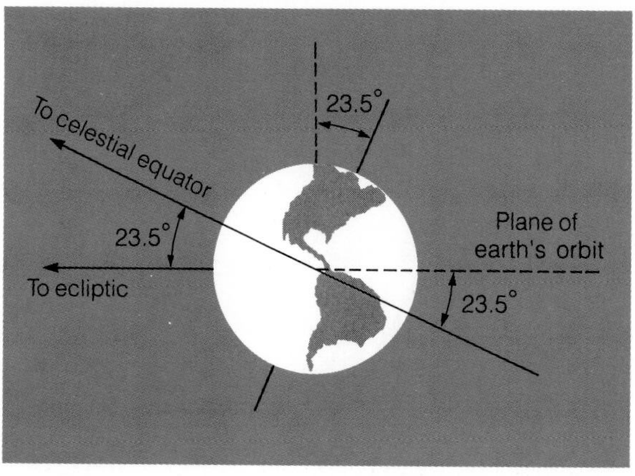

FIGURE 2–13 Because the earth's equator is tipped 23.5° to its orbit, the ecliptic is tipped 23.5° to the celestial equator.

and the constellations in the evening sky. As the earth moves through space, we observe the sun from a different part of earth's orbit, and the sun appears to be in Capricornus in February. Thus, as the earth moves along its orbit, the sun seems to move eastward through the constellations, taking a year to circle the sky one time.

The apparent path of the sun around the sky is called the **ecliptic**. Because the apparent motion of the sun is due to the orbital motion of the earth, it is easy to see that the ecliptic is the projection of the earth's orbit on the sky. If the celestial sphere were a great screen illuminated by the sun at the center, the shadow cast by earth's orbit would be the ecliptic (Figure 2-11).

Because of the rotation of the earth, this slow eastward motion of the sun is not easy to visualize. Since the earth spins on its axis once each day, we see the sun, stars, moon, and planets rise in the east and set in the west. While this daily motion is taking place, the sun is moving slowly eastward along the ecliptic about 1° per day, which is about twice its own angular diameter.

An additional complication is that the earth's axis is not perpendicular to the plane of its orbit. Its axis of rotation is tipped 23.5° from perpendicular. The spinning earth, like a spinning top, holds its axis fixed in space as it moves around the sun (Figure 2-12). (Precession alters this only very slowly.)

Recall that the celestial equator is the projection of the earth's equator, and that the ecliptic is the projection of the earth's orbit. Because the earth is tipped 23.5°, its equator is tipped 23.5° from the plane of Earth's orbit. When we project this on the sky, we find that the ecliptic and celestial equator meet at an angle of 23.5° (Figure 2-13).

This angle is very important to us who live on Earth. Because the ecliptic is tipped with respect to the celestial equator, the earth passes through a yearly cycle of seasons.

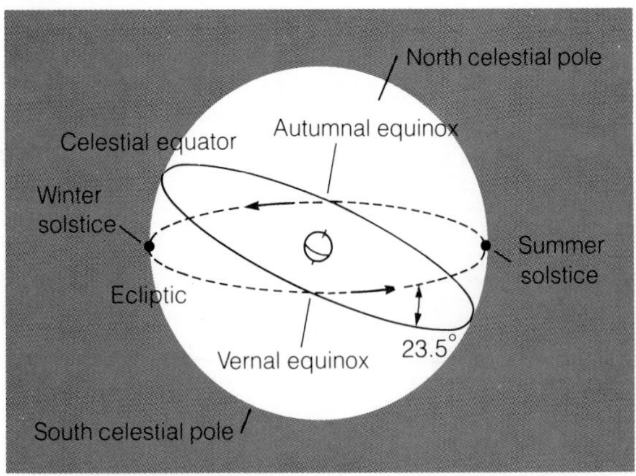

FIGURE 2–14 The ecliptic (dashed line), the sun's apparent path through the sky, crosses the celestial equator at the equinoxes. The solstices mark the most northerly and most southerly points.

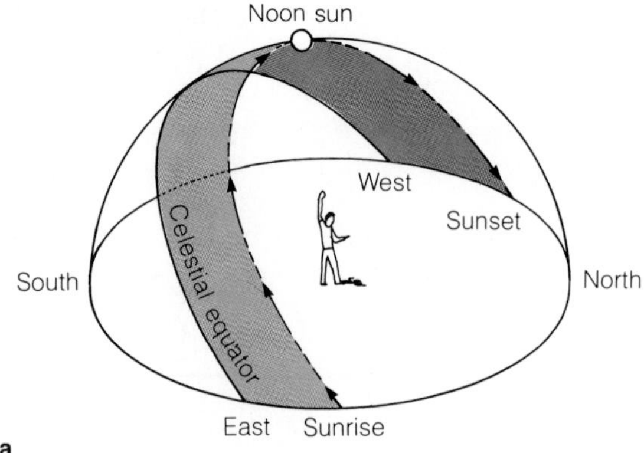

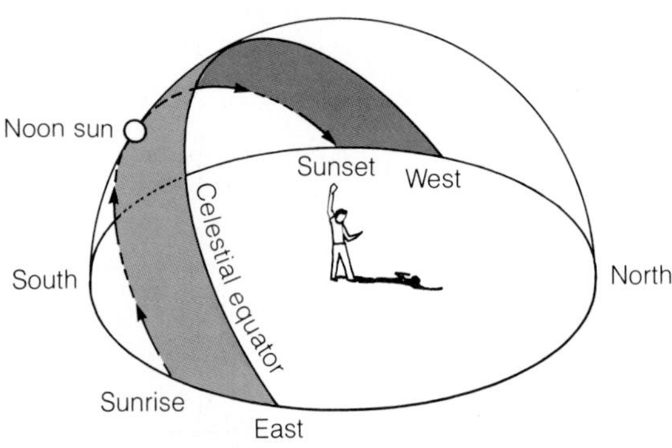

FIGURE 2–15 (a) The path of the sun across the sky at the summer solstice. ((b) The sun's path at the winter solstice.

THE SEASONS

The seasonal temperature depends on the amount of heat we receive from the sun. For the temperature of North America to remain constant, there must be a balance between the amount of heat we gain and the amount we radiate to space. If we receive more heat than we lose, we get warmer; if we lose more than we gain, we get cooler.

The motion of the sun around the ecliptic tips the heat balance one way in summer and the opposite way in winter. Because the ecliptic is inclined with respect to the celestial equator, the sun spends half the year in the northern celestial hemisphere and half the year in the southern celestial hemisphere (Figure 2-14). When the sun is in the northern celestial hemisphere, the northern half of the earth receives more direct sunlight—and therefore more heat—than the southern half. This makes North America, Europe, and Asia warmer.

The seasons are reversed in the southern half of the earth (see Figure 2-12). While the sun is in the northern celestial hemisphere warming North America, South America becomes cooler. Southern Chile has warm weather on New Year's Day and cold in July.

Four locations along the ecliptic help define the beginnings of the seasons (Figure 2-14). The **vernal equinox** is the place in the sky where the sun crosses the celestial equator moving northward. "Vernal," like "verdant" and "Vermont," comes from the word for green. "Equinox" comes from the word for equal and refers to the fact that we have equal amounts of daylight and darkness when the sun is at the equinox. The **summer solstice** is the point on the ecliptic where the sun is farthest north. "Solstice" comes from words that mean "sun stationary" and refers to the fact that the sun pauses there in its northward travel before beginning to move south. The **autumnal equinox** is the point where the sun crosses the equator moving southward, and the **winter solstice** is the place on the sky where the sun is farthest south.

According to our modern calendars, the seasons begin at the moment the sun crosses these four places on the ecliptic. The sun crosses the vernal equinox on or about March 21, and we say spring has begun. Summer begins about June 22 when the sun reaches the summer solstice, and autumn begins about September 22 when the sun crosses the autumnal equinox. Winter begins about December 22 when the sun reaches the winter solstice. (These dates vary slightly because of leap year and other factors.)

To see how we can get more heat from the summer sun, think about the path the sun takes across the sky between sunrise and sunset. Figure 2-15 shows these paths when the sun is at the summer solstice and at the winter solstice as seen by a person living at latitude 40°, an average latitude for most of the United States. Notice in Figure 2-15a that at the summer solstice the sun rises in the northeast, moves high across the sky, and sets in the northwest. But at the winter solstice, Figure 2-15b, the sun rises in the southeast, moves low across the sky, and sets in the southwest. Two features of these paths tip the heat balance.

First, the summer sun is above the horizon for more hours of each day than the winter sun. Summer days are long, and winter days are short. Because the sun is above our horizon longer in summer, we receive more energy each day.

Second, the sun stands high in the sky at noon on a summer day. It shines almost straight down, as shown by our small shadows. On a winter day, however, the noon sun is low in the southern sky. The ground gains less heat from the winter sun because the sunlight strikes the ground at an angle and spreads out, as shown by our longer shadows. These two effects work together to tip the heat balance and produce the seasons.

Notice that the seasons are not caused by changes in the distance from the earth to the sun. Earth's orbit is slightly elliptical, but the total variation in the earth–sun distance is only about 2 percent. The earth is actually at **perihelion** (closest to the sun) in the first week of January, when it is winter in the Earth's northern hemisphere. The earth passes **aphelion** (farthest from the sun) in early July, when it is summer in the northern hemisphere. Thus spring begins, not because the earth is drawing closer to the sun, but because the sun is moving into the northern half of the sky.

Of course, the weather does not turn warm the instant spring begins. The ground, air, and oceans are still cool from winter, and they take a while to warm up. Likewise, in the fall the earth slowly releases the heat it has stored through the summer. Because of this thermal lag, the average daily temperatures lag behind the solstices by about 1 month. Although the sun crosses the summer solstice on about June 22, the hottest months are July and August. The coldest months are January and February, even though the sun passes the winter solstice earlier, about December 22.

In ancient times the solstices and equinoxes were celebrated with rituals and festivals. Shakespeare's play *A Midsummer Night's Dream* describes the enchantment of the summer solstice night. (In Shakespeare's time, the equinoxes and solstices were taken to mark the midpoint of the seasons.) Many North American Indians marked the summer solstice with ceremonies and dances. Early church officials placed Christmas in late December to coincide with an earlier pagan celebration of the winter solstice. Some writers have speculated that the secular popularity of Easter, Thanksgiving, and Christmas stems from their proximity to the solstices and equinoxes.

2.4 THE MOTION OF THE PLANETS

The ecliptic is not only important to our daily lives because of its connection with the seasons but also because it is the path followed by the moon and planets. In this section we will discuss the way the planets parade along the ecliptic, and we will save the moon's motions for the next chapter.

THE MOVING PLANETS

Most of the planets of our solar system are visible to the unaided eye, though they produce no light of their own. We see them by reflected sunlight. Mercury, Venus, Mars, Jupiter, and Saturn are all visible to the naked eye, but Uranus is usually too faint to be seen and Neptune is never bright enough. Pluto is even fainter, and we need a large telescope to find it. (See the supplement "Observing the Sky" at the back of this book for a simple method of locating the planets.)

All the planets of our solar system move in nearly circular orbits around the sun. If we were looking down on the solar system from the north celestial pole, we would see the planets moving in the same counterclockwise direction around their orbits (Chapter 1). The farther from the sun, the more slowly the planets move.

When we look for planets in the sky, we always find them near the ecliptic because their orbits lie in nearly the same plane as the orbit of the earth. As they orbit the sun, they appear to move generally eastward along the ecliptic. In fact, the word *planet* comes from the Greek word meaning "wanderer." Mars moves completely around the ecliptic in slightly less than 2 years, but Saturn, being farther from the sun, takes nearly 30 years.

As seen from Earth, Venus and Mercury can never move far from the sun because their orbits are inside Earth's orbit. They sometimes appear near the western horizon just after sunset or near the eastern horizon just before sunrise. Venus is easier to locate because its larger orbit carries it higher above the horizon than Mercury (Figure 2-16). Mercury's orbit is so small that it can never get far from the sun. Consequently, it is usually hard to see against the sun's glare, and is often hidden in the clouds and haze near the horizon. On the other hand, at certain times when it is farthest from the sun, Mercury shines brightly and can be located near the horizon in the evening or morning sky. (See the supplement "Observing the Sky" at the back of this book.)

By tradition, any planet visible in the evening sky is called an **evening star**, although planets are not stars. Any planet visible in the sky shortly before sunrise is a **morning star**. Perhaps the most beautiful is Venus, which can become as bright as minus fourth magnitude. As Venus moves around its orbit, it can dominate the western sky each evening for many weeks, but eventually its orbit carries it back toward the sun, and it is lost in the haze near the horizon. In a few weeks it reappears in the dawn sky, a brilliant morning star.

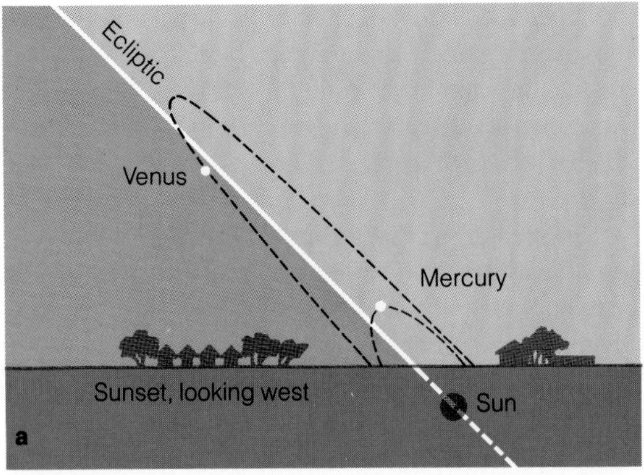

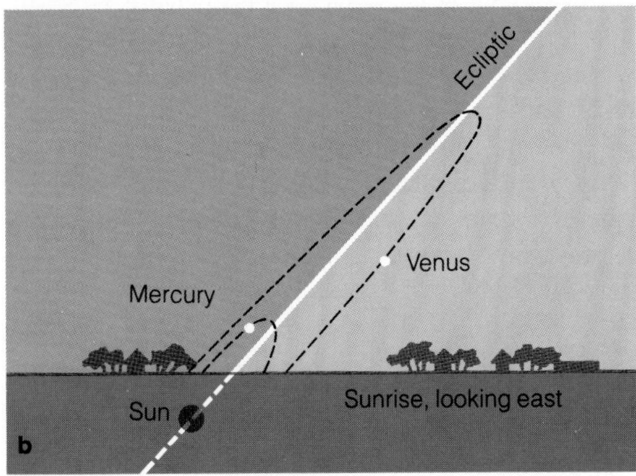

FIGURE 2–16 The orbits of Venus and Mercury sometimes carry them far enough from the sun to be visible (a) in the evening sky soon after sunset or (b) in the morning just before sunrise.

ASTROLOGY

The stately motions of the visible planets along the ecliptic have become a part of our culture through astrology, a superstition that originated in Babylonian religion roughly 1000 BC. The positions of the planets supposedly influenced the fate of kings and great nations. When these beliefs were adopted by the Greeks, they were extended to all individuals, each person's destiny being determined by the positions of the planets along the ecliptic at the moment of the person's birth.

Because the planets do not move exactly along the ecliptic, ancient astrologers defined a **zodiac**, a band 18° wide and centered on the ecliptic (Figure 2-17). They divided this band into 12 signs, each 30° across and named for one of the constellations on the ecliptic. The first sign, Aries, begins at the vernal equinox and runs eastward one-twelfth of the way around the sky, where it adjoins the next sign, Taurus, and so on.

A **horoscope** is a chart that shows the location of the heavenly bodies among these zodiacal signs at the moment of a person's birth. An individual's astrological sign is determined by the sign occupied by the sun at the moment of birth. The exact relationships of the planets to the signs, the horizon, and to each other are also important to the astrologer. A horoscope that properly shows all of this celestial information must be quite complex; it must incorporate the exact time, latitude, and longitude of the person's birth. Given this complexity, it is obvious that the horoscopes printed in newspapers and magazines cannot be based on real computations of planetary positions.

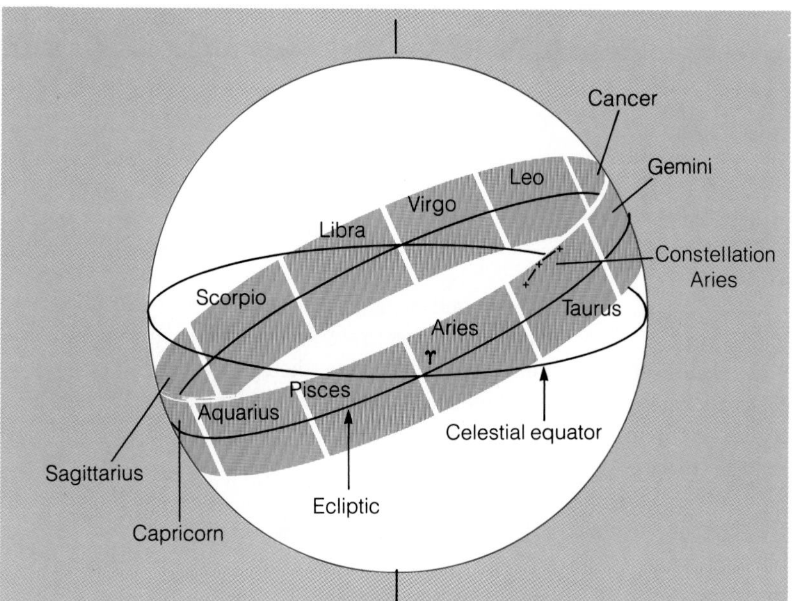

FIGURE 2–17 The signs of the zodiac. The zodiac is a band 18° wide and centered on the ecliptic. It is divided into 12 equal signs, and the positions of the sun, moon, and planets among these signs are the basis of astrology. The stars of the constellation Aries are plotted to show how precession has moved the stars of the constellations one entire sign along the zodiac. There is no evidence that astrological predictions are valid, and the subject is considered a superstition.

Other problems further undermine the validity of horoscopes. For instance, precession shifts the vernal equinox westward by about one full constellation every 2100 years, and this shift moves the astrological signs against the background of the constellations. When astrology began more than 3000 years ago, the vernal equinox was in the constellation Aries, but about the time of Christ, it entered the constellation Pisces. (Note the location of the constellation Aries in Figure 2-17.) Thus the signs no longer match the constellations, and a person born under the sign of Cancer, for instance, began life when the sun was in the constellation Gemini. About AD 2600, the vernal equinox will enter the constellation Aquarius, heralding the "age of Aquarius," and throwing the signs of the zodiac two full constellations out of place.

The interpretation of the horoscope results in predictions that are notoriously vague, easy to accept in any situation, but difficult to test. Nevertheless, hundreds of tests of astrology have been made. Astrologers, astrology magazines, and tabloid newspapers have touted tests showing that astrology works, but when the same tests are repeated using rigidly unbiased testing procedures, astrology fails the test. One of the best and fairest tests was conducted by a team of scientists at the University of California at Berkeley working with the assistance and approval of professional astrologers. This experiment showed to high statistical significance that even professional astrologers could not match a subject's personality with the subject's astrological interpretation.

Astronomers view astrology with irritation, embarrassment, and a special kind of respect. Astrology irritates astronomers because many people confuse astronomy with astrology and expect astronomers to help interpret horoscopes. Astronomers sometimes express embarrassment for people who still believe in astrology, but at the same time, astronomers respect the historical importance of astrology. The first astronomers were astrologers, and the science and superstition parted company only a few hundred years ago. Perhaps it is best to treat astrology as an interesting part of human history, the seed from which modern astronomy grew, but of no more practical significance than the advice in a fortune cookie.

PERSPECTIVE

CLIMATE AND ICE AGE

EARTH'S CLIMATE

Weather is what happens today, but climate is the average of what happens over tens of years. Occasional hot summers, floods, droughts, and other variations in weather are usually only random, short-term changes, whereas changes in climate are slow.

Some climatic changes are caused by humans. We release gases into the atmosphere and change its properties, causing a slow warming called the greenhouse effect. Some waste gases are attacking the ozone layer in our atmosphere. We will discuss these problems in Chapter 21 when we discuss the planets and their atmospheres, but here we are interested in periodic, long-term changes in earth's climate—the ice ages.

LONG-TERM VARIATION IN CLIMATE

The earth has gone through periods called ice ages, when the worldwide climate was cooler and dryer and thick layers of ice covered the higher latitudes. The earliest known ice age occurred roughly 570 million years ago, and the next occurred about 280 million years ago. The latest began only 3 million years ago and may not have ended yet. That is, we may even now be in an ice age. Because dating these periods is difficult, the timing of the ice ages is uncertain. Nevertheless, some earth scientists believe they occur about every 250 million years.

From plant and animal fossils, paleontologists conclude that the normal climate between ice ages is about 10°C warmer than it is today. Tropical forests extended about 1100 km (700 miles) farther north than they do today, and the temperature forests of North America reached up to 2200 km (1400 miles) farther north. In addition, the chemical composition of fossil shells suggests that in the past the oceans were warmer. The most recent variation began roughly 45 million years ago, when the earth's climate began to cool, and culminated in the beginning of an ice age a few million years ago.

GLACIAL PERIODS

During an ice age, water freezes to form large polar caps, and the atmosphere becomes cooler and dryer. Ice sheets may cover as much as 30 percent of the land, but the remainder of the earth remains ice free.

In the course of an ice age, the ice alternately advances and melts back. A **glacial period** is an interval when the ice sheets engulf huge areas of the land; an **interglacial period** is the time when the ice sheets melt back. Glacial and interglacial periods alternate during an ice age, in a cycle lasting roughly 40,000 years.

We are living in an interglacial period. Geological evidence, such as large boulders transported hundreds of miles from their place of origin, deposits of rock debris washed off of glaciers, and striations in rock surfaces cut by moving ice all show that Canada and much of the northern United States were repeatedly coverd by sheets of ice up to 2 km thick (Figure 2-18). The last of these ice sheets melted about 20,000 years ago. If the advance of the glaciers is really periodic, and if the current ice age is not coming to an end, we have another 20,000 years before the ice begins its next crushing advance.

The advance and retreat of the glaciers seems to depend on how warm the summers are in the northern half of the earth. If the summers are not warm enough, the previous winter's snow and ice fail to melt completely and accumulate year after year, building into advancing glaciers. However, if the summers are

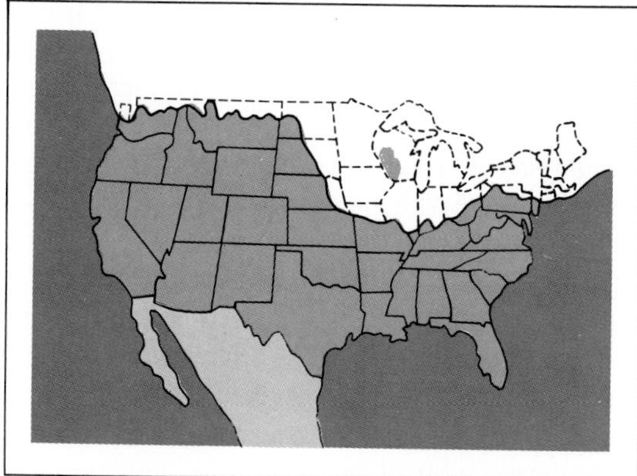

FIGURE 2–18 Only 20,000 years ago much of the northern United States was covered by glacier ice. The solid line indicates the farthest advance of ice.

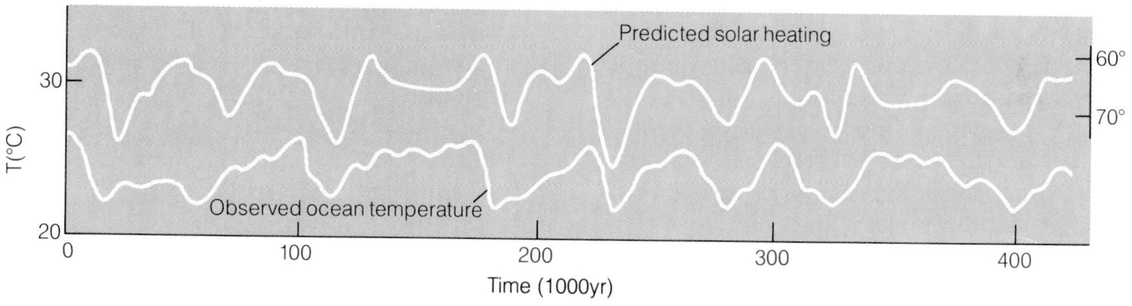

FIGURE 2–19 The Milankovitch theory predicts periodic changes in solar heating (shown here as the equivalent summer latitude of the sun). Over the last 400,000 years these changes seem to vary in step with ocean temperatures measured from fossils in sediment layers. (Adapted from Cesare Emiliani.)

warm enough to melt the previous winter's snow and ice, the ice sheets do not grow.

The northern half of the earth is more important to the worldwide climate than the southern half, because most of the land mass is north of the Earth's equator. Cooling or warming this land mass directly affects the ice sheets and the global climate.

In 1920, the Yugoslavian meteorologist, Milutin Milankovitch, proposed that small changes in the rotation and revolution of the earth affect the heat balance and modify the world climate. This theory was long disregarded by earth scientists. Then, in the 1960s, studies of ocean sediment revealed that earth's past climate has varied in step with the predictions of the Milankovitch theory (Figure 2-19). The theory is now widely accepted among earth scientists as the primary cause of long-term changes in the earth's climate.

THE CHANGING SHAPE OF THE EARTH'S ORBIT

Because the earth's orbit is ellipitcal, its distance from the sun varies slightly. Its average distance of 1 astronomical unit equals 1.5×10^8 km (93×10^6 miles), but in the first days of January each year it passes perihelion, the point in its orbit where it is closest to the sun. It is then about 2 percent closer to the sun than average, but the slight increase of heat received from the sun is wiped out by the seasonal temperature var-

Perspective *(continued)*

iation from the inclination of the earth's, even though the earth is slightly close to the sun. In the same way, the earth passes aphelion, the farthest point from the sun, in early July.

By studying the motion of the earth and the other planets, astronomers have discovered that the shape of the earth's orbit changes with a period of about 93,000 years. When the orbit gets more elliptical the variation in the distance from earth to sun is more than 2 percent and may be enough to make our winters milder and the summers cooler. If the summers in the northern half of the earth are cooler, ice and snow accumulate and glaciers advance. When the orbit becomes more circular, summers are warmer and warmer and ice sheets melt back.

This sounds like a good theory, but glacial periods occur every 40,000 years not every 93,000 years. There must be more to the ice ages than just variation in the shape of the earth's orbit.

PRECESSION

When we discussed the rotation of the earth, we saw that the earth's axis precesses like a spinning top. This precession is slowly changing the seasons. The earth is now tipped toward the sun in June, producing summer in the northern half of the earth. But in 13,000 years precession will have tilted the earth the other way, away from the sun in June, producing winter in the northern half of the planet. Of course, we will adjust our calendar to move the months with the seasons, keeping June a summer month. The important point is not the month but the place in the earth's orbit where winter occurs.

It happens that winter in the northern half of the earth now occurs when the earth is near perihelion. Because it is 2 percent closer to the sun, our winters are very slightly less severe than they would be if the earth's orbit were circular. However, in 13,000 years precession will have moved winter to the other side of the earth's orbit, and we will be 2 percent farther from the sun in winter, and it will be slightly colder. Summer will occur near perihelion where we will be 2 percent closer to the sun, and summers will be warmer. If summer in the earth's northern half is warmer, the ice should melt faster, and the glaciers should recede. Thus, we might expect the earth's climate to change with the same period as precession—about 26,000 years.

It is not that simple, however. The effects of precession combine with the effects of the changing shape of the earth's orbit. If the orbit is almost circular, then it doesn't matter when perihelion occurs because the earth won't be significantly closer to the sun. But if the orbit is more elliptical, the time of perihelion is important. The fact that the two variations work together with different periods complicates the problem. A third variation in the motion of the earth adds another complication.

THE INCLINATION OF THE EARTH'S AXIS

Not only can astronomers study the changing shape of the earth's orbit and its axial precession, they can also study its changing inclination. The earth has not always been tipped at an angle of 23.5°. The inclination varies with a period of about 41,000 years from 22° to 28°. Because the ecliptic is the projection of the earth's orbit onto the sky, a change in the earth's axial tilt changes the inclination of the ecliptic to the celestial equator (see Figure 2-13). If this inclination becomes smaller, the sun does not travel as far north in the sky during summer, producing cooler summers and favoring the accumulation of ice sheets.

It is tempting to point to the 40,000-year glacial cycle and identify it with the 41,000-year variation in the earth's inclination, but we would probably be wrong to do so. We must remember that there are at least three factors working to change the climate, each with a different period. When the three processes work together to produce cool summers, ice sheets may accumulate, producing a glacial period. But the pattern of the advance and retreat of the glaciers is very complex because of the three effects at work.

SUMMARY

Astronomers divide the sky into 88 areas called constellations. Although the constellations originated in Greek mythology, the names are Latin. Even the modern constellations, added to fill in between the ancient figures, have Latin names. The names of stars usually come from ancient Arabic, though modern astronomers often refer to a star by constellation and Greek letters assigned according to brightness within each constellation.

The magnitude system is the astronomer's brightness scale. First-magnitude stars are brighter than second-magnitude stars, which are brighter than third-magnitude stars, and so on. The magnitude we see when we look at a star in the sky is its apparent visual magnitude.

The celestial sphere is a model of the sky, carrying the celestial objects around the earth. Because the earth rotates eastward, the celestial sphere appears to rotate westward on its axis. The northern and southern celestial poles are the pivots on which the sky appears to rotate. The celestial equator, an imaginary line around the sky above the earth's equator, divides the sky in half.

Because the earth orbits the sun, the sun appears to move eastward around the sky following the ecliptic. Because the ecliptic is tipped 23.5° to the celestial equator, the sun spends half the year in the northern celestial hemisphere and half the year in the southern celestial hemisphere, producing the seasons. The seasons are reversed south of the earth's equator. That is, while the northern hemisphere experiences a warm season, the southern hemisphere experiences a cold season.

Of the nine planets in our solar system, Mercury, Venus, Mars, Jupiter, and Saturn are visible to the naked eye. Their orbital motion around the sun carries them along the zodiac, a band 18° wide and centered on the ecliptic. The positions of the sun, moon, and these five planets are the basis of astrology, an ancient superstition that originated in Babylonia about 1000 BC.

The motion of the earth may change in ways that can affect the climate. Changes in orbital shape, in precession, and in axial tilt can alter the planet's heat balance and may be responsible for the ice ages and glacial periods.

NEW TERMS

constellation
magnitude scale
apparent visual magnitude (m_V)
celestial sphere
diurnal motion
north and south celestial poles
circumpolar constellation
celestial equator
minute of arc
second of arc
precession
ecliptic
vernal equinox
summer solstice
autumnal equinox
winter solstice
perihelion
aphelion
morning and evening stars
zodiac
horoscope
glacial period
interglacial period

QUESTIONS

1. Why are most modern constellations composed of faint stars or located in the southern sky?
2. What does a star's Greek letter designation tell us that its ancient Arabic name does not?
3. From your knowledge of star names and constellations, which of the following pairs of stars is the brighter and which is the fainter? Explain your answers.
   a. α Ursae Majoris; θ Ursae Majoris
   b. λ Scorpii; β Pegasus.
   c. β Ursae Minoris; β Orionis
4. Give two reasons why the magnitude scale might be confusing.
5. Why do modern astronomers continue to use the celestial sphere when they know that stars are not all at the same distance?
6. How do we define the celestial poles and the celestial equator?
7. From what locations on earth is the north celestial pole not visible? The south celestial pole? The celestial equator?
8. If the earth did not turn on its axis, could we still define an ecliptic? Why or why not?
9. Give two reasons why winter days are colder than summer days.
10. What would our seasons be like on earth if the earth were tipped 35° instead of 23.5°? What would they be like if the earth's axis were perpendicular to its orbit?
11. How do the seasons in earth's southern hemisphere differ from those in the northern hemisphere?
12. Why don't the planets move exactly along the ecliptic?
13. Why can we be sure the astrological predictions printed in newspapers and magazines are not based on true calculations of the positions of celestial bodies in a horoscope?
14. Why should the eccentricity of the earth's orbit make winter in the earth's northern hemisphere different from winter in earth's southern hemisphere?
15. How might small changes in the inclination of earth's axis to the plane of its orbit affect the growth of glaciers?

PROBLEMS

1. If one star is 40 times brighter than another star, how many magnitudes brighter is it?
2. If two stars differ by 8.6 magnitudes, what is their brightness ratio?
3. If star A is fourth magnitude and star B is sixth magnitude, which is brighter? By what factor?
4. By what factor is the sun brighter than the full moon? (HINT: See Figure 2-5.)
5. Sketch the celestial sphere and label the poles, equator, and horizon.
6. Draw a diagram like Figure 2-15 and show the path of the sun across the sky at the time of the vernal equinox.

ACTIVITY: THE CYCLE OF SEASONS

Most of the human race lives on this planet unaware of the passing cycle of the seasons. Hidden in our cities, we notice the changing weather of the seasons but not the changing motion of the sun that causes the seasons. Yet it is quite possible to make simple observations of the sun that will reveal the cycle of the seasons.

Horizon Observations If you are an early riser, you may enjoy observing the rising point of the sun each morning, or you may prefer to note the setting point each evening. We will proceed as if you were observing sunsets, but the same method will work for watching sunrises.

Choose a window, balcony, rooftop, or hilltop to use as your regular observing position, and then make a sketch of the horizon from the southwest to the northwest. Make note of the position of reference marks such as distant mountain peaks, buildings, towers, and so on. Then, about once a week at sunset note the position of the setting sun on your diagram.

In just a few days you will notice that the sunset point moves along the horizon, southward in the fall and northward in the spring. Observe the sunset on the day of the equinox, September 22 or March 21, and the sun will be setting directly in the west. In the summer, you will see the sun setting somewhere along the northwestern horizon, and in the winter, it will set somewhere along the southwestern horizon.

If at all possible, continue to chart the sunset position through a solstice, December 22 or June 22. As a solstice approaches, you will notice that the motion of the sun along the horizon slows, and then seems to stop for a few days around the solstice. Then you will notice the sun reversing its motion as the solstice passes.

If you didn't know when the solstices and equinoxes were, you could find out from your observations. You could find the dates of the solstices (sun stationary) to within a day or two, and then you could count days halfway to the next solstice to find the equinox. You could even discover the length of the year using this method. In fact, many different cultures did exactly that. American Indians, ancient Chinese, Egyptians, the Maya, and the builders of Stonehenge, to name a few, used horizon observations to follow the cycle of the sun.

RECOMMENDED READING

ALLEN, RICHARD HINCKLEY *Star Names: Their Lore and Meaning.* New York: Dover, 1963.

BARNARD, MARY *Time and the White Tigress.* Portland, Oregon: Breitenbush Books, 1986.

BOK, BART J., and LAWRENCE JEROME *Objections to Astrology.* Buffalo, N.Y.: Prometheus Books, 1975.

CARLSON, SHAWN "A Double Blind Test of Astrology." *Nature 318* (Dec. 5, 1985), p. 419.

EVANS, D. L., and H. J. FREELAND "Variations in the Earth's Orbit: Pacemaker of the Ice Ages?" *Science 198* (Nov. 4, 1977), p. 528.

GINGERICH, OWEN "Ancient Egyptian Sky Magic." *Sky and Telescope 65* (May 1983), p. 418.

GOLDBERG, S. "Is Astrology Science?" *The Humanist 39* (March 1979), p. 9.

GRIBBIN, J. "Why Does Earth's Climate Change?" *Astronomy 6* (Feb. 1978), p. 18.

HAYS, D. D., J. IMBRIE, and N. J. SHACKLETON "Variations in the Earth's Orbit: Pacemaker of the Ice Ages." *Science 194* (Dec. 10, 1976), p. 1121.

HOLZINGER, J. R., and M. A. SEEDS *Laboratory Exercises in Astronomy.* Ex. 5, 17, 23, and Appendix A. New York: Macmillan, 1976.

IMBRIE, J., and K. P. IMBRIE *Solving the Mystery.* Short Hills, N.J.: Enslow, 1979.

KELLY, IVAN "The Scientific Case Against Astrology." *Mercury 9* (Nov./Dec. 1980), p. 135.

KIDWELL, PEGGY ALDRICH "Elijah Burritt and the 'Geography of the Heavens.'" *Sky and Telescope 69* (Jan. 1985), p. 26.

KUNITZSCH, PAUL "How We Got Our 'Arabic' Star Names." *Sky and Telescope 65* (Jan. 1983), p. 20.

Mag 6 Star Atlas. Barrington, N.J.: Edmund Scientific, 1982.

MECHLER, GARY, CYNDI McDANIEL, and STEVEN MULLOY "Response to the National Enquirer Astrology Study." *The Skeptical Enquirer* (Winter 1980–81), p. 34.

MENZEL, D. H. *A Field Guide to the Stars and Planets.* Boston: Houghton Mifflin, 1964.

NORTON, A. P. *A Star Atlas.* Cambridge, Mass.: Sky Publishing, 1964.

Seasonal Star Charts. Northbrook, Ill.: Hubbard Press, 1972.

WARNER, DEBORAH JEAN "Blaeu's Failed Celestial Globe." *Sky and Telescope* 69 (April 1985), p. 294.

WHITNEY, CHARLES A. *Whitney's Star Finder.* New York: Alfred A. Knopf, 1980.

WILLIAMSON, ROY A. *Living the Sky: The Cosmos of the American Indian.* Boston: Houghton Mifflin, 1984.

CHAPTER 3

LUNAR PHASES, TIDES, AND ECLIPSES

Even the best of men
May turn to a wolf at night
When the wolfbane blooms
And the moon is bright.

Proverb
from old Wolfman movies

The moon has tremendous fascination for us. It is the brightest object in the night sky, and its monthly cycle of phases has served lovers and predators since the first humans left the forests to live in the grasslands. Human culture is filled with traditions and superstitions connected with the moon.

Some people still believe that moonlight causes insanity. "Don't stare at the moon. You'll go crazy," more than one child has been warned.* The word "lunatic" comes from a time when even doctors thought that the insane were "moonstruck." A "mooncalf" is someone who has been crazy since birth, and the word is probably also related to the belief that moonlight can harm unborn children.

You have probably heard that people act less rationally when the moon is bright. Nurses, police, and teachers will sometimes make that claim, but careful studies of hospital and police records show that there is no real relation between the moon and erratic behavior. The moon is so impressive that we *expect* it to influence us.

In fact, the moon does have an influence on the earth. The gravitational attraction of the moon causes the ocean tides and slows the earth's rotation. But that force is much too small to affect humans.

*When I was very small, my grandmother told me if I gazed at the moon, I might go crazy. But it was too beautiful, and I ignored her warning. I secretly watched the moon from my window, became fascinated by the sky; and became an astronomer.

FIGURE 3–1 (a) A twelfth-century Mayan symbol believed to represent a solar eclipse. The black-and-white sun symbol hangs from a rectangular sky symbol while a voracious serpent approaches from below. (b) Chinese representation of a solar eclipse as a dragon devouring the sun. (c) A wall carving from the ruins of a temple at Vijayanagara in southern India symbolizes a solar eclipse as two snakes approaching the disk of the sun. (b, Yerkes Observatory photograph; c, T. Scott Smith.)

If the moon has any effect on us, it is only because lunar phenomena are so dramatic. The most dramatic of those phenomena are eclipses, those events that turn the full moon red or bring sudden darkness on a sunny day.

The ancient Chinese greeted solar eclipses by using noisemakers and by shooting arrows toward the heavens. The noise and arrows were intended to scare off the great dragon that was slowly devouring the sun (Figure 3–1). The ceremony never failed, and the dragon always retreated.

One story tells of Hsi and Ho, two Chinese astronomers who got drunk and either failed to predict the solar eclipse of October 22, 2137 BC, or were unable to conduct the proper ceremonies to scare away the dragon. Once the emperor recovered from the terror of the eclipse, he had the unfortunate astronomers beheaded.

This chapter is about lunar phenomena—phases, tides, and eclipses. We will try to put aside the ancient associations with evil and enjoy the earth's good luck in having such a beautiful satellite like the moon.

3.1 THE PHASES OF THE MOON

THE MOON'S ORBIT

The moon orbits around the earth at an average distance of 384,401 km. Its orbit is slightly elliptical so its distance can vary by about 6 percent. Seen from a point far above the earth's North Pole, the moon orbits counterclockwise (eastward) with a period of 27.321661 days. This is called the **sidereal period** because it is referenced to the stars. The moon takes 27.321661 days to circle the sky once and return to the same place among the stars. We see this motion as the moon drifts eastward in the night sky. It moves over 13° in a day or about its own diameter (0.5°) in an hour.

In its eastward motion, the moon stays near the ecliptic. Its orbit is inclined by 5° 8′ 43″ to the plane of earth's orbit, so its path around the sky is inclined to the ecliptic by the same angle. It can never wander farther than 5° 8′ 43″ north or south of the ecliptic.

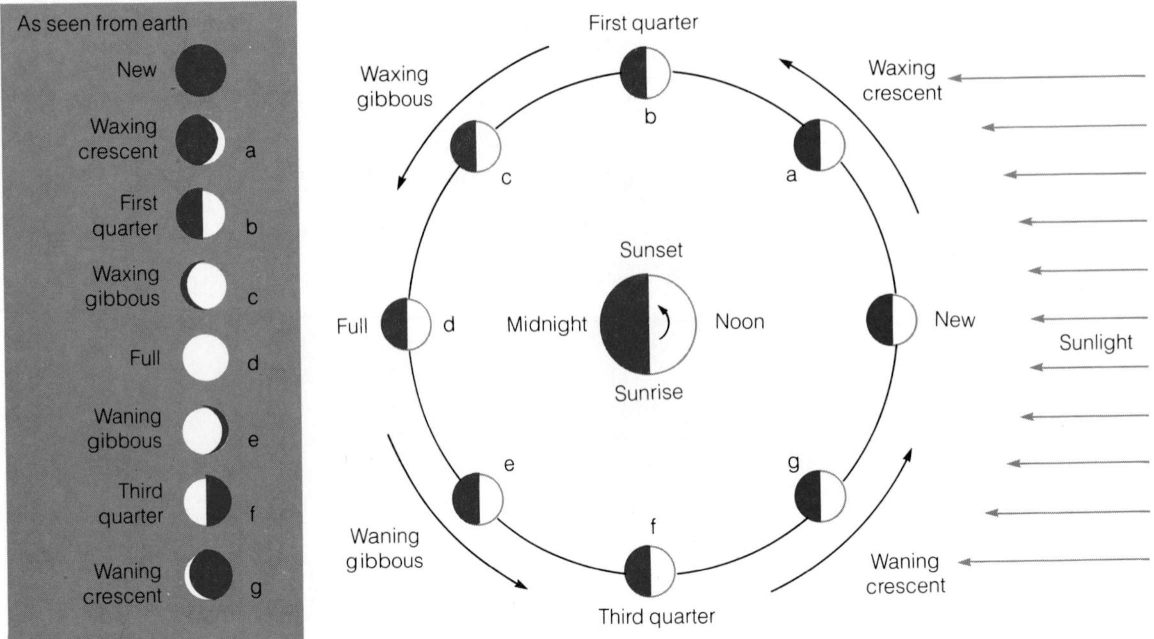

FIGURE 3–2 The phases of the moon are produced by the varying amounts of the illuminated surface we can see. The box shows the moon as it appears from the earth. Lettered phases a to g refer to Figure 3–3.

If we watch the moon for a few nights, we will notice its phases, its cycle of illuminated shapes. The moon shines by reflected sunlight, and as the moon orbits the earth, we see different portions of the sunlit side. Thus, the moon's shape appears to change (Figures 3–2 and 3–3). As it orbits the earth, it goes through a cycle of phases.

THE PHASE CYCLE

The first half of the lunar cycle extends from new moon to full moon. When the moon is approximately between the earth and sun, the side toward us is in darkness. The moon is invisible, and we refer to it as new moon. A few days after new moon, it has moved far enough along its orbit to allow the sun to illuminate a small sliver of the side toward us, and we see a thin crescent. Night by night this crescent moon waxes (grows), until we see half of the side toward us illuminated by sunlight and refer to it as first quarter. The moon continues to wax, becoming gibbous or protuberant. Gibbous comes from a Latin word meaning hunchbacked and is pronounced Gib'es—"Gib" as in Gibson and "es" as in estimate. When it is nearly opposite the sun, the side toward earth is fully illuminated, and we see a full moon.

The second half of the lunar cycle reverses the first half. After reaching full, the moon wanes (shrinks) through gibbous phase to third quarter, then through crescent to new moon. To distinguish between the gibbous and crescent phases of the first and second half of the cycle, we refer to gibbous waning and crescent waning when the moon is shrinking, and gibbous waxing and crescent waxing when it is growing.

This cycle of phases takes longer than the moon's orbital period. Imagine that we begin watching the moon when it is near the sun in the sky. The moon circles the sky in one sidereal period and returns to the same place among the stars, but the sun is not there. During the 27.321661 days the moon needed to circle the sky, the sun moved eastward along the ecliptic. The moon must travel a bit over 2 more days to catch up with the sun and complete the cycle of phases. Thus, the lunar phases repeat with a period of 29.5305882 days. This is called the **synodic period**—the period with respect to the sun. "Synodic" comes from the Greek word *sunodos*, which is a combination of words for "together" and "journey."

To summarize, let us follow the moon through one cycle of phases (Table 3–1). At new moon, the moon is nearly in line with the sun and sets in the west with the sun. Therefore, we see no moon at new moon. A few days

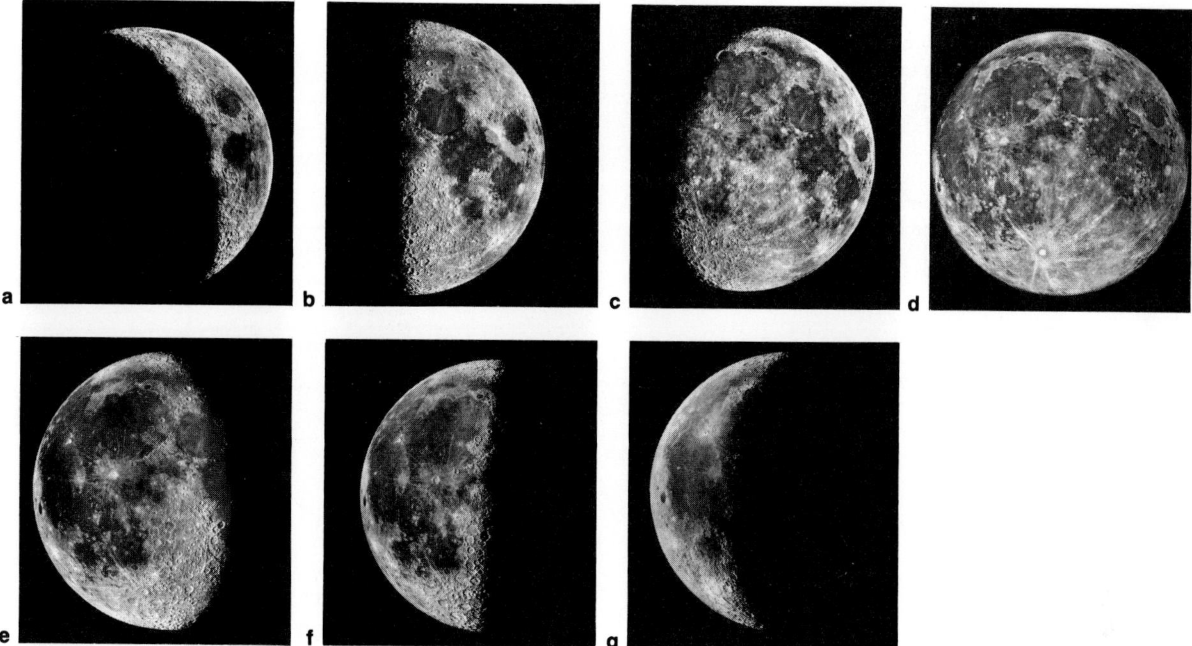

FIGURE 3–3 The lunar phases. Compare with Figure 3–2. (Lick Observatory photographs.)

Table 3–1 Times of moon rise and moon set.

Phase	Moon Rise	Moon Set
New	Dawn	Sunset
First quarter	Noon	Midnight
Full	Sunset	Dawn
Third quarter	Midnight	Noon

after new moon, we see the waxing crescent above the western horizon soon after sunset, and each evening it is fatter and higher above the horizon, until, about 1 week after new moon, it reaches first quarter and stands high in the southern sky at sunset. The first-quarter moon does not set until about midnight. In the days following first quarter, the moon waxes fatter, becoming gibbous waxing. Each evening we find it farther east among the stars and it sets later and later. About 2 weeks after new moon, the moon reaches full, rising in the east as the sun sets in the west. The full moon is visible all night long, setting in the west at sunrise.

The waning phases of the moon may be less familiar because the moon is not visible in the early evening sky. As it wanes through gibbous, it rises later and later. By the time it reaches third quarter, it does not rise until midnight. The waning crescent does not rise until even later, and if we wish to see the thin waning crescent just before new moon, we must get up before sunrise and look for the moon above the eastern horizon.

3.2 THE TIDES

THE CAUSE OF THE TIDES

The earth's gravity draws us downward with a force we refer to as our weight, but that is not the only force acting on us. The moon is less massive and is farther away than the center of the earth is, but its gravity measurably affects the earth. The side of the earth facing the moon is about 6400 km (4000 miles) closer to the moon than the center of the earth is, and the moon's gravity pulls on it more

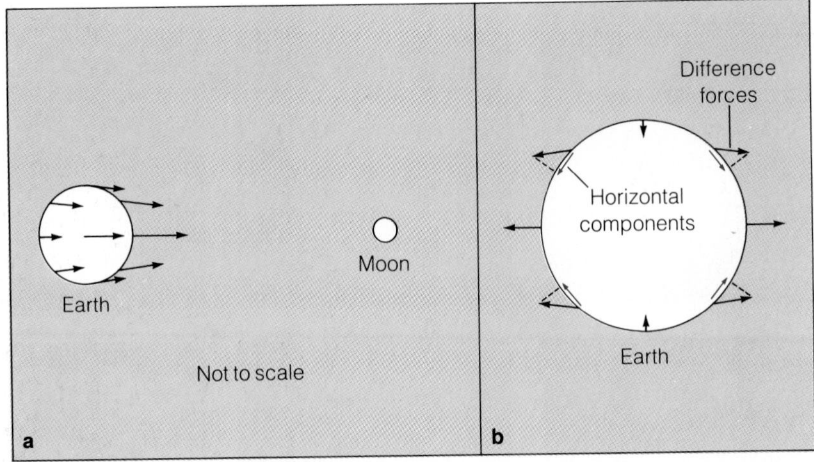

FIGURE 3–4 The moon's gravity does not act equally on all parts of the earth's surface (a). Subtracting the force acting on earth's center reveals small difference forces (b). The horizontal components of these forces make the ocean waters flow into tidal bulges on the near and far sides of the earth.

strongly than on the earth's center. We have tides because the moon's gravity does not pull equally strongly or in exactly the same direction on all parts of the earth.

The tides are caused by these differences between the forces acting on the surface of the earth and the force acting on its center. In Figure 3–4a we represent the forces acting on different parts of the earth's surface as arrows pointing toward the moon. When we subtract the force acting on the earth's center, we find small difference forces that are not all perpendicular to the earth's surface (Figure 3–4b). The horizontal components of these forces make the ocean waters flow into the tidal bulges.

Note that the ocean waters are not pulled up away from the earth. Rather the horizontal part of the difference forces of the moon's gravity makes the ocean waters flow over earth's surface into bulges on the near and far sides of the earth.

TIDAL EFFECTS

We can see dramatic evidence of tidal forces if we watch the ocean shore for a few hours. Although the earth rotates on its axis, the tidal bulges remain fixed along the earth–moon line. As the turning earth carries us into a tidal bulge, the ocean water deepens, and the tide crawls up the beach. Later when the earth carries us out of the bulge, the water becomes shallower, and the tide falls. Because there are two bulges on opposite sides of the earth, the tides rise and fall roughly twice a day, although the times change with the phases of the moon.

The sun, too, produces tidal bulges on the earth. At new moon and at full moon, the moon and sun produce tidal bulges that add together (Figure 3–5a) and produce extreme tidal changes; high tide is very high, and low tide is very low. Such tides are called **spring tides**, even though they occur at every new and full moon and not just in the spring. **Neap tides** occur at first- and third-quarter moon, when the moon and sun pull at right angles to each other (Figure 3–5b). Then the tides do not add together and are less extreme than usual.

Tidal forces can have surprising effects. The friction of the ocean waters with the seabeds slows the rotation of the earth by 0.001 second per day per century. Fossils of marine animals confirm that only 400 million years ago the earth's day was 22 hours long. In addition, the earth's gravitational field exerts tidal forces on the moon, and, although there are no bodies of water on the moon, friction within the flexing rock has slowed the moon's rotation to the point that is now keeps the same face toward the earth.

Tidal forces can also affect orbital motion. Friction with the rotating earth drags the tidal bulges eastward out of a direct earth–moon line (Figure 3–6). These tidal bulges contain a large amount of mass, and their gravitational field pulls the moon forward in its orbit. As a result, the moon's orbit is growing larger, and it is receding from the earth at about 3 cm per year, an effect that astronomers can measure by bouncing laser beams off reflectors left on the lunar surface by the Apollo astronauts.

These and other tidal effects are important in many areas of astronomy. In later chapters we will see how tidal forces can pull gas away from stars, rip galaxies apart, and

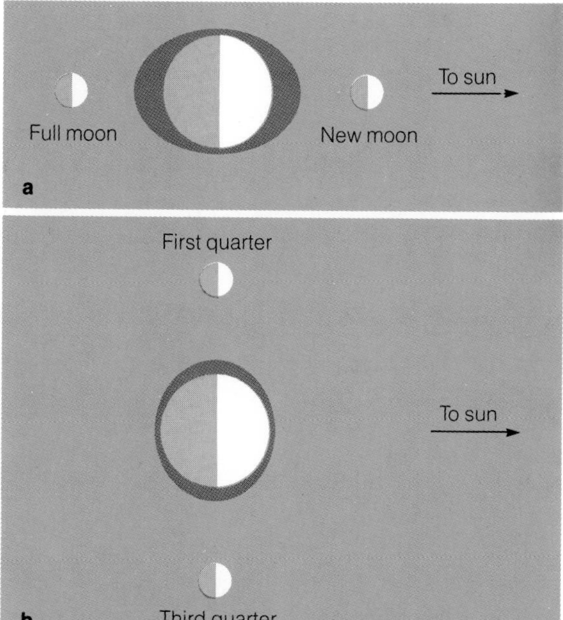

FIGURE 3–5 (a) When the moon and the sun pull in the same direction, their tidal forces add, and the tidal bulges are larger. Thus, spring tides occur at new moon and full moon. (b) When the moon and sun pull at right angles, their tidal forces do not add, and the tidal bulges are smaller. Such neap tides occur at first- and third-quarter moon.

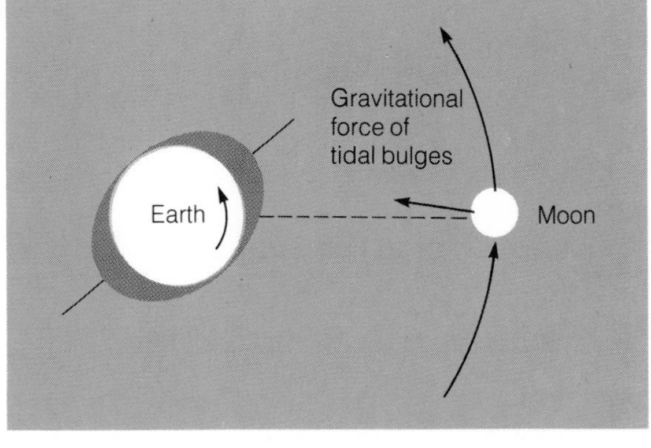

FIGURE 3–6 The rotation of the earth drags the tidal bulges ahead of the earth–moon line (exaggerated here). The gravitational attraction of these masses of water pulls the moon forward in its orbit, forcing its orbit to grow in size.

melt the interiors of satellites orbiting near massive planets. For now, however, we must consider yet another kind of lunar phenomena—eclipses.

3.3 LUNAR ECLIPSES

A lunar eclipse occurs at full moon when the moon moves through the shadow of the earth. Because the moon shines only by reflected sunlight, we see the moon gradually darken as it enters the shadow.

EARTH'S SHADOW

The earth's shadow consists of two parts. The **umbra** is the region of total shadow. If we were floating in space in the umbra of the earth's shadow, we would see no portion of the sun. However, if we moved into the **penumbra**, we would be in partial shadow and would see part of the sun peeking around the edge of the earth. In the penumbra the sunlight is dimmed but not extinguished.

We can construct a model of this by pressing a map tack into the eraser of a pencil and holding the tack between a light bulb a few feet away and a white cardboard screen (Figure 3–7). The light bulb represents the sun, and the map tack represents the earth. When we hold the screen close to the tack, we see that the umbra is nearly as large as the tack and that the penumbra is only slightly larger. However, as we move the screen away from the tack, the umbra shrinks and the penumbra expands. Beyond a certain point, the shadow has no dark core at all, showing that the screen is beyond the end of the umbra.

The umbra of the earth's shadow is about 1.4 million km (860,000 miles) long and points directly away from the sun. A giant screen placed in the shadow at the average distance of the moon would reveal a dark umbra about 9000 km (5700 miles) in diameter, and the faint outer edges of the penumbra would mark a circle about 16,000 km (10,000 miles) in diameter. For comparison,

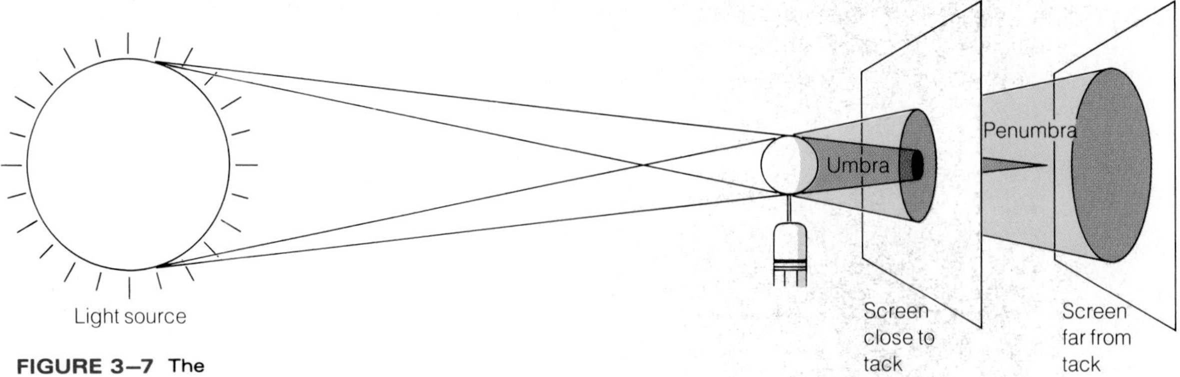

FIGURE 3–7 The shadows cast by a map tack resemble those of the earth and moon. The umbra is the region of total shadow; the penumbra is the region of partial shadow.

the moon's diameter is only 3476 km (2160 miles). Thus, when the moon's orbit carries it through the umbra, it has plenty of room to become completely immersed in shadow.

TOTAL LUNAR ECLIPSES

If the orbit of the moon carries it entirely into the umbra, so that no part of the moon protrudes into the partial sunlight of the penumbra, we say the eclipse is total (Figures 3–8 and 3–9 and Color Plate 1).

A **total lunar eclipse** occurs gradually, although the exact timing depends on how the moon's orbit crosses the earth's shadow. Because the moon moves eastward a distance equal to its own diameter each hour, it takes about an hour to enter the outer edge of the penumbra. As it moves deeper into the penumbra, it grows dimmer, and, about an hour after entering the penumbra, it reaches the edge of the umbra. The moon takes about an hour to enter the umbra and become totally eclipsed. (See Table 3–2.)

Although the moon is totally eclipsed, it does not disappear completely. Sunlight, bent by the earth's atmosphere, leaks into the umbra and bathes the moon in a faint glow. Because blue light is scattered by the earth's atmosphere more easily than red light, it is red light that penetrates to illuminate the moon in a coppery glow. (See Color Plate 2.) If we were on the moon during totality and looked back at the earth, we would not see any part of the sun because it would be entirely hidden behind the earth. However, we would see the earth's atmosphere

Table 3–2 Eclipses of the moon 1990 to 2003.

Year	Date	Partial or Total	Best Observing Location
1990	Feb. 9	T	India
	Aug. 6	P	Australia
1991	Dec. 21*	P	Hawaii
1992	June 15*	P	Chile
	Dec. 9*	T	North Africa
1993	June 4	T	South Pacific
	Nov. 29*	T	Mexico
1994	May 25*	P	Brazil
1995	April 15	P	South Pacific
1996	April 4*	T	South Atlantic
	Sept. 27*	T	Brazil
1997	March 24*	P	Brazil
	Sept. 16	T	Indian Ocean
1999	July 28*	P	South Pacific
2000	Jan. 21*	T	West Indies
	July 16	T	Australia
2001	Jan. 9	T	Saudi Arabia
	July 5	P	South Pacific
2003	May 16*	T	Brazil
	Nov. 9*	T	West Africa

*Visible in at least part from the United States.

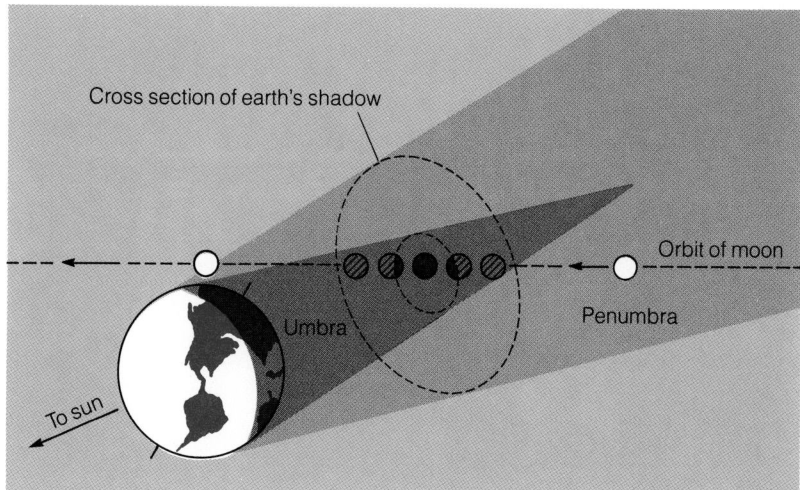

FIGURE 3–8 During a total lunar eclipse the orbit of the moon carries it through the penumbra and completely into the umbra. Compare with Figure 3–9.

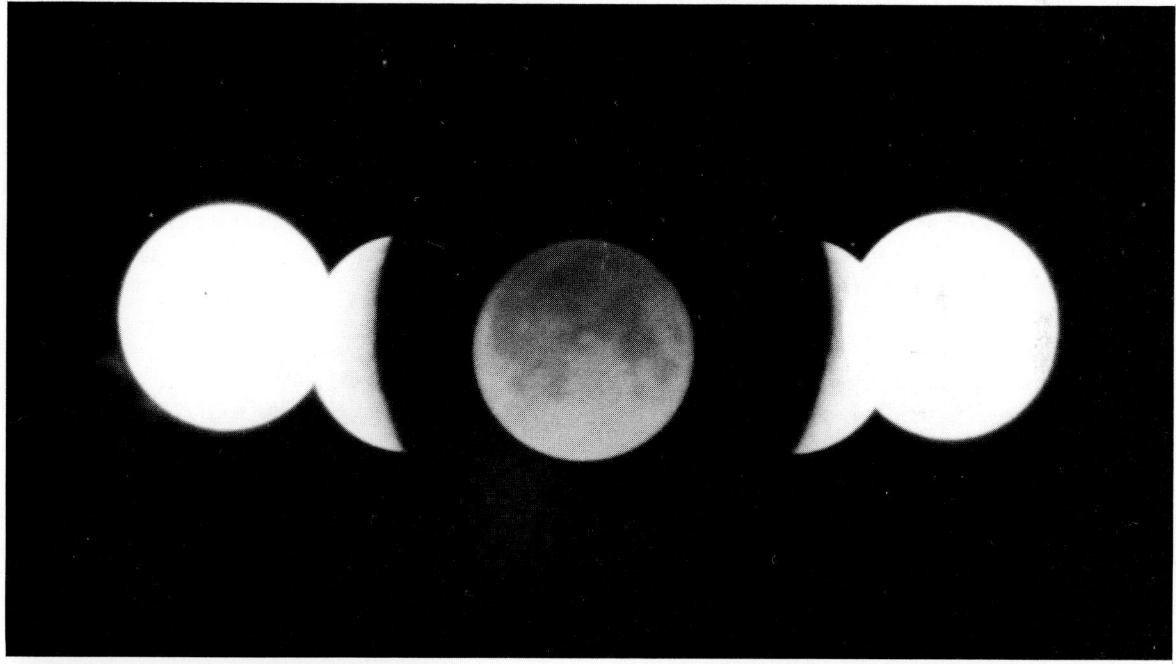

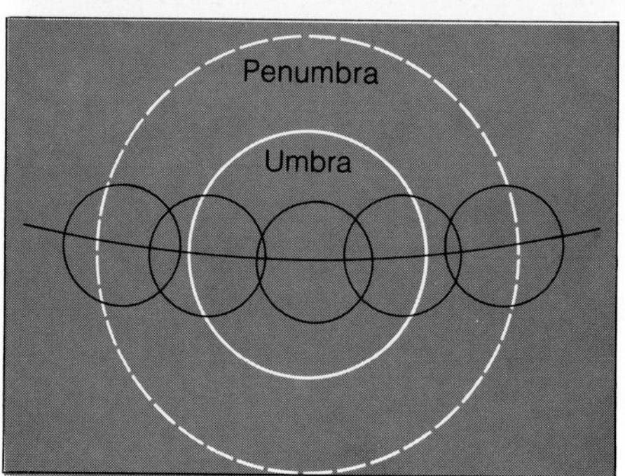

FIGURE 3–9 This multiple-exposure photo of a lunar eclipse spans 5 hours and shows the moon passing through earth's umbra and penumbra (left) from the right. The totally eclipsed moon (center) was 10,000 times dimmer than the full moon, so the exposure was lengthened to record the fainter image. Photographic effects make the moon's orbit appear curved here. Compare with Figure 3–8. (© 1982 Dr. Jack B. Marling.)

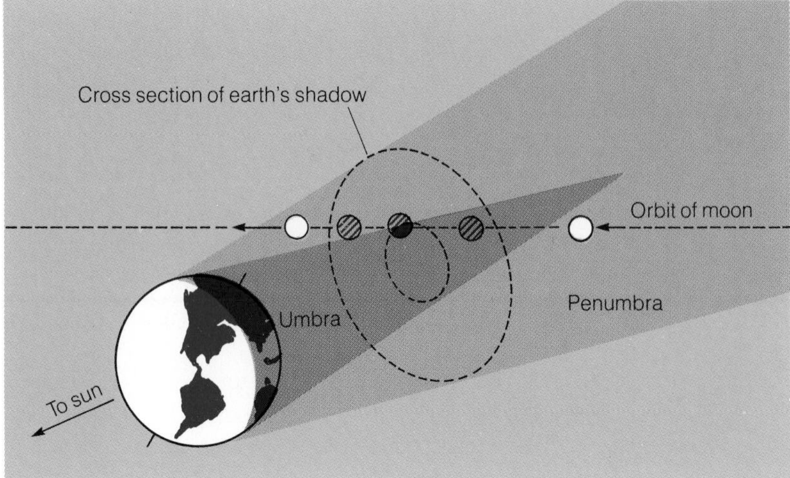

FIGURE 3–10 During a partial eclipse the orbit of the moon carries it through the penumbra and only partially into the umbra.

illuminated from behind by the sun in a spectacular sunset completely ringing the earth. It is the red glow from this sunset that gives the totally eclipsed moon its reddish color.

How dim the totally eclipsed moon becomes depends on a number of things. If the earth's atmosphere is especially cloudy in those regions that must bend light into the umbra, the moon will be darker than usual. An unusual amount of dust in the earth's atmosphere, from volcanic eruptions, for instance, also causes a dark eclipse. Also, total lunar eclipses tend to be darkest when the moon's orbit carries it through the center of the umbra.

As the moon moves through the earth's umbral shadow, we can see that the shadow is circular. From this the Greek philosopher Aristotle (384–322 BC) concluded that the earth had to be a sphere, because only a sphere could cast a shadow that was always circular.

Depending on the geometry of the eclipse, the moon can take as long as 1 hour and 40 minutes to cross the umbra and another hour to emerge into the penumbra. Still another hour passes as it emerges into full sunlight. A total eclipse of the moon can take 6 hours from start to finish.

Not all eclipses of the moon are total. If the moon only partially enters the umbra, the eclipse is termed a **partial lunar eclipse** (Figure 3–10). Partial eclipses usually are not as beautiful as total eclipses because the faint coppery glow of the eclipsed moon is lost in the glare of the uneclipsed portions. If the moon does not enter the umbra at all, but only passes through the penumbra, the eclipse is termed a **penumbral eclipse**. The partial dimming of the moon in the penumbra is often difficult to detect.

3.4 SOLAR ECLIPSES

We on earth are very lucky because our moon has the same angular diameter as our sun, so it can cover the sun and produce spectacular solar eclipses. Of course, the sun is 400 times larger in diameter than the moon, but it is also 400 times farther away. The sun and moon are both about 0.5° in angular diameter. It is this coincidence that allows us to see solar eclipses. (Box 3–1 describes a simple rule called the small-angle formula, which allows us to calculate angular diameters.)

A solar eclipse happens when the moon moves between the earth and sun. If the moon covers the sun completely, the eclipse is a **total solar eclipse**. If the moon covers only part of the sun, the eclipse is a **partial solar eclipse**.

During a single solar eclipse, people in one place on earth may see a total eclipse, while people only a few hundred kilometers away see a partial eclipse. To understand how that can happen, we must consider the moon's shadow.

THE MOON'S SHADOW

Like the earth's shadow, the moon's shadow consists of a central umbra of total shadow and a penumbra of partial shadow. What we see when the moon crosses in front of the sun depends on where we are in the moon's shadow.

The moon's umbral shadow produces a spot of darkness only about 269 km (167 miles) in diameter on the earth's surface (Figure 3–11). If we are in this spot of total shadow, we see a total solar eclipse. If we are just outside

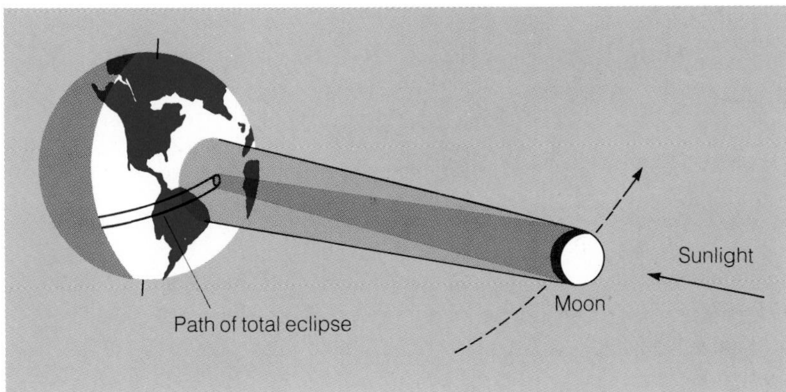

FIGURE 3–11 Observers in the path of totality see a total solar eclipse when the umbral shadow sweeps over them. Those in the penumbra see a partial eclipse.

the umbral shadow, but in the penumbra, we see part of the sun peeking around the moon, and the eclipse is partial. Of course, if we are outside the penumbra, we see no eclipse at all. Because of the orbital motion of the moon, its shadow sweeps across the earth at speeds of at least 1700 km/h (1060 mph). To be sure of seeing a total solar eclipse, we must place ourselves in the **path of totality**, the path swept out by the umbral spot.

TOTAL SOLAR ECLIPSES

A total solar eclipse begins when we first see the edge of the moon encroaching on the sun. This is also the moment when the edge of the penumbra sweeps over our location.

During the partial phase, part of the sun remains visible, and it is hazardous to look at the eclipse without protection. Even dense filters and exposed film are not necessarily safe because some do not block the invisible heat radiation (infrared) that can burn the retina of our eyes. This has led officials to warn the public not to look at solar eclipses, and it has frightened some people into locking themselves and their children into windowless rooms during eclipses. In fact, the sun is a bit less dangerous than usual during an eclipse because part of the bright surface is covered by the moon. But an eclipse is dangerous in that it can tempt us to look at the sun without proper precautions.

The safest and simplest way to observe the partial phases of a solar eclipse is to use pinhole projection. Poke a small pinhole in a sheet of cardboard. Hold the sheet with the hole in the sunlight and allow light to pass through the hole to a second sheet of cardboard (Figure 3–12).

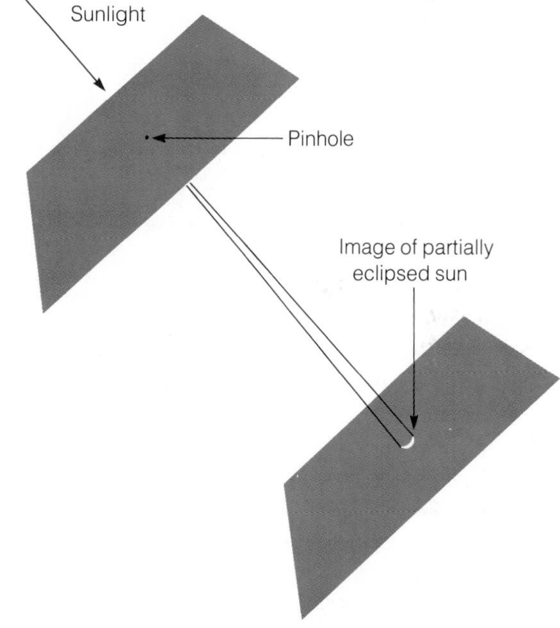

FIGURE 3–12 A safe way to view the partial phases of a solar eclipse. Use a pinhole in a card to project an image of the sun on a second card. The greater the distance between the cards the larger and fainter the image will be.

On a day when there is no eclipse, the result is a small round spot of light that is an image of the sun. During the partial phases of a solar eclipse, the image shows the dark silhouette of the moon obscuring part of the sun. These pinhole images of the partially eclipsed sun can also be seen in the shadows of trees as the sunlight peeks

BOX 3–1

The Small-Angle Formula

The angular diameter of an object is related by a simple formula to its linear diameter and its distance (Figure 3–13). Linear diameter is the distance between an object's opposite sides. The linear diameter of the moon, for instance, is 3476 km. The angular diameter of an object is the angle formed by lines extending from opposite sides of the object and meeting at our eye. Clearly, the farther away an object is, the smaller its angular diameter.

In the small-angle formula, we will always measure angular diameter in seconds of arc, and we will always use the same units for distance and linear diameter.

$$\frac{\text{angular diameter}}{206{,}265} = \frac{\text{linear diameter}}{\text{distance}}$$

We can use this formula to find any of these three quantities if we know the other two.

Consider an earthly example. Suppose that we see an automobile 1 km away and note that its angular diameter is about 13 minutes of arc. Remember that 1 km is 1000 m and that 1 minute of arc equals 60 seconds of arc. Thus, 13 minutes of arc equals 780 seconds of arc. We can find its linear diameter from the formula.

$$\frac{780}{206{,}265} = \frac{\text{linear diameter}}{1000}$$

So the linear diameter of the car is 780,000/206,265, which equals 3.8 m; it must be a limousine.

We can also find angular diameter. The moon has a linear diameter of 3476 km and a distance of about 384,000 km. Then its angular diameter is

$$\frac{\text{angular diameter}}{206{,}265} = \frac{3476}{384{,}000}$$

through the tiny openings between the leaves and branches. This can produce an eerie effect just before totality as the remaining sliver of sun produces thin crescents of light on the ground under trees.

Totality begins as the last sliver of the sun's bright surface disappears behind the moon. This is the same as the moment when the edge of the umbra sweeps over our location. So long as any of the sun is visible, the countryside is bright; but, as the last of the sun disappears, dark falls in a few seconds. Automatic streetlights come on, car drivers switch on their headlights, and birds go to roost. The darkness of totality depends on a number of factors, including the weather at the observing site, but it is usually dark enough to make it difficult to read the settings on cameras.

The totally eclipsed sun is a spectacular sight. With the moon covering the bright disk of the sun, called the **photosphere**, we can see the sun's faint outer atmosphere, the **corona**, glowing with a pale, white light so faint we can safely look at it directly. This corona is made of low-density, hot gas, which is given a wispy appearance by the solar magnetic field (Figure 3–14). Also visible just above

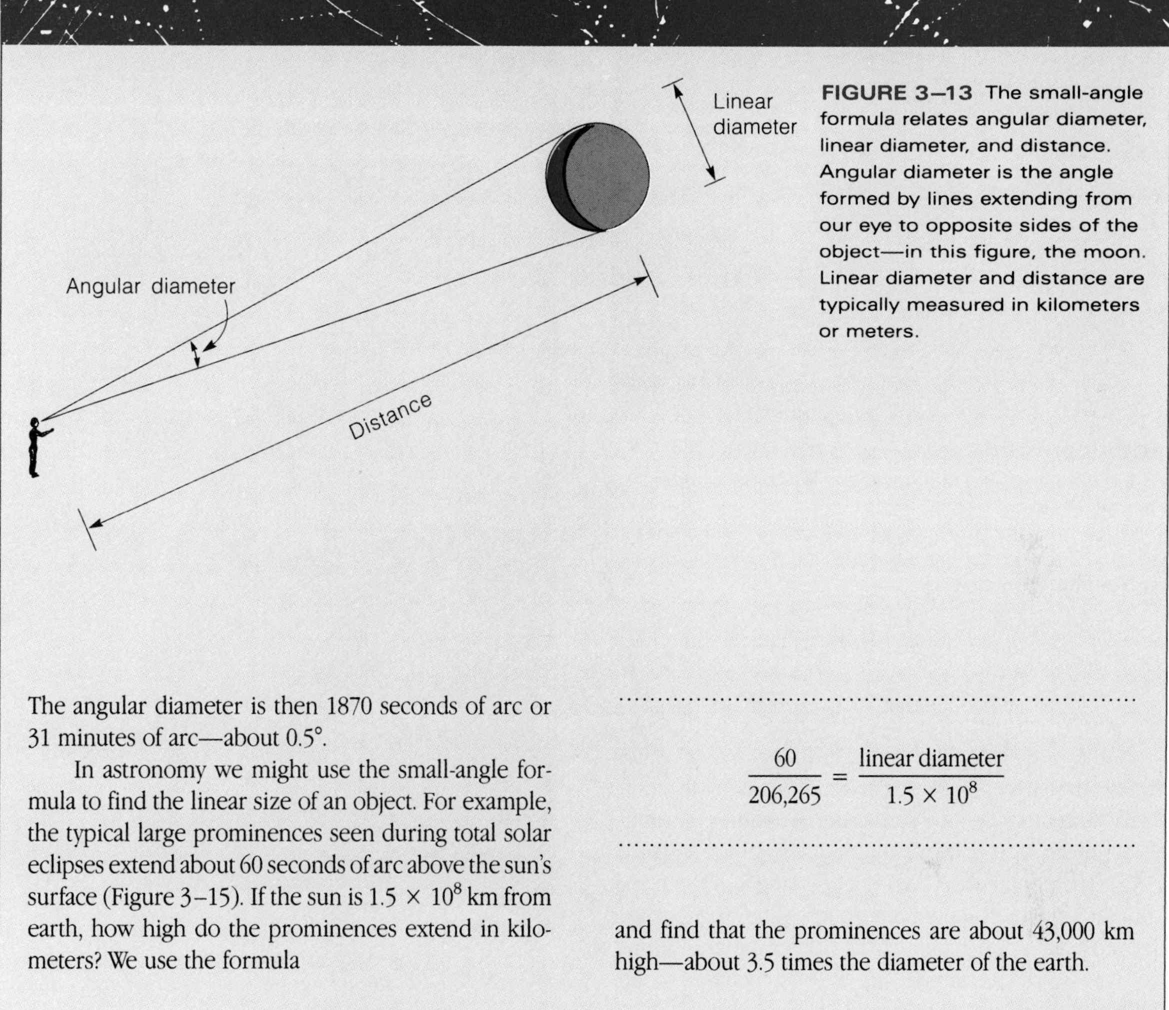

FIGURE 3–13 The small-angle formula relates angular diameter, linear diameter, and distance. Angular diameter is the angle formed by lines extending from our eye to opposite sides of the object—in this figure, the moon. Linear diameter and distance are typically measured in kilometers or meters.

The angular diameter is then 1870 seconds of arc or 31 minutes of arc—about 0.5°.

In astronomy we might use the small-angle formula to find the linear size of an object. For example, the typical large prominences seen during total solar eclipses extend about 60 seconds of arc above the sun's surface (Figure 3–15). If the sun is 1.5×10^8 km from earth, how high do the prominences extend in kilometers? We use the formula

$$\frac{60}{206{,}265} = \frac{\text{linear diameter}}{1.5 \times 10^8}$$

and find that the prominences are about 43,000 km high—about 3.5 times the diameter of the earth.

the photosphere is a thin layer of bright gas called the **chromosphere**. The chromosphere is often marked by eruptions on the solar surface called **prominences** (Figure 3–15), which glow with a clear, pink color due to the high temperature of the gases involved. (See Color Plate 3.) The small-angle formula (Box 3–1) tells us that a large prominence is about 3.5 times the diameter of the earth.

Totality cannot last longer than 7.5 minutes under any circumstances, and the average is only 2–3 minutes. Totality ends when the sun's bright surface reappears at the trailing edge of the moon. This corresponds to the moment when the trailing edge of the moon's umbra sweeps over the observer.

Once totality is over, daylight returns quickly, and the corona and chromosphere vanish. Astronomers travel great distances to place their instruments in the path of totality to study the faint outer corona and make other measurements only possible during the few minutes of a total solar eclipse.

But not all solar eclipses are total. In more than half of all eclipses, the moon, following its slightly elliptical orbit, is too far from the earth, and its umbral shadow

FIGURE 3–14 The sun's extended atmosphere is visible during a total solar eclipse. This photograph was taken November 12, 1966, with a special filter to enhance the outer portions of the corona. The planet Venus is visible near the left edge. (G. Newkirk, Jr., High Altitude Observatory.)

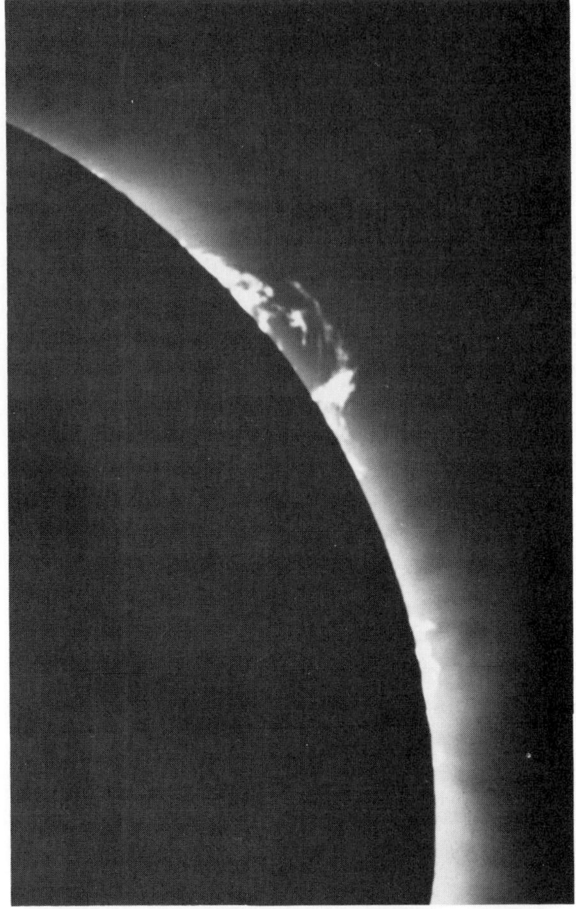

FIGURE 3–15 During a total solar eclipse, great eruptions on the solar surface, called prominences, are visible protruding beyond the moon's edge. (Lick Observatory photograph.)

does not reach earth's surface (Figure 3–16). In these cases, the moon's angular diameter is slightly too small to cover the sun completely. At mideclipse, when the moon is exactly centered on the sun, a ring or annulus of bright photosphere is visible surrounding the moon (Figure 3–17). The countryside does not get dark, and the glare from the exposed ring of photosphere hides the corona, chromosphere, and prominences. Such **annular eclipses** are not nearly as interesting as total eclipses. (See Table 3–3.)

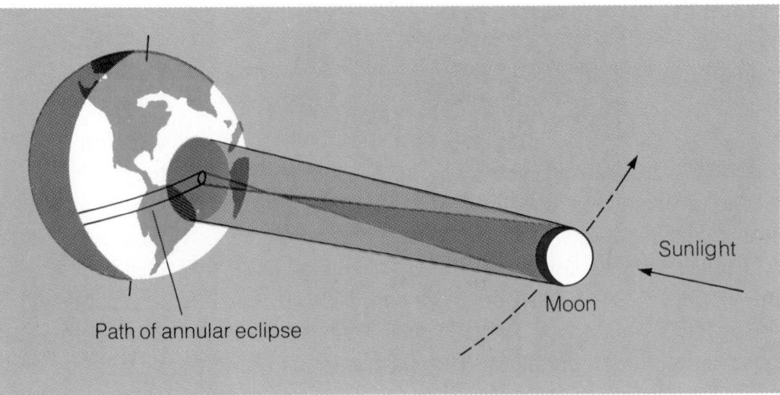

FIGURE 3–16 If the moon is near the farther part of its orbit, the umbral shadow does not reach the earth, resulting in an annular eclipse.

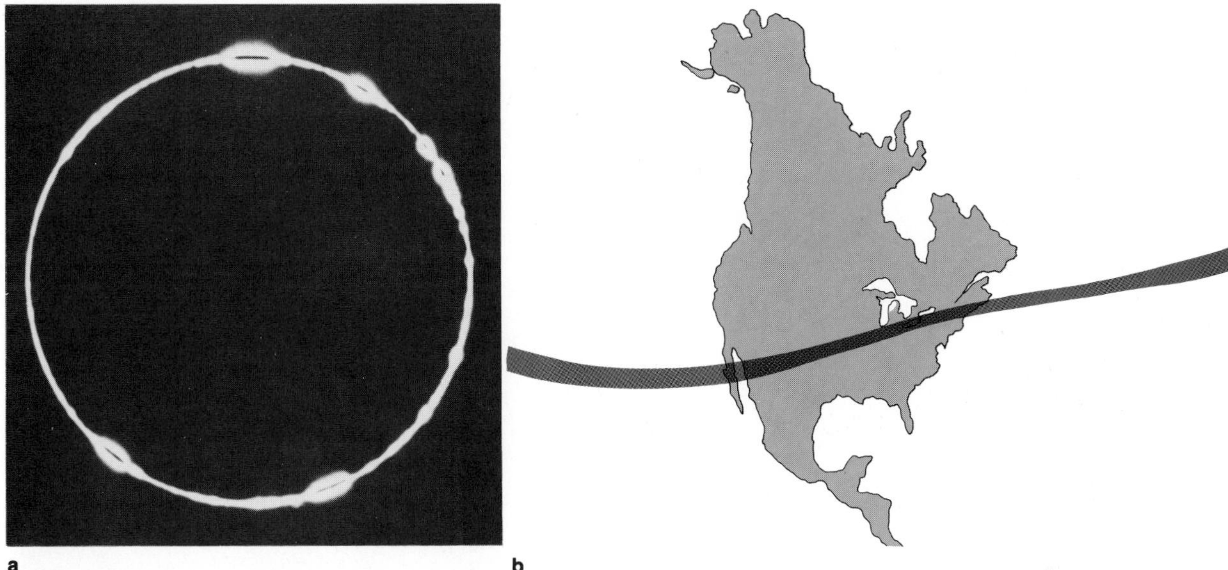

a b

FIGURE 3–17 (a) The annular eclipse of April 28, 1930. A bright ring of photosphere remains visible at the edge of the moon. (Lick Observatory photograph.) (b) The eclipse of May 10, 1994 will be annular as seen from within the path shown and partial as seen from the rest of the United States, Canada, and Mexico.

Table 3–3 Total and annular eclipses 1988 to 2000.

Year	Date	Total or Annular	Best Observing Location
1990	Jan. 26	A	Antarctica, South Atlantic
	July 22	T	Finland, northern Siberia
1991	Jan. 15	A	Australia, New Zealand
	July 11	T	Hawaii, Mexico, South America
1992	Jan. 4	A	Pacific
	June 30	T	South America, Africa
1994	May 10	A	Pacific, central U.S.
	Nov. 3	T	South America, Atlantic
1995	April 29	A	South Pacific, South America
	Oct. 24	T	Iran, India, Southeast Asia
1997	March 9	T	Asia, Siberia, Arctic
1998	Feb. 26	T	Pacific, South America
	Aug. 22	A	Indian Ocean, Indonesia
1999	Feb. 16	A	Indian Ocean, Australia
	Aug. 11	T	Europe, Asia, India
2001	June 21	T	Atlantic, Africa, Indian Ocean
	Dec. 14	A	Pacific, Central America

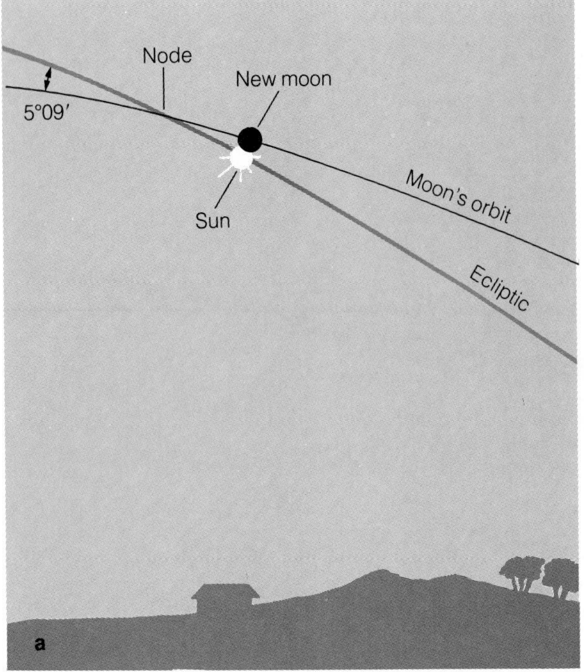

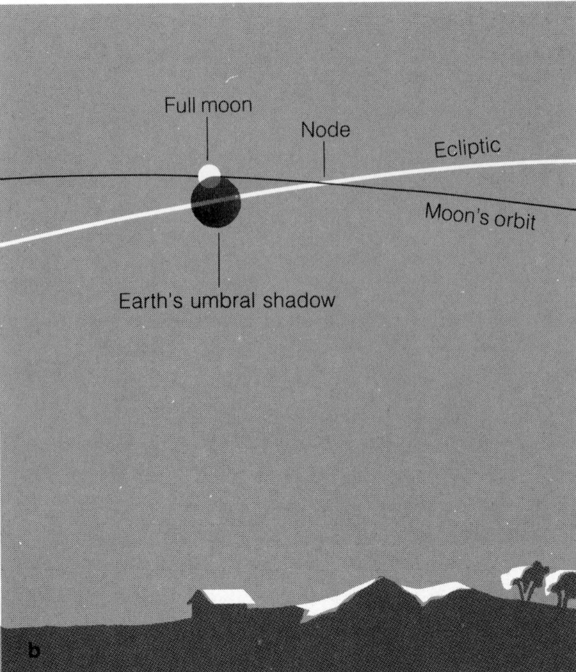

FIGURE 3–18 Eclipses can occur only near the nodes of the moon's orbit. (a) A solar eclipse occurs when the moon meets the sun near a node. (b) A lunar eclipse occurs when the sun and moon are at opposite nodes. Partial eclipses are shown here for clarity.

3.5 PREDICTING ECLIPSES

We should not be surprised that modern astronomers, armed with computers, can predict eclipses of the sun and moon. But, in fact, it is not particularly difficult to predict eclipses. Anyone with a pocket calculator can predict an eclipse in just a few hours of calculation.* What is surprising is that primitive astronomers, without computers, calculators, or even a knowledge of simple algebra could predict eclipses. They did it by noticing the pattern in the dates of eclipses.

We will discuss eclipse prediction for a number of reasons. First, it is an important part of the history of astronomy. But it also shows how apparently complex phenomena in the sky can be understood in terms of cycles. In addition, eclipse prediction will exercise our mental muscles and force us to see the earth, moon, and sun as objects moving through space.

CONDITIONS FOR AN ECLIPSE

We can predict eclipses by understanding the conditions that make them possible. As we begin to think about these conditions, we must be sure we understand our point of view. Later we will change our point of view, but to begin we will imagine that we can look up into the sky from our home on earth and see the sun moving along the ecliptic and the moon moving along its orbit.

The orbit of the moon is tipped 5°8′43″ to the plane of the earth's orbit, so we see the moon follow a path tipped by that angle to the ecliptic. Each month the moon crosses the ecliptic at two points called **nodes** (Figure 3–18). At one node it crosses going southward, and 2 weeks later it crosses at the other node going northward.

Eclipses can only occur when the sun is near a node. A solar eclipse is caused by the moon passing in front of the sun, and this can only happen when the sun is near

*See Peter Duffett-Smith, *Practical Astronomy with Your Calculator*, 2nd ed. (Cambridge, England: Cambridge University Press, 1981) or Peter Duffett-Smith, *Astronomy with Your Personal Computer* (Cambridge, England: Cambridge University Press, 1985).

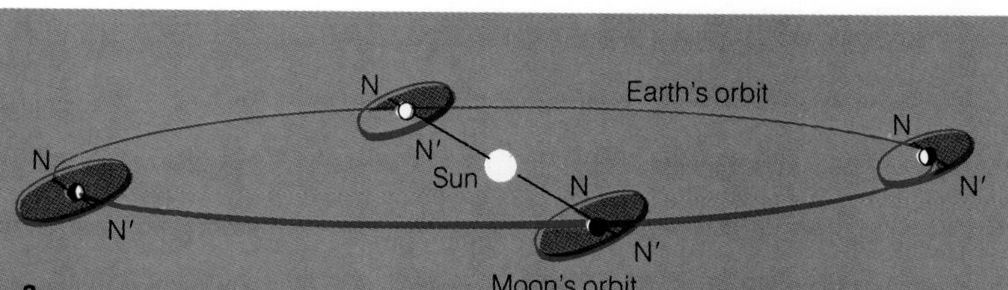

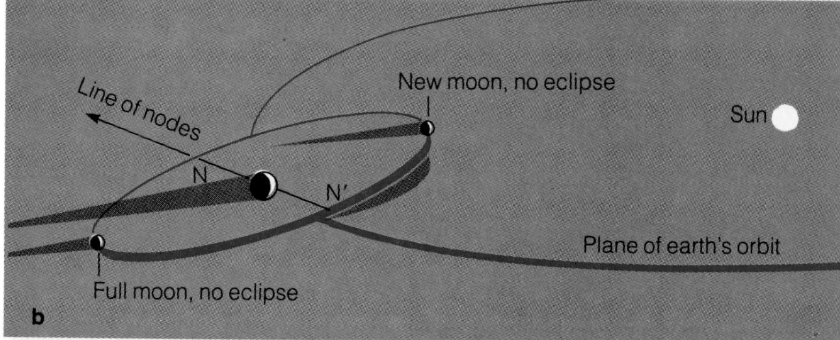

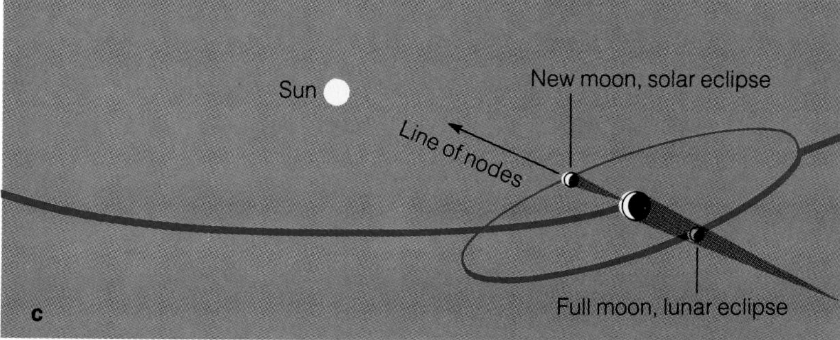

FIGURE 3–19 (a) The moon's orbit is tipped about 5° to Earth's orbit. The nodes N and N′ are the points where the moon passes through the plane of Earth's orbit. (b) If the line of nodes does not point at the sun, the shadows miss and there are no eclipses at new moon and full moon. (c) At those parts of Earth's orbit where the line of nodes points toward the sun, eclipses are possible at new moon and full moon.

a node. A lunar eclipse occurs when the moon enters the earth's shadow, but that shadow always follows the ecliptic exactly opposite the sun. Thus, the moon can only enter the shadow when the shadow is at a node, and that means the sun must be on the opposite side of the sky at the other node.

Thus, there are two conditions for an eclipse. The sun must be crossing a node, and the moon must be crossing the same node (solar eclipse) or the other node (lunar eclipse). Clearly, solar eclipses can only occur when the moon is new, and lunar eclipses can only occur when the moon is full.

An **eclipse season** is the period of time during which the sun is close enough to a node for an eclipse to occur. For a solar eclipse, an eclipse season is about 32 days. Any new moon during this period will produce a solar eclipse. For lunar eclipses, the eclipse seaason is a bit shorter, about 22 days. Any full moon in this period will be eclipsed.

This makes eclipse prediction easy. We simply keep track of where the moon crosses the ecliptic, and when the sun is near one of these nodes we predict that the nearest new moon will cause a solar eclipse and the nearest full moon will cause a lunar eclipse. This system works fairly well, and ancient astronomers such as the Maya may have used such a system. But we could do better if we changed our point of view.

THE VIEW FROM SPACE

Let us change our point of view and imagine that we are looking at the orbits of the earth and moon from a point far away in space. We see the moon's orbit as a smaller disk tipped at an angle to the larger disk of the earth's orbit (Figure 3–19a). As the earth orbits the sun, the moon's orbit remains fixed in direction. The nodes of the moon's orbit are the points where it passes through

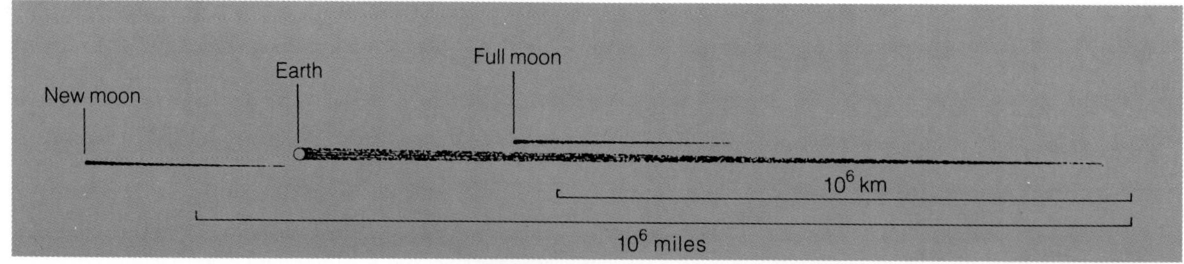

FIGURE 3–20 This scale drawing of the umbral shadows of the earth and moon shows how easy it is for the shadows to miss their mark at full moon and new moon. Diameter of the earth and moon are exaggerated by a factor of two for clarity.

the plane of the earth's orbit; an eclipse season occurs each time the line connecting these nodes, the **line of nodes**, points toward the sun (Figure 3–19c).

The shadows of the earth and moon, seen from space, are very long and thin (Figure 3–20). Only at the time of an eclipse season, when the line of nodes points toward the sun, do the shadows produce eclipses.

From our point of view in space, we would see the orbit of the moon precess like a hubcap spinning on the ground. This precession is caused mostly by the gravitational influence of the sun, and it makes the line of nodes rotate once every 18.6 years. People back on Earth see the nodes slipping westward along the ecliptic 19.4° per year, and the sun takes only 346.62 days (an **eclipse year**) to return to a node. This means that, according to our calendar, the eclipse seasons begin about 19 days earlier every year (Figure 3–21).

The cyclic pattern of eclipses shown in Figure 3–21 makes eclipse prediction simple. Because the moon's orbit precesses, the eclipse seasons occur 19 days earlier each year. Any new moon during this period will cross in front of the sun and cause a solar eclipse, and any full moon will enter earth's shadow and cause a lunar eclipse.

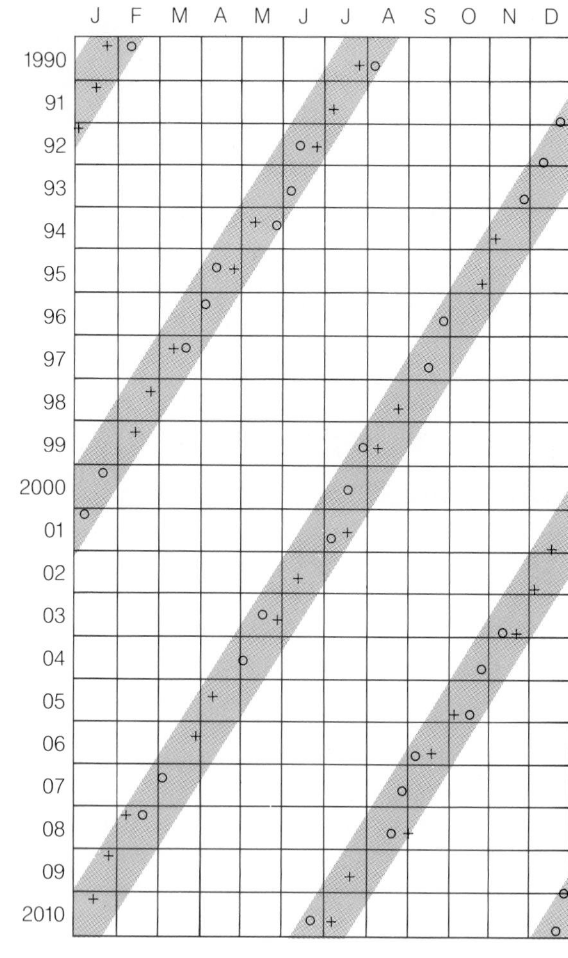

FIGURE 3–21 A calendar of eclipse seasons. Each year the eclipse seasons begin about 19 days earlier. Any new moon or full moon that occurs during an eclipse season results in an eclipse. Not all eclipses are shown here.

THE SAROS CYCLE

Ancient astronomers could predict eclipses in a crude way using the eclipse seasons, but they could have been much more accurate if they recognized that eclipses occur following certain patterns. The most important of these is the **Saros cycle**. After one Saros cycle of 18 years 11⅓ days, the pattern of eclipses repeats. In fact, *Saros* comes from a Greek word that means "repetition."

The eclipses repeat because, after one Saros cycle, the moon and the nodes of its orbit return to the same

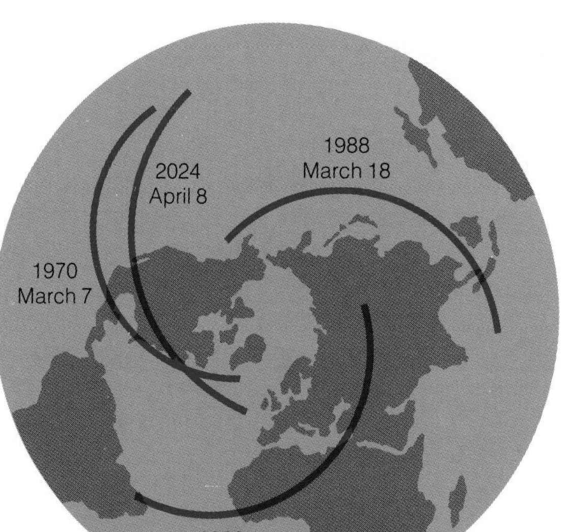

FIGURE 3–22 The Saros cycle at work. The total solar eclipse of March 7, 1970, will recur after 18 years 11 days over the Pacific Ocean. After another interval of 18 years 11 days the same eclipse will be visible from Asia and Africa. After a similar interval, the eclipse will again be visible from the United States.

place with respect to the sun. One Saros contains 6585.321 days, which is equal to 223 lunar months. Therefore, after one Saros cycle the moon is back to the same phase it had when the cycle began. But one Saros is nearly equal to 19 eclipse years. After one Saros cycle, the sun has returned to the same place it occupied with respect to the nodes of the moon's orbit when the cycle began. If an eclipse occurred on a given day, then 18 years 11⅓ days later the sun, moon, and the nodes of the moon's orbit return to nearly the same relationship and the eclipse happens all over again.

Although the eclipse repeats almost exactly, it is not visible from the same place on earth. The Saros cycle is one-third of a day longer than 18 years 11 days. When the eclipse recurs, the earth will have rotated one-third of a turn farther east, and the eclipse will occur 8 hours of longitude west of its earlier location (Figure 3–22). Thus, after three Saros cycles, a period of 54 years and 1 month, the same eclipse occurs in the same part of the earth.

One of the most famous predictors of eclipses was Thales of Miletus (about 640–546 BC) who supposedly learned of the Saros cycle from the Chaldeans who discovered it. No one knows which eclipse Thales predicted, but some scholars suspect the eclipse of May 28, 585 BC. In any case, the eclipse occurred at the height of a battle between the Lydians and the Medes, and the mysterious darkness in the midafternoon so startled the two Greek factions that they concluded a truce.

In fact, many historians doubt that Thales actually predicted the eclipse. The year 585 BC is very early for the Greeks to have known of the Saros cycles. The important point is not that Thales did it, but that he could have done it. If he had had records of past eclipses of the sun visible from Greece, he could have discovered that they tended to recur with a period of 54 years 1 month: Indeed, he could have predicted the eclipse without ever understanding what the sun and moon were or how they moved.

The phases of the moon, the tides, solar and lunar eclipses may seem to be complex events, but they are, in fact, simple patterns that repeat cycle after cycle. We have analyzed these lunar phenomena and found those cyclic patterns, which is a good illustration of what scientists do. They study nature searching for the patterns that give meaning to the universe.

SUMMARY

Because we see the moon by reflected sunlight, its shape appears to change as it orbits the earth. The lunar phases wax from new moon to first quarter to full moon, and wane from full moon to third quarter to new moon. A complete cycle of lunar phases takes 29.53 days.

The moon's gravitational field exerts tidal forces on the earth that pull the ocean waters up into two bulges, one on the side of the earth facing the moon and the other on the side away from the moon. As the rotating earth carries the continents through these bulges of deeper water, the tides ebb and flow. Friction with the seabeds slows the earth's rotation, and the gravitational force the bulges exert on the moon force its orbit to grow larger.

A lunar eclipse occurs when the moon enters the earth's shadow. If the moon becomes completely immersed in the umbra, the central shadow, the eclipse is termed total. The moon glows coppery red due to the sunlight bent as it passes through earth's atmosphere. If the moon enters the umbral shadow only partly, the eclipse is termed partial, and the reddish glow is not visible. A penumbral eclipse occurs when the moon passes through the penumbra, the region of partial shadow. Penumbral eclipses are not very noticeable.

A solar eclipse occurs when the earth passes through the moon's shadow. To see a total solar eclipse, observers must place themselves in the path of totality, the path swept out by the umbra of the moon. As the umbra sweeps over the observers, they see the bright surface of the sun, the photosphere, blotted out by the moon. Then the fainter chromosphere and corona, higher layers of the sun's atmosphere, become visible. Eruptions on the solar surface, called prominences, may be visible peeking around the edge of the moon.

The corona, chromosphere, and prominences are not visible to observers outside the path of totality. If the observers are in the path of the penumbra of the moon, they will see a partial solar eclipse, but the photosphere will never be completely hidden.

If an eclipse of the sun occurs when the moon is not in the nearer part of its orbit, the moon's umbra does not reach earth's surface. In these cases, the eclipse is not total. The moon's angular diameter is too small to cover the sun completely. Even at maximum eclipse, therefore, a bright ring or annulus of the photosphere is visible around the edge of the moon. This annular eclipse is not as dramatic as a total eclipse.

Ancient astronomers could predict eclipses because they occur not randomly, but in a pattern. Two conditions must be met if an eclipse is to occur. First, the moon must be on or near the ecliptic. The two points where the moon crosses the ecliptic are called the nodes of the moon's orbit. The second condition is that the sun must be at or near one of the nodes. This means that eclipses can occur only at new moon or full moon during two eclipse seasons that are about 32 days long and that occur almost 6 months apart.

Because the moon's orbit precesses, the nodes slip westward along the ecliptic, and the eclipse seasons begin 19 days earlier each year. The moon, sun, and nodes return to the same relative positions every 18 years 11⅓ days in what is called the Saros cycle. After the passage of a Saros cycle, the pattern of eclipses begins to repeat. This means that ancient astronomers could predict eclipses just by examining the dates of previous eclipses. The Chaldeans discovered the Saros cycle, and the ancient Greeks used it. Many other primitive cultures also used this method.

NEW TERMS

sidereal period
synodic period
spring and neap tides
umbra
penumbra
total eclipse (lunar or solar)
partial eclipse (lunar or solar)
penumbral eclipse
path of totality
photosphere
corona
chromosphere
prominences
annular eclipse
node
eclipse season
line of nodes
eclipse year
Saros cycle

QUESTIONS

1. Which lunar phases would be visible in the sky at dawn? At midnight?
2. If you looked back at the earth from the moon, what phase would you see when the moon was full? New? A first-quarter moon? A waxing crescent?
3. If the moon did not exist, would the earth's oceans still have tides? Why or why not?
4. Give examples to show how tides can alter the rotation and revolution of celestial bodies. (HINT: Recall from Chapter 2 the difference between rotation and revolution.)
5. Could a solar-powered spacecraft generate any electricity while passing through the earth's umbral shadow? The penumbral shadow?
6. Draw the umbral and penumbral shadows onto Figure 3–2. Explain why lunar eclipses can only occur at full moon and solar eclipses can only occur at new moon.
7. If you were on the moon while people on the earth saw a total lunar eclipse, what would you see when you looked back at the earth? Why would that make the moon look red?
8. How did lunar eclipses lead Aristotle to conclude that the earth was round?
9. Why isn't the corona visible during partial or annular solar eclipses?
10. Why can't the moon be eclipsed when it is halfway between the nodes of its orbit?
11. Why aren't eclipses separated by one Saros cycle visible from the same location on earth?
12. How could Thales of Miletus have predicted the date of a solar eclipse without observing the location of the moon in the sky?

PROBLEMS

1. Identify the phases of the moon if on March 21 the moon is located at (a) the vernal equinox, (b) the autumnal equinox, (c) the summer solstice, (d) the winter solstice.
2. Identify the phases of the moon if at sunset the moon is (a) near the eastern horizon, (b) high in the southern sky, (c) in the southeastern sky, (d) in the southwestern sky.
3. About how many days must elapse between first-quarter moon and third-quarter moon?

4. How many hours should elapse between successive high tides, if tides at a given location were caused only by the sun's gravity? Only by the moon's gravity? Why is there a difference?
5. How many times larger than the moon is the diameter of the earth's umbral shadow at the moon's distance? (HINT: See Figure 3–9).
6. Use the small-angle formula to calculate the angular diameter of the sun as seen from the earth.
7. During solar eclipses, large solar prominences are often seen extending 5 minutes of arc from the edge of the sun's disk. How far is this in kilometers? In earth diameters?
8. If a solar eclipse occurs on October 3, why can't there be a lunar eclipse on October 13? Why can't there be a solar eclipse on December 28?
9. A total eclipse of the sun was visible from Canada on July 10, 1972. When will this eclipse occur again? From what part of the earth will it be total?
10. When will the eclipse described in Problem 9 next be total as seen from Canada?
11. When will the eclipse seasons occur during the current year? What eclipse will occur?

ACTIVITY: DO-IT-YOURSELF ECLIPSE PREDICTION

To predict eclipses yourself, you need the star charts at the end of this book, and a calendar that shows the date and time of new and full moons. You should be able to predict the date and time of eclipses and the approximate location on earth for best viewing.

Begin by observing the location of the moon among the stars on as many nights as possible. Mark the location on a star chart that shows the ecliptic. Figure S–3 of the supplement would do, but you may want to use a larger chart for better accuracy. After observing for ten nights or so, sketch a smooth line through the points. This is the orbit of the moon, and where it crosses the ecliptic is a node of the moon's orbit. Continue observing until you can locate at least one of the nodes. The other node is exactly halfway around the sky.

Find the dates when the sun will cross the two nodes. (See Table S–3 in the supplement.) These dates fix the centers of the eclipse seasons. Consult your calendar to find the date and time of new moons and full moons that occur during the eclipse seasons. These will mark the dates and times of eclipses.

Look at the time of day of new moon or full moon to find the best location on earth from which to see the eclipse. A solar eclipse that occurs at 1:00 AM by your clocks will be visible from the far side of the earth. On the other hand, a lunar eclipse that occurs at 1:00 AM will be easily visible from your location.

Check your predictions by looking in a farmers almanac or in *The Astronomical Almanac* for the current year published by the U.S. Government Printing Office. Better yet, observe the eclipse yourself.

RECOMMENDED READING

AVENI, ANTHONY F. *Skywatchers of Ancient Mexico.* Austin, Texas: University of Texas Press, 1980, pp. 67–82.

BRACHER, KATHERINE "Getting to the 1860 Solar Eclipse." *Sky and Telescope 61*(Feb. 1981), p. 120.

DUFFET-SMITH, PETER *Astronomy with Your Personal Computer.* Cambridge, England: Cambridge University Press, 1985.

———. *Practical Astronomy with Your Calculator*, 2nd ed. Cambridge, England: Cambridge University Press, 1981.

HAWKINS, G. S. *Stonehenge Decoded.* Garden City, N.Y.: Doubleday, 1965.

KUHLMAN, HENRY "The Shadows Know." *Sky and Telescope 61* (April 1981), p. 296.

MARSCHALL, LAURENCE A. "A Tale of Two Eclipses." *Sky and Telescope 57* (Feb. 1979), p. 116.

———. "Shadow Bands—Solar Eclipse Phantoms." *Sky and Telescope 67* (Feb. 1984), p. 116.

———. "The Maximum Eclipse." *Sky and Telescope 59* (May 1980), p. 383.

MEEUS, JEAN "Solar Eclipse Diary: 1985–1995." *Sky and Telescope 68* (Oct. 1984), p. 296.

MITCHELL, SAMUEL A. *Eclipses of the Sun*, 5th ed. Westport, Conn.: Greenwood, 1951.

PERCY, JOHN R., ed. *The Observer's Handbook* (current year). Toronto: Royal Astronomical Society of Canada.

SHIPMAN, H. L. "Megaliths and the Moon: Eclipse Prediction in the Stonehenge Era." *The Griffith Observer 38* (Feb. 1974), p. 7.

STEPHENSON, F. RICHARD "Historical Eclipses." *Scientific American 247* (Oct. 1982), p. 170.

WHITEMAN, MARK "Eclipse Predictions on Your Computer." *Astronomy 14* (Nov. 1986), p. 67.

See also current issue of *Sky and Telescope* and *Astronomy* for information about coming eclipses.

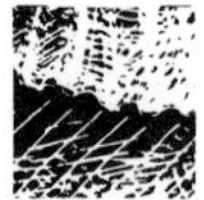

C H A P T E R 4

THE ORIGIN OF MODERN ASTRONOMY

Oh, my dear Kepler, how I wish that we could have one hearty laugh together!

From a letter by Galileo Galilei

The history of astronomy is more like a soap opera than a situation comedy. In a simple half-hour sitcom, only a few characters carry the story, but the plot of a soap opera is so complex, so filled with characters and subplots only tenuously related to each other, that the action jumps from character to character, from subplot to subplot, with almost no connection. The history of astronomy is similarly complex, filled with brilliant people who lived at different times and worked in different parts of the world.

The most important character in our story is Nicolaus Copernicus, the sixteenth-century astronomer who revolutionized astronomy by proposing that the sun, not the earth, is the center of the heavens. This was so controversial a theory that one of its defenders was burnt at the stake. Most of this chapter will deal with Copernicus, his theory, its implications, and its reception in the century that followed his death. However, Copernicus is not the only character in our story. We will also meet an astronomer who was condemned by the Inquisition, an astronomer whose mother was tried for witchcraft, and another astronomer who wore false noses.

Though much of our story will take place in the sixteenth and seventeenth centuries, we must recognize that astronomy had its beginnings when the first semihuman creatures looked up at the moon and stars and wondered what they were. Since then, hundreds of generations of talented astronomers have lived and worked on this planet and left almost no record of their accomplishments. Only by studying the archaeological ruins of their temples and observatories can we get a hint of their insights. Still other cultures, such as ancient Greece, have left some written records from which we can reconstruct the sophisticated astronomy of the ancient world.

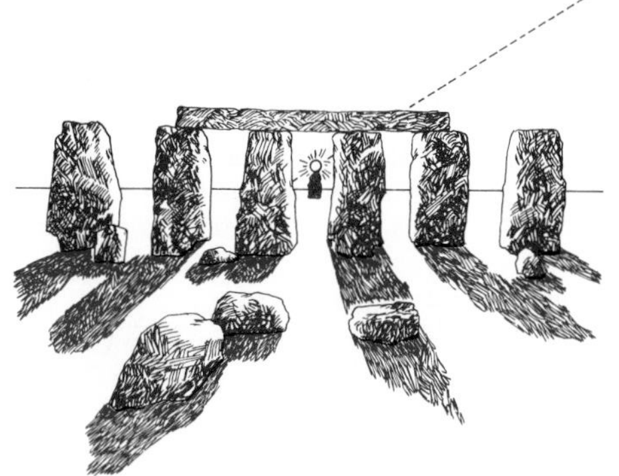

FIGURE 4–1 Stonehenge, the second most popular tourist attraction in England (after the Tower of London), is a Stone Age observatory containing more than a dozen astronomical alignments. The most important alignment is visible from the center of the monument. Looking northeast, we would see the sun rise over the Heelstone on the morning of the summer solstice (left). Path of rising sun shown as dashed line. (Joseph R. Holzinger.)

4.1 THE ROOTS OF ASTRONOMY

Astronomy has its origin in that most noble of all human traits, curiosity. Just as modern children ask their parents what the stars are and why the moon changes, so did ancient humans ask themselves these questions. The answers, often couched in mythical or religious terms, reveal a great reverence for the order of the heavens.

ARCHAEOASTRONOMY

Archaeoastronomy, the study of the astronomy of ancient, primitive peoples, came to public attention in 1965 when Gerald Hawkins published a book called *Stonehenge Decoded.* In that book, he reported that Stonehenge, the Stone Age arrangement of massive stones standing on Salisbury Plain in southern England (Figure 4–1), was an astronomical observatory.

Stonehenge was built in stages from about 2800 BC to about 1075 BC, a period extending from the late Stone Age into the Bronze Age. As early as 1740 AD, the English scholar Dr. W. Stukely suggested that the heelstone and avenue pointed toward the rising sun at the summer solstice, but few accepted the idea. Only in recent times have scholars begun to consider possible astronomical sight lines at Stonehenge. The axis of Stonehenge, represented by the avenue, points toward the northeastern horizon where the sun rises on the day of the summer solstice (Figure 4-1). Other sight lines seem to point toward the most northerly and most southerly risings of the moon and similar points on the horizon (Figure 4–2).

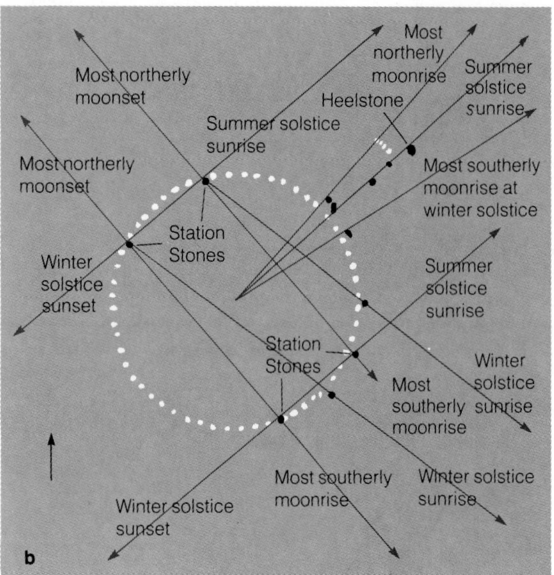

FIGURE 4–2 The major known or proposed astronomical alignments at Stonehenge involve smaller outlying stones and not the tall upright stones that are most prominent in photographs.

FIGURE 4–3 The stone ring at Callanish in Scotland's Outer Hebrides dates from the Stone Age and has astronomical alignments. About 900 Stone Age rings, most less impressive than Stonehenge and Callanish, dot the British Isles and many have astronomical significance. (Courtesy Gerald and Margaret Ponting.)

It does seem possible that the primitive peoples who built Stonehenge used it to observe the sun and moon and thus regulate their calendar. Calendars are very important in agricultural societies, and nearly all calendars are based on the sun and moon. But some astronomers contend that the people of Stonehenge used the 56 Aubrey Holes as counters to keep track of the locations of the nodes of the moon's orbit and were thus able to predict eclipses of the moon.

Although Stonehenge may be the most sophisticated Stone Age observatory, it is not the only one. Much of England and northern Europe is dotted by stone rings and monuments (Figure 4–3), many of which seem to have astronomical alignments, particularly toward the summer solstice sunrise.

The early inhabitants of North America were also interested in astronomy. The Big Horn Medicine Wheel* on a 3000-m shoulder of the Big Horn Mountains of Wyoming was used by the Plains Indians about AD 1500–1750. This arrangement of rocks marks a 28-spoke wheel about 27 m in diameter. Piles of rocks, called cairns, mark the center and six locations on the circumference. Astronomer John Eddy has discovered that these mark sight lines toward a number of important points on the eastern horizon and could have been used as a calendar to help schedule hunting, planting, harvesting, and celebrations (Figure 4–4).

Many other American Indian sites have astronomical alignments. More than three dozen medicine wheels are known, although most do not have the sophistication of the Big Horn Medicine Wheel. The Moose Mountain Wheel, for instance, seems to have been in use as early as AD 100. Alignments at some Mound Builder sites of the Midwest show that these peoples were familiar with the sky. In the Southwest, the Pueblo Indian ruins in Chaco Canyon, New Mexico, among others, have alignments that point toward the rising and setting of the sun at the summer and winter solstices.

Primitive astronomy also flourished in Central and South America. The Maya and Aztec empires built many temples aligned with the solstice rising of the sun and with the extreme points where Venus rose and set. The Caracol Temple in the Yucatán is a good example (Figure

*Medicine is used here to mean magical power.

a

b

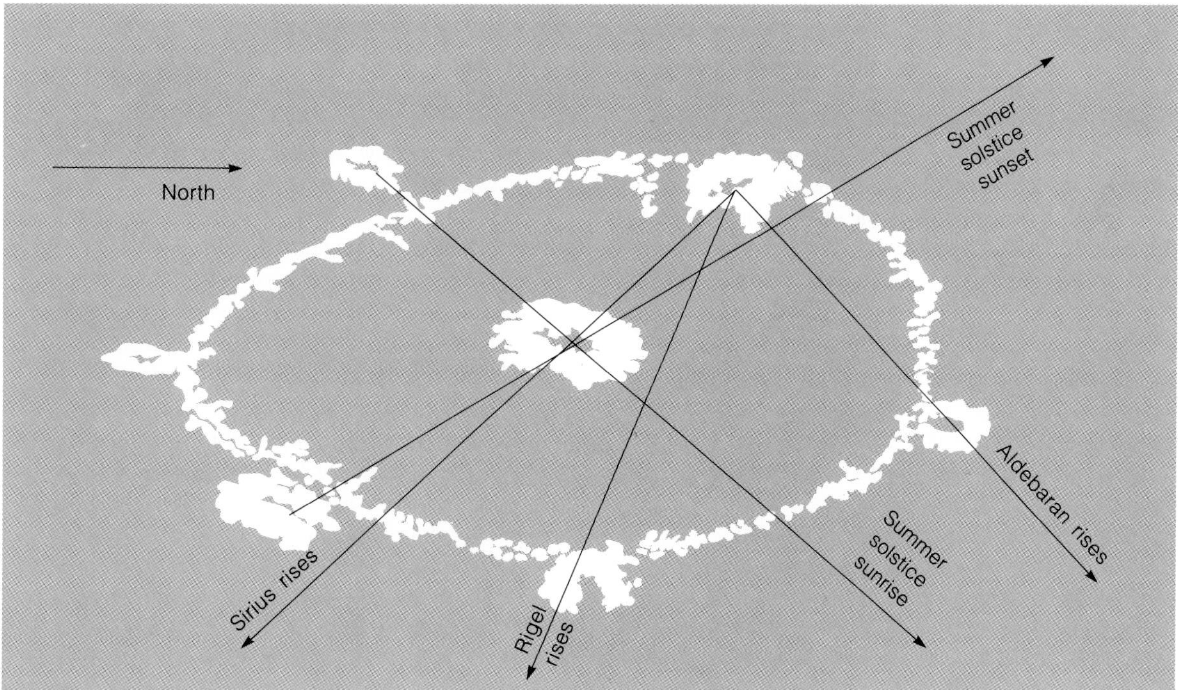

FIGURE 4–4 The Big Horn Medicine Wheel in Wyoming was built by American Indians before the westward spread of European settlers. (a) Its rock cairns are aligned with important events along the horizon. (b) The alignment suggests that the Indians used the medicine wheel as an observatory to regulate their calendar. (Courtesy John A. Eddy.)

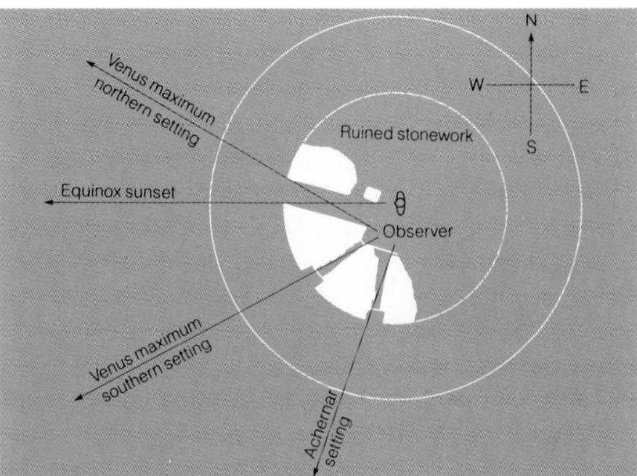

FIGURE 4–5 The Caracol at the Mayan city of Chichén Itzá is a 1000-year-old observatory. The remaining windows in the partially ruined tower contain sight lines that point toward the equinox sunset, the setting places of Venus at its most northerly and most southerly positions, and the setting place of the star Achernar. Various lines of evidence suggest that the Caracol was directly associated with the worship of the Venus god, Quetzalcóatl. (Courtesy Anthony Aveni.)

4–5). It is a circular tower containing complicated passageways and a spiral staircase (thus the name *Caracol*—"the snail's shell"). The tubelike windows at the top of the tower point toward the equinox sunset point, and the most northerly and most southerly setting points of Venus. Unfortunately, only about one-third of the tower top survives, so the directions of any other windows are forever lost.

Archaeoastronomers are uncovering the remains of ancient astronomical observatories around the world. Some temples in the jungles of Southeast Asia, for instance, are believed to have astronomical alignments. Unfortunately, their exploration is presently impossible because of the political instability of the region.

Other scholars are looking not at temples, but at small artifacts from thousands of years ago. Scratches on certain bone and stone implements seem to follow a pattern and may be an attempt to keep a record of the phases of the moon (Figure 4–6). Some scientists contend that humanity's first attempts at writing were stimulated by a desire to record and predict lunar phases.

Archaeoastronomy is uncovering the earliest roots of astronomy and simultaneously revealing some of the first human efforts at systematic inquiry. The most important lesson of archaeoastronomy is that humans don't have to be technologically sophisticated to admire and study the universe.

One thing about archaeoastronomy is especially sad. Although we are learning how primitive people observed the sky, we may never know what they thought about their universe. Many had no written language. In other cases, the written record has been lost. Dozens of beautiful Mayan manuscripts, for instance, were burned by Spanish missionaries who believed they were the work of Satan. That is one reason why our history of astronomy really begins with the Greeks. Some of their manuscripts have survived, and we can discover what they thought about the shape and motion of the heavens.

THE ASTRONOMY OF GREECE

Greek astronomy was based on the astronomy of Babylonia and Egypt, but these astronomies were heavily influenced by religion or astrology. Those ancient astronomers studied the motions of the heavens as a way of

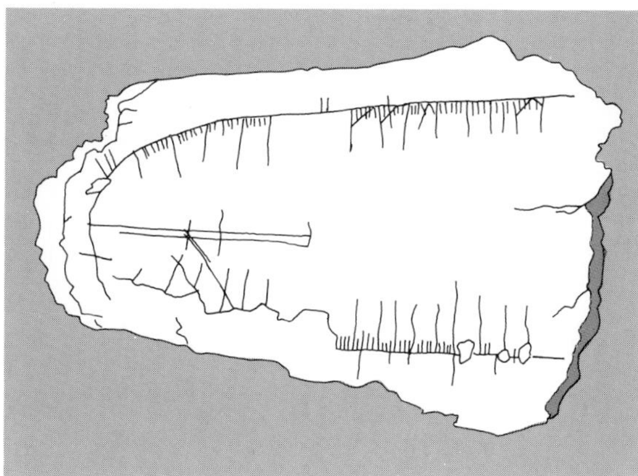

FIGURE 4–6 A fragment of a 27,000-year-old mammoth tusk found at Gontzi in the Ukraine contains scribe marks on its edge, simplified in this drawing. These markings have been interpreted as a record of four cycles of lunar phases. Although controversial, such finds suggest that some of the first human attempts at recording events in written form were stimulated by astronomical phenomena.

worship and divination. The Greek astronomers studied astronomy in an entirely new way—they tried to understand the universe.

This new attitude toward the heavens, a truly scientific attitude, was made possible by two early Greek philosophers. Thales of Miletus (c. 624–547 BC) lived and worked in what is now Turkey. He taught that the universe is rational, and that the human mind can understand why the universe works the way it does. This view contrasts sharply with those of earlier cultures, which believed that the ultimate causes of things are mysteries beyond human understanding. To Thales and his followers, the mysteries of the universe are mysteries because they are unknown, not because they are unknowable.

The second philosopher who made the new scientific attitude possible was Pythagoras (c. 570–500 BC). He and his students noticed that many things in nature seem to be governed by geometrical or mathematical relations. Musical notes, for example, are related in a regular way to the lengths of plucked strings. This led Pythagoras to propose that all nature was underlain by musical principles, by which he meant mathematics. One result of this was the later belief that the harmony of the celestial movements produced actual music, the music of the spheres. But at a deeper level, the teachings of Pythagoras made Greek astronomers look at the universe in a new way. Thales said that the universe could be understood, and Pythagoras said that the underlying rules were mathematical.

In trying to understand the universe, Greek astronomers did something that Babylonian astronomers had never done—they tried to construct descriptions based on geometrical forms. Anaximander (c. 611–546 BC) described a universe made up of wheels filled with fire: The sun and moon are holes in the wheels through which we see the flames. Philolaus (fifth century BC) argued that the earth moves in a circular path around a central fire (not the sun), which is always hidden behind a counter earth located between the fire and the earth. This was the first theory to suppose that the earth is in motion.

Plato (428–347 BC) was not an astronomer, but his teachings influenced astronomy for 2000 years. Plato argued that the reality we see is only a distorted shadow of a perfect, ideal form. If our observations are distorted, then observation can be misleading, and the best path to truth is through pure thought on the ideal forms that underlie nature. This led later astronomers to try to understand the universe without extensive observation.

Plato also argued that the most perfect form was the circle, and that therefore all motions in the heavens should be made up of combinations of circular motion. Of course, the most perfect motion is uniform motion, so later astronomers tried to describe the motions of the heavens using the principle of **uniform circular motion.**

Pythagoras had taught that the earth is a sphere and that the stars, sun, moon, and five planets are carried by rotating spheres. But Eudoxas of Cnidus (409–356 BC), a student of Plato, expanded this universe of nested spheres to include uniform circular motion. A total of 27 spheres rotated on pivots to produce the observed motions of the heavens (Figure 4–7).

Aristotle (384–322 BC) taught and wrote on philosophy, history, politics, ethics, poetry, drama, and so on. Because of his sensitivity and insight, he became the great authority of antiquity, and astronomers for almost 2000 years used the Greek model of the universe as described by Aristotle.

Aristotle believed that the universe is divided into two parts—the earth, corrupt and changeable, and the heavens, perfect and immutable. Like most of his predecessors, he believed that the earth was the center of the universe, so his model was **geocentric** (earth-centered).

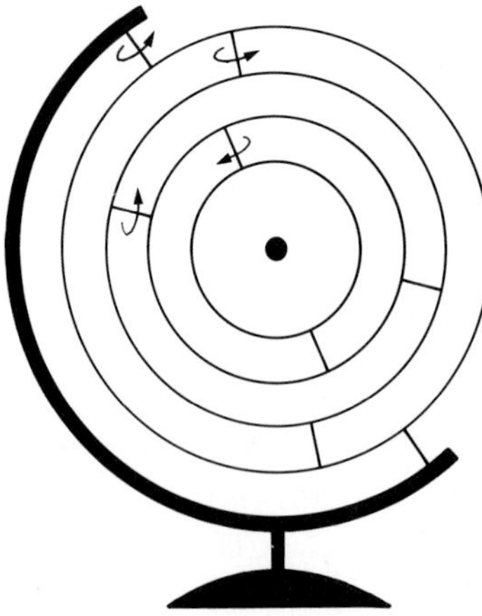

FIGURE 4–7 The spheres of Eudoxus explained the motions in the heavens by means of nested spheres rotating about various axes at different rates. The earth is located at the center. (Eudoxus used 27 spheres; Aristotle used 56.)

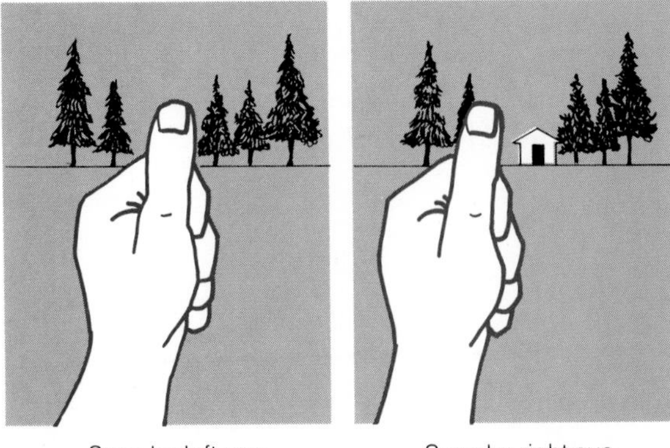

FIGURE 4–8 To demonstrate parallax, close one eye and cover a distant object with your thumb held at arm's length. Look with the other eye and your thumb appears to have shifted position. (See Figure 9–3.)

The heavens surrounded the earth, and Aristotle added more crystalline spheres to bring the total to 56.

Because he believed the earth is immobile, he had to make these spheres whirl westward around the earth each day, while they move with respect to each other to produce the motions of the sun, moon, and planets. Like most Greek philosophers, Aristotle viewed the universe as a perfect heavenly machine that was not many times larger than earth itself.

About a century after Aristotle, the Alexandrian philosopher Aristarchus proposed a theory that the earth rotated on its axis and revolved around the sun. This theory is, of course, correct, but most of the writings of Aristarchus were lost, and the theory was not well known. Later astronomers rejected any suggestion that the earth could move because it conflicted with Aristotle's theory and because the astronomers saw no parallax.

Parallax (p) is the apparent change in the position of an object due to a change in the location of the observer. It is actually an old friend, although you may not have known its name, for you use parallax to judge the distance to things. To see how this works, close your right eye and use your thumb, held at arm's length, to cover some distant object, a building perhaps. Now look with your right eye. Your thumb seems to move to the left, uncovering the building (Figure 4–8). This apparent shift is parallax, and your brain uses it to estimate distances to objects around you.

Ancient astronomers reasoned that if the earth moves around the sun, we should be viewing the stars from different positions in earth's orbit at different times. Then we should see parallax shifting the apparent positions of the stars. In the most extreme cases, this should distort the shapes and sizes of the constellations. That is, constellations should look largest when the earth is nearest that part of the sky. Because they did not see this parallax, they concluded that the earth does not move. Actually, they saw no parallax because the stars are much farther away than they supposed, and the parallax is much too small to be visible to the naked eye.

Aristotle had taught that the earth had to be round because it always casts a round shadow during lunar eclipses, but he could only estimate its size. About 200 BC Eratosthenes, working in the great library in Alexandria, found a way to calculate the earth's radius. He learned from travelers that the city of Syene (Aswan) in southern Egypt contained a well into which sunlight shone vertically on the day of the summer solstice. Thus, the sun was at the zenith at Syene, but on that same day in Alexandria, he noted that the sun was 1/50th of the circumference of the sky (about 7°) south of the zenith. Because sunlight comes from such a great distance, its rays arrive at earth traveling almost parallel. Thus, Eratosthenes could conclude from simple geometry (Figure 4–9) that the distance from Alexandria to Syene was 1/50th the circumference of the earth.

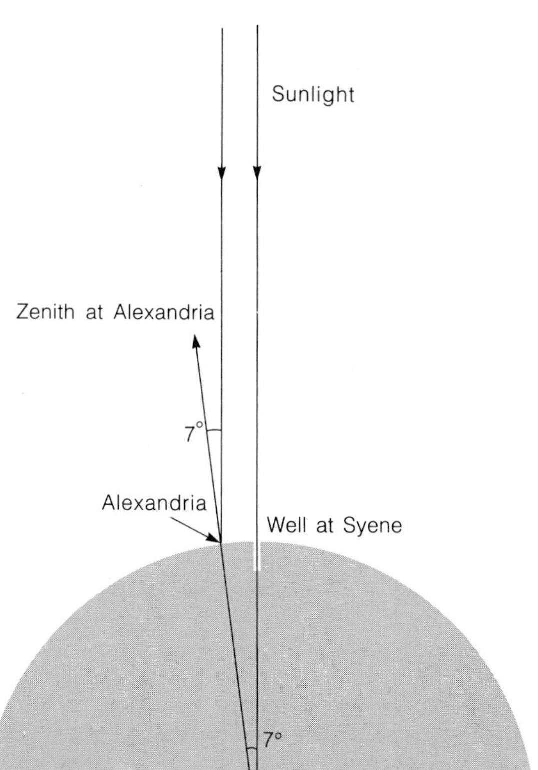

FIGURE 4–9 On the day of the summer solstice, sunlight fell to the bottom of a well at Syene, but on the same day the sun was about 1/50th of a circle (7°) south of the zenith at Alexandria. This told Eratosthenes that the distance from Syene to Alexandria was 1/50th of the circumference of the earth. Thus, Eratosthenes was able to find the radius of the earth.

To find the circumference of the earth, Eratosthenes had to know the distance from Alexandria to Syene. Travelers told him it took 50 days to cover the distance, and he knew that a camel can travel about 100 stadia per day. Thus, the total distance was about 5000 stadia. If 5000 stadia is 1/50th the earth's circumference, then the earth must be 250,000 stadia around, and dividing by 2π, Eratosthenes found the radius of the earth to be 40,000 stadia.

We don't know how accurate Eratosthenes was. The stadium (singular of stadia) had different lengths in ancient times. If we assume 6 stadia to the kilometer, then Eratosthenes' result was too big by only 4 percent. If he used the Olympic stadium, his result was 14 percent too big.

FIGURE 4–10 The muse of astronomy guides Ptolemy (about AD 140) in his study of the heavens. (Courtesy Owen Gingerich and The Houghton Library.)

In any case, this was a much better measurement of the radius of the earth than Aristotle's estimate, which was only about 40 percent of the true radius.

Despite the theories and discoveries of such philosophers as Aristarchus and Eratosthenes, the Aristotelian model of the universe was accepted as correct for almost two millennia. One reason was Aristotle's authority in all branches of learning, but another reason has to do with the last great astronomer of classical times, Ptolemy.

THE PTOLEMAIC UNIVERSE

Claudius Ptolemaeus (Figure 4–10) was one of the great astronomer–mathematicians of antiquity. His nationality and birth date are unknown, but he lived and worked in the Greek settlement at Alexandria about AD 140. He assured the survival of Aristotle's universe by fitting to it a sophisticated mathematical model.

In agreement with Aristotle, the Ptolemaic model was geocentric (earth-centered). In addition, it incorporated

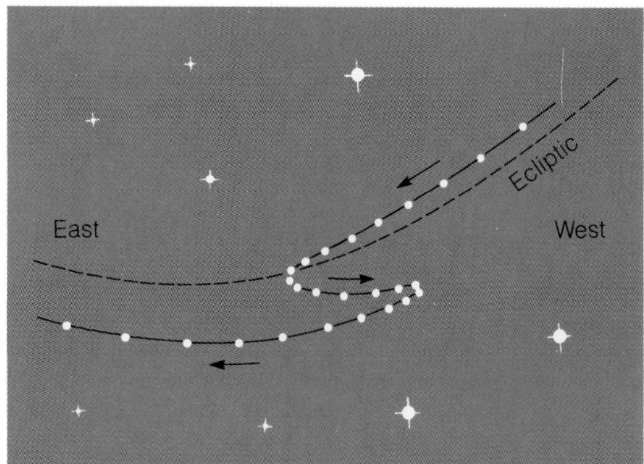

FIGURE 4–11 The motion of Mars along the ecliptic is shown at 10-day intervals. Though it usually moves eastward, it sometimes slows to a stop and moves westward in retrograde motion.

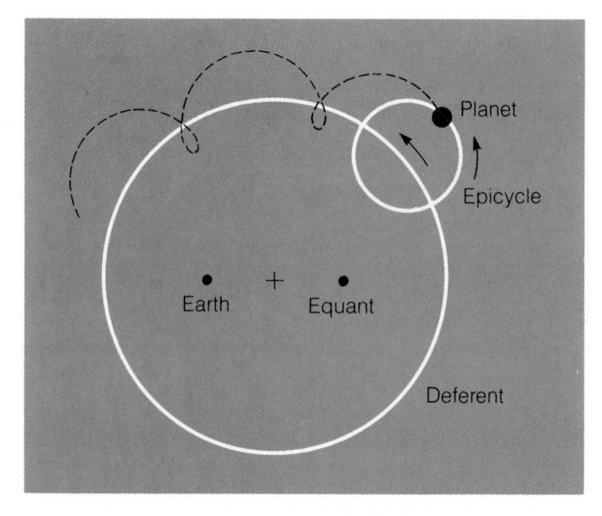

FIGURE 4–12 Ptolemy accounted for a planet's motion by placing it on a small circle (epicycle) that moved along a larger circle (deferent). Viewed from the equant, the center of the epicycle would have moved at constant speed.

the Greek belief that the heavenly bodies move perfectly with uniform circular motion. But simple, circular paths centered on the earth do not account for the motions of the planets in the sky. The planets sometimes move faster and sometimes slower, and occasionally they appear to slow to a stop and move backward over a period of months, tracing a **retrograde loop** (Figure 4–11).

To describe the complicated planetary motions and yet preserve a geocentric model with uniform circular motion, Ptolemy adopted a system of wheels within wheels. The planet moves in a small circle called an **epicycle**, and the center of the epicycle moves along a larger circle around the earth called a **deferent** (Figure 4–12). By adjusting the size of the circles and the rate of their motion, Ptolemy could account for most planetary movement. But as a final adjustment, he placed the earth off-center in the deferent circle and specified that the center of the epicycle would only appear to move at constant speed if viewed from a point called the **equant** located on the other side of the deferent's center. Thus, by using a few dozen circles of various sizes rotating at various rates, Ptolemy's system predicted the positions of the planets (Figure 4–13).

About AD 140 Ptolemy included this work in a book now known as *Almagest*. Ptolemy never knew that title; he called his work *Mathematical Syntaxis*. With Ptolemy's death classical astronomy ended forever. Western civilization was slipping into the shadows of the Dark Ages, and the invading Islamic nations dominated astronomy for almost 1000 years. Arabian astronomers translated, studied, and preserved many classical manuscripts, and in Arabic Ptolemy's book became *Al Magisti* (Greatest). Beginning in the 1200s, Europeans began recovering their classical heritage through Arabic translations.* In Latin, Ptolemy's book became *Almagestum* and thus our modern *Almagest*. For 1000 years Arab astronomers studied and preserved Ptolemy's work, but they made no significant improvement in his theory.

At first the Ptolemaic system predicted the positions of the planets well, but as centuries passed, errors accumulated. A watch that gains only 1 second a year will keep time well for many years, but the error gradually accumulates. After a century the watch would be 100 seconds fast. So too did the errors in the Ptolemaic system gradually accumulate as the centuries passed. Arabian and later European astronomers tried to update the system, computing new constants and sometimes adding more epicycles. In the middle of the thirteenth century a team of

*One such book was Ptolemy's *Geography*. In it he gives Earth's circumference as 18,000 km, nearly agreeing with Aristotle's estimate and ignoring the much larger result obtained by Eratosthenes. From Ptolemy's book, Christopher Columbus obtained the circumference of the earth and concluded he could easily reach Asia by sailing west. Columbus may have discovered America because Ptolemy respected the authority of Aristotle.

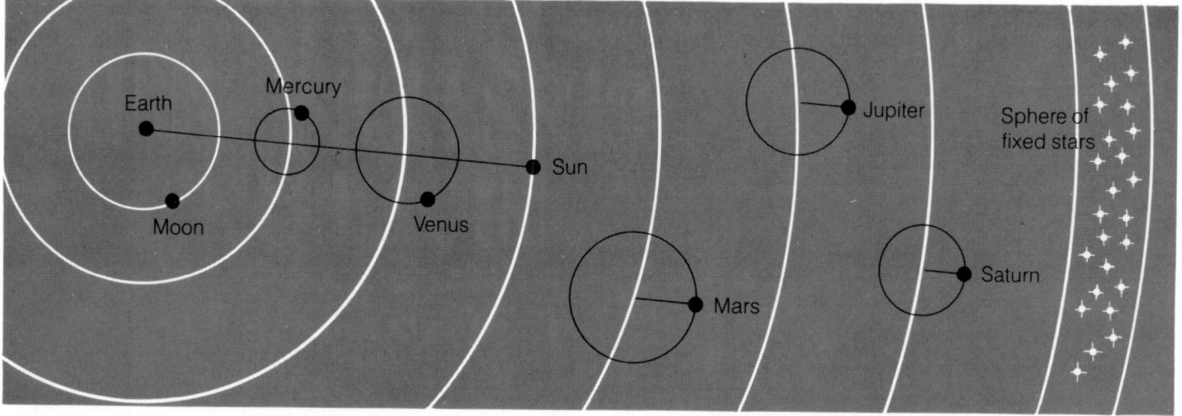

FIGURE 4–13 The Ptolemaic system was geocentric (earth-centered) and based on the uniform circular motion of epicycles. Note that the centers of the epicycles of Mercury and Venus must lie on the earth–sun line.

astronomers supported by King Alfonso X of Castile worked for 10 years revising the Ptolemaic system and publishing the result as the *Alfonsine Tables*. It was the last great adjustment of the Ptolemaic system.

4.2 THE COPERNICAN REVOLUTION

The work of Nicolaus Copernicus was revolutionary in that it dethroned the Ptolemaic theory (Figure 4–14). For more than 1600 years, astronomers had believed that the earth was fixed at the center of the universe. The Copernican theory, which maintains that the sun is at the center and that the earth is merely one of the planets that rotate on their axes and revolve around the sun, was a startling contention. Its gradual acceptance was a revolution that overthrew past authority. Indeed our use of the word *revolutionary* to mean "one who challenges accepted authority" comes from *De Revolutionibus Orbium Coelestium* (On the Revolution of the Celestial Orbs), the book in which Copernicus challenged the authority of Ptolemy and Aristotle.

COPERNICUS THE REVOLUTIONARY

By 1543, the year Copernicus died, the Ptolemaic system, in spite of many revisions, was still a poor predictor of planetary positions. Yet, because of the authority of Aristotle, it was the officially accepted theory of the universe. The Catholic Church had adopted the teachings of Aristotle as part of church dogma, and anyone who questioned the Ptolemaic system risked a charge of heresy.

Throughout his life Copernicus had been associated with the church. His uncle, by whom he was raised and educated, was an important bishop in Poland, and after studying canon law and medicine in some of the major universities in Europe, Copernicus became a canon of the church at Frauenburg at the age of 24. He served as secretary and personal physician to his powerful uncle for 15 years. When his uncle died, Copernicus returned to live in quarters adjoining the cathedral in Frauenburg. Because of this long association with the church and his fear of persecution, he hesitated to publish his revolutionary ideas in astronomy.

In fact, his theory was already being discussed long before the publication of his book. His interest in astronomy had begun during his college days, and he apparently doubted the Ptolemaic system even then. Some time well before 1530, he wrote a short pamphlet that discussed the motion of the sky and outlined his theory that the sun, not the earth, is the center of the universe and that the earth rotates on its axis and revolves around the sun. To avoid criticism and possible charges of heresy, he distributed his pamphlet in handwritten form to scientific friends. By 1515 he was well known, and by 1530 church officials were asking about his theory.

Copernicus worked on his book *De Revolutionibus Orbium Coelestium* over a period of many years and was still refining the calculations at the time of his final illness. He did not want to publish an incomplete work. In addition, he was hesitant to publish his work until he was safely beyond the reach of his critics. This was a time of

FIGURE 4–14 Nicolaus Copernicus (1473–1543) proposed that the sun, not the earth, is the center of the universe. These stamps were issued in 1973 to commemorate the 500th anniversary of his birth. Note the sun-centered model on the United States stamp and the pages from his book *De Revolutionibus* on the East German stamp (DDR).

NICOLAI COPERNICI

net, in quo terram cum orbe lunari tanquam epicyclo contineri diximus. Quinto loco Venus nono mense reducitur. Sextum denique locum Mercurius tenet, octuaginta dierum spacio circũ currens. In medio uero omnium residet Sol. Quis enim in hoc

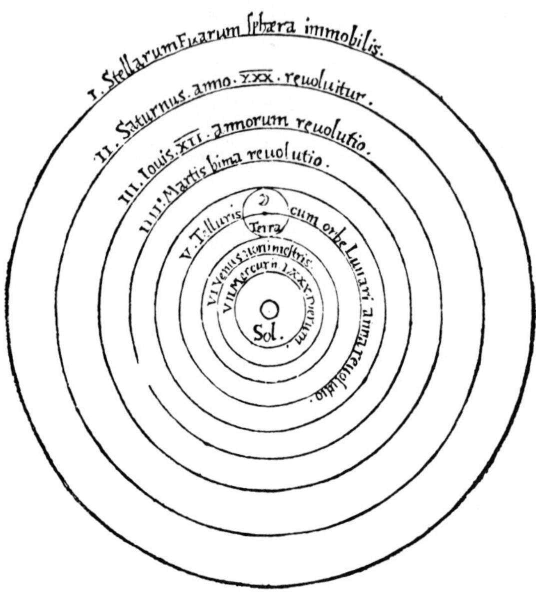

pulcherrimo templo lampadem hanc in alio uel meliori loco poneret, quàm unde totum simul possit illuminare? Siquidem non inepte quidam lucernam mundi, alij mentem, alij rectorem uocant. Trimegistus uisibilem Deum, Sophoclis Electra intuentẽ omnia. Ita profecto tanquam in solio regali Sol residens circum agentem gubernat Astrorum familiam. Tellus quoque minime fraudatur lunari ministerio, sed ut Aristoteles de animalibus ait, maximam Luna cum terra cognationẽ habet. Cõcipit interea à Sole terra, & impregnatur annno partu. Inuenimus igitur sub hac

FIGURE 4–15 The Copernican universe as reproduced in *De Revolutionibus*. The earth and all of the known planets moved in separate orbits centered on the sun, surrounded by an outer, immobile sphere of fixed stars. Compare the simplicity of this diagram with the Ptolemaic system in Figure 4–13. (Yerkes Observatory photograph.)

rebellion in the church—Martin Luther was speaking harshly about fundamental church teachings; others, both scholars and scoundrels, were questioning the authority of the church. It is understandable that the church did not welcome more criticism, even on things as abstract as astronomy. Remember also, the penalty for heresy could have been burning at the stake. Thus, Copernicus did not approve the publication of his book until he realized he was dying. Although he did see printed proof pages, he never saw the book in its final bound form.

De Revolutionibus did not prove that the Ptolemaic theory was wrong. It placed the sun at the center of the universe and thus reduced the earth to a mere planet orbiting the sun with the others. Copernicus quoted a number of criticisms of the Ptolemaic theory while defending his own, but his arguments were not conclusive. His theory did not predict the positions of the planets more accurately than Ptolemy's. Because Copernicus retained epicycles and deferents much like Ptolemy's, the *Prutenic Tables* (1551), tables of planetary position based on the Copernican system, were no more accurate than the *Alfonsine Tables.* Both could be in error by as much as 2°, four times the angular diameter of the moon.

The Copernican *system* was wrong, but the Copernican *theory,* that the universe is **heliocentric** (sun-centered), was right. Why that theory gradually won acceptance despite the inaccuracy of the epicycles and deferents is a question historians still debate. There are probably a number of reasons, including the revolutionary temper of the times, but the most important factor may be the elegance of the theory. Placing the sun at the center of the universe produced a symmetry among the motions of the planets that was pleasing to the eye as well as to

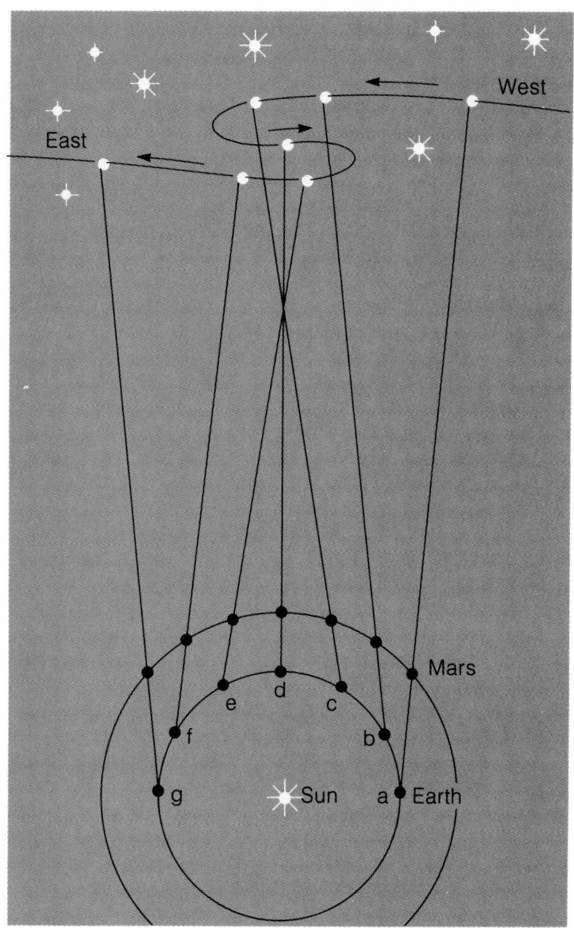

FIGURE 4–16 The Copernican explanation of retrograde motion. As Earth overtakes Mars (a-c), it appears to slow its eastward motion. As Earth passes Mars (d), it appears to move westward (retrograde). As Earth draws ahead of Mars (e-g), it resumes its eastward motion against the background stars. Compare with Figure 4–11. The positions of Earth and Mars are shown at equal intervals of 1 month.

the intellect (Figure 4–15). No longer did Venus and Mercury revolve around empty points located between Earth and sun. Now they, like the rest of the planets, move in orbits around the sun.

In addition, the Copernican theory explained the retrograde (westward) motion of the planets in a straightforward way. The earth moves faster along its orbit than the planets that lie farther from the sun. Consequently, the earth periodically overtakes and passes these planets. Imagine that you are riding in a race car driving rapidly along the inside lane of a circular race track. As you pass slower cars driving in the outer lanes, they fall behind, and, if you did not know you were moving, it would seem that the cars in the outer lanes occasionally slowed to a stop and then backed up for a short interval. The same thing happens as Earth passes a planet such as Mars. Although Mars moves steadily along its orbit, as seen from Earth, it appears to slow to a stop and move westward (retrograde) as Earth passes it (Figure 4–16). Because the planetary orbits do not lie in precisely the same plane, a planet does not resume its eastward motion in precisely the same path it followed earlier. Consequently, it describes a loop whose shape depends on the angle between the orbital planes. This simple explanation of retrograde motion did not prove that the Copernican theory was correct, but it was a much simpler explanation than that provided by the Ptolemaic theory.

However interesting astronomers found *De Revolutionibus,* it did not become an instant best seller. It was written in Latin, the language of science at that time, and it included mathematical and philosophical arguments that were beyond the grasp of most people. Thus the church took little notice at first, probably assuming that with the author dead the theory would wither in a few years. The book was not officially condemned until 1616.

GALILEO THE DEFENDER

Copernicus was a mathematician and astronomer and is remembered for his improvement of astronomical theory. Galileo Galilei (Figure 4–17) is most famous as the defender of Copernicanism. This is, to a certain extent, unfair because Galileo was a mathematician and scientist of deep insight. He made significant discoveries in the physics of motion and invented a calculating device that was a forerunner of the slide rule. However, the astronomical thrust of his life was directed at the defense of Copernicanism, not at the improvement of astronomical theory.

Galileo was born in Pisa in 1564. At the university he studied medicine, although his true interest was mathematics, mechanics, and astronomy. We can be sure he was talented in mathematics because, even though family finances forced him to leave school without a degree, he obtained a position as a professor of mathematics at the University of Pisa only 4 years later.

In Pisa his teaching assignment included astronomy, and he must have taught the Ptolemaic system. He may even then have been a Copernican, but he would have been foolish to try to teach it. Italy was no place to introduce unorthodox ideas that might challenge church teachings. Perhaps, while he taught astronomy, he became convinced that the universe is Copernican, not Ptolemaic.

In 1609 Galileo obtained some lenses and built a telescope (Figure 4–18), following descriptions of similar instruments built by Dutch lens makers. He did not invent the telescope, nor was he the first to turn one toward the sky, but he was the first to study, night after night, the telescopic appearance of the heavenly bodies, record his observations, and report them publicly. In 1610 he published a book called *Sidereus Nuncius* (The Sidereal Messenger), in which he described what he saw through his telescope and showed how his observations supported the Copernican theory.

One of his important discoveries was that Venus goes through phases like the moon. In the Ptolemaic theory Venus revolves around an epicycle located between Earth and the sun (Figure 4–13). Thus an observer on Earth would never see the planet fully illuminated by the sun. That is, it would always be seen as a crescent. But Galileo saw Venus go through a complete set of phases, which proves that it goes around the sun (Figure 4–19).

In addition, he discovered four satellites orbiting the planet Jupiter. Some critics of the Copernican theory had said Earth could not move because the moon would be left behind. But Jupiter moved yet kept its satellites, so Galileo's discovery proved that Earth could move and keep its moon. Also, the accepted teaching was that everything revolves around Earth. Galileo's telescope revealed that some objects orbit Jupiter, proving there could be other centers of motion.

Finally, the telescope challenged Aristotle's contention that the heavens are perfect and unchanging. Galileo observed spots on the sun, raising the suspicion that the sun is less than perfect. By noting the movement of the spots, he concluded that the sun rotates on its axis, just as Copernicus said the earth does. When he observed the moon, he found not a polished perfect sphere, but a craggy, mountainous terrain surrounding flat areas that were later mistaken for seas.

Sidereus Nuncius was a popular book, and it made Galileo famous. He soon left his teaching position and went to Florence where he became personal philosopher and mathematician to the Grand Duke of Tuscany. When Galileo visited Rome in 1611, he gave well-attended lectures and had long and friendly discussions with the powerful Cardinal Barberini. But he also made enemies. Personally, Galileo was outspoken, forceful, sometimes tactless, and he enjoyed debate. Most of all he enjoyed being right. In the years following publication of *Sidereus Nuncius,* in lectures, debates, and letters, he offended important people who questioned his telescopic discoveries.

FIGURE 4–17 Galileo Galilei (1564–1642), remembered as the great defender of Copernicanism, also made important discoveries in the physics of motion. He is honored here on an Italian 2000-lira note. The reverse side shows one of his telescopes at lower right and a modern observatory above.

By 1616 Galileo was the center of a storm of controversy. Some said he was mistaken (or worse), and others refused even to look through a telescope. Pope Paul V decided to end the disruption, and so, when Galileo visited Rome in 1616, Cardinal Bellarmine interviewed him and ordered him to end his astronomical work. Books relevant to Copernicanism were banned, including *De Revolutionibus,* and Galileo rode back to Florence.

In some respects the years that followed gave Galileo hope that he could resume his work. His interviews with the pope and Cardinal Bellarmine in 1616 had been stern but not disrespectful, and in a last interview before he left Rome the pope had shown open friendship. Then in 1621 Pope Paul V died and his successor, Pope Gregory XV, died in 1623. The next pope was Galileo's friend Cardinal Barberini, who took the name Urban VIII. Galileo

FIGURE 4–18 Galileo's telescopic discoveries generated intense interest and controversy. Some critics refused to look through a telescope lest it deceive them. (Yerkes Observatory photograph.)

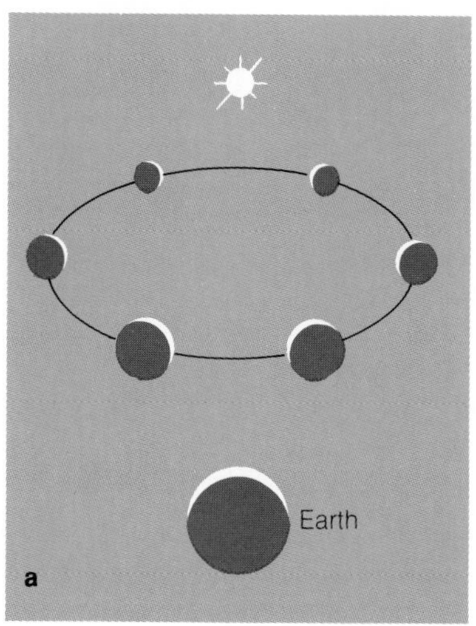

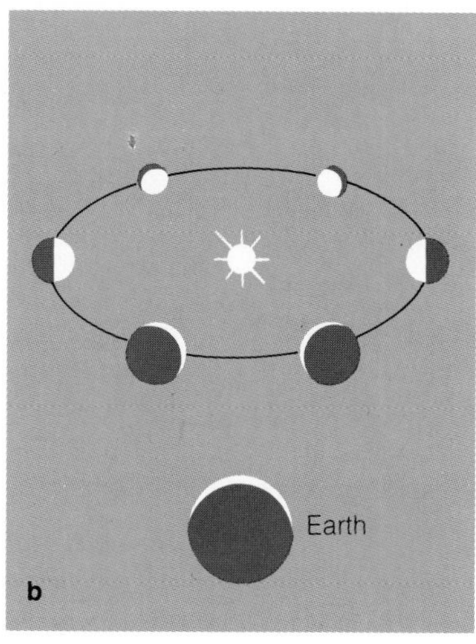

FIGURE 4–19 (a) In the Ptolemaic universe Venus moves around an epicycle between Earth and sun and would always appear as a crescent. (b) Galileo's telescope showed that Venus goes through a full set of phases, proving that it must orbit the sun as in the Copernican universe. Compare with Figure 4–13.

visited the new pope and, though the order of 1616 was not revoked, Urban VIII was friendly.

In 1624 Galileo began to write a book defending the Copernican theory. It was a long project that was finally completed in the last days of 1629. After some delay the book was approved by the local censor in Florence and by the head censor of the Vatican in Rome. It was printed in 1632.

The book was called *Dialogo Dei Due Massimi Sistemi* (Dialogue Concerning the Two Chief World Systems) (Figure 4–20). It was written as a conversation between three friends who debate the Copernican and Ptolemaic theories. Salviati, a swift-tongued defender of Copernicus, dominates the book. Sagredo is a reasonably intelligent but uninformed believer in the Ptolemaic theory who is gradually convinced to adopt Copernicanism. Simplicio, the third character in the book, is the dismal defender of Ptolemy. In fact, Simplicio does not seem very bright.

Galileo seriously misjudged his safety in writing the book. For one thing, the book was not an unbiased account of the two systems, although he later claimed it was. More importantly, either intentionally or unintentionally, he exposed the pope's authority to possible ridicule. Pope Urban VIII was fond of arguing that, as God was omnipotent, He could construct the universe in any form while making it appear to us to have a different form. Galileo placed this argument in the mouth of Simplicio, and the enemies of Galileo showed the passage to the pope as an example of Galileo's disrespect. The pope suddenly ordered Galileo to Rome to meet the Inquisition.

Galileo was interrogated by the Inquisition four times. He was not tortured, but he was threatened with torture. At first Galileo tried to maintain his pride and defend his ideas, but he was soon cowed by the Inquisition. He must have thought often of Giordano Bruno, who had been tried, condemned, and burned at the stake in Rome in 1600. Bruno had been a bitter critic of the church on a number of theological questions, but one of his offenses was Copernicanism.

In the end, because Galileo's book had been approved by the censors, the Inquisition had to return to the orders given Galileo in 1616. The official record of the interview between Galileo and Cardinal Bellarmine included the statement that Galileo was "not to hold, teach, or defend in any way" the principles of Copernicus.* Galileo's assertion that his dialogue was an unbiased discussion, not a defense of any one theory, was worthless even if it had been true.

On June 22, 1633, at the age of 70, Galileo knelt before the Inquisition and read a recantation admitting his errors. Tradition has it that as he rose he whispered *"Eppur si muove"* (still it moves) referring to the earth, but it is unlikely that anyone with any sense would risk such defiance.

Galileo was sentenced to life imprisonment. Perhaps through the intervention of the pope, he was held in confinement at his villa, where he could meet his family, though other visitors were forbidden. During these years Galileo studied mechanics and physics and even wrote a book on the subject that was published in Holland in 1638.

During his last few years he was allowed a few visitors, and two young scientists came to stay with him. As he was blind by then, he no doubt enjoyed their discussions. At last, on January 8, 1642, after 10 years' imprisonment, 99 years after the death of Copernicus, Galileo died.

The church's condemnation of Galileo has been a subject of heated debate for 350 years. Many have argued that it shows that religion is incompatible with science and is incapable of incorporating new discoveries into religious teachings. In 1979, however, Pope John Paul II ordered a reexamination of the case against Galileo. Referring specifically to Galileo, whom he called "the founder of modern physics," Pope John Paul II said, "We cannot but deplore certain attitudes that have led many to conclude that faith and science are mutually opposed." It appears that Galileo has been forgiven.

*Because the document is unsigned, some scholars suspect it was forged and that Galileo never received those instructions in 1616.

FIGURE 4–20 Aristotle, Ptolemy, and Copernicus discuss astronomy in this frontispiece from Galileo's book *Dialogue Concerning the Two Chief World Systems.* (Courtesy Owen Gingerich and The Houghton Library.)

4.3 PLANETARY MOTION

While Galileo defended Copernicanism in Florence and Rome, two other astronomers worked in northern Europe beyond the sway of the Inquisition. One made precise observations of the planetary positions, and the other analyzed those positions mathematically. Their lives culminated in conclusive proof that the Copernican system

FIGURE 4–21 Tycho Brahe (1546–1601), a Danish nobleman, established an observatory at Hveen and measured planetary positions with high accuracy. His artificial nose is suggested in this engraving.

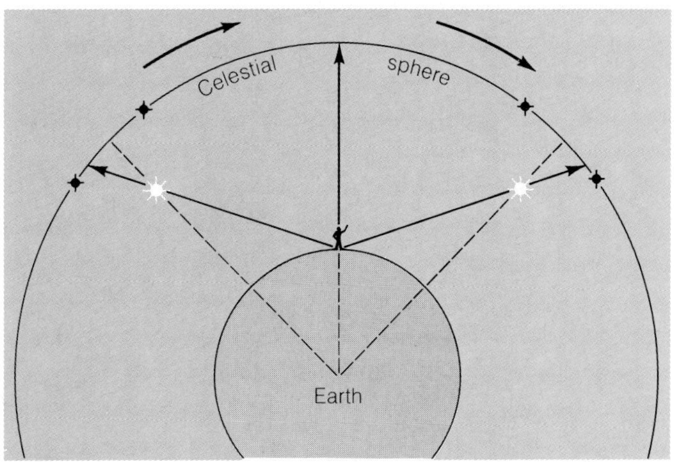

FIGURE 4–22 Daily parallax from a stationary earth. Looking down on the earth and celestial sphere from the north celestial pole, we see the celestial sphere rotating westward. The supernova is shown here closer to the earth than the stars of the celestial sphere. Because the observer is not at the earth's center, the supernova would appear to move against the background stars through the night. Tycho could not detect this daily parallax, and so he concluded the new star was no closer to earth than the moon and probably lay on the sphere of the stars.

could work and eventually led to a new understanding of planetary motion and the nature of the universe.

TYCHO THE OBSERVER

The great observational astronomer of our story is a Danish nobleman, Tycho Brahe (Figure 4–21), born December 14, 1546, only 3 years after the publication of *De Revolutionibus.* Great figures in history are often referred to by their last name, but Tycho Brahe is usually called Tycho. Were he alive today, he would no doubt object to such familiarity from his obvious inferiors. He was well known for his vanity and lordly manners.

Tycho's college days were eventful. He was officially studying law with the expectation that he would enter Danish politics, but he made it clear to his family that his real interest was astronomy and mathematics. It was also during his college days that he became involved in a duel and received a wound that disfigured his nose. For the rest of his life he wore false noses made of gold and silver and stuck on with wax. The disfigurement probably did little to improve his disposition.

Tycho's first astronomical observations were made while he was a student. In 1563 Jupiter and Saturn passed very near each other in the sky, nearly merging into a single point on the night of August 24. Tycho found that the *Alfonsine Tables* were a full month in error and that the *Prutenic Tables* were in error by a number of days. These discrepancies dismayed Tycho and sparked his interest in the motions of the planets.

In 1572 a brilliant "new star" (now called Tycho's supernova) appeared in the sky. Such changes in the stars puzzled classically trained astronomers. Aristotle had argued that the starry sphere was perfect and unchanging, and therefore such new stars had to lie closer to the earth than the moon. Layers below the moon were thought to be less perfect and thus more changeable.

Tycho observed that the new star moved westward through the night, keeping pace with the stars, so he concluded that its position should shift slightly through the night as seen by an observer on earth (Figure 4–22). This

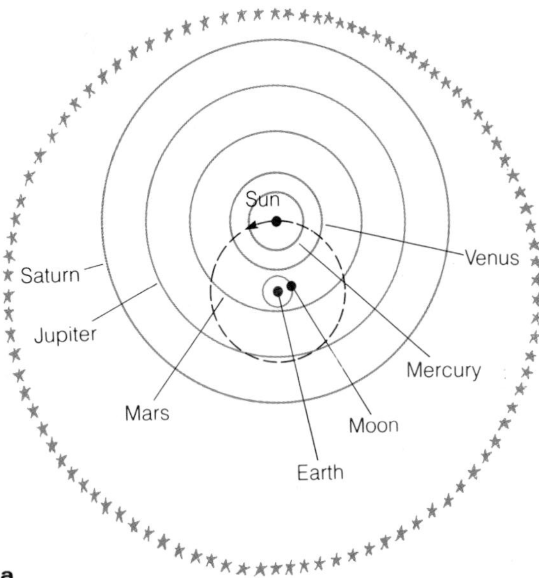

FIGURE 4–23 (a) Tycho Brahe's model of the universe held that the earth was fixed at the center of the starry sphere. The moon and sun circled the earth while the planets circled the sun. (b) Much of Tycho's success was due to his skill in designing large, accurate instruments. In this engraving of his mural quadrant, the figure of Tycho, his dog, and the background scene are painted on the wall within the arc of the quadrant. The observer (Tycho himself) at the extreme right peers through a sight out the loophole in the wall at the upper left to measure a star's altitude above the horizon. (Yerkes Observatory photograph.)

change in position, called daily parallax, would make the star appear slightly east of its average position when it was in the eastern sky and slightly west of its average position when it was in the western sky. Tycho had great confidence in his talents for astronomy and mathematics and was sure he could detect this daily parallax, if it existed, by carefully measuring the position of the new star against background stars.

Tycho saw no daily parallax in the new star and concluded that it was farther away than the moon and was probably among the stars of the celestial sphere. This was a dramatic discovery because Aristotle held that the heavens above the moon were perfect and unchanging. Classical astronomers had always believed that changing objects such as comets and new stars were sublunar (below the moon). The appearance of a new star in the supposedly unchanging heavens beyond the moon forced Tycho to question the Ptolemaic system. He summarized his results in a small book, *De Stella Nova* (The New Star), published in 1573.

The book attracted the attention of astronomers throughout Europe, and soon Tycho was summoned to the court of the Danish king Frederik II and offered funds to build an observatory on the island of Hveen just off the Danish coast. Tycho also received a steady source of income as landlord of a coastal district from which he collected rents. (He was not a popular landlord.) On Hveen, Tycho constructed a luxurious home with four towers especially equipped for astronomy and populated it with servants, assistants, and a dwarf to act as jester. Soon Hveen was an international center of astronomical study.

Tycho's great contribution to the birth of modern astronomy was not theoretical. In fact, his grand theory of the universe was wrong. Because he could measure no parallax for the stars, he concluded the earth had to be stationary, thus rejecting the Copernican theory. How-

FIGURE 4–24 Johannes Kepler (1571–1630) derived three laws of planetary motion from Tycho Brahe's observations of the positions of the planets. This Romanian stamp commemorates the 400th anniversary of Kepler's birth. Ironically it contains an error—the orientation of the moon.

ever, he also rejected the Ptolemaic theory because of its inaccurate predictions. Instead, he devised a complex theory in which the earth is the immobile center of the universe around which the sun and moon move. The remaining planets orbit the sun (Figure 4–23a). This system is nearly the same as the Copernican system except that the earth is held fixed instead of the sun.

The true value of Tycho's work was observational. Because he was able to devise new and better observing instruments, he was able to make highly accurate observations of the positions of the stars, sun, moon, and planets. Tycho had no telescopes—they were not invented until after his death—so his observations were made by the naked eye peering along sights on his large instruments (Figure 4–23b). Despite these limitations, he measured the positions of 777 stars to better than 4 minutes of arc and measured the positions of the sun, moon, and planets almost daily for the 20 years he stayed on Hveen.

Unhappily for Tycho, King Frederik II died in 1588, and his young son took the throne. Suddenly, Tycho's temper, vanity, and noble presumptions threw him out of favor. In 1596, taking most of his instruments and books of observations, he went to Prague, the capital of Bohemia, and became imperial mathematician to the Holy Roman Emperor Rudolph II. His assignment there was to revise the *Alfonsine Tables* and publish the revision as a monument to his new patron. It would be called the *Rudolphine Tables.*

Tycho intended to base the *Rudolphine Tables* not on the Ptolemaic system, but rather on his own Tychonic system, proving once and for all the validity of his theory. To assist him, he hired a few mathematicians and astronomers, including one Johannes Kepler. Then in November 1601, Tycho overate at a nobleman's home, suffered a serious attack, and collapsed. Before he died 9 days later, he asked Rudolph II to make Kepler imperial mathematician. Thus, the newcomer Kepler became Tycho's replacement (though at half Tycho's salary).

KEPLER THE ANALYST

No one could have been more different from Tycho Brahe than Johannes Kepler (Figure 4–24). He was born December 27, 1571, to a poor family in a region now included in southwest Germany. His father was unreliable and shiftless, principally employed as a mercenary soldier, fighting for whoever paid enough. He finally failed to return from a military expedition, either because he was killed or because he found circumstances more to his liking. Kepler's mother was apparently an unpleasant and unpopular woman. She was accused of witchcraft in later years, and Kepler had to defend her in a trial that dragged on for 3 years. She was finally acquitted but died the following year.

Kepler was the oldest of six children, and his childhood was no doubt unhappy. The family was not only poor and often lacked a father, but it was also Protestant in a predominantly Catholic region. In addition, Kepler was never healthy, even as a child, so it is surprising that he did well in the pauper's school he attended, eventually winning a scholarship to the university at Tübingen, where he studied to become a Lutheran pastor.

During his last year of study, Kepler accepted a job in Graz teaching mathematics and astronomy. Evidently, he was not a good teacher. He had few students his first year, and none at all his second. His superiors put him to work teaching a few introductory courses and preparing an annual almanac that contained astronomical, astrological, and weather predictions. Through good luck in 1595 some of his weather predictions were fulfilled and he gained a reputation as an astrologer and seer, and even in later life he earned money from his almanacs.

While still a college student, Kepler had become a believer in the Copernican theory, and at Graz he used his extensive spare time to study astronomy. By 1596, the same year Tycho left Hveen, Kepler was ready to solve the mystery of the universe. That year he published a book called *The Forerunner of Dissertations on the Universe, Containing the Mystery of the Universe.* Like nearly all scientific works, the book was in Latin, and is now known as *Mysterium Cosmographicum.*

The book begins with a long appreciation of Copernicanism and then goes on to speculate on the spacing of the planetary orbits. Kepler knew of the five planets, Mercury, Venus, Mars, Jupiter, and Saturn, with the moon considered a sixth planet. According to his theory, the spheres containing the six planets were separated and their relative sizes fixed by the five regular solids*—the cube, tetrahedron, dodecahedron, icosahedron, and octahedron (Figure 4–25). Kepler went on to give astrological, numerological, and even musical arguments for his theory.

The second half of the book is no better than the first, but it has one virtue. As Kepler tried to fit the five solids to the planetary orbits, he demonstrated that he was a talented mathematician and that he was well versed in astronomy. He sent copies to Tycho and to Galileo, and both recognized his talent in spite of the mystical content of the book.

Life was unsettled for Kepler because of the persecution of Protestants in the region, so when Tycho invited him to Prague in 1600, he came readily, anxious to work with the famous astronomer. Tycho's sudden death in 1601 left Kepler in a position to use the observations from Hveen to analyze the motions of the planets and complete the *Rudolphine Tables.* Tycho's family, recognizing that

*A regular solid is a three-dimensional body each of whose faces is the same. A cube is a regular solid, each of whose faces is a square.

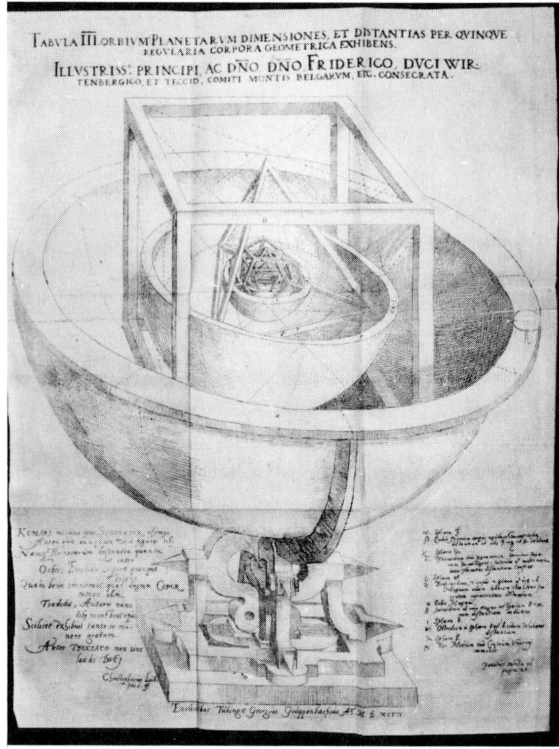

FIGURE 4–25 Kepler believed the five regular solids were the spacers between the spheres containing the planetary orbits. His book *Cosmographicum Mysterium* contained this foldout illustration of the spheres and spacers. (Courtesy Owen Gingerich and The Houghton Library.)

Table 4–1 Kepler's laws of planetary motion.

I. The orbits of the planets are ellipses with the sun at one focus.

II. A line from the planet to the sun sweeps over equal areas in equal intervals of time.

III. A planet's orbital period in years squared is equal to its average distance from the sun in astronomical units cubed:

$$P_y^2 = a_{AU}^3$$

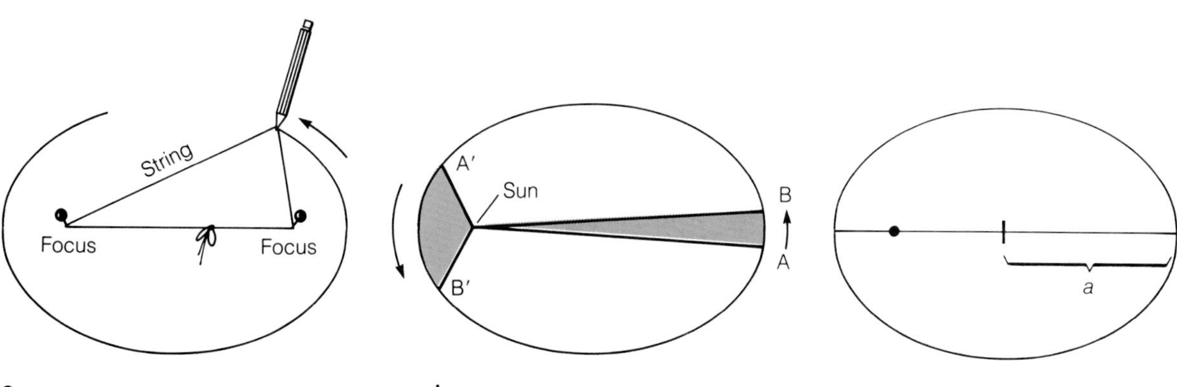

FIGURE 4–26 (a) Drawing an ellipse with two tacks and a loop of string. (b) "A line from a planet to the sun sweeps over equal areas in equal intervals of time." (c) The average distance from a planet to the sun equals *a*, the semimajor axis of its orbit.

Kepler was a Copernican and guessing that he would not follow the Tychonic system in completing the *Rudolphine Tables,* sued to recover the instruments and books of observations. The legal wrangle went on for years. They did recover the instruments Tycho had brought to Prague, but Kepler had the books and he kept them.

Whether Kepler had any legal right to Tycho's records is debatable, but he put them to good use. He began by studying the motion of Mars, trying to deduce from the observations how the planet moves. By 1606 he had solved the mystery, this time correctly. The orbit of Mars (and all planets) is an ellipse with the sun at one focus. Thus, he abandoned the 2000-year-old belief in circular motion. But the mystery is even more complex. The planets do not move at constant speed along their orbits—they move faster when close to the sun and slower when farther away. Kepler therefore abandoned both uniform motion and circular motion.

Kepler published his results in 1609 in a book called *Astronomia Nova* (New Astronomy). Like Copernicus's book, *Astronomia Nova* did not become an instant success. It is written in Latin for other scientists and is highly mathematical. In some ways the book is surprisingly advanced. For instance, Kepler discusses the force that holds the planets in their orbits and comes within a paragraph of discovering the principle of mutual gravitation.

Despite the abdication of Rudolph II in 1611, Kepler continued his astronomical studies. He wrote about a supernova that had appeared in 1604 (now known as Kepler's supernova) and about comets, and he wrote a textbook about Copernican astronomy. In 1619 he published *Harmonice Mundi* (The Harmony of the World) in which he returned to the cosmic mysteries of *Mysterium Cosmographicum.* The only thing of note in *Harmonice Mundi* is his discovery that the radii of the planetary orbits are related to the planet's orbital period. That and his two previous discoveries are now recognized as Kepler's three laws of planetary motion (Table 4–1).

Kepler's first law states that the orbits of the planets around the sun are ellipses with the sun at one focus. An **ellipse** is defined as a figure drawn around two points called the **foci** (singular **focus**) such that the distance from one focus to any point on the ellipse back to the other focus equals a constant, the longest diameter of the ellipse. This makes it very easy to draw ellipses with two thumbtacks and a loop of string. Press the thumbtacks into a board, loop the string about them, and place a pencil in the loop as in Figure 4–26a. If you keep the string taut as you move the pencil, it traces out an ellipse. The closer together the thumbtacks, the more nearly circular the ellipse. Though Kepler was able to determine the elliptical shape of the planetary orbits, they are nearly circular. Of the planets known to Kepler, Mercury has the most elliptical orbit, but even it differs only slightly from a circle (Figure 4–27).

Kepler's second law states that a line from the planet to the sun sweeps over equal areas in equal intervals of time. This means that when the planet is closer to the sun and the line connecting it to the sun is shorter, the planet must move more rapidly if the line is to sweep over the same area. Thus, the planet in Figure 4–26b would move from point A′ to point B′ in 1 month sweeping out the area shown. But when the planet is farther from the sun, 1 month's motion would carry it from A to B.

Kepler's third law states that a planet's orbital period squared is proportional to its average distance from the sun cubed. Because of the way the planet moves along its orbit, its average distance from the sun is equal to half of the long diameter of the elliptical orbit, the so-called **semimajor axis** *a* (Figure 4–26c). If we measure this quantity in astronomical units and the orbital period in years, we can summarize the third law as:

$$P_y^2 = a_{au}^3$$

For example, Jupiter's average distance from the sun is roughly 5 AU. The semimajor axis cubed would be 125, so the period must be the square root of 125, about 11 years.

It is important to notice that Kepler's three laws are empirical. That is, they describe the phenomenon without explaining why it occurs. Kepler derived them from Tycho's extensive observations, not from any fundamental assumption or theory. In fact, Kepler never knew what holds the planets in their orbits or why they continue to move around the sun. His books are a fascinating blend of careful observation, mathematical analysis, and mystical theory.

In spite of Kepler's recurrent affairs with astrology and numerology, he continued to work on the *Rudolphine Tables*. At last in 1628 they were ready, and he financed their printing himself, dedicating them to the memory of Tycho Brahe. In fact, Tycho's name appears in larger type on the title page than Kepler's own. This is especially surprising when we recall that the tables were based on the heliocentric theory of Copernicus and the elliptical orbits of Kepler, not on the Tychonic system. The reason for Kepler's care was Tycho's family, still powerful and still intent on protecting the memory of Tycho. They even demanded a share of the profits and the right to censor the book before publication, although they changed nothing but a few words on the title page.

The *Rudolphine Tables* was Kepler's masterpiece, the final proof that the Copernican theory was an accurate description of the heavens. It was the proof that Copernicus had sought but failed to find.

Kepler died November 15, 1630. During his life he had been an astrologer, mystic, numerologist, and seer, but he became one of the great astronomers. His work paved the way for the rise of modern astronomy.

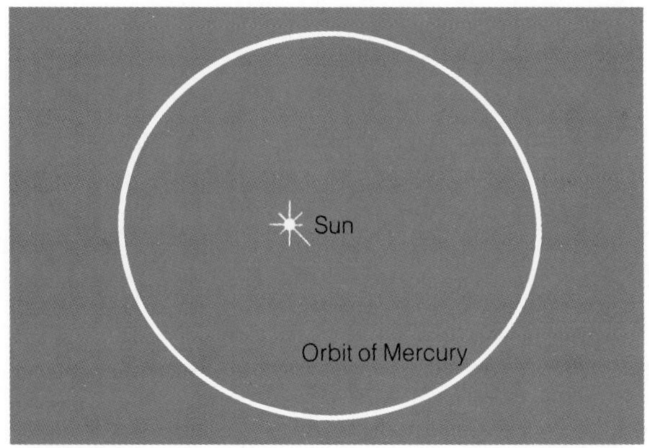

FIGURE 4–27 The orbits of the planets are nearly circular. Of the planets known to Kepler, Mercury has the most elliptical orbit as shown in this scale drawing. Pluto's orbit is only slightly more elliptical.

4.4 MODERN ASTRONOMY

We date the origin of modern astronomy from the 99 years between the deaths of Copernicus and Galileo because it was an age of transition. That period marked the transition between the Ptolemaic theory and the Copernican, but it also marked a transition in the nature of astronomy in particular and science in general. Before the events of our story, scientific principles were drawn not from observation but from philosophical judgments of what the universe should be like. Thus, Aristotle believed the heavenly bodies are perfect because he felt they should be. In such an atmosphere scientific discoveries and observations had to be bent to fit expectations.

The discoveries of Kepler and Galileo were accepted in the 1600s because the world was in transition. Astronomy was not the only thing changing during this period. The Renaissance is commonly taken to be the period between 1300 and 1600, and thus the 99 years of this history lie at the culmination of the reawakening of learning in all fields (Figure 4–28). The world was open to new ideas and new observations. Martin Luther remade religion, and other philosophers and scholars reformed their areas of human knowledge. Had Copernicus not published his theory, someone else would have suggested that the universe is heliocentric. History was ready to shed the Ptolemaic theory.

In addition, this period marks the beginning of the modern scientific method. Beginning with Copernicus,

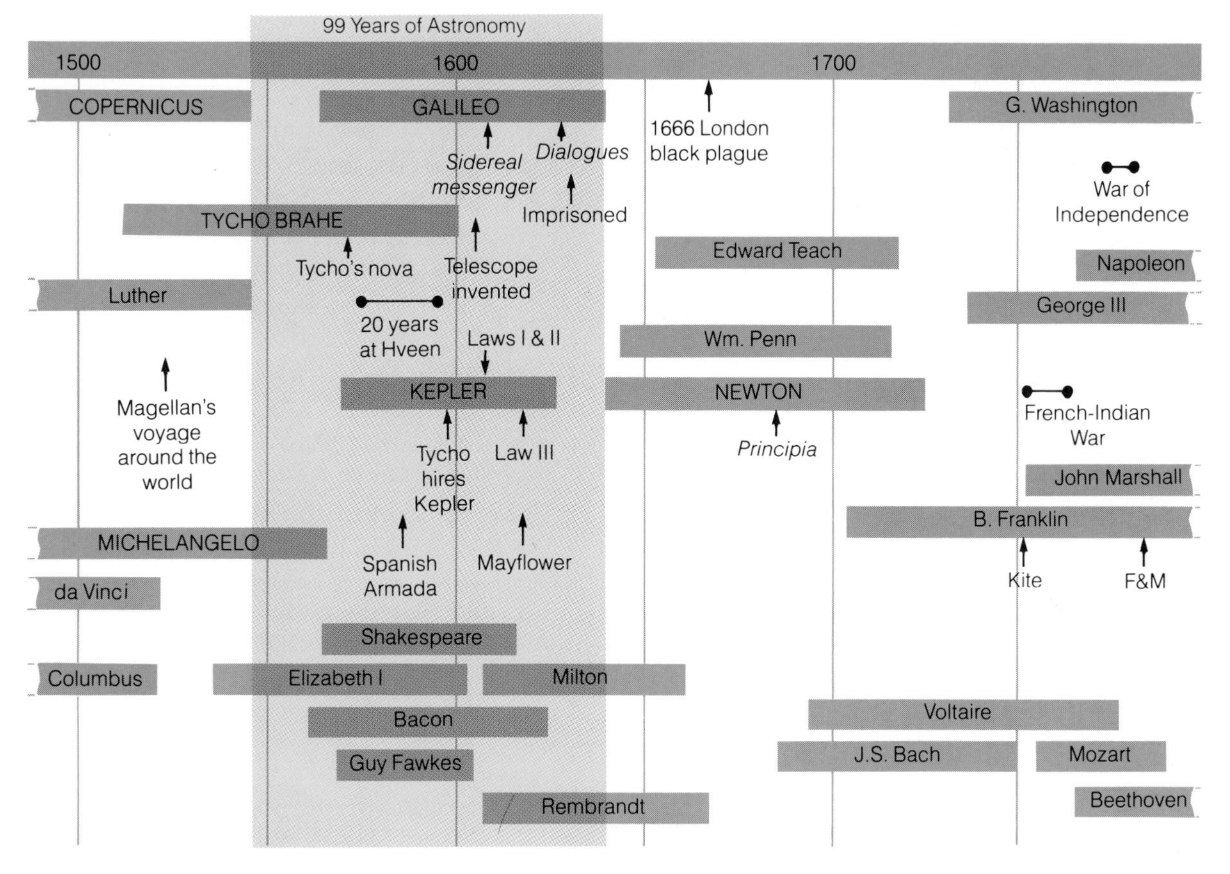

FIGURE 4–28 The 99 years between the deaths of Copernicus and Galileo (1543–1642) marked the transition from the ancient astronomy of Ptolemy and Aristotle to the revolutionary theory of Copernicus. This period saw the birth of modern scientific astronomy.

Tycho, Kepler, and Galileo, scientists depended more and more on evidence, observation, and measurement. This, too, is coupled to the Renaissance and its advances in metalworking and lens making. Before our story began, no astronomer had looked through a telescope because one could not be made. By 1642, not only telescopes, but also other sensitive measuring instruments made science into something new and precise. Also, the growing number of scientific societies increased the exchange of observations and theories among scientists and stimulated more and better work. However, the most important advance was the application of mathematics to scientific questions. Kepler's work demonstrated the power of mathematical analysis, and as the quality of these numerical tools improved, the progress of science accelerated. Our story, therefore, is the story of the birth not just of modern astronomy, but of modern science as well.

We will continue this story in the next chapter with the life of Isaac Newton. We will see how his accomplishments had their origins in the work of Galileo and Kepler.

SUMMARY

Archaeoastronomy, the study of the astronomy of ancient cultures, is revealing that primitive astronomers could make sophisticated astronomical observations. The Stone Age people of Britain who built Stonehenge aligned its long avenue and Heelstone to point toward the rising sun at the summer solstice. Other stones mark sight lines toward the rising moon. It seems clear that Stonehenge was a calendrical device, and some believe that its builders used it to predict eclipses.

Other archaeoastronomy sites span the world. A great many stone circles dot England and northern Europe, and related pat-

terns, called medicine wheels, show that North American Indians also practiced astronomy. The Mound Builders of the Midwest, the Indians of the Southwest, and the Aztecs, Maya, and Incas of Central and South America also studied astronomy.

Traditional histories of astronomy usually begin with the Greeks. The Greek philosopher Aristotle held that the earth is fixed at the center of the universe, and Ptolemy based a mathematical model of the moving planets on this geocentric universe. To preserve the concept of uniform circular motion, he used epicycles, deferents, and equants for each of the planets, sun, and moon.

Near the end of his life, in 1543, Nicolaus Copernicus published his theory that the sun is the center of the universe. Because the teachings of Aristotle had been adopted as official church doctrine, the heliocentric theory of Copernicus was considered radical.

Galileo Galilei was the first astronomer to use the newly invented telescope for a systematic study of the heavens, and he found that the appearance of the sun, moon, and planets supported the heliocentric theory. After he described his findings in a book, the Church, in 1616, ordered him to stop teaching the theory. When he published another book on the subject in 1632, he was tried by the Inquisition and ordered to recant. He was imprisoned until his death in 1642.

In northern Europe, beyond the power of the church, Tycho Brahe rejected the Ptolemaic theory and the Copernican theory. In Tycho's own theory, the earth is fixed at the center of the universe and the sun and moon revolve around the earth. The planets orbit the moving sun. Although his theory is wrong, he was a brilliant observer and accumulated 20 years of precise measurements of the positions of the sun, moon, and planets.

At Tycho's death, one of his assistants, Johannes Kepler, became his successor. Analyzing the observations, Kepler finally discovered how the planets move. His three laws of planetary motion denied uniform circular motion. The first law says the orbits of the planets are elliptical rather than circular, and the second law says the planets do not move at a uniform rate but move faster when near the sun and slower when more distant. The third law relates orbital period to orbital size.

NEW TERMS

archaeoastronomy	deferent
uniform circular motion	equant
geocentric universe	heliocentric universe
parallax (p)	ellipse
retrograde loop	focus
epicycle	semimajor axis

QUESTIONS

1. What evidence do we have that early human cultures observed astronomical phenomena?
2. Why did Plato propose that all heavenly motion was uniform and circular?
3. How do the epicycles of Mercury and Venus differ from those of Mars, Jupiter, and Saturn?
4. Suggest a number of reasons why Copernicus may have delayed publication of his book.
5. Why did Copernicus have to keep epicycles and deferents in his system?
6. Explain how each of Galileo's telescopic discoveries supported the Copernican theory.
7. Galileo was condemned but Kepler, also a Copernican, was not. Why not?
8. What connections do you see between the Vatican's recent decision to forgive Galileo and the controversy over teaching evolution in public schools?
9. When Tycho observed the new star of 1572, he could detect no parallax. Why did that result undermine belief in the Ptolemaic theory?
10. Does Tycho's theory of the universe explain the phases of Venus that Galileo observed? Why or why not?
11. How do the first two of Kepler's three laws overthrow one of the basic beliefs of classical astronomy?

PROBLEMS

1. Draw and label a diagram of the eastern horizon from northeast to southeast and label the rising point of the sun at the solstices and equinoxes. (See Figures 2–15 and 4–3b.)
2. If you lived on Mars, which planets would describe retrograde loops? Which would never be visible as crescent phases?
3. Galileo's telescope showed him that Venus has a large angular diameter (61 seconds of arc) when it is a crescent and a small angular diameter (10 seconds of arc) when it is nearly full. Use the small-angle formula to find the ratio of its maximum distance to its minimum distance. Is this ratio compatible with the Ptolemaic universe shown in Figure 4–13?
4. Galileo's telescopes were not of high quality by modern standards. He was able to see the rings of Saturn, but he

never reported seeing features on Mars. Use the small-angle formula to find the angular diameter of Mars when it is closest to Earth. How does that compare with the maximum angular diameter of Saturn's rings?

5. If a planet has an average distance from the sun of 4 AU, what is its orbital period?
6. If a space probe is sent into an orbit around the sun, which brings it as close as 0.5 AU and as far away as 5.5 AU, what will be its orbital period?
7. Pluto orbits the sun with a period of 247.7 years. What is its average distance from the sun?

RECOMMENDED READING

ARMITAGE, A. *Copernicus: The Founder of Modern Astronomy.* New York: A. S. Barnes, 1962.

AVENI, A. F. "Tropical Archeoastronomy." *Science 213* (10 July 1981), p. 161.

BERRY, A. *A Short History of Astronomy.* 1898. Reprint. New York: Dover, 1961.

BOORSTIN, DANIAL J. *The Discoverers.* New York: Vintage Books, 1983.

BROAD, WILLIAM J. "A Bibliophile's Quest for Copernicus." *Science 218* (12 Nov. 1982), p. 661.

BRONOWSKI, J. *The Ascent of Man.* Chapters 6 and 7. Boston: Little, Brown, 1973.

BULLOUGH, VERN L. (Ed.). *The Scientific Revolution.* New York: Krieger Publishing, 1970.

CHRISTIANSON, J. "The Celestial Palace of Tycho Brahe." *Scientific American 204* (Feb. 1961), p. 118.

DICKS, D. R. *Early Greek Astronomy to Aristotle.* Ithaca, N.Y.: Cornell University Press, 1970.

DREYER, J. L. E. *A History of Astronomy from Thales to Kepler.* 1906. Reprint. New York: Dover, 1981.

DURHAM, FRANK, and ROBERT D. PURRINGTON *Frame of the Universe.* New York: Columbia University Press, 1983.

FERMI, L., and G. BERNARDINI *Galileo and the Scientific Revolution.* Greenwich, Conn.: Fawcett Publications, 1965.

GINGERICH, O. "Johannes Kepler and the Rudolphine Tables." *Sky and Telescope 42* (Dec. 1971), p. 328.

———. "Copernicus and Tycho." *Scientific American 229* (Dec. 1973), p. 86.

———. "Tycho Brahe and the Great Comet of 1577." *Sky and Telescope 54* (Dec. 1977), p. 452.

———. "Great Conjunctions, Tycho, and Shakespeare." *Sky and Telescope 61* (May 1981), p. 394.

———. "The Galileo Affair." *Scientific American 247* (Aug. 1982), p. 133.

———. "Laboratory Exercises in Astronomy—The Orbit of Mars." *Sky and Telescope 66* (Oct. 1983), p. 300.

HAWKINS, GERALD S. *Stonehenge Decoded.* Garden City, N.Y.: Doubleday, 1965.

HOYLE, FRED *On Stonehenge.* San Francisco: W. H. Freeman, 1977.

KOESTLER, A. *The Watershed: A Biography of Johannes Kepler.* Garden City, N.Y.: Doubleday, 1960.

KOYRE, A. *From the Closed World to the Infinite Universe.* New York: Harper & Row, 1958.

KRUPP, E. C. *Echoes of the Ancient Skys.* Cambridge, Mass.: Harper & Row, 1983.

KUHN, THOMAS S. *The Copernican Revolution.* Cambridge, Mass.: Harvard University Press, 1957.

———. *The Structure of Scientific Revolutions.* 2nd ed. Chicago: University of Chicago Press, 1970.

MARSHACK, ALEXANDER *The Roots of Civilization.* New York: McGraw-Hill, 1972.

MOORE, P. *Watchers of the Sky.* New York: G. P. Putnam's Sons, 1973.

ROSEN, EDWARD "Render Not Unto Tycho That Which is Not Brahe's." *Sky and Telescope 61* (June 1981), p. 476.

SHEA, W. R. *Galileo's Intellectual Revolution.* New York: Macmillan, 1972.

WALLACE, WILLIAM A. (Ed.) *Reinterpreting Galileo.* Washington, D.C.: Catholic University of America Press, 1986.

WILSON, C. "How Did Kepler Discover His First Two Laws?" *Scientific American 226* (March 1972), p. 93.

C H A P T E R 5

NEWTON, EINSTEIN, AND GRAVITY

Nature and Nature's laws
lay hid in night:
God said, "Let Newton be!"
and all was light.
Alexander Pope

Isaac Newton was born in Woolsthorpe, England on December 25, 1642, and on January 4, 1643. This was not a biological anomaly but a calendrical quirk. Most of Europe, following the lead of the Catholic countries, had adopted the Gregorian calendar, but Protestant England continued to use the Julian calendar. Thus, December 25 in England was January 4 in Europe. If we take the English date, then Newton was born in the same year that Galileo Galilei died.

Newton went on to become one of the greatest scientists who ever lived (Figure 5–1), but even he admitted the debt he owed to those who had studied nature before him. He said, "If I have seen farther than other men, it is because I stood on the shoulders of giants."

One of those giants was Galileo (Figure 5–2). Although we remember Galileo as the defender of the Copernican theory, he was also a trained scientist who studied the motions of falling bodies.

Newton based his own work on the discoveries of Galileo and others. During his life he studied optics, invented calculus, developed three laws of motion, and discovered the principle of mutual gravitation. Of these, the last two are the most important to us in this chapter because they make it possible to understand the orbital motion of the moon and planets.

Newton's laws of motion and gravity were astonishingly successful in describing the heavens. A friend of Newton, Edmund Halley, used Newton's laws to calculate the orbits of comets and discovered that certain comets that had been seen throughout recorded history were actually a single object returning every 75 years, now known as Comet Halley.

FIGURE 5–1 Isaac Newton (1642–1727) was the founder of modern physics. He made important discoveries in optics, developed three laws of motion, invented differential calculus, and discovered the law of mutual gravitation. (Grundy Observatory.)

FIGURE 5–2 Galileo Galilei (1564–1642), usually remembered as the defender of Copernicanism, was also a talented scientist. He studied the motions of falling bodies and discovered the law of inertia. (Grundy Observatory.)

For over two centuries following the publication of Newton's works, astronomers used his laws to describe the universe. Then, early in this century, Albert Einstein proposed a new way to describe gravity. The new theory did not replace Newton's laws but rather showed that they were only approximately correct and could be seriously in error under special circumstances. We will see how Einstein's theories further extend our understanding of the nature of gravity. Just as Newton had stood on the shoulders of Galileo, Einstein had stood on the shoulders of Newton.

5.1 GALILEO AND NEWTON

Johannes Kepler discovered three laws of planetary motion, but he never understood why the planets move along their orbits. At one place in his writings he wonders if they are pulled along by magnetic forces emanating from the sun. In another place, he speculates that they are pushed along their orbits by angels beating their wings.

Isaac Newton refined Kepler's model of planetary motion but did not perfect it. In science a **model** is an intellectual conception of how nature works. No model is perfect; although Kepler's model was better than Aristotle's, it still included angels to push the planets along their orbits. Newton improved Kepler's model by expanding it into a general model of motion and gravity, but Newton's model was not perfect either. Newton never understood what gravity was. It was as mysterious as an angel pushing the moon inward toward the earth instead of forward along its orbit.

To understand science, we must understand the importance of models. The scientist studies nature by creating new models or refining old models. Yet a model can never be perfect, because it can never represent the universe in all its intricacies. Instead, a model must be a limited approximation of a single phenomenon, such as orbital motion. It is fitting that Newton's discoveries all began with Kepler's fellow Copernican, Galileo Galilei.

GALILEO AND MOTION

Even before Galileo built his first telescope, he had begun studying the motion of freely moving bodies. After the Inquisition condemned and imprisoned him in 1633, he continued his study of motion. He seems to have realized that he would have to understand motion before he could truly understand the Copernican theory. That he was eventually able to formulate principles that later led Newton to the laws of motion and the theory of gravity is a tribute to Galileo's ability to set aside authority and think for himself.

The authority of the age was Aristotle, whose ideas on motion were hopelessly confused. Aristotle said that the world is made up of four elements: earth, water, air, and fire. According to this idea, all earthly things—wood, rock, flesh, bone, metals, and so on—are made up of mixtures of earth and water. The motions of bodies are determined by their natural tendencies to move toward their proper places in the cosmos. Earth and water, and all things composed of them, move toward the center of the cosmos, which, in Aristotle's geocentric universe, is the center of the earth (Figure 5–3). Thus objects fall downward because they are moving toward their proper place.* On the other hand, fire and air move upward toward the heavens, their proper place in the cosmos.

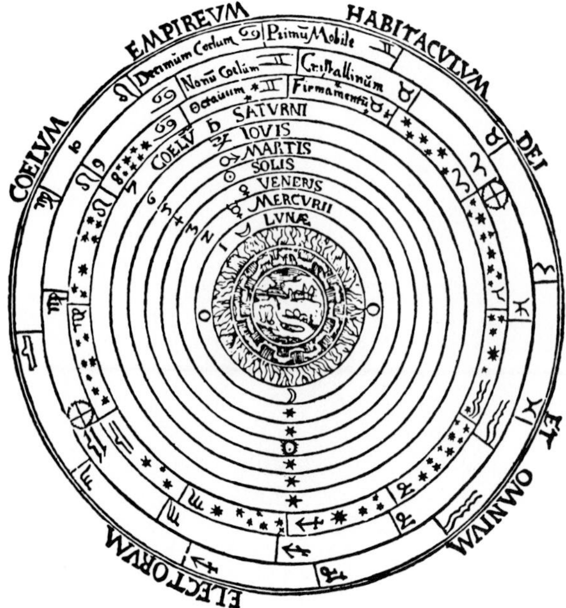

FIGURE 5–3 In Aristotle's universe the earth is located at the center. Of the four elements, earth and water tend to move toward the center of the universe. Fire and air move up toward the sphere of stars. (From *Cosmographia*, Peter Apian, 1539.)

Aristotle called these motions **natural motions** to distinguish them from **violent motions** produced, for instance, when we push on an object and make it move other than toward its proper place. According to Aristotle, such motions stop as soon as the force is removed. To explain how an arrow could continue to move upward even after it had left the bowstring, he said currents in the air around the arrow carried it forward even though the bowstring no longer exerted a force on it.

These ideas about the proper places of objects, natural and violent motion, and the necessity of a force to preserve motion were still the accepted theory in Galileo's time. In fact, in 1590 when Galileo was 26, he wrote a short work called *De Motu* (On Motion) that deals with the proper places of objects and their natural motions.

In Galileo's time, and for the two preceding millennia, scholars commonly tried to resolve problems of science by referring to authority. To analyze the flight of a cannonball, for instance, they would turn to the writings of Aristotle and other classical philosophers and try to deduce what those philosophers would have said on the subject. This generated a great deal of discussion but little real progress. Galileo broke with this tradition and conducted experiments of his own.

He began by studying the motions of falling bodies, but he quickly discovered that the velocities were so great and the times so short, he could not measure them accurately. Consequently, he turned to polished bronze balls rolling down gently sloping inclines. In that instance, the velocity is lower and the time longer. Using an ingenious water clock, he was able to measure the time the balls took to roll given distances down the incline, and he correctly recognized that these times are proportional to the times taken by falling bodies.

He found that falling bodies do not fall at constant rates, as Aristotle had said, but are accelerated. That is, they move faster with each passing second. Near the surface of the earth, a falling object will have a velocity of 9.8 m/sec (32 ft/sec) at the end of 1 second, 19.6 m/sec (64 ft/sec) after 2 seconds, 29.4 m/sec (96 ft/sec) after 3 seconds, and so on. Each passing second adds 9.8 m/sec (32 ft/sec) to the object's velocity (Figure 5–4). In modern terms, this is called the **acceleration of gravity** at the earth's surface.

*This is one reason why Aristotle had to have a geocentric universe. If the center of the earth had not also been the center of the cosmos, his theory of gravity would not have worked.

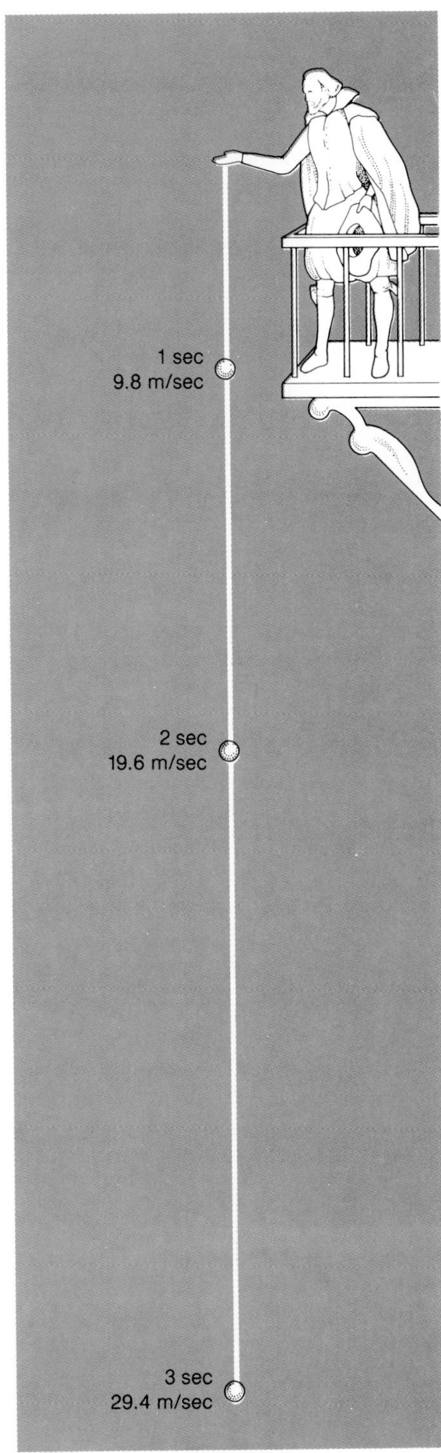

FIGURE 5–4 Galileo found that a falling object is accelerated downward. Each second its velocity increases by 9.8 m/sec (32 ft/sec).

Galileo also discovered that the acceleration does not depend on the weight of the object. This, too, is contrary to the teachings of Aristotle, who believed that heavy objects, containing more earth and water, fall with higher velocity. Galileo found that the acceleration of a falling body is the same whether it is heavy or light.* According to some accounts, he demonstrated this by dropping balls of iron and wood from the top of the Leaning Tower of Pisa to show that they would fall together and hit the ground at the same time (Figure 5–5a). In fact, he probably didn't perform the experiment. It would not have been conclusive anyway, because of air resistance. More than 300 years later, Apollo 15 astronaut David Scott, standing on the airless lunar surface, demonstrated Galileo's discovery by dropping a feather and a steel geologist's hammer. They hit the lunar surface at the same time (Figure 5–5b).

Having described natural motion, Galileo turned his attention to violent motion—that is, motion directed other than toward an object's proper place in the cosmos. He pointed out that an object rolling down an incline is accelerated toward the earth and that an object rolling up the same incline is decelerated. If the surface were perfectly horizontal and frictionless, he reasoned, there could be no acceleration or deceleration to change the object's velocity, and, in the absence of friction, the object would continue to move forever. In his own words, "any velocity once imparted to a moving body will be rigidly maintained as long as the external causes of acceleration or retardation are removed."

This is contrary to Aristotle's belief that motion can only continue if a force is present to maintain it. In fact, Galileo's statement is a perfectly valid summary of the law of inertia, which became Newton's first law of motion.

Galileo published his work on motion in 1638, 2 years after he had become entirely blind and only 4 years before his death. The book was called *Mathematical Discourses and Demonstrations Concerning Two New Sciences, Relating to Mechanics and to Local Motion.* It is known today as *Two New Sciences.*

The book is a brilliant achievement for a number of reasons. To understand motion, Galileo had to abandon the authority of the age, make his own experiments, and draw his own conclusions. In a sense, this was the first

*In 1986 a team of physicists reported evidence for a fifth force, which would cause a tiny difference in the acceleration of two objects made of different materials. Since then a number of sensitive experiments have tried to detect this fifth force with mixed results. Physicists now question its existence. If it were confirmed, such a new fundamental force would be an exciting addition to modern physics, but its effects would be small, and it would not affect our discussion.

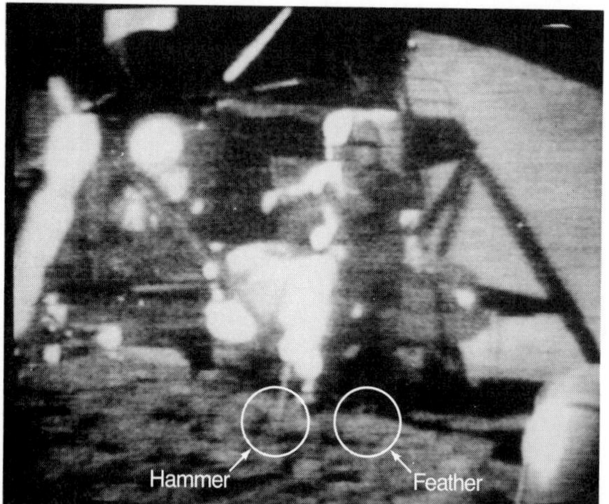

FIGURE 5–5 (a) According to tradition, Galileo demonstrated that the acceleration of a falling body is independent of its weight by dropping balls of iron and wood from the Leaning Tower of Pisa. In fact, air resistance would have confused the result. (b) In a historic television broadcast from the moon on August 2, 1971, Apollo 15 Astronaut David Scott dropped a hammer and a feather at the same instant. In the vacuum of the lunar surface, they fell together (NASA).

example of experimental science. But Galileo had also to generalize his experiments to discover how nature worked. Though his apparatus was finite and plagued by friction, he was able to imagine an infinite, totally frictionless plane on which a body moves at constant velocity. In his workshop the law of intertia was obscure, but in his imagination, it was clear and precise.

NEWTON AND THE LAWS OF MOTION

Newton's three laws of motion (Table 5–1) are critical to our understanding orbital motion. They apply to any moving object, from an automobile driving along a highway to galaxies colliding with each other.

The first law is really a restatement of Galileo's law of inertia. An object continues at rest or in uniform motion in a straight line unless acted upon by some force (Figure 5–6a). An astronaut drifting in space will travel at a constant rate in a straight line forever if no forces act.

Newton's first law also explains why a projectile continues to move after all forces have been removed—for instance, why an arrow continues to move after leaving the bowstring. The object continues to move because it has momentum. An object's **momentum** is a measure of the amount of motion.

Table 5–1 Newton's three laws of motion.

I. A body continues at rest or in uniform motion in a straight line unless acted upon by some net force.

II. The acceleration of a body is inversely proportional to its mass, directly proportional to the net force, and in the same direction as the net force.

III. To every action, there is an equal and opposite reaction.

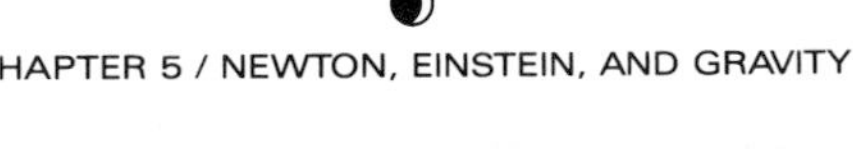

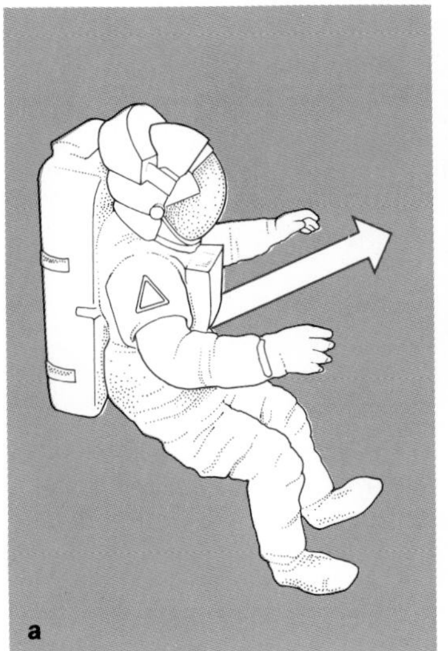

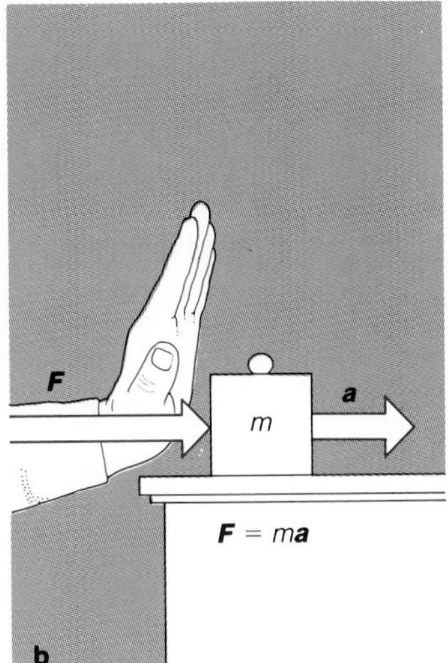

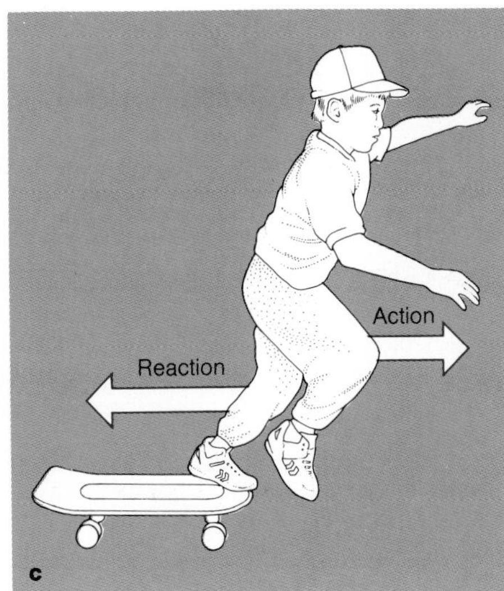

FIGURE 5–6 Newton's three laws of motion.

An object's momentum depends on its velocity and its mass.* A low-velocity object such as a paper clip tossed across a room, has little momentum, and we could easily catch it in our hand. But the same paper clip fired at the speed of a rifle bullet would have tremendous momentum, and we would not dare try to catch it.

But momentum also depends on the mass of an object. **Mass** is a measure of the amount of matter in an object and is *not* the same as the object's weight. Our weight is the force that earth's gravity exerts on the mass of our bodies. Floating in space we would have no weight at all, but we would still have mass. Using the metric system, we will measure mass in kilograms. (See Appendix B.)

To see how momentum depends on mass, imagine that, instead of tossing us a paper clip, someone tosses us a bowling ball. A bowling ball contains much more mass than a paper clip and therefore has much greater momentum at the same velocity.

Newton's first law explains the consequences of the conservation of momentum. When we say that momentum is conserved, we mean it remains constant until something acts to change it. A moving object has a given amount of momentum. To change that momentum, either by changing the speed or the direction, we must exert some force on the object. Newton's first law and the concept of momentum came from the work of Galileo.

*Mathematically, momentum is the product of mass and velocity.

Newton's second law of motion discusses forces, and Galileo did not talk about forces. He spoke instead of accelerations. Newton saw that an acceleration is the result of a force acting on a mass (Figure 5–6b). Newton's second law is commonly written as

$$F = ma$$

Once again, we must carefully define terms. An **acceleration** is a change in velocity, and a **velocity** is a directed rate of motion. By rate of motion, we mean, of course, a speed, but the word *directed* has a special meaning. Speed itself does not have any direction associated with it, but velocity does. If you drive a car in a circle at 55 mph, your speed is constant, but your velocity is changing. Thus, an object experiences an acceleration if its speed changes or if its direction of motion changes. Every automobile has three accelerators—the gas pedal, the brake pedal, and the steering wheel. All three change the auto's velocity.

The acceleration of a body is proportional to the force applied. This is reasonable. If we push gently against a grocery cart with frictionless wheels, we do not expect a large acceleration. The second law of motion also says that the acceleration is inversely proportional to the mass of the body. This too is reasonable. If the cart were filled with bricks and we pushed it gently, we would expect very little result. If it were full of table tennis balls, however, it might move easily in response to a gentle push. Finally, the second law says the resulting acceleration is in the direction of the force. This is also what we would expect. If we push on a cart which is not moving, we expect it to begin moving in the direction we push.

The second law of motion is important because it establishes a precise relationship between cause and effect. Objects do not just move. They accelerate due to the action of a force. Moving objects do not just stop. They decelerate due to a force. Also, moving objects don't just change direction for no reason. Any change in direction is a change in velocity and requires the presence of a force. Aristotle said that objects move because they have a tendency to move. Newton said objects move due to a specific cause, a force.

Newton's third law of motion specifies that for every action there is an equal and opposite reaction. In other words, forces must occur in pairs directed in opposite directions. For example, if you stand on a skateboard and jump forward, the skateboard will shoot away backward. As you jump, your feet exert a force against the skateboard, which accelerates it toward the rear. But forces must occur in pairs, so the skateboard must exert an equal but opposite force on your feet, which accelerates your body forward (Figure 5–6c).

MUTUAL GRAVITATION

The three laws of motion led Newton to consider the force that causes objects to fall. The first and second laws tell us that falling bodies accelerate downward because some force must be pulling downward on them. Newton wondered what that force could be. Whatever the force is, the third law of motion says that if there is a force pulling the object downward, there has to be a matching force directed upward. If it is the earth that pulls the object downward, then the object should pull the earth upward. That is, the attraction must be mutual.

Newton was also aware that some force has to act on the moon. The moon follows a curved path around the earth, and motion along a curved path is accelerated motion. The second law says that a force is required to make the moon follow that curved path.

Newton wondered if the force that holds the moon in its orbit could be the same force that causes rocks to roll downhill—gravity. He was aware that gravity extends at least as high as the tops of mountains, but he did not know if it could extend all the way to the moon. He believed that it could, but he thought it would be weaker at greater distances. He also guessed that its strength would decrease as the square of the distance increases.

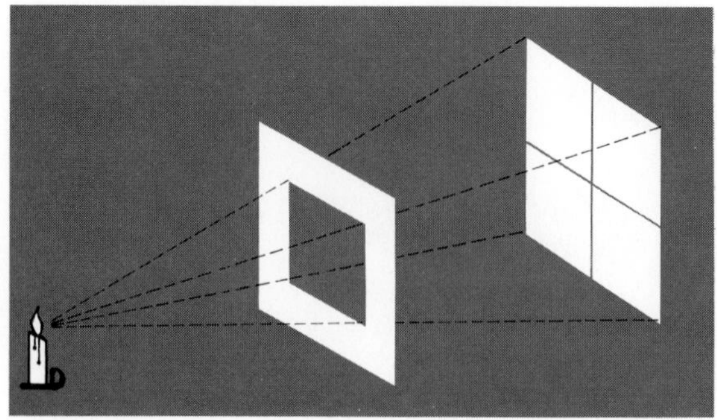

FIGURE 5–7 The inverse square law. At twice the distance from a candle flame, the light is spread over four times the area and is thus only one-fourth as intense.

This relationship, the **inverse square law**, was familiar from Newton's work on optics where it applied to the intensity of light. A screen set up one unit from a candle flame receives a certain amount of energy on each square meter. However, if that screen is moved to a distance of two units, the light that originally illuminated 1 m^2 must cover 4 m^2 (Figure 5–7). Thus the intensity of light is inversely proportional to the square of the distance to the screen.

Newton made two assumptions that enabled him to predict the strength of earth's gravity at the distance of the moon. He assumed that the strength of gravity follows the inverse square law and that the critical distance is not the distance from the surface of the earth, but the distance from the center of the earth. Because the moon is about 60 earth radii away, earth's gravity at the distance of the moon should be about 60^2 times less than at earth's surface. Instead of being 9.8 m/sec^2, it should be about 0.0027 m/sec^2.

Now, Newton thought, could this acceleration keep the moon in orbit? He knew the moon's distance and its orbital period, so he could calculate the actual acceleration needed to keep it moving along its curved path. The answer is 0.0027 m/sec^2. To the accuracy of Newton's data for the radius of the earth, it was exactly what his assumptions predicted. The moon is held in its orbit by gravity, and gravity obeys the inverse square law.

Newton's third law says that forces always occur in pairs, and this quickly led Newton to realize that gravity

FIGURE 5–8 A cannon on a high mountain could put its projectile into orbit if it could achieve a high enough velocity. Newton published a similar figure in *Principia*.

is mutual. If the earth pulls on the moon, then the moon must pull on the earth. Gravitation is a general property of the universe. The sun, planets, and all their moons must also attract each other by mutual gravitation. In fact, every particle in the universe must attract every other particle.

Because we do not find ourselves attracting particles of mass—books, rocks, passing birds, and so on—drawn to us by our personal gravity, Newton concluded that the gravitational force depends on the mass of the object. Large masses, like the earth, have strong gravity, but smaller masses, like people, have weaker gravity. Combining this mass dependence with the inverse square law led to the famous formula for the gravitational force between masses M and m:

$$F = -G\frac{Mm}{r^2}$$

The constant G is the gravitational constant, and r is the distance between the masses. The negative sign tells us the force is attractive, pulling the masses together and making r decrease. In plain English this is Newton's law of gravitation: The force of gravity between two masses M and m is proportional to the product of the masses and inversely proportional to the square of the distance between them.

Newton's model of gravity was a difficult idea for physicists of his time to accept because it is an example of action at a distance. The earth and moon exert forces on each other although there is no physical connection between them. Modern scientists resolve this problem by referring to gravity as a **field**. The presence of the earth produces a gravitational field directed toward the center of the earth. The strength of the field decreases according to the inverse square law. Any particle of mass in that field experiences a force that depends on the mass of the particle and the strength of the field at the particle's location. The resulting force is directed toward the center of the field.

The field is an elegant way to describe gravity, but it does not tell us what gravity is. For that we must wait until later in this chapter when we will discuss Einstein's theory of curved space-time.

5.2 ORBITAL MOTION

Newton's laws of motion and gravitation make it possible to understand why the planets move along their orbits. We can understand how they are held in their curved paths, and we can even discover why Kepler's laws work.

ORBITING THE EARTH

To illustrate the principle of orbital motion, imagine that we position a large cannon at the top of a mountain, point the cannon horizontal, and fire it (Figure 5–8). The cannonball falls to earth some distance from the foot of the mountain. The more gunpowder we use, the faster the ball travels, and the farther from the foot of the mountain it falls. If we use enough powder, the ball travels so fast it never strikes the ground. Earth's gravity pulls it toward the earth's center, but the earth's surface curves away from it at the same rate at which it falls. We say it is in orbit. If the cannonball is high above the atmosphere, where there is no friction, it will fall around the earth forever. Real earth satellites do fall back to earth sometimes, but that is caused by friction with the tenuous upper layers of earth's atmosphere.

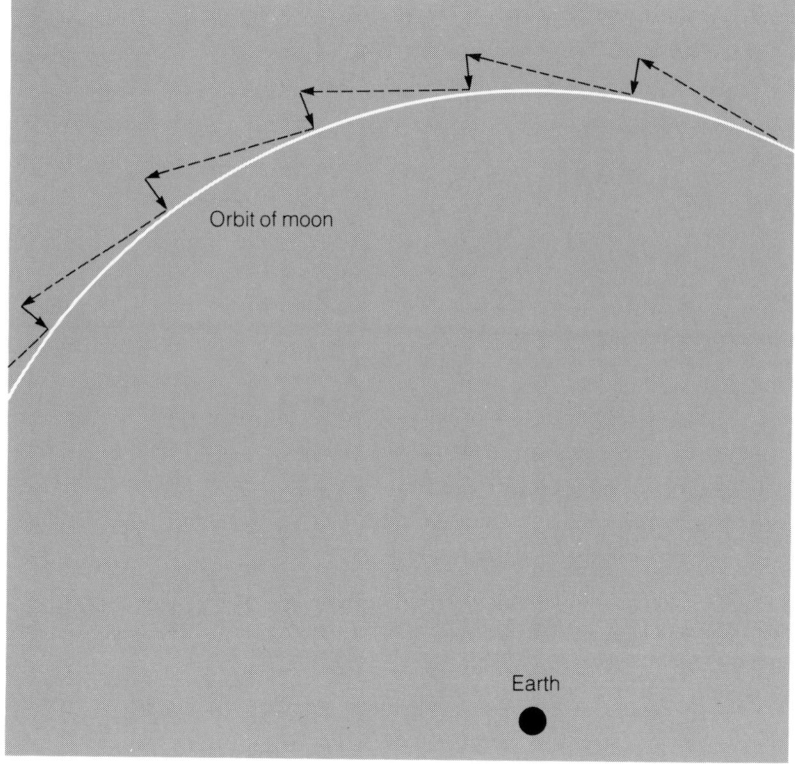

FIGURE 5–9 In the absence of forces, the moon would follow a straight-line path (dashed arrows). It follows a curved path because earth's gravity accelerates it toward the earth (solid arrows).

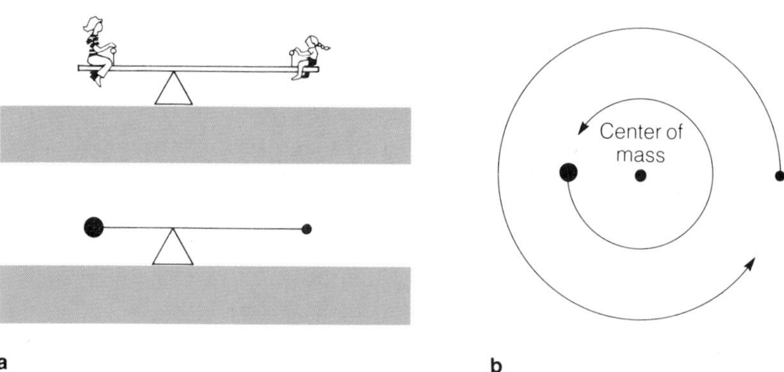

FIGURE 5–10 (a) If two bodies were connected by a massless rod and placed in a uniform gravitational field like that at the earth's surface, they would balance at their center of mass. (b) If the bodies orbit each other, the center of mass remains fixed, and the objects move around it. The center of mass of the earth–moon system lies inside the earth.

The first law of motion says that an object in motion tends to stay in motion in a straight line unless acted upon by some force. Thus, the cannonball in our example travels in a curve around the earth only because earth's gravity acts to pull it away from its straight line motion.

Similarly, the moon should move in a straight line at constant speed (dashed lines in Figure 5–9) were it not for gravity accelerating it toward the earth. Each second the moon moves 1043 m (3420 ft) eastward along its orbit and falls about 1.6 mm (about 1/16 inch) toward earth. The combination of these two motions results in a closed orbit around the earth. An object's **circular velocity** is the lateral velocity the object must have to remain in a circular orbit. (See Box 5–1.)

A space capsule orbiting the earth is not "beyond earth's gravity" to use a phrase common in old science fiction movies. Like the moon, the space capsule is accelerated toward the earth. That, combined with its lateral motion, places it in a closed orbit. Astronauts inside are weightless, not because they are beyond earth's gravity,

BOX 5–1
Orbital Velocity

How fast must we travel to stay in orbit around another body? If we assume we are to stay in a circular orbit and that our mass is small compared to the object we orbit, then the answer is V_c, the circular velocity.

$$V_c = \sqrt{\frac{GM}{r}}$$

In this formula, M is the mass of the central body, r is the radius of the orbit, and G is the gravitational constant, 6.67×10^{-11} N · m²/kg², where N stands for newton, a measure of force.

For example, how fast does the moon travel in its orbit? The mass of the earth is 5.98×10^{24} kg, and the radius of the moon's orbit is 3.84×10^{8} m. Then the velocity is

$$V_c = \sqrt{\frac{6.67 \times 10^{-11} \times 5.98 \times 10^{24}}{3.84 \times 10^{8}}}$$
$$= 1020 \text{ m/sec}$$

The same kind of arithmetic would tell us that a space capsule just above earth's atmosphere [and thus about 6500 km (4000 miles) from earth's center] would have a circular velocity of about 8000 m/sec (about 18,000 mph).

but because they are falling at the same rate as their capsule. Rather than say the astronauts are weightless, it is more accurate to say they are in free-fall.

To be precise, we should avoid saying that the moon orbits the earth. In fact, the moon and earth orbit each other. Gravitation is mutual, and if the earth pulls on the moon, then the moon pulls on the earth. The two bodies revolve around their common **center of mass**, the balance point of the system. If the earth and moon could be connected by a massless rod and placed in a uniform gravitational field such as that near the earth's surface, the system would balance at its center of mass like a child's seesaw (Figure 5–10). If the system revolves, the moon and the earth each describe an orbit around the center of mass.

As we might expect from experience on a seesaw, the balance point, or center of mass, is closest to the more massive body. The center of mass of the earth–moon system lies only 4708 km (2926 miles) from the earth's center. This places the center of mass inside the earth. As the moon revolves around its orbit, the earth swings about the center of mass. (We learned in Chapter 3 that this motion is one of the causes of the tides.)

CONIC SECTIONS

Kepler's laws refer only to elliptical orbits. But Newton's work showed that a number of different kinds of orbits are possible. These orbits are called **conic sections** because they can be constructed by cutting or sectioning a cone (Figure 5–11).

If a cone is cut by a plane perpendicular to the cone's axis, the section is circular. If the plane is tipped slightly, the section is elliptical. Both of these kinds of orbits are called **closed orbits** because they return back on themselves. If you were in a spacecraft orbiting earth in a closed orbit, you would return to the same place in the orbit time after time.

If a cone is cut by a plane parallel to one side of the cone, the section is a parabola; if the plane is tipped even farther, the section is a hyperbola. Such orbits are called **open** or **escape orbits*** because they never return to the same point. If you were in a spacecraft following an open

*Open orbits are also called open trajectories.

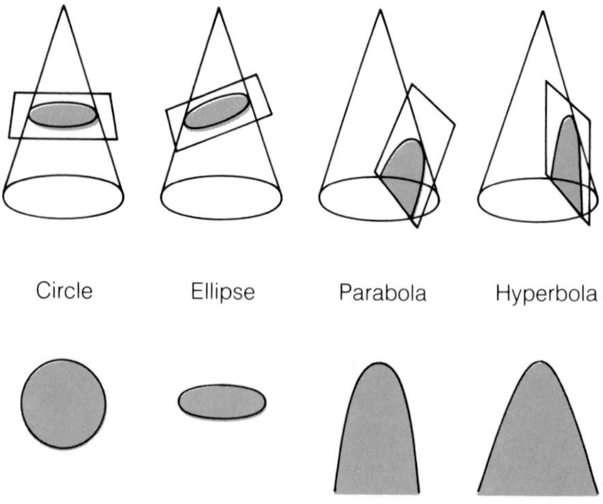

FIGURE 5–11 Conic sections are curves produced by cutting (sectioning) a cone with a plane.

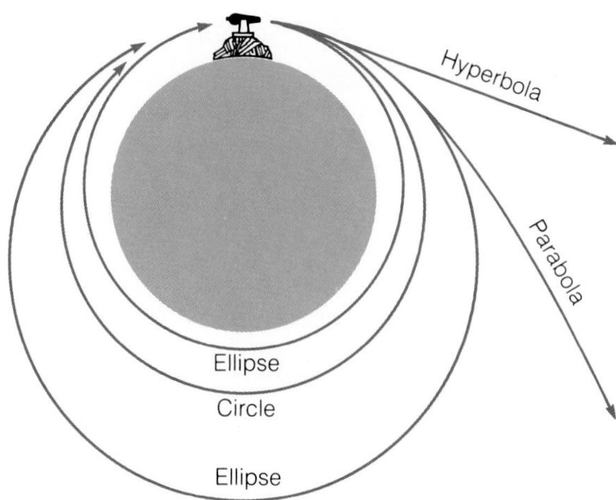

FIGURE 5–12 If an object's orbital velocity is less than escape velocity, it will follow an elliptical or circular orbit. If its velocity equals or exceeds escape velocity, it will follow a parabolic or hyperbolic orbit and escape from the earth.

orbit, you would be leaving the earth along a path that did not return.

In any kind of orbit, the center of mass of the system is located at one of the foci of the orbit. In the case of the Space Shuttle orbiting the earth, the center of mass is essentially the center of the earth. But if the two masses are more nearly equal, the center of mass is not the center of either body. This is true of the earth–moon system.

We can analyze these open and closed orbits by thinking again of the cannon in Figure 5–8. If the cannonball travels at the circular velocity, it will fall in a circular orbit around the earth. Of course, if the velocity is much too small, the ball will hit the earth. However, if the velocity of the ball is only slightly less than the circular velocity, the ball will follow an elliptical orbit with its highest point, the *apogee,* at the cannon (Figure 5–12). If the velocity of the ball is slightly greater than the circular velocity, it will again follow an elliptical path, but with the lowest part of its orbit, the *perigee,* at the cannon. All of these orbits are closed—the cannonball returns to its starting point. If we really performed this experiment, the cannonball would circle the earth in about 90 minutes, returning to its starting point and smashing the cannon to fragments.

Suppose the ball's velocity is even greater. If it equals or exceeds **escape velocity** (Box 5–2), the velocity needed to escape from the surface of a body, then it will follow an open orbit. If the velocity equals escape velocity, the orbit is parabolic. If it is greater than escape velocity, the orbit is hyperbolic. In both cases, the cannonball leaves the earth never to return.

KEPLER'S LAWS REEXAMINED

Now that we understand Newton's laws, gravity, and orbital motion, we can understand Kepler's laws of planetary motion in a new way.

Kepler's first law says the orbits of the planets are ellipses with the sun at one focus. Kepler wondered why the planets keep moving along these orbits, and now we know the answer. They move because there is nothing to slow them down. Newton's first law says that a body in motion stays in motion unless acted on by some force. The gravity of the sun accelerates the planets inward toward the sun and holds them in their orbits, but it doesn't pull backward on the planets, so they don't slow to a stop. With no friction, they must continue to move.

The orbits of the planets must be ellipses because gravity follows the inverse square law. If gravity depended on $1/r^3$ or $1/r^4$ or anything other than $1/r^2$, the orbits of the planets could not be ellipses.

Kepler's second law says that a planet moves faster when it is near the sun and slower when it is farther away. Once again, Newton's discoveries explain why. Earlier we saw that a body moving on a frictionless surface will continue to move in a straight line until it is acted on by some force. We said the object had momentum. But an object

BOX 5–2

Escape Velocity

How fast must a rocket travel to escape from the earth? We know, of course, that no matter how far it travels, it can never escape from earth's gravity. The effects of earth's gravity extend to infinity. However, it is possible for a rocket to travel so fast that gravity can never slow it to a stop. Thus, it could leave the earth.

The escape velocity is the velocity required to escape from the surface of an astronomical body. In this example we will consider the earth, but in later chapters we will discuss the escape velocity of other planets, stars, and even black holes.

The escape velocity V_e is given by a simple formula:

$$V_e = \sqrt{\frac{2GM}{R}}$$

where G is the gravitational constant, 6.67×10^{-11} N $\cdot$ m^2/kg^2, M is the mass of the astronomical body, and R is its radius. For the earth $M = 5.98 \times 10^{24}$ kg and $R = 6.38 \times 10^6$ m. Then the escape velocity from earth's surface is

$$V_e = \sqrt{\frac{2 \times 6.67 \times 10^{-11} \times 5.98 \times 10^{24}}{6.38 \times 10^6}}$$
$$\approx \sqrt{\frac{12.5 \times 10^7 \text{m}^2}{\text{sec}^2}}$$
$$\approx 1.12 \times 10^4 \text{ m/sec}$$
$$= 11.2 \text{ km/sec}$$

This is equal to about 25,000 mph.

Compare the escape velocity with the circular velocity given in Box 5–1. The escape velocity is just $\sqrt{2}$ times larger than the circular velocity.

Notice that the formula says that escape velocity depends on both mass and radius. A massive body might have a low escape velocity if it has a very large radius. We will meet such objects when we talk about giant stars. On the other hand, a rather low mass body can have a very large escape velocity if it has a very small radius, a condition we will discuss when we meet black holes.

set rotating on a frictionless surface will continue rotating until something acts to speed it up or slow it down. Such an object has **angular momentum**, a measure of the rotation of the body about some point. A planet circling the sun has a given amount of angular momentum, and with no outside influences to speed it up or slow it down, it must conserve its angular momentum. That is, its angular momentum must remain constant.

Mathematically, a planet's angular momentum is the product of its mass, velocity, and distance from the sun. This explains why a planet must speed up as it comes closer to the sun along an elliptical orbit. Its angular momentum is conserved, so as its distance from the sun decreases, its velocity must increase. In the same way, the planet's velocity must decrease as its distance from the sun increases.

This conservation of angular momentum is actually a common human experience. Skaters spinning slowly can draw their arms and legs closer to their axes of rotation and, through conservation of angular momentum, spin faster (Figure 5–13). To slow their rotation, they again extend their arms. Similarly, divers can spin rapidly in the tuck position and slow their rotation by stretching into the extended position.

Kepler's third law is also explained by a conservation law, but in this case, it is the law of conservation of energy. A planet orbiting the sun has a specific amount of energy that depends only on its average distance from the sun.

FIGURE 5–13 Skaters demonstrate conservation of angular momentum when they spin faster by drawing their arms and legs closer to their axes of rotation.

FIGURE 5–14 Isaac Newton and his book *Principia* are honored on this English one-pound note. The telescope and prism beside him represent his discoveries in optics, and the diagram in the background symbolizes his work on motion and gravity.

That energy can be divided between energy of motion and energy stored in the gravitational attraction between the planet and the sun. The energy of motion depends on how fast the planet moves, and the stored energy depends on the size of its orbit. The relation between these two kinds of energy is fixed by Newton's laws. That means there has to be a fixed relationship between the rate at which a planet moves around its orbit and the size of the orbit—between its orbital period P and the orbit's semimajor axis a. This is just Kepler's third law.

ASTRONOMY AFTER NEWTON

Isaac Newton published his work in July 1687 in a book called *Philosophiae Naturalis Principia Mathematica* (Mathematical Principles of Natural Philosophy), now known as *Principia* (Figure 5–14). It is one of the most

FIGURE 5–15 Albert Einstein at his desk in the Swiss patent office in Bern, 1905. (Einstein Archives. Courtesy AIP Niels Bohr Library.)

important books ever written. The principles changed astronomy, science, and the way we think about nature.

Principia changed astronomy and ushered in the age of gravitational astronomy. No longer did astronomers appeal to the whim of the gods to explain things in the heavens. No longer did they speculate on why the planets move. They knew that the motions of the heavenly bodies are governed by simple, universal rules that describe the motions of everything from planets to falling apples. Suddenly the universe was understandable in simple terms.

Newton's laws of motion and gravity made it possible for astronomers to calculate the orbits of planets and moons. Not only could they explain how the heavenly bodies move, they could predict future motions. This subject, known as gravitational astronomy, dominated astronomy for almost 200 years and is still important. It included the calculation of the orbits of comets and asteroids, and the theoretical prediction of the existence of two planets, Neptune and Pluto.

Principia also changed science. The works of Copernicus and Kepler had been mathematical, but no book before had so clearly demonstrated the power of mathematics as a language of precision. Newton's arguments were couched in geometrical terms instead of the new analytical methods developed by European mathematicians, but *Principia* was so powerful an illustration of the quantitative study of nature, that scientists around the world adopted mathematics as their most powerful tool.

Also, *Principia* changed the way we think about nature. Newton showed that the rules that govern the universe are simple. Particles move according to three rules of motion and attract each other with a force called gravity. These motions are predictable, and that makes the universe a vast machine based on a few simple rules. It is complex only in that it contains a vast number of particles. In Newton's view, if he knew the location and motion of every particle in the universe, he could, in principle, derive the past and future of the universe in every detail. This mechanical determinism has been undermined by modern quantum mechanics, but it dominated science for more than two centuries as scientists thought of nature as a beautiful clockwork that would be perfectly predictable if we knew how all the gears meshed.

Most of all, Newton's work broke the last bonds between science and formal philosophy. Newton did not speculate on the good or evil of gravity. He did not debate its meaning. Not more than a hundred years before, scientists would have argued over the "reality" of gravity. Newton didn't care for these debates. He wrote, "It is enough that gravity exists and suffices to explain the phenomena of the heavens."

Newton made other discoveries of importance to astronomy, but we will reserve them for later chapters. Now we must move forward two centuries to see how Albert Einstein described gravity in a new and powerful way.

5.3 EINSTEIN AND RELATIVITY

In the early years of this century, Albert Einstein, then a clerk in the Swiss patent office (Figure 5–15), began thinking about how motion and gravity interact. He soon gained international fame by showing that Newton's laws of motion and gravity were only partially correct. The revised theory became known as the theory of relativity. As we will see, there are really two theories of relativity.

SPECIAL RELATIVITY

Einstein began by thinking about how moving observers see events around them. His analysis led him to the first postulate of relativity, also known as the principle of relativity.

First Postulate (The Principle of Relativity) Observers can never detect their *uniform* motion except relative to other objects.

You may have experienced the first postulate while sitting on a train in a station. You suddenly notice that the train on the next track has begun to creep out of the station. However, after several moments you realize that it is your own train that is moving and that the other is still motionless on its tracks.

Consider another example. Suppose you are floating in a spaceship in interstellar space and another spaceship comes coasting by (Figure 5–16a). You might conclude that it is moving and you are not, but someone in the other ship might be equally sure that you are moving and they are not. The principle of relativity says that there is no experiment you could perform to decide which ship is moving and which is not. This means there is no such thing as absolute rest—all motion is relative.

Because neither you nor the people in the other spaceship could perform any experiment to detect your absolute motion through space, the laws of physics must have the same form in both spaceships. Otherwise, experiments would give different results in the two ships and you could decide who is moving. Thus, a more general way of stating the first postulate refers to these laws of physics.

First Postulate (Alternate Version) The laws of physics are the same for all observers, no matter what their motion, so long as they are not *accelerated.*

The words *uniform* and *accelerated* are important. If either of the spaceships were to fire its rockets, then its velocity would change. The crew of that ship would know it because they would feel the acceleration pressing them into their couches. Accelerated motion, therefore, is different—we can always tell which ship is accelerating and which is not. The postulates of **special relativity** apply only to observers in uniform motion. That is why it is called *special* relativity.

The first postulate fit with Einstein's conclusion that the speed of light must be constant for all observers. No matter how you move, your measurement of the speed of light has to give the same result (Figure 5–16b). This became the second postulate of special relativity.

Second Postulate The velocity of light is constant and will be the same for all observers independent of their motion relative to the light source.

Once Einstein had accepted the basic postulates of relativity, he was led to some startling discoveries. Newton's laws of motion and gravity worked well so long as distances were small and velocities were low. But when we begin to think of very large distances or very high velocities, Newton's laws are no longer adequate to describe what happens. Instead, we must use relativistic physics. For example, special relativity shows that the observed mass of a moving particle depends on its velocity. The higher the velocity the greater the mass of the particle. This is not significant at low velocities, but it becomes very important as the velocity approaches the velocity of light.

This discovery led to yet another insight. The relativistic equations that describe the energy of a moving particle predict that the energy of a motionless particle is not zero. Rather its energy at rest is m_0c^2. This is of course the famous equation

$$E = m_0c^2$$

The m_0 is the rest mass of the particle, and c is the speed of light. This suggests that mass and energy are related, and we will see in later chapters how nature can convert one into the other inside stars.

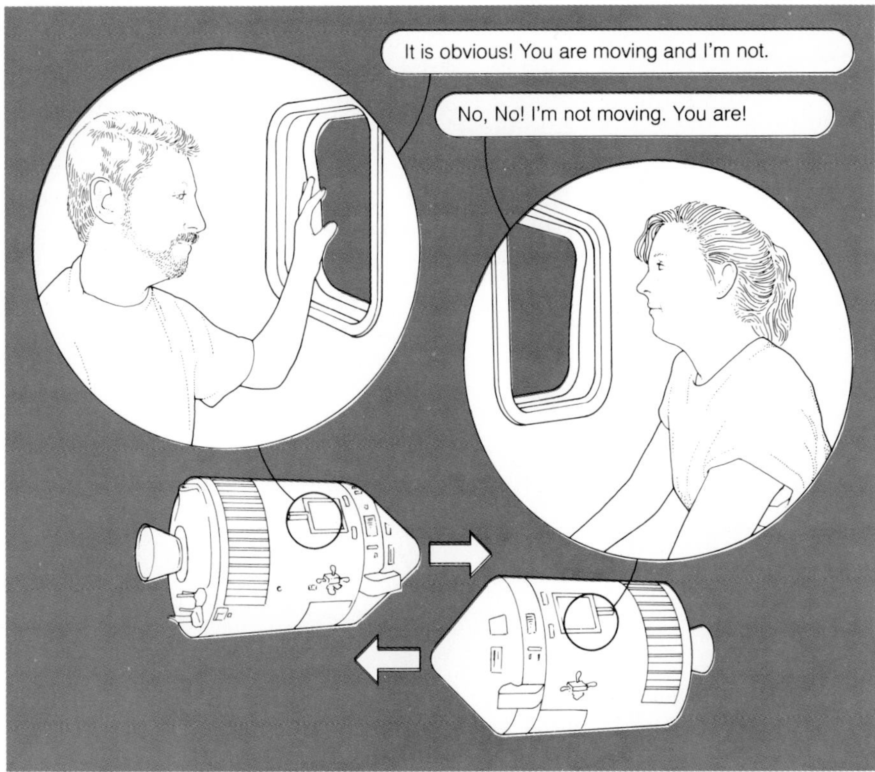

a

b

FIGURE 5–16 (a) The principle of relativity says that observers can never detect their uniform motion, except relative to other observers. Thus, neither of these travelers can decide who is moving and who is not. (b) If the principle of relativity is correct, then the velocity of light must be a constant for all observers. If the velocity of light did depend on the motion of the observer through space, then the observers above could decide who was moving and who was not.

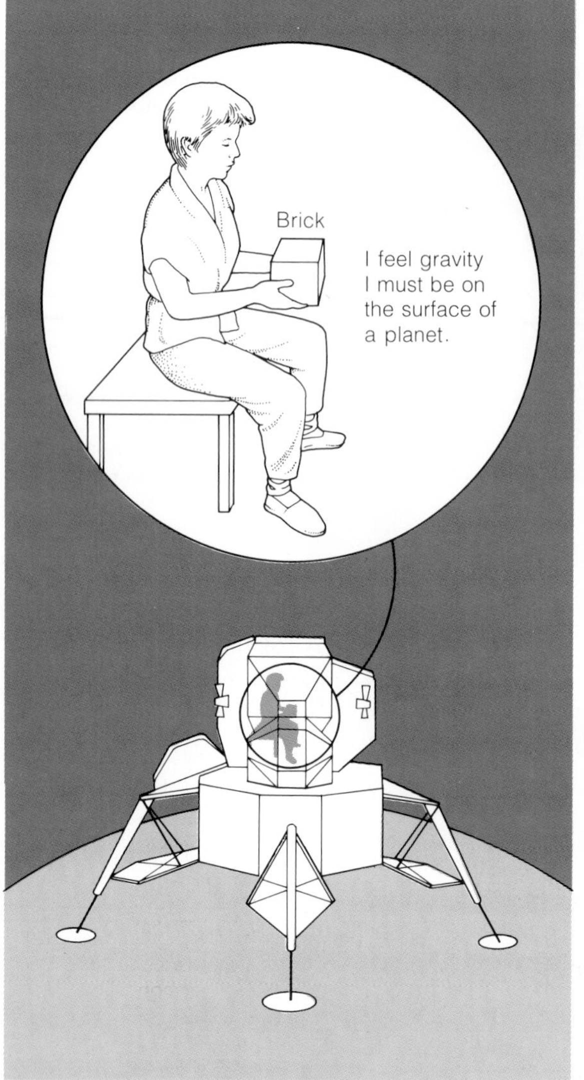

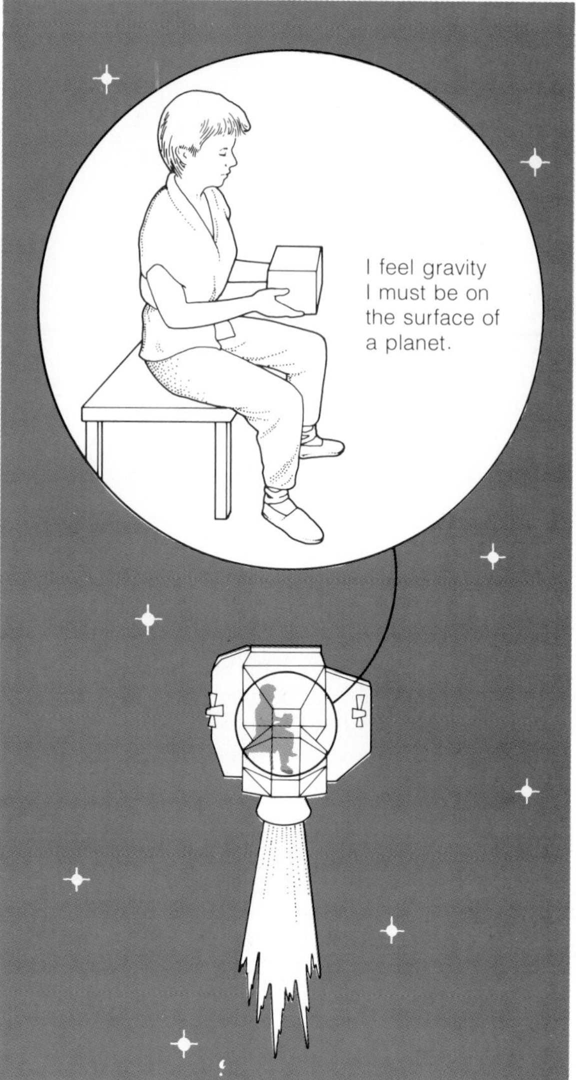

FIGURE 5–17 (a) An observer in a closed spaceship on the surface of a planet feels gravity. (b) In space, with the rockets firing and accelerating the spaceship, the observer feels inertial forces that are equivalent to gravitational forces.

For example, suppose that we convert 1 kg of matter into energy. We must express the velocity of light as 3×10^8 m/sec, and our result is 9×10^{16} joules (J) (approximately equal to a 20 megaton bomb). (A **joule** is a unit of energy roughly equivalent to the energy given up when an apple falls from a table to the floor.) Our simple calculation shows that the energy equivalent of even a small mass is very large.

A detailed discussion of the major consequences of the special theory of relativity is beyond the scope of this book. Instead, we must consider Einstein's second advance, the general theory.

THE GENERAL THEORY OF RELATIVITY

In 1916 Einstein published a more general version of the theory of relativity that dealt with accelerated as well as uniform motion. This **general theory of relativity** contained a new description of gravity.

Einstein began by thinking about observers in accelerated motion. Imagine an observer sitting in a spaceship (Figure 15–17). Such an observer cannot distinguish between the force of gravity and the inertial forces produced by the acceleration of the spaceship. This led Einstein to conclude that gravity and motion through space–time are related, a conclusion now known as the equivalence principle.

The Equivalence Principle Observers cannot distinguish locally between inertial forces due to acceleration and uniform gravitational forces due to the presence of a massive body.

The importance of the theory of general relativity lies in its description of gravity. Einstein concluded that gravity, inertia, and acceleration are all associated with the way space and time are related. This relation is often referred to as curvature, and a one-line description of general relativity explains a gravitational field as a curved region of space-time.

Gravity According to General Relativity Mass tells space-time how to curve, and the curvature of space-time (gravity) tells mass how to accelerate.

Thus, we feel gravity because the mass of the earth causes a curvature of space-time. The mass of our bodies responds to that curvature by accelerating toward the center of the earth. According to general relativity, all masses cause curvature, and the larger the mass the more severe the curvature.

CONFIRMATION OF THE CURVATURE OF SPACE TIME

Einstein's general theory of relativity has been confirmed by a number of experiments, but two are worth mentioning here because they were among the first tests of the theory. One involves Mercury's orbit and the other involves eclipses of the sun.

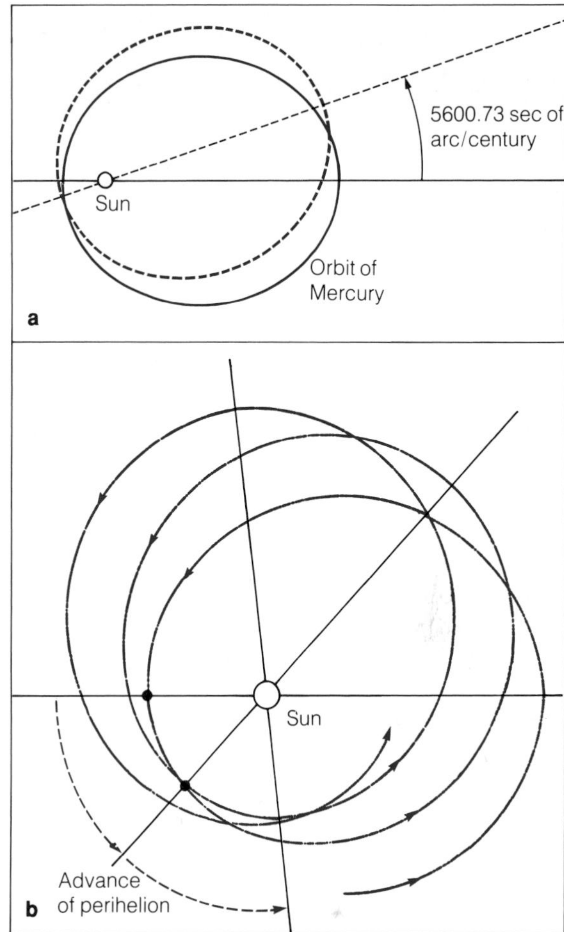

FIGURE 5–18 (a) Mercury's orbit precesses 5600.73 seconds of arc per century—43.11 seconds of arc per century faster than predicted by Newton's laws. (b) Even when we ignore the influences of the other planets, Mercury's orbit is not a perfect ellipse. Curved space-time near the sun distorts the orbit from an ellipse into a rosette. The advance of Mercury's perihelion is exaggerated about a million times in this figure.

Johannes Kepler understood that the orbit of Mercury is elliptical, but only since 1859 have astronomers known that the long axis of the orbit sweeps around the sun in a motion called precession (Figure 5–18). The total observed precession is 5600.73 seconds of arc per century (as seen from Earth), or about 1.5° per century. This precession is produced by the gravitation of Venus, Earth, and the other planets. However, when astronomers used Newton's description of gravity, they calculated that the precession should amount to only 5557.62 seconds of arc per century. Thus, Mercury's orbit is advancing 43.11 sec-

Table 5–2 Orbital precession.

Planet	Observed Precession (seconds of arc)	Predicted Precession (seconds of arc)
Mercury	43.11 ± 0.45	43.03
Venus	8.4 ± 0.48	8.6
Earth	5.0 ± 1.2	3.8
Icarus	9.8 ± 0.8	10.3

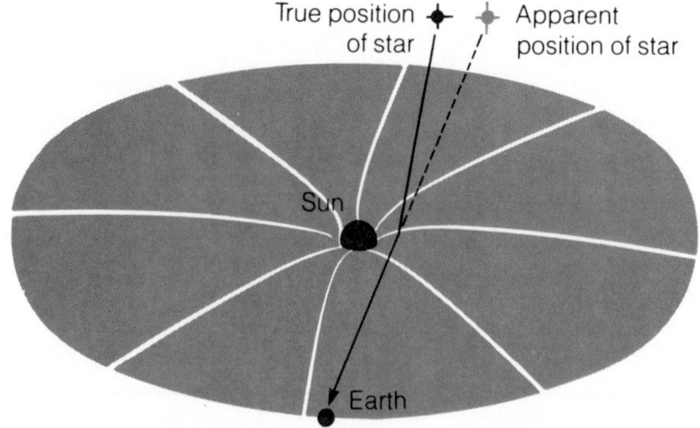

FIGURE 5–19 Like a depression in a putting green, the curved space-time near the sun deflects light from distant stars and makes them appear to lie in slightly different positions.

onds of arc per century faster than predicted by Newton's laws.

This is a tiny effect. Each time Mercury returns to perihelion, its closest point to the sun, it is about 29 km (18 miles) past the position predicted by Newton's laws. This is such a small distance compared with the planet's diameter of 4850 km that it could never have been detected had it not been cumulative. Each orbit Mercury gains 29 km, and in a century it gains over 12,000 km—more than twice its own diameter. Thus this tiny effect, called the advance of perihelion of Mercury's orbit, accumulated into a serious discrepancy in the Newtonian description of the universe.

The advance of perihelion of Mercury's orbit was one of the first problems to which Einstein applied the principles of general relativity. First, he calculated how much the sun's mass curves space-time in the region of Mercury's orbit, and then he calculated how Mercury moves through the space-time. The theory predicted that the curved space-time should cause Mercury's orbit to advance by 43.03 seconds of arc per century, well within the observational accuracy of the excess.

Einstein was elated with this result, and he would be even happier with modern studies that have shown that Mercury, Venus, Earth, and even Icarus, an asteroid that comes close to the sun, have orbits observed to be slipping forward due to the curvature of space-time near the sun (Table 5–2 and Figure 5–18b).

This same effect has been detected in pairs of stars that orbit each other. In some cases the advance of perihelion agrees with general relativity; in many cases, the sizes and masses of the stars are not well enough known to be certain of the theoretical rate of advance we should expect. But in a few cases, the star's orbits appear to be changing faster than predicted. This may be a critical test for Einstein's theory. (See, for example, the article by David H. Smith in the recommended readings for this chapter.)

A second test of the curvature of space-time was directly related to the motion of light through the curved space-time near the sun. The equations of general relativity predicted that light would be deflected by curved space-time just as a rolling golf ball is deflected by undulations in a putting green (Figure 5–19). Einstein predicted that starlight grazing the sun's surface would be deflected by 1.75 seconds of arc. Starlight passing near the sun is normally lost in the sun's glare, but during a total solar eclipse stars beyond the sun could be seen. As soon as Einstein published his theory, astronomers rushed to observe such stars and thus test the curvature of space-time.

The first solar eclipse following Einstein's announcement in 1916, was June 8, 1918. It was cloudy. The next occurred on May 29, 1919, only months after the end of World War I, and was visible from Africa and South America. British teams went to both Brazil and Príncipe, an island off the coast of Africa. First, they photographed that part of the sky where the sun would be located during the eclipse and measured the positions of the stars on the plates. Then during the eclipse they photographed the same star field with the eclipsed sun located in the middle. After measuring the plates, they found slight changes in the positions of the stars. During the eclipse, the positions of the stars on the plates were shifted outward, away from the sun (Figure 5–20). If a star had been located at

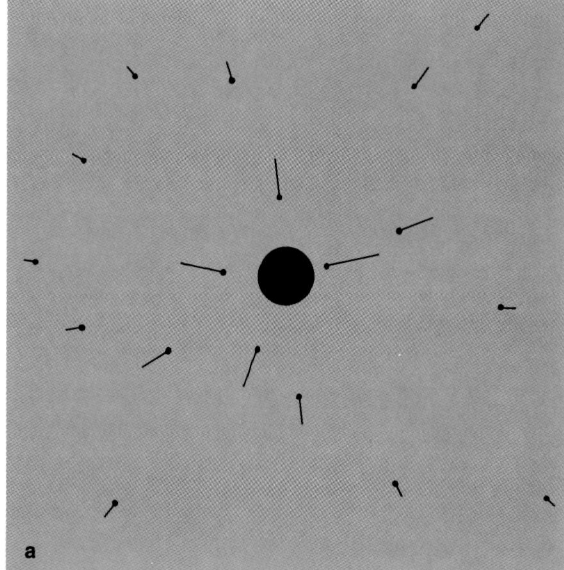

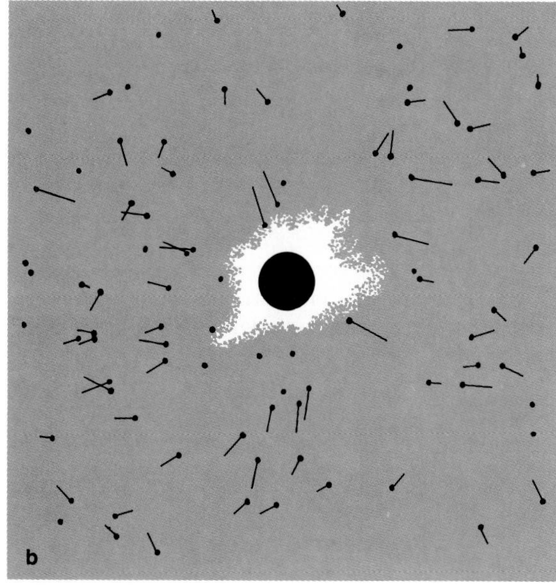

FIGURE 5–20 (a) Schematic drawing of the deflection of starlight by the sun's gravity. Dots show the true positions of the stars as photographed months before. Lines point toward the positions of the stars during the eclipse. (b) Actual data from the eclipse of 1922. Random errors of observation cause some scatter in the data, but in general the stars appear to move away from the sun by 1.77 seconds of arc at the edge of the sun's disk. The deflection of stars is magnified by a factor of 2300 in both (a) and (b).

the edge of the solar disk, it would have been shifted outward by about 1.8 seconds of arc. This represents good agreement with the theory's prediction.

This test has been repeated at many total solar eclipses since 1919 with similar results. The most accurate results were obtained in 1973 when a Texas–Princeton team measured a deflection of 1.66 ± 0.18 seconds of arc—good agreement with Einstein's theory.

The general theory of relativity is critically important in modern astronomy. We will discuss it again when we meet black holes, distant galaxies, and the big bang universe. The theory revolutionized modern physics by providing a theory of gravity based on the geometry of curved space-time. Thus, Galileo's inertia and Newton's mutual gravitation are shown to be fundamental properties of space and time.

SUMMARY

Galileo Galilei took the first steps toward understanding motion and gravity when he began to study falling bodies. He found that a falling object is accelerated. That is, it falls faster and faster with each passing second. The rate at which it accelerates, termed the acceleration of gravity, is 9.8 m/sec^2 (32 ft/sec^2) at the earth's surface and does not depend on the weight of the object. According to tradition, Galileo demonstrated this by dropping balls of iron and wood from the Leaning Tower of Pisa to show that they would fall together. Finally, Galileo stated the law of inertia. In the absence of friction, a moving body on a horizontal plane will continue moving forever.

Newton adopted Galileo's law of inertia as his first law of motion. (See Table 5–1.) The second law of motion established the relationship between the force acting on a body, its mass, and the resulting acceleration. The third law says that forces occur in pairs acting in opposite directions.

Newton also developed an explanation for the accelerations that Galileo had discovered—gravity. By considering the motion of the moon, Newton was able to show that objects attract each other with a gravitational force that is proportional to the product of their masses and inversely proportional to the square of the distance between them.

If we understand Newton's laws of motion and gravity, we can better understand orbital motion. An object in space would move along a straight line and quickly leave the earth were it not for earth's gravity accelerating the object toward earth's center and forcing it to follow a curved path—an orbit. If there is no friction, the object will fall around its orbit forever.

Newton's laws also illuminate the meaning of Kepler's three laws of planetary motion. (See Table 4–1.) The planets follow elliptical orbits because gravity follows the inverse square law. The planets move faster when closer to the sun and slower when farther away because they conserve angular momentum. The same law makes ice skaters spin faster when they draw their arms and legs nearer their body. The planets' orbital period squared is proportional to their orbital radii cubed because the moving planets conserve energy.

In fact, Newtonian gravity and motion show that an ellipse is only one of a number of orbits that a celestial body can follow. These shapes are called conic sections because they can be formed by slicing, or sectioning, a cone. The circle and the ellipse are closed orbits that return to their starting points. If an object moves at circular velocity, V_c, it will follow a circular orbit. If its velocity equals or exceeds the escape velocity, V_e, it will follow a parabola or hyperbola. These orbits are termed open because the object never returns to its starting place.

Newton's laws changed astronomy and our view of nature. They made it possible for astronomers to predict the motions of the heavenly bodies using the analytic power of mathematics. Newton's laws also showed that the apparent complexity of the universe is based on a few simple principles, natural laws.

Albert Einstein published two theories that extended Newton's laws of motion and gravity. The special theory of relativity, published in 1905, applied to observers in uniform motion. The theory held that the speed of light is a constant for all observers and that mass and energy are related by the expression $E = m_0c^2$.

The general theory of relativity, published in 1916, held that a gravitational field was a curvature of space-time caused by the presence of a mass. Thus, the mass of the earth curves space-time and the mass of our bodies responds to that curvature by accelerating toward the center of the earth. This curvature of space-time was confirmed by the slow rotation of the orbit of Mercury and by the deflection of starlight observed during a 1919 total solar eclipse.

NEW TERMS

model	circular velocity
natural motion	center of mass
violent motion	conic sections
acceleration of gravity	closed orbit
momentum	open (escape) orbit
mass	escape velocity
acceleration	angular momentum
velocity	special relativity
inverse square law	joule (J)
field	general relativity

QUESTIONS

1. Why wouldn't Aristotle's explanation of gravity work if the earth was not the center of the universe?
2. According to the principles of Aristotle, what part of the motion of a baseball pitched across home plate is natural motion? What part is violent motion?
3. If we drop a feather and a steel hammer at the same moment, they should hit the ground at the same instant. Why doesn't this work on Earth and why does it work on the moon?
4. How did Galileo idealize his inclines to conclude that an object in motion stays in motion until it is acted on by some force?
5. Give an example to illustrate each of Newton's laws.
6. What is the difference between mass and weight? Between speed and velocity?
7. Why did Newton conclude that some force had to pull the moon toward the earth?
8. Why did Newton conclude that gravity had to be mutual and universal?
9. How does the concept of a field explain action at a distance? Name another kind of field also associated with action at a distance.
10. Why can't a space capsule go "beyond earth's gravity"?
11. What is the center of mass of the earth–moon system? Where is it?
12. How do skaters and planets orbiting the sun conserve angular momentum?
13. Why is the period of an open orbit undefined?
14. How does the first postulate of special relativity imply the second?
15. When we ride a fast elevator upward we feel slightly heavier as the trip begins and slightly less heavy as the trip ends. How is this related to the equivalence principle?
16. From your knowledge of general relativity, would you expect radio waves from distant galaxies to be deflected as they pass near the sun? Why?

PROBLEMS

1. Compared to the strength at the earth's surface, how much weaker is the earth's gravity at a distance of 10 earth radii from the center of the earth? At 20 earth radii?
2. If a lead ball falls from a high tower on earth, what will be its velocity after 2 seconds? After 4 seconds?

3. What is the circular velocity of an earth satellite 1000 km above earth's surface? (HINT: The radius of the earth is 6380 km. See Box 5–1.)
4. What is the circular velocity of an earth satellite 36,000 km above earth's surface? What is its orbital period?
5. Describe the orbit followed by the slowest cannonball in Figure 5–8 on the assumption that the cannonball could pass freely through the earth. (Newton got this problem wrong the first time he tried it.)
6. If we could launch rockets from a tower 100 km high, how would the escape velocity compare with that at the earth's surface?

RECOMMENDED READING

BROAD, W. J. "Priority War: Discord in Pursuit of Glory." *Science 211*(30 Jan. 1981), p. 465.

BRONOWSKI, J. *The Ascent of Man.* Boston: Little, Brown, 1973.

COHEN, I. B. *The Newtonian Revolution.* London: Cambridge University Press, 1980.

———. "Newton's Discovery of Gravity." *Scientific American 244* (March 1981), p. 167.

DRAKE, S. "Galileo's Discovery of the Laws of Free Fall." *Scientific American 228* (May 1973), p. 85.

———. "The Role of Music in Galileo's Experiments." *Scientific American 232* (June 1975), p. 98.

———. *Galileo at Work.* Chicago: University of Chicago Press, 1978.

———. "Newton's Apple and Galileo's Dialogue." *Scientific American 243* (Aug. 1980), p. 150.

FERMI, L., and G. BERNARDINI *Galileo and the Scientific Revolution.* Greenwich, Conn.: Fawcett Publications, 1965.

FRANKLIN, ALLAN *The Principle of Inertia in the Middle Ages.* Boulder, Colo.: Colorado Associated University Press, 1976.

HALL, A. RUPERT *Philosophers at War.* London: Cambridge University Press, 1980.

HARRISON, EDWARD "Newton and the Infinite Universe." *Physics Today 39* (Feb. 1986), p. 24.

MANUEL, F. A. *A Portrait of Isaac Newton.* Washington, D.C.: New Republic Press, 1979.

MOORE, P. *Watchers of the Sky.* New York: G. G. Putnam's Sons, 1973.

SCHWARTZ, JOE, and MICHAEL McGUINNESS *Einstein for Beginners.* New York: Pantheon, 1979.

SHEA, W. R. *Galileo's Intellectual Revolution.* New York: Macmillan, 1972.

SMITH, DAVID H. "Testing Relativity with DI Herculis." *Sky and Telescope 71*(March 1986), p. 236.

WESTFALL, R. S. "Newton and the Fudge Factor." *Science 179* (23 Feb. 1973), p. 751. See also *Science 180* (15 June 1973), p. 1118.

———. *Never at Rest.* Cambridge, Mass.: Cambridge University Press, 1980.

ZIRKER, JACK B. "Testing Einstein's General Relativity During Eclipses of the Sun." *Mercury 14* (July/Aug. 1985), p. 98.

CHAPTER 6

LIGHT AND TELESCOPES

He burned his house down for the fire insurance
And spent the proceeds on a telescope.

Robert Frost
"THE STAR-SPLITTER"

What do fleas living on rats have to do with modern astronomy? That may sound like part of a bad joke, but it is actually related to the subject of this chapter. We will discuss the tools that modern astronomers use, and those tools are connected by an interesting sequence of events to rats and their fleas.

The most horrible disease in history—the black plague—was spread by flea bites, and the fleas lived on the rats that infested the cities. Plague broke out in London in 1665, and, although no one knew how the disease spread, those people who lived in the country were less likely to get the plague than city dwellers. All who could left the cities. When the plague reached Cambridge, the universities were closed, and both students and faculty fled to the English countryside. One was the young Isaac Newton. From the summer of 1665 to 1667, he spent most of his time in his mother's cottage in the small village of Woolsthorpe. While he was there, he conducted an experiment that changed the history of science.

Boring a hole in a shutter, he admitted a thin beam of sunlight into his darkened room. A glass prism placed in the beam threw a rainbow of color across the wall—a spectrum (Color Plate 4). When he used a second prism to recombine the colors, they produced white light. From this and other experiments conducted in his bedroom, he concluded that white light was made up of a mixture of all the colors of the rainbow.

When the plague abated, Newton returned to the university, where he began experimenting with telescopes. He discovered that telescopes made of lenses produced colored fringes around bright objects in the field of view because the glass lenses broke the light into colors just as his prism broke up the sunlight. To solve the problem, Newton designed and built a telescope containing a mirror instead of a lens. Although the first model

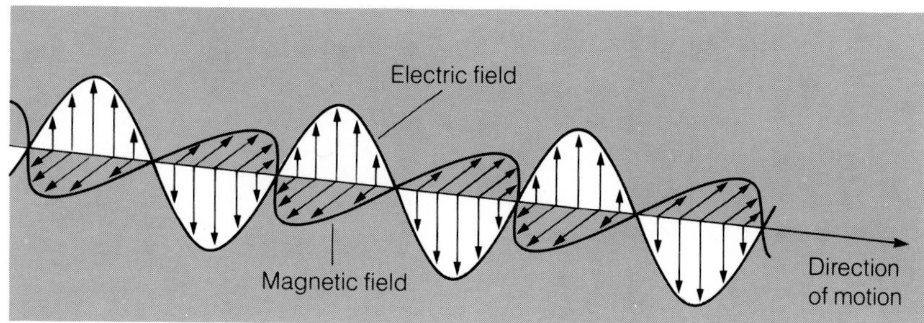

FIGURE 6–1 Electric and magnetic fields travel together through space as electromagnetic radiation.

hardly exceeded 1 inch in diameter, when Newton presented it to the Royal Society in 1671, it established his reputation as a scientist.

For a century after Newton's first telescope, astronomers did little with such devices, but as instrument makers grew more skilled at making large telescopes, they became the principal tool of the astronomer. Telescopes are important in astronomy because they gather light and concentrate it for study. The larger the telescope, the more light it gathers, and thus astronomers are still striving to build bigger telescopes to gather more light from the objects in the sky. Like Newton's original telescope, almost all modern telescopes use mirrors rather than lenses to avoid spreading the light into its component colors.

6.1 RADIATION: INFORMATION FROM SPACE

Astronomers are in the light business. Almost everything we know about the universe, we learn by analyzing the light gathered by telescopes. Thus to understand astronomy, we must understand light.

ELECTROMAGNETIC RADIATION

Light is merely one form of radiation called **electromagnetic** because it is associated with changing electric and magnetic fields that travel through space and transfer energy from one place to another. When light enters our eye, the fluctuating electric and magnetic fields carry energy that stimulates nerve endings, and we see what we call light.

The oscillating electric and magnetic fields that constitute electromagnetic radiation move through space at about 300,000 km/sec (186,000 miles/sec). This speed is commonly referred to as the speed of light *c,* but it is in fact the speed of all such radiation. If we represent the fluctuating electric and magnetic fields as arrows, electromagnetic radiation might resemble Figure 6–1.

Electromagnetic radiation is a wave phenomenon—that is, it is associated with a periodically repeating disturbance, or wave. We are familiar with waves in water. If we disturb a quiet pool of water, waves spread across the surface. Imagine that we use a meter stick to measure the distance between the successive peaks of a wave. This distance is the **wavelength**, usually represented by the Greek letter lambda (λ). If we were measuring ripples in a pond, we might find the wavelength is a few centimeters, whereas the wavelength of ocean waves might be a hundred meters or more. There is no restriction on the wavelength of electromagnetic radiation. Wavelengths can range from smaller than an atom to larger than the earth.

It is incorrect, or at least incomplete, to say that electromagnetic radiation is a wave, because it has sometimes the properties of a wave and sometimes the properties of a particle. For instance, the beautiful colors in a soap bubble arise from the wave nature of light. On the other hand, when light strikes the photoelectric cell in a camera's light meter, it behaves like a stream of particles carrying specific amounts of energy. Throughout his life, Newton believed that light is made up of particles, but we now recognize that light can behave as both particle and wave. Our model of light is thus more complete than Newton's. We will refer to "a particle of light" as a **photon**, and we can recognize its dual nature by thinking of it as a bundle of waves.

The amount of energy a photon carries depends on its wavelength. The shorter the wavelength, the more energy the photon contains; the longer the wavelength, the less energy it contains.

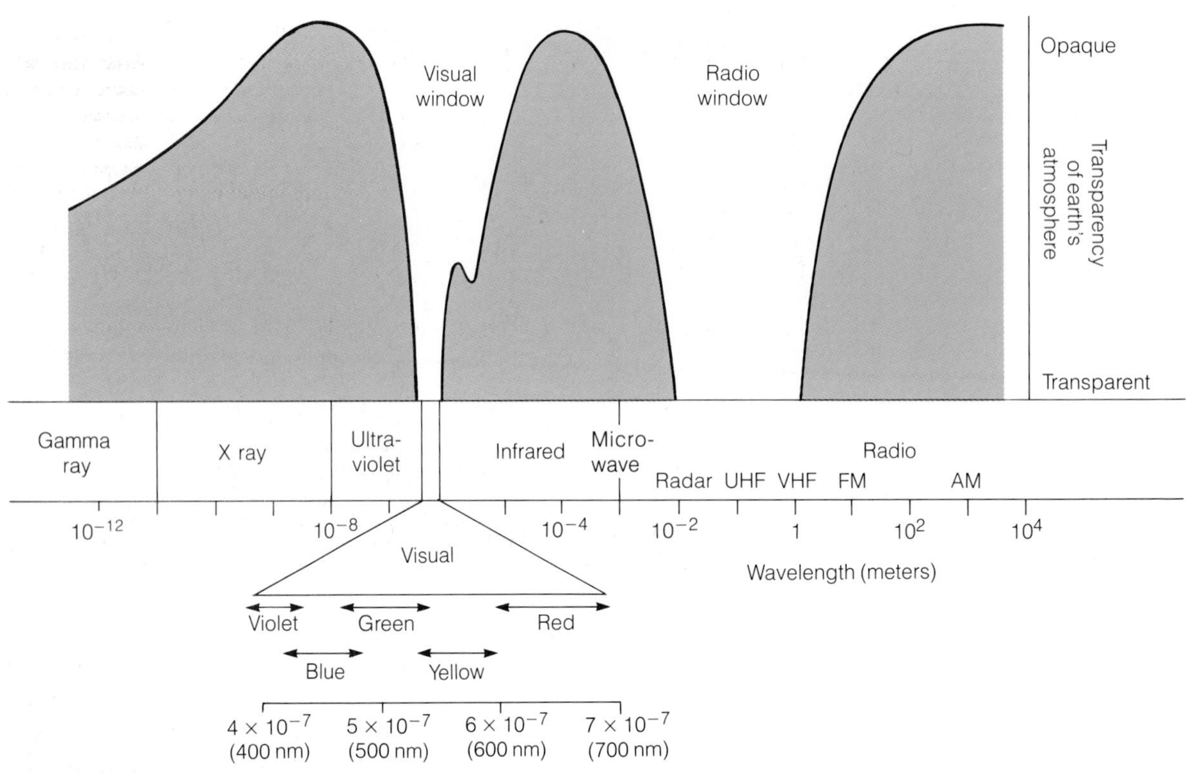

FIGURE 6–2 The electromagnetic spectrum includes all wavelengths of electromagnetic radiation. Earth's atmosphere is relatively opaque at most wave-lengths. Visual and radio windows allow light and some radio waves to reach earth's surface.

THE ELECTROMAGNETIC SPECTRUM

A spectrum is an array of electromagnetic radiation in order of wavelength. We are most familiar with the spectrum of visible light, which we see in rainbows for instance, but the visible spectrum is merely a small segment of the much larger electromagnetic spectrum (Figure 6–2).

The average wavelength of visible light is about 0.00005 cm. We could put 50 light waves end to end across the thickness of a sheet of household plastic wrap. It is too awkward to measure such short distances in centimeters, so we will measure the wavelength of light in **nanometers** (nm). One nanometer is 10^{-9} m. The wavelength of visible light ranges from about 400 nm to 700 nm. (See Box 6–1.)

Just as we sense the wavelength of sound as pitch, we sense the wavelength of light as color. Light near the short wavelength end of the visible spectrum (400 nm) looks violet to our eyes, and light near the long wavelength end (700 nm) looks red (Figure 6–2).

Beyond the red end of the visible spectrum lies **infrared radiation**, where wavelengths range from 700 nm to about 0.1 cm. Our eyes are not sensitive to this radiation, but our skin senses it as heat. A "heat lamp" is just a bulb that gives off principally infrared radiation.

Beyond the infrared part of the electromagnetic spectrum lie radio waves. The radio radiation used for AM radio transmissions has wavelengths of a few kilometers, whereas FM, television, military, governmental, and ham radio transmissions have wavelengths that range down to a few meters. Microwave transmissions, used for radar and long-distance telephone communication for instance, have wavelengths from a few centimeters down to about 0.1 cm.

The distinction between the wavelength ranges is not sharp. Long-wavelength infrared radiation and the shortest microwave radio waves are the same. Similarly, there is no clear division between the short-wavelength infrared and the long-wavelength part of the visible spectrum. It is all electromagnetic radiation.

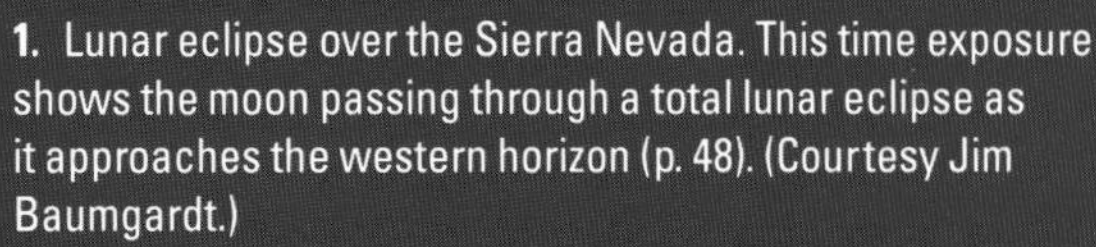

1. Lunar eclipse over the Sierra Nevada. This time exposure shows the moon passing through a total lunar eclipse as it approaches the western horizon (p. 48). (Courtesy Jim Baumgardt.)

2. During a total lunar eclipse the moon turns coppery red as sunlight, refracted by the earth's atmosphere, illuminates the moon in a red, sunset glow (p. 48). (Celestron International.)

3. A total solar eclipse. The pale white corona and pink prominences are visible when the moon covers the sun's brilliant photosphere (pp. 53, 178). (William P. Sterne, Jr.)

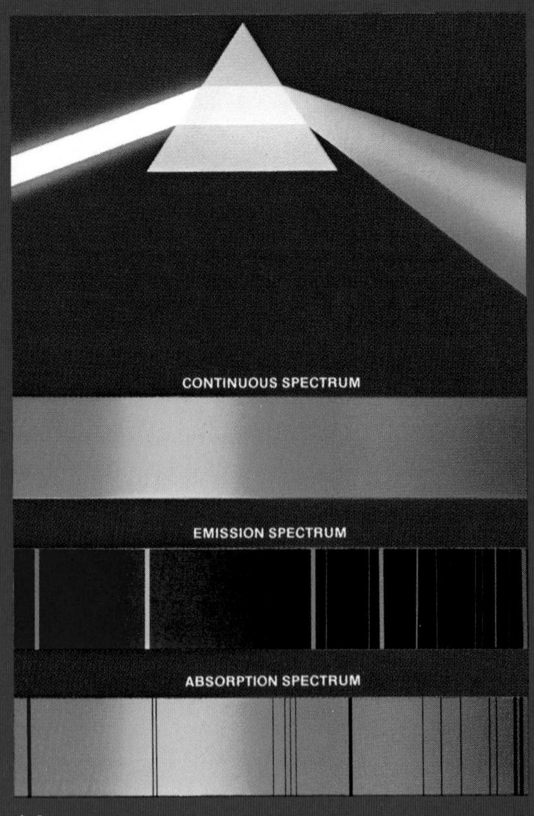

▲4

▲5

4. The spectrograph may be the most common instrument used at observatories. Here, white light passing through a prism is dispersed into a spectrum. A continuous spectrum is featureless, but an emission spectrum, such as that of helium, contains bright lines. The sun's spectrum is an absorption spectrum (pp. 108, 121). (Courtesy *Astronomy Magazine.*)

5. Modern observatories such as Kitt Peak National Observatory near Tucson are located on high mountains where the air is clear and calm and the night sky is dark (p. 118). (National Optical Astronomy Observatories.)

6. The Trifid nebula in Sagittarius illustrates three kinds of nebulae. The pink glow is produced by ionized hydrogen in an emission nebula around hot stars. The blue glow is reflected starlight producing a reflection nebula. The dark lanes are thick dust clouds (p. 114). (Celestron International.)

7. Bright enough to see with the naked eye, the Orion nebula is a glowing emission nebula surrounding a small cluster of newly formed stars. One or more of these stars is hot enough to emit ultraviolet radiation capable of ionizing the surrounding hydrogen gas. The pink color is typical of ionized hydrogen (pp. 114, 221, 240). (Celestron International.)

▼6

▼7

9. Auroral displays are triggered by the arrival of a shock wave in the solar wind. This ultraviolet image of the earth was obtained by the Dynamics Explorer 1 satellite. The oval shape of the display shows the influence of the earth's magnetic field (p. 482). (L. A. Frank, University of Iowa/NASA.)

▼9

▲8

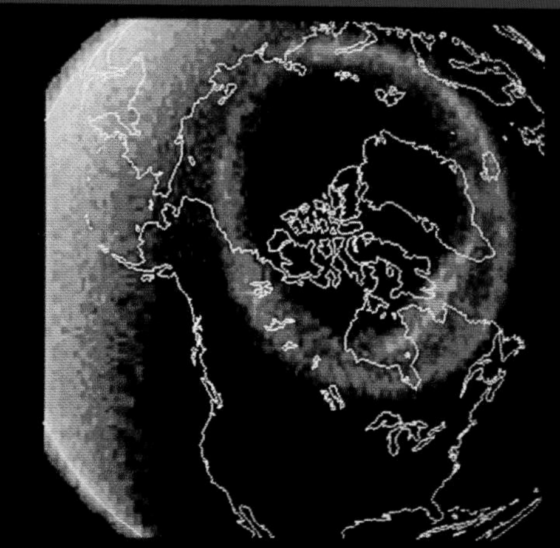

8. Streamers in the sun's corona. An image from the Solar Maximum Mission satellite (inset) has been computer processed to produce a false-color image that reveals subtle variations in brightness (p. 178). (JPL/NASA.)

10. Just as the sun has sunspots, other stars seem to have starspots. Sophisticated mathematical analysis of spectral lines was used to generate this map showing the location of silicon-deficient regions (green) on the surface of the star Gamma-2 Arietis. Regions overabundant in silicon are mapped in red (p. 172). (Courtesy Steven Vogt, Artie Hatzes, and Dan Penrod, Lick Observatory/UC Santa Cruz.)

11. Helioseismology is the study of the modes of vibration of the sun. This computer-generated image shows one of nearly 10 million possible modes of oscillation. Red represents receding regions and blue represents approaching regions. Comparison of observations with such models can reveal details of the sun's internal structure (p. 163). (National Optical Astronomy Observatories.)

▼10

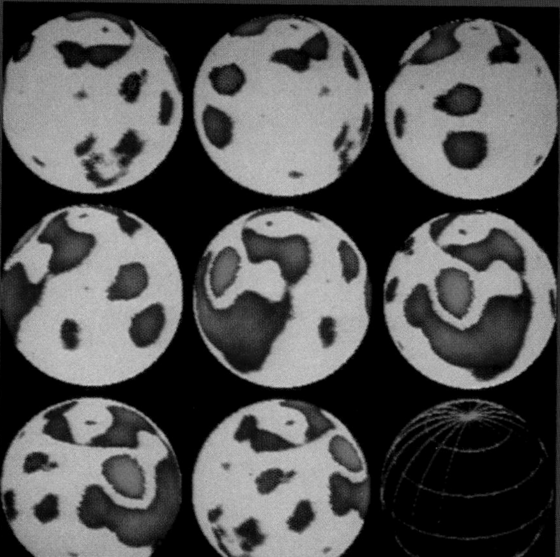

▼11

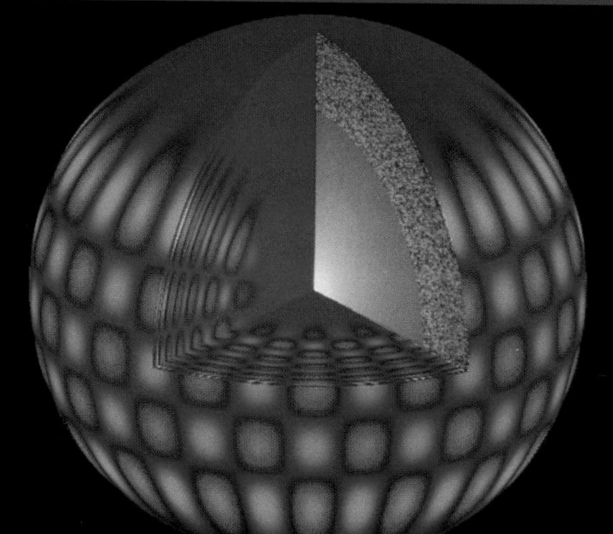

12. The Lagoon nebula in Sagittarius is a cloud of gas and dust about 60 ly in diameter, excited by the ultraviolet radiation of the hot, young stars within (pp. 114, 121). (National Optical Astronomy Observatories.)

13. The birth of a star. The nebula S106 surrounds a young star expelling ionized gas in a bipolar flow to the upper left and lower right. The star, located at the bright dot, is presumably surrounded by a rotating disk of gas and dust. This radio image was obtained at a wavelength of 6 cm and computer processed to produce a color-coded radio map (p. 230). (Courtesy John Bally.)

▼13

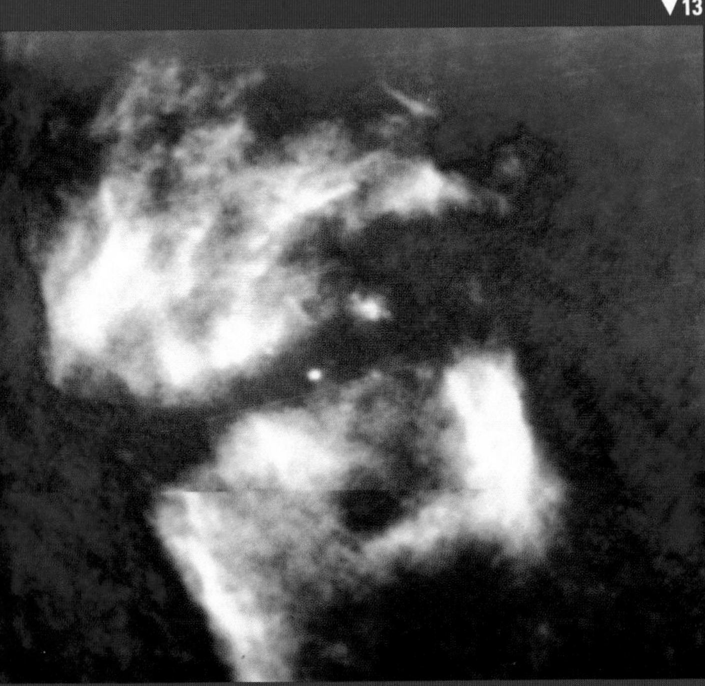

▲12

14. An infrared view of the entire sky obtained by the Infrared Astronomy Satellite shows the strong infrared emission from the Milky Way as the white strip running from left to right. Much of this emission comes from dust and gas warmed by stars. Orion lies at the extreme right. Many of the small sources far above and below the plane are distant galaxies (pp. 309, 324). (NASA.)

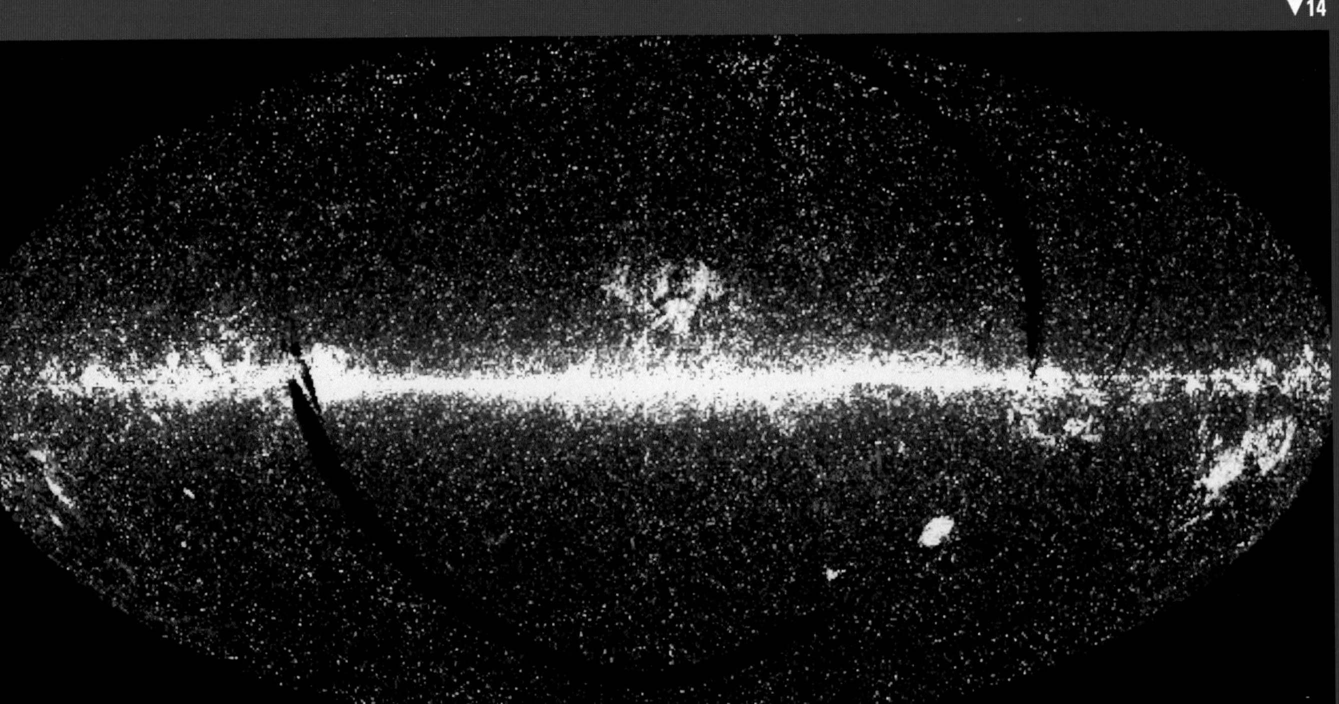

▼14

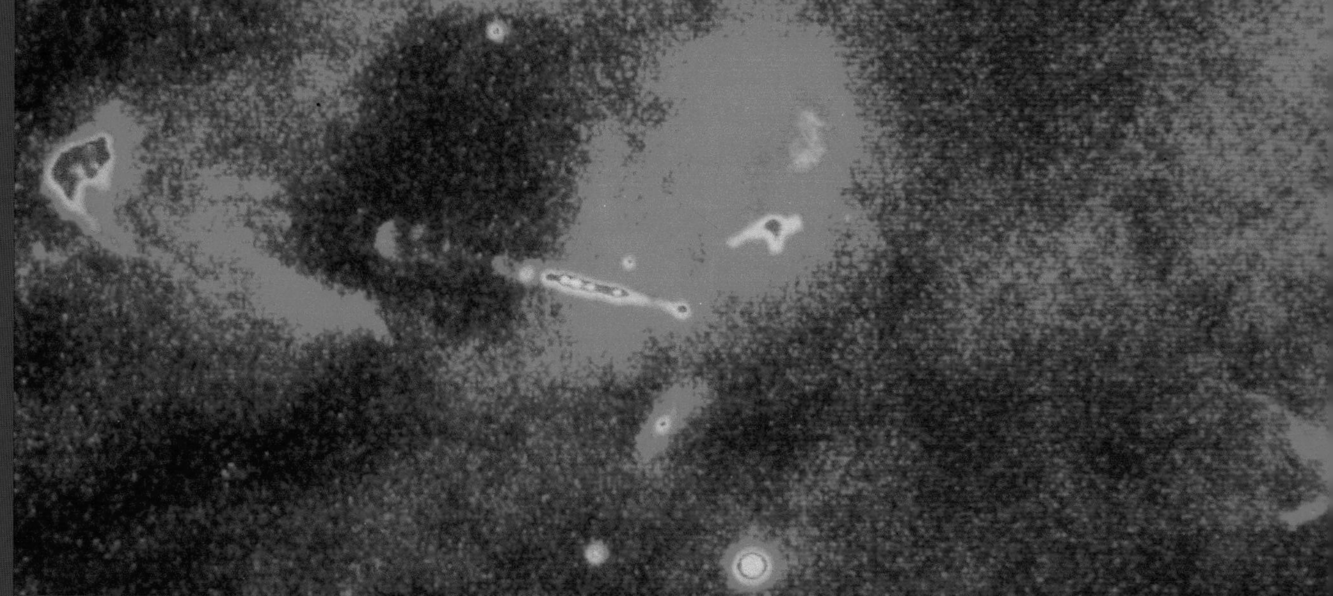

▲15

15. Bipolar flow can excite Herbig-Haro objects. This false color CCD image was originally made in the light of once ionized sulfur. It reveals a newly formed star (at center) emitting a jet which strikes the interstellar medium and creates the Herbig-Haro object HH34S (left). A similar jet not visible here extends to the right and excites the Herbig-Haro object HH34N (extreme right) (p. 229). (Obtained by Reinhard Mundt at the Calar Alto 3.5 m telescope.)

16. Discovered by the Infrared Astronomy Satellite, the infrared cirrus appears to be a wispy distribution of very cold dust clouds scattered through interstellar space (p. 225). (NASA.)

▼16

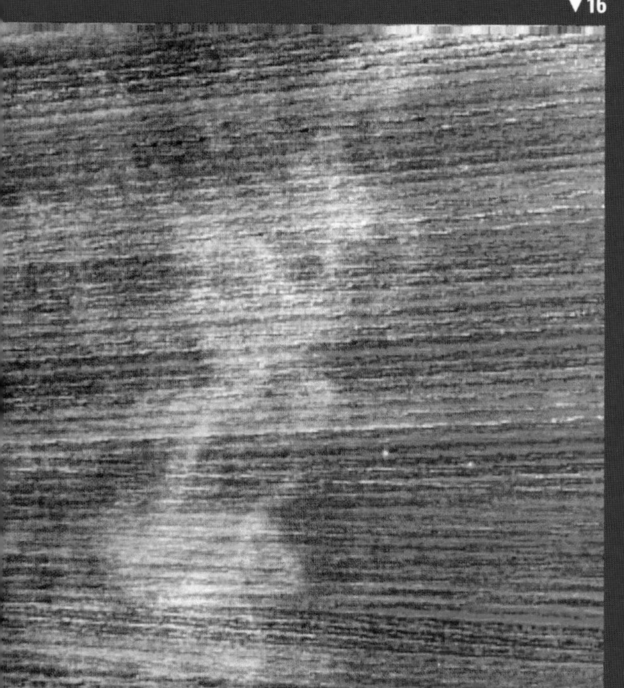

17. Cold dust clouds fill the constellation Orion in this false-color image recorded by the Infrared Astronomy Satellite. The bright region at the center bottom is the Orion molecular cloud, which includes the Orion nebula itself. The circular dust cloud at the top surrounds Lambda Orionis. Betelgeuse is the blue dot to the lower left of the dust ring (pp. 240, 321). (NASA/JPL courtesy Z. Souras.)

▼17

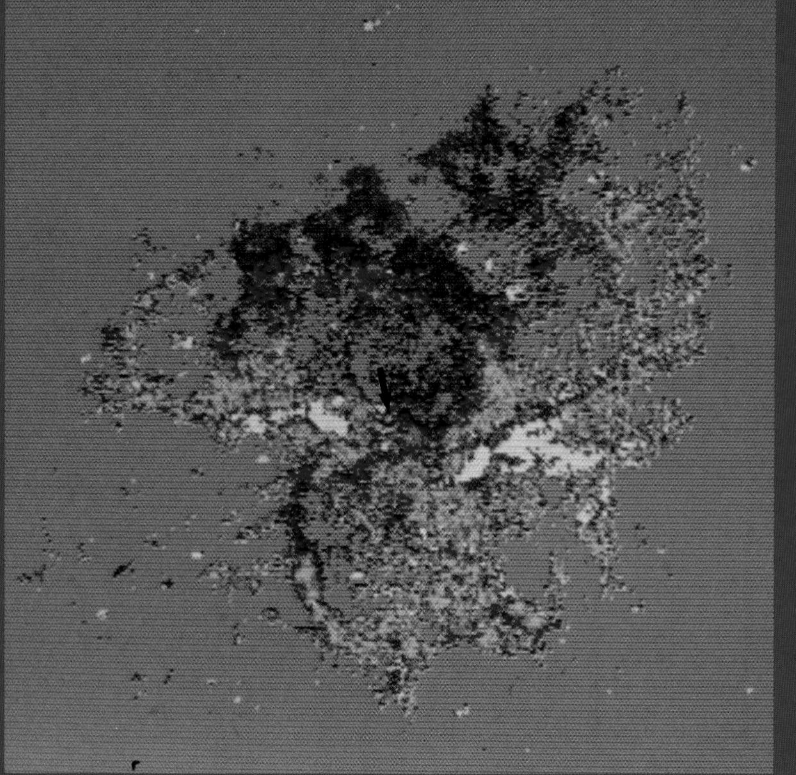

▲18

18. This false color image of the Crab nebula maps the ratio of helium to hydrogen. The filaments in the nebula are rich in helium and the white areas contain over 90 percent helium. Presumably these atoms were made in the massive star before it exploded as a supernova to produce the Crab nebula and its pulsar (arrow) (p. 271). (Courtesy Gordon M. MacAlpine and Alan K. Uomoto.)

▼19

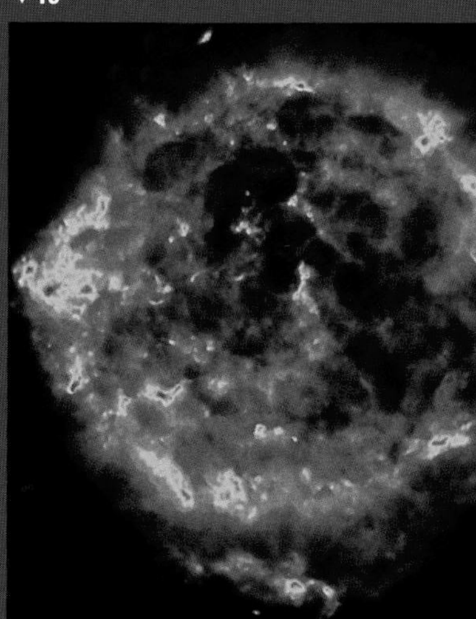

19. Cassiopeia A is a supernova remnant, an expanding shell of gas created by the explosive death of a massive star about 300 years ago. VLA radio telescope data from three different observing runs were combined to produce this false-color radio map (p. 272). (Courtesy NRAO/AUI, Observers R. J. Tuffs, R. A. Perley, M. T. Brown, and S. F. Gull.)

20. Image analysis of the galaxy M81. The original image from the Mt. Palomar 1.2 m Schmidt telescope is digitized for computer processing (a) and stretched to "untip" the galaxy (b). Further processing produces a false-color image (c) that reveals subtle details near the nuclear bulge. Compare with Color Plate 22 (pp. 120, 322). (Courtesy D. M. Elmegreen, B. G. Elmegreen, and P. E. Seiden, IBM Research Center.)

▼20a

▼20b

▼20c

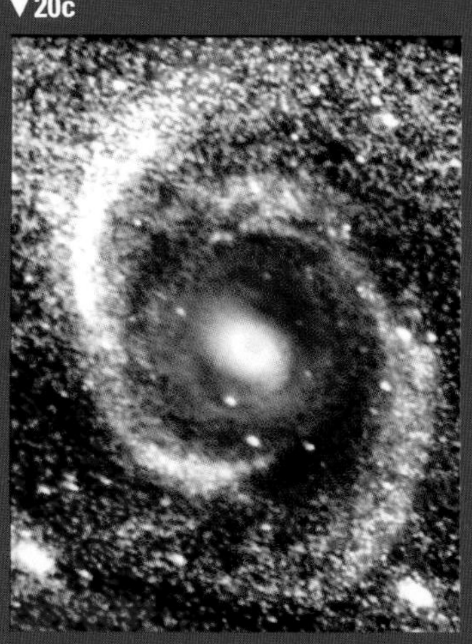

▲21

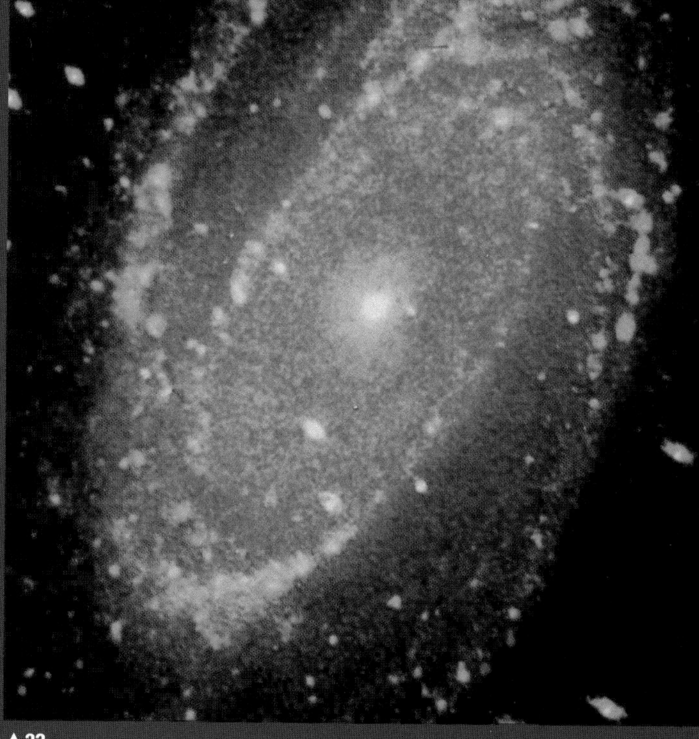

▲22

21. Photographic color enhancement of the barred spiral galaxy NGC 1097 shows that the arms are blue and the nuclear bulge is red. The faint linear feature to the upper left is real and appears to be associated with the galaxy (pp. 123, 322). (Courtesy Halton Arp.)

22. Color enhancement of the relatively nearby galaxy M81 shows that the disk of the galaxy is red. The green knots along the arms are produced by emission from oxygen atoms in ionized gas clouds around hot stars. Compare with Color Plate 20 (pp. 123, 322). (Courtesy Halton Arp.)

23. A false-color radio map of the active galaxy NGC 1265 shows that it is ejecting twin beams of matter as it moves through the intergalactic medium (pp. 123, 356, 357). (Courtesy NRAO/AUI, observers C. P. O'Dea and F. N. Owen.)

24. The double-lobed radio galaxy 3C 219 appears in this false-color map as an active core (red) ejecting a narrow jet pointing toward one of the two radio lobes containing hot spots (yellow) (pp. 123, 356). (Courtesy NRAO/AUI, observers A. H. Bridle and R. A. Perley.)

25. The radio galaxy 3C 75 contains two active nuclei (red) and two pairs of radio jets twisted by the relative motion of the intergalactic medium and the orbital motion of the two active cores (pp. 123, 356). (Courtesy NRAO/AUI, observers F. N. Owen, C. P. O'Dea, and M. Inoue.)

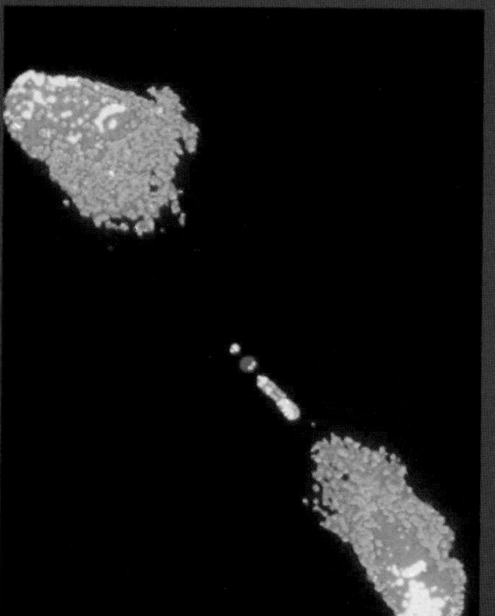

▼25

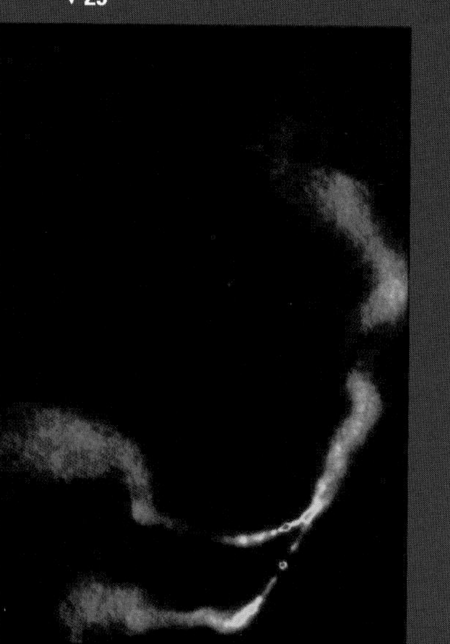

▼24

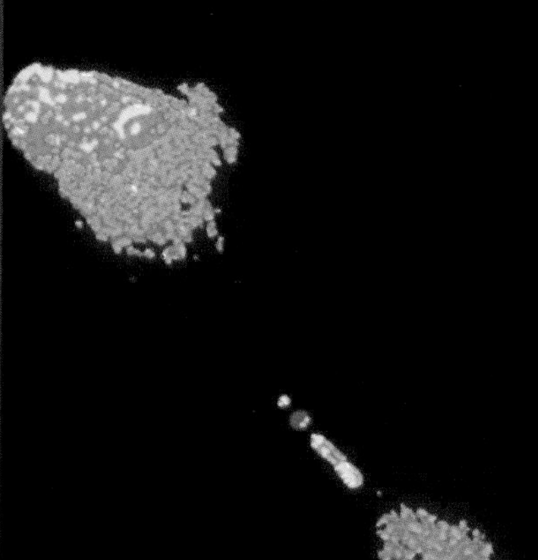

▼23

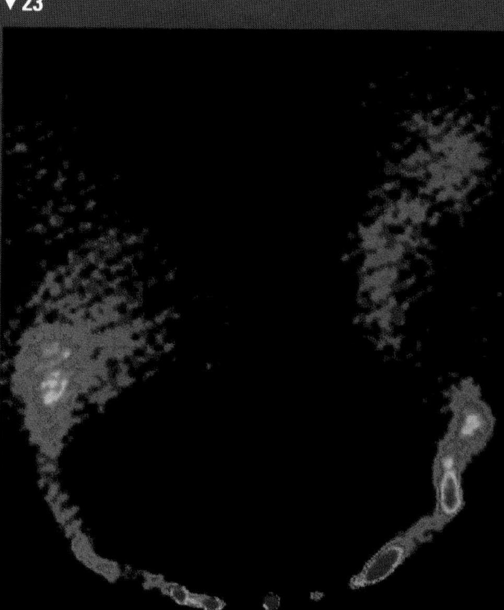

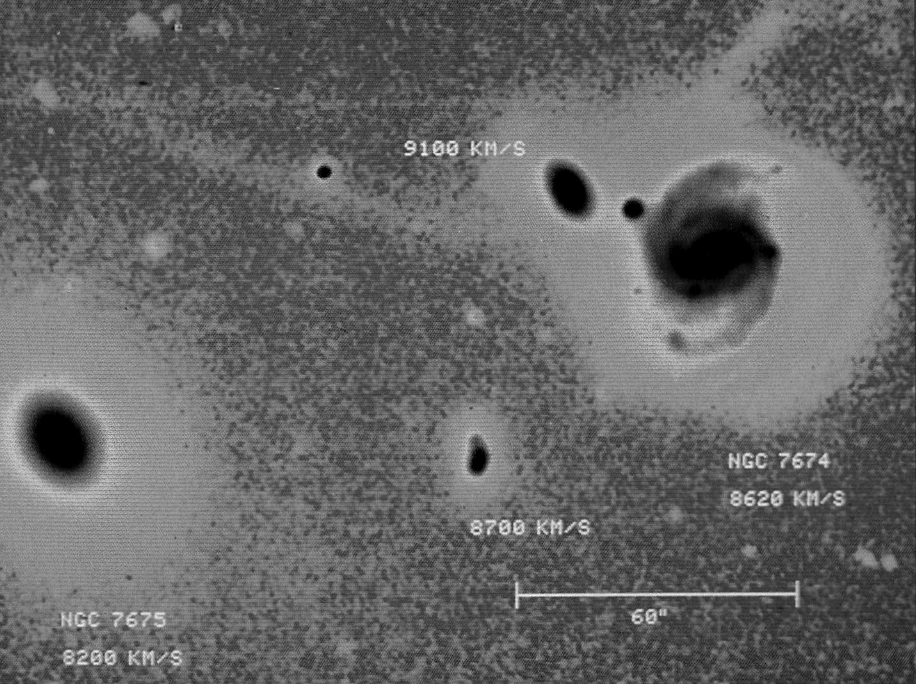

▲26

26. The Seyfert galaxy NGC 7674 (upper right) appears to be interacting with another galaxy, as shown by the tails. Studies show that Seyfert galaxies occur in such interacting pairs more often than we would expect from chance. Perhaps interactions can trigger Seyfert outbursts (p. 359). (Courtesy John W. Mackenty, Institute for Astronomy, University of Hawaii.)

29. A radio wavelength gravitational lens of the sort called an Einstein Ring. A massive foreground object, which is not luminous enough to be visible in this VLA map, is imaging radio energy from a more distant radio galaxy to form a nearly perfect ring with two compact sources. Such Einstein Rings can only occur when the lensing body is nearly in line with the distant object (p. 367). (Courtesy NRAO/AUI, J. N. Hewitt, and E. L. Turner.)

▼29

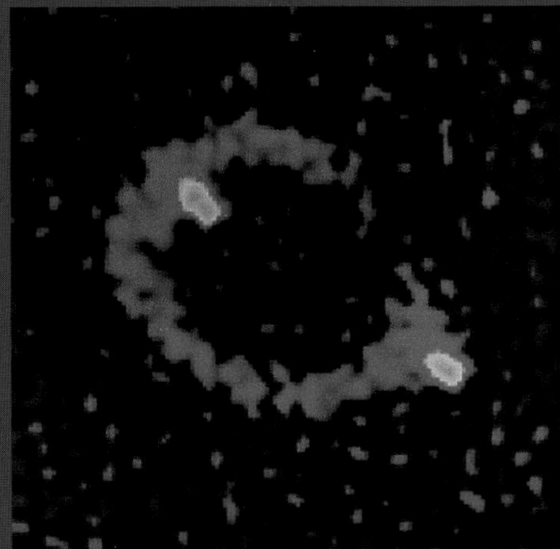

27, 28. False-color images of NGC 4319 and Markarian object 205 reveal what some believe is a link between the galaxy and the quasar. These computer enhancements are specially processed to reveal subtle intensity differences (p. 364). (Peter Wehinger.)

▼27

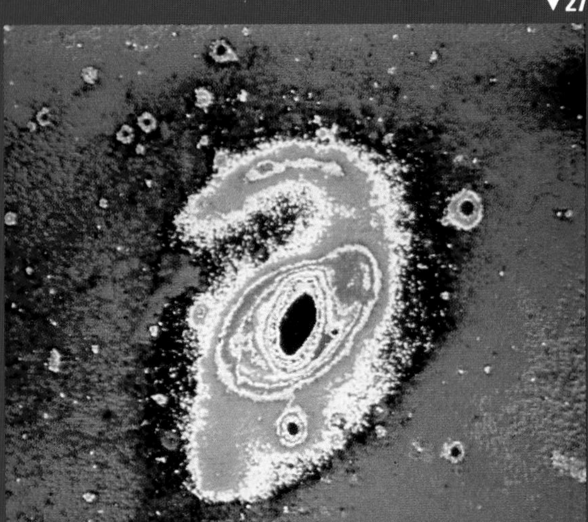

▼28

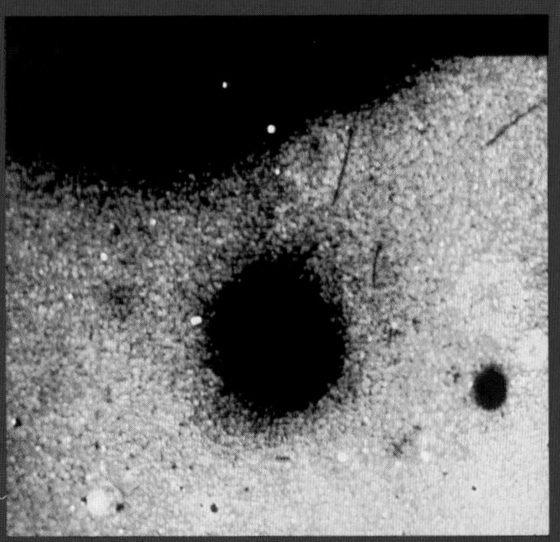

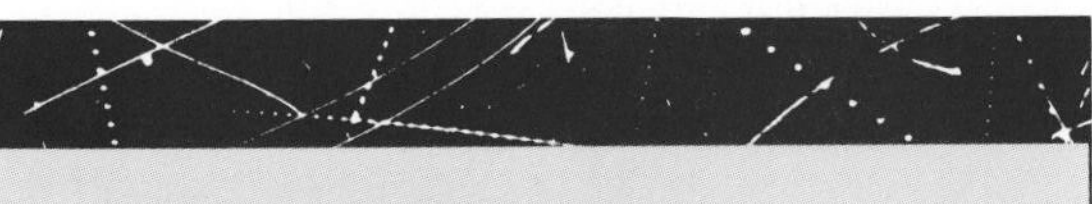

BOX 6–1
Measuring the Wavelength of Light

The wavelength of light is so short we must use more convenient units. Here we will use nanometers because it is consistent with the International System. One nanometer is 10^{-9} meter.

Another unit that astronomers use commonly, and a unit that you will see in other references on astronomy, is the **Angstrom** (Å). One Angstrom is 10^{-10} m, and visible light has wavelengths between 4000 Å and 7000 Å. One nanometer contains 10 Angstroms.

You may also find radio astronomers describing wavelengths in centimeters or millimeters, and infrared astronomers often refer to wavelengths in micrometers (or microns). One micrometer (μm) is 10^{-6} meter.

At wavelengths shorter than violet, we find **ultraviolet radiation**, with wavelengths ranging from 400 nm down to about 10 nm. At even shorter wavelengths lie X rays and gamma rays. Again, the boundaries between these wavelength ranges are not clearly defined.

X rays and gamma rays can be dangerous, and even ultraviolet photons have enough energy to do us harm. Small doses produce a suntan, sunburn, and skin cancers. Contrast this to the lower-energy infrared photons. Individually they have too little energy to affect skin pigment, a fact that explains why you can't get a tan from a heat lamp. Only by concentrating many low-energy photons in a small area, as in a microwave oven, can we transfer significant amounts of energy.

We are interested in electromagnetic radiation because it brings us clues to the nature of stars, planets, and other celestial objects. However, only a small part of this radiation can get through the earth's atmosphere. Only visible light and some radio waves can reach the surface of the earth; other wavelengths are absorbed. The highest parts of the atmosphere absorb X rays, gamma rays, and some radio waves, and a layer of ozone (O_3) at an altitude of about 30 km absorbs ultraviolet radiation. In addition, water vapor in the lower atmosphere absorbs infrared radiation. The wavelength regions in which our atmosphere is transparent are called **atmospheric windows.** Obviously, if we wish to study the sky from the earth's surface, we must look out through one of these windows.

Having described the nature of electromagnetic radiation and the electromagnetic spectrum, we can now study the tools astronomers use to analyze radiation.

6.2 ASTRONOMICAL TELESCOPES

Most astronomical telescopes are designed to observe visible light from the earth's surface. Later in this chapter we will discuss telescopes that observe radio waves and other telescopes that observe from space. Here, however, we must discuss the two kinds of earth-based telescopes—those that use lenses and those that use mirrors.

REFRACTING TELESCOPES

A **refracting telescope** works by refracting, or bending, light with a lens. Because of the shape of the lens, light striking the edge bends more than light striking the central area, and the light that enters the lens from some distant object comes together to form a small, inverted image (Figure 6–3). The **focal length** of a lens is the distance from the lens to the point where it focuses parallel rays of light (Figure 6–4). The focal length depends on the curvature of the glass surfaces, not on the diameter of the lens.

Notice that the image is inverted. This is true of all such optical systems including cameras and the human eye. The image could be re-inverted by an extra lens, but this is an unnecessary expense and a possible source of distortion. Consequently, astronomical telescopes, like microscopes, produce inverted images.

To build a refracting telescope, we need a lens of relatively long focal length to form an image of the object we wish to view. This lens is often called the **objective lens** because it is closest to the object. To view the image we add a lens of short focal length called an **eyepiece** to enlarge the image and make it easy to see (Figure 6–5). Thus the eyepiece acts as a magnifier. By changing eyepieces we can easily change the magnification of the telescope.

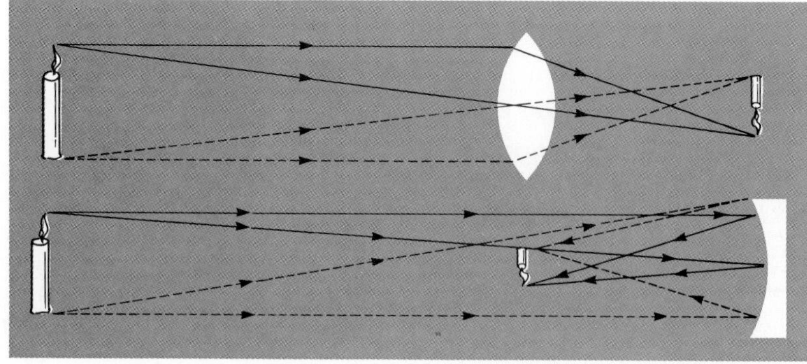

FIGURE 6–3 A lens forms an image by refracting (bending) light. A curved mirror can form an image by reflecting light. In both cases, the image of distant objects is inverted.

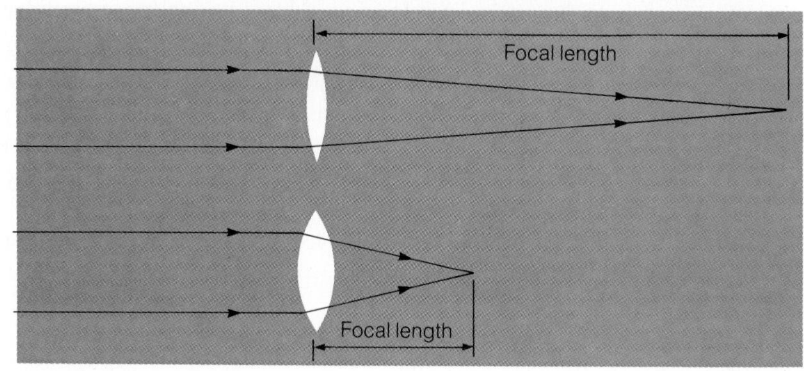

FIGURE 6–4 The focal length of a lens is the distance from the lens to the point where parallel rays of light come to a focus. The lens above has a longer focal length than the lens below.

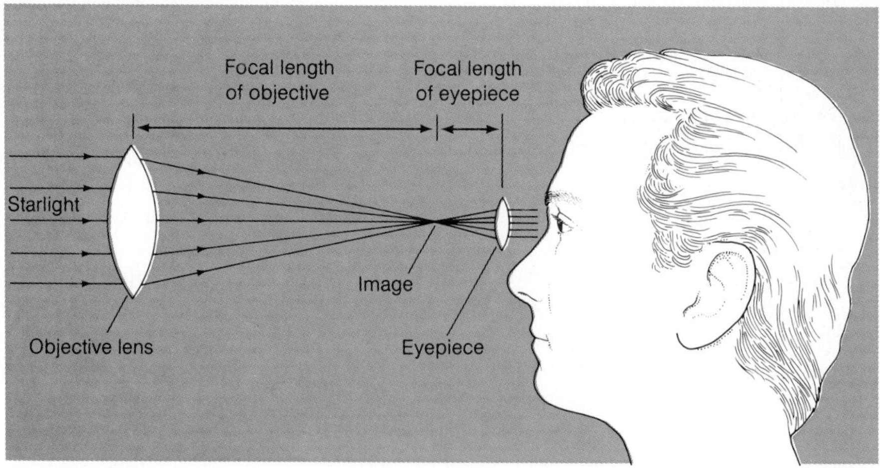

FIGURE 6–5 In a refracting telescope, the objective lens forms an image that is magnified by the eyepiece.

Refracting telescopes suffer from a serious optical distortion (aberration) that limits what we can see through them. When light is refracted through glass, shorter wavelengths bend more than longer wavelengths, and blue light comes to a focus closer to the lens than does red light (Figure 6–6). If we focus the eyepiece on the blue image, the red light is out of focus, producing a red blur around the image. If we focus on the red image, the blue light blurs. This color separation is called **chromatic aberration.**

A telescope designer can partially correct for this by replacing the single objective lens with one made of two lenses, ground from different kinds of glass. Such lenses, called **achromatic lenses,** can be designed to bring any

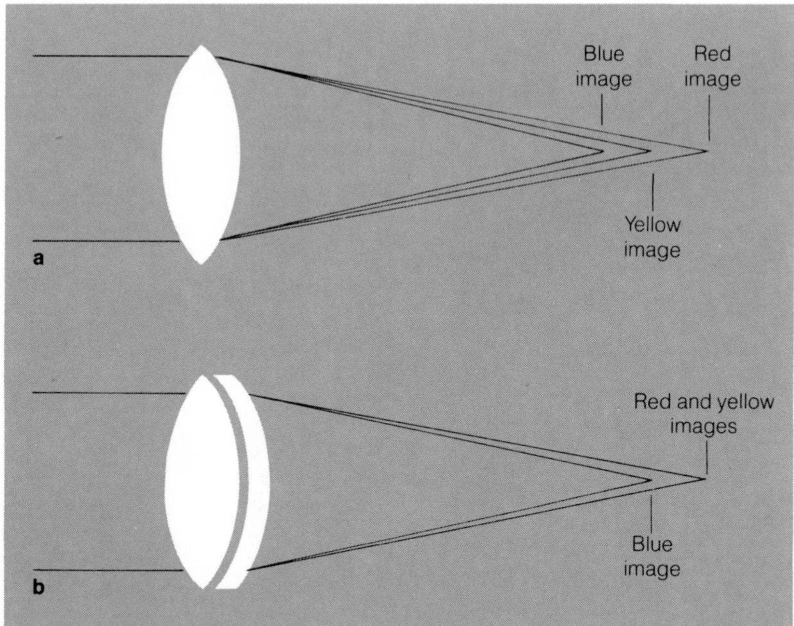

FIGURE 6–6 (a) A normal lens suffers from chromatic aberration because short wavelengths bend more than long wavelengths. (b) An achromatic lens made in two parts can bring any two colors to the same focus, but other colors remain slightly out of focus.

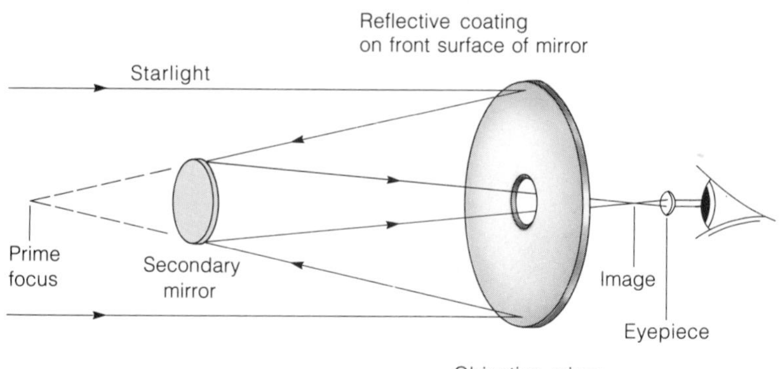

FIGURE 6–7 In a Cassegrain reflecting telescope, the objective mirror forms an image that is magnified by the eye-piece. For convenient viewing a secondary mirror reflects the light through a hole in the objective.

two colors to the same focus (Figure 6–6). Because our eyes are most sensitive to red and yellow light, we might bring these two colors to the same focus, but blue and violet would still be out of focus, producing a hazy blue fringe around bright objects.

Refracting telescopes were popular through the nineteenth century, but they are no longer economical for professional astronomy. A large achromatic lens is very expensive because it contains four matched optical surfaces and must be made of high-quality glass. Refractors can't be made larger than about 1 m in diameter because such large lenses sag under their own weight. Also, large refractors have very long telescope tubes that require large observatory domes. Modern telescopes, just as Newton's first telescope, focus light with mirrors.

REFLECTING TELESCOPES

A **reflecting telescope** uses a concave mirror, the **objective mirror**, to focus starlight into an image. Objective mirrors are usually made of special kinds of glass or quartz covered with a thin layer of aluminum to act as a reflecting surface.

The objective mirror forms an image at the location called the **prime focus** at the upper end of the telescope tube (Figure 6–7). Because it is usually inconvenient to

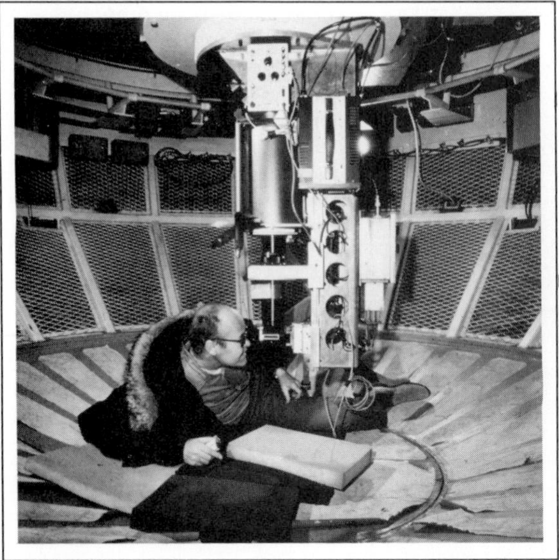

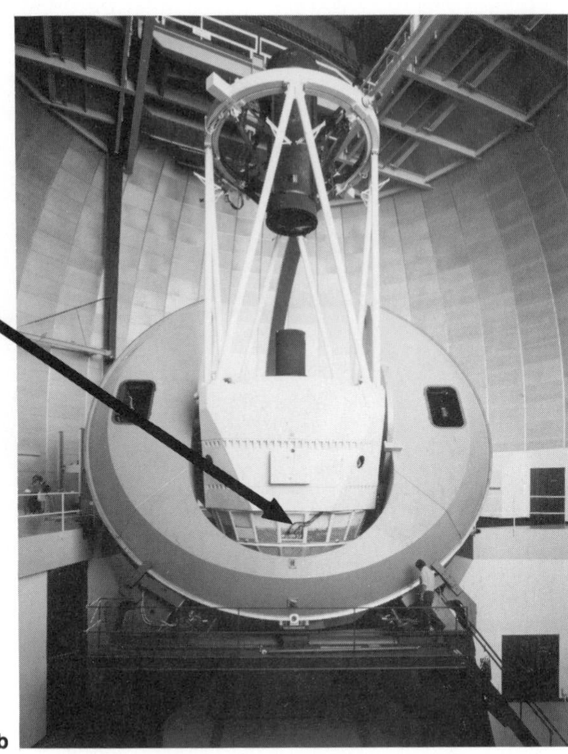

FIGURE 6–8 (a) Astronomer C. R. Lynds works in the Cassegrain observing cage beneath the objective mirror of the 4-m Mayall telescope at Kitt Peak National Observatory. (b) The 4-m (158-inch) Mayall telescope. (© Association of Universities for Research in Astronomy, Inc., Kitt Peak National Observatory.)

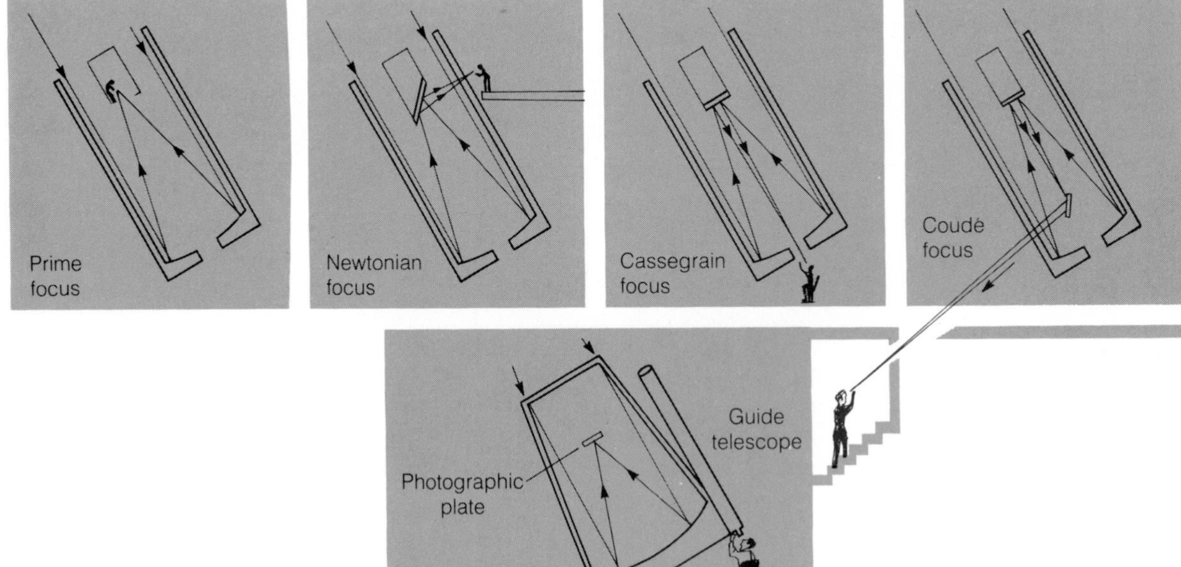

FIGURE 6–9 Many large telescopes can be used at a number of focal positions. Schmidt cameras, however, are used only for photography.

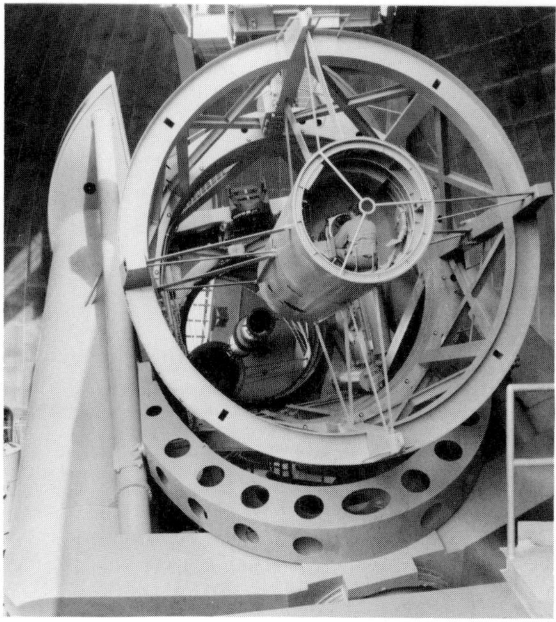

FIGURE 6–10 The prime focus cage of the 5-m (200-inch) Hale telescope on Mount Palomar. Note the objective mirror below. (Palomar Observatory photograph.)

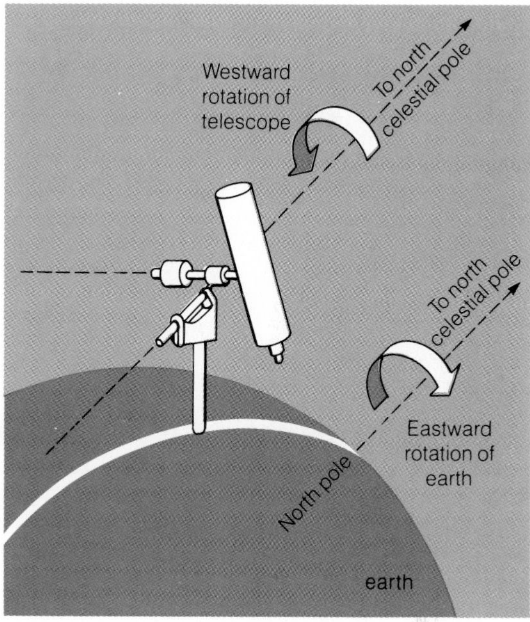

FIGURE 6–11 Westward motion around the polar axis of an equatorial mounting counters the eastward rotation of the earth and keeps the telescope pointed at a given star.

view the image there, a smaller **secondary mirror** reflects the light to a more accessible location. In one popular arrangement, the secondary mirror reflects the light back down the telescope tube through a hole in the center of the objective. This kind of telescope is called a **Cassegrain telescope** (Figures 6–7 and 6–8).

Other focal arrangements are also common (Figure 6–9). The largest telescopes usually have prime focus cages where the astronomer can ride inside the telescope tube to observe very faint objects (Figure 6–10). The **Newtonian focus**, named after Newton's first telescope, is common for smaller telescopes, but awkward for large instruments. The **coudé focus** (from the French word for "elbow") bends the light path and sends it to a separate observing room where very large or very delicate instruments such as spectrographs can be used. Most large telescopes can be used in more than one focal position, but **Schmidt cameras** are especially designed for wide-field photography and can't be changed to other arrangements.

Nearly all recently built telescopes are reflectors. Because the light does not enter the glass, there is no chromatic aberration, the glass need not be of perfect optical quality, and the mirror can be supported over its back to reduce sagging. Also, reflecting telescopes tend to be shorter and thus require smaller mountings and observatory buildings.

Telescope mountings can be more expensive than the telescope itself. The mounting must support the optics, protect against vibration, move accurately to designated objects, and then compensate for the earth's rotation. Because the earth rotates eastward, the telescope mounting must contain a **sidereal drive** (literally, a star drive) that can move the telescope smoothly westward to follow the object being studied. To simplify this motion, most telescope mountings are **equatorial mountings** with one axis of rotation, the **polar axis**, parallel to the earth's axis. Rotation around the polar axis moves the telescope parallel to the celestial equator (Figure 6–11). Advances in technology are changing the way astronomers build and mount telescopes.

NEW GENERATION TELESCOPES

The largest telescope in the world is the 6-m (236-inch) reflector on Mount Pastukhov in the Soviet Union, and

a

b

FIGURE 6–12 (a) The world's largest traditional telescope is the 6-m reflector atop Mt. Pastukhov in the Caucasus Mountains of the southern USSR. It uses a single thick mirror. (b) The six 1.8-m mirrors of the Multiple Mirror Telescope (MMT) move on a single mounting and are the equivalent of a 4.5-m telescope. Both telescopes have alt-azimuth mountings. (b, Courtesy of the Whipple Observatory, a joint facility of the University of Arizona and the Smithsonian Institution.)

the second largest is the 5-m (200-inch) Hale telescope on Mount Palomar. Both are traditional telescopes in that they use single, thick mirrors. The 5-m mirror weighs 14.5 tons, and its mounting weighs 500 tons. New technology is making such massive, expensive telescopes obsolete.

One way to reduce the cost of a telescope is to build the mirror in segments. Smaller mirror segments are less expensive to make and sag less under their own weight. The Multiple Mirror Telescope (MMT) uses six round mirrors, each 1.8 m (72 inches) in diameter on one mounting. The mirrors bring their light to a single image, producing the equivalent of a 4.5-m (176-inch) telescope. The Keck telescope now under construction atop the volcano Mauna Kea in Hawaii will use 36 hexagonal mirror segments. Their alignment will be controlled by a computer to form a mirror 10 m (400 inches) in diameter (Figure 6–13b).

Thin telescope mirrors weigh less and are cheaper to make and support, but such "floppy mirrors" cannot hold their shape alone. Computerized support systems must control their alignment as the telescope moves. The mirrors for the MMT were built by fusing thin sheets of glass with an egg crate lattice of glass to make thin mirrors. The original design called for a laser–computer system to control the alignment of the six mirrors, but that proved unnecessary. The Keck mirrors are single sheets of glass only about 8 cm (3 inches) thick. They are hinged at the edges, and computers will monitor their position and adjust them as needed.

Lightweight mirrors and computer controls make possible economies in the mounting. The MMT and Keck telescopes are on **alt-azimuth mountings.** Like cannons, they can be moved up or down (altitude) and left or right (azimuth). This type of mounting is not convenient for a sidereal drive, but it is very strong, and a computer can continuously move such a telescope to follow a star. Such small mountings make possible smaller, less expensive observatories. The buildings for the MMT and the Keck telescopes actually rotate with the telescopes.

Grinding large mirrors can take years and cost millions of dollars, but a new technique speeds the process. The mold for the mirror, filled with molten glass, is placed in a spinning oven. The rotation forces the liquid glass toward the outer rim of the mold forming a concave surface; once cooled, the glass is pre-formed to the approximate shape.

Astronomers are rushing to build giant telescopes with spun-cast, thin mirrors. Plans call for the six mirrors of the MMT to be replaced by a single large spun-cast mirror. The National Optical Astronomy Observatory now plans to build an 8-m telescope. One consortium of universities is building a 3.5-m telescope, and another consortium is planning an 8-m telescope for Chile. Japanese astronomers expect to build a 7-m telescope, and the European Southern Observatory proposes to build four 8-m telescopes that can be operated as a single instrument. This Very Large Telescope would be the equivalent of a 16-m telescope.

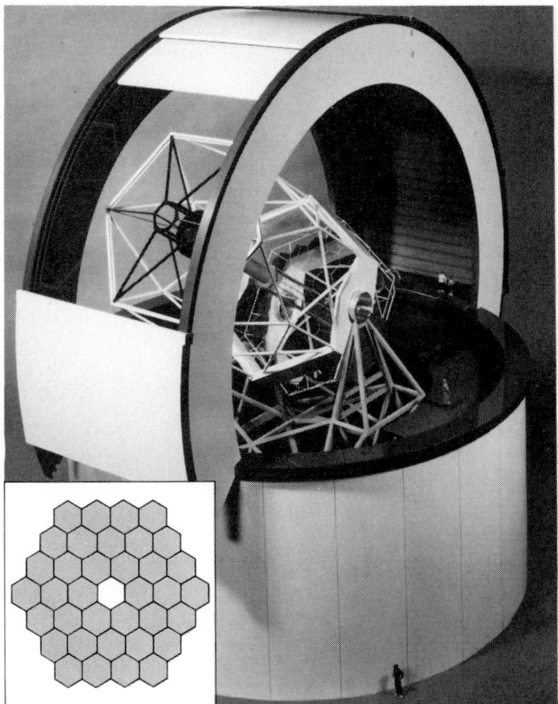
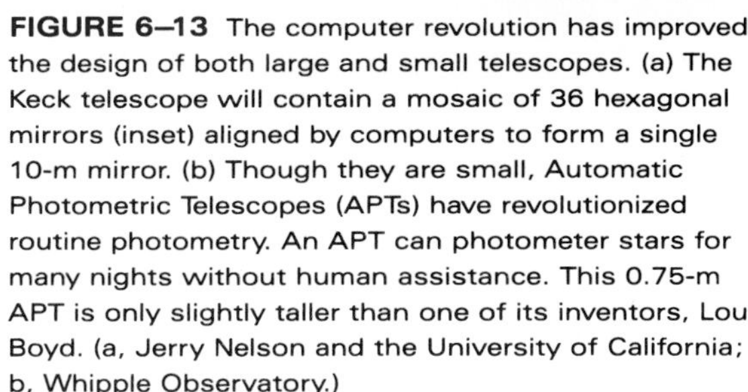

FIGURE 6–13 The computer revolution has improved the design of both large and small telescopes. (a) The Keck telescope will contain a mosaic of 36 hexagonal mirrors (inset) aligned by computers to form a single 10-m mirror. (b) Though they are small, Automatic Photometric Telescopes (APTs) have revolutionized routine photometry. An APT can photometer stars for many nights without human assistance. This 0.75-m APT is only slightly taller than one of its inventors, Lou Boyd. (a, Jerry Nelson and the University of California; b, Whipple Observatory.)

Not all new generation telescopes are large. Telescopes as small as 25 cm (10 inches) have become powerful research instruments thanks to computer control. Six Automatic Photometric Telescopes (APTs) now observe from the APT Service Observatory atop Mt. Hopkins near Tucson, Arizona. Every clear night of the year, desktop computers at the observatory guide the telescopes to photometer selected variable stars. During the day, the computers report their observations and receive further instructions from astronomers over phone lines. Such APTs can be quite small (Figure 6–13b) and inexpensive, so a number of others are under construction. The tireless APTs are changing the way astronomers study variable stars, a revolution made possible by the development of small computers.

THE POWERS OF A TELESCOPE

Refractor or reflector, the telescope aids our eyes in three ways: light-gathering power, resolving power, and magnifying power.

The most important thing a telescope does is gather light. Most interesting celestial objects are faint sources of light, so we need a telescope that can gather large amounts of light to produce a bright image. **Light-gathering power** refers to the ability of a telescope to collect light. Catching light in a telescope is like catching rain in a bucket—the bigger the bucket, the more it catches (Figure 6–14). This is the main reason why astronomers use large telescopes.

The second telescopic power, **resolving power**, refers to the ability of the telescope to reveal fine detail. Whenever light is focused to form an image, a small blurred fringe surrounds the image (Figure 6–15). Because this **diffraction fringe** surrounds every point of light in the image, we cannot see fine detail. There is nothing we can do to eliminate diffraction fringes; they are produced by the wave nature of light as it passes through the telescope. However, if we use a large-diameter telescope, the fringes are smaller and we can see smaller details. Thus the larger the telescope, the better its resolving power.

In addition to resolving power, two other factors—lens quality and atmospheric conditions—limit the detail we can see through a telescope. A telescope must contain high quality optics to achieve its full potential resolving power. Even a large telescope shows us little detail if its optics are marred with imperfections. Also, when we look through a telescope, we are looking through miles of turbulent air in earth's atmosphere, which makes the image dance and blur, a condition called **seeing**. On a night when the atmosphere is unsteady and the images are blurred, the seeing is bad. Even under good seeing conditions, the detail visible through a large telescope is limited, not by its diffraction fringes, but by the air through which the observer must look. A telescope performs better on a high mountaintop where the air is thin and steady, but even there the earth's atmosphere limits the detail the telescope can reveal. (See Color Plate 5.)

The third and least important power of a telescope is **magnifying power**, the ability to make the image bigger. Because the amount of detail we can see is limited by the seeing conditions and the resolving power, very high magnification does not necessarily show us more detail. Also, we can change the magnification by changing the eyepiece, but we cannot alter the telescope's light-gathering power or resolving power.

If you visit a department store to shop for a telescope, you will probably find telescopes described according to magnification. One may be labeled an "80 power telescope" and another a "40 power telescope." However, the magnifying power really tells us little about the telescopes. Astronomers identify telescopes by diameter because that determines both light-gathering power and resolving power.

We have described the three powers of the telescope in general terms. Box 6–2 gives them mathematical form.

6.3 SPECIAL INSTRUMENTS

Looking through a telescope doesn't tell us much. To use an astronomical telescope to learn about stars, we must be able to analyze the light the telescope gathers. Special instruments attached to the telescope make that possible.

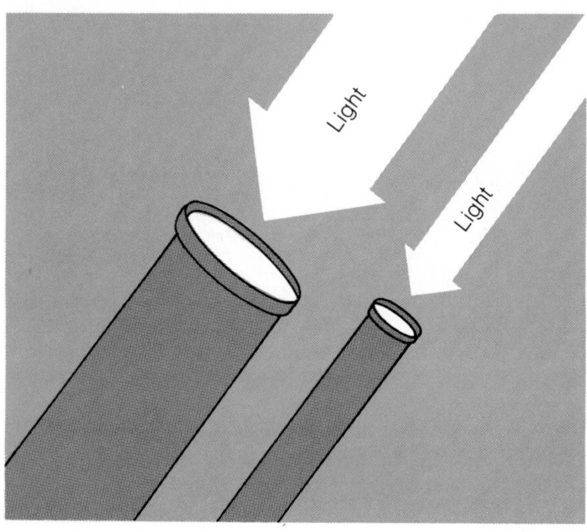

FIGURE 6–14 Gathering light is like catching rain in a bucket. A large-diameter telescope gathers more light and has a brighter image than a smaller telescope of the same focal length.

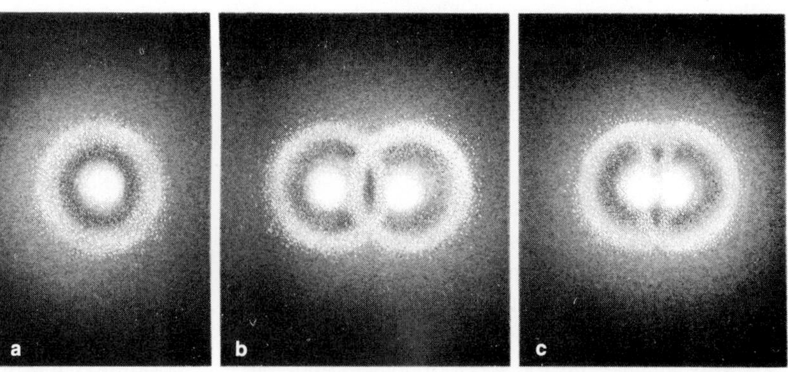

FIGURE 6–15 (a) Diffraction fringes surround every star image. (b) Two stars close to each other have overlapping fringes. (c) Two stars very close to each other blend together.

BOX 6–2

The Powers of the Telescope

Light-gathering power is proportional to the area of the telescope objective. A lens or mirror with a large area gathers a large amount of light. Because the area of a circular lens or mirror of diameter D is $\pi(D/2)^2$, we can compare the areas of two telescopes, and therefore their relative light-gathering powers, by comparing the square of their diameters. That is, the ratio of the light-gathering powers of two telescopes A and B is equal to the ratio of their diameters squared:

$$\frac{LGP_A}{LGP_B} = \left(\frac{D_A}{D_B}\right)^2$$

For example, suppose we compare a 4-cm telescope with a 24-cm telescope. How much brighter would a star look with the 24-cm telescope than with the 4-cm?

$$\frac{LGP_{24}}{LGP_4} = \left(\frac{24}{4}\right)^2 = 6^2 = 36 \text{ times brighter}$$

Our eye acts like a telescope with a diameter of about 0.8 cm (⅓ inch), the diameter of the pupil. How much brighter will stars look if we use a 24-cm telescope to aid our eyes?

$$\frac{LGP_{24}}{LGP_{eye}} = \left(\frac{24}{0.8}\right)^2 = (30)^2 = 900 \text{ times brighter}$$

The resolving power of a telescope is the angular distance between two stars that are just barely visible through the telescope as two separate images. For optical telescopes, the resolving power α, in seconds of arc, equals 11.6 divided by the diameter of the telescope in centimeters:

$$\alpha = \frac{11.6}{D}$$

For example, a 25-cm telescope has a resolving power of 0.46 seconds of arc. If the lenses are of good quality and if the seeing is good, we should be able to distinguish as separate points of light any pair of stars farther apart than 0.46 seconds of arc. If the stars are any closer together, diffraction fringes blur the stars together into a single image (Figure 6–15). Obviously, we would like to use large telescopes to make α as small as possible

The magnification of a telescope is the ratio of the focal length of the objective lens or mirror F_o divided by the focal length of the eyepiece F_e:

$$M = \frac{F_o}{F_e}$$

For instance, if the focal length of a telescope is 80 cm and we use an eyepiece with a focal length of 0.5 cm, the magnification is 80/0.5 or 160 times.

IMAGING SYSTEMS

The original imaging device in astronomy was the photographic plate. It could record faint objects in long-time exposures and could be stored for later analysis. But photographic plates have been largely replaced in astronomy by electronic imaging systems.

Low-light level television cameras were the first electronic replacements for the photographic plate, but the newest and most common system used in astronomy is a **charge-coupled device** (CCD) (Figure 6–16a)—typically a quarter million, microscopic, light-sensitive diodes in an array about the size of a postage stamp. These can be used like a small photographic plate, but they have dramatic advantages. They can detect both bright and faint objects in a single exposure, are much more sensitive than photographic plates, and can be read directly into computer memory for later analysis.

However an image is recorded, astronomers often reproduce the image in exaggerated ways to bring out subtle details. If an image of a faint object is reproduced as a negative, the sky is white and the stars are dark (Figure 6–16b). This makes the faint parts of the image easier to see. With a computer, astronomers can easily manipulate an image to produce **false color** images. Examine the color plates in this book for numerous examples. Remember that the colors in such images are merely codes to intensity and are not related to the true colors of the objects. To measure the true colors, we need a photometer.

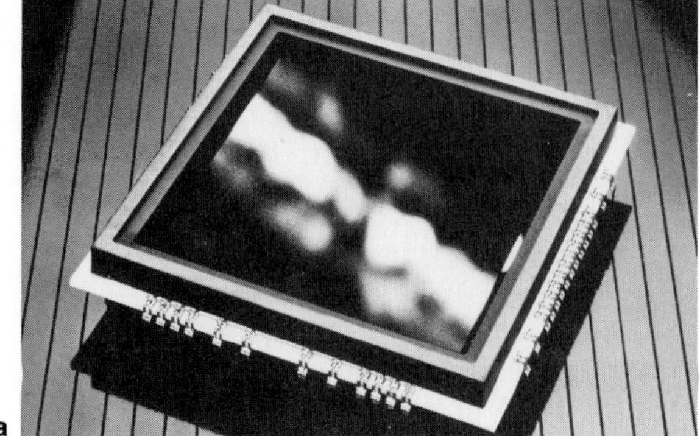
a

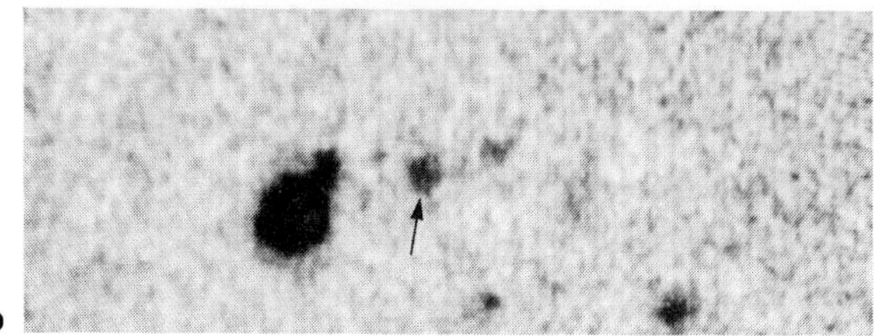
b

FIGURE 6–16 (a) This large CCD chip contains over 4 million light-sensitive diodes in an array 55.3 mm square. (b) An electronic image of a giant galaxy (arrow) so distant that the light took 10 billion years to reach us. This image is reproduced as a negative. The sky is light and the stars and galaxy are black. (a, Copyright © 1986 Tektronix, Inc. All rights reserved. Reprinted by permission of Tektronix, Inc.; b, Courtesy Hyron Spinrad.)

THE PHOTOMETER

A **photometer** is nothing more than a sensitive light meter that can measure the brightness and color of stars. Such a photometer contains a sensitive detector that produces an electric current when struck by light. The telescope focuses starlight onto the detector, and the resulting current can be measured.

To measure the color of a star, we pass the light through filters. A filter transmits only those wavelengths in a certain range—for instance, blue light between 400 nm and 480 nm. The difference between the brightness of the star through a red and a blue filter tells us the color of the star.

We also use a photometer to make measurements in the near ultraviolet or near infrared. With the proper detector and filters, the photometer measures the brightness at wavelengths we cannot see. Of course, if we go

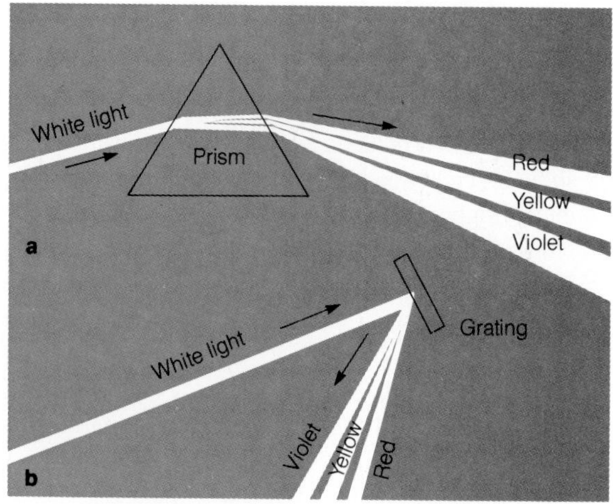

FIGURE 6–17 (a) A prism bends light by an angle that depends on the wavelength of the light. Short wavelengths bend most. (b) White light can be spread into a spectrum by reflection from a grating, producing the same effect as a prism.

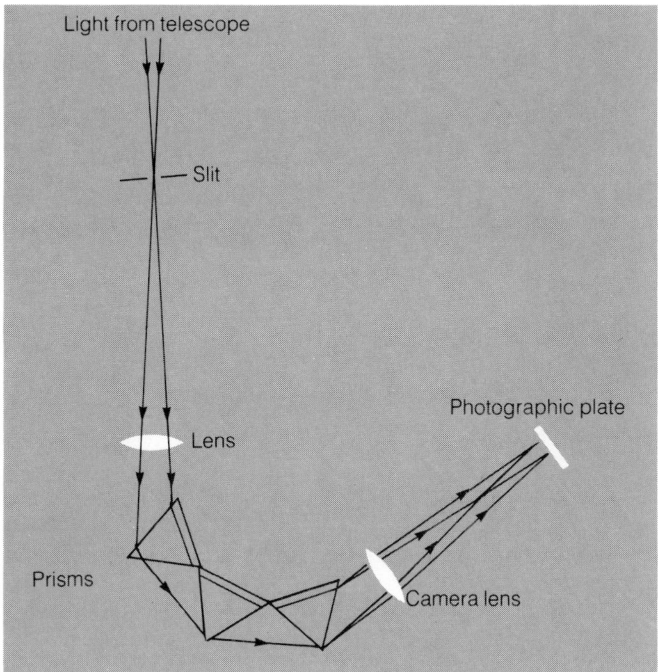

FIGURE 6–18 In a three-prism spectrograph, light from the telescope passes through the slit and through three prisms before being focused on the photographic plate.

too far into the ultraviolet or infrared, the earth's atmosphere is not transparent, and we cannot make such observations from earth's surface.

THE SPECTROGRAPH

A **spectrograph** is a device that separates starlight according to wavelength to produce a spectrum. We can see how this works by reproducing Newton's original experiment. In place of a hole in a shutter, we might use a narrow slit to produce a thin beam of light. When that beam passes through a prism, the angle through which it is bent depends on wavelength—violet bends most and red least—so the light leaving the prism is spread into a spectrum (Figure 6–17 and Color Plate 4). A typical prism spectrograph contains more than one prism to spread the light further and lenses to guide the light into the prism and to focus the light onto a photographic plate (Figure 6–18).

Nearly all modern spectrographs use a grating in place of a prism. A **grating** is a piece of glass with thousands of microscopic parallel lines scribed onto its surface. Different wavelengths of light reflect from the grating at slightly different angles, so white light is spread into a spectrum.

6.4 RADIO TELESCOPES

All the telescopes and instruments we have discussed look out through the visible light window in earth's atmosphere, but there is another window running from a wavelength of 1 cm to about 1 m (Figure 6–2). By building the proper kinds of instruments, we can study the universe through this radio window.

OPERATION OF A RADIO TELESCOPE

A radio telescope usually consists of four parts: a dish reflector, an amplifier, a receiver, and a recorder (Figure 6–19). The dish reflector gathers incoming radio waves and focuses them on the antenna, which absorbs the energy. The signal is amplified, and its intensity is recorded usually under the direction of a computer.

Like the mirror of an optical telescope, the dish reflector must be of large diameter to gather as much radio energy as possible, but because radio photons have such long wavelengths, the surface of the dish need not be as smooth as a mirror. In fact, wire mesh is a very good reflector of all but the shortest radio waves (Figure 6–19 inset).

The dish also acts to exclude all radio energy except that coming from a small region of the sky. By scanning

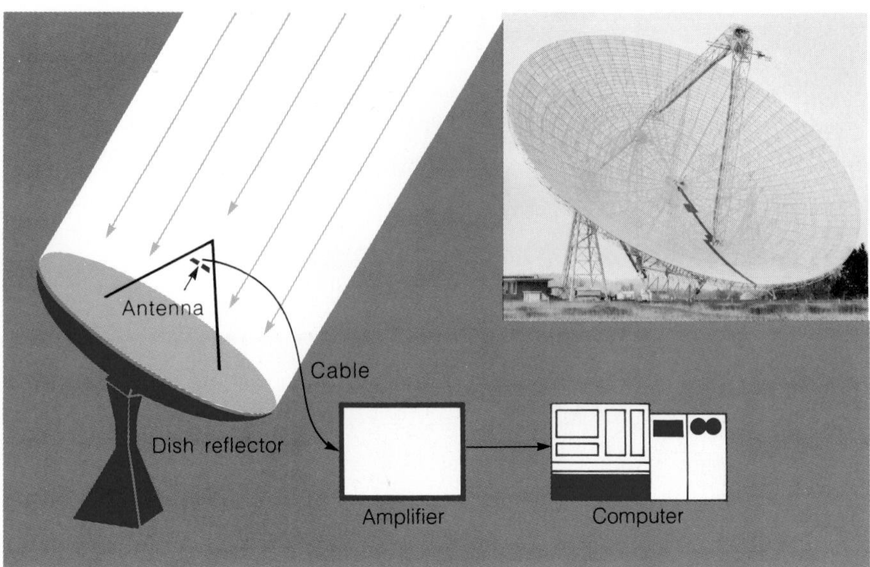

FIGURE 6–19 In most radio telescopes, a dish reflector concentrates the radio signal on the antenna. The signal is then amplified and recorded. For all but the shortest radio waves, wire mesh is an adequate reflector (inset). (National Radio Astronomy Observatory, operated by Associated Universities, Inc. under a contract with the National Science Foundation.)

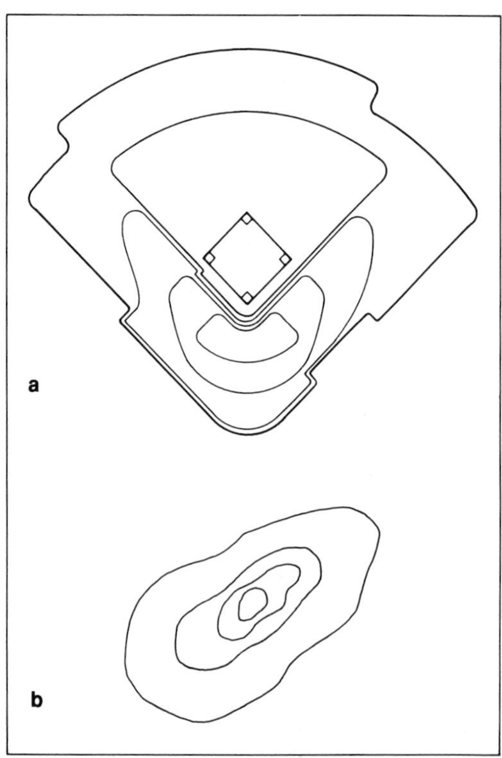

FIGURE 6–20 Contour maps. (a) A contour map of a baseball stadium shows regions of similar admission prices. The most expensive seats are behind home plate. (b) A contour map of a gas cloud in space shows regions of similar radio intensity. The radio signals are strongest near the center.

FIGURE 6–21 The largest radio telescope in the world is the 300-m dish suspended in a valley in Arecibo, Puerto Rico. The antenna hangs above the dish on cables stretched from towers. (Cornell University.)

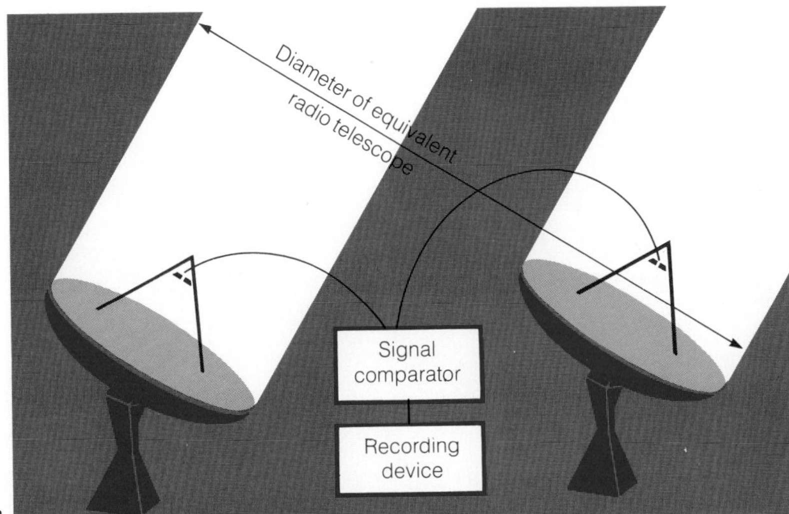

FIGURE 6–22 (a) A radio interferometer consists of two or more radio telescopes whose signals are combined to give the resolving power of a much larger instrument. (b) The Very Large Array (VLA) radio interferometer uses up to 27 dishes along the 20-km-long arms of a Y to simulate a radio telescope 40 km in diameter. (National Radio Astronomy Observatory, operated by Associated Universities, Inc. under a contract with the National Science Foundation.)

from spot to spot, the radio astronomer can construct a map showing the intensity of the radio energy coming from different regions of the object being studied. These maps are usually drawn as contour maps, which mark areas of constant radio intensity (Figure 6–20). Such maps can also be manipulated in a computer to produce false color maps. See Color Plates 21–25 for examples.

Because they focus much longer wavelength photons, radio telescopes have much worse resolving power than optical telescopes. A dish 30 m in diameter receiving radiation at a wavelength of 21 cm has a resolving power of about 0.5°. Such a radio telescope would be unable to show us any details in the sky smaller than the angular diameter of the moon. The larger the diameter of the dish, the better the resolving power (and the stronger the signal), so radio telescopes tend to be very large. The largest radio dish in the world is 300 m (1000 ft) in diameter. It is built into a mountain valley in Arecibo, Puerto Rico (Figure 6–21), and the operators depend on the rotation of the earth to point the telescope at different regions of the sky.

THE RADIO INTERFEROMETER

To improve the resolving power of their telescopes, radio astronomers have linked radio telescopes together to form **radio interferometers**—two or more radio telescopes that combine their signals to simulate one big telescope (Figure 6–22a). The system has the resolving power of a telescope whose diameter is equal to the separation between the two smaller telescopes.

Radio astronomers have built radio interferometers of various sizes. The Very Large Array (VLA) is a Y-shaped pattern of 27 radio dishes spread across the New Mexico desert (Figure 6–22b). They simulate a radio telescope 32 km (20 mi) in diameter and can produce radio maps ten times as detailed as the best photographs taken from earth. (See Color Plates 21–25).

Radio astronomers have also linked radio telescopes on opposite sides of the world to simulate a radio telescope thousands of miles in diameter. Connecting such widely separated radio telescopes is called **Very Long Baseline Interferometry** (VLBI), and a new system of ten

FIGURE 6–23 At 4150 m, the top of Mauna Kea is home to (clockwise from the bottom) NASA–University of Hawaii 3-m infrared telescope, Canada–France–Hawaii 3.6-m telescope, Hawaii 0.6-m telescope, Hawaii l2.24-m telescope, United Kingdom 3.8-m Infrared Telescope, and a Hawaiian 0.6-m telescope. (Dale P. Cruikshank, Institute for Astronomy.)

radio dishes is being constructed from the Caribbean to Hawaii. When this Very Long Baseline Array (VLBA) is completed in the 1990s, radio astronomers will be able to make radio maps 100 times more detailed than the best photographs taken from earth.

ADVANTAGES OF A RADIO TELESCOPE

The radio telescope has three advantages over an optical telescope. First, radio telescopes can detect cold clouds of gas in space. These gas clouds are important because they give birth to stars, but the gas is so cold (about 50 K) that it emits no visible light. In Chapter 7 we will see how cold hydrogen can emit photons with a wavelength of 21 cm and how other gases such as carbon monoxide can emit other wavelengths.

A radio telescope can also detect extremely hot gas. In many processes in the universe, gas is heated so hot it emits powerful radio signals. Thus radio telescopes can map the gas clouds associated with exploding stars and galaxies.

The third advantage of radio telescopes is that they can see through the dust clouds that block our view at visible wavelengths. Because radio photons have longer wavelengths, they are not affected by the microscopic particles of dust that float in space. But those dust specks scatter visual wavelength photons and make the dust clouds opaque. Radio signals from far across the galaxy pass unhindered through the dust, giving us an unobscured view.

6.5 SPACE ASTRONOMY

Celestial bodies emit radiation at many different wavelengths, but earth's atmosphere absorbs most such photons. From the earth's surface we can observe at visual wavelengths, in the near infrared, the near ultraviolet, and at some radio wavelengths. But if we wish to observe at other wavelengths, we must get above earth's atmosphere, and that means we must put telescopes into space.

INFRARED ASTRONOMY

Some infrared radiation penetrates partially open windows in our atmosphere scattered from 1.2 μm (microns) to 40 μm. In this range, called the near infrared, most of the radiation is absorbed by water vapor in the earth's atmosphere, so it is an advantage to place telescopes on mountains where the air is thin and dry. Two major infrared telescopes, a 3.0-m (118-in) and a 3.8-m (150-in), observe from the 13,600-foot high top of Mauna Kea in Hawaii. At this altitude they are above most of the water vapor in the atmosphere (Figure 6–23).

The far infrared—wavelengths longer than 40 μm—can tell us about planets, comets, forming stars, and other cool objects, but photons of these wavelengths are strongly absorbed by the atmosphere. To observe in the far infrared, telescopes must venture high in the atmosphere. Remotely operated, infrared telescopes suspended under balloons have reached altitudes as high as 41 km (25 mi).

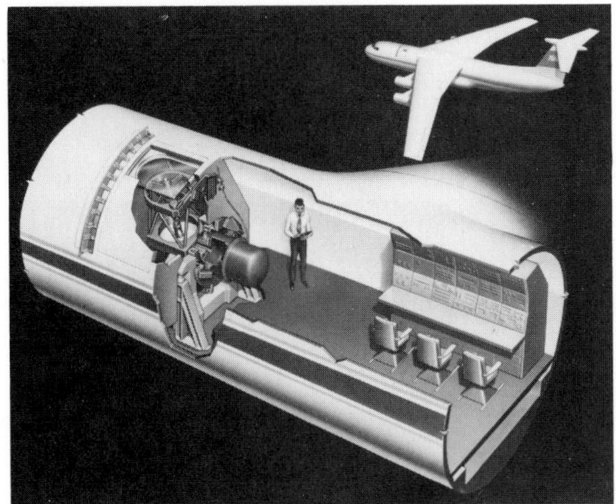

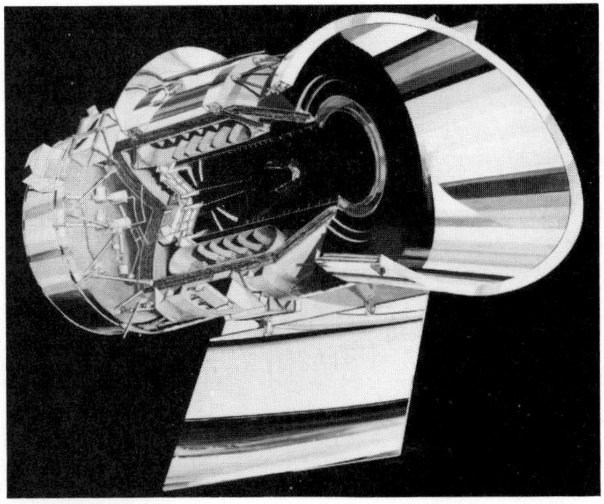

FIGURE 6–24 (a) The Gerard P. Kuiper Airborne Observatory carries a 91-cm infrared telescope and astronomers to altitudes of 40,000 feet to get above 99 percent of the water vapor in the earth's atmosphere. (b) The Infrared Astronomical Satellite (IRAS) mapped the entire sky at wavelengths of 10, 25, 50, and 100 μm. (c) The Space Infrared Telescope Facility (SIRTF), now in the planning stage, would observe in the far infrared. (NASA.)

Another solution is to fly the infrared telescope to high altitudes in an airplane. NASA has modified a Lockheed C-141 jet transport to carry a 91-cm infrared telescope and a crew of a dozen astronomers to altitudes of 40,000 feet to get above 99 percent of the water vapor in the earth's atmosphere (Figure 6–24a). The rings of Uranus, for example, were discovered by astronomers flying in this aircraft.

The ultimate solution to the problem of atmospheric absorption is to place the telescope in orbit above the atmosphere. The Infrared Astronomical Satellite (IRAS) (Figure 6–24b) carried a 56-cm (22-in) telescope cooled to nearly absolute zero by liquid helium. It observed from an orbit 900 km high through most of 1983 before its helium coolant was exhausted. Among its many discoveries, IRAS found a disk of cold matter around the bright star Vega. This is the first clear evidence of planet-like material orbiting a star other than our sun. NASA is now planning the Space Infrared Telescope Facility (SIRTF) (Figure 6–24c), which would become a major infrared observatory in space.

IRAS, like all infrared telescopes, has to be cooled to very low temperatures. Infrared radiation is heat, and if the telescope is warm it will emit many times more infrared than that coming from a distant object. Imagine trying to look at a dim, moonlit scene through binoculars that were glowing brightly. In the near infrared, only the detector, the element on which the infrared radiation is focused, must be cooled. To observe in the far infrared, however, the entire telescope must be cooled.

ULTRAVIOLET ASTRONOMY

Some infrared telescopes can observe from mountaintops, but ultraviolet radiation beyond about 300 nm, the far ultraviolet, is completely absorbed by the ozone layer extending from 20 km to above 40 km in our atmosphere. To get above this layer, ultraviolet telescopes must go into space.

The first far ultraviolet observations of a celestial body were made in 1946 when a captured German V-2 rocket lifted instruments to an altitude of 100 km and recorded

FIGURE 6–25 (a) In the control room of the IUE, astronomers send commands up to the satellite. (b) A sample IUE spectrum of the star AO Cas shows absorption due to silicon atoms at 139.3 nm. (a, NASA; b, Courtesy George E. McCluskey and *The Astrophysical Journal,* published by the University of Chicago Press; © 1981 The American Astronomical Society.)

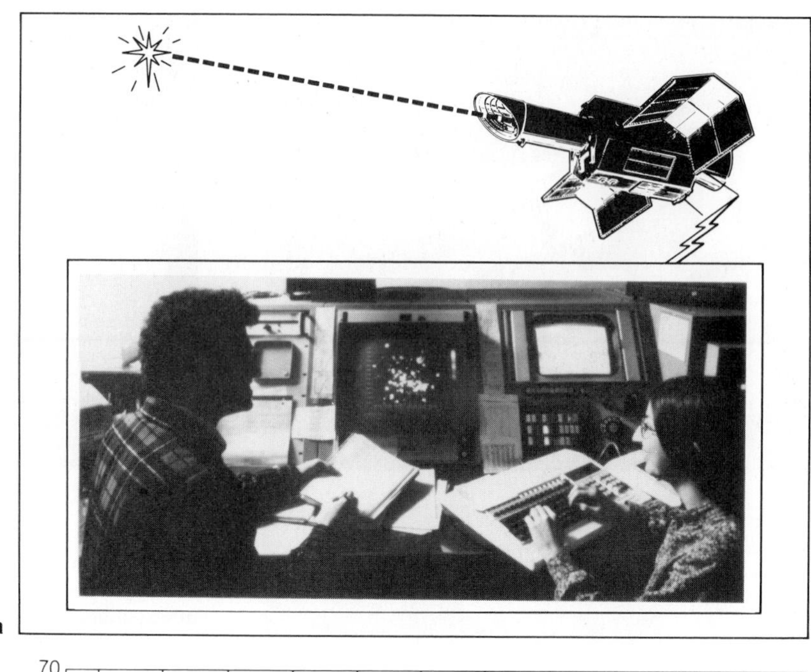

a

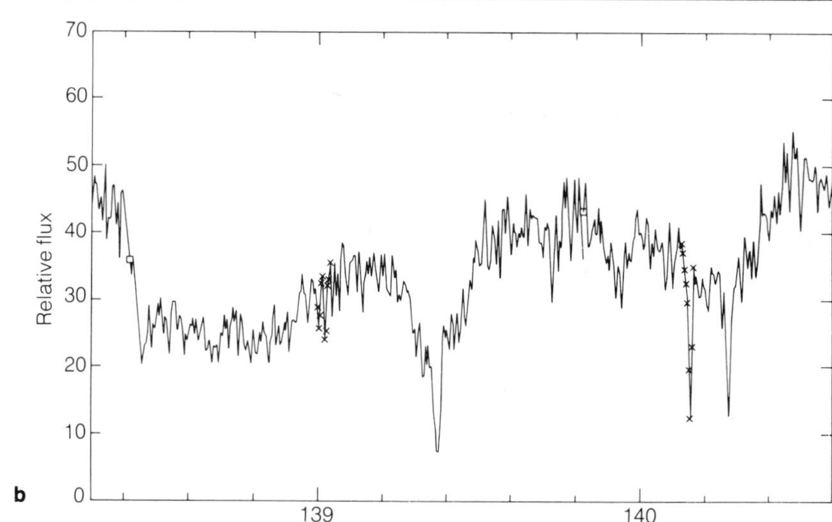

b

the spectrum of the sun. Since then, many rockets have carried ultraviolet instruments on short trips into space, but extended studies are possible only from satellites.

One of the most successful ultraviolet observatories is the International Ultraviolet Explorer (IUE) (Figure 6–25). Launched in January 1978, it carries a 45-cm (18-in) telescope with attached spectrographs. Spectra are later transmitted to an earth-based computer system, which controls the satellite.

IUE has become an important observatory available to any astronomer with a good research proposal, but it has long outlived its projected life. The satellite has had a number of mechanical and electronic failures over the years. Only the creative genius of NASA engineers and programmers have kept it operating. The world's astronomers fear it will not survive much longer.

What is it like to use the IUE? First, we submit a proposal explaining the kind of observations we want to make and justifying the project. If our proposal is approved, we go to the Goddard Space Flight Center and, at our appointed time, enter the IUE control room, a rather small cubicle filled with control consoles. With expert advice

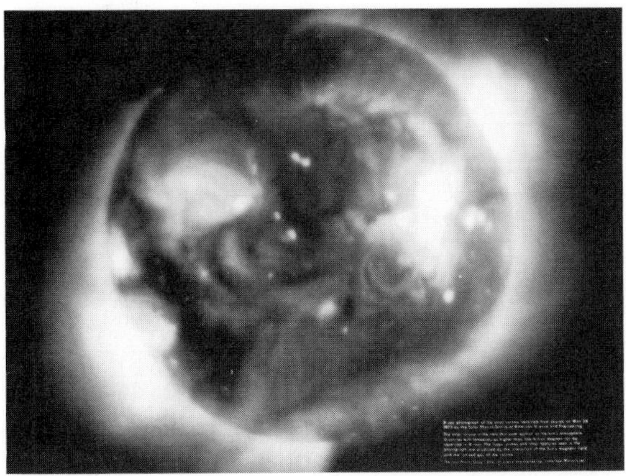

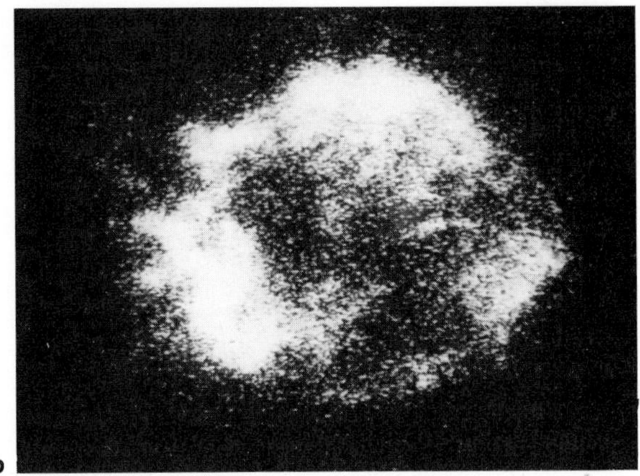

a b

FIGURE 6–26 (a) An X-ray image of the sun taken from Skylab reveals regions of X-ray emission in the corona. (b) A photograph of the X-ray image of Cassiopeia A, a supernova remnant obtained with the HEAO-2 orbiting observatory. The X rays reveal a shell of gas expanding away from the site of a stellar explosion that occurred about 310 years ago. (a, NASA; b, reprinted courtesy of S. S. Murray, G. Fabbiano, A. Fabian, A. Epstein, and R. Giacconi, from *The Astrophysical Journal,* published by the University of Chicago Press; ©1979 The American Astronomical Society.)

from NASA technicians, we command the IUE to move to the first object on our observing list and wait as the satellite moves to the proper coordinates (Figure 6–25). We use star charts to be certain the satellite is pointed at the correct object, and then we command the satellite to begin recording a spectrum. An exposure might last a few minutes or many hours, but when it is done, we see the spectrum on the television screen and can analyze it quickly with computer facilities. The spectrum is also recorded on magnetic tape for more detailed analysis later.

X-RAY ASTRONOMY

Beyond the far ultraviolet, at wavelengths from 10 nm to 0.01 nm, lie the X rays. These very high-energy photons can be produced only by violent, high-energy events, so they tell us about high-energy violence in the universe. We detect X rays coming from exploding stars, from colliding galaxies, and from matter smashing into the surface of a neutron star or falling into a black hole—phenomena we will be examining in greater detail in later chapters.

Although early X-ray observations of the sky were made from balloons and small rockets in the 1960s, the age of X-ray astronomy did not really begin until 1970 when an X-ray telescope named Uhuru (Swahili for "freedom") was put into orbit. Uhuru detected nearly 170 separate sources of celestial X rays. Later the three High Energy Astronomy Observatories (HEAO) satellites, carrying more sensitive and more sophisticated equipment, pushed the total to many hundreds. The second HEAO satellite, named the Einstein Observatory, used special optics to produce X-ray images. Similar X-ray optics aboard Skylab produced X-ray images of the sun (Figure 6–26).

The European Space Agency launched an X-ray observatory (Exosat) in May 1983, and it was very successful in studying sites of high-energy violence in the universe. Unfortunately, Exosat failed in April 1986, so earth's astronomers currently have no major X-ray observatory in space.

New X-ray observatories are in the planning stages. German and British astronomers are now working on a satellite called Rosat that will map the sky at X-ray wavelengths. NASA is planning a new X-ray observatory called the Advanced X-ray Astrophysics Facility (AXAF). With a 1.2-m (47-in) telescope operating from 6 nm to 0.15 nm, AXAF will have about 100 times the sensitivity of the best X-ray observatories yet launched. Plans called for Rosat to be launched in 1987, and AXAF in the early 1990s, but problems with the Space Shuttle will delay these projects.

GAMMA-RAY TELESCOPES

Gamma rays have wavelengths even shorter than X rays, and that means they have even higher energies. This high energy makes them doubly difficult to detect. First, gamma rays cannot be focused or detected like X rays. Thus,

PERSPECTIVE

THE HUBBLE SPACE TELESCOPE

When the Space Shuttle Challenger exploded during launch in January 1986, space astronomy suffered a major setback. The Space Shuttles are needed to carry space observatories into orbit. SIRTF, AXAF, Rosat, and other space observatories are designed for shuttle launch, and in some cases, for repair by astronauts from a shuttle. The delay in the shuttle program will delay these and many other space programs, but perhaps the most important project delayed by the shuttle disaster is the Hubble Space Telescope (Figure 6–27).

For over 10 years, American astronomers and engineers have worked to develop the Edwin P. Hubble Space Telescope (HST). Named for the man who discovered the expansion of the universe, it is the largest space telescope ever built. When HST is finally placed into orbit, it will become one of the most important observatories in history.

HST will carry a 2.4-m (96-in) telescope with spectrographs and cameras capable of detecting objects fifty times fainter than can be seen by any telescope on the earth. It will be able to see objects of a given brightness out to seven times farther away. From above the earth's atmosphere it will be able to observe in the near infrared, visible, and far ultraviolet, and with no atmospheric turbulence, it will be able to see detail ten times smaller than can earth-based telescopes.

The list of problems that HST should be able to resolve is impressive. It will be able to improve measurements of the distances to nearby stars and the furthest galaxies, thus helping pin down the age and size of the universe. It will show us the details of how stars form and how they die, and it might be able to detect planets orbiting nearby stars. Within our own solar system, it will monitor the weather on Venus, Mars, Jupiter, and Saturn; study the nuclei of comets; and search for satellites and rings around the outer planets.

Operating the space telescope will be much like running the IUE. The control center is at the Space Telescope Science Institute at Johns Hopkins University. Astronomers from around the world are proposing research projects for the telescope, and the cream

gamma-ray telescopes are more complex and can see less detail than similar X-ray telescopes. Also, gamma rays are such high-energy photons that natural processes produce only a few. Gamma-ray telescopes must count gamma rays one at a time. For example, a gamma-ray telescope observed an intense source for a month and counted only 3000 gamma rays. That is one every 15 minutes.

In addition, gamma rays are almost totally absorbed by our atmosphere, so gamma-ray telescopes must observe from orbit. NASA has operated three important gamma-ray telescopes aboard satellites including one on the Small Astronomy Satellite 2 (SAS 2). The Soviet Union has flown two. The most recent gamma-ray telescope was aboard the satellite COS B launched in 1975 by the European Space Agency. It surveyed the sky and detected gamma rays from clouds of gas in space, from the center of our galaxy, and from the remains of exploded stars. Although gamma rays are very difficult to detect, they tell us about important processes in the universe, including the birth and death of stars.

SUMMARY

Electromagnetic radiation is an electric and magnetic disturbance that transports energy at the speed of light. The electromagnetic spectrum includes radio waves, infrared radiation, visible light, ultraviolet radiation, X rays, and gamma rays.

We can think of "a particle of light," a photon, as a bundle of waves that sometimes acts as a particle and sometimes as a wave. The energy a photon carries depends on its wavelength. The wavelength of visible light, usually measured in nanometers

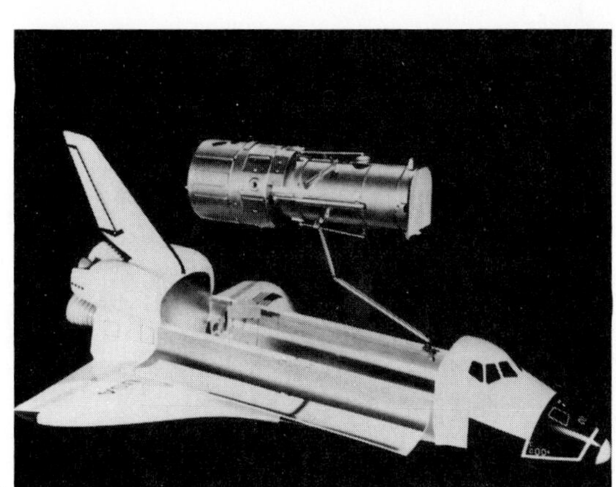

FIGURE 6–27 (a) The Space Telescope will be carried into orbit by the Space Shuttle. (b) Once deployed, it will function as a remote control observatory obeying commands from astronomers on earth. Plans call for the Space Telescope to be maintained by Space Shuttle visits every few years. (NASA.)

of the crop will be selected. Operators will communicate with the satellite through a network of data satellites as instructions flow up to HST and data in the form of images and spectra flow down to the astronomers on earth.

Now that the Space Shuttles have resumed operation, the world's astronomers are waiting for the HST to be launched. It is scheduled to go into orbit in early 1990 and begin routine operations by mid 1990.

Despite the difficulty and expense, astronomers will continue to send telescopes into space. The advantages of observing from above the atmosphere are too impressive to ignore, and the puzzles of the universe are too tempting to forget.

(10^{-9} m), ranges from 400 nm to 700 nm. Infrared and radio photons have longer wavelengths and carry less energy. Ultraviolet, X-ray, and gamma-ray photons have shorter wavelengths and carry more energy.

Astronomical telescopes are of two types, refractor and reflector. A refractor uses a lens to bend the light and focus it into an image. Because of chromatic aberration, refracting telescopes cannot bring all colors to the same focus, resulting in color fringes around the images. An achromatic lens partially corrects for this, but such lenses are expensive and cannot be made larger than about 1 m in diameter.

Reflecting telescopes use a mirror to focus the light and are less expensive than refracting telescopes of the same diameter. In addition, reflecting telescopes do not suffer from chromatic aberration. Thus, most recently built telescopes are reflectors.

The largest telescopes in the world are the 6-m telescope in Russia, the 5-m telescope in California, and the Multiple Mirror Telescope (equivalent to a 4.5-m telescope) in Arizona. New telescopes are being designed and built using new technology to achieve diameters as large as 15 m.

Special instruments attached to a telescope analyze the light the telescope gathers. The photographic plate records vast amounts of detail for later analysis, but electronic imaging systems such as CCDs have largely replaced photographic plates. CCDs are more sensitive, record both faint and bright objects, and can be read directly into computer memory. A photometer measures the brightness and color of the light entering a telescope. Spectrographs, using prisms or gratings, break the starlight into a spectrum, which then can be photographed or electronically recorded.

The powers of a telescope are light-gathering power, resolving power, and magnifying power. The first two of these depend on the telescope's diameter; thus, astronomical telescopes often have large diameters.

To observe radio signals from celestial objects, we need a radio telescope, which usually consists of a dish reflector, an antenna, an amplifier, and a recorder. Such an instrument measures the intensity of radio signals over the sky and constructs radio maps. The poor resolution of the radio telescope can be improved by combining it with another radio telescope to make a radio interferometer. The three principal advantages of radio telescopes are that they can detect the very cold gas clouds in space, can detect regions of very hot gas produced by exploding stars or erupting galaxies, and can look through the dust clouds that block our view at optical wavelengths.

Observations at some wavelengths in the near infrared are possible from high mountaintops or from high-flying aircraft. At these altitudes the air is thin and dry, and the infrared radiation can reach the telescope. For instance, infrared telescopes can operate at 4150 m (13,600 ft) atop the volcano Mauna Kea in Hawaii.

At other infrared wavelengths, the telescope must observe from above earth's atmosphere—from orbit. The Infrared Astronomy Satellite observed from its orbit about 900 km high. It carried a 56-cm telescope and was cooled by liquid helium.

To observe at these short wavelengths, the telescope must be above the ozone layer in earth's atmosphere. That means the telescope must be in orbit. The International Ultraviolet Explorer is an ultraviolet astronomy satellite operated jointly by United States' and European astronomers. X-ray and gamma-ray telescopes must also be placed in orbit above the absorbing layers of earth's atmosphere.

The launch of the Hubble Space Telescope (HST) has been delayed by difficulties with the Space Shuttle. When HST is eventually launched in early 1990, it will become the most powerful astronomical facility in use. Compared with present earth-based telescopes, it will be able to detect objects fifty times fainter, seven times more distant, and ten times smaller.

NEW TERMS

electromagnetic radiation
wavelength
photon
nanometer (nm)
Angstrom (Å)
infrared radiation
ultraviolet radiation
atmospheric window
refracting telescope
focal length
objective lens
eyepiece
chromatic aberration
achromatic lens
coudé focus
Schmidt camera
sidereal drive
equatorial mounting
polar axis
alt-azimuth mounting
light-gathering power
resolving power
diffraction fringe
seeing
magnifying power
charge-coupled device (CCD)
false color
reflecting telescope
objective mirror
prime focus
secondary mirror
Cassegrain telescope
Newtonian focus
photometer
spectrograph
grating
radio interferometer
Very Long Baseline Interferometry (VLBI)

QUESTIONS

1. Almost all the animals on earth have developed eyes sensitive to visual wavelengths. How is earth's atmosphere responsible for this similarity? Why don't any animals see at radio wavelengths? (HINT: Consider the diameter of eye needed to obtain adequate resolving power.)
2. Why do nocturnal animals usually have large pupils in their eyes? How is that related to astronomical telescopes?
3. Astronomers have not constructed a single, major refracting telescope during this century. Why not?
4. How have computers and new technology changed the design of astronomical telescopes and their mountings?
5. Small telescopes are often advertised as "200 power" or "magnifies 200 times." As someone knowledgeable about astronomical telescopes, how would you improve such advertisements?
6. An astronomer recently said, "Some people think I should give up photographic plates." Why might she change to something else?
7. What purpose do the colors in a false color image or false color radio map serve?
8. Some radio astronomers hope to place radio telescopes in very large orbits around the earth and use them with earth-based radio telescopes. Why would this be an advantage?
9. Why are we able to observe in the infrared from high mountains and aircraft, but must go into space to observe in the far ultraviolet?
10. The moon has no atmosphere at all. At what wavelengths could we observe if we had an observatory on the lunar surface?

PROBLEMS

1. The thickness of the plastic in plastic bags is about 0.001 mm. How many wavelengths of red light is this?

2. Measure the actual wavelength of the wave in Figure 6–1. In what portion of the electromagnetic spectrum would it belong?
3. Compare the light-gathering powers of the 5-m telescope and a 0.5-m telescope.
4. How does the light-gathering power of the largest telescope in the world compare with that of the human eye. (HINT: Assume that the pupil of your eye can open to about 0.8 cm.)
5. What is the resolving power of a 25-cm telescope? What do two stars, 1.5 seconds of arc apart, look like through this telescope?
6. Most of Galileo's telescopes were only about 2 cm in diameter. Should he have been able to resolve the two stars mentioned in Problem 5?
7. How does the resolving power of the 5-m telescope compare with that of the Hubble Space Telescope? Why will the Space Telescope outperform the 5-m telescope?
8. If we build a telescope with a focal length of 1.3 m, what focal length should the eyepiece have to give a magnification of 100 times?
9. Astronauts observing from a space station need a telescope with a light-gathering power 15,000 times that of the human eye, capable of resolving detail as small as 0.1 seconds of arc, and having a magnifying power of 250. Design a telescope to meet their needs. Could you test your design by observing stars from Earth?
10. A spy satellite orbiting 400 km above the earth is supposedly capable of counting individual people in a crowd. What minimum diameter telescope must the satellite carry? (HINT: Use the small-angle formula.)

RECOMMENDED READING

BISHOP, ROY L. "Newton's Telescope Revealed." *Sky and Telescope 59* (March 1980), p. 207.

BOK, BART J. "The Promise of the Space Telescope." *Mercury 12* (May/June 1983), p. 66.

CAMERON, R. M. "NASA's 91-cm Airborne Telescope." *Sky and Telescope 52* (Nov. 1976), p. 327.

CLASSEN, J., and NORMAN SPERLING "Telescopes for the Record." *Sky and Telescope 61* (April 1981), p. 303.

COHEN, MARTIN *In Quest of Telescopes.* Cambridge, Mass.: Sky Publishing, 1981.

CORDOVA, FRANCE A., and KEITH O. MASON "Exosat: Europe's New X-Ray Satellite." *Sky and Telescope 67* (May 1984), p. 397.

CORNELL, JAMES "Six New Eyes Peer from Mount Hopkins." *Sky and Telescope 58* (July 1979), p. 23.

FEDERER, C. A., JR. "The VLA: Ears on the Universe." *Sky and Telescope 60* (Dec. 1980), p. 472.

GENET, RUSSELL, KENNETH KISSELL, and GEORGE ROBERTS "Our Turn at Kitt Peak." *Sky and Telescope 63* (March 1982), p. 240.

GIACCONI, RICCARDO "The Einstein X-ray Observatory." *Scientific American 242* (Feb. 1980), p. 80.

GORDON, MARK A. "VLBA—A Continent-Size Radio Telescope." *Sky and Telescope 69* (June 1985), p. 487.

KING, G. C. *The History of the Telescope.* Cambridge, Mass.: Sky Publishing, 1955.

KRAUS, J. *Big Ear.* Powell, Ohio: Cygnus-Quasar Books, 1976.

KRISTIAN, JEROME, and MORLEY BLOUKE "Charge-Coupled Devices in Astronomy." *Scientific American 247* (Oct. 1982), p. 67.

LONGAIR, MALCOLM "The Scientific Challenge of Space Telescope." *Sky and Telescope 69* (April 1985), p. 306.

MAMMANA, DENNIS L. "The Incredible Spinning Oven." *Sky and Telescope 70* (July 1985), p. 7.

MARX, SIEGFRIED, and WERNER PFAU *Observatories of the World.* New York: Van Nostrand Reinhold, 1982.

MOOD, STEPHANIE, and JOHN MOOD "Palomar and the Politics of Light Pollution." *Astronomy 13* (Nov. 1985), p. 6.

PANKONIN, VERNON "Protecting Radio Windows for Astronomy." *Sky and Telescope 61* (April 1981), p. 309.

ROBINSON, LEIF J. "Update: Telescopes of the Future." *Sky and Telescope 72* (July 1986), p. 23.

ROSCH, JEAN, and JEAN DRAGESCO "The French Quest for High Resolution." *Sky and Telescope 59* (Jan. 1980), p. 6.

SCHIELDS, JOHN POTTER "Backyard Radio Astronomy." *Astronomy 11* (March 1983), p. 75.

SMITH, DAVID H. "MERLIN: A Wizard of a Telescope." *Sky and Telescope 67* (Jan. 1984), p. 31.

———. "King of the Space Telescopes." *Sky and Telescope 70* (Sept. 1985), p. 216.

———. "Cologne's Submillimeter Dwarf." *Sky and Telescope 72* (Sept. 1986), p. 216.

STROM, STEPHEN E. "Infrared and Submillimeter Astronomy from Space." *Sky and Telescope 65* (April 1983), p. 312.

TUCKER, WALLACE, and RICCARDO GIACCONI "The Birth of X-Ray Astronomy." *Mercury 14* (Nov./Dec. 1985), p. 178.

TUCKER, WALLACE, and KAREN TUCKER *The Cosmic Inquirers.* Cambridge, Mass.: Harvard University Press, 1986.

WILSON, TOM "The New Improved Palomar Sky Survey." *Astronomy 14* (Oct. 1986), p. 16.

WOLFF, S. "The Search for Aperture: A Selective History of Telescopes." *Mercury 14* (Sept./Oct. 1985), p. 139.

UNIT 2

THE STARS

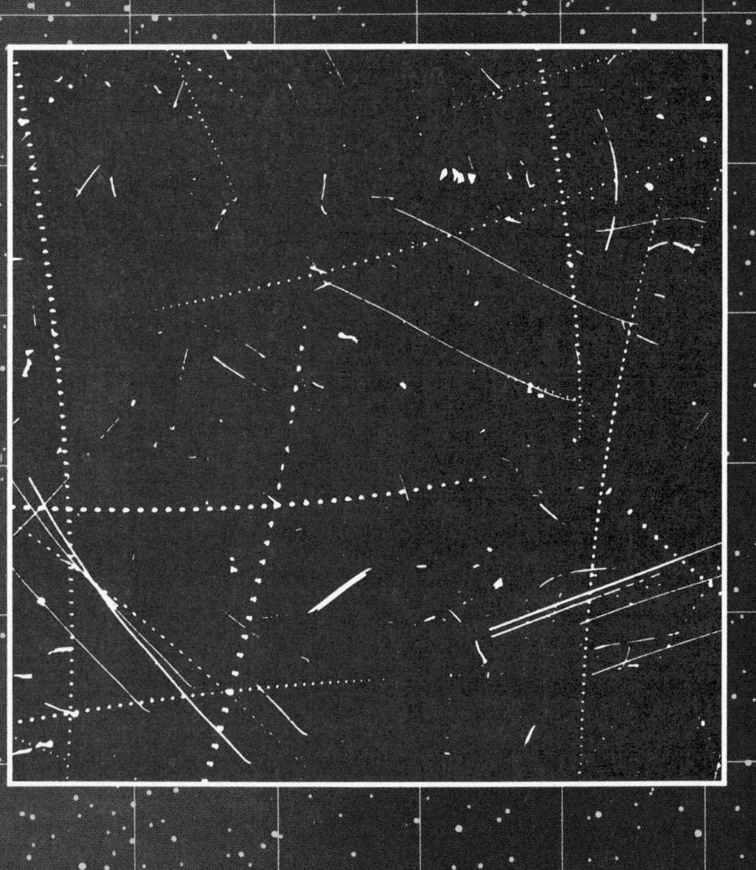

CHAPTER 7

STARLIGHT AND ATOMS

Awake!
for Morning in the Bowl of
Night Has flung the Stone
that puts the Stars to Flight:
And Lo! the Hunter of the
East has caught
The Sultan's Turret in a Noose
of Light.

trans. Edward Fitzgerald
THE RUBÁIYÁT OF OMAR KHAYYÁM

Sunlight is nothing but bright starlight—the sun is about 10 billion times brighter than the brightest star, but that is only because it is so close. The sun is, in fact, an ordinary star whose light differs in no detail from the light of hundreds of millions of similar stars scattered across our galaxy. If we want to understand what stars are, we have only to look at the sun, the nearest star.

However close the sun may seem in comparison with other stars, it is far beyond our reach. The sun is 93 million miles away—a distance so great a commercial jet could not travel that far in 5 years. Thus we cannot probe and sample the sun. No laboratory jar on earth holds a sample labeled "sun stuff," and no instrument has ever descended into the sun to tell us the temperature, pressure, composition, and so on. As with all stars, the only information we can obtain about the sun comes to us hidden in light. Whatever we want to know about stars, we must catch in a "noose of light."

The stars are so far away that it would not be surprising if earthbound humans knew almost nothing about them. But we can learn about the stars by analyzing the light we receive. We begin by considering the bulk properties of that light—how much red and how much blue are present in a star's light. This will tell us the approximate temperature of the star, but to obtain more detail we must examine the starlight more closely.

Starlight spread into a spectrum contains dark lines that can tell us a great deal about the gases that make up the star. But to analyze those lines, we must know how atoms interact with light. We begin with the hydrogen atom because it is the most common atom in the universe and the simplest.

This chapter is the foundation of our study of stars because it describes how we get information from starlight. In the next chapter, we will apply these techniques to our own sun to see close up a star in action. In later chapters, we will apply the same techniques to other stars and discover the surprising variety among the stars.

7.1 STARLIGHT

If you look at the stars in the constellation Orion, you will notice that they are not all the same color. Betelgeuse, in the upper left corner, is quite red; Rigel, in the lower right corner is blue. These differences in color arise from the way the stars produce light and give us our first clue to the temperatures of stars.

THE ORIGIN OF STARLIGHT

The starlight we see comes from the gases in the outer surface of the star—the photosphere. The gases deep inside the star also emit light, but it is absorbed before it can escape, and the low-density gas above the photosphere is too thin to emit significant amounts of light. Thus, the photosphere of a star is that layer of gases dense enough to emit significant amounts of light but thin enough so the photons can escape. (Recall that we discussed the photosphere of the sun in Chapter 3.)

To see how the photosphere of a star can produce light, we must consider two things—how photons are produced and how the temperature of a material is related to motion among its atoms.

A photon can be produced by a changing electric field. An **electron** is a negatively charged subatomic particle, and if we disturb the motion of an electron the sudden change in the electric field around it can cause the emission of a photon. For example, if you run a comb through your hair while standing near an AM radio, you can produce popping noises on the radio. The moving comb disturbs the electrons in both the comb and hair, building static electricity. The sudden sparks of static electricity produce electromagnetic waves that the radio picks up as noise. This illustrates an important principle: When we change the motion of an electron, we generate an electromagnetic wave.

The second thing we must discuss is the nature of heat. When we say that something is hot, we mean that the atoms in the material are moving rapidly. Heat energy (also called thermal energy) is present in a material as the energy of motion of the atoms or molecules of the material. In general, the hotter a material is, the more motion among its atoms or molecules.

With these two ideas in mind we can understand how hot objects emit photons. Rapidly moving atoms collide with electrons in the material, and the sudden changes in the motion of the electrons can produce photons. Thus, we can expect a hot object to emit electromagnetic radiation. Such radiation is **black body radiation** and is quite common. The light we see coming from the hot filament in an incandescent light bulb is black body radiation.

In the heated filament, gentle collisions produce low-energy photons with long wavelengths, and violent collisions produce high-energy photons with short wavelengths. If we graph the energy emitted at different wavelengths, we get a curve like those shown in Figure 7–1. The curve shows that gentle collisions and violent collisions are rare. Most collisions are intermediate in violence, producing photons of intermediate wavelength.

Now we can understand how the photosphere of a star produces light. The light is black body radiation produced by the rapidly moving atoms in the gas.

One characteristic of this black body radiation is that hot stars look bluer than red stars. Consider a hot star and a cool star. Because the gases at the surface of the hotter star are hot, collisions between atoms and electrons will be, on average, more violent, and that star will radiate higher-energy photons. Higher-energy photons have shorter wavelengths, so the hotter star will tend to emit bluer light. In contrast, the cooler star will, on average, have less violent collisions between atoms and electrons and so will radiate lower-energy, longer wavelength photons. That is, the cooler star will be redder.

A second characteristic is that hot stars emit more radiation per second from 1 square meter of their surface than do cooler stars. In a hot gas, the atoms move more rapidly and collide more often. Thus a hot gas will emit a larger number of photons per second than a cooler gas. (Box 7–1 gives these two principles mathematical form.)

Black body radiation is important in astronomy because stars radiate approximately like black bodies. In later sections of this chapter we will see how atoms modify the star's radiation, but we can estimate a star's surface temperature quite accurately by thinking of it as a black body and measuring its color.

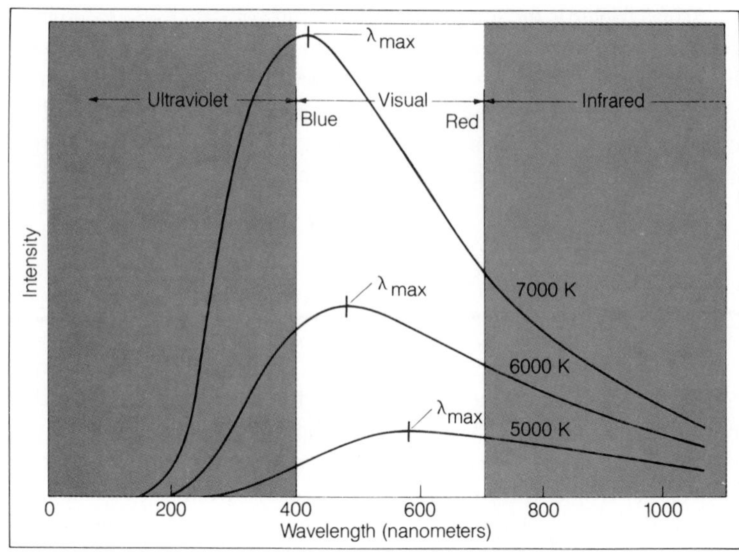

Figure 7–1 The intensity of radiation emitted by a heated body depends on wavelength. λ_{max} designates the wavelength of maximum intensity. Hotter objects radiate more energy and have shorter λ_{max} than cooler objects. A hot object radiates more blue light than red and therefore looks blue. A cool object radiates more red than blue and therefore looks red.

COLOR INDEX

Hot stars look blue, but if we want to know the temperature of the star's surface, we must have a way of measuring color. In astronomy the **color index** is a measure of color.

To measure the color index of a star we need a photometer and a set of standard filters. The most commonly used filters are the blue (B) and visual (V). Each filter isolates a specific part of the spectrum. The B filter, for example, will not allow any light through except those photons with wavelengths between about 400 nm and 480 nm. The V filter is transparent to photons with wavelengths between about 500 nm and 600 nm, roughly approximating the sensitivity of the human eye.

If we measure the magnitude of a star through each of these filters, then the difference between the two magnitudes is a number related to the color of the star. This is commonly written B-V and is referred to as the B-V color index.

For example, a hot star radiates much more blue light than red and will be brighter through the B filter. That means that the B magnitude will be smaller and the B-V color index will be negative. The bluest stars have a B-V color index of about −0.4 and have surface temperatures of about 50,000 K. Red stars radiate much more red light than blue, so they will be fainter through the blue filter. Then B-V for a red star will be positive. The reddest stars have a B-V color index of about 2 and surface temperatures of about 2000 K.

Many filter systems have been designed by astronomers to measure different parts of the spectrum. Some are transparent in the ultraviolet and some in the infrared, but in each case, the filters sample broad bands of the spectrum. That is enough to tell us the temperatures of the stars, but to get more detailed information, we must look at the spectrum of starlight much more closely.

7.2 ATOMS

The atoms in the surface layers of stars leave their marks on the light the stars emit. By understanding what atoms are and how they interact with light, we can decode the spectra of the stars. We begin in this section by constructing a working model of an atom.

A MODEL ATOM

In Chapter 2 we devised a model of the sky, the celestial sphere, to help us think about the nature and motion of the heavens. In the case of the atom, we again need a model.

Our model of the atom consists of a small central **nucleus** surrounded by a cloud of whirling electrons. The nucleus has a diameter of about .0000016 nm, and the cloud of electrons has a diameter of about .1 to .5 nm. (Recall from Chapter 6 that 1 nm is 10^{-9} m.) Household

BOX 7–1

Radiation from a Heated Object

All objects radiate some electromagnetic radiation because the atoms of all objects have some motion, and this thermal energy of motion is more violent for hotter objects. When moving atoms collide with electrons, some of the energy can be radiated away as a photon. We can graph the intensity of radiation given off at different wavelengths as shown in Figure 7–1. Note that an object emits very short wavelength and very long wavelength photons much less often than moderate wavelength photons.

The wavelength at which an object emits the maximum amount of energy, called the **wavelength of maximum** (λ_{max}), depends on the object's temperature. If we heat the object, the average photon emitted will have a shorter wavelength. This is commonly expressed as Wien's law.

$$\lambda_{max} = \frac{3{,}000{,}000}{T}$$

This is a powerful tool in astronomy because it means we can find the temperature of an object from its wavelength of maximum and vice versa. Suppose, for example, we found that a star's light was most intense at a wavelength of 1000 nm. Then its temperature would have to be about (3,000,000/1000) or about 3000 K.

Later we will meet objects much hotter than most stars, and such objects radiate most of their energy at very short wavelengths. The hottest stars, for instance, radiate most of their energy in the ultraviolet. When we discuss neutron stars and black holes, we will discover clouds of gas with temperatures of over 1,000,000 K. Such objects radiate tremendous amounts of X rays.

The total amount of radiation emitted at all wavelengths depends on the number of collisions per second. If an object's temperature is high, there are many collisions and it emits more light than a cooler object of the same size. We measure energy in units called joules (J); one joule is about the energy of an apple falling from a table to the floor. The total radiation given off by 1 m^2 of the object in joules per second equals a constant number, represented by σ, times the temperature raised to the fourth power.* This is often called the Stefan–Boltzmann law.

$$E = \sigma T^4 (\text{J/sec/m}^2)$$

If we doubled an object's temperature, for instance, it would radiate 2^4, or 16, times more energy. Thus, we can expect hot stars to radiate large amounts of energy from each square meter, and most of their radiation is at short wavelengths.

*For the sake of completeness we should note that the constant σ equals 5.67×10^{-8} J/m^2 · sec · degree^4).

plastic wrap is about 100,000 atoms thick. This makes the atom seem very small, but the nucleus is 100,000 times smaller.

Atoms are so small that any scale model we might make would have to be magnified tremendously. Suppose that we could make a hydrogen atom bigger by a factor of 10^{12} (1 million million). Only then would it be big enough to examine.

The nucleus of a hydrogen atom is a proton whose diameter is about .0000016 nm, or 1.6×10^{-13} cm. Multiplying by a factor of 10^{12} magnifies it to 0.16 cm, about the size of a grape seed. The electron cloud* has a diam-

*For a representative diameter, we take the size of the atom's second orbit (Figure 7–4).

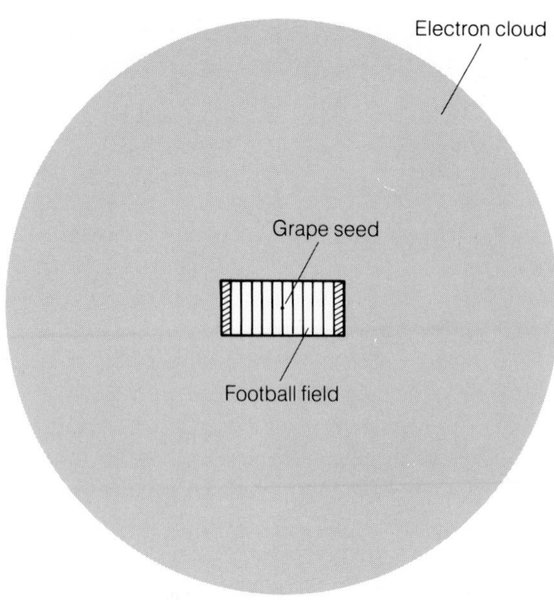

Figure 7–2 Magnifying a hydrogen atom by 10^{12} makes the nucleus the size of a grape seed and the outer electron cloud about 5½ times bigger than a football field. The electron itself is still too small to see.

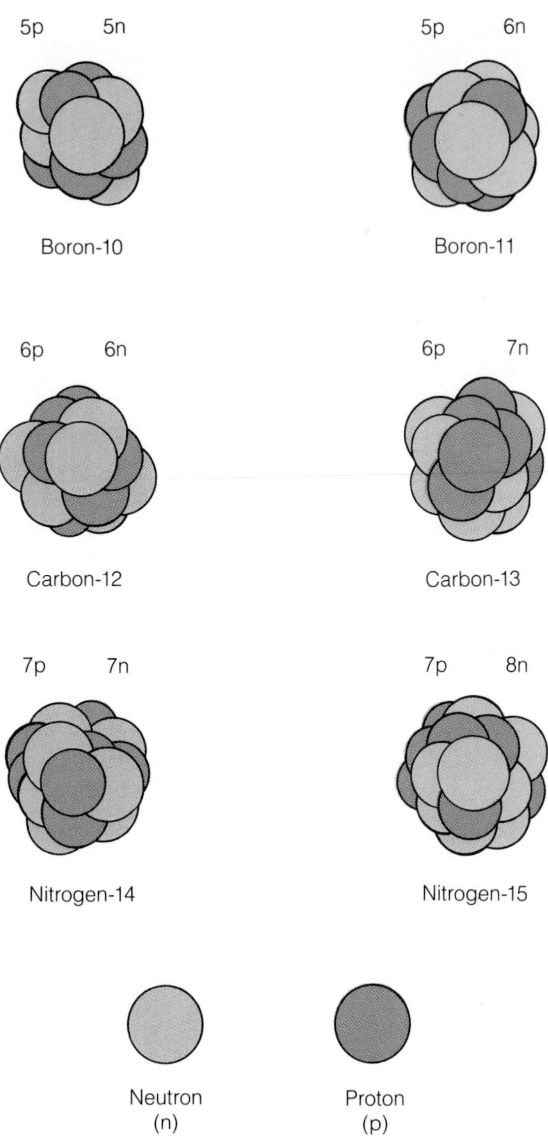

Figure 7–3 Atomic nuclei of the isotopes of boron, carbon, and nitrogen.

eter of about .5 nm, or 5×10^{-8} cm. When we magnify the atom by 10^{12}, this becomes 500 m, or about 5½ football fields laid end to end (Figure 7–2). When you imagine a grape seed in the midst of 5½ football fields, orbited by one magnified electron still too small to be visible, you see that an atom is mostly empty space.

The mass of a hydrogen atom is very small—about 2×10^{-24} g. For comparison, a paper clip has a mass of about 1 g. Individual atoms have such small masses that only by assembling vast numbers can nature build such massive objects as the stars. The sun, for instance, contains about 10^{57} atoms.

The nucleus of our model atom contains nearly all of the mass. The mass of the proton is about 1836 times the mass of the electron. Thus, the electrons in an atom never represent more than about 0.05 percent of the atom's mass.

The nucleus of a typical atom is more complicated than that of hydrogen in that it contains two different kinds of particles, protons and neutrons. **Protons** carry a positive electrical charge, and **neutrons** have no charge. Consequently, an atomic nucleus, made of protons and neutrons, has a net positive charge.

The electrons surrounding the nucleus carry a charge equal to but opposite from that on the nucleus. In a neutral atom, the number of electrons equals the number of protons. Thus, the positive charge on each proton is balanced by the negative charge on an electron, and the atom is electrically neutral.

There are over a hundred kinds of atoms, called chemical elements. The kind of element an atom represents depends only on the number of protons in the nucleus. For example, carbon has six protons and six

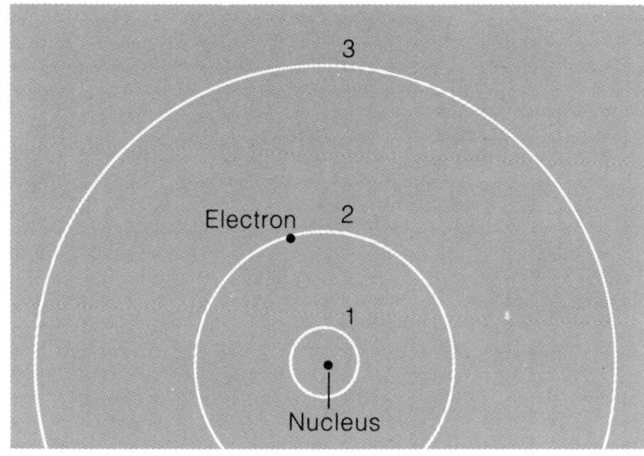

Figure 7–4 The electron in a hydrogen atom may occupy any permitted energy level in the atom but not energy levels in between. Different atoms, because of the different charge on their nuclei and because of the different numbers of electrons present, have different patterns of energy levels.

neutrons in its nucleus. Adding a proton produces nitrogen, and subtracting a proton produces boron.

However, we can change the number of neutrons in an atom's nucleus without changing the atom significantly. For instance, if we add a neutron to the carbon nucleus, we still have carbon, but it is slightly heavier than the more common form of carbon. Atoms that have the same number of protons but a different number of neutrons are **isotopes**. Carbon has two stable isotopes. One form contains six protons and six neutrons, making a total of twelve particles, and is thus called carbon-12. Carbon-13 has six protons and seven neutrons in its nucleus. Figure 7–3 shows schematically the nuclei of a few isotopes.

Protons and neutrons are bound tightly into the nucleus, but the electrons are held loosely in the electron cloud. Running a comb through your hair creates a static charge by removing a few electrons from their atoms. This process is called **ionization**, and the atom that has lost one or more electrons is an **ion**. The neutral carbon atom, with six protons and six neutrons in its nucleus, has six electrons, which balance the positive charge of the nucleus. If we ionize the atom by removing one or more electrons, the atom is left with a net positive charge. Under some circumstances, an atom may capture one or more extra electrons, giving it a net negative charge. Such a negatively charged atom is also an ion.

Atoms form bonds with each other by exchanging or sharing electrons. Two or more atoms bonded together form a **molecule**. Few atoms can form chemical bonds in stars. The high temperatures produce such violent collisions between atoms that most molecules would quickly break up. Only in the coolest stars are the collisions gentle enough to permit chemical bonds. We will see later that the presence of molecules such as titanium oxide (TiO) in a star is a clue that the star is cool. In later chapters we will also see that molecules can form in cool gas clouds in space and in the atmospheres of planets.

ELECTRON SHELLS

So far we have described the electron cloud only in a general way, but the specific way electrons behave within a cloud is very important in astronomy.

The electrons are bound to the atom by the attraction between their negative charge and the nucleus's positive charge. If we wish to ionize the atom, we need a certain amount of energy to pull an electron away from its nucleus. This energy is the electron's **binding energy**, the energy that holds it to the atom.

The rules that govern the motion of an electron within an atom are the same rules that govern our lives, but, because the atom is very small, peculiar effects show up. Called **quantum mechanics**, these rules specify that an electron can have only certain amounts of binding energy. Physicists refer to these possible amounts of binding energy as **energy levels** (Figure 7–4). An electron with a certain amount of binding energy is said to be "in" the corresponding energy level.

These energy levels are like steps in a staircase: You can stand on the number one step or the number two step, but not on the number one and one-quarter step. Similarly, the electron can be in any of the permitted energy levels but may not occupy any level between (Figure 7–4). That is, the electron may have any of the permitted energies but no other energy.

The arrangement of permitted energy levels depends mainly on the charge on the nucleus and the number of electrons in the atom. Thus, each kind of element has its own pattern of energy levels. Isotopes of the same element have nearly the same pattern because they have the same number of protons. However, ionized atoms have energy level patterns that differ from their un-ionized forms. Thus, the arrangement of energy levels differs for every kind of atom and ion.

The properties of energy levels are important because the electrons in the levels can interact with light, our major clue from celestial objects. By understanding this interaction, we can interpret the spectra of stars and learn such things as their composition and temperature.

7.3 THE INTERACTION OF LIGHT AND MATTER

We begin our study of light and matter by considering the hydrogen atom. As we noted earlier, hydrogen is both simple and common. Roughly 90 percent of all atoms in the universe are hydrogen.

THE EXCITATION OF ATOMS

The hydrogen atom in Figure 7–5 has its electron in the lowest permitted energy level, where it is tightly bound to the atom. We can move the electron to a higher level by supplying some energy. This is like moving a flowerpot from a low shelf to a high shelf; the higher we move the pot, the more energy we must expend. The amount of energy needed to move the electron from one level to a higher level is just the energy difference between the levels.

If we move the electron from a low-energy level to a higher level, we say the atom is **excited.** That is, we have added energy to the atom in moving its electron. If the electron falls back to a lower-energy level, the energy is returned.

An atom can become excited if it collides with another atom. During the collision, some of the energy of motion of the atoms can be absorbed by one or both of the atoms, leaving them in an excited state. This is very common in a hot, dense gas where the atoms move rapidly and collide often.

An atom can also become excited if it absorbs a photon. Only a photon with exactly the right energy can move the electron from one energy level to another. Too much or too little energy, and the photon cannot be absorbed. Because the energy of a photon depends on its wavelength, only photons of certain wavelengths can be absorbed by a given kind of atom. The atoms in Figure 7–6, for example, can absorb any of three different wavelength photons, moving the electron up to any of three permitted energy levels. Any other wavelength photon has too much or too little energy to be absorbed.

Atoms, like humans, cannot remain excited forever. The excited atom is unstable and must eventually (usually within 10^{-6} to 10^{-9} seconds) give up the energy it has absorbed, and the electron returns to the lowest-energy level. Because the electrons eventually tumble down to this bottom level, it is known as the **ground state.**

When the electron drops from a higher-energy level to a lower level, it must give up the excess energy—

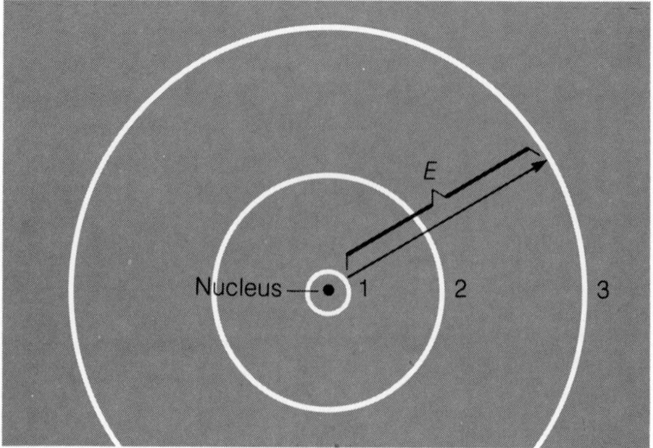

Figure 7–5 An atom with an electron in the lowest-energy level can absorb energy *E* and move its electron to a higher-energy level.

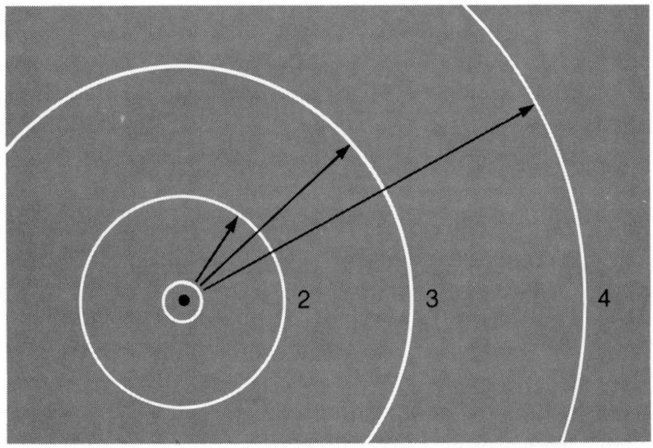

Figure 7–6 This atom can absorb one of three different wavelength photons and move the electron to one of the higher-energy levels.

usually by emitting a photon. Study the sequence of events in Figure 7–7 to see how an atom can absorb and emit certain wavelength photons.

Because only certain energy levels are permitted in an atom, only certain energy differences can occur. Each type of atom or ion has its unique set of energy levels, so each atom or ion absorbs and emits photons with a unique set of wavelengths. Thus, we can identify the elements in a gas by studying the characteristic wavelengths of light absorbed or emitted.

This process of excitation and emission produces a common sight. The gas in a neon sign glows because a high voltage forces electrons to flow through the gas, exciting the atoms by collisions. Almost as soon as an

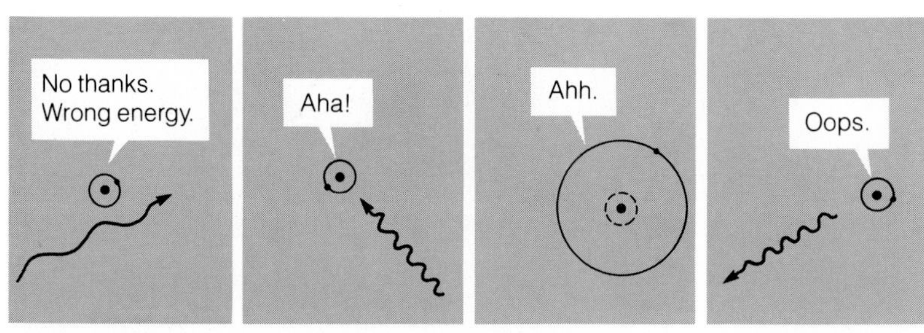

Figure 7–7 An atom can absorb a photon only if the photon has the correct energy. The excited atom is unstable and within a fraction of a second returns the electron to a lower-energy level, reradiating the photon in a random direction.

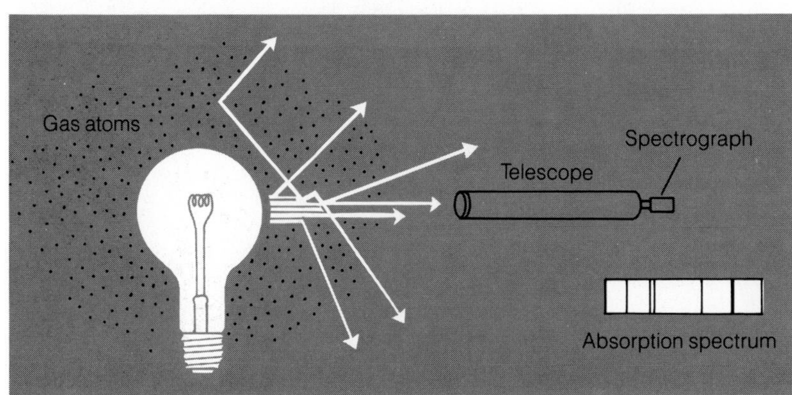

Figure 7–8 Photons of the proper wavelengths can be absorbed by the gas atoms and reradiated in random directions. Because these particular photons do not reach the telescope, the spectrum is dark at the wavelengths of the missing photons.

atom is excited, its electron drops back to lower-energy levels, emitting the surplus energy as a photon of a certain wavelength. Neon is a popular gas for such signs because it emits a large number of red, orange, and yellow wavelength photons, which we see as a rich reddish orange. So-called neon signs of other colors contain other gases or mixtures of gases selected to produce other colors.

THE FORMATION OF A SPECTRUM

To see how an astronomical object can produce a spectrum, imagine a cloud of hydrogen floating in space with an incandescent light bulb glowing inside it. The bulb glows because its filament is hot, producing black body radiation as described in Box 7–1. Thus, it emits photons of all wavelengths, and a spectrum of its light would reveal an uninterrupted band of color called a **continuous spectrum.**

However, the light from this bulb must pass through the hydrogen gas before it can reach our telescope (Figure 7–8). Most of the photons will pass through the gas unaffected because they have wavelengths the hydrogen atoms cannot absorb, but a few photons will have the right wavelengths. These photons cannot pass through the gas because they are absorbed by the first atom they meet. The atom is excited for a fraction of a second, and the electron then drops back to a lower orbit and a new photon is emitted. The original photon was traveling through the gas toward our telescope, but the new photon is emitted in some random direction. Very few of these new photons leave the cloud in the direction of our telescope, so the light that finally enters the telescope has very few photons at the wavelengths the atoms can absorb. When we form a spectrum from this light, photons of these wavelengths are missing and the spectrum has dark lines at the positions these photons would have occupied. These dark lines are called **absorption lines** because the atoms absorbed the photons. A spectrum containing absorption lines is an **absorption spectrum** (also called a **dark line spectrum**). (See Color Plate 4.)

What happens to the photons that were absorbed? They bounce from atom to atom, being absorbed and emitted over and over until they escape from the cloud. If, instead of aiming our telescope at the bulb, we swing it to one side so that no light from the bulb enters the telescope, we can photograph a spectrum of the light emitted by the gas atoms (Figure 7–9). In that case, the

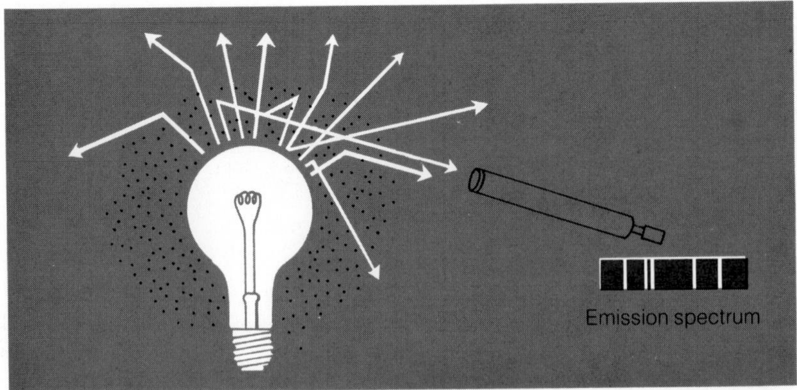

Figure 7–9 Pointing the telescope away from the bulb, we can receive only those photons the atoms can absorb and reradiate, producing emission lines in the spectrum.

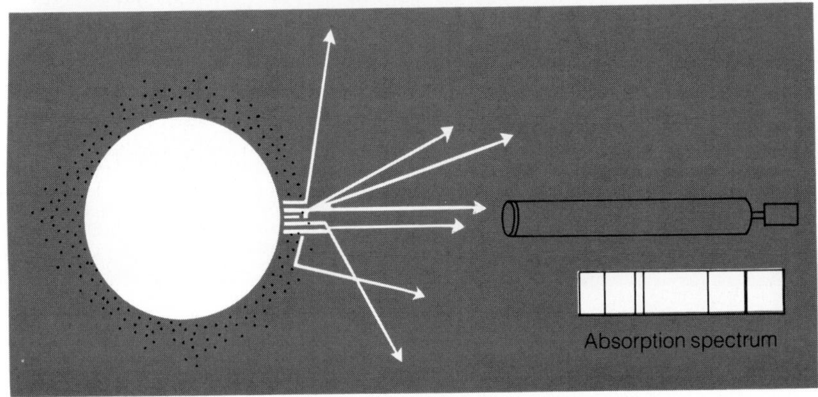

Figure 7–10 A star produces an absorption spectrum because its atmosphere absorbs certain wavelengths in the spectrum.

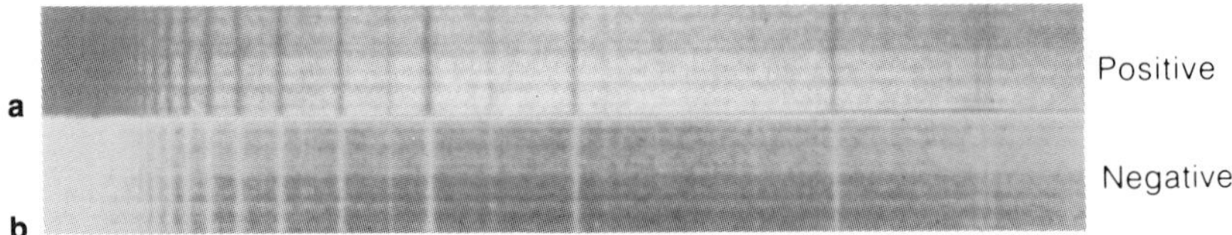

Figure 7–11 Astronomers often reproduce stellar spectra as negative images. In the positive image (a) we see the starlight spread into a band of light with dark lines at certain wavelengths. In a negative reproduction of the same absorption spectrum (b), we see the spectrum as it would look on a photographic plate, and subtler detail is usually visible. (Adapted from H. A. Abt, A. B. Meinel, W. W. Morgan, and J. W. Tapscott, *An Atlas of Low-Dispersion Grating Stellar Spectra,* Kitt Peak National Observatory, 1968.)

only photons entering the telescope are photons that were absorbed and reemitted. A spectrum of this light is almost entirely dark except for the wavelengths corresponding to the photons the gas can absorb and reemit. Thus, we will see a spectrum containing only bright lines on a dark background. These bright lines are called **emission lines,** and a spectrum with emission lines is an **emission spectrum** (also called a **bright line spectrum**). The spectrum of a neon sign, for instance, is an emission spectrum. Also, the bluish purple color of mercury-vapor streetlights and the pinkish orange color of sodium-vapor streetlights are produced by the emission lines of those elements. (See Color Plate 4.)

The light bulb and hydrogen gas cloud produce a spectrum in the same way a star does. A star is all gas, however. Its outer layers, its photosphere, act as a hot,

BOX 7–2

Kirchhoff's Laws

The Heidelberg scientist Gustav Robert Kirchhoff (1824–1887) lived long before twentieth century physicists came to understand atomic structure. Kirchhoff knew nothing of electrons in energy levels, but from his experiments during the 1850s with the spectra of gases and solids, he was able to summarize the nature of spectra in three statements now known as Kirchhoff's laws.

Law I: The Continuous Spectrum

A solid, liquid, or dense gas excited to emit light will radiate at all wavelengths and thus produce a continuous spectrum.

Law II: The Emission Spectrum

A low-density gas excited to emit light will do so at specific wavelengths and thus produce an emission spectrum.

Law III: The Absorption Spectrum

If light comprising a continuous spectrum is allowed to pass through a cool, low-density gas, the resulting spectrum will have dark lines at certain wavelengths. That is, it will be an absorption spectrum.

Kirchhoff experimented in his laboratory and found that sodium vapor heated in a flame produced two bright emission lines close together in the yellow part of the spectrum. But when he passed light of a continuous spectrum through the same sodium vapor, he saw two absorption lines *at the same wavelengths.* Other atoms produced lines at other wavelengths. Thus, he understood that sodium somehow was responsible for producing lines at these two wavelengths.

Kirchhoff used his understanding of spectra to perform one of the first analyses of a stellar spectrum. Because he saw two absorption lines in the spectrum of sunlight at the same wavelengths as the emission lines produced by sodium vapor, he concluded that the sun's atmosphere contained sodium. By comparing dark lines in the solar spectrum with bright lines produced by excited gases in his laboratory, Kirchhoff was eventually able to identify some half-dozen elements in the sun. Generations of astronomers have used Kirchhoff's laws to analyze spectra of everything from the moon to the farthest galaxies.

bright surface emitting radiation at all wavelengths much like the filament in a light bulb. Above these layers lies the thinner gas of the star's atmosphere. As the light travels upward through the star's atmosphere, photons of certain wavelengths are absorbed by atoms and so never reach us. The spectrum of a star is an absorption spectrum whose dark lines indicate which wavelengths the atoms absorbed (Figures 7–10 and 7–11).

We have discussed the three types of spectra—continuous, emission, and absorption. These three kinds of spectra are summarized in three statements now known as **Kirchhoff's laws** (Box 7–2).

The absorption lines in stellar spectra provide a windfall of data about the star's surface layers. By studying the spectral lines, we can identify the elements in the stellar atmosphere and find the temperature of the atoms. To see how to get all this information, we need to look carefully at the way the hydrogen atom produces lines in a star's spectrum.

THE HYDROGEN SPECTRUM

As you must have gathered by now, each element has its own spectrum, unique as a human fingerprint, and it can be recognized by its spectrum across trillions of miles. To see how hydrogen produces its spectrum, we must draw a scale diagram of its permitted energy levels, making the radius of the level proportional to its energy (Figure 7–12). Then we can examine the way such atoms interact with light.

A **transition** occurs in an atom when an electron changes energy levels. In our diagram of a hydrogen atom we can represent transitions by arrows pointing from one

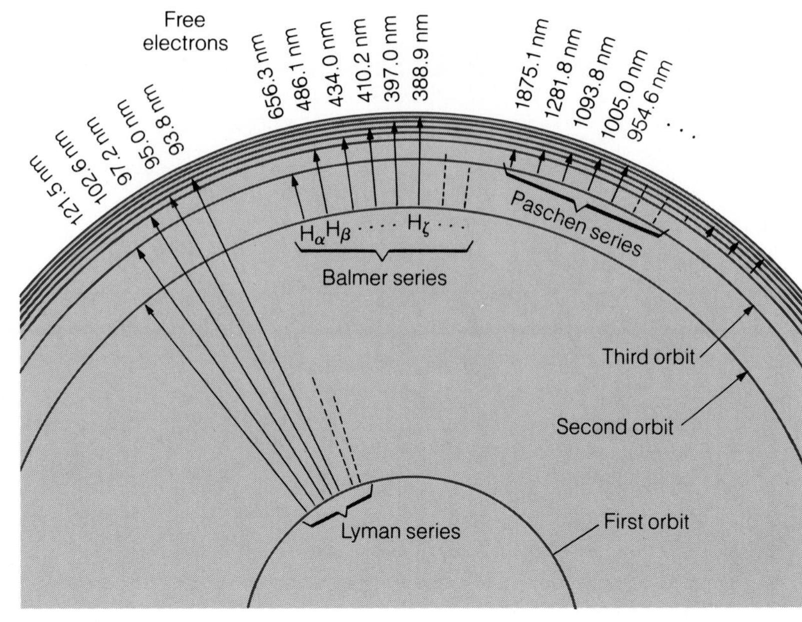

Figure 7–12 The levels in this diagram are spaced to represent the energy the hydrogen atom's electron can have. The transitions, drawn as arrows, can be grouped into series according to the lowest level. This drawing shows only a few of the infinity of transitions and series possible. Note that the arrows point upward and thus represent transitions that would absorb energy.

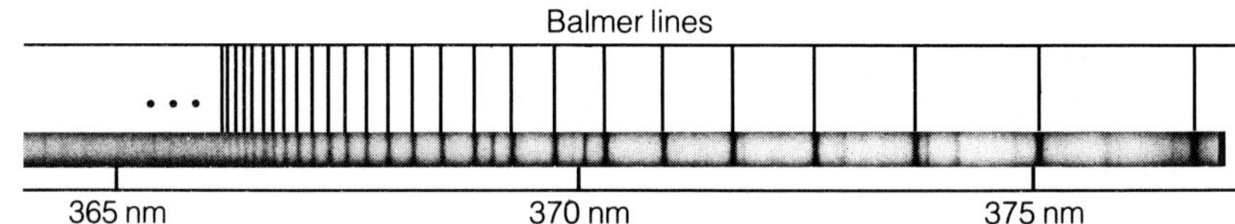

Figure 7–13 The Balmer lines photographed in the near ultraviolet. (Mount Wilson and Las Campanas Observatories, Carnegie Institution of Washington.)

level to another. If the arrow points upward, the atom must absorb energy, and if the arrow points downward, the atom must emit energy.

If the transition results in the absorption or emission of a photon, the length of the arrow tells us its energy. Long arrows represent large amounts of energy and thus short-wavelength photons. Short arrows represent smaller amounts of energy and longer-wavelength photons.

We can divide the possible transitions in a hydrogen atom into groups called series, according to their lowest energy level. Those arrows whose lower ends rest on the ground state represent the **Lyman series**; those resting on the second energy level, the **Balmer series**; and those resting on the third, the **Paschen series.** In principle, each series contains an infinite number of transitions, and there are an infinite number of series. Figure 7–12 shows only the first few transitions in the first few series.

The Lyman series transitions involve large energies, as shown by the long arrows in Figure 7–12. These energetic transitions produce lines in the ultraviolet part of the spectrum where they are invisible to the human eye. Nor do any Paschen lines lie in the visible part of the spectrum. These transitions involve small energies, and thus produce spectral lines in the infrared.

Balmer series transitions produce the only spectral lines of hydrogen in the visible part of the spectrum. Figure 7–12 shows that the first few Balmer series transitions are intermediate between the energetic Lyman transitions and the low-energy Paschen transitions. These Balmer lines are labeled by Greek letters for easy identification. H_α is a red line, H_β is blue, and H_γ and H_δ are violet. These four lines blend together to produce the purple-red color characteristic of glowing clouds of hydrogen in space. (See Color Plates 6, 7 and 12.) The remaining Balmer lines have wavelengths too short to see, but they are in the near ultraviolet and are not absorbed by the earth's atmosphere. They can be photographed easily from earth-based observatories (Figure 7–13).

Balmer series lines are important because, as we have seen, they are the only hydrogen lines we can study from the earth's surface. In the next section, we will see how Balmer series lines can tell us a star's temperature.

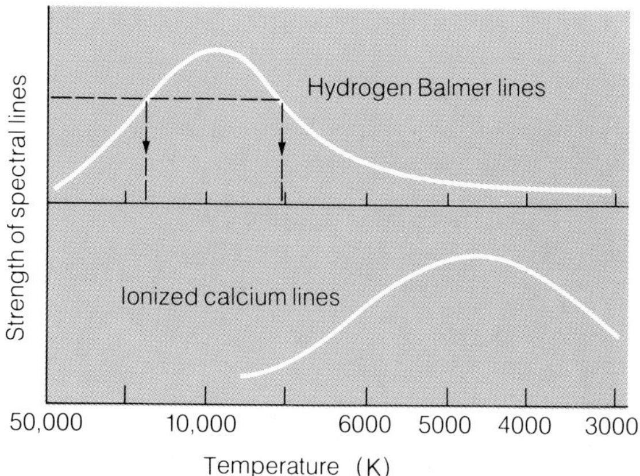

Figure 7–14 The strength of the Balmer lines in a stellar spectrum depends on the temperature of the star. A star with medium strength Balmer lines could have one of two possible temperatures. Other elements, such as once ionized calcium, behave similarly, but reach maximum strength at different temperatures. This can help us choose the correct temperature.

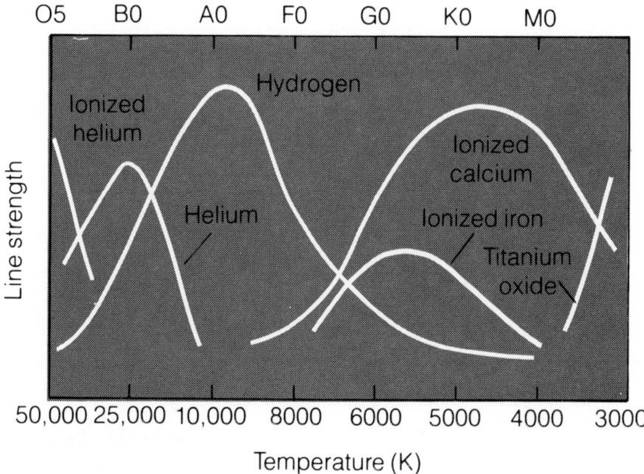

Figure 7–15 The strengths of spectral lines can tell us the temperature of a star. This relationship is the basis for the spectral classification system.

7.4 STELLAR SPECTRA

In later chapters we will use spectra to study galaxies and planets, but we begin here by studying the spectra of stars. Such spectra are the easiest to understand, and the nature of stars is central to our study of all celestial objects.

THE BALMER THERMOMETER

We can use the Balmer absorption lines as a thermometer to find the temperatures of stars. Chapter 6 showed how to estimate temperature from color, but the strengths of the Balmer lines in a star's spectrum give a much more accurate estimate of the star's temperature.

The Balmer thermometer works because the Balmer absorption lines are produced only by atoms whose electrons are in the second energy level (Figure 7–12). If the star is cool, there are few violent collisions between atoms to excite the electrons, and most atoms have their electron in the ground state. If most electrons are in the ground state, they can't absorb photons in the Balmer series. As a result, we should expect to find weak Balmer absorption lines in the spectra of cool stars.

In hot stars, on the other hand, there are many violent collisions between atoms, exciting electrons to high energy levels or knocking the electron clear out of some atoms. That is, some atoms are ionized. Thus, few atoms have electrons in the second orbit to form Balmer absorption lines, and we should expect hot stars, like cool stars, to have weak Balmer absorption lines.

At some intermediate temperature the collisions are just right to excite large numbers of electrons into the second energy level. With many atoms excited to the second level, the gas absorbs Balmer-wavelength photons strongly and thus produces strong Balmer lines.

To summarize, the strength of the Balmer lines depends on the temperature of the star's surface layers. Both hot and cool stars have weak Balmer lines, but medium-temperature stars have strong Balmer lines.

Theoretical calculation can predict just how strong the Balmer lines should be for stars of various temperatures. The details of these calculations are not important to us, but the results are. Figure 7–14 shows the strength of the Balmer lines for various stellar temperatures. We could use this as a temperature indicator except that the curve gives us two answers. A star with Balmer lines of a certain strength might have either of two temperatures, one high and one low. We must examine other spectral lines to choose the correct temperature.

We have seen how the strength of the Balmer lines depends on temperature. The same process affects the spectral lines of other elements, but the temperature at which they reach maximum strength differs for each element. If we add these elements to our graph, we get a handy tool for taking the stars' temperatures (Figure 7-15).

We can determine a star's temperature by comparing the strengths of its spectral lines with our graph. For instance, if we photographed a spectrum of a star and found medium-strength Balmer lines and strong helium lines, we could conclude it had a temperature of about 20,000 K. But if the star had weak hydrogen lines and

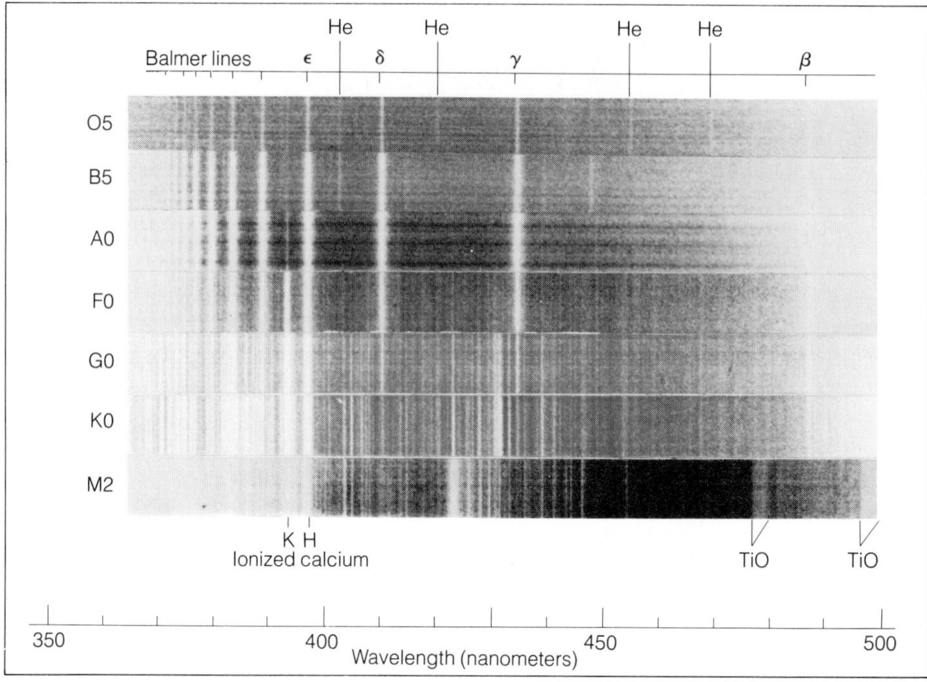

Figure 7–16 The spectra of stars of various classes illustrate how the strengths of lines change with temperature. These spectra are reproduced as the astronomer sees them on the photographic plates. Thus, the dark spectral lines appear light on the plates. (Adapted from H. A. Abt, A. B. Meinel, W. W. Morgan, and J. W. Tapscott, *An Atlas of Low-Dispersion Grating Stellar Spectra,* Kitt Peak National Observatory, 1968.)

strong lines of ionized iron, we would assign it a temperature of about 5800 K, similar to the sun.

The spectra of the coolest stars contain dark bands produced by molecules such as titanium oxide (TiO). Because of their structure, molecules can absorb photons at many wavelengths, producing numerous, closely spaced spectral lines that blend together to form bands. These molecular bands appear only in the spectra of the coolest stars because, as mentioned before, only there can molecules avoid the collisions that would break them up in hotter stars. Thus, the presence of dark bands in a star's spectrum indicates that the star is very cool. (See, for example, the star labeled M2 in Figure 7–16.)

By analyzing stellar spectra, astronomers have found that the hottest stars have surface temperatures above 40,000 K. The coolest have temperatures of about 2000 K. Compare these with the surface temperature of the sun, about 5800 K.

SPECTRAL CLASSIFICATION

We have seen that the strengths of spectral lines depend on the surface temperature of the star. From this we can predict that all stars of a given temperature should have similar spectra. If we learn to recognize the pattern of spectral lines produced by a 6000 K star, for instance, we need not use Figure 7-15 every time we see that kind of spectrum. In other words, we can save time by classifying stellar spectra rather than by analyzing each one individually.

The first widely used classification system was made by astronomers at Harvard during the 1890s and 1900s. One of them, Annie J. Cannon, personally inspected and classified the spectra of over 250,000 stars. The spectra were first classified in groups labeled A through Q, but some groups were later dropped, merged with others, or reordered. The final classification includes the seven

Table 7–1 Spectral classes.

Spectral Classes	Approximate Temperature (K)	Hydrogen Balmer Lines	Other Spectral Features
O	40,000	Weak	Ionized helium
B	20,000	Medium	Neutral helium
A	10,000	Strong	Ionized calcium weak
F	7,500	Medium	Ionized calcium weak
G	5,500	Weak	Ionized calcium medium
K	4,500	Very Weak	Ionized calcium strong
M	3,000	Very Weak	TiO strong

spectral classes, or **types**, still used today: O, B, A, F, G, K, M.*

This sequence of spectral types, called the **spectral sequence**, is important because it is a temperature sequence. The O stars are the hottest, the B stars next hottest, and so on. The temperature continues to decrease down to the M stars, the coolest of all.

We can classify a star by examining features in its spectrum, as described in Table 7–1. For example, if it has weak Balmer lines and lines of ionized helium, it must be an O star. This table is based on the same information we used in Figure 7–15.

The spectra shown in Figure 7–16 illustrate how spectral lines change from class to class. Study the Balmer line H_γ in the center of the spectra. In the spectrum of the O5 star, H_γ is not very strong. (Notice that these spectra are reproduced as negatives, so the dark lines are white.) As we run our eye down the spectra, H_γ becomes stronger, reaching maximum strength about A0, and then becomes weaker in the cooler stars. The hotter O5 star has weak Balmer lines because most of the hydrogen atoms are excited to energy levels above the second and thus cannot absorb Balmer photons. The cooler stars (F through M) have weak Balmer lines because most of the hydrogen atoms are not excited out of the ground state.

The spectral lines of other atoms also change from class to class. Helium is visible only in the spectra of the hottest classes, and the titanium oxide bands only in the coolest. The two lines of ionized calcium, labeled H and K, increase in strength from A to K and then decrease from K to M. Because the strength of these spectral features depends on temperature, it requires only a few minutes to compare a star's spectrum with Table 7–1 or Figure 7–16 and determine its temperature.

We can be more precise if we divide each spectral class into ten subclasses. For example, spectral class A consists of the subclasses A0, A1, A2 . . . A8, A9. Next comes F0, F1, F2, and so on. Thus A5 lies halfway between A0 and F0. This finer division, of course, demands that we look carefully at a spectrum, but it is worth the effort, for the subclasses give us a star's temperature with an accuracy of about 5 percent. The sun, for example, is not just a G star, but a G2 star, with a temperature of about 5800 K.

*Generations of astronomers have remembered the spectral sequence using the mnemonic, "Oh, Be A Fine Girl, Kiss Me." Nationwide contests to find a less sexist mnemonic have failed to displace this traditional sentence.

RADIAL VELOCITY

A stellar spectrum can tell us many things besides temperature. One important thing we can learn from a spec-

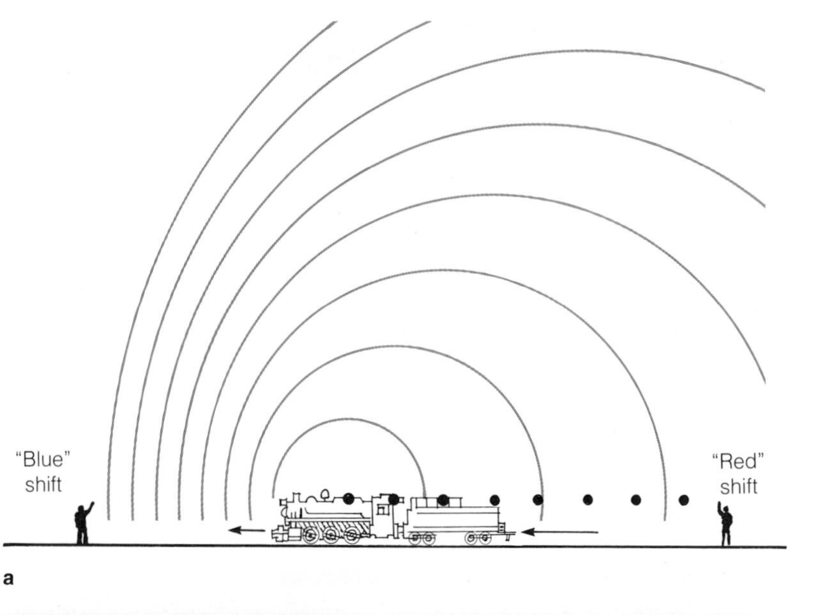

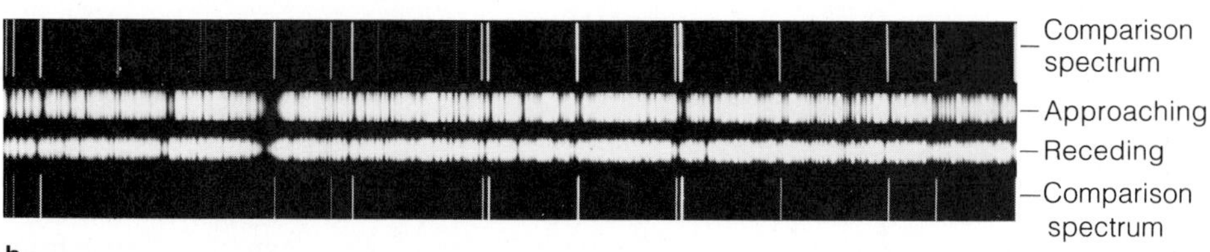

Figure 7–17 (a) Successive clangs of the engine bell (marked by dots) occur closer to the observer ahead, decreasing the distance the sound must travel. Thus, the observer hears the bell ring more often than it really does. The observer behind the train hears the bell ring less often. This is an example of the Doppler effect. (b) The upper spectrum of Arcturus was taken when the earth's orbital motion carried it toward the star. The lower spectrum was taken 6 months later when the earth was receding from Arcturus. The difference in the wavelengths of the lines is due to the Doppler shift. The bright lines above and below the stellar spectra are comparison spectra added within the spectrograph to provide reference wavelengths unaffected by the Doppler shift. (b, Mount Wilson and Las Campanas Observatories, Carnegie Institution of Washington.)

trum is related to the motion of the star relative to the earth.

If a star is approaching the earth, all of the wavelengths in the star's spectrum are shifted slightly toward the blue end of the spectrum. This is sometimes called a **blue shift** (Figure 7–17). Similarly, if a star is receding from the earth, all the wavelengths in the spectrum are shifted slightly toward the red end of the spectrum in what is termed a **red shift**.

This is known as the Doppler effect (Box 7–3). The red shift or blue shift in a stellar spectrum is very small and does not affect the color of the star, but we can use it to calculate the radial velocity of the star. The **radial velocity** (V_r) is the velocity with which the star is traveling away (or toward) the earth (Figure 7–18). Motion perpendicular to this radial velocity is called the **transverse velocity** and cannot be detected by the Doppler effect.

We will use the Doppler effect later to study such things as the motions of stars, the masses of galaxies, and the expansion of the universe.

BOX 7–3

The Doppler Effect

The **Doppler effect** is the change in wavelength of radiation due to the relative motion of the source and observer. To see how this works, imagine standing on a railroad track as a train approaches with the engine bell ringing once each second (Figure 7–17). When the bell rings, the sound travels ahead of the engine to reach your ears. One second later the bell rings again, but not at the same place. During that one second the engine moved closer to you, so the bell is closer at its second ringing. Now the sound has a shorter distance to travel and reaches your ears a little sooner than it would have if the engine had not moved. The third time the bell rings it is even closer. By timing the ringing of the bell, you would observe that the bell seemed to be ringing more often than once each second, all because the engine was approaching.

Standing behind the engine would give the opposite effect. You would find that each successive ring takes place farther away from you and the rings would sound more than 1 second apart. These apparent changes in the rate of the ringing bell are an example of the Doppler effect.

We can think of the peaks of the electromagnetic waves leaving a star as a series of clangs from a bell. If a star is moving toward us, we see the peaks of the light waves closer together than expected, making the wavelengths slightly shorter than they would have been if the star were not moving. If the star is going away from us, the peaks of the light waves are slightly farther apart and the wavelengths are longer. Thus, the lines in a star's spectrum are shifted slightly toward the blue if the star is approaching and toward the red if it is receding.

For convenience, we have assumed that the earth is stationary and the star is moving, but the Doppler effect depends only on relative motion. Thus, we cannot say that either the earth or star is stationary—only that there is relative motion between them. In addition, the Doppler effect depends only on the radial velocity, that part of the velocity directed away from or toward earth (Figure 7–18). The Doppler effect cannot reveal relative motion to the right or left—the transverse velocity.

How much the spectral lines change depends on the radial velocity. This can be expressed as a simple ratio relating the radial velocity V_r, divided by the speed of light c, to the change in wavelength $\Delta\lambda$ divided by the unshifted wavelength λ_0:

$$\frac{V_r}{c} = \frac{\Delta\lambda}{\lambda_0}$$

This expression is quite accurate for the low velocities of stars, but we will need a better version later when we discuss objects moving with very high velocities.

Suppose that we observe a line in a star's spectrum with a wavelength of 600.1 nm. Laboratory measurements show that the line should have a wavelength of 600 nm. That is, its unshifted wavelength is 600 nm. What is the star's radial velocity? First we note that the change in wavelength is 0.1 nm.

$$\frac{V_r}{c} = \frac{0.1}{600} = 0.000167$$

Multiplying by the speed of light, 3×10^5km/sec, gives the radial velocity, 50 km/sec. Because the wavelength is shifted to the red (lengthened), the star must be receding from us.

You may be quite familiar with this method of speed measurement. Police radar uses the Doppler shift in reflected radio waves to determine the velocity of approaching cars.

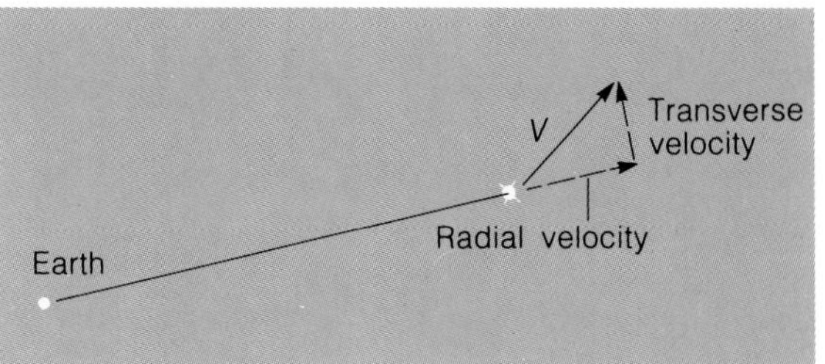

Figure 7–18 The radial velocity of a star is the part of the star's velocity *V*, which is directed away from (or toward) the earth. The Doppler effect can tell us a star's radial velocity but cannot tell us the transverse velocity—the part of the velocity perpendicular to the radial direction.

COMPOSITION

On Earth many branches of science and industry use spectrographs to analyze samples to determine their chemical composition. Astronomers too can use spectra in this way to tell which chemical elements are present in a celestial body and in what proportion. But recall that stellar spectra are produced by the surface layers of the star—the photosphere. Thus, stellar spectra tell us the composition of the gases at the star's surface.

Identifying the elements in a star by identifying the lines in the star's spectrum is relatively straightforward. For example, two dark absorption lines appear in the yellow region of the solar spectrum at the wavelengths 589 nm and 589.6 nm. The only atom that can produce this pair of lines is sodium, so we must conclude that the sun contains sodium. Over 90 elements have been found this way in the sun.

However, just because the spectral lines characteristic of an element are missing, we cannot conclude that the element itself is absent. For example, the hydrogen Balmer lines are weak in the sun's spectrum, yet the sun is about 75 percent hydrogen by mass. The reason for this apparent paradox is that the sun is too cool to produce strong Balmer lines. Similarly, an element's spectral lines may be absent from a star's spectrum because the star is too hot or too cool to excite those atoms to the orbits that produce visible spectral lines.

Detailed spectral analysis taking the star's temperature into consideration can reveal the abundance of the chemical elements in the star. The results of such studies show that nearly all stars have compositions similar to the

Table 7–2 The most abundant elements in the sun.

Element	Percentage by Number of Atoms	Percentage by Mass
Hydrogen	92.0	73.4
Helium	7.8	25.0
Carbon	0.03	0.3
Nitrogen	0.008	0.1
Oxygen	0.06	0.8
Neon	0.008	0.1
Magnesium	0.002	0.05
Silicon	0.003	0.07
Sulfur	0.002	0.04
Iron	0.004	0.2

Source: Adapted from C. W. Allen, *Astrophysical Quantities*, London: The Athlone Press, 1976.

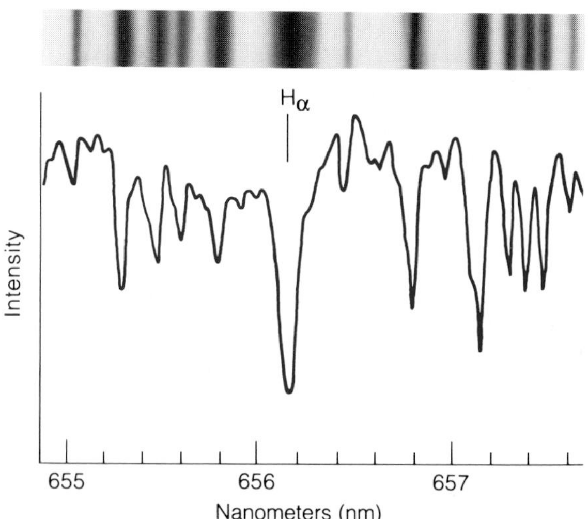

Figure 7–19 A CCD (p. 120) recorded the spectrum of the K2 star HD 80715 from which this short segment was taken. The data have been converted from an absorption spectrum (represented schematically above) to a graph of intensity versus wavelength. Dips in the curve show where the spectrum was darker. The profile of the H_α absorption line is visible at the center of the figure. (Adapted from data obtained by Harold Nations and Sam Barden.)

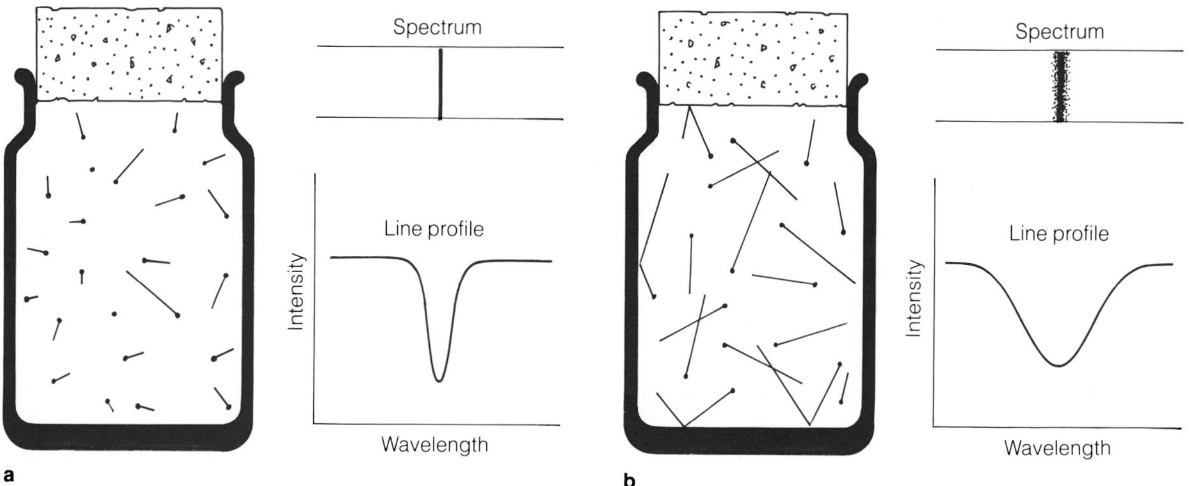

Figure 7–20 Doppler broadening. The atoms of a gas are in constant motion. Photons emitted by atoms moving toward the observer will have slightly shorter wavelengths, and those emitted by atoms moving away will have slightly longer wavelengths. This broadens the spectral line. If the gas is cool (a), the atoms do not move very fast, the Doppler shifts are small, and the line is narrow. If the gas is hot, (b), the atoms move faster, the Doppler shifts are larger, and the line is broader.

sun's—about 92 percent of the atoms are hydrogen, 7.8 percent are helium, and a small percentage are heavier elements. (See Table 7–2.)

So far we have seen how astronomers can extract information from the presence and strength of lines in a star's spectrum. Now we turn to another source of information—the shapes of spectral lines.

THE SHAPES OF SPECTRAL LINES

When astronomers refer to the shape of a spectral line, they mean the variation of intensity across the line. An absorption line, for instance, is darkest at the center and fades to either side. This shape is sometimes called a **line profile.**

A line profile is a graph that shows the brightness of the starlight at different wavelengths across the spectral line. If a spectrum is recorded on a glass plate, then the darkening of the plate must be measured at many different points along the spectrum to construct a line profile. But most modern astronomical spectrographs record the spectrum with a CCD or some other computer-readable device. Thus, the brightness of the light at different wavelengths is recorded directly in the computer and can be plotted to display a line profile (Figure 7–19).

The exact shape of a line can tell us a great deal about a star, but the most important characteristic is the width of the line. Spectral lines are not perfectly narrow; if they were, we could not see them. They have a natural width because nature allows an atom some leeway in the energy it may absorb or emit. The quantum mechanics behind this effect is beyond the scope of this discussion, but the result is very simple. In the absence of all other effects, spectral lines have a natural width of about .001 to .00001 nm—very narrow indeed.

The natural widths of spectral lines are not important in most branches of astronomy because other effects smear out the lines and make them much broader. For example, if a star spins rapidly the Doppler effect will broaden the spectral lines. As the star rotates, one side will recede from us and the other side will approach. Light from the receding side will be red-shifted, and light from the approaching side will be blue-shifted, so any spectral lines will be broadened.

Another important process is called **Doppler broadening.** To consider this process, let us imagine that we photograph the spectrum of a jar full of hydrogen atoms (Figure 7–20). Because the gas has some thermal energy (it is not at absolute zero), the gas atoms are in motion. Some will be coming toward our spectrograph, and some will be receding. Most, of course, will not be traveling very fast, but some will be moving very quickly. The photons emitted by the atoms approaching us will have slightly shorter wavelengths because of the Doppler effect, and photons emitted by atoms receding from us will have slightly longer wavelengths. Thus, the Doppler shifts due to the motions of the individual atoms will smear the spectral line out and make it broader. We have described the Doppler broadening of an emission line, but the effect is the same for absorption lines.

The extent of Doppler broadening depends on the temperature of the gas. If the gas is cold, the atoms travel at low velocities, and the Doppler shifts are small (Figure 7–20a). If the gas is hot, however, the atoms travel faster, Doppler shifts are larger, and the lines will be wider (Figure 7–20b).

Another form of broadening is called **collisional broadening** because it involves collisions between atoms. If an atom is in the process of emitting or absorbing a photon and it collides with another atom, ion, or electron, the energy levels can be disturbed, and the wavelength of the photon affected. Collisional broadening is most effective when the gas is hot and relatively dense, as it is in the sun. The high temperature ensures that atoms will move rapidly and collide violently, and the relatively high density ensures the atoms will collide often. In a later chapter we will see how collisional broadening affects the widths of spectral lines that are formed by low-density gas in giant stars and in interstellar space.

This chapter has described some of astronomy's most powerful techniques. By understanding how atoms and light interact, how spectra are produced, and how the spectral lines acquire their shapes, astronomers can discover the composition, temperature, density, and motions of gas anywhere in the universe. All they need is a glimmer of light to spread out into a spectrum.

In the next chapter, we will apply these tools to the sun, our own star. In subsequent chapters, we will explore farther into the universe, analyzing the spectra of other stars, nebulae, and distant galaxies.

SUMMARY

Stars emit black body radiation from the dense gases of their photospheres, and as that radiation passes through the less dense gas of the star's atmosphere, the atoms absorb certain wavelength photons to produce dark lines in the spectrum.

A heated solid, liquid, or dense gas emits black body radiation that contains all wavelengths and thus produces a continuous spectrum. Black body radiation is most intense at the wavelength of maximum λ_{max}, which depends on the temperature of the radiating body. Hot objects emit mostly short-wavelength radiation, whereas cool objects emit mostly long-wavelength radiation. This effect gives us clues to the temperatures of stars—hot stars are blue, and cool stars are red.

A low-density gas excited to emit radiation will produce an emission (or bright line) spectrum. Light from a source of a continuous spectrum passing through a low-density gas will produce an absorption (or dark line) spectrum. These three kinds of spectra are described by Kirchhoff's laws.

The lines in spectra are produced by the electrons that surround the nucleus of the atom. The electrons may occupy only certain permitted energy levels, and photons may be absorbed or emitted when electrons move from one energy level to another. Because each kind of atom and ion has a different set of energy levels, each kind of atom and ion can absorb or emit only certain wavelength photons. The hydrogen atom can produce the Lyman series lines in the ultraviolet, the Balmer series in the visual and near ultraviolet, the Paschen series in the infrared, and many more.

In cool stars the Balmer lines are weak because most atoms are not excited out of the ground state. In hot stars the Balmer lines are weak because most atoms are excited to higher orbits or ionized. Only at medium temperatures are the Balmer lines strong. We can use this effect as a thermometer for determining the temperature of a star. In its simplest form this amounts to classifying the stars' spectra in the spectral sequence: O, B, A, F, G, K, M.

When a source of radiation is approaching us, we observe shorter wavelengths, and when it is receding, we observe longer wavelengths. This Doppler effect makes it possible for the astronomer to measure a star's radial velocity, that part of its velocity directed toward or away from earth.

The widths of spectral lines are affected by a number of processes. With no outside influences, a spectral line has a natural width that is very small. Doppler broadening, due to the thermal motion of the gas atoms, depends on the temperature of the gas. Collisional broadening occurs when the atoms emitting or absorbing photons collide with other atoms, ions, or electrons. The collisions disturb the atomic energy levels and slightly alter the wavelengths the atoms can absorb or emit. In a dense, hot gas, where the gas atoms collide often, collisional broadening is important.

NEW TERMS

electron

black body radiation

wavelength of maximum (λ_{max})

color index

nucleus

absorption spectrum (dark line spectrum)

emission line

emission spectrum (bright line spectrum)

Kirchhoff's laws

proton
neutron
isotope
ionization
ion
molecule
binding energy
quantum mechanics
energy level
excited atom
ground state
continuous spectrum
absorption line
transition
Lyman, Balmer, and Paschen series
spectral class or type
spectral sequence
blue and red shifts
radial velocity (V_r)
transverse velocity
Doppler effect
line profile
Doppler broadening
collisional broadening

QUESTIONS

1. Why can a good blacksmith judge the temperature of a piece of heated iron by its color?
2. Why do hot stars look bluer than cool stars?
3. Use black body radiation to explain why you could hold a cigarette between your lips and light it with a match, but you couldn't hold the cigarette in your lips while you light it by sticking the tip into a bonfire. The match and bonfire have about the same temperature. (This is just one of the hazards of smoking.)
4. What is the difference between a neutral atom, an ion, and an excited atom?
5. How do the energy levels in an atom determine which wavelength photons it can absorb or emit?
6. What kind of spectrum would you expect to record if you observed molten lava? In practice, to view molten lava you must look through gases boiling out of the lava. What kind of spectrum might you see in that case?
7. Why do the strength of the Balmer lines depend on the temperature of the star?
8. Why does a stellar spectrum tell us about the surface layers but not the deeper layers of the star?
9. Explain the similarities between Figure 7–15, Figure 7–16, and Table 7–1.
10. Why do we not see TiO bands in the spectra of hot stars?
11. How could Kirchhoff have identified sodium in the solar spectrum in the 1850s, long before physicists understood how atoms interact with light?
12. In observing a star's spectrum, we note that the lines of a certain element are not present. Are we safe in concluding that the star does not contain this element? Why or why not?
13. If a star moves exactly perpendicular to a line connecting it to the earth, will the Doppler effect change its spectrum? Why or why not?
14. Why would we expect a star that rotates very rapidly to have broad spectral lines? What else could broaden the lines?

PROBLEMS

1. Human body temperature is about 310 K (98.6°F). At what wavelength do humans radiate the most energy? What kind of radiation do we emit?
2. If a star has a surface temperature of 20,000 K, at what wavelength will it radiate the most energy?
3. Infrared observations of a star show that it is most intense at a wavelength of 2000 nm. What is the temperature of the star's surface?
4. If we double the temperature of a black body, by what factor will the total energy radiated per second per square meter increase?
5. If one star has a temperature of 6000 K and another star has a temperature of 7000 K, how much more energy per second will the hotter star radiate from each square meter of its surface?
6. Transition A produces light with a wavelength of 500 nm. Transition B involves twice as much energy as A. What wavelength light does it produce?
7. Determine the temperatures of the following stars based on their spectra. Use Figure 7–16.
   (a) Medium-strength Balmer lines, strong helium lines
   (b) Medium-strength Balmer lines, weak ionized calcium lines
   (c) TiO bands very strong
   (d) Very weak Balmer lines, strong ionized calcium lines
8. To which spectral classes do the stars in Problem 7 belong?
9. In a laboratory the Balmer beta line has a wavelength of 486.1 nm. If the line appears in a star's spectrum at 486.3 nm, what is the star's radial velocity? Is it approaching or receding?
10. The highest-velocity stars an astronomer might observe have velocities of about 400 km/sec. What change in wavelength would this cause in the Balmer gamma line? (HINT: Wavelengths are given in Figure 7–12.)
11. Use a ruler to measure the wavelength of the Balmer alpha line in Figure 7–19. What is the radial velocity of this star?

RECOMMENDED READING

ALLER, L. H. *Atoms, Stars, and Nebulae.* Cambridge, Mass.: Harvard University Press, 1971.

BOORSE, H., and L. MOTZ *The World of the Atom.* New York: Basic Books, 1966.

BOOTH, V. *The Structure of Atoms.* New York: Macmillan, 1964.

FEINBERG, G. *What Is the World Made of?* Garden City, N.Y.: Anchor Press/Doubleday, 1977.

GIBSON, E. G. *The Quiet Sun.* Washington, D.C.: NASA SP-303, 1973.

GOLDBERG, L. "Ultraviolet Astronomy." *Scientific American 220* (June 1969), p. 92.

HOLZINGER, J. R., and M. A. SEEDS *Laboratory Exercises in Astronomy.* Ex. 25, 28, 29. New York: Macmillan, 1976.

KALER, JAMES B. "Origins of the Spectral Sequence." *Sky and Telescope 71.* (Feb. 1986), p. 129.

PASACHOFF, J. M., J. L. LINSKY, B. M. HAISCH, and A. BOGGESS "IUE and the Search for a Lukewarm Corona." *Sky and Telescope 57* (May 1979), p. 438.

SNOW, T. P. "Ultraviolet Spectroscopy with Copernicus." *Sky and Telescope 54* (Nov. 1977), p. 371.

THACKERAY, A. D. *Astronomical Spectroscopy.* New York: Macmillan, 1961.

WELTHER, B. "Annie Jump Cannon: Classifier of Stars." *Mercury 13 (Jan./Feb. 1984), p. 28.*

CHAPTER 8

THE SUN—OUR STAR

A wit once remarked that solar astronomers would know a lot more about the sun if it were farther away. This contains a grain of truth; the sun is just a humdrum star and there are billions like it in the sky, but the sun is the only one close enough to show surface detail. Solar astronomers can see so much detail, swirling currents of gas and arching bridges of magnetic force, that present theories seem inadequate. Yet the sun is not a complicated body. It is just a star.

In their general properties, stars are very simple. They are great balls of hot gas held together by their own gravity. Their gravity would make them collapse into small, dense bodies were they not so hot. The tremendously hot gas inside stars has such a high pressure that the stars would surely explode were it not for their own confining gravity. Thus stars, like soap bubbles, are simple structures balanced between opposing forces that would destroy them.

Although stars are simple, and the sun is one of the simplest kinds of stars, we can see complex detail on the sun. The spots, eruptions, and storms visible on its face are important because they tell us what an average star is like. By understanding some of the complex processes we can see on the sun, we may better understand the more distant stars.

Another reason that this solar activity is important is that it affects the earth. The sun's atmosphere of very thin gas reaches out past the orbit of the earth, and thus any change in the sun can have a direct effect on the earth. Also, we get nearly all of our energy from the sun—oil and coal are merely stored sunlight—and our pleasant climate on earth is maintained by energy from the sun. Should the sun's energy output vary by even a small amount, life on earth might vanish.

All cannot live
on the piazza, but everyone
may enjoy the sun.
ITALIAN PROVERB

As we discuss the sun in this chapter, we will think of it in three ways. The detail we can see makes the sun an interesting body in itself, and we can enjoy it for its own complexity and beauty. However, the sun is also interesting because of its influences on earth. Finally, and most importantly, we will look at the sun as a representative star.

8.1 THE STRUCTURE OF THE SUN AND OTHER STARS

The sun is a spherical ball of hot gas about 1.4×10^6 km (860,000 miles) in diameter, making it about 109 times the diameter of the earth (Data File 1a). Roughly 75 percent of the sun's mass is hydrogen, 25 percent is helium, and a tiny fraction is heavier elements. (See Table 7–2). The surface of the sun has a temperature of about 5800 K—hot enough to vaporize iron. In all these respects the sun is characteristic of stars. A few are larger, most are smaller, some are poorer in heavy elements, and some are hotter, but they are all built from the same master blueprint.

To understand the structure of the sun and other stars, we must answer three questions. First, how does the sun manufacture its energy? Second, how does that energy get to the sun's surface? Finally, what keeps the sun stable—that is, why doesn't it expand or contract? These questions will lead our imaginations into a region where our bodies can never go—the heart of the sun.

SOLAR ENERGY GENERATION

The sun constantly radiates light and heat into space, so it would cool significantly in only a few thousand years if it could not replace the energy it loses. Somehow it has to make energy.

Not until the 1930s did astronomers realize how the sun generates its energy. The energy is produced deep in the interior of the sun where the temperature is so high the atoms have lost their electrons. The matter is a stormy sea of high-speed atomic nuclei and free electrons. The sun's energy is produced by fusion reactions that occur when the nuclei collide. These reactions, much like those that occur in hydrogen bombs, are called fusion reactions because they fuse nuclei together.

In the case of the sun, the reactions fuse four hydrogen nuclei to make a single helium nucleus. Because one helium nucleus has 0.7 percent less mass than four hydrogen nuclei, it seems that some mass vanishes in the process.

$$\begin{aligned} \text{4 hydrogen nuclei} &= 6.693 \times 10^{-27}\ \text{kg} \\ \text{1 helium nucleus} &= 6.645 \times 10^{-27}\ \text{kg} \\ \hline \text{difference in mass} &= 0.048 \times 10^{-27}\ \text{kg} \end{aligned}$$

However, this mass does not actually vanish; it merely changes form. The equation $E = m_0c^2$ reminds us that mass and energy are related, and under certain circumstances, mass may become energy and vice versa. Thus the 0.048×10^{-27} kg do not vanish, but merely become energy. To see how much, we use Einstein's equation:

$$\begin{aligned} E &= m_0c^2 \\ &= (0.048 \times 10^{-27}\ \text{kg})\,(3 \times 10^8\ \text{m/sec})^2 \\ &= 0.43 \times 10^{-11}\ \text{J} \end{aligned}$$

This a very small amount of energy, hardly enough to raise a housefly one-thousandth of an inch. Because one reaction produces such a small amount of energy, it is obvious that many reactions are necessary to supply the energy needs of a star. The sun, for example, needs 10^{38} reactions per second, transforming 5 million tons of mass into energy every second, just to stay hot enough to resist its own gravity.

Two nuclei can fuse only if they come close together, but atomic nuclei have positive charges and repel each other with an electrostatic force. To overcome this force, atomic nuclei must collide at high velocity. If the particles in a material are moving fast, we say that the material is hot. Thus, atomic fusion reactions can occur only if the gas is very hot—at least 10^7 K.

Fusion reactions also require that the gas be very dense—denser than solid lead. We know that the sun requires 10^{38} reactions per second to manufacture sufficient energy. But fusion occurs in only a small percent of all collisions, so the sun requires many collisions between

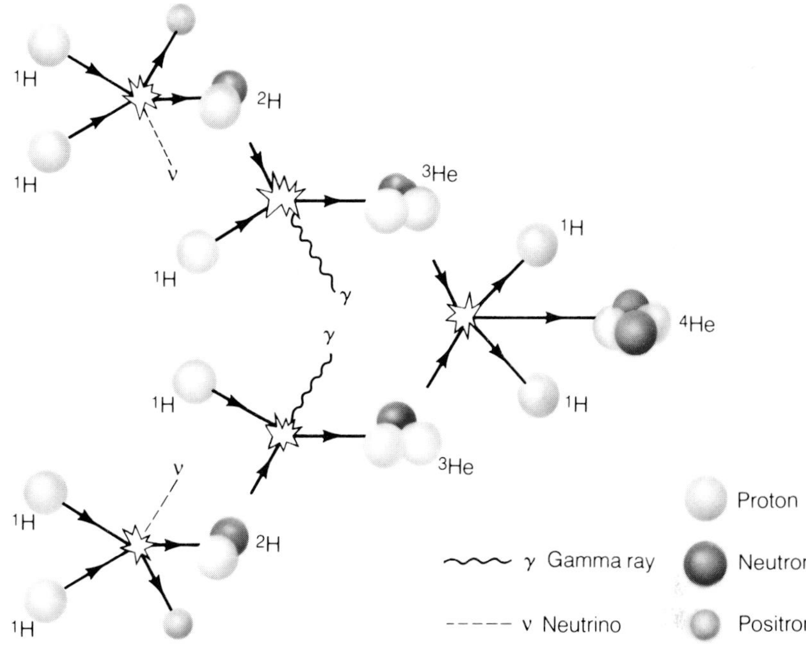

Figure 8–1 The proton–proton chain combines four protons (at left) to produce one helium nucleus (at right) plus energy.

nuclei each second. Only where the gas is very dense are there enough collisions to meet the sun's energy needs.

We can symbolize this process with a simple nuclear reaction:

$$4\,{}^{1}H \rightarrow {}^{4}He + \text{energy}$$

In this equation, ^{1}H represents a proton, the nucleus of the hydrogen atom, and ^{4}He represents the nucleus of a helium atom. The superscripts indicate the approximate weight of the nuclei (the number of protons plus the number of neutrons). The actual steps in the process are more complicated than this convenient summary suggests. Instead of waiting for four hydrogen nuclei to collide simultaneously, a highly unlikely event, the process can proceed step by step in a chain of reactions—the proton–proton chain.

The **proton–proton chain** is a series of three nuclear reactions that builds a helium nucleus by adding together protons. This process is efficient at temperatures above 10,000,000 K. The sun, for example, manufactures over 90 percent of its energy in this way.

The three steps in the proton–proton chain entail these reactions:

$$
\begin{aligned}
{}^{1}H + {}^{1}H &\rightarrow {}^{2}H + e^{+} + \nu \\
{}^{2}H + {}^{1}H &\rightarrow {}^{3}He + \gamma \\
{}^{3}He + {}^{3}He &\rightarrow {}^{4}He + {}^{1}H + {}^{1}H
\end{aligned}
$$

In the first step, two hydrogen nuclei (two protons) combine to form a heavy hydrogen nucleus, emitting a particle called a positron (a positively charged electron) and a **neutrino** (a subatomic particle originally thought to have zero mass and to travel at the velocity of light). In the second reaction, the heavy hydrogen nucleus absorbs another proton, and, with the emission of a gamma ray, becomes a lightweight helium nucleus. Finally, two light helium nuclei combine to form a common helium nucleus and two hydrogen nuclei. Because the last reaction needs two ^{3}He nuclei, the first and second reactions must occur twice (Figure 8–1). The net result of this chain reaction is the transformation of four hydrogen nuclei into one helium nucleus plus energy.

DATA FILE 1

The Sun

Average Distance from the earth	1.00 AU (1.495979×10^8 km)
Maximum Distance from the earth	1.0167 AU (1.5210×10^8 km)
Minimum Distance from the earth	0.9833 AU (1.4710×10^8 km)
Average diameter seen from the earth	0.53° (32 minutes of arc)
Period of rotation	25 days at equator
Radius	6.9599×10^5 km
Mass	1.989×10^{30} kg
Average density	1.409 g/cm^3
Escape Velocity at surface	617.7 km/sec
Luminosity	3.826×10^{26} J/sec
Surface temperature	5800 K
Central temperature	15×10^6 K
Spectral type	G2 V
Apparent visual magnitude	−26.74
Absolute visual magnitude	4.83

Convective zone
Radiative zone
Core energy generation
Moon
Earth

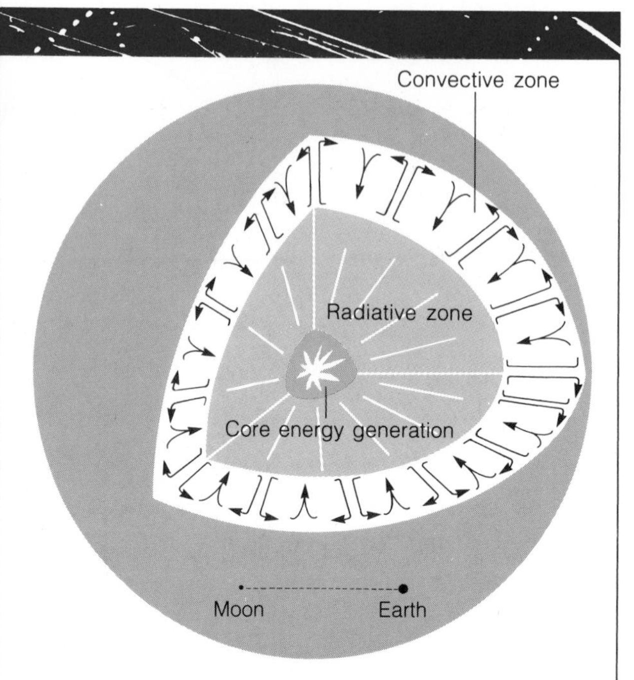

a. The sun and its interior drawn to the same scale as the earth-moon system.

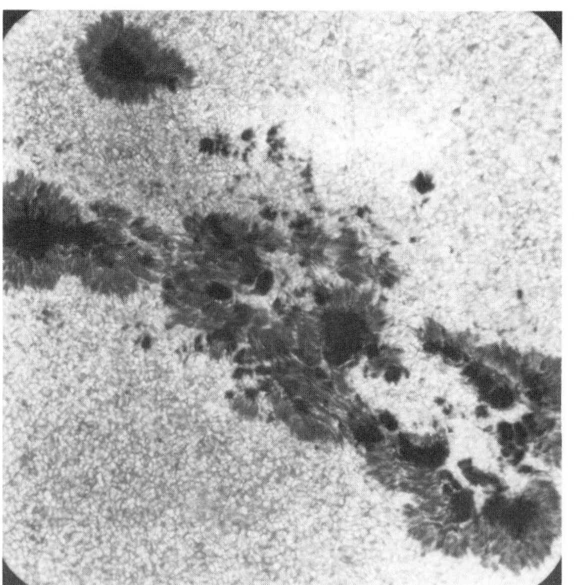

b. Sunspots are cooler and darker than the photosphere. Note the granulation in the photosphere. (NOAO.)

c. Photographed at the wavelength of the hydrogen Balmer alpha line, the surface of the sun is blotched by active regions and dark filaments. (©1979 Sacramento Peak Observatory.)

d. A prominence at the edge of the sun consists of ionized gases trapped in the sun's magnetic field. (©Sacramento Peak Observatory.)

The energy appears in the form of gamma rays, positrons, the energy of motion of the nuclei, and neutrinos. The gamma rays are photons that are absorbed by the surrounding gas before they can travel more than a few centimeters. This heats the gas and helps maintain the pressure. The positrons produced in the first reaction combine with free electrons and both particles vanish, converting their mass into gamma rays. Thus, the positrons also help keep the center of the star hot. In addition, when fusion reactions produce new nuclei, they fly apart at high velocity. This energy of motion helps raise the temperature of the gas. The neutrinos, however, resemble photons except that they almost never interact with other particles. The average neutrino could pass unhindered through a lead wall 1 light-year thick. Thus, the neutrinos do not help heat the gas, but race out of the star at the speed of light, carrying away roughly 2 percent of the energy produced.

The sun's energy comes almost entirely from the proton–proton chain, but other kinds of stars manufacture their energy using other fusion reactions. Stars more massive than the sun fuse hydrogen using a different chain of reactions, and some stars fuse helium and heavier elements. We will discuss these reactions in Chapter 11.

The sun's supply of hydrogen fuel is not unlimited. It has been converting hydrogen into helium for about 5 billion years, and thus helium is more abundant in the sun's core where the reactions occur than near the surface. As this helium ash accumulates in the solar core and as the hydrogen there is used up, the sun must change. Eventually, in about 5 billion years, the fuel will be exhausted and the sun will die, and of course so will life on earth. This too is a universal trait of stars. Nothing lasts forever, not even the stars.

Figure 8–2 The solar neutrino experiment consists of 100,000 gallons of cleaning fluid held in a tank nearly a mile underground. Solar neutrinos trapped in the cleaning fluid convert chlorine atoms into argon atoms that can be counted by their radioactivity. (Brookhaven National Laboratory.)

THE SOLAR NEUTRINO MYSTERY

We cannot see into the sun's interior, so we cannot directly confirm this theoretical description of the sun's energy source. But there is one way to measure the rate at which the nuclear fires are burning in the sun's core—by trapping neutrinos.

During the moment it takes to read this sentence, approximately 10^{12} solar neutrinos pass through your body. If we could catch and study some of these particles, we could learn about conditions at the sun's center. But because neutrinos almost never react with other particles, they are very difficult to detect.

Chemist Raymond Davis, Jr. devised an experiment to count solar neutrinos. He knew that some of the neutrinos coming from the interior of the sun had the proper energy to transform chlorine nuclei into radioactive argon nuclei. Because the neutrinos only rarely interact with chlorine nuclei, Davis had to include a great many chlorine nuclei in his experiment, so he filled a tank with 100,000 gallons of a cleaning fluid rich in chlorine atoms (Figure 8–2). By counting the number of radioactive argon atoms that appeared in the tank and by knowing the probability that a neutrino would change a chlorine nucleus into an argon nucleus, Davis measured the number of neutrinos coming from the sun's interior.

The tank of cleaning fluid and all of its associated equipment are buried nearly a mile deep in a South Dakota gold mine flooded with water to protect the chlorine from cosmic rays. A cosmic ray could produce a radioactive argon nucleus and thus distort the count of solar neutrinos.

From its inception in 1970, this solar neutrino experiment has generated one of the biggest controversies in twentieth-century astronomy. The tank catches only about one-third as many neutrinos as predicted by theory. Davis

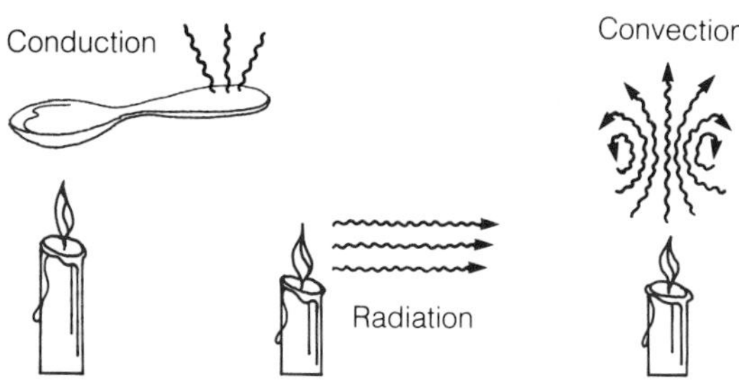

Figure 8–3 Conduction, radiation, and convection are the principal modes of energy transport within a star.

has calibrated the equipment with great care, and others have repeated the theoretical calculations, but the discrepancy remains. How do we explain the missing neutrinos?

The most obvious suggestion is that the sun's center is not as hot as we think. But if the central temperature is lower, the nuclear reactions would not produce as much energy, and the sun would begin to contract. Some measurements of the sun's radius suggest that it may be contracting very slowly. If so, the sun may be temporarily producing less nuclear energy and thus fewer neutrinos. However, no one knows why the sun's nuclear reactions should fluctuate in this way, and most astronomers do not accept this explanation.

A number of explanations have been proposed. One theory suggested that neutrinos oscillate among three different types and that Davis's tank can detect only one type. This has been a very popular proposal, but recent attempts to detect neutrino oscillation have failed, and studies of the neutrinos from the supernova that was discovered in February 1987 (Chapter 13) have also raised doubts about neutrino oscillation.

Another explanation proposed that weakly interacting massive particles (WIMPs) could spread the thermal energy outward from the core of the sun. This would lower the central temperature and thus explain the small number of solar neutrinos. The total amount of energy the sun generates would not fall, because the nuclear reactions would occur over a larger part of the sun's interior. WIMPs are predicted by certain theories of subatomic particle structure but have never been detected in the laboratory. We will discover later (page 163) that other evidence suggests WIMPs exist in the sun.

The solar neutrino problem is one of the great controversies of modern astronomy because it is a direct conflict between observation and theory. Whether neutrino oscillations prove to be the solution or not, we need better measurements of the rate at which the sun is producing neutrinos. Raymond Davis and others have proposed a new detector that would use tons of the rare metal gallium, but funding ($15 to $25 million) was recently denied. Soviet astronomers are building a 50-ton gallium detector, and Europeans are planning a similar detector in Italy. Detectors using argon or heavy water are also being planned. In the next few years two new neutrino detectors may be in operation, giving us a better look into the regions of solar energy generation.

ENERGY TRANSPORT

The sun and other stars generate nuclear energy in their deep interiors, but that energy must flow outward to their surfaces to replace the energy radiated into space as light and heat. If this outward flow of energy in the sun were suddenly shut off, the sun's surface would gradually cool and dim, and the sun would begin to shrink. Thus, stars can exist only so long as energy can move from their cores to their surfaces. In the material of which stars are made, energy may move by conduction, radiation, or convection.

Conduction is the most familiar form of heat flow. If you hold the bowl of a spoon in a candle flame, the handle of the spoon grows warmer. Heat, in the form of motion of the particles in the metal, is conducted from molecule to molecule up the handle, until the molecules of metal under your fingers begin to move faster and you sense heat (Figure 8–3). Thus, conduction requires close contact between the molecules. Because matter in most stars is gaseous, conduction is unimportant. Conduction is significant only in peculiar stars that have tremendous internal densities.

The transport of energy by radiation is another familiar experience. Put your hand beside a candle flame, and you can feel the heat. What you actually feel are infrared photons radiated by the flame (Figure 8–3). Because photons are packets of energy, your hand grows warm as it absorbs them.

Radiation is the principal means of energy transport in the sun's interior. Photons are absorbed and reemitted in random directions over and over as they work their way outward. It takes about 1 million years for the average photon to travel from the sun's center to its surface.

The flow of energy by radiation depends on how difficult it is for the photons to move through the gas. If the gas is cool and dense, the photons are more likely to be absorbed or scattered, and thus the radiation does not penetrate the gas very well. We would call such a gas opaque. In a hotter, lower-density gas, the photons can penetrate more easily; such a gas is less opaque. The **opacity** of a gas—its resistance to the flow of radiation—depends strongly on its temperature.

If the opacity of a gas is high, radiation cannot flow through it easily. Energy moving outward from the center of a star first moves through the hotter gas of the deep interior. Because the gas is so hot, it is transparent, and the energy moves as radiation. But the outer portions of the star are cooler and therefore more opaque. Like water behind a dam, energy builds up raising the temperature until the gas begins to churn. Hot gas, being less dense, rises, and cool gas, being more dense, sinks. This is convection, the third way energy can move in a star.

Convection is a common experience; the wisp of smoke rising above a candle flame travels upward in a small convection current (Figure 8–3). If you hold your hand above the flame, you can feel the rising current of hot gas. In stars, energy may be carried upward by rising currents of hot gas hundreds or thousands of miles in diameter.

Convection is important in stars because it carries energy and mixes the gas. Convection currents flowing through the layers of a star tend to homogenize the gas, giving it a uniform composition throughout the convective zone. As you might expect, this mixing affects the fuel supply of the nuclear reactions, just as the stirring of a campfire makes it burn more efficiently.

The convection in the sun stirs a zone about 200,000 km deep, just below the visible surface (Data File 1a). Below that, the energy moves as radiation, and the gas does not get mixed. Deep in the interior of the sun there is no convection. Hydrogen nuclei are fusing to form helium, and nothing removes the helium ashes or brings fresh hydrogen down into the core. Thus, the sun is like a great pot of mashed potatoes burning at the bottom and stirred only slightly at the top. We will see in later chapters that most stars have interiors like the sun's.

WHAT SUPPORTS THE SUN?

From its surface to its interior, the sun is gaseous. On the earth a puff of gas, vapor from a smokestack perhaps, dissipates rapidly, driven by the motion of the air and by the random motions of the atoms in the gas. However, the sun differs from a puff of gas in a very important way—its mass. The sun is over 300,000 times more massive than the earth, and that mass produces a tremendous gravitational field that draws the sun's gases into a sphere.

With such a strong gravitational field it might seem that the gases of the sun should compress into a tiny ball, but a second force balances gravity and prevents the sun from shrinking. The gas of which the sun is made is quite hot, and because it is hot, it has a high pressure. The pressure pushes outward; indeed, without the restraint of the sun's gravity, it would blast the sun apart. Thus, the sun is balanced between two forces—gravity trying to squeeze it tighter and gas pressure trying to make it expand.

To discuss the forces inside the sun, we can imagine that the sun's interior is divided into concentric shells like those in an onion. We can then discuss the temperature, density, pressure, and so on in each shell. Keep in mind, however, that these helpful shells do not really exist. The sun and stars are not composed of separable layers. The layers are only a convenience in our discussion.

The gravity–pressure balance that supports the sun is a fundamental part of stellar structure known as the law of **hydrostatic equilibrium.** It says that, in a stable star like the sun, the weight of the material pressing downward on a layer must be balanced by the pressure of the gas in that layer. *Hydro* implies we are discussing a fluid—the gases of the star. *Static* implies that the fluid is stable—neither expanding nor contracting.

The law of hydrostatic equilibrium can prove to us that the interior of the sun must be very hot. Near the sun's surface there is little weight pressing down on the gas, so the pressure must be low, implying a low temperature. But as we go deeper into the sun, the weight becomes larger, so the pressure, and therefore the temperature, must also increase (Figure 8–4). Near the sun's surface the temperature is only about 5800 K, but at the sun's center the temperature is believed to be about 15,000,000 K.

Of course, the interior of the sun is kept hot by the nuclear reactions occurring at the core, and the outward flow of energy keeps each layer in the sun hot enough to support the weight pressing down from above. In a sense, the sun is supported by the flow of energy from its center to its surface. Turn off that energy, and gravity would gradually force the sun to collapse into its center.

HELIOSEISMOLOGY

The center of the sun is forever hidden from our sight, but we are beginning to explore the solar interior through **helioseismology**, the study of the way the sun vibrates.

Just as geologists can study the earth's interior by observing how sound waves produced by earthquakes are reflected and transmitted by the layers of the earth's

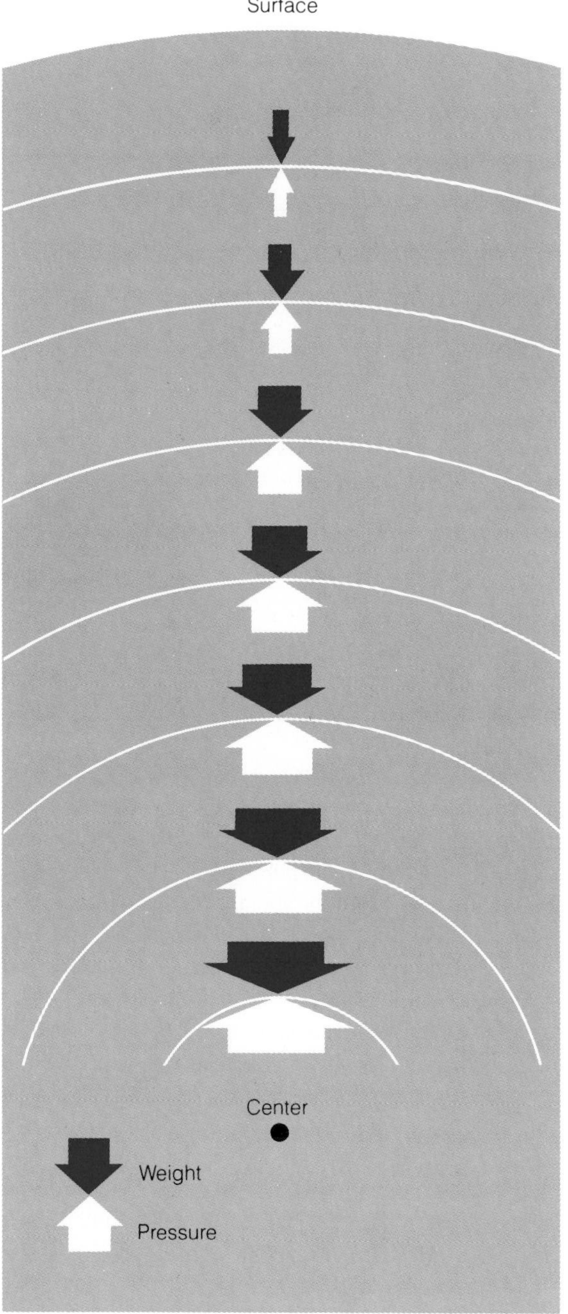

Figure 8–4 The law of hydrostatic equilibrium says the pressure in each layer must balance the weight on that layer. As a result, pressure and temperature must increase from the surface of a star to its center.

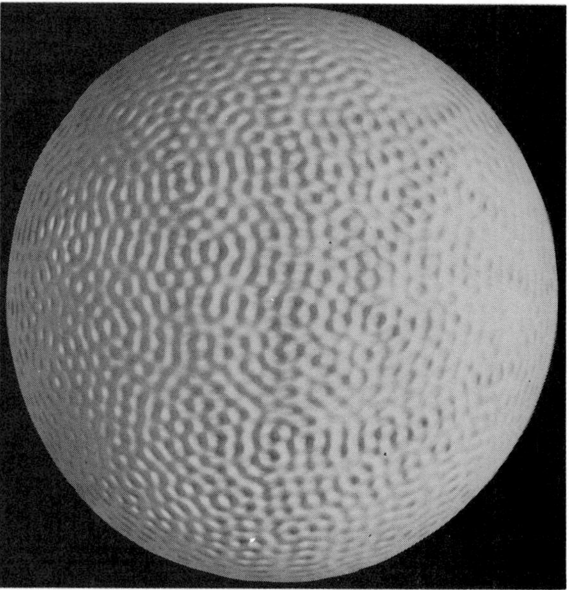

Figure 8–5 Like a ringing bell, the sun vibrates in as many as 10 million different modes. For each mode, the peaks and valleys of the vibration are spaced differently over the solar surface. This simulation shows the complex pattern of surface displacement caused by the combination of 100 possible modes. By studying how such vibrations travel through the sun's interior, solar astronomers can explore the sun's inner layers just as seismologists can explore the earth's interior. (David Hathaway/NASA.)

interior, so too can helioseismologists explore the sun's interior. We can't put detectors on the sun, but we can measure the motions of the sun's surface using the Doppler effect. As the vibrations pass through the sun's interior, the solar surface we see—the photosphere—moves up and down in a complicated pattern (Figure 8–5 and Color Plate 11). The Doppler shifts caused by the moving surface, although very small, can tell solar astronomers which frequency vibrations are present. By comparing these with models of the sun, astronomers can determine the structure and motions of the sun's internal layers.

These studies have shown that the standard model of the sun's interior is wrong in some details. The interior seems to be slightly cooler than previously thought, and the chemical composition may be slightly richer in helium. Also, the core of the sun seems to rotate faster than the exterior, and very large currents of gas move in banana-shaped giant cells oriented north-south deep inside the convective region.

Some of these discoveries are especially exciting because they relate to the solar neutrino problem. Any suggestion that the sun's center is cooler than supposed would help explain the deficiency in observed neutrinos. In addition, recent studies suggest that the presence of WIMPs in the sun would not only lower the sun's central temperature (and thus explain the missing neutrinos) but also would explain the vibration frequencies observed in the sun.

At least two teams of astronomers are building observatories specifically for helioseismology. As these begin to produce data and as computers become faster and better able to analyze large masses of data, we may be able to draw a detailed map of the sun's interior.

When we look at the sun, we see nothing of the interior. We look down through low-density gas layers to the photosphere. These layers—the sun's atmosphere—are complex, fascinating, and dramatically beautiful.

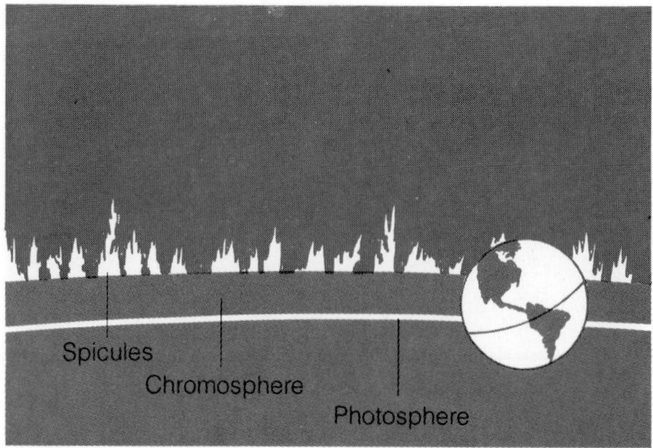

Figure 8–6 A cross section at the edge of the sun shows the relative thickness of the photosphere and chromosphere. The photosphere is only a few hundred kilometers thick. Above it is the lower-density chromosphere. The earth is shown for scale.

8.2 THE SOLAR ATMOSPHERE

The sun's atmosphere is a layer of low-density gas extending from the visible surface out to about 5×10^6 km. There is no sharp upper boundary. Rather, the upper layers blend into the breeze of gas flowing away from the sun and past the planets. Thus, we could say that the sun's atmosphere extends to envelope the earth.

THE PHOTOSPHERE

The visible surface of the sun, the photosphere, is a layer of gas only about 500 km deep from which we receive most of the sun's light. Below the photosphere, the gas is denser and hotter and therefore radiates more light. However, that light cannot escape from the sun because of the outer layers of gas. Thus, we cannot detect light directly from these deeper layers. Above the photosphere, the gas is less dense and thus is unable to radiate much light. The photosphere is the layer in the sun's atmosphere that is dense enough to emit plenty of light, but of low enough density to allow the light to escape.

The photosphere is actually a very narrow layer. If the sun were to shrink to the size of a bowling ball, the photosphere would be no thicker than a layer of tissue paper wrapped without wrinkles around the ball (Figure 8–6).

One reason the photosphere is so shallow is related to the hydrogen atom. Because the temperature of the photosphere is sufficient to ionize some atoms, there is a large number of free electrons in the gas. Neutral hydrogen atoms can add an extra electron and become an H^- (H minus) ion, but this extra electron is held so loosely that almost any photon has energy enough to free it. In the process, of course, the photon is absorbed. Thus the H^- ions are very good absorbers of photons and make the gas of the photosphere very opaque. Light from below cannot escape easily, and we see a well-defined surface—the thin photosphere.

Most of the light we see comes from a region of the photosphere with a temperature of about 6000 K, roughly the temperature of a hot welding arc. We receive smaller amounts of light from deeper regions of the photosphere where the temperature is about 8000 K and from higher regions where the temperature is about 4000 K. The light we see coming from the sun is therefore a mixture of photons emitted by gases at various temperatures.

Although the photosphere appears to be substantial, it is really a very low-density gas. The density in the middle of the photosphere is only 0.1 percent that of air at sea level. This is a rather good vacuum. To find gases as dense as the air we breathe, we would have to descend about 48,000 km below the photosphere, about 7 percent of the way to the sun's center. If someone could invent a fantastically efficient insulation, we could fly a spaceship right through the photosphere.

Good photographs of the photosphere show that it is not uniform, but mottled by a pattern of bright cells called **granulation** (Data File 1b). Each granule is about 1500 km in diameter—slightly larger than Texas—and is separated from its neighbors by a dark boundary. A granule lasts for about 10 minutes before it dissipates or merges with neighboring granules. Measurements of Doppler shifts show that the centers of granules are rising gas slightly hotter than the sinking gas in the dark boundary regions.

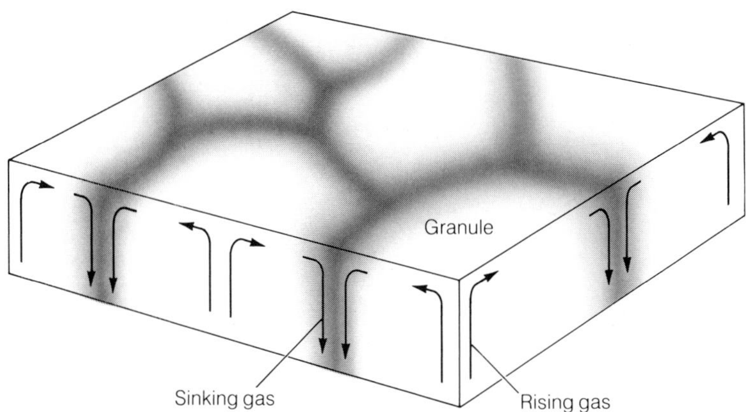

Figure 8–7 Rising convection currents in the sun heat areas of the surface, producing bright granules. Cooler material sinks at the darker edges of the granule.

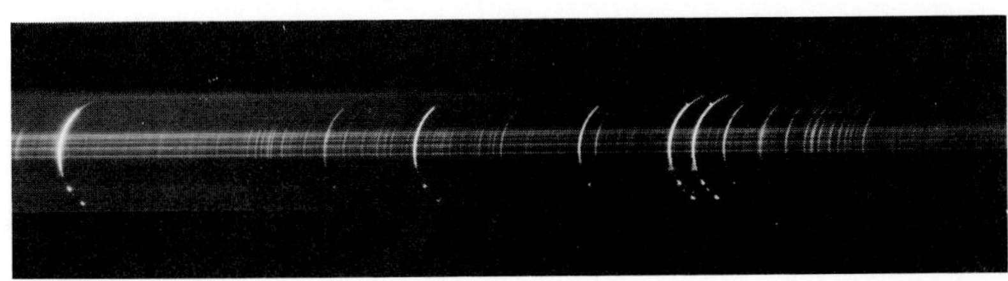

Figure 8–8 The flash spectrum of the chromosphere shows emission lines of high ionization. The lines are curved because the thin crescent of chromosphere acts as the slit of the spectrograph. (Mount Wilson and Las Campanas Observatories, Carnegie Institution of Washington.)

Solar astronomers believe the granules are the tops of rising currents in the sun's convection zone (Figure 8–7). Rising currents of hot gas reach the photosphere and heat it, making it slightly brighter above the rising gas. As the gas cools, it sinks, and the region above the sinking gas is slightly darker because it is slightly cooler. The granulation in the photosphere confirms that the sun's outer layers are convective.

Recent spectroscopic studies of the solar surface have revealed another kind of granulation. **Supergranules** are regions about 30,000 km in diameter (about 2.3 times the diameter of the earth) and include about 300 granules. These supergranules are regions of very slowly rising currents that last about a day. They may be the surface traces of larger convection cells deeper under the photosphere. Observations based on helioseismology suggest even larger cells, called giant cells, deeper in the convection zone.

THE CHROMOSPHERE

Above the photosphere lies a nearly invisible layer of gas about 10,000 km thick (Figure 8–6). It is about 1000 times fainter than the photosphere, so we can see it only during a total solar eclipse when the moon covers the brilliant photosphere. Then, for a few seconds, the chromosphere flashes into view as a thin line of pink just above the photosphere. The term *chromosphere* comes from the Greek word *chroma* meaning "color."

If astronomers observe the spectrum of the sun during a total solar eclipse, they see the usual absorption spectrum so long as any part of the photosphere is visible. However, when the moon covers the photosphere, the chromosphere is visible for a few seconds extending beyond the edge of the moon in a thin crescent, and the solar absorption spectrum suddenly flashes into an emission spectrum produced by the hot, low-density gas of the chromosphere (Figure 8–8). The lines of this **flash spectrum** curve because the spectrograph does not contain a slit to isolate the light but merely uses the narrow crescent of chromosphere. The lines in the emission spectrum are generally the same as those in the solar absorption spectrum, with the addition of the lines of un-ionized helium, which are normally invisible in the sun's spectrum.

As the moon covers more and more of the chromosphere, the lines in the flash spectrum change in peculiar ways. Let's analyze these changes in detail. They will not only reveal the structure of the sun's atmosphere, but

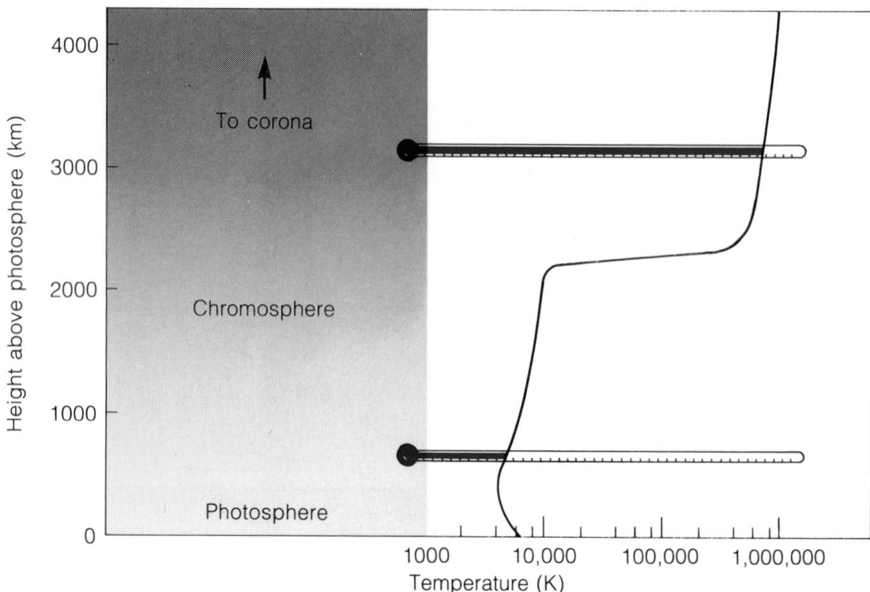

Figure 8–9 If we could place thermometers in the sun's atmosphere, we would discover that the temperature increases from 6000 K at the photosphere to 10^6 K in the corona. Not shown in this figure are the irregularities in the chromosphere (such as spicules), which extend as high as 12,000 km above the photosphere.

also the analysis will illustrate how we can extract information from a spectrum if we know how atoms interact with light.

The presence of emission lines in the flash spectrum is itself important. Recall from Figure 7–8 our example of a gas cloud surrounding a light bulb. There we saw an absorption spectrum when we looked at the bulb through the gas, and an emission spectrum when we looked at the gas alone. In the case of the sun, the photosphere plays the role of the glowing bulb, and the chromosphere plays the part of the gas cloud. The atoms in the chromosphere absorb photons as they leave the photosphere below. This forms the absorption lines we see when we look at the photosphere. But these same atoms emit photons in random directions. When the moon blots out the brighter photosphere, we can see the fainter light emitted by the atoms of the chromosphere. Thus, we see emission lines where we saw absorption lines before. This accounts for the beautiful color of the chromosphere—it is glowing like a giant neon sign. (See Color Plate 3.)

When the flash spectrum first appears, we receive light from the top, middle, and bottom of the chromosphere. Because the bottom is brightest, it dominates the spectrum. In the spectrum of this lowest layer, we see the Balmer lines plus lines of neutral helium. For neutral helium to radiate, the temperature must be at least 10,000 K. But if it is much higher, the Balmer lines would be weak because the hydrogen atoms would be excited to higher energy levels. So the temperature of the lowest layers of the chromosphere must be about 10,000 K.

When the moon conceals these lower layers, we can see the emission of the middle layers dominating the flash spectrum. The Balmer lines start to fade, and lines of ionized helium appear. To ionize helium, the temperature must be at least 20,000 K, so this tells us the temperature at this layer. To confirm our estimate of the temperature, we also find lines of ionized iron and titanium that require high temperatures.

When the moon moves on and covers all but the top of the chromosphere, we see weak lines of very highly ionized atoms such as calcium, iron, and strontium. One line, for example, is produced by iron atoms that have lost 13 electrons. The temperature must be very high indeed to produce such extreme ionization.

Notice the similarity between our analysis of the flash spectrum and our classification of stars in the previous chapter. In each case, we used the excitation of different elements to judge temperature.

The flash spectrum, analyzed in this way, can tell us the temperature at each layer in the sun's atmosphere (Figure 8–9). At the photosphere the temperature is about 6000 K. It decreases slightly as we go upward, reaching a minimum of about 4000 K just above the photosphere. Above that, in the chromosphere, the temperature increases rapidly to 1,000,000 K at a height of 10,000 km, the beginning of the corona.

The flash spectrum can even tell us how dense the sun's atmosphere is. The emission lines fade away not

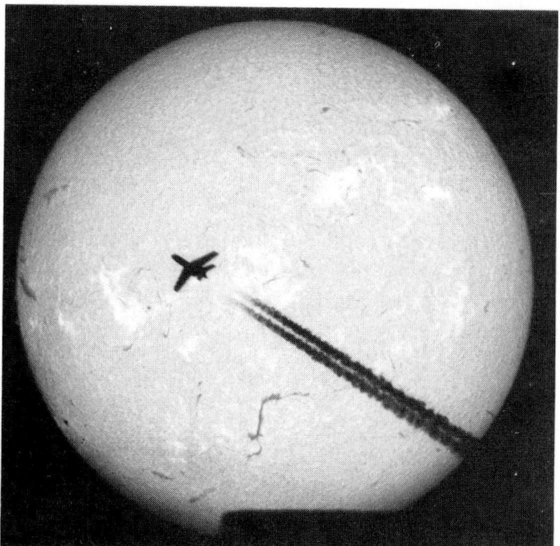

Figure 8–10 An H-alpha filtergram is a photograph recording photons at the wavelength of the Balmer alpha line of hydrogen. Such photons originate in the chromosphere, so filtergrams reveal structure in the chromosphere. This H-alpha filtergram also recorded a passerby in the earth's atmosphere. (Pennsylvania State University.)

Figure 8–11 Spicules are flamelike spikes extending up to 12,000 km above the photosphere. (©Association of Universities for Research in Astronomy, Inc., Sacramento Peak Observatory.)

only because the temperature increases with height, but also because the density of the gas decreases. Near the photosphere the gas is rather dense, only about 10^4 times thinner than the air we breathe, but at the top of the chromosphere it is nearly a vacuum, about 10^{13} times thinner than air.

Although the chromosphere is not visible to the naked eye outside of solar eclipses, it can be photographed if special filters are used to admit only those photons easily absorbed by certain atoms and ions. Such photographs are called **filtergrams.** An H-alpha filtergram (Figure 8–10), for instance, is formed by photons with wavelengths in the Balmer alpha line of hydrogen. Because hydrogen can absorb these photons so readily, they cannot have come from deep in the chromosphere. Photons with these wavelengths could only have escaped from the upper layers of the chromosphere. Thus, the H-alpha filtergram shows detail in the uppermost layers of the chromosphere. By tuning the filter to the wavelengths of photons slightly less likely to be absorbed, the solar astronomer can photograph different depths in the solar atmosphere.

Filtergrams of the chromosphere reveal **spicules**—flamelike structures 100–1000 km in diameter extending up to 12,000 km above the photosphere and lasting from 5 to 15 minutes (Figures 8–6 and 8–11). These spicules appear to be cool regions (about 10,000 K) extending up into the much hotter corona (about 500,000 K). Seen at the edge of the solar disk, these spicules blend together and look like flames covering a burning prairie, but filtergrams of spicules located near the center of the solar disk show that they spring up around the edges of supergranules like weeds around flagstones. Some astronomers suggest that the spicules are channels through which energy flows from the convective zone into the corona.

THE CORONA

The sun's atmosphere extending above the chromosphere is termed the corona after the Greek word for "crown." Although these outermost layers of the sun are far removed from the sun's surface, they are closely coupled with the solar convective zone and events in the chromosphere.

The corona is visible to the naked eye only during total solar eclipses when the moon covers the bright photosphere (Figure 3–14). Then the corona shines with a milky glow not quite as bright as the full moon. Eclipse photographs taken from the ground can trace the corona out to a distance of about 10 solar radii, and photographs taken from high-flying balloons or aircraft can trace the corona out to 30 solar radii.

By using a special telescope called a **coronagraph**, earthbound astronomers can see the corona at times when there are no eclipses. The coronagraph uses a disk to cover the brilliant photosphere and light baffles to reduce scattered light in the telescope. Also, coronagraphs are generally placed on high mountains to reduce light scattered from earth's atmosphere. Good coronagraphs can detect the corona out to about 1.3 solar radii.

The spectrum of the corona consists of a continuous spectrum with superimposed emission lines. The continuous spectrum is produced by sunlight scattered from dust and free electrons in the corona. Because of the very high temperatures (about 1,000,000 K), the electrons travel at high velocities and the resulting Doppler shifts smear out any absorption lines in the sunlight to produce a continuous spectrum. The emission lines are produced by very low-density, highly ionized gases. In the lower corona, atoms such as twice-ionized oxygen emit photons, but in the outer corona, the atoms are more highly ionized. Emission from these ions is clear evidence of low density and high temperature.

The temperature in the corona rises as we travel outward. Just above the chromosphere, in the region called the transition region, the temperature increases 500,000 K in only 300 km. In the lower corona, the temperature is about 500,000 K and in the outer corona it may be as high as 3,500,000 K. The density of this gas must be very low indeed or it would emit a great deal of light. In fact, the density of the outer corona is only about 1 to 10 atoms/cm^3.

How can the corona be hotter than the photosphere? This question has puzzled solar astronomers for many years. If we could safely visit the sun, we might hear (and feel) an extremely low-pitched rumble caused by the turbulence in the convection zone; most solar astronomers assumed that these sound waves were the source of heat in the corona. As the sound waves traveled upward through the photosphere and chromosphere, the falling density of the gas would convert them into **shock waves** (the astrophysical equivalent of sonic booms). Energy from the shock waves would agitate the gas atoms, thus raising the temperature.

In the 1970s astronomers found reason to doubt the sound wave hypothesis. Theoretical calculations suggested that the shock waves would dissipate in the lower chromosphere and never reach the corona. Observations made from space at ultraviolet and X-ray wavelengths showed no shock waves in the lower corona, but they did reveal that many stars had emission lines in their spectra, which could only have come from coronae. Even the K and M main-sequence stars, which have very gentle convection and should not generate shock waves, have hot coronae.

A newer theory supposes that the corona is heated by the sun's magnetic field. As we will see later in this chapter, the sun's magnetic field is very complicated and can store large amounts of energy. Apparently some of that energy heats the corona.

The hot gases of the corona blow away from the sun as the **solar wind**—the moving outer extension of the corona. It contains mostly ionized hydrogen (protons with free electrons) but also heavier elements and a trapped magnetic field. Containing only a few particles per cubic centimeter, it blows past the earth at about 400 km/sec and interacts with earth's magnetic field in complex ways. (See Chapter 20.)

Because of the gas carried away by the solar wind, the sun is slowly losing mass. This is a minor loss for the sun, amounting to some 10^7 tons per year, only about 10^{-14} of a solar mass per year. Other stars, however, at other stages in their lives, can lose mass rapidly.

Do other stars have chromospheres, coronae, and stellar winds like the sun's? Ultraviolet spectra taken by orbiting space telescopes such as the IUE (Figure 6–25) suggest that the answer is yes. The spectra of many stars contain emission lines in the far ultraviolet that could have been formed only in the low-density, high-temperature gases of the upper chromosphere and lower corona. Thus the sun, for all its complexity, seems to be a normal star.

8.3 SOLAR ACTIVITY

So far we have described the sun as if it were a static, unchanging object. But, in fact, the sun is a highly variable body whose appearance is constantly changing. The most obvious features of this solar activity are the sunspots.

SUNSPOTS

A **sunspot** is a cool, dark area of the solar surface (Figure 8–12). The center of the spot, called the umbra, is darker than the outer border, the penumbra. The average spot is about twice the diameter of the earth and may last for a week or so. Sunspots tend to form in groups, and a large

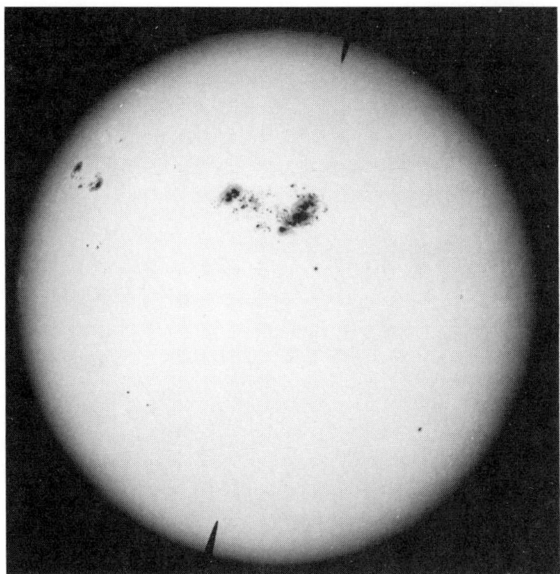

Figure 8–12 Sunspots look like dark areas on the solar disk because they are about 1500 K cooler than the photosphere. Although still very hot, the cooler spots radiate less light and look dark in comparison to the photosphere. Note that sunspots tend to occur in large groups containing two concentrations of spots. (Mount Wilson and Las Campanas Observatories, Carnegie Institution of Washington.)

group may contain up to 100 individual spots and may last as long as 2 months or more.

Observing the sun with the unprotected eye can be very dangerous (Box 8–1), but sometimes large sunspots are visible to the naked eye at sunset or sunrise when the sun's brilliance is dimmed. The Chinese observed sunspots this way as early as the fifth century BC. In 1610 Galileo turned his telescope toward the sun and discovered that sunspots are quite common. He used his observations to show that the sun rotates on its axis with a period, as seen from the earth, of about 27 days.

Sunspots look dark because they are cooler than the photosphere. The center of a large sunspot is about 4240 K. The temperature of the photosphere is about 5800 K, so the cooler spot looks dark in contrast. In fact, a sunspot emits quite a bit of radiation. If the sun were magically removed and only an average-size sunspot were left behind, it would glow a brilliant orange-red and would be brighter than the full moon.

A clue to the origin of sunspots appeared in 1908 when the American astronomer George Ellery Hale discovered magnetic fields in sunspots. Hale was able to do this because a magnetic field affects the permitted energy levels in an atom. With no magnetic field present, a particular atom might absorb photons of a certain wavelength, producing a spectral line. If the atom is located in a magnetic field, however, the energy levels are split into multiple levels, and the single spectral line could appear as three or more spectral lines at slightly different wavelengths. This is known as the **Zeeman effect**. The separation of the spectral lines depends on the strength of the magnetic field, so Hale could measure the strength of magnetic fields on the sun by looking for this effect. He found that the field in a sunspot is about 1000 times stronger than the sun's average field (Figures 8–13 and 8–14).

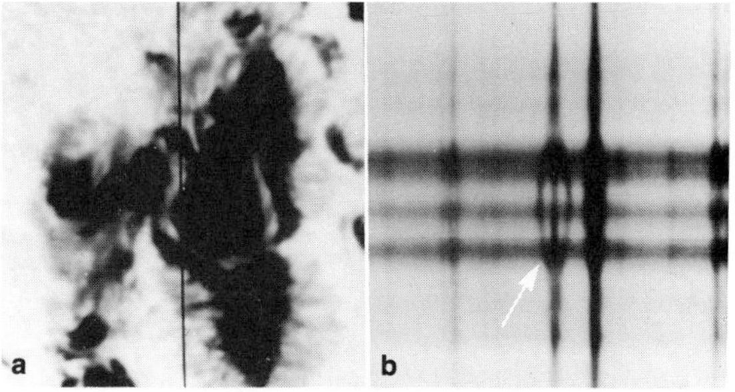

Figure 8–13 If we place the slit of our spectrograph across a sunspot (a), we admit light from the photosphere where the magnetic field is weak and from the spot where the field is strong. The resulting spectrum (b) shows a single line in the regions outside the sunspot, but the Zeeman effect splits that line into three components (arrow) inside the spot. This would allow us to measure the magnetic field inside sunspots. (©Association of Universities for Research in Astronomy, Inc., Kitt Peak National Observatory.)

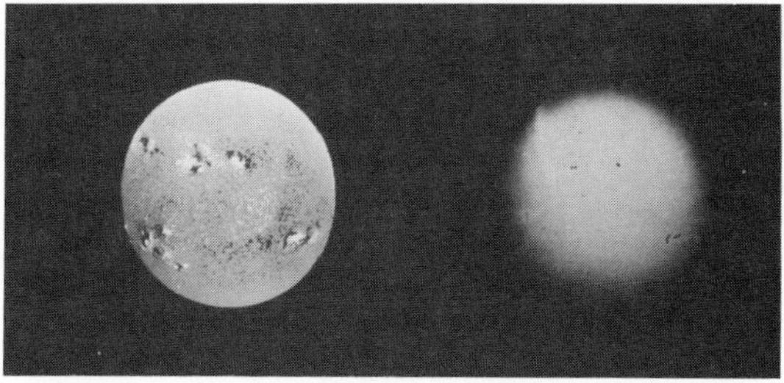

Figure 8–14 The white light image of the sun (right) shows the location of a few sunspots. The magnetic map, or magnetogram (left), was constructed using the Zeeman effect to show the location of strong magnetic fields. Regions of strong magnetic north are dark and magnetic south are light. Both images were recorded on January 3, 1978. (National Optical Astronomy Observatories.)

BOX 8–1

Observing the Sun with a Small Telescope

The visible and infrared radiation coming from the sun is intense enough to burn the retina of the human eye and cause blindness. This can happen to someone staring at the sun with the naked eye as well as to observers using telescopes. The added light-gathering power of telescopes, binoculars, and even the view-finders of cameras can convert bright sunlight into an intense beam of energy. One professor regularly impressed his students by lighting his pipe in the beam of sunlight emerging from the eyepiece of an 11-inch telescope aimed at the sun. A few puffs were enough to set the tobacco blazing.

Some commercially available telescopes come with solar filters. However, those filters that are made of dark glass to be fitted over or inside the eyepiece of the telescope are not safe even if they sufficiently reduce the intensity of the visible and infrared radiation. The full intensity of the sunlight enters the telescope and focuses on the filter. The heat can crack the filter, and if you are observing when the filter breaks, intense sunlight can burn your eye. Such eyepiece filters are not to be trusted.

Yet there are safe ways to observe the sun with a telescope. If a solar filter is a silvery-looking sheet of glass to be placed over the front of the telescope, it is probably safe. Such filters admit only a tiny fraction of the sunlight to the telescope, and there is little heat buildup. If such a filter is damaged, by a scratch for instance, it may not be safe.

Even if a telescope has no filter, it can still be used to observe the sun by the projection method (Figure 8–15). Without looking through the main telescope or finder telescope, point the telescope at the sun and then focus the image of the sun on a sheet of white

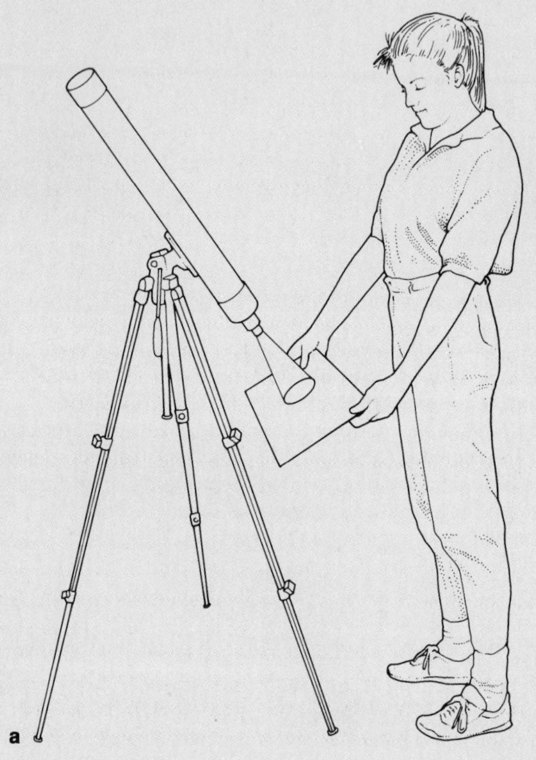

Figure 8–15 Looking through a telescope at the sun is dangerous, but you can always view the sun safely with a small telescope by projecting its image on a white screen (a). If you sketch the location and structure of sunspots on successive days (b) you will see the rotation of the sun and gradual changes in the size and structure of sunspots.

This suggests that the powerful magnetic field causes a sunspot by inhibiting convection. Ionized gas is made up of electrically charged particles, and such particles cannot move freely in a magnetic field. Astronomers often say the magnetic field is "frozen into" the ionized gas. This simply means the gas and magnetic field are locked together. Convection currents just under the photosphere might be reduced by the magnetic fields in sunspots, causing a decrease in the temperature and producing a dark sunspot.

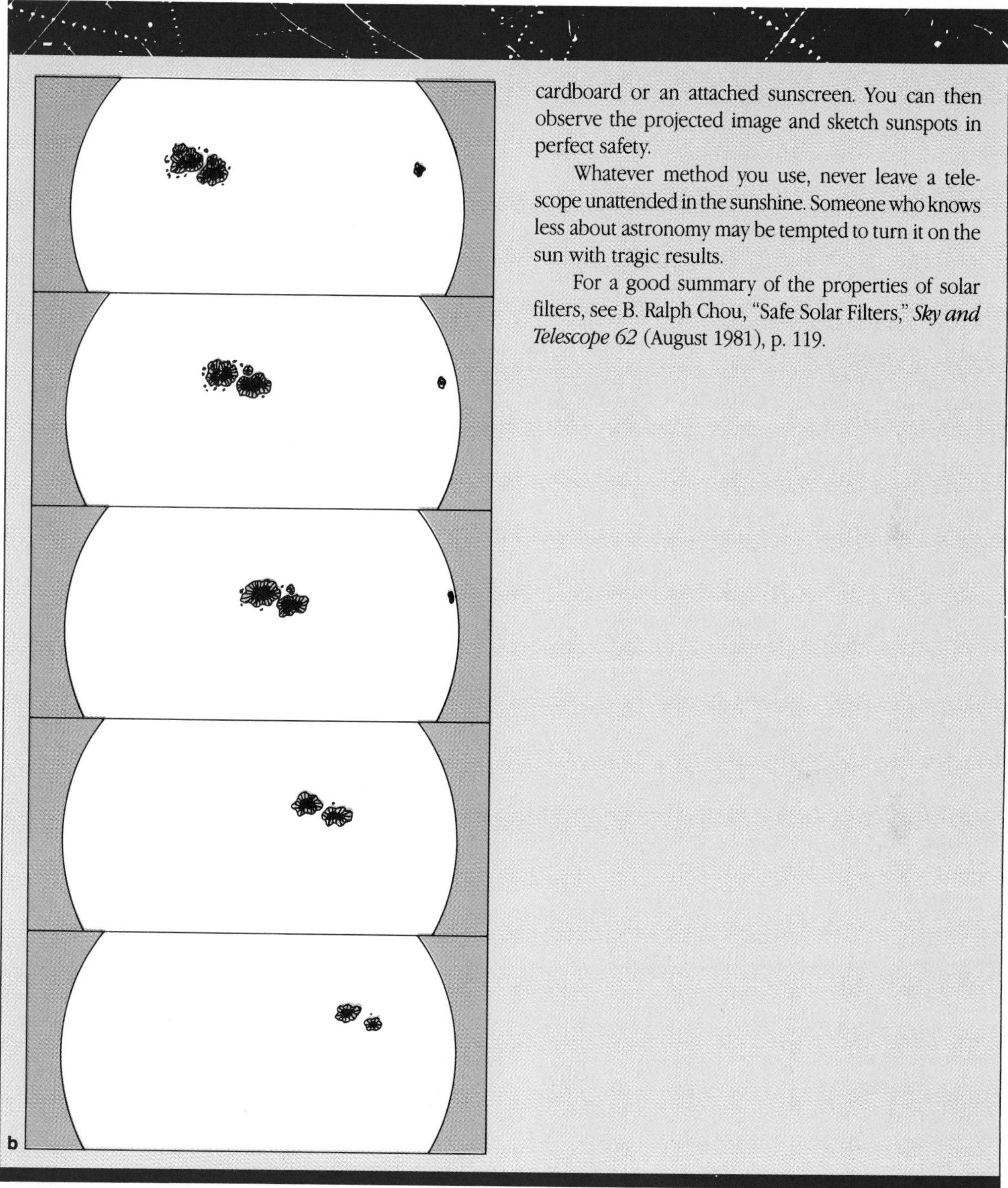
b

cardboard or an attached sunscreen. You can then observe the projected image and sketch sunspots in perfect safety.

Whatever method you use, never leave a telescope unattended in the sunshine. Someone who knows less about astronomy may be tempted to turn it on the sun with tragic results.

For a good summary of the properties of solar filters, see B. Ralph Chou, "Safe Solar Filters," *Sky and Telescope 62* (August 1981), p. 119.

Do other stars have sunspots, or rather "starspots," on their surfaces? This is a difficult question because, except for the sun, the stars are so far away no surface detail is visible. Some stars, however, vary in brightness in ways that suggest they are mottled by randomly placed, dark spots. As the star rotates, its total brightness changes slightly, depending on the number of spots facing in our direction. This has been suggested to explain the variation of the RS Canum Venaticorum stars whose spots may cover as much as 25 percent of the surface.

Also, some stars show spectral features that suggest the presence of magnetic fields and starspots. Ultraviolet observations reveal stars whose spectra contain emission lines commonly produced by the regions around spots on the sun. This suggests that these stars, too, have spots. One team of astronomers has used the Doppler shifts in the spectrum of the rotating star HR1099 to construct a map showing the distribution of dark spots on its surface (Figure 8–16 and Color Plate 10). Such results suggest that the sunspots we see on our sun are not unusual.

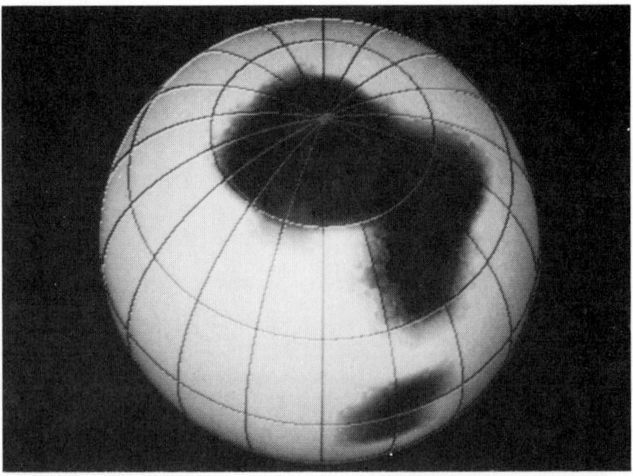

Figure 8–16 This computer-generated map shows the distribution of dark spots over the surface of the star HR1099. The map was constructed by analyzing the Doppler shifts in the spectrum of the rotating star. Although more extensive than sunspots, these dark regions are believed to be related to magnetic activity similar to that responsible for sunspots. (Courtesy Steven Vogt, Artie Hatzes, and Don Penrod.)

THE SUNSPOT CYCLE

The total number of sunspots visible on our sun is not constant. In 1843 the German amateur astronomer Heinrich Schwabe noticed that the number of sunspots varies with a period of about 11 years. This is now known as the sunspot cycle (Figure 8–17a). At sunspot maximum there are often as many as 100 spots visible at any one time, but at sunspot minimum there are only a few small spots. The next sunspot maximum will occur in the fall of 1991.

At the beginning of each sunspot cycle the spots begin to appear in the sun's middle latitudes about 35° above and below the sun's equator. As the cycle proceeds, the spots appear at lower latitudes, until, near the end of the cycle, they are appearing within 5° of the sun's equator. If we plot the latitude of the appearance of sunspots versus time, the diagram takes on the appearance of butterfly wings (Figure 8–18). Such diagrams are now known as **Maunder butterfly diagrams,** named after E. Walter Maunder of the Royal Greenwich Observatory, who first discovered the effect.

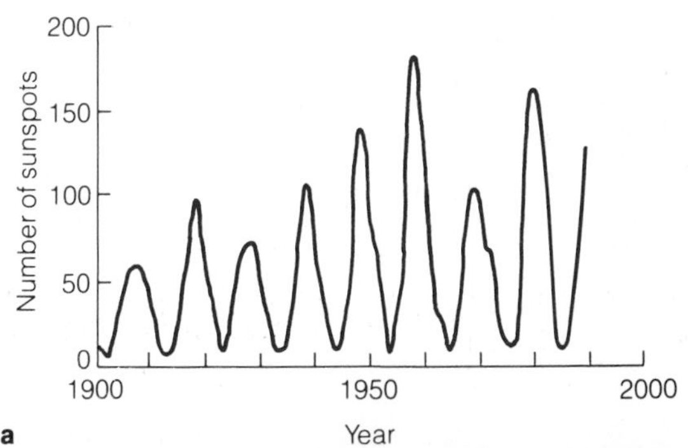

THE MAGNETIC CYCLE

The sunspot cycle seems to be related to an overall magnetic cycle in the sun. Alternate sunspot cycles feature spots of reversed polarity. For example, if we used the Zeeman effect to study the magnetic fields in sunspots, we might find that every sunspot pair is made up of a magnetic pair (Figure 8–19a and b). One spot would be a magnetic north pole and one a south pole. We might also discover that the leading spot of each pair in the northern hemisphere is a north pole and that each trailing spot is a south pole. South of the sun's equator, we would find the polarities reversed (Figure 8–19c). Even more mysterious, if we watched for an entire sunspot cycle, we would discover that the overall polarity reverses from cycle to cycle.

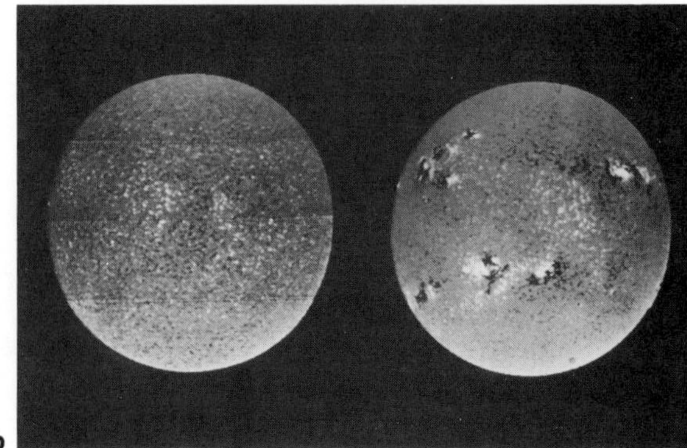

Figure 8–17 (a) The number of sunspots varies with an 11-year period. (b) Magnetograms of the sun show the magnetic field over the sun's disk at sunspot minimum (left) and at sunspot maximum (right). (National Optical Astronomy Observatory.)

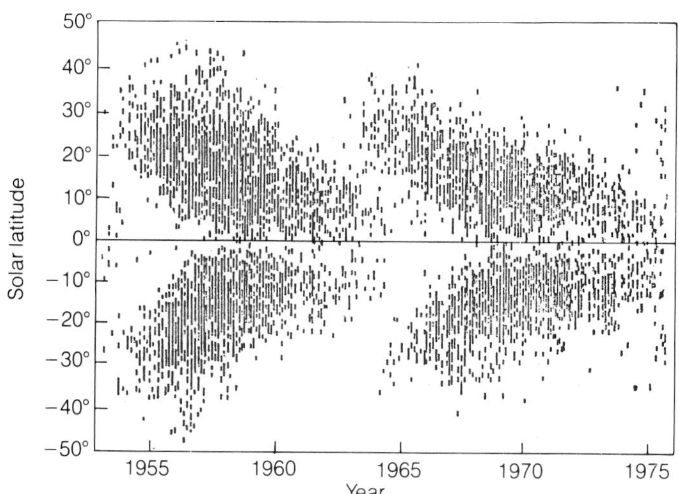

Figure 8–18 The Maunder butterfly diagram is a plot of the latitude on the sun where sunspots first appear. Early in a sunspot cycle, the spots appear at higher latitudes. Later in the cycle, they appear nearer the equator.

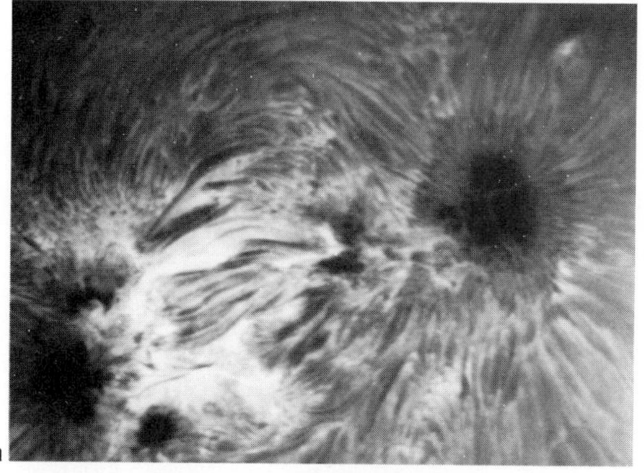

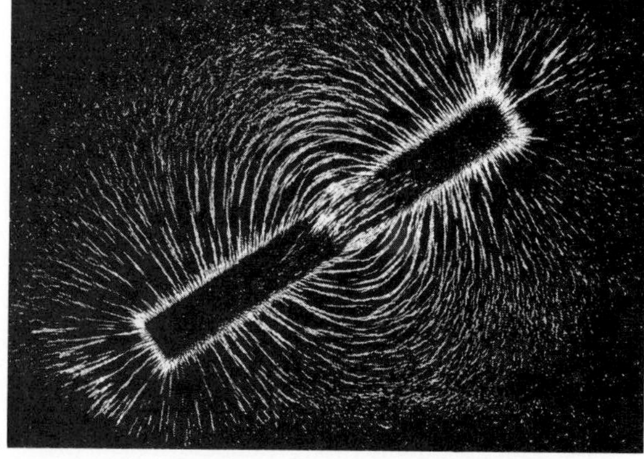

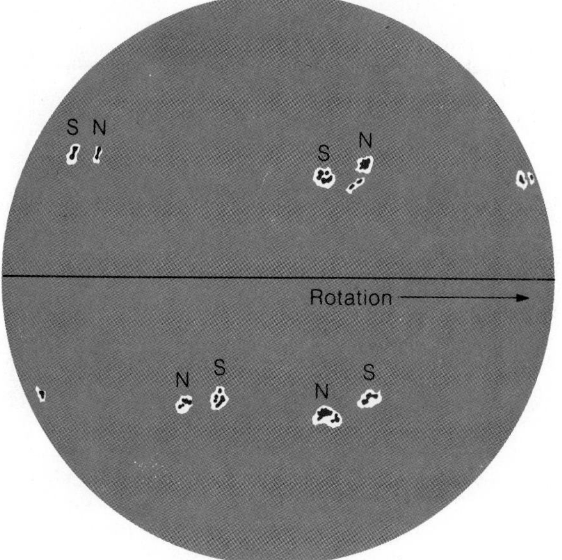

Figure 8–19 An H-alpha filtergram of a sunspot group (a) shows that structures in the chromosphere over the spots follow a pattern much like iron filings sprinkled over a bar magnet (b). The polarity of sunspot groups in the sun's southern hemisphere is the reverse of that in the northern hemisphere (c). (a, ©Association of Universities for Research in Astronomy, Inc., Sacramento Peak Observatory; b, Grundy Observatory.)

Although this magnetic cycle is not fully understood, a model proposed in 1961 by Mount Wilson astronomer Horace Babcock explains the cycle as an interaction between the sun's magnetic field and the sun's rotation.

The earth, being solid, rotates as a rigid body, but the sun is a sphere of gas, and some parts of it rotate faster than other parts. The sun's equator rotates once in about 25 days, while the higher latitudes take up to 29 days to

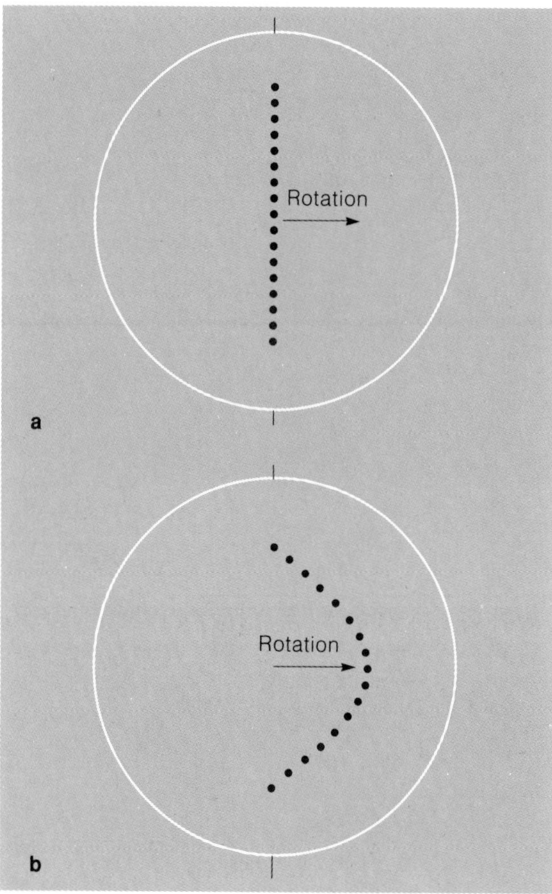

Figure 8–20 Differential rotation. The sun's equator rotates with a shorter period than the higher latitudes. If we could arrange a line of sunspots running from north to south (a), those near the equator would pull ahead. A month later (b) we would find the line of sunspots distorted into a curve.

rotate once. This means that objects near the equator pull ahead of objects farther north or south (Figure 8–20). Such rotation is termed **differential rotation.**

The differential rotation of the sun drags the equatorial parts of the magnetic field around the sun. The field winds up like a loop of thread on a spool. Turbulence in the convection zone further twists the field into ropelike

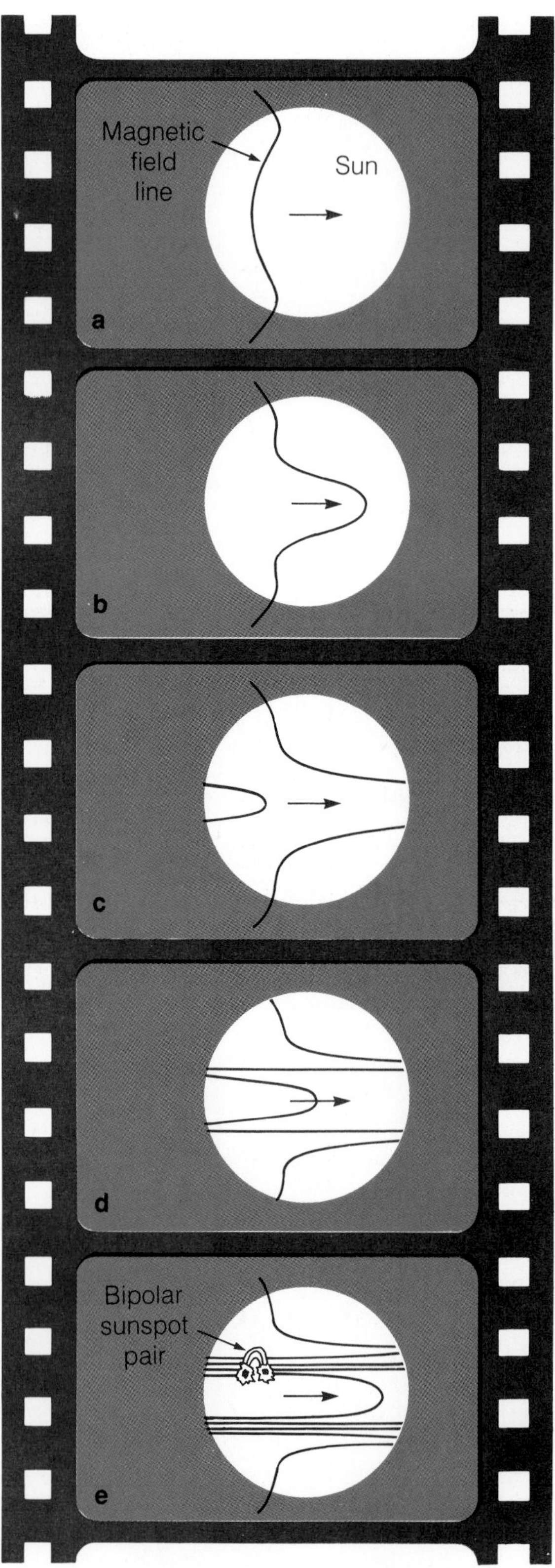

Figure 8–21 The Babcock model suggests that the differential rotation of the sun winds up the magnetic field. When the field becomes tangled and bursts through the surface, it forms sunspot pairs. For the sake of clarity, only one line in the sun's magnetic field is shown here.

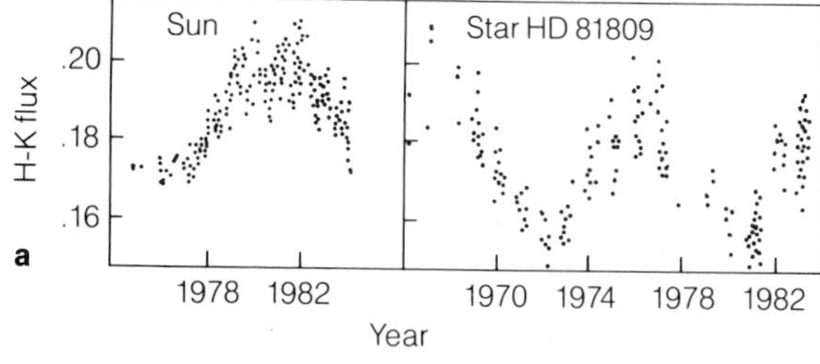

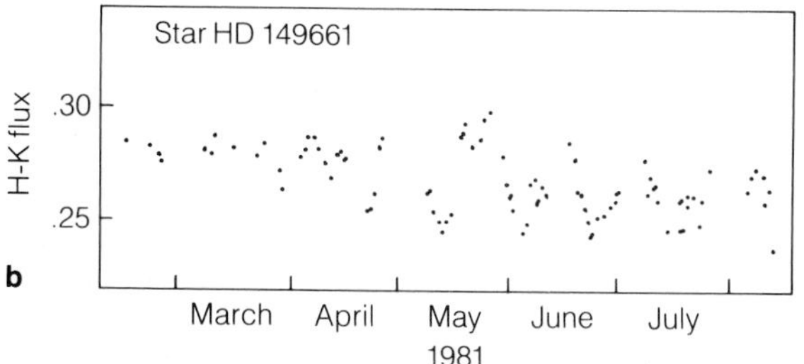

Figure 8–22 The average amount of emission in the H and K lines of calcium (mean H-K flux) is related to magnetic activity. (a) In the sun, this emission is stronger when sunspot activity is higher. The sunlike star HD 81809 appears to have a 9-year starspot cycle. (b) Rapid variation in H and K emission is caused by the rotation of the star, carrying groups of spots into and out of sight. The star HD 149661, for example, rotates in about 21 days. (Adapted from data by Livingston and Vaughan.)

tubes of magnetic field, which tend to float upward. Where these magnetic tubes burst through the sun's surface, sunspot pairs occur (Figure 8–21).

Babcock's theory explains the butterfly diagram. The magnetic field becomes tangled first at higher latitudes and later at lower latitudes. Thus, the sunspots appear at higher latitudes early in the cycle and at progressively lower latitudes as the magnetic field becomes more tightly wound.

The Babcock theory even explains the reversal of the sun's magnetic field from cycle to cycle. When the magnetic field becomes severely tangled, it breaks and reorders itself into a simpler pattern and the differential rotation begins winding it up again. This marks the beginning of a new cycle, but because of the way the field is reordered, it is reversed, and the new cycle begins with magnetic north replaced by magnetic south.

We have said that the period of the sunspot cycle is 11 years, but the overall magnetic cycle in the sun repeats every 22 years. For 11 years the sun's magnetic field maintains its orientation, and the sun goes through a sunspot cycle. Then the magnetic field reverses, and the sun goes through another sunspot cycle. Thus, the overall period of the Babcock cycle in the sun is 22 years.

Notice that Babcock's theory is a model of the sun's magnetic field. Because no model can perfectly represent nature, we should not ask whether the model is correct, but rather how well it explains the sun's magnetic field. In fact, the Babcock model is very successful at explaining the sun's magnetic cycles.

Because we believe that the sun is a representative star, we should expect other stars to have similar cycles of starspots. We can't see individual spots from earth, of course, but certain features in stellar spectra are associated with magnetic fields. Regions of strong magnetic fields on the solar surface emit strongly at the central wavelengths of the H and K lines of ionized calcium. This calcium emission appears in the spectra of other sunlike stars and tells us that those stars too have strong magnetic fields on their surfaces. These stars presumably have starspots as well.

In 1966, astronomers began measuring the strength of this H and K emission in the spectra of stars. The stars to be observed were selected to be similar to the sun. Their temperatures range from 1000 K hotter than the sun to 3000 K cooler, and they are all generating their energy by hydrogen fusion. These stars are thus most likely to have sunlike magnetic activity on their surfaces.

The observations show that the strength of the emissions in the spectra of these stars varies from year to year. The H and K emission averaged over the sun's disk varies with the sunspot cycle, and similar periodic variations can be seen in the spectra of the stars studied (Figure 8–22a). The star HD 81809, for instance, appears to have a starspot cycle lasting 9 years. Thus, we can be sure that stars like the sun do have magnetic fields and are subject to magnetic cycles.

The rotation periods of these stars are also apparent from these observations. If a star has a major region of magnetic activity and rotates every 20 days, we see the emission appear for 10 days and then disappear for 10 days as the rotation of the star carries the region to the far side of the star (Figure 8–22b).

These observations confirm our belief that the sun is an average sort of star, that is, not peculiar. Most other stars like our sun have magnetic fields, starspots, and go through magnetic cycles. More important, these studies are helping us understand our sun better.

Although the connection between the magnetic cycle in the sun and the sunspot cycle is not clearly understood, it does seem clear that the sun's magnetic field governs the production of sunspots. It also controls other forms of solar activity such as prominences and flares.

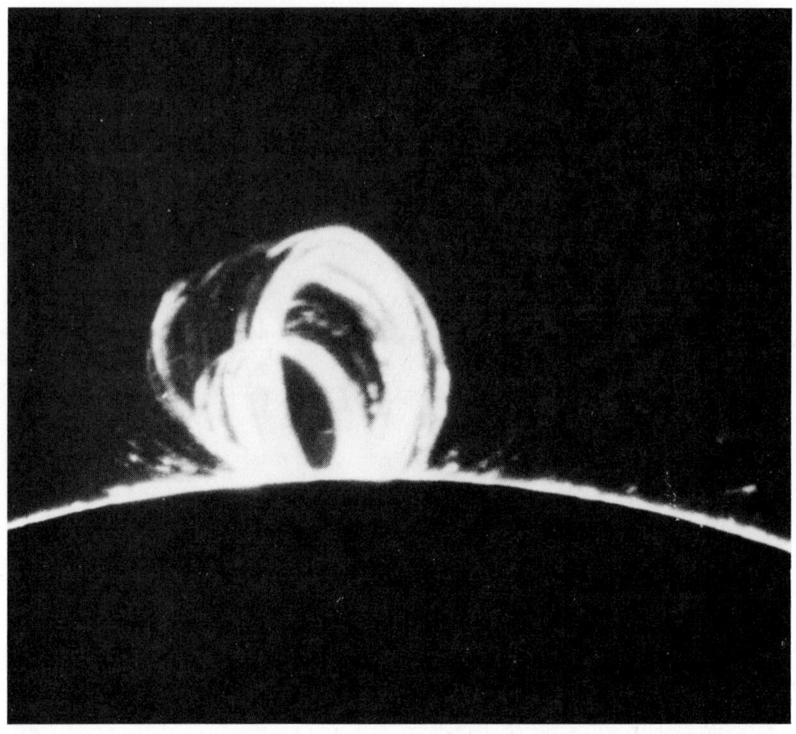

Figure 8–23 Prominences, which are chromospheric eruptions, are dominated by the complex magnetic fields above solar active regions. (©Association of Universities for Research in Astronomy, Inc., Sacramento Peak Observatory.)

PROMINENCES AND FLARES

Prominences are visible during total solar eclipses as red protrusions at the edge of the solar disk. (See Color Plate 3.) The red color is the same as the red color of the chromosphere and comes from the emission lines of hydrogen.

Prominences seem to be controlled by magnetic fields. Many are arch-shaped, looking much like the patterns made by iron filings sprinkled over a magnet (Figures 8–19b and 8–23). In filtergrams of the sun, prominences show as dark filaments that wind through the magnetically active regions around sunspots. Eruptive prominences burst out of these complex magnetic fields and may shoot upward 500,000 km in a few hours. Quiescent prominences may develop as graceful arches over sunspot groups and can last weeks or even months. Eruptive or quiescent, prominences are clearly ionized gases trapped in the twisted magnetic fields of active regions.

Flares are much more violent than prominences (Figure 8–24). A **flare** is an eruption on the solar surface that rises to maximum in a few minutes and decays in an hour or less. During that time it emits vast amounts of X-ray, ultra-violet, and visible radiation, and streams of high-energy protons and electrons. A large flare can release 10^{25} J, the equivalent of 2 billion megatons of TNT.

Solar flares seem clearly linked to the magnetic field. They almost always occur near sunspot groups and may recur over and over at the same place. A large spot group may experience 100 small flares a day, although only one flare a year may be bright enough to see in visible light. Various theories suggest that flares occur when sharp twists in the magnetic field store up great quantities of energy and then release it all at once.

Solar flares can have important effects on earth. X-ray and ultraviolet radiation reaches the earth in only 8 minutes and increases the ionization in the earth's upper atmosphere. This alters the reflection of shortwave radio signals and can absorb them completely, thereby interfering with communications. Flares can eject high-energy particles at a third the speed of light, but most particles ejected from flares have lower velocities and reach the earth hours or days after the flare as gusts in the solar wind. Such gusts interact with the earth's magnetic field and generate tremendous electrical currents (as much power as a million megawatts), which flow down into the earth's atmosphere near the magnetic poles. There they can excite the atoms of the upper atmosphere to glow in

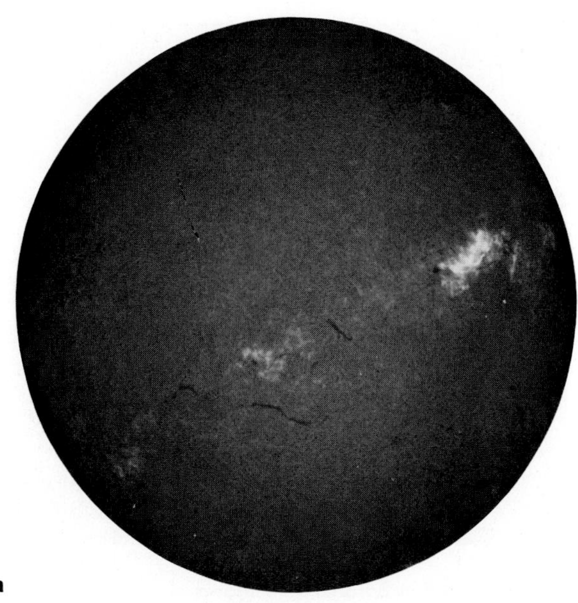

a

displays called **auroras** (Figure 8–25 and Color Plate 9) at altitudes of 100–400 km. These currents can also disturb the earth's magnetic field and cause magnetic storms in which compasses behave erratically.

Other effects of solar flares include surges in high-voltage power lines and radiation hazards to passengers in supersonic transports and in spacecraft. The U.S. Air Force watches for flares from observatories around the world, and other solar astronomers have developed ways of predicting which flares will affect earth.

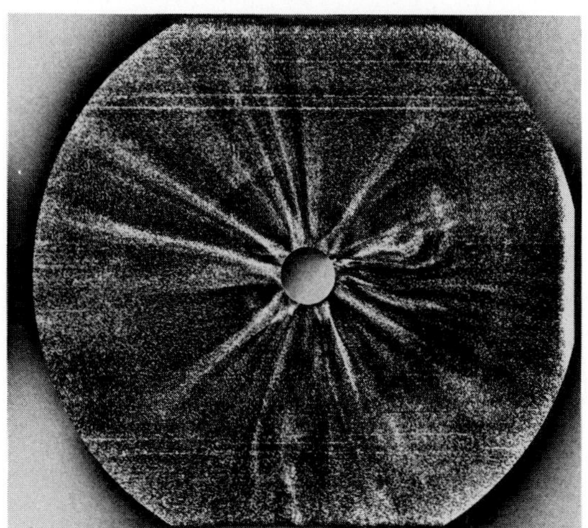

b

Figure 8–24 (a) This H-alpha filtergram shows a solar flare erupting near the right edge of the sun's disk. This flare of April 24, 1984 was about eight times the diameter of the earth and caused a major disruption in radio communication. (b) Events on the sun's surface can affect the outer corona. The solar eclipse of February 16, 1980 was photographed from a jet 11 km (36,000 ft) above the Indian Ocean. Computer processing reveals twisted streamers extending beyond 12 $R_{\odot}$. (a, Donald F. Neidig, National Solar Observatory/Sacramento Peak; b, C. Keller, Los Alamos National Laboratory.)

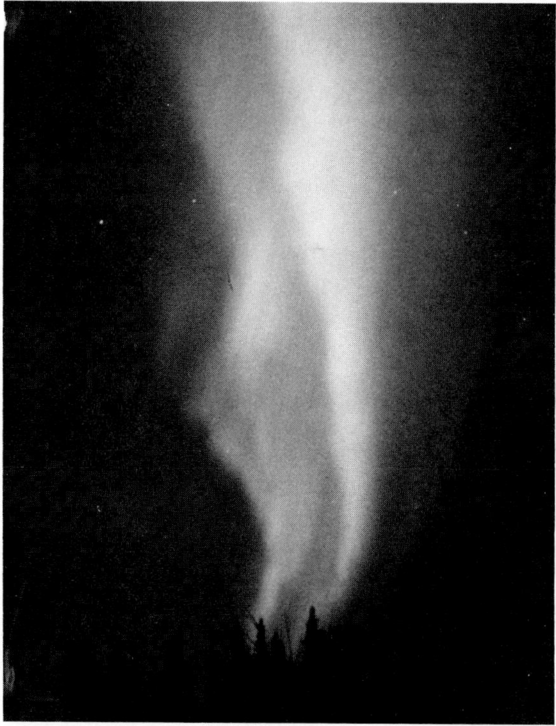

Figure 8–25 The wavering curtains of light called aurora are caused by excited atoms in earth's upper atmosphere. (Mario Grassi.)

CORONAL ACTIVITY

The corona of the sun also takes part in the solar activity cycle. At sunspot minimum, observers of total solar eclipses see a small, slightly flattened corona. But at sunspot maximum, eclipse observers are treated to a blazing corona that is nearly circular. (Compare Figure 3–14 and Color Plate 3.)

However, observing eclipses from earth's surface is not the best way to study the corona. Scientists from Los Alamos National Laboratory have mounted equipment in an Air Force NC-135 and flown at nearly 12,200 m (40,000 ft) to photograph the eclipsed sun from above 80 percent of the earth's atmosphere. They could detect the corona out to 20 solar radii (Figure 8–24b).

The results of such studies show that the corona is not the uniform halo of gas that earlier astronomers imagined. It is actually composed of streamers shaped by the solar magnetic field (Color Plate 8). The corona looks uniform during some eclipses because we see the streamers in projection at the sun's edge. Photographs made by the Los Alamos team can trace coronal streamers out to 12 solar radii.

The gradual changes in the streamers can be interrupted by coronal transients, disturbances that erupt outward at about 500 km/sec. Although some transients become as large as the sun itself, they represent very little mass and are clearly related to the complex magnetic fields in active regions.

The streamers seem to draw their bulbous shapes from loops of magnetic fields that extend to a few solar radii, where the charged particles trapped in the fields are able to escape in long thin streams. However, in some regions of the solar surface, the magnetic fields do not loop back, and the particles stream away from the sun unimpeded. These regions appear in X-ray images of the corona as cooler, lower-density regions called **coronal holes** (Figure 8–26). Although there are permanent coronal holes at the sun's north and south poles, the distribution of coronal holes over the rest of the sun depends on the solar activity cycle, further evidence that it is a magnetic phenomenon. Coronal activity is important because it affects the solar wind and thus affects the earth. Many solar astronomers believe that the solar wind is composed of particles streaming away from coronal holes. Understanding the solar wind, therefore, seems to depend on our understanding coronal activity.

The solar activity we see in the photosphere, chromosphere, and corona is complex and beautiful, and some solar activities are important for what they tell us about the sun and stars. A few are important because they can affect the earth. But one last aspect of solar activity is critically important to life on earth—the constancy of the sun.

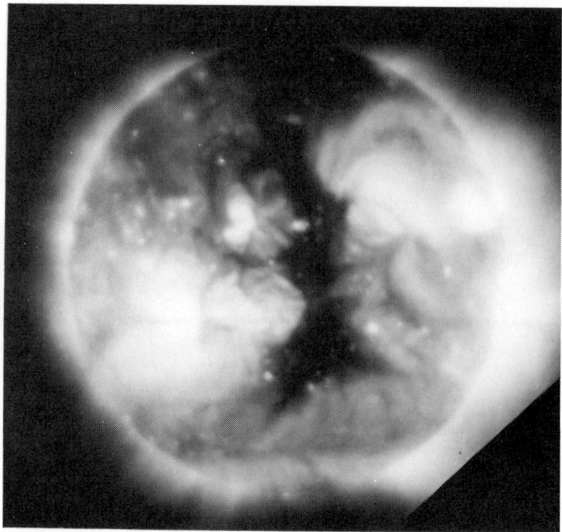

Figure 8–26 A coronal hole shows as a dark area near the center of the solar disk in the Skylab photo of the sun made at X-ray wavelengths. (American Science and Engineering.)

THE SOLAR CONSTANT

If the sun's energy output varies by even a small amount, life on earth might end. The continued existence of our civilization and our species depends on the constancy of our sun, but we know very little about the variation of the sun's energy output.

The energy production of the sun can be measured by adding up all of the energy falling on 1 sq m of the earth's surface during 1 second. Of course, some correction for the absorption of earth's atmosphere is necessary, and we must count all wavelengths from X-rays to radio waves. The result, which is called the **solar constant**, amounts to about 1360 J per square meter per second. A change in the solar constant of only 1 percent could change the average temperature of the earth by 1 to 2°C (about 1.8 to 3.6°F). For comparison, during the last ice age, the average temperature on Earth was about 5°C cooler than it is now.

Some of the best measurements of the solar constant were made by instruments aboard the Solar Maximum Mission satellite (repaired in orbit in 1984). These have shown variations in the energy received from the sun of

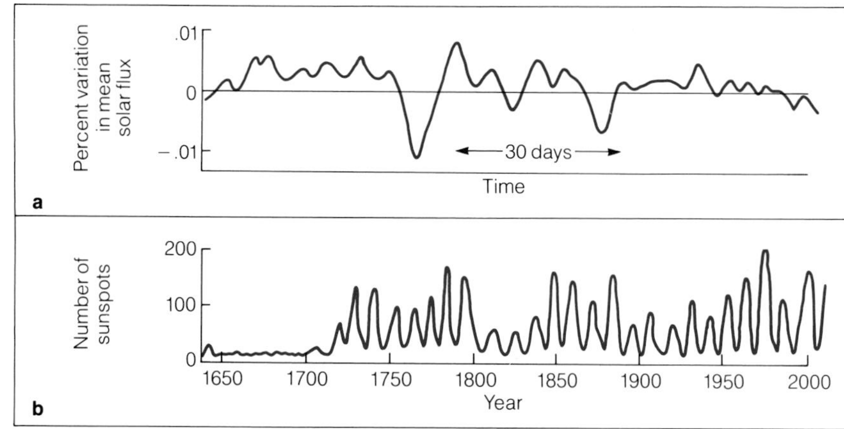

Figure 8–27 The Inconstant Sun: (a) Variations in the solar constant of roughly 0.01 percent are caused by the passage of sunspots across the disk. This data from the Solar Maximum Mission satellite spans 153 days. (b) The Maunder Minimum was a period between 1645 and 1715 when there were very few sunspots. (a, Adapted from data by Willson, Duncan, and Geist.)

about 0.01 percent that lasted for days or weeks (Figure 8–27a). Superimposed on that random variation is a long-term decrease of about 0.018 percent per year. Observations made from the NIMBUS 7 satellite find a 0.015 percent per year decrease, and observations made from sounding rockets and balloons detect a 0.016 percent per year decrease. This long-term decrease may be related to a cycle of activity on the sun with a period longer than the 22-year magnetic cycle.

Small random fluctuations will not affect our climate, but a long-term decrease over a decade or more could cause worldwide cooling. History contains some evidence that the solar constant may have varied in the past. The "little ice age" was a period of unusually cool weather in Europe and America that lasted from about 1430 to 1850. The average temperature worldwide was about 1 K cooler than it is now. Although there are many things that affect climate, this could have been caused by a decrease in the solar constant by only 1 percent.

This period of cool weather included an era of few sunspots now known as the **Maunder Minimum*** (Figure 8–27b). Between 1645 and 1715 there is almost no record of sunspots, even though the telescope had been perfected and was being actively used by astronomers. Reports of total solar eclipses during this period make no mention of the corona or chromosphere, and there is almost no record of auroral displays. The Maunder Minimum seems to have been a period of reduced solar activity. Solar astronomer John Eddy believes the historical record shows traces of at least 11 similar periods extending back to 3000 BC. If that is true, then our sun is not a perfectly stable star, and the survival of humanity may depend on our ability to adapt to changes in the solar constant.

*Ironically the Maunder Minimum coincides with the reign of Louis XIV, the "Sun King."

SUMMARY

The sun is a typical star, and, like all stars, it is held together by its own gravity and supported by the high pressure of the hot gases in its interior. This balance is called hydrostatic equilibrium. The sun keeps its interior hot by nuclear fusion reactions that fuse hydrogen into helium—the proton–proton chain. Although the experiment measuring the number of neutrinos coming from the sun raises questions about the nuclear reactions going on there, no one doubts that the sun's energy comes from nuclear fusion.

The nuclear reactions occur in the central core of the sun, and the energy moves outward as radiation, and, in the outer layers, as convection currents. Conduction, a third means of energy transport, is important in only a few stars.

The atmosphere of the sun consists of three layers: the photosphere, chromosphere, and corona. The photosphere, or visible surface, is a thin layer of low-density gas that is the level in the sun from which visible photons most easily escape. It is marked by granulation, a pattern produced by solar convection.

The chromosphere is most easily visible during total solar eclipses when it flashes into view for a few seconds and produces the flash spectrum. This spectrum shows that the chromosphere is hot, ionized gas. Filtergrams taken in the light emitted by specific atoms show that the chromosphere is filled with large jets called spicules.

The corona is the sun's outermost atmospheric layer. It is composed of very low-density, very hot gas extending at least 20 solar radii from the sun. Although it can be studied with a special telescope called a coronagraph, its outer layers are visible only during total solar eclipses. Its high temperature—up to 3.5×10^6 K—is believed to be maintained by interaction with the solar magnetic field. The outer parts of the corona merge with the solar wind, a breeze of low-density ionized gas streaming away from the sun. Thus the earth, which is bathed in the solar wind, is orbiting within the sun's outer atmosphere.

The sunspots are the most prominent example of solar activity. A sunspot appears to have a dark center called the umbra and a slightly lighter border called the penumbra. A sunspot seems dark because it is slightly cooler than the rest of the photosphere. The average sunspot is about twice the size of the earth and contains magnetic fields about 1000 times stronger than the sun's average field. Sunspots are thought to form because the magnetic field inhibits convection. The average number of sunspots visible varies with a period of about 11 years and appears to be related to the solar magnetic cycle.

Alternate sunspot cycles have reversed magnetic polarity, and this has been explained by the Babcock model of the magnetic cycle. In this theory the differential rotation of the sun winds up the magnetic field. Tangles in the field rise to the surface and cause sunspot pairs. When the field becomes strongly tangled, it reorders itself into a simpler but reversed field, and the cycle starts over.

Prominences and flares are other examples of solar activity. Prominences occur in the chromosphere; their arch shapes show that they are formed of ionized gas trapped in the magnetic field. Flares too seem to be related to the magnetic field. They are sudden eruptions of X-ray, ultraviolet, and visible radiation and high-energy particles that occur among the twisted magnetic fields around sunspot groups. Flares are important because they can have effects on Earth such as communications blackouts and auroras.

Activity in the corona is also guided by the magnetic field. The corona seems to be composed of streamers of thin, hot gas escaping from the magnetic field. Coronal transients are eruptions that balloon outward and distort the streamers. In some regions of the corona, the magnetic field does not loop back to the sun, and the gas escapes unimpeded. These regions are called coronal holes and are believed to be the source of the solar wind.

Because the sun is so active, it would not be surprising if its total energy output varies. Such variations could have important consequences for the earth. The solar constant, a measure of the total energy coming from the sun, has been measured by instruments in spacecraft and found to vary by about 0.1 percent. Such short-term variations are not dangerous. However, the observations also show that the sun is currently fading by about 0.018 percent per year. This is probably a cyclic phenomenon, but if it continues for a decade or more it could affect climate. Historical records show that from 1645 to 1715 there were almost no sunspots. This Maunder Minimum occurred during a period of cool weather in Europe and America called the little ice age, which suggests that the solar constant may change occasionally and alter earth's climate.

NEW TERMS

proton–proton chain
neutrino
opacity
hydrostatic equilibrium
helioseismology
granulation
supergranule
flash spectrum
filtergrams
spicule
coronagraph
shock wave
solar wind
sunspot
Zeeman effect
Maunder butterfly diagram
differential rotation
prominence
flare
aurora
coronal hole
solar constant
Maunder Minimum

QUESTIONS

1. Why is it wrong to say that the sun makes its energy by fusing hydrogen atoms together, and why is it correct to say that the sun makes its energy by fusing hydrogen nuclei together?
2. Why can't hydrogen fusion occur in the sun's convection zone?
3. Why was the neutrino experiment buried nearly a mile underground? How can that help detect neutrinos?
4. How do helioseismology and neutrino detectors give us information about the sun's interior?
5. Give everyday examples of the transport of energy by conduction, convection, and radiation in addition to those in Figure 8–3.
6. If the sun is gaseous and has a strong gravitational field, why doesn't it contract into its center?
7. What does granulation tell us about the sun's interior?
8. Why does the flash spectrum contain spectral lines in emission that are not present as absorption lines in the spectrum of the photosphere?
9. Why do we think the sunspot cycle is controlled by the magnetic cycle of the sun?
10. How can flares on the sun affect the earth?
11. In a best-selling novel of a few years ago, astronauts on the moon were killed by high-energy protons from a solar flare that reached the moon 8 minutes after the flare occurred. What is wrong with this plot device?
12. What evidence do we have that other stars have stellar winds, chromospheres, coronas, starspots, and magnetic cycles?

PROBLEMS

1. Circle all of the ^{1}H and ^{4}He nuclei in Figure 8–1 and explain how the proton–proton chain can be summarized by $4\ ^1\text{H} \rightarrow\ ^4\text{He}$ + energy.
2. How much energy is produced when the sun converts 1 kg of mass into energy?
3. How much energy is produced when the sun converts 1 kg of hydrogen into helium? (HINT: How does this problem differ from Problem 2?)
4. A 1-megaton nuclear weapon produces about 4×10^{15} J of energy. How much mass must vanish when a 5-megaton weapon explodes?
5. The center of the sun is believed to have a temperature of about 10^7 K. What wavelength photons does this gas emit most commonly? What part of the spectrum do they occupy?
6. If the smallest detail we can see with the Hubble Space Telescope is 0.05 seconds of arc in diameter, what is the smallest diameter feature we will be able to see on the surface of the sun? (HINT: Use the small-angle formula.)
7. What is the angular diameter of a star like the sun located 5 ly from the earth? Will the Hubble Space Telescope be able to detect detail on the surface of such a star?
8. If a sunspot has a temperature of 4200 K and the solar surface has a temperature of 5800 K, how many times brighter is the surface compared to the sunspot? (HINT: See Box 7–1.)
9. A solar flare can release 10^{25} J. How many megatons of TNT would be equivalent?
10. The United States consumes about 2.5×10^{19} J of energy in all forms in a year. How many years could we run the United States on the energy released by the solar flare in Problem 9?
11. Neglecting energy absorbed or reflected by our atmosphere, the solar energy hitting 1 sq m of earth's surface is 1360 J/sec (the solar constant). How long does it take a baseball diamond (90 feet on a side) to receive 1 megaton of solar energy? (HINT: See Problem 4.)

RECOMMENDED READING

BOHM-VITENSE, ERIKA "Chromospheres, Transition Regions, and Coronas." *Science 223* (24 Feb. 1984), p. 777.

DELANCY, MARY MARTIN "The Case of the Missing Sunspots." *Astronomy 9* (Feb. 1981), p. 66.

Fire of Life: The Smithsonian Book of the Sun. Washington, D.C.: Smithsonian Books, 1981.

FRANCO, ANNA and DAVID H. SMITH "Vanishing Solar Neutrinos." *Sky and Telescope 73* (Feb. 1987), p. 149.

GIOVANELLI, R. *Secrets of the Sun.* Cambridge, Mass.: Cambridge University Press, 1984.

GOLUB, LEON "Solar Magnetism: A New Look." *Astronomy 9* (March 1981), p. 66.

———. "What Heats the Solar Corona?" *Astronomy 10* (Sept. 1982), p. 74.

HARVEY, J., M. POMERANTZ, and T. DUVALL "Astronomy on Ice." *Sky and Telescope 64* (Dec. 1982), p. 520.

HALL, DOUG "Starspots." *Astronomy 11* (Feb. 1983), p. 66.

KALER, JAMES B. "Cousins of Our Sun: The G Stars." *Sky and Telescope 72* (Nov. 1986), p. 450.

MARAN, STEPHEN P. "Coronal Revisionism." *Natural History 92* (Jan. 1983), p. 74.

———. "The Jewel in the Satellite." *Natural History 95* (Jan. 1986), p. 84.

MAXWELL, ALAN "Solar Flares and Shock Waves." *Sky and Telescope 66* (Oct. 1983), p. 285.

MULLAN, DERMOT J. "Tuning into the Interior of a Star." *Astronomy 12* (Dec. 1984), p. 66.

NEIDIG, DONALD F. and JACQUE M. BECKERS "Observing White Light Flares." *Sky and Telescope 65* (March 1983), p. 226.

NORMAN, ERIC B. "Neutrino Astronomy: A New Window on the Universe." *Sky and Telescope 70* (Aug. 1986), p. 101.

O'LEARY, BRIAN "The Stormy Sun." *Sky and Telescope 60* (Sept. 1980), p. 199.

RUST, DAVID M. "Solar Flares, Proton Showers, and the Space Shuttle." *Science 216* (28 May 1982), p. 939.

SCHWARZSCHILD, BERTRAM "Conversion in Matter May Account for Missing Solar Neutrinos." *Physics Today 39* (June 1986), p. 17.

SOFIA, SABATINO, PIERRE DEMARQUE, and ANDREW ENDAL "From Solar Dynamo to Terrestrial Climate." *American Scientist 73* (July/Aug. 1985), p. 326.

STAHL, PHILIP A. "Prominences." *Astronomy 11* (Jan. 1983), p. 66.

WALLENHORST, STEVEN G. "Sunspot Numbers and Solar Cycles." *Sky and Telescope 64* (Sept. 1982), p. 234.

WILLIAMS, GEORGE E. "The Solar Cycle in Precambrian Time." *Scientific American 255* (Aug. 1986), p. 88.

WILSON, OLIN C., ARTHUR H. VAUGHN, and DIMITRI MIHALAS "The Activity Cycles of Stars." *Scientific American 244* (Feb. 1981), p. 104.

WILSON, RICHARD C., HUGH HUDSON, and MARTIN WOODARD "The Inconstant Solar Constant." *Sky and Telescope 67* (June 1984), p. 501.

WOLFSON, RICHARD "The Active Solar Corona." *Scientific American 248* (Feb. 1983), p. 104.

CHAPTER 9

MEASURING STARS

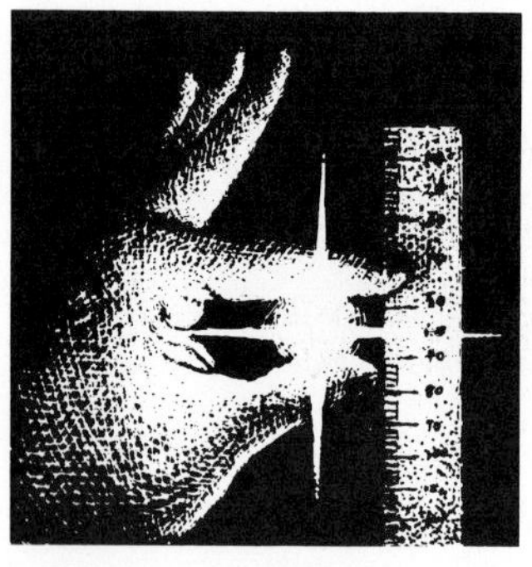

If it can't be expressed in figures, it is not science; it is opinion.

Robert Heinlein
THE NOTEBOOKS OF LAZARUS LONG

The stars are unimaginably remote. The nearest star is our sun, only 1.5×10^8 km (93 million miles) away, so close that light takes only 8 minutes to reach the earth. The next nearest star α Centauri is nearly 300,000 times farther away, about 4 light-years. Recall from Chapter 1 that a light-year is the distance light travels in 1 year—about 9.5×10^{12} km (5.8 trillion miles).

Despite their great distances, the stars are the key to astronomy. The universe is filled with stars, and if we are to understand the universe, we must discover how stars are born, live, and die. We begin our study in this chapter by gathering data about the intrinsic properties of the stars—those properties inherent in the nature of the stars. In later chapters, we will use these data to deduce the life cycles of different kinds of stars.

Unfortunately, determining a star's intrinsic properties is quite difficult. When we look at a star through a telescope, we see only a point of light that tells us nothing about the star's energy production, temperature, diameter, or mass. Because we cannot visit stars, we can only observe from the earth and unravel the properties of stars through the analysis of starlight. One of the reasons astronomy is interesting is that it contains so many such puzzles, each demanding a different method of solution.

To simplify our task, this chapter will concentrate on two intrinsic stellar properties. Our goals will be to find out how much energy stars emit and how large they are. In the next chapter, we will discuss the masses of stars. These three parameters, combined with stellar temperatures—an intrinsic stellar property discussed in Chapter 7—will give us an overview of the nature of stars and provide us with the data we need to consider the lives of stars.

Although we begin this chapter with two goals firmly in mind, we immediately meet a short detour. To find out how much energy a star emits,

FIGURE 9–1 We must know how far away a source of light is before we can judge its intrinsic brightness. In this case, both light sources could look the same brightness to the observer.

we must know how far away it is. If at night we see bright lights approaching on the highway, we cannot tell whether the lights are the intrinsically bright headlights of a distant truck or the intrinsically faint lights on a nearby bicycle (Figure 9–1). Only when we know the distance to the lights can we judge their intrinsic brightness. In the same way, to find the intrinsic brightness of a star and thus the amount of energy it emits, we must know its distance. Our short detour will provide us with a method of measuring stellar distances.

Once we know the distance to a star, finding its total energy output and estimating its diameter are simple. When we do this for many stars, we discover that there are different kinds of stars, and we will conclude this chapter by considering the frequency of stellar types—that is, we will try to decide which kinds of stars are common and which are rare. These family relations among the stellar types will be an important clue to the life stories of the stars.

9.1 MEASURING THE DISTANCES TO STARS

Determining the distance to a star is difficult because astronomers cannot journey to the star. They must, instead, measure the distance indirectly, much as surveyors measure the distance across a river they cannot cross. We will begin by reviewing this method and then apply it to stars.

THE SURVEYOR'S METHOD

To measure the distance across a river, a team of surveyors begins by driving two stakes into the ground. The distance between the stakes is the baseline of the measurement. The surveyors then choose a landmark on the opposite side of the river, a tree perhaps, thus establishing a large triangle marked by the stakes and the tree. Using surveyors' instruments, they sight the tree from the two ends of the baseline and measure the two angles on their side of the river (Figure 9–2).

Knowing two angles and the length of the side between them, the baseline, the surveyors can find the distance across the river by trigonometry or by constructing a scale drawing. For example, if the baseline was 50 m and the angles were 66° and 71°, they could draw a line 50 mm long to represent the baseline. Using a protractor, they could construct angles of 66° and 71° at each end of the baseline, and then extend the two sides until they met at C, the location of the tree. Measuring the height of the triangle in the drawing, they would find it was 64 mm high, and thus conclude that the distance across the river to the tree was 64 m.

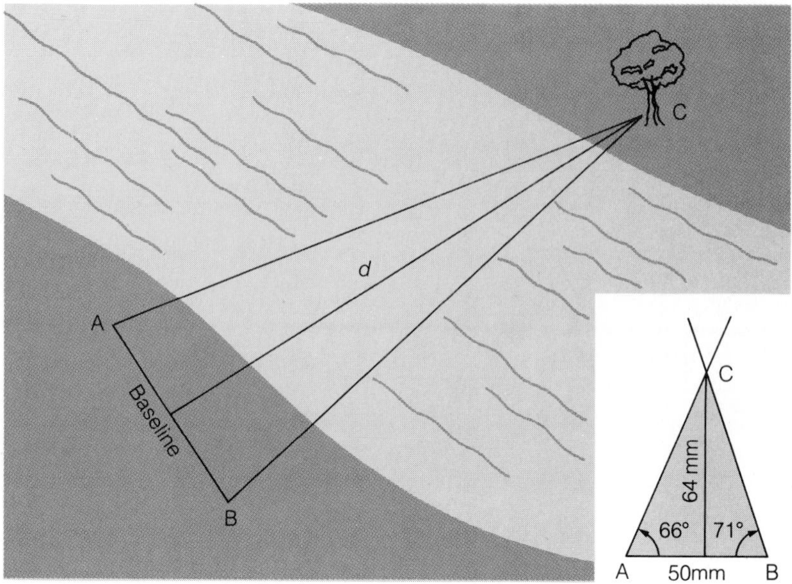

FIGURE 9–2 Surveyors can find the distance *d* across the river by measuring the baseline and the angles A and B and then constructing a scale drawing of the triangle.

The more distant an object is, the longer the baseline we must use to measure the distance to the object. We could use a baseline 50 m long to find the distance across a river, but to measure the distance to a mountain on the horizon, we might need a baseline 5 km long.

THE ASTRONOMER'S METHOD

No baseline we could draw on the earth would be long enough to allow us to measure the distance to the stars. Instead, we must use the diameter of the earth's orbit—the longest baseline possible at present. We can use this baseline because the earth carries us from one side of its orbit to the other as it circles the sun.

If we took a photograph of a nearby star and then waited 6 months, the earth would have moved halfway around its orbit. We could then take another photograph of the star at a point in space 2 AU (astronomical units) from the point where the first photograph was taken. Thus, our baseline would equal the diameter of earth's orbit, or 2 AU.

We would then have two photographs of the same part of the sky taken from slightly different locations in space. If we examined the two photographs, we would discover that the nearby star was not in exactly the same place as seen against the background of more distant stars (Figure 9–3). This apparent shift in the position of the star is called parallax. Parallax is the apparent change in the position of an object due to a change in the location of the observer. We saw in Chapter 4 that parallax is an everyday experience. Our thumb, held at arm's length, appears to shift position against a distant background when we look with first one eye and then the other (Figure 4–8).

The larger the distance to a star, the smaller the star's parallax. The distant stars in the background of our photographs are so distant they hardly shift at all, and we can use them as references against which to measure the parallax of nearby stars.

The quantity that astronomers call the **stellar parallax** (*p*) is half of the total shift of the star (Figure 9–3). Once this angle and the baseline of the observations are known, astronomers can calculate the distance to the star. Astronomers measure parallax and surveyors measure the angles at the ends of the baseline, but both measurements tell us the same thing—the shape of the triangle and thus the distance to the object in question.

Measuring the small angle *p* is very difficult (Figure 9–4). The nearest star, α Centauri, has a parallax of only 0.76 seconds of arc, and the more distant stars have even smaller parallaxes. To see how small these angles are, hold a piece of paper edgewise at arm's length. The thickness of the paper covers an angle of about 30 seconds of arc.

We cannot use a scale drawing to find the distance to a star because the distance is so large and the angle is so small. Even for the nearest star, the triangle would have

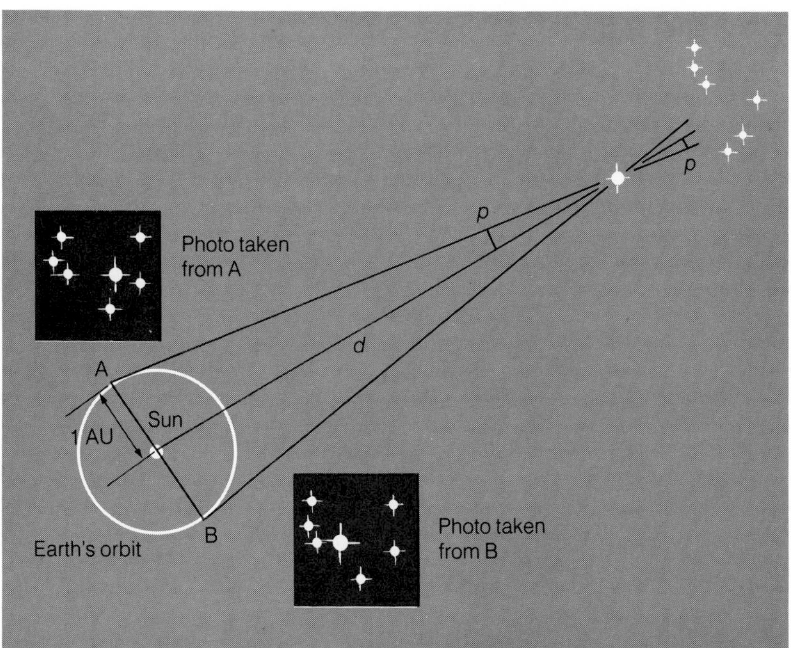

FIGURE 9–3 We can measure a nearby star's parallax by photographing it from two points along the earth's orbit, A and B. Half of the star's total change in position from one photograph to the other is its parallax *p*.

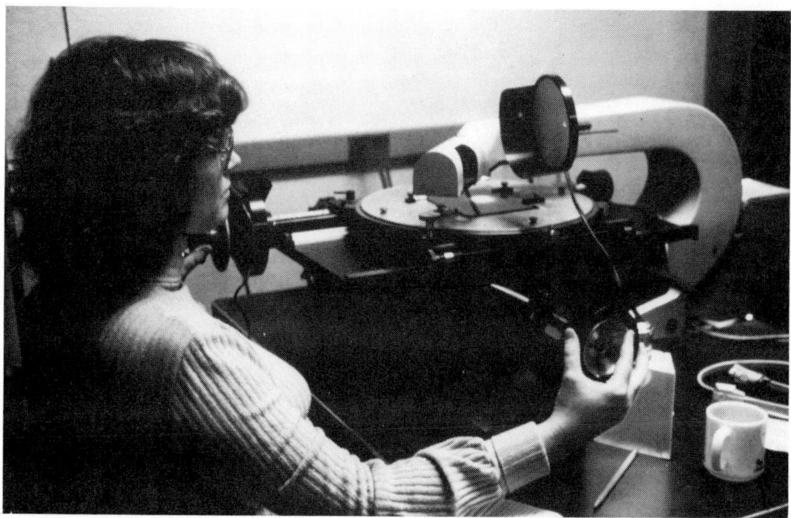

FIGURE 9–4 Parallax plates must be measured with great care on precision measuring machines. (Courtesy of Arthur Upgren.)

to be 300,000 times longer than it was wide. If the baseline in our drawing was 1 inch, the triangle would have to be about 5 miles long. Box 9–1 describes how we could find the distance from the parallax without drawing scale triangles.

The distances to the stars are so large it is not convenient to use astronomical units. As Box 9–1 explains, when we measure distance via parallax, it is convenient to use the unit of distance called a parsec (abbreviated pc). Parsec is derived from the definition: One **parsec** is the distance a star would have if its **par**allax was 1 **sec**ond of arc. One parsec is 206,265 AU, roughly 3.26 light-years.*

*The parsec is used throughout astronomy because it simplifies the calculation of distance. However, there are instances where the light-year is also convenient. Consequently, the chapters that follow use either parsecs or light-years as convenience and custom dictate.

BOX 9–1

Parallax and Distance

We wish to find the distance to a star from its measured parallax. To see how this is done, imagine that we observe the earth from the star. Figure 9–5 shows that the angular distance from the sun to the earth would equal the star's parallax *p*. To find the distance, we recall that the small-angle formula in Box 3–1 relates an object's angular diameter, its linear diameter, and its distance. In this case, the angular diameter is *p*, and the linear diameter is 1 AU. Then the small-angle formula, rearranged slightly, tells us that the distance to the star in AU is equal to 206,265 divided by the parallax in seconds of arc.

$$d = \frac{206{,}265}{p}$$

Because the parallaxes of even the nearest stars are less than 1 second of arc, the distances in AU are inconveniently large numbers. To keep the numbers manageable, astronomers have defined the **parsec** as their unit of distance in a way that simplifies the arithmetic. One parsec equals 206,265 AU, so the equation becomes

$$d = \frac{1}{p}$$

Thus, a parsec is the distance to an imaginary star whose parallax is 1 second of arc.

For example, the star Altair has a parallax of 0.20 seconds of arc. Then its distance is

$$d = \frac{1}{0.2} = 5\,\text{pc}$$

Because 1 parsec equals about 3.26 light-years, Altair is about 16.3 light-years away.

Because of the blurring caused by the earth's atmosphere, we can't measure parallaxes smaller than about 0.01 seconds of arc. This means that the farthest star whose parallax we can measure is about 100 pc away. Less than 1000 stars have accurately measured parallaxes.

Two space telescopes yet to be launched should improve the measurement of parallax. Once placed in orbit by the Space Shuttle, the Hubble Space Telescope will be able to measure parallax ten times more accurately than ground-based telescopes. Thus, the Space Telescope should be able to measure the parallax of stars out to 1000 pc. This tenfold increase in distance will produce a thousandfold increase in the number of stars whose parallax can be measured. In 1989, the European Space Agency launched the satellite Hipparcos. Although the satellite was stranded in a low orbit, its designers expect it to complete at least part of its mission to measure the parallaxes of 120,000 stars to very high precision.

Parallax can tell us the distance to the nearer stars, but how can we find out which stars are nearby? The answer is to watch how the stars move over the years.

PROPER MOTION

All the stars in the sky, including our sun, are moving along orbits around the center of our galaxy. We don't notice that motion over periods of years, but over the centuries it can significantly distort the shape of constellations (Figure 9–6).

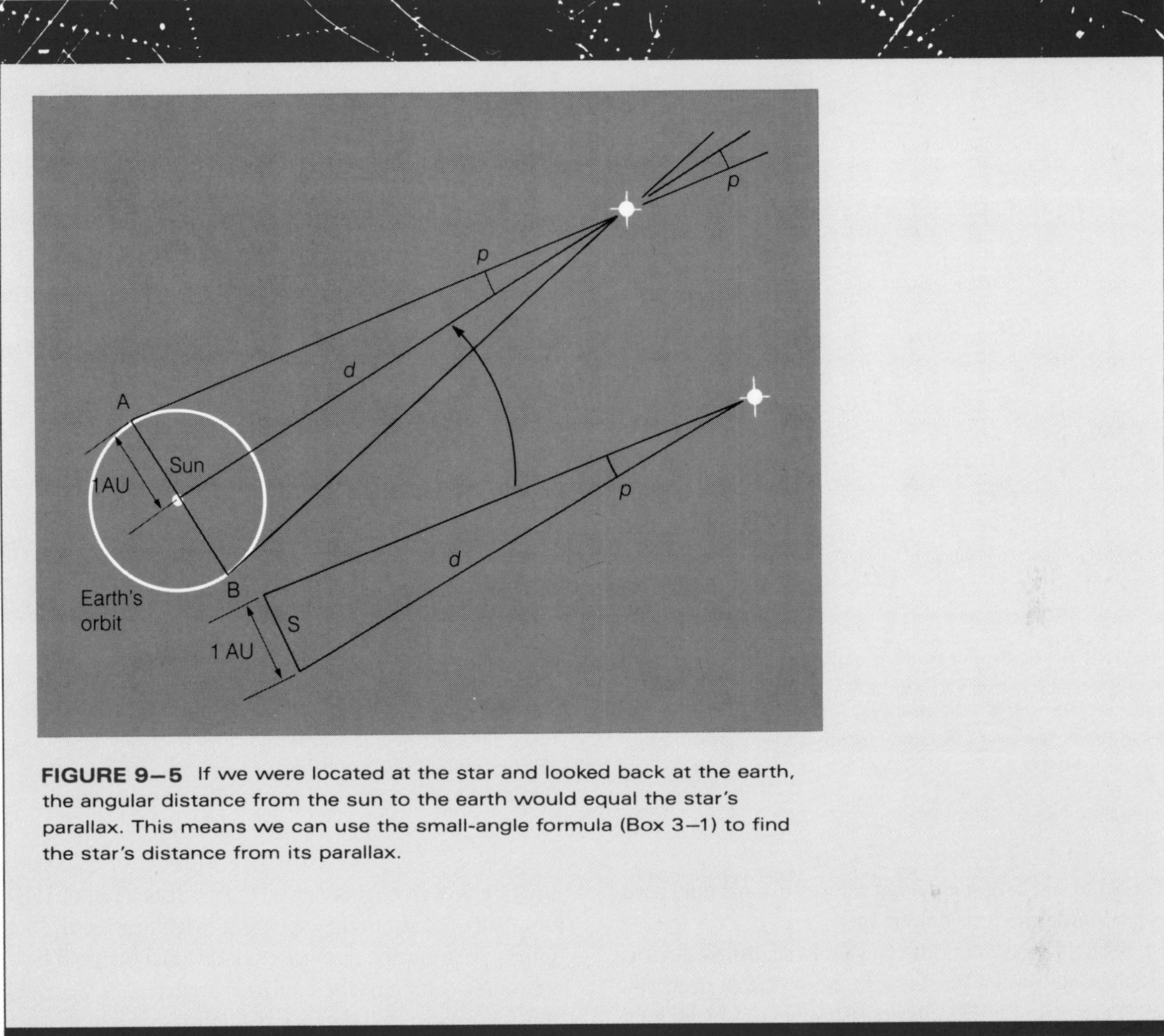

FIGURE 9–5 If we were located at the star and looked back at the earth, the angular distance from the sun to the earth would equal the star's parallax. This means we can use the small-angle formula (Box 3–1) to find the star's distance from its parallax.

If we photograph an area of the sky on two dates separated by 10 years or more, we can notice that some of the stars on the photograph have moved very slightly against the background stars. This motion, expressed in units of seconds of arc per year, is the **proper motion** of the stars. For example, the star Altair (α Aquilae) has a proper motion of 0.662 seconds of arc per year, but Albireo (β Cygni) has a proper motion of only 0.002 seconds of arc per year.

Why should one star have a larger proper motion than another star? For one thing, a star might be moving almost directly toward or away from us, and thus its position on the sky would change very slowly. That is, its transverse velocity (Chapter 7) could be quite low, resulting in a small proper motion (Figure 9–7a).

Another reason why a star might have a small proper motion is that it could be quite far away from us. Then even a large transverse velocity would not produce a large proper motion (Figure 9–7b). Thus Albireo, at a distance of 116 pc, has a smaller proper motion, and Altair, at a distance of only 4.9 pc, has a larger proper motion.

We can use proper motion to look for nearby stars. If we see a star with a small (or zero) proper motion, it is probably a distant star. We cannot be sure because it could be a nearby star traveling almost directly toward or away from us. But on the average, stars with small proper motions are distant. Similarly, stars with large proper motions are usually nearby stars. In fact, stars with proper motions above a certain limit must be nearby. A distant star traveling fast enough to have such a large proper

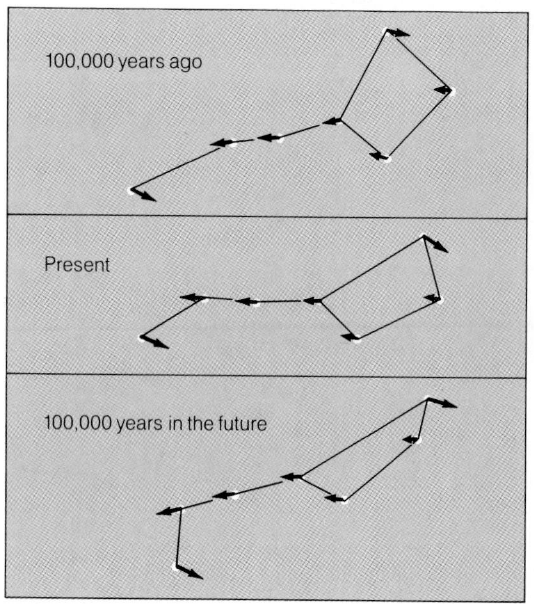

FIGURE 9–6 Proper motion slowly alters the positions of the stars in the sky. The Big Dipper looked quite different 100,000 years ago, and the proper motions of the stars (arrows) continue to change the Dipper's shape.

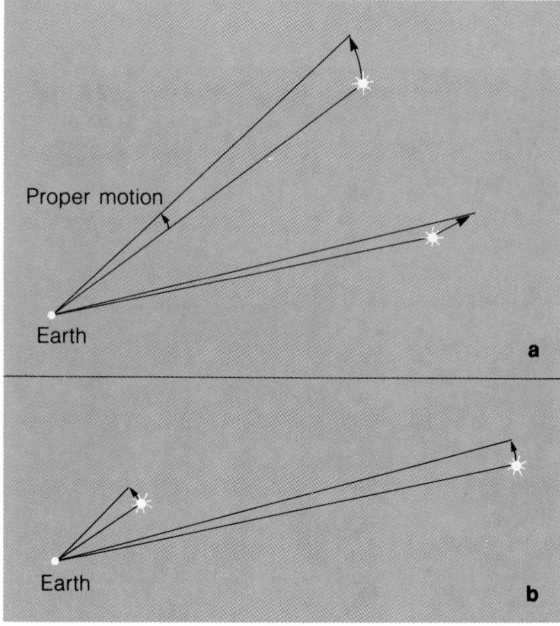

FIGURE 9–7 The proper motion of a star is the angular distance it moves across the sky in 1 year. Two stars with the same velocity through space can have different proper motions if one of them is moving approximately toward or away from the earth (a), or if one star is much more distant than the other (b).

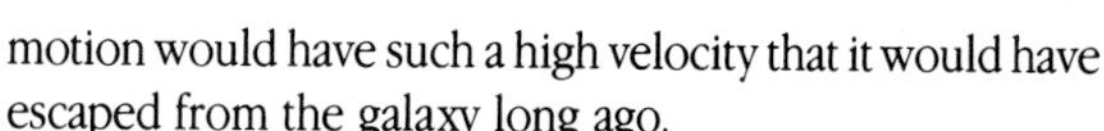

motion would have such a high velocity that it would have escaped from the galaxy long ago.

Thus, proper motions can give us statistical clues to distance. We can locate stars that are probably nearby—good candidates for parallax measurements—by looking for stars with large proper motions. Other statistical techniques can use proper motions to find the average distance to selected groups of stars.

Having found a way to locate nearby stars and measure their distances, we are ready to discuss the first of the three stellar parameters—brightness. Our goal is to find out how much energy stars emit.

9.2 INTRINSIC BRIGHTNESS

If we view a streetlight from nearby, it may seem quite bright, but if we view it from a hilltop miles away, it appears faint. Its apparent brightness depends on its distance, but its intrinsic brightness, the amount of light it emits, is independent of distance. When we look at stars, we face the same problem we might face trying to judge the brightness of city lights viewed from a distant hilltop (Figure 9–8). We can judge apparent brightness easily, but unless we know the distances to individual points of light we cannot determine their intrinsic brightnesses. We could not, for instance, tell distant streetlights from dimmer, but nearer light bulbs. Once an astronomer determines the distance to a star, however, it is simple to calculate its intrinsic brightness from its apparent brightness and distance.

We will use two terms to refer to a star's intrinsic brightness. One, related to the magnitude system, is common in astronomy because its use simplifies calculations involving distance. A second term refers directly to the amount of energy the star emits in 1 second.

ABSOLUTE VISUAL MAGNITUDE

Judging the intrinsic brightness of stars would be easier if they were all at the same distance. Although astronomers can't move stars about and line them up at some standard distance, they can calculate how bright a star of known distance would appear at any other distance. In

FIGURE 9–8 The valley below Mount Wilson sparkles with the lights of Los Angeles, Pasadena, Hollywood, and more than 40 other communities. Without knowing the distance to a light on the valley floor, we cannot estimate its intrinsic brightness. Similarly, we must know a star's distance before we can find its intrinsic brightness. (Mount Wilson and Las Campanas Observatories, Carnegie Institution of Washington.)

this way, they refer to the intrinsic brightness of a star as its **absolute visual magnitude** M_v—the apparent visual magnitude the star would have if it were 10 pc away (Box 9–2).

The symbol for absolute visual magnitude is a capital *M* with a subscript v. The subscript tells us it is visual magnitude based only on the wavelengths of light we can see. Other magnitude systems are based on other parts of the electromagnetic spectrum such as the infrared, ultraviolet, and so on. Yet another magnitude system refers to the total energy emitted at all wavelengths. We will limit our discussions to visual magnitudes.

The intrinsically brightest stars known have absolute magnitudes of about −8, and the faintest about +19. The nearest star to the sun, α Centauri, is only 1.4 pc away, and its apparent magnitude is 0.0, indicating that it looks bright in the sky. However, its absolute magnitude is 4.39, telling us it is not intrinsically very bright. Because we know the distance to the sun and can measure its apparent magnitude, we can find its absolute magnitude: about 4.78. If the sun were only 10 pc away from us, it would look no brighter than the faintest star in the handle of the Little Dipper.

LUMINOSITY

The **luminosity** (*L*) of a star is the total amount of energy the star radiates in 1 second—not just visible light, but all wavelengths. To find a star's luminosity, we begin with its absolute magnitude, make a small correction, and compare the star with the sun.

The correction we must make adjusts for the light emitted at wavelengths we cannot see. Absolute visual magnitude includes only visible light. The absolute magnitudes of hot stars and cool stars will underestimate their total luminosities because those stars radiate significant amounts of radiation in the ultraviolet or infrared parts of the spectrum. We can correct for the missing radiation because the amount of missing energy depends on only the star's temperature. For hot and cool stars the correction can be large, but for medium-temperature stars like the sun, the correction is small. Adding the proper correction to the absolute visual magnitude changes it into the **absolute bolometric magnitude**—the absolute magnitude the star would have if we could see all wavelengths.

Once we know a star's absolute bolometric magnitude, we can find its luminosity by comparing it with the

BOX 9–2

Absolute Magnitude

Apparent magnitude tells us how bright a star looks (Box 2–1), but absolute magnitude tells us how bright the star really is. The absolute visual magnitude M_v of a star is the apparent visual magnitude of the star if it were 10 pc away. If we know a star's apparent magnitude and its distance, we can calculate its absolute magnitude. The equation that allows this calculation relates apparent magnitude m, distance in parsecs d, and absolute magnitude M_v:

$$m - M_v = -5 + 5\log_{10}(d)$$

Sometimes it is convenient to rearrange the equation and write it as

$$d = 10^{\frac{m - M_V + 5}{5}}$$

It is the same equation, so we can use whichever form is most convenient in a given problem. This equation shows that the difference between apparent and absolute magnitude depends only on the distance to the star.

The quantity $m - M_v$ is called the **distance modulus**, a measure of how far away the star is. If the star is very far away, the distance modulus is large; if the star is close, the distance modulus is small. We could use the preceding equation to make a table of distance moduli (Table 9–1).

Table 9–1 Distance moduli.

$m - M_V$	d(pc)
0	10
1	16
2	25
3	40
4	63
5	100
6	160
7	250
8	400
9	630
10	1000
•	•
•	•
•	•
15	10,000
•	•
•	•
•	•
20	100,000
•	•
•	•
•	•

If we know the distance to a star, we can find its distance modulus from the table. If we subtract the distance modulus from the apparent magnitude, we get the star's absolute magnitude. For example, Deneb is 490 pc away and has an apparent magnitude of 1.26. From the table we find that its distance modulus is about 8.5. Then its absolute magnitude is about 1.26 − 8.5 or about −7.2. Deneb is intrinsically a very bright star. If it were only 10 pc away, it would dominate the night sky, shining 5.8 magnitudes or over 200 times brighter than Sirius.

sun. The absolute bolometric magnitude of the sun is +4.7. For every magnitude a star is brighter than 4.7, it is 2.512 times more luminous than the sun. (See Box 2–1.) Thus, a star with an absolute bolometric magnitude of 2.7 is 2 magnitudes brighter than the sun and 6.3 times more luminous (6.3 is approximately 2.512 × 2.512).

Canopus, for example, has an absolute bolometric magnitude of −0.7. That makes it 5 magnitudes brighter than the sun. A difference of 5 magnitudes is defined to be a factor of 100 in brightness, so the luminosity of Canopus is 100 times the sun's luminosity or 100 $L_{\odot}$.

The symbol $L_{\odot}$ represents the luminosity of the sun, a number we can calculate in a direct way. Because we know how much solar energy hits 1 square meter in 1 second just above the earth's atmosphere (the solar constant defined in the previous chapter) and how far it is from the earth to the sun, it is a simple matter to calculate how much energy the sun must radiate in all directions to provide the earth with the energy it receives per second. (See Problem 9.) Thus we find that the luminosity of the sun is about 4×10^{26} Joules/sec.

Let's review: If we can measure the parallax of a star, we can find its distance, calculate its absolute visual magnitude, correct for the light we can't see to find the absolute bolometric magnitude, and then find the luminosity in terms of the sun. If we want to know the luminosity in joules per second, we need only multiply by the sun's luminosity.

In this way we have reached our first goal. We have found a way to discover the energy emitted by stars. The range of stellar luminosities is very large. The most luminous stars have luminosities of about $10^5 L_{\odot}$, and the least luminous are roughly $10^{-4} L_{\odot}$. Knowing the luminosities of stars helps us toward our second goal, finding their diameters.

9.3 THE DIAMETERS OF STARS

Two factors influence an object's luminosity: temperature and surface area. For example, you can eat dinner by candlelight because the candle flame has a small surface area and consequently a small luminosity. However, if the flame were 12 ft tall, it would have a very large surface area from which to radiate, and, although it was no hotter than a normal candle flame, its luminosity would drive you away from the table.

In a similar way, a hot star may not be very luminous if it has a small surface area, although it could be highly luminous if it were larger. Even a cool star could be luminous if it had a large surface area. (See Box 9–3.) Because of this dependence on both temperature and surface area, we can use stellar luminosities to determine the diameters of stars if we can separate the effects of temperature and surface area.

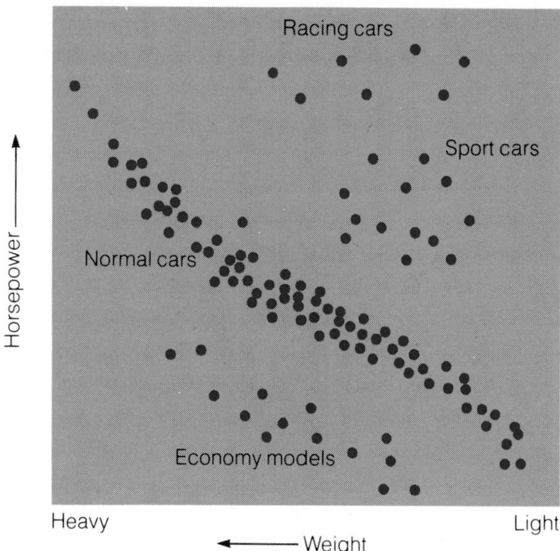

FIGURE 9–9 We could analyze automobiles by plotting their horsepower versus their weight and thus reveal relationships between various models. Most would lie somewhere along the main sequence of "normal" cars.

THE H–R DIAGRAM

The **Hertzsprung–Russell diagram**, named after its discoverers Ejnar Hertzsprung and Henry Norris Russell, is a graph that separates the effects of temperature and surface area on stellar luminosities and enables us to sort the stars according to their diameters. The **H–R diagram** (as it is often called) is the most important diagram in astronomy. It appears in this book 26 times. Referring to it sometimes as "the diagram," astronomers use it as a graphic way of thinking about the different kinds of stars. The diagram is a powerful tool because it sorts the stars by their diameters.

To see how a simple diagram can reveal hidden relationships among similar objects, to see how the H–R diagram can sort the stars into family groups according to size, let us begin with a similar diagram we might use to sort automobiles.

We could plot a diagram such as Figure 9–9 to show horsepower versus weight for various makes of cars. We

BOX 9–3

Luminosity, Radius, and Temperature

The luminosity L of a star depends on two things—its size and temperature. If the star has a large surface area from which to radiate, it can radiate a great deal. Recall from our discussion of black body radiation in Box 7–1 that the amount of energy emitted per second from each square meter of the star's surface is σT^4. Thus, the star's luminosity can be written as its surface area in square meters times the amount it radiates per square meter.

That is

$$L = \text{area} \times \sigma T^4$$

Because a star is a sphere, we can use the formula Area = $4\pi R^2$. Then the luminosity is

$$L = 4\pi R^2 \sigma T^4$$

This seems complicated, but if we divide by the same quantities for the sun we can cancel out the constants:

$$\frac{L}{L_\odot} = \left(\frac{R}{R_\odot}\right)^2 \left(\frac{T}{T_\odot}\right)^4$$

Here the symbol $\odot$ stands for the sun.

This is a very handy formula. Suppose we knew that a star was 10 times the sun's radius but only half as hot. How luminous would it be?

$$\frac{L}{L_\odot} = \left(\frac{10}{1}\right)^2 \left(\frac{1}{2}\right)^4 = \frac{100}{1}\frac{1}{16} = 6.25$$

The star would be 6.25 times the sun's luminosity.

We can also use this formula to find the diameters of stars. Suppose we found a star whose absolute magnitude was +1. It is 4 magnitudes brighter than the sun, and we recall from Box 2–1 that each magnitude represents a factor of 2.512. Then 4 magnitudes represents a factor of 2.512^4, or about 40. The star's luminosity is therefore 40 $L_\odot$. From its spectrum we can find its temperature—say twice the sun's temperature. What is its radius? Using our formula, we find

$$\frac{40}{1} = \left(\frac{R}{R_\odot}\right)^2 \left(\frac{2}{1}\right)^4$$

Solving for the radius we get

$$\left(\frac{R}{R_\odot}\right)^2 = \frac{40}{2^4} = \frac{40}{16} = 2.5$$

So the radius is

$$\frac{R}{R_\odot} = \sqrt{2.5} = 1.58$$

The star is 58 percent larger in radius than the sun.

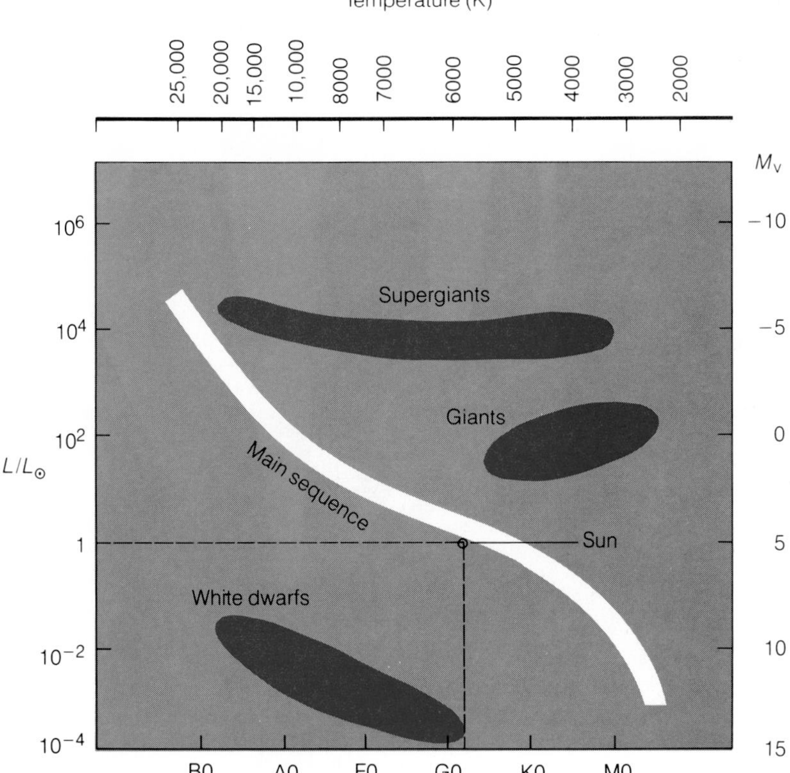

FIGURE 9–10 An H–R diagram. Roughly 90 percent of all stars lie on the main sequence.

would find that in general the more a car weighs, the more horsepower it has. Most cars would fall somewhere along the sequence of cars running from heavy, high-powered cars to lightweight, low-powered models. We could call this the main sequence of cars. But some would have much more horsepower than normal for their weight—the sport or racing models—and the economy models would have less power than normal for cars of the same weight. Just as this diagram would help us understand the different kinds of autos, the H–R diagram helps us understand the different kinds of stars.

The H–R diagram relates the intrinsic brightness of stars to their surface temperatures (Figure 9–10). We may plot either absolute magnitude or luminosity on the vertical axis of the graph, because both refer to intrinsic brightness. As you will remember from Chapter 7, spectral type is related to temperature, so we may plot either spectral type or temperature on the horizontal axis. Technically, only graphs of absolute magnitude versus spectral type are H–R diagrams. However, we will refer to plots of luminosity versus either spectral type or surface temperature by the generic term *H–R diagram.*

A point on an H–R diagram shows a star's luminosity and surface temperature. Points near the top of the diagram represent very luminous stars, and points near the bottom represent very faint stars. Points on the left represent hot stars, and points on the right represent cool stars. Notice that the location of a star in the H–R diagram has nothing to do with its location in space. Two stars near each other in the H–R diagram have similar properties but need not be near each other in space. Also, as a star ages, its luminosity and surface temperature change and the point which represents it moves in the H–R diagram, but this has nothing to do with the star's actual motion through space.

GIANTS, SUPERGIANTS, AND DWARFS

The **main sequence** is the region of the H–R diagram running from upper left to lower right (narrow area of Figure 9–10), which includes roughly 90 percent of all stars. These are the "ordinary" stars. As we might expect, the hot main-sequence stars are brighter than the cool

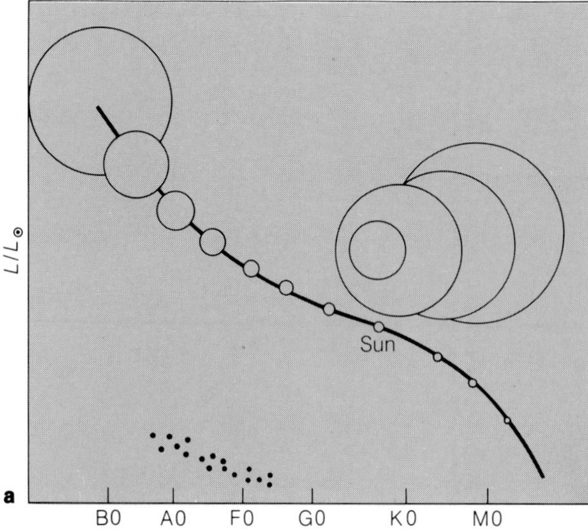

FIGURE 9–11 (a) This H–R diagram shows the relative sizes of stars. Giant stars are 10 to 100 times larger than the sun, and white dwarfs are about the size of the earth. (The dots representing white dwarfs here are much too large.) Supergiants are too large for this diagram. (b) To compare the size of the largest stars imagine that the sun is the size of one of your eyeballs. Then the largest supergiants would be the size of a hot air balloon.

main-sequence stars. The sun is a medium-temperature main-sequence star.

Just as sports cars do not fit in with the normal cars in Figure 9–9, some stars do not fit in with the main-sequence stars in Figure 9–10. The **giant stars** lie at the upper right of the H–R diagram. These stars are cool, radiating little energy per square meter. Nevertheless, they are highly luminous because they have enormous surface areas; hence the name *giant stars.* In fact, we can estimate the size of these giants with a simple calculation. Notice from the H–R diagram that they are about 100 times more luminous than the sun although they have about the same surface temperature. Thus, they must have about 100 times more surface area than the sun, indicating that their diameters must be about 10 times larger than the sun's (Figure 9–11).

Near the top of the H–R diagram we find a few stars called **supergiants.** These exceptionally luminous stars are 10 to 1000 times the sun's diameter. The stars Betelgeuse and Rigel (α and β Orionis) are both supergiants. The largest known supergiant may be μ Cephei, an M2 Ia star whose radius is believed to be about 3700 $R_{\odot}$. If μ Cephei were to replace the sun at the center of our solar system, it would engulf Mercury, Venus, Earth, Mars, Jupiter, Saturn, and nearly reach the orbit of Uranus (Figure 9–11).

In the lower left of the H–R diagram are the economy models, stars that are very faint although they are hot. Clearly such stars must be small. These are the **white dwarf stars,** dying stars that have collapsed to about the size of the earth and are slowly cooling off (Figure 9–12).

LUMINOSITY CLASSIFICATION

We can tell from a star's spectrum what kind of star it is. Recall from Chapter 7 that collisional broadening can make spectral lines wider when the density of a gas is relatively high and the atoms collide with each other often.

Main-sequence stars are relatively small and have dense atmospheres where the gas atoms collide often and distort their electron energy levels. Thus, lines in the spectra of main-sequence stars are broad. On the other hand, giant stars are larger, their atmospheres are less dense, and the atoms disturb each other relatively little. Collisional broadening is not as important in the atmospheres of giant stars, and their spectra contain sharp lines. As we

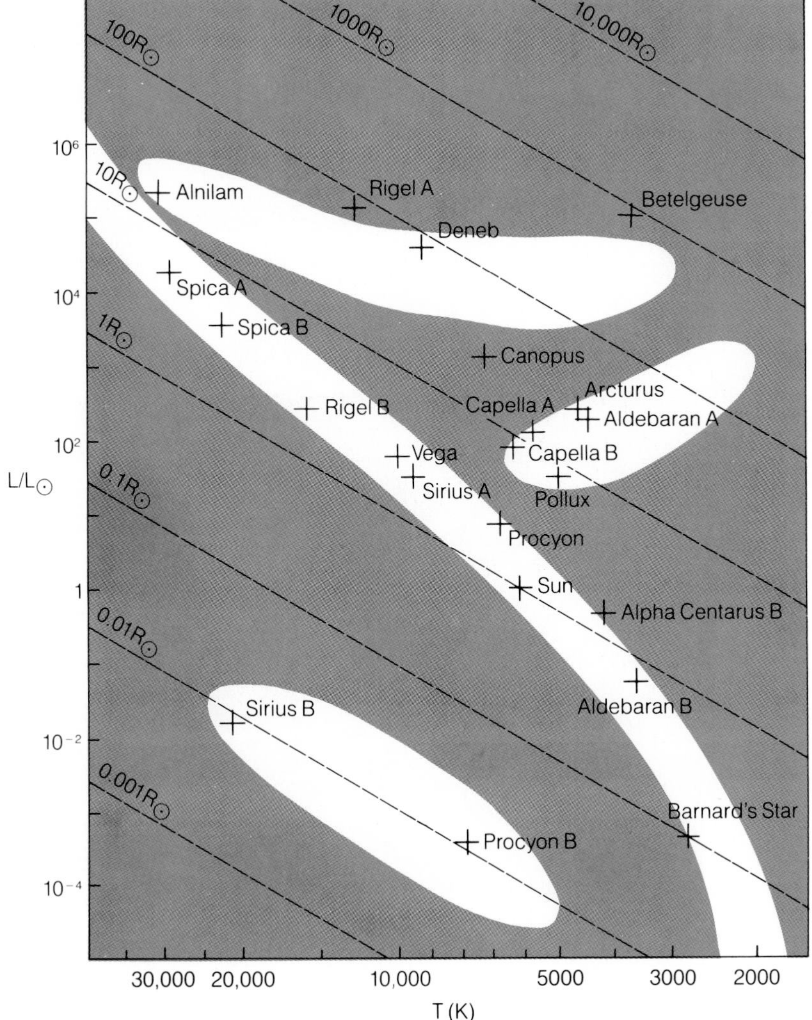

FIGURE 9–12 An H–R diagram drawn with luminosity versus surface temperatures. Diagonal lines are lines of constant radius. White areas mark approximate location of supergiants, giants, the main sequence, and white dwarfs. Note that giant stars are 10 to 100 times larger than the sun and that white dwarfs are about 100 times smaller than the sun—that is, about the size of the earth.

might expect, the lines in the spectra of supergiants are even sharper.

Thus, we can look at a star's spectrum and classify its luminosity (Figure 9–13). We can tell whether it is a supergiant, a bright or ordinary giant, a subgiant, or a main-sequence star. Although these are the **luminosity classes**, the names refer to the sizes of the stars because size is the dominating factor in determining luminosity. Supergiants, for example, are very bright because they are very large.

These luminosity classes are represented by the roman numerals I through V, as shown in this list:

- Ia Bright supergiant
- Ib Supergiant
- II Bright giant
- III Giant
- IV Subgiant
- V Main-sequence star

Using letters for subclasses, we can distinguish between the bright supergiants (Ia) such as Rigel (β Orionis) and the regular supergiants (Ib) such as Polaris, the North Star. The star Adhara (ε Canis Majoris) is a bright giant (II), Capella (α Aurigae) is a giant (III), and Altair (α Aquilae) is a subgiant (IV). The sun is a main-sequence (V). The luminosity class usually appears after the spectral type, as in G2 V for the sun. White dwarf stars don't enter into this classification because their spectra are quite unlike those of other stars.

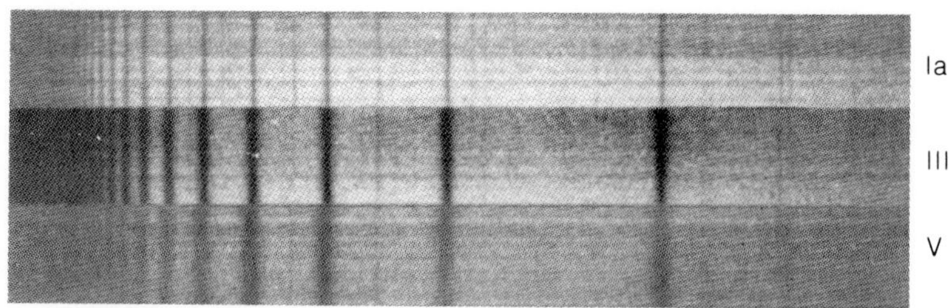

FIGURE 9–13 Differences in widths and strengths of spectral lines distinguish the spectra of supergiants, giants, and main-sequence stars, thus making the luminosity classification possible. (Adapted from H. A. Abt, A. B. Meinel, W. W. Morgan, and J. W. Tapscott. (*An Atlas of Low-Dispersion Grating Stellar Spectra,* Kitt Peak National Observatory, 1968.)

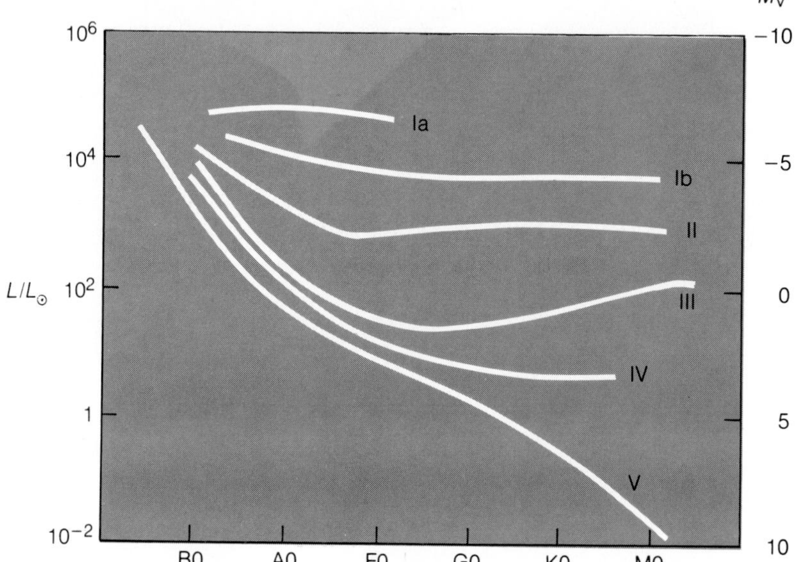

FIGURE 9–14 The approximate location of the luminosity classes on the H–R diagram.

We can, as in Figure 9–14, plot the locations of the luminosity classes on the H–R diagram. Remember that these are rather broad classifications. A star of luminosity class III may lie slightly above or below the line labeled III. The lines are only approximate.

The luminosity classes are an important tool because the luminosity of a star can give us a clue to its distance. This method of finding distance is called **spectroscopic parallax** (Box 9–4), and we can use it to estimate the distance to stars that are too far away to have measurable parallaxes. As long as we can photograph a star's spectrum, we can estimate its distance.

The two goals of this chapter have been achieved; we now know how to find the luminosities and diameters of stars. We have discovered that there are four different kinds of stars—main-sequence stars, giants, supergiants, and white dwarfs—and that these types differ in their intrinsic properties, luminosity, and diameter.

A third intrinsic property of a star is its mass, but the determination of stellar masses is such an important and interesting aspect of astronomy, we will consider it in a separate chapter. First, however, we pause in the Perspective on page 198 to consider the frequency of the stellar types—that is, how common are the different types of stars? The problems in answering this question illustrate the importance of distance and luminosity in astronomy and give us clues to the evolution and death of the stars.

SUMMARY

We can measure the distance to the nearer stars by observing their parallaxes. The more distant stars are so far away that their parallaxes are unmeasurably small. To find the distances to these stars, we must use spectroscopic parallax. Stellar distances are commonly expressed in light-years or parsecs. One light-year is

BOX 9–4

Spectroscopic Parallax

Driving along the highway at night, we often see lights dotting the countryside. We cannot immediately tell how far away these lights are unless we know how bright they really are. If we know that one of the lights is an airport searchlight and another is the headlight on a bicycle, we can judge their distances. The same is true of stars. If we can discover the luminosity of a star, we can use its apparent brightness to estimate its distance.

The method of **spectroscopic parallax** lets us find the distance to a star by classifying its spectrum according to spectral type and luminosity class. We can then look it up on an H–R diagram such as Figure 9–14 and read off its absolute magnitude M_V. Once we measure its apparent magnitude *m,* we can calculate the distance using either the equation (left) from Box 9–2, or a distance modulus table such as Table 9–1.

$$d = 10^{\frac{m - M_V + 5}{5}}$$

For example, Spica is classified as B1 V, and its apparent magnitude is +1. From Figure 9–14 we can estimate that a B1 V star should have an absolute magnitude of about −3. Therefore, its distance modulus is 4, and the distance (taken from the table) is about 63 pc.

This method is not very accurate because there is some uncertainty in Figure 9–14 due to individual differences between stars. Consequently, when we classify a star's spectrum, we can't be sure of its exact absolute magnitude. It might be a little brighter or fainter than the diagram predicts. If the star is just one magnitude fainter than we expect, the distance we calculate is 37 percent too small. Although this method is not very accurate, spectroscopic parallax is often the only method available to measure distance.

the distance light travels in 1 year, about 9.5×10^{12} km or about 5.8 trillion miles. One parsec is about 206,265 AU—the distance to an imaginary star whose parallax is 1 second of arc. One parsec equals 3.26 light-years.

One way to locate nearby stars is to look for stars with high proper motions. Proper motion is the angular motion of a star across the sky and is usually expressed in seconds of arc per year. Nearby stars are likely to have larger proper motions than more distant stars.

Once we know the distance to a star, we can find its intrinsic brightness, expressed as its absolute visual magnitude. We can add a small correction to account for the light at wavelengths we cannot see and thus transform absolute visual magnitude into absolute bolometric magnitude. From this we can find the luminosity of the star, the total energy radiated in 1 second. Luminosity is often expressed in terms of the sun, as in $10\ L_{\odot}$.

The H–R diagram is a graph in which we plot stars according to their intrinsic brightness and their surface temperature. In the diagram, roughly 90 percent of all stars fall on the main sequence, the more massive being hotter, larger, and more luminous. However, the giants and supergiants are much larger and lie above the main sequence; they are more luminous than main-sequence stars of the same temperature. The white dwarfs are hot stars, but they fall below the main sequence because they are so small.

Because the atmospheres of giant and supergiant stars have low density, collisional broadening does not widen the lines in their spectra very much. Thus, giants and supergiants have sharper spectral lines than do main-sequence stars. In fact, it is possible to assign stars to luminosity classes by the widths of their spectral lines. Class V stars are main-sequence stars with broad spectral lines. Giant stars (III) have sharper lines, and supergiants (I) have extremely sharp spectral lines.

A survey in the neighborhood of the sun shows us that the most common kind of stars are the lower main-sequence stars. The hot stars of the upper main sequence are very rare. Giants and supergiants are also rare, but white dwarfs are quite common, although they are faint and hard to find.

PERSPECTIVE

A NEIGHBORHOOD SURVEY

SURVEYING A REPRESENTATIVE SAMPLE

Suppose you took a survey in your neighborhood to find how many people have gray eyes. If you knew the area of your neighborhood in square miles, you could then say, "In my neighborhood, x people per square mile have gray eyes." Next you might think of extending your conclusion to the country as a whole: "In America x people per square mile have gray eyes." Of course if you did not live in an average neighborhood, your result would be wrong, but if you had sampled a truly representative neighborhood, your survey would give valid results about the entire population.

We can do the same thing with the stars. We can ask how many stars of each spectral type are whirling through each million cubic parsecs of space. We can't count every star in our galaxy, but we can take a survey in the region of space near the sun. Because we think the solar neighborhood is a fairly average sort of place, we can use this local survey to reach conclusions about the entire population of stars.

UNITS OF STELLAR DENSITY

The result of such a survey of stars is called a **stellar density function**, a description of the abundance of different types of stars in space. A simple form of the stellar density function appears in Figure 9–15, giving the abundance of the stars of each spectral type in terms of the number of each type we would expect to find in 1,000,000 pc^3.

Clearly, the most common kind of star is an M star. Every million cubic parsecs of space contains 50,000 M stars, but only about 2000 B stars. One of the questions we must resolve in the next few chapters is, Why are M stars so common?

But notice that the stellar density function does not tell us whether these M stars are main-sequence stars, giants, or supergiants. It lumps all M stars into one category. To distinguish among the luminosity classes, we must consider both spectral type and absolute magnitude when we make our survey. Unfortunately, certain problems stand in our way.

THREE PROBLEMS IN COUNTING STARS

The astronomer could make these surveys in the neighborhood of the sun by counting all the stars within a given distance. A sphere of radius 62 pc contains 1,000,000 pc^3. Thus, if we could count all of the M stars, for example, within 62 pc of the sun, we would have the frequency of M stars per million cubic parsecs.

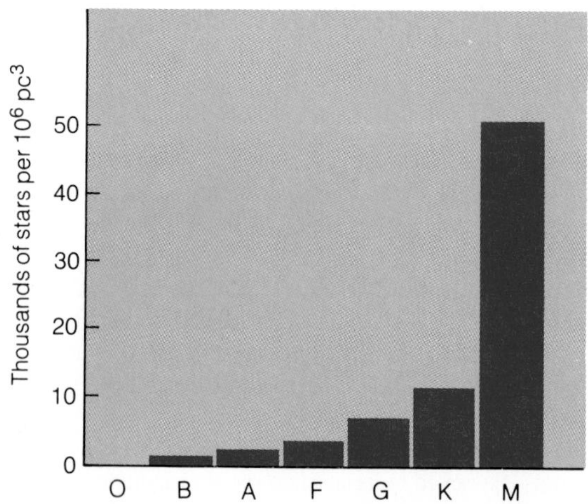

FIGURE 9–15 The stellar density function shows that M stars are the most common stars in space and that O stars are very rare.

But this survey of the stars poses three problems. First, to determine which stars are within 62 pc of the sun, we must measure their distances. However, stars near the outer edge of a sphere 62 pc in radius have small parallaxes that are difficult to measure accurately, and the method of spectroscopic parallax is not accurate enough for this purpose. We could count stars in a smaller sphere, but some stars are so rare that we might not find any in such a small volume of space.

A second problem for the stellar surveyor is the intrinsic faintness of stars such as M stars and white dwarfs. These are so faint that they are very hard to see if they are only a few dozen parsecs away. For example, a white dwarf 62 pc away is over 1500 times fainter than the faintest star visible to the naked eye—very hard to find, indeed.

The third problem with these surveys is that the hottest stars are rare. There are no O stars at all within 62 pc of the sun. We must extend our survey to great distances before we find many of these hot, luminous stars. At such distances the parallaxes are too small to be measured, and we have to find distances in other, less accurate ways, such as spectroscopic parallax.

THE FREQUENCY OF STELLAR TYPES

Despite these difficulties, astronomers have been able to piece together the stellar density function of the stars in our galaxy. Using statistical methods, they can now tell us not just the abundance of stars of a given spectral type, but also how many stars of that type are likely to be giants, how many supergiants, and how many white dwarfs. This detailed version of the stellar density function is shown in Figure 9–16.

Notice how common the main-sequence stars are. There are about 50,000 in every million cubic parsecs.

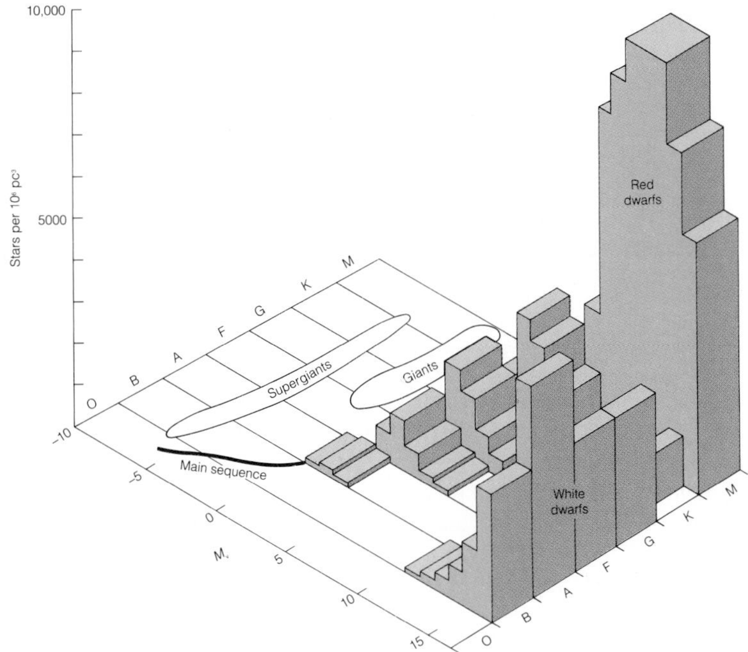

FIGURE 9–16 In this histogram, bars rise from an H-R diagram to represent the number of stars of given spectral type and absolute magnitude found in one million cubic parsecs. The main-sequence M stars are the most common stars, followed closely by the white dwarfs. The upper-main-sequence stars, giants, and supergiants are so rare their bars are not visible in this diagram.

Perspective: A Neighborhood Survey *(continued)*

Notice also how few M giants there are, only about 20 per million cubic parsecs. The white dwarfs and the main-sequence M stars are the most common kinds of star (Figure 9–17). Fortunately for us, these stars are very faint, for if they were as bright as supergiants, the sky would be filled with their glare and we would hardly see anything else. As it is, these most common stars are so faint they are hard to find even with large telescopes.

The most luminous stars are very rare. On the average there are only 0.03 main-sequence O stars per million cubic parsecs. That is, we would have to search through about 30,000,000 pc^3 to find an O star. Put yet another way, only one star in 4 million is an O star. The giants are slightly more abundant than this. There are a few hundred giants and subgiants in every million cubic parsecs, but the supergiants are very rare. There are about 0.07 supergiants per million cubic parsecs. Luckily, these very rare kinds of stars are also very luminous. We can see them from great distances. If they were as faint as the sun, we might never know they existed.

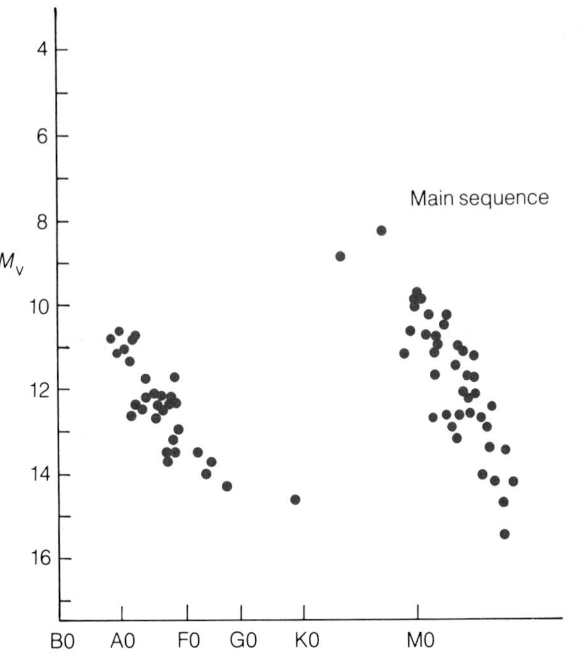

FIGURE 9–17 An H–R diagram of stars within 100 pc of the sun shows how common main-sequence stars are. A few of the more luminous stars, such as Sirius, are not plotted. (Adapted from official U.S. Naval Observatory photograph.)

NEW TERMS

stellar parallax (p)
parsec (pc)
proper motion
absolute visual magnitude (M_v)
distance modulus ($m - M_v$)
luminosity (L)
absolute bolometric magnitude
H–R diagram
main sequence
giant stars
supergiant stars
white dwarf stars
luminosity class
spectroscopic parallax
stellar density function

QUESTIONS

1. Why are parallax measurements limited to the nearest stars?
2. How would having an observatory on Mars help astronomers measure parallax?
3. How would having an observatory in orbit above the earth's atmosphere help astronomers measure parallax?
4. If someone asked you to compile a list of the nearest stars to the sun based on your own observations, what measurements would you make and how would you analyze them to detect nearby stars?

5. For which stars does absolute visual magnitude differ least from absolute bolometric magnitude? Why?
6. How can a cool star be more luminous than a hot star? Give some examples.
7. How can we be certain that the giant stars are actually larger than the sun?
8. Describe the steps in using the methods of spectroscopic parallax. Do we really measure a parallax?
9. What are the approximate radii and intrinsic brightnesses of stars in the following classes: G2 V, G2 III, G2 Ia.
10. What is the most common kind of star?

PROBLEMS

1. If a star has a parallax of 0.050 seconds of arc, what is its distance in parsecs? In light-years? In AU?
2. If a star has a parallax of 0.016 seconds of arc and has an apparent magnitude of 6, how far away is it and what is its absolute magnitude?
3. Complete the following table.

m	M_v	d (pc)	p (sec of arc)
—	7	10	—
11	—	1000	—
—	−2	—	0.025
4	—	—	0.040

4. If a main-sequence star has a luminosity of 400 $L_{\odot}$, what is its spectral type? (HINT: See Figure 9–10.)
5. If a star has an apparent magnitude equal to its absolute magnitude, how far away is it in parsecs? In light-years?
6. If a star has an absolute bolometric magnitude that is 8 magnitudes brighter than the sun, what is the star's luminosity?
7. If a star has an absolute bolometric magnitude that is 1 magnitude fainter than the sun, what is the star's luminosity?
8. An O8 V star has an apparent magnitude of +1. Use the method of spectroscopic parallaxes to find the distance to the star. Why might this distance be inaccurate?
9. Find the luminosity of the sun given the radius of the earth's orbit and the solar constant (Chapter 8). Make your calculation in two steps. First, use $4\pi R^2$ to calculate the area in square meters of a sphere surrounding the sun with a radius of 1 AU. Second, multiply by the solar constant to find the total solar energy passing through the sphere in 1 second. That is the luminosity of the sun. Compare your result with that in Appendix C.
10. In the following table, which star is brightest in apparent magnitude? Most luminous in absolute magnitude? Largest? Farthest away?

Star	Spectral type	m
a	G2 V	5
b	B1 V	8
c	G2 Ib	10
d	M5 III	19
e	white dwarf	15

11. Using the data in Appendix C, plot H–R diagrams for the brightest stars and for the nearest stars. Explain why they are different.

RECOMMENDED READING

DeVORKIN, D. H. "Steps Toward the Hertzsprung–Russell Diagram." *Physics Today 31* (March 1978), p. 32.

EVANS, D. S., T. G. BARNES, and C. H. LACY "Measuring Diameters of Stars." *Sky and Telescope 58* (August 1979), p. 130.

GETTS, JUDY A. "Decoding the Hertzsprung–Russell Diagram." *Astronomy 11* (Oct. 1983), p. 16.

HACK, M. "The Hertzsprung–Russell Diagram Today." *Sky and Telescope 31* (May/June 1966), pp. 260 and 333.

IRWIN, J. B. "The Case of the Degenerate Dwarf." *Mercury 7* (Nov./Dec. 1978), p. 125.

PHILIP, A. G. D., and L. C. GREEN "Henry Norris Russell and the H–R Diagram." *Sky and Telescope 55* (April 1978), p. 306.

———. "The H–R Diagram as an Astronomical Tool." *Sky and Telescope 55* (May 1978), p. 395.

REDDY, FRANCIS "How Far the Stars." *Astronomy 11* (June 1983), p. 6.

STRUVE, O., and V. ZEBERGS *Astronomy of the 20th Century.* New York: Crowell, Collier and Macmillan, 1962.

Verschuur, G. L. "Measuring Star Diameters." *Astronomy 2* (Dec. 1974), p. 36.

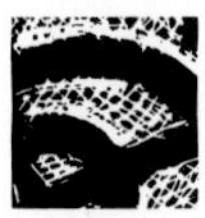

CHAPTER 10

BINARY STARS

LQMZISF LQMZISF SMLLSF JLOH KCQ M QCZAFH QKOL TCN OHF ND OBCPF LKF QCHSA JC KMGK SMIF O AMOECZA MZ LKF JIT*

Everybody loves hidden messages. They can be cryptograms like the one at the left or secret messages in invisible ink. There can be meaning in ancient hieroglyphics or systems of cosmology in ancient monuments like Stonehenge. Perhaps that is one reason why astronomers find **binary stars**—pairs of stars that orbit around each other—so fascinating. Properly decoded, their message tells us something we can learn in no other way, the masses of the stars.

If we look at a star through a telescope, there is nothing to tell us the star's mass. There is no clue buried in its luminosity, temperature, color, or spectrum. Only by studying pairs of stars that orbit each other can we find the masses of the stars. The orbits are governed by the gravity of the stars, and the gravity depends on the amount of mass in the stars. Thus, the orbits contain clues to the masses.

The behavior of certain binary stars also contains clues to the diameters of the stars, and this, combined with the masses, will give us important hints about the formation and evolution of stars. When we look at the masses of stars, we will discover a pattern concealed in the H–R diagram, and when we look at the density of matter in stars, we will find a different pattern. We would never unravel the puzzle of stellar evolution without this information from binary stars.

*This cryptogram contains two messages. One is easy to find, but the other is hidden deeper.

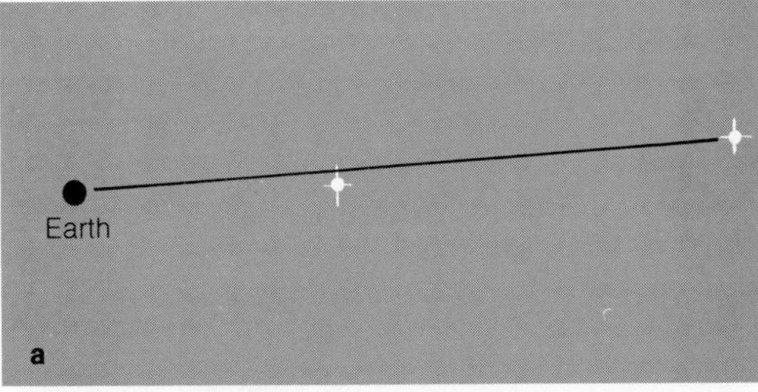

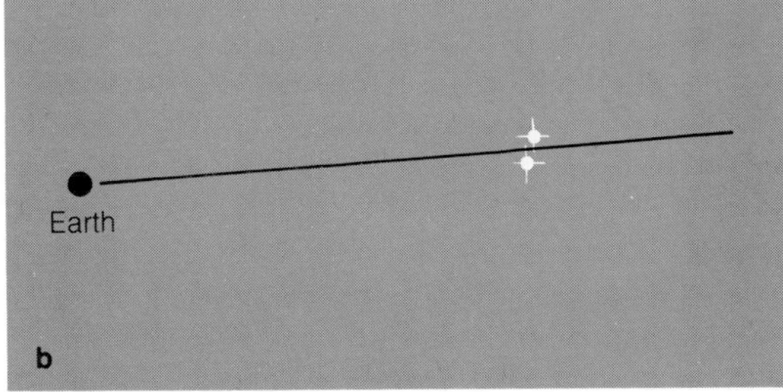

FIGURE 10–1 (a) An optical binary is a pair of stars that appear along the same line of sight but are not related. That is, they do not orbit each other. (b) A visual binary is a pair of stars that orbit each other.

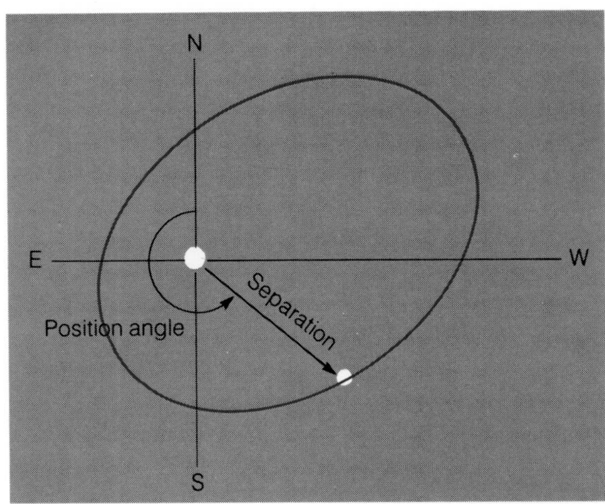

FIGURE 10–2 Measuring a visual binary. Separation is the angular distance between stars in seconds of arc, and position angle is measured from north toward the east around to the fainter star.

In this chapter we will discover a number of different kinds of binary stars. Some are easier to study, some give better results, and some are more common. But remember, in every case, the secret is hidden in the orbital motion.

10.1 VISUAL BINARIES

Through even a small telescope, we can see hundreds of **double stars**—close pairs of stars. Some of these are **visual binaries,** pairs of stars that are physically associated with each other. That is, they orbit around each other. Rarely are double stars composed of a nearby star and a much more distant star that only seem to be associated. Clearly, these **optical binaries** do not orbit each other (Figure 10–1), and although they are no longer important in modern astronomy, they sparked the discovery of binary stars almost two centuries ago.

Starting about 1782, William Herschel began looking for optical binaries. He hoped to detect the parallax of the nearby star by comparing it with the more distant star. Although he catalogued hundreds of pairs of stars, none were optical binaries. By 1804, he had accumulated enough measurements of the pair of stars known as Castor to show that they were in orbit around each other. This was the first orbital motion detected outside the solar system, and it marked the beginning of the observation of binary stars.

OBSERVING VISUAL BINARIES

In a visual binary the stars are usually only a few seconds of arc apart, and they orbit around each other with a period that is usually a few decades long. Because the stars are so close to each other, it is difficult to measure their positions precisely. Nevertheless, because their periods are so long, astronomers can make many measurements over the years and average the results. Some of the errors of measurement average out.

If we were to begin observing a visual binary, we would want to measure the separation between the stars in seconds of arc and the position of the stars with respect to a north-south line. We could do this by drawing a line from the brighter star to the fainter star and measuring the angle this line makes with a north-south line. This is called the **position angle** and is always measured from north toward the east to the fainter star (Figure 10–2).

FIGURE 10–3 Filar micrometers were once used regularly to measure the separation and position angle for visual binaries. Most such measurements are now made photographically. (Grundy Observatory photograph.)

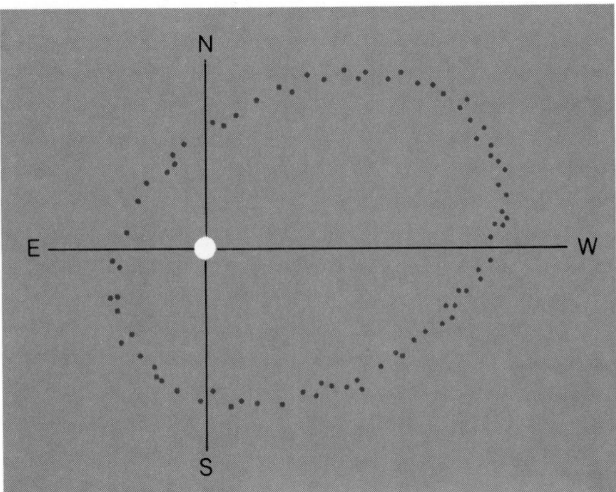

FIGURE 10–4 Plotting the separations and position angles on polar graph paper reveals the apparent relative orbit of the fainter star about the brighter star.

Note from Figure 10–2 that east is to the left in diagrams of the sky. On a map of the earth, we place north at the top and east to the right, but sky charts are made to be held overhead. If you face south and hold Figure 10–2 overhead, you will see that east must be to the left in such diagrams.

Not many decades ago astronomers measured position angles and separations directly at the telescope using a device known as a **filar micrometer** (Figure 10–3). This instrument carried very fine cross hairs, often made from spider silk, that could be moved with high precision to line up with the stars. The separation and position angle could be read from dials on the micrometer. At about the turn of the century, many observatories had such instruments, and astronomers spent long hours at their telescopes making precise, difficult observations of visual binaries.

Modern astronomers measure most visual binaries by photographing the stars and then measuring the photographic plate. They can take photographs faster than they could make the measurements directly, and the plates can be measured repeatedly by high-precision instruments to determine separations and position angles. Yet a few talented observers continue to use filar micrometers to measure visual binaries that are too close together to be measured photographically.

If we were studying a visual binary, making the observations might take many years, but that is only the first step. We would next plot the data to produce a chart of the orbit (Figure 10–4). Of course, both stars are moving around the center of mass of the system (Chapter 5). However, because of the way we measure separations and position angles, the bright star appears to be stationary in our data. Our graph would show the motion of the fainter star relative to the brighter star, and thus we call it the **apparent relative orbit**.

The orbit shown in Figure 10–4 is an apparent relative orbit. The word *apparent* refers to the fact that we see the orbit tipped at some unknown angle (Figure 10–5). To find the **true relative orbit**, we must untip the apparent relative orbit. The shape of the apparent relative orbit contains clues to the shape of the true orbit. For example, the center of the apparent orbit must be the same as the center of the true orbit; tipping an orbit at an angle distorts its shape but does not move the center. We also know that the two stars obey Kepler's laws, so the bright star must be at one focus of the true orbit. Then a line through the bright star and the center of the apparent orbit must be the major axis of the true ellipse. Various methods exist for solving the puzzle and discovering the shape of

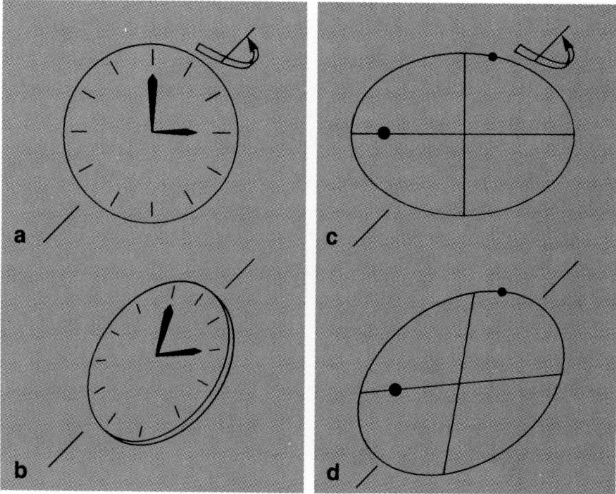

FIGURE 10–5 Just as the circular dial of a clock (a) tipped at an angle is distorted into an ellipse (b), so the elliptical orbit of a binary star tipped at an angle is distorted into different ellipse. To find the mass of a binary system, astronomers must observe the apparent relative orbit (d) and mathematically untip it to find the true relative orbit (c).

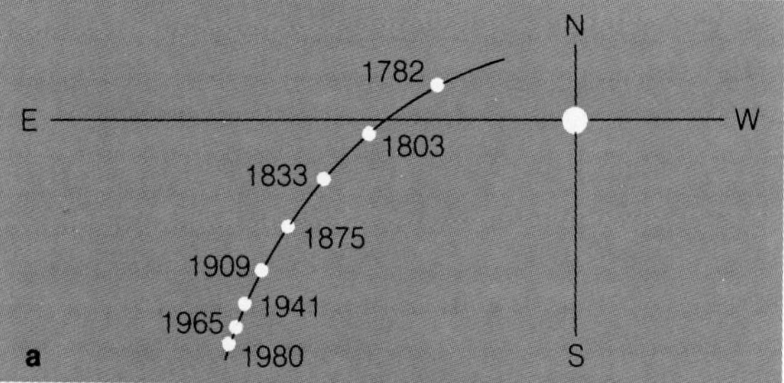

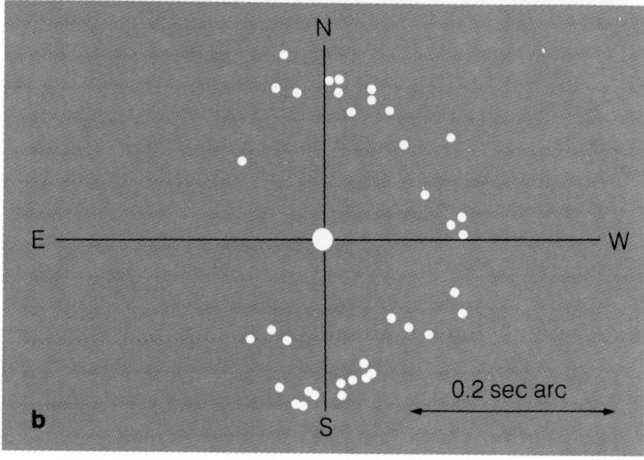

FIGURE 10–6 (a) The orbit of the visual binary Gamma Leonis has a very long period. Attempts to analyze the partial orbit have given results for the period that range from 400 to 620 years. (b) The visual binary A88 has such a small separation it is very difficult to measure accurately, although its period is only 22 years. (From data by Robert G. Aitken.)

the true orbit. Once we know the true relative orbit, we are almost ready to calculate the masses of the stars.

Unfortunately, not all visual binaries can be observed well enough to analyze. If the stars were far enough apart so we could measure their separations accurately, they would be so far apart in space they would have very long orbital periods, often thousands of years. Humans have not been accumulating observations long enough to plot the apparent orbits of such long period visual binaries, and binaries with partial orbits cannot be analyzed accurately (Figure 10–6a). On the other hand, if the orbital period were short enough to observe conveniently, the stars would probably be so close together we couldn't observe their separations with any accuracy (Figure 10–6b).

THE MASSES OF VISUAL BINARIES

If we observed a visual binary system and found its true relative orbit, we could find the total mass of the system using simple physics. We would need two pieces of information: the orbital period in years and the average separation between the stars in astronomical units (AU). Finding the orbital period from the orbit is simple, but finding the separation is a bit more complicated.

The average distance between the two stars equals the semimajor axis of the true relative orbit, but we know this distance as an angle in seconds of arc, not in AU. To convert an angular measurement into a linear measurement, we need to know the distance to the star. Then we can use the small-angle formula (Box 3–1) to find the average separation in AU.

If we know the period and separation, the mass of the system is easy to find. Newton's first law of motion says that an object in motion tends to stay in motion in a straight line unless acted upon by some force. The two stars in Figure 10–7 should move at constant speed in the straight-line paths shown by the dashed lines unless forces act upon them. In a binary star system the stars move around each other in orbits because their mutual gravitation pulls them away from straight-line motion. We could ask ourselves how much gravity has to be present to hold the two stars at their average separation while they whirl around each other at their given period. The amount of gravity present depends only on the amount of mass in the system, so this would tell us the sum of the masses of the two stars. This is actually an application of Kepler's third law of planetary motion as shown in Box 10–1.

To find the individual masses of the stars, we have to know their relative distances from the center of mass. Finding the center of mass is simple because it must always be a point on the line connecting the two stars. As we observe the stars revolving around each other, the line connecting them always passes through the center of mass (Figure 10–7).

The distances of the stars from the center of mass give us a clue to their relative masses. The more massive star will be closer, and the less massive will be farther away. In Figure 10–7, for instance, star A is about twice as far from the center of mass as is star B. This tells us that star A is only half the mass of star B. Thus, the relative distances of the stars from the center of mass tell us the ratio of the stellar masses. If we know the sum of the masses and their ratio, we can find their individual masses.

Only a few dozen visual binaries have been analyzed to give good masses. Many systems are too close together or have periods that are too long. Some are particular in that only one of the stars is bright enough to see.

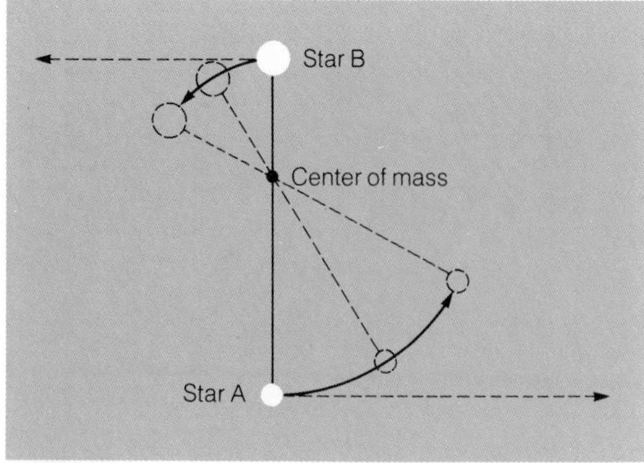

FIGURE 10–7 Without gravity the two stars would follow the straight-line paths shown by dashed lines. Gravity pulls the stars into curved orbits about their common center of mass. The line connecting the stars always passes through their center of mass.

ASTROMETRIC BINARIES

A binary is called an **astrometric binary** when one of the stars is too faint to be seen. In that case we can tell that the system is a binary only by noticing the motion of the visible star around the center of mass. This requires highly precise measurements of the position of the star.

All stars move through space, and over periods of years or decades we can detect the motion of many of these stars as their proper motion (Chapter 9). Single stars follow paths that are, over periods as short as decades or centuries, straight lines, but binary stars follow more complicated paths.

The center of mass of a binary system follows a straight line, but the stars follow wavy paths as they orbit the center of mass (Figure 10–8). Thus, we can recognize astrometric binaries by the proper motions of the stars about their center of mass, but we cannot find the masses unless both stars are visible. Often only one star is bright enough to be visible.

It is interesting to note that this may be one way to search for planets orbiting nearby stars. A massive planet like Jupiter is much less massive than a star, but it could affect the motion of its star. Such a variation in the motion of a nearby star may eventually give us our first direct evidence that other stars have planets just like the sun.

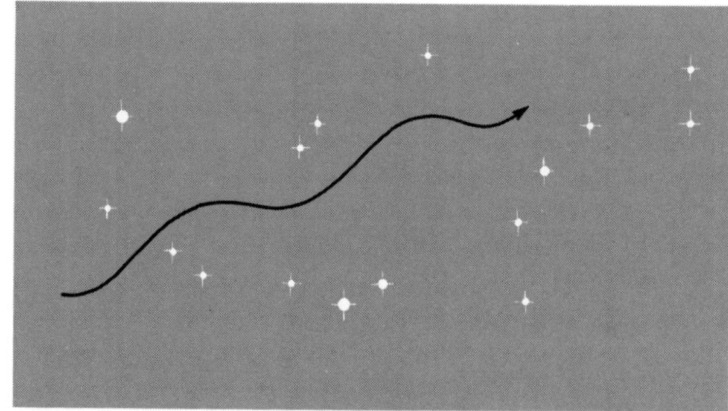

FIGURE 10–8 An astrometric binary is a binary in which one star is too faint to be visible. The binary nature of the system can be recognized because the visible star follows a wavy path across the sky due to its motion around the center of mass of the system.

FAMOUS VISUAL BINARIES

Sirius, the brightest star in the sky, is an important example of a visual binary. In 1844 the German astronomer Friedrich Wilhelm Bessel (1784–1846) discovered that Sirius is an astrometric binary. That is, Sirius does not follow a straight path across the sky, but moves along a wavy path. In 1862, while testing a new telescope lens, the American telescope maker Alvan Clark discovered the companion to Sirius. Dubbed Sirius B, it is about 9 mag-

BOX 10–1

The Masses of Binary Stars

According to Newton's laws of motion and gravity, the total mass of two stars orbiting each other is related to the average distance between them, a, and their orbital period, P. If the masses are M_A and M_B, then

$$M_A + M_B = \frac{a^3}{P^2}$$

In this formula, we will measure a in AU, P in years, and the mass in solar masses.

Notice that this formula is related to Kepler's third law of planetary motion. Almost all of the mass of the solar system is in the sun. If we apply this formula to any planet in our solar system, the total mass is 1 solar mass. Then the formula becomes $P^2 = a^3$, which is Kepler's third law.

In other star systems the total mass is not necessarily 1 solar mass, and this gives us a way to find the masses of binary stars. If we can find the average distance between the two stars in AU and their orbital period in years, the sum of the masses of the two stars is just a^3/P^2.

For example, if we observe a visual binary star with a period of 32 years and an average separation of 16 AU, then

$$M_A + M_B = \frac{16^3}{32^2} = \frac{4096}{1024} = 4 \text{ solar masses}$$

To find the individual masses, we must know how far the stars are from their center of mass. If star A is 12 AU away from the center of mass and star B is 4 AU away, then the ratio of the masses must be 12⁄4 with star B the more massive. What two numbers add up to 4 and have the ratio 12⁄4? Star B must be 3 solar masses, and star A must be 1 solar mass.

nitudes fainter than Sirius A and is never farther away than 11.5 seconds of arc. Nevertheless, careful observations over the years have shown that both stars are moving around the center of mass as it moves through space (Figure 10–9).

Observations of the masses of Sirius A and Sirius B show that Sirius A is a fairly normal star of about 2.35 solar masses. However, Sirius B has a mass of 0.98 solar mass but is not much larger than the earth. It is one of the first white dwarfs ever discovered, and its high density is characteristic of such objects.

Cygnus (the Swan) contains a number of interesting binary stars (Figure 10–10). Viewed through a small telescope, Albireo (β Cygni) is a beautiful sight, appearing as a golden yellow K3 star and a sapphire blue B8 star. For years astronomers thought Albireo was a binary with a period as long as 100,000 years, but it shows no orbital motion, and a recent study suggests that the stars lie at different distances from earth and are thus not a real binary system. Another visual binary in Cygnus is 61 Cygni (Figure 10–11). Its period is 653 years, which is too long for convenient analysis, but it is interesting in a number of ways. It is the eleventh closest star (3.4 pc), and in 1838 it became the first star to have its parallax measured. Also, spectroscopic analysis suggests that one of its stars may have a planetlike companion about eight times the mass of Jupiter.

Binary stars are very common, but many are so close to each other that they cannot be resolved individually. Then other methods must be used to find their orbital period and separation. Though the methods differ, the goal is the same—discovering the masses of the stars.

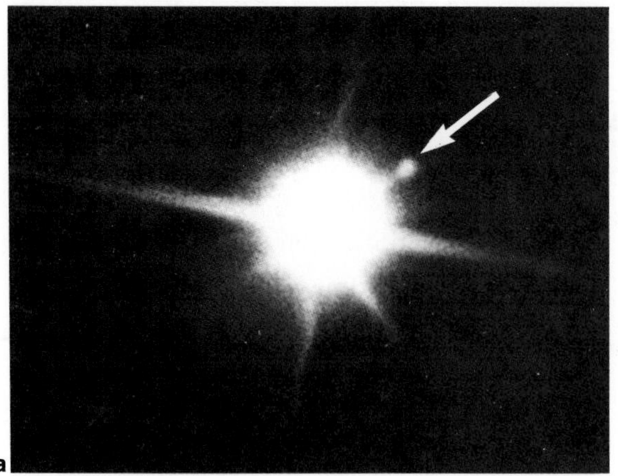

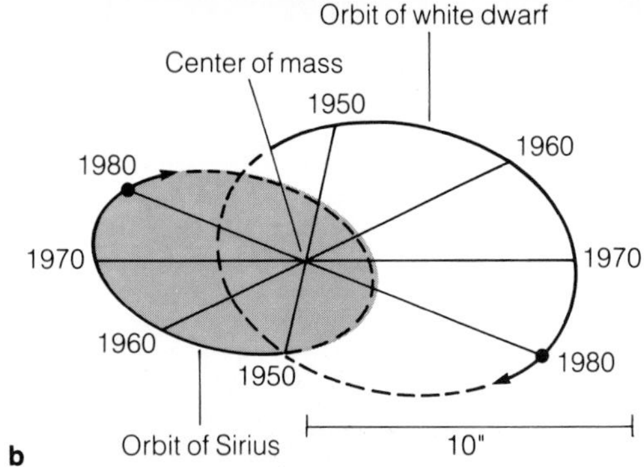

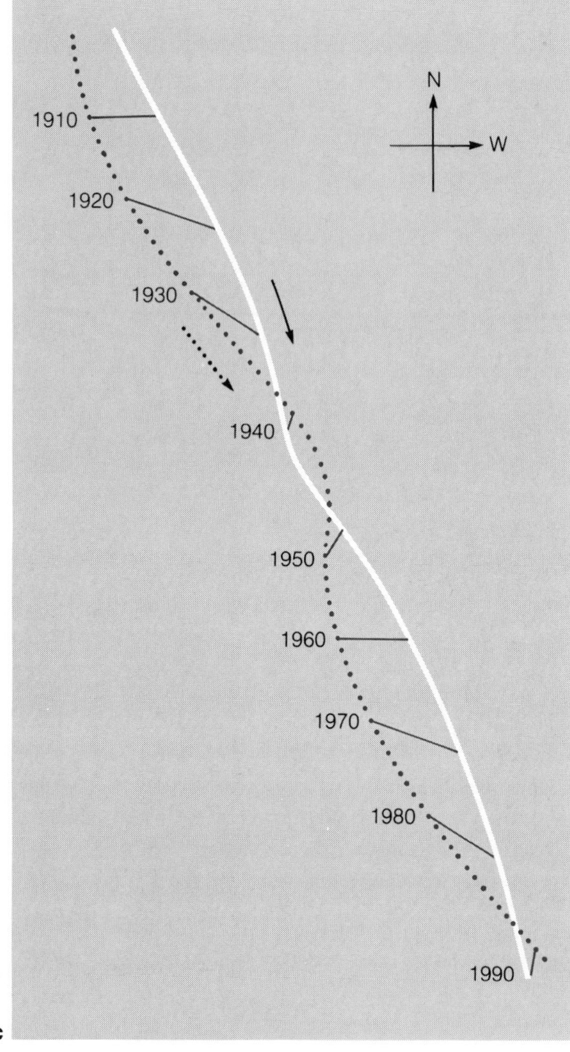

FIGURE 10–9 (a) Sirius and its white dwarf companion (arrow), a visual binary system, photographed in 1960. The size of the images is related to the brightness of the stars, not to their diameter. (b) The orbits of the two stars relative to the center of mass. (c) The motion of Sirius A (solid line) and Sirius B (dotted line) as they move southward across the sky. (Lick Observatory photograph.)

10.2 SPECTROSCOPIC BINARIES

If the stars of a binary system are too close together to be visible separately, the telescope shows us a single point of light. Only by taking a spectrum, which is formed by light from both stars, can we sometimes tell that there are two stars, not one, present. Such a system is called a **spectroscopic binary.**

DOUBLE-LINE SPECTROSCOPIC BINARIES

If the stars of a spectroscopic binary have about the same brightness, we will be able to see spectral lines from both stars. Such a binary is called a **double-line spectroscopic binary.**

Because the stars in a spectroscopic binary orbit each other, they alternately approach and recede from us, as in Figure 10–12. As one star comes toward us, its spectral lines are Doppler-shifted toward the blue. The other star is moving away from us, and its spectral lines are shifted toward the red. Half an orbit later the star that was approaching is receding. As we watch the spectrum of the binary system, we see the spectral lines split into two parts that move apart and then move together as the stars follow their orbits (Figures 10–12 and 10–13).

If we convert these Doppler shifts into radial velocities, we can plot them in a graph called a **radial velocity**

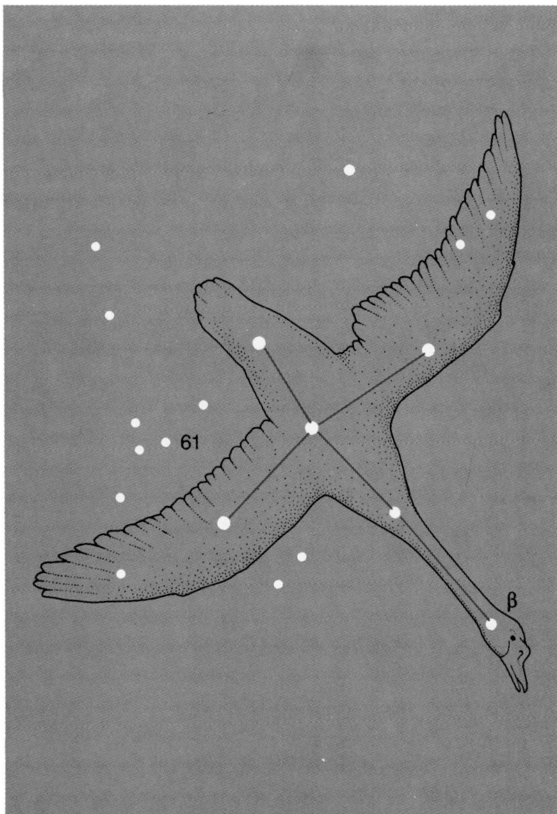

FIGURE 10–10 The constellation Cygnus the Swan contains many interesting binary stars. Albireo (β Cygni) is a beautiful sight through a small telescope. The binary 61 Cygni is one of the closest star systems to the sun and the first star to have its parallax measured. The brighter stars of Cygnus make up the asterism known as the Northern Cross.

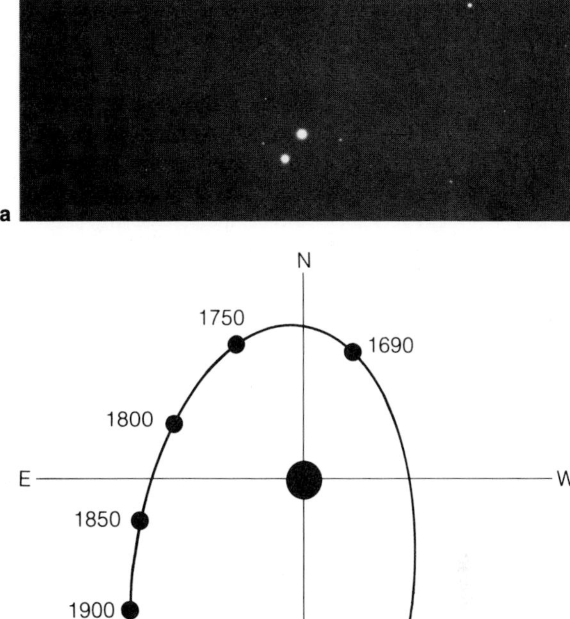

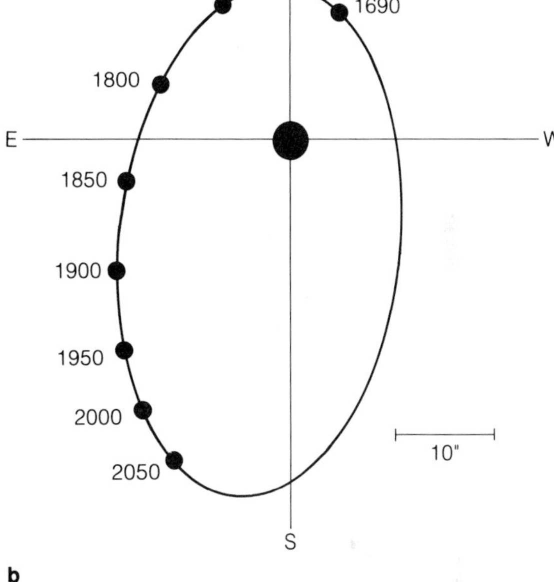

FIGURE 10–11 The visual binary 61 Cygni (a) photographed about 1960, and its relative apparent orbit (b). (Lick Observatory photograph.)

curve. For a double-line spectroscopic binary, this shows the radial velocity of both stars. The curves in Figure 10–14 are drawn for a hypothetical binary in which the orbits are circular. If the orbits are elliptical, the curves are more complicated, but properly analyzed the radial velocity curves can tell us the eccentricity of the elliptical orbits.

Except for the complication of elliptical orbits, analyzing a double-line spectroscopic binary seems straightforward. The radial velocity curve in Figure 10–14 shows that the two stars are moving about their common center of mass, and that the center of mass has a radial velocity of its own. We would expect this, of course; it would be unusual to find a star system that is not moving with respect to our solar system. From the curves we can tell that the velocity of star A never differs from the velocity of the center of mass by more than 40 km/sec, and that the velocity of star B never differs by more than 20 km/sec. If we are seeing the orbits edge-on, the orbital velocity of star A with respect to star B must be 60 km/sec.

We can find the mass if we can find the radius of the orbit, and we can do that because we know the orbital velocity and the period. Multiplying the velocity times the period, we find the total distance around the orbit, its circumference. In our idealized example shown in Figure 10–14, we assumed a circular orbit, so it is easy to find the radius of the orbit from its circumference. The radius of the orbit is just the circumference divided by 2π. Once we know the radius and the period, we can find the mass as shown in Box 10–1.

We can even find the individual masses. The velocity of star A relative to the center of mass is twice that of star B, so star B must be twice as massive as star A.

This sounds easy, but we have skipped over one important fact. We don't know that the orbits are edge-on as seen from earth. They could be inclined at any angle, and we can never find that inclination. We can find the inclination of a visual binary because we can see the stars moving along their orbits. In a spectroscopic binary, however, we cannot see the individual stars nor find the inclination nor untip the orbits. The velocities we observe are not the true orbital velocities but only the part of that velocity directed radially toward or away from earth. Because we cannot find the inclination, we cannot correct these radial velocities to their true orbital velocities. Therefore we cannot find the true masses. All we can find from a spectroscopic binary is a lower limit to the masses.

SINGLE-LINE SPECTROSCOPIC BINARIES

If one of the stars in a spectroscopic binary is too faint, we will not be able to see its spectral features in the spectrum of the system. Such a binary is called a **single-line spectroscopic binary** because only a single set of lines is visible in its spectrum.

If we measure the wavelengths of the visible lines, we find them shifting first to the red and then to the blue, as the brighter star orbits the center of mass. We could convert these Doppler shifts into radial velocities and plot them to produce a radial velocity curve for the system, but it would show the motion of only one star (Figure 10–15). We cannot find the orbital velocity of one star around the other and so cannot even find the lower limit to the masses of the individual stars. All we can find is the **mass function**, a number that is related to the mass of the invisible star, the total mass of the system, and the inclination. There is no way to unscramble these quantities.

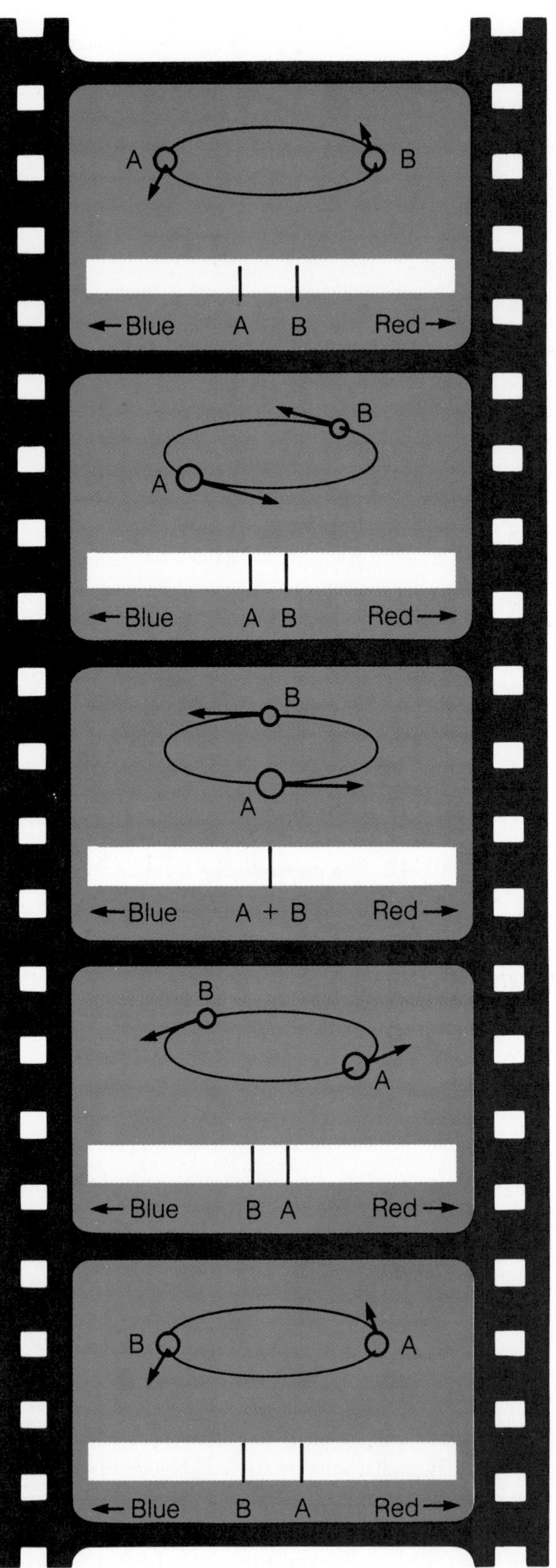

FIGURE 10–12 As the stars of a spectroscopic binary revolve around their common center of mass, they alternately approach and recede from the earth. The Doppler shifts cause their spectral lines to move back and forth across each other. The variation of a single spectral line is shown here. Compare with Figure 10–13.

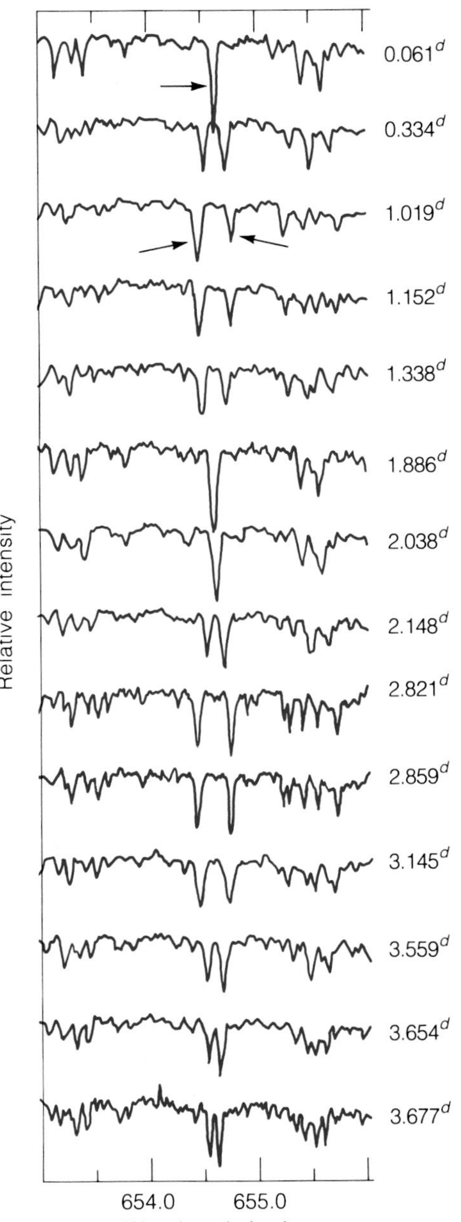

FIGURE 10–13 Fourteen spectra of the spectroscopic binary HR 80715 show how the single line of iron (arrow in first spectrum) is split into two components (arrows in third spectrum) by the orbital motion of the stars. The laboratory wavelength of the line is 654.315 nm. Times of each spectrum are given in days from an arbitrary starting time. (Adapted from data courtesy of Samuel C. Barden and Harold L. Nations.)

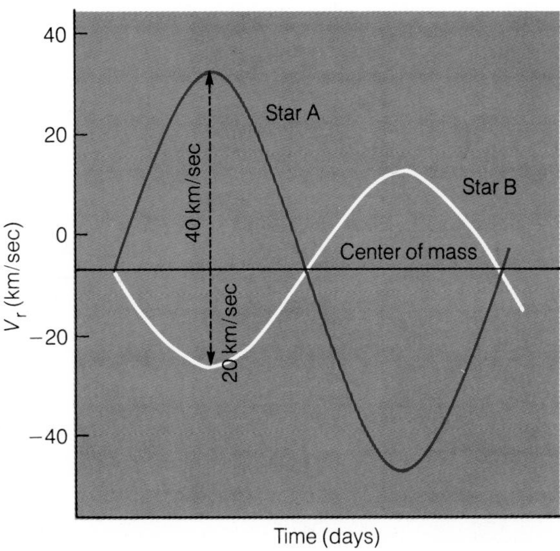

FIGURE 10–14 The radial velocity curve for a hypothetical spectroscopic binary with circular orbits.

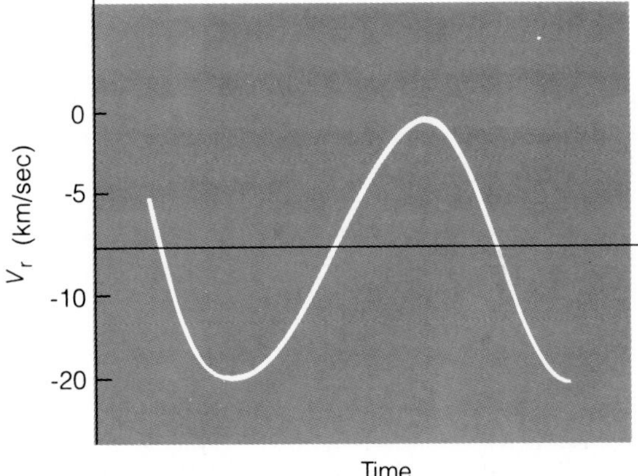

FIGURE 10–15 The radial velocity curve for a single-line spectroscopic binary shows the motion of only one star.

THE USE OF SPECTROSCOPIC BINARIES

It might seem that spectroscopic binaries are of no use because neither double-line nor single-line systems can tell us stellar masses. However, statistical analysis of spectroscopic binaries can give useful results.

For instance, if we had data from a dozen spectroscopic binaries containing A stars, we might assume that

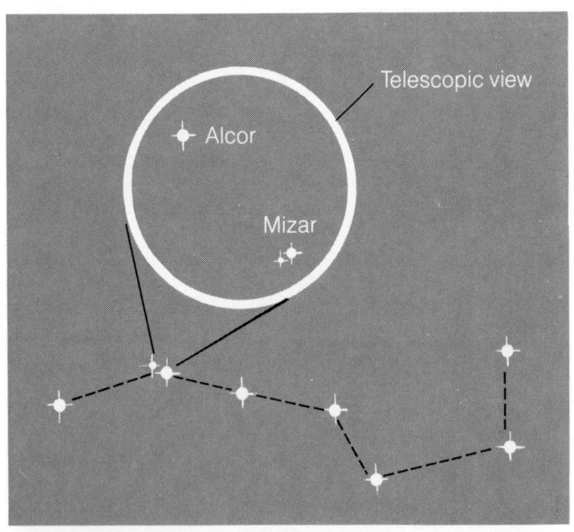

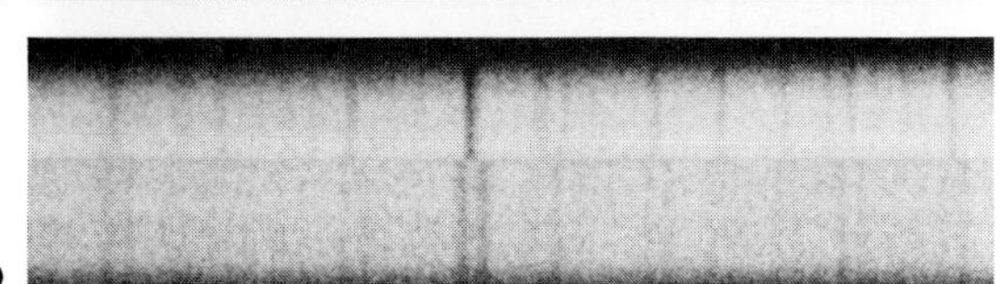

FIGURE 10–16 At the bend of the handle of the Big Dipper lies Mizar, a visual binary (a). Mizar, its companion (visible through small telescopes), and the nearby star Alcor are all spectroscopic binaries. Two spectra of Mizar (b) recorded at different times show how a spectral line is separated into two components by the orbital motion of the stars. (Compare with Figure 10–13.) (Spectra Courtesy Mount Wilson and Las Campanas Observatories, Carnegie Institution of Washington.)

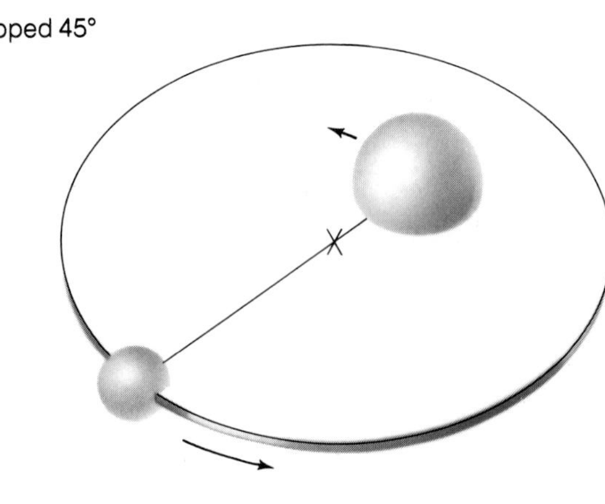

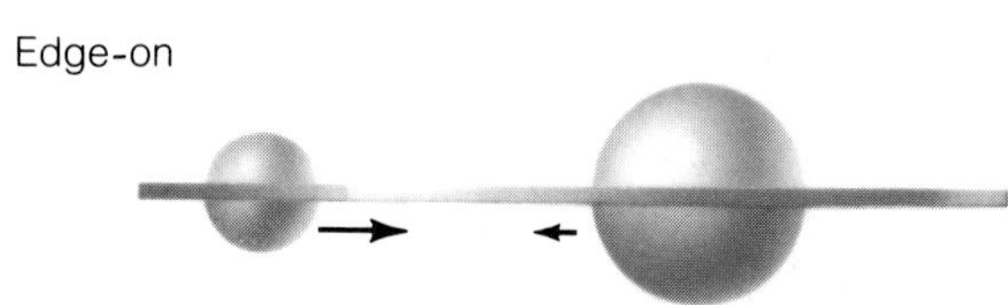

FIGURE 10–17 Imagine a model of a binary system with balls for stars and a disk of cardboard for the plane of the orbits. Only if we view the system edge-on do we see the stars cross in front of each other. These eclipsing binary systems are rare.

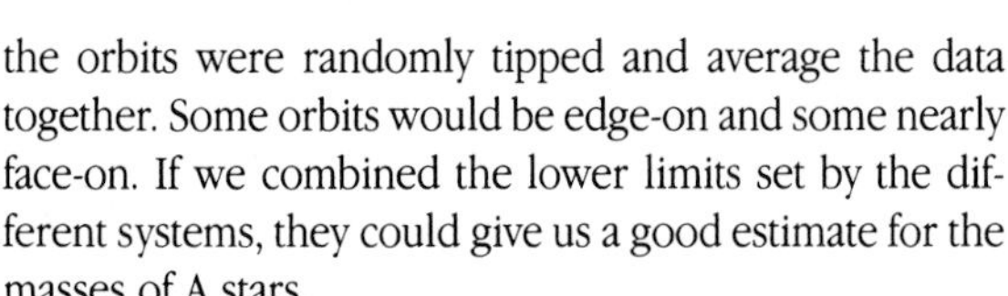

the orbits were randomly tipped and average the data together. Some orbits would be edge-on and some nearly face-on. If we combined the lower limits set by the different systems, they could give us a good estimate for the masses of A stars.

This kind of statistical analysis is important because visual binaries that can be analyzed are very rare, and in some cases the only way to get data on stellar masses is to turn to spectroscopic binaries. Also, spectroscopic binaries are very common, so there are large numbers of them that can be analyzed.

In fact, spectroscopic binaries are so common that many of the familiar stars in the sky are spectroscopic binaries. Capella (α Aurigae), for instance, is a spectroscopic binary with a period of 104 days. A small telescope shows that Mizar, the star at the bend of the handle of the Big Dipper, is a visual binary (Figure 10–16). Spectroscopic observations show that both of the stars in the visual binary are themselves spectroscopic binaries, making Mizar a "double double star." Near Mizar is Alcor, a fainter star just visible to the naked eye. It too is a spectroscopic binary.

Spectroscopic binaries are common but do not give good results individually. Visual binaries give good results when they can be analyzed, but few have short enough periods and large enough separations. Yet another kind of binary star can give us information about stellar masses and can also tell us the diameters of the stars.

10.3 ECLIPSING BINARIES

Rare among binary stars are those with orbits inclined so the stars cross in front of each other. Seen from the earth, the two stars are not resolvable; that is, they look like a single point of light. However, when one star crosses in front of the other, part of the light is eclipsed, and we recognize the system as an **eclipsing binary.**

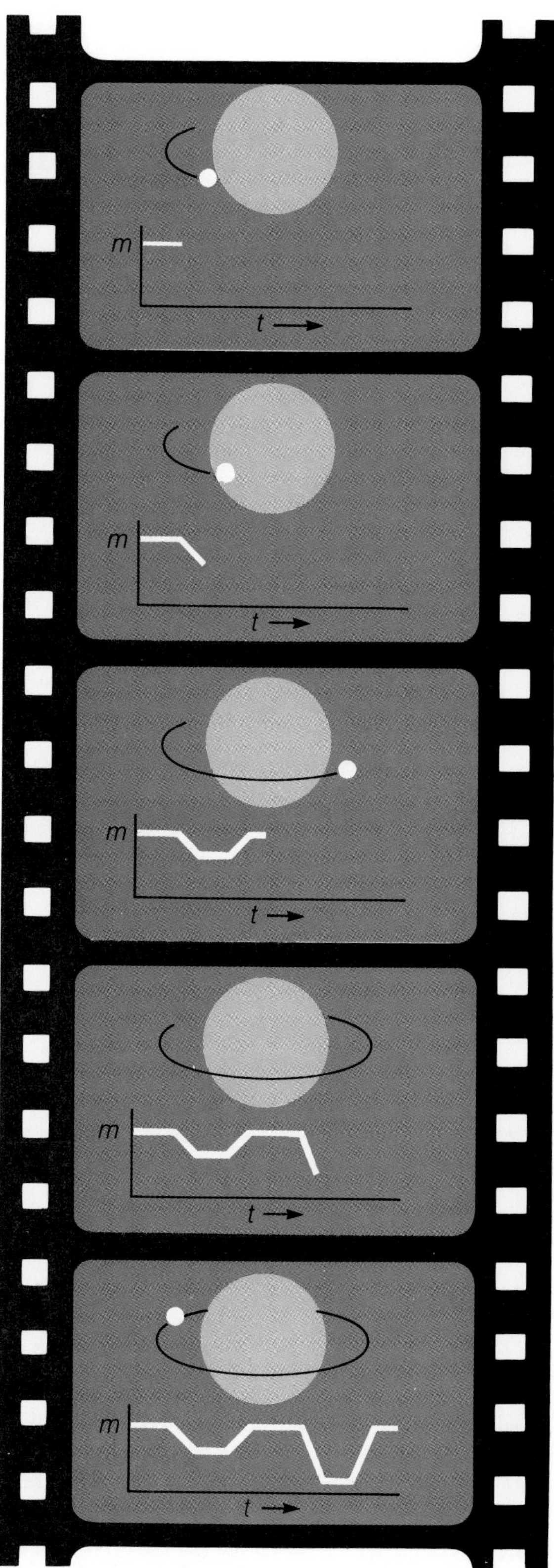

MASSES AND DIAMETERS

Imagine a model of a binary star system in which a cardboard disk represents the orbital plane, as in Figure 10–17. If the orbits are seen edge-on from earth, the two stars cross in front of each other. The small star crosses in front of the large star, and then, half an orbit later, the large star crosses in front of the small star. Each time one star crosses in front of the other, the total brightness of the system decreases in an eclipse (Figure 10–18).

There are two eclipses during each orbital revolution. The deeper eclipse is called **primary minimum,** and the shallower is called **secondary minimum.** If we observe such a system and plot the changing magnitude versus time, we get a graph called a **light curve.** This would show the eclipses as in Figure 10–18.

This graph can tell us a great deal about the stars. For instance, we can find the ratio of the temperatures of the stars. During both eclipses the same amount of area is hidden from our view. First the small star hides an area of the large star that is equal to its own cross section. Then when the small star is eclipsed, the same amount of area is hidden. Any difference in the amount of light lost during the two eclipses must arise from the differences in the temperature of the two stars, not from differences in the area hidden.

For example, study the system shown in Figure 10–18. When the small star crosses in front of the larger cooler star, the total brightness of the system declines slightly because a small area of the cooler star is hidden from view. When the small star is eclipsed behind the larger star, the same amount of area is hidden, but it is much hotter than that hidden before, so the resulting eclipse is much deeper.

We can find the masses of the stars in an eclipsing system if we can get spectra showing the Doppler shifts of the two stars. We can't analyze the Doppler shifts alone to find masses because we don't know the inclination, but the light curve can tell us how the orbits are tipped. We know we are seeing the orbits nearly edge-on or we would not see eclipses at all, and the shape of the eclipses can tell us if the orbits are tipped slightly from exactly edge-

FIGURE 10–18 As the stars in an eclipsing binary cross in front of each other, the total brightness of the system declines. The light curve is a graph of magnitude versus time. This example shows a small, hot star orbiting a large, cool star.

on. If we know the inclination, we can use the radial velocity curve to get the orbital velocities and analyze the system like a double-line spectroscopic binary. In this case, because we know the inclination, we can find the true masses of the stars.

Eclipsing binaries are especially important because their light curves enable us to measure the diameters of the stars. From the light curve we can tell how long it took for the small star to cross the large star. Multiplying this time interval by the orbital velocity of the small star gives us the diameter of the larger star. We could also determine the diameter of the small star by noting how long it took to disappear behind the edge of the large star. For example, if it took 300 seconds for the small star to disappear while traveling 500 km/sec relative to the large star, it must be 150,000 km in diameter.

Of course there are complications. Although many eclipsing binary star systems have circular orbits due to the influences of tides between the stars, some have elliptical orbits, and these affect the shape of the light curves. In some cases, the orbits are tipped slightly, so the stars do not cross directly in front of each other, and neither star is completely eclipsed (Figure 10–19). These are called partial eclipses. In some cases, the stars are so close together they distort each other from perfect spheres, and the rotation of these nonspherical stars causes confusing variations in the brightness of the system. It is even possible for the stars to be so close that the hotter star heats up one side of the cooler star and thus further confuses the light variation.

Some of these complications can be accounted for, and some can make a system so complex it cannot be analyzed. The systems that can be solved give us important information, not only about stellar masses, but also about stellar diameters.

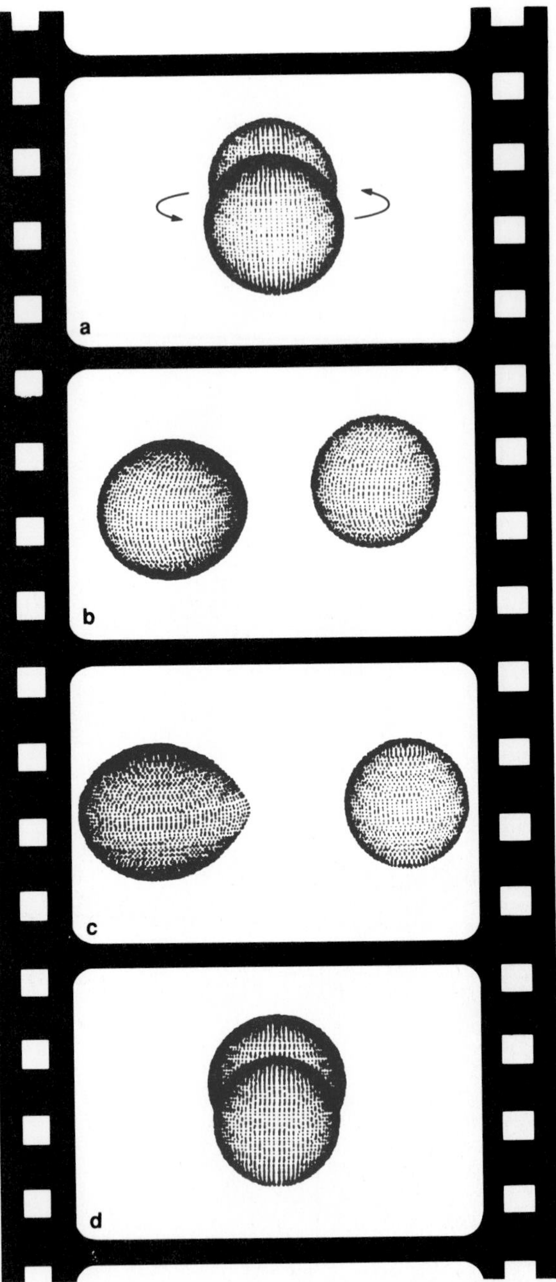

FIGURE 10–19 Four computer-generated views of the eclipsing binary Algol. Because the orbits of the stars are tipped slightly from our line of sight, we see partial eclipses (a) and (d). The less massive star is tidally distorted from a spherical shape (c). (Courtesy R. E. Wilson, University of Florida and *Mercury*.)

ALGOL AND OTHER ECLIPSING BINARIES

Algol (β Persei) is one of the best-known eclipsing binaries because its eclipses are visible to the naked eye. Normally about magnitude 2.1, its brightness drops to magnitude 3.4 in eclipses lasting about 10 hours (Figure 10–20). This is a decrease of about 68 percent and is easily visible to the naked eye.

The eclipses were first reported in 1669 by Geminiano Montanari, mathematics professor at the universities of Bologna and Padua, but the star's variation was not explained until 1783. In that year, the English astronomer John Goodricke realized that the variations of Algol are periodic, occurring every 68¾ hours, and he offered two possible explanations. It might be a rotating star with dark spots on its surface, or it might be a star orbited by a dark

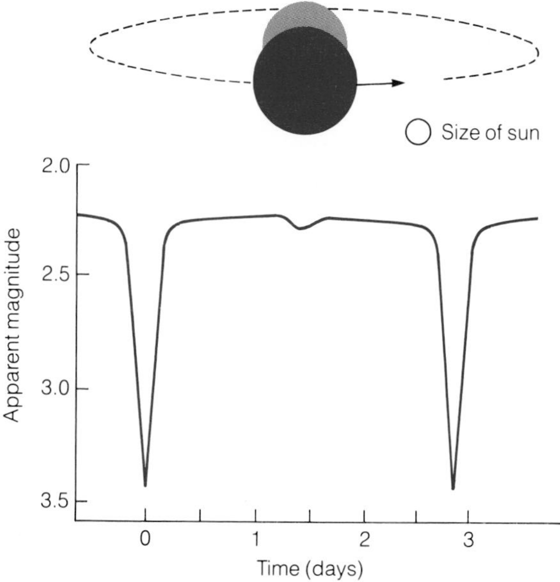

FIGURE 10–20 Algol is a famous eclipsing binary because its eclipses are deep enough to be visible to the naked eye.

FIGURE 10–21 The eclipsing binary Algol is the star on the demon's head in this drawing showing Perseus and the head of the gorgon Medusa. *Algol* comes from the Arabic for "the demon's head." (Adapted from Duncan Bradford, *The Wonders of the Heavens,* Boston: John B. Russell, 1837.)

companion that periodically passes between it and the earth. Of course, the latter explanation has proved correct, and the dark companion is merely the fainter of the two stars.

Because the variation of Algol is so easy to observe, some have speculated that it was known three millennia ago when the oldest constellations originated. *Algol* comes from the Arabic for "the demon's head," and it is associated in constellation mythology with the severed head of Medusa, the sight of whose serpentine locks turned mortals to stone (Figure 10–21). Indeed, in some accounts, Algol is the winking eye of the demon.

A number of other bright stars are eclipsing binaries. The star β Lyrae is a peculiar eclipsing binary with a period of 12.9 days. Its eclipses are about 1 magnitude deep, and its light curve is complicated by the distorted shape of the stars and by glowing clouds of gas floating between the two stars. Another star, ε Aurigae, has an orbital period of 27.06 years, and its eclipses last about 700 days. Its light curve is so distorted that astronomers cannot agree on the characteristics of the stars.

Compare the very long period of ε Aurigae with the 12-hour period of the faint star catalogued as +16°516. The stars that compose it orbit so quickly that its eclipses last only about 45 minutes, and it drops from maximum brightness to minimum in only 48 seconds.

The diverse properties of eclipsing binaries make them fascinating puzzles in their own right, but they are most valuable when they are decoded to reveal the true masses and diameters of stars. In the next section we will assemble such data from all types of binaries and look for relationships among the stars.

10.4 STELLAR MASS AND DENSITY

All of our discussion of binary stars has been aimed at a single goal: We want to know the masses of stars. Binary stars are entertaining puzzles, and many astronomers specialize in their study. But entertaining or not, binaries are critical to our discussion because they give us a way to find stellar masses.

Knowing the mass and diameter of a single star is of little use. Only when we assemble data for many stars do we begin to see a pattern. That pattern will eventually lead us to understand how stars are born, age, and die. In the rest of this chapter, we will assemble the stellar masses determined from binary star systems and look for the pattern.

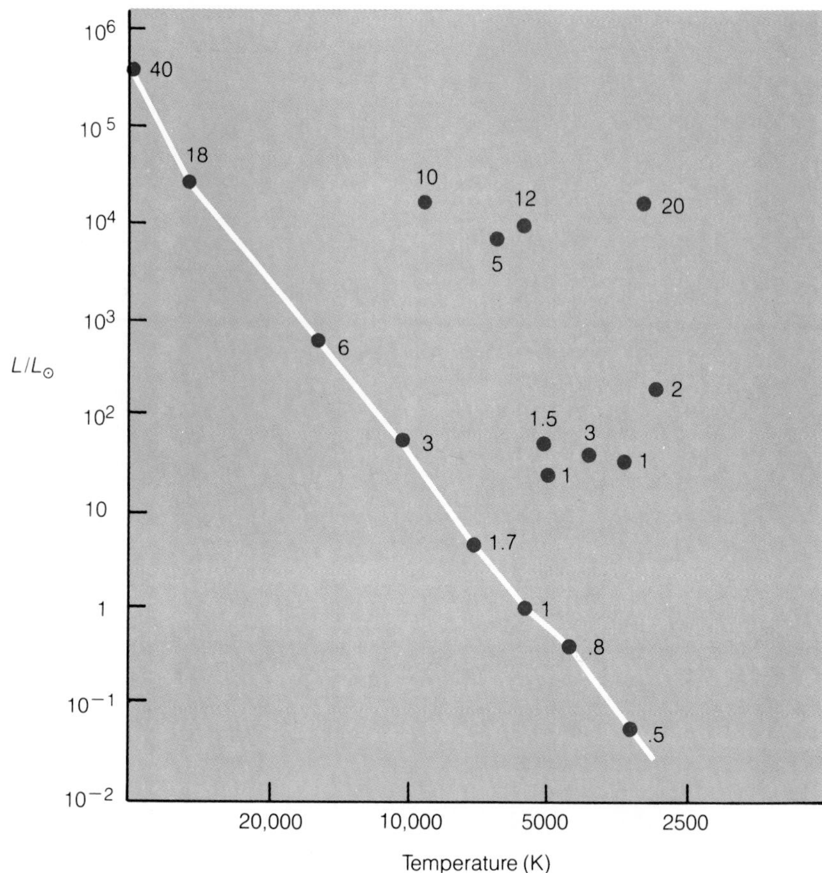

FIGURE 10–22 Stellar masses plotted in the H–R diagram. Main-sequence stars are ordered. That is, the more massive a star is, the more luminous it is. But giants and supergiants are not ordered. In the giant and supergiant region of the diagram, different masses are jumbled together with no pattern.

STELLAR MASSES

If we began plotting stars in the H–R diagram and writing their mass beside each dot, we would discover that the supergiants tend to be a bit more massive than the sun, but that there is no pattern to their masses (Figure 10–22). That is, massive stars and lower mass stars are all jumbled together in the supergiant region. The same is true for the giants. Some giants are more massive than the sun, and some are about the same mass as the sun, but they are all jumbled together in the giant region.

When we begin to plot main-sequence stars, however, we find a pattern. The farther up the main sequence a star is located, the more massive it is. Stars near the top of the main sequence, the O and B stars, are quite massive. The F and G stars in the middle of the main sequence are about the mass of the sun, and the M stars at the lower end are much less massive than the sun. This ordering of main-sequence stars by their mass is a clue to how stars work.

In fact, this pattern among the stars can be expressed as a direct relationship between mass and luminosity. The more massive a main-sequence star is, the more luminous it is. If we plot all of the stars with well-determined masses according to their luminosities, we get a diagram that illustrates this mass–luminosity relation (Figure 10–23). The only stars that don't obey the rule are the giants, supergiants, and white dwarfs. This too is a clue to how stars work.

Notice from the mass–luminosity diagram how large the range in luminosity is. The range of masses extends from about 0.08 solar masses to about 50 solar masses—a factor of 600. But the range of luminosities extends from about 10^{-6} to about 10^6 solar luminosities—a factor of 10^{12}. Clearly, a small difference in mass causes a large difference in luminosity.

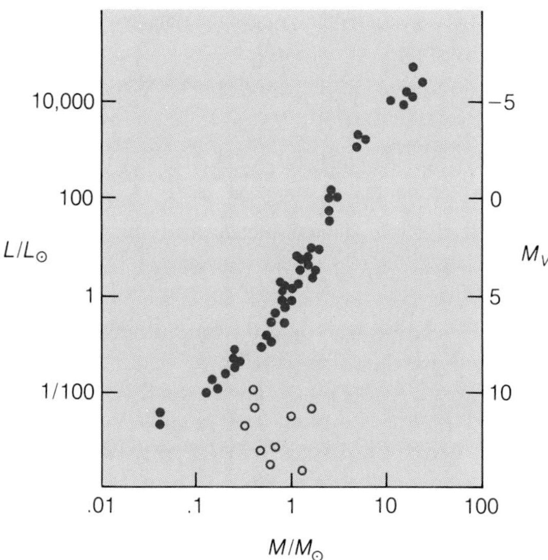

FIGURE 10–23 The mass–luminosity relation shows that the more massive a main-sequence star is, the more luminous it is. The open circles represent white dwarfs, which do not obey the relation.

STELLAR DENSITIES

The data we get from binary stars can be used in another way. We can combine it with stellar diameters to find the average densities of the stars. All we have to do is divide the mass of a star by the volume it occupies. This gives us its average density. We can be sure that stars are denser near their centers and less dense near their surfaces, but these average densities are sufficient to reveal another important pattern in the H–R diagram.

The average densities of main-sequence stars are about like that of the sun. Some stars are a bit denser, and some are a bit less dense, but in general, main-sequence stars are about as dense as water.

However, binary stars tell us that giant stars and supergiants are not tremendously massive stars, yet we know from their positions in the H–R diagram that they must be very large. If we divide their ordinary masses by their gigantic volumes, we find they have very low densities. Giant stars have densities ranging from 0.1 to 0.01 g/cm^3. The enormous supergiants have still lower densities, ranging from 0.001 to 0.000001 g/cm^3. This is thinner than the air we breathe.

On the other hand, binary stars such as Sirius show that the white dwarfs have masses of about 1 solar mass. Yet they have to be small stars not much larger than the earth. This means the average density of a white dwarf is about 2,000,000 g/cm^3 or more. On earth, a cubic centimeter would weigh as much as a limousine.

Density divides stars into three groups. Most stars are main-sequence stars with densities like the sun's. Giants and supergiants are very low-density stars, and white dwarfs are high-density objects. We will see in later chapters that these densities reflect different stages in the evolution of stars.

SUMMARY

The only way to get information about the masses of stars is to find them in binary systems—that is, in systems of two stars orbiting each other. The size and period of the orbits depend on the masses of the stars. If we can find the average separation of the two stars in AU and the orbital period in years, the total mass in solar masses is a^3/P^2.

The individual masses can be found by studying the individual motions of the stars. The more massive star will be located closer to the center of mass of the system, and the less massive star will be farther away. Thus, the ratio of the distances of the stars from the center of mass tells us the ratio of their masses. If we know the sum of the masses and their ratio, we can find the individual masses.

Visual binaries are those systems in which both stars are visible. They often have periods of tens or hundreds of years, though only the shorter period visual binaries can be analyzed. The advantage to studying visual binaries is that we can see the actual shape of the orbit and thus can discover the angle at which it is tipped. This means the true masses of the stars can be found.

In an astrometric binary, only one of the stars is bright enough to see, and we can recognize it as a binary star system only by watching the star wobble back and forth as it orbits the center of mass. These do not yield stellar masses. Sirius was originally discovered to be an astrometric binary, though its faint, white dwarf companion was later detected visually.

If the two stars in a binary system are close together, we may not be able to see them individually from earth. The system may look like a single point of light. However, we may be able to study the orbital motion of the system by measuring the Doppler shifts of the spectral lines of the two stars. If lines are visible from both stars, the system is called a double-line spectroscopic binary. If only one set of lines is visible, it is called a single-line spectroscopic binary. In both cases we cannot calculate true masses because we cannot find the inclination of the orbit. Thus, the results of spectroscopic binaries are most useful in statistical studies where the effects of orbital inclination can be averaged out.

If the two stars cross in front of each other, part of the light is periodically eclipsed. We can recognize the binary nature of the star from its light curve. The shape of the light curve can tell us the surface brightness of the two stars, their diameters, and the inclination of the orbit. If we can find the inclination, we can find the masses from the radial velocity curve.

These data on masses and diameters reveal important patterns in the H–R diagram. The giants and supergiants consist of stars of various masses jumbled together in no apparent order. However, the stars on the main sequence are ordered according to mass. The most massive stars lie on the upper main sequence and the least massive on the lower main sequence. This relationship is reflected in the mass–luminosity diagram.

If we divide the mass of a star by the volume it occupies, we get its average density. These average densities divide the stars in the H–R diagram into three groups. The giants and supergiants are very low-density stars—some are thinner than air. Main-sequence stars are all about as dense as water. White dwarfs are all about as massive as the sun, but about the size of the earth, so they are very dense—about 2,000,000 g/cm^3.

These patterns hidden in the H–R diagram are important clues to how stars work. We will follow this trail through the next chapters.

NEW TERMS

binary stars
double stars
visual binary
optical binary
position angle
filar micrometer
apparent relative orbit
true relative orbit
astrometric binary
spectroscopic binary
double-line spectroscopic binary
radial velocity curve
single-line spectroscopic binary
mass function
eclipsing binary
primary minimum
secondary minimum
light curve

QUESTIONS

1. What did William Herschel hope to discover from his study of double stars?
2. Why don't any visual binaries have very short periods such as a few days or less?
3. Why do we have to know the distance to a visual binary before we can compute the masses of the stars?
4. How do we find the ratio of the masses in a visual binary system? In a spectroscopic binary system?
5. Explain how you can be certain that the orbit shown in Figure 10–4 is not a true relative orbit.
6. Why can't we find the masses of the stars in a spectroscopic binary system?
7. How do astrometric binaries resemble single-line spectroscopic binaries?
8. Why can we find the masses of the stars in an eclipsing binary but not in a spectroscopic binary?
9. How could we find the diameters of the stars in an eclipsing binary (assuming that the orbit is exactly edge-on)?
10. What would the light curve in Figure 10–18 look like if the stars had the same temperature?
11. What is the mass–luminosity diagram?
12. How does average density divide stars into three categories?

PROBLEMS

1. Draw in the center and the major axis of the true orbit in Figure 10–4.
2. What is the total mass of a visual binary system if its average separation is 8 AU and its period is 20 years?
3. What is the mass ratio of the stars in Figure 10–7? In Figure 10–9b?
4. Assume that the stars in Figure 10–7 have a separation of 18 AU and a period of 30 years. What is the total mass? What is the mass ratio? What are their individual masses?
5. Measure the orbit of 61 Cygni in Figure 10–11 and compute the semimajor axis in astronomical units. The distance to the system is 3.4 pc. (Disregard the slight distortion of the orbit caused by its inclination.)
6. Use the result of Problem 5 and the given period and orbit of 61 Cygni to compute the total mass of the stars.
7. If the period of the spectroscopic binary in Figure 10–14 is 67 days and the orbit is edge-on, what is the total mass of the system? What are the masses of the two stars?
8. Measure the wavelengths of the iron lines in Figure 10–13 and plot a radial velocity curve for this system. What is the period? Assuming that the orbit is circular and edge-on, what is the total mass? The individual masses?
9. If the eclipsing binary in Figure 10–18 has a period of 32 days, an orbital velocity of 153 km/sec, and an orbit which is nearly edge-on, what is the circumference of the orbit? The radius of the orbit? The mass of the system?
10. If the orbital velocity of the eclipsing binary in Figure 10–18 is 153 km/sec and the smaller star becomes completely eclipsed in 2½ hours, what is its diameter?

ACTIVITY: OBSERVING ALGOL

Most eclipsing binary stars are too faint to see with the naked eye, but a few are quite bright. Algol, the winking eye of the demon, goes through eclipses every 2.8673 days, and its changes in brightness are easy to detect.

Observations Use the star charts at the back of the book to locate the constellation Perseus. It is visible in the evening sky from September through May. It is approximately overhead at midnight in early November and at sunset in early February.

Once you have located Perseus, use the star chart below to find Algol (β Persei). Estimate the brightness of Algol to a tenth of a magnitude by comparing it to other stars in the constellation.

Eclipses Normally Algol is about 2.1 magnitude but once every 2.8673 days it fades to about 3.4 magnitude. These eclipses last only about 10 hours, so you may have to look at Algol on a number of evenings before you catch it in eclipse. Dates and times of mid-eclipse are published regularly in *Sky and Telescope.*

If you see Algol entering eclipse, estimate its brightness every half hour or so, making careful note of the time of each observation. Later, you can plot these data on graph paper to produce a light curve.

If you are observing Algol with binoculars or a small telescope, don't fail to look at the Double Cluster in Perseus.

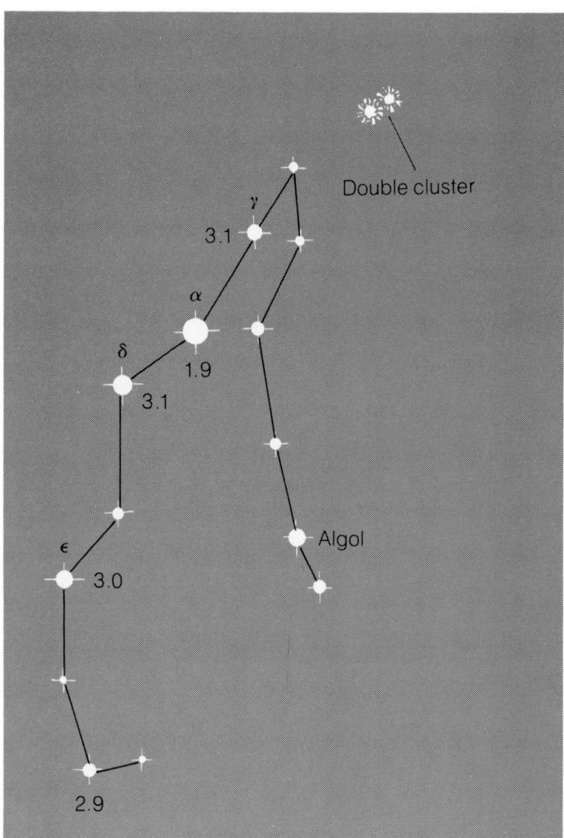

Estimate the brightness of Algol by comparing it with the other stars in the constellation of Perseus. Magnitudes are given here without decimal points. For example, α Persei has a magnitude of 1.9.

RECOMMENDED READING

ASHBROOK, JOSEPH "Visual Double Stars for the Amateur." *Sky and Telescope 60* (Nov. 1980), p. 379.

BATTEN, ALAN H. "The Story of OΣ 341." *Sky and Telescope 59* (March 1980), p. 200.

CHRISTY, JAMES W. "Searching for Dim Companions." *Natural History 90* (June 1981), p. 70.

CROSWELL, KEN "When Stars Coalesce." *Astronomy 13* (May 1985), p. 67.

DARLING, DAVID "Mystery Star." *Astronomy 11* (Aug. 1983), p. 66.

EVANS, ANEURIN "Laboratory Exercises in Astronomy—The Orbit of a Visual Binary." *Sky and Telescope 60* (Sept. 1980), p. 195.

GLASBY, JOHN STEPHEN *Variable Stars.* Cambridge, Mass.: Harvard University Press, 1969.

HOPKINS, JEFFREY L. and ROBERT E. STENCEL "Epsilon Aurigae." *Astronomy 14* (Feb. 1986), p. 6.

LETTVIN, JEROME Y. "The Gorgon's Eye" in *Astronomy of the Ancients,* ed. Kenneth Brecher and Michael Feirtag. Cambridge, Mass.: MIT Press, 1979.

MacROBERT, ALAN "The Puzzle of Epsilon Aurigae." *Sky and Telescope 70* (Dec. 1985), p. 527.

MATTEI, JANET AKYÜZ, ERNST H. MAYER, and MARVIN E. BALDWIN "Variable Stars and the AAVSO." *Sky and Telescope 60* (Sept. 1980), p. 180.

———. "Observing Variable Stars." *Sky and Telescope 60* (Oct. 1980), p. 285.

PITHER, COLIN M. "Measuring Double Stars with a Grating Micrometer." *Sky and Telescope 59* (June 1980), p. 519.

PLAVEC, MIREK J. "IUE Looks at the Algol Paradox." *Sky and Telescope 65* (May 1983), p. 413.

SEEDS, MICHAEL A. "The Wink in the Demon's Eye." *Astronomy 8* (Dec. 1980), p. 66.

STROHMEIER, W. *Variable Stars.* New York: Pergamon Press, 1972.

TRIMBLE, VIRGINIA "A Field Guide to Close Binary Stars." *Sky and Telescope 68* (Oct. 1984), p. 306.

CHAPTER 11

THE FORMATION OF STARS

Jim he allowed [the stars] was made, but I allowed they happened. Jim said the moon could'a laid them; well, that looked kind of reasonable, so I didn't say nothing against it, because I've seen a frog lay most as many, so of course it could be done.

Mark Twain
THE ADVENTURES OF HUCKLEBERRY FINN

Stars exist because of gravity. They form because gravity makes clouds of gas contract, and they generate nuclear energy because gravity squeezes their cores to unearthly densities and temperatures. In the end stars die because they exhaust their fuel supply and can no longer withstand the force of their own gravity.

In Chapter 8, we saw that a star can remain stable only by maintaining great pressure in its interior. Gravity tries to make it contract. However, if the internal temperature is high enough, pressure pushes outward just enough to balance gravity. Thus a star is a battlefield, where pressure and gravity struggle for dominance.

Only by generating tremendous amounts of energy can a star maintain the gravity–pressure balance. The sun, for example, generates 6×10^{13} times more energy per second than all of the coal, oil, natural gas, and nuclear power plants on the earth. Like the sun, most stars generate their power by fusion reactions that consume hydrogen and produce helium. However, we will discover in this chapter that not all stars fuse their hydrogen in the same way as does the sun.

The overall anatomy of a stable star is elegant in its simplicity—four laws of stellar structure suffice to describe how mass and energy behave inside a star. Using these laws, astrophysicists can create mathematical models of stars that tell them what internal conditions are like and how the star will change over billions of years. Nearly everything we know about the internal structure and evolution of stars comes from such models.

In this chapter we will see how the simple laws of stellar structure rule the formation of stars, the ignition of their nuclear fuels, and their long, stable lives on the main sequence. In the next two chapters, we will follow the life story of stars from the main sequence to their deaths.

FIGURE 11–1 An emission nebula is a cloud of gas excited by hot stars. Its spectrum is an emission spectrum. This nebula also contains dark clouds of dust and a young star cluster, NGC 6611, near the center of the nebula. (© Association of Universities for Research in Astronomy, Inc., Kitt Peak National Observatory.)

11.1 THE INTERSTELLAR MEDIUM

Stars are born out of the gases of the **interstellar medium**, the thinly scattered matter between the stars. Starting in the 1930s, astronomers realized that space was not empty and perfectly transparent. It is filled with a low-density mixture of atoms, specks of dust, and molecules. Concentrated in some instances into large dark clouds, this is the material from which stars form.

Observations suggest that the interstellar medium consists mostly of hydrogen gas with other atoms present in much smaller amounts, roughly similar to the composition of the sun. The dust seems to represent only about 1 microscopic speck per 10^6 m^3. That is about the same as two dust specks per Astrodome of volume. In most regions the gas has a very low density, about 1 to 10 atoms per cubic centimeter. The air we breathe contains 10^{19} atoms per cubic centimeter.

NEBULAE

Some evidence of the existence of an interstellar medium is obvious to the naked eye. Seen in a dark sky, the Great Nebula in Orion is easily visible to the naked eye as a patch of haze in the sword of Orion (Figure 2–4). A small telescope reveals even more such **nebulae**—clouds of gas and dust.

One kind of nebula is called an **emission nebula** (also known as diffuse or bright nebulae) because its spectrum is an emission spectrum. Recalling Kirchhoff's second law (Box 7–2), we would expect an emission spectrum to be produced by an excited cloud of low-density gas. Thus, emission nebulae glow for the same reason that neon signs glow. Both produce emission spectra when the gas is excited.

The gas in emission nebulae is ionized by short-wavelength photons emitted by a nearby hot star (Figure 11–1). As the ionized atoms recapture an electron, the electron cascades down through the energy levels emitting photons. These emission nebulae have a characteristic red color (Color Plates 7 and 12) because of the strong Balmer alpha-emission line of hydrogen at 656.3 nm.

Only hot stars emit enough ultraviolet radiation to ionize the surrounding gas to any great distance. An O5 star, for example, can ionize the gas out to a distance of about 50 pc. Because the gas is ionized, such regions are often called **H II regions**. H II is shorthand notation for ionized hydrogen. (H I is neutral hydrogen.) The Orion Nebula discussed in the Perspective at the end of this chapter is an H II region.

FIGURE 11–2 The dust around the stars of the Pleiades star cluster reflects the light of the stars and produces a reflection nebula. The spectrum of the nebula is just the reflected spectrum of the stars. The cluster, visible to the naked eye in Taurus, contains about 500 stars. (Lick Observatory photograph.)

Another kind of nebula is called a **reflection nebula**. The spectrum of a reflection nebula is simply a reflected spectrum of the stars that illuminate the nebula. This tells us that the nebula does not contain excited gas but must contain solid specks of dust capable of reflecting light. A well-known example of a reflection nebula is that around the stars of the Pleiades star cluster (Figure 11–2). Although the stars are just visible to the naked eye, the dusty nebula is only visible on time-exposure photographs.

Dark nebulae are thick dusty clouds that block our view of distant stars. We see them as great dark patches in the sky where there seem to be few stars. Of course, the clouds contain gas and dust, but in the dark nebulae, we can see no hot star exciting the gas (Figure 11–3).

The smallest of the dark nebulae are the **Bok globules**,* small dark clouds less than 1 pc in diameter, containing 10 to 1000 solar masses. Bok globules are seen silhouetted against bright nebulae (Figure 11–4). Some globules contain infrared sources at their centers and appear to be collapsing clouds that will eventually become stars; others may be fragments of larger clouds that were shredded by the violence of the star-formation process. In either case, Bok globules appear to be associated with the birth of new stars.

*Named after astronomer Bart Bok.

EXTINCTION AND REDDENING

One way we know that dust is present in the interstellar medium is that it makes distant stars fainter than they would be if space were perfectly transparent. This is called **extinction**, and, in the neighborhood of the sun, it amounts to about 1.9 magnitudes per 1000 pc.

Another way we can detect the presence of dust is the effect it has on the colors of stars. An O star should be blue, but some stars whose spectra are that of an O star seem much redder than they should be. Termed **interstellar reddening**, this effect is caused by dust particles scattering light. Because the dust specks have diameters roughly the same as the wavelength of blue light, the blue photons can be scattered easily. Red photons, due to their longer wavelengths, are less likely to be scattered (Figure 11–5). Thus, the light that reaches us from a distant star is relatively rich in long-wavelength photons, making it seem redder.

This kind of scattering is quite common. Sunlight passing through our atmosphere is scattered by molecules and dust in the air with blue photons being scattered more than red. Whichever direction we look in the sky we see these scattered blue photons, and the sky looks blue. At sunset or sunrise, sunlight travels through such a long path in our atmosphere that it loses many of its bluer photons, and thus the setting or rising sun looks red.

FIGURE 11–3 A dark nebula consisting of dust and gas blocks our view of more distant stars and gas. This example, known as the Horsehead Nebula, is about 500 pc from earth and measures, from nose to ear, about 0.55 pc. (Palomar Observatory photograph.)

FIGURE 11–4 Bok globules are small black clouds of interstellar matter. They may be collapsing clouds about to form stars or the fragments of clouds that have already formed stars. The smallest globules visible in this photograph are roughly 100 times the diameter of our solar system. (Lick Observatory photograph.)

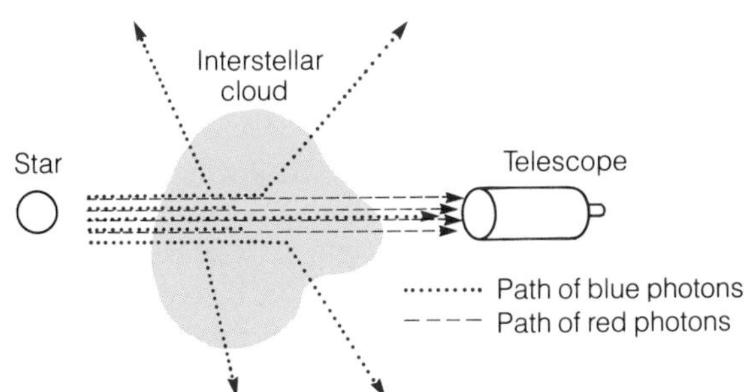

FIGURE 11–5 Seen through a dust cloud, a star appears redder because the blue photons, having shorter wavelengths, are more likely to be scattered by the dust grains.

Studies of interstellar reddening show that the dust specks must have diameters of a few millionths of a centimeter—about the size of the specks in cigarette smoke. They must contain only about 10^8 atoms. The centers of the dust particles must be made of silicates (rocky materials) or carbon, but they must also have a coating of frozen ices including water, carbon dioxide, methane, and ammonia. The silicate and carbon cores are believed to condense like soot in the cool atmospheres of some giant stars, and the icy mantles are believed to condense on the cores inside cold clouds of gas and dust.

INTERSTELLAR ABSORPTION LINES

Extinction and reddening give us a way to study the dust in space, but we need other observations to study the gaseous component of the interstellar medium. One way to detect this gas is to look for the absorption lines it forms in the spectra of distant stars—the **interstellar absorption lines**.

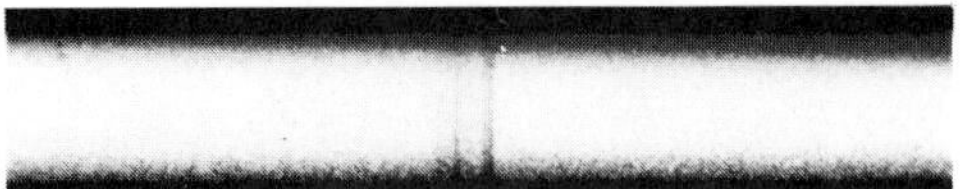

FIGURE 11–6 Interstellar lines of calcium in the spectrum of the hot star ϵ Orionis show that the light has passed through interstellar gas clouds. Hot stars do not produce calcium absorption lines in their own spectra. Multiple lines indicate multiple clouds with different velocities. (Mount Wilson and Las Campanas Observatories, Carnegie Institution of Washington.)

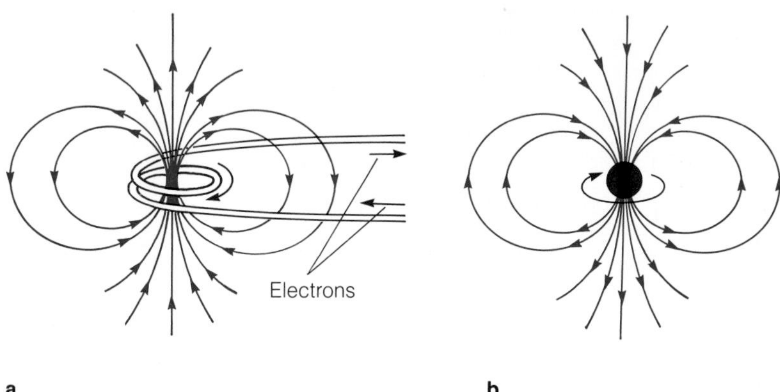

FIGURE 11–7 (a) Electrons flowing through a coil of wire produce a magnetic field, as in an electromagnet. (b) The rotation of the positively charged proton produces a similar magnetic field.

We can observe these lines best in the spectra of hot stars. Hot stars have few lines to confuse the spectrum and are very bright so they can be seen at great distance. Thus, we can record the spectrum of ϵ Orionis, a B0 supergiant, and detect very narrow lines of once ionized calcium (Figure 11–6). The lines of Ca II should not be present in the spectrum of a hot star, so we can conclude that they must be produced by the interstellar medium.

The nature of these lines tells us two things about the interstellar gas. First, the lines are exceedingly narrow. Clearly, the gas is not very dense or collisional broadening would widen the lines. Also, the gas cannot be very hot or Doppler broadening would be significant. Thus, we know the gas is cold and of very low density. In many cases, we see multiple components of the lines slightly displaced. This must occur when the starlight, on its way to earth, passes through clouds of gas with slightly diferent radial velocities. Then the Doppler effect displaces the individual spectral lines slightly in wavelength.

Interstellar lines of calcium, sodium, and iron have been detected, but no hydrogen lines have been seen from the earth's surface. The gas is so cold, no warmer than 100 K, that the hydrogen atoms are all in the ground state and cannot absorb Balmer wavelength photons. They can, however, absorb Lyman wavelength photons; these lines were eventually detected when telescopes such as the IUE were able to observe in the ultraviolet from above earth's atmosphere.

RADIO OBSERVATIONS OF THE INTERSTELLAR MEDIUM

As mentioned in Chapter 6, radio telescopes are important because they can detect the 21-cm wavelength radiation emitted by clouds of cool hydrogen in space. This radiation is emitted when a hydrogen atom's electron changes its energy by changing the direction of its spin.

This **21-cm radiation** was predicted in the mid-1940s by H. C. van de Hulst. (It was not actually detected until 1951.) We can understand how such a theoretical prediction was made if we think of the structure of a hydrogen atom. It consists of a proton and an electron, both of which must spin like tiny tops. Spinning a charged top is similar to making electricity flow through a coil of wire. It produces a magnetic field (Figure 11–7).

We have all played with small magnets and noticed that they repel each other in one orientation and attract each other if reversed. In the same way, the small magnetic fields produced by the spinning proton and electron can repel or attract each other. In one orientation the electron is slightly less tightly bound, and in the other orientation is slightly more tightly bound.

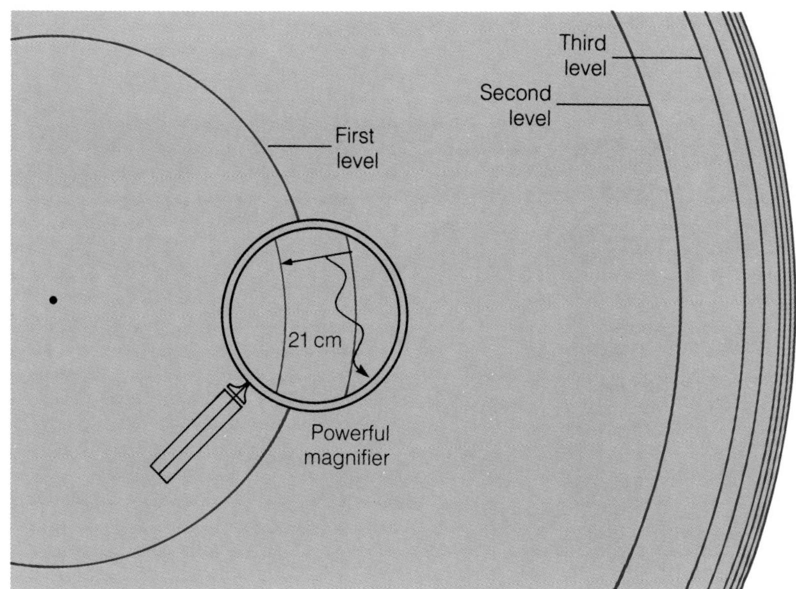

FIGURE 11–8 The magnetic fields of the proton and electron split the ground state of the hydrogen atom into two very close energy levels. A transition from the upper to the lower level emits a 21-cm photon.

Because of these magnetic fields, the ground state of the hydrogen atom is really two energy levels of very slightly different energy (Figure 11–8). If the electron is in the higher orbit, it can spontaneously flip over and spin the other way, dropping to the lower orbit and emitting a photon. The two energy levels are so close together that the photon emitted in the transition must have a very low energy—corresponding, in fact, to a wavelength of 21 cm.

Only very cold, low-density clouds of atomic hydrogen will emit 21-cm radiation. Once a hydrogen atom is excited into the upper of the two levels, it will, on average, stay there for 11 million years before spontaneously dropping to the lower level and emitting a 21-cm photon. The atoms of hot dense hydrogen in stars collide much too often, disturbing the electron before it can radiate a 21-cm photon. But atoms in the interstellar medium collide about once every million years, so a few do manage to radiate 21-cm photons. Also, hydrogen atoms linked into molecules cannot radiate 21-cm photons because the energy levels for a molecule are different from those for an atom.

Unfortunately for astronomers, molecular hydrogen does not emit photons efficiently but other molecules do, and over 60 of these have been detected in the interstellar medium (Table 11–1). Some are quite complex, and it is not clear how they form, although most astronomers suspect they originate on the surfaces of dust grains. Some of these molecules have not been synthesized yet on earth, but some are common. Ethyl alcohol, the alcohol some humans drink, has been detected, which once led a team of radio astronomers to report the mass of a particular cloud using units of billions of fifths of whiskey.

The most important of these molecules may be carbon monoxide (CO). Although it is about 10,000 times less common than molecular hydrogen, it radiates at a wavelength of 2.6 mm, and is used as a tracer of molecular hydrogen. Where we see CO we can be certain H_2 is common.

Such molecules occur in dusty dense clouds. Outside the clouds the ultraviolet radiation in starlight would break them apart, but inside the denser clouds, protected by the dust in the cloud, the molecules can survive.

In 1983 the Infrared Astronomy Satellite (IRAS) observed a complex of wispy clouds that became known as the **infrared cirrus** because of its overall resemblance to cirrus clouds in the earth's atmosphere. The infrared cirrus consists of dusty clouds of interstellar matter with temperatures of about 30 K. (See Color Plate 16.) Recent studies of CO emission from molecular clouds have shown that at least some of the infrared cirrus is associated with molecular clouds within a few hundred parsecs of the sun. This suggests that the gas and dust in the molecular clouds is not uniform but very patchy.

The largest of these molecular clouds, the **giant molecular clouds**, are clearly associated with star formation. The typical giant molecular cloud is a few hundred parsecs in diameter and contains up to 0.5 million solar masses. Although we say they are very dense at about 10^6 particles/cm^3, they are about 10 trillion times less dense

Table 11–1 Molecules detected in the interstellar medium.

Number of Atoms									
2	3	4	5	6	7	8	9	11	13
H_2	H_2O	NH_3	C_4H	CH_3OH	CH_3CCH	CH_3COOH	CH_3OCH_3	HC_9N	$HC_{11}N$
CO^+	O_3	H_2C_2	CH_2NH	CH_3SH	CH_3CHO	CH_3C_3N	CH_3CH_2OH		
CO	N_2H	H_2CO	CH_2CN	CH_3CN	CH_3NH_2		CH_3CH_2CN		
CH	HCO^+	H_2CS	HCOOH	NH_2CHO	CH_2CHCN		CH_3C_4H		
CH^+	HCS^+	HNCO	CH_2CO		HC_5N		HC_7N		
CN	HCN	HNCS	HC_3N		CH_3C_2H				
CS	C_2H	C_3H	H_2C_2O						
C_2	SO_2	C_3O							
CH^+	H_2S	C_3N							
OH	HCO	$HOCO^+$							
NO	HCO^+								
NS	OCS								
SO	HNO								
SiO	HOC^+								
SiS	NaOH								

than air. Nevertheless, the extinction in these clouds is so great that we cannot see through them at visual wavelengths. Radio observations tell us that the gas is very cold, about 10 K near the center. Yet the clouds also contain infrared sources that are believed to be stars in the process of forming, and bright H II regions excited by young hot stars are often found near their outer edges, as in the case of the Orion Nebula. (See Color Plates 14, 15, and 17.) Evidently, at least some giant molecular clouds are sites of active star formation.

11.2 THE BIRTH OF STARS

Stars have been forming continuously since our galaxy took shape over 10 billion years ago. We know this for two reasons. First, the sun is only about 5 billion years old, a relative newcomer compared to the older stars in our galaxy. Second, we can see hot, blue stars such as Spica (α Virginis), a B1 main-sequence star. As we will see in Box 11–2, such massive stars have very short lives. In fact, a star like Spica can last only 10 million years and thus must have formed recently.

The key to understanding star formation is the correlation between young stars and clouds of gas. Where we find the youngest groups of stars we also find large clouds of gas illuminated by the hottest and brightest of the new stars (Figure 11–1). This leads us to suspect that stars form from such clouds, much as raindrops condense from the water vapor in a thundercloud. The central problem for our discussion of star formation is how the large, low-density, cold clouds of gas become comparatively small, high-density, hot stars. Gravity is the key.

CONTRACTION AND HEATING

The combined gravitational attraction of the atoms in a cloud of gas squeezes the cloud, pulling every atom toward

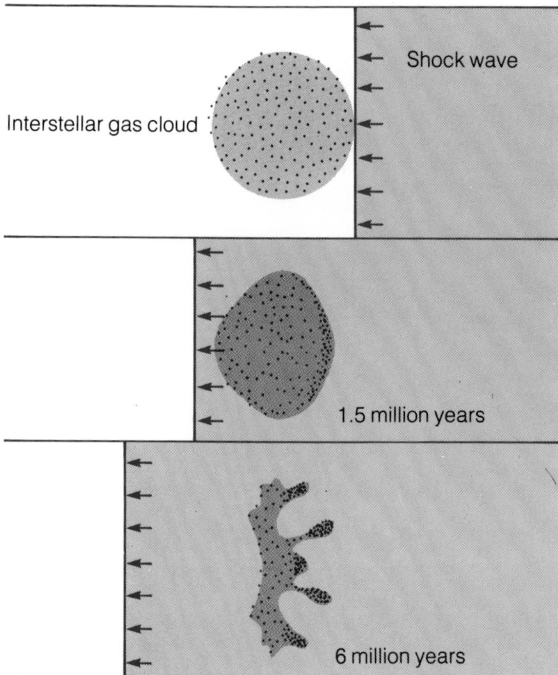

FIGURE 11–9 An idealized interstellar gas cloud is struck by a passing shock wave and compressed to densities that may trigger star formation. (Adapted from computer models by Paul R. Woodward.)

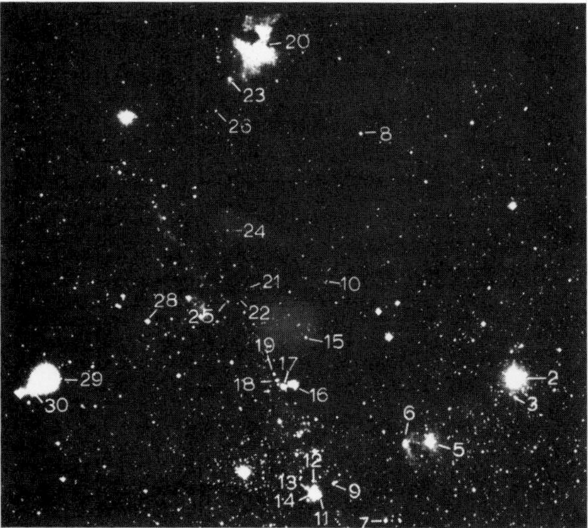

FIGURE 11–10 Canis Major R1 is a dusty nebula shining by the reflection of light from young stars (numbered). The arc-shaped nebula is about 100 light-years long and lies at the edge of an expanding shell of hot gas believed to be the remains of a supernova explosion. (Courtesy William Herbst.)

the center. Thus, we might expect every cloud eventually to collapse and become a star, but two factors oppose gravity and resist collapse. First, the atoms of the gas move quickly, even at temperatures of 10 K. The average hydrogen atom in such a cloud moves about 0.5 km/sec (1100 mph). This thermal motion would make the cloud drift apart if gravity were too weak to hold it together. The second factor that resists collapse is turbulence. Gas in the cloud may be churned by the heating effects of nearby stars, collisions with other gas clouds, or currents of gas pushing through the interstellar medium from the explosion of distant stars. Whatever the cause, these turbulent motions resist the collapse of the cloud.

If we measured the diameter and density of an interstellar cloud, we could find its mass and thus its gravity. Comparing its gravity to its temperature and estimating the effects of turbulence would tell us whether the cloud is dominated by gravity and will collapse, or whether it is dominated by thermal and turbulent motions and will drift apart. The disappointing result of such measurements is that many clouds are not dominated by gravity. That is, few should collapse spontaneously due solely to their own gravity.

Until recently this finding left astronomers with no mechanism to explain the new stars we see all around us, but recent mathematical models of gas clouds colliding with shock waves show that the clouds are compressed and disrupted into fragments (Figure 11–9). Some of these fragments probably become dense enough to collapse and form stars.

At least four different processes can trigger star formation. Supernovae (exploding stars described in Chapter 13) can produce powerful shock waves. Also, the ignition of very hot stars can release such a burst of radiation that the surrounding interstellar matter is pushed back into an expanding shell. A third trigger may be the collision of molecular clouds. Because the clouds contain magnetic fields, they cannot pass through each other, and a collision can compress the gas and trigger star formation. Yet another trigger may be the spiral pattern in our Milky Way galaxy. One theory suggests that the spiral arms contain shock waves which compress the interstellar medium and trigger star formation.

All of these processes have been observed. For example, young stars have been found at the outer edge of nebulae expanding away from the sites of ancient supernova explosions (Figure 11–10). In another study,

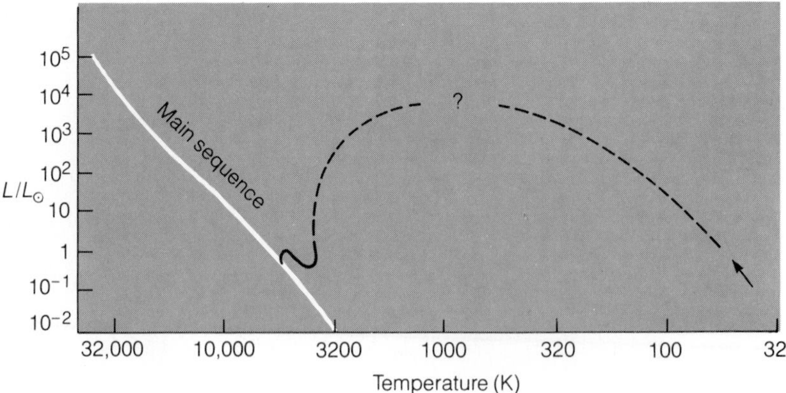

FIGURE 11–11 To follow the contraction of a collapsing gas cloud (dashed line), we must extend the H–R diagram to very low temperatures. The exact behavior of such a cloud is uncertain because of the dust.

x-ray astronomers found a giant bubble or shell of hot gas lying behind the dust clouds in Cygnus. At a distance of 6000 pc the shell is about 400 pc in diameter. Apparently the passage of a spiral arm triggered star formation on the far side of the dust clouds about 3 million years ago, and the successive generations of stars have inflated the shell through supernovae and the ignition of hot stars. The expansion of the shell is triggering the collapse of gas clouds to form still more new stars.

Once a cloud is triggered into collapse, its gravity draws each atom toward the center, and the atoms gather speed as they fall. As the cloud contracts, the atoms collide with each other more often; the increased velocity, which the atoms picked up in falling inward, is converted to random motion among the atoms. Because the temperature of a gas is related to the average velocity of the atoms, the temperature of the gas increases. Thus, the contraction of an interstellar gas cloud heats the gas by converting gravitational energy into thermal energy.

A small gas cloud could thus contract, grow denser and hotter, and, presumably, become a star. But a larger cloud cannot contract into a single object; because of instabilities, it must fragment, producing a cluster of 10 to 1000 stars.

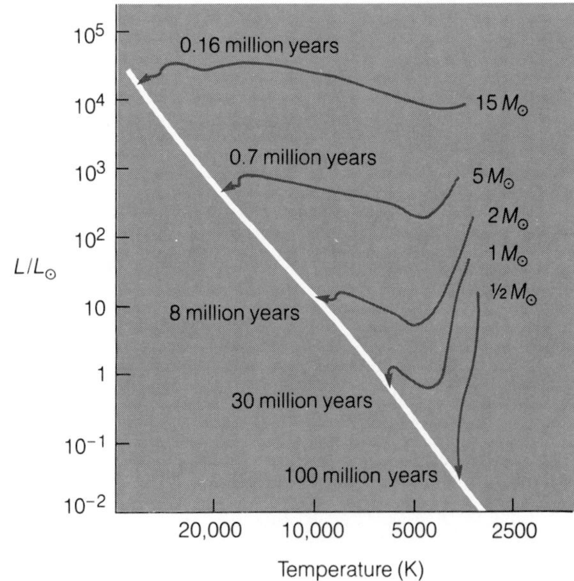

FIGURE 11–12 The more massive a protostar is, the faster it contracts. A 1 $M_\odot$ star requires 30 million years to reach the main sequence. (Adapted from Michael A. Seeds, "Stellar Evolution," *Astronomy*, Feb. 1979.)

PROTOSTARS

To follow the story of star formation further, we must concentrate on a single fragment of a collapsing cloud as it contracts, heats up, and begins to behave like a star. Although the term **protostar** is used rather loosely by astronomers, we will define it to be a prestellar object that is hot enough to radiate infrared radiation but not hot enough to initiate nuclear fusion.

We must extend the H–R diagram to very low temperatures to follow this process (Figure 11–11). The details are poorly known because of the complex effects of dust in the gas. The object begins as a cold gas cloud, but as it contracts, it grows warmer by converting its gravitational energy into thermal energy. Half of this energy radiates away, but the remaining energy heats the center of the protostar closer to the ignition temperature for nuclear reactions—a few million degrees at least. Finally,

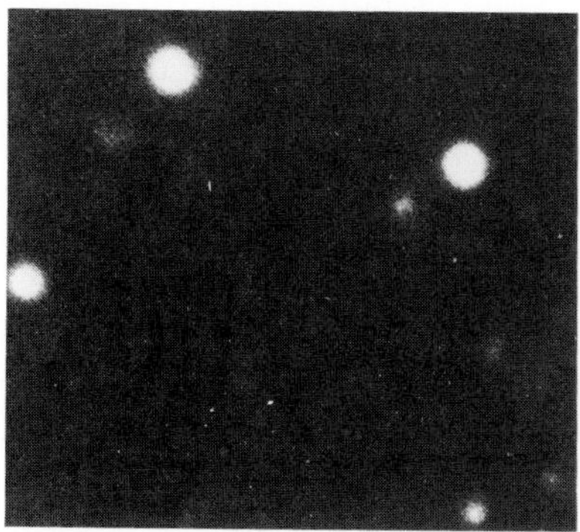

FIGURE 11–13 A star just emerging from its cocoon. A 1964 photograph (left) shows no star, but a 1983 photograph (right) shows the image of a star (arrow). Spectra show that the star is surrounded by an expanding shell of gas and dust. (©1960 National Geographic Society, Palomar Sky Survey; ©1983 AURA, Inc., National Optical Astronomy Observatories, CTIO.)

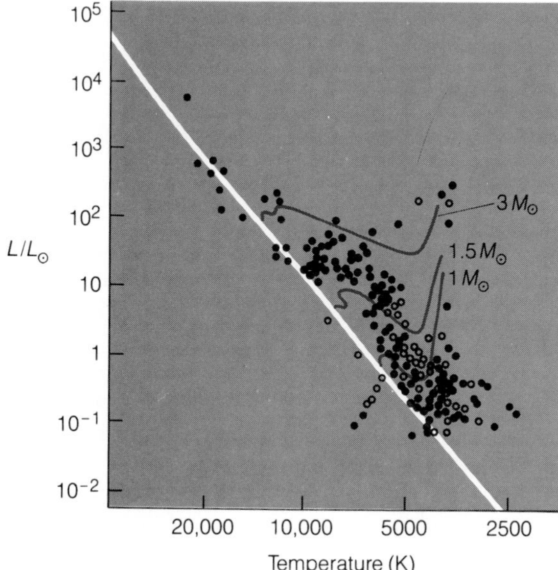

FIGURE 11–14 An H–R diagram of the star cluster NGC 2264 reveals many low mass stars that have not yet reached the main sequence. Many of these stars are T Tauri stars (open circles). The cluster is only a few million years old. (Adapted from a diagram by M. Walker.)

as hydrogen fusion begins at the protostar's center, the contraction slows, and the object, now a star, settles onto the main sequence where it remains most of its life.

How long a time this contraction takes depends on the mass of the star. More massive stars have stronger gravity and contract rapidly (Figure 11–12). The sun took about 30 million years to reach the main sequence, but a 30 $M_\odot$ star requires only 30,000 years. A 0.2 $M_\odot$ star takes 1 billion years to reach the main sequence.

STAR FORMATION CONFIRMED

The preceding scenario of star formation is based on theory. To test the theory and its supporting assumptions, we must look for real protostars.

Unfortunately, a protostar is not easy to find. The protostar stage is short compared to a star's lifetime. Also, the protostar is surrounded when it forms by a cloud of gas and dust called a cocoon. This cocoon absorbs all of the light from the protostar and radiates the energy away as infrared radiation. Thus, young protostars are invisible at optical wavelengths. The Infrared Astronomy Satellite found many infrared sources that are probably protostars within cocoons.

Eventually, the protostar must become hot enough to blow away this cocoon. Herbig–Haro objects, small nebulae that vary irregularly in brightness, were originally thought to be protostars in the early stages of clearing their cocoons. New observations, however, suggest that many of these objects are small gas clouds being excited by intense stellar winds from newly formed stars (Color Plate 15). On the other hand, some Herbig–Haro objects do seem to be emerging protostars. In March 1983 astronomers at Cerro Tololo Inter-American Observatory in Chile discovered a new star near the edge of Herbig–Haro object Number 57 (Figure 11–13). The star was not visible on photographs taken some years before.

T Tauri stars, named after variable star T in Taurus, do seem to be stars emerging from their cocoons. We know they are no older than 100 million years because they occur in some young star clusters such as NGC 2264. This cluster of stars formed so recently that the less massive stars, contracting slower than the massive stars, have not yet reached the main sequence (Figure 11–14). The temperature and luminosity of the T Tauri stars in the

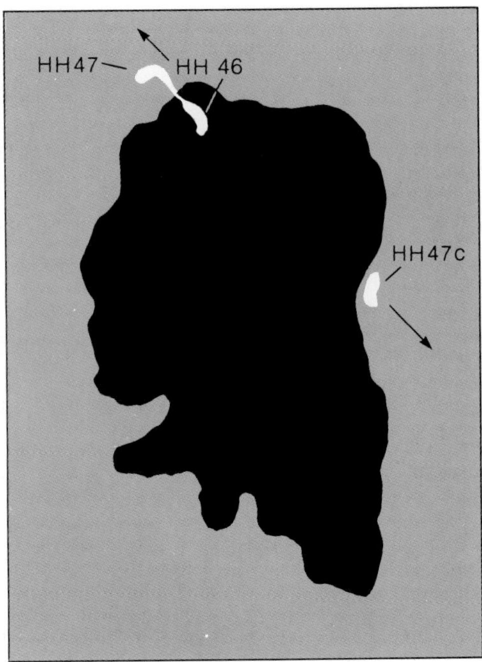

FIGURE 11–15 The Herbig–Haro objects HH46 and HH47 are moving toward us while the fainter HH47c is receding. They appear to have been driven outward in opposite directions from a young star located near HH46 within the dark globule. (©1982 AURA, Inc., National Optical Astronomy Observatories, CTIO.)

cluster place them to the right of the main sequence, just where we would expect to find protostars.

Other characteristics of the T Tauri stars support our belief that they are clearing their cocoons. Some of these stars are strong infrared sources, showing that they are still associated with clouds of dust. But the remnants of their cocoons are detectable in their spectra, where Doppler-shifted lines show that the stars are surrounded by expanding clouds of gas.

Optical, infrared, and radio observations reveal streams of gas blasting away from many forming stars in two oppositely directed beams or jets called **bipolar flows** (Figure 11–15 and Color Plate 13). Some T Tauri stars are sources of such flows, and Herbig–Haro objects are often involved where the jets interact with surrounding gas.

The cause of bipolar flows is not well known, but they seem to be a result of powerful stellar winds blowing away from the protostar. One theory suggests that gas falling into a disk of gas rotating around the star drags in some of the magnetic field with it. The combination of the rotation of the disk, the stellar wind, and the magnetic field could eject gas along the axis of rotation (Figure 11–16). We will discover in later chapters that both disks and bipolar flows are common—they occur in dying stars and exploding galaxies and young stars.

Infrared observations show that many stars are surrounded by clouds or disks of dusty material. The Infrared Astronomy Satellite found dusty clouds around a number of stars including Vega and Fomalhaut. Earth-based infrared observations have found other such stars. Modern astronomers believe that the planets of our solar system formed in such a disk of dust when the sun formed. One star, β Pictoris, is clearly at the center of an edge-on disk of particles extending out ten times farther than the distance from the sun to Pluto (Color Plates 30 and 31). Such a disk of material could be forming planets.

The evidence is accumulating rapidly that bipolar flows are related to disks of material around young stars. If that interpretation proves correct, then it would be a very common event and is probably a stage through which all stars pass as they clear their cocoons. This has great philosophical significance because, if planets do form in such disks of material, planetary systems must be very common. No planets have yet been detected orbiting other

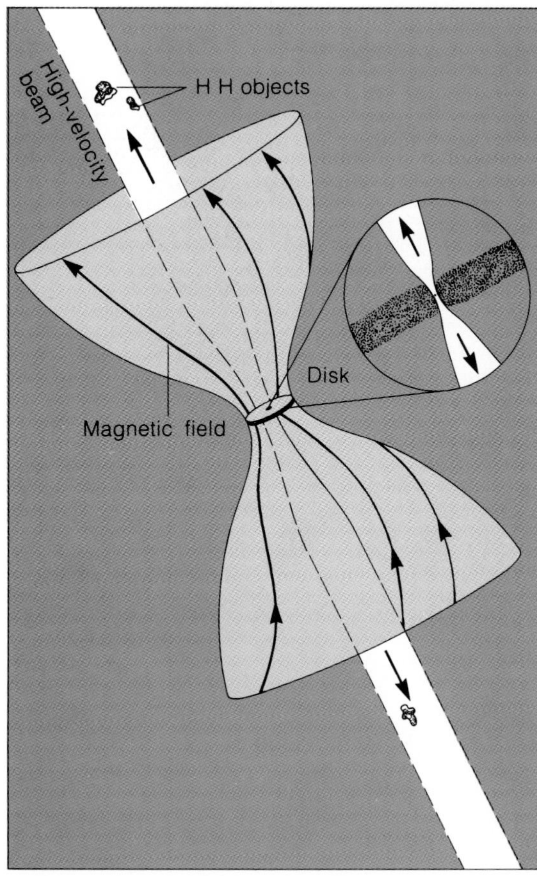

FIGURE 11–16 A theoretical model of bipolar flow shows how it may be related to in-falling gas dragging the magnetic field into a spinning disk around a forming star (inset). The resulting gas flows may travel at 300 km/sec and excite Herbig–Haro objects when small clouds get caught in the beams. (Adapted in part from diagrams by Ralph E. Pudritz and Colin A. Norman.)

stars, but observations by the Hubble Space Telescope may detect them. We will discuss the formation of planets in detail in Chapter 19. Here we are interested in how the stars themselves form.

Stars form because gravity forces interstellar gas clouds to collapse. The contraction of the cloud fragments heats the gas until the center of the protostar can ignite its thermonuclear fires and halt its collapse. Clearly, internal nuclear reactions are as important to a star's life as gravity is to its formation. Before we can follow the evolution of stars further, we must examine these reactions—the sources of stellar energy.

11.3 NUCLEAR ENERGY SOURCES

In Chapter 8 we learned that the sun generates its energy from nuclear fusion reactions that combine hydrogen nuclei to build helium nuclei. We also found ample evidence that the sun is a typical star, so we can have confidence that all stars generate their energy through nuclear fusion reactions. Not all stars, however, use exactly the same reactions and the same fuels as does the sun.

HYDROGEN FUSION

The nuclear reactions that fuse hydrogen into helium are critically important to the formation of stars. The contraction of the protostar continues until the central temperature and pressure become high enough to ignite nuclear reactions, generate energy, and halt the contraction. The first reactions to ignite and produce useful amounts of energy are those that fuse hydrogen. Thus, hydrogen fusion is important because it halts the contraction of protostars.

Hydrogen fusion is also the longest lasting source of energy in a star. Although some stars can fuse other fuels at later stages, all stars fuse hydrogen for about 90 percent of their total lifetime. Once a contracting protostar ignites hydrogen, it becomes a stable star and remains stable for a very long time.

About 90 percent of all stars (excluding white dwarfs) are now fusing hydrogen. The sun is a hydrogen-fueled star and has been fusing hydrogen for about 5 billion years. It has enough hydrogen fuel left to last another 5 billion years. We can see evidence that a great many other stars are fusing hydrogen when we look at the H–R diagram. All main-sequence stars generate their energy from hydrogen. The numbers of main-sequence stars that we found in the survey of the neighborhood around the sun (Figures 9–16 and 9–17) show how common hydrogen-powered stars are.

Most hydrogen-fusing stars generate their energy in the same way the sun does—by the proton–proton chain. We discussed this sequence of three nuclear reactions in Chapter 8 and discovered that it combines four hydrogen nuclei to make one helium nucleus and in the process converts a small amount of mass into energy. The proton–proton chain, operating in the sun's core, generates nearly all of the sun's energy.

However, about 10 percent of the sun's energy comes from a different set of reactions. This process, the CNO cycle, is common in some main-sequence stars, but it is merely a different way of fusing hydrogen.

Much like the proton–proton chain, the **CNO (carbon–nitrogen–oxygen) cycle** is a series of nuclear reactions that produce energy through hydrogen fusion. However, the CNO cycle uses carbon as a catalyst; that is, the carbon nucleus makes the reaction possible but is not altered in the end. Although it seems quite different from the proton–proton chain, the CNO cycle combines four hydrogen nuclei to make one helium nucleus plus energy, which we can verify by counting protons in Figure 11–17.

Because the carbon nucleus has a charge six times that of hydrogen, the electrostatic force of repulsion between a carbon nucleus and a proton is high, and much higher temperatures are necessary to force the proton into the carbon nucleus. Thus, the CNO cycle is dominant only in stars more massive than about 1.1 $M_\odot$. These stars have central temperatures hotter than 16,000,000 K. Stars less massive than this, such as the sun, are cooler at their centers and are dominated by the proton–proton chain.

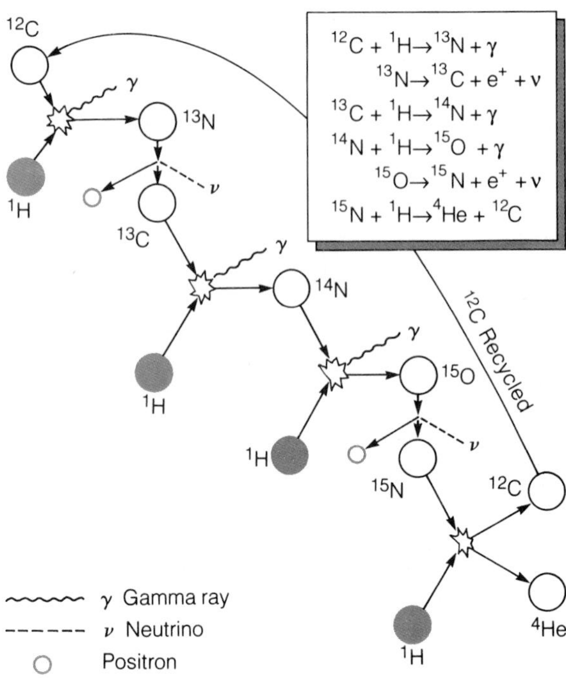

FIGURE 11–17 The CNO cycle uses ^{12}C as a catalyst to combine four hydrogen atoms (^1H) to make one helium atom (^4He) plus energy. The carbon atom reappears at the end of the process ready to start the cycle over.

THE PRESSURE–TEMPERATURE THERMOSTAT

Nuclear reactions in stars manufacture energy and heavy nuclei under the supervision of a built-in thermostat that keeps the reactions from erupting out of control. That thermostat is the relation between pressure and temperature discussed at the beginning of this chapter.

In a star the nuclear reactions generate just enough energy to balance the inward pull of gravity. Consider what would happen if the reactions begin to produce too much energy. The extra energy would raise the internal temperature of the star, and, because the pressure of the gas depends on its temperature, the pressure would also rise. The increased pressure would make the star expand. Expansion would cool the gas and lower its density, slowing the nuclear reactions. Thus, the star has a built-in regulator that keeps the nuclear reactions from going too fast.

The same thermostat also keeps the reactions from slowing. Suppose the nuclear reactions begin to produce too little energy. Then the inner temperature would fall, lowering the pressure and allowing gravity to squeeze the star slightly. Compression would heat the gas and make it denser, thus increasing nuclear energy generation until the star was stable again.

The stability of a star depends on this relation between pressure and temperature. If an increase or decrease in temperature produces a corresponding change in pressure, the thermostat functions correctly and the star is stable. We will see in this chapter how the thermostat accounts for the mass–luminosity relation. In Chapter 13 we will see what happens to a star when the thermostat breaks down completely and the nuclear fires burn unregulated.

INSIDE STARS

In Chapter 10 we discovered that the stars on the main sequence are ordered according to mass with the upper

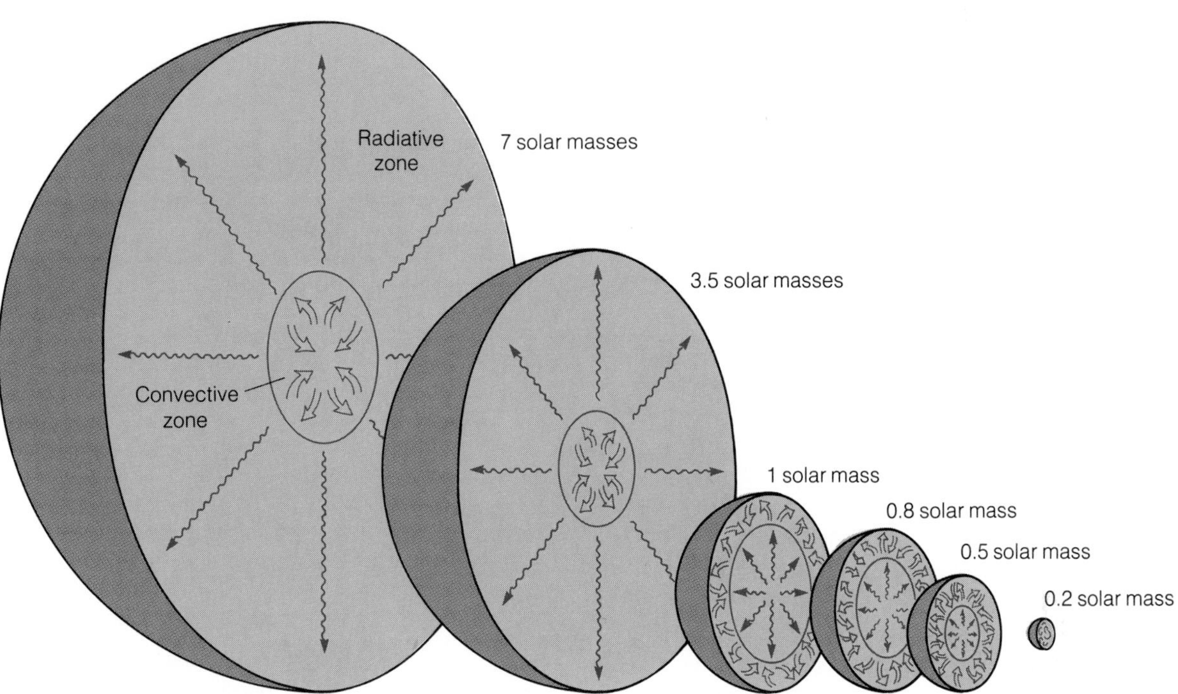

FIGURE 11–18 Inside stars. The more massive stars have small convective cores and radiative envelopes. Stars like the sun have radiative cores and convective envelopes. The lowest mass stars are convective throughout. (Adapted from Michael A. Seeds, "Stellar Evolution," *Astronomy*, Feb. 1979.)

main-sequence stars being most massive and the lower main-sequence stars being least massive. Combining this with what we know about hydrogen burning tells us that there are two kinds of main-sequence stars: upper main-sequence stars that fuse hydrogen on the CNO cycle and lower main-sequence stars that fuse hydrogen on the proton–proton chain. From the outside these stars differ only in size, temperature, and luminosity, but inside they are quite different.

The upper main-sequence stars are more massive and thus must have higher central temperatures to withstand their own gravity. These high central temperatures permit the star to fuse hydrogen on the CNO cycle and that affects the internal structure of the star.

The CNO cycle is very temperature-sensitive. If the central temperature of the sun rose by 10 percent, energy production by the proton–proton chain would rise by about 46 percent, but energy production by the CNO cycle would shoot up by 350 percent. This means that the more massive stars generate almost all of their energy in a tiny region at their very centers where the temperature is highest. A 10 solar mass star, for instance, generates 50 percent of its energy in its central 2 percent of mass.

This concentration of energy production at the very center of the star causes a "traffic jam" as the energy tries to flow away from the center. Transport of energy by radiation can't drain the energy away fast enough, and the central core of the star churns in convection as hot gas rises upward and cooler gas sinks downward. Farther from the center, the traffic jam is less severe, and the energy can flow outward as radiation. Thus, massive stars have convective cores at their centers and radiative envelopes extending from their cores to their surfaces (Figure 11–18).

Main-sequence stars less massive than about 1.1 solar masses cannot get hot enough to fuse much hydrogen on the CNO cycle. They generate most of their energy by the proton–proton chain, which is not as sensitive to temperature, and thus the energy generation occurs in a larger region in the star's core. The sun, for example, generates

Table 11–2 The four laws of stellar structure.

1. Hydrostatic equilibrium	The weight on each layer is balanced by the pressure in that layer.
2. Energy transport	Energy moves from hot to cool by radiation, convection, or conduction.
3. Continuity of mass	Total mass equals sum of shell masses. No gaps allowed.
4. Continuity of energy	Total luminosity equals sum of energies generated in each shell.

50 percent of its energy in a region that contains 11 percent of its mass. Because the energy generation is not concentrated at the very center of the star, no traffic jam develops and the energy flows outward as radiation. Only near the surface, where the gas is cooler and therefore more opaque, does convection stir the gas. Thus, the less massive stars on the main sequence have radiative cores and convective envelopes.

The lowest mass stars have a slightly different kind of structure. For stars less than about 0.4 solar mass, the gas is relatively cool compared to the inside of more massive stars, and the radiation cannot flow outward easily. Thus, the entire bulk of these low-mass stars is stirred by convection.

The energy generation inside stars not only stops their contraction, but it also determines the internal structure of the star. In the next section, we will learn still more about the long, peaceful lives of main-sequence stars.

11.4 MAIN-SEQUENCE STARS

If a protostar could think, it would face a number of urgent questions as it contracted toward the main sequence: How should I generate energy? How luminous should I be? How long can I remain stable? We might summarize the protostar's quandary in a single question: How am I supposed to know what kind of star to be? Of course, protostars can't think, and nature sees to it that each protostar becomes the right kind of main-sequence star. In this section, we will analyze nature's star-making methods and answer the protostar's questions.

As humans we might have a question of our own at this point: How can mere humans know what happens inside a star or how that star will change over billions of years? We will answer that question first by discussing stellar models.

STELLAR MODELS

The structure of a star is based on four simple laws of physics. We have previously discussed the law of hydrostatic equilibrium, which says the weight on each layer in a star must be balanced by the pressure in that layer. We have also discussed the law of energy transport, which says that energy must flow from hot to cool regions by radiation, convection, or conduction.

To these laws we add two very simple laws that are both laws of continuity. They tell us that the distribution of matter and energy inside the star must vary smoothly from surface to center. No gaps are allowed. In fact, both laws may seem familiar because we have all heard of the conservation of mass and energy.

The **continuity of mass law** says that the total mass of the star must equal the sum of the masses of its shells and that the mass must be distributed smoothly throughout the star. In a sense, the continuity of mass law is the law of conservation of mass with the added proviso that the mass be distributed smoothly.

The **continuity of energy law** says that the amount of energy flowing out the top of a shell must equal the

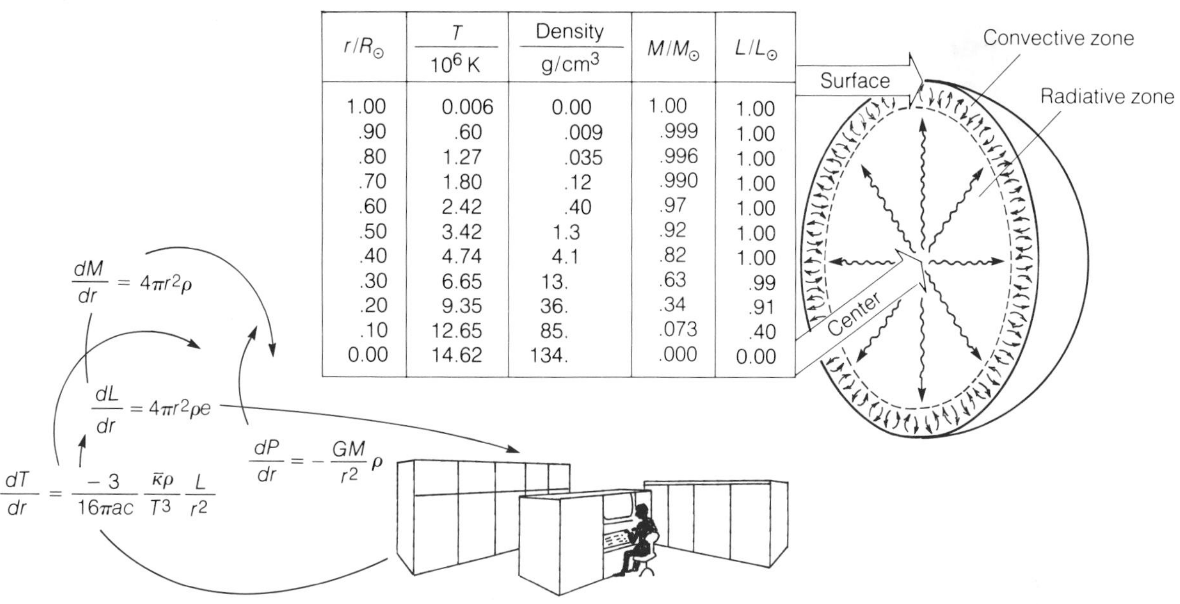

$r/R_\odot$	T (10^6 K)	Density (g/cm^3)	$M/M_\odot$	$L/L_\odot$
1.00	0.006	0.00	1.00	1.00
.90	.60	.009	.999	1.00
.80	1.27	.035	.996	1.00
.70	1.80	.12	.990	1.00
.60	2.42	.40	.97	1.00
.50	3.42	1.3	.92	1.00
.40	4.74	4.1	.82	1.00
.30	6.65	13.	.63	.99
.20	9.35	36.	.34	.91
.10	12.65	85.	.073	.40
0.00	14.62	134.	.000	0.00

FIGURE 11–19 A stellar model is a table of numbers that represents conditions inside a star. This table describes the interior of the sun. Each line in the table tells us the radius, temperature, and density of a shell in the model star, plus the mass it encloses and the energy flowing through it. (Adapted from Michael A. Seeds, "Stellar Evolution," *Astronomy*, Feb. 1979.)

amount coming in at the bottom plus whatever energy is generated within the shell. Further, it says that the energy leaving the surface of the star—the luminosity—must equal the sum of the energies generated in the shells. Thus, the energy law is really a version of the law of conservation of energy.

The four laws of stellar structure are much more precise than the four statements in Table 11–2. In fact, the four laws can be expressed as four equations (Figure 11–19). What these equations look like is not important to us, but applying them to build mathematical models of stars can tell us how stars work, how they are born, and how they die.

If we wanted to build a model of a star, we would have to divide the star into about 100 concentric shells and then write down the four equations of stellar structure for each shell. We would then have 400 equations that would have 400 unknowns—namely, the temperature, density, mass, and energy flow in each shell. Solving 400 equations simultaneously is not easy, and the first such solutions, done by hand before the invention of the electronic computer, took months of work. Now a properly programmed computer can solve the equations in a few seconds and print a table of numbers that represents the conditions in each shell of the star. Such a table, shown in Figure 11–19, is a **stellar model**.

Notice that these models are quantitative—that is, properties have specific numerical values. Earlier in this book, we discussed models that were qualitative—our model of the sunspot cycle, for instance. Both kinds of models are useful, but a model is much more useful when it is quantitative. This reflects the power of mathematics as a way of thinking quantitatively.

The model shown in Figure 11–19 represents our sun. As we scan the table from top to bottom, we descend from the surface of the sun to its center. The temperature increases rapidly as we move downward, reaching a maximum of about 15,000,000 K at the center. At this temperature, the gas is not very opaque, and the energy can flow outward as radiation. In the cooler outer layers, the gas is more opaque and the outward-flowing energy forces these layers to churn in convection. This model of the sun, like all stellar models, lets us study the otherwise inaccessible layers inside a star.

BOX 11–1

The Mass–Luminosity Relation

We can calculate the approximate luminosity of a star from a simple equation. A star's luminosity in terms of the sun's luminosity equals its mass in solar masses raised to the 3.5 power:

$$L = M^{3.5}$$

This is the mathematical form of the mass–luminosity relation.

We can do simple calculations with this equation if we remember that raising a number to the 3.5 power is the same as cubing it and then multiplying by its square root. For example, suppose that we observe a star whose mass is four times the mass of the sun. Thus:

$$L = M^{3.5} = 4^{3.5} = 4 \cdot 4 \cdot 4\sqrt{4} = 64 \cdot 2 = 128$$

The star emits 128 times as much energy as the sun.

Stellar models also let us look into the star's past and future. In fact, we can use models as time machines to follow the evolution of stars over billions of years. To look into a star's future, for instance, we use a stellar model to determine how fast the star consumes its fuel in each shell. As the fuel is consumed, the chemical composition of the gas changes and the amount of energy generated declines. By calculating the rate of these changes, we can predict what the star will look like at any point in the future.

THE ENDS OF THE MAIN SEQUENCE

One of the things that mathematical models of stars can tell us is why the main sequence has ends. If a protostar has too small a mass, it cannot become a star, and another protostar with too much mass may also have trouble stealing onto the main sequence.

The low-mass end of the main sequence is set by the ignition temperature required by the hydrogen-fusing reactions. To produce reasonable amounts of energy by the proton–proton chain, the center of a star must have a temperature of at least 10,000,000 K. But a protostar heats up by contracting, and, if its mass is too low (less than about 0.08 $M_{\odot}$), it will not be able to contract fast enough or far enough to heat up to this temperature. A protostar with the mass of the earth, for instance, would never get hot enough to ignite hydrogen and would continue to contract until it became a small, cool, dense body—a planet. If Jupiter had been about 100 times more massive, it could have been a star.

Protostars that are not massive enough to ignite nuclear fusion have been termed **brown dwarfs**. Such objects would cool slowly as they radiated their internal energy as infrared photons. Their infrared radiation and low luminosities would make them very difficult to find, and none have been identified with certainty as yet.

The rules of star making are not clear at the massive end of the main sequence. Stellar models predict that the most massive star cannot exceed a limit of about 60–100 solar masses—more massive than this and the star would be unstable. In addition, massive clouds of gas contracting

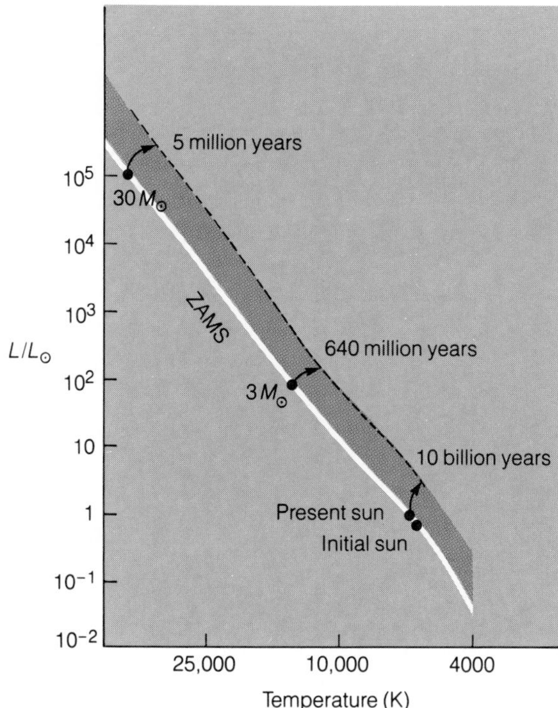

FIGURE 11–20 The main sequence is not a line but a band (shaded). Stars begin their main-sequence lives on the lower edge, which is called the zero age main sequence (ZAMS). As hydrogen fusion changes their composition, the stars slowly move across the band.)

to become protostars may break up into lower-mass fragments and thus make the formation of extremely massive stars highly unlikely.

If a contracting protostar has a mass between about 0.08 solar masses and about 60–100 solar masses, it can become a common main-sequence star. However, the protostar might still have a question: How luminous should I be? The answer to that question will tell us more about how a star works.

THE MASS–LUMINOSITY RELATION

In Chapter 10 we saw that the more massive stars are also more luminous (Box 11-1). This mass–luminosity relation can be explained by the theory of stellar structure, giving us direct observational confirmation of the theory.

The key to the mass–luminosity relation is the law of hydrostatic equilibrium, which says that pressure must balance weight, and the pressure–temperature thermostat, which regulates energy production. We have seen that a star's internal pressure stays high because generation of thermonuclear energy keeps its interior hot. Because more massive stars have more weight pressing down on the inner layers, their interiors must have high pressures and thus must be hot. For example, the temperature at the center of a 15 $M_\odot$ star is about 34,000,000 K, more than twice the central temperature of the sun.

Because massive stars have hotter cores, their nuclear reactions burn more fiercely. That is, their pressure–temperature thermostat is set higher. The nuclear fuel at the center of a 15 $M_\odot$ star fuses over 3000 times more rapidly than the fuel at the center of the sun. The rapid reactions in massive stars make them more luminous than the lower-mass stars. Thus, the mass–luminosity relation results from the requirement that a star support its weight by generating nuclear energy.

THE LIFE OF A MAIN-SEQUENCE STAR

A main-sequence star supports its weight by fusing hydrogen, but its supply of hydrogen is limited. Thus, it is inevitable that the star change when it exhausts its hydrogen fuel. However, long before the star runs out of hydrogen, its changing composition forces small alterations in its structure. When these small changes occur in the sun, life on Earth will vanish.

Hydrogen fusion combines four nuclei into one. Thus, as a main-sequence star consumes its hydrogen, the total number of nuclei in its interior decreases. Each newly made helium nucleus can exert the same pressure as a hydrogen nucleus, but because the gas has fewer nuclei, its total pressure is less. This unbalances the gravity–pressure stability, and gravity squeezes the core of the star more tightly. As the core contracts, its temperature increases, and the nuclear reactions burn faster, releasing more energy and making the star more luminous. This additional energy flowing outward through the envelope forces the outer layers to expand and cool, so the star becomes slightly larger, brighter, and cooler.

As a result of these gradual changes in main-sequence stars, the main sequence is not a sharp line across the H–R diagram but rather a band (shaded in Figure 11–20). Stars begin their stable lives fusing hydrogen on the lower

Table 11–3 Main-sequence stars.

Spectral Type	Mass (Sun = 1)	Luminosity (Sun = 1)	Approximate Years on Main Sequence
O5	40	405,000	1×10^6
B0	15	13,000	11×10^6
A0	3.5	80	440×10^6
F0	1.7	6.4	3×10^9
G0	1.1	1.4	8×10^9
K0	0.8	0.46	17×10^9
M0	0.5	0.08	56×10^9

edge of this band, the **zero-age main sequence** (ZAMS), but gradual changes in luminosity and surface temperature move the stars upward and to the right. By the time they reach the upper edge of the main sequence (the dashed line in Figure 11–20), they have exhausted nearly all of the hydrogen in their centers. Thus, we find main-sequence stars scattered throughout this band at various stages of their main-sequence lives.

These gradual changes in the sun will spell trouble for earth. When the sun began its main-sequence life about 5 billion years ago, it was only about 60 percent of its present luminosity. This, by the way, makes it difficult to explain how the earth has remained at roughly its present temperature for at least 3 billion years. Some experts suggest that the earth's atmosphere has gradually changed and thus compensated for the increasing luminosity of the sun.

By the time the sun leaves the main sequence in 5 billion years, the sun will have twice its present luminosity. This will raise the average temperature on earth by at least 19°C (34°F). As this happens over the next few billion years, the polar caps will melt, the oceans will evaporate, and much of the atmosphere will vanish into space. Clearly the future of the earth as the home of life is limited by the future evolution of the sun.

Once a star leaves the main sequence, it evolves rapidly and dies. The average star spends 90 percent of its life burning hydrogen on the main sequence. This explains why 90 percent of all stars are main-sequence stars. We are most likely to see a star during that long, stable period while it is on the main sequence.

The number of years a star spends on the main sequence depends on its mass (Table 11–3). Massive stars use fuel rapidly and live short lives, but low-mass stars conserve their fuel and shine for billions of years. For example, a 25 $M_\odot$ star will exhaust its hydrogen and die in only about 7 million years. This means that life is very unlikely to develop on planets orbiting massive stars. The stars do not live long enough for life to get started and evolve into complex creatures. We will discuss this problem in detail in Chapter 25.

Very low-mass stars, the red dwarfs, use their fuel so slowly they last for 200–300 billion years. Because the universe seems to be only 10–20 billion years old, red dwarfs must still be in their infancy. Box 11–2 explains how we can quickly estimate the life expectancies of stars from their masses.

Nature makes more low-mass stars than massive stars, but this fact is not sufficient to explain the vast numbers of low-mass stars that fill the sky. An additional factor is the stellar life-times. Because low-mass stars live long lives, there are more of them in the sky than massive stars. Look at Figure 9–16 and notice how much more common the lower main-sequence stars are than the massive O and B

BOX 11.2

Life Expectancies of the Stars

Because main-sequence stars consume their fuel at an approximately constant rate, we can estimate the amount of time a star spends on the main sequence—its life expectancy T—by estimating the amount of its fuel and its rate of fuel consumption:

$$T = \frac{\text{fuel}}{\text{rate of consumption}}$$

The amount of fuel a star has is proportional to its mass, and the rate at which it burns its fuel is proportional to its luminosity. Thus, its life expectancy must be proportional to M/L. But we can simplify this further because, as we saw earlier, (Box 11–1) the luminosity of a star depends on its mass raised to the 3.5 power ($L = M^{3.5}$). Thus the life expectancy is

$$T = \frac{M}{M^{3.5}}$$

or

$$T = \frac{1}{M^{2.5}}$$

If we express the mass in solar masses, the lifetime will be in solar lifetimes. For example, a 4 solar mass star will last for

$$T = \frac{1}{4^{2.5}} = \frac{1}{4 \cdot 4\sqrt{4}} = \frac{1}{32} \text{ solar lifetime}$$

Studies of solar models show that the sun, presently 5 billion years old, can last another 5 billion years. Thus, a solar lifetime is approximately 10 billion years, and a 4 solar mass star will last for about

$$T = \frac{1}{32} \times (10 \times 10^9 \text{y}) = 310 \times 10^6 \text{ years}$$

stars. The main-sequence K and M stars are so faint they are difficult to locate, but they are very common. The O and B stars are luminous and easy to locate, but, because of their fleeting lives, there are never more than a few on the main sequence at any one time.

When a star finally exhausts its hydrogen fuel, it can no longer resist the pull of its own gravity. Contraction resumes, and the star collapses. As the star dies, it can delay its end by burning other fuels, but as we will discover in Chapter 13, nothing can steal gravity's final victory.

Before we examine the deaths of stars, we should look once more at regions where stars are forming. Recent observations in the ultraviolet, infrared, and radio parts of the spectrum have revealed that a nebula familiar to generations of astronomers is actually a region of intense star formation.

PERSPECTIVE

THE ORION NEBULA

The fuzzy wisp of nebula visible to the naked eye in Orion's sword has attracted the attention of astronomers and casual stargazers throughout history. Commonly referred to as the Great Nebula in Orion, it is a striking sight through binoculars or a small telescope, and through a large telescope it is breathtaking. At the center lie four brilliant blue-white stars known as the Trapezium, and surrounding them are the glowing filaments of a nebula over 8 pc across (Color Plate 17 and Figure 11–21). Like a great thundercloud illuminated from within, the churning currents of gas and dust testify to the violence of the mechanisms that created them. However, a deeper significance lies hidden, figuratively and literally, behind the Great Nebula. In the last decade radio and infrared astronomers have discovered a vast dark cloud lying just beyond the visible nebula—a cloud in which stars are now being created.

FIGURE 11–21 The Great Nebula in Orion is a glowing cloud of hot gas over 8 pc in diameter. It is excited by a cluster of hot stars, the Trapezium, that lies at its center.(©1980 Royal Observatory Edinburgh.)

STAR FORMATION IN ORION

It should not surprise us that Orion is a site of star formation. The stars that make up the constellation are mostly hot, bright, main-sequence stars. These massive stars have short lifetimes and therefore must have formed recently. In addition, the region contains associations of T Tauri stars, which are believed to be stars in the later stages of contraction to the main sequence. These new stars most likely formed from the large gas and dust clouds that fill the Orion region.

The Great Nebula represents a late stage of star formation. The four stars we see as the Trapezium are only the brightest of a cluster of a few hundred stars that are no more than a million years old. The most massive, a star of about 40 solar masses, must burn its fuel at a tremendous rate to support its large mass. Thus, it is very hot and luminous. Its surface temperature is about 30,000 K, and it is about 300,000 times more luminous than the sun. The other stars are too cool to affect the surrounding gas very much, but the 40 solar mass star is so hot it radiates large amounts of ultraviolet radiation (Figure 11–22). These ultraviolet photons have enough energy to ionize the hydrogen gas in the region near the Trapezium, creating the glowing clouds we see as the Great Nebula.

Although the nebula looks impressive, it is nearly a vacuum, containing a mere 600 atoms/cm^3. For comparison, the interstellar medium has an average density of about 1 atom/cm^3, and the density of air at sea level is about 10^{19} atoms/cm^3. Infrared observations show that the thin, ionized gas is mixed with sparsely

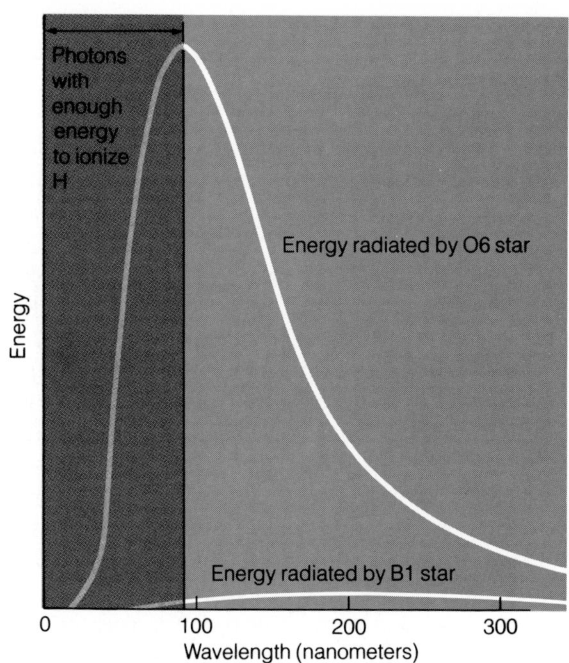

FIGURE 11–22 Photons with wavelengths shorter than 91.2 nm have enough energy to ionize hydrogen. The O6 star is the only star in the Trapezium hot enough to produce appreciable ionization.

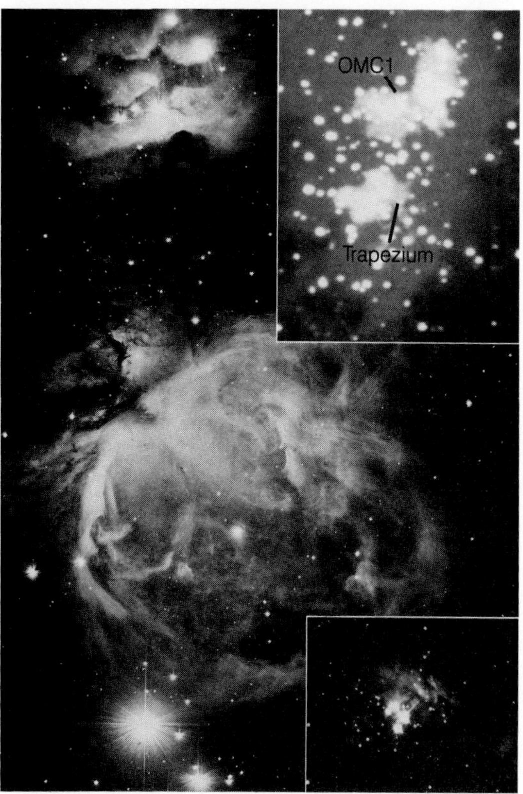

FIGURE 11–23 The Great Nebula in Orion and the young star cluster at its center called the Trapezium (inset lower right) are beautiful even through a small telescope. But they are only the visible portion of a much larger cloud of gas, dust, and forming stars. The Orion molecular clouds (OMC1 and OMC2) can be mapped at radio and infrared wavelengths. An infrared image of the region (upper right) shows the Trapezium cluster and OMC1 containing the BN object and KL nebula. (Lick Observatory photographs. Upper inset courtesy Ian McLean, Joint Astronomy Center, Hawaii.)

scattered dust, heated by the central stars to a temperature of about 70 K.

THE MOLECULAR CLOUD

The importance of the Orion region was established when observations at wavelengths longer than visible light revealed a dense cloud of gas lying just beyond the Great Nebula. These observations spanned the region of the spectrum that includes infrared and short-wavelength radio waves. At these wavelengths, hot stars and ionized gas are invisible, but cool, dense gas is detectable. The gas cloud beyond the Great Nebula is so dense that molecules such as carbon monoxide have formed where the dense dust protects them from ultraviolet radiation. These molecules emit the radio signals that make the cloud detectable. As shown in Figure 11–23, the cloud contains two dense regions, named the Orion Molecular Clouds 1 and 2 (OMC1 and OMC2).

These clouds are significantly different from the Great Nebula. The molecular clouds contain at least 10^6 atoms/cm^3 and large amounts of dust. In addition, observations at different wavelengths show that the

Perspective: The Orion Nebula
(continued)

clouds are warmest near their centers. Astronomers conclude that the clouds are growing hotter at their centers because they are contracting.

THE INFRARED CLUSTERS

Infrared observations reveal that clusters of warm objects lie at the centers of both OMC1 and OMC2. Invisible at optical wavelengths, these objects are evidently stars in pre-main-sequence stages, wrapped deep in dust cocoons. A number of interesting objects lie near the center of OMC1.

The Becklin–Neugebauer object (BN object), named after its discoverers, is probably a single star nearing the main sequence. Athough the star may have a surface temperature of about 30,000 K, it is hidden inside a cloud of cold dust roughly 300 AU in diameter. The dust, at a mere 600 K, is a strong source of infrared radiation. Although the BN object was originally thought to be a protostar, recent radio observations have detected ionized gas at the center. Thus, the BN object is probably a hot, young star that has formed within the last 100,000 years and has not yet cleared away its cocoon.

The second infrared object in OMC1 is the Kleinmann–Low (KL) nebula about 2000 AU in diameter and containing a number of compact sources. Some of these are dense knots of gas and dust, and some are very young stars still hidden from view (at visual wavelengths) within the dusty KL nebula. For example, one of the objects, Infrared Compact Source 2 (IRc2), is dimmed by 100 magnitudes at visual wavelengths; only one out of every 10^{40} photons of visible light it emits escapes from the dusty nebula. Yet IRc2 has been observed in detail at infrared and short radio wavelengths.

Infrared and radio maps reveal powerful bipolar flows centered on IRc2, and one model suggests that it is a massive protostar at the center of an expanding disk less than 0.1 pc across. Presumably the disk focuses the gas blasting away from the young star into two jets along the axis of the disk. Where the expanding disk pushes into the surrounding nebula, clumps of gas no bigger than earth's orbit are compressed and and molecules of water form. Radio emission from these water molecules has been detected at a wavelength of 13 mm.

The object at the center of IRc2 must be very young. The bipolar flow carries away mass at the tremendous rate of 0.01 to 0.001 $M_{\odot}$ per year. At this rate, even a massive star would waste away in 1000 to 10,000 years.

THE FUTURE OF THE ORION CLOUD

All evidence points to star formation within the molecular cloud just behind the Great Nebula in Orion. To understand the past and predict the future of this cloud,

SUMMARY

The interstellar medium consists mostly of hydrogen with traces of heavier gases and some dust. We can see the hydrogen where it is excited by hot young stars to form emission nebulae or H II regions. We can see the dust in reflection nebulae where it reflects starlight. We can also detect the interstellar gas by the narrow absorption lines it forms in the spectra of distant stars and by the 21-cm radiation emitted by cold hydrogen atoms. We can detect the dust by the extinction and reddening of starlight.

Radio observations show that the interstellar medium is very patchy with the largest and densest clouds called giant molecular clouds. When such a cloud is compressed by shock waves, parts of the cloud can collapse to form stars.

A contracting protostar is large and cool and would appear in the red giant region of the H–R diagram were it not surrounded by a cocoon of dust and gas. The dust in the cocoon absorbs the protostar's light and reradiates it as infrared radiation. Many infrared sources are probably protostars. T Tauri stars may be cocoon stars in later stages as the cocoon is absorbed by

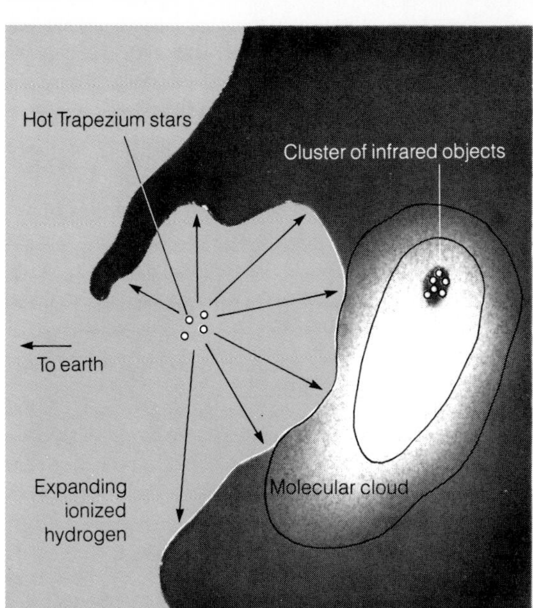

FIGURE 11–24 A side view of the Great Nebula would show the newly formed Trapezium stars heating and driving away the surrounding gas as new stars form in OMC1.

we have only to look at the ionized nebula around the Trapezium stars.

A million years ago the Trapezium stars were contracting protostars buried within a molecular cloud. As they approached the main sequence, their temperatures increased, and their radiation drove away their cocoons. But not until the 40 $M_{\odot}$ star turned on was there sufficient ultraviolet radiation to ionize the gas. The ionization transformed the cloud from an opaque shield into a transparent gas and forced it to expand, an expansion that continues to this day. We can see evidence of this expansion in the Doppler shifts of spectral lines and in the twisted filaments of gas within the nebula.

Once the hot star ionized the gas, star formation in the Trapezium region stopped. In more distant parts of the cloud, the contraction of the gas into protostars continues. Indeed, the ionized gases pushing against the remains of the cloud may have triggered the collapse of more protostars. We now see the front of the cloud torn by the expanding gases around the Trapezium cluster, while deeper within the cloud more protostars are forming (Figure 11–24).

In the next few thousand years, the familiar outline of the Great Nebula will change, and a new nebula may form as the protostars in the molecular clouds reach the main sequence and the most massive become hot enough to ionize the surrounding gas. Thus, the Great Nebula in Orion and its parent molecular cloud is an example of how the formation of stars continues even today.

the protostar or driven into space. Very young star clusters, like NGC 2264, contain large numbers of T Tauri stars.

As a protostar's center grows hot enough to fuse hydrogen, it settles onto the main sequence to begin its long hydrogen-fusion phase. If it is a low-mass star, it cannot become very hot and must generate energy by the proton–proton chain. But if it is a massive star, it will become hot enough to use the CNO cycle. In either case, the nuclear reactions are regulated by the pressure–temperature thermostat in the star's core.

Because of the temperature sensitivity of the CNO cycle, most of the energy generated by stars more massive than about 1.1 solar masses is produced right at the star's center. The resulting traffic jam as this energy flows outward causes the core to be convective, whereas the outer part of the star transports energy by radiation. In stars less massive than about 1.1 solar masses, the energy is generated on the proton–proton chain in a larger volume of the core. Thus, there is no traffic jam, and these stars have radiative cores and convective envelopes.

Almost everything we know about the internal structure of stars comes from mathematical stellar models. The models are based on four simple laws of stellar structure. Two laws say that mass and energy must be conserved and spread smoothly through the star. Another, the law of hydrostatic equilibrium, says the star must balance the weight of its layers by its internal pressure. Yet another says energy can only flow outward by conduction, convection, or radiation.

The mass–luminosity relation is explained by the requirement that a star support the weight of its layers by its internal pressure. The more massive a star is, the more weight it must support, and the higher its internal pressure must be. To keep its pressure high, it must be hot and generate large amounts of energy. Thus, the mass of a star determines its luminosity. The massive stars are very luminous and lie along the upper main sequence. The less massive stars are fainter and lie lower on the main sequence.

How long a star can stay on the main sequence depends on its mass. The more massive a star is, the faster it uses up its hydrogen fuel. A 25 $M_{\odot}$ star will exhaust its hydrogen and die in only about 7 million years, but the sun can last for 10 billion years.

The Orion nebula is an example of a star-forming region. The bright stars we now see are ionizing part of the gas and forcing it to expand, and this gas is pushing into cooler regions of the cloud. Radio and infrared observations show that the compressed regions of the gas cloud are giving birth to a new generation of stars.

NEW TERMS

interstellar medium	protostar
nebula	cocoon
emission nebula	Herbig–Haro objects
H II region	T Tauri star
reflection nebula	bipolar flows
dark nebula	CNO cycle
Bok globules	continuity of mass law
extinction	continuity of energy law
interstellar reddening	stellar model
interstellar absorption lines	brown dwarf
21-cm radiation	zero-age main sequence (ZAMS)
infrared cirrus	
giant molecular cloud	

QUESTIONS

1. What evidence do we have that the spaces between the stars are not totally empty?
2. What evidence do we have that the interstellar medium contains both gas and dust?
3. If starlight on its way to the earth passed through a cloud of interstellar gas that was very hot instead of very cold, would you expect the resulting interstellar absorption lines to be broader or narrower than usual? Why?
4. When we see distant streetlights through smog, they look dimmer and redder than they do normally. But when we see the same streetlights through fog or falling snow, they look dimmer but not redder. Use your knowledge of the interstellar medium to discuss the relative sizes of the particles in smog, fog, and snowstorms compared to the wavelength of light.
5. How does a protostar convert gravitional energy into thermal energy?
6. What observational evidence do we have that (a) star formation is a continuous process? (b) protostars really exist? (c) the Orion region is actively forming stars?
7. How do the proton–proton chain and the CNO cycle resemble each other? How do they differ?
8. How does the pressure–temperature thermostat control the nuclear reactions inside stars?
9. How do more massive main-sequence stars differ from lower-mass main-sequence stars?
10. Why is there a lower limit to the mass a star can have?
11. Why is there a mass–luminosity relation?
12. Why does a star's life expectancy depend on its mass?

PROBLEMS

1. A small Bok globule in the Rosette nebula shown in Figure 11–4 has a diameter of 20 seconds of arc. If the nebula is 1000 pc from the earth, what is the diameter of the globule?
2. Extinction dims starlight by about 1.9 magnitudes/1000 pc. What fraction of photons survive a trip of 1000 pc? (HINT: See Box 2–1.)
3. The density of air in a child's balloon 20 cm in diameter is roughly the same as the density of air at sea level, 10^{19} particles/cm^3. To how large a diameter would you have to expand the balloon to make the gas inside the same density as the interstellar medium, about 1 particle/cm^3? (HINT: The volume of a sphere is $\frac{4}{3}\pi R^3$.)
4. If a giant molecular cloud has a diameter of 30 pc and drifts relative to neighboring clouds at 20 km/sec, how long will it take to travel its own diameter?

5. A typical giant molecular cloud has a diameter of 30 pc and a density of 300 hydrogen molecules/cm^3. What is the mass of such a cloud in solar masses? (HINT: The mass of a hydrogen atom and the mass of the sun are given in Appendix C.)
6. If a typical giant molecular cloud could radiate only from its surface (100 K), what would its total luminosity be in terms of the sun? (HINT: See Box 9–3.) Actual clouds are more luminous than this because they do not radiate only from their surfaces.
7. In the model shown in Figure 11–19, how much of the sun's mass is hotter than 12,000,000 K?
8. The gas in a bipolar flow can travel as fast as 300 km/sec. How long would it take to travel 1 light-year?
9. If brown dwarfs exist they are about the size of Jupiter and have surface temperatures of about 1400 K. What is the total luminosity of such an object? What is its absolute bolometric magnitude? (HINT: See Box 9–3.)
10. What is the luminosity of a 4 $M_\odot$ star? Of a 9 $M_\odot$ star?
11. What is the life expectancy of a 16 $M_\odot$ star? Of a 50 $M_\odot$ star?
12. The hottest star in the Orion nebula has a surface temperature of 30,000 K. At what wavelength does it radiate the most energy? (HINT: See Box 7–1.)

RECOMMENDED READING

ALLEN, DAVID, JEREMY GAILEY, and A. R. HYLAND "Infrared Images of the Orion Nebula." *Sky and Telescope 67* (March 1984), p. 222.

BALLEY, JOHN "Bipolar Gas Jets in Star-Forming Regions." *Sky and Telescope 66* (Aug. 1983), p. 94.

———. "Interstellar Molecular Clouds." *Science 232* (11 April 1986), p. 185.

BOK, B. J. "Early Phases of Star Formation." *Sky and Telescope 61* (April 1981), p. 284.

BOSS, ALAN P. "Collapse and Formation of Stars." *Scientific American 251* (Jan. 1985), p. 40.

CASH, WEBSTER, and PHILIP CHARLES "Stalking the Cygnus Superbubble." *Sky and Telescope 59* (June 1980), p. 455.

CHURCHWELL, ED, and KEVIN J. ANDERSON "The Anatomy of a Nebula." *Astronomy 13* (June 1985), p. 66.

COHEN, M. "The Shocking Truth About Herbig–Haro Objects." *Griffith Observer 45* (Jan. 1981), p. 9.

CROSWELL, KEN "Stars Too Small to Burn." *Astronomy 12* (April 1984), p. 16.

DARLING, DAVID "The Molecules Between the Stars." *Astronomy 10* (March 1982), p. 82.

DAVIES, JOHN K. "The Herbig Stars." *Astronomy 13* (Dec. 1985), p. 90.

GEHRZ, ROBERT D., DAVID C. BLACK, and PHILIP M. SOLOMON "The Formation of Stellar Systems from Interstellar Molecular Clouds." *Science 224* (25 May 1984), p. 823.

HERBST, WILLIAM, and GEORGE E. ASSOUSA "Supernova and Star Formation." *Scientific American 241* (Aug. 1979), p. 138.

LADA, CHARLES J. "Energetic Outflows from Young Stars." *Scientific American 247* (July 1982), p. 82.

———. "Star in the Making." *Sky and Telescope 72* (Oct. 1986), p. 334.

LITTLE, LESLIE "How Astronomers Watch the Birth of Stars." *New Scientist 101* (March 1984), p. 12.

MARSCHALL, LAURENCE "Secrets of Interstellar Clouds." *Astronomy 10* (March 1982), p. 6.

MURRAY, MARY "The Search for Unlit Stars." *Science News 130* (1 Nov. 1986), p. 282.

PARESCE, FRANCESCO, and STUART BOWYER "The Sun and the Interstellar Medium." *Scientific American 255* (Sept. 1986), p. 93.

REIPURTH, B. "Bok Globules." *Mercury 13* (March/April 1984), p. 50.

RODRIGUEZ, LUIS "Cosmic Jets: Bipolar Outflows in the Universe." *Astronomy 12* (June 1984), p. 66.

ROBINSON, LEIF J. "Orion's Stellar Nursery." *Sky and Telescope 64* (Nov. 1982), p. 430.

SCHORN, R. A. "Good-bye to Supermassive Stars." *Sky and Telescope 71* (Jan. 1986), p. 13.

SEEDS, M. "Stellar Evolution." *Astronomy 7* (Feb. 1979), p. 6. Reprinted in *Astronomy: Selected Readings*, ed. M. A. Seeds. Menlo Park, Calif.: Benjamin/Cummings, 1980.

SMITH, DAVID H. "Reflection Nebulae: Celestial Veils." *Sky and Telescope 70* (Sept. 1985), p. 207.

SPITZER, L. "Interstellar Matter and the Birth and Death of Stars." *Mercury 12* (Sept./Oct. 1983), p. 142.

WALDROP, M. MITCHELL "Stellar Nurseries." *Science 83 4* (May 1983), p. 40.

WELCH, W. J., S. N. VOGEL, R. L. PLAMBECK, M. C. H. WRIGHT, and J. H. BIEGING "Gas Jets Associated with Star Formation." *Science 228* (21 June 1985), p. 1389.

CHAPTER 12

STELLAR EVOLUTION

We should be unwise to trust scientific inference very far when it becomes divorced from opportunity for observational test.

Sir Arthur Eddington
THE INTERNAL CONSTITUTION OF THE STARS

The stars are going out. Slowly but surely they are using up their fuel, and, when they can no longer fuse hydrogen at their centers, they will have to leave the main sequence. But they don't wink out like embers falling on the hearth. They swell rapidly into giant stars 10 to 1000 times the size of the sun. For a short time, a few million to a billion years, they are the majestic beacons that are visible across thousands of parsecs.

Although we say a star swells rapidly and spends a short time as a giant, these changes happen slowly compared to a human life. How can we be sure stars really do evolve? Stellar evolution and the formation of giant stars might be an astronomer's daydream. What evidence do we have that stars evolve?

We will look for evidence of the evolution of stars in two places. We will discover that certain kinds of stars fluctuate in brightness like great beating hearts. These **variable stars** will not only give us insight into how giant stars are constructed but also will provide us with clear evidence that the stars are evolving.

We also will look at clusters of stars. Because all the stars in a given cluster are formed at about the same time, the present state of the cluster can reveal how the stars have evolved since the cluster formed. In fact, we will even be able to estimate the age of a cluster by combining our theories of stellar evolution with observations of the cluster's H–R diagram. Clusters of stars may be the single, most important kind of observational evidence we have against which to test our theories of stellar evolution.

From beginning to end, this chapter is the story of the evolution of stars from main sequence to giant. We will trace the passage of stars through this phase of existence and see how they generate their energy and why they swell so large.

12.1 POST MAIN-SEQUENCE EVOLUTION

In the previous chapter we learned that main-sequence stars generate their energy by fusing hydrogen at their centers and that they have a limited amount of hydrogen available. Thus, they can remain stable for a limited time. This leads us to an important question: What happens to a star after it exhausts the hydrogen fuel at its center?

The answer to that question will help answer three other questions that we asked in earlier chapters. Why are giant stars so large? Why are they so uncommon? And, why do they have such low densities? Now we are ready to answer those questions by discussing the evolution of stars after they leave the main sequence.

EXPANSION INTO A GIANT

To understand how stars evolve, we must recognize that they are not well-mixed. That is, their interiors are not stirred. The centers of stars like the sun are radiative and thus are not mixed at all. More massive stars have convective cores that mix the central regions (see Figure 11–18), but these regions are not very large, and thus, for the most part, these stars too are not mixed.

In this respect stars are like a campfire that is not stirred; the ashes accumulate at the center, and the fuel in the outer parts never gets used. Nuclear fusion consumes hydrogen nuclei and produces helium nuclei, the "ashes," at the star's center. Nothing mixes the interior of the star, so the helium nuclei remain where they are in the center of the star, and the hydrogen in the outer parts of the star is not mixed down to the center where it can be fused.

The helium ashes that accumulate in the star's core cannot fuse into heavier elements because the temperature is too low. As a result, the helium accumulates, the hydrogen is used up, and the core becomes an inert ball of helium. As this happens, the energy production in the core falls, and the weight of the outer layers forces the core to contract.

Although the contracting helium core cannot generate nuclear energy, it does grow hotter because it converts gravitational energy into thermal energy (see Chapter 11). The rising temperature heats the unprocessed hydrogen just outside the core, hydrogen that was never before hot enough to fuse. When the temperature of the surrounding hydrogen becomes high enough, it ignites in a shell of fusing hydrogen. Like a grass fire burning outward from an exhausted campfire, the hydrogen-fusing shell burns outward, leaving helium ash behind and increasing the mass of the helium core.

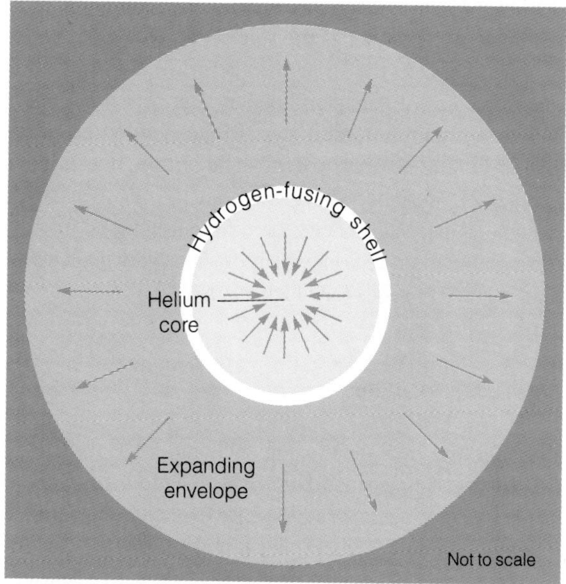

FIGURE 12–1 When a star runs out of hydrogen at its center, it ignites a hydrogen-fusing shell. The helium core contracts and heats, while the envelope expands and cools. (For a scale drawing see Figure 12–3.)

At this stage in its evolution, the star overproduces energy—that is, it produces more than it needs to balance its own gravity. The helium core, having no nuclear energy sources, must contract, and that contraction converts gravitational energy into thermal energy—the contraction heats the helium core. Some of that heat leaks outward through the star. At the same time, the hydrogen-fusing shell produces energy as the contracting core brings fresh hydrogen closer to the center of the star and heats it to high temperature. The result is a flood of energy flowing outward through the outer layers of the star, forcing them to expand and swelling the star into a giant (Figure 12–1).

The expansion of the envelope dramatically changes the star's location in the H–R diagram. As the outer layers expand outward, the surface cools, and the star moves quickly to the right into the red giant region. As the radius continues to increase, the enlarging surface area makes the star more luminous, moving it upward in the H–R diagram (Figure 12–2). Aldebaran (α Tauri), the glowing red eye of Taurus the bull, is such a red giant, having a diameter 25 times that of the sun, but with a surface temperature only half that of the sun.

This explains the large diameters, low densities, and mixed masses of giants and supergiants. They are mid- and upper-main-sequence stars that expanded to large size and low density as hydrogen-shell fusion began (Figure 12–2). Most become giants, but the most massive become supergiants. Stars of different masses funnel through the giant region, which explains why giant stars have assorted masses (see Figure 10–22).

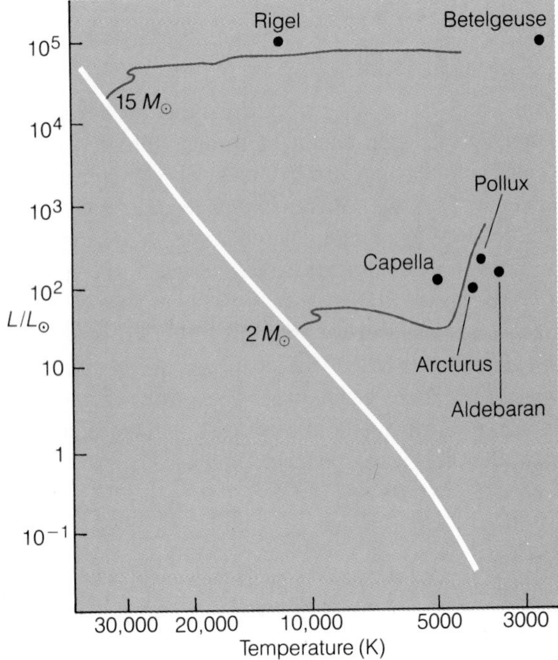

FIGURE 12–2 Many of the bright, red stars in the night sky are giants. Not long ago they were main-sequence stars, but they expanded when their cores ran out of hydrogen. Supergiants like Rigel and Betelgeuse are more massive stars from the upper main sequence.

HELIUM FUSION

Although the hydrogen-fusing shell can force the envelope of the star to expand, it cannot stop the contraction of the helium core. Because the core has no energy source, gravity squeezes it tighter, and it becomes very small. If we represent the helium core of a 5 $M_\odot$ star with a quarter, the outer envelope of the star would be about the size of a baseball diamond. Yet the core would contain about 12 percent of the star's mass.

The extreme contraction of the helium core causes the temperature to rise very high. When the helium reaches 100,000,000 K, it begins to fuse in nuclear reactions that convert it to carbon.

We can summarize the helium-fusing process in two steps:

$$^4He + {}^4He \rightarrow {}^8Be + \gamma$$
$$^8Be + {}^4He \rightarrow {}^{12}C + \gamma$$

This process is complicated by the fact that beryllium-8 is very unstable and may break up into two helium nuclei before it can absorb another helium nucleus. Three helium nuclei can also form carbon directly, but such a triple collision is unlikely. Because a helium nucleus is called an alpha particle, astronomers often refer to helium fusion as the **triple alpha process**.

Some stars begin helium fusion gradually, but stars in a certain mass range begin with an explosion called the **helium flash**. This explosion is caused by the density of the helium, which can reach 15,000 g/cm^3. On Earth a teaspoon of this material would weigh as much as a small automobile. At these densities, the gas is **degenerate** (Box 12–1), and its pressure no longer depends on temperature. Thus the pressure–temperature thermostat that controls the nuclear fusion reactions no longer works.

When the helium ignites, it generates energy, which raises the temperature. Because the pressure–temperature thermostat is not working in the degenerate gas, the core does not respond to the higher temperature by expanding. Rather, the higher temperature forces the reactions to go faster, which makes more energy, which raises the temperature, which makes the reactions go faster, which. . . . Thus the ignition of helium fusion in a degenerate gas results in a runaway explosion so violent that for a few minutes the helium core generates more energy than an entire galaxy. At its peak, the core makes 10^{14} times more energy per second than the sun.

Although the helium flash is sudden and powerful, it does not destroy the star. In fact, if you were observing a giant star as it experienced a helium flash, you would see no outward evidence of an eruption. The helium core

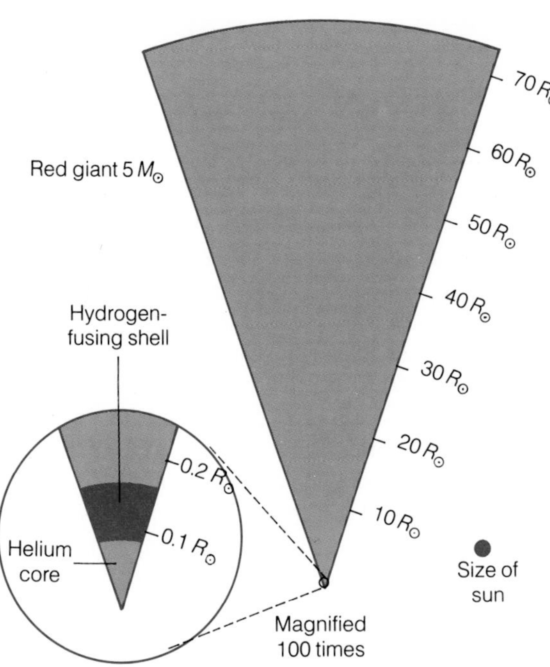

FIGURE 12–3 A plug cut from a 5 $M_\odot$ red giant would be almost all low-density envelope. The magnified image shows the hydrogen-fusing shell and the helium core. The core contains roughly 12 percent of the star's mass. (Adapted from Michael A. Seeds, "Stellar Evolution," *Astronomy*, Feb. 1979.)

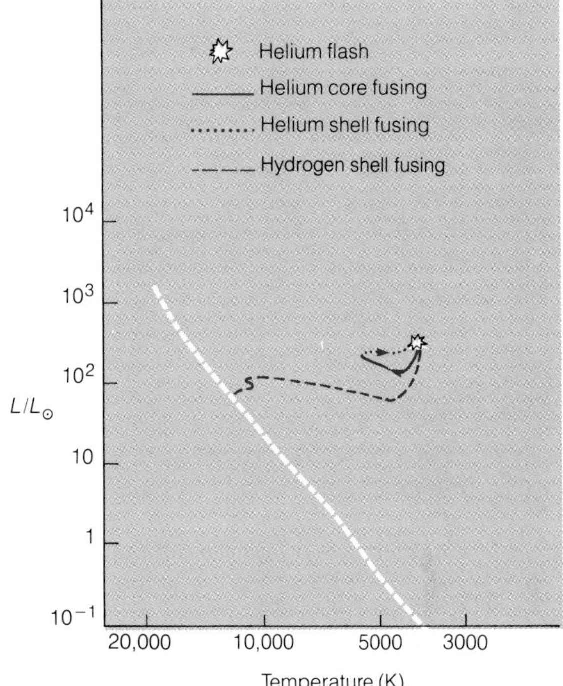

FIGURE 12–4 After the helium flash at the tip of the red giant track, the star moves through a loop in the diagram as it fuses helium in its core and then in a shell.

is quite small (Figure 12–3), and all of the energy of the explosion is absorbed by the distended envelope. Also, the helium flash is a very short-lived event. In a matter of minutes, the core of the star becomes so hot it is no longer degenerate, the pressure–temperature thermostat brings the helium fusion under control, and the star proceeds to fuse helium steadily in its core (Figure 12–4). We will discuss this post-helium-flash evolution later.

Not all stars experience a helium flash. Stars less massive than about 0.4 $M_\odot$ can never get hot enough to ignite helium, and stars more massive than about 3 $M_\odot$ ignite helium before their cores become degenerate. In such stars, pressure depends on temperature, so the pressure–temperature thermostat keeps the helium fusion under control.

If the helium flash occurs only in some stars and is a very short-lived event, which is not visible from outside the star, why should we worry about it? The answer is that it prevents us from understanding stellar evolution with confidence. The medium-mass stars, which experience the helium flash, are those that we must consider most carefully in studying stellar evolution. Massive stars are not very common; low-mass stars evolve so slowly that we cannot see much evidence of low-mass stellar evolution. Thus, studies of stellar evolution must concentrate on moderate-mass stars, which do experience the helium flash. But the helium flash occurs so rapidly and so violently that computer programs cannot follow the changes in the star's internal structure in detail. To follow the evolution of moderate-mass stars like the sun past the helium flash, astronomers must make assumptions about the way the helium flash affects stars' internal structure.

Post-helium-flash evolution is generally understood. As the core temperature rises, the gas ceases to be degenerate, and helium fusion proceeds under the control of the pressure–temperature thermostat. Throughout these events the hydrogen-fusion shell continues to produce energy, but the new helium-fusion energy produced in the core makes the core expand. That expansion absorbs energy previously used to support the outer layers of the

BOX 12–1
Degenerate Matter

Normally the pressure in a gas depends on its temperature, but under certain circumstances nature may break that rule, often with catastrophic consequences for a star. To see how this works, we must consider the energy of the free electrons in an ionized gas.

The gas inside a star is completely ionized. That is, it consists of the bare nuclei of the atoms mixed with free electrons. At high densities and temperatures the pressure of the gas is due almost entirely to these electrons.

If the density is very high, the particles of the gas are forced close together, and two laws of quantum mechanics become important. First, quantum mechanics says that the moving electrons confined in the star's core can have only certain amounts of energy just as the electron in an atom can occupy only certain energy levels (see Chapter 7). We can think of these permitted energies as the rungs in a ladder. An electron can occupy any rung, but not the spaces between.

The second quantum mechanical law (called the Pauli Exclusion Principle) says that two identical electrons may not occupy the same energy level. Because electrons spin in one direction or the other, two electrons can occupy an energy level if they spin in opposite directions. Such a level is completely filled and a third electron cannot enter because, whichever way it spun, it would be identical to one or the other of the two electrons already in the level. Thus, no more than two electrons can occupy the same energy level.

In a low-density gas, there are few electrons per cubic centimeter, so there are plenty of empty energy levels (Figure 12–5). However, if a gas becomes very dense, nearly all of the lower energy levels may be occupied, and the gas is termed **degenerate**. In such matter, an electron cannot slow down because there are no open energy levels for it to drop down to. It can speed up only if it can absorb enough energy to leap to the top of the energy ladder, where there are empty energy levels.

This has two important effects in stars. First, the degenerate gas resists compression. To compress the gas, we must push against the moving electrons, and

star. As a result the outer layers contract, and the surface of the star grows slightly hotter. In the H–R diagram the star moves down toward lower luminosity and to the left toward higher temperature (Figure 12–4).

Helium fusion produces carbon and oxygen as "ash," and these accumulate at the center of the star. They do not fuse because the core is not hot enough. Thus, the helium in the core is gradually converted into carbon–oxygen ashes. As this happens the core contracts, grows hotter, and ignites a helium-fusion shell. At this stage the star has two shells producing energy. A hydrogen-fusion shell eats it way outward leaving behind helium ash; a helium-fusion shell eats its way outward into the helium leaving behind carbon–oxygen ash. Unable to generate energy in its carbon–oxygen core, the star expands its outer layers and moves back toward the right in the H–R diagram, completing a loop (Figure 12–4).

THE FUSION OF ELEMENTS HEAVIER THAN HELIUM

The evolution of a star after helium exhaustion is uncertain, but the general plan is clear. The inert carbon–oxygen core contracts and becomes hotter. Stars more massive than about 3 $M_{\odot}$ can reach temperatures of 600,000,000 K and ignite carbon. Subsequent contraction may be able to fuse oxygen, silicon, and other heavy elements (Table 12–1).

At temperatures above 600,000,000 K, carbon fuses rapidly in a complex network of reactions illustrated in Figure 12–6, where each arrow represents a different nuclear reaction. The process is complicated because nuclei can react by adding a proton, a neutron, or a helium nucleus or by combining directly with other nuclei.

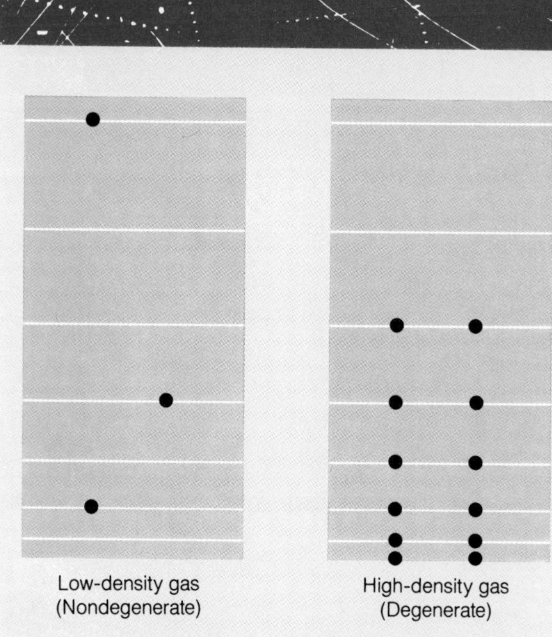

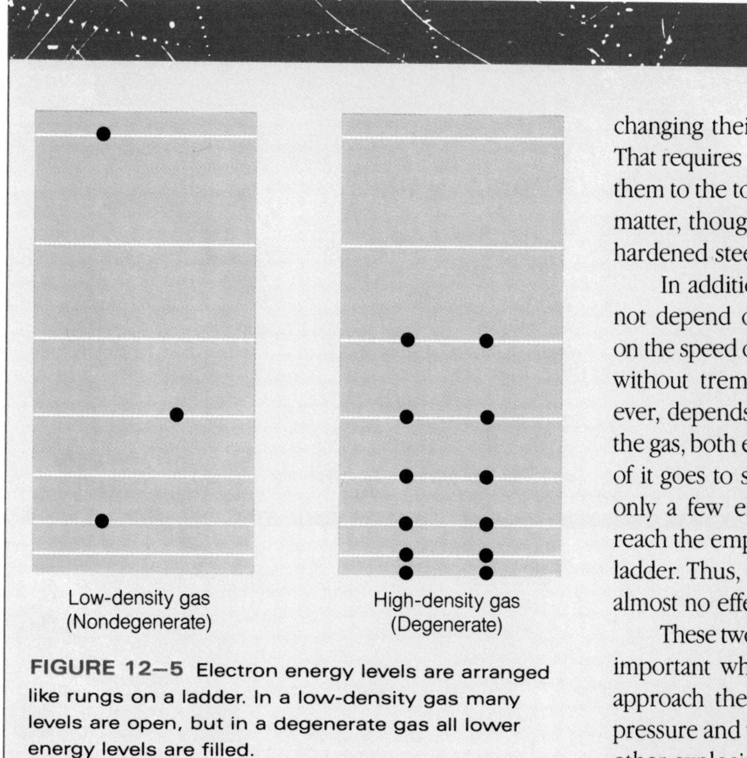

FIGURE 12–5 Electron energy levels are arranged like rungs on a ladder. In a low-density gas many levels are open, but in a degenerate gas all lower energy levels are filled.

changing their motion means changing their energy. That requires tremendous effort because we must boost them to the top of the energy ladder. Thus degenerate matter, though still a gas, takes on the consistency of hardened steel.

In addition, the pressure of degenerate gas does not depend on temperature. The pressure depends on the speed of the electrons, which cannot be changed without tremendous effort. The temperature, however, depends on the motion of all of the particles in the gas, both electrons and nuclei. If we add heat, most of it goes to speed up the motions of the nuclei, and only a few electrons can absorb enough energy to reach the empty energy levels at the top of the energy ladder. Thus, changing the temperature of the gas has almost no effect on the pressure.

These two properties of degenerate matter become important when stars leave the main sequence and approach their final collapse. The independence of pressure and temperature causes the helium flash and other explosive effects such as supernova explosions, and the hardened steel effect supports white dwarfs and neutron stars.

Unstable nuclei can decay by ejecting an electron, a positron, or a helium nucleus or by splitting into fragments. The complexity of this process makes it difficult to determine exactly how much energy will be generated and how many heavy nuclei will be produced.

Reactions at still higher temperatures can convert magnesium and silicon into yet heavier nuclei, and the star can continue to balance its high internal pressure against its own gravity. These stars become complexly structured, as shells fusing different elements lie one on top of the other like the layers in a spherical cake.

Eventually, all stars face collapse. Less massive stars cannot get hot enough to ignite the heavier nuclei. The more massive stars can ignite heavy nuclei, but these stars consume their fuels at a tremendous rate, so they run through their available fuel quickly. No matter what the star's mass, it will eventually run out of usable fuels, and gravity will win the struggle with pressure. We will discuss the ultimate deaths of stars in the next chapter.

The time a star spends as a giant or supergiant is small compared with its life on the main sequence. The sun, for example, will spend a total of about 10 billion years as a main-sequence star, but only about a billion years as a giant. The more massive stars pass through the giant stages even more rapidly. Because of the short time a star spends as a giant, we are unlikely to see many such stars. This illustrates an important principle in astronomy: The shorter the time a given evolutionary stage takes, the less likely we are to see stars in that particular stage. Thus, we see a great many main-sequence stars near the sun, but few giants (see Figure 9–16).

Although we say that some stages of a star's life are short, they are long compared to our human life spans. Thus, we do not see stars evolve; no one has ever seen a

Table 12–1 Nuclear reactions in massive stars.

Nuclear Fuel	Nuclear Products	Minimum Ignition Temperature	Main-Sequence Mass Needed to Ignite Fusion	Duration of Fusion in a $25M_\odot$ Star
H	He	4×10^6 K	$0.1\,M_\odot$	7×10^6 yr
He	C, O	120×10^6 K	$0.4\,M_\odot$	0.5×10^6 yr
C	Ne, Na, Mg, O	600×10^6 K	$4\,M_\odot$	600 yr
Ne	O, Mg	1.2×10^9 K	$\sim 8\,M_\odot$	1 yr
O	Si, S, P	1.5×10^9 K	$\sim 8\,M_\odot$	~0.5 yr
Si	Ni to Fe	2.7×10^9 K	$\sim 8\,M_\odot$	~1 day

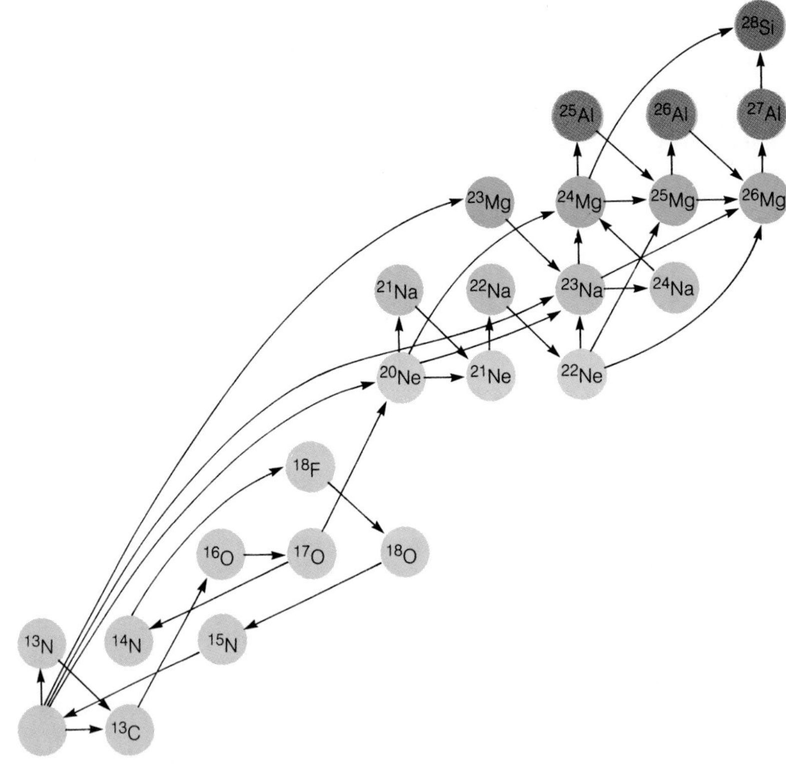

FIGURE 12–6 Carbon fusion involves many possible reactions that build numerous heavy nuclei.

main-sequence star swell into a giant. In the rest of this chapter, we will search for observational confirmation of the theory of stellar evolution. To paraphrase the quotation at the beginning of this chapter, we shouldn't believe any theory until we compare it with reality, or seeing is believing.

12.2 VARIABLE STARS

For millennia poets and scientists believed that the stars are constant and unchanging. Shakespeare refers to "the fixed stars." But we know now that the stars are changing, slowly evolving as they consume their fuels. Here we

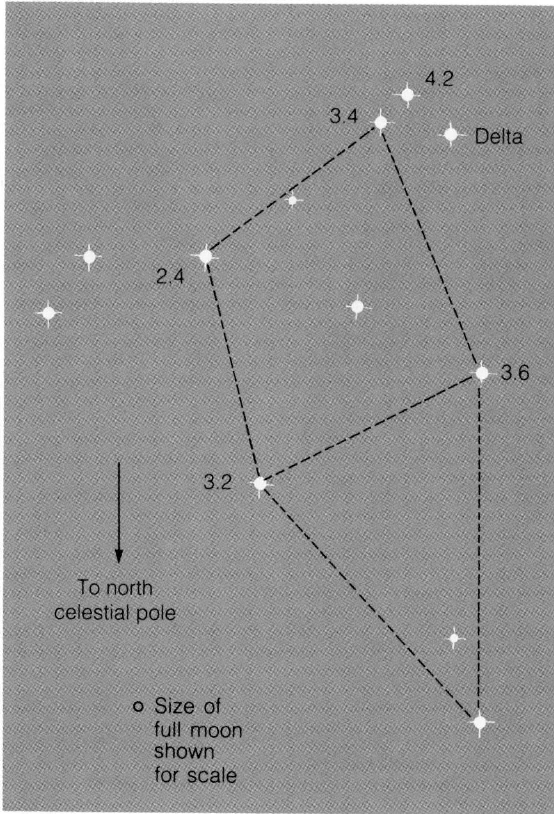

FIGURE 12–7 Cepheid variable stars are named after star delta in Cepheus. It varies from 3.6 magnitude to 4.3. The magnitudes of nearby stars are given here. The constellation is shown as it appears on October evenings.

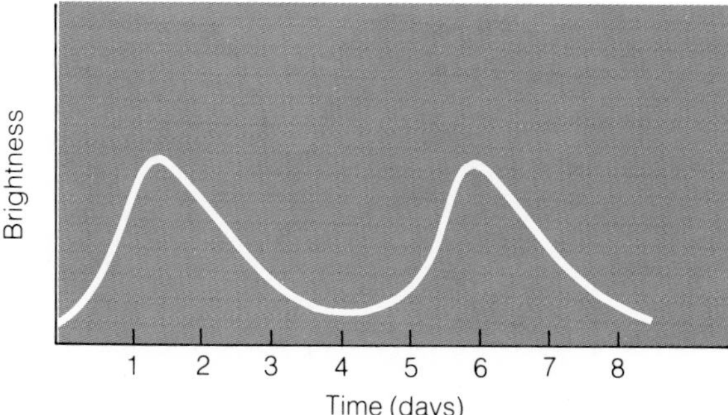

FIGURE 12–8 A typical light curve for a Cepheid variable.

will consider still more active stars—variable stars that change their brightness rapidly and periodically.

While Shakespeare was busy writing plays, the Dutch astronomer David Fabricius was studying the sky. On August 13, 1596, he discovered that the Star o Ceti is variable in brightness, sometimes fading from third magnitude to invisibility. Not until almost 1640 did astronomers realize that its variations are periodic, repeating with a period of 332 days. That star, now known as Mira, meaning "wonderful," was the first-known variable star. (See "Observational Activity: Naked-Eye Variable Stars" at the end of this chapter.)

CEPHEID VARIABLES

In 1784 the deaf and dumb English astronomer John Goodricke, then 19 years old, discovered that the star δ Cephei is variable, but unlike Mira, it has a short period, only 5.4 days (Figure 12–7). Goodricke died at the age of 21, but his discovery ensures his place in the history of astronomy. Since 1784 almost 1000 stars like δ Cephei have been discovered. Known as **Cepheids**, they are the most important kind of variable star.

Cepheid variables are giant stars with spectral types between F and K. The most rapidly varying complete a cycle—bright, faint, bright—in about 2 days, whereas the slowest take as long as 60 days. A plot of a variable's magnitude versus time, called a light curve, illustrates this periodic variation (Figure 12–8). Some Cepheids change their brightness by only 0.1 magnitude, whereas others change by as much as 2 magnitudes.

Giant stars are rare, and giants that are Cepheids are even more rare. Nevertheless, some familiar stars are Cepheids. Polaris, for instance, is a Cepheid with a period of 3.9696 days and an amplitude (difference between brightest and faintest) of only 0.1 magnitude.

Cepheids became important stars in 1912 when Henrietta S. Leavitt (1868–1921), an astronomer at the Harvard College Observatory, published the results of her study of variable stars in the Small Magellanic Cloud. The Large and Small Magellanic clouds are star systems now known to be companions to our Milky Way galaxy, and Leavitt searched plates to locate variable stars and determine their periods. She discovered that the longer period Cepheids always have brighter average magnitudes. Because all of the stars in the Small Magellanic Cloud are at about the same distance from earth, this means the long period Cepheids are more luminous.

Unfortunately, Leavitt had no way of knowing the distance to any of these stars, so she could not find their absolute magnitudes. Her diagram was a plot of apparent magnitude versus period (Figure 12–9). Nor could she find the distance to any of the Cepheids near the sun. Cepheids are rare, and it happens that none are near

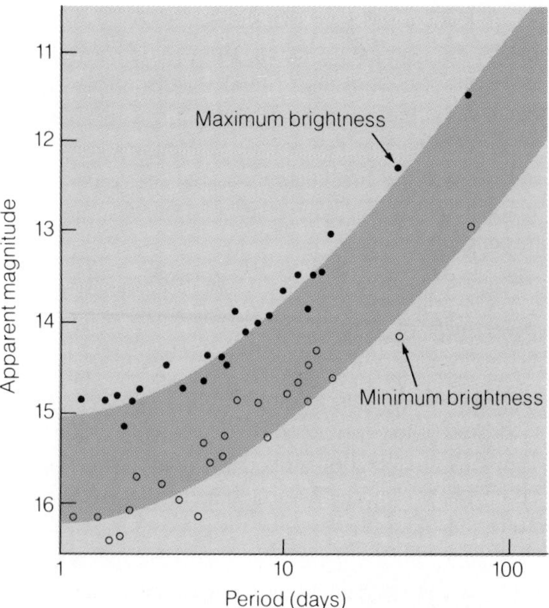

FIGURE 12–9 The first period–luminosity diagram showed the maximum brightness (filled circles) and minimum brightness (open circles) of Cepheids in the Small Magellanic Cloud. This suggested a relation between period and intrinsic brightness (shaded band).

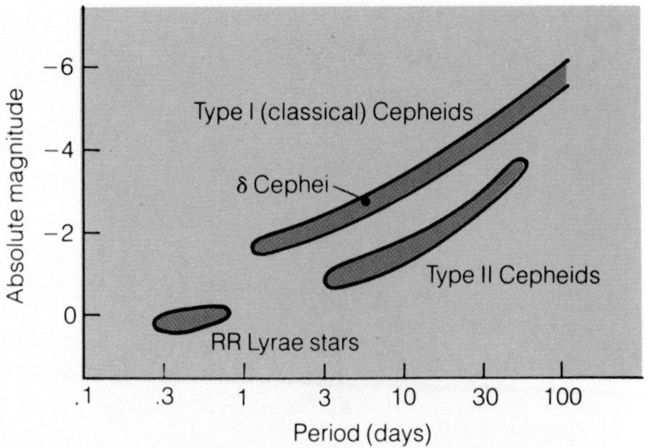

FIGURE 12–10 The modern period–luminosity diagram includes two types of Cepheids and the RR Lyrae stars.

enough to have measurable parallaxes. Soon, however, the Mount Wilson astronomer Harlow Shapley used statistical methods to find the average absolute magnitudes of a few Cepheids near the sun. Thus, in one of the most seminal papers in twentieth-century astronomy, he was able to replace the apparent magnitudes on Leavitt's graph with absolute magnitudes.

This diagram, shown in modern form in Figure 12–10, is now known as the **period–luminosity diagram**, although it is usually drawn as period and absolute magnitude. Modern astronomers know of two types of Cepheids that differ slightly in luminosity. Type I Cepheids are richer in heavy nuclei than Type II. A related type of star is the **RR Lyrae variables**, named after variable star RR in the constellation Lyra.

The stars in the period–luminosity diagram are important because we can use them to find distances even beyond the limit for accurate parallaxes. Suppose we discovered a Type I Cepheid in a distant galaxy. By measuring its brightness on a sequence of photographic plates, we could find its period and apparent magnitude. From the period–luminosity diagram we could find its absolute magnitude. The difference between its apparent magnitude and its absolute magnitude is $(m - M_v)$, its distance modulus. Then we could find the distance to the galaxy by referring to a table such as Table 9–1. (See Box 12–2.)

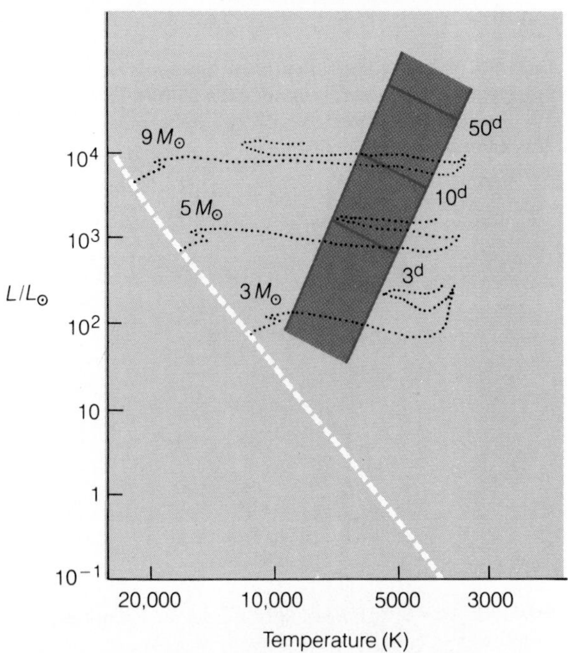

FIGURE 12–11 The instability strip (shaded) is a region of the H–R diagram in which stars are unstable and pulsate as variable stars. Massive stars cross the strip at higher luminosities than lower-mass stars, and, because of their higher mass and larger radius, they pulsate with longer periods.

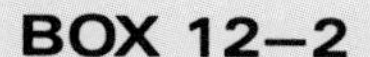

BOX 12–2
Cepheid Distance Indicators

Suppose we discovered a Type I Cepheid in the Small Magellanic Cloud with a period of 10 days and an average apparent magnitude of 16.5. How could we find the distance?

If we look at the period–luminosity diagram, we find that a Type I Cepheid with a period of 10 days has an absolute magnitude of about −3. The distance modulus is

$$m - M_V = 16.5 - (-3) = 19.5$$

We could estimate the distance from a table such as Table 9–1, or we could solve the magnitude–distance formula

$$d = 10^{\frac{m - M_V + 5}{5}}$$

From this we would find that the distance is 80,000 pc.

This is too large. The accepted distance is about 60,000 pc. We might be more accurate if we used an average distance from a number of stars. But we should also consider the effects of interstellar dust that makes the stars look too faint and thus too far away. With proper care for these details, we could find accurate distances using Cepheid variables.

The Cepheids are important because they can tell us the distances to galaxies, but they are also important for what they tell us about the evolution of stars.

PULSATING STARS

Why do Cepheids pulsate? Why is there a period–luminosity relation? The answers to these questions start with the evolution of stars in the H–R diagram. After a star leaves the main sequence, it moves back and forth in the giant region of the H–R diagram. If it passes through a region called the **instability strip** (Figure 12–11), it becomes unstable and pulsates as a variable star.

Variable stars in the instability strip pulsate like beating hearts because of energy-absorbing layers in their outer envelopes. These layers are the regions where hydrogen and helium are partially ionized. Like springs, these ionization zones can absorb energy when they are compressed and release it when they expand. If these layers are near the surface of the star, it will oscillate like a weight bouncing on a spring.

Imagine the surface layers of such a pulsating star as the surface expands. The ionization zones also expand, and as they expand they release stored energy, which makes the surface layers expand even faster. The surface of the star overshoots its equilibrium position—it expands too far—and eventually falls back. As the surface layers contract, the ionization zones are compressed and absorb energy that would otherwise help slow the contraction. This allows the contraction to go even faster, and again the layers overshoot until the pressure in the star slows them to a stop and makes them expand again. Thus, the surface of these stars is not in hydrostatic equilibrium, but rather pulsates in and out.

Cepheids change their radius by 5–10 percent as they pulsate, and other kinds of pulsating stars can change by 20–30 percent. Although this seems like a large change, it affects only the surface layers. The center of the star is too dense to be affected.

Only stars in the instability strip have the ionization zones in the right place to cause pulsation. If the star is too hot, the zones are too near the surface where the gas is thin. If the star is too cool, the zones are buried too deep to affect the envelope.

The instability strip also explains the period–luminosity relation. The more massive a star, the larger it becomes as a giant; the larger a star, the more slowly it pulsates—just as large bells vibrate more slowly than small bells. Thus, period of pulsation depends on mass. But luminosity is also related to mass, and when two things depend on a third thing, the two things must depend on each other. Thus, there must be a relationship between period and luminosity.

The evolution of a giant star may carry it through the instability strip a number of times, and each time it can become a variable star. If we could watch a star as it entered the instability strip, we could see it begin to pulsate, and this would confirm our belief that the stars are evolving. Unfortunately, they evolve so slowly we cannot expect the motion of a star into or out of the instability strip to be common.

Nevertheless, some astronomers believe that the Type II Cepheid known as RU Camelopardalis is leaving the instability strip. In 1966 University of Toronto astronomers discovered that it had stopped pulsating. Over the decade that followed, RU Camelopardalis seems to have recovered some of its variability, so it may be at the edge of the instability strip.

Clearer evidence of evolution can be found in the slowly changing periods of many Cepheids. Like a clock that gains a second every day, Cepheids with slowly changing periods gradually run "fast" or "slow" by an appreciable amount. These changes are believed to occur because the evolution of the star across the instability strip gradually changes its radius and thus the period of its pulsation.

Other kinds of variable stars are not understood as well as Cepheids. **Long-period variables** are very cool giants that vary with periods of months or years. Some, like Mira, are quite periodic, but others are irregular. Many of these stars are known to eject large amounts of gas and dust into the interstellar medium, and this may be related to their pulsation. Another type of variable star is the **β Canis Majoris variable**, which has a short period and a high surface temperature. They are B stars and seem to be just leaving the main sequence. None of these kinds of stars lies in the instability strip, and the causes of their pulsation are poorly understood (Figure 12–12).

If we can understand why the different kinds of variable stars pulsate, we may better understand how stars

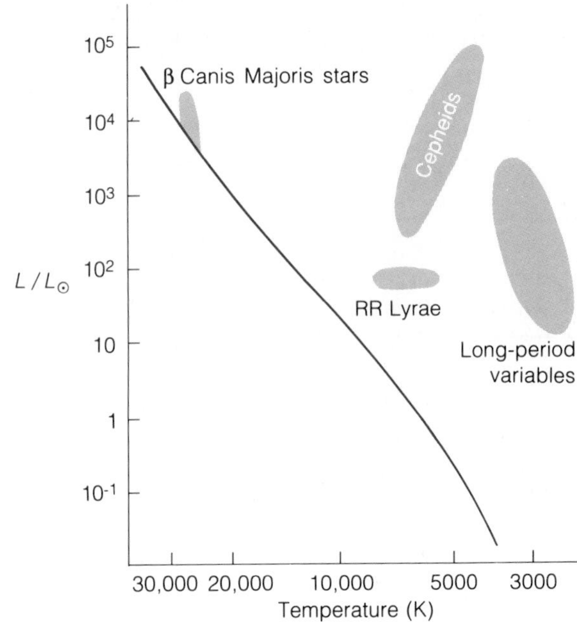

FIGURE 12–12 The Cepheid and the RR Lyrae variable stars lie in the instability strip, but the long-period variables are much cooler stars. The β Canis Majoris stars are also pulsating variables, but the reason for their pulsation is not well understood.

evolve. But there is another way to confirm that stars really do evolve. We can look at the groups of stars called star clusters.

12.3 STAR CLUSTERS

The theory of stellar evolution is so complex and involves so many assumptions that astronomers would have little confidence in it were it not for H–R diagrams of clusters of stars. By observing the properties of star clusters of different ages, we can see clear evidence of the evolution of stars.

To grasp the difficulty of understanding stellar evolution, consider an analogy suggested by William Herschel. Suppose a visitor who had never seen a tree wandered through a forest for an hour looking at the falling seeds, rising sprouts, young saplings, mature trees, and fallen logs. Could such an observer understand the life cycle of the trees? Astronomers face the same problem when they try to understand the life story of the stars.

We do not live long enough to see stars evolve. We see only a momentary glimpse of the universe, a snapshot,

FIGURE 12–13 The open star cluster NGC 2682 in Cancer. All of the stars in such clusters have roughly the same age, chemical composition, and distance. (Palomar Observatory photograph.)

in which all of the stages in the life cycle of the stars are represented. Unscrambling these stages and putting them in order is an almost impossible task. Only by looking at selected groups of stars, star clusters, can we see the pattern.

OBSERVING STAR CLUSTERS

Three things make star clusters useful. All the stars in a cluster have about the same age and about the same chemical composition. In addition, they all lie at about the same distance from the earth. These family relationships among the stars make the beautiful star clusters (Figure 12–13) objects of value.

All of the stars in a cluster presumably formed from the same cloud of gas, and they all began contracting at about the same time. Although the massive stars reached the main sequence sooner than the low-mass stars, the spread in ages is not significant compared with the age of most clusters. Thus, we can study the evolution of stars of different masses without worrying about differences in age.

Another important factor is chemical composition. Because all of the stars formed from the same cloud of gas, we might expect them to have similar chemical compositions. Although there are certainly exceptions to this, it is probably true for most stars in most clusters. This means the stars in a cluster differ from each other only because of the mass they contain, not because of differences in composition.

Clusters are also convenient laboratories for the study of stellar evolution because all of the stars lie at about the same distance from Earth. Stars on the near side of the cluster are slightly closer to us, of course, but compared to the distance to the cluster, this difference is not important. With all of the stars at the same distance, we can easily arrange the stars in order of luminosity. The stars that are brightest in apparent magnitude are also the most luminous. Thus, we can study the relative luminosities of the stars without worrying about the distances to the individual stars.

We can also study the surface temperatures of the stars in a number of ways. We might take spectra of the stars and classify them according to spectral type. But that would be very time consuming; some clusters contain a thousand stars. We could use a photometer and measure the brightness of the stars through a blue and a yellow filter to find their color. (See Chapter 6.) This too would be time consuming. The most efficient way would be to photograph the cluster through the yellow and the blue filters and then determine the colors from the plates. From the colors we could deduce surface temperatures or spectral types.

This would be enough information to draw an H–R diagram for the cluster. We could plot color, spectral type, or surface temperature on the horizontal axis—all three

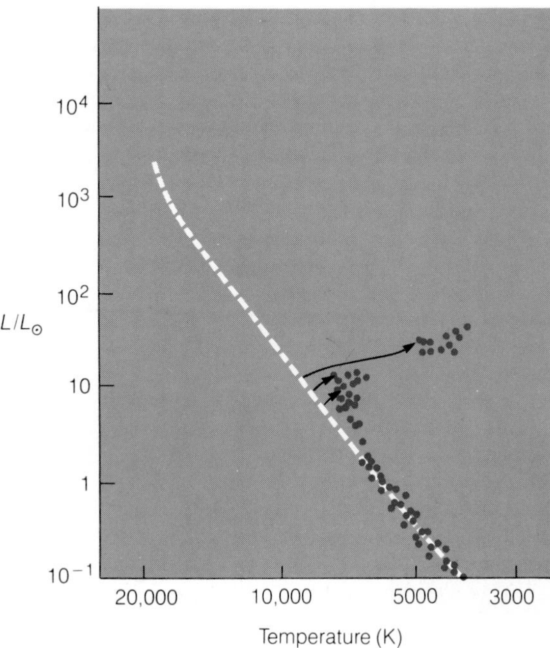

FIGURE 12–14 The H–R diagram of a hypothetical cluster of stars would reveal that the less-massive stars are still on the main sequence, that some are just beginning to leave the main sequence, and that some have become giants.

are related to the star's temperature. If we didn't know the distance to the cluster, we could not convert apparent magnitude into absolute magnitude, but we could still plot an H–R diagram using apparent magnitudes on the vertical axis. All of the stars are at about the same distance and thus have the same distance modulus. Plotting apparent magnitude instead of absolute magnitude would not change the distribution of dots in the diagram.

If we made such observations of a star cluster, we might get a diagram such as that in Figure 12–14. We would discover that most of the stars are on the main sequence but some have evolved off of the main sequence to become giants. The massive stars on the upper main sequence would be missing entirely because they live such short lives. Properly interpreted, such a diagram can tell us a great deal about stellar evolution.

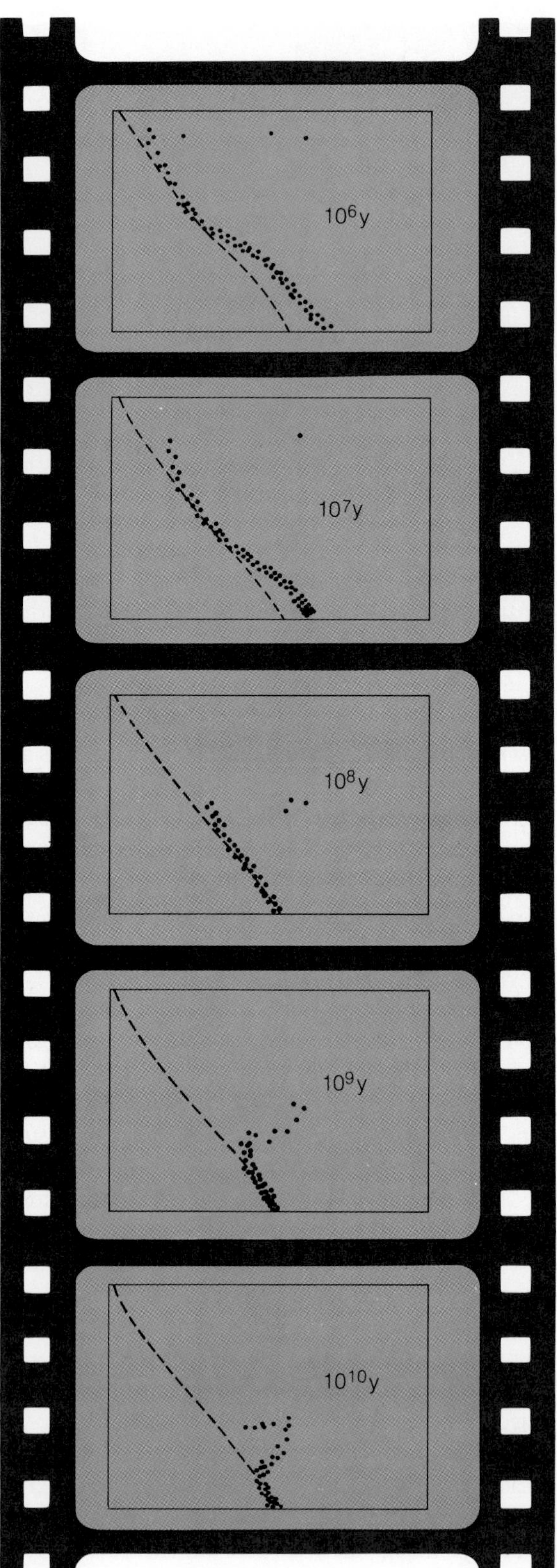

FIGURE 12–15 A series of H–R diagrams, like frames in a film, illustrate the evolution of a cluster of stars. Massive stars approach the main sequence faster, live shorter lives, and die sooner than lower-mass stars. Compare with Figure 12–16.

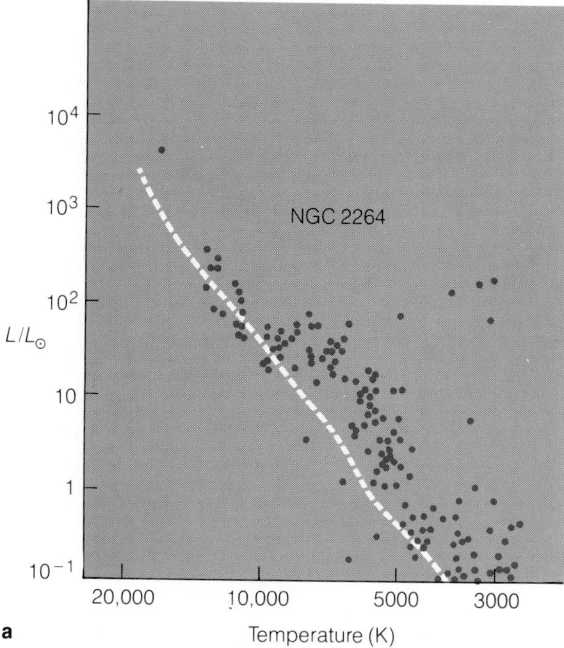

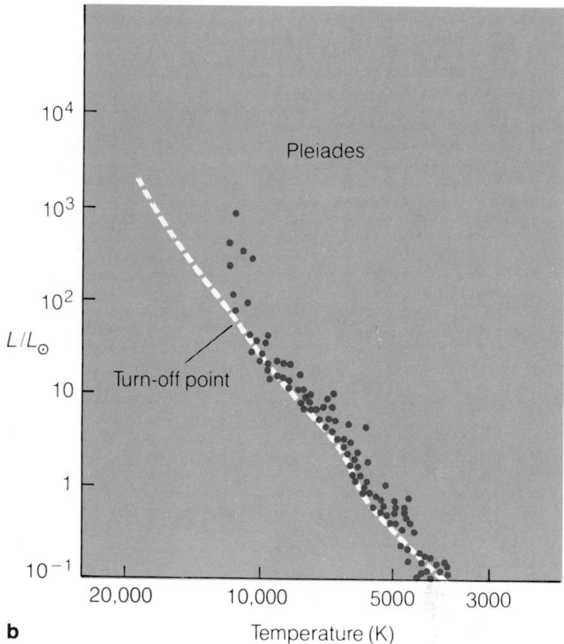

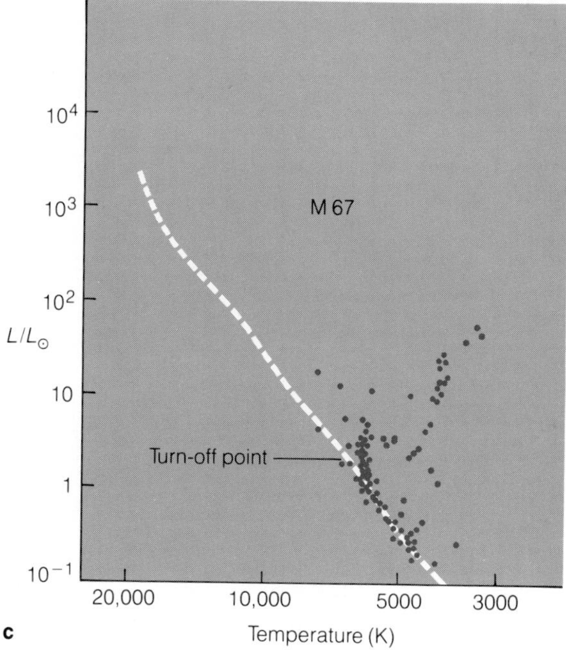

FIGURE 12–16 (a) NGC 2264 is a cluster only a few million years old. Its lower-mass stars are still approaching the main sequence. (b) The Pleiades is an older star cluster, but most of its stars are still on the main sequence. (c) M 67 is about 5 billion years old, and all of its more massive stars have died.

STELLAR EVOLUTION CONFIRMED

To interpret H–R diagrams of star clusters, we must imagine how the diagram changes with time. Suppose we follow the evolution of a star cluster by making H–R diagrams like frames in a film (Figure 12–15). Our first frame shows the cluster only 10^6 years after it began forming, and already the most massive stars have reached the main sequence, consumed their fuel, and moved off to become supergiants. However, the medium- to low-mass stars have not yet reached the main sequence.

Because evolution is such a slow process, we cannot make the time step between frames equal or we would fill more than 1000 pages with nearly identical diagrams. Instead, we increase the time step by a factor of 10 with each frame. Thus the second frame shows the cluster after 10^7 years and the third after 10^8 years.

By the third frame, all massive stars have died, and stars slightly more massive than the sun are beginning to leave the main sequence. Notice that the lowest-mass stars have finally begun to fuse hydrogen. Only after 10^{10} years does a star of the sun's mass begin to swell into a giant.

These five frames were made from theoretical models of stellar evolution, but they compare very well with H–R diagrams of real star clusters. NGC 2264 is only a few million years old and still has many of its lower-mass stars contracting toward the main sequence (Figure 12–16a). The Pleiades, a cluster visible to the naked eye in Taurus, is older than NGC 2264, dating back about 100 million

FIGURE 12–17 The globular cluster M 13 in Hercules contains about 10^6 stars in a region only about 25 pc in diameter. This cluster is at least 10 billion years old. (Palomar Observatory photograph.)

years (Figure 12–16b). Compare these younger clusters with M 67, a faint cluster of stars about 5 billion years old (Figure 12–16c).

We can estimate the age of a star cluster by noting the point where its stars turn off the main sequence and move toward the red giant region. The masses of the stars at this **turn-off point** tell us the age of the cluster, because those stars are on the verge of exhausting their hydrogen-fusing cores. Thus, the life expectancy (Box 11–2) of the stars at the turn-off point equals the age of the cluster.

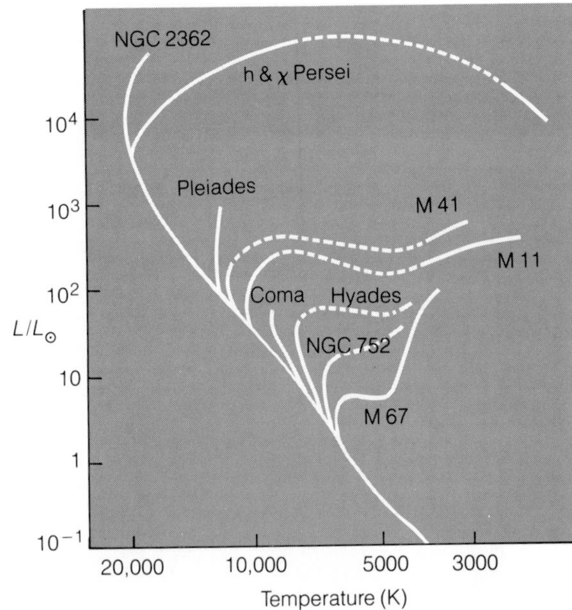

FIGURE 12–18 The combined H–R diagrams of nine open star clusters illustrate how clusters of different ages have different turn-off points.

TWO KINDS OF CLUSTERS

Even a glance through a small telescope will convince us that there are two kinds of star clusters in the sky. The H–R diagrams of these two kinds of clusters confirm that they differ dramatically in a number of ways.

One kind of star cluster is known as an **open cluster** because its stars are uncrowded, and the cluster has an open, transparent appearance. Figure 12–13 shows an open cluster. Such a cluster can contain from ten to a few thousand stars in a region about 25 pc in diameter.

The other kind of cluster is known as a **globular cluster*** because the cluster is shaped like a globe (Figure 12–17). Such clusters can contain 10^5 to 10^6 stars in a region 10–30 pc in diameter. The stars in a globular cluster are crowded a thousand times more densely than the stars near the sun. The average distance between the stars

*"Glob" in globular cluster is pronounced like "glob of butter," not like "globe" meaning a sphere.

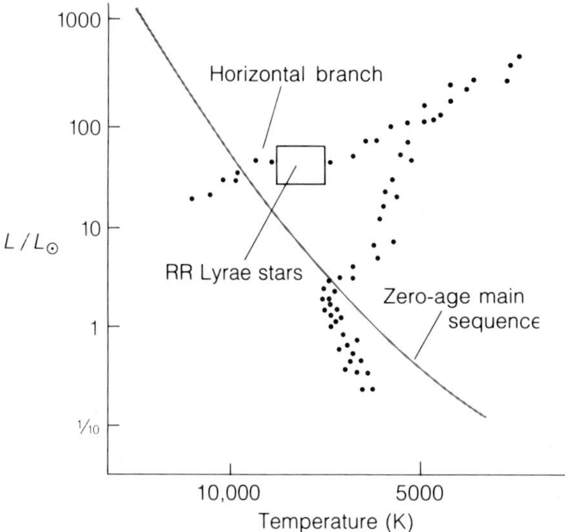

FIGURE 12–19 The H–R diagrams of globular clusters show by their turn-off points that they are very old. The main sequences of such clusters lie below the zero-age main sequence because the stars are poor in elements heavier than helium. The horizontal branch contains a gap (boxed) where RR Lyrae stars are located. Such stars are not plotted here because their brightnesses are variable.

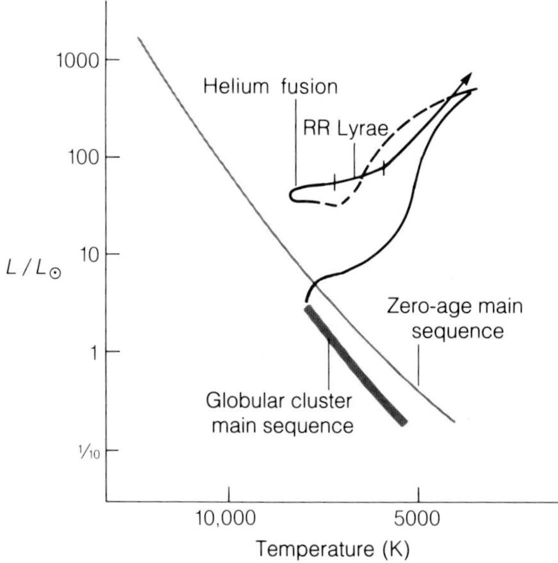

FIGURE 12–20 The schematic evolutionary track of a low-mass, globular cluster star shows how the helium-fusion loop carries the star through the instability strip and along the horizontal branch. Portions of the evolution that occur especially rapidly are shown dashed.

at the center of a globular cluster is only a few light-months. If the earth were located inside a globular cluster, we would see thousands of stars brighter than first magnitude in our sky.

The H–R diagrams of these clusters tell us that they are quite different from open clusters. The open clusters tend to be young to medium age. We can combine their H–R diagrams (Figure 12–18) and see a range of turn-off points demonstrating a range of ages.

But the H–R diagrams of globular clusters reveal very red turn-off points, showing that these clusters are very old (Figure 12–19). Calculated ages range from 10–15 billion years or more, making them the oldest clusters of stars known. We will discuss the ages of globular clusters again in Chapter 15 because they give us clues to the age of our galaxy.

Another thing we see in the H–R diagrams of globular clusters is that their main sequence is slightly fainter than the zero-age main sequence. Theoretical models of stars predict that this could be caused by a low abundance of elements heavier then helium. The spectra of these stars reveal that they are, indeed, deficient in these heavy elements. Why these old clusters are poor in the heavier atoms is an important clue to how our galaxy formed (Chapter 15).

Finally, we notice that globular cluster H–R diagrams have stars located along a **horizontal branch** running toward the blue side of the diagram. Studies of the evolution of low-mass stars show that those stars loop far to the left in the H–R diagram as they fuse helium in their interiors. Compare Figure 12–20 with Figure 12–19. The H–R diagrams of globular clusters trace out the evolutionary tracks followed by low-mass stars and once again confirm that our theories of stellar evolution are, in general at least, correct.

This chapter has discussed the evolution of stars, but it has ignored the most interesting question of all. What happens to a star when it uses up the last of its nuclear fuel? Clearly, stars have to die. The question is how. We will discuss that in the next chapter.

SUMMARY

When a main-sequence star exhausts its hydrogen, it does not just wink out. Its core contracts, and it begins to fuse hydrogen in a shell around its core. The outer parts of the star, its envelope, swell and the star becomes a giant. Because of this expansion, the surface of the star cools, and it moves toward the right in the H–R diagram. The most massive stars move across the top of the diagram as supergiants.

As the core of the star continues to contract, it finally ignites helium fusion, a process that converts helium into carbon. If the core becomes degenerate before helium ignites, the pressure of the gas does not depend on its temperature, and, when helium ignites, it explodes in the helium flash. Although the helium flash is violent, the star absorbs the extra energy and quickly brings the helium-fusing reactions under control.

After helium is exhausted in the core, it can fuse in a shell around the core. Then, if the star is massive enough, it can contract and ignite other fuels such as carbon.

As the giant star burns additional fuels, it moves about in the giant region of the H–R diagram. If it enters the instability strip, it can become unstable as ionization zones of hydrogen and helium act as valves in the outer layers of the star and force it to expand and contract. We see such pulsations from earth as periodic changes in the brightness of the star.

Cepheid stars, named after the prototype star δ Cephei, are important because they can be used as distance indicators. We can measure a Cepheid's period of variation and then read its absolute magnitude from the period–luminosity diagram. Comparing absolute magnitude with apparent magnitude tells us the distance to the star.

Pulsating stars confirm that stars are evolving. At least one star has been found to stop its pulsation temporarily, as if it were approaching the edge of the instability strip. Other stars have periods that change slowly, showing that the evolution of the star is changing its radius and thus its natural period.

Another confirmation of stellar evolution comes from star clusters. Because all the stars in a cluster have about the same distance, composition, and age, we can see the effects of stellar evolution in the H–R diagram of a cluster. Massive stars evolve faster than low-mass stars, so in a given cluster, the most massive stars leave the main sequence first. We can even judge the age of such a cluster by looking at the turn-off point, the location on the main sequence where the stars turn off to the right and become giants. The life expectancy of a star at the turn-off point equals the age of the cluster.

There are two types of star clusters. Open clusters contain 10 to 1000 stars and have an open, transparent appearance. Globular clusters contain 10^5 to 10^6 stars densely packed into a spherical shape. The open clusters tend to be young to middle-aged, but globular clusters tend to be very old. The ages of globular clusters range from 10–15 billion years or more. Also, globular clusters tend to be poor in elements heavier than helium.

NEW TERMS

variable star

triple alpha process

helium flash

instability strip

long-period variable

β Canis Majoris variable

degenerate matter

Cepheid variable star

period–luminosity diagram

RR Lyrae variable

turn-off point

open star cluster

globular cluster

horizontal branch

QUESTIONS

1. Why do we say that stars are not mixed when some have convective cores?
2. Why can't the core of a star remain stable if it is too cool to ignite a nuclear fuel?
3. Why do stars move to the right in the H–R diagram when they expand?
4. What causes the helium flash? Why does it make it difficult for astronomers to understand the later stages of stellar evolution?
5. How do some stars avoid the helium flash?
6. What determines if a star will be a giant or a supergiant?
7. Why are giant stars so low in density?
8. Why are lower-mass stars unable to ignite more massive nuclear fuels such as carbon?
9. How do some variable stars prove that stars are evolving?
10. How can we estimate the age of a star cluster?
11. How do star clusters confirm that stars evolve?
12. Why can't globular clusters contain B stars?

PROBLEMS

1. If the sun expands to a radius 100 times its present radius, what would its density be? (HINT: The volume of a sphere is $\frac{4}{3}\pi R^3$.)
2. What is the absolute magnitude of a Type I Cepheid with a period of 30 days?
3. δ Cephei has an apparent magnitude of about 4 and a period of 5.4 days. What is its absolute magnitude, distance modulus, and distance? (HINT: See Figure 12–10.)
4. A star cluster contains RR Lyrae stars with an apparent magnitude of about 10. How far away is the cluster?
5. The brightest variable stars in a distant galaxy have apparent magnitudes of 19. How far away is the galaxy?
6. If the stars at the turn-off point in a star cluster have masses of about 4 $M_{\odot}$, how old is the cluster? (HINT: See Box 11–2.)

7. If an open cluster contains 500 stars and is 25 pc in diameter, what is the average distance between the stars? (HINTS: What share of the volume of the cluster surrounds the average star? The volume of a sphere is $\frac{4}{3}\pi R^3$.)

8. Repeat Problem 7 for a typical globular cluster containing 10^6 stars in a sphere 25 pc in diameter.

OBSERVATIONAL ACTIVITY: NAKED-EYE VARIABLE STARS

Of the few thousand variable stars known, most are much too faint to see with the naked eye. However, a few very interesting stars are quite bright and we can see their brightness vary without the aid of a telescope. In fact, the two stars we will discuss were among the first variable stars discovered.

Mira the Wonderful The variation of Omicron Ceti was first seen by the Dutch astronomer David Fabricius on August 13, 1596, and it has become known as Mira meaning "The Wonderful." It is a long-period variable whose brightness ranges from about 3.5 to 9.3 with an irregular period of about 330 days.

To find Mira, use the accompanying star chart to locate Cetus the whale to the west of Orion. Cetus is prominent in the evening sky from October to January. You may not be able to see Mira the first time you look. At its faintest, it is much too faint to see with the naked eye. If you don't see Mira at first, check its location at each opportunity. Dates when Mira is at maximum are sometimes published in magazines such as *Sky and Telescope*.

δ Cephei δ Cephei is easy to locate in the northern sky. Use the star charts at the end of this book to find the constellation Cepheus and then use the chart in Figure 12–7 to locate δ Cephei.

The light variations of δ Cephei are easily visible to the naked eye. Estimate the brightness of the star by comparing it with other stars in the constellation. The magnitudes of these stars are given in Figure 12–7 without decimal points to avoid confusion with faint stars. Keep careful records of your observations, including the date, time, brightness of δ Cephei, and the condition of the sky. After a number of nights you can graph these observations to produce a light curve. The period of δ Cephei is 5.36634 days.

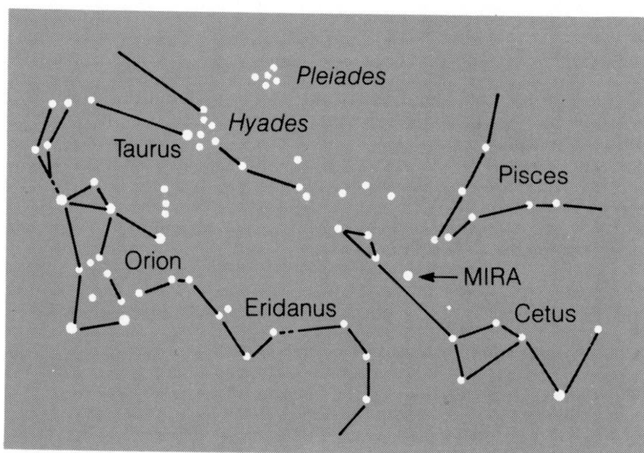

Mira is located in Cetus west of Orion.

RECOMMENDED READING

CLARK, GAIL O. "Ancients of the Universe." *Astronomy 13* (May 1985), p. 6.

———. "Stellar Populations: Key to the Clusters." *Astronomy 14* (Oct. 1986), p. 106.

GINGERICH, OWEN "Harlow Shapley and the Cepheids." *Sky and Telescope 70* (Dec. 1985), p. 540.

HOLZINGER, J. R., and M. A. SEEDS *Laboratory Exercises in Astronomy.* Ex. 30, 31, 34. New York: Macmillan, 1976.

LESH, J. R. "Swarms of Stars: Cosmic Calibrators." *Astronomy 6* (March 1978), p. 6. Reprinted in *Astronomy: Selected Readings.* ed. M. A. Seeds. Menlo Park, Calif.: Benjamin/Cummings, 1980, p. 111.

MARSCHALL, LAURENCE A., LIANG-TAI GEORGE CHIU, and WILLIAM F. VAN ALTENA "Star Cluster Membership: Separating Sheep from Goats." *Sky and Telescope 62* (Aug. 1981), p. 112.

MATTEI, JANET AKYÜZ, ERNST H. MAYER, and MARVIN E. BALDWIN "Variable Stars and the AAVSO." *Sky and Telescope 60* (Sept. 1980), p. 180.

———. "Observing Variable Stars." *Sky and Telescope 60* (Oct. 1980), p. 285.

PASACHOFF, J. M., and R. W. GOEBEL "Laboratory Exercises in Astronomy—Cepheid Variables and the Cosmic Distance Scale." *Sky and Telescope 57* (March 1979), p. 241.

PELTIER, L. C. "Hunting Variable Stars." *Astronomy 3* (Feb. 1975), p. 51.

PERCY, J. R. "Pulsating Stars." *Scientific American 232* (June 1975), p. 66.

———. "Cepheids: Cosmic Yardsticks, Celestial Mysteries." *Sky and Telescope 68* (Dec. 1984), p. 517.

SANNER, FRED "Betelgeuse: Searching for Supergranules." *Astronomy 11* (May 1983), p. 66.

SEEDS, MICHAEL A. "Stellar Evolution." *Astronomy 7* (Feb. 1979), p. 6. Reprinted in *Astronomy: Selected Readings.* ed. M. A. Seeds. Menlo Park, Calif.: Benjamin/Cummings, 1980, p. 95.

STROHMEIER, W. *Variable Stars.* Oxford, England: Pergamon Press, 1972.

CHAPTER 13 THE DEATHS OF STARS

Natural laws have no pity.

Robert Heinlein
THE NOTEBOOKS OF LAZARUS LONG

Gravity is patient—so patient it can kill stars. A star is a ball of gas caught in a battle between gravity trying to make it contract and nuclear fusion trying to make it expand. So long as the star can fuse nuclear fuels and keep its interior hot, it can withstand its own gravity and shine in the darkness. But no star has unlimited fuel, and, when the fuel is gone, gravity wins, and the star dies.

How a star dies depends on its mass. Low-mass stars die with a whimper, slowly shrinking away as they consume their fuels. Medium-mass stars like the sun puff out elegant spherical nebulae and then collapse into hot embers. But the massive stars, the rarest and fastest living members of the stellar community, die in tremendous explosions that destroy the star.

In Chapter 11 we decided that a protostar becomes a true star when it ignites its nuclear fuels. According to that definition, a star's life ends when its nuclear fires go out, and that is the subject of this chapter—the processes by which stars die.

13.1 LOWER MAIN-SEQUENCE STARS

The stars of the lower main sequence share a common characteristic—they have relatively low masses. That means that they face similar fates as they exhaust their nuclear fuels.

When a star exhausts one nuclear fuel, its interior contracts and grows hotter until the next nuclear fuel ignites. The contracting star heats up by converting gravitational energy into thermal energy, so low-mass stars cannot get very hot, and this limits the fuels they can ignite. The lowest-mass stars, for example, cannot get hot enough to ignite helium fusion.

Structural differences divide the low-mass stars into two subgroups—very low-mass stars, and medium-mass stars such as the sun (see Figure 11–18). The critical difference between the two groups is the extent of interior convection. If the star is convective, fuel is constantly mixed, and the resulting evolution is drastically altered.

RED DWARFS

The stars on the lowest part of the main sequence are called **red dwarfs** because they are very cool and very small. We saw in Chapter 11 that the stars less massive than 0.4 $M_\odot$ are convective from their centers to their surfaces. The gas is constantly mixed, so hydrogen is consumed, and helium accumulates uniformly throughout the star. Because the star cannot develop an inert helium core surrounded by a shell of unprocessed hydrogen, it never ignites a hydrogen shell and cannot become a giant.

As nuclear reactions convert hydrogen to helium, the star slowly contracts and heats up. Because it has a low mass, it cannot get hot enough to ignite its helium. Contraction continues until the gas becomes degenerate and, like solid steel, resists further compression.

Thus red dwarfs contract, heat up, and move to the left side of the H–R diagram to become white dwarfs. As we will see, white dwarfs are small, degenerate stars, unable to ignite their remaining nuclear fuels.

SUNLIKE STARS

Stars with masses between roughly 4 $M_\odot$ and 0.4 $M_\odot$,* including the sun, evolve in the same way. They can ignite hydrogen and helium and become giants, but they cannot get hot enough to ignite carbon, the next fuel in the sequence (see Table 12–1). When they reach that impasse, they collapse.

To understand the collapse of these stars, we must consider two concepts: mixing and expansion. The interiors of these sunlike stars are not well-mixed (see Figure 11–18). As we learned in Chapter 11, stars of 1.1 $M_\odot$ or less, including the sun, have no convection near their centers so they are not mixed at all. Stars more massive than 1.1 $M_\odot$ have small zones of convection at their centers, but this mixes no more than about 12 percent of the star's mass. Thus medium-mass stars, whether they have convective cores or not, are not mixed and the helium accumulates in an inert helium core surrounded by unprocessed hydrogen. When this core contracts, the unprocessed hydrogen just outside the core ignites in a shell and swells the star into a giant.

As a giant, the star fuses helium in its core and then in a shell surrounding a core of carbon and oxygen. This core contracts and grows hotter, but cannot become hot enough to ignite the carbon. Thus, the carbon-oxygen core is a dead end for these medium-mass stars.

Because no nuclear reactions can begin in the carbon core, it cannot resist the weight pressing down on it. In addition, just outside the carbon core, the helium-fusion shell converts helium into carbon and thus increases the mass of the carbon core. Thus, the core must contract. The energy released by the contracting core, plus the energy generated in the helium- and hydrogen-fusing shells, flows outward and makes the envelope of the star expand.

This forces the star to become a very large giant. Its radius may become as large as the radius of Earth's orbit, and its surface becomes as cool as 2000 K. Such a star can lose large amounts of mass from its surface.

MASS LOSS FROM STARS

We know that stars can lose mass because we can see mass streaming away from the sun. The solar wind is a breeze of gas flowing out of the sun's hot corona and escaping into space.

Other stars like the sun are also losing mass. Observations in the ultraviolet and X-ray parts of the spectrum have been made with space telescopes such as the International Ultraviolet Explorer (IUE) and the High Energy Astrophysical Observatories (HEAO). Such observations show strong emission from many main-sequence stars, emission that must originate in hot chromospheres and coronas like the sun's. If these stars have outer atmospheres like the sun, they presumably have similar winds of hot gas.

This kind of mass loss does not appear to be sufficient to alter the evolution of the star. The sun, for example, loses about 0.001 solar mass per billion years. Even over its entire lifetime of 10 billion years, the sun will not lose any appreciable fraction of its mass.

However, the sun will not always be a simple main-sequence star. It will eventually become a giant star, and the spectra of some giants contain distorted spectral lines caused by a rapid flow of gas away from the star. Because

*This mass limit is uncertain, as are many of the masses quoted here. The evolution of stars is highly complex, and such parameters are not well known.

the spectra of some of these stars do not show the characteristic emission, we must assume that the stars do not have hot coronas, but other processes could drive mass loss. The stars are so large that gravity is weak at their surfaces and convection in the cool gas can drive shock waves outward and power mass loss. In addition, some giants are so cool that specks of carbon dust condense in their atmospheres, just as soot can condense in a fireplace. The pressure of the star's radiation can push this dust and any atoms that collide with the dust completely out of the star.

Whatever drives this mass loss from giants, it can affect the mass of the star appreciably in a short time. A star expanding as its carbon core contracts could lose an entire solar mass in only 10^5 years, which is not a long time in the evolution of a star. Thus, a star that began its existence on the main sequence with a mass of 8 $M_\odot$ might reduce its mass to only 3 $M_\odot$ in half a million years.

Stellar mass loss confuses our story of stellar evolution. We would like to say that stars more massive than a certain limit will evolve one way and stars less massive will evolve another way. But stars may lose enough mass to alter their own evolution. Thus, we must consider the initial mass a star has on the main sequence and the mass it retains after mass loss. Because we don't know exactly how effective mass loss is, it is difficult for us to be exact about the mass limits in our discussion of stellar evolution.

The mass loss we have discussed so far has been a gradual process, but a sunlike star with a contracting carbon core expands so far that it can lose all of its outer layers in a sudden expulsion. This surface expulsion is helped by the growing pressure of the radiation inside the star and by instabilities in the helium-fusion shell, which can cause periodic bursts of energy production. The expanding layers are visible as small, spherical nebulae.

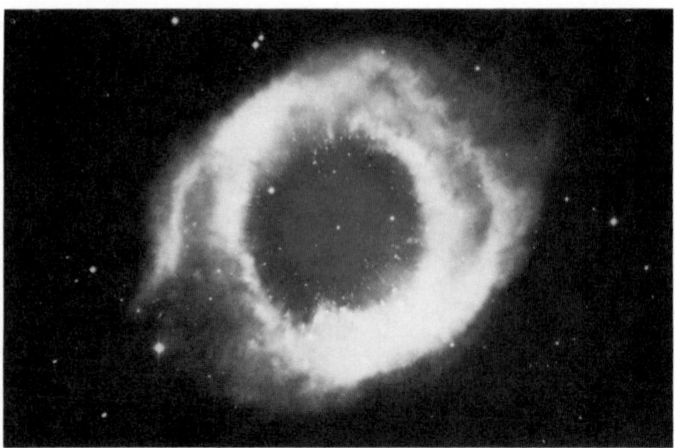

FIGURE 13–1 A planetary nebula is the ejected surface of a medium-mass star. The hollow sphere of gas appears as a ring when seen from a distance. The stellar remains collapse into a hot, faint object (visible at the center of the nebula), which eventually will become a white dwarf. This nebula, known as the Helical Nebula, is half as large in angular diameter as the full moon. (Palomar Observatory photograph.)

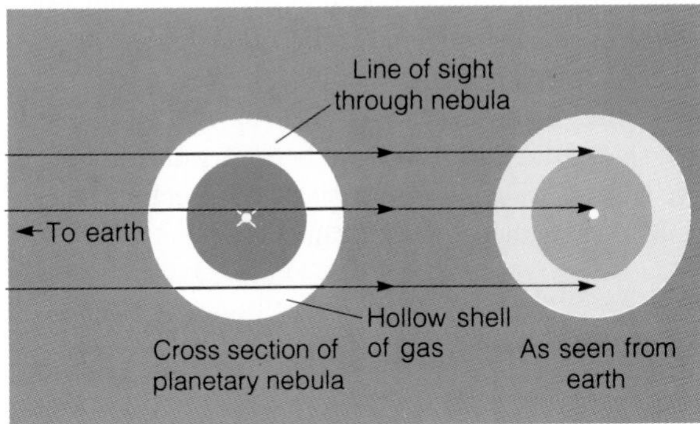

FIGURE 13–2 Planetary nebulae are hollow spheres of excited gas, but they often look like rings as seen from Earth. Through the center of the nebula, we see only two thicknesses of the shell, the front and the back. But near the edge of the nebula our line of sight passes through a larger amount of gas, and the nebula looks brighter. Thus, we see a bright ring with a fainter glow inside.

PLANETARY NEBULAE

Even a small telescope will reveal small circles of haze scattered among the stars. These are named **planetary nebulae** because they look like the small, greenish disk of Uranus, but they are actually the expelled outer layers of giant stars. As these layers overtake the lower-density gas lost by the star when it was a giant, they concentrate the gas just as a snowplow concentrates snow. Ionized by the ultraviolet radiation from the stellar remains at the center, the hollow sphere of gas glows with an emission spectrum.

Larger telescopes reveal about 1500 of these nebulae with diameters ranging from 1000 AU to a parsec. In some cases the exposed, hot interior of the star is visible at the center of the nebula (Figure 13–1). In many cases, because of a geometrical effect, the nebula looks like a smoke ring but is in fact a hollow sphere. When we look at the nebula near its center, we look through the near side and the far side. But when we look near the edge of the nebula, we look through a greater mass of glowing gas, and the nebula looks brighter. Thus, we see the hollow sphere as a ring (Figure 13–2).

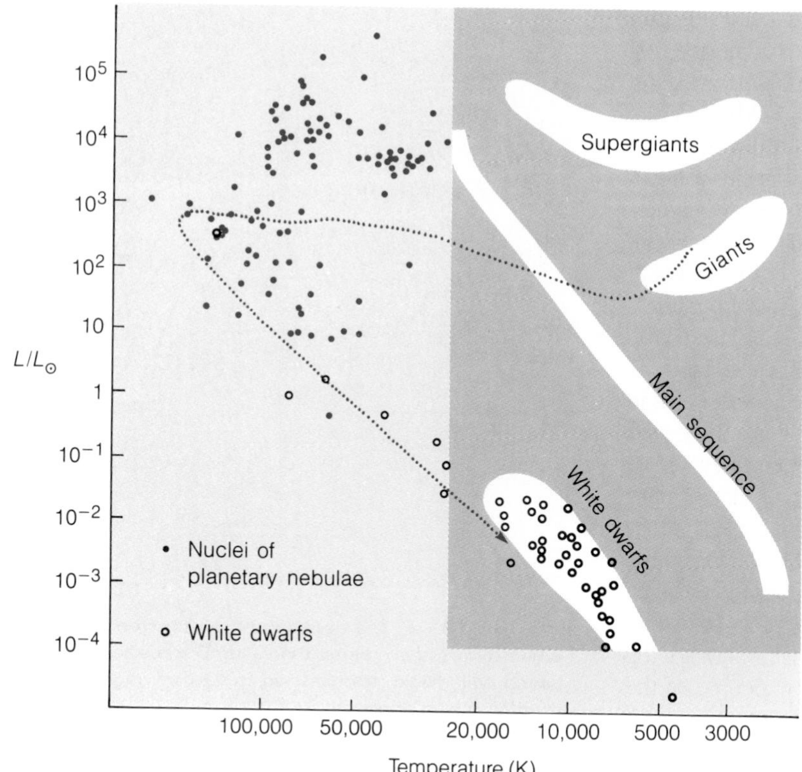

FIGURE 13–3 Our customary H–R diagram (shaded as in Figure 9–10) must be extended to higher temperatures to show the evolution of a star after it ejects a planetary nebula. Central stars of planetary nebulae (filled circles) are hotter and more luminous than most white dwarfs (open circles). The dotted line shows the evolution of a 0.8 $M_\odot$ star as it collapses toward the white dwarf region. Presumably the central stars of planetary nebulae cool to become white dwarfs long after their nebulae have dissipated. (Adapted from diagrams by C. R. O'Dell and S. C. Vila.)

The spectra of these nebulae are emission spectra with lines of such elements as H, He, N, O, C, Ne, S, Ar, Cl, Fe, and so on. The gases glow green because of strong emission lines of OIII (twice-ionized oxygen) in the green part of the spectrum. Doppler shifts show that the nebulae are expanding at 10–20 km/sec.

Although the ejected gas is only about 10 percent of the star's total mass, it is an important part of a stable star—namely, the insulating blanket that confines the star's internal heat. When the surface puffs away into space, the white-hot core is exposed, and the star emits intense ultraviolet radiation. It is this radiation that ionizes and drives away the gases of the nebula.

If we follow the evolution of such a star as it produces a planetary nebula, we see it move rapidly to the left of the H–R diagram. At this point it consists of an inert carbon core surrounded by helium- and hydrogen-fusing shells and topped by a shallow atmosphere of hydrogen and helium. The surface temperatures of these planetary nebula nuclei range from 25,000–100,000 K or more. But with the insulating surface layers gone, the star loses heat rapidly and continues to contract until it becomes degenerate and enters the white dwarf region (Figure 13–3).

WHITE DWARFS

Both low-mass red dwarfs and medium-mass stars eventually become white dwarfs. Our survey of neighboring stars (Chapter 9) showed that most stars have masses less than that of the sun, but these are very long-lived stars. Our galaxy is probably not old enough for many of these red dwarfs to have become white dwarfs. However, white dwarfs are very numerous—our galaxy probably contains billions—so they must be the remains of stars with masses similar to the sun's.

The first white dwarf discovered was the faint companion to Sirius. In that visual binary system, the bright star is Sirius A. The white dwarf, Sirius B, is 10,000 times fainter than Sirius A. The orbital motions of the stars (shown in Figure 10–9) tell us that the white dwarf's mass is about 1 $M_\odot$, and its blue-white color tells us that its surface is hot, about 32,500 K. Because its luminosity is low, it must have a small surface area—in fact, it is about 76 percent earth's diameter. The mass and size imply that its average density is over 3×10^6 g/cm^3. On earth a teaspoonful of Sirius B material would weigh more than 15 tons (Figure 13–4).

A normal star is supported by energy flowing outward from its core, but a white dwarf has no internal energy source, so there is nothing to oppose gravity and the gas becomes degenerate (see Box 12–1). Thus, a white dwarf is supported not by energy flowing outward but by the refusal of its electrons to pack into a smaller volume.

Not all of the white dwarf is degenerate. Computer models predict that a crust about 50 km (30 miles) thick forms at the surface. The bottom of this crust consists of atoms locked in a rigid crystalline lattice; the upper layers blend slowly into a very hot atmosphere of ionized gas.

The tremendous surface gravity of white dwarfs—100,000 times that of earth—affects their atmospheres in strange ways. The heavier atoms in the atmosphere tend to sink, leaving the lightest gases at the surface. We see some white dwarfs with atmospheres of almost pure hydrogen, whereas others have atmospheres of nearly pure helium. Still others, for reasons not well understood, have atmospheres that contain traces of heavier atoms. In addition, the powerful surface gravity pulls the white dwarf's atmosphere down into a very shallow layer. If the earth's atmosphere were equally shallow, people on the top floors of skyscrapers would have to wear oxygen masks.

Clearly, a white dwarf is not a normal star. It generates no nuclear energy, is almost totally degenerate, and, except for a thin layer at its surface, contains no gas. Instead of calling a white dwarf a "star," we call it a **compact object.** Later in this chapter we will discuss two other kinds of compact objects—neutron stars and black holes.

A white dwarf's future is bleak. As it radiates energy into space, its temperature gradually falls, but it cannot shrink any smaller because its degenerate electrons cannot get closer together. This degenerate matter is a very good thermal conductor, so heat flows to the surface and escapes into space, and the white dwarf gets fainter and cooler, moving downward and to the right in the H–R diagram. Because the white dwarf contains a tremendous amount of heat, it needs billions of years to radiate that heat through its small surface area. Eventually, such objects may become cold and dark, so-called **black dwarfs.** Our galaxy is probably not old enough to contain many.

Perhaps the most interesting thing about white dwarfs appears in mathematical models. The equations predict that if we added mass to a white dwarf, its radius would *shrink* because added mass would increase its gravity and squeeze it tighter. If we added enough to raise its total mass to about 1.4 $M_{\odot}$, its radius would shrink to zero (Figure 13–5). This is called the **Chandrasekhar limit** after the astronomer who discovered it.

The Chandrasekhar limit tells us that no white dwarf can exist with a mass greater than 1.4 $M_{\odot}$. It seems quite likely that stars more massive than this can ultimately collapse into white dwarfs if they lose enough mass when they are giants. We have seen in an earlier section that mass loss is common and very effective. A 6 $M_{\odot}$ star should be able to reduce its mass to 1.4 $M_{\odot}$ before it collapses. Some astronomers have suggested that stars as massive as 10 $M_{\odot}$ may ultimately collapse into white dwarfs with final masses less than the Chandrasekhar limit.

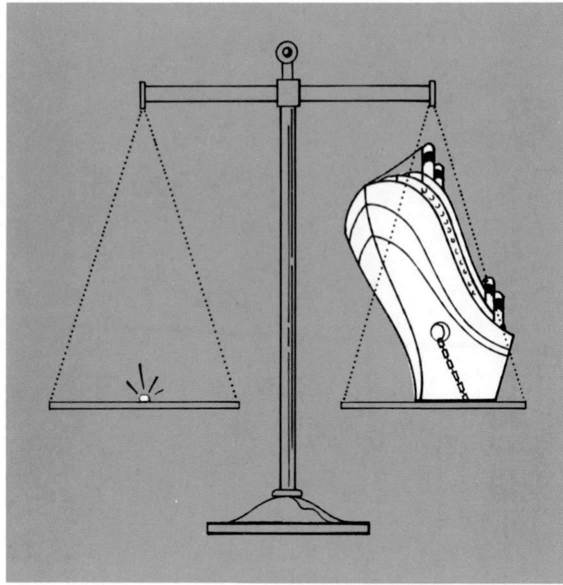

FIGURE 13–4 The degenerate matter from inside a white dwarf is so dense that a lump the size of a beach ball would, transported to earth, weigh more than an ocean liner.

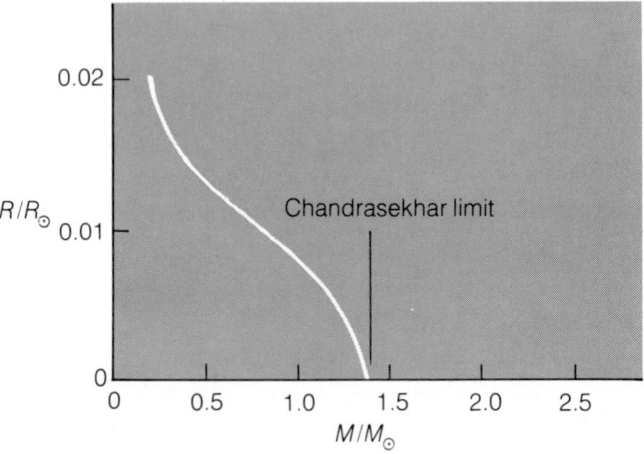

FIGURE 13–5 The more massive a white dwarf, the smaller its radius. Stars more massive than the Chandrasekhar limit of 1.4 $M_{\odot}$ cannot be white dwarfs.

But what happens to more massive stars? How can a 20 $M_{\odot}$ or 40 $M_{\odot}$ star die? Such massive stars evolve so rapidly that mass loss cannot reduce their mass below the Chandrasekhar limit before they collapse. To resolve this problem, we must consider the collapse of these massive stars in detail.

13.2 THE UPPER MAIN SEQUENCE

We have seen that the low-mass red dwarfs and the sunlike, medium-mass stars die relatively quietly as they exhaust their nuclear fuels. The most violent event in the evolution of these stars is the ejection of surface layers to form planetary nebulae. Massive stars, however, end their days in spectacular explosions, which destroy the star.

HYDROGEN, HELIUM, AND CARBON

The evolution of an upper main-sequence star begins like the evolution of a sunlike star. It exhausts the hydrogen in its core, and, when the core contracts, a hydrogen-fusing shell ignites, swelling the star into a giant. While it is a giant, it fuses helium in its core and then in a shell, leaving behind a carbon–oxygen core, which contracts and grows hotter.

Mass loss will reduce the mass of the star, but if it still has a mass between 3 $M_{\odot}$ and 9 $M_{\odot}$ when its carbon core begins to contract, it develops a potentially lethal problem. The contracting core becomes degenerate, turning off the pressure–temperature thermostat that normally would keep nuclear reactions under control. The carbon–oxygen core in such a star is a bomb, destined to explode when the temperature reaches 600,000,000 K, the ignition temperature for carbon. The resulting explosion is called the **carbon detonation.**

The carbon detonation may be powerful enough to blow some stars apart, and it could be responsible for some of the violent stellar explosions known as supernovae. As we will see later, supernovae are rare and poorly understood. Other stars may be able to survive carbon detonation because the rising temperature eventually forces the core to expand and the gas stops being degenerate.

Stars more massive than about 9 $M_{\odot}$ do not face the carbon detonation because their cores are so hot carbon ignites before the gas becomes degenerate. Thus, the pressure–temperature thermostat is in working order and carbon fusion turns on gradually. Stars that survive carbon ignition continue their lives for a short while as they fuse heavier elements (Table 12–1), but they too eventually face an explosive end. The development of an iron core spells their absolute and final collapse.

THE IRON CORE

Some stars do survive carbon ignition, but we cannot be certain of their evolution. Nuclear reactions involving heavier elements are very complex (see Figure 12–6), and stellar models are unable to follow the behavior of stars as they fuse these fuels. Nevertheless, we can predict the general plan of their evolution.

As the star fuses heavy elements, each new fuel ignites first in the core and then in a shell, building layer after layer of heavy elements inside the star. Because fusion makes a more massive atomic nucleus by sticking little nuclei together, each successive stage involves fewer particles. But higher temperatures are required for each new fuel. Thus, the reactions are faster and more violent for each new fuel (Table 12–1).

The accumulation of iron atoms in the star's core forbodes the end of heavy-element fusion and the death of the star. Nuclear fusion reactions can release energy if the nucleus produced is bound more tightly than the lower-mass nuclei that were fused. Because iron is the most tightly bound nucleus of all (Figure 13–6), no nuclear fusion reactions can combine iron nuclei and release energy. In fact, any nuclear reactions that do fuse iron into heavier nuclei absorb rather than release energy.

When a massive star develops an iron core, nuclear fusion cannot occur in the core to produce energy, and the core contracts and grows hotter. The shells around the core burn outward, fusing lighter elements and leaving behind more iron, which increases the mass of the core further. When the mass of the iron core exceeds a limiting mass of 1.3 $M_{\odot}$ to 2 $M_{\odot}$ (depending on the mass of the star), the core must collapse.

As the core contracts, two processes can make it contract even faster. Heavy nuclei in the core can capture high-energy electrons, thus removing thermal energy from the gas. In more massive stars, temperatures are so high that many photons have gamma-ray wavelengths and can break more massive nuclei into less massive nuclei. This reversal of the nuclear fusion process absorbs energy and allows the core to collapse even faster.

Although a massive star may live for millions of years, its iron core—only about 500 km in diameter—collapses in only a few thousandths of a second. This collapse happens so rapidly that our most powerful computers are not capable of following the details. One thing is clear, however. The collapse of the iron core of a massive star triggers a star-destroying explosion—a **supernova.**

SUPERNOVA EXPLOSIONS

Modern theory predicts that the collapse of a massive star can eject the outer layers of the star to produce a supernova explosion while the core of the star collapses to form a neutron star or a black hole. We will discuss neutron stars and black holes in detail in the next chapter, but here we will concentrate on the process that triggers the supernova explosion.

We can't be sure how a supernova explodes, but recent advances in mathematical-modeling techniques and increases in computer power and speed are giving us clues. The collapse of the innermost part of the degenerate core allows the rest of the star's interior to fall inward, creating a tremendous "traffic jam" as all the nuclei fall toward the center. It is as if everyone in Indiana suddenly tried to drive their car into downtown Indianapolis. Not only is there a traffic jam downtown but also in the suburbs; as more cars arrive, the traffic jam spreads outward. Similarly, as the inner core falls inward, a shock wave (a traffic jam) develops and begins to move outward. Containing about 100 times the energy necessary to destroy the star, such a shock wave was thought to be the cause of the supernova explosion.

Recent models, however, show that this shock wave stalls within 25–40 thousandths of a second. Matter flows inward just as fast as the shock wave spreads outward. If this happens to a star, then it presumably will collapse without any visible explosion. Some theorists believe that neutrinos generated in the breakup of heavy nuclei can spread energy across the shock wave, and, about a quarter of a second after the collapse begins, reaccelerate the stalled shock wave. If that is the case, then the outward moving shock wave blasts the envelope of the star into space, producing the supernova explosion.

The supernova we see is the brightening of the star as its distended outer layers are blasted outward. As the cloud of gas expands and thins, it begins to fade. But the way it fades in some cases suggests that nuclear reactions in the compressed outer layers have enriched it with short-lived radioactive nuclei such as nickel-56. The gradual decay of these nuclei can keep the gas hot and prevent it from fading rapidly. Thus, the supernova explosion may be violent enough to trigger nuclear fusion in the outer layers of the dying star.

New super computers may eventually tell us more about how supernovae explode, but whatever the mechanism, it must generate tremendous energy to account for supernovae. A single supernova explosion is equivalent to the explosion of 10^{28} megatons of TNT. (For comparison, this much TNT would amount to about a trillion times the mass of the earth.)

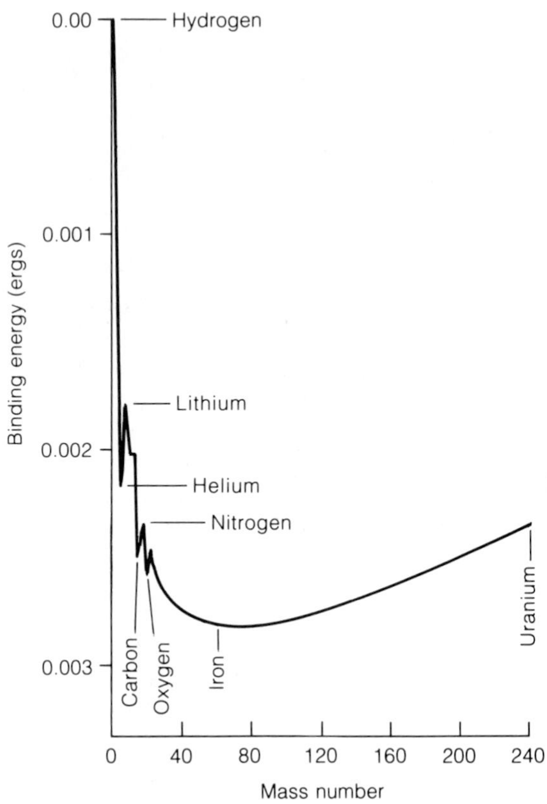

FIGURE 13–6 Binding energy is the energy that holds an atomic nucleus together. When hydrogen fuses into helium, it releases the excess binding energy. Helium fusing into carbon releases less energy. Because iron is the most tightly bound of all atomic nuclei, it and heavier atoms cannot release energy by fusion reactions.

OBSERVATIONS OF SUPERNOVAE

In 1054 Chinese astronomers saw a "guest star" appear in the constellation we know as Taurus the Bull. The star quickly became so bright it was visible in the daytime. After a month's time, it slowly faded, taking almost 2 years to vanish from sight. When modern astronomers turned their telescopes to the location of the guest star, they found a cloud of gas about 1.35 pc in radius, expanding at 1400 km/sec. Projecting the expansion back in time,

FIGURE 13–7 The Crab nebula is a supernova remnant, the remains of a supernova observed by the Chinese in AD 1054. (Palomar Observatory photograph.)

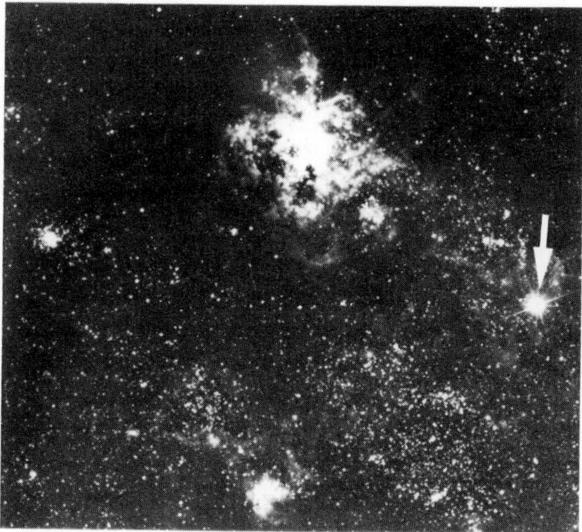

FIGURE 13–8 Supernova 1987A was discovered February 23, 1987. The Tarantula nebula lies at the top of this photo with the supernova at lower right (arrow). Visible from earth's southern hemisphere, the supernova brightened through the spring and summer of 1987 reaching a peak brighter than 3rd magnitude. It is the first naked-eye supernova since the invention of the astronomical telescope. (National Optical Astronomy Observatories.)

they concluded the expansion must have begun about 900 years ago, just when the guest star made its visit. Thus, we think the nebula, now called the Crab nebula because of its shape (Figure 13–7 and Color Plate 18), marks the site of the 1054 supernova.

Supernovae are rare. Only a few have been seen with the naked eye in recorded history. Arab astronomers saw one in 1006, and the Chinese saw one in 1054. European astronomers observed two—one in 1572 (Tycho's supernova) and one in 1604 (Kepler's supernova). Also, the guest stars of 185, 386, 393, and 1181 may have been supernovae.

In the centuries following the invention of the astronomical telescope in 1609, no supernova was seen bright enough to be visible to the naked eye. The ones seen were in distant galaxies and are thus fainter and harder to study. Then in the early hours of February 23, 1987, astronomer Ian Shelton discovered a 5th magnitude object in the southern sky—a supernova still growing in brightness (Figure 13–8). Shelton's supernova, known officially as SN 1987A, is located only 53,000 pc away in the Large Magellanic Cloud, a small satellite galaxy to our own Milky Way. The first naked-eye supernova in 383 years gives us ringside seats at the most spectacular event in astronomy.

For example, at 2:35 AM EST on February 23, 1987, 18 hours before the supernova was seen, a blast wave of neutrinos swept through the earth.* Instruments buried in a salt mine near Cleveland and a similar detector in Japan recorded the passage of the neutrinos coming from the direction of SN1987A and thus confirmed the theory that core collapse triggers supernovae. The neutrinos were apparently produced during the sudden disruption of heavy nuclei in the core of the collapsing star and the conversion of the resulting protons into neutrons.

In some ways SN1987A is peculiar. From the study of supernovae in other galaxies, astronomers have found two types of supernovae. **Type I supernovae** become about 4 billion times more luminous than the sun, decline rapidly at first, and then more slowly (Figure 13–9). **Type II supernovae** become only about 0.6 billion times the sun's luminosity and decline in a more irregular way. Spectra of Type II supernovae show hydrogen lines, but spectra of Type I supernovae do not.

Type I supernovae seem to occur when a white dwarf in a binary system gains mass from its companion (a process we will discuss in detail later in the chapter) and exceeds

*Within a few seconds of that moment, roughly 20 trillion neutrinos passed harmlessly through your body.

the Chandrasekhar limit. This causes the white dwarf to collapse, triggering a star-destroying explosion. Spectra of Type I supernovae lack hydrogen lines because white dwarfs contain little hydrogen.

Type II supernovae are believed to occur when the iron core of a massive star collapses. These massive stars were generally thought to be red supergiants, but observations made before the explosion of SN1987A show that the star that exploded was a blue supergiant. Apparently SN1987A was a Type II supernova, but the star that collapsed was not as distended and cool as was thought usual. That may explain why the light curve of SN1987A does not resemble the typical light curves of either type (Figure 13–9).

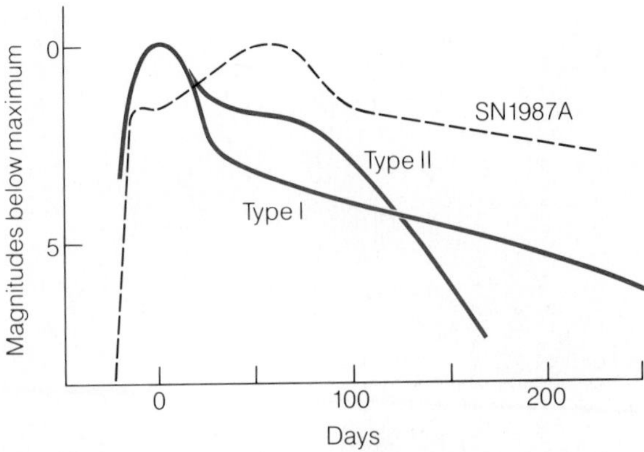

FIGURE 13–9 Type I supernovae decline rapidly at first and then more slowly, but Type II supernovae pause for about 100 days before beginning a steep decline. These light curves have been adjusted to the same maximum brightness. Generally, Type II supernovae are about 2 magnitudes fainter than Type I. SN1987A shown for comparison.

One reason SN1987A does not fit either of the two classes may be that the classification of supernovae is incomplete. Not many supernovae have been studied well, and very few have been seen in their early stages. There may be types of supernovae that are so rare they have never been seen by earth's astronomers. New types and subtypes may be added as more supernovae are discovered by automated searches now underway. For example, one subtype, termed "peeled" supernovae, seems to be produced by the collapse of massive stars that have lost their outer layers. We will see later how a star could be peeled by its companion in a binary system.

Continuing observations of SN1987A are revealing how a supernova explosion evolves. The steady decline of SN1987A's brightness is probably powered by the decay of short-lived radioactive isotopes nickel-56 and cobalt-56. Such nuclei must have been created during the supernova explosion. Nine months after the explosion, the supernova began to fade at a faster rate, apparently because the expanding shell of gas has become transparent enough to allow gamma rays to escape from the interior. Infrared observations made from the Kuiper Airborne Observatory show that the cooling gas is beginning to form dust particles.

Although the supernova explosion fades to obscurity in a year or two, an expanding shell of gas marks the explosion site. The gas, originally expelled at 10,000–20,000 km/sec may carry away one-fifth of the mass of the star. The collision of that expanding gas with the surrounding interstellar medium can sweep up even more gas and excite it to produce a **supernova remnant**, the nebulous remains of a supernova explosion.

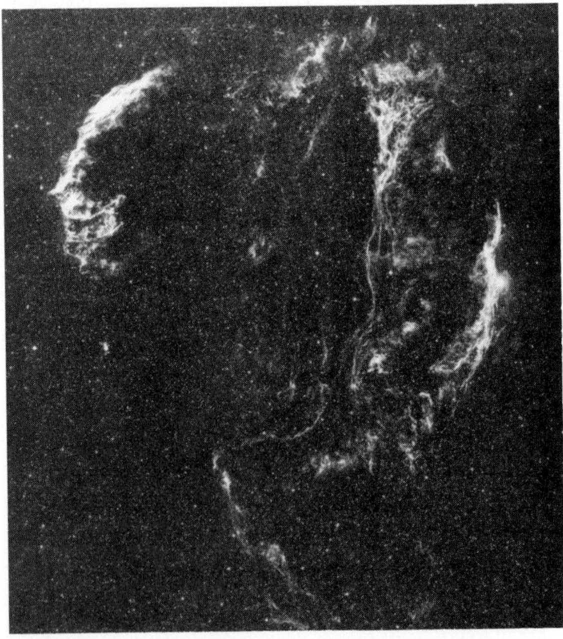

FIGURE 13–10 The Veil nebula, a supernova remnant about 50,000 years old, is about 40 pc in diameter. (Palomar Observatory photograph.)

Supernova remnants look quite delicate and do not survive very long—a few tens of thousands of years—before they gradually mix with the interstellar medium and vanish. The Crab nebula is a young remnant, only 900 years old. The Veil nebula in Cygnus (Figure 13–10) is larger and more diffuse, having originated in a supernova about 50,000 years ago.

Some supernova remnants are visible only at radio wavelengths (Color Plate 19). Such supernovae remnants do not emit detectable light, but rather they radiate radio

BOX 13–1
Synchrotron Radiation

When a charged particle moves through a magnetic field, its path is bent. This process is used to guide beams of electrons inside television picture tubes and draw images on the face of the tube. When the path of an electron if changed by a magnetic field, it converts some of its energy of motion into a photon of electromagnetic radiation. This is known as synchrotron radiation because it was first observed in a particle accelerator called a synchrotron.

The energy radiated by deflected electrons in a television picture tube is not very strong because (a) the electrons do not move very rapidly and (b) there are not many electrons involved. But excited nebulae such as supernova remnants can contain large numbers of high-energy electrons and thus radiate strong synchrotron radiation. In the Crab nebula, for example, high-speed electrons spiral around lines of the nebula's magnetic field (Figure 13–11). As each electron follows a curved path, it radiates away some of its energy of motion as synchrotron radiation. Low-speed electrons radiate at longer wavelengths, and higher-speed electrons radiate at shorter wavelengths, so synchrotron radiation is spread over a wide range of wavelengths.

As any given electron radiates, it loses energy, slows down, and spirals along a smaller radius. Thus, the high-speed electrons producing synchrotron radiation in celestial bodies must be continuously resupplied, or the synchrotron radiation will diminish.

Some of the light we see coming from the Crab nebula is synchrotron radiation produced by very high-speed electrons, but the electrons in the nebula could not have kept it shining for its entire 900-year history. They lose energy too fast. In the Crab nebula, the high-speed electrons are continuously resupplied by the collapsed remains of the star at the center of the nebula—a neutron star.

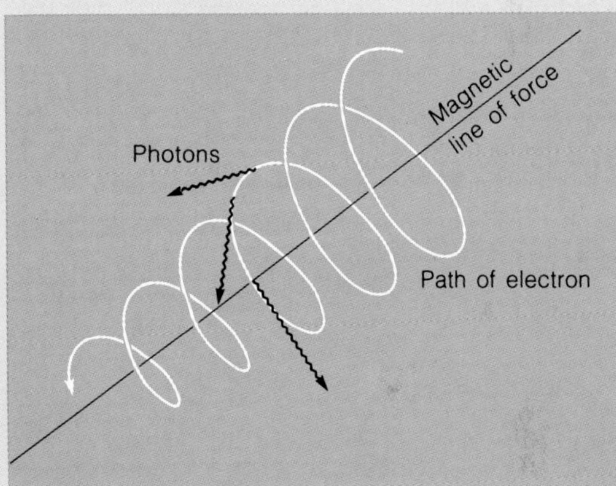

FIGURE 13–11 Synchrotron radiation is emitted when an electron spirals around a magnetic field line and radiates its energy away as photons. Only a few photons are shown here for clarity.

energy called **synchrotron radiation** (Box 13–1). Such radiation is produced by high-speed electrons spiraling through a magnetic field. The Crab nebula is a strong source of synchrotron radiation, and the electrons there are so energetic that some of the synchrotron radiation is emitted at optical wavelengths. The production of synchrotron radiation in supernova remnants tells us that tremendous energy is present to accelerate the electrons to high speed.

Orbiting X-ray telescopes such as HEAO-2 and Exosat have observed a number of supernova remnants at X-ray wavelengths. These images show that the supernova remnants are spherical shock waves expanding outward and heating the shocked gas to very high temperatures (Figure 13–12). The Crab nebula is quite a young remnant and is filled with hot gas that is bright at X-ray wavelengths. In the next chapter we will discover that the gas in the Crab nebula is heated by a neutron star—the remains of the original star that exploded.

Supernovae are important in astronomy because they are associated with birth and death. We saw in Chapter 11 that the compression of the interstellar medium by supernova remnants can trigger star formation. Of course, supernovae mark the deaths of massive stars, and supernovae exploding near the earth might even cause the deaths of living things on the earth (Box 13–2). But supernovae explosions also mark the birth of neutron stars and black holes, objects we will discuss in the next chapter.

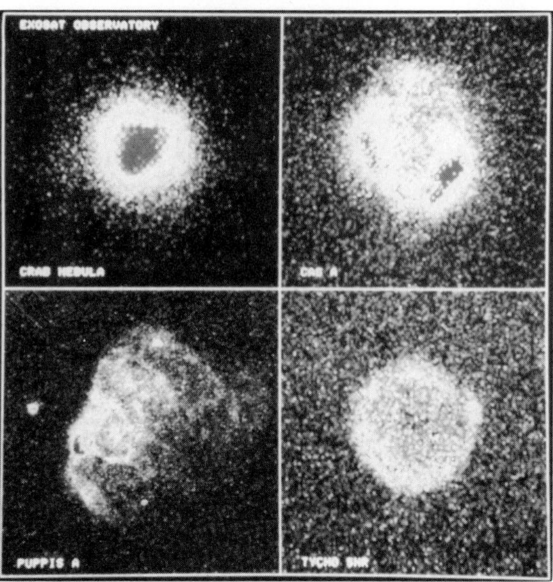

FIGURE 13–12 X-ray images of supernova remnants obtained with Exosat. Most supernova remnants are more or less spherical shock waves pushing into the interstellar medium. The Crab nebula, perhaps because it is young, is still filled with high-temperature gas. (Courtesy W. Brinkman and B. Aschenbach of the Max-Planck Institute, J. Davelaar of Exosat Observatory, and the European Space Agency.)

13.3 THE EVOLUTION OF BINARY STARS

So far we have discussed the deaths of stars as if they were all single objects that never interact. But more than half of all stars are members of binary star systems. Most such binaries are far apart, and one of the stars can swell into a giant and eventually collapse without affecting the companion star. Some systems, however, are close together. When the more massive star begins to expand, it interacts with its companion star in peculiar ways.

These interacting binary stars are interesting objects themselves, but they are also important because they help us explain observed phenomena such as nova explosions. In the next chapter, we will use them to help us find black holes.

MASS TRANSFER

Binary stars can sometimes interact by transferring mass from one star to the other. Of course, the gravitational field of each star holds its mass together, but the gravitational fields of the two stars, combined with the rotation of the binary system, define a teardrop-shaped surface around each star called the **Roche surface** (Figure 13–13). Matter inside a star's Roche surface is gravitationally bound to the star. The size of the Roche surface depends on the mass of the star and the separation between the stars.

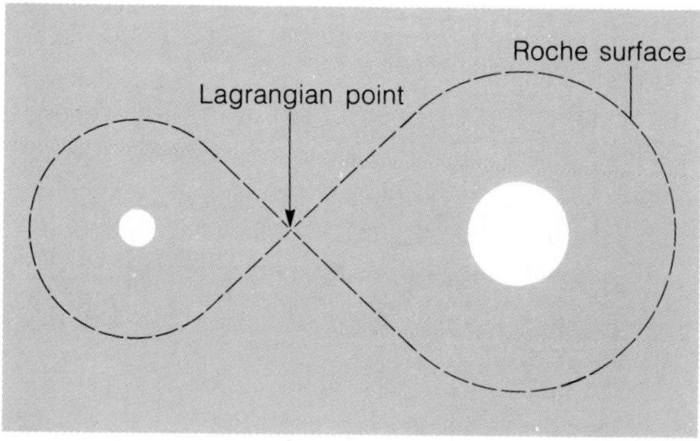

FIGURE 13–13 The Roche surface around a pair of binary stars outlines the two volumes of space that the stars control gravitationally. The Roche surfaces touch at the Lagrangian point where matter can flow from one star to the other.

Like the lobes of a giant dumbell, the Roche surfaces in a binary star meet at the **Lagrangian point** somewhere between the stars (Figure 13–13). The Lagrangian point connects the two Roche surfaces (often termed Roche lobes) and allows matter to flow from one lobe to the other.

In general, there are only two ways matter can escape from a star and reach the Lagrangian point. First, if a star has a strong stellar wind, some of the gas blowing away from the star can pass through the Lagrangian point and be captured by the other star. Second, if an evolving star expands so far that it fills its Roche lobe, then matter will flow through the Lagrangian point onto the other star. Mass transfer driven by a stellar wind tends to be slow, but mass transfer driven by an expanding star can occur rapidly.

BOX 13–2

Beware of Local Supernovae

Although supernovae are rare events, they are very powerful and could affect life on planets orbiting nearby stars. In fact, supernovae explosions long ago may have affected earth's climate and the evolution of life.

If a supernovae occurred within about 50 ly of the earth, the human race would have to abandon the surface and live below ground for at least a few decades. The burst of gamma rays and high-energy particles from the supernova explosion could kill many life-forms and could cause serious genetic damage in others. The only way we could avoid this radiation would be to move our population into tunnels below the earth's surface. Of course, if a supernova did occur, we would not have time to dig enough tunnels.

Even if we could survive in tunnels long enough for the radioactivity on the earth's surface to subside, we might not like the earth when we emerged. Genetic mutation induced by radioactivity could alter plant and animal life so seriously that we might not be able to support our population. After all, we humans depend almost totally on grass for food. The basic human foods—milk, butter, eggs, wheat, tomatoes, lettuce and meat (Big Macs, in other words)—are grass and similar vegetation processed into different forms. Seafood is merely processed ocean plankton and plant life, which might also be altered or damaged by a local supernova explosion. Even if surface life survived the radiation, damage to the delicate upper layers of our atmosphere might alter the climate dramatically.

Local supernovae have been suggested as a possible cause for occasional climate changes and extinctions in Earth's past. It is possible that a supernova does occur near Earth every few hundred million years. Such an explosion has been suggested as a speculative explanation of the extinction of the dinosaurs.

However, we seem to be fairly safe for the moment. No star within 50 ly is known to be a massive giant capable of exploding as a supernova. One candidate, the peculiar variable star η Carinae (Figure 13–14), is poorly understood, but some astronomers suggest it is a massive star losing its envelope and approaching final collapse sometime in the next few million years. At a distance of about 3700 ly, we should have a ringside, although presumably safe, seat. Of course, a cooling white dwarf teetering on the edge of the Chandrasekhar limit could collapse and produce a Type I supernova explosion. Even a nearby white dwarf would be quite faint and thus might go unnoticed until it collapsed.

FIGURE 13–14 The η Carinae nebula is a spectacular site of star formation. The gas is excited by a number of hot bright stars, one of which, η Carinae itself, is apparently massive and highly variable. Some astronomers suspect that it will become a supernova within the next few million years. (©1969 National Optical Astronomy Observatories/Cerro Tololo.)

RECYCLED STELLAR EVOLUTION

Mass transfer between stars can affect the evolution of the stars in surprising ways. In fact, this is the explanation of a problem that puzzled astronomers for many years.

In some binary systems, the less massive star has become a giant, while the more massive star is still on the main sequence. If more massive stars evolve faster than lower-mass stars, how does the low-mass star in such binaries manage to leave the main sequence first? This is called the Algol paradox after the binary system Algol. (See Figures 10–20 and 10–21).

Mass transfer explains how this could happen. Imagine a binary system that contains a 5 $M_{\odot}$ star and a 1 $M_{\odot}$ companion (Figure 13–15). The two stars formed at the same time, so the more massive star will evolve faster and leave the main sequence first. When it expands into a giant, it can fill its Roche lobe and transfer matter to the low-mass companion. Thus, the massive star could evolve into a lower-mass star, and the companion could gain mass and become a massive star still on the main sequence. Thus, we might find a system such as Algol containing a 5 $M_{\odot}$ main-sequence star and a 1 $M_{\odot}$ giant.

The evolution of close binary stars could result in one of the stars having its outer layers "peeled" away. A massive star expanding to become a giant could lose its outer layers to its companion and then collapse to form a lower-mass peculiar star. A few such peculiar stars are known. Of course, if the massive star explodes as a supernova, we would see the peculiar type of supernova called a peeled supernova.

Another exotic result of the evolution of close binary systems is the merging of the stars. We see many binaries in which both stars have expanded to fill their Roche surfaces and spill mass out into space. If the stars are close enough together and if they expand rapidly enough, theorists believe that the two stars could merge into a single, rapidly rotating giant star. Most giants rotate slowly because they conserved angular momentum as they expanded, but examples of rapidly rotating giants are known. Inside

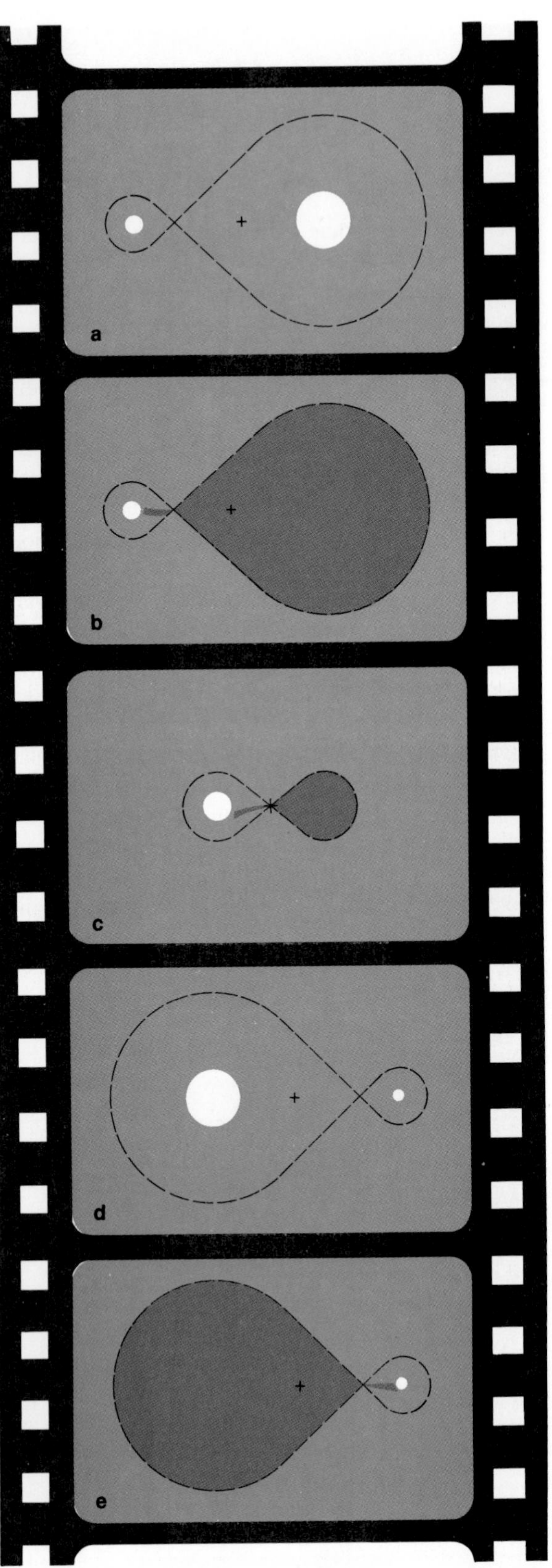

FIGURE 13–15 The evolution of a close binary system. As the more massive star evolves (a), it fills its Roche lobe (b) and begins to transfer mass through the Lagrangian point to its companion. The companion grows more massive (c), and the star losing mass collapses (in this example) into a white dwarf (d). The companion, now a massive star, evolves into a giant and begins transferring mass back to the white dwarf (e).

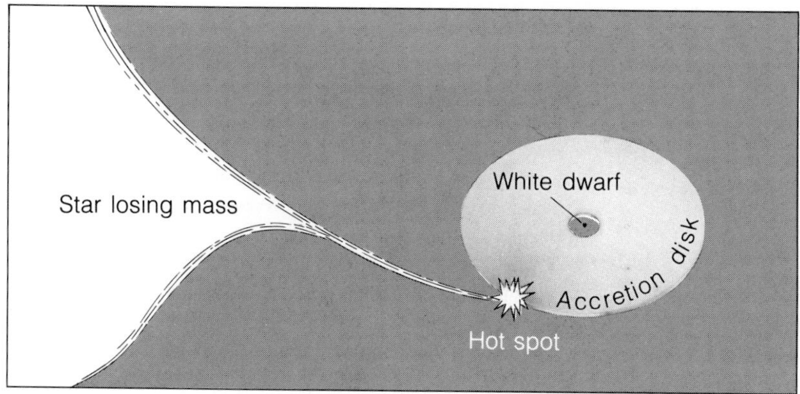

FIGURE 13–16 Matter flowing into a small star like a white dwarf cannot fall directly into the star. It must form a whirling accretion disk. The impact of matter flowing into the disk can create a hot spot on the disk.

the distended envelope of such a star, the cores of the two stars could continue to orbit each other until friction slows them down and they sink to the center.

Yet another exotic possibility can arise if the massive star in a binary system transfers mass to its companion and then collapses to form a white dwarf. Such a system can later become a site of tremendous explosions when the companion begins to transfer mass back to the white dwarf. To see how these explosions occur, we must consider how mass falls into a star.

ACCRETION DISKS

We say that matter flows through the Lagrangian point and falls onto the white dwarf, but the matter cannot fall directly into a white dwarf because of the conservation of angular momentum. Instead, it falls into a whirlpool around the white dwarf. For a common example, consider a bathtub full of water. Gentle currents in the water give it some angular momentum, but its slow circulation is not apparent until we pull the stopper. Then, as the water rushes toward the drain, conservation of angular momentum forces it to form a whirlpool. This same effect forces gas falling into a white dwarf to form a whirling disk of gas called an **accretion disk** (Figure 13–16).

Two important things happen in an accretion disk. First, the gas in the disk grows very hot due to friction and tidal forces. The disk also acts as a brake, ridding the gas of its angular momentum and allowing it to fall into the white dwarf. The temperature of the gas in the inner parts of an accretion disk can exceed 1,000,000 K and emit intense X-rays. In addition, the matter falling inward from the accretion disk can cause a violent explosion, if it accumulates on a white dwarf.

NOVAE

Nova is Latin for "new," and in astronomy it refers to the appearance of what seems to be a new star. A nova can appear in the sky and brighten in a few days, and then fade back to obscurity during the next few months (Figure 13–17). A nova, however, is not a new star, but the eruption of an old star, a white dwarf.

Nova explosions appear to be caused by the transfer of matter from a normal star, through an accretion disk, onto the surface of a white dwarf. Because the matter comes from the surface of a normal star it is rich in unfused fuel, mostly hydrogen, and when it accumulates on the surface of the white dwarf, it forms a layer of unprocessed fuel. As the layer deepens, it becomes denser and hotter until the hydrogen fuses in a sudden explosion.

The nova explosion blows off the surface of the white dwarf in a shell of gas traveling thousands of kilometers per second. Although it contains only about 0.0001 $M_{\odot}$, this expanding shell can become 100,000 times more luminous than the sun. As the shell grows larger and less dense, it cools, and the nova fades.

The explosion hardly disturbs the white dwarf and its companion star. Mass transfer quickly resumes, and a new layer of fuel begins to accumulate. How fast the fuel builds up depends on the rate of mass transfer. According to this theory, some novae might need 1000–100,000 years to accumulate another explosive layer. Others might need only a few weeks. **Recurrent novae** are stars that erupt every few dozen years, and **dwarf novae** undergo small novalike explosions every few days or weeks. Although there are important differences between these types of stars, they appear to be related to the same process—accretion into a disk. For example, observations of dwarf novae such as Z Camelopardalis show irregular flickering

FIGURE 13–17 Nova Cygni 1975 photographed near maximum when it was second magnitude (top) and later when it had declined to about eleventh (bottom). (Lick Observatory photographs.)

believed to be due to the impact of matter on the hot spot at the edge of its accretion disk (Figure 13–16).

Not all novae have accretion disks. For example, Nova Cygni 1975 reached second magnitude in August 1975 (Figure 13–17) and for a short time was visible to the naked eye. Although it has faded now, continuing studies show that the white dwarf has a magnetic field more than 20 million times stronger than earth's. The field is so strong it forces the ionized gas from the companion star to flow directly into the magnetic poles, located near the white dwarf's equator, where the impact heats it to 10^8 K. The rapid rotation of the white dwarf makes it slightly variable as these hot spots cross the visible side of the star. No doubt, nature has other surprises waiting for astronomers studying the evolution of binary stars.

In this chapter we have traced the evolution of stars, both as single objects and as members of binary systems. We have found that all stars end in one of three final states—white dwarfs, neutron stars, and black holes. We have considered white dwarfs, but we have not discussed neutron stars and black holes. Those objects will be the subject of the next chapter.

SUMMARY

When a star's central hydrogen-fusing reactions cease, its core contracts and heats up, igniting a hydrogen-fusing shell and swelling the star into a cool giant. The contraction of the star's core ignites helium first in the core and later in a shell. If the star is massive enough, it can eventually fuse carbon and other elements.

If a star's mass lies between about 0.4 and 3 $M_{\odot}$, its helium core becomes degenerate before the helium ignites. In degenerate gas, pressure does not depend on temperature so there is no pressure–temperature thermostat to control the reactions. As a result, the core explodes in a helium flash. All of the energy produced is absorbed by the star. A similar event happens when carbon ignites in stars between about 3 and 9 $M_{\odot}$, except that it is much more powerful. This carbon detonation is one possible cause of supernovae.

How a star evolves depends on its mass. Stars less massive than about 0.4 $M_{\odot}$ are completely mixed and have very little hydrogen left when they die. They cannot ignite a hydrogen shell or a helium core, so they become white dwarfs—degenerate stars about the size of the earth—containing no nuclear reactions. Our galaxy does not seem old enough for any of these stars to have produced white dwarfs.

Medium-mass stars between about 0.4 $M_{\odot}$ and 3 $M_{\odot}$ become giants and fuse helium but cannot fuse carbon. They swell very large as their carbon cores contract and eventually eject their surface layers to form planetary nebulae. Such nebulae are expanding hollow shells of gas ionized by the high-temperature core of the star left at the center. These planetary nebula nuclei eventually collapse to form white dwarfs.

Because stars lose mass through stellar winds, the mass of a star can decrease as it evolves. Thus, it seems possible for a star with a main-sequence mass of 6 $M_{\odot}$ to lose enough mass to produce a planetary nebula and a white dwarf. No white dwarf can exist if it is more massive than the Chandrasekhar limit of 1.4 $M_{\odot}$, so the most massive stars cannot collapse into white dwarfs. Presumably, they collapse to form other compact objects—neutron stars or black holes.

The most massive stars can get hot enough to ignite carbon and other nuclear fuels. Eventually they develop iron cores, which cannot release energy through fusion reactions, and collapse to produce supernova explosions. The reason a collapsing star explodes is not well understood, but it seems to be blasted apart by a shock wave rising from the interior. Type II supernovae appear to be caused by the collapse of a massive star, but Type I supernovae may be caused by the collapse of a white dwarf, which has gained enough mass to exceed the Chandrasekhar limit. The ejected envelope of the star produces a supernova remnant.

Evolving stars in close binary systems can transfer matter from one star to the other, producing an accretion disk around the star gaining mass. This can dramatically alter the evolution of the stars. If the star gaining mass is a white dwarf, the added fuel on its surface can explode as a nova explosion.

NEW TERMS

red dwarf
planetary nebula
compact object
black dwarf
Chandrasekhar limit
carbon detonation
supernova, Type I and Type II
supernova remnant
synchrotron radiation
Roche surface
Lagrangian point
accretion disk
nova
recurrent nova
dwarf nova

QUESTIONS

1. Why can't the lowest-mass stars become giants?
2. Presumably all the white dwarfs we see in our galaxy were produced by sunlike stars of medium mass. Why couldn't any of the white dwarfs we see have been produced by the deaths of the lowest-mass stars?
3. What leads us to believe that stars can lose mass?
4. What kind of spectrum does the gas in a planetary nebula produce? Where does it get the energy to radiate?
5. The coolest stars we see at the center of planetary nebulae are about 25,000 K. Why don't we see planetary nebulae containing cooler central stars? (HINT: What kind of photons excite the gas in a planetary nebula?)
6. As white dwarfs cool, they move toward the lower right in the H–R diagram (Figures 13–3 and 9–12), maintaining constant radius. Why don't they contract as they cool?
7. All white dwarfs are about the same mass—somewhere around 1 $M_{\odot}$. Why?
8. Why does a massive star explode when it develops an iron core?
9. What could cause a Type I supernova explosion?
10. How do supernova remnants produce synchrotron radiation?
11. Why is a star in a close binary system unable to swell into a full size giant star? Why can a star become a giant if it is in a binary system with a large separation?
12. How could mass transfer onto a white dwarf produce a nova? A Type I supernova?

PROBLEMS

1. How long would it take for a star of 0.4 $M_{\odot}$ to produce a white dwarf? (HINT: see Box 11–3.)
2. The Ring nebula in Lyrae is a planetary nebula with an angular diameter of 72 seconds of arc and a distance of 5000 ly. What is its linear diameter? (HINT: See Box 3–1.)
3. If the Ring nebula is expanding at a velocity of 15 km/sec, typical of planetary nebulae, how old is it? (HINT: 1 ly = 9×10^{12} km and 1 y = 3.15×10^{7} sec.)
4. Suppose that a planetary nebula is 1 pc in diameter and the Doppler shifts in its spectrum show that it is expanding at 30 km/sec. How old it is it? (HINT: See Problem 3.)
5. A planetary nebula photographed 20 years ago and photographed today has increased its radius by 0.6 seconds of arc. If Doppler shifts in its spectrum show that it is expanding at a velocity of 20 km/sec, how far away is it? (HINT: First figure out how many parsecs it has increased in radius in 20 years. Then see Box 3–1.)
6. The Crab nebula is now 1.35 pc in radius and is expanding at 1400 km/sec. About when did the supernova occur?
7. Doppler shifts in the spectra of the Veil nebula (Figure 13–10) show that it is now expanding at 70 km/sec. If it is increasing its radius by 0.03 seconds of arc per year, how far away is it? (HINT: How many kilometers will its radius increase in 1 year? See Box 3–1.)
8. The Veil nebula is now 2.6° in diameter and lies about 500 pc distant. If it is 50,000 years old, what was its average velocity of expansion? (HINT: Find its radius in parsecs first.)
9. The supernova remnant Cassiopia A (Color Plate 19) is expanding in radius at a rate of about 0.5 seconds of arc per year. Doppler shifts show the velocity of expansion is about 5700 km/sec. How far away is the nebula?
10. Cassiopia A has a radius of about 2.5 minutes of arc. If it is expanding at 0.5 seconds of arc per year, when did the supernova explosion occur? (There is no record of a supernova being seen at that time.)

RECOMMENDED READING

BALICK, BRUCE "The Shaping of Planetary Nebulae." *Sky and Telescope 73* (Feb. 1987), p. 125.

CROSWELL, KEN "FG Sagittae: One Piece of the Puzzle." *Astronomy 11* (Oct. 1983), p. 74.

DE VAUCAULEURS, G. "The Supernova of 1885 in Messier 31." *Sky and Telescope 70* (Aug. 1985), p. 115.

KAFATOS, MINAS, and ANDREW G. MICHALITSIANOS "Symbiotic Stars." *Scientific American 251* (July 1984), p. 84.

KALER, JAMES B. "Planetary Nebulae and Stellar Evolution." *Mercury 10* (July/Aug. 1981), p. 114.

———. "Bubbles from Dying Stars." *Sky and Telescope 63* (Feb. 1982), p. 129.

———. "Planetary Nebulae and the Death of Stars." *American Scientist 74* (May/ June 1986), p. 244.

KWOK, SUN "Not with a Bang but a Wimper." *Sky and Telescope 63* (May 1982), p. 449.

LEMONICK, MICHAEL D. "Supernova." *Time 129* (23 March 1987), p. 60.

MALIN, DAVID F. "A Look at Some Unstable Stars." *Sky and Telescope 63* (Jan. 1982), p. 221.

MARAN, S. P. "A Nonconforming Supernova." *Natural History 90* (May 1981), p. 78.

———. "The Origin of the Crab Nebula." *Natural History 91* (Oct. 1982), p. 20.

———. "Star Burst." *Natural History 92* (July 1983), p. 26.

MORRISON, NANCY D., and STEPHEN GREGORY "What Makes Massive Stars Explode." *Mercury 15* (May/June 1986), p. 77.

MULLAN, DERMOTT "Caution! High Winds Beyond this Point." *Astronomy 10* (Jan. 1982), p. 74.

NICASTRO, ANTHONY J. "White Dwarfs: Big Things in Small Packages." *Astronomy 12* (July 1984), p. 6.

PACZYNSKI, BOHDAN "Binary Stars." *Science 225* (20 July 1984), p. 275.

REDDY, FRANCIS "Supernova: Still a Challenge." *Sky and Telescope 66* (Dec. 1983), p. 485.

SCHORN, RONALD A. "A Supernova in Our Backyard." *Sky and Telescope 73* (April 1987), p. 382.

SEEDS, M. A. "Stellar Evolution." *Astronomy 7* (Feb. 1979), p. 6. Reprinted in *Astronomy: Selected Readings.* ed. M. A. Seeds. Menlo Park, Calif.: Benjamin/Cummings, 1980, p. 95.

———. "The Wink in the Demon's Eye." *Astronomy 8* (Dec. 1980), p. 66.

SEWARD, FREDERICK, D. PAUL GORENSTEIN, and WALLACE H. TUCKER "Young Supernova Remnants." *Scientific American 253* (Aug. 1985), p. 88.

STENCEL, ROBERT E. "Mass Loss from Stars." *Astronomy 7* (Nov. 1979), p. 78.

STEPHENSON, F. R., and D. H. CLARK "Historical Supernovas." *Scientific American 234* (June 1976), p. 100.

THEOKAS, ANDREW "The Dance of the Double Stars." *New Scientist 107* (19 Sept. 1985), p. 48.

TRIMBLE, VIRGINIA "White Dwarfs: The Once and Future Suns." *Sky and Telescope 72* (Oct. 1986), p. 348.

———. "Exploding Stars, Superbubbles, and the HEAO Observations." *Mercury 13* (Sept./Oct. 1984), p. 130.

TUCKER, W. "Supernovae, Dinosaurs, and Us." *Mercury 9* (July/ Aug. 1980), p. 95.

WALI, KAMESHWAR C. "Chandrasekhar vs. Eddington—An Unexpected Confrontation." *Physics Today 35* (Oct. 1982), p. 33.

WALLERSTEIN, G., and S. WOLFF "The Next Supernova." *Mercury 10* (March/April 1981), p. 44.

WILLIAMS, ROBERT E. "The Shells of Novas." *Scientific American 244* (April 1981), p. 120.

WINGET, DONALD E. "ZZ Ceti Stars: Variable White Dwarfs." *Sky and Telescope 64* (Sept. 1982), p. 216.

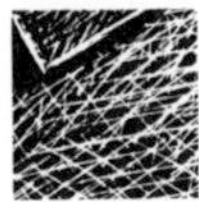

CHAPTER 14

NEUTRON STARS AND BLACK HOLES

However a star dies, gravity assures us that its last remains must eventually reach one of three final states—white dwarf, neutron star, or black hole. These objects, often called compact objects, are small, high-density monuments to the power of gravity. Every star faces an ultimate collapse into such an object.

We discussed white dwarfs in the preceding chapter. In this chapter, we will compare the theoretical predictions of the existence of neutron stars and black holes to observations. The question is, Do neutron stars and black holes really exist? The search for neutron stars and black holes has been one of the greatest adventures of modern astronomy, and it isn't over yet.

Almost anything is easier to get into than out of.

Agnes Allen

14.1 NEUTRON STARS

A **neutron star** is the core of a star that has collapsed to a radius of only 10 km and to a density so high only neutrons can exist. The proof that these theoretical neutron stars actually exist has combined theoretical astrophysics, radio astronomy, and optical astronomy.

PREDICTING THE PROPERTIES OF NEUTRON STARS

The existence of neutron stars and their general properties were predicted theoretically over 50 years ago. The neutron was detected in the laboratory

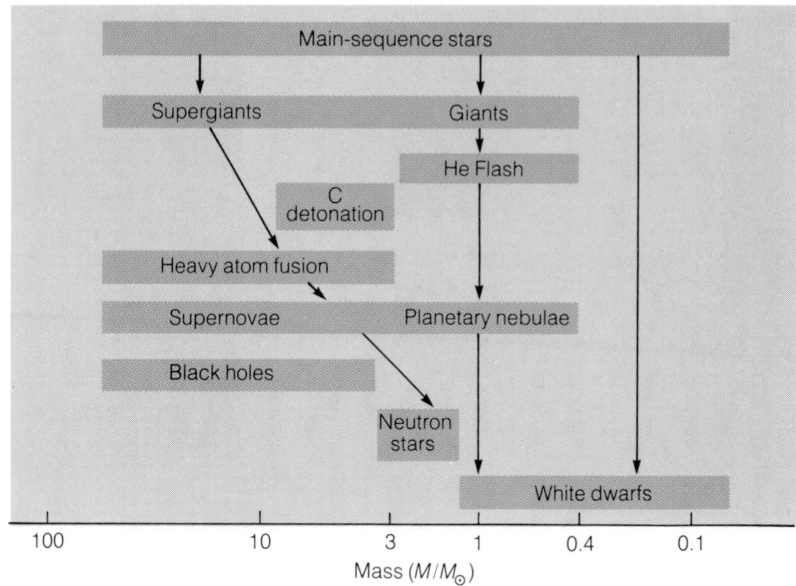

FIGURE 14–1 How a star evolves depends on its mass. The low- and medium-mass stars become white dwarfs. More massive stars can eventually collapse to form neutron stars or black holes, depending on the amount of mass left. Mass loss during the star's evolution can move it to the right in this diagram. (Adapted from Michael A. Seeds, "Stellar Evolution," *Astronomy*, Feb. 1979, p. 6.)

in 1932, and the following year Walter Baade and Fritz Zwicky predicted that supernova explosions might leave behind neutron stars.

If a supernova explosion leaves behind a mass between 1.4 $M_\odot$ and 2–3 $M_\odot$, then gravity will be strong enough to force it to collapse into a neutron star (Figure 14–1). The object will collapse to the density of a white dwarf, which normally would be supported by degenerate electrons, but the gravity is strong enough to overcome the pressure of the degenerate electrons. The object collapses to even higher density, atomic nuclei break up, and electrons are forced to combine with protons to form neutrons.

The neutrons become degenerate and halt the collapse at a radius of 10–15 km. The entire mass of the collapsed star is so compressed that its density is about 10^{14} g/cm^3. On Earth a sugar cube–size lump of this material would weigh 100 million tons. This is roughly the density of an atomic nucleus, and, to a certain extent, we could think of the neutron star as a gigantic atomic nucleus.

Degenerate neutrons can support the weight of a neutron star up to a mass of 2–3 $M_\odot$. This limit is not well-known because the theoretical properties of pure neutron matter are not well understood. Nevertheless, theory predicts that a mass greater than 2–3 $M_\odot$ would not remain a neutron star but would collapse into a black hole. Thus, theory predicts that neutron stars may not be more massive than 2–3 $M_\odot$.

The sudden collapse of a star down to a radius of 10 km can give the star a rapid rotation, a strong magnetic field, and a high temperature. All stars rotate, and as a star collapses it must rotate faster, conserving angular momentum, just as spinning ice skaters rotate faster as they draw in their arms. If the sun collapsed to a radius of 10 km, its rotation would increase from once every 25 days to 1000 times a second. A star collapsing into a neutron star would eject part of its mass and thus lose some angular momentum, but we could expect a newly formed neutron star to rotate 100 times a second.

The collapse to neutron star densities would also produce a powerful magnetic field. The magnetic field in a star is trapped in the ionized gases, and the field would be concentrated and strengthened as the star collapses. We might expect the neutron star to have a magnetic field a billion times stronger than the star had. We will measure the strengths of magnetic fields using the unit **Gauss** (G). The average magnetic field at the surface of the earth is about 0.5 G, and the sun's average field is about 1 G. Some stars have fields as strong as 1000 G, so a newly formed neutron star might have a magnetic field of 10^{12} G, about 3 million times stronger than the strongest field that can be produced in laboratories on Earth.

We have seen in earlier chapters that the contraction of a star converts gravitational energy into thermal energy and heats the star. The extreme contraction of a star collapsing into a neutron star should heat it to very high temperatures. A surface temperature of 10^6 K would be typical.

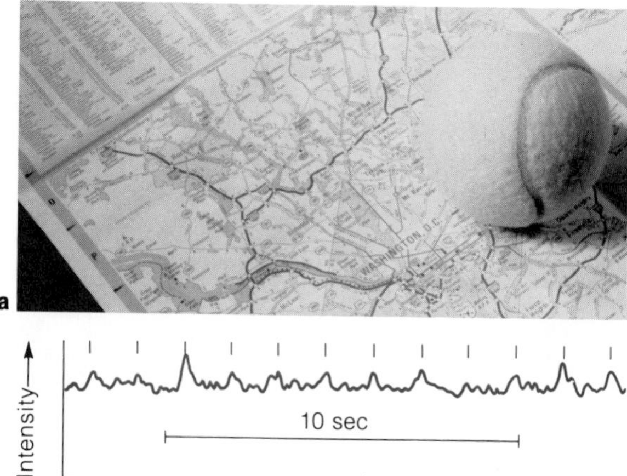

FIGURE 14–2 (a) A tennis ball and a road map illustrate the size of a neutron star. Such a star would fit inside the beltway around Washington, D.C. (b) The first pulsar CP 1919 was discovered November 28, 1967 when its regularly spaced pulses (marked by tics) were noticed in the output of a radio telescope.

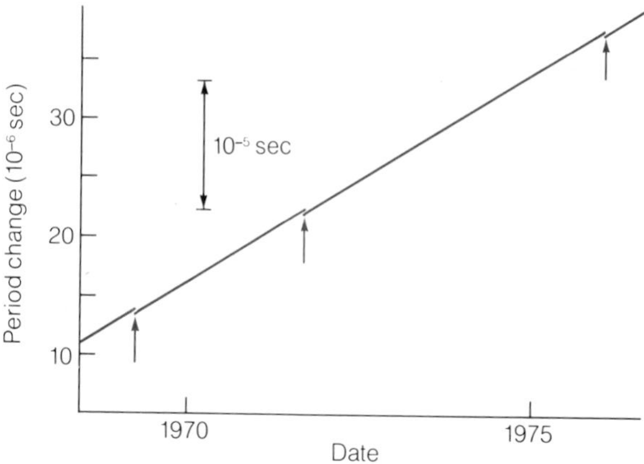

FIGURE 14–3 Glitches occur when a pulsar, gradually slowing down, suddenly speeds up by a small amount. Here three glitches in the pulsation of the Vela pulsar occurred over an 8-year interval.

We might expect such a hot object to be easily visible, but theory predicts otherwise. Neutron stars are small, and their small surface cannot radiate much energy even at such a high temperature. A typical neutron star 10 km in radius with a surface temperature of 10^6 K would radiate only about 15% as much black body radiation as the sun (Boxes 7–1 and 9–3). Also, most of the energy radiated would be in the X-ray part of the spectrum and thus not visible to earth-based telescopes. In the mid-1960s astronomers did not expect to be able to detect neutron stars. Progress came not from theory but from observation.

THE DISCOVERY OF PULSARS

In November 1967, Jocelyn Bell, a graduate student at the University of Cambridge, England, found a peculiar pattern on a paper chart from a radio telescope. Unlike other radio signals from celestial objects, this was a series of pulses (Figure 14–2) with a highly regular period of 1.33730119 seconds. Bell and Anthony Hewish, the director of the experiment, investigated further and found that the signals could not be local. Day after day, they came from the same place among the stars.

Another possibility—that it was a signal from a distant civilization—led them to consider naming it LGM for Little Green Men. But within a few weeks the team found three more objects in other parts of the sky, pulsing with different periods. The objects were clearly natural, and the team dropped the name LGM in favor of **pulsar**—a pulsing radio source.

As more pulsars were found, astronomers argued over their nature. Although the periods were almost as constant as an atomic clock, months of observation showed that many of the pulsars were slowing. Periods were growing longer by a few billionths of a second per day.

In early 1969 radio astronomers were surprised to discover that a pulsar had changed its period suddenly. The Vela pulsar has a period of about 0.089 seconds and is gradually slowing. But in early 1969 the pulsar suddenly sped up. It decreased its period by about 10^{-7} seconds (Figure 14–3). Known as **glitches**, these sudden changes in period have been seen in a small number of pulsars. Something had to regulate the pulsation precisely, had to slow down gradually, but might be subject to occasional glitches.

Pulsars cannot be stars. A normal star is too big to pulse that fast. Nor can a star with a hot spot on its surface spin fast enough to produce the pulses. Some astronomers suggested spinning white dwarfs, but the fastest pulsar then known blinked 30 times a second. Even a white dwarf, the smallest known star, would fly apart if it spun 30 times a second.

The pulses themselves last only about 0.001 second, and that is a clue. If a white dwarf blinked on then off in that interval, we would not see a 0.001 second pulse. The near side of the white dwarf would be about 6000 km closer to us, and light from the near side would arrive 0.022 second before the light from the bulk of the white dwarf. Thus, its short blink would be smeared out into a longer pulse. This is an important principle in astronomy: An object cannot change its brightness appreciably in an interval shorter than the time light takes to cross its diameter. If pulses from pulsars are no longer than 0.001 second, then the object cannot be larger than 300 km in diameter.

Only a neutron star is small enough to be a pulsar. In fact, a neutron star is so small, it can't vibrate slowly enough, but it can spin as fast as 1000 times a second without flying apart. Thus, astronomers began to suspect that pulsars might be spinning neutron stars.

The missing link between pulsars and neutron stars was found in October 1968 when radio astronomers found a pulsar at the heart of the Crab nebula (Figure 13–7).

The Crab nebula is the remnant of the supernova of 1054, and theory predicts that some exploding stars may leave behind neutron stars. The Crab nebula pulsar is evidently such an object.

If we reconsider the theoretical properties of neutron stars and combine them with the observed properties of pulsars, we can devise a model of a pulsar.

A MODEL OF A PULSAR

Pulsar is a misnomer. The periodic flashing of pulsars is linked to rotation, not pulsation. The spinning neutron star emits beams of radiation that sweep around the sky. When one of these beams sweeps over us, we detect a pulse, just as sailors see a pulse of light when the beam from a lighthouse sweeps over their ship. In fact, this model is called the **lighthouse theory.**

The lighthouse theory is generally accepted, and astronomers are becoming more confident of the mechanism that produces the beams. The theory suggests that the neutron star spins so fast and its magnetic field is so strong that it acts like a generator and creates an electric field around itself. This field is so intense that it rips charged particles, mostly electrons, out of the surface near the magnetic poles and accelerates them to high velocity. These accelerated electrons emit photons traveling in the same direction as the electrons. Thus, the photons leave the neutron star in narrow beams shining out of the magnetic poles. If the magnetic axis is inclined with respect to the axis of rotation, as is the case with the earth and most of the planets in the solar system that have magnetic fields,

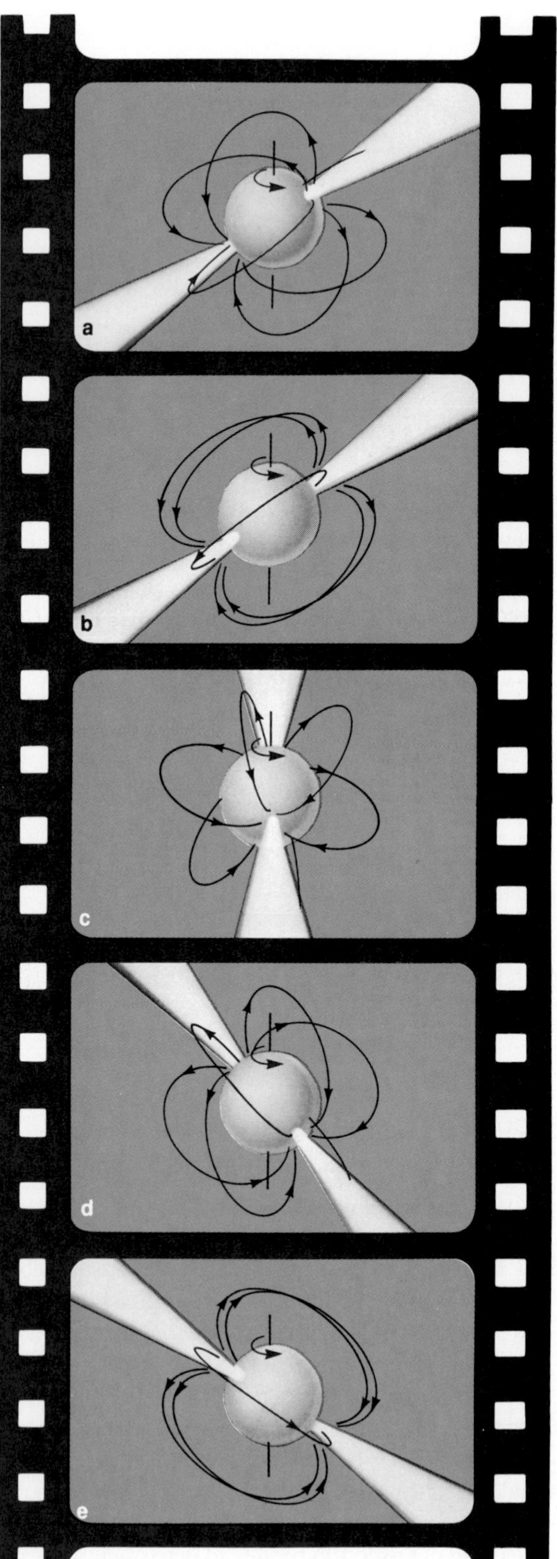

FIGURE 14–4 Beams emerging from the magnetic poles of a neutron star sweep around the sky like beams of light from a lighthouse.

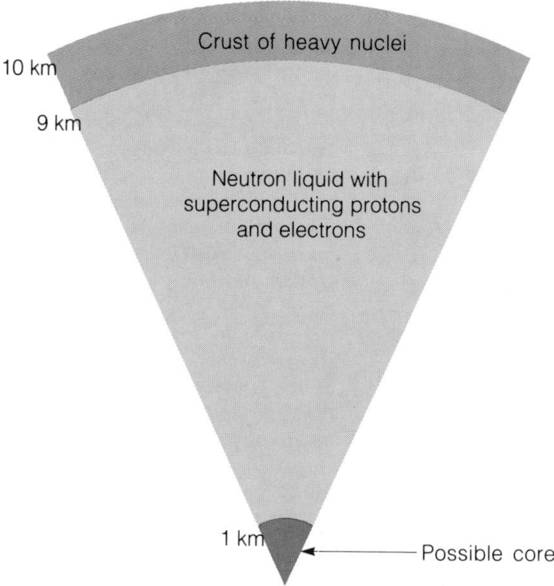

FIGURE 14–5 Theoretical models of a neutron star predict a crystalline crust about 1 km thick and 10,000 trillion times more rigid than steel. Deeper, the pressure produces a neutron liquid with some free protons and electrons. The magnetic field is rooted in this superconducting liquid. A solid core may exist in slightly more massive neutron stars.

then the neutron star will sweep the beams around the sky (Figure 14–4).

About 500 pulsars are known, but there are probably many more. Only when a pulsar's beams sweep over the earth do we detect its presence. In most cases, the beams of radio energy never point to the earth, and the pulsar remains invisible.

Two properties of pulsars support the lighthouse theory. First, many pulsars are slowing down. Their periods are increasing by a few billionths of a second each day. The amount of energy that a pulsar radiates into space per day, about 10^5 times more than the sun, is approximately equal to the amount of energy a spinning neutron star would lose by increasing its period by a few billionths of a second. This explains the source of the pulsar's energy. The spinning neutron star and its intense magnetic field convert energy of rotation into radiation.

Consider the case of the Crab nebula, for instance. In Chapter 13 we saw that the nebula could not continue to emit synchrotron radiation for very long because the high-speed electrons gradually lose energy. Evidently, all the energy to keep the nebula excited and to keep the electrons traveling rapidly enough to generate synchrotron radiation comes from the gradual slowing of the neutron star at the center of the nebula.

The glitch is the second property of pulsars that supports the lighthouse theory. To see how the model could produce glitches, we must consider the internal structure of a neutron star. Theoretical models of neutron stars are difficult to compute because no one knows exactly how pure neutron matter behaves.

Most theorists agree, however, on the three principal layers in a neutron star (Figure 14–5). Near the surface the pressure is less, and atomic nuclei can exist as a rigid crystal layer roughly 1 km thick. This crystalline crust is roughly 10^{16} times more rigid than steel. Although this material is very strong, the tremendous gravity at the surface would prevent neutron star mountains from being more than a few millimeters high.

Below the crust the pressure is high enough to force the material into a liquid state made up mostly of neutrons with some free protons and electrons. Because the protons and electrons can move through the neutron liquid with almost no friction at all, the material is an almost perfect conductor of electricity—a **superconductor.** Apparently, the neutron star's magnetic field is anchored in this spinning mass of superconducting liquid.

Some theories of nuclear particle physics suggest that a neutron star might contain a solid core of massive particles. Whether such a core actually exists depends on details of quantum mechanics and on the behavior of gravity at very high strength. As these details are not understood yet, we cannot be sure that neutron stars have solid cores.

So far as glitches are concerned, the crust is the important layer. The rapid spin of a neutron star tends to flatten it slightly, but as its rotation slows, gravity tries to squeeze it into a more spherical shape. Stresses build in the crust until it breaks in a "starquake," the neutron star equivalent of an earthquake. Once the crust is broken, the neutron star can contract, becoming slightly more spherical, and its rotation increases as it conserves angular momentum. This speed-up decreases the pulse length and we detect a glitch.

This could explain the glitches in the Crab pulsar, but those in the Vela pulsar seem too big and too common. A recent study suggests that, unlike the Crab pulsar, the Vela pulsar has a solid core and that circulation in the liquid interior couples the rotating core and the crust. Thus, astronomers may eventually be able to use glitches to explore the interior of neutron stars just as geologists use earthquakes to probe the interior of the earth.

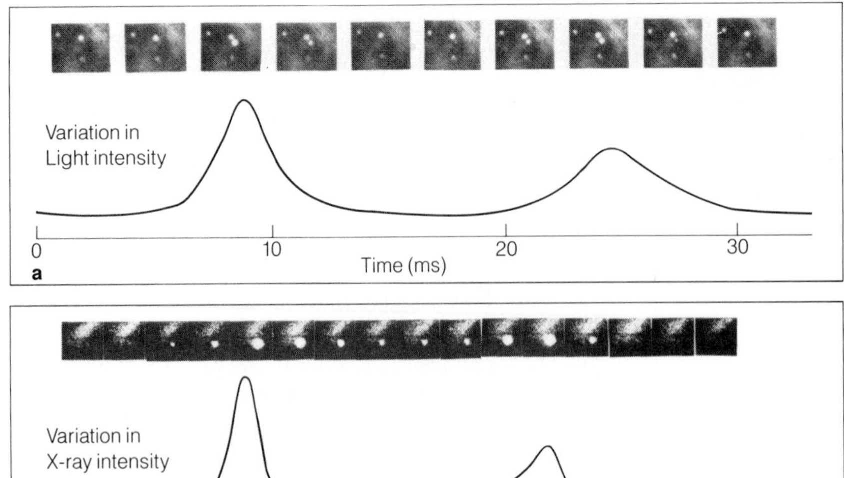

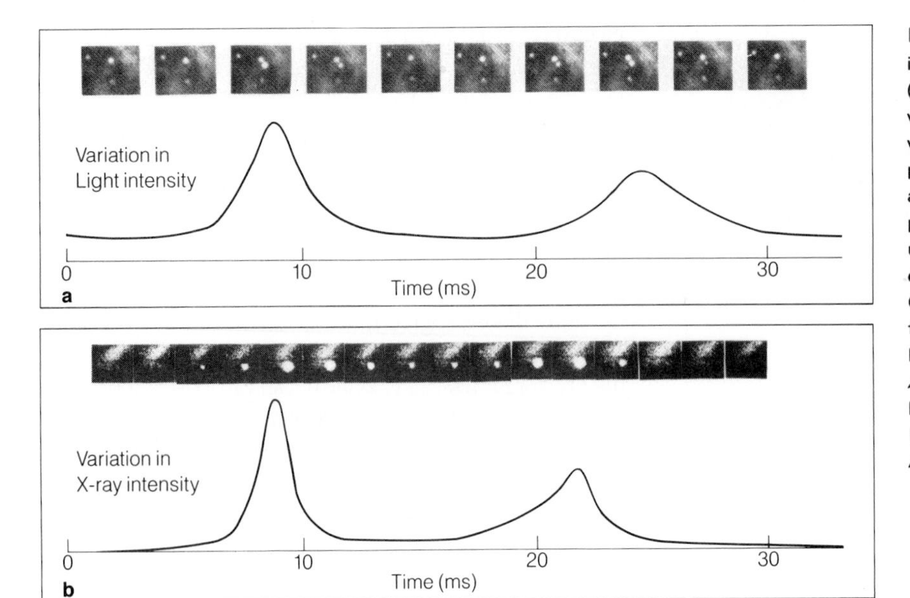

FIGURE 14–6 High-speed images of the Crab nebula pulsar (arrow) show it pulsing at visual wavelengths (a) and at X-ray wavelengths (b). The period of pulsation is 33 milliseconds (ms), and each cycle includes two pulses as its two beams of unequal intensity sweep over the earth. (a, ©AURA, Inc., National Optical Astronomy Observatories, KPNO; b, Courtesy F. R. Harnden, Jr., from *The Astrophysical Journal*, published by the University of Chicago Press, ©1984 The American Astronomical Society.)

THE EVOLUTION OF PULSARS

Theorists are beginning to understand how a pulsar ages. When it first forms, it is spinning very fast, perhaps nearly 100 times a second, and it contains a strong magnetic field. As it converts its energy of rotation into radiation, it gradually slows, and its magnetic field grows weaker. The average pulsar is apparently about 2×10^6 years old, and the oldest is about 10^7 years. Presumably, by the time a pulsar gets older than that, it is rotating too slowly to generate detectable beams.

If a pulsar contains a strong magnetic field and spins very fast, then it is capable of emitting very strong beams of radiation. In addition, it is capable of emitting shorter wavelength photons than older, slower pulsars. The Crab nebula pulsar, the youngest known, emits pulses at radio, infrared, visible, X-ray, and gamma-ray wavelengths (Figure 14–6). In fact, the pulsar has been identified as a star at the center of the nebula long thought to be the remains of a supernova. No one knew it blinked on and off 30 times a second because the blinks blended together when viewed through a telescope or on photographic plates. Not until the star was observed electronically in 1969 were the blinks detected.

At least four pulsars are known that produce visible pulses. All are fast-spinning neutron stars (Table 14–1), and three appear to be young neutron stars located inside supernova remnants. In addition to the Crab nebula, the Vela pulsar produces optical pulses with a period of 0.089 seconds. It has an age of about 11,000 years—young for a pulsar. Another pulsar in the Large Magellanic Cloud (LMC) blinks 20 times a second and produces visible pulses. Its age is unknown but it is located in a supernova remnant. A fourth visible light pulsar, PSR 1937 + 21, pulses very fast but appears to be an old pulsar. (More about PSR 1937 + 21 follows.)

Evidently, we should expect to find the youngest pulsars inside supernova remnants, as in the case of the Crab nebula pulsar. However, not every supernova remnant contains a pulsar, and not every pulsar is located inside a supernova remnant. Many supernova remnants probably contain pulsars whose beams never sweep over the earth. It will be difficult to detect such pulsars. Also, some pulsars have high proper motions, which suggests that a supernova explosion can occur slightly off center or it can disrupt a binary system. Either would give a pulsar a high velocity, and it could leave its supernova remnant quickly. Of course, supernova remnants do not survive more than 50,000 years or so before they mix into the interstellar medium. Because the average pulsar is about 2×10^6 years old, its supernova remnant was lost long ago.

Of course, the best place to look for a very young pulsar would be at the heart of SN1987A. That supernovae ejected large amounts of matter in an expanding shell, which was at first too thick to allow light to escape from the interior. Thus astronomers could not see the neutron star that they believed had been produced by the supernova. But as the shell of gas expanded and cooled it became transparent, and astronomers around the world watched

Table 14–1 Optical pulsars.

Location	Identification	Period (sec)	Age	In Supernova Remnant
Crab	PSR 0531 + 21	0.033	900 years	Yes
Vela	PSR 0833 − 45	0.089	11,000 years	Yes
LMC	PSR 0540 − 69.3	0.050	Unknown	Yes
Vulpecula	PSR 1937 + 21	0.0016	Old	No

to catch the first glimpse of the central neutron star. Finally, on January 18, 1989, astronomers at Cerro Tololo Inter-American Observatory in Chile detected what may be the youngest known pulsar.

Astronomers hoped they would see a pulsar when the debris cleared, but they were unprepared for what their analysis revealed. The brightness of the source was pulsing at 1968.63 pulses per second. One of the scientists said, "We were waiting for the egg to hatch, but when it did, instead of a chicken we got a Ferrari turning over at 120,000 r.p.m.!"

If the object is a true pulsar, the neutron star must be spinning nearly fast enough to fly into pieces. In fact, Doppler shifts observed over the 7-hour observation suggest the object may have a low mass companion. Some think the companion may be material ejected from the rapidly spinning neutron star. Others suggest the pulses are caused, not by the spinning of a neutron star, but by its vibration.

The pulses have not been seen since that night, perhaps because we glimpsed the object through a temporary break in the expanding shell. Thus the existence of the youngest known pulsar remains unconfirmed.

The possible pulsar in SN1987A presents astronomers with a problem. Theory predicts that pulsars are born spinning at no more than 100 times a second and slow as they age. If the object in SN1987A is a true pulsar, it is hard to understand how it got such a rapid spin.

Some of the oldest pulsars are also difficult to explain. An old pulsar should spin slowly and have a weak magnetic field, but PSR 1937 + 21 is a pulsar that pulses 642 times a second and is slowing down only slightly. It has been called the millisecond pulsar because its period is roughly a millisecond. If it slows only slightly, then its magnetic field must be weak, a sign of age. The problem is how an old pulsar could spin so rapidly. It could not have been born spinning so fast because it would have slowed as it aged.

It now appears that the millisecond pulsar is an old neutron star that has gradually lost most of its magnetic field but was once in a binary system. Mass flowing from its companion, like water hitting a mill wheel, spun the old neutron star up to 642 rotations per second. With its old, weak magnetic field, it slows down only gradually and will continue to spin for a very long time.

Observations suggest that the millisecond pulsar is not now part of a binary system. How it could have escaped from a binary and its evolution in general are not well understood.

At least two other very fast pulsars, also called millisecond pulsars, have been found. Both PSR 1855 + 09 (186 pulses per second) and PSR 1953 + 29 (163 pulses per second) are members of binary systems (Table 14–2). To understand such peculiar pulsars, we must consider the binary pulsars.

BINARY PULSARS

Of the hundreds of pulsars now known, a few are located in binary systems. These pulsars are of special interest because we can learn more about them by studying their orbital motion. In a few cases, mass transferred from their companion stars can generate X-rays.

Table 14–2 Binary and millisecond pulsars.

Pulsar (PSR)	Pulse Period (sec)	Orbital Period (days)	Orbital Eccentricity	Mass of Companion ($M_\odot$)	Comment
1937 + 21	0.0016	—	—	—	First millisecond pulsar
1855 + 09	0.0054	12.33	0.00002	0.2–0.4	White dwarf companion detected
1953 + 29	0.0061	117.35	0.0003	0.2–0.4	Binary millisecond pulsar
1913 + 16	0.059	0.32	0.6171	1.4	First binary pulsar
0655 + 64	0.1956	1.03	<0.00005	0.7–1.3	White dwarf companion detected
1831 − 00	0.5209	1.81	<0.005	0.06–0.13	
0820 + 02	0.8649	1232.40	0.0119	0.2–0.4	White dwarf companion detected
2303 + 46	1.0664	12.34	0.6584	1.2–2.5	

The first binary pulsar was discovered in 1974 when astronomers noticed that the pulse period of the pulsar PSR 1913 + 16, normally 0.059 seconds, was changing (Figure 14–7). When plotted against time, the period changes repeated every 7.75 hours.

Thinking of the Doppler shifts seen in the spectra of spectroscopic binaries, the radio astronomers realized that the pulsar had to be in a binary system with a period of 7.75 hours. When the orbital motion of the pulsar carries it away from the earth, we see the period slightly lengthened, just as the wavelength of light emitted by a receding source is lengthened. Then, when the pulsar rounds its orbit and approaches the earth, we see the period slightly shortened. Thus, the Doppler changes in the period of the pulsar allow us to plot a radial velocity curve (Figure 14–7). The shape of this curve resembles that of a single-line spectroscopic binary with an elliptical orbit (Chapter 10).

When we try to analyze the orbital motion of PSR 1913 + 16, we confront the same problems we encounter when we analyze a single-line spectroscopic binary. We cannot detect the companion, so we cannot include its orbital motion, and we cannot find the inclination of the orbit. We do not see any eclipses—the pulsar never disappears behind the companion star—so we can be sure that the orbit is not edge-on. Reasonable assumptions about the size and mass of the companion, combined with detailed analysis of the orbital motion, suggest that the mass of the neutron star is about 2 $M_\odot$.

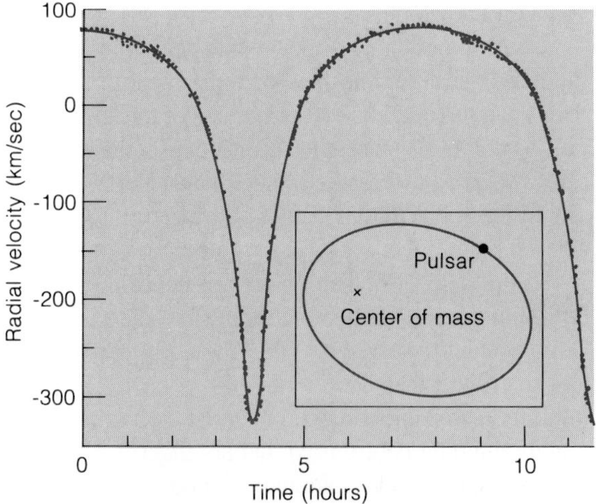

FIGURE 14–7 The period of pulsation of pulsar PSR 1913 + 16 varies over an interval of 7.75 hours. This is caused by the Doppler effect as the pulsar orbits in a binary system. Here, points represent observations of the period converted to radial velocity using the Doppler effect, and the solid line represents a solution based on the elliptical orbit shown in the inset. (Adapted from data by Hulse and Taylor.)

In the years since the discovery of this binary pulsar, astronomers have found that the orbital period is growing shorter. Evidently, the two stars are losing orbital energy and spiraling closer together. This loss of energy is exactly

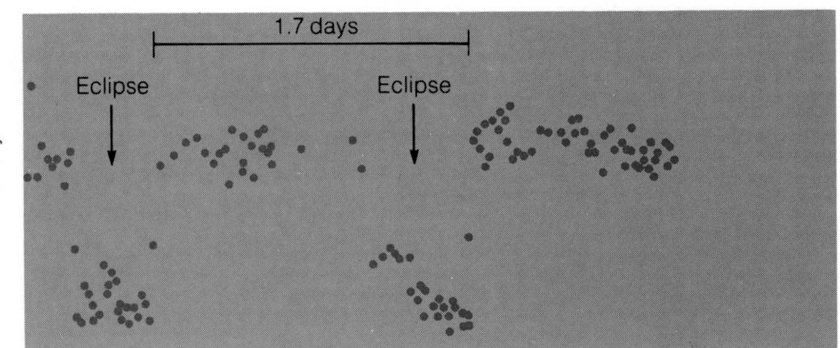

FIGURE 14–8 The X rays from Hercules X-1 disappear as the X-ray pulsar is eclipsed behind its stellar companion.

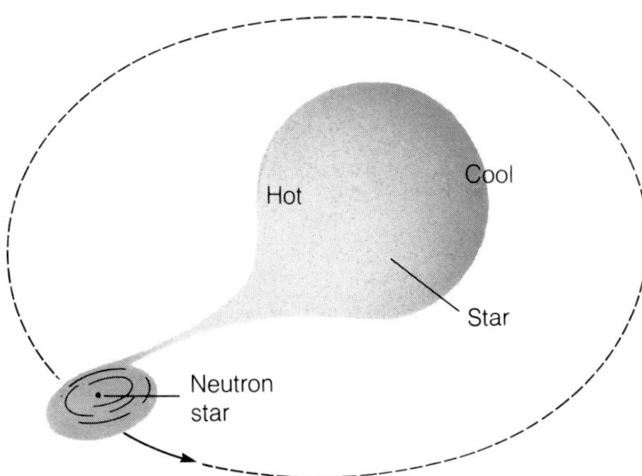

FIGURE 14–9 In Hercules X-1, X rays from the X-ray pulsar heat the near side of the star to 20,000 K. The rotating system shows us the hot side and the cool side of the star alternately, varying the visual brightness.

that predicted by general relativity. The energy lost is apparently carried away by gravitational waves (Box 14–1). Thus, the existence of gravitational waves seems to have been confirmed although laboratory instruments have not yet detected such waves.

The binary pulsar PSR 1913 + 16 is one of seven know binary pulsars (Table 14–2). Such systems are now relatively quiet, but they may have once been violently active. The companion star was probably once a giant star losing mass to the neutron star, and that matter falling into the neutron star would have liberated tremendous energy. A single marshmallow dropped onto the surface of a neutron star from a distance of 1 AU would hit with an impact equivalent to a 3-megaton nuclear warhead. Even a small amount of matter transferred to a neutron star can generate high temperatures and release X rays and gamma rays.

Hercules X-1 is such a system. X rays from Hercules X-1 arrive in pulses with a period of 1.2372253 seconds. This suggests a spinning neutron star. But the pulses vanish for 5.8 hours every 1.7 days (Figure 14–8), and that suggests an eclipsing binary. To understand Hercules X-1, we must combine its X ray properties with its appearance at visible wavelengths.

Optical astronomers know this object as the peculiar variable star HZ Herculis, a complex binary system. The normal star in the system has expanded to fill its Roche surface and is now transferring matter to its neutron star companion at about 10^{-9} $M_{\odot}$ per year. Theorists do not agree on whether the matter forms an accretion disk, but it eventually becomes trapped by the neutron star's powerful magnetic field and falls in at the magnetic poles. The impact of this matter creates temperatures as high as 10^8 K and generates beams of radiation, mostly X rays, shining out of the poles. The spinning neutron star sweeps these beams around the sky once every 1.2372253 seconds.

HZ Herculis varies in brightness with a period of 1.7 days, and its spectrum shows Doppler shifts that change with that period. It appears to be a 2 $M_{\odot}$ star with a surface temperature of about 7000 K. The star is variable because X rays from the neutron star heat one side to about 20,000 K. As the system revolves, we see first the hot side and then the cool side of the star (Figure 14–9). This explains the disappearance of the pulses every 1.7 days. The system is an eclipsing binary, and the neutron star disappears behind the companion once each orbit.

Hercules X-1 is a very complex system not yet well-understood. For example, sudden outbursts of gamma rays have been detected. They may be related to uneven flow of matter into the accretion disk.

Irregular flow of matter into accretion disks may be very common in such systems. The **quasi-periodic objects** (QPOs) are X-ray sources that pulse irregularly (perhaps

BOX 14–1

Gravitational Radiation

Einstein's theory of general relativity, published in 1916, explained that a gravitational field was actually a curvature of space-time induced by the presence of a mass. From this Einstein predicted that any change in the distribution of the mass would change the curvature of space-time and that change would move at the velocity of light like ripples across a pond. These waves are now known as **gravitational waves.**

University of Maryland physicist Joseph Weber made the first attempts to detect gravitational waves in the 1960s. He used strain gauges attached to a large metal cylinder (Figure 14–10a). Passing gravitational waves should set the cylinder vibrating by tiny amounts, which should be detected by the strain gauges. To avoid confusion from local vibrations, he placed another cylinder in a laboratory near Chicago. Only simultaneous events in both cylinders could be gravitational waves. His results were exciting but controversial, and nearly a dozen similar experiments have failed to detect gravitational waves with certainty.

Many groups around the world are now building gravitational wave detectors. A team at Stanford has built a detector similar to Weber's. A 5-ton aluminum bar 10 feet long and 3 feet across is suspended on delicate springs inside a tank of liquid helium that cools it to near absolute zero. It must be cooled because the motion of a single atom at room temperature is 3000 times more energetic than a gravitational wave. At low temperatures, atoms in the bar move more slowly and gravitational waves are easier to detect. The Stanford experiment is still being developed and will be paired with a similar detector at Louisiana State University. Other cooled bars are being built around the world.

A team at Caltec is building a detector that uses laser beams instead of massive bars. In its final form, two evacuated tunnels 1.2 m in diameter and 4 km long will be built in the Mohave Desert. A laser beam split into two parts is sent down each tunnel and reflected back to the starting point. If the light waves in the laser beam start out in step, and if the lengths of the two tunnels are not disturbed, the light waves will reach the finish line in step. But a passing gravitational wave will slightly alter the length of one tunnel or the other, and that will throw the light waves slightly out of step at the finish line. The detector in the Mojave Desert will be linked to a similar detector in Maine to form LIGO, the Laser Interferometer Gravitational Wave Observatory. Similar laser beam detectors are under construction or being planned in Europe and Asia.

Any event that changes the distribution of mass can produce gravitational waves—snapping your fin-

flicker would be a better word) with periods of about 0.03 seconds. One theory is that blobs of matter in the accretion disk gradually approach the rapidly rotating neutron star until its magnetic field begins to rip gas out of the rapidly orbiting blobs. As the matter flows into the neutron star, the pulsation dies away until a new irregularity in the disk spirals closer to the spinning neutron star.

Binary systems containing neutron stars can evolve in peculiar ways. For example, the X-ray source 4U1820-30 is a binary system in which a neutron star orbits a white dwarf with a period of only 11 minutes. The separation between the stars is about a third the distance between the earth and moon. To explain how such a very close pair could originate, theorists suggest that a neutron star collided with a giant star and went into an orbit *inside* the star. The neutron star ate away the giant's envelope from the inside, leaving the white dwarf behind. Matter still flows from the white dwarf into an accretion disk around the neutron star.

Objects known as **bursters** produce irregularly spaced bursts of X rays (Figure 14–11) and appear to be related to accretion onto a neutron star. X-ray bursters are thought to be neutron stars gaining mass from a companion star. According to one model, as the gas—mostly hydrogen—trickles onto the surface of the neutron star, it fuses into helium. When the helium layer reaches a depth of about 1 m, it explosively fuses into carbon and produces an X-ray burst. If the explosion is delayed slightly, the helium layer grows slightly deeper and, when the burst does

FIGURE 14–10 Joseph Weber works on an early gravitational wave detector weighing 3 tons. Strain gauges around the middle of the aluminum cylinder are intended to detect vibrations stimulated by passing gravitational waves. (Courtesy Joseph Weber.)

gers, for example—but only rapid events involving large masses produce gravitational waves strong enough to detect. For example, binary neutron stars should gradually spiral into each other, and their merger would emit a burst of gravitational waves. Such events occur in our galaxy every few hundred years, and the new detectors should be able to detect these mergers in nearby galaxies. Thus experimenters expect to detect several events per year.

Supernovae are another source of gravitational waves. In the final collapse of a massive star's core, as much as 0.1 percent of its mass may be radiated away as gravitational waves. Once operational, the new instruments should detect a number of supernovae each year among neighboring galaxies. A network of detectors in different parts of the world would be able to pinpoint the direction of the gravitational waves and warn optical astronomers where to train their telescopes to catch the supernova eruption in its earliest stages.

The new generation of gravitational wave detectors are not built yet, but the experimenters are now confident that gravitational waves do exist. The gradual loss of orbital energy by the binary pulsar PSR 1913 + 16 confirms that gravitational waves are real. When the new detectors begin operation, they will give us an entirely new way to observe the universe.

occur, it is larger than average. Thus the model explains why delayed bursts tend to be larger (Figure 14–11). Notice the similarity between this model and the accreting white dwarf model of novae.

14.2 BLACK HOLES

If an object with a mass greater than 2–3 $M_{\odot}$ collapses, no known force in nature can stop it. The object reaches white dwarf density, but the degenerate electrons cannot support the weight, and the collapse continues. When the object reaches neutron star density, the degenerate neutrons cannot support the weight, and the collapse goes on. The object quickly becomes smaller than an electron. No one knows of anything that can stop gravity from squeezing the object to zero radius and infinite density.

As an object shrinks, its density and the strength of its gravity at its surface increases, and when an object shrinks to zero radius, its density and gravity become infinite. Mathematicians call such a point a **singularity**. Physically we have difficulty thinking about infinite density and zero radius objects, but, even if such objects exist, they may not be visible to us. Theory predicts they will be hidden inside a region of space called a **black hole**.

Although black holes are difficult to discuss without general relativity and sophisticated mathematics, we can use common sense and some simple physics to see why they form. Finding the velocity we need to escape from

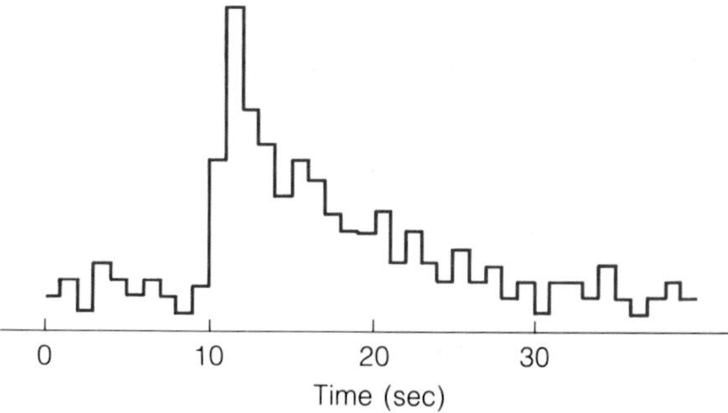

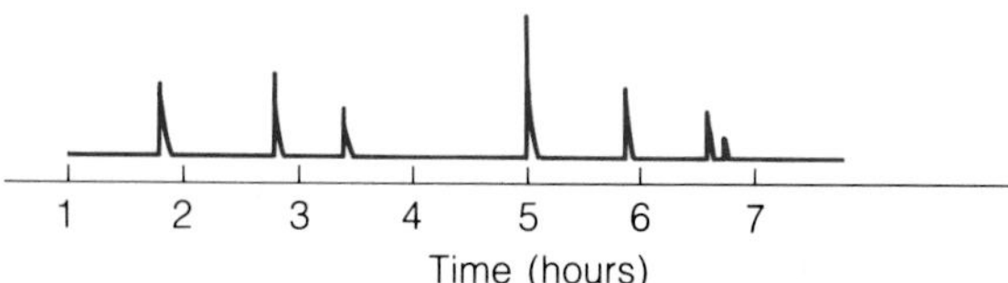

FIGURE 14–11 (a) A typical X-ray burst rises to full intensity in a few seconds and then fades in about 20 seconds. (b) The longer the interval since the preceding pulse, the brighter the next pulse will be. This suggests the continuous accumulation of energy, which is occasionally released in bursts.

the gravity around a celestial body will help explain how black holes were first predicted theoretically and how they might be detected.

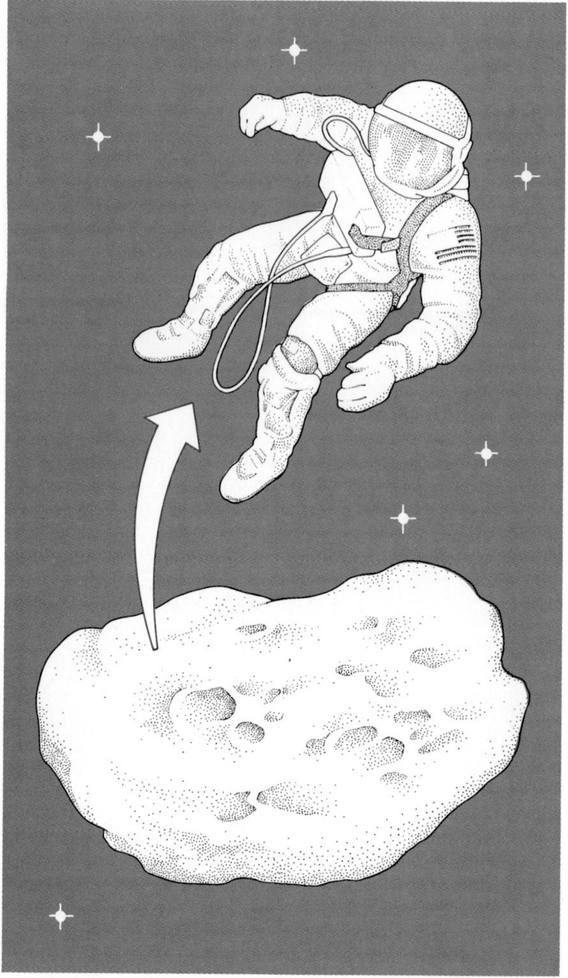

FIGURE 14–12 Escape velocity, the velocity needed to escape from a celestial body, depends on mass. The escape velocity at the surface of a very small body would be so low we could jump into space. The escape velocity of the earth is much larger, about 11 km/sec (25,000 mph).

ESCAPE VELOCITY

Suppose we threw a baseball straight up. How fast must we throw it if it is not to come down? Of course gravity would pull back on the ball, slowing it, but if the ball were traveling fast enough to start with, it would never come to a stop and fall back. Such a ball would escape from the earth. As we discovered in Box 5–2, the escape velocity is the initial velocity an object needs to escape from a celestial body (Figure 14–12).

Whether we are discussing a baseball leaving the earth or a photon leaving a collapsing star, the escape velocity depends on two things: the mass of the celestial body and the distance from the center of mass to the escaping object. If the celestial body had a large mass, its gravity would be strong and we would need a high velocity to escape, but if we began our journey farther from the center of mass, the velocity needed would be less. For example, to escape from earth a spaceship would have to leave its surface at 11 km/sec (25,000 mph), but if we could launch spaceships from the top of a tower 1000 miles high, the escape velocity would be only 8.8 km/sec (20,000 mph). If we could make an object massive enough and/or small enough, its escape velocity could be greater than the speed of light. Such an object could never be seen because light could never leave it.

Rev. John Mitchell, a British amateur astronomer, was the first person to realize that Newton's laws of gravity and motion contained this implication. In 1783 he pointed out that an object 500 times the radius of the sun but of the same density would have an escape velocity greater

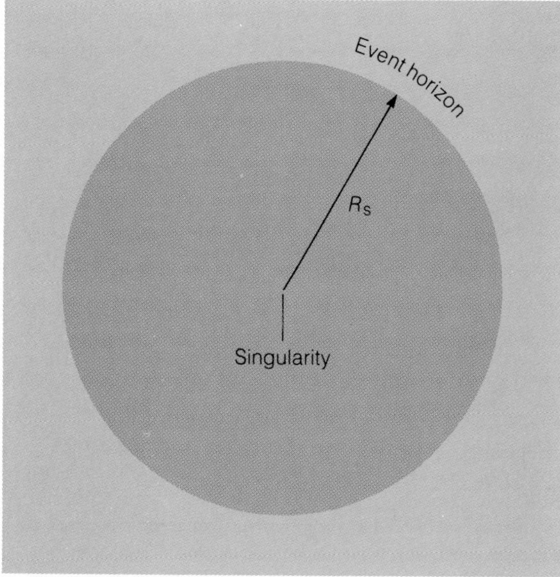

FIGURE 14–13 A black hole forms when an object collapses to a small size (perhaps to a singularity) and the escape velocity in its neighborhood is so great light cannot escape. The boundary of this region is called the event horizon because any event that occurs inside is invisible to outside observers. The radius of the region is R_s the Schwarzschild radius.

than the speed of light. Then, "all light emitted from such a body would be made to return towards it." Mitchell had discovered the black hole.

SCHWARZSCHILD BLACK HOLES

If the core of a star collapses and contains more than 2–3 $M_\odot$, it will continue to collapse to a singularity and form a black hole. Some theorists believe that a singularity is impossible and that when we better understand the laws of physics, we will discover that the collapse halts before diameter zero.

It makes little difference to us, however. If the object becomes small enough, the escape velocity nearby is so large that no light can escape. We can receive no information about the object or about the volume of space near it, and we refer to this volume as a black hole. The boundary of this region is called the **event horizon** (Figure 14–13), because any event that takes place inside the surface is invisible to an outside observer. To see how such a region can exist we must consider general relativity.

In Chapter 5 we saw how general relativity explained a gravitational field as a curvature of space-time. That is, the presence of mass curves space-time in a way that we experience as gravity. Einstein published the equations that described his theory in 1916, and almost immediately the astronomer Karl Schwarzschild found a way to solve the equations to describe the gravitational field around a single, nonrotating, electrically neutral lump of matter. That solution contained the first general relativistic description of a black hole; nonrotating, electrically neutral black holes are now known as Schwarzschild black holes.

Schwarzschild's solution showed that if matter is packed into a small enough volume, then space-time is curved back on itself. Objects can still follow paths that lead into the region, but no path leads out, so nothing can escape, not even light. Thus, the inside of the black hole is totally beyond the view of an outside observer. The event horizon is the boundary between the isolated volume of space-time and the rest of the universe, and the radius of the event horizon is called the **Schwarzschild radius R_s**—the radius within which an object must shrink to become a black hole.

The Schwarzschild radius depends on only the mass of the object. A 1 $M_\odot$ black hole would have a Schwarzschild radius of about 3 km, a 2 $M_\odot$ object a radius of 6 km, and so on.

Any object could be a black hole if it were smaller than its Schwarzschild radius. For example, if we could squeeze the earth to a radius of about 1 cm, its gravity would be so strong it would become a black hole. Fortunately, the earth will not collapse spontaneously into a black hole because its mass is less than the critical mass of 2–3 $M_\odot$. Only exhausted stellar cores more massive than this can form black holes under the sole influence of their own gravity. In this chapter, we are interested in black holes that might originate from the deaths of massive stars. These would have masses larger than 3 $M_\odot$. In the following chapters, we will meet black holes whose masses might exceed 10^6 $M_\odot$.

Do not think of black holes as giant vacuum cleaners that will pull in everything in the universe. A black hole is just a gravitational field, and at a reasonably large distance its force is quite small. If the sun were replaced by a 1 $M_\odot$ black hole, the orbits of the planets would not change at all.

BLACK HOLES HAVE NO HAIR

Theorists who study black holes are fond of saying, "Black holes have no hair." By that they mean that once matter forms a black hole, it loses almost all of its normal properties. A black hole made of a collapsed star will be indistinguishable from a black hole made from peanut butter or fake-fur mittens. Once the matter is inside the event horizon, it may retain only three properties—mass, angular momentum, and electrical charge.

The Schwarzschild black hole is represented by a solution to Einstein's equations for the special case where the object has only mass. Schwarzschild black holes do not rotate or have charge. The solutions for rotating or charged black holes (or for rotating, charged black holes) are more difficult and have been found in only the last few decades. Generally, rotating, charged black holes are similar to Schwarzschild black holes.

It does not seem that astronomers need worry about charged black holes because stars, whose collapse presumably forms black holes, cannot have large electrostatic charges. Suppose that we could give the sun a large positive charge. It would begin to repel protons in its corona and attract electrons, and it would soon return to neutral charge. Thus, we should expect black holes to be electrically neutral.

But everything in the universe seems to rotate, and collapsing stars spin rapidly as they conserve angular momentum. Thus, we should probably expect black holes to have angular momentum. In 1963 the New Zealand mathematician Roy P. Kerr found a solution to Einstein's equations that describes a rotating black hole. This is now known as the **Kerr black hole**.

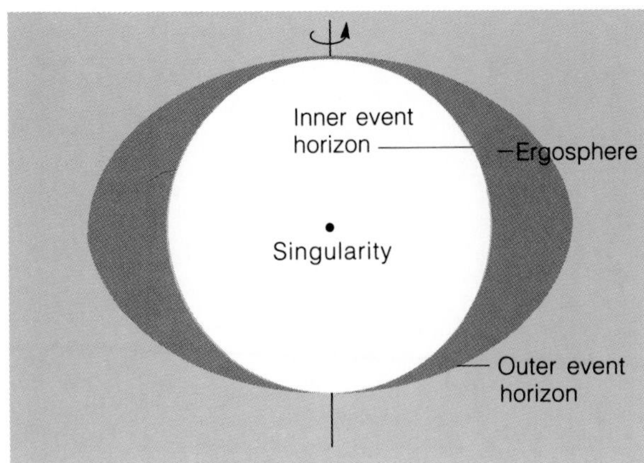

FIGURE 14–14 The Kerr black hole consists of a rotating mass at the center, surrounded by two event horizons. The outer event horizon marks the boundary within which an observer cannot resist being dragged around the black hole with space-time. The inner event horizon marks the boundary from within which an observer cannot escape. The volume between the event horizons is known as the ergosphere.

The rotation of a Kerr black hole splits the event horizon into two concentric surfaces, which touch at the poles (Figure 14–14). The region between these surfaces is known as the **ergosphere**. To understand the nature of the ergosphere, we must imagine that we approach a spinning black hole in a spaceship. Far away from the event horizon, all is well, but as we draw closer we begin to feel as if we were spinning around the black hole. We might feel dizzy, although our spaceship is keeping us from rotating when compared to the distant stars. What we feel is the space-time near the black hole being dragged around with the spinning mass.

The outer event horizon marks the boundary beyond which we may not go without being dragged around with the rotating black hole. We might still be able to escape falling into the black hole, but we could not resist being dragged along with space-time within the ergosphere. Of course, the inner event horizon marks the boundary from within which we could never escape.

The word *ergosphere* comes from the Greek word *ergo* meaning "work" because the rotating space-time in the ergosphere can do work on a particle—that is, the particle can gain energy. In particular, the Kerr solution shows that a particle that enters the ergosphere can break into two pieces, one falling into the black hole and the other escaping with more energy than it had when it entered. Thus, energy can be extracted from a rotating black hole, and, as a result, the black hole slows its rotation very slightly.

The Kerr solution is a fascinating bit of theoretical physics, but it has an important application in astronomy. Almost certainly black holes rotate, and matter falling into black holes must pass through the ergosphere. Thus, we may eventually find situations where energy is extracted from rotating black holes.

LEAPING IN

Because we cannot observe a black hole up close, let us examine their peculiar properties by using our imaginations. Let's imagine that we leap, feet first, into a Schwarzschild black hole.

If we were to leap into a black hole of a few solar masses from a distance of an astronomical unit, the gravitational pull would not be very large, and we would fall slowly at first. Of course, the longer we fell and the closer we came to the center, the faster we would travel. Our wristwatches would tell us that we fell for about 65 days before we reached the event horizon.

Our friends who stayed behind would see something

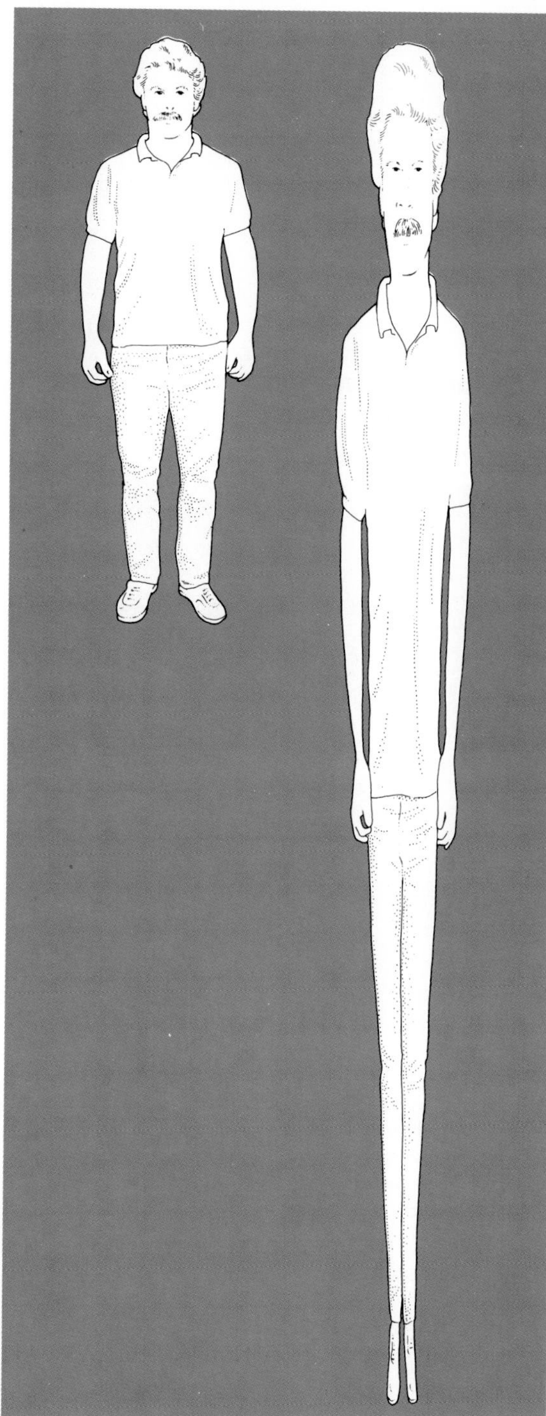

FIGURE 14–15 Leaping feet first into a black hole. A person of normal proportions (left) would be distorted by tidal forces (right) long before reaching the event horizon around a typical black hole of stellar mass. Tidal forces would stretch the body lengthwise while compressing it horizontally. Friction from this distortion would heat the matter to high temperatures.

different. They would see us falling slower as we came closer to the event horizon because, as explained by general relativity, clocks slow down in curved space-time. This is known as **time dilation.** In fact, our friends would never actually see us cross the event horizon. To them we would fall slower and slower until we seemed hardly to move. Generations later, our descendents could focus their telescopes on us and see us still inching closer to the event horizon. We, however, would have sensed no slowdown and would conclude that we had crossed the event horizon after only about 65 days.

Another relativistic effect would make it difficult to see us with normal telescopes. As light travels out of a gravitational field, it loses energy, and its wavelength grows longer. This is known as the **gravitational red shift.** Although we would notice no effect as we fell toward the black hole, our friends would need to observe at longer and longer wavelengths in order to detect us.

These relativistic effects seem peculiar, but other effects would be quite unpleasant. Imagine again that we are falling feet first toward the event horizon of a black hole. We would feel our feet, which would be closer to the black hole, being pulled in more strongly than our heads. This is a tidal force, and at first it would be minor. But as we fell closer, the tidal force would become very large (Figure 14–15). Another tidal force would compress us as our left side and our right side both fell toward the center of the black hole. For any black hole with a mass like that of a star, the tidal forces would crush us laterally and stretch us longitudinally long before we reached the event horizon. The friction from such severe distortions of our bodies would heat us to millions of degrees, and we would emit X rays and gamma rays.

Some years ago a popular book suggested that we could travel through the universe by jumping into a black hole in one place and popping out of another somewhere far across space. That makes good science fiction, but tidal forces would make it an unpopular form of transportation even if it worked.

Our imaginary leap into a black hole is not entirely frivolous. We now know how to find a black hole. Look for a strong source of X rays. It may be a black hole into which matter is falling.

THE SEARCH FOR BLACK HOLES

Do black holes really exist? Earlier in this chapter we saw how theory predicted the existence of neutron stars and how the discovery of pulsars confirmed that neutron stars do exist. Can we do the same thing for black holes? Theory

Table 14–3 Three black hole candidates.

Object	Location	Companion Star	Orbital Period	Mass of Compact Object
Cyg X-1	Cygnus	O supergiant	5.6 days	6–11 $M_\odot$
LMC X-3	Dorado	B3 main-sequence	1.70 days	10 $M_\odot$
A0620-00	Monoceros	K main-sequence	7.75234 hours	More than 3.18 $M_\odot$

predicts that they exist, but can we find one or more objects that are obviously black holes?

If matter were falling into a black hole, it should become very hot and radiate X rays. Of course, an isolated black hole will not have much matter falling inward, but a black hole in a binary system could gain mass from its companion, form a very hot accretion disk, and emit intense X rays. Thus, to find possible black holes we must look at X-ray binary systems. Some, like Hercules X-1 will contain neutron stars, but some may contain black holes.

We can tell the difference between a neutron star and a black hole in an X-ray binary if we can find the mass. Neutron stars must be less than 3 $M_\odot$. Any compact object in an X-ray binary with a mass of 3 $M_\odot$ or more must be a black hole.

So far astronomers have suggested three X-ray binaries as strong candidate black holes (Table 14–3). In each case, the X rays are generated by matter flowing into a hot accretion disk around a compact object more massive than we would expect for a neutron star.

Cygnus X-1, the first X-ray object discovered in Cygnus, is apparently a hot supergiant O star and a compact object orbiting each other with a period of 5.6 days. The X rays are generated when matter in the supergiant's stellar wind flows into an accretion disk around the compact object (Figure 14–16). Although the compact object is invisible, we can estimate its mass from the Doppler shifts in the spectrum of the visible supergiant. If we assume that the supergiant has a mass of about 30 $M_\odot$, typical for a star of that temperature and luminosity, then the compact object must have a mass of 6–11 $M_\odot$ with 9 $M_\odot$ being most probable. This is well above the maximum mass for a neutron star.

Nevertheless, some observers have argued that the supergiant is not a normal star, that a third star may be present in the system, and so on. These possibilities mean that the compact object could, under certain peculiar circumstances, be a neutron star. Thus, Cyg X-1 is a good candidate for a black hole but is not conclusive.

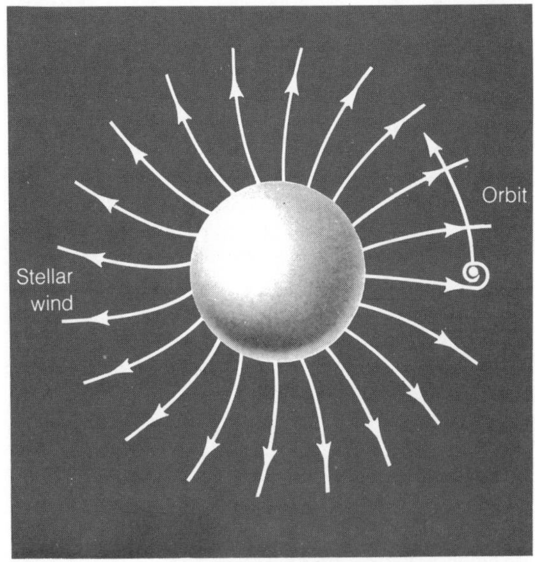

FIGURE 14–16 The X-ray source Cygnus X-1 is generally believed to be a supergiant O star in a binary system with a black hole. Some of the gas from the O star's stellar wind flows into an accretion disk around the black hole, and the hot accretion disk emits the X rays we detect.

LMC X-3 is another black hole candidate. (The letters LMC refer to the Large Magellanic Cloud, a small galaxy near our own Milky Way galaxy.) The compact object in LMC X-3 has a mass of at least 7 $M_\odot$ and probably 10 $M_\odot$, twice the mass of its companion. The system is a powerful source of X rays, and the visible light from the companion star is variable because the gravity of the compact object distorts the star into an egg shape. As the distorted star rotates, its total brightness varies. Yet the compact object is not detectable at visual wavelengths. Any reasonably normal star of that mass would be visible, so most astronomers now accept LMC X-3 as a black hole.

A third candidate is the X-ray system A0620-00 in the constellation Monoceros. It is only 1000 pc away and contains an ordinary main-sequence K star. The star and compact object orbit each other once every 7.75234 hours, and the light of the K star is variable because it is distorted into an egg shape by its companion. Analysis shows that the mass of the compact object must be greater than 3.18 $M_{\odot}$ and may be more than 7.3 $M_{\odot}$. Although this is slightly lower than the mass of the compact objects in Cyg X-1 and LMC X-3, the K star seems normal in every way, and thus it is easier to analyze the system and find the mass of the compact object. Thus, A0620-00 seems a likely candidate for a black hole.

Theory appears to be confirmed. Black holes do exist. In later chapters, we will discuss the possibility that black holes containing millions, perhaps billions, of solar masses lie in the cores of galaxies. For now, however, we will consider an object that is of interest, not because it might contain a black hole, but because it shows how a compact object can produce tremendous energy when matter falls in at a high rate.

PERSPECTIVE

SS 433—THE PROTOTYPE ENERGY MACHINE

SS 433* is a peculiar X-ray binary system, but it is unique in that it is converting gravitational energy into oppositely directed jets of radiation and matter, blasting away from the system at one-fourth the speed of light. Astronomers commonly refer to the unknown energy sources that lie at the centers of galaxies as **energy machines.** Here, we will study SS 433 as a prototype energy machine, which may give us clues to an entire family of objects that produce energy in oppositely directed jets of radiation and matter.

We have already seen such low-energy phenomena in the bipolar flows and disks around protostars (see Chapter 11) and higher-energy effects in the accretion disks discussed in this chapter. We will discover similar, though larger and more violent, events when we discuss peculiar galaxies (Chapter 17). Yet again we will see the same mechanism at work when we discuss the origin of the solar system (Chapter 19). In a sense, SS 433 is a prototype that illustrates how nature solves the same problem, the accretion of matter into a gravitational field, in the same way each time it arises.

SS 433 is a rather faint star that is also a source of X rays. Astronomers studying the optical spectrum found three sets of spectral lines. One set is blue-shifted, one set is red-shifted, and one set is almost unaffected. If these are interpreted as Doppler shifts, they indicate that one part of the object is approaching Earth at about 15 percent of the speed of light while another part is receding at a similar velocity. The stationary lines indicate that a third part of the object is not moving at a high velocity.

Over a number of months, astronomers found that the red- and blue-shifted lines move back and forth across each other with a period of 164 days, much like the lines of a spectroscopic binary (Figures 14–17 and 10–13). Careful measurements of the so-called stationary lines showed that they too move, but with a small velocity and with a period of about 13 days. This lower velocity variation suggests that the object may be part of a binary system. *(continued)*

*Object number 433 is in the Stephenson–Sanduleak catalog of stars showing H-alpha emission.

Perspective: SS433—the Prototype Energy Machine *(continued)*

Three facts hint that SS 433 contains a compact object. Astronomers found that the object is a source of radio, infrared, X-ray, and gamma-ray radiation—clear evidence of the high energies we expect from compact objects. Also, the object lies near the center of a spherical shell of gas called W50. This shell is about 100 pc in diameter and seems to be a supernova remnant. Finally, SS 433 changes in brightness rapidly and irregularly. All of this suggests a compact object—a neutron star or a black hole.

At this point we can't be sure what the compact object is, but most astronomers now agree on the following general model for the system (Figure 14–18). A compact object and a star orbit each other with a period of 13.1 days. Mass flows in a torrent from the normal star into a hot accretion disk around the compact object. The high-temperature disk emits powerful beams of radiation perpendicular to the plane of the disk. As the in-falling matter overloads the disk, the pressure of radiation inside the disk blasts small blobs of gas out of the disk along the beams at about 25 percent of the velocity of light. (Because the disk ejects blobs of gas repeatedly each time the disk overloads, some theorists have slyly dubbed it the "burping disk hypothesis.") Thus, the system produces bipolar jets.

This model explains the observed emission lines and Doppler shifts. The emission lines in the spectrum are apparently produced by these gas blobs caught in the beams. Blobs in the beam that is directed away from the earth produce red-shifted lines, and blobs in the beam directed toward the earth produce blue-shifted lines.

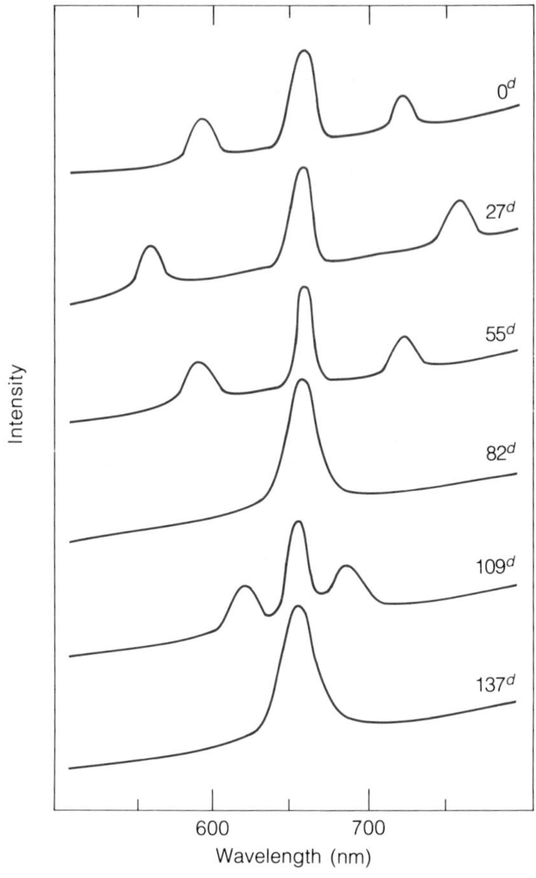

FIGURE 14–17 Six spectra of SS 433 recorded at intervals of about 27 days show three emission lines. The central line is unshifted H-alpha and the other two lines are red- and blue-shifted H-alpha. Compared with Figure 10-13. (These spectra have been simplified for clarity.)

The compact object and its accretion disk precess just as the spinning earth does, and this precession sweeps the beams around a conical path with a period of 164 days (Figure 14–18). This explains the cyclical motion of the emission lines shown in Figure 14–17. The beams never point exactly toward the earth, but when a beam is pointed roughly in our direction, it produces blue-shifted spectral lines. As precession swings the beam away from the earth, the blue shift decreases and eventually becomes a red shift. Thus, the emission lines in the spectrum cross over each other with a period of 164 days.

The unshifted emission lines are believed to come from the companion star or from the hot accretion disk. Thus, the unshifted lines are affected only by the lower-velocity, 13.1 day orbital motion of the system.

Observations at many wavelengths support this model (Figure 14–19). Regular radio maps of W50 show that it has two "ears" projecting on either side. Very long baseline radio maps made with telescopes

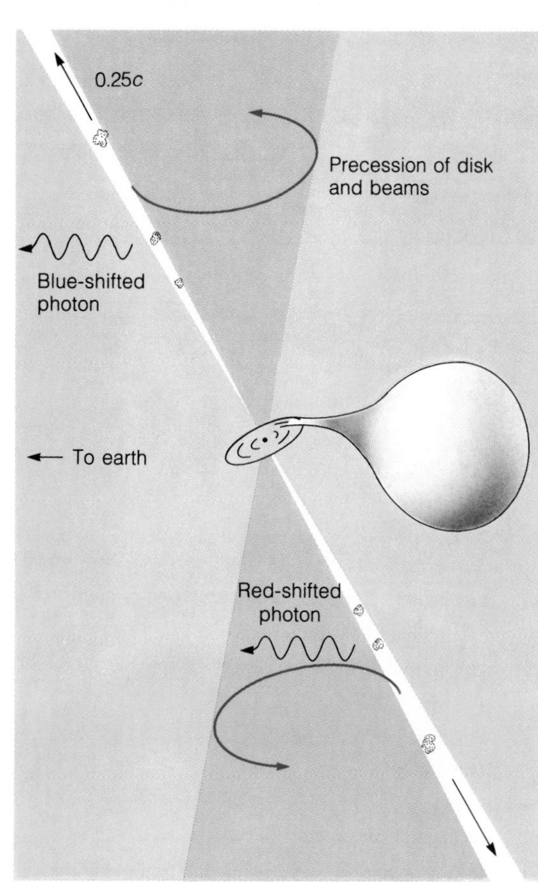

FIGURE 14–18 The generally accepted model of SS 433 includes a compact object with a very hot accretion disk producing beams of radiation with embedded blobs of gas traveling at 25 percent the velocity of light. The precession of the disk swings the beams around a conical path every 164 days.

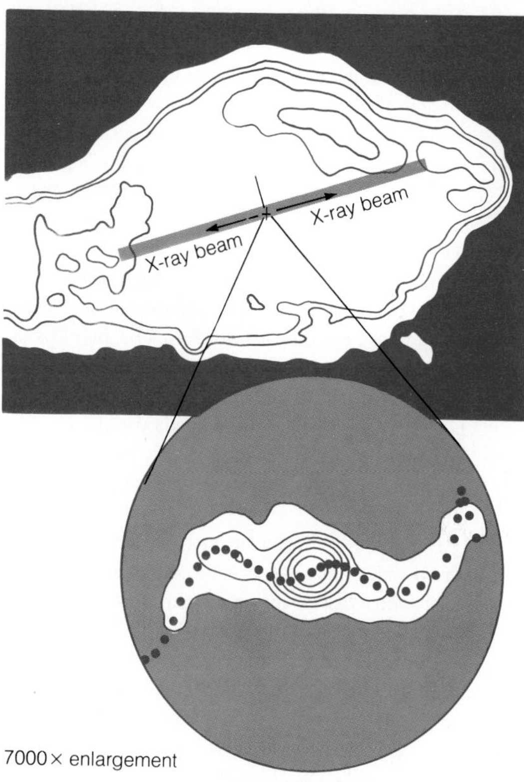

FIGURE 14–19 The radio map at the top shows SS 433 at the center of the radio source W50. The "ears" on W50 extend to east and west and correspond to the direction of the beams detected at X-ray wave-lengths. The inset, about 80 light-days in diameter, is a high-resolution radio map showing the curvature of the beams caused by the precession of the accretion disk. Black dots are the paths of ejected gas clouds predicted by theory. (Adapted from maps by B. Geldzahler and R. Spencer.)

on opposite sides of the earth show that the central source is less than 0.002 second of arc across and that it has elongations pointing toward the ears in W50. X-ray studies made with the Einstein satellite can trace the beams at least 1° from the center, and they too point toward the ears in W50. High-resolution maps made with the MERLIN radio telescope in England have been able to follow individual blobs of gas as they flow outward along the beams.

Thus it is clear. SS 433 is somehow converting the energy of the in-falling matter into beams of radiation and gas of tremendous power. Yet these phenomena are weak compared with the radiation and jets of matter we will discover in the hearts of galaxies (Chapter 17). The mechanism may be the same, with the amount of energy generated being the main difference. If so, SS 433 is a small prototype of nature's greatest energy machines.

SUMMARY

If the remains of a star collapse with a mass greater than the Chandrasekhar limit of 1.4 $M_{\odot}$, then the object cannot reach stability as a white dwarf. It must collapse to the neutron star stage with a radius of about 10 km and a density equal to that of an atomic nucleus. Such a neutron star can be supported by the pressure of its degenerate neutrons. But if the mass is greater than 2–3 $M_{\odot}$, then the degenerate neutrons cannot stop the collapse, and the object must become a black hole.

Theory predicts that a neutron star should rotate very fast, be very hot, and have a strong magnetic field. Such objects have been identified as pulsars, sources of pulsed radio energy. Pulsars are evidently spinning neutron stars, which emit beams of radiation from their magnetic poles. As they spin, they sweep the beams around the sky, and, if the beams sweep over the earth, we detect pulses. This is known as the lighthouse theory. The spinning neutron star slows as it radiates energy into space, and glitches in its pulsation appear to be caused by starquakes in the crust of the neutron star.

A few pulsars have been found in binary systems. One such binary pulsar is losing orbital energy at the rate predicted by the theory of gravitational waves. In some binary systems such as Hercules X-1, mass flows into a hot accretion disk around the neutron star and causes the emission of X rays. In other systems called X-ray bursters, the accumulation of fuel on the neutron star causes periodic outbursts of X rays. SS 433 is an extreme case of such a system; its accretion disk is hot enough to drive blobs of gas away in its beams of radiation. SS 433 is considered a small prototype of the powerful energy machines found in the centers of some galaxies.

If a collapsing star has a mass greater than 2–3 $M_{\odot}$, then it must contract to a very small size—perhaps to a singularity, an object of zero radius. Near such an object, gravity is so strong not even light can escape, and we term the region a black hole. The surface of this region, called the event horizon, marks the boundary of the black hole. The Schwarzschild radius is the radius of this event horizon, amounting to only a few kilometers for black holes of stellar mass.

If we were to leap into a black hole, we would experience peculiar effects. Our friends who stayed behind would see two relativistic effects. Our clocks would slow down relative to our friends' clocks because of time dilation in the strong gravitational field. Also, the gravitational red shift would cause the light we emitted to be shifted to longer wavelengths. As we fell into the black hole, we would feel tidal forces that would deform and heat our mass until we grew hot enough to emit X rays. Any X rays emitted before we crossed the event horizon can escape.

To search for black holes, we must look for binary star systems in which mass flows into a compact object and emits X rays. If the mass of the compact object is greater than 2–3 $M_{\odot}$, then the object is presumably a black hole. Three such objects have been located.

NEW TERMS

neutron star	singularity
Gauss (G)	black hole
pulsar	event horizon
glitch	Schwarzschild radius (R_s)
lighthouse theory	Kerr black hole
superconductor	ergosphere
gravitational wave	time dilation
quasi-periodic object (QPO)	gravitational red shift
burster	energy machine

QUESTIONS

1. Why is there an upper limit to the mass of neutron stars? Why is that upper limit not well known?
2. Explain in detail why we expect neutron stars to be hot, spin fast, and have strong magnetic fields.
3. Why can't we locate neutron stars with visual wavelength telescopes?
4. Why does the short length of pulsar pulses eliminate normal stars as possible pulsars?
5. What observations support the lighthouse theory?
6. According to our model of a pulsar, if a neutron star formed with no magnetic field at all, could it be a pulsar? Why or why not?
7. Why did astronomers first assume that the millisecond pulsar was very young?
8. Why do we suspect that only very fast pulsars can emit visible pulses?
9. How does an X-ray burster resemble a nova?
10. If the sun were replaced by a 1$M_{\odot}$ black hole, how would the earth's orbit change?
11. What do we mean when we say, "Black holes have no hair"?
12. In this chapter we described what would happen if we jumped into a Schwarzschild black hole. From what you have read, what would happen if we jumped into a Kerr black hole?
13. What evidence do we have that black holes exist?
14. What evidence do we have that SS 433 is a binary system?

PROBLEMS

1. If a neutron star has a radius of 10 km and rotates 642 times a second, what is the speed of the surface at the neutron star's equator in terms of the speed of light?
2. Suppose that a neutron star has a radius of 10 km and a temperature of 10^6 K. How luminous is it? (HINT: See Box 9–3.)
3. A neutron star and a white dwarf have been found orbiting each other with a period of 11 minutes. If their masses are typical, what is their average separation? (HINT: See Box 10–1.)
4. If the accretion disk around a neutron star has a radius of 2×10^5 km, what is the orbital velocity of a particle at its outer edge? (HINT: See Box 5–1.)
5. What is the escape velocity from the surface of a typical neutron star? How does that compare with the speed of light? (HINT: See Box 5-2.)
6. If the earth's moon were replaced by a typical neutron star, what would the angular diameter of the neutron star be as seen from the earth? (HINT: See Box 3–1.)
7. If the inner accretion disk around a black hole has a temperature of 10^6 K, at what wavelength will it radiate the most energy? What part of the spectrum is this in? (HINT: See Box 7–1.)
8. What is the orbital period of a bit of matter in an accretion disk 2×10^5 km from a $10\ M_\odot$ black hole? (HINT: See Box 5–1 or 10–1.)
9. If SS 433 consists of a $20\ M_\odot$ star and a neutron star orbiting each other every 13.1 days, then what is their average separation? (HINT: See Box 10–1.)

RECOMMENDED READING

ANDERSON, L. "X-Rays from Degenerate Stars." *Mercury 5* (Sept./Oct. 1976), p. 6.

CAMERON, A. G. W. ed. *Astrophysics Today.* New York: American Institute of Physics, 1984.

CLARK, DAVID H. *The Quest for SS 433.* New York: Viking Penquin Inc., 1986.

DARLING, DAVID "The Quest for Black Holes." *Astronomy 11* (July 1983), p. 6.

———. "Space, Time, and Black Holes." *Astronomy 8* (Oct. 1980), p. 66.

GREENSTEIN, GEORGE "Neutron Stars and the Discovery of Pulsars." *Mercury 14* (March/April 1985), p. 34, and *14* (May/June 1985), p. 66.

———. *Frozen Star.* New York: New American Library, 1984.

HALL, DONALD E. "The Hazards of Encountering a Black Hole." *The Physics Teacher 23* (Dec. 1985), p. 540, and *24* (Jan. 1986), p. 29.

MARAN, S. P. "The Origin of the Crab Nebula." *Natural History 91* (Oct. 1982), p. 20.

MARGON, BRUCE "Relativistic Jets in SS 433." *Science 215* (15 Jan. 1982), p. 247.

PARKER, BARRY "In and Around Black Holes." *Astronomy 14* (Oct. 1986), p. 6.

PACZYNSKI, BOHDAN "Binary Stars." *Science 225* (20 July 1984), p. 275.

PINES, DAVID "Accreting Neutron Stars, Black Holes, and Degenerate Dwarf Stars." *Science 207* (8 Feb. 1980), p. 597.

SCHORN, RONALD A. "Binary Pulsars: Back from the Grave." *Sky and Telescope 72* (Dec. 1986), p. 588.

———. "SS 433—Enigma of the Century." *Sky and Telescope 62* (Aug. 1981), p. 100.

SCHWARTZ, RICHARD "A Hunt for Flashing Stars." *Sky and Telescope 72* (Dec. 1986), p. 560.

SEEDS, M. A. "Stellar Evolution." *Astronomy 7* (Feb. 1979), p. 6. Reprinted in *Astronomy: Selected Readings.* ed. M. A. Seeds. Menlo Park, Calif.: Benjamin/Cummings, 1980, p. 95.

SEWARD, FREDERICK D. "Neutron Stars in Supernova Remnants." *Sky and Telescope 71* (Jan. 1986), p. 6.

SEXL, R., and H. SEXL "Curved Space-Time Near a Neutron Star." *Mercury 9* (March/April 1980), p. 38.

———. *White Dwarfs-Black Holes: An Introduction to Relativistic Astrophysics.* New York: Academic Press, 1979.

SMITH, DAVID H. "The Melodious Pulsar." *Sky and Telescope 66* (Oct. 1983), p. 311.

SPETZ, GARY W. "The Detection of Gravity Waves." *The Physics Teacher 22* (May 1984), p. 282.

STOKES, G. M., and J. J. MICHALSKY. "Cygnus X-1," *Mercury 8* (May/June 1979), p. 60.

SCIENCE FICTION

FORWARD, ROBERT L. *Starquake.* New York: Ballantine Books, 1985.

———. *Dragons Egg.* New York: Ballantine Books, 1986.

NIVEN, LARRY *Neutron Star.* New York: Ballantine Books, 1968.

UNIT 3

THE UNIVERSE

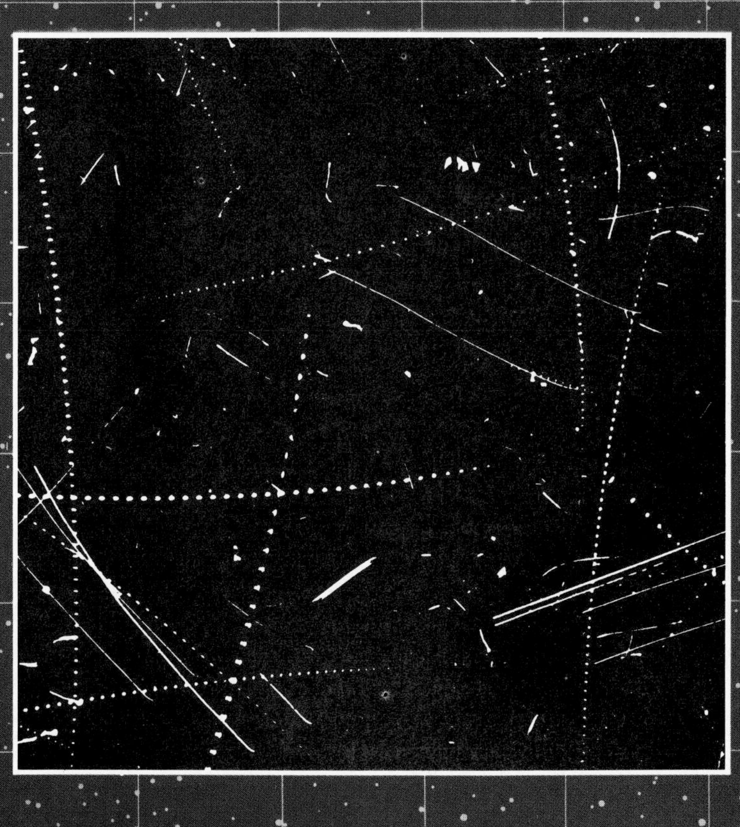

CHAPTER 15

THE MILKY WAY

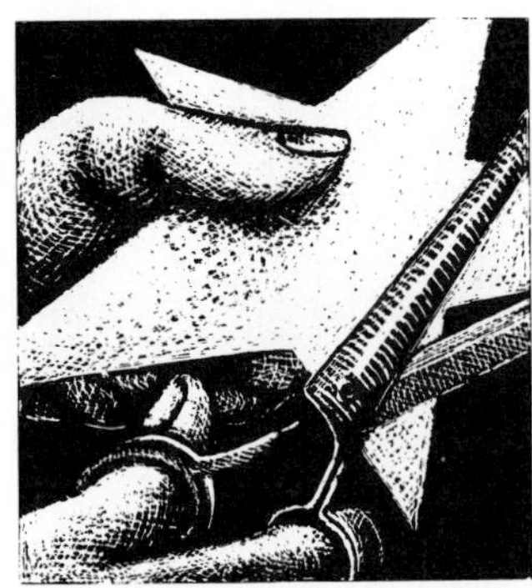

Jane was watching Mrs. Corry splashing the glue on the sky and Mary Poppins sticking on the stars. . . . "What *I* want to know," said Jane, "is this: Are the stars gold paper or is the gold paper stars?" There was no reply to her question and she did not expect one. She knew that only someone very much wiser than Michael could give her the right answer. . . .

P. L. Travers
MARY POPPINS

P. L. Travers wrote her story about the supercalifragilisticexpialidocious nanny in 1934, and at that time no one could answer the question that Jane asked in the quotation above. But by the 1960s, astronomers could assure her that the gold paper was stars, or rather, the elements of which gold paper is made were created inside generation after generation of stars. The story of how the stars made the chemical elements is one of the keys to understanding our Milky Way galaxy.

The ancient Greeks named the faint band of light that stretches around the sky *galaxies kuklos,* the "milky circle." The Romans changed the name a bit to *via lactea,* the "milky road" or "milky way." Today we can recognize the Milky Way as the glow of 200 billion stars whirling in a great wheellike system that includes our sun. Millions of other such systems dot the sky and, drawing on the Greek word for milk, we call them galaxies.

Almost every celestial object visible to our naked eyes is a member of our Milky Way galaxy. Two exceptions are the **Magellanic Clouds**, small irregular galaxies located in the southern sky. Another exception is the Andromeda galaxy, just visible to our unaided eyes as a faint patch of light in the constellation of Andromeda.* Our galaxy probably looks much like the Andromeda galaxy (Figure 15–1).

Unfortunately, we cannot see much of our own galaxy from our position inside it. Clouds of dust block our view, like dark thunderclouds against a sunset (Figure 15–2). Photographs taken from earth can explore only about

*Consult the star charts at the end of this book to locate the Milky Way and the Andromeda galaxy in your night sky.

FIGURE 15–1 The Great Galaxy in Andromeda is very similar to our own Milky Way galaxy. (Lick Observatory photograph.)

FIGURE 15–2 Vast dust clouds (black on this photograph) block our view of the Milky Way. This photograph of the region of Sagittarius looks toward the center of our galaxy (marked by the small box). (Mount Wilson and Las Campanas Observatories, Carnegie Institution of Washington.)

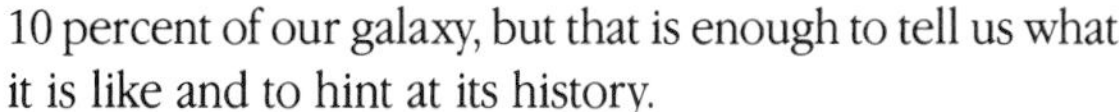

10 percent of our galaxy, but that is enough to tell us what it is like and to hint at its history.

The key to understanding our galaxy is the separation of its stars into two regions. Most stars, including the sun and all the gas and dust, lie in a thin, rotating disk. Some stars and star clusters, however, are scattered through a spherical volume that completely encloses the galactic disk. The motions, compositions, and ages of the stars in these two distributions can help us visualize the formation of our galaxy.

15.1 PROPERTIES OF THE MILKY WAY GALAXY

Before we can consider the origin and evolution of our Milky Way galaxy, we must survey it to see what its overall characteristics are. We must discover what kinds of objects are present and where they are located in the galaxy. Such data will later help us speculate on the history of the galaxy. Our first task, however, is to answer a difficult question: How big is the galaxy?

SIZE

Humanity has known of the Milky Way since before recorded history, but only recently have astronomers understood what the Milky Way is and how large it is. Copernicus, Kepler, Galileo, and their contemporaries limited their speculations to the sun and planets. Later astronomers supposed the stars are either confined to a sphere at great distance or scattered into limitless space. Most of these theories were tainted by philosophical and theological preconceptions.

The English astronomer Sir William Herschel (1738–1822) (Figure 15–3) and his sister Caroline (1750–1848) were the first to study the extent of the Milky Way scientifically. They assumed that the sun is located inside a great cloud of stars and that they could see to the edges of the cloud. By counting the number of stars visible through his telescope in a given direction, Herschel believed he could gauge the relative distance to the edge. If he saw few stars, the edge was nearby; if he saw many, the edge was very distant. Calling his method "star-gauges," he and Caroline counted stars in 683 preselected regions of the sky and outlined a model of the galaxy. According to this model, the galaxy is a great, irregular grindstone of stars with the sun near the center (Figure 15–4).

The grindstone model of the galaxy was confirmed by Jacobus C. Kapteyn (1851–1922) during the first decades of this century. He analyzed the brightness, number, and proper motions of stars and concluded that the system is about 10 kpc in diameter and about 2 kpc thick with the sun near the center. (A kpc is a **kiloparsec** and equals 1000 pc.)

In the years following World War I, a young astronomer named Harlow Shapley (1885–1972) proved that our star system is much larger than anyone had supposed. Shapley's study was one of the turning points of modern astronomy and is worth retracing in detail.

Shapley began by noticing that open clusters are located along the band of the Milky Way, but that most of the globular clusters are located on one side of the sky (Figure 15–5). In fact, more than half of all known globular clusters in our galaxy lie in or near the constellation Sagittarius (Figures 15–6 and 15–7). This suggested to Shapley that our sun is not located at the center of the star system.

To estimate the distance to the center of the star system, Shapley had to find the distances to the globular clusters. They are much too far away to have measurable parallaxes, but they do contain variable stars. These stars gave Shapley the clue he needed to find their distance.

a

b

FIGURE 15–3 William Herschel made many discoveries with the telescopes he built. He mapped the "grindstone" shape of our galaxy, and, as commemorated on the paper he holds in this engraving (a), he discovered the planet Uranus. (b) His great telescope was supported by scaffolding with the observer riding in the cage just below the lip of the tube. The telescope was moved by assistants. (Grundy Observatory.)

Some years earlier Henrietta S. Leavitt had discovered variable stars in the Small Magellanic Cloud and had found that there was a relation between the period of pulsation and the apparent brightness of the stars. Leavitt did not know the distance to the Small Magellanic Cloud, so she could not calculate absolute magnitudes. She drew a period–luminosity diagram for the variable stars with apparent magnitude on the vertical axis (Figure 12–9).

Knowing about Leavitt's work, Shapley made a very important assumption: that the variable stars he saw in the globular clusters were the same kinds of stars as the shorter period variables in Leavitt's period–luminosity diagram, stars we know today as RR Lyrae stars. Thus, Shapley needed to know the absolute magnitudes of the stars in the period–luminosity diagram. That would tell him how luminous the RR Lyrae stars were, and he could use that to find the distances to the globular clusters.

Astronomers had found variable stars scattered throughout our star system, but none of those variables was close enough to have a measurable parallax. A number of the brighter variables did have measured proper motions. We saw in Chapter 9 that nearby stars tend to have larger proper motions than more distant stars. Shapley used the average proper motion of only eleven nearby Cepheid variables to estimate their average distance, and thus their average absolute magnitude (Box 9–2). That allowed him to erase the apparent magnitudes from the period–luminosity diagram and replace them with a scale of absolute magnitudes (Figure 12–10).

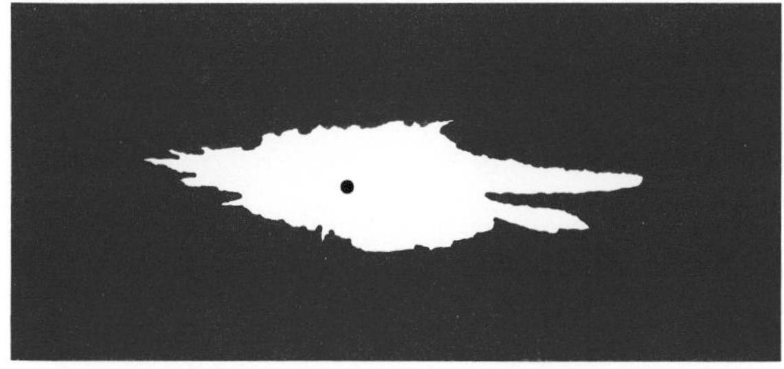

FIGURE 15–4 An edge-on view of Herschel's grindstone model of the Milky Way galaxy. The dot marks the location of the sun.

In this way, Shapley calibrated the variable stars for use in finding distances. From the diagram he could see that RR Lyrae stars have absolute magnitudes of about zero magnitude absolute. He could then subtract this from the apparent magnitude of the variable stars in a cluster to find the distance modulus and thus the distance to the cluster (Box 9–2).

This worked well for the nearer globular clusters, but the variable stars in the more distant clusters are too faint to detect. Shapley estimated the distances to these more distant clusters by calibrating the diameters of the clusters. For the clusters whose distance he knew, he could use their angular diameters and the small-angle formula to calculate linear diameters in parsecs (Box 3–1). He found that the nearby clusters are about 25 pc in diameter, which he assumed is the average diameter of all globular clusters. He then used the angular diameters of the more distant clusters to find their distances.

Notice how in this study, Shapley calibrated variable stars and then cluster diameters for use in finding diameters. This technique, the calibration of one parameter to find another, is common in astronomy. We used it earlier

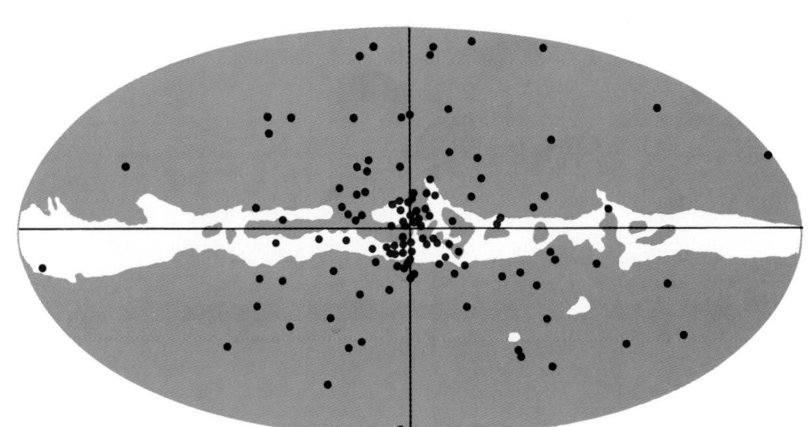

FIGURE 15–5 If the globe of the sky were split open and pressed flat, we could easily see the concentration of globular clusters (dots) toward the center of our galaxy. The unshaded area is the Milky Way with the Large and Small Magellanic Clouds at the lower right.

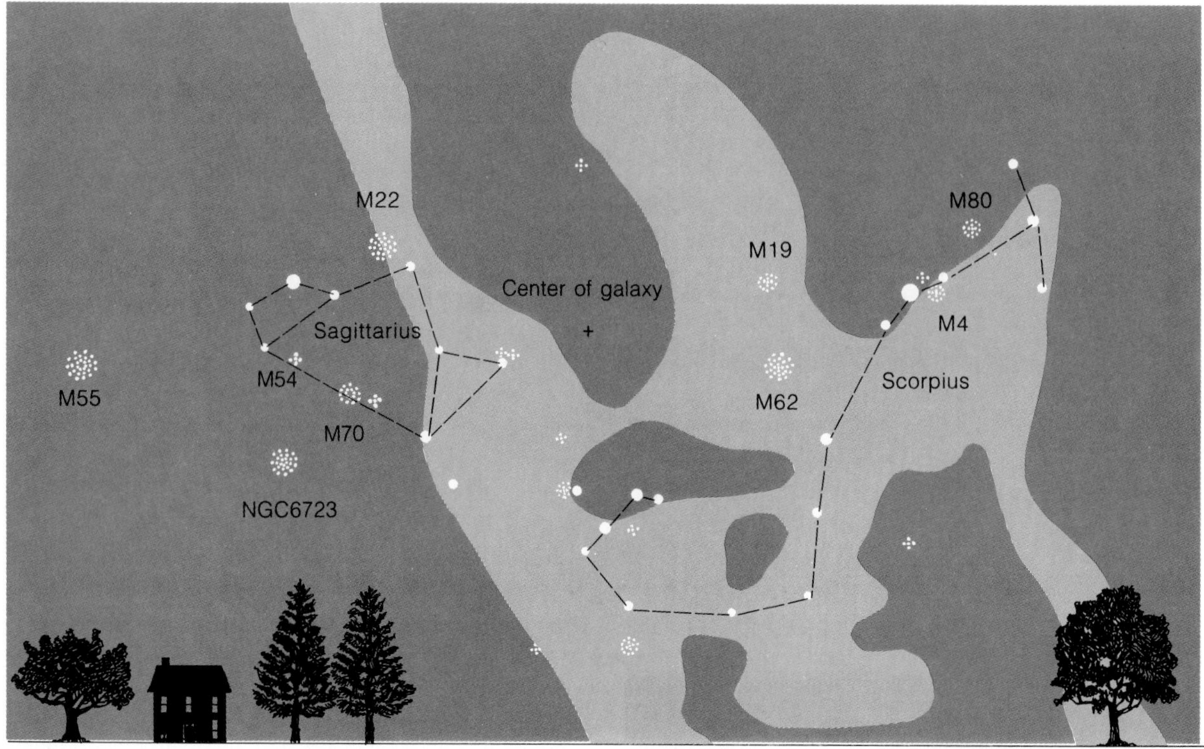

FIGURE 15–6 A few of the brighter globular clusters in the Scorpius–Sagittarius region. The brightest are visible through good binoculars on a dark night. The position of the constellations is shown as they appear above the southern horizon on a summer night as seen from a latitude of 40°N, typical for most of the United States.

when we calibrated spectral features against the temperatures of stars (Chapter 7), and we will use it again to find the distances to other galaxies (Chapters 16 and 17).

When he plotted the direction and distance to the globular clusters, he found they are distributed in a swarm centered not on the sun but on a point thousands of parsecs away in the direction of Sagittarius (Figure 15–8b). He reasoned that because the motions of the clusters are dominated by the gravity of the entire galaxy, the center of the swarm should coincide with the center of the galaxy. Measuring the distance to this center revealed that the galaxy is much bigger than anyone had supposed.

Shapley's first estimate for the diameter of the galaxy was 100,000 pc, about ten times the diameter of the grindstone model. Earlier astronomers had underestimated the size of the galaxy because great clouds of dust block our view in the plane of the Milky Way. Thus, Herschel was not counting stars out to the edge of the star system as he assumed but only out to the first dense cloud of dust. Shapley's work avoided this limitation because he looked at globular clusters out in the halo of the galaxy, and no dust clouds block our view there.

Nevertheless, Shapley's result was not quite correct; his result was about three times too big. Like all astronomers at the time, Shapley assumed that space was empty and transparent. In fact, our galaxy is filled with interstellar gas and dust that makes distant stars look fainter than they would be otherwise and thus makes them look farther away. Shapley overestimated the size of the galaxy because the variable stars he saw were slightly dimmed by the interstellar gas and consequently looked farther away than they really were.

Although his final result was a bit larger than the true distance to the center of the galaxy, Shapley's dis-

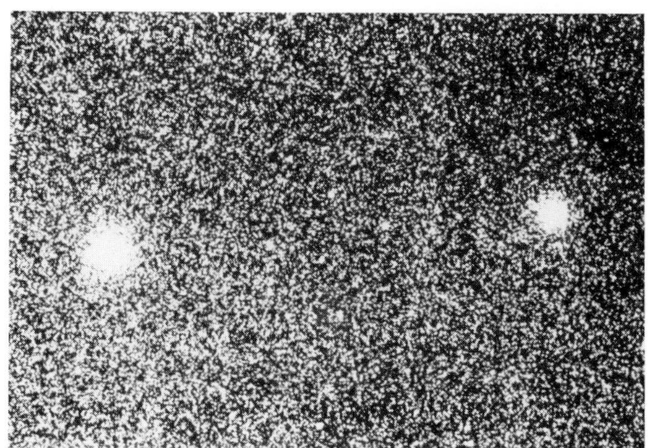

FIGURE 15–7 The two globular clusters NGC 6528 (left) and 6522 (right) are seen against the background of crowded stars toward the center of our galaxy. In Figure 15-6, these clusters are located just above the tip of the spout of the Sagittarius teapot. (National Optical Astronomy Observatories.)

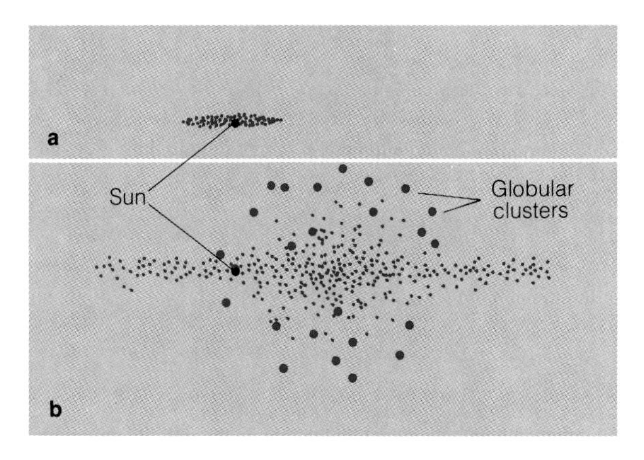

FIGURE 15–8 (a) Before Shapley studied globular clusters, astronomers thought the galaxy to be only about 10 kpc in diameter. (b) Shapley's work determining the distribution of the globular clusters showed that the galaxy is much larger and that the sun is not located at the center.

covery was a turning point in modern astronomy. It overthrew the small, sun-centered model and showed that the galaxy is much larger than what we see from the earth.

Like Copernicus, Shapley helped change the way we think of our place in nature. Previous to Shapley, our solar system appeared to be the center of the galaxy, perhaps the center of the universe. Shapley's result placed us in the suburbs of a very large galaxy. As he said, "[It] is a rather nice idea because it means that man is not such a big chicken. He is incidental—my favorite term is 'peripheral.' "*

COMPONENTS

Modern astronomers recognize that the galaxy in which we live consists of two parts—a disk and a halo. The nature of these two components contains clues to the origin and evolution of the galaxy.

The **disk component** of our galaxy consists of all matter confined to the plane of the disk—that is, everything in the disk itself. This includes stars, open star clusters, and nearly all of the galaxy's gas and dust.

*Harlow Shapley, *Through Rugged Ways to the Stars.* New York: Scribner's, 1969, p. 60.

We cannot quote a single number for the thickness of the disk for two reasons. First, the disk does not have sharp boundaries. The stars become gradually less crowded as we move away from the plane of the galaxy. Second, the thickness of the disk depends on the kind of object we study. The O stars and dense dust clouds lie within 100 pc of the galactic plane, but stars like the sun are much less confined: The disk defined by such stars is roughly 1000 pc thick (Figure 15–9).

Vast dust clouds are scattered throughout the disk. We can see these on visual-wavelength photographs as dark clouds (Figure 15–2), and the Infrared Astronomy Satellite found the disk glowing brightly in the infrared radiation emitted by the dust and cool stars (Color Plate 14). These dust clouds block our view of distant stars and make it difficult to estimate the diameter of the disk and the location of the sun. Astronomers usually quote a diameter of 30 kpc (about 100,000 ly), but recent research suggests that the galaxy may be smaller. The most recent estimates of the distance to the center of the galaxy, based on the distances to globular clusters and the motions of dense gas clouds near the center, range from 6.8–8.5 kpc. In November 1985 members of the International Astronomical Union passed a resolution urging astronomers to use a value of 8.5 kpc (with an uncertainty of ±1 kpc). Many books and articles, however, continue to use 10 kpc as a matter of convenience. Whatever the exact value, the

FIGURE 15–9 The Milky Way galaxy seen face-on and edge-on illustrates the shape and location of the disk, halo, and nucleus. Note the position of the sun.

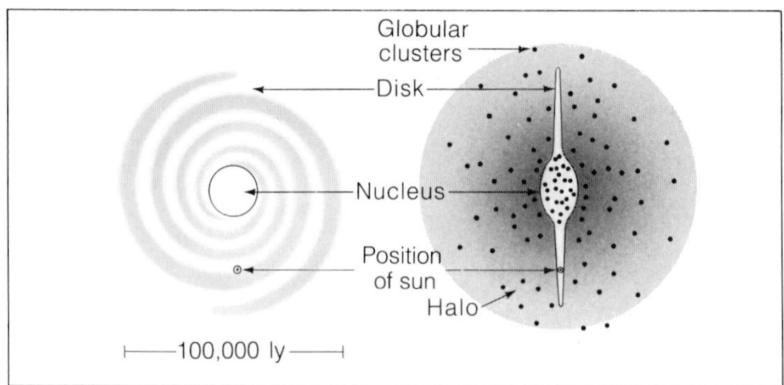

sun is roughly ⅔ of the way from the center of the disk to the edge.

Spread throughout the galaxy is a very weak magnetic field averaging 7×10^{-6} G. For comparison the magnetic field at the surface of the earth is about 0.5 G. Although the galaxy's magnetic field is weak, it is strong enough to trap **cosmic rays** and confine them to our galaxy (Box 15–1). These cosmic rays are not waves but high-energy particles, mostly protons and electrons, at least some of which originate in supernova remnants. During the last second, about a hundred of these particles have passed through your head doing little or no damage. Thus, the disk of our galaxy is filled by a sea of high-energy particles, confined by a weak magnetic field.

The most striking features of the disk component are the **spiral arms**, long spiral patterns of bright stars, star clusters, gas, and dust. Such spiral arms are easily visible in photographs of other galaxies, and we will see later in this chapter that our own galaxy contains a similar spiral pattern.

The disk component contains two types of star clusters—associations and open clusters. **Associations** are groups of 10 to 100 stars so widely scattered in space their mutual gravity cannot hold the association together. We find the stars moving together through space (Figure 15–10) because they formed from a single gas cloud and have not yet wandered apart. However, associations are short-lived, and the stars eventually go their separate ways. Two types of young associations—O and B associations and T Tauri associations (see Chapter 11)—are located along the spiral arms.

Open clusters (Figure 15–11) contain 100 to 1000 stars in a region about 25 pc in diameter. Because they have more stars and occupy less space than associations, open clusters are more firmly bound by their gravity. Such clusters do lose stars occasionally as close encounters between cluster members eject stars from the group. However, it takes tens of millions of years for this process to affect the cluster. Thus, open clusters are much more stable and longer-lived than associations.

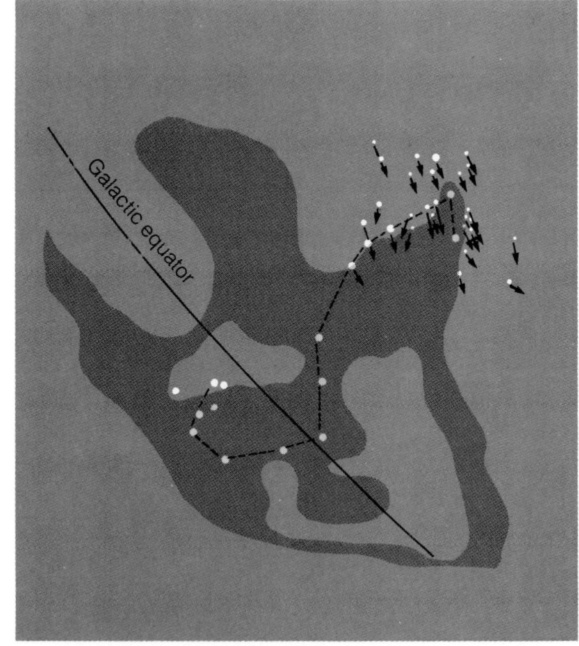

FIGURE 15–10 Many of the stars in the constellation Scorpius are members of an O and B association. They have formed recently from a single gas cloud and are moving together southwest along the Milky Way (shaded).

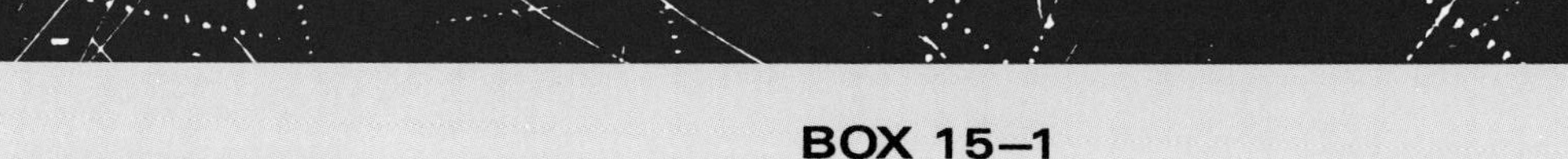

BOX 15–1
Cosmic Rays

If you play table tennis, you may have an advantage in understanding the origin of cosmic rays. Austrian physicist Victor Hess (1883–1964) discovered them during a balloon ascent in 1912, and they have been a puzzle for astronomers ever since.

We know today that cosmic rays are not waves as their name implies but subatomic particles traveling at nearly the speed of light and entering our atmosphere from all directions. Many collide with the nuclei of gas atoms in the upper atmosphere and shower us with subatomic fragments, but others reach the surface and pass through our bodies at the rate of about 0.5/sec/cm^2.

About 92 percent of cosmic rays are protons, 6.3 percent are helium nuclei, 0.9 percent are electrons, and the remaining 0.7 percent are made up by nuclei of the heavier elements. This is roughly the abundance of elements in the sun and stars and suggests that cosmic rays originate in stars.

Observations of the direction of arriving cosmic rays cannot reveal their origin because they are charged particles and are deflected by magnetic fields in space. The earth's magnetic field protects us from the lower-energy cosmic rays, and the sun's magnetic field and solar wind deflects some others. In fact, we receive only about half as many cosmic rays at sunspot maximum when the solar magnetic field is more extended as compared with sunspot minimum.

The galactic magnetic field is much weaker but the cosmic rays travel great distances through space. A typical cosmic ray has its direction of travel changed by 90° for every 0.1 pc it travels. Thus, when we observe cosmic rays, we find them arriving from all directions uniformly, just as we see photons arriving from all directions when we stand in a heavy fog bank. Consequently, we cannot tell where cosmic rays come from.

Some of the lower-energy cosmic rays originate in the sun and a few of the highest-energy particles must come from outside the galaxy, but most come from sources within the galaxy. Theorists suppose that most cosmic rays originate in supernova explosions and are further accelerated in supernova remnants. Observations of supernova remnants show that some are emitting gamma rays, which are presumably produced by cosmic rays striking gas atoms. This confirms our suspicion that supernova remnants are associated with cosmic rays.

Once a cosmic ray is produced inside a supernova, it can be accelerated by bouncing back and forth between the tangled magnetic fields in the remnant. We see the same effect when we bounce a table tennis ball and then pin it beneath a paddle. As the paddle descends on the ball, it bounces faster and faster, gaining energy each time it collides with the approaching paddle. Similarly, a cosmic ray bouncing between approaching magnetic fields in a supernova remnant can gain tremendous energy before it escapes.

The cosmic rays that continuously smash down through our atmosphere are samples of matter from dying stars. We are just beginning to be able to use the information in these particles to study the deaths of stars and the nature of our galaxy.

The second component of our galaxy is the **spherical component**, which includes all matter in the galaxy scattered in a spherical distribution around the center. This includes the halo and the nuclear bulge.

The **halo** contains a thin scattering of stars, star clusters, and almost no gas and dust. It contains no young bright stars, but rather cool, lower main-sequence stars and giants. These stars are scattered so thinly that it is difficult to gauge the extent of the halo, but observations suggest that it must extend well beyond the edge of the disk.

The 100 or so star clusters in the halo are globular clusters (Figure 15–7 and 12–17). These clusters contain 100,000 to 1,000,000 stars crowded into a sphere about 10–30 pc in diameter. Because globular clusters contain so many stars in such small regions, the clusters are very stable and have survived for billions of years. We saw in Chapter 12 that the turn-off points in their H–R diagrams

FIGURE 15–11 The open cluster NGC 457 lies in the Cassiopeia segment of the Milky Way. At a distance of 2800 pc, the young cluster contains a few thousand stars although only the brightest 100 or so are visible at the center of this photo. The rest of the stars shown are Milky Way stars not associated with the cluster. (National Optical Astronomy Observatories.)

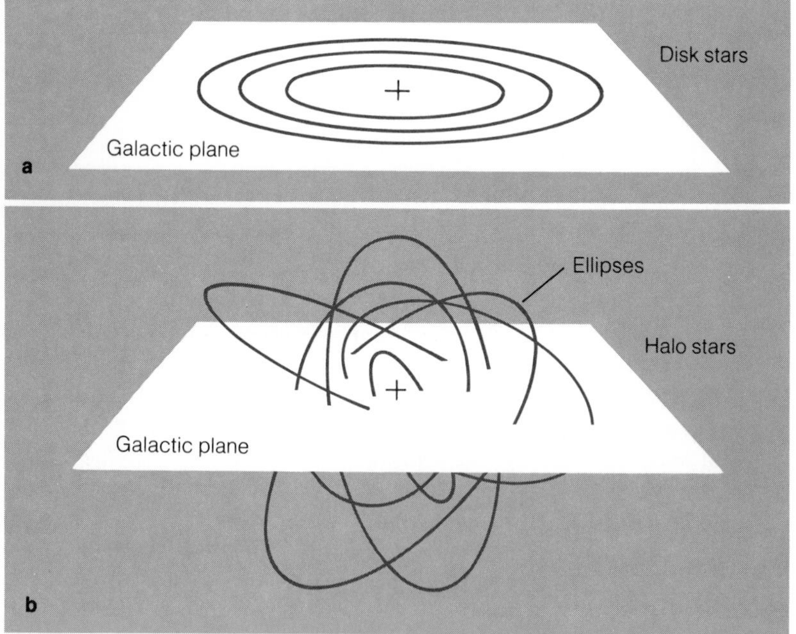

FIGURE 15–12 (a) Stars in the galactic disk have nearly circular orbits that lie in the plane of the galaxy. (b) Stars in the halo have randomly oriented, elliptical orbits.

(Figure 12–19) tell us that their ages range from 10 billion years for the youngest to 15 billion years for the oldest. This makes globular clusters the oldest known objects in our galaxy and fixes the age of our galaxy as at least 15 billion years.

The cloud of stars that lies at the center of the galaxy is the **nuclear bulge**, the most crowded part of the spherical component. The stars of the nuclear bulge are similar to the halo stars, although evidence suggests that the center of the bulge contains some young hot stars. The dust clouds in the disk and the tremendous crowding of stars in the central regions hopelessly obscure the galactic center at visual wavelengths. Later in this chapter we will review the radio and infrared observations of the galaxy's core.

So far our inventory has been static, merely listing the contents and sizes of the disk and spherical components. Now we must add motion and study galactic rotation.

BOX 15-2

The Mass of the Galaxy

Although the galaxy is immense, a surprisingly simple calculation reveals its mass. We will treat the galaxy and the sun as a binary star system. In Chapter 10, we discovered that the total mass of a binary star system depends only on the separation of the stars a and their orbital period P.

$$M = \frac{a^3}{P^2}$$

If we express a, the average radius of the orbit, in astronomical units (AU) and P in years, the total mass is in solar masses.

The radius of the sun's orbit is about 8500 pc. To convert this to AU, recall that a parsec equals 206,265 AU. Then a equals 1.75×10^9 AU.

The sun's orbital period equals the circumference of its orbit divided by its velocity. The circumference is just 2π times the radius, and we know the sun's orbital velocity, at least approximately, from studies of the Doppler shifts in the spectra of stars and other galaxies. Simple arithmetic shows that P is about 240×10^6 years.

Now we can find the total mass.

$$M = \frac{(1.75 \times 10^9)^3}{(240 \times 10^6)^2} = 0.93 \times 10^{11} M_\odot$$

Although the arithmetic is simple, the measurements of a and the sun's orbital velocity are very difficult, so this mass is only approximate. Also, it includes only the mass inside the sun's orbit. Matter in the galaxy outside the sun's orbit does not affect its orbital period. Thus, we should increase our estimate to include this matter. The total mass of the galaxy is probably $2 \times 10^{11} M_\odot$ or more.

THE ROTATION OF THE GALAXY

The galaxy must rotate; if it did not, all the stars would fall into the center. That does not happen because each star follows its own orbit around the center of the galaxy.

Disk stars move in nearly circular orbits that lie in the plane of the galaxy (Figure 15–12). The sun, for example, is a disk star and moves about 220 km/sec in the direction of Cygnus, carrying the earth and other planets along with it. Because its orbit is a circle with a radius of 8.5 kpc, it takes about 240×10^6 years to make one circuit around the galaxy. As shown in Box 15–2, this tells us that the mass of our galaxy is at least $2 \times 10^{11} M_\odot$.

The motion of stars near the sun shows that the disk does not rotate as a solid body (Figure 15–13). Stars orbiting 1 kpc closer to the center of the galaxy move faster, and stars 1 kpc outside the sun's orbit move more slowly. We thus say that the disk of the galaxy rotates differentially. (Recall that we defined differential rotation when we discussed the rotation of the sun in Chapter 8.)

Contrast this with the rigid rotation of a record on a turntable. Three spots lined up on the record would stay together as the record turned. Because the outermost spot had the farthest to go, it would move fastest, and the innermost spot would move slowest. In the galaxy, three stars lined up near the sun as in Figure 15–13 do not stay aligned. Because the galaxy rotates differentially, the innermost star pulls ahead and the outermost star falls behind. A graph of orbital velocity versus radius is called a **rotation curve**.

This rotation has been used to tell us the mass of the galaxy, but recent observations of the rotation curve suggest that the galaxy may be much more massive than we suspect. At least five different teams of astronomers have studied the motions of distant globular clusters, clouds of gas, individual stars, clusters of stars, and emission nebulae. They all find evidence that the rotation curve does not continue to decrease beyond the location of the sun. In fact, it seems to increase. That is, matter located farther from the center of the galaxy than the sun may have higher orbital velocities than the sun (Figure 15–14).

A rising rotation curve implies that the galaxy is more massive than we thought. If all of the mass of the galaxy were inside the orbit of a star 15 kpc from the center of the galaxy, increasing the size of its orbit would mean it moved more slowly. This is how the planets orbit the sun; the farther away a planet is, the slower it moves. This kind of motion is called **Keplerian motion**, referring to Kepler's laws of planetary motion. However, if more distant stars move along their orbits more quickly, their orbits must be enclosing more mass, so the galaxy must have appreciable mass in its outer parts. This is supported by studies of the rotation curves of other galaxies and by theoretical studies of the stability of our own galaxy.

Many astronomers now suspect that the halo of our galaxy extends much farther into space than was generally accepted. This so-called **galactic corona** could extend out as far as 100,000 pc—seven times the presently accepted extent of the galaxy—and it could contain up to 10^{12} solar masses. Such an extensive corona might interact with nearby galaxies such as the Magellanic Clouds.

Whereas the motions within the disk are regular, motions within the halo are much more random. Each halo star and globular cluster follows its own randomly inclined elliptical orbit (Figure 15–12). These orbits carry the stars and clusters far out into the spherical halo, where they move slowly, but when they fall back into the inner part of the galaxy, their velocities increase. Those presently passing through the disk of the galaxy are called **high-velocity stars.**

The differences between the disk component and the spherical component illuminate our galaxy's past. We have already seen that the two components differ in number of stars, amount of gas and dust, and orbital shape and orientation. In the next section we will discover that they also differ in chemical composition. That clue, combined with our knowledge of stellar evolution, will suggest a model for the formation of the galaxy.

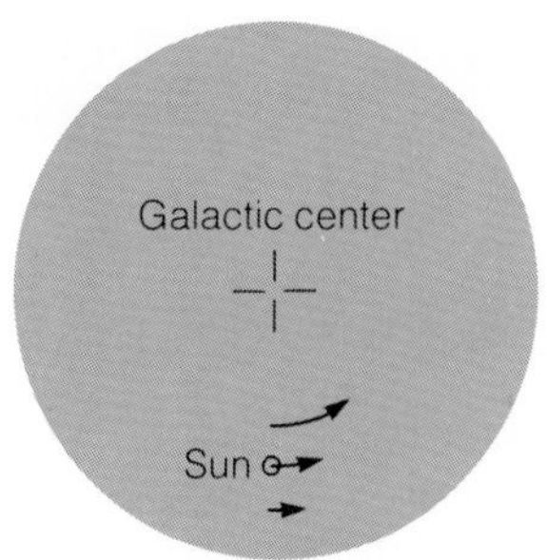

FIGURE 15–13 The differential rotation of the galaxy's disk means that not all stars orbit the galaxy with the same period. Three stars lined up near the sun will not remain aligned as they follow their orbits. The inner star will pull ahead, and the outer star will fall behind.

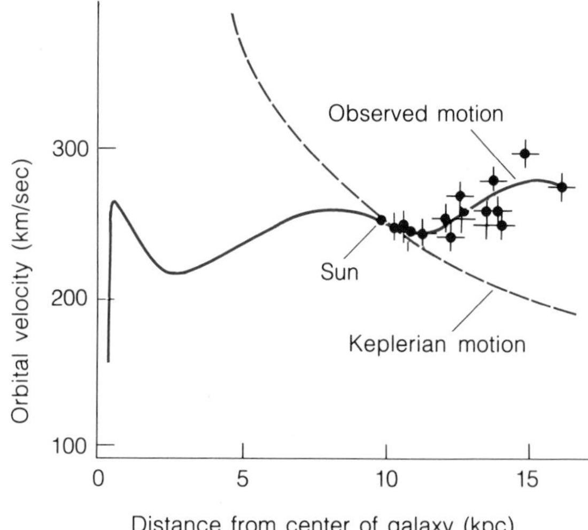

FIGURE 15–14 If most of the galaxy's mass were inside the sun's orbit, we would expect the rotation curve to obey Kepler's third law outside the sun's orbit. But the observed orbital velocities in the outer disk are higher than expected. This suggests that the galaxy contains significant amounts of mass outside the orbit of the sun. Points with error bars represent the orbital velocities of dense gas clouds in the disk of the galaxy. Note that these data have been plotted using a radius of 10 kpc for the sun's orbit. (Adapted from data by Leo Blitz.)

15.2 THE ORIGIN OF THE MILKY WAY

Just as paleontologists reconstruct the history of life on earth from the fossil record, astronomers can reconstruct the past of our galaxy from the fossil it left behind as it evolved. That fossil is the spherical component of the galaxy. The stars we see now in the nuclear bulge and halo formed when the galaxy was young. By studying those stars, their chemical composition, and their distribution in the galaxy, we can devise a theory to explain how the galaxy formed.

Table 15–1 Stellar populations.

	Population I		Population II	
	Extreme	**Intermediate**	**Intermediate**	**Extreme**
Location	Spiral arms	Disk	Nuclear bulge	Halo
Metals	3%	1.6%	0.8%	Less than 0.8%
Shape of orbit	Circular	Slightly elliptical	Moderately elliptical	Highly elliptical
Ave. age (y)	100 million and younger	0.2–10 billion	2–10 billion	10–14 billion

STELLAR POPULATIONS

Near the end of World War II, astronomers realized that there are two types of stars in the galaxy. The type they were accustomed to studying were those located in the disk, like those near the sun. These they called **population I stars.** The second type, called **population II stars**, are usually found in the halo, in globular clusters, or in the central bulge. In other words, the two stellar populations are associated with the two components of the galaxy.

The stars of the two populations are very similar. They burn nuclear fuels and evolve in nearly identical ways. They differ only in the abundance of atoms heavier than helium, atoms that astronomers refer to collectively as **metals.** (Note that this is not the way the word *metal* is commonly used by nonastronomers.) Population I stars are metal-rich, containing 2–3 percent metals, whereas population II stars are metal-poor, containing only about 0.1 percent metals. The metal content of the star defines its population.

Population I stars belong to the disk component of the galaxy and are sometimes called disk population stars. They have circular orbits in the plane of the galaxy and are relatively young stars that formed within the last few billion years. The sun is a population I star, as are the Type I Cepheids discussed in Chapter 12.

Population II stars belong to the spherical component of the galaxy and are sometimes called the halo population stars. These stars have randomly tipped, elliptical orbits, and are old stars. The metal-poor globular clusters are part of the halo population, as are the RR Lyrae and Type II Cepheids.

Since the discovery of stellar populations, astronomers have realized that there is a gradation between populations (Table 15–1). Extreme population I stars, like the young, hot stars in the constellation Orion, are found only in the spiral arms. Slightly less metal-rich population I stars, called intermediate population I stars, are located throughout the disk. The sun is such a star. Stars even less metal-rich, such as stars in the nuclear bulge, belong to the intermediate population II. The most metal-poor stars are those in the halo and in globular clusters. These are extreme population II stars.

Why do the disk and halo stars have different metal abundances? The answer to this question is the key to the history of our galaxy. We must begin by discussing the cycle of element building in the Milky Way galaxy.

NUCLEOSYNTHESIS: THE ELEMENT-BUILDING CYCLE

Of the 81 stable elements that occur on the earth, all but hydrogen and helium were made in stars in a process known generally as **nucleosynthesis.** In Chapter 13 we saw how the nuclei heavier than hydrogen and helium up to iron were synthesized by nuclear fusion in massive stars, and how nuclei more massive than iron were made during the short-lived violence of supernova explosions. Gold, silver, platinum, and many other elements heavier than iron are rare, and therefore precious, because they are only made in supernovae (Figure 15–15).

The gas from which our galaxy formed should have been almost free of metals (atoms heavier than helium) because no stars had yet formed. The original gas must have been almost pure hydrogen atoms (90 percent) and helium atoms (10 percent). (Hydrogen and helium were produced by the big bang that began the universe, a subject we will discuss in Chapter 18.)

Thus, the first generation of stars in our galaxy was metal-poor, and now, 10–16 billion years later, their spectra still show few metal lines. Only the most massive of these stars would have made metals (elements heavier than helium), and those metals would have stayed locked in the stars' centers. They could have been released only if the stars exploded as supernovae. The oldest stars we see today, the population II stars, have weak metal lines in their spectra (Figure 15–16) because they formed before many metals could be made.

The most massive of the first generation of stars died as supernovae and enriched the interstellar gas with metals. Successive generations of stars formed from this enriched gas and, in turn, further enriched the gas through the supernovae of the massive stars. By the time the sun formed, about 5 billion years ago, the element-building cycle had added about 1.6 percent metals. Since then, supernovae have further enriched the interstellar gas, and stars forming now, such as those in the Orion nebula, contain 2–3 percent metals. We call these stars population I stars. Thus, metal abundance varies between populations because of the production of heavy atoms in successive generations of stars.

Globular-cluster stars do contain a few metals, and that fact raises one of the interesting mysteries of modern astronomy: Where did the metals come from? We will see in Chapter 18 that the universe should have made very few metals when it began, so the gas from which our galaxy formed should have been nearly metal-free. Some theories suggest that there was an age of star formation before our galaxy took shape and that these stars manufactured some metals that were later incorporated in the extreme population II stars. Other theories suggest that our galaxy began with a burst of star formation that created massive, highly luminous stars that lived short lives and enriched the forming galaxy with metals. In any case, the presence of metals in globular-cluster stars may someday tell us about conditions before the formation of our galaxy.

The population II stars are in the halo and the population I stars are in the disk because the population II stars formed long ago when the galaxy was young and drastically different from its present form. By studying the distribution of the stellar populations, we can visualize a process by which our galaxy could have formed.

THE FORMATION OF THE GALAXY

A number of theories have been proposed to try to account for the origin of the Milky Way galaxy, and it is not possible at present to choose the correct version. However, we can summarize these theories by discussing a traditional hypothesis popular for some decades and a newer idea only a few years old.

The traditional theory supposes that the Milky Way galaxy began 10–15 billion years ago as a swirl of hydro-

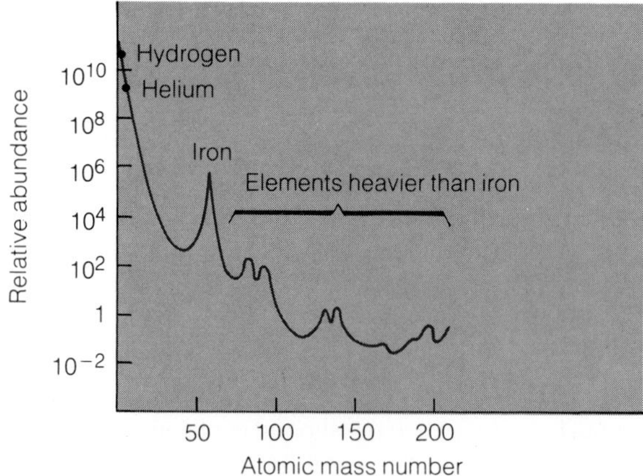

FIGURE 15–15 The abundance of the elements. Because they are made only during supernova explosions, the elements heavier than iron are rare.

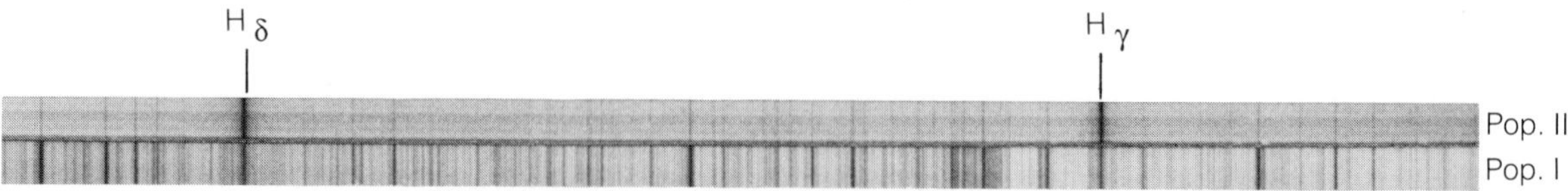

FIGURE 15–16 Spectra of population II (top) and population I (bottom) stars of similar spectral type show hydrogen lines of equal strength, but the lines of heavier atoms are conspicuously weak in the population II star's spectrum. (Lick Observatory photograph.)

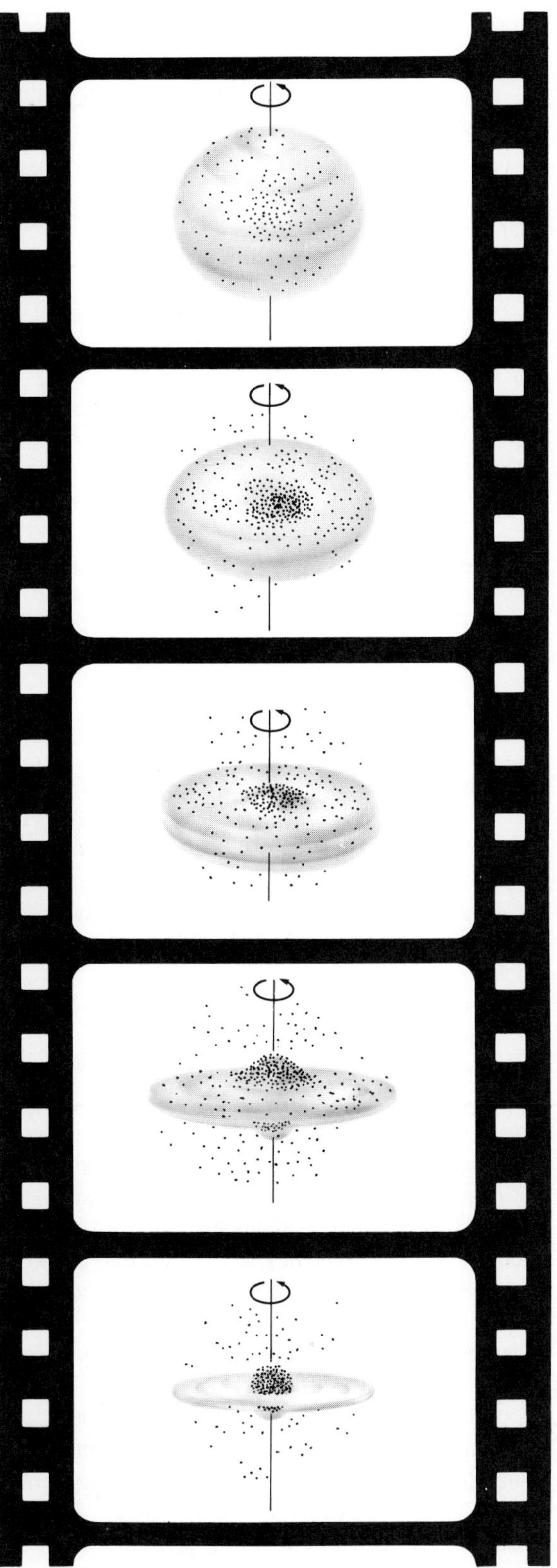

gen and helium gas that began to contract under the influence of its own gravity. As it contracted, density grew until the gas began to fragment into individual clouds. Because the original gas was turbulent, the clouds formed with random velocities. When the gas density grew high enough, some of the gas began to form clusters of metal-poor stars. Because the galaxy was approximately spherical at this time, these first star clusters formed in a spherical distribution that we see today as the halo.

These star clusters may not have looked like globular clusters. Many probably contained too few stars and were too scattered through space to hold themselves together. They gradually dissociated, freeing stars that wandered through the spherical cloud. Clusters that had more stars packed in a smaller volume survived, although they, too, lost members occasionally. Today, more than 10 billion years later, the surviving clusters are the highly stable globular clusters.

Although the galaxy began as a spherical distribution of gas, it immediately began to collapse into a rotating disk. Randomly moving eddies in the cloud collided with each other and the turbulent motions canceled out, leaving the galaxy with a uniform rotation. A low-density cloud of gas rotating uniformly around an axis cannot maintain a spherical shape. A star can remain spherical because it has high internal pressure to balance the weight of the gas, but in a cloud where the density is low, this pressure is not effective. Like a blob of pizza dough spun into the air, the cloud must flatten into a rotating disk (Figure 15–17).

According to the traditional theory, this collapse into a disk took billions of years and did not alter the orbits of the stars that had formed when the galaxy was spherical. The halo population stars were left behind as a fossil of the early galaxy. Subsequent generations of stars formed in flatter distributions. The intermediate population I stars, for instance, are scattered 100 parsecs above and below the plane of the disk. The gas distribution in the galaxy has now become so flat that the newest stars, the extreme population I stars, are confined to a disk less than 100 pc thick. In proportion to its diameter, this is comparable to the crust on a very thin, large pizza.

FIGURE 15–17 According to the traditional theory, the galaxy began as a spherical cloud of gas (shaded) in which stars and star clusters (dots) formed. As the rotating gas cloud collapsed into a disk, the halo stars were left behind as a fossil of the early galaxy.

In addition to a change in distribution, the shapes of stellar orbits also changed as the galaxy flattened. When the galaxy was young, the turbulent gas moved at random, and the stars that formed then took up orbits with random orientations and random shapes. As the galaxy collapsed into a disk, the random gas motion canceled out, and the gas took up more circular orbits, as did the forming stars. This explains why the oldest stars in the galaxy have the most elliptical orbits.

Simultaneously with these changes in distribution and orbital shape, the metal abundance in the stars grew with every generation, producing the stellar populations summarized in Table 15–1. The populations and their properties provide a permanent record of our galaxy's past.

This traditional view of the formation of the Milky Way sounds good, but new theories differ from it in some important respects. We will summarize these as a single hypothesis.

Some studies suggest that the hydrogen and helium gas produced large star clusters some time before the formation of the Milky Way galaxy. These star clusters may have contained 10^8 stars, some of which were massive and exploded as supernovae, thus slightly enriching the remaining gas with metals. Collisions between these star clusters began to build a larger cluster, which attracted more clusters and eventually grew into a galaxy-size swarm of stars of approximately spherical shape. As this cluster grew massive, the remaining gas in the neighborhood fell into the galaxy and gradually collapsed to form a disk. Like the traditional theory, the new theory explains the distribution of populations by the gradual change in the shape of the galaxy. The important difference is in the process by which the galaxy first took form.

In the next two chapters we will find evidence that galaxies occasionally merge with each other and that such mergers affect their evolution. Thus, we can suspect that our own galaxy's history includes the merging of small systems of stars.

Even if we knew how the galaxy formed and evolved, there would still be mysteries hidden in the Milky Way's star clouds. In the next sections, we will consider two of these problems: spiral arms and the galactic nucleus.

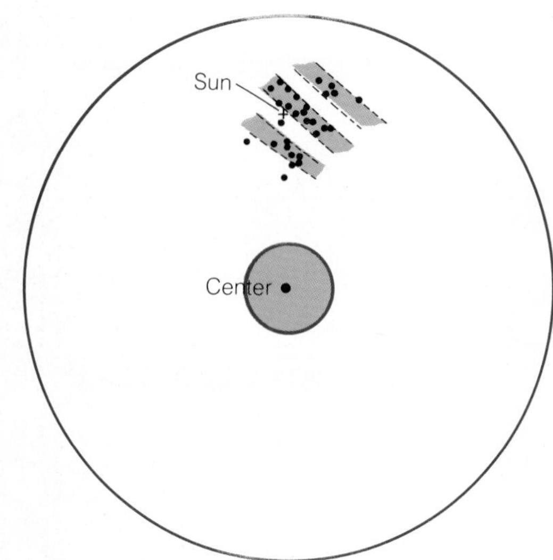

FIGURE 15–18 The O and B associations near the sun lie along three bands (shaded). These are segments of spiral arms.

15.3 SPIRAL ARMS

The most striking feature of galaxies like the Milky Way is the system of spiral arms that wind outward through the disk. These arms contain swarms of hot blue stars, clouds of dust and gas, and young star clusters. The spiral pattern presents astronomers with two problems. First, how can we study the spiral arms of our galaxy when our view is obscured by dense dust clouds? Second, what are the spiral arms? In this section we will see that mapping the location of the arms is easier than explaining their origin.

TRACING THE SPIRAL ARMS

Studies of other galaxies show us that spiral arms contain hot blue stars. Thus, one way to study the spiral arms of our own galaxy is to locate these stars. Fortunately, this is not difficult because O and B stars are often in associations, and, being very bright, they are easy to detect across great distances. Unfortunately, at these great distances their parallax is too small to measure, so their distances must be found by other means, usually by spectroscopic parallax (see Box 9–4).

O and B associations near the sun are not located randomly (Figure 15–18). They form three bands, indicating that there are three segments of spiral arms near the sun. If we could penetrate the dust clouds, we could locate other O and B associations and trace the spiral arms farther, but, like a traveler in a fog, we can see only the region near us.

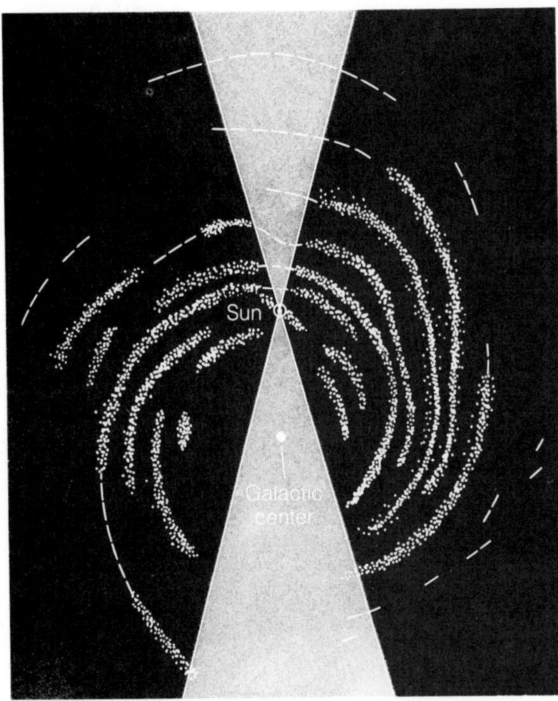

FIGURE 15–19 A radio map of our galaxy at a 21-cm wavelength shows traces of a spiral pattern in the distribution of neutral hydrogen. The radio emission from clouds within the two wedge-shaped regions cannot be unscrambled to show discrete arms. (Adapted from data by Gerrit Verschuur.)

Objects used to map spiral arms are called **spiral tracers.** O and B associations are good spiral tracers because they are bright and easy to see at great distances. Other tracers include young open clusters, clouds of hydrogen ionized by hot stars (emission nebulae), and certain kinds of variable stars.

Notice that all spiral tracers are young objects. O stars, for example, live only a few million years. If their orbital velocity is about 250 km/sec, they cannot have moved more than about 500 pc since they formed. This is less than the width of a spiral arm. Because they don't live long enough to move away from the spiral arms, they must have formed there.

The youth of spiral tracers gives us an important clue to the nature of the arms. Somehow the arms are associated with star formation. Before we can follow this clue, however, we must extend our map of spiral arms to show the entire galaxy.

RADIO MAPS OF SPIRAL ARMS

The dust clouds that block our view at visual wavelengths are transparent at radio wavelengths because radio waves are much longer than the diameter of the dust particles. When we point a radio telescope at a section of the Milky Way, we receive 21–cm radio signals (Chapter 11) coming from cold clouds of neutral hydrogen in a number of spiral arms at various distances across the galaxy. In most areas of the disk, the signals can be unscrambled using the Doppler shifts of the 21–cm radiation, and thus we can map the distribution of neutral hydrogen gas throughout the disk of the galaxy (Figure 15–19). Only where we look toward the center of the galaxy or away from the center (the two wedge shapes in Figure 15–19) are we unable to separate emission from clouds at different distances. Within these wedge-shaped regions the gas clouds all move perpendicular to our line of sight and have minimal Doppler shifts.

Radio astronomers can also use the strong radio emission from carbon monoxide (CO) to map the location of giant molecular clouds in the plane of the galaxy. Two such maps are shown in Figure 15–20. These studies show that the giant molecular clouds are located in large structures that resemble segments of spiral arms.

Radio maps reveal a number of things. First, the spiral pattern we see near the sun continues throughout the disk. Second, the spiral arms are rather irregular and interrupted by bends, spurs, and gaps. The stars we see in Orion, for example, appear to be a detached segment of a spiral arm. There are significant sources of error in the radio-mapping method, but many of the irregularities along the arms seem real and photographs of nearby spiral galaxies show similar features. In fact, a few astronomers argue that our galaxy has no overall spiral pattern, only a distribution of broken segments. Other astronomers argue that the spiral pattern is four-armed rather than the "grand design" two-armed pattern usually assumed.

If we combine optical and radio data, we can construct a general map of our galaxy (Figure 15–21a). From this it seems that we live in a spiral galaxy and that the sun is now on the inner edge of a spiral arm segment known as the Orion arm because the bright stars of Orion are part of that arm. But it is not possible to connect all the arm segments in the map into a single, grand design of two arms. A "best guess" of what our galaxy looks like is shown in Figure 15–21b.

The most important feature of the spiral arms is easily overlooked—spiral arms are regions of higher gas density, richly populated by bright young stars. The spiral arms are somehow associated with star formation.

THE DENSITY WAVE THEORY

Just what are spiral arms? We can be sure they are not physically connected structures like bands of magnetic field holding the gas in place. If they were, the strong differential rotation of the galaxy would destroy them within a billion years. They would get wound up and torn apart like paper streamers caught on the wheel of a speeding car. Yet spiral arms are common in galaxies and must last many billions of years.

Most astronomers believe that spiral arms are dynamically stable—they retain the same appearance although the gas, dust, and stars in them are constantly changing. To see how this works, think of the traffic jam behind a slow-moving truck. Seen from an airplane, the traffic jam would be stable, moving slowly down the highway. But any given car would approach from behind, slow down, work its way to the front of the jam, pass the truck, and resume speed. The individual cars in the jam are constantly changing, but the traffic jam itself is dynamically stable.

In the **density wave theory**, the spiral arms are dynamically stable regions of compression that move slowly around the galaxy, just as the truck moves slowly down the highway. Gas clouds, moving at orbital velocity around the galaxy, overtake the slow moving arms from behind and slam into the gas already in the arms. The sudden compression of the gas can trigger the collapse of the gas

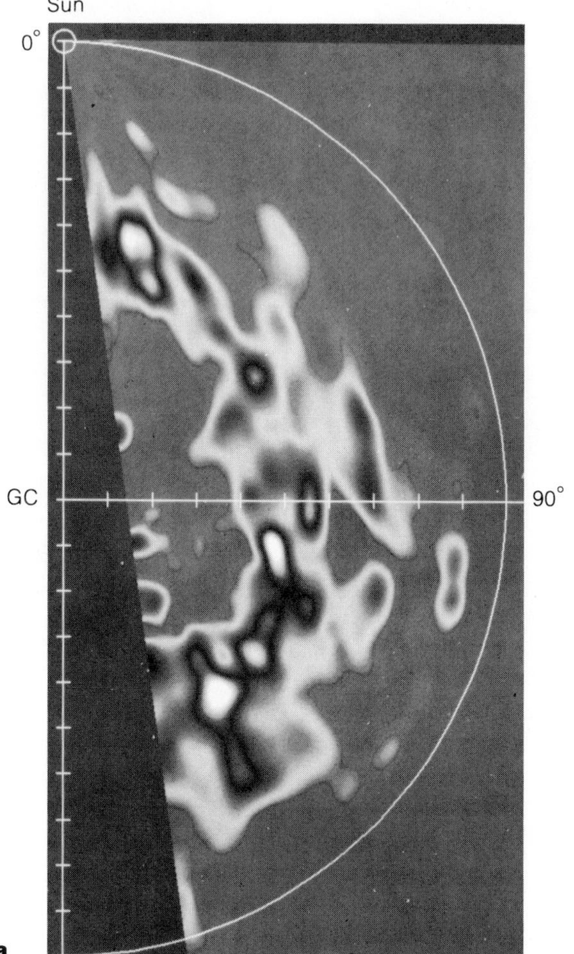

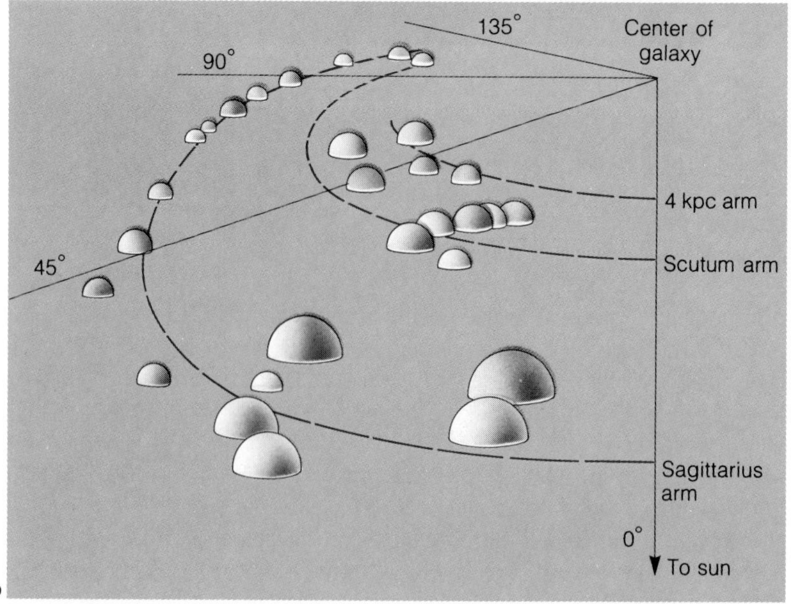

FIGURE 15–20 (a) The CO distribution in the galactic plane as seen from directly above the galactic center (GC). (b) The location of giant molecular clouds in our galactic plane are shown here as seen from a location 2 kpc directly above the sun. Angles in these diagrams are galactic longitudes measured clockwise from the sun. (a, Dan P. Clemens; b, adapted from a diagram by T. M. Dame, B. G. Elmegreen, R. S. Cohen, and P. Thaddeus.)

clouds and the formation of new stars (see Chapter 11). The newly formed stars and the remaining gas eventually move on through the arm and emerge from the front of the slow-moving arm to resume their travels around the galaxy (Figure 15–22). Thus, whenever we look we see a spiral arm, although the stars, gas, and dust are constantly changing.

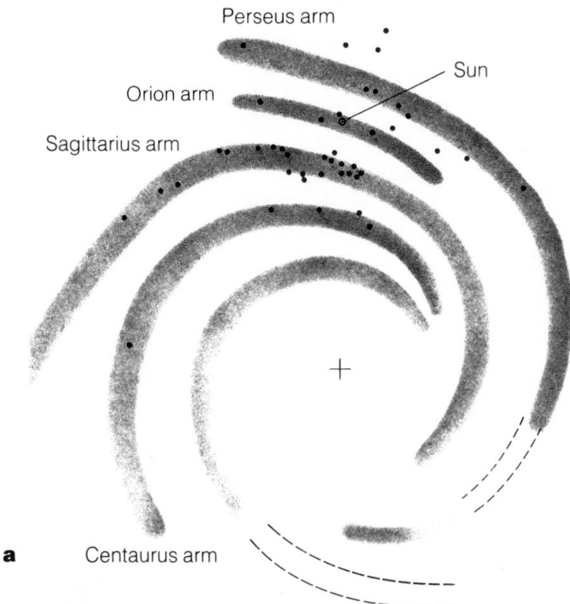

FIGURE 15–21 (a) Combined optical and radio data reveal the overall spiral pattern of the Milky Way. (b) An observer 1 million pc away might have this view of our galaxy. The cross marks the location of the sun. (b, Painting by M. A. Seeds based on a study by G. De Vaucouleurs and W. D. Pence.)

Stars of all masses form in the spiral arms, but the O and B stars are the brightest. These massive stars have very short lifetimes and cannot live long enough to move out of the spiral arm. They illuminate the spiral arms and make them stand out against the background of the rest of the galaxy. Lower-mass stars, like the sun, live longer and have time to travel around the galaxy many times. The sun probably formed in a star cluster 5 billion years ago when a gas cloud smashed into a spiral arm. Since that time the sun has escaped from its cluster and made roughly 20 trips around the galaxy, passing through spiral arms many times.*

The giant molecular-cloud complexes seem to be associated with spiral arms (Figure 15–20b), which is not surprising. They may form from the compression of gas clouds overtaking the arms. They are certainly active regions of star formation, with as much as 30 percent of their mass going into new stars. The Orion complex is a giant molecular-cloud complex (Color Plate 17) in which many infrared protostars are embedded, and the Ophiuchus dark cloud is a similar region with at least 20 infrared sources. Such a giant molecular-cloud complex may form stars for 10–100 million years before it is disrupted by the energy from the new stars. This is about how long the cloud would need to cross a spiral arm.

What we know about star formation seems to fit the spiral density wave theory very well, but the theory has two problems. First, how is the complicated spiral disturbance started and sustained? Spiral density waves should slowly die out in a billion years or so. Something must regenerate the spiral wave. One possibility is that the galaxy is naturally unstable to certain disturbances, just as a guitar string is unstable to certain vibrations. Any sudden disturbance—the rumble of a passing truck, for example—can set the string vibrating at its natural frequencies. Similarly, minor fluctuations in the galaxy's disk such as supernovae explosions might regenerate the waves. Another suggestion is that collisions between galaxies or the tidal forces produced by a passing galaxy might sustain the wave pattern.

The second problem for the density wave theory involves the spurs and branches in the arms of our own and other galaxies. Computer models of density waves produce regular, two-armed spiral patterns. Some galaxies, called grand-design galaxies, do indeed have sym-

*Some scientists have suggested that the periodic passage of the sun through the denser gas in spiral arms could affect the amount of sunlight reaching the earth and thus cause ice ages. This is interesting but highly speculative.

metric two-armed patterns but others do not (Color Plates 20–22). Other galaxies have a great many short spiral segments, giving them a fluffy appearance. These galaxies have been termed **flocculent**, meaning "woolly." Our galaxy is probably an intermediate galaxy. How can we explain these variations if the density wave theory always produces two-armed, grand-design spiral patterns? Perhaps the solution lies with a new theory that involves the compressional triggering of star formation.

SELF-SUSTAINING STAR FORMATION

An alternative theory to explain the spiral arms supposes that they are sustained by a kind of chain reaction of star formation. This **self-sustaining star formation** theory, like the density wave theory, is consistent with what we know about star formation, although it explains spiral arms in quite a different way from the spiral density wave theory.

Suppose a gas cloud forms a star cluster. The most massive star would be very hot, and the blast of radiation could compress nearby parts of the gas and trigger further star formation (Figure 15–23a). Alternatively, the most massive star could compress the neighboring gas when it explodes as a supernova. Massive stars live such short lives that their birth and their supernova death are separated by a mere instant in the life of a galaxy. Either way, star formation in a large gas cloud can be self-sustaining. Examples of such star formation have been found and studied, so we know it works. The Orion complex is such a region. The question is, does it make spiral arms?

The self-sustaining star formation theory depends on the differential rotation of the galaxy to produce spiral arms. Suppose that the galaxy had no spiral arms and self-sustaining star formation made stars in clumps scattered throughout the disk. Differential rotation would tend to drag the inner edges of the patches ahead of the outer edges, and each clump would begin to look like a segment of a spiral arm. If the star formation were truly self-sustaining, each clump would tend to grow larger until

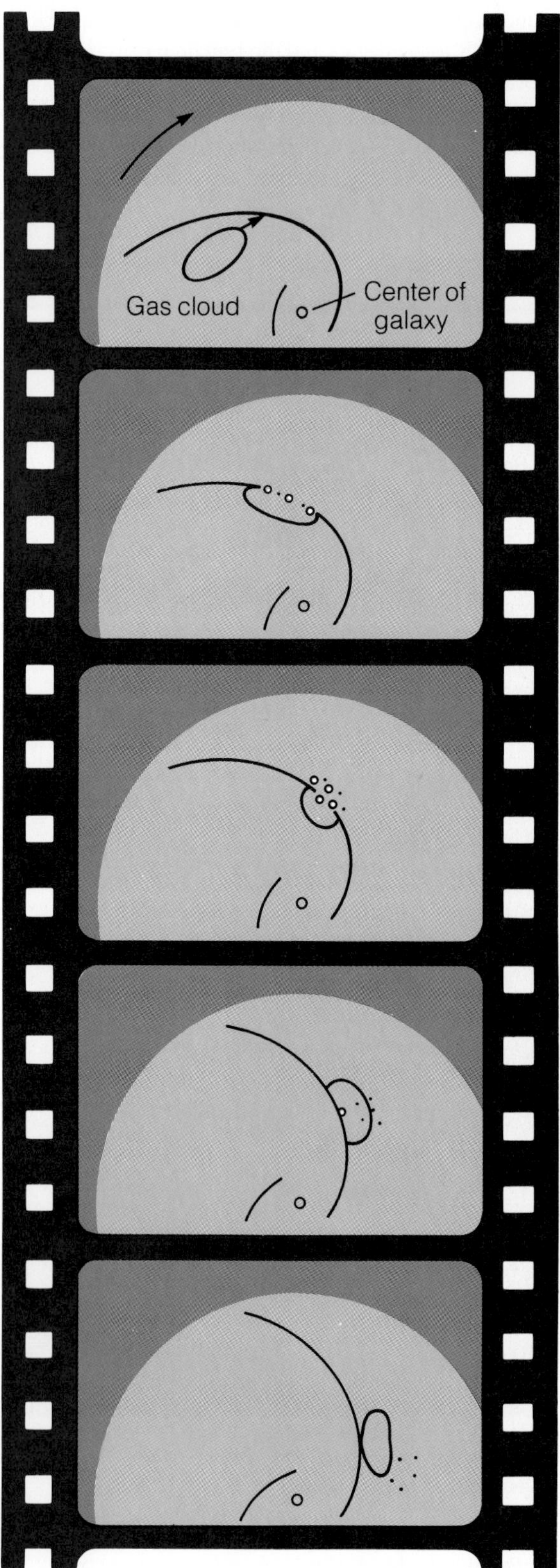

FIGURE 15–22 According to the density wave theory, gas clouds overtake the spiral arm from behind and smash into the density wave. The compression triggers the formation of stars. The massive stars (open circles) are so short-lived that they die before they can leave the spiral arm. The less massive stars (dots) emerge from the front of the arm with the remains of the gas cloud.

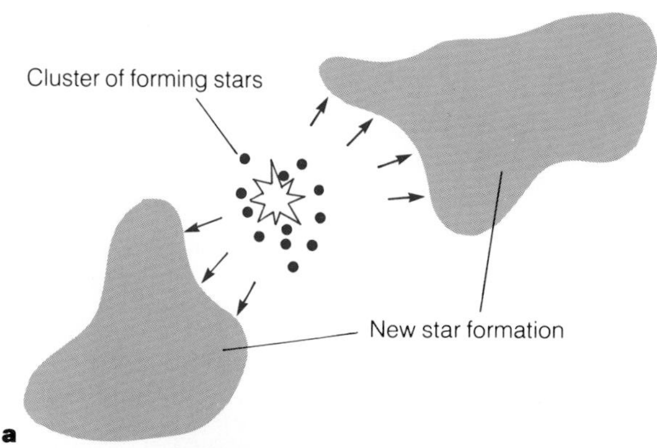

FIGURE 15–23 (a) Self-sustaining star formation may occur when a cluster of forming stars develops a massive member. The energetic turn-on or sudden supernova can compress nearby gas clouds and trigger new star formation. (b) A computer model of spiral arms produced by self-sustaining star formation is superimposed on galaxy M 81. The model has been tipped to match the inclination of the galaxy. (b, Reprinted courtesy of Humberto Gerola and *The Astrophysical Journal*, published by the University of Chicago Press, ©1978 The American Astronomical Society.)

the gas was exhausted or driven away. Then the galaxy would be littered with growing spiral segments.

Computer simulations of self-sustaining star formation do produce patterns that have spiral features (Figure 15–23b), but they do not have the symmetrical, grand-design patterns that we see in some galaxies. It seems likely that some galaxies are dominated by density waves and develop into beautiful, two-armed spiral galaxies. Other galaxies may be dominated by self-sustaining star formation and thus lack grand-design patterns. Instead, these galaxies may be flocculent. The difference may lie in the presence or absence of a trigger for the spiral-density wave. A galaxy that does not interact with other galaxies may not have a density wave generated and may consequently be flocculent.

Our galaxy seems intermediate; it seems to have a strong spiral pattern but also contains branches and spurs. It may have a density wave producing a grand-design pattern with self-sustaining star formation creating the spurs and branches (Figure 15–21b).

Spiral arms are graceful streamers of light in spiral galaxies like the Milky Way, and the processes that create them are fascinating in their complexity. Another region of our galaxy, the nucleus, is equally fascinating and complex, but it substitutes raw power for grace.

15.4 THE NUCLEUS

Forever hidden from human eyes, the center of our galaxy lies behind thick clouds of gas and dust. We can observe it, however, at radio, infrared, X-ray, and gamma-ray wavelengths. These photons penetrate the dust clouds and tell us of stars crowded together, a spinning disk of gas, a vast magnetic field, and of gas rushing outward.

OBSERVATIONS

On a dark night far from city lights, the Milky Way is dramatic, but even then we might notice only a slight thickening toward Sagittarius, the direction of the galaxy's center (Figure 15–24). Even Shapley's study of globular clusters identified the direction of the center only approximately. When radio astronomers turned their telescopes toward Sagittarius, they found a collection of radio sources, the most powerful of which, **Sagittarius A**, is believed to be the galactic core (Figure 15–25).

No visible light from the nucleus reaches us, but we can get information at other wavelengths. Before we

FIGURE 15–24 This photomosaic shows the Milky Way from Cassiopeia (left) to Sagittarius (right). The galactic center lies in Sagittarius within the box. See Figure 15-25. (Palomar Observatory photograph.)

explore the core in detail, let us survey the kinds of information we can get. Radio, infrared, X-ray, and gamma-ray photons have different energies and originate in different ways, so they can tell us about different areas of the core.

Radio telescopes aimed at the core can detect radiation from hot gas, most of which lies in clouds of ionized hydrogen around hot stars. Because hot stars live short lives and must have formed recently, this radiation is evidence that some star formation is still going on in the nucleus of the galaxy.

Of course, radio telescopes can detect 21–cm radiation from neutral hydrogen, but the nucleus is also a source of synchrotron radiation (Box 13–1). This radiation is produced when high-energy electrons move through a magnetic field. Thus, the synchrotron radiation is evidence of a magnetic field and some source of energy powerful enough to accelerate electrons to very high speeds.

Infrared radiation with a wavelength of 2 μm (micrometers) tells us about stars. Dust particles in space are not hot enough to emit many photons of so short an infrared wavelength. Instead, this radiation comes from cool giant stars. To produce so much of this radiation, the stars must be very crowded.

The Infrared Astronomy Satellite (Chapter 6) was able to image the entire Milky Way at wavelengths of 12, 25, 60, and 100 μm (Color Plate 14). Such long-wavelength photons could not have originated from even the coolest stars. Instead, they are emitted by interstellar dust grains with temperatures of a few tens of degrees Kelvin. Presumably, the dust is warmed by the closely packed stars in the core.

Orbiting space telescopes have detected X rays and gamma rays coming from the galactic core. The X rays must originate in very high-temperature objects—perhaps neutron stars or black holes in X-ray binaries scattered through the core. The gamma rays may be produced by collisions between high-energy particles or by the total annihilation of electrons and their antiparticles, positrons. More about that later.

These are the wavelengths we must use to explore the heart of our galaxy. To see what this radiation can tell us, let's take an imaginary journey from the sun to the center of the galaxy.

CONDITIONS IN THE CENTER

If we could suspend the laws of physics and journey from the earth to the center of our galaxy at the speed of light, it would take us 26,000–32,000 years. Remember, we are not sure of the distance. Along the way we would find some telling clues to the nature of the core.

We know from radio observations at 2.6 mm, the wavelength of emission from CO, that we would find a ring of giant molecular-cloud complexes from 4–6 kpc from the center, near the outer edge of the nuclear bulge. Inside the bulge we would find a dense aggregate of old stars, as shown by infrared observations, and a thin mixture of gas and dust. Some astronomers believe that the gas and dust is sparse there because it has been used up by star formation.

A little farther on our journey, 3 kpc from the center, we would find what seems to be a ring of cool hydrogen

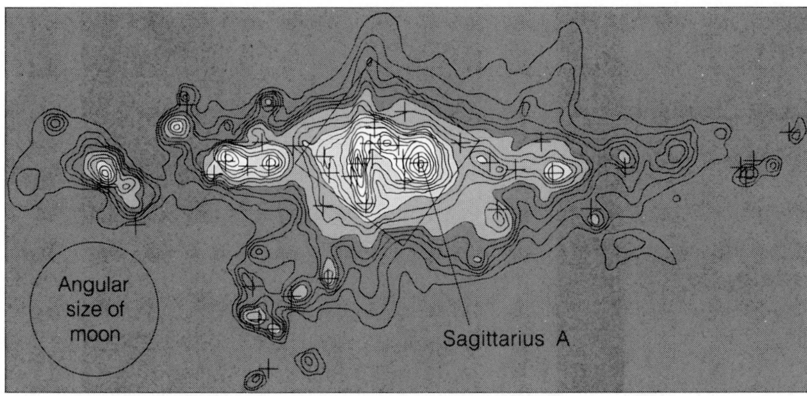

FIGURE 15–25 A radio map of the Sagittarius region of the Milky Way reveals an intense radio source, Sagittarius A, at the expected location of the center. Crosses mark far infrared sources associated with star formation. Box marks boundaries of Figure 15-26a. Angular size of the moon shown for comparison. (Adapted from observations by W. L. Altenhoff, D. Downes, T. Pauls, and J. Schraml; and by S. F. Odenwald and G. G. Fazio.)

expanding outward. Earthbound astronomers can detect it at 21 cm and know that the near side is approaching at 53 km/sec. The far side is receding at 135 km/sec. Radio astronomers have named these the 3-kpc arm (near side) and the 135-km/sec arm (far side), although no one is sure that they are actually part of the spiral pattern. When first discovered in 1964, they were explained as part of a giant gas ring puffed out of the center in an eruption about 30 million years ago.

Drawing closer to the very center of the galaxy, only a few hundred parsecs away, we would find a mixture of molecular clouds, dust, and both neutral and ionized hydrogen. The ionized gas is located around hot, young stars that have formed only recently.

As we approached the very center we would find a ring of ionized gas a few parsecs in radius rotating on an axis tipped at about 45°. We know this because infrared astronomers have detected an emission line of ionized neon at a wavelength of 12.8 μm. The Doppler shifts show that the gas is revolving around the core at high velocity.

In this region the stars are very crowded. In the region of space near the sun, light from one star can travel to the next star in about 5 years. Near the core of the galaxy, the stars are so close together that light can travel between stars in only 4 days. That is, the stars are only about 800 AU (0.004 pc) apart.

The orbital velocities of the gas clouds, evidenced by the emission from ionized neon, and presumably the velocities of the stars are very high in the inner few parsecs. At the very center is a mysterious source of radio, infrared, and gamma-ray radiation less than 10 AU (0.00005 pc) in diameter. Its mass must be very large, perhaps a few million solar masses, to keep the rapidly moving gas and stars in their orbits. Astronomers now speculate that the central object is a massive black hole, but whatever it is, it must be a tremendous source of energy.

THE ENERGY MACHINE

We know that the center of our galaxy is a powerful source of energy. One important piece of evidence surfaced when astronomers observed the center of the galaxy and detected gamma rays of a certain energy.

Although gamma rays are electromagnetic radiation, astronomers rarely refer to their wavelengths but rather specify their energy in **electron volts** (eV). One electron volt is a very small amount of energy. A housefly crawling 1 cm up a wall expends energy equivalent to 3 trillion eV. A photon of visible light has an energy of about 2 eV, but the particular gamma rays detected coming from the center of the galaxy have energies of 511 keV (kiloelectron volts). This is equivalent to a wavelength of about 0.0024 nm.

Gamma rays of 511 keV are produced by the annihilation of electrons and positrons (antielectrons). But what could produce high enough energies at the center of the galaxy to create positrons? Whatever it is, it must be small. As observed by the HEAO 3 satellite and later balloon flights, the 511-keV gamma-ray source decreased in intensity around 1980. To make such a sudden change, it must be smaller than 0.3 pc. At present only one hypothesis seems to fit all of the facts: matter flowing into a massive black hole. That hypothesis works because accre-

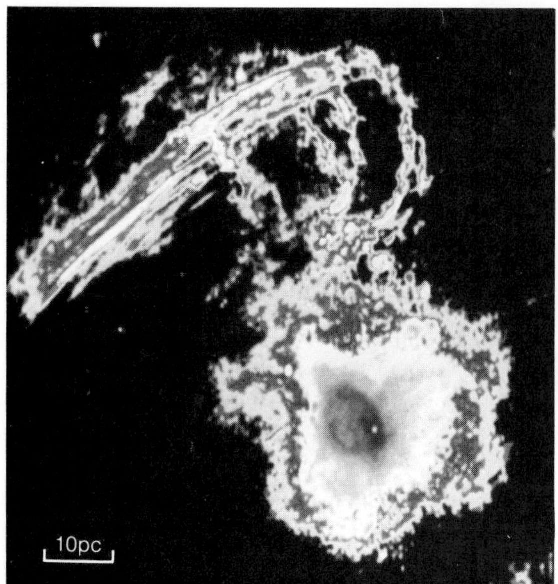

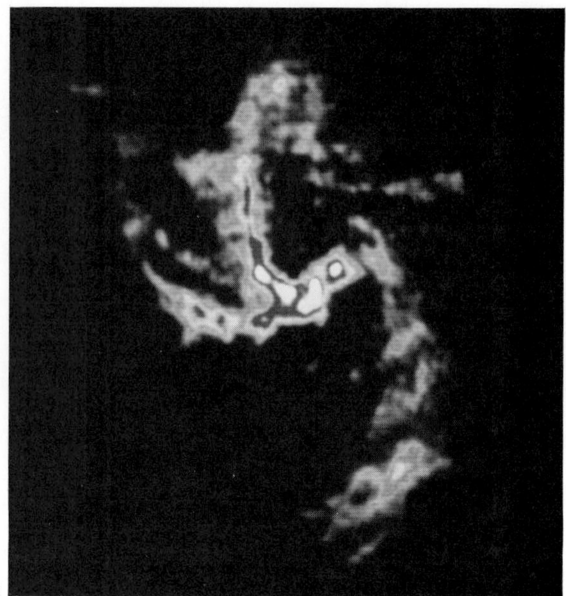

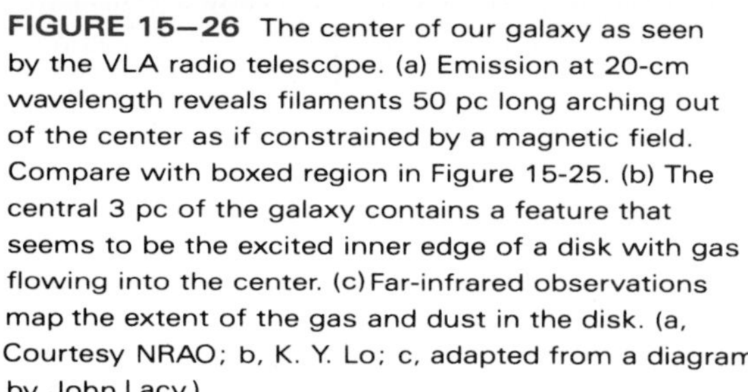
FIGURE 15–26 The center of our galaxy as seen by the VLA radio telescope. (a) Emission at 20-cm wavelength reveals filaments 50 pc long arching out of the center as if constrained by a magnetic field. Compare with boxed region in Figure 15-25. (b) The central 3 pc of the galaxy contains a feature that seems to be the excited inner edge of a disk with gas flowing into the center. (c) Far-infrared observations map the extent of the gas and dust in the disk. (a, Courtesy NRAO; b, K. Y. Lo; c, adapted from a diagram by John Lacy.)

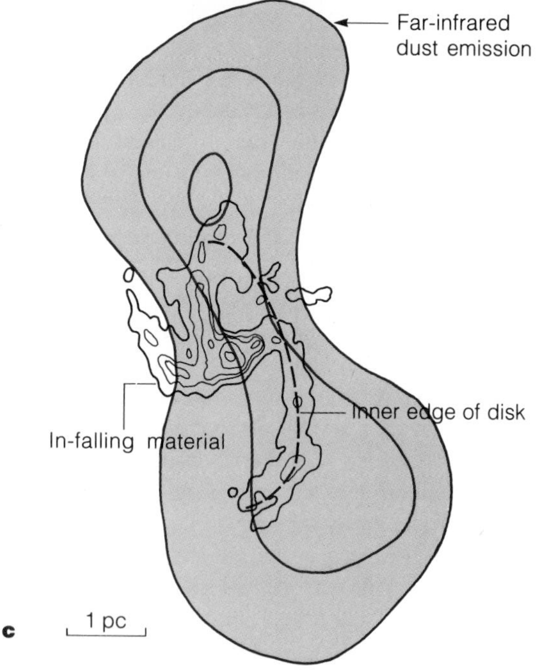

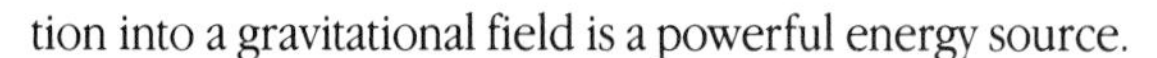

tion into a gravitational field is a powerful energy source.

Matter falling into a gravitational field can supply unlimited amounts of energy because there is no known limit to the amount of matter that can occupy a small space. If the center of our galaxy contained a collapsed object with a mass of 1 million solar masses or more, mass falling in would release tremendous energy. The more massive the central object, the more energy an infalling bit of matter releases.

Other processes have limits on their total energy production. For example, a supernova converts very little of its mass into energy because it blows itself apart so quickly. We could suppose many stars exploding as supernovae, but there is a limit to the number of stars we could pack into a small space without having them merge to form a massive black hole.

To make an energy machine at the center of our galaxy, we would need a collapsed object of a few million solar masses. Supermassive stars have been considered, but theory predicts that they would not be stable and would explode or collapse into a black hole. A supermassive neutron star would surely become a black hole. Such arguments are convincing astronomers that a supermassive black hole at the center of our galaxy could explain the observations.

We will see in Chapter 17 that many galaxies have peculiar energy sources at their cores. Some are much more powerful than the one at the center of our galaxy. Some of these energy machines are producing more energy than the galaxy itself and ejecting great jets of matter in opposite directions out of the galaxy.

VLA radio maps of the central core of the galaxy (Figure 15–26), combined with infrared observations, suggest a model for the central energy machine. A large-scale radio map of the central 50 pc (Figure 15–26a) shows the bright core with thin filaments 50 pc long and only 1 pc wide arching out of the core. These filaments seem to be constrained by a magnetic field, and some theorists now suggest that the field is generated by a rotating disk of matter around the core.

A higher-resolution radio map spanning only the inner few parsecs (Figure 15–26b) reveals a peculiar spiral feature. By comparing this with far-infrared maps of the gas and dust distribution near the core, we can model this spiral as the glowing inner edge of a disk of gas only a few parsecs in radius. Matter appears to be flowing into some energy source at the center. This suggests accretion into a massive black hole.

Other radio observations have detected a faint jet of matter flowing southward out of the nuclear region. Such jets are reminiscent of the powerful SS 433 with its pair of jets of radiation and matter (Chapter 14). An SS 433–style energy machine at the center of our galaxy would have to be more powerful, and therefore more massive than SS 433, but the process could be very similar. We will see in Chapter 17 that many galaxies have peculiar cores emitting jets of matter.

We cannot yet be sure that there is a black hole at the galactic center. To make further progress in our discussion, we must compare our galaxy with others. Many of those resemble the Milky Way, and by understanding how our galaxy compares with other galaxies, we may be better able to understand how our galaxy formed and evolved.

Disk stars are metal-rich population I stars moving in circular orbits that lie in the plane of the disk. Because stars farther from the center move slower in their orbits, the disk rotates differentially.

The spherical component consists of a nuclear bulge at the center and a halo of thinly scattered stars and globular clusters that completely envelops the disk. Halo stars are metal-poor population II stars moving in random, elliptical orbits.

The distribution of populations through the galaxy suggests a way the galaxy could have formed from a spherical cloud of gas that gradually flattened into a disk. The younger the stars, the more metal-rich they are and the more circular and flat their orbits are.

The very youngest objects lie along spiral arms within the disk. They live such short lives they don't have time to move from their place of birth in the spiral arms. Maps of the spiral tracers and cool hydrogen clouds reveal the spiral pattern of our galaxy. The spiral arms are also outlined by giant molecular-cloud complexes detectable by radio emission from carbon monoxide molecules. These clouds are sites of star formation.

The spiral-density wave theory suggests that the spiral arms are regions of compression that move through the disk. When an orbiting gas cloud smashes into the compression wave, the gas cloud forms stars. Another process, self-sustaining star formation, may act to modify the arms as the birth of massive stars triggers the formation of more stars by compressing neighboring clouds.

The nucleus of the galaxy is invisible at optical wavelengths, but radio, infrared, X-ray, and gamma-ray radiation can penetrate the dust clouds. These wavelengths reveal crowded central stars and heated clouds of dust. Some of the central features are expanding outward. The 3-kpc arm and the 135-km/sec arm are both moving outward from the center, as is a ring of dense molecular clouds. These expanding features suggest the nucleus may have exploded millions of years ago.

The very center of the Milky Way is marked by a radio source, Sagittarius A, which is also a source of infrared radiation. The core of the source is only 10 AU in diameter and contains about $5 \times 10^6 M_\odot$. Theorists suspect the core contains a massive black hole generating energy by accreting matter.

SUMMARY

The Milky Way galaxy contains two components, the disk component and the spherical component. Harlow Shapley determined the size of the galaxy and the peripheral location of the sun by studying the distribution of globular clusters. The traditional description of the galaxy includes a disk about 30 kpc in diameter with the sun about two-thirds of the way to the edge. However, studies of the rotation curve of the galaxy hint that it may be much larger and much more massive than previously believed.

NEW TERMS

Magellanic Clouds
kiloparsec (kpc)
disk component
cosmic ray
spiral arm
association
spherical component
high-velocity star
population I
population II
metals
nucleosynthesis
spiral tracers
density wave theory

halo
nuclear bulge
rotation curve
Keplerian motion
galactic corona
flocculent
self-sustaining star formation
Sagittarius A
electron volt (eV)

QUESTIONS

1. Why is it diffucult to specify the dimensions of the disk and halo?
2. Why didn't astronomers before Shapley realize how large the galaxy is?
3. If supernovae produce cosmic rays, why don't we observe cosmic rays coming from supernovae remnants? What do we observe?
4. What evidence do we have that our galaxy has a galactic corona?
5. Explain why some star clusters lose stars more slowly than others.
6. Contrast the motion of the disk stars and halo stars. Why do their orbits differ?
7. Why do high-velocity stars have lower-metal abundance than the sun?
8. Why are metals less abundant in older stars than in younger?
9. Why are all spiral tracers young?
10. Why couldn't spiral arms be physically connected structures? What would happen to them?
11. Why does self-sustaining star formation produce clouds of stars that look like segments of spiral arms?
12. Describe the kinds of observations we can make to study the galactic nucleus.
13. What evidence do we have that the nucleus of the galaxy contains an energy source that is very small?

PROBLEMS

1. Make a scale sketch of our galaxy in cross section. Include the disk, sun, nucleus, halo, and some globular clusters. Try to draw the globular clusters to scale size.
2. Because of dust clouds, we can see only about 5 kpc into the disk of the galaxy. What percentage of the galactic disk can we see? (HINT: Consider the area of the entire disk and the area we can see.)
3. If the fastest passenger aircraft can fly 0.45 km/sec (1000 mph), how long would it take to reach the sun? The galactic center? (HINT: 1 pc $= 3 \times 10^{13}$ km.)
4. If a typical halo star has an orbital velocity of 250 km/sec, how long does it take to pass through the disk of the galaxy? Assume that the disk is 1000 pc thick.
5. The brightest star in Figure 15–11 is ϕ Cassiopeia, an F0 star with an apparent magnitude of 5.0. If it is a member of the cluster, it is 2800 pc away. What is its absolute magnitude? What kind of star is it?
6. If the RR Lyrae Stars in a globular cluster have apparent magnitudes of 14, how far away is the cluster? (HINT: See Box 12–2.)
7. If interstellar dust makes an RR Lyrae variable star look 1 magnitude fainter than it should, by how much will we overestimate its distance? (HINT: See Box 9–2.)
8. If a globular cluster is 10 minutes of arc in diameter and 8.5 kpc away, what is its diameter? (HINT: Use the small-angle formula, Box 3–1.)
9. If we assume that a globular cluster 4 minutes of arc in diameter is actually 25 pc in diameter, how far away is it? (HINT: Use the small-angle formula, Box 3–1.)
10. If the sun is 5 billion years old, how many times has it orbited the galaxy?
11. If the true distance to the center of the galaxy is found to be 7 kpc and the orbital velocity of the sun is 220 km/sec, what is the minimum mass of the galaxy? (HINT: See Box 15–2.)
12. What temperature would interstellar dust have to have to radiate most strongly at 100 μm? (HINTS: 1 μm = 1000 nm. See Box 7–1.)
13. Infrared radiation from the center of our galaxy with a wavelength of about 2 μm (2×10^{-6}) comes mainly from cool stars. Use this wavelength as λ_{max} and find the temperature of the stars.
14. If an object at the center of our galaxy has a linear diameter of 10 AU, what will its angular diameter be as seen from the earth? (HINT: See Box 3–1.)

RECOMMENDED READING

BERENDZEN, RICHARD, RICHARD HART, and DANIEL SEELEY *Man Discovers the Galaxy*. New York: Science History Publications, 1976.

BHAT, CHAMAN, CHRIS MAYER, and ARNOLD WOLFENDALE "In Search of the Source of Cosmic Rays." *New Scientist 108* (6 Feb. 1986), p. 48.

BLITZ, LEO "Giant Molecular-Cloud Complexes in the Galaxy." *Scientific American 246* (April 1982), p. 84.

BOK, B. "The Milky Way Galaxy." *Scientific American 244* (March 1981), p. 92.

———. "A Bigger and Better Milky Way Galaxy." *Astronomy 12* (Jan. 1984), p. 6.

BOK, BART J., and PRISCILLA F. BOK *The Milky Way*. 5th ed. Cambridge, Mass.: Harvard University Press, 1981.

CLARK, GAIL O. "Stellar Populations: Key to the Clusters." *Astronomy 14* (Oct. 1986), p. 106.

FRIEL, EILEEN "A Symposium on Stellar Populations." *Mercury 13* (Nov./Dec. 1984), p. 165.

HASLAM, GLYN, RICHARD WIELEBINSKI, and WOLFGANG PRIESTER "Radio Maps of the Sky." *Sky and Telescope 63* (March 1982), p. 230.

HERBST, W., and G. E. ASSOUSA "Supernovas and Star Formation." *Scientific American 241* (Aug. 1979), p. 138.

HIRSCHFELD, ALAN "How Far Is the Galactic Center?" *Sky and Telescope 68* (Dec. 1984), p. 498.

KILLIAN, ANITA "Galactic Center Update." *Sky and Telescope 71* (March 1986), p. 255.

KRAUS, JOHN "The Center of Our Galaxy." *Sky and Telescope 65* (Jan. 1983), p. 30.

LESH, J. R. "Swarms of Stars: Cosmic Calibrators." *Astronomy 6* (March 1978), p. 6. Reprinted in *Astronomy: Selected Readings*. ed. M. A. Seeds. Menlo Park, Calif.: Benjamin/Cummings, 1980, p. 111.

MALIN, DAVID F. "Dust Clouds of Sagittarius." *Sky and Telescope 63* (March 1982), p. 255.

MEWALDT, RICHARD A., EDWARD C. STONE, and MARK E. WIEDENBECK "Samples of the Milky Way." *Scientific American 247* (Dec. 1982), p. 108.

PASACHOFF, J. M., and R. W. GOEBEL "Laboratory Exercises in Astronomy—Cepheid Variables and the Cosmic Distance Scale." *Sky and Telescope 57* (March 1979), p. 241.

ROBINSON, LEIF J. "The Black Heart of the Milky Way." *Sky and Telescope 64* (Aug. 1982), p. 133.

SMITH, DAVID H., and LEIF J. ROBINSON "Dissecting the Hub of Our Galaxy." *Sky and Telescope 68* (Dec. 1984), p. 494.

TWAROG, B. "The Chemical Evolution of Our Galaxy." *Mercury 14* (July/Aug. 1985), p. 119.

WHITNEY, C. A. *The Discovery of Our Galaxy*. New York: Alfred A. Knopf, 1971.

CHAPTER 16

GALAXIES

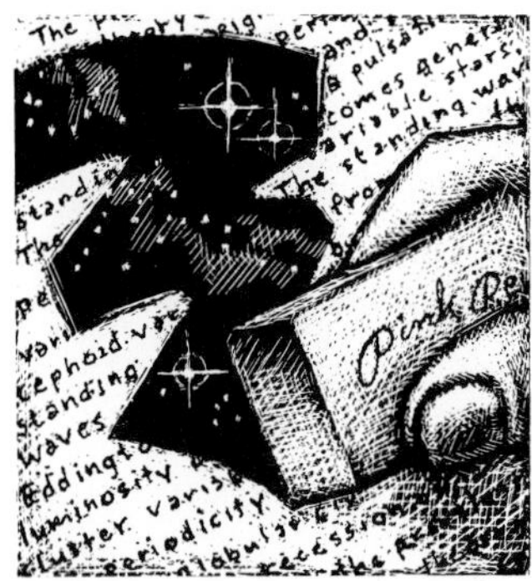

A hypothesis or theory is clear, decisive, and positive, but it is believed by no one but the man who created it. Experimental findings, on the other hand, are messy, inexact things which are believed by everyone except the man who did that work.

Harlow Shapley
THROUGH RUGGED WAYS TO THE STARS

If an evil wizard turned off all the stars of the Milky Way galaxy, our sky would be very black. Only the Great Galaxy in Andromeda and the two Magellanic Clouds—small, nearby galaxies—would be visible to the naked eye. Yet the universe is filled with roughly 100 billion galaxies, which cover the entire sky except along the plane of the Milky Way (Figure 16–1). There, in what is known as the **zone of avoidance**, the thick dust clouds of our Milky Way block our view of the distant galaxies.

Previous chapters took us to the stars, out into the spiral arms of our galaxy, and finally into its mysterious core. Yet we have hardly begun our exploration of the universe. If our journey to the farthest galaxies were compared to a flight around the world, our airplane would not yet have traveled 1000 feet. In this chapter we leave behind the familiar Milky Way and penetrate deep into space, out among the vast clouds of galaxies.

We do not, however, leave behind the tools and methods of previous chapters. The tools we developed to determine the properties of stars (Chapter 9) are useful for determining the properties of galaxies. The most important properties are distance, size, luminosity, mass, and motion. As in the case of stars, distance is the key. Once we know the distance to a galaxy, its size and luminosity are easy to find. In some cases, we can even find the masses of galaxies.

Although millions of galaxies are visible on long-exposure photographs of the sky, there are only a few types of galaxies. The shapes of these galaxy types, the kinds of stars they contain, and their distribution in clusters of galaxies are hints to their origin and evolution.

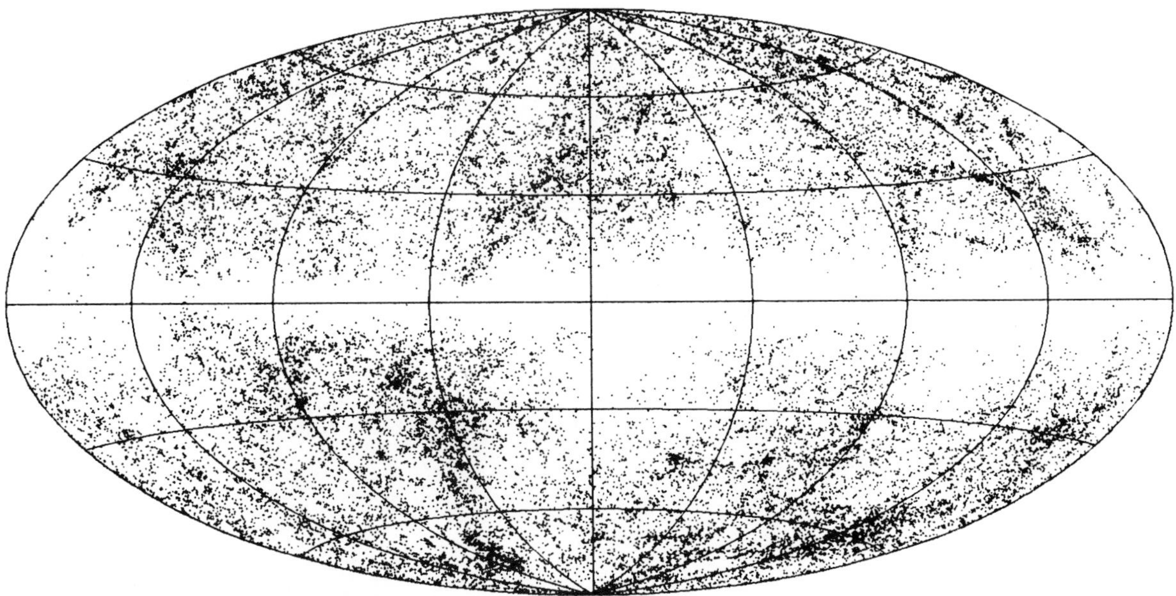

FIGURE 16–1 A total of 34,729 galaxies are plotted in this map of the entire sky. The plane of the Milky Way runs along the center from right to left. The dust clouds in our galaxy block our view of other galaxies, producing the region of missing galaxies known as the zone of avoidance. Note the clumpy distribution of galaxies. (Courtesy of Nigel A. Sharp.)

16.1 MEASURING THE PROPERTIES OF GALAXIES

Perhaps the basic problem in astronomy is finding the distance to objects. Until we know the distance, we can discover very few of an object's properties. Since astronomers first identified galaxies as objects outside the Milky Way, they have sought to find the distances to individual galaxies. In this section we will see how the distances to the galaxies unlock their basic properties.

DISTANCE, DIAMETER, AND LUMINOSITY

The distances to galaxies are so large it is not convenient to express them in light-years, parsecs, or even kiloparsecs. Instead we will use the unit **megaparsec** (Mpc), or 1,000,000 pc. One Mpc equals 3.26 million ly, or approximately 2×10^{19} miles.

To find the distance to a galaxy, we must search among its stars for a familiar object whose luminosity or diameter is known. Such objects are called **distance indicators.**

Because their period is related to their luminosity, Cepheid variable stars are fairly reliable distance indicators. Chapter 12 explained how a giant star evolving through the instability strip in the H–R diagram could pulsate with a period between 1–60 days. Because the period of pulsation and the star's average luminosity depend on its mass, a period–luminosity relation exists (Figure 12–10). If we know the star's period, we can use the period–luminosity diagram to learn its average absolute magnitude. By comparing its apparent magnitude with its absolute magnitude, we can find its distance (Box 12–2).

Cepheids can reveal the distance to any galaxy in which individual Cepheids are visible (Figure 16–2). Unfortunately, only about a dozen galaxies are close enough to have visible Cepheids. Generally galaxies beyond 6 Mpc are too distant to have detectable Cepheids, but during 1986 astronomers used CCD images to identify

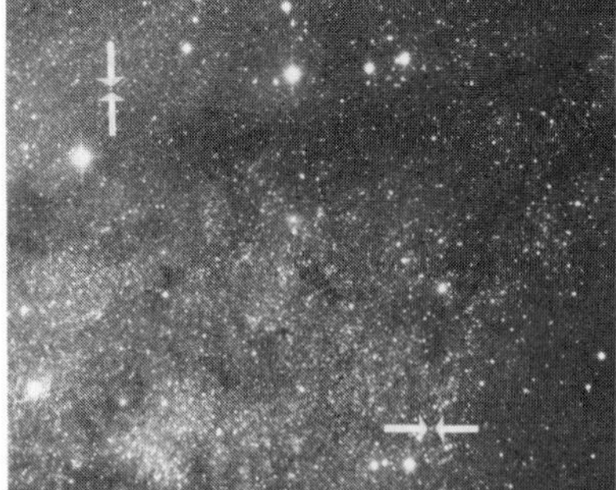

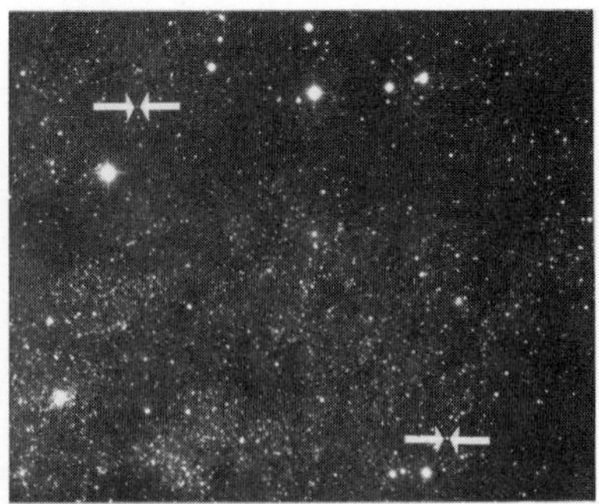

FIGURE 16–2 The Andromeda galaxy is relatively close and can be resolved into individual stars. In these two views of a section of the galaxy, variable stars are visible. A determination of the stars' periods and average apparent magnitudes can establish the distance to the galaxy. (Palomar Observatory photograph.)

FIGURE 16–3 H-II regions are visible in the spiral arms of this galaxy as small, glowing patches of gas. Because H-II regions have known diameters, we can measure their angular diameters and use the small angle formula to find the distance to the galaxy. (Palomar Observatory.)

Cepheids in the galaxy M101 at a distance of 6.3–7.9 Mpc. Such new techniques may be able to extend the limits slightly, but the Hubble Space Telescope (HST) will be able to search for Cepheids out to about 40 Mpc.

Studies of the few galaxies whose distances are known from Cepheids reveal large numbers of bright giants and supergiants, large globular clusters, and occasional novae at maximum brightness. These objects have absolute magnitudes of about -9, so they too can be used as distance indicators—that is, they can be calibrated. If we found bright giants and supergiants in a galaxy, we could measure their apparent magnitude and find the distance to the galaxy. The same is true for novae and large globular clusters. These distance indicators are good out to about 25 Mpc, beyond which objects of this brightness are invisible.

Another distance indicator is the cloud of ionized hydrogen around hot stars—H II regions (Figure 16–3). The Great Nebula in Orion is an H II region. Studies of the nearest galaxies, whose distances are known from other distance indicators, show that the largest of these H II regions have predictable diameters. If we detect H II regions in a distant galaxy, we can measure their angular diameter and assume they have the same diameter as the H II regions in other galaxies. Then we can find the distance from the small-angle formula (Box 3–1). The H II regions are important distance indicators because they are easy to detect and because they give us distances to

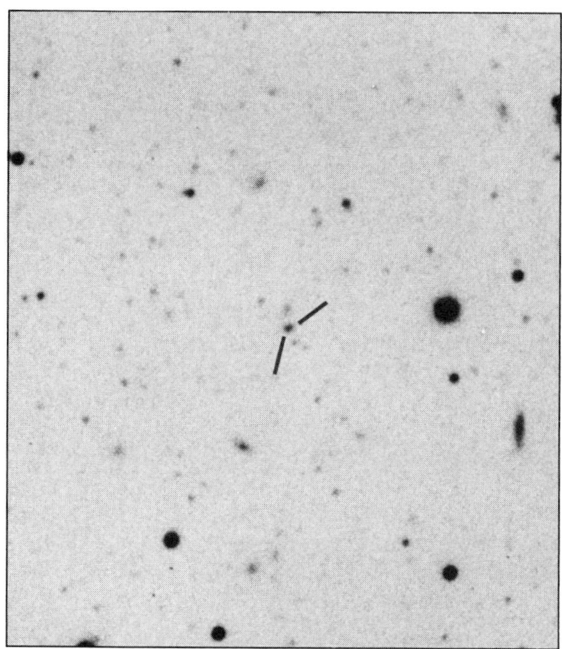

FIGURE 16–4 One of the most distant galaxies known—the visible galaxy associated with radio source 3C 13—is shown on this negative reproduction of a photographic plate. The galaxy is roughly 10 billion light-years away. (Hyron Spinrad.)

more distant objects. Unfortunately, the size of the largest H II regions may differ slightly from galaxy to galaxy, and this variation limits the reliability of the method.

Supernovae can also give us distances, if we assume that supernovae in other galaxies reach the same maximum brightness as supernovae in nearby galaxies. This method will work to great distances, but supernovae are rare, so we might wait 50–100 years to see one in a particular galaxy.

To measure distances to the farthest galaxies, we must use the galaxies themselves as distance indicators. Studies of nearby galaxies, whose distances are known by more reliable methods, tell us the brightness of the brightest galaxies. If we see a very distant group of galaxies, we might assume that the brightest galaxies in the group were of the same brightness as the brightest nearby galaxies. Then we could find the distance to the group. Averaging the brightest galaxies in the group and other statistical methods help, but this is not a very accurate technique.

In spite of the uncertainties, it is clear that galaxies are far apart and scattered through the universe to tremendous distances. The nearest large galaxy, the Andromeda galaxy, is 0.66 Mpc (2.2 million ly) distant. One of the most distant galaxies presently known is 3C13, at a distance of about 10 billion ly. Little is visible at such distances (Figure 16–4). Radio telescopes can detect faint sources of radio energy that are evidently distant galaxies much more luminous in radio energy than in visible light. Thus, radio telescopes can detect galaxies beyond the limit of the largest optical telescopes.

When the HST is launched by the Space Shuttle, it will greatly improve our knowledge of the distances to galaxies. It will be able to detect Cepheids out to 40 Mpc—as far as the very important group of galaxies known as the Virgo cluster. Not only will this better determine the distances to nearby galaxies, but it will also improve the calibration of other methods of finding distances. In addition, the HST will be able to detect galaxies at even greater distances than can present telescopes.

The vast distances between the galaxies produce an effect akin to time travel. When we look at a galaxy billions of light-years away, we see it as it was billions of years ago when its light began the journey toward earth. Thus, when we look at a distant galaxy we look into the past by an amount called the **look-back time**, a time in years equal to the distance to the galaxy in light-years. Although we have expressed most distances in parsecs, it is convenient to express the distances to galaxies in light-years because of this relationship between distance and look-back time.

The look-back time for nearby galaxies is not significant because galaxies change very slowly. But when we look at more distant galaxies, the look-back time becomes an appreciable part of the age of the universe. We will see evidence in Chapter 18 that the universe began 10–20 billion years ago. Thus, we see the most distant visible galaxies as they were when the galaxies and the universe were much younger. This effect will be important in this and the next chapter.

Clearly the distance to a galaxy is the key to finding its diameter and luminosity. If we can measure its distance by some reliable indicator, we can use its apparent brightness to find its luminosity (Box 9–2). This will work so long as we did not estimate the galaxy's luminosity in the first place to find its distance. In addition, once we know the distance, we can measure a galaxy's angular diameter and then use the small-angle formula (Box 3–1) to find its diameter in parsecs.

Finding a galaxy's diameter and luminosity is easy when the distance is known, but finding a galaxy's mass is a challenging puzzle.

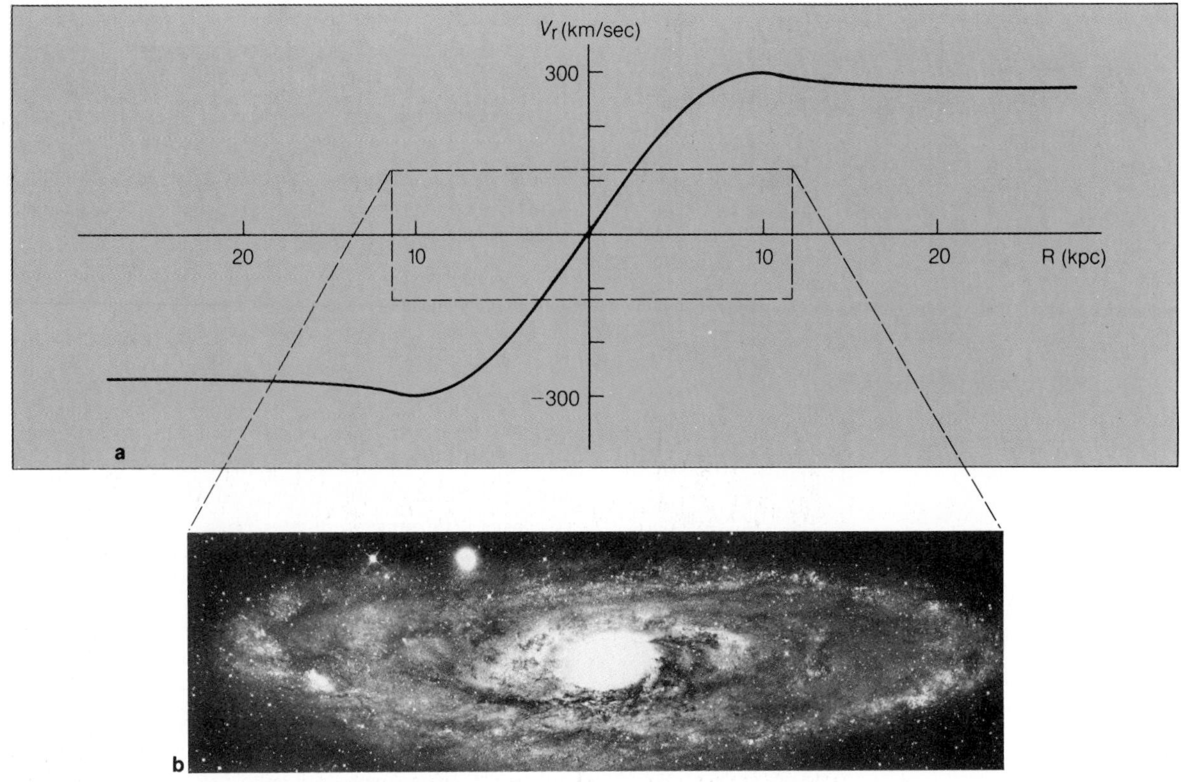

FIGURE 16–5 The rotation curve of the Andromeda galaxy (a) shows that orbital velocity in the outer regions does not decline. This implies that the galaxy is surrounded by a corona of dark matter. Compare with Figure 15–14. The photo (b) shows only the brighter part of the galaxy, which has a diameter of about 23 kpc. If the entire galaxy were visible in the night sky, it would span over 5°—ten times the diameter of the full moon. (Diagram adapted from 21-cm observations by Morton S. Roberts and Robert N. Whitehurst; Lick Observatory photograph.)

MASS

Although the mass of a galaxy is difficult to determine, it is an important quantity. It tells us how much matter the galaxy contains, which gives us clues to the galaxy's origin and evolution. In this section we will examine four ways to find the masses of galaxies.

One way is called the **rotation curve method.** We begin by photographing the galaxy's spectrum at different points along its diameter and plotting the Doppler-shift velocities in a rotation curve like that in Figure 16–5. This tells us how fast the galaxy is rotating. The sizes of the orbits the stars follow around the galaxy's center are related to the size of the galaxy, easily found from its angular diameter and its distance. We then ask how massive the galaxy must be to hold stars in orbits of that size at that velocity. This is the same method we used in Box 15–2 to find the mass of our own galaxy from the size of the sun's orbit and its orbital velocity.

The rotation curve method may be the best way to find the masses of galaxies, but it suffers from two shortcomings. First, it can be applied only to the nearer galaxies. More distant galaxies look so small that the astronomer cannot photograph their spectrum at different points along their diameter and thus cannot determine their rotation curves. Second, the rotation curve method measures the mass only in the luminous part of the galaxy, the part we can see, and astronomers are finding increasing evidence that galaxies have extended coronas. For example, the rotation curves of some galaxies may not turn downward in Keplerian motion far from the center (Figure 15–14). As we saw in the case of the Milky Way galaxy, this tells us that the outer parts of the galaxy contain significant amounts of mass. A recent study of the spiral galaxy NGC 3067 shows that 94 percent of the mass lies outside the visible part of the galaxy. The rotation curve method misses this mass, so we must use masses found in this way with some care.

It is not unusual to find pairs of galaxies orbiting each other. In such cases we analyze them as if they were a pair of binary stars. The only difficulty with this **double galaxy method** is that the galaxies take hundreds of millions of years to orbit once, so over periods of thousands of years we would see no motion. Because we can't observe the shapes of the orbits, we can't determine the orbital inclination, and we can't find their true masses. The solution to this problem is to average the results for many pairs of galaxies. Different orbital inclinations should average out and give good average masses.

The **cluster method** of finding the masses of galaxies depends on the motions of galaxies within a cluster. If we measure the radial velocities of many galaxies in a cluster, we find that some velocities are larger than others because of the orbital motions of the individual galaxies in the cluster. Given the range of velocities and the size of the cluster, we can ask how massive a cluster of this size must be to hold itself together with this range of velocities. Dividing the total mass of the cluster by the number of galaxies in the cluster yields the average mass of the galaxies. This method contains the built-in assumption that the cluster is gravitationally bound; that is, it is not coming apart. If it is, our result is too large. Because it seems likely that most clusters are bound, the method is probably valid.

However, some large clusters do not appear to contain enough mass to hold themselves together. Assuming that the clusters are not coming apart, we are led to the conclusion that the clusters must contain large amounts of mass, which we cannot see. This has been called the **missing mass** problem, but it might be better to think of it as the hidden mass problem. The mass must be there, but we haven't been able to detect it.

The missing mass may represent as much as 90 percent of the mass of the clusters, but no one knows what that matter could be. It can't be hot gas or we would see X rays or ultraviolet radiation. Some clusters do contain hot gas that emits X rays, but it is not enough mass to hold the cluster together. Similarly, the missing mass can't be cool stars, dust clouds, or cold gas or we would detect it at infrared and radio wavelengths. In recent years theorists have begun referring to this mass as **dark matter.** They suggest that dark matter is made up of exotic atomic particles trapped in extended galaxy coronae much like the corona that must surround our own galaxy. Whatever this dark matter is, it seems clear that we see only the most luminous parts of a galaxy.

The fourth way of measuring a galaxy's mass is the **velocity dispersion method.** It is really a version of the cluster method. Instead of looking at the motions of galaxies in a cluster, we look at the motions of stars in the galaxy and estimate the mass needed to hold the galaxy together. This method, like the rotation curve method, measures only the inner, visible parts of a galaxy and gives generally lower masses.

Measuring the masses of galaxies tells us two things. First, the range of masses is large—from 10^{-6} as much as the Milky Way galaxy to 50 times more. And second, there is more to a galaxy than meets the eye.

MOTION

Because many galaxies look like great whirlpools of stars and Doppler shifts prove that they are rotating, we might hope to see some motion. Unfortunately, galaxies are too large and too distant and move too slowly to show any visible change over hundreds or thousands of years.

Although we cannot watch galaxies move, we can measure radial velocities by the Doppler effect. Such measurements led to a discovery made in the years following World War I when astronomers began to accumulate spectra of many faint nebulae. As it became evident that many of these were galaxies, a few astronomers noticed that the spectra of galaxies contained lines shifted toward the red. Assuming that this was due to the Doppler effect, they concluded that nearly all of the galaxies were receding from us.

In 1929 the American astronomer Edwin Hubble announced a general law of red shifts now known as the **Hubble law.** This law says that a galaxy's velocity of recession equals a constant times its distance. Thus, the more distant a galaxy is, the faster it recedes from us (Figure 16–6). The constant H, now known as the **Hubble constant,** is very difficult to determine.

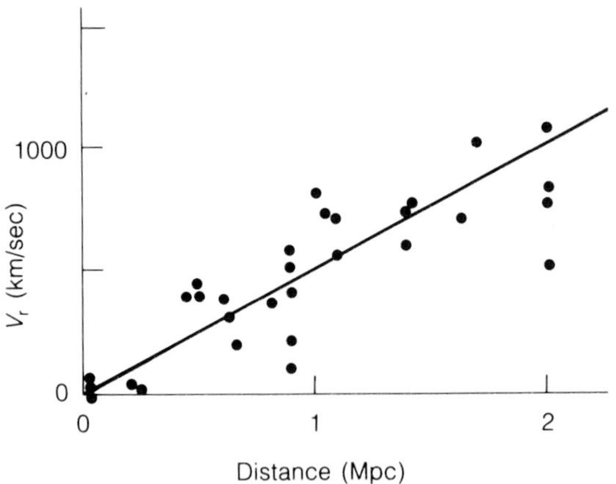

FIGURE 16–6 Edwin Hubble's first diagram of the velocities and distances of galaxies did not probe very deeply into space. It did show, however, that the galaxies are receding from each other. Because of errors in his distances, Hubble's first estimate of the Hubble constant was too large.

BOX 16–1
The Hubble Law

The Hubble law relates a galaxy's radial velocity V_r in kilometers per second to its distance D in megaparsecs. We can visualize this relation as a graph in which we plot radial velocity and distance for a number of galaxies (Figure 16–6). Points that represent galaxies fall along a straight line, showing that the more distant a galaxy, the faster it recedes from us. A galaxy's radial velocity in kilometers per second equals the Hubble constant times its distance in megaparsecs.

$$V_r = HD$$

The best measurements of distance and velocity suggest that H is approximately equal to 70 km/sec/Mpc. This tells us that for every million parsecs that separate two galaxies, they recede from each other at 70 km/sec. Thus, two galaxies 10 Mpc apart move away from each other at 700 km/sec.

If we can measure the radial velocity of a galaxy, we can estimate its distance from our galaxy by dividing by the Hubble constant. For example, the Virgo cluster of galaxies has a radial velocity of 1180 km/sec. To find its distance, we divide by H, giving 16.8 Mpc.

One important study of the recession of the galaxies suggests that H equals about 50 km/sec/Mpc.* However, other studies yield values as high as 100 km/sec/Mpc. The uncertainty arises from the difficulty of determining the distances to galaxies. Different groups of astronomers have found distances in different ways and have arrived at different values of H. In Chapter 18 we will see that this has important consequences for our understanding of the history of the universe, but here it is sufficient to recognize that the Hubble constant is poorly known and to adopt a provisional value of about 70 km/sec/Mpc.

The Hubble law is important because it is commonly interpreted to show that the universe is expanding. In Chapter 18 we will discuss the implications of this expansion, but here we will use the Hubble law as a practical way to estimate the distance to a galaxy. As Box 16–1 explains, a galaxy's velocity of recession divided by the Hubble constant equals its distance. This makes it relatively easy to find galactic distances, because large telescopes can photograph the spectrum of a distant galaxy and reveal its red shift even though distance indicators such as variable stars are totally invisible.

However, we cannot abandon distance indicators and use the Hubble law exclusively. In Chapter 17 we will discuss peculiar galaxies that, according to some astronomers, may not obey the Hubble law. In addition, Chapter 18 will describe how distant galaxies may recede faster than the Hubble law predicts. To detect such departures from the law, we must measure distances with distance indicators and use the red shifts only for estimates.

It is clearly impossible to find the distance, diameter, luminosity, and mass of each of the millions of galaxies in the sky. Instead, we will classify the galaxies into categories and then study the properties of the galaxies in each category. Understanding the differences between these types of galaxies will lead us to theories of their origin and evolution.

*H has the units of a velocity divided by a distance. These are usually written as km/sec/Mpc, meaning km/sec per Mpc.

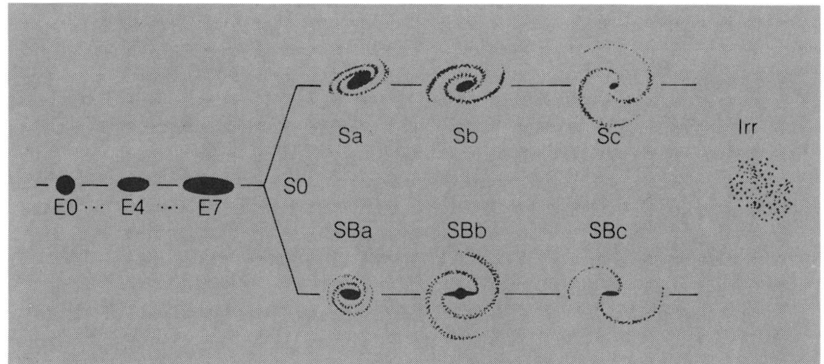

FIGURE 16–7 The Hubble tuning fork diagram organizes the galaxy types in an orderly way. The farther to the right a galaxy is located in the diagram, the richer it is in gas, dust, metals, and young stars.

a

b

FIGURE 16–8 (a) M 87 is a giant elliptical galaxy surrounded by a swarm of more than 500 globular clusters. (b) NGC 185 is a dwarf elliptical peculiar for the dust cloud seen against its bright center. It is a satellite of the Andromeda galaxy and is resolved into individual stars. (a, National Optical Astronomy Observatories; b, Lick Observatory photograph.)

16.2 THE MORPHOLOGY OF GALAXIES

Morphology, meaning form or structure, is often used by biologists to refer to the variation in structure among related organisms. Because we must compare different types of galaxies that appear to be related but differ in form, we refer to our study as galactic morphology. Just as the morphology of organisms may tell a biologist how a species evolved, galactic morphology may give us clues to how galaxies evolve.

We will begin our study of galaxy shapes by classifying them into three broad classes: elliptical, spiral, and irregular. We can then subdivide these classes to account for small variations in form among similar galaxies. To organize these classes in an easily remembered system, astronomers usually arrange them in a **tuning fork diagram** as first devised by Edwin Hubble (Figure 16–7).

ELLIPTICAL GALAXIES

About 70 percent of all galaxies are **elliptical** (Figure 16–8). They are round or elliptical in shape, have almost no visible gas or dust, lack hot bright stars, and have no spiral pattern. The stars in elliptical galaxies are more crowded toward the center, and the outer parts of some larger ellipticals are peppered by hundreds of globular clusters (Figure 16–8a).

The largest ellipticals are about 5 times larger in diameter than our galaxy, and the smallest are 100 times smaller. This wide range is also reflected in their mass, with the most massive being 50 times the mass of our galaxy and the least massive 10^{-6} times (Table 16–1). The largest of these galaxies are called giant ellipticals and the smallest, dwarf ellipticals (Figure 16–8b).

Elliptical galaxies are identified by the letter *E* followed by a number ranging from 0 to 7, indicating the apparent shape of the galaxy. Galaxies that look circular are classified E0; the more elliptical the galaxy's outline, the larger the number. No elliptical galaxy is known to be more elliptical than E7.

The spectra of elliptical galaxies indicate that they are limited to a single generation of stars. A few studies have detected small dust and gas clouds where a few new stars are forming but for the most part all of their stars are old. Lack of extensive gas and dust means that few new stars can form, so the only stars we see are old, lower-mass stars like those in our galaxy's spherical component. The massive stars would have died long ago.

It is tempting to think of an elliptical galaxy, with its swarm of globular clusters, as a galaxy that lacks a disk component, but that may be wrong. Until recently, astronomers thought that E galaxies are shaped like **oblate spheroids**—that is, like hamburger buns (Figure 16–9a). This would be the shape we would expect if the galaxies were flattened by their own rotation. However, photographs of E galaxies show only their projected outlines, not their true three-dimensional shapes. Recent evidence suggests that E galaxies are not necessarily oblate spheroids. Some may be **prolate spheroids**—shaped like melons (Figure 16–9b). In fact, some may be **triaxial ellipsoids**—hot dog buns that are longer than they are wide and wider than they are thick (Figure 16–9c).

Another clue to the origin of these galaxies may lie in the dwarf ellipticals. These are small systems, containing as few as 10 million stars (about 0.01 percent the number in our galaxy), and extending only a few hundred parsecs in diameter. In fact, it is not clear if there is a difference between a small dwarf elliptical galaxy and a large globular cluster.

Because these galaxies are small and contain few stars, they are not very luminous and are thus hard to find. Of the approximately two dozen galaxies near the

Table 16–1 The properties of galaxies.*

	Elliptical	Spiral	Irregular
Mass	0.0001–50	0.005–2	0.0005–0.15
Diameter	0.01–5	0.2–1.5	0.05–0.25
Luminosity	0.00005–5	0.005–10	0.00005–0.1

*In units of the mass, diameter, and luminosity of the Milky Way.

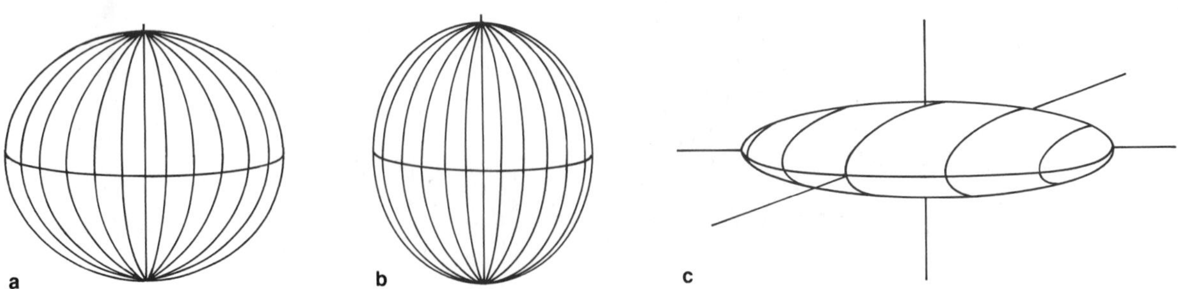

FIGURE 16–9 Possible shapes for elliptical galaxies: (a) oblate spheroid, (b) prolate spheroid, (c) triaxial ellipsoid.

Milky Way galaxy, a dozen are dwarf ellipticals. If dwarf ellipticals are that common throughout the universe, they are the most common form of galaxy.

SPIRAL GALAXIES

Although **spiral galaxies** make up only 15 percent of all galaxies, they are the most striking. Their distinguishing characteristic is an obvious disk component that contains gas, dust, and hot bright stars.

FIGURE 16–10 NGC 5866 is an S0 galaxy with a disk of stars but little gas and dust. (Mount Wilson and Las Campanas Observatories, Carnegie Institution of Washington.)

The gas and dust clouds in spiral galaxies support active star formation, which makes spiral galaxies bright with the light of newly formed stars. Spiral galaxies are so bright that they are easy to see at great distances; about two-thirds of all *known* galaxies are spirals although they make up only about 15 percent of all galaxies. Most galaxies are probably fainter elliptical galaxies, which we do not see as easily as the gaudy spirals.

The largest spirals are about 1.5 times larger in diameter than our galaxy, and the smallest are about five times smaller. The most massive are slightly more massive than the Milky Way, and the least massive are less than 1 percent of that mass. Our Milky Way galaxy is larger and more massive than average.

Among spiral galaxies we identify three distinct types: S0 galaxies, normal spirals, and barred spirals. Unlike our galaxy, the S0 galaxies show no obvious spiral arms, have very little gas and dust, and contain very few hot, bright stars (Figure 16–10). However, they have an obvious disk component with a large nucleus at the center. They appear to be intermediate between elliptical and spiral galaxies.

Normal spiral galaxies can be further subclassified into three groups according to the size of their nuclear bulge and the degree to which their arms are wound up (Figure 16–11). Spirals that have little gas and dust, larger nuclear bulges, and tightly wound arms are classified Sa. Sc galaxies have large clouds of gas and dust, small nuclear bulges, and very loosely wound arms. The Sb galaxies are intermediate between Sa and Sc. Because there is more gas and dust in the Sb and Sc galaxies, we find more

FIGURE 16–11 Normal spirals are classified according to the size of the nuclear bulge and the tightness of the arms. (Mount Wilson and Las Campanas Observatories, Carnegie Institution of Washington.)

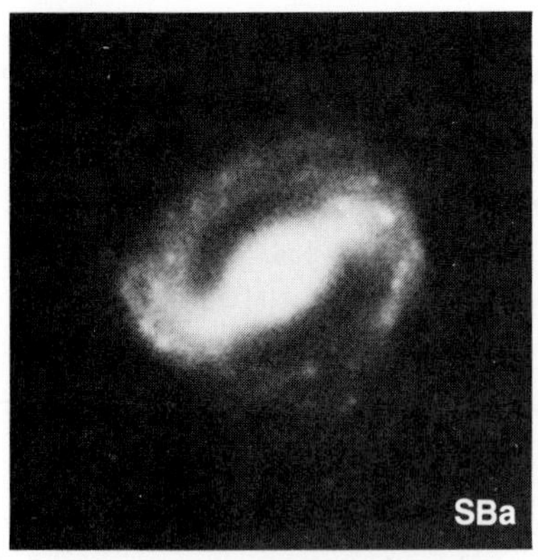

FIGURE 16–12 Barred spiral galaxies have an elongated nuclear bulge from which the arms spring. (Mount Wilson and Las Campanas Observatories, Carnegie Institution of Washington.)

young, hot, bright stars along their arms. The Andromeda galaxy (Figure 16–5) and our own Milky Way galaxy are Sb galaxies.

About 20 percent of spirals have an elongated nuclear bulge with spiral arms springing from the ends of the bar. These **barred spiral galaxies** are classified SBa, SBb, or SBc according to the same criteria listed for normal spirals (Figure 16–12). The elongated shape of the nuclear bulge is not well-understood, but some astronomers working with computer models have succeeded in imitating the rotating bar structure (Figure 16–13). It appears to occur when an instability develops in the stellar distribution within the rotating galaxy. The gravitational field of the bar alters the orbits of the inner stars and generates a stable, elongated nuclear bulge. Thus, except for the peculiar rotation in the nucleus, barred spirals are similar to normal spiral galaxies.

IRREGULAR GALAXIES

About 15 percent of all galaxies are **irregular** (Figure 16–14). Irregular galaxies have a chaotic appearance with large clouds of gas and dust mixed with both young and old stars. They have no obvious spiral arms or nuclei. Because they are small, ranging from 5–25 percent the diameter of our galaxy, they are difficult to detect. The Magellanic Clouds are the best-studied examples of irregular galaxies.

First clearly described by the navigator on Magellan's voyage around the world in 1521, the Magellanic Clouds appear to be small galaxies near the Milky Way galaxy. They are small, 7 kpc and 3 kpc in diameter, and have low masses, about 20 billion solar masses for the Large Cloud and about 10 percent of that for the Small Cloud. About 15 percent of their mass is present as gas and dust, and this is responsible for the numerous emission nebulae and the hot young stars that dot the clouds.

We know a great deal about the Magellanic Clouds because they are nearby, only 50–60 kpc away. However, at greater distances the small irregular galaxies, along with the dwarf ellipticals, are difficult to see. From a survey of nearby galaxies, we can estimate that most galaxies are elliptical and that irregular galaxies are about as common as spirals.

Although typical irregular galaxies show no spiral pattern, some astronomers believe that the Large Magellanic Cloud does contain a barred spiral pattern. This has been mapped on long-exposure photographs. Are the Magellanic Clouds really irregular galaxies? The situation is confused by the interaction of the Magellanic Clouds with each other and with our own galaxy. Radio maps show that the clouds are enveloped in a common cloud of neutral hydrogen, and a bridge of gas connects them to the Milky Way galaxy (Figure 16–15a). It seems possible that they recently passed through the disk of our galaxy. They are being distorted by tidal forces and are distorting the disk of our galaxy, bending one edge down and the other edge up like the brim of a hat (Figure 16–15b).

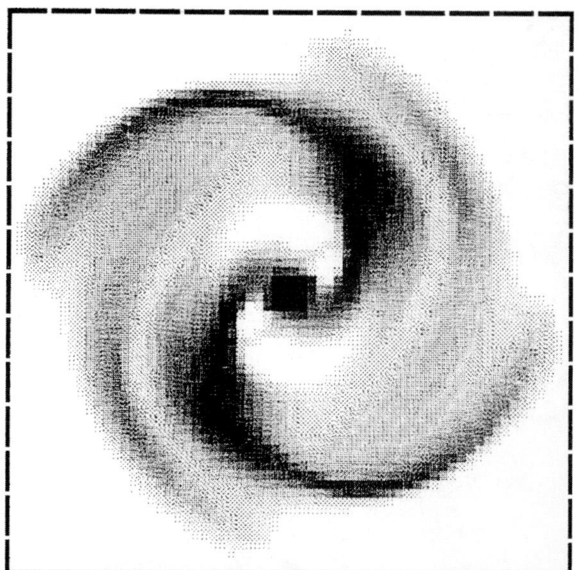

FIGURE 16–13 A computer model of a rotating disk galaxy reproduces the bar-shaped nuclear bulge and spiral arms common in real galaxies. (Reprinted courtesy of William V. Schempp and *The Astrophysical Journal,* published by the University of Chicago Press; © 1982 The American Astronomical Society.)

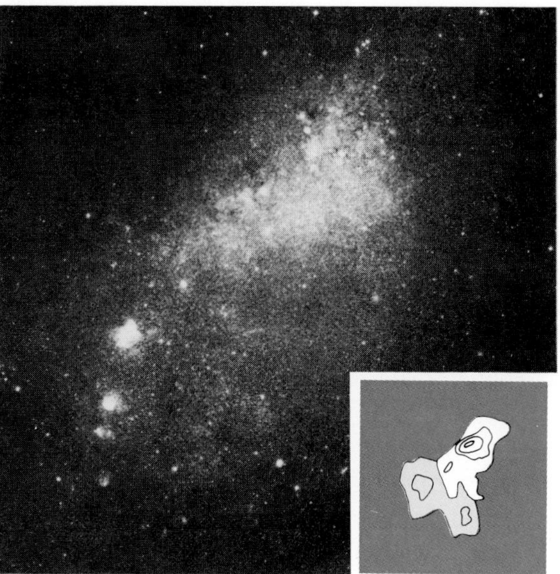

FIGURE 16–14 The Small Magellanic Cloud (SMC) is an irregular galaxy. Radio and optical studies show that the SMC is actually two clouds (inset). The Small Magellanic Cloud remnant (white) is about 20,000 ly closer to us than the mini-Magellanic Cloud (shaded). The two clouds appear to be connected by a long streamer of gas and stars. (Photo courtesy National Optical Astronomy Observatories; diagram adapted from data by Don S. Mathewson and Vincent L. Ford.)

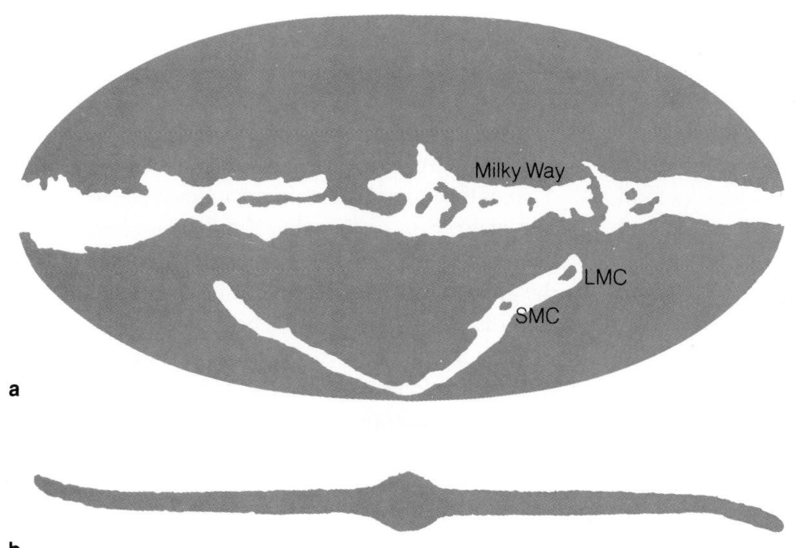

FIGURE 16–15 The interaction of the Magellanic Clouds and the Milky Way galaxy. (a) If we split the celestial sphere open and pressed it flat, we would find the Magellanic Clouds just south of the Milky Way. Radio telescopes show that they are enveloped in a common cloud of neutral hydrogen and connected to the Milky Way by a gaseous trail. (b) Seen from outside, our galaxy is distorted like the brim of a hat because of the tidal influence of the Magellanic Clouds.

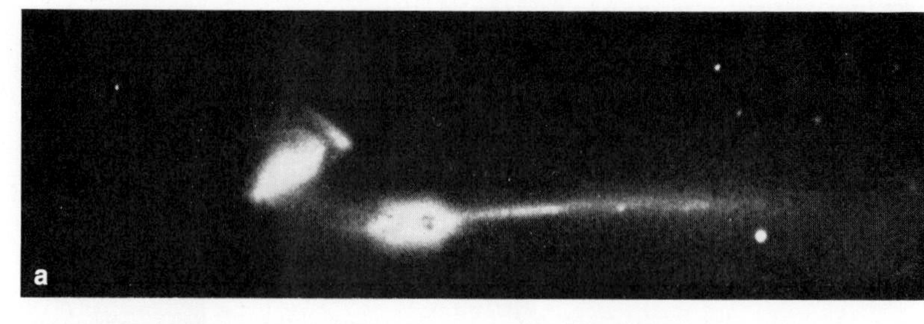

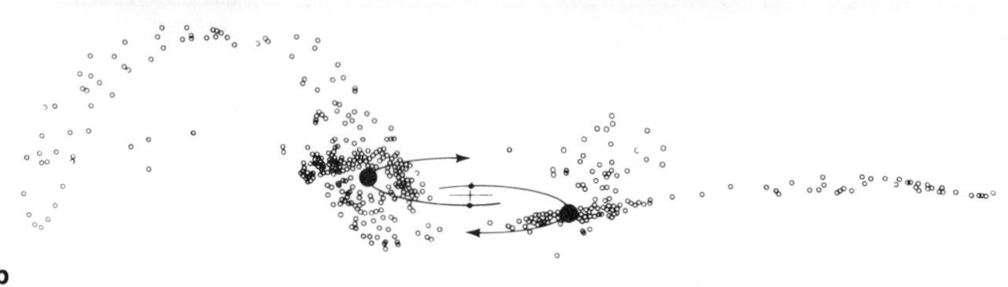

FIGURE 16–16 (a) The Mice are a pair of interacting galaxies with peculiar tails. (b) A computer simulation of a close encounter between normal galaxies produces similar tails. (From "Violent Tides Between Galaxies" by A. Toomre and J. Toomre. Copyright © 1973 by *Scientific American, Inc.* All rights reserved. Photograph courtesy of Halton Arp/Mount Wilson and Las Campanas Observatories.)

Recent radio and optical studies of the Small Magellanic Cloud suggest that it is not a single object. When we look at the Small Magellanic Cloud we see two smaller clouds almost exactly in front of each other (Figure 16–14). The Small Magellanic Cloud remnant is about 20,000 ly closer to us than the so-called mini-Magellanic Cloud. These two cloudlets appear to be the denser parts of a long tendril of gas, dust, and stars, which is pointed almost exactly away from us. When we look at the Small Magellanic Cloud, we look along that tendril and see what appears to be a single cloud. It seems very likely that the small galaxy was disrupted by a recent interaction with the large cloud and with the disk of our galaxy.

The interaction of the Magellanic Clouds with our galaxy illustrates how difficult it is to discuss the structure of galaxies in isolation. Galaxies occur in clusters. They interact and distort each other, and sometimes they even pass through each other. To understand how galaxies form and evolve, we must study them in their natural communities.

16.3 THE LIVES OF THE GALAXIES

When we discussed the origin of our own Milky Way galaxy in Chapter 15, we compared two theories. The first theory supposes that the galaxy formed in isolation from a single cloud of gas. Astronomers have generally accepted that idea since Harlow Shapley's discovery of the true extent of the galaxy. The second theory, seriously considered only since the early 1970s, supposes that the galaxy formed over time by the merger of a number of smaller clouds of stars and gas.

These two theories for the origin of the Milky Way reflect the basic question we must answer concerning the origin of galaxies. Did the galaxies form separately out of isolated clouds of gas, or did they interact with each other as they developed? In this section we will see evidence that interactions, collisions, and mergers between galaxies affect their structure and evolution.

COLLIDING GALAXIES

Galaxies should collide fairly often. The average separation between galaxies is only about 20 times their diameter. Like two elephants blundering about at random under a circus tent, galaxies should bump into each other once in a while. Stars, on the other hand, almost never collide with each other. In the region of the galaxy near the sun, the average separation between stars is about 10^7 times their diameter. Thus, collision between two stars is about as likely as collision between two gnats flitting about at random in a football stadium.

Large telescopes reveal hundreds of galaxies that appear to be colliding with other galaxies. One of the most famous pairs of colliding galaxies, NGC 4676A and NGC 4676B, is called The Mice because of the taillike deformities (Figure 16–16). In addition to tails, some interacting galaxies seem to be connected by a bridge of matter.

When two galaxies collide, they actually pass through each other, and there is no direct contact between the stars of one galaxy and the stars of the other. The gas in the galaxies does interact, but it is mainly the galaxies' gravitational fields that twist and deform the galactic shapes, producing peculiar tails and bridges. Because this interaction may last hundreds of millions of years, it is impossible for us to watch peculiar features develop in real galaxies. But computer models can simulate such a collision and display it on a TV screen. In a large modern computer, such a simulation takes only 5 minutes or so. Although highly simplified, these models produce tails and bridges very similar to those seen in peculiar galaxies.

By repeated adjustments of the masses of the galaxies, their velocities, the angle of their encounter, and other parameters, the computer can produce a convincing imitation of The Mice. The tails visible in the photograph are apparently due to tidal interactions as the galaxies swing past each other along narrow orbits.

With different initial parameters, collision forms a bridge that seems to connect the galaxies. A striking example of this is M 51, the Whirlpool galaxy (Figure 16–17). Although the Whirlpool is often used as an example of a normal spiral, it is peculiar in that its long arm appears to connect with a smaller galaxy. The computer model of this collision suggests that the two galaxies have collided recently. We now see the small galaxy beyond its larger neighbor, and the spiral arm that seems to connect them only passes in front of the smaller galaxy.

Because collisions between galaxies are fairly common, we should expect to see an occasional head-on collision. Computer models suggest that such an interaction could produce a **ring galaxy** such as that shown in Figure 16–18. The low expected frequency of head-on collisions agrees with the rarity of ring galaxies.

So far we have considered collisions in which the galaxies interact and separate, but something different can happen if the galaxies collide at lower velocity. The tidal distortion converts some of the energy of motion of the galaxies into random energy of motion of stars. Thus, the galaxies are slowed by the collision, and, if they were

FIGURE 16–17 A computer simulation of a collision between galaxies produces a pattern very similar to M 51, the Whirlpool galaxy (bottom) and its small companion. (From "Violent Tides Between Galaxies" by A. Toomre and J. Toomre. Copyright © 1973 by *Scientific American, Inc.* All rights reserved. Lick Observatory photograph.)

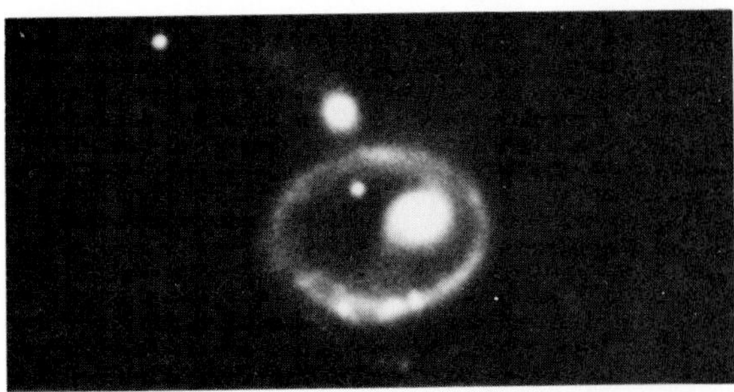

FIGURE 16–18 Ring galaxies are rare and sometimes lack nuclei. In this case, however, the nucleus is visible as the bright object within the rings. Such systems are believed to be the product of collisions between galaxies. (National Optical Astronomy Observatories.)

going slow enough to start with, they may not have escape velocity after the collision. They will pass each other, slow to a stop, fall back, and collide again and again, eventually merging into a single galaxy.

This merging of galaxies can be more dramatic if one of the galaxies is much larger than the other. The larger galaxy first rips away the small galaxy's outer stars and then begins to pull apart the denser core. The core is quite stable and settles toward the center of the larger galaxy even while the larger galaxy continues to "digest" it. This has been called **galactic cannibalism.***

We know that galactic cannibalism happens because we can see it occurring. Computer models can simulate the process, and, in that way, we can "see" step by step how a large galaxy disrupts a smaller companion. The actual process occurs so slowly that we cannot see such changes in real galaxies, but many interacting pairs of galaxies are known and share many of the characteristics that appear in the computer simulations (Figure 16–19).

*A few radio astronomers now refer to the smaller galaxies, which fall victim to galactic cannibalism, as "missionary galaxies."

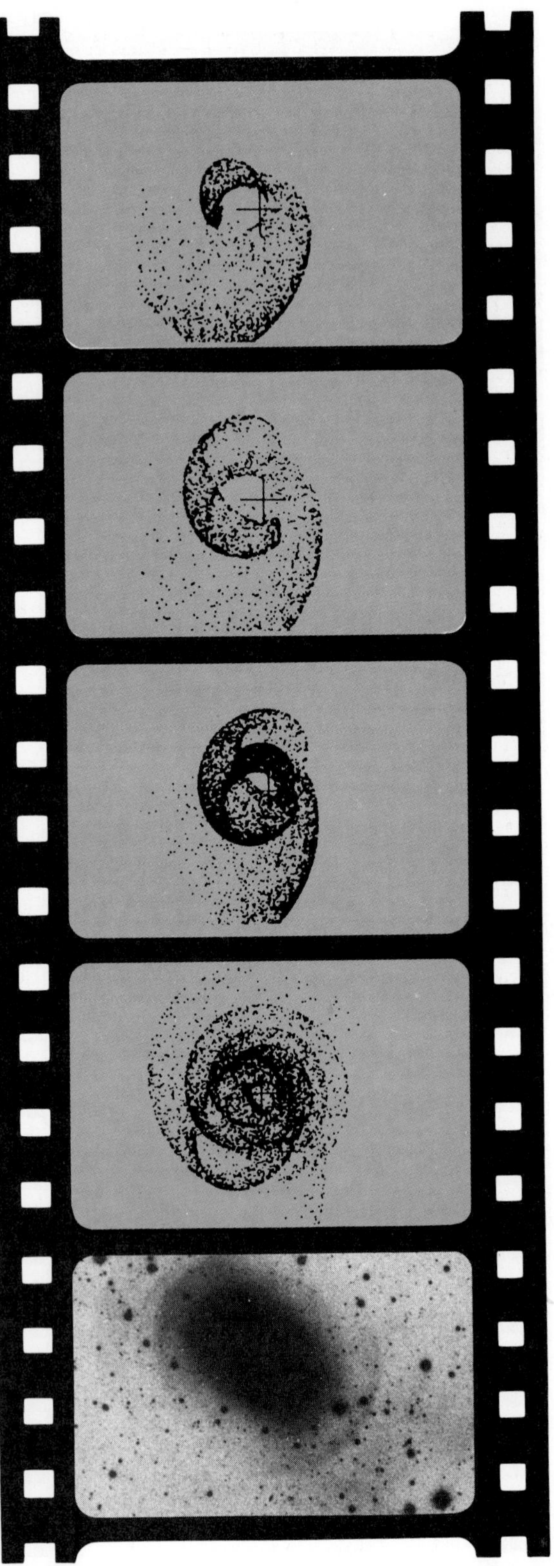

FIGURE 16–19 Computer simulations of merging galaxies suggest that they can produce shells of stars as the two galaxies whirl around their common center of mass. Only one galaxy is shown in this sequence. Such shells are seen around galaxies such as NGC 3923 (bottom) shown here as a negative image specially enhanced to reveal low-contrast features. (Model courtesy Francois Schweizer and Alar Toomre; photo courtesy David Malin, Anglo-Australian Observatory.)

FIGURE 16–20 The Hercules cluster contains about 300 galaxies and is approximately 175 Mpc distant. (Palomar Observatory photograph.)

Until the mid-1980s, astronomers argued that certain giant elliptical galaxies, which have multiple nuclei, were cannibal galaxies. One such galaxy is known with eight different nuclei. These multiple nuclei were supposedly the undigested bits of the galaxy's most recent meals—that is, the cores of smaller galaxies, which the larger galaxy had recently absorbed. Newer studies suggest that these giant ellipticals are not as voracious as first supposed. They may have absorbed no more than 10 percent of their present mass through cannibalism, and their multiple nuclei may be smaller galaxies seen in front of or behind the larger galaxy. Nevertheless, clear examples of cannibal galaxies are known.

Even our own galaxy shows evidence of cannibalism. One team of astronomers has found young population I stars several kiloparsecs south of the galaxy's disk, where only population II stars should lie. They suggest that these stars formed when our galaxy cannibalized a smaller galaxy and the compression triggered star formation in the gas of the victim. Even if this tentative theory proves incorrect, we can see clearly that our galaxy is now beginning to digest the Magellanic Clouds and has probably consumed other small galaxies in the past.

Collisions between galaxies are clearly important, not only because collisions alter the nature of galaxies, but because galaxies are located in groups and clusters where collisions occur regularly.

CLUSTERS OF GALAXIES

Single galaxies are rare. Most occur in clusters containing a few to a few thousand galaxies in a volume 1–10 Mpc across (Figure 16–20). Our Milky Way galaxy is a member of a cluster containing slightly over two dozen galaxies, and surveys have catalogued over 2700 other clusters within 4 billion ly.

We can sort clusters of galaxies into rich clusters and poor clusters. **Rich clusters** contain a thousand or more galaxies, many elliptical, scattered through a volume roughly 3 Mpc (10^7 ly) in diameter. Such a cluster is nearly always condensed; that is, the galaxies are concentrated toward the cluster center. And at the center, such clusters often contain one or more giant elliptical galaxies.

The Virgo cluster is an example of a rich cluster. It contains over 2500 galaxies located about 20 Mpc (65 million ly) away. Earth-based telescopes cannot see Cepheid variables at this distance, so the distance to the cluster is known from supernovae, globular clusters, and so forth. The Hubble Space Telescope will be able to find Cepheids in these galaxies and will give us a more accurate distance. The Virgo cluster, like most rich clusters, is centrally condensed and contains the giant elliptical galaxy M 87 at its center (Figure 16–8a). Collisions must be common in such a crowded cluster, and some of the material ripped from galaxies by tidal interactions presumably sinks toward the center, increasing the mass of the giant elliptical at its center.

X-ray observations have found that at least 40 of these rich clusters are filled with a hot gas—an intracluster medium. This gas cannot be left over from galaxy formation because it is rich in metals. It must have come from supernovae and stellar winds in the galaxies. A galaxy moving through such a medium would feel a tremendous wind stripping the galaxy of its remaining gas and dust. This may explain why these clusters contain more elliptical and S0 galaxies and fewer spirals. A spiral in such a cluster could lose its gas and dust and stop making new stars to illuminate its spiral arms.

Stripping and collisions are much less important in the **poor clusters**. They contain less than a thousand (and often only a few) galaxies and are not condensed toward the center. With fewer galaxies, there is less gas in the intracluster medium, so stripping does not rob the galaxies of their gas and dust. Collisions are also less common. This may explain why such clusters usually contain a larger proportion of spiral galaxies, which we recall are rich in dust and gas.

Our Milky Way galaxy is a member of a poor cluster known as the Local Group (Figure 16–21). It contains a little more than two dozen galaxies scattered irregularly through a volume roughly 1 Mpc in diameter. Of the brighter galaxies, 14 are elliptical, 3 are spiral, and 4 are irregular.

We can't be sure how many galaxies are in the Local Group because the dust in the Milky Way could be hiding a few. For instance, Maffei I and Maffei II are galaxies hidden behind the dust of the Milky Way. Maffei I is a large elliptical galaxy that would be visible to the naked eye if there were no dust, and Maffei II is a spiral. They were discovered by the Italian astronomer Paolo Maffei studying the area in the infrared. Although difficult to determine, estimates of the distances to these galaxies put Maffei I at or just beyond the edge of the Local Group. Maffei II is more distant, about 5 Mpc away.

The Local Group illustrates the subclustering found in poor galaxy clusters. The two largest galaxies, the Milky Way and the Andromeda galaxy, are the centers of two subclusters. The Milky Way is accompanied by the Magellanic Clouds and seven other dwarf galaxies. The Andromeda galaxy is attended by more dwarf elliptical galaxies (two of which are visible in Figure 15–1 on page 305) and a small spiral, M33. The two large galaxies orbit each other while maintaining gravitational control over the small galaxies.

Just as the clustering of galaxies is repeated on a smaller scale in subclustering, it is repeated on a larger scale with superclustering. Clusters of galaxies seem to be associated with each other in groupings called **superclusters**. The Local Group is a part of the Local Supercluster, an approximately disk-shaped swarm of galaxy clusters including the Virgo cluster near its center. It is 25–50 Mpc in diameter (Figure 16–22).

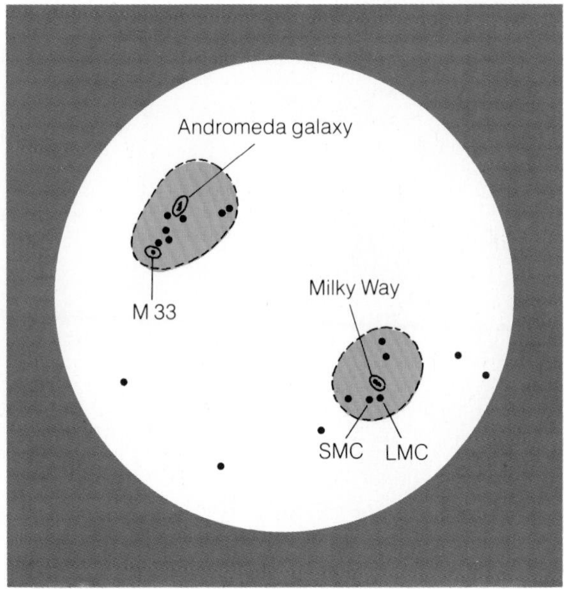

FIGURE 16–21 A map of the Local Group of galaxies showing relative distances. Note the two subclusters. The outer circle is 4 million ly in diameter.

Until the last decade, astronomers had pictured a universe filled with isolated superclusters, but a number of recent studies have shown that the superclusters are linked together in a filamentary network with great voids between. This structure is visible in Figure 16–23, one of the first diagrams to show the effect, and in more recent studies. It is even visible in plots of the brighter galaxies such as shown in Figure 16–1. Clearly, the galaxies, clusters, and superclusters are not scattered uniformly through space; they are linked in a complex structure like Swiss cheese.

The study of this distribution of superclusters is especially difficult because the galaxies are so faint that only the largest telescopes can be helpful in measuring red shifts and estimating distances. In addition, even a small area of the sky contains vast numbers of galaxies, each of which must be studied separately. Surveys have explored a few percent of the sky and have found a number of superclusters and intervening voids at typical distances of 75–150 Mpc. This structure raises a critical ques-

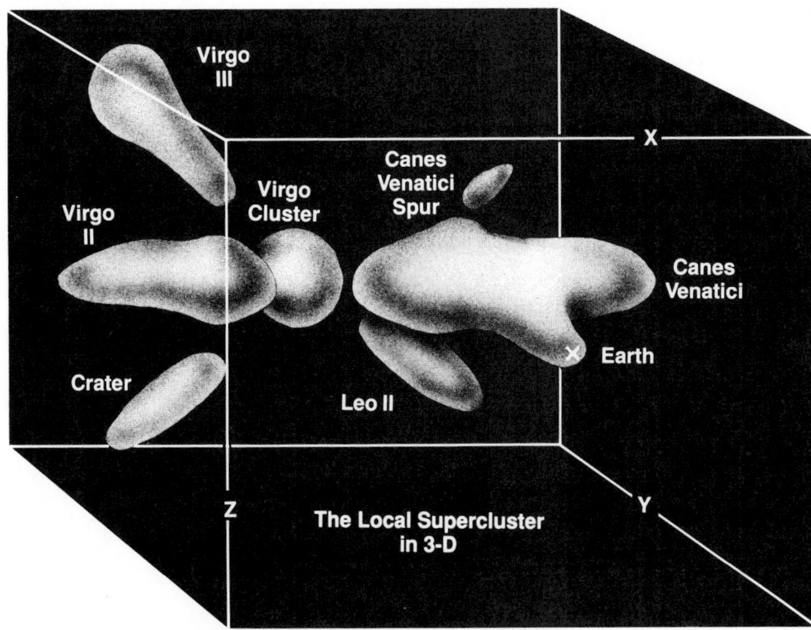

FIGURE 16–22 The local supercluster in three dimensions. It is composed of a number of clusters spread through a volume of space 25–50 Mpc in diameter. (The actual dimensions of the supercluster depend on the value of *H*.) (Courtesy R. Brent Tully and *Sky and Telescope*.)

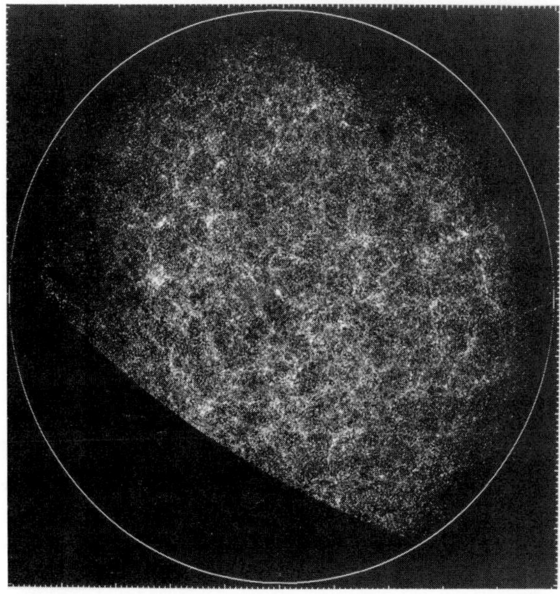

FIGURE 16–23 The filamentary appearance of the distribution of galaxies was first seen in this plot of over a million galaxies. The dust clouds of the Milky Way are located around the outer circle, and the dark area to the lower left is the southern part of the sky not visible from northern latitudes. The significance of the filamentary structure seen here is being debated on statistical grounds, but the existence of filaments and voids has been confirmed by other studies. (Courtesy M. Seldner, B. L. Siebers, E. J. Groth, and P. J. E. Peebles, *The Astronomical Journal 82*, 1977, p. 249.)

tion: How did the distribution of galaxies become so clumpy? The answer to that question is related to the origin and evolution of galaxies.

A MODEL OF GALAXY EVOLUTION

Galaxies are apparently shaped by both internal and external influences. The morphology of galaxies gives us clues to the internal influences produced by different initial conditions such as mass, rotation, and so forth. The classification of the galaxies is a first step. However, we must also consider how external influences such as collisions and stripping can affect a galaxy's shape. Clusters of galaxies can tell us about these external influences.

Before we try to build a model of galactic evolution, we must pause to eliminate a tempting but discredited idea. When the tuning fork diagram first appeared, astronomers theorized that it was an evolutionary diagram showing how elliptical galaxies evolved into spiral and then into irregular galaxies. This simple theory does not work. Elliptical galaxies contain little or no gas and are not making many new stars. Thus, an ellliptical galaxy can never become a spiral.

In addition, evolution can't go from irregular to spiral to elliptical. Only four irregular galaxies, those in the Local Group, are close enough to study in detail, but all four contain both young and old stars. Thus, they cannot be young galaxies.

The tuning fork diagram is not an evolutionary diagram. Galaxies do not evolve from elliptical to spiral or vice versa, any more than cats evolve into dogs. The tuning fork is a star-formation diagram showing that different types of galaxies have had different histories of star formation.

If we accept the older belief that galaxies formed and evolved in isolation, then we might suppose that elliptical galaxies used up their gas and dust in a sudden burst of star formation. If they formed from gas clouds with little rotation or turbulence, they might have contracted rapidly and triggered star formation suddenly. Spiral galaxies, we might suppose, formed from more rapidly rotating clouds, contracted more slowly, and conserved their gas and dust.

However, we now know that galaxies did not form suddenly from isolated gas clouds. We see clear evidence that galaxies have developed gradually through interactions, mergers, and collisions. These interactions may be the dominant influence of the structure of galaxies.

Collisions between galaxies will trigger star formation and drive the galaxies to consume their gas and dust, making new stars. Such galaxies undergoing a burst of star formation would have been brilliant with young, hot stars. In fact, a number of galaxies have been identified that appear to be undergoing a burst of star formation. These **starburst galaxies** are very luminous in the infrared due to dust warmed by newly formed stars, and they seem to be the products of collisions and mergers. The energy released by supernovae among the young stars could drive the remaining gas and dust out of the galaxy. Thus, a galaxy involved in a collision could be driven to consume much of its gas in a starburst and then lose the rest of its gas as it was blown away.

In addition to collisions, the motion of a galaxy through the gas of an intracluster medium would tend to strip the galaxy's gas away in only a few hundred million years. This appears to be happening in the Virgo cluster, where X-ray observations show a hot (about 100,000,000 K) intracluster medium concentrated toward the center.

The characteristics of elliptical and spiral galaxies suggest a model for the evolution of galaxies. Elliptical galaxies appear to be the products of collisions and mergers. The average bright elliptical galaxy may be the merged remains of from four to ten spiral galaxies. Such mergers can easily explain the various shapes of elliptical galaxies, including the triaxial ellipsoids.

But we do not have to suppose that all elliptical galaxies have been formed by mergers. At least some of the smaller elliptical galaxies and the S0 galaxies, which we see in rich clusters of galaxies, have been stripped of gas and dust by their rapid motion through the intracluster medium. Also, the dwarf ellipticals cannot be merged spirals because the average dwarf elliptical contains much less mass than a single spiral galaxy. Perhaps the dwarf ellipticals have lost their gas and dust due to stripping and through close encounters with larger galaxies.

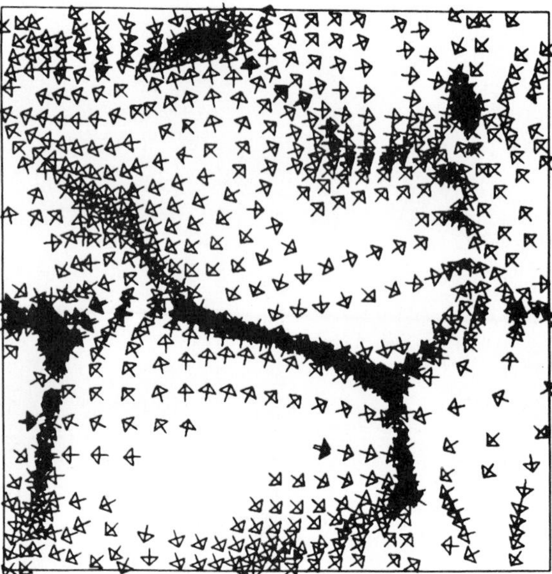

FIGURE 16–24 Filaments and pancakes resembling the observed distribution of clusters of galaxies develop in this computer model showing the motions of particles of mass (arrows) in a universe dominated by massive neutrinos. (Courtesy Paul R. Shapiro, Curtis Struck-Marcell, Adrian L. Melott, and *The Astrophysical Journal,* published by the University of Chicago Press; © 1983 The American Astronomical Society.)

According to this model, the spiral galaxies are systems that have not experienced major collisions since their formation. They may have interacted with a number of smaller galaxies, cannibalizing some, stripping gas and stars from others, just as the Milky Way is presently interacting with the Magellanic Clouds. But the spiral galaxies we see today cannot have encountered other galaxies of comparable mass very often.

Although this new model of galactic evolution seems successful in explaining the characteristics of galaxies, it omits one important factor. It does not account for the origin of galaxies. In fact, the origin of galaxies, clusters, filaments, and voids is poorly understood. We do not know whether the filaments formed first and the galaxies formed from the gas in the filaments, or whether the galaxies formed first in clusters and some process created the voids as the universe expanded. Some theoretical calculations predict that clusters will occur in long filaments or in pancakes—thin walls of matter separating two voids. Such models are very sophisticated and require special assumptions. The model shown in Figure 16–24, for instance, assumes that the motion of matter is dominated by the gravity of massive neutrinos. At present, however, we have no conclusive evidence that neutrinos have mass.

The filaments and voids represent the largest known structure in the universe. If we eventually understand how these structures originated, it surely will be linked to an improved understanding of how the universe began and how it has evolved. The study of the universe as a whole is the subject of Chapter 18.

Before we can consider the universe as a whole, however, we must examine the galaxies from a different perspective. Some galaxies are suffering tremendous eruptions as their cores blast radiation and matter outward. In the next chapter, we will discover that these peculiar galaxies are closely related to collisions and mergers between galaxies.

SUMMARY

To measure the properties of galaxies, we must first find their distances. For the nearer galaxies, we can judge distances using distance indicators, objects whose luminosity or diameter is known. The most accurate distance indicators are Cepheid variable stars. Other distance indicators are bright giants and supergiants, globular clusters, and novae. Another type of distance indicator, the largest H II regions, have known diameters. To use them, we compare their angular diameter with their known linear diameter. In addition, we can estimate the distance to the farthest galaxy clusters using the average luminosity of the brightest galaxies.

The Hubble law shows that the radial velocity of a galaxy is proportional to its distance. Thus we can use the Hubble law to estimate distances. The galaxy's radial velocity divided by the Hubble constant is its distance in megaparsecs.

The masses of galaxies can be measured in four ways—the rotation curve method, the double galaxy method, the cluster method, or the velocity dispersion method. The first method is the most accurate, but it is applicable only to nearby galaxies.

If we assume that the galaxy clusters are not coming apart—that their total gravity is sufficient to hold their most rapidly moving galaxies—then we must conclude that the clusters contain missing mass. This mass must be present, perhaps as massive galaxy coronae, but it is not visible. The nature of this dark matter is unknown.

We can divide galaxies into three classes—elliptical, spiral, and irregular—with subclasses giving the galaxy's shape or the amount of gas and dust present. The galaxy types appear to reflect different histories of star formation. The elliptical galaxies contain little gas and dust and cannot make many new stars. The spiral and irregular galaxies contain large amounts of gas and dust and are still forming stars.

The rotation and turbulence of the gas clouds from which the galaxies formed may have influenced their final form, but modern studies suggest that collisions between galaxies are also important. At high velocities colliding galaxies can distort each other through tidal forces, creating tails and bridges. Such collisions can drive a galaxy to consume its gas and dust in rapid star formation. At lower velocities colliding galaxies can merge to become a single galaxy. Such galaxy cannibalism may be common; our own Milky Way may be consuming the Magellanic Clouds.

Stripping may also affect the evolution of galaxies. When two galaxies collide or when a galaxy moves rapidly through the gas of an intracluster medium, the galaxy can be stripped of its gas and dust. Such galaxies might resemble elliptical and S0 galaxies more than spirals. Thus, we can understand why spirals are less common in rich clusters where collisions are more common.

Normal elliptical galaxies are apparently formed by the merger of spiral galaxies. The average, bright elliptical galaxy may contain the merged remains of from four to ten spiral galaxies. Dwarf ellipticals are different systems and appear to be small galaxies stripped of gas and dust by their motion through the intracluster gas and by interactions with larger galaxies. According to this theory, the spiral galaxies have not experienced collisions with large galaxies very often since they were formed.

The clusters of galaxies appear to be united in superclusters, which are, in turn, united in a network of filaments surrounding empty voids. How such filaments and voids formed is not well-understood.

NEW TERMS

zone of avoidance
megaparsec (Mpc)
distance indicator
elliptical galaxy
oblate spheroid
prolate spheroid

look-back time
rotation curve method
double galaxy method
cluster method
missing mass
dark matter
velocity dispersion method
Hubble law
Hubble constant (H)
tuning fork diagram
triaxial ellipsoid
spiral galaxy
barred spiral galaxy
irregular galaxy
ring galaxy
galactic cannibalism
rich galaxy cluster
poor galaxy cluster
supercluster
starburst galaxy

QUESTIONS

1. If a civilization lived on a planet in an E0 galaxy, do you think they would have a zone of avoidance? Why or why not?
2. Why wouldn't white dwarfs make good distance indicators?
3. Why isn't the look-back time important among nearby galaxies?
4. Explain how the rotation curve method of finding a galaxy's mass is similar to the method used to find the masses of binary stars.
5. Explain how the Hubble law permits us to estimate the distances to galaxies.
6. Draw and label a tuning fork diagram. Why can't evolution go from elliptical to spiral? From spiral to elliptical?
7. If all elliptical galaxies were triaxial, then we would never see an elliptical galaxy with a circular outline on a photograph. True or false? Explain your answer. (HINT: Can a football ever cast a circular shadow?)
8. What is the difference between an Sa and Sb galaxy? Between an S0 and an Sa galaxy? Between an Sb and an SBb galaxy? Between an E7 and an S0 galaxy?
9. How can collisions affect the shape of galaxies?
10. What evidence do we have that galactic cannibalism really happens?
11. Describe the future evolution of a galaxy that we now see as a starburst galaxy. What will happen to its interstellar medium?
12. Why does the intracluster medium help determine the nature of the galaxies in a cluster?

PROBLEMS

1. If a galaxy contains a Type I (classical) Cepheid with a period of 30 days and an apparent magnitude of 20, what is the distance to the galaxy? (HINT: See Box 12–2.)
2. If you find a galaxy that contains globular clusters that are 2 seconds of arc in diameter, how far away is the galaxy? (HINTS: Assume a globular cluster is 25 pc in diameter. See Box 3–1).
3. If a galaxy contains a supernova that at brightest has an apparent magnitude of 17, how far away is the galaxy? (HINT: Assume that the absolute magnitude of the supernova at brightest is −19.)
4. A galaxy has been found that is 10 kpc in diameter and whose outer stars orbit the center with a period of 200 million years. What is the mass of the galaxy? On what assumptions does this result depend? (HINT: See Box 15–2.)
5. Use the data in Figure 16–5 to calculate a minimum mass for the Andromeda galaxy. (HINT: See Box 15–2.)
6. Among the globular clusters orbiting a distant galaxy, the fastest is traveling 420 km/sec and is located 11 kpc from the center of the galaxy. Assuming the globular cluster is gravitationally bound to the galaxy, what is the mass of the galaxy? (HINTS: The galaxy had a slightly faster globular cluster, but it escaped some time ago. See Box 5–2.)
7. If a galaxy has a radial velocity of 4000 km/sec, how far away is it? (HINT: See Box 16–1.)
8. Sketch the Milky Way galaxy and the Andromeda galaxy to scale in Figure 16–21 and the Local Group in Figure 16–22.

RECOMMENDED READING

BERENDZEN, RICHARD, RICHARD HART, and DANIEL SEELEY *Man Discovers the Galaxies.* New York: Columbia University Press, 1984.

BERTOLA, FRANCESCO "What Shape Are Elliptical Galaxies?" *Sky and Telescope 61* (May 1981), p. 380.

BURNS, JACK O. "Dark Matter in the Universe." *Sky and Telescope 68* (Nov. 1984), p. 396.

DE BEER, KLASS S., and BLAIR D. SAVAGE "The Coronas of Galaxies." *Scientific American 247* (Aug. 1982), p. 54.

FIELD, GEORGE B. "The Hidden Mass in Galaxies." *Mercury 11* (May/June 1982), p. 74.

FINKBEINER, ANN K. "Cold Dark Matter and the Origin of Galaxies." *Astronomy 13* (April 1985), p. 67.

———. "Fossils of Something Interesting: The Large Scale Structure of the Universe." *Astronomy 12* (Nov. 1984), p. 18.

GINGERICH, OWEN, and BARBARA WELTHER "Harlow Shapley and the Cepheids." *Sky and Telescope 70* (Dec. 1985), p. 540.

GREGORY, S., and N. MORRISON "The Formation of Galaxies and Clusters." *Mercury 14* (May/June 1985), p. 85.

GREGORY, STEPHEN A., and LAIRD A. THOMPSON "Superclusters and Voids in the Distribution of Galaxies." *Scientific American 246* (March 1982), p. 106.

HODGE, PAUL W. *Galaxies.* Cambridge, Mass.: Harvard University Press, 1986.

KAUFMAN, MICHELE "Tracing M81[1] Spiral Arms." *Sky and Telescope 73* (Feb. 1987), p. 135.

KRAUSS, LAWRENCE M. "Dark Matter in the Universe." *Scientific American 255* (Dec. 1986), p. 58.

LEA, S. "M-87." *Mercury 12* (Jan./Feb. 1983). p. 25.

MACROBERT, ALAN "No Missing Mass?" *Sky and Telescope 70* (July 1985), p. 22.

MARAN, S. P. "Ring Galaxies." *Natural History 86* (Nov. 1977), p. 106.

MARSCHALL, LAURENCE "Superclusters: Giants of the Cosmos." *Astronomy 12* (April 1984), p. 6.

———. "Galactic Coronas." *Astronomy 10* (Nov. 1982), p. 74.

MATHEWSON, DON "The Mini-Magellanic Cloud." *Mercury 13* (March/April 1984), p. 57.

———. "The Clouds of Magellan." *Scientific American 252* (April 1985), p. 107.

MORRISON, N., and S. GREGORY "Luminous Galaxies with Extended Halos." *Mercury 13* (July/Aug. 1984), p. 115.

———. "More About Galaxies with Extended Halos." *Mercury 13* (Sept./Oct. 1984), p. 157.

PARKER, BARRY "Mystery of the Missing Mass." *Astronomy 12* (Nov. 1984), p. 6.

———. "Celestial Pinwheels: Spiral Galaxies." *Astronomy 13* (May 1985), p. 14.

PERATT, ANTHONY L. "Simulating Spiral Galaxies." *Sky and Telescope 68* (Aug. 1984), p. 118.

REDDY, FRANCIS "To Sculpt the Galaxies." *Astronomy 11* (Jan. 1983), p. 6.

RICHSTONE, DOUGLAS "Encounters of the Galactic Kind." *Natural History 91* (July 1982), p. 60.

ROBINSON, LEIF "Galactic Cannibalism." *Sky and Telescope 61* (Feb. 1981), p. 108.

RUBIN, VERA C. "Dark Matter in Spiral Galaxies." *Scientific American 248* (June 1983), p. 96.

SCHWEIZER, FRANCOIS "Colliding and Merging Galaxies." *Science 231* (17 Jan. 1986), p. 227.

SILK, JOSEPH "Formation of Galaxies." *Sky and Telescope 72* (Dec. 1986), p. 582.

STROM, K. M., and S. E. STROM "Galactic Evolution: A Survey of Recent Progress." *Science 216* (7 May 1982), p. 571.

TOOMRE, A., and J. TOOMRE "Violent Tides Between Galaxies." *Scientific American 233* (Dec. 1973), p. 38.

TULLEY, R. BRENT "Unscrambling the Local Supercluster." *Sky and Telescope 63* (Jan. 1982), p. 550.

VAN DEN BERGH, S. "Golden Anniversary of Hubble's Classification System." *Sky and Telescope 52* (Dec. 1976), p. 410.

CHAPTER 17
PECULIAR GALAXIES

Virtually all scientists have the bad habit of displaying feats of virtuosity in problems in which they can make some progress and leave until the end the really difficult central problems.

M. S. Longair
HIGH ENERGY ASTROPHYSICS

Astronomers get used to odd phone calls like this one that came an hour before dawn.*

Voice: Hello ... Dr. Seeds? Sorry to wake you up. You don't know me but you spoke to my church group last week, and I just got up to go to the bathroom, and there is a real bright thing out over my barn and it's flashing colors and its really bright, and ...

Seeds: Which direction's your barn?

Voice: Out back ... uh ... east mostly.

Seeds: Venus ... it's Venus ...

Voice: Are you sure? It's awful bright, and it looks like it's right above my barn.

Seeds: Venus ... been rising before the sun for the last few weeks ... twinkles when it's low on the horizon ... not above your barn ... 'bout 100 million miles away.

Voice: You sure? It looks like it's over my barn.

If we don't know how far away an astronomical object is, it is almost impossible to figure out what it is from its appearance. Modern astronomers had this problem in the early 1960s when they discovered a number of small, starlike objects with totally unrecognizable spectra. Only after a number of years did they realize that the objects were not nearby stars but the most distant galaxies in the visible universe.

Over the last two decades astronomers have been able to connect these very distant, very luminous galaxies with other peculiar galaxies,

*This is a true story. See Figure 2–16b.

which are less luminous but closer to us. They all appear to be suffering from tremendous eruptions occurring in their cores—eruptions that are blasting jets of excited gas outward at high velocity, creating vast clouds of gas emitting radio energy, and illuminating the cores of the galaxies with intense electromagnetic radiation at all wavelengths.

In the quotation that opens this chapter, Longair speaks of "the really difficult central problems" of science. So far as galaxies are concerned, the really difficult central problem lies at the very centers of the galaxies. What is the energy machine? In this chapter we will link together our theories of the origin of the Milky Way, the origin and evolution of spiral and elliptical galaxies, the interactions of merging galaxies, and the accretion of matter into black holes, and we will construct a tentative theory to explain the peculiar galaxies.

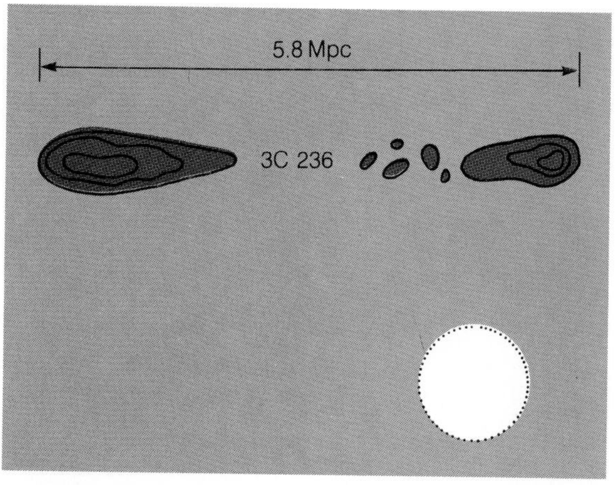

FIGURE 17–1 The radio lobes of 3C 236, one of the largest radio galaxies known, span 5.8 Mpc. The circle illustrates the approximate size of the Local Group.

17.1 ACTIVE GALAXIES

In the 1950s astronomers began to use radio telescopes to map the sky, and they soon noticed that some galaxies were sources of unusually intense radio energy. All galaxies, including our Milky Way galaxy, emit some radio energy from neutral hydrogen, molecular clouds, pulsars, and so on, but some galaxies emitted 10^7 times more radio energy. These galaxies became known as **radio galaxies.**

Later observations showed that many of these galaxies were also emitting at infrared, ultraviolet, and X-ray wavelengths. Therefore, the term radio galaxy has been replaced gradually by **active galaxy**, referring to a galaxy that produces much more energy than a normal galaxy. Because this energy is always produced in the cores of galaxies, these energy sources have become known as **active galactic nuclei (AGN)**.

In our search for the nature of the AGN, we will begin by studying the radio emissions of certain of these galaxies. The pattern of these radio emissions is, like a footprint at the scene of the crime, a tell-tale clue.

DOUBLE-LOBED RADIO SOURCES

Beginning in the 1950s radio astronomers found that some radio sources in the sky consisted of pairs of radio-bright lobes. When optical telescopes studied the locations of these **double-lobed radio sources**, they found galaxies located between the lobes. Apparently, the galaxies produced the radio lobes, although in some cases the central galaxies were not emitting detectable radio energy.

Radio lobes are generally much larger than the galaxy they accompany. Many are as large as 60 kpc (200,000 ly) in diameter, twice the size of the Milky Way galaxy. From tip to tip the radio lobes span hundreds of kiloparsecs. Records for the largest, smallest, brightest, and so on are broken almost daily in astronomy, but one of the largest known radio galaxies is 3C 236 (the 236th source in the *Third Cambridge Catalogue of Radio Sources*). Its radio lobes span 5.8 Mpc (19×10^6 ly) (Figure 17–1).

Radio lobes have two properties that hint at their origin: They radiate synchrotron radiation and often have hot spots (regions of intense radio emission) at the edge of the lobe farthest from the central galaxy.

Synchrotron radiation, remember, is produced when high-speed electrons move through a magnetic field (see Box 13–1). Although the field in a radio lobe is at least 1000 times weaker than the earth's, it fills a tremendous volume and represents vast stored energy. In addition, the high-speed electrons must be traveling near the speed of light and so must be highly energetic. Such electrons are usually referred to as relativistic electrons. Thus, the total energy stored in a radio lobe is very large, about 10^{53} J—approximately the energy we would get if we converted the mass of a million suns entirely into energy. Clearly, the process that creates radio lobes must involve tremendous power.

The second hint to the nature of the radio lobes is that many contain **hot spots**, regions of high intensity, on the edge of the lobes farthest from the galaxy (Figure

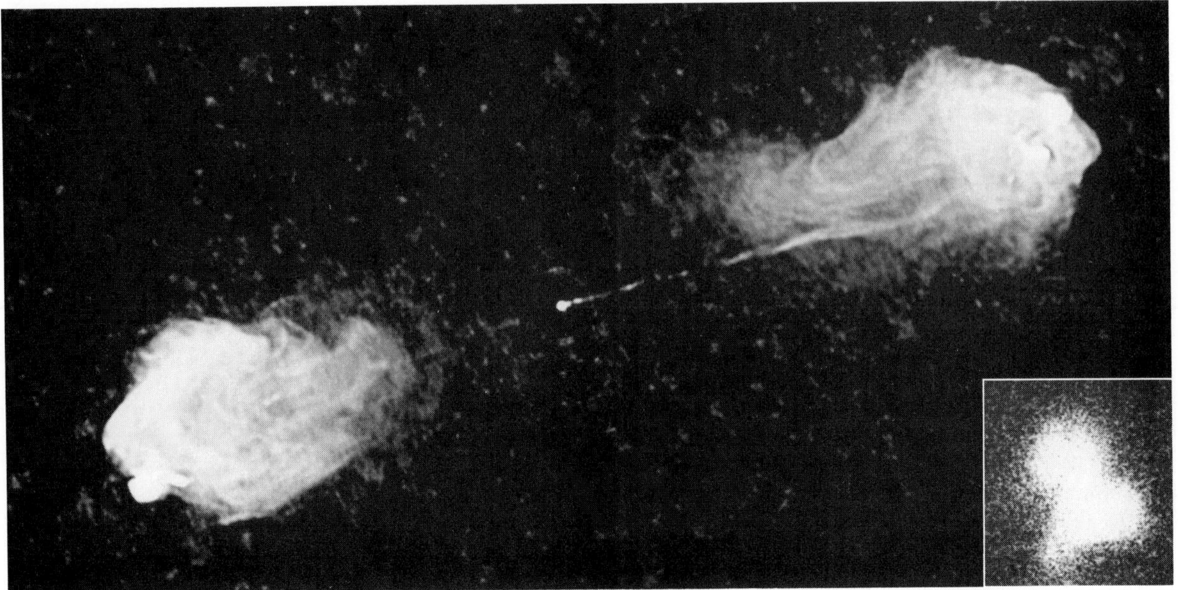

FIGURE 17–2 Cygnus A is a powerful, double-lobed radio galaxy. This radio map produced by the VLA radio telescope shows hot spots as brightenings on the outer edges of the lobes. Note the long thin jet of matter flowing from the galaxy to the righthand lobe. The jet is at least 50 kpc long—about 50 percent longer than the diameter of our galaxy. A fainter, counter jet is also visible leading to the left lobe. In a visible-light photograph the deformed central galaxy (inset) is roughly the size of the region between the radio lobes. (National Radio Astronomy Observatory; Palomar Observatory photograph.)

17–2). This suggests that the lobes are expanding away from the galaxy and that their leading edges are colliding with the intergalactic gas, which would compress the magnetic field and the hot gas, producing stronger radio emission from that region. In fact, the hot spots can even give us a clue to the source of energy that excites the lobes.

The presence of synchrotron radiation and hot spots plus the finely detailed radio images produced by such radio telescopes as the VLA have convinced astronomers that the radio lobes are produced by gas ejected from the central galaxy. This central galaxy is usually a giant elliptical and is often located in a crowded cluster. Many of these galaxies are deformed or otherwise peculiar in their visible appearance, and their spectra contain emission lines of highly excited low-pressure gas. All these properties suggest that the galaxies have suffered some eruption.

Cygnus A (Figure 17–2), the brightest radio source in Cygnus and the second brightest in the sky, was the first double-lobed source discovered (in 1953). It radiates about 10^7 times more radio energy than the Milky Way galaxy from two lobes containing hot spots. The central galaxy is oddly distorted, but at its distance of 225 Mpc (730 million ly) we cannot see it well. Recent studies have revealed a thin jet of high-energy matter flowing from the core of the galaxy all the way into the northwest lobe. Such jets are very common in these galaxies.

Centaurus A is another double-lobed radio source with a jet (Figure 17–3). Its outer lobes span 10° in the sky. (Recall that the full moon spans only 0.5°!) Located between the lobes is NGC 5128, a giant elliptical galaxy encircled by a mysterious ring of dust in which new stars appear to be forming. Long-exposure photographs reveal a faint optical jet extending 40 kpc (130,000 ly) along the axis of rotation toward the northern lobe.

Very "deep" photographs (reaching very faint stars) show that this galaxy is surrounded by concentric shells of stars (Figure 17–4). Computer simulations of mergers between galaxies have produced such features (Figure 16–19). A different study reveals that the bright, spherical component of the galaxy is rotating slowly along an axis that lies in the plane of the dust ring. The dust ring, of course, is rotating normally around an axis perpendicular

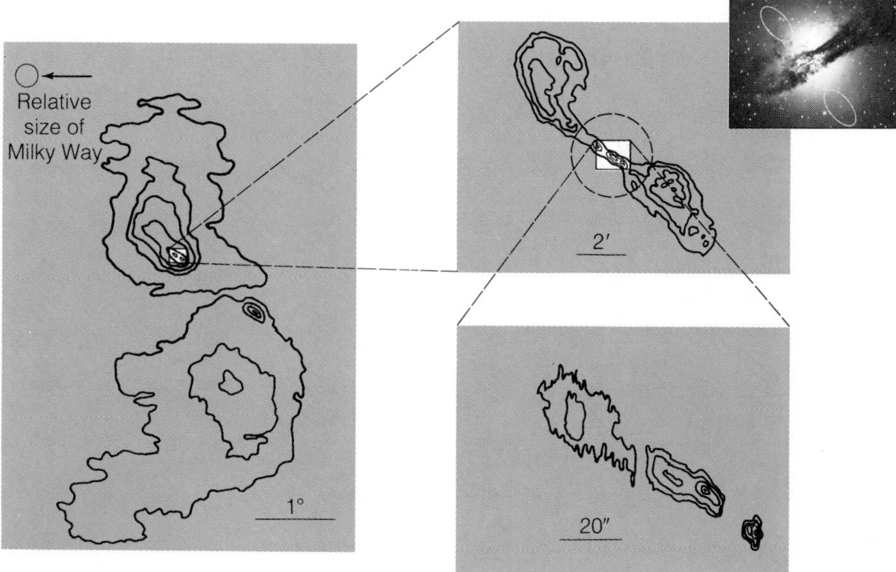

FIGURE 17–3 Centaurus A consists of a very large pair of radio lobes within which higher-resolution radio maps reveal a smaller pair. The size of the visible galaxy NGC 5128 is shown by the dashed circle. At the highest resolution, radio maps reveal a jet of matter about 3 kpc long springing from the nucleus (lower right). (Adapted from VLA maps by Jack O. Burns, Eric D. Feigelson, and Ethan J. Schreier; Palomar Observatory photograph.)

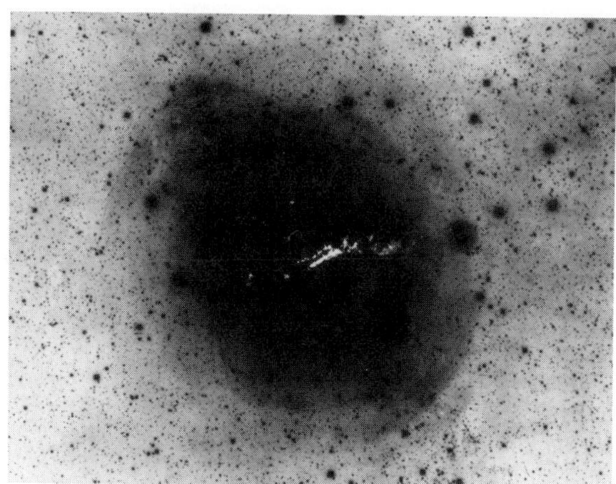

FIGURE 17–4 Faint shells of stars surround the peculiar galaxy NGC 5128 in this negative print (white sky—dark stars). Such shells are believed typical of merged galaxies and suggest that the activity of this galaxy could be related to a merger that occurred about 1 billion years ago. (Compare with Figures 17–3 and 16–19.) (Four-meter CTIO image courtesy National Optical Astronomy Observatories.)

to the plane of the ring. Thus, the dust ring is following an orbit that passes over the poles of the main galaxy. Other such galaxies are known and are believed to be the result of collisions. Both the concentric shells and the peculiar rotation suggest that NGC 5128 is the product of a merger between an elliptical and a spiral galaxy, which took place within the last billion years. We will see later in this chapter how such a merger might have triggered the galaxy's outbursts.

High-resolution radio maps show that an inner pair of lobes is centered on the galaxy, and, at the highest resolution, we can see a radio jet (Figure 17–3, lower right). These maps suggest that an eruption at the center of the galaxy has been active sporadically for a long time, ejecting gas in jets along its axis of rotation to produce the various lobes we see. These active galaxies were once referred to as "exploding," but an explosion is a sudden event that is over quickly. Perhaps *erupting* is a better term. In any case, it is clear from galaxies such as NGC 5128 that the power source at the center of the galaxy is capable of sustained activity.

JETS FROM AGN

Now that we have discussed a few examples of double-lobed radio galaxies, can we construct a model that explains their characteristics? We will not try to guess the identity of the energy source at the center—we will save that until the end of this section. Let's begin by trying to find a way for such a powerhouse to create radio lobes.

Our working hypothesis is that the lobes are created and sustained by jets of hot gas coming from the AGN. The **double-exhaust model** supposes that two jets exist, which bore through the interstellar medium forming tunnels leading out of the galaxy. The jets push into the intergalactic medium and inflate cavities of hot gas on either side of the galaxy. We see these cavities as radio lobes, and the points where the jets push back the intergalactic medium we see as hot spots (Figure 17–5).

This explains how the lobes hold themselves together. Their magnetic and gravitational fields are much too weak, but they are evidently confined by their impact with the intergalactic medium. Their gas cools, of course, and mixes with the intergalactic medium, but as long as the jet is on, it will keep pumping new hot gas into the cavity.

Thanks to the newest high-resolution radio telescopes and orbiting X-ray telescopes, we can see this happening (Color Plates 23, 24, and 25). The jets are very narrow and very hot and are bright at radio and X-ray wavelengths. The gas in some jets travels about 5000 km/sec (in NGC 5128, for instance), while the gas velocity in other jets (such as Cygnus A) may be a sizable fraction of the speed of light. Such jets have been found in 90 percent of all low-luminosity radio galaxies and may exist in all of them. We will see later that such jets occur in other, higher-luminosity objects.

This double-exhaust model is not entirely consistent with observations, however. For example, the beams tunneling through the galaxy could develop instabilities that would spread the beams. No one knows exactly why the beams are so narrow. Most jets, when viewed at a high enough resolution, are made up of a series of blobs or knots. This may be at least partially due to turbulence, but it may also be due to individual outbursts by the energy source at the center of the galaxy.

Another problem is that nearly all active galaxies with visible jets have double lobes but only one jet. One team of astronomers has suggested that the energy source is unstable and ejects matter in first one direction and then the other along the axis. Thus, there are two jets, but only

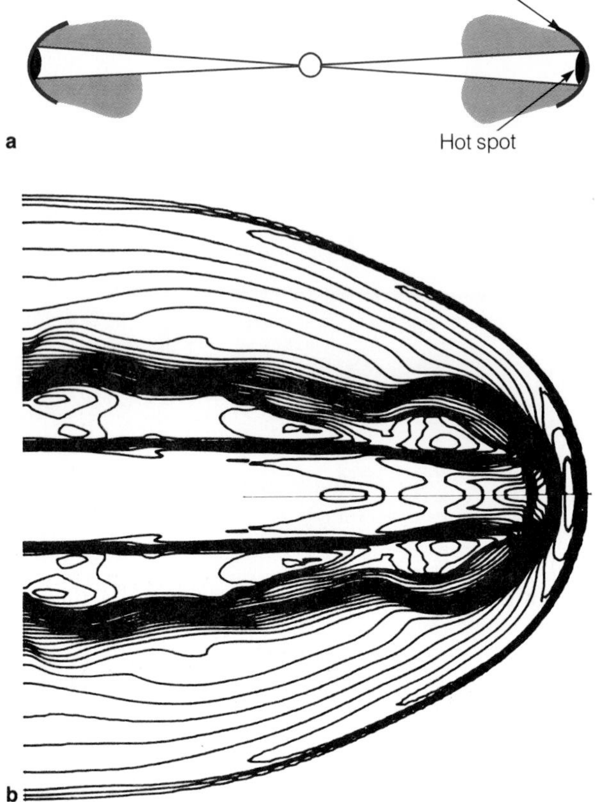

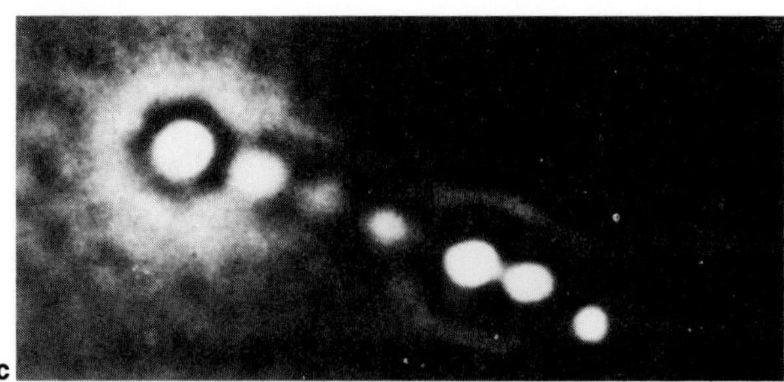

FIGURE 17–5 (a) In the double-exhaust model, the hot spots are produced by jets of high-speed gas pushing into the intergalactic medium. (b) In this computer simulation of a galactic jet, contours show the density of the hot gas of the jet, which travels 19.2 times faster than the speed of sound in the cooler intergalactic medium. Note the high-density cap formed at the end of the jet. (c) A computer-enhanced image of the jet in the giant elliptical galaxy M87 shows irregularities along the jet. (b, Courtesy Clifford N. Arnold and W. David Arnett; c, Courtesy Halton Arp.)

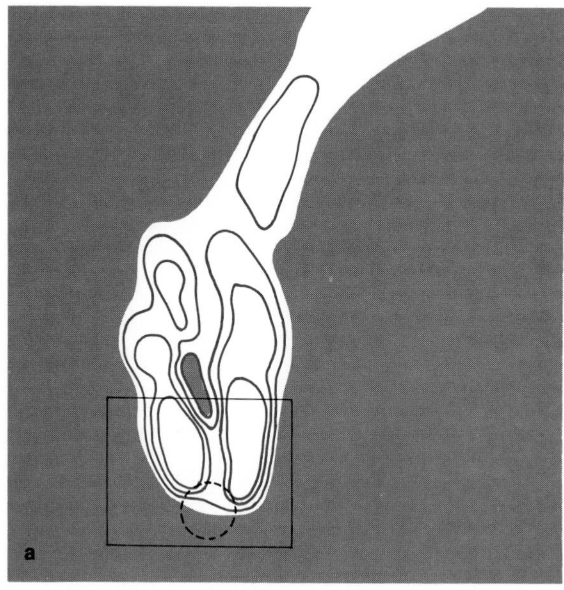

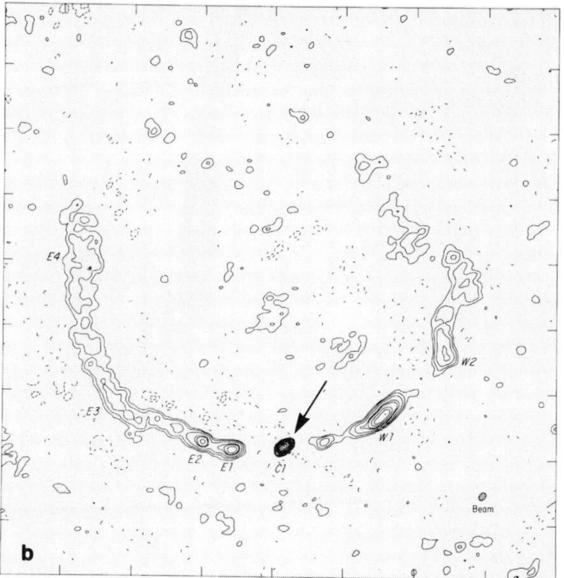

FIGURE 17–6 (a) Schematic radio map of the head–tail galaxy NGC 1265. The dashed circle marks the location of the visible galaxy, and the box marks the region shown in part b. (b) A high-resolution radio map shows that the galaxy contains an AGN (arrow), ejecting jets in opposite directions. The galaxy is moving about 2000 km/sec in the direction of the arrow, and the relative motion through the intergalactic medium blows the jets back to form a tail. (Courtesy Frazer N. Owen and the National Radio Astronomy Observatory.)

one is active at any given time. Another possibility is that relativistic effects permit us to see only those jets pointed more or less toward us. The jets pointing away from us would be significantly fainter and might not be detectable. In the case of Cygnus A, for example, a very faint jet has been detected pointing counter to the main jet.

The double-exhaust model depends on the existence of an intergalactic medium. We can be sure such a medium exists because of related active galaxies originally called **head–tail radio galaxies.** When they were first discovered, at fairly low resolution, they looked like speedboats racing across a lake and leaving a wake behind (Figure 17–6a). Higher-resolution observations resolved the wake into two streams of material trailing away from the galaxy (Figure 17–6b). Apparently these head–tail radio galaxies are ejecting material that would normally form radio lobes, but the galaxy is moving relative to the intergalactic medium. The ejected material is swept away in a long tail, like the plume from a twin smokestack carried away by the breeze. This interpretation is supported by the observation that such galaxies occur in clusters of galaxies where the intergalactic medium is relatively dense.

A few head–tail galaxies have complex structures such as blobs or spirals among the two streams (Figures 17–7 and 17–8). The blobs suggest that the jets do not remain "on" all the time, and the spirals suggest that the axis along which the beams shine may be precessing and spewing relativistic particles out of the galaxy in swirling streams.

One interesting object illustrates the power in such a jet. NGC 541 is an elliptical galaxy with a known jet. Minkowski's object, long known as a peculiar blue object nearby, has recently been recognized as a small, irregular galaxy passing through the jet (Figure 17–9). The tremendous impact of the high-speed gases of the jet has triggered a burst of star formation in the little galaxy, making it very bright and blue with hot, young stars.

THE ENERGY MACHINES

Clearly, galaxies can be powerful sources of energy, but so far we have ignored the "really difficult central problem." What object at the center of a galaxy could produce such energy? That is, what is the energy machine?

The radio lobes associated with many active galaxies are important clues, but not all active galaxies have these features. Some galaxies radiate tremendous energy directly

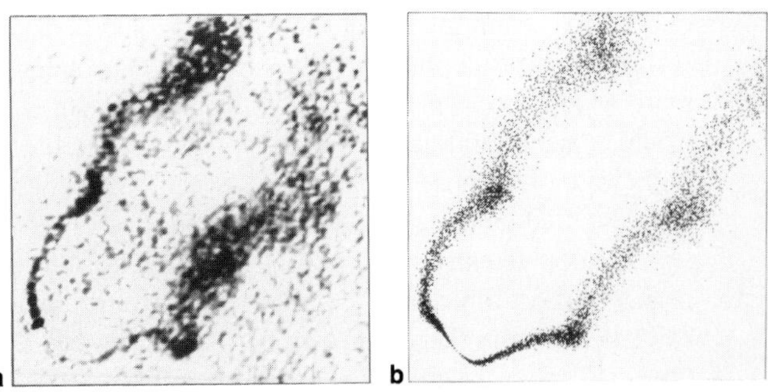

FIGURE 17–7 (a) VLA radio map of 3C 129 showing the spiral form of the streams produced by this head–tail galaxy. (b) Computed stream pattern from a model using a precessing-beam source. (Map courtesy of Laurence Rudnick [U. Minn.] and Jack Burns [U. New Mexico]. Model courtesy of Vincent Icke [U. Minn.] and *The Astrophysical Journal*, published by the University of Chicago Press, © 1981 The American Astronomical Society.)

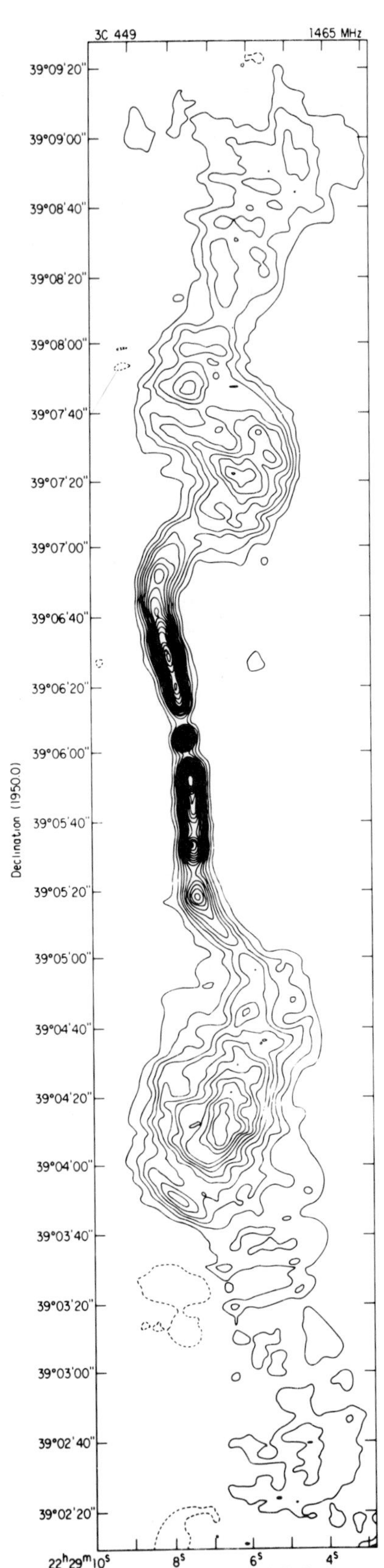

FIGURE 17–8 The active galaxy 3C 449 has spiral patterns in its radio lobes. They may be produced by a source of jets that orbits the nucleus at a distance of about 1 kpc. An alternate model suggests that the jets (each about 20 kpc long) precess much like the beams produced by the much smaller system SS 433. At visual wavelengths the galaxy appears to be an elliptical galaxy about the size of a quarter laid over the central spot in this diagram. (Courtesy Richard A. Perley.)

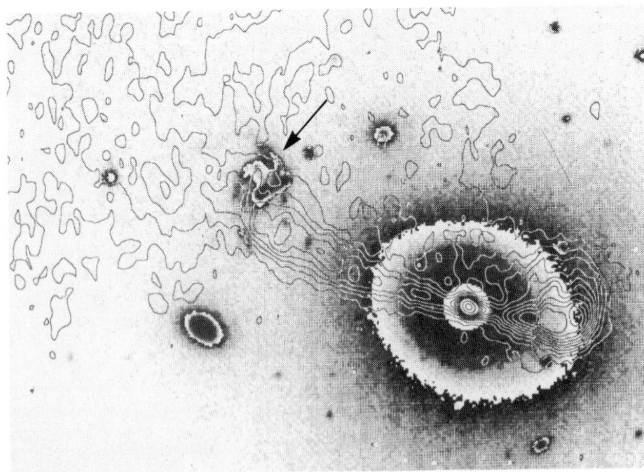

FIGURE 17–9 A computer has combined an optical photograph (negative) and a radio map to produce this illustration of the interaction of a jet and a galaxy. The jet, shown by contours, originates in the elliptical galaxy NGC 541 at the right. The jet strikes Minkowski's object (arrow), now recognized as a small irregular galaxy passing through the jet. The impact stimulates a burst of star formation in the small galaxy. (Courtesy Dr. Wil J. M. van Breugel.)

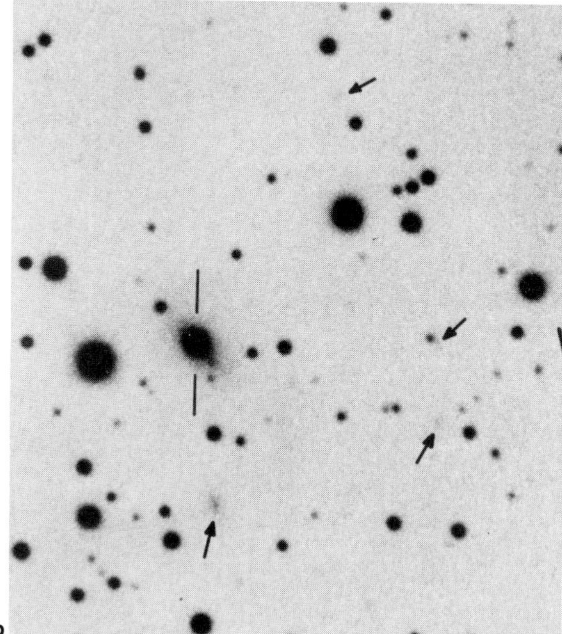

FIGURE 17–10 Two galaxies with bright cores. (a) The radio source Perseus A coincides with a Seyfert galaxy NGC 1275, a peculiar spiral with an intensely bright, small nucleus. The distorted appearance suggests a collision between galaxies. (b) On this negative photo, BL Lac is the starlike image between the bars. The fuzz around the image appears to come from a giant elliptical galaxy that contains BL Lac. Arrows point to other faint galaxies. (a, National Optical Astronomy Observatories; b, Courtesy T. D. Kinman, Kitt Peak National Observatory Photo Lab No. 10221.)

from their centers. Such active nuclei are very common; our own galaxy may have a mildly active nucleus.

Seyfert galaxies are spiral galaxies with unusually bright, tiny cores that fluctuate in brightness (Figure 17–10a). Spectra reveal emission from ionized gas moving at high velocities, and most such galaxies are bright infrared sources. Some are radio sources, and at least one (NGC 1275) is an X-ray source.

A recent study shows that Seyfert galaxies are three times more common in interacting pairs of galaxies than in isolated galaxies and that roughly 25 percent have peculiar shapes suggesting tidal interactions with other galaxies. This suggests that galaxy nuclei may be triggered into activity. (See Color Plate 26.)

Another interesting statistic is that 2 percent of all spiral galaxies are Seyferts. Either galaxy nuclei are active 2 percent of the time, or 2 percent of all galaxies have active nuclei. The peculiar heart of our own galaxy suggests that all galaxies can occasionally develop an active nucleus.

Observations of the core of the Seyfert galaxy NGC 6814 show that it fluctuates in brightness in a time as short as 100 seconds. Recall that an astronomical body cannot change its brightness in a time shorter than the time light takes to cross its diameter. Thus, the energy machine at the center of NGC 6814 must be very small—no bigger than 100 light-seconds in diameter (about 0.1 the diameter of the earth's orbit.)

Many other galaxies have small, bright cores. One extreme group is known as **BL Lac objects** or Lacertids. They are named after their prototype BL Lac, which was thought to be a conventional variable star (Figure 17–10b), and, being located in the constellation of Lacerta, was given the appropriate variable star designation BL Lac. However, in 1968 astronomers discovered that the object is a powerful radio source. In the early 1970s a team of astronomers was able to blot out the brilliant central point of light and photograph the fuzz surrounding the object. The spectrum of this fuzz is typical of a giant elliptical galaxy. Thus, BL Lac appears to be a powerful energy source, presumably the core, in a distant galaxy. A few dozen BL Lac objects are now known.

The giant elliptical galaxy M 87, located at the center of the Virgo cluster, also has a peculiar bright core visible

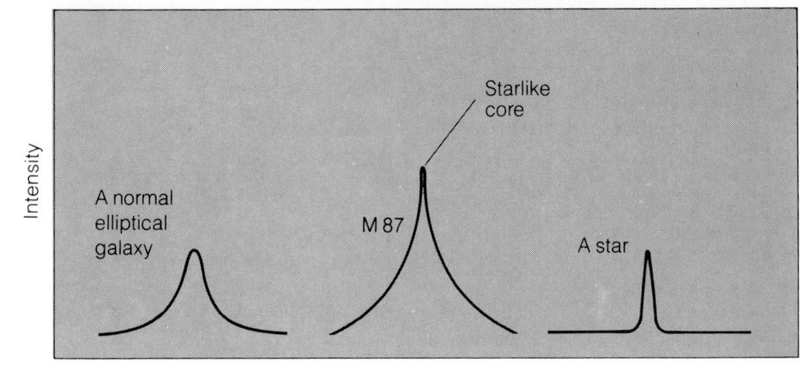

FIGURE 17–11 (a) M 87 is a giant elliptical galaxy known to have an active core. The inset is a short-exposure photo with the same scale and orientation as the main photo. It shows the bright core and a 1500 pc long jet pointing to the upper right. (b) A graph of intensity across the center of M 87 reveals that its bright nucleus is very small, quite unlike that of a normal eliptical galaxy. (National Optical Astronomy Observatories.)

at optical, radio, and X-ray wavelengths (Figure 17–11). The impressive feature of M 87, however, appears only on short-exposure photographs, which reveal a tremendous jet of matter 1500 pc long squirting out of the nucleus. The jet emits synchrotron radiation, contains knots, and has been detected at X-ray wavelengths by the Einstein Observatory.

M 87 provided the first observational clue to what could cause such activity in galaxies. Astronomers studying the light from the nucleus found that the stars there were tremendously crowded, producing a starlike point of light at the center (Figure 17–11b). Another team found that velocities at the center were very high. If the rapidly moving stars are held in orbit by gravity, the core must be very massive—$5 \times 10^9 M_\odot$. (This is an application of the velocity dispersion method described in Chapter 16.)

This mass could be a dense swarm of stars, but the astronomers point out that their results are consistent with a 5-billion solar mass black hole. If this is true, then matter falling into the black hole could grow hot enough to provide the energy for the radio galaxy.

Not all astronomers agree that M 87 contains a black hole, but the proposal that active galactic nuclei contain massive black holes has grown more popular with each passing year. A scientific hypothesis can never be proved correct, only disproved, so when a hypothesis survives many tests over a number of years, scientists grow more confident of it. Thus, we cannot prove that any galaxy contains a massive black hole, but the hypothesis is now an acceptable working model.

One test of the hypothesis is the size of the active region. As radio astronomers have improved the resolving power of their telescopes, they have found the basic structure of active galaxies dictated by events in a tiny region at their center. For example, the elliptical galaxy NGC 6251 is bracketed by radio lobes spanning 2 Mpc. At higher resolution, we see a jet 200 kpc long pointing toward one of the lobes. Using an international network

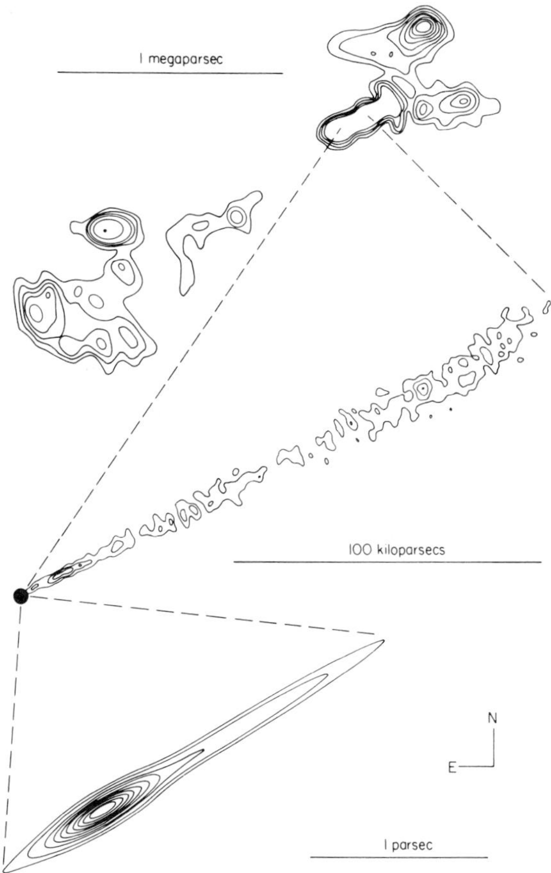

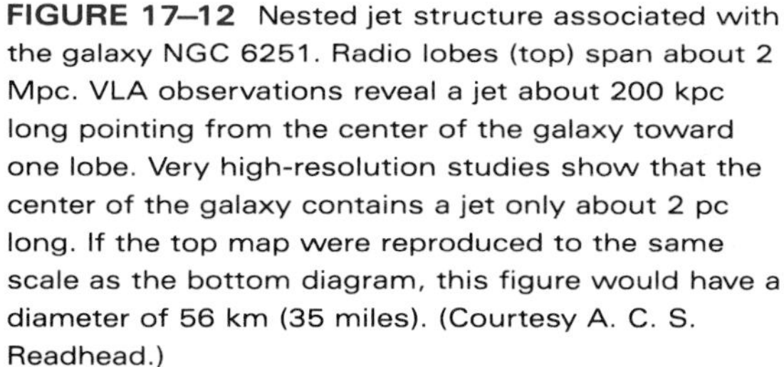

FIGURE 17–12 Nested jet structure associated with the galaxy NGC 6251. Radio lobes (top) span about 2 Mpc. VLA observations reveal a jet about 200 kpc long pointing from the center of the galaxy toward one lobe. Very high-resolution studies show that the center of the galaxy contains a jet only about 2 pc long. If the top map were reproduced to the same scale as the bottom diagram, this figure would have a diameter of 56 km (35 miles). (Courtesy A. C. S. Readhead.)

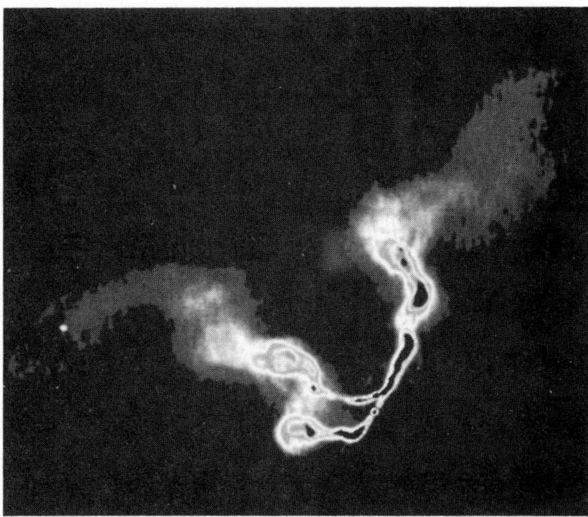

FIGURE 17–13 The twin-tail radio galaxy 3C 75 appears to consist of jets emitted from two twin-jet radio sources within the binary nucleus of an active galaxy (Color Plate 25). This galaxy is the central galaxy in the cluster Abell 400. The jets and radio lobes are twisted by the motion of the nuclei around each other and by their motion through the intra-cluster medium. The radio features span a region about 1 million ly in diameter. The field of view is 20 × 15 minutes of arc, slightly smaller than the angular diameter of the full moon. (National Radio Astronomy Observatory.)

of radio telescopes, radio astronomers have found a central jet only 3 pc long pointing along the axis of the larger jet (Figure 17–12). Clearly, the source of the energy is very small.

If our black hole model is correct, how could it produce jets along a preferred axis? Astronomers now compare active galactic nuclei with the peculiar binary star system SS 433 (Chapter 14). That system may consist of a 10-$M_{\odot}$ black hole pulling matter from its companion star and producing a very hot accretion disk. The hot disk somehow ejects beams of radiation and hot gas in opposite directions, sweeping in spirals as the disk precesses. To create a radio galaxy with its jets, we would have to build an SS 433 energy source about 100 times more powerful. That would take a black hole of 10^6–10^9 $M_{\odot}$.

A study of the Seyfert galaxy NGC 4151 seems to have detected mass flowing into the core. The observations were made by analyzing the shapes of spectral lines as distorted by Doppler shifts. The results show mass flowing into a region about 2 light-days in diameter at a rate consistent with a central mass of 60 million solar masses. If the central mass is a black hole, the inflowing mass could easily power the Seyfert galaxy.

What could cause an inflow of matter? That process may be related to tidal interactions between galaxies. When two galaxies interact, their gravitational fields tend to distort their orbital motions, and astronomers now believe that this will feed the black holes at their centers (Figure 17–13). This distortion would explain why active galaxies are more common in crowded clusters and close pairs and why Seyfert galaxies are more likely to occur in pairs. Theoretical simulation of collisions between galaxies show that such a collision does throw enough matter into the central region to power an active galactic nucleus (Figure 17–14). Perhaps every galaxy carries at its heart a black hole waiting to be fed. If so, then activity in galactic nuclei could be common.

The evidence we see is consistent with such a power source. It would explain the jets, the radio lobes, the precession, and the alignment of activity along an axis maintained by a very small central engine. Thus, we see that although double-lobed radio sources seem different from active core sources, they may be related by some process common to all galaxies. In fact, many astronomers now believe that the same phenomena on a larger scale can explain the quasars.

17.2 QUASARS

Quasars (also called quasi-stellar objects or QSOs) are small, powerful objects that seem to lie far away. Some quasars are radio sources, and astronomers are coming to think of them as very powerful, very distant peculiar galaxies. Their significance, however, goes far beyond this interpretation. First, they are so distant that their look-back times are tremendous. They are, therefore, stepping-stones that lead us from a study of individual galaxies to a study of the origin and evolution of the universe. In addition, the discovery of quasars in the 1960s revolutionized astronomy. The objects were so unbelievable that scientists had to reexamine the validity of the most basic natural laws. Although no longer at a boil, that controversy still simmers, so we must discuss the debate stimulated by the discovery of quasars.

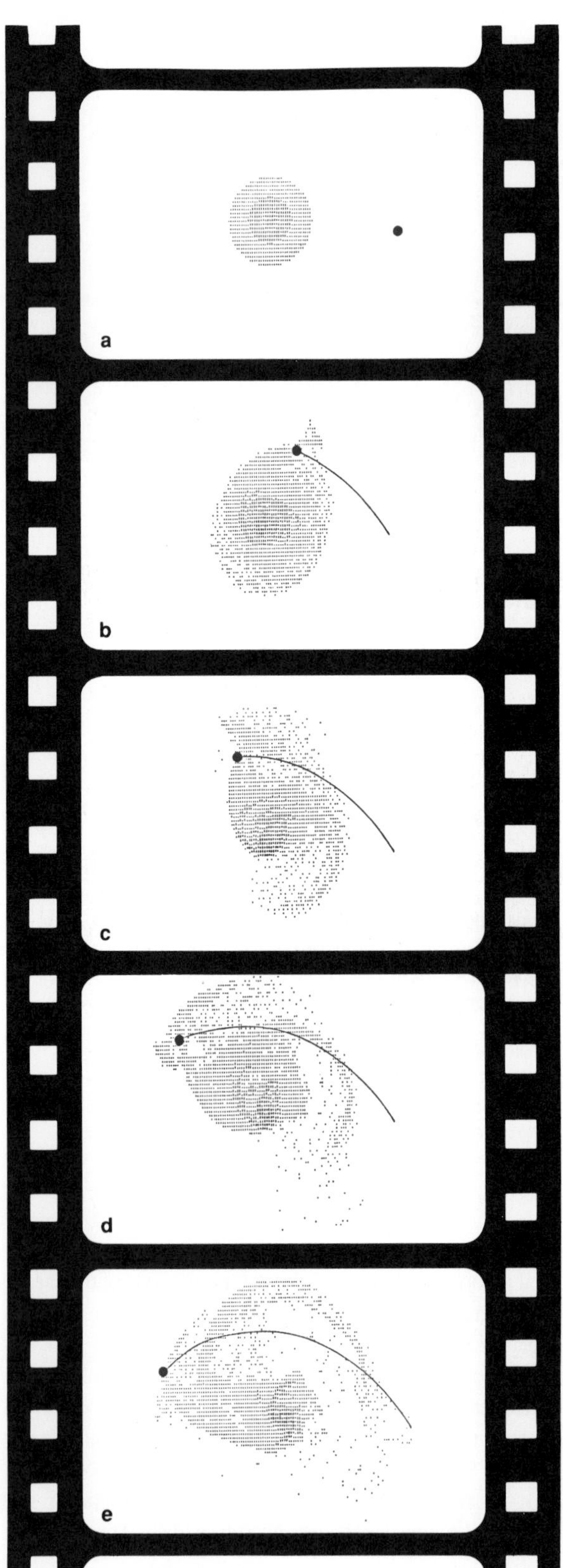

FIGURE 17–14 This computer simulation shows a disk galaxy during an encounter with a smaller galaxy, represented by a black dot. Not only is the disk galaxy distorted by the encounter but also is made unstable, and significant amounts of matter flow into the core. This study suggests that encounters can force enough matter into a galaxy's nucleus to power an active-core energy source. (Courtesy G. Byrd, M. Valtonen, B. Sundelius,and L. Valtaoja.)

DISCOVERY

The existence of quasars became apparent gradually over a period of a few years. Not until the mid-1960s did astronomers realize that the quasars were a dramatically puzzling phenomenon.

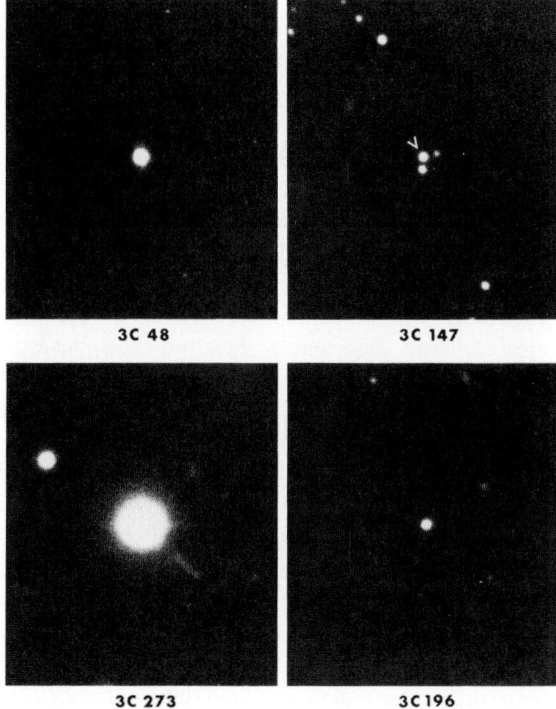

FIGURE 17–15 Quasars look starlike in photographs and are not obviously normal radio galaxies. Note the jet extending away from 3C 273. (Palomar Observatory photograph.)

Radio astronomers first detected quasars as sources of radio energy. A large radio telescope can detect myriads of very distant galaxies, many of which are not visible at optical wavelengths. Radio interferometers (Chapter 6) could measure the angular diameters of these objects. Many were double-lobed radio sources. But some radio sources had angular diameters so small they could not be measured. We now know that only 10 percent of the quasars are radio sources, but it was the radio astronomers who first called attention to them because of their small size.

When optical astronomers photographed the locations of these pointlike radio sources, they did not find the broad fuzzy images of galaxies, but rather small starlike points of light (Figure 17–15). The first to be located optically was 3C 273, and 3C 48 was identified soon after. Because the objects looked starlike at optical wavelengths, they were called quasi-stellar objects, or quasars.

Recall from the introduction to this chapter that we cannot judge the true nature of an astronomical body when we do not know its distance. Many astronomers assumed that the quasars were nearby, in our own back yard. The key to the nature of the quasars was hidden in their spectra.

The spectra of quasars were a mysterious combination of a continuous spectrum plus a few unidentifiable emission lines. In the last months of 1963 Maarten Schmidt at Hale Observatories tried red-shifting the hydrogen Balmer lines to see if they could be made to agree with the lines in the spectrum of 2C 273. At a red shift of 15.8 percent, three lines clicked into place (Figure 17–16). Other quasar spectra quickly yielded to this approach, revealing even larger red shifts. The red shift of 3C 48, for instance, was 0.37.

To an astronomer, the red shift Z is the change in wavelength $\Delta\lambda$ divided by the unshifted wavelength λ_0:

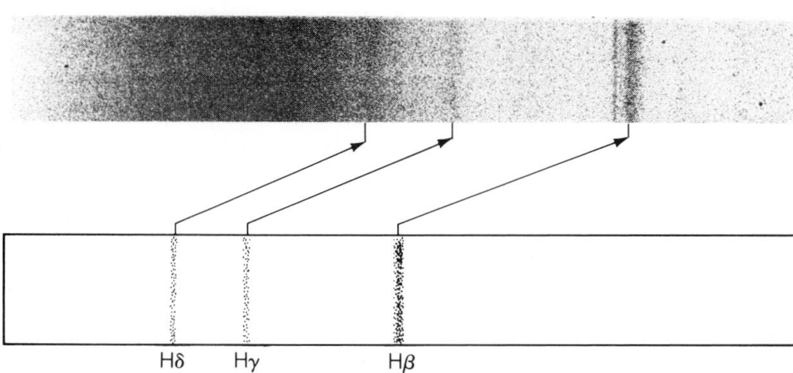

FIGURE 17–16 The spectrum of 3C 273 (reproduced here as a negative image) contains three hydrogen Balmer lines red-shifted by 15.8 percent. The drawing shows the unshifted positions of the lines. (Courtesy Maarten Schmidt.)

$$\text{red shift} = Z = \frac{\Delta\lambda}{\lambda_0}$$

Quasars have large red shifts. If these shifts arise from the expansion of the universe, then the Hubble law predicts large distances. Thus the large red shift of a quasar can be interpreted to mean great distance.

The implications of the large red shifts were startling. At the time, the largest observed galaxy red shifts were about 1, but quasars were soon found with much larger red shifts and, presumably, much larger distances. Normal galaxies at such large distances would be invisible, but the quasars were clearly visible on photographs. They had to be 10 to 1000 times more luminous than a large galaxy.

As astronomers tried to explain the super luminosity of the quasars, another observation complicated the problem. Teams of astronomers discovered that the light from quasars was fluctuating rapidly and erratically. In some cases, a quasar could change its brightness by a large factor in a month, a week, or perhaps even a day.

In Chapter 14 we decided that an object cannot change its brightness appreciably in a time less than the time that light takes to cross its diameter. If the quasars could fluctuate rapidly, they had to be small, not more than a light-week or a light-day in diameter. How can quasars generate 10 to 1000 times more energy than a galaxy in a volume only a few light-weeks in diameter?

In the 20 years since quasars were discovered, astronomers have learned a bit more. Roughly 3000 quasars have been found, most of which have red shifts larger than 1. (As explained in Box 17–1, these large red shifts require that we use the relativistic red shift.) Red shifts are now known as large as 4.1. This implies that the most distant quasars are more than 15.5 billion ly away.

If quasars are as far away as their large red shifts indicate, then their look-back times are very large, and we see them as they were when the universe was significantly younger. In the next chapter we will see that the universe is believed to be 15–20 billion years old. Thus, we see the most distant quasars as they were when the universe was only a few billion years old.

About 10 percent of quasars have absorption lines as well as emission lines in their spectra. These absorption lines have a smaller red shift than the emission lines, and in some cases, multiple sets of absorption lines are observed, each with its individual red shift. The red shift of the absorption lines is less than that of the emission lines, and astronomers have argued that the absorption lines are formed as the light from the quasar passes through normal galaxies on its way to the earth. The quasars are so distant that their light could pass through galaxies that are still too far away to be visible from the earth. Thus, the quasars must be much farther away than these very distant galaxies.

The quasar puzzle quickly became the quasar controversy. A majority of astronomers believed the quasars were very distant, superluminous objects, but a minority suggested that they were not as far away as they seemed.

THE LOCAL HYPOTHESIS

How could a very small object generate 10 to 1000 times more energy than a galaxy? No one could solve the puzzle of the quasar energy source, so some turned the problem around. They proposed that the quasars were not at the great distances suggested by their red shifts, but were local objects, scattered among the galaxies. If they really were that close, they would not be superluminous, and it would not be so difficult to explain their energy source.

This local hypothesis didn't really solve the quasar puzzle, however. It only traded the energy puzzle for a red-shift puzzle. If the quasars are not at great distances, then what produces their fantastic red shifts?

A number of versions of the local hypothesis were proposed, but none could account for the very large red shifts. We refer to red shifts produced by the expansion of the universe as cosmological red shifts, drawing on the word *cosmology,* which refers to the study of the universe as a whole. If a local hypothesis was to succeed, it had to involve noncosmological red shifts. Although the laws of nature did not account for such phenomena, a few astronomers argued that undiscovered laws of nature might be responsible, and they attempted to find examples of noncosmological red shifts—that is, a red shift that was clearly not produced by the expansion of the universe. A single conclusive example would demonstrate that a local hypothesis was possible.

One of the most studied examples was first discussed in 1971. The quasar Markarian 205 is located very near the galaxy NGC 4319, and some astronomers claimed to detect a bridge of matter connecting the two (Figure 17–18a). These observations are shown dramatically in Color Plates 27 and 28. If they are linked together, then they are at the same distance. However, the galaxy has a red shift of 0.006 and the quasar has a red shift of 0.07. If they are at the same distance, then the red shift

BOX 17–1

The Relativistic Red Shift

The classical red-shift equation discussed in Box 7–3 is only an approximation. It works quite well so long as the radial velocity of the star or galaxy is much smaller than the speed of light, and because the velocities of planets, binary stars, and clusters of stars in our galaxy are hardly ever more than a few hundred kilometers per second, the classical red-shift equation is accurate enough for most astronomical problems.

However, when the velocity of an object is an appreciable fraction of the speed of light, we must use the correct equation. Because this equation is derived by the theory of relativity, it is called the **relativistic red-shift** equation. It relates the radial velocity V_r to the speed of light c and the red shift Z.

$$\frac{V_r}{c} = \frac{(Z+1)^2 - 1}{(Z+1)^2 + 1} \qquad \text{where} \quad Z = \frac{\Delta\lambda}{\lambda_0}$$

For example, suppose a quasar has a red shift of 2. What is its velocity? First, Z + 1 equals 3, and thus $(Z + 1)^2$ equals 9. Then the velocity in terms of the speed of light is

$$\frac{V_r}{c} = \frac{9-1}{9+1} = \frac{8}{10} = 0.8$$

Thus, the quasar has a radial velocity of 80 percent of the speed of light.

Figure 17–17 illustrates how V_r/c depends on Z. For small red shifts the velocity is the same in both the classical and relativistic case, but as Z gets larger, the difference between the classical approximation and the true velocity increases. No matter how large Z becomes, the velocity can never quite equal the speed of light.

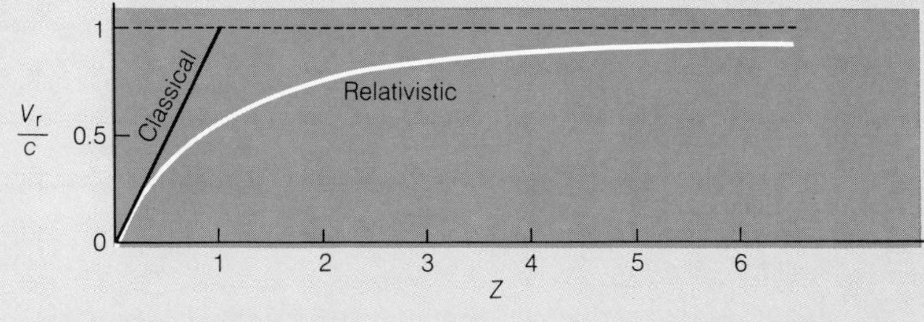

FIGURE 17–17 At high velocities the relativistic red shift must be used in place of the classical approximation. Note that no matter how large Z gets, the speed can never equal the speed of light.

of the quasar cannot arise from the expansion of the universe.

Many astronomers believed that the quasar was much farther away than the galaxy and that they only appeared to be linked together. If that was true, Markarian 205 was not an example of a noncosmological red shift. The question of Markarian 205 may be nearing resolution. Observations published in 1985 show that the link between the galaxy and the quasar is actually a distorted galaxy at the same distance as the quasar. Apparently, the quasar and the distorted galaxy are near each other and only appear to be linked to the much closer galaxy NGC 4319.

Other such examples have been proposed but none is conclusive. An entirely different kind of evidence is shown in Figure 17–18b. A few astronomers have found that quasars tend to be located around nearby galaxies

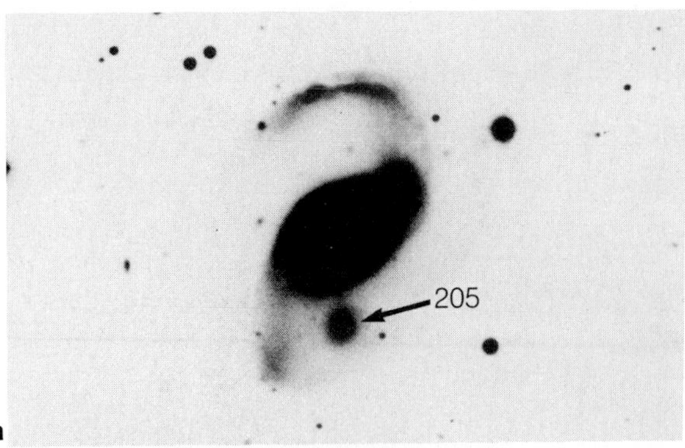

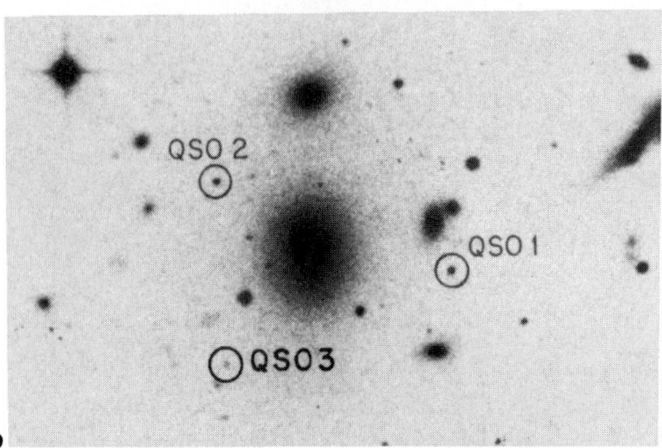

FIGURE 17–18 Proposed examples of noncosmological red shifts. (a) The quasar Markarian 205 appears to be connected to the galaxy NGC 4319 in this negative photograph. (b) This negative image of the galaxy NGC 3842 (center) reveals three quasars within 73 seconds of arc of the galaxy. Some astronomers argue that such groupings occur too often to be chance alignments of distant quasars and nearby galaxies. If the quasars and galaxies are at the same distance, then the large red shifts of the quasars cannot arise from the expansion of the universe. (Courtesy Halton Arp.)

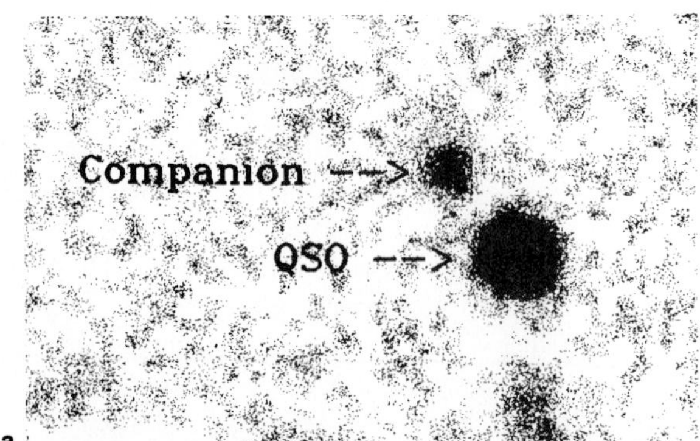

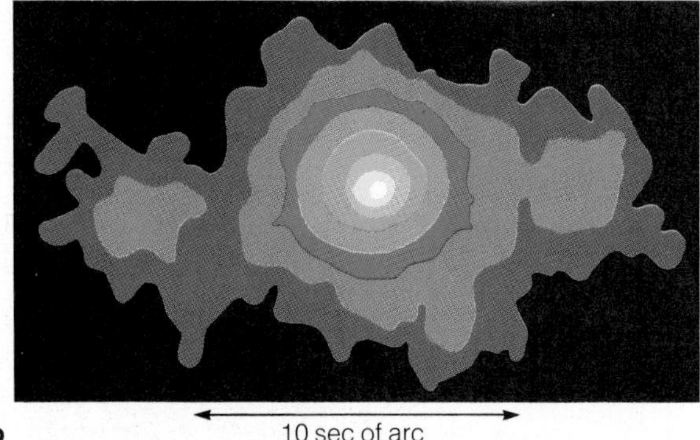

FIGURE 17–19 (a) Quasar PKS 1614 + 051, like a number of other quasars, is known to be associated with a distant cluster of galaxies with the same red shift as the quasar. One of the companion galaxies is shown in this negative image. At about 14.5 billion ly, this is one of the most distant galaxies ever detected. (b) This intensity map of quasar 3C 206 was produced by scanning a photographic plate. The "ears" to right and left are galaxies in a cluster with the same red shift as the quasar. The "fuzz" surrounding the quasar itself has a spectrum like that of a galaxy. (a, S. Djorgovski; b, Adapted from observations by S. Wyckoff, G. Wehinger, and T. Gehren.)

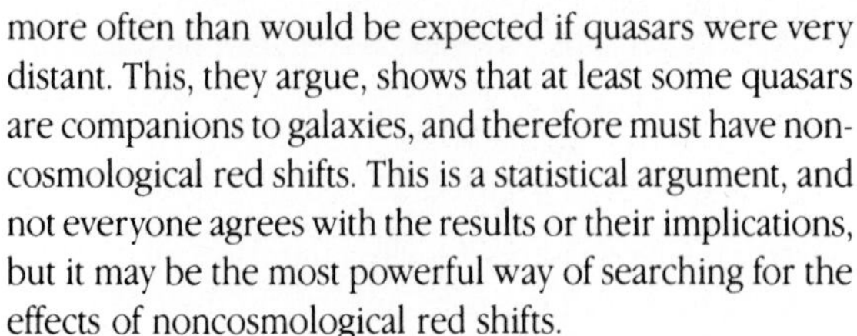

more often than would be expected if quasars were very distant. This, they argue, shows that at least some quasars are companions to galaxies, and therefore must have noncosmological red shifts. This is a statistical argument, and not everyone agrees with the results or their implications, but it may be the most powerful way of searching for the effects of noncosmological red shifts.

The local hypothesis has been generally abandoned, not because of any astonishing discovery, but because of the lack of any conclusive evidence that quasars are nearby and because of a slow accumulation of evidence that they are very distant and may be associated with active galactic nuclei.

THE QUASAR–GALAXY CONNECTION

A number of lines of evidence suggest that quasars are objects in very distant galaxies. Unlike the local hypothesis, this requires that they be superluminous. We will discuss the energy source later.

As electronic imaging such as CCDs (Chapter 6) developed in the 1970s and 1980s, astronomers found that quasars with the smaller red shifts, and presumably smaller distances, are often found in clusters of galaxies (Figure 17–19). These galaxies are very faint, but they appear to be typical of the kinds of galaxies we would expect to find in a crowded cluster. The red shifts of these

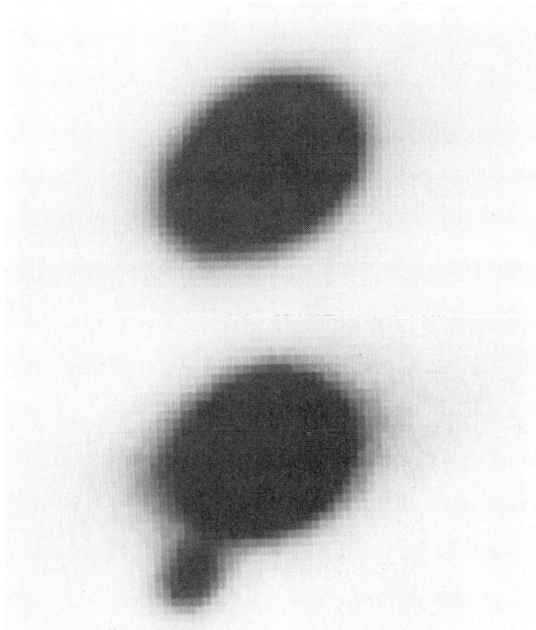

FIGURE 17–20 These two negative photographs show the quasar 1059 + 730. The lower photograph contains an extra image—a supernova occurring in the galaxy containing the quasar. The apparent magnitude of the object and the distance to the quasar based on its red shift give an absolute magnitude for the supernova of −17.6, a value consistent with the supernova explanation. (Courtesy Bruce Campbell.)

galaxies are the same as that of the quasar they accompany, implying that the cluster and quasar both lie at the same great distance.

These same observation techniques have allowed the discovery of quasar fuzz. The best images of the low-red-shift quasars often show that they are surrounded by fuzz (Figure 17–19b). The spectrum of this fuzz is typical of a galaxy with the same red shift as the quasar. The quasar 3C 273, for example, appears to lie in a giant elliptical galaxy.

Faint galaxies accompanying quasars and quasar fuzz are visible only for low-red-shift quasars. Higher-red-shift quasars are so distant that galaxies near them would be invisible to our best telescopes and instruments. The Hubble Space Telescope will no doubt discover more.

Recently astronomers studying the quasar QSO 1059 + 730 discovered an image very near the quasar that did not appear on other plates. After eliminating other possibilities, they concluded that the object was a supernova explosion in the galaxy that hosts the quasar (Figure 17–20). This seems to confirm the view that otherwise normal galaxies can be the site of quasars.

A discovery made in 1979 further supports the belief that quasars are very distant and not local. The object 0957 + 561 lies just a few degrees west of the bowl of the Big Dipper and consists of two quasars separated by only 6 seconds of arc. The optical spectra of these objects proved to be nearly identical with the same red shift (1.40) and the same relative strengths of lines. Quasar spectra are as different as fingerprints, so when two quasars so close together proved to have the same spectra and the same red shift, it was evident that they were separate images of the same quasar.

The two images are formed by an intervening galaxy. The gravitational field of the galaxy deflects the light of the quasar and focuses it into two images (Figure 17–21a). The galaxy itself is much too far away to be easily visible against the glare of the quasar, but Dr. Alan Stockton used computer-image processing to reveal the hidden image of the galaxy. The left portion of Figure 17–21b and c shows a negative reproduction of a photograph of the two quasars. In the right portion, the image of the upper quasar has been subtracted from the image of the lower quasar. What remains is the small image (arrow) of a galaxy, probably a giant elliptical.

The discovery of the double nature of 0957 + 561 is significant for two reasons. First, the effect, known as the **gravitational lens effect**, was predicted in 1936 by Albert Einstein but had never been seen before. Thus, the discovery further confirms general relativity (Color Plate 29). Second, the gravitational lens effect is important because it shows that the quasar cannot be local. It must be much farther away than the galaxy, and the galaxy is a giant elliptical galaxy that is just barely bright enough to detect. The galaxy must be fairly distant itself. Thus, the quasar cannot be local.

Since 1979 half a dozen gravitational lenses have been found. In each case, light from a distant quasar is deflected by the gravitational field of a nearer galaxy and multiple images are produced. The quasar PG 1115 + 08 in Leo, for example, consists of three images within 3 seconds of arc of each other, all with the same red shift.

The discovery of gravitational lenses has been a serious blow to the local hypothesis because it shows that at least some quasars are very distant, far beyond the more distant galaxies. If any quasar is at the distance indicated by its red shift, then it is superluminous, and we must try to explain its energy source. It is possible that some quasars are local and that noncosmological red shifts do occur, but they no longer help us solve the quasar puzzle.

A MODEL QUASAR

Assuming that the great majority of astronomers are right and that quasars are at very great distances, can we combine the observed facts and produce a model of a quasar? Our model must explain what we see and be consistent with what we know about the rest of the universe.

At the center of our model an unknown source of energy must accelerate electrons to high speeds. These electrons, spiraling through a weak magnetic field, emit synchrotron radiation, both light and radio. This accounts for the continuous spectrum. Such a region can be no larger than a few light-weeks in diameter to allow the rapid fluctuations observed. The quasar 1525 + 227 is a source of X rays, and statistical studies show that it is flickering in a time as short as 200 seconds. If the X rays are produced by the quasar's energy source, it cannot be larger in diameter than the earth's orbit.

The emission lines in a quasar spectrum do not seem to fluctuate rapidly, so our model can create them in a larger region of gas ionized by the synchrotron radiation streaming out of the core. The absorption lines we see in quasar spectra could be produced as light from the quasar passes through the galaxies on its way to the earth. Although there do not seem to be enough galaxies to produce all the lines we see, we must remember the evidence that galaxies are surrounded by extensive coronae. If galaxies are that big, there may be enough to produce such effects.

Our model has explained the spectrum, but can it explain the energy source? In this case, we might use the same energy source we suggested for active galactic nuclei—a massive black hole. To make a quasar, we have only to throw mass down the black hole faster than we do in an active galaxy. A black hole gobbling down 10 solar masses per year through a hot accretion disk could power a quasar.

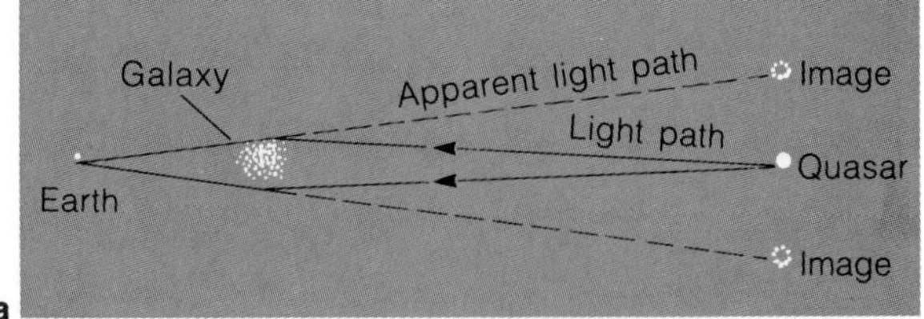

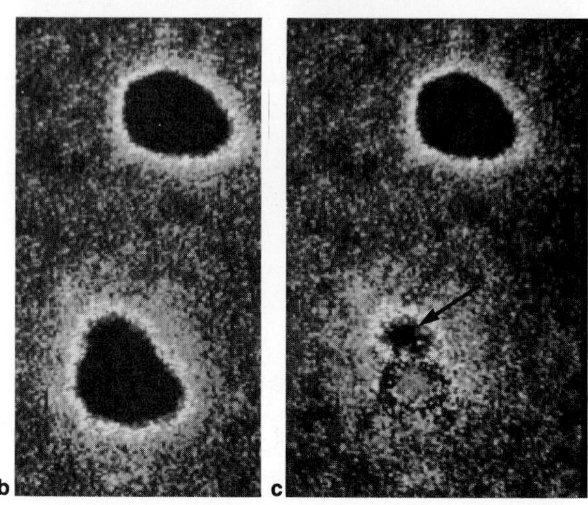

FIGURE 17–21 The gravitational lens effect. (a) A distant quasar can appear to us as multiple images if its light is deflected and focused by the mass of an intervening galaxy. The galaxy, being much fainter than the quasar, is not easily visible from the earth. (b) The two images of quasar 0957 + 561 appear as black blobs in this computer-enhanced image. When the upper image is subtracted from the lower image (c), we are able to detect the faint image of the giant elliptical galaxy (arrow) whose gravity produces the gravitational lens effect. (Courtesy Alan Stockton, Institute for Astronomy, University of Hawaii.)

In some ways quasars resemble active galaxies. For example, images of 3C 273 show a jet of material (Figure 17–15) much like the jets we see being ejected from active galaxies, and quasar 2300 – 189 appears to be emitting jets of matter (Figure 17–22a) as it precesses, rather like the powerful SS 433 (Chapter 14). A few quasars are flanked by radio lobes (Figure 17–22b), and many, perhaps a third, appear to be interacting with neighboring objects. At least two quasars have been found that have curved tails characteristic of those formed during collisions and mergers between galaxies. In addition, quasar energy sources may be as small as the energy sources in the cores of Seyfert galaxies, so many astronomers believe that the active galactic nuclei seen in Seyfert galaxies are lower-luminosity versions of quasars.

If quasars are powered by black holes, then we can draw interesting inferences from their association with distant galaxies. Quasars seem to occur in regions where galaxies are crowded, and the galaxies near them are often distorted. This might mean that the quasar was triggered into existence when interacting galaxies fed matter into a massive black hole in one of the galaxies. But we must also recall the enormous look-back times to quasars. If they are associated with galaxies, those galaxies are young. We do not see quasars near us, at small look-back times. Quasars seem to have formed when the universe

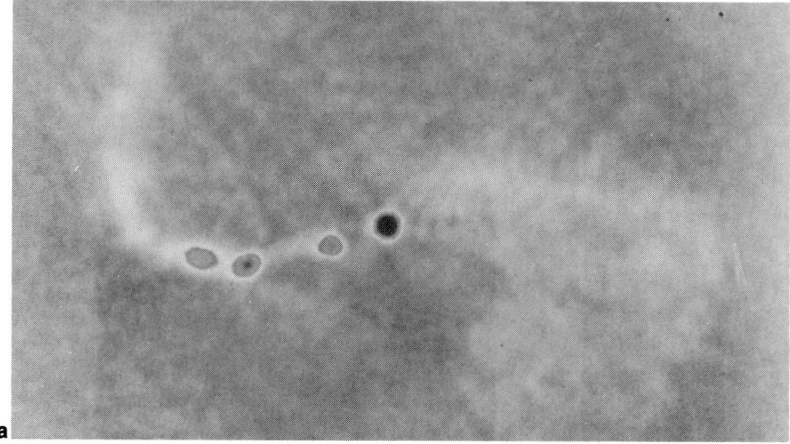

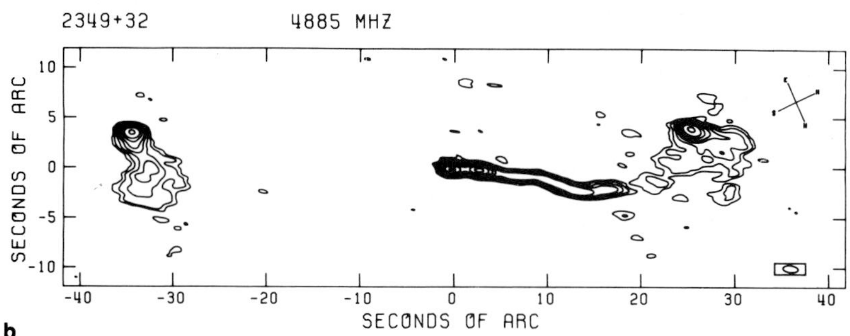

FIGURE 17–22 (a) A VLA radio image of quasar 2300 − 189 reveals precessing jets suggestive of the much smaller and nearer X-ray binary SS 433 (Chapter 14). (b) The quasar 4C 32.69 consists of a bright energy source at the center of this map with a jet extending toward the radio lobe at the right. Another radio lobe lies to the left. (a, Courtesy James J. Condon and Hugh S. Murdoch, National Radio Astronomy Observatory; b, Courtesy Robert Potash.)

and galaxies were young, and this may be related to the early formation of the galaxies.

Another question is, Where are all the dead quasars? If quasars formed naturally when galaxies formed, or when young galaxies in crowded clusters interacted, why don't we see them now? Some astronomers believe we do see them in the form of active galactic nuclei. These may be black holes that long ago gobbled mass at a great rate and created a quasar, but now are reduced to a lesser diet and power the center of an active galaxy. If that is true, then quasars may be the ancestors of both active and normal galaxies.

SUPERLUMINAL EXPANSION

A few astronomers have objected that the quasars cannot be located in such very distant galaxies because, if the quasars are really that far away, then a few examples seem to be expanding at speeds greater than the velocity of light.

Radio astronomers discovered this **superluminal expansion** by using Very Long Baseline Interferometry (VLBI)—the interconnection of radio telescopes in different parts of the world. Such a network of antennae can resolve details to about 0.002 second of arc in diameter. At the distance of quasar 3C 273, for example, this corresponds to a few parsecs.

For a few quasars these radio maps show small blobs near the quasars, and maps made over a few years show these blobs moving away from the quasar. The quasar 3C 273, for example, has a blob very near the quasar, which is moving about 0.0008 seconds of arc farther from the quasar each year (Figure 17–23). If we believe that the quasar's red shift of 0.16 arises from the expansion of the universe, then the quasar is about 960 Mpc distant and the small-angle formula tells us that the blob is moving away from the quasar at a speed of 3.8 pc per year. This is the same as 12 ly per year, which is impossible. The theory of relativity clearly states that nothing can travel faster than the velocity of light.

One way to solve this problem is to assume that the quasar is local. If it is only 10 Mpc away, then the blob is moving at a velocity less than 15 percent the velocity of light. But, of course, if quasars are local, we can't explain their large red shifts.

The most commonly discussed theory to explain this superluminal expansion without making the quasar a local object is the **relativistic jet model**. It supposes that the quasar is ejecting a jet of matter at nearly the velocity of light and that the jet is pointed nearly toward the earth. Then a blob of matter in the jet could appear to move away from the quasar at velocities greater than the velocity of light.

To see how this can happen, consider Figure 17–24. A blob of matter in the jet emits radiation that will reach the earth in 1977. In the next 35 years that radiation travels 35 ly toward the earth, but the blob, traveling at 98 percent the velocity of light, travels 34 ly closer to earth. Radiation that it emits from this second position is only 1 ly behind the radiation it emitted 35 years before. The newly emitted radiation will arrive on earth in 1978, and we will see the blob move in only 1 year a distance that actually took 35 years. This is a simplified explanation that omits relativistic effects, but it clearly shows that a blob of gas in a relativistic jet can appear to travel faster than light if the jet points nearly toward the earth.

This jet hypothesis is not completely successful. One problem is that there may be too many cases of superluminal expansion to be explained by relativistic jets. In a survey of 12 quasars, 6 had superluminal expansion. We might expect a few jets to be directed toward the earth, but surely not 50 percent. The explanation may lie in the fact that these 12 quasars were not selected randomly, so they may not be representative of the population of quasars in general.

Although most astronomers accept the relativistic jet explanation, for now at least, they are still uneasy. That is characteristic of the quasars; since they were first discovered, they have made astronomers uneasy.

Our study of peculiar galaxies has led us far out in space and back in time to quasars. The light now arriving from the most distant quasars left when the universe was

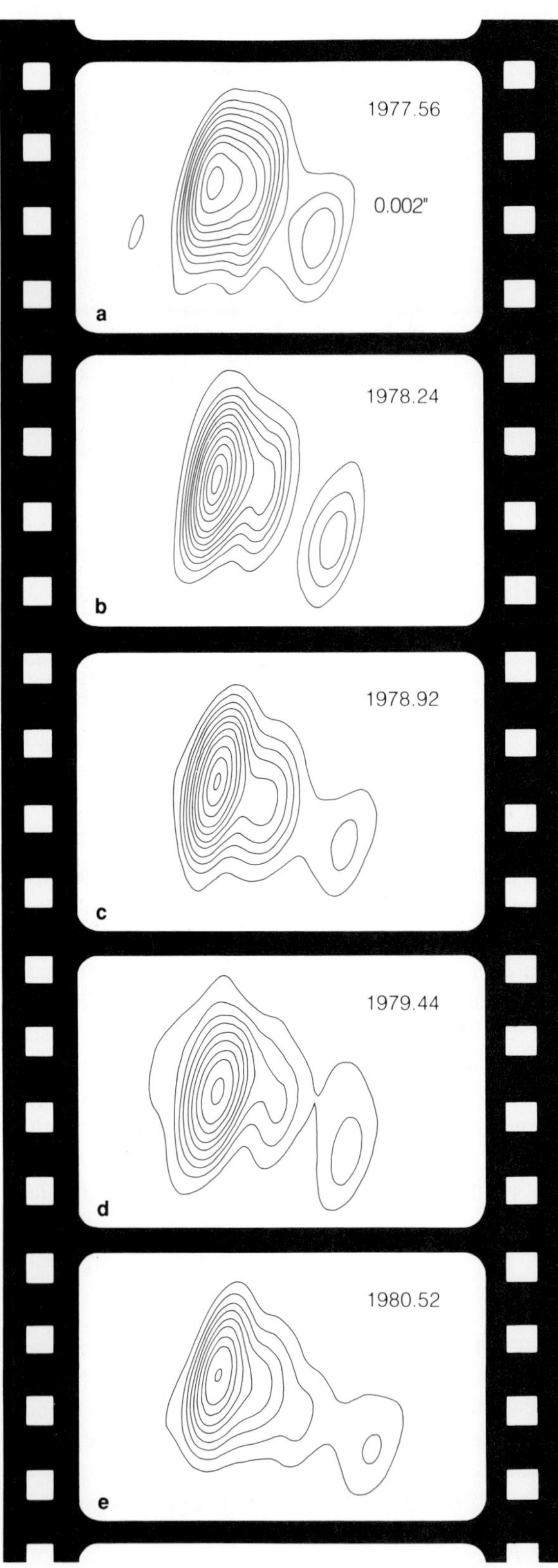

FIGURE 17–23 Superluminal expansion. Very high-resolution radio maps of quasar 3C 273 show a blob separating from the quasar at a rate of 0.0008 seconds of arc per year. If the quasar is at the distance indicated by its red shift, the blob must be moving over ten times the velocity of light. (Courtesy Timothy Pearson and *Nature*.)

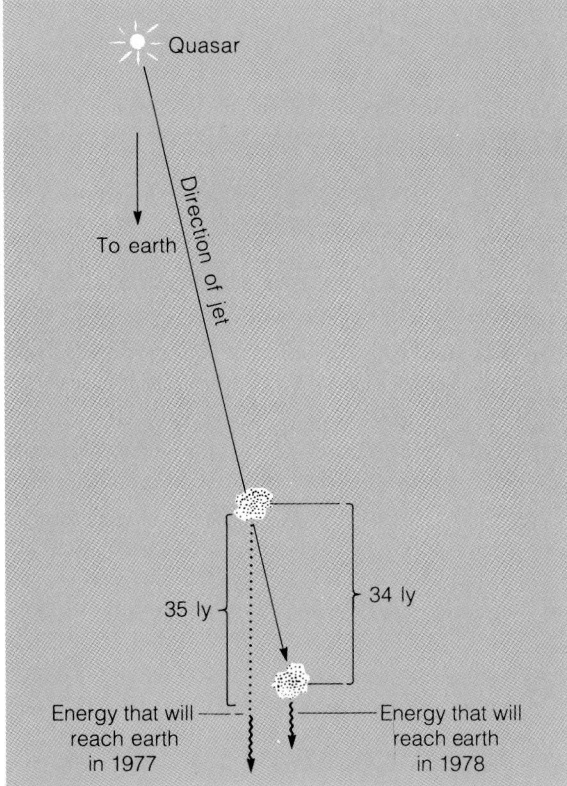

FIGURE 17–24 The relativistic jet model explains superluminal expansion by supposing that the quasar emits a jet that points nearly at the earth. If a blob of gas in the jet, traveling at nearly the velocity of light, emits radio energy toward the earth at two times separated by 35 years, the first bundle of energy will have only a 1 ly lead. As the energy arrives at the earth, the blob will appear to move in 1 year a distance that actually took 35 years.

only about one-fourth its present age. The quasars stand at the very threshold of the study of the universe itself.

SUMMARY

Some galaxies have peculiar properties. Double-lobed radio galaxies emit radio energy from the lobes on either side of the galaxies. Some are known to have jets extending from the galaxies out into the radio lobes. Other galaxies are known to have active nuclei. Seyfert galaxies are spirals with small, highly luminous cores; some giant ellipticals have small, energetic cores, some of which have jets of matter rushing outward.

The proposed explanation for double-lobed radio galaxies is called the double-exhaust model. This theory assumes that the galaxy ejects jets of relativistic particles in two beams that push into the intergalactic medium and blow up the radio lobes like balloons. The impact of the beams on the intergalactic medium produces the hot spots detected on the leading edges of many radio lobes. We know that many active galaxies produce jets, and we can see instances where these jets are blown back by the intergalactic medium to produce head–tail galaxies.

The source of energy in the active galactic nuclei is unknown, but it may involve gravity. Observations have hinted that M 87, a giant elliptical galaxy with an active core and a jet, may contain a supermassive black hole. Although the observations are far from conclusive, theorists are now considering such objects as the power sources for active galaxies. Matter falling into such an object could release tremendous energy and, by analogy with the less massive SS 433, might eject beams.

Active galactic nuclei seem to occur in galaxies involved in collisions or mergers. Seyfert galaxies, for example, are three times more common in interacting galaxies than in isolated galaxies. This suggests that an encounter between galaxies can throw matter into the central regions of the galaxies where it can feed a black hole and release energy. Galaxies not recently involved in collisions will not have matter flowing into their central black hole and will not have active galactic nuclei.

The quasars may be related objects. Their spectra show emission lines with large red shifts. Their large red shifts imply they are very distant, and that they are visible at all implies that they are superluminous. Because they fluctuate rapidly, they must be no larger than a few hundred light-seconds in diameter.

Because they lie at great distances, we see them as they were long ago. The look-back time to the most distant quasar is about 15.5 billion years. They may be young interacting galaxies distorting each other and funneling mass into black holes, or they may be young galaxies in the early stages of formation. In any case, heavy mass flow into a central black hole could heat the gas and produce the observed synchrotron radiation and emission lines.

A growing mass of evidence suggests that the quasars are the active cores of distant galaxies. The discovery of the gravitational lens effect shows that at least a few quasars lie at cosmological distances. Also, the spectrum of the fuzz that surrounds some quasar images looks like the spectrum of a normal galaxy with the same red shift as the quasar.

Less easy to explain are the cases of superluminal expansion discovered in some quasars. Blobs of material appear to be rushing away from some quasars at speeds as high as 12 times the speed of light, but this may be an illusion caused by a jet of relativistic material that happens to be directed almost exactly at the earth.

Because of the quasar controversy, astronomers are reexamining the red shift–distance relation. Searches have found a number of objects with red shifts that do not agree with their presumed distances. None of these, however, is conclusive. They could be chance alignments of distant objects with large red shifts and nearby objects with small red shifts. Until some unambiguous example is found, astronomers will continue to depend on the Hubble law.

NEW TERMS

radio galaxy
active galaxy
active galactic nucleus (AGN)
double-lobed radio source
hot spot
double-exhaust model
head–tail radio galaxy
Seyfert galaxy
BL Lac object
quasar
relativistic red shift
gravitational lens effect
superluminal expansion
relativistic jet model

QUESTIONS

1. What is the difference between the terms "radio galaxy" and "active galaxy"?
2. What evidence do we have that radio lobes are inflated by jets from active galactic nuclei (AGN)?
3. Why is it significant that most head–tail galaxies occur in clusters of galaxies?
4. What evidence do we have that AGN contain black holes?
5. What properties of SS 433 resemble the energy source in AGN?
6. If you located a galaxy that had not interacted with another galaxy recently, would you expect it to be active or inactive? Why?
7. Seyfert galaxies are three times more common in close pairs of galaxies than in isolated galaxies. What does that suggest?
8. Why do we conclude that quasars must be small? How do quasars resemble the AGN in Seyfert galaxies?
9. How does our model quasar account for the different components in a quasar's spectrum?
10. Why have most astronomers abandoned the local hypothesis?
11. What kind of observations would be needed to confirm the local hypothesis?
12. Why are some astronomers concerned about the proportion of quasars that have superluminal expansion?

PROBLEMS

1. The total energy stored in a radio lobe is about 10^{53} J. How many solar masses would have to be converted into energy to produce this energy? (HINTS: Use $E = m_0 c^2$. One solar mass equals 2×10^{30} kg.)
2. If the jet in NGC 5128 is traveling 5000 km/sec and is 40 kpc long, how long will it take for gas to flow from the core of the galaxy out to the end of the jet?
3. Cygnus A is 225 Mpc away, and its jet is about 50 seconds of arc long. What is the length of the jet in parsecs? (HINT: See Box 3–1.)
4. If the average giant elliptical galaxy is 50 kpc in diameter, calculate an estimate of the distance to the galaxy in Figure 17–8. (HINT: First estimate the angular diameter of the elliptical galaxy from the information in the caption.)
5. Use the small-angle formula to find the linear diameter of a radio source with an angular diameter of 0.0015 seconds of arc and a distance of 3.25 Mpc.
6. If the active core of a galaxy contains a black hole of 10^6 $M_\odot$, what will the orbital period be for matter orbiting the black hole at a distance of 0.33 AU? (HINT: See Box 10–1.)
7. If a quasar is 1000 times more luminous than an entire galaxy, then what is the absolute magnitude of such a quasar? (HINT: The absolute magnitude of a bright galaxy is about -21.)
8. If the quasar in Problem 7 were located at the center of our galaxy, what would its apparent magnitude be? (HINT: See Box 9–2.)
9. What is the radial velocity of 3C 48 if its red shift is 0.37? (HINT: See Box 17–1.)
10. The hydrogen Balmer line Hβ has a wavelength of 486.1 nm. It is shifted to 563.9 nm in the spectrum of 3C 273. What is the red shift? (HINT: What is $\Delta\lambda$?)
11. Plot the red shifts and velocities of quasars 3C 48, 3C 273, and the most distant known quasar (red shift 4.01) in Figure 17–17.
12. The quasar images in Figure 17–21b are separated (center to center) by 6 seconds of arc. What is the angular diameter of the giant elliptical galaxy? Assume, conservatively, that this galaxy has a linear diameter of about 30 kpc and use the small-angle formula to estimate its distance.

RECOMMENDED READING

ARP, HALTON C. "Related Galaxies with Different Redshifts." *Sky and Telescope 65* (April 1983), p. 307.

BALICK, B. "Quasars with Fuzz." *Mercury 12* (May/June 1983), p. 31.

BLANDFORD, ROGER D., MITCHELL C. BEGELMAN, and MARTIN J. REES "Cosmic Jets." *Scientific American 246* (May 1982), p. 124.

BURNS, JACK O., and R. MARCUS PRICE "Centaurus A: The Nearest Active Galaxy," *Scientific American 249* (Nov. 1983), p. 56.

CAMPBELL, B. "A Supernova Explosion in a Galaxy Containing a Quasar." *Mercury 14* (Nov./Dec. 1985), p. 184.

CAPRIOTTI, EUGENE R. "Seyfert Galaxies." *Natural History 90* (Feb. 1981), p. 82.

DARLING, DAVID "The Quest for Black Holes." *Astronomy 11* (July 1983), p. 6.

DOWNS, ANN "Radio Galaxies." *Mercury 15* (March/April 1986), p. 34.

DURIC, NEBOJSA "A New Source of Cosmic Rays." *Astronomy 12* (July 1984), p. 66.

FEIGELSON, ERIC D., and ETHAN J. SCHREIER "The X-Ray Jets of Centaurus A and M 87." *Sky and Telescope 65* (Jan. 1983), p. 6.

GORENSTEIN, MARC V. "Charting Paths Through Gravity's Lens." *Sky and Telescope 66* (Nov. 1983), p. 390.

HOFF, DARREL B. "Laboratory Exercises in Astronomy—Quasars." *Sky and Telescope 63* (Jan. 1982), p. 20.

LAWRENCE, J. "Gravitational Lenses and the Double Quasars." *Mercury 9* (May/June 1980), p. 66.

LONGAIR, M. S. *High Energy Astrophysics.* New York: Cambridge University Press, 1981.

MARAN, STEPHEN P. "Seeing Double and Seeing Triple." *Natural History 90* (July 1981), p. 24.

———. "The Quasar Controversy Continues." *Natural History 91* (Jan. 1982), p. 85.

MORRISON, N., and S. GREGORY "Centaurus A: The Nearest Active Galaxy." *Mercury 13* (May/June 1984), p. 75.

———. "A Remarkable Image of Cygnus A." *Mercury 14* (March/April 1985), p. 55.

SCHENDEL, JACK "Are Blazars Quasars?" *Astronomy 8* (Feb. 1980), p. 67.

———. "Looking Inside Quasars." *Astronomy 10* (Nov. 1982), p. 6.

SCHILD, RUDOLPH E. "Arp 220 Revealed." *Sky and Telescope 69* (Jan. 1985), p. 24.

SHAWCROSS, WILLIAM E. "Multiple Quasars and Gravitational Lenses." *Sky and Telescope 60* (Dec. 1980), p. 486.

SMITH, DAVID H. "Mysteries of Cosmic Jets." *Sky and Telescope 69* (March 1985), p. 213.

SULENTIC, JACK W. "Are Quasars Far Away?" *Astronomy 12* (Oct. 1985), p. 66.

TYSON, TONY, and MARC GORENSTEIN "Resolving the Newest Gravitational Lens." *Sky and Telescope 70* (Oct. 1985), p. 319.

WYCKOFF, SUSAN, and PETER A. WEHINGER "Are Quasars Luminous Nuclei of Galaxies?" *Sky and Telescope 61* (March 1981), p. 200.

CHAPTER 18

COSMOLOGY

The Universe, as has been observed before, is an unsettlingly big place, a fact which for the sake of a quiet life most people tend to ignore.

Douglas Adams

THE RESTAURANT AT THE END OF THE UNIVERSE

What is the biggest number? A billion? How about a billion billion? That is only 10^{18}. How about 10^{100}? That very large number is called a googol, a number that is at least a billion billion times larger than the total number of atoms in all of the galaxies we can observe. Can you name a number bigger than a googol? Try a billion googols, or better yet, a googol of googols. You can go even bigger than that, ten raised to a googol. (That's a googolplex.) No matter how large a number you name, we can immediately name a bigger number. That is what infinity means—big without limit.

If the universe is infinite in size, then you can name any distance you want, and the universe is bigger than that. Try a googol to the googol light-years. If the universe is infinite, there is more universe beyond that distance.

Is the universe infinite or finite? In this chapter we will try to answer that question and others, not by playing games with big numbers, but through **cosmology**, the study of the universe as a whole.

If the universe is not infinite, then we face the edge-center problem. Suppose that the universe is not infinite, and you journey out to the edge of the universe. What do you see: a wall of cardboard; a great empty space; nothing, not even space? What happens if you try to stick your head out beyond the edge? These are almost nonsense questions, but they illustrate the problem we have when we try to think about an edge to the universe. Modern cosmologists believe that the universe cannot have an edge. In this chapter, we will see how the universe might be finite but have no edge.

If the universe has no edge, then it can have no center. We find the centers of things—globular clusters, sheets of paper, bowling balls, and pizzas—by referring to their edges.

Finally, we must answer a third question. How old is the universe? If it is not eternal (infinite in age), then it had a beginning, and we must think about how the universe began. Of course, we must also think about the possibility that the universe may eventually end.

18.1 THE STRUCTURE OF THE UNIVERSE

We begin our study of cosmology by considering the most basic property of the universe—its geometry. By understanding the geometry of space-time, we will see how we might discover whether the universe is infinite or finite.

WHY DOES IT GET DARK AT NIGHT?

We have all noticed that the night sky is dark. However, reasonable assumptions about the geometry of the universe can lead us to the conclusions that the night sky should glow as brightly as a star's surface. This conflict between observation and theory is called **Olbers' paradox** after Heinrich Olbers, a Viennese physician and astronomer, who discussed the paradox in 1826.

However, Olbers' paradox is not Olbers', and it isn't a paradox. The problem of the dark night sky was first discussed by Thomas Digges in 1576 and was further analyzed by such astronomers as Johannes Kepler in 1610 and Edmund Halley in 1721. Olbers gets the credit through an accident of scholarship on the part of modern cosmologists who did not know of previous discussions. What's more, Olbers' paradox is not a paradox. We will be able to understand why the night sky is dark by revising our assumptions about the nature of the universe.

To begin, let's state the so-called paradox. Suppose we assume that the universe is static, infinite, and eternal. Suppose we also assume that it is uniformly filled with stars. (The aggregation of stars into galaxies makes no difference to our argument.) If we look in any direction, our line of sight must eventually reach the surface of a star (Figure 18–1). Consequently, every point on the surface of the sky should be as bright as the surface of a star, and it should not get dark at night.

Of course, the most distant stars would be much fainter than the nearer stars, but there would be more distant stars than nearer stars. The intensity of the light from a star decreases according to the inverse square law, so distant stars would not contribute much light. However, the farther we look in space, the larger the volume we survey. Thus, the number of stars we see at any given distance increases as the square of the distance. The two effects cancel out, and the stars at any given distance contribute as much total light as the stars at any other distance. Then given our assumptions, every spot on the sky must be occupied by the surface of a star, and it should not get dark at night.

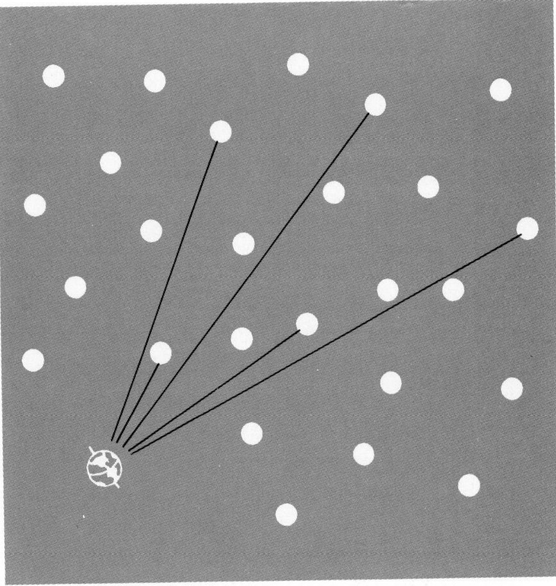

FIGURE 18–1 If the universe is infinitely large and infinitely old and uniformly filled with stars, any line from the earth should eventually reach the surface of a star. This predicts that it should not get dark at night, a puzzle commonly referred to as Olbers' paradox.

Imagine the entire sky glowing with the brightness of the surface of the sun. The glare would be overpowering. In fact, the radiation would rapidly heat the earth and all other celestial objects to the average temperature of the surface of the stars, 1000 K at least. Thus, we can pose Olbers' paradox in another way: "Why is the universe so cold?"

Kepler and many other astronomers of his time saw no paradox in the dark night sky because they believed in a finite universe. Once they looked beyond the sphere of stars, there was nothing to see except, perhaps, the dark floor of heaven. But soon western astronomers accepted that the universe was infinite, and thus they had to explain why the sky is dark between the stars.

Olbers assumed that the sky was dark because clouds of matter in space absorb the radiation from distant stars. But this interstellar medium would gradually heat up to the average surface temperature of the stars, and the gas and dust clouds should be glowing as brightly as the stars.

Today cosmologists believe they understand why the sky is dark. Olbers' paradox makes the incorrect prediction that the sky should be bright because it is based on two incorrect assumptions. The universe is neither static nor infinitely old.

In Chapter 16 we saw that the galaxies are receding from us. The distant stars in these galaxies are receding from the earth at high velocity, and their light is Doppler-shifted to long wavelengths. We can't see the light from these stars because their light is red-shifted, and the energy of the photons is reduced to levels we cannot detect. Expressed in another way, the universe is cold because the photons from very distant stars arrive with such low energy they cannot heat up objects. Although this explains part of the problem, the red shifts of the distant galaxies are not enough to make the sky as dark as it appears.

The second part of the explanation was first stated by Edgar Allan Poe in 1848. He proposed that the night sky was dark because the universe was not infinitely old but had been created at some time in the past. The more distant stars are so far away that light from them has not reached us yet. That is, if we look far enough, the look-back time is greater than the age of the universe, and we look back to a time before stars began to shine. Thus, the night sky is dark because the universe is not infinitely old.

This is a powerful idea because it clearly illustrates the difference between the universe and the observable universe. The universe is everything that exists, and it could be infinite. But the observable universe is the part that we can see. We will learn later that the universe is 15–20 billion years old. In that case, the observable universe has a radius of 15–20 billion ly. Do not confuse the observable universe, which is finite, with the universe as a whole, which could be infinite.

The assumptions that we made when we described Olbers' paradox were at least partially in error. This illustrates the importance of assumptions in cosmology and serves as a warning that our common sense expectations are not dependable. All of astronomy is reasonably unreasonable—that is, reasonable assumptions often lead to unreasonable results. That is especially true in cosmology, so we must examine our assumptions with special care.

BASIC ASSUMPTIONS

Although we could make many assumptions, three are basic—homogeneity, isotropy, and universality.

Homogeneity is the assumption that matter is uniformly spread throughout space. Obviously this is not true on the small scale, because we can see matter concentrated in planets, stars, and galaxies. Homogeneity refers to the large-scale distribution. If the universe is homogeneous, we should be able to ignore individual galaxies and think of matter as an evenly spread gas in which each particle is a cluster of galaxies. Recent observations suggest that the universe is homogeneous only on the largest scales.

Isotropy is the assumption that the universe looks the same in every direction, that it is isotropic. On the small scale this is not true, but if we ignore local variations like galaxies and clusters of galaxies, the universe should look the same in any direction. For example, we should see roughly the same number of galaxies in every direction. Again, observations suggest that the universe is isotropic on the largest scales.

The most easily overlooked assumption is **universality**, which holds that the physical laws we know on earth apply everywhere in the universe. Although this may seem obvious at first, some astronomers challenge universality by pointing out that when we look out in space we look back in time. If the laws of physics change with time, we may see peculiar effects when we look at distant galaxies. For now we will assume that the physical laws observed on earth apply everywhere in the universe.

The assumptions of homogeneity and isotropy lead to an assumption so fundamental it is called the **cosmological principle.** According to this principle, any observer in any galaxy sees the same general features of the universe. For example, all observers should see the same kinds of galaxies. As in previous assumptions, we ignore local variations and consider only the overall appearance of the universe, so the fact that some observers live in galaxies in clusters and some live in isolated galaxies is only a minor irregularity.

Evolutionary changes are not included in the cosmological principle. If the universe is expanding and the galaxies are evolving, observers living at different times may see galaxies at different stages. The cosmological principle says that once observers correct for evolutionary changes, they should see the same general features.

The cosmological principle is actually an extension of the Copernican principle. Copernicus said that the earth is not in a special place; it is just one of a number of planets orbiting the sun. The cosmological principle says that there are no special places in the universe. Local irregularities aside, one place is just like another. Our location in the universe is typical of all other locations.

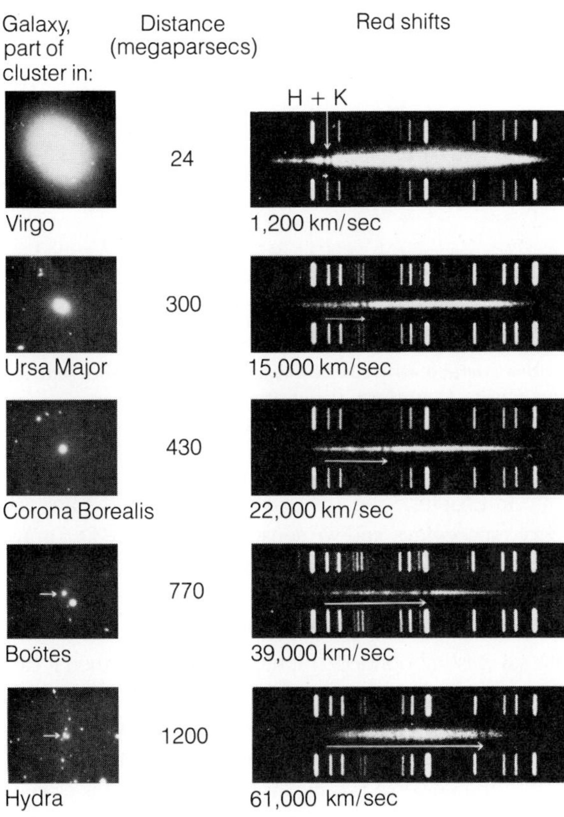

FIGURE 18–2 The red shifts in these galaxy spectra appear as displacements of the H and K lines of calcium toward the red (arrows). The red shifts are assumed to be Doppler shifts due to the expansion of the universe. (Palomar Observatory photograph.)

If we accept the cosmological principle, then we may not imagine that the universe has an edge or a center. Such locations would not be the same as all other locations.

Once we establish our basic assumptions, we face a choice between two methods of attack. We can observe the universe and try to deduce its properties from its appearance, or we can build a theoretical universe based on our assumptions and the laws of nature and then compare our theory with reality. Both methods are valid.

THE EXPANSION OF THE UNIVERSE

All of cosmology is based on a single fact. The spectra of galaxies contain red shifts that are proportional to their distances (Figure 18–2). These red shifts are commonly referred to as Doppler shifts due to the recession of the galaxies, which is why we say the universe is expanding.

Edwin P. Hubble discovered the velocity–distance relationship in 1929 using the spectra of only 46 galaxies (Figure 16–6). Since then the Hubble law, as it has come to be known, has been confirmed for hundreds of galaxies out to great distances. This law (Box 16–1) is clear evidence that the universe is expanding uniformly and has no center.

To see how the Hubble law implies uniform, centerless expansion, imagine that we make a loaf of raisin bread (Figure 18–3). As the dough rises, the expansion pushes the raisins away from each other. Two raisins that were originally separated by only 1 cm move apart rather

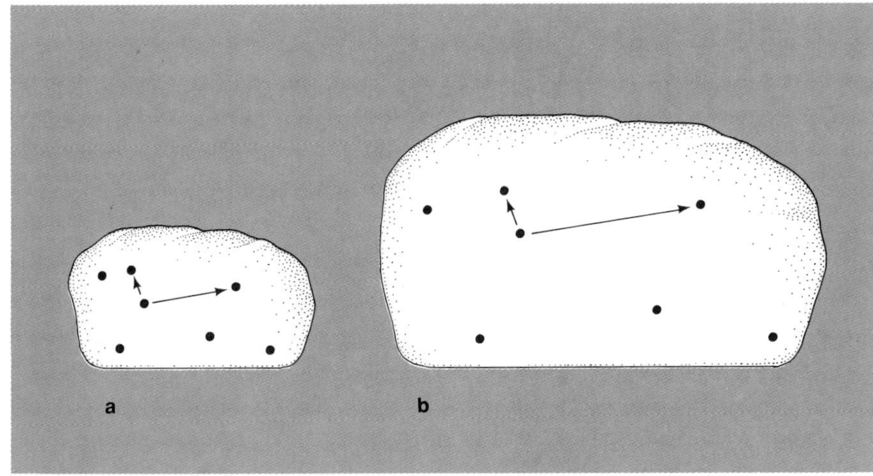

FIGURE 18–3 The uniform expansion of the universe can be represented by raisins in a loaf of raisin bread (a). As the dough rises (b), raisins originally near each other move apart more slowly, and raisins originally farther apart move away from each other more rapidly. A colony of bacteria living on any raisin would find that the velocities of recession of the other raisins are proportional to their distances.

slowly, but two raisins that were originally separated by 4 cm of dough are pushed apart more rapidly. Thus, the uniform expansion of the dough causes the raisins to move away from each other at velocities proportional to their distances. According to the Hubble law, the larger the distance between two galaxies, the faster they recede from each other. This is exactly the result we expect from uniform expansion.

The raisin bread also illustrates that the expansion has no identifiable center. A colony of bacteria living on one of the raisins would see themselves surrounded by a universe of receding raisins. (Here we must assume that they cannot see to the edge of the loaf—that is, they must not be able to see to the edge of their universe.) It does not matter on which raisin the bacteria live. The bacterial astronomers would measure the distances and velocities of recession and derive a bacterial Hubble law showing that the velocities of recession are proportional to the distances. So long as they cannot see the edge of the loaf, they cannot identify any raisin as the center of the expansion.

Similarly, we see galaxies receding from us, but we cannot identify any galaxy or point in space as the center of the expansion. Any observer in any galaxy should see the same expansion that we see.

Although astronomers and cosmologists commonly refer to these red shifts as Doppler shifts and often speak of the recession of the galaxies, relativity provides a more elegant explanation. Einstein's theory of general relativity explains the expansion of the universe as an expansion of space-time itself. A photon traveling through this space-time is stretched as space-time expands, and the photon arrives with a longer wavelength than it had when it left. Photons from distant galaxies travel for a longer time and are stretched further than photons from nearby galaxies. Thus, the expansion of the universe is an expansion of the geometry and not just a simple recession of the galaxies. In fact this relativistic explanation will make no difference to our discussion, but it reminds us that nature is often more elegant than our nonmathematical language.

THE GEOMETRY OF SPACE-TIME

How can the universe expand if it does not have any extra space to expand into? How can it be finite if it doesn't have an edge? The properties we have ascribed to the universe seem to violate common sense, but we can understand some of these if we apply a few ideas from general relativity.

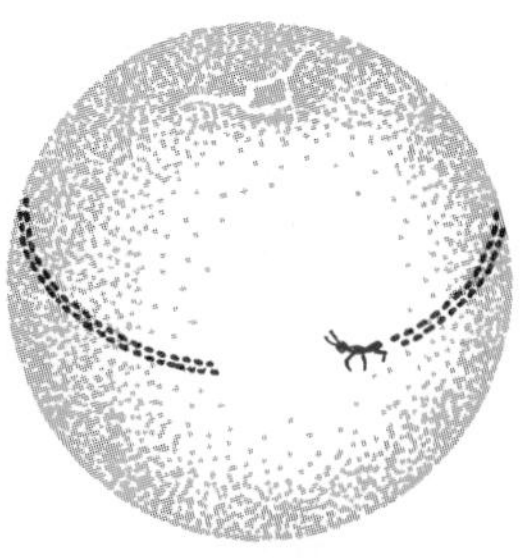

FIGURE 18–4 An ant confined to the two-dimensional surface of an orange could explore the entire surface without coming to an edge. Were it to leave dirty footprints, it might realize that its two-dimensional universe was finite but unbounded.

In Chapter 5 we saw that the presence of mass can curve space-time, and we sense that curvature as a gravitational field. Einstein's theory also predicts that the universe as a whole can be curved (Box 18–1).

If the universe is infinite, then it has no edge, and we say it is unbounded. If it has no boundary, then it can have no center because we define the centers of things by reference to their edges. In terms of curvature, such possible universes are said to be open or flat.

If the universe is finite, however, we say it is closed. Only a finite number of cubic centimeters exist in a closed universe. Yet such a universe need not have a boundary. To see why, think of an ant living on an orange (Figure 18–4). The ant can walk all over the orange, and, if it leaves dirty footprints, it might eventually say, "I have visited every square centimeter in my two-dimensional universe, so it must be finite. But I can't find any edge. My universe is finite but unbounded." Our universe could similarly be of finite volume if it is curved such that it is closed.

We must not try to visualize the expansion of the universe as an outer edge moving into previously unoccupied space. Open, flat, or closed, the universe has no edge, so it does not need additional room to expand. The universe contains all of the volume that exists, and the expansion is a change in the nature of space-time that causes that volume to increase.

According to all the evidence and consistent with general relativity, we live in an expanding universe. If the universe were a videotape, we could run it backward and see the universe contracting. Gradually, the galaxies would approach each other until they began to merge. The total volume of the universe would decrease until all the matter and energy that exists would be trapped in a very high-temperature, high-density state—a state called the big bang.

BOX 18–1
The Curvature of Space

Most modern cosmological theories are based on Einstein's general theory of relativity and its main feature, curved space-time. This curvature of space-time is important in cosmology because the mass of the universe can produce a general curvature in space-time that determines the motion of the universe.

To discuss curvature, we can use a two-dimensional analogy of our three-dimensional universe. Suppose an ant was confined to a two-dimensional surface. Not only must the ant stay on the surface, but the ant's light cannot travel perpendicular to the surface. This surface could be flat (zero curvature), spherical (positive curvature), or saddle-shaped (negative curvature) (Figure 18–5). Because the ant cannot leave the surface, it might be unaware of the true curvature of its universe.

One way the ant could detect the curvature would be to draw circles and measure their areas. On the flat surface a circle would always have an area of exactly πr^2, but on the positively curved surface its area would be less than πr^2, and on the negatively curved surface, more. Drawing small circles would not suffice because the area of small circles would not differ noticeably from πr^2, but if the ant could draw big enough circles, it could actually measure the curvature of its two-dimensional universe.

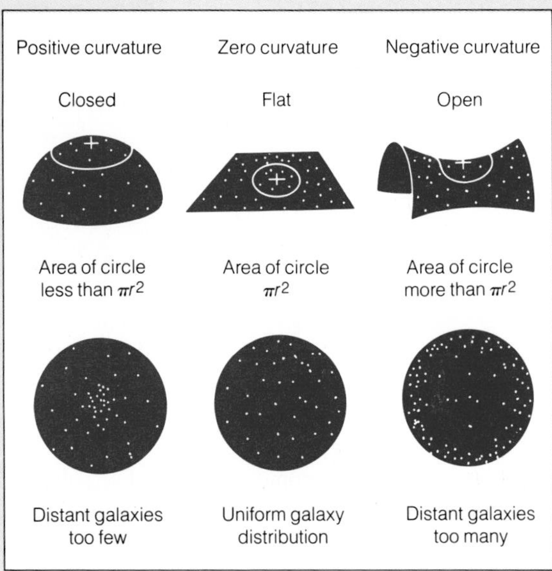

FIGURE 18–5 In two-dimensional space, curvature distorts the area of a circle; in three-dimensional space, it distorts the volume of a sphere. We can measure distortion and detect curvature by counting galaxies.

We are three-dimensional creatures, but our universe might still be curved, and we could measure that curvature by measuring the volume of spheres. If space-time is flat, then no matter how big a sphere we measure, its volume will always be $\frac{4}{3}\pi r^3$. But in positively curved space-time, spheres would have less volume than this; in negatively curved space-time, they would have more. The spheres must be many megaparsecs in radius for the difference to be detectable.

We could measure the volume of such spheres by counting the number of galaxies within a certain distance r from the earth. If galaxies are homogeneously scattered through space, then the number within distance r should be proportional to the volume of the sphere. If, as we count to greater and greater distances, we find the number of galaxies increasing proportional to $\frac{4}{3}\pi r^3$, space-time is flat. However, if we find an excess of distant galaxies, space-time is negatively curved, and if we find a deficiency, space-time is positively curved (Figure 18–4).

By thinking of the two-dimensional analogy shown in Figure 18–4, astronomers usually refer to a universe with positive curvature as a **closed universe.** A zero-curvature universe is said to be a **flat universe**, and a universe with negative curvature is an **open universe.** Notice that flat and open universes are infinite, but closed universes are finite.

A closed universe has a finite volume but no edge. Like the ant in Figure 18–5, we might explore our universe and visit every location but never find an edge. Of course, an open universe, being infinite, also has no edge. Thus, we can answer our question about the edge of the universe. Whether it is open or closed, the universe can have no boundary.

FIGURE 18–6 The city limits sign for Cosmos, Minnesota, reminds us of the big bang theory. Evidently, the population of Cosmos is in flux. The number has been painted out. (Courtesy Richard Fluck.)

18.2 THE BIG BANG

The **big bang theory** has become a part of modern culture (Figure 18–6), but the details are less well-known. The theory supposes that the universe began a finite time ago in a high-temperature, high-density state. The expansion of the universe since that time has allowed the matter to cool and form galaxies, and the universe continues to expand today.

THE BEGINNING

Although our imaginations try to visualize the big bang as a localized event, we must keep firmly in mind that the big bang did not occur at a single place but filled the entire volume of the universe. The matter of which we are made was part of that big bang, so we are inside the remains of that event, and the universe continues to expand all around us. We cannot point in any particular direction and say, "The big bang occurred over there." The big bang occurred everywhere (Figure 18–7).

Although the big bang occurred long ago, the theory predicts that we can detect it because of the finite speed of light. When we look at a distant galaxy, we do not see it as it is now, but as it was long ago. We see the most distant galaxies as they were soon after they formed from the chaos of the big bang. It doesn't matter what direction we look because we are inside the explosion and galaxies surround us on all sides.

Suppose that we look even further, past the most distant objects, backward in time to the fiery clouds of matter in the exploding fireball (Figure 18–7c). From these great distances we receive light that was emitted by the hot gas soon after the explosion began. Again, it does not matter what direction we look in space because we are inside the explosion surrounded by a distant wall of fire.

The radiation that comes to us from this distance has a tremendous red shift. The farthest visible galaxies have red shifts of about 1.0, and the farthest quasars about 4, but the radiation from the big bang fireball has a red shift of about 1000. Thus, the light emitted by the fireball arrives at the earth as infrared radiation and short radio waves. The theory predicts that we should be able to detect this radiation.

PRIMORDIAL BACKGROUND RADIATION

In the mid-1960s two Bell Laboratories physicists, Arno Penzias and Robert Wilson, were using a horn antenna to measure the radio brightness of the sky (Figure 18–8). Their measurements showed a peculiar noise in the system. After rebuilding the receiver and still detecting the noise, they decided the problem lay with a pair of pigeons living inside the antenna. They trapped the pigeons, relocated them, and began cleaning out the droppings. Perhaps they would have enjoyed the job more if they had known they would win the 1978 Nobel Prize for physics for the discovery they were about to make.

When the antenna was cleaned, they measured the brightness of the sky at radio wavelengths and again found the low-level radio noise. The pigeons were innocent, but what was causing the signal?

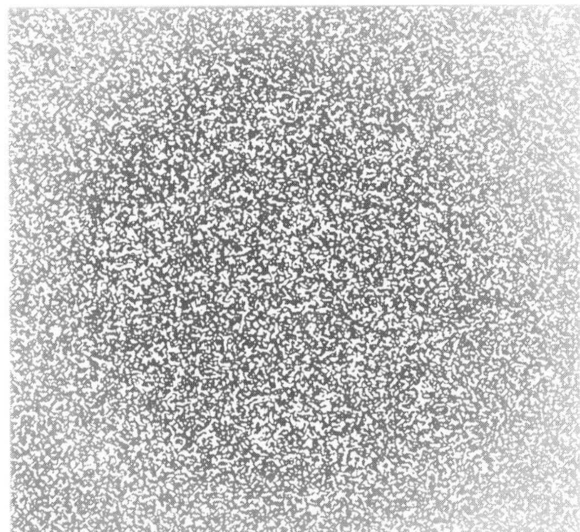

a

A region of the universe during the big bang

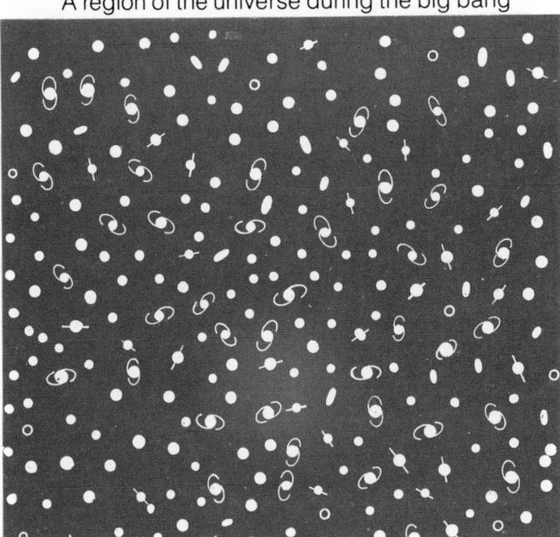

b

A region of the universe now

c

The present universe as it appears from our galaxy

FIGURE 18–7 Three views of a small region of the universe centered on our galaxy. (a) During the big bang explosion the region is filled with hot gas and radiation. (b) Later the gas forms galaxies, but we can't see the universe this way because the look-back time distorts what we see. (c) Near us we see galaxies, but farther away we see young galaxies (dots), and at great distance we see radiation (arrows) coming from the hot clouds of the big bang explosion.

FIGURE 18–8 Arno Penzias (right) and Robert Wilson pose before the horn antenna with which they discovered the primordial background radiation. (Bell Laboratories.)

The explanation of the radio noise goes back to 1948 when George Gamow predicted that the early stages of the big bang would be very hot and would emit large amounts of black body radiation. A year later the physicists Ralph Alpher and Robert Herman pointed out that the very large red shift would lengthen the wavelengths and make the gas clouds of the big bang seem very cold. In the late 1940s there was no way to detect this radiation, but in the mid-1960s Robert Dicke at Princeton developed his own theories and concluded that the radiation was just strong enough to detect. Dicke and his team began building a receiver to search for it.

When Penzias and Wilson heard of Dicke's work in a chance phone conversation, they recognized the noise they had detected as radiation coming from the big bang, the **primordial background radiation**.

Since its discovery the primordial background radiation has been measured at many wavelengths (Figure

18–9). Ground-based radio telescopes have measured its strength in the radio and microwave parts of the spectrum, and rocket and balloon-borne instruments have measured it in the far infrared. These tests confirm that it is black body radiation with an apparent temperature of about 2.7 K.

It may seem strange that the hot gas of the big bang seems to have a temperature of only 2.7 K, but recall the tremendous red shift. When we look back to the big bang, we see light that has a red shift of about 1000—that is, the wavelengths of the photons are about 1000 times longer than when they were emitted. The gas clouds that emitted the photons had a temperature of about 3000 K, and they emitted black body radiation with a λ_{max} of about 1000 nm (Box 7–1). Although this is in the near infrared, the gas would also have emitted enough visible light to glow orange-red. But the red shift has made the wavelengths about 1000 times longer, so λ_{max} is about 1 million nm (equivalent to 1 mm). Thus, the hot gas of the big bang seems to be 1000 times cooler, about 2.7 K.

Although the radiation is almost perfectly isotropic, recent observations show a small departure from complete isotropy. The primordial background radiation seems to have slightly shorter wavelengths—it seems hotter—in the direction of the constellation Leo. In the opposite direction, the radiation seems slightly cooler. This difference is evidently due to the motion of our galaxy through space. The Milky Way galaxy is moving about 540 km/sec in the direction of Leo, causing a slight blue shift in the background radiation that makes it appear slightly hotter. Conversely, radiation from the opposite side of the universe is slightly red-shifted and looks cooler.

The isotropy of the background radiation implies that the early universe was very homogeneous. Today, however, we see galaxies grouped together in clusters and superclusters, and recent observations have suggested that there are great voids between the superclusters, voids as large as 100 Mpc in diameter (Figure 18–10). If the early universe was as homogeneous as the background radiation suggests, how did it get to be so clumpy today? This problem is related to the formation of the galaxies and is not understood at present.

Further research involving the primordial background radiation may reveal how irregularities in the early universe stimulated the formation of superclusters of gal-

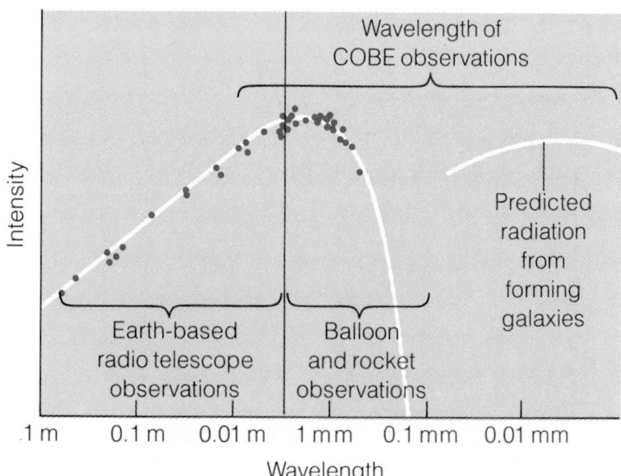

FIGURE 18–9 Current observations suggest that the primordial background radiation is black body radiation with a temperature of about 2.7 K. The COBE satellite will observe it at a wide range of wavelengths and will also search for infrared radiation from the first galaxies to form soon after the big bang.

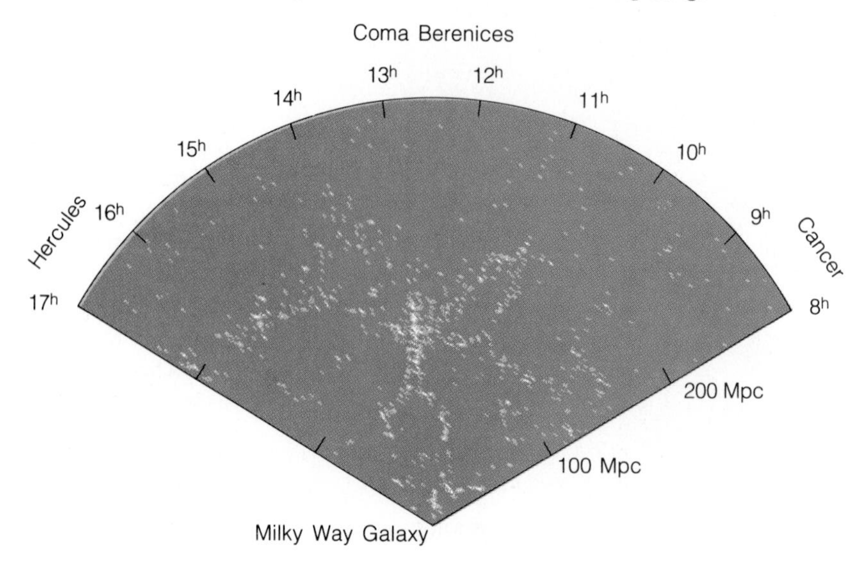

FIGURE 18–10 A slice of the universe from Hercules to Cancer plots galaxies according to their distance from our galaxy. Galaxies appear to be located on the surfaces of great voids like the soap in soap bubbles. (Courtesy John P. Huchra. Harvard-Smithsonian Center for Astrophysics.)

Coma Berenices
Hercules
Cancer
Milky Way Galaxy
100 Mpc
200 Mpc

axies. A satellite called the Cosmic Background Explorer (COBE) carries three far infrared telescopes cooled by a year's supply of liquid helium. It will observe from 1 micron (10^{-6} m) to 1 cm and will search for radiation from dust, young stars, and forming galaxies back to within 1 million years after the big bang. It will also look for irregularities in the background radiation that are the precursors of forming galaxies (Figure 18–9).

The greatest importance of the primordial background radiation lies in its interpretation as radiation from the big bang. If that interpretation is correct, the radiation is evidence that a big bang did occur. That evidence is so strong that the **steady state theory**, which held that the universe was eternal and unchanging, has been abandoned (Box 18–2). Nearly all astronomers now accept that the universe evolves and probably began with a big bang.

BOX 18–2
The Steady State Theory

Through the 1950s and most of the 1960s astronomers had an alternative to the big bang theory. The steady state theory proposed that the universe did not evolve, that it always had the same general properties. Stars and galaxies might age and die, but new stars and new galaxies would be born to take their place and thus preserve the general properties of the universe.

One of the most distinctive features of the steady state theory was caused by the expansion of the universe. As the universe expands its average density should decrease. The steady state theory, however, held that the universe did not change, so it proposed that matter was created continuously to maintain the density of the universe. This newly created matter, presumably in the form of hydrogen atoms, collected in great clouds between the receding galaxies and eventually gave birth to new galaxies to take the place of aging galaxies no longer able to make new stars.

The steady state theory remained a controversial although exciting idea in cosmology until the late 1960s. By then, the primordial background radiation had been widely accepted as evidence that a big bang had occurred, and the steady state theory was gradually abandoned.

A HISTORY OF THE BIG BANG

Modern cosmologists have been able to reconstruct the history of the early universe to reveal how energy and matter interacted as the universe began. As we recount that history, remember that the big bang did not occur in a specific place. The big bang filled the entire volume of the universe from the first moment.

We cannot begin our history at time zero, because we do not understand the physics of matter and energy under such extreme conditions, but we can come close. At a time earlier than 0.0001 seconds after the beginning, the universe was filled with high-energy photons having a temperature over 10^{12} K and a density greater than 5×10^{13} g/cm^3. When we say the photons had a given temperature, we mean that the photons were the same as black body radiation emitted by an object of that temperature. Thus, the photons in the early universe were gamma rays of very short wavelength and therefore very high energy. When we say that the radiation had a certain density, we refer to Einstein's equation $E = m_0c^2$. We can express a given amount of energy in the form of radiation as if it were matter of a given density. In the early universe, the radiation had a density roughly equal to the density of an atomic nucleus.

If two photons have enough energy, they can collide and convert their energy into a pair of particles—a particle of normal matter and a particle of antimatter. In the early universe, photons had sufficient energy to produce proton–antiproton pairs and neutron–antineutron pairs. However, when a particle collides with its antiparticle, the particles are annihilated, and the mass is converted into a pair of gamma-ray photons. Thus, the early universe was a soup of energy continuously switching from photons to particles and back again.

While all of this went on, the universe was expanding, and the wavelengths of the photons were lengthened by the expansion. This lowered the energy of the gamma rays, and the universe cooled. By the time the universe was 0.0001 seconds old, its temperature had fallen to 10^{12} K. By this time the average energy of the gamma rays had fallen below the energy equivalent to the mass of a proton or a neutron, so the gamma rays could no longer produce such heavy particles. The particles combined with their antiparticles and quickly converted most of the mass into photons.

It would seem from this that all the protons and neutrons should have annihilated with their antiparticles, but for quantum mechanical reasons a small excess of normal particles existed. For every billion protons annihilated by antiprotons, one survived with no antiparticles to destroy it. Thus, we live in a world of normal matter, and antimatter is very rare.

Although the gamma rays did not have enough energy to produce protons and neutrons, they could produce electron–positron pairs, which are about 1800 times less massive than protons and neutrons. This continued until the universe was about 4 seconds old, at which time the expansion had cooled the gamma rays to the point where they could no longer create electron–positron pairs. Thus the protons, neutrons, and electrons of which our universe is made were produced during the first 4 seconds in the history of the universe.

This soup of hot gas and radiation continued to cool and eventually began to form atomic nuclei. High-energy gamma rays can break up a nucleus, so the formation of such nuclei could not occur until the universe had cooled somewhat. By the time the universe was 3 minutes old, protons and neutrons could link to form deuterium, the nucleus of a heavy hydrogen atom, and, by the end of the next minute, further reactions began converting deuterium into helium. But no heavier atoms could be built because no stable nuclei exist with atomic weights of 5 or 8 (in units of the hydrogen atom). Cosmic element building during the big bang had to proceed rapidly, step by step, like someone hopping up a flight of stairs (Figure 18–11). The lack of stable nuclei at atomic weights of 5 and 8 meant there were missing steps in the stairway, and the step-by-step reactions could not jump over these gaps.

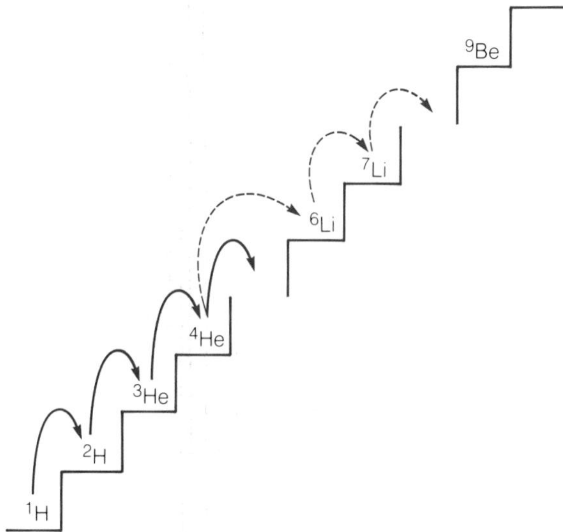

FIGURE 18–11 Cosmic element building. During the first few minutes of the big bang, temperatures and densities were high, and nuclear reactions built heavier elements. Because there are no stable nuclei with atomic weights of 5 or 8, the process built very few atoms heavier than helium.

By the time the universe was 30 minutes old, it had cooled sufficiently that nuclear reactions had stopped. About 25 percent of the mass was helium nuclei, and the rest was in the form of protons—hydrogen nuclei. This is the cosmic abundance we see today in the oldest stars. (The heavier elements, remember, were built by nucleosynthesis inside many generations of massive stars.) The cosmic abundance of helium was fixed during the first minutes of the universe.

For the first million years, the universe was dominated by radiation (Figure 18–12). The gamma rays interacted continuously with the matter, and they cooled together as the universe expanded. The gas was ionized because it was too hot for the nuclei to capture electrons to form neutral atoms, and the free electrons made the gas very opaque. A photon could not travel very far before it collided with an electron and was deflected. Thus, radiation and matter were locked together.

When the universe reached an age of about 1 million years, it was cool enough for nuclei and electrons to form neutral atoms. The free electrons were captured by atomic nuclei, the gas became transparent, and the radiation was free to travel through the universe. The temperature at the time of this **recombination** was about 3000 K. We see these photons arriving now as primordial background radiation.

After recombination the universe was no longer dominated by radiation. Instead, matter was free to move under the influence of gravity, so we say the universe after recombination is dominated by matter. How the matter cooled and collected into clouds and eventually gave birth to galaxies in clusters and superclusters is not well-understood.

THE END OF THE UNIVERSE: A QUESTION OF DENSITY

Will the universe ever end? Will it go on expanding forever with stars burning out, galaxies exhausting their star-forming gas and becoming cold, and dark systems

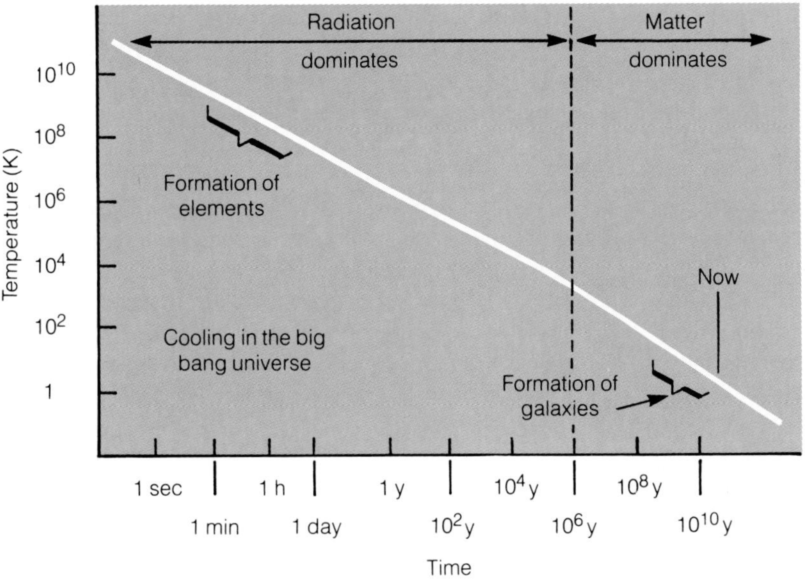

FIGURE 18–12 During the first few minutes of the big bang, some hydrogen became deuterium and helium and a few heavier atoms. Later, when radiation was no longer dominant, galaxies formed, and nuclear reactions inside stars made the rest of the chemical elements.

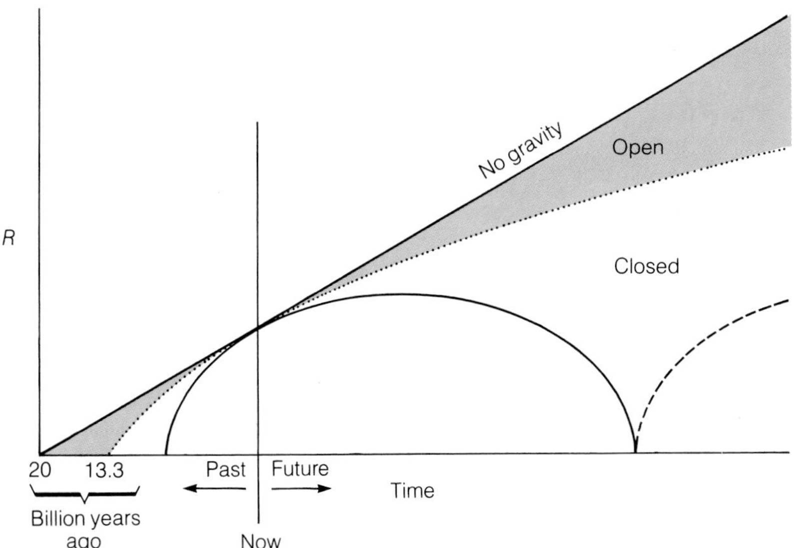

FIGURE 18–13 The expansion of the universe as a function of time. *R* is some measure of the size of the universe. Open universe models expand without end (shaded region), but closed models pass through repeated big bangs (curved solid line). Dotted line marks the dividing line between open and closed models. Note that the estimated age of the universe depends on the rate at which the expansion is slowing down.

expanding forever through an endless, dark universe? How else could the universe end?

Earlier we decided that the universe could be open, flat, or closed. The geometry of the universe determines how it will end, so we must consider these three possible universes.

The general curvature of the entire universe is determined by its density. If the average density of the universe is equal to the **critical density** of 4×10^{-30} g/cm^3, space-time will be flat. If the average density of the universe is less than the critical density, the universe is negatively curved and open. If the average density is more, the universe is positively curved and closed (Box 18–1).

We decided earlier that an open universe and a flat universe are both infinite. Both of these universes will expand forever. The gravitational field of the material in the universe, present as a curvature of space-time, will cause the expansion to slow; but, if the universe is open, it will never come to a stop (Figure 18–13). If the universe is flat, it will slow to a stop after an infinite time. Thus, if the universe is open or flat, it will expand forever, and the galaxies will eventually become black, cold, solitary islands in a universe of darkness.

If the universe is closed, however, its fate will be quite different. In a closed universe the gravitational field, present as curved space-time, is sufficient to slow the expansion to a stop and make the universe contract. Eventually the contraction will compress all matter and energy back into the high-energy, high-density state from which the universe began. This end to the universe, a big bang in reverse, has been called the big crunch. Nothing in the universe could avoid being destroyed in such a universe.

Some theorists have suggested that the big crunch will spring back to produce a new big bang and a new expanding universe. This theory, called the **oscillating universe theory**, predicts that the universe undergoes alternate stages of expansion and contraction. Recent theoretical work, however, suggests that successive bounces of an oscillating universe would be smaller and smaller until the oscillation ran down. In addition, we have no theoretical reason to explain how a big crunch could be converted into a new big bang.

The big question, of course, is, What is the density of the universe?

It is quite difficult to measure the density of the universe. We could count galaxies in a given volume, multiply by the average mass of a galaxy, and divide by the volume, but there are uncertainties. We are not sure of the average mass of a galaxy, and many galaxies are too small to see even nearby. We would also have to include the mass equivalent to the energy present in the universe, because mass and energy are related. The best attempts yield a density of about 5×10^{-31} g/cm^3, about 10 percent of the mass needed to close the universe.

However, we have seen clear evidence that the universe contains undetected mass. In Chapter 16 we saw that some galaxy clusters seem to contain more mass than we can see, and our studies of our own galaxy and others suggest that galaxies are surrounded by massive coronae. Whatever this matter is, it is not luminous or we would see it, so astronomers refer to it as **dark matter**.

It does not appear that dark matter can be the normal kind of matter with which we are familiar. Hot gas would emit X rays, and cold gas would emit infrared and radio energy. Faint cool stars or large numbers of planets would emit infrared radiation. We detect none of this, so the dark matter seems not to be normal material.

One suggestion is that dark matter is made up of neutrinos. Neutrinos were originally thought to be massless, but work during the late 1970s hinted that they might have a very small mass. If the neutrino does have a mass, it could provide the missing mass because there are about 10^8 neutrinos in the universe for every normal particle. A number of teams have attempted to measure the mass of the neutrino and have not succeeded. Cosmologists presently view the neutrino as an unlikely possibility.

Theorists have suggested a wide range of subatomic particles that could make up dark matter. Axions, photinos, and WIMPs (weakly interacting massive particles) are only some of the particles that have been proposed. None of these has been detected in the laboratory yet, so cosmologists are presently waiting to see if the existence of any of these particles can be confirmed experimentally.

If dark matter is not sufficient to close the universe, then it must be open. In fact, isotopes such as deuterium and lithium-7 suggest that the universe is not closed. Deuterium is an isotope of hydrogen in which the nucleus contains a proton and a neutron. One out of every 6000 hydrogen atoms in the earth's oceans is actually deuterium. This element is so easily converted into helium that none can be produced in stars. In fact, stars destroy what deuterium they have to start with. Thus, the deuterium in the earth's oceans must have been made in the first few minutes of the big bang explosion. If the universe now contained enough normal mass to be closed, then during the big bang it should have been so dense it would have destroyed most of the deuterium and would have produced isotopes such as lithium-7. That we have deuterium in our oceans and little lithium-7 seems to show that the universe cannot contain enough normal matter to be closed.

Whether the universe is open or closed, the big bang theory has been phenomenally successful in explaining the origin and evolution of the universe. At present, no other theory is a serious contender. But there are serious questions yet to be resolved concerning the big bang. Future cosmologists will need to examine both observation and theory to find the answers.

18.3 REFINING THE BIG BANG: THEORY AND OBSERVATION

Many questions remain to puzzle cosmologists, but the three we will examine will illustrate some important trends in future research. We can phrase these problems as questions. Why was there a big bang? How old is the universe? How does the universe evolve?

THE QUANTUM UNIVERSE

A newly developed theory, combining general relativity and quantum mechanics, may be able to tell us how the big bang began and why the universe is in the particular state that we observe. Some theorists claim that the new theory will explain why there was a big bang in the first place. To introduce the new theory, we can consider two unsolved problems in the current big bang theory.

One of the problems is called the **flatness problem.** The universe seems to be balanced near the boundary between an open and a closed universe. That is, it seems nearly flat. Given the vast range of possibilities, from zero to infinite, it seems peculiar that the density of the universe is within a factor of 10 of the critical density that would make it flat. If dark matter is as common as it seems, the density may be even closer than a factor of 10.

Even a small departure from critical density when the universe was young would be magnified by subsequent expansion. To be so near critical density now, the density of the universe during its first moments must have been within 1 part in 10^{49} of the critical density. So the flatness problem is: Why is the universe so nearly flat?

Another problem with the big bang theory is the isotropy of the primordial background radiation. Once we correct for the motion of our galaxy, we see the same background radiation in all directions to at least 1 part in 1000. Yet when we look at the background radiation coming from two points in the sky separated by more than a degree, we look at two parts of the big bang that were not causally connected when the radiation was emitted. That is, when recombination occurred and the gas of the big bang became transparent to the radiation, the universe was not old enough for any signal to have traveled from one of these regions to the other. Thus, the two spots we look at did not have time to exchange heat and even out their temperatures. Then how did every part of the entire big bang universe get to be so precisely the same temperature by the time of recombination? This is called the **horizon problem** because the two spots are said to lie beyond their respective light-travel horizons.

The key to these two problems and to others involving subatomic physics may lie with a new theory. It has been called the **inflationary universe** because it predicts a sudden expansion when the universe was very young, an expansion even more extreme than that predicted by the big bang theory.

To understand the inflationary universe, we must recall that physicists know of only four forces—gravity, the electromagnetic force, the strong force, and the weak force. We are familiar with gravity, and the electromagnetic force is responsible for making magnets stick to refrigerator doors and cat hair stick to wool sweaters charged with static electricity. The strong force holds atomic nuclei together, and the weak force is involved in certain kinds of radioactive decay. (In 1986, physicists announced the discovery of a fifth force. A number of sensitive tests have yielded mixed results, so this exciting discovery remains controversial.)

For many years theorists have tried to unify these forces—that is, they have tried to describe the forces with a single mathematical law (Figure 18–14). A century ago James Clerk Maxwell showed that the electric force and the magnetic force were really the same effect, and we now count them as a single electromagnetic force. About 15 years ago theorists succeeded in unifying the electromagnetic force and the weak force in what they called the electroweak force, effective only for processes at very high energy. At lower energies, the electromagnetic force and the weak force behave differently. Now theorists have found ways of unifying the electroweak force and the strong force at even higher energies. These new theories are called **Grand Unified Theories** or **GUTs.**

Studies of GUTs suggest that the universe expanded and cooled until about 10^{-35} seconds after the big bang when it became so cool that the forces of nature began to separate from each other. This released tremendous energy, which suddenly inflated the universe by a factor between 10^{20}–10^{30}. At that time the part of the universe that we can see now, the entire observable universe, was no larger than the volume of an atom, but it suddenly inflated to the volume of a cherry pit and then continued its slower expansion to its present extent. (Remember that we do not accept an edge in our model of the universe. That is why we speak of the volume of a cherry pit and not its diameter.)

That sudden inflation can solve the flatness problem and the horizon problem. The sudden inflation of the universe would have forced whatever curvature it had toward zero, just as inflating a balloon makes a small spot on its surface flatter. Thus, we now see the universe nearly flat because of that sudden inflation long ago. In addition, because the part of the universe we can see was once no larger in volume than an atom, it had plenty of time to equalize its temperature before the inflation occurred. Now we see the same temperature for the background radiation in all directions.

The inflationary universe is based, in part, on quantum mechanics, and a slightly different aspect of quantum mechanics may explain why there was a big bang at all.

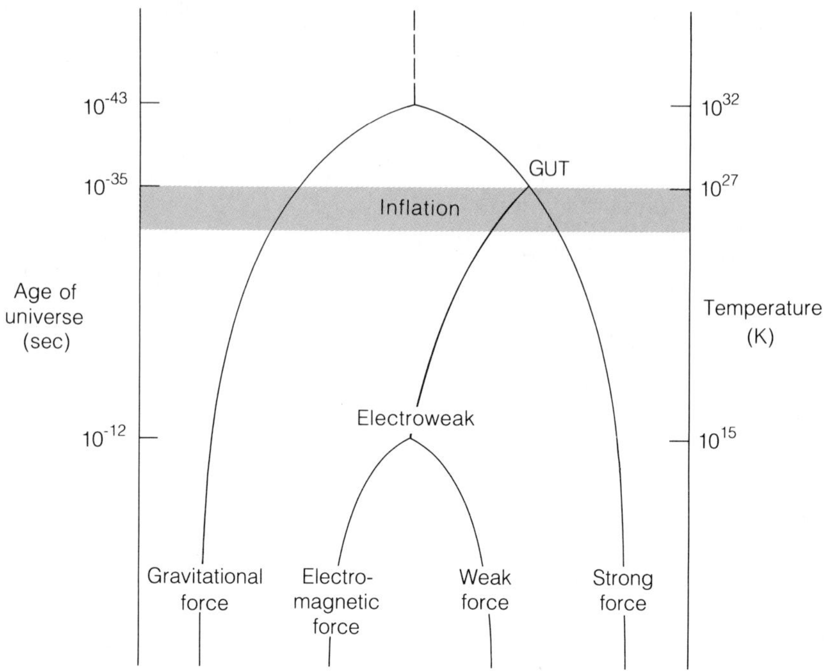

FIGURE 18–14 When the universe was very young and hot (top), the four forces of nature were indistinguishable. As the universe began to expand and cool, the forces separated and triggered a sudden inflation in the size of the universe.

Theorists believe that a universe totally empty of matter could be unstable and decay spontaneously by creating pairs of particles until it was filled with the hot, dense state we call the big bang. This theoretical discovery has led some cosmologists to believe that the universe could have been created by a chance fluctuation in space-time. In the words of physicist Frank Wilczyk, "The reason there is something instead of nothing, is that 'nothing' is unstable."

The inflationary universe and other theoretical phenomena predicted by quantum mechanics have become an important and exciting part of modern cosmology. But another area of excitement and controversy is centered around an aspect of traditional observational cosmology that began with Edwin Hubble's discovery of the expanding universe.

THE AGE OF THE UNIVERSE

In 1929 Edwin Hubble announced that the universe was expanding and reported the rate of that expansion, the number now known as the Hubble constant (*H*). Within a few years astronomers had proposed the idea that the universe began in a big bang and had used *H* to extrapolate backward to find the age of the universe. But when they used the constant that Hubble had determined, 530 km/sec/Mpc, they found that the universe was only about half as old as the earth. Something was wrong.

If we know *H,* we can estimate the age of the universe. The Hubble constant tells us how fast the galaxies are receding from each other, and, assuming that they have always traveled at their present rate, we simply ask how long it took for the galaxies to reach their present separations (Box 18–3). Hubble's original value for *H* gave an age of about 2 billion years, but we now know that his distances were too small. The modern, corrected value of *H* gives an age of 10–20 billion years. For comparison, the earth is no older than 4.6 billion years.

When we use *H* to make these simple calculations of the age of the universe, we are assuming that the galaxies have always traveled at their present rate. But we saw earlier that gravity slows the expansion of the universe, so the galaxies traveled faster in the past, and the estimate for the age of the universe must be an upper limit. That is, it would be the age of the universe if there were no gravity slowing the expansion (solid line in Figure 18–13). To find the true age, we must know the extent to which gravity has slowed the expansion, and that depends on the average density of the universe. As we have seen earlier in this chapter, the average density is highly uncertain.

As an example, let us assume that *H* is 50 km/sec/Mpc. Then, if gravity has not slowed the expansion of the

BOX 18–3

The Hubble Constant and the Age of the Universe

The Hubble constant H is a measure of the rate of expansion of the universe. It contains enough information to estimate the age of the universe.

To discover how long ago the universe began expanding, divide the distance D to a galaxy by the velocity V_r with which that galaxy recedes. The result is the time the galaxy took to travel the distance, assuming that it has maintained constant velocity.

Because D is measured in megaparsecs and V_r in kilometers per second, we must convert D into kilometers by multiplying by 3.085×10^{19}, the number of kilometers in 1 Mpc. Then dividing by V_r yields the age of the universe in seconds. To convert to years, we must divide by 3.15×10^7, the number of seconds in a year. Thus the age of the universe in years is

$$T = \frac{D}{V_r} \times \frac{3.085 \times 10^{19}}{3.15 \times 10^7}$$

or approximately

$$T \approx \frac{D}{V_r} \times 10^{12}\text{y}$$

However, we don't have to measure D and V_r if we know the Hubble constant. To see why, recall that H is a galaxy's radial velocity divided by its distance:

$$H = \frac{V_r}{D}$$

So we can simplify our formula for the age of the universe.

$$T \approx \frac{1}{H} \times 10^{12}y$$

If H is 50 km/sec/Mpc, the universe is about 20×10^9 years old, assuming there is no gravity to slow the expansion. To find the true age of the universe, we must know the extent to which gravity has slowed the expansion, and that depends on the average density of the universe.

universe, it is 20 billion years old. But we know the universe contains some matter. If the average density of the universe is equal to the critical density, the density needed to close the universe (dotted line in Figure 18–13), the universe must be two-thirds as old as the estimate just presented—about 13.3×10^9 years. If the density is less than that, the universe is open, and its age is between 13.3×10^9 and 20×10^9 years. If the density is more than the critical value, the universe is younger than 13.3×10^9 years.

Other evidence suggests that the universe is older than this. Measured ages for globular clusters based on the turn-off points in their H–R diagrams range from 14.8–20.4 billion years with uncertainties as large as ± 3.9 billion years. Another study suggests that the universe must be between 16.8–23.8 billion years old to allow time for nucleosynthesis in massive stars to build the present amount of heavy elements. Such estimates of the age of the universe suggest that H cannot be much larger than 70 km/sec/Mpc.

Measuring the value of H is difficult for two reasons. First, we must find the distances of galaxies that are so far away we cannot see such dependable distance indicators as the Cepheid variables. Hubble's distances were much too small, and his value of H was 530 km/sec/Mpc, much larger than modern estimates, which range from 50–100 km/sec/Mpc.

A second reason it is difficult to measure H is related to the motions of the galaxies themselves. Our Milky Way galaxy and the Local Group of galaxies in which we live seems to be falling toward the massive Virgo and Hydra-Centaurus superclusters of galaxies. Different teams of astronomers correct for this motion differently when they calculate the radial velocities of other galaxies. Also, recent studies have begun to suggest that streaming motions occur among the galaxies. Our Local Group, the Virgo supercluster and the Hydra-Centaurus supercluster, may all be streaming in the same direction, perhaps due to the forces that produced the voids and filaments in the distribution galaxies. If these streaming motions do occur, then the velocities of distant galaxies may be affected, thus distorting estimates of H.

The current controversy over the value of H and the age of the universe is of fundamental importance. Some astronomers believe H is about 50 km/sec/Mpc, and others believe it is as large as 100 km/sec/Mpc. Still other studies yield numbers between these extremes. A recent study using distances derived from the rate of expansion of supernovae remnants in other galaxies gives a value of 65 km/sec/Mpc. In this book we adopt a provisional value of 50 km/sec/Mpc because it is most consistent with other estimates of the age of the universe. We may eventually obtain a more accurate value of H when the Hubble Space Telescope begins observing from orbit. It will be able to detect distance indicators in galaxies about seven times farther away than earth-based telescopes.

THE EVOLUTIONARY UNIVERSE

The big bang theory is widely accepted in astronomy, but it leaves us with a number of possibilities. The universe may be open, closed, or flat. How can we decide which version is correct?

A **cosmological test** is a measurement or observation whose result can help us choose between rival cosmological theories. Almost all such tests are searches for evolution in the general properties of the universe. When we detect such evolution, we confirm that we live in a universe that is changing, and knowing in detail how the universe is evolving should eventually allow us to choose the right version of the big bang theory.

FIGURE 18–15 A cluster of galaxies slightly more than 4 billion ly away shows little detail. (Palomar Observatory photograph.)

Although galaxies are visible at optical wavelengths out to distances of about 10 billion ly, astronomers must concentrate on galaxies within about 2 billion ly to see any detail (Figure 18–15). Thus, present optical telescopes can look back no more than about 2 billion years, and what they show us implies that galaxies have not changed drastically in that short interval. To see the effects of evolution in the universe, we must look much farther.

Radio telescopes can probe much farther than optical telescopes, mainly because some galaxies are much more luminous at radio wavelengths than at optical wavelengths. Consequently, radio astronomers can perform a cosmological test by counting the number of these radio galaxies at various distances. If distant regions of space contain more galaxies than nearby regions, it would show that the universe has changed.

Unfortunately, this process requires that we know the distance of the galaxies, a quantity that radio astronomers can only estimate. However, if we assume that all radio galaxies radiate the same power, we can assume that the weaker signals come from those more distant.

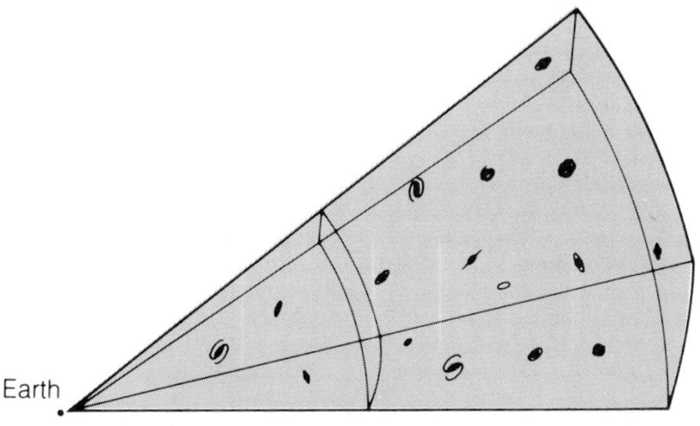

FIGURE 18–16 The volume we sample increases as we look farther away. The number of galaxies we count in a flat universe should be proportional to the distance cubed.

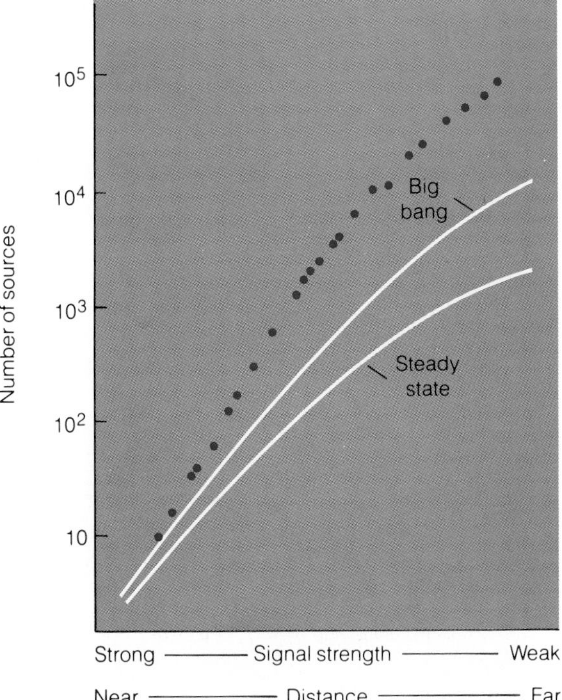

FIGURE 18–17 The excess of faint radio sources shows that the universe has changed. Observations do not agree with any current model, suggesting that the evolution of galaxies is distorting the counts.

If the universe were not expanding, the number of galaxies per unit volume must remain the same, and as we look out into space to large look-back times, we should see the same density of galaxies. However, the volume in which we count increases as the cube of the distance (Figure 18–16), so we should see more faint radio galaxies even if the density is constant. The critical question is, Do we see more than we should?

In 1968 the English radio astronomer Sir Martin Ryle published the results of extensive counts of radio sources, most of them presumably radio galaxies. He found that the number of sources increased with distance as expected, but that there were too many faint sources (Figure 18–17). Apparently there is an excess of distant radio galaxies. This implies that radio galaxies were more common in the distant past, or that they were more powerful and thus easier to detect. In either case, the universe was different in the past.

The excess of faint radio sources shows that the universe is evolving, but it does not agree with any version of the big bang—open, closed, or flat. This does not mean that none of these is correct, only that we don't know how galaxies themselves evolve. If young galaxies are more luminous at radio wavelengths than older galaxies, then we will see more faint sources because when we look at faint sources, we look to great look-back times and see galaxies when they were young.

Of course, we suspect that quasars are related to young, distant, active galactic nuclei. If quasars really are at the distances predicted by their red shifts, as most astronomers now believe, their distribution in space can tell us about the history of the universe. As we look out to greater distances, we find more quasars per unit volume of space. This shows that quasars were more common in the past, and it is clear evidence that the universe has changed. But once we look beyond a red shift of about 2, we find fewer quasars (Figure 18–18). A few quasars with red shifts over 4.0 are known, but none have been found with larger red shifts. This appears to mean that when the universe was young, it contained few quasars. Sometime between 1.3–4 billion years after the big bang, quasars began to appear in large numbers and then they began to disappear. This is consistent with the belief that quasars are the active cores of forming galaxies, and it is clear evidence that the universe is evolving. But once again, we do not know how quasars themselves evolve, so we cannot use this information to choose among open, closed, and flat universes.

Because we don't understand how galaxies evolve with time, counts of galaxies are not conclusive. And direct measurements of the density of the universe are confused by the evidence that the universe contains large amounts of dark matter. At present our story of the evolution of the universe is incomplete, and we can only say that the present evidence slightly favors an open universe.

Yet another cosmological test may help us choose. The open, closed, and flat universe versions of the big bang theory predict that the universe is expanding, but they differ in how its expansion changes with time. The open big bang theory says gravity is slowing the expansion slightly, and the closed big bang theory says gravity is slowing the expansion rapidly. If we could measure a change in the rate of expansion, we might be able to choose the correct theory.

The Hubble constant H describes the rate of expansion of the universe, and if the universe is decelerating, H must be decreasing. This decrease would be much too slow to detect in hundreds or even thousands of years, but it may be detectable if we extend the velocity–distance diagram to great distance.

Galaxies plotted according to their velocity and distance fall along a line indicating that the more distant galaxies are receding faster than those nearby. The line's slope (steepness) equals H (Figure 18–19). Thus, the faster the universe expands, the steeper the line will be. For nearby galaxies the line has a slope of about 50 km/sec/Mpc.* If we look at distant galaxies, we look back in time to an earlier age when H may have been larger. If that is the case, the slope of the line should get steeper as we look farther into space.

If the universe were not slowing, the line should be straight, but if gravity is slowing the expansion, the line should curve upward, indicating faster expansion in the past. Because the open universe theory supposes that gravity has slowed the universe only slightly, it predicts a small increase in H as we look back in time. But the closed universe theory says gravity is strong and the universe has been slowed down drastically. Thus, it predicts a much larger departure from a straight line (Figure 18–19). Clearly we can choose the correct theory if we can plot galaxies out to the point where the lines diverge.

Unfortunately, getting the distances to these galaxies is difficult because they are too far away to have visible distance indicators such as Cepheid variables, luminous supergiants, or globular clusters. Instead, the astronomer must judge the distances by calibrating the properties of the galaxies themselves. Nearby galaxies, whose distances can be found from the indicators just mentioned, provide the luminosities of different kinds of galaxies. By observ-

*Recall that some studies claim this slope is near 100 km/sec/Mpc

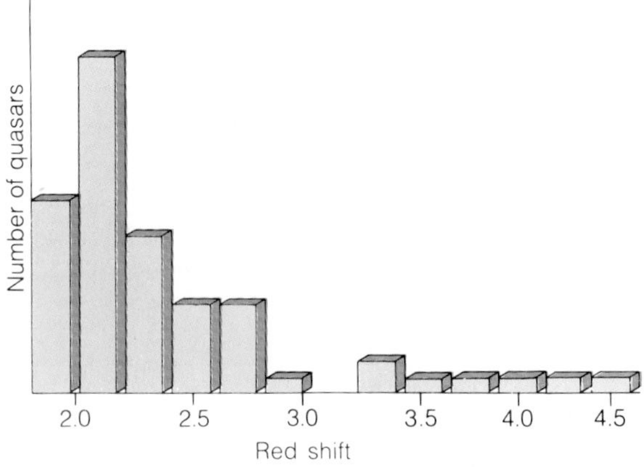

FIGURE 18–18 The distribution of quasars with red shift. The number of quasars declines beyond a red shift of about 2. (Adapted from data by Patrick S. Osmer.)

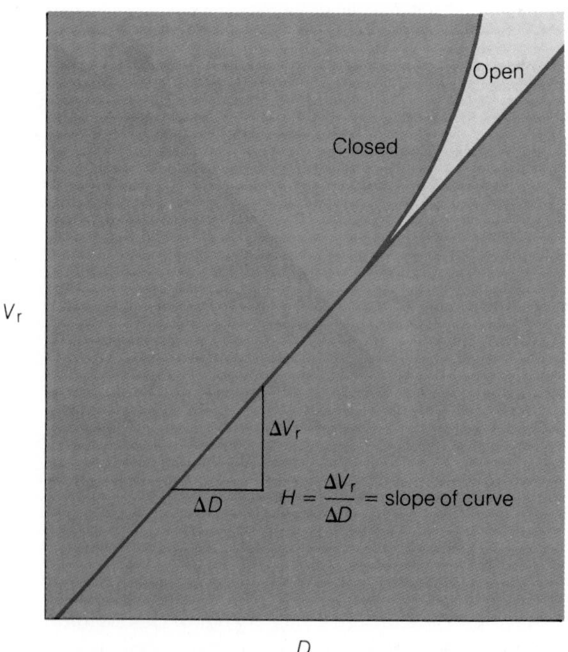

FIGURE 18–19 The slope of the relation in the velocity–distance diagram is equal to the Hubble constant H. Here ΔV_r and ΔD represent differences in velocity and distance. The shaded area represents the region occupied by models of open universes. Models of closed universes would be represented by lines curving above the shaded region.

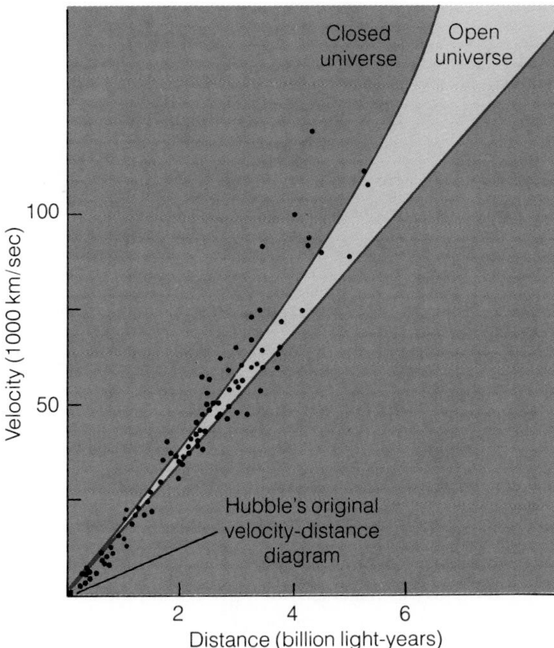

FIGURE 18–20 Hubble's original velocity–distance diagram (lower left) has now been extended to great distance. The data suggest that the universe was expanding faster in the past. It is not possible to choose between the open and closed models on the basis of these data. (Adapted from data by J. Kristian, A. Sandage, and J. A. Westphal.)

ing the apparent magnitude of similar, more distant galaxies, the astronomer can calculate distances.

These distances are uncertain because we don't know how galaxies evolve. We might suspect that long ago, when galaxies had more gas and dust, they could make more hot, bright stars and were more luminous. To judge the distance to a galaxy on the basis of its luminosity, we must correct for this excess of bright stars and for other evolutionary effects. Unfortunately, the exact form of this correction is unknown, and thus the distances are uncertain.

One of the most extensive studies of this kind has been made by Allan Sandage and his co-workers. They have revised the distance calibration, starting with the luminosity of the variable stars used as distance indicators in the nearest galaxies and working their way outward to more distant galaxies. These results, plotted in Figure 18–20, show the velocities of the galaxies falling above a straight line, indicating that the universe is slowing down. However, the uncertainties in the data do not permit a choice between the open big bang and the closed big bang theories. In the future, better observations made from the Space Telescope and a better understanding of the luminosities of young galaxies may reduce the uncertainties and extend the data to greater distances.

The evidence is inconclusive. The distances to the farthest galaxies are not known well enough to allow us to use the increase in H at great distances to show that the universe is open, closed, or flat.

SUMMARY

The fact that the night sky is dark shows that the universe is not infinitely old. If it were infinite in extent and age, then every spot on the sky would glow as brightly as the surface of a star. This problem, known incorrectly as Olbers' paradox, illustrates how important assumptions are in cosmology.

The basic assumptions of cosmology are homogeneity, isotropy, and universality. Homogeneity says the matter is spread uniformly through the universe. Isotropy says the universe looks the same in any direction. Both deal only with general features. Universality assumes that the laws of physics known on earth apply everywhere. In addition, the cosmological principle asserts that the universe looks the same from any location.

The Hubble law implies that the universe is expanding. Tracing this expansion backward in time we come to an initial high-density state commonly called the big bang, the explosion that started the expansion. From the Hubble constant we can conclude that the expansion began 10–20 billion years ago.

The universe can be infinite or finite, but it cannot have an edge or a center. Such regions would violate the cosmological principle. Instead, we assume that the universe occupies curved space-time and thus could be finite but unbounded. Depending on the average density of the universe, it could be open, flat, or closed. Measurements of the density of the universe are uncertain because of the presence of dark matter, which we cannot detect easily.

During the first few minutes of the big bang, about 25 percent of the matter became helium, and the rest remained hydrogen. Very few heavy atoms were made. As the matter expanded, instabilities caused the formation of clusters of galaxies, which are still receding from each other.

Whether the universe expands forever or slows to a stop and falls back depends on the amount of matter in the universe. If the average density is greater than the critical density of 4×10^{-30} g/cm^3, it will provide enough gravity to slow the expansion and force the universe to collapse. The collapse will smash all matter back to a high density from which a new big bang may emerge. Such a universe is termed closed because the curvature of space is positive. If the average density is less than 4×10^{-30} g/cm^3, gravity will be unable to stop the expansion, and it will continue forever. Such a universe is termed open because the space curvature is negative.

The inflationary universe is a new theory that combines quantum mechanics, general relativity, and cosmology. It predicts that the universe underwent a sudden inflation when it was only 10^{-35} seconds old. This seems to explain why the universe is so flat and why the primordial background radiation is so isotropic.

We can estimate an upper limit to the age of the universe from the Hubble constant, but we need to know the average density of the universe to find a true age. Also, the value of the Hubble constant is not well-known. It is believed to be between 50–100 km/sec/Mpc. If it is more than about 70 km/sec/Mpc, the age of the universe will be less than the ages determined for some star clusters.

Many pieces of evidence show that we live in an evolving universe. Primordial background radiation, which is detected in the far-infrared and short-radio region of the electromagnetic spectrum, is actually red-shifted light emitted by the hot clouds of the big bang. This is clear evidence that there was a big bang. The excess of faint radio sources and the deficiency in quasars with red shifts greater than 2 shows that galaxies have evolved since the origin of the universe.

Observations show that the distant galaxies are receding from each other faster than the nearer galaxies. That is, the Hubble constant was larger in the past. This shows that the universe is slowing its expansion and is not in a steady state, but the data are not good enough to distinguish between a closed universe and an open universe.

NEW TERMS

cosmology	steady state theory
Olbers' paradox	recombination
homogeneity	critical density
isotropy	oscillating universe theory
universality	dark matter
cosmological principle	flatness problem
closed universe	horizon problem
flat universe	inflationary universe
open universe	grand unified theories (GUTs)
big bang theory	cosmic string
primordial background radiation	cosmological test

QUESTIONS

1. Would the night sky be dark if the universe were only 1 billion years old and was contracting instead of expanding? Explain your answer.
2. How can we be located at the center of the observable universe if we accept the Copernican principle?
3. Would Copernicus have accepted the cosmological principle? Explain your opinion.
4. Why can't an open universe have a center? Why can't a closed universe have a center?
5. What evidence do we have that the universe is expanding? That it began with a big bang?
6. Why couldn't atomic nuclei exist when the universe was younger than 3 minutes?
7. Why is it difficult to determine the present density of the universe?
8. How does the inflationary universe resolve the flatness problem? The horizon problem?
9. If the Hubble constant is really 100 km/sec/Mpc, then much of what we understand about the evolution of stars and star clusters must be wrong? Explain why.
10. What evidence do we have that the universe is evolving?
11. If the expansion of the universe is slowing down, why should the Hubble constant be larger for galaxies seen at great distances?
12. What observations would you recommend that the Hubble Space Telescope make to help us choose between open, flat, and closed models of the universe?

PROBLEMS

1. Use the data in Figure 18–2 to plot a velocity–distance diagram, find *H,* and determine the approximate age of the universe.
2. If a galaxy is 8 Mpc away from us and recedes at 456 km/sec, how old is the universe assuming that gravity is not slowing the expansion? How old is the universe if it is flat?
3. If the temperature of the big bang had been 10^6 K at the time of recombination, what wavelength of maximum would the primordial background radiation have as seen from the earth?
4. If the average distance between galaxies is 2 Mpc and the average mass of a galaxy is $10^{11}\ M_{\odot}$, what is the average density of the universe? (HINTS: The volume of a sphere is $\frac{4}{3}\pi r^3$. The mass of the sun is 2×10^{33} g.)

5. If the value of the Hubble constant were found to be 60 km/sec/Mpc, how old would the universe be if it was not slowed by gravity? If it were flat?
6. Hubble's first estimate of the Hubble constant was 530 km/sec/Mpc. If his distances were too small by a factor of 7, what answer should he have obtained?
7. What is the maximum age of the universe predicted by Hubble's first estimate of the Hubble constant?
8. High-resolution radio observations show that a galaxy contains a supernova remnant that is expanding 0.00004 seconds of arc per year, and spectra show the remnant is expanding at 10,000 km/sec. How far away is the galaxy?

RECOMMENDED READING

ANTHONY, SIMON "Superstrings: A Theory of Everything?" *New Scientist 105* (29 Aug. 1985), p. 34.

BARROW, JOHN D., and JOSEPH SILK "The Structure of the Early Universe." *Scientific American 242* (April 1980), p. 118.

BAUER, CLAUDE "COBE's Quest." *Astronomy 14* (Aug. 1986), p. 74.

BURNS, JACK "Very Large Structures in the Universe." *Scientific American 255* (July 1986), p. 38.

———. "Dark Matter in the Universe." *Sky and Telescope 68* (Nov. 1984), p. 396.

CLOSE, FRANK *The Cosmic Onion.* New York: American Institute of Physics, 1986.

DAVIS, PAUL "The Arrow of Time." *Sky and Telescope 72* (Sept. 1986), p. 239.

———. "Relics of Creation." *Sky and Telescope 69* (Feb. 1985), p. 112.

DE VAUCOULEURS, GERARD "The Distance Scale of the Universe." *Sky and Telescope 66* (Dec. 1983), p. 511.

DISNEY, M. *The Hidden Universe.* New York: Macmillan, 1984.

FERRIS, T. "The Radio Sky and the Echo of Creation." *Mercury 13* (Jan./Feb. 1984), p. 2.

GIBBONS, GARY "Quantum Gravity: The Last Frontier." *New Scientist 105* (31 Oct. 1985), p. 45.

GUTH, ALAN H., and PAUL J. STEINHARDT "The Inflationary Universe." *Scientific American 250* (May 1984), p. 116.

HARRISON, EDWARD R. *Masks of the Universe.* New York: Macmillan, 1985.

———. "The Dark Night Sky Riddle: A 'Paradox' that Resisted Solution." *Science 226* (23 Nov. 1984), p. 941.

———. *Cosmology: The Science of the Universe.* New York: Cambridge University Press, 1981.

———. "The Paradox of the Dark Night Sky." *Mercury 9* (July/Aug. 1980), p. 83.

HODGE, PAUL "The Cosmic Distance Scale." *American Scientist 72* (Sept./Oct. 1984), p. 474.

KIPPENHAHN, RUDOLF "Light from the Depths of Time." *Sky and Telescope 73* (Feb. 1987), p. 140.

LINDE, ANDREI "The Universe: Inflation Out of Chaos." *New Scientist 105* (7 March 1985), p. 14.

MACROBERT, ALAN "Beyond the Big Bang." *Sky and Telescope 65* (March 1983), p. 211.

MATHER, JOHN "A Look at the Primeval Explosion." *New Scientist 106* (16 Jan. 1986), p. 48.

MELOTT, ADRIAN "Cosmology on a Computer." *Astronomy 11* (June 1983), p. 66, and *11* (July 1983), p. 66.

ODENWALD, STEN "The Planck Era." *Astronomy 12* (March 1984), p. 6.

———. "The Decay of the False Vacuum." *Astronomy 11* (Nov. 1983), p. 66.

OSMER, PATRICK S. "Quasars as Probes of the Distant and Early Universe." *Scientific American 246* (Feb. 1982), p. 126.

PAGE, D., and R. McKEE "The Future of the Universe." *Mercury 12* (Jan./Feb. 1983), p. 17.

PARKER, BARRY "Discovery of the Expanding Universe." *Sky and Telescope 72* (Sept. 1986). p. 227.

SCHNEIDER, STEPHEN E., and YERVANT TERZIAN "Between the Galaxies." *American Scientist 72* (Nov./Dec. 1984), p. 574.

SCHRAMM, DAVID N. "The Early Universe and High-Energy Physics." *Physics Today 36* (April 1983), p. 27.

SHU, FRANKL "The Expanding Universe and the Large-Scale Geometry of Space-Time." *Mercury 12* (Nov./Dec. 1983), p. 162.

SMITH, DAVID H. "The Rise and Fall of a Cosmic String." *Sky and Telescope 72* (Aug. 1986), p. 115.

———. "The Inflationary Universe Lives?" *Sky and Telescope 65* (March 1983), p. 207.

VAN DEN BERGH, SIDNEY "The Size and Age of the Universe." *Science 213* (21 Aug. 1981), p. 825.

WEINBERG, STEVEN *The First Three Minutes.* New York: Basic Books, 1977.

———. "Origins." *Science 230* (4 Oct. 1985). p. 15.

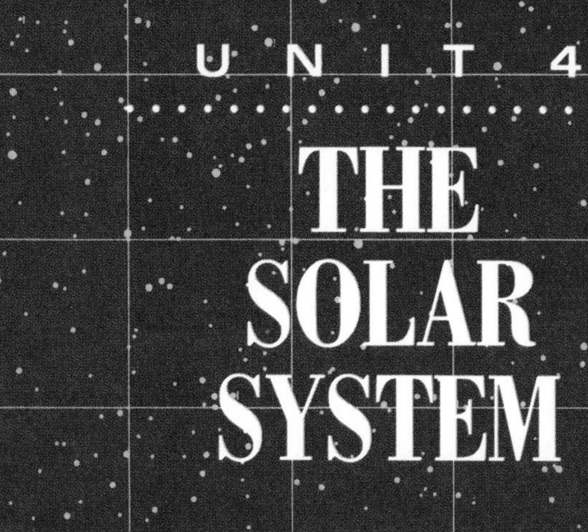

UNIT 4

THE SOLAR SYSTEM

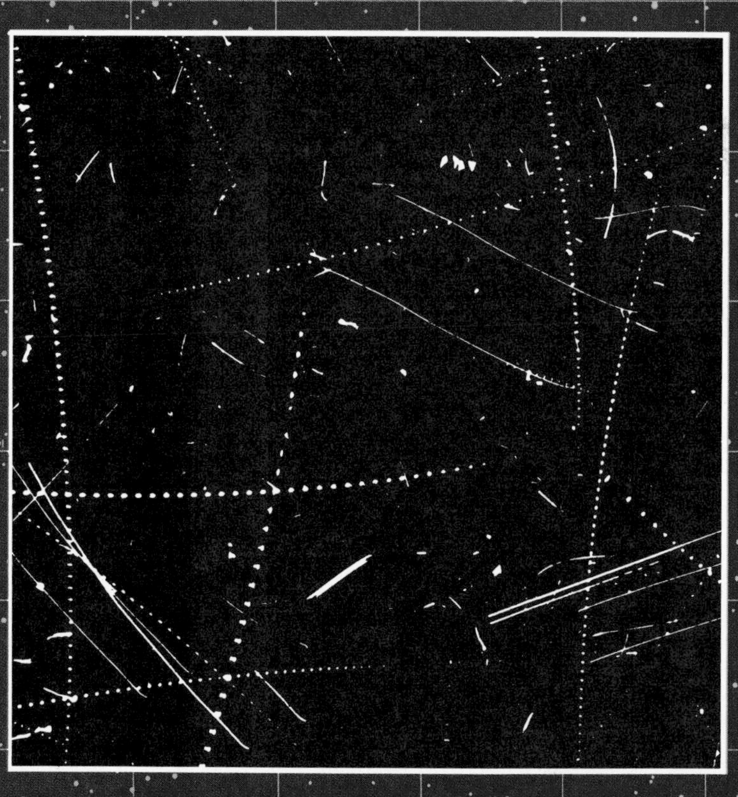

CHAPTER 19

THE ORIGIN OF THE SOLAR SYSTEM

What place is this? Where are we now?

Carl Sandburg
GRASS

Microscopic creatures live in the roots of your eyelashes. Don't worry. Everyone has them, and they are harmless.* They hatch, fight for survival, mate, lay eggs, and die in the tiny spaces around the roots of our eyelashes without doing us any harm. Some live in renowned places—the eyelashes of the Queen of England, of a glamorous movie star, of a famous athlete—but the tiny beasts are not self-aware; they never stop to say, "Where are we?" Humans are much more intelligent; we can wonder where we are in the universe and how we came to be here.

In this chapter we begin exploring our solar system: its sun, nine planets, and some scattered gas, dust, and debris. Our planetary system occupies no more than 10^{-30} of the volume of the observable universe. If the words in this book represent the volume of the universe, the solar system would be less than a trillionth of a trillionth of one letter. However, the significance of the solar system—for us at least—far exceeds its relative size.

There are many reasons for studying the solar system. Not the least of these reasons is that understanding how planets form and evolve teaches us more about our own planet. Our planet now has so many people that we are altering our environment in dramatic ways, and it is possible we could make earth uninhabitable. Understanding the origins of the barren wastes of Mars and the sulfuric acid fogs of Venus may teach us how to preserve our more comfortable climate.

Astronomers need no justification for studying the solar system, for it is the only collection of nonluminous bodies in the universe that we can

**Demodex folliculorum* has been found in 97 percent of individuals with healthy skin.

study in detail. If there are planets orbiting other stars, they are too far away and reflect too little light to be visible. Yet there may be more planets in the universe than stars, so, if only to complete our study of celestial objects, we must examine the solar system.

An intriguing reason for studying the planets of our solar system is the search for life beyond the earth. As if enchanted, some matter on the earth's surface lives and is aware of its existence. To search for other living beings we must examine the surfaces of planets—stars are too hot, and space is too cold for the evolution of life forms. So far as we know, life seems to require the moderate conditions found only on the surfaces of some planets. To focus our search, we must understand how planets form and evolve.

Above all, we study the solar system because it is our home in the universe. Because we are an intelligent species, we have the right and the responsibility to wonder what we are and where we are. Our kind have inhabited this solar system for at least a million years, but only within our lifetime have we begun to understand what a solar system is. Like sleeping passengers on a train, we waken, look out at the passing scenery, and mutter to ourselves, "What place is this? Where are we now?"

19.1 THE GREAT CHAIN OF ORIGINS

We are linked through a great chain of beginnings that leads backward through time to the first instant of the big bang. The gradual discovery of the links in that chain is one of the most exciting adventures of the human intellect. In earlier chapters we discussed some of that story: the origin of the universe, the galaxies, the stars, and the elements. Here we will explore further to consider the formation of planets.

A REVIEW OF THE ORIGIN OF MATTER

The mass of which your body is made came into existence within minutes of the big bang explosion. The protons, neutrons, and electrons were created by the energy present as gamma rays; by the time the universe was only 3 minutes old, these subatomic particles had combined to form atomic nuclei.

About 25 percent of the mass became helium, an inert chemical element, which is not present in your body. Most of the matter in the big bang, however, remained in the form of single protons, the nuclei of hydrogen atoms. Today roughly 75 percent of the mass in the universe is still hydrogen, including the hydrogen atoms in the water molecules in your body. These hydrogen nuclei have survived unchanged since the big bang.

Within a billion years or so after the big bang, matter began to collect into galaxies and form stars. Nuclear fusion reactions inside the stars arranged the protons and neutrons into new partnerships and created heavier elements. The carbon, nitrogen, and oxygen so important throughout your body, the calcium in your bones, and all of the most common heavy elements up to the iron in your blood have been cooked up by nuclear fusion inside many generations of stars.

The elements heavier than iron were created by rapid nuclear fusion, which occurred during supernovae explosions. These elements are much less common than elements lighter than iron, but some of them, such as the iodine in your thyroid gland, play important roles in your body chemistry. The supernovae explosions also spread these newly created elements back into the interstellar medium where they could be incorporated in the formation of new stars. Indeed, the expanding supernovae remnants sometimes trigger the formation of new stars.

The sun is one of the stars that formed from these clouds of gas enriched in heavy elements. How a tiny fraction of that matter came together to form the earth, the oceans, and your body is a story we will explore in the rest of this book. As we explore we must remember the great chain of origins that produced the chemical elements. As geologist Preston Cloud remarked, "Stars have died that we might live."

THE ORIGIN OF PLANETS

In earlier chapters we saw how stars can form from the gravitational contraction of gas and dust clouds. In at least some cases, that contraction must be triggered by compression of the gas cloud, perhaps by a nearby supernova explosion.

As stars form in these contracting clouds, they remain surrounded by cocoons of dust and gas, and the rotation of the cloud causes that dust and gas to form a spinning disk around the protostar. When the center of the star grows hot enough to ignite nuclear reactions, its surface quickly heats up, becomes more luminous, and blows away the gas and dust cocoon.

Modern theories suppose that planets form in the rotating disks of gas and dust around young stars (Figure 19–1). Infrared observations of T Tauri stars, for instance,

FIGURE 19–1 (a) Because the solar nebula was rotating, (b) it contracted into a disk, and (c) the planets formed with orbits lying in nearly the same plane.

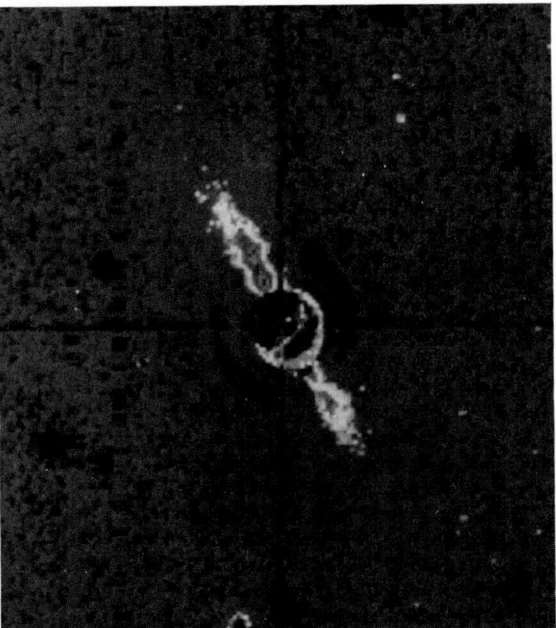

FIGURE 19–2 A computer-enhanced image of the ring of dust surrounding the star β Pictoris. (The glare of the star is hidden behind the mask at the center.) The ring is nearly edge-on to earth, and its radius is about ten times the radius of Pluto's orbit. Planet formation may have occurred recently or be presently occurring in the disk. (Courtesy of R. J. Terrile, JPL, and B. A. Smith, University of Arizona.)

show that some are surrounded by gas clouds rich in dust, and spectra show that these stars are blowing away their nebulae at speeds up to 200 km/sec. The presence of bipolar flows from young stars confirms that such systems contain gas and dust distributed in a disk-shaped cloud (Chapter 11). Our own planetary system probably formed in such a disk-shaped cloud around the sun. When the sun became luminous enough, the remaining gas and dust was blown away into space leaving the planets orbiting the sun. This is known as the **solar nebula theory** because the planets form from the nebula around the protosun (Figure 19–1). If the theory is right, then planets form as a by-product of star formation, and most stars should have planetary systems.

We cannot check our theory observationally because even the nearest stars are so distant that telescopes cannot detect the faint light reflected from planets that might be orbiting the star. But infrared and radio observations do suggest that some stars are surrounded by planet-building material. The Infrared Astronomy Satellite did detect shells of warm dust around two dozen nearby stars, including the bright star Vega. Earth-based observations have detected disks of gas and dust around three different stars, including β Pictoris (Figure 19–2 and Color Plates 30 and 31).

Spectroscopic observations also suggest that other stars have planets. When a star and planet revolve around their common center of mass, the star will have a slightly variable radial velocity. Precision observations reveal a number of stars that have companions with masses too small to be stars. For example, the star HD 114762 appears to have a companion of about 10 Jupiter masses in an 84-day-long orbit. Another, GLS 846 appears to have a 5-Jupiter mass planet orbiting it every 24.3 days. Lower mass planets, because they affect the motion of their star

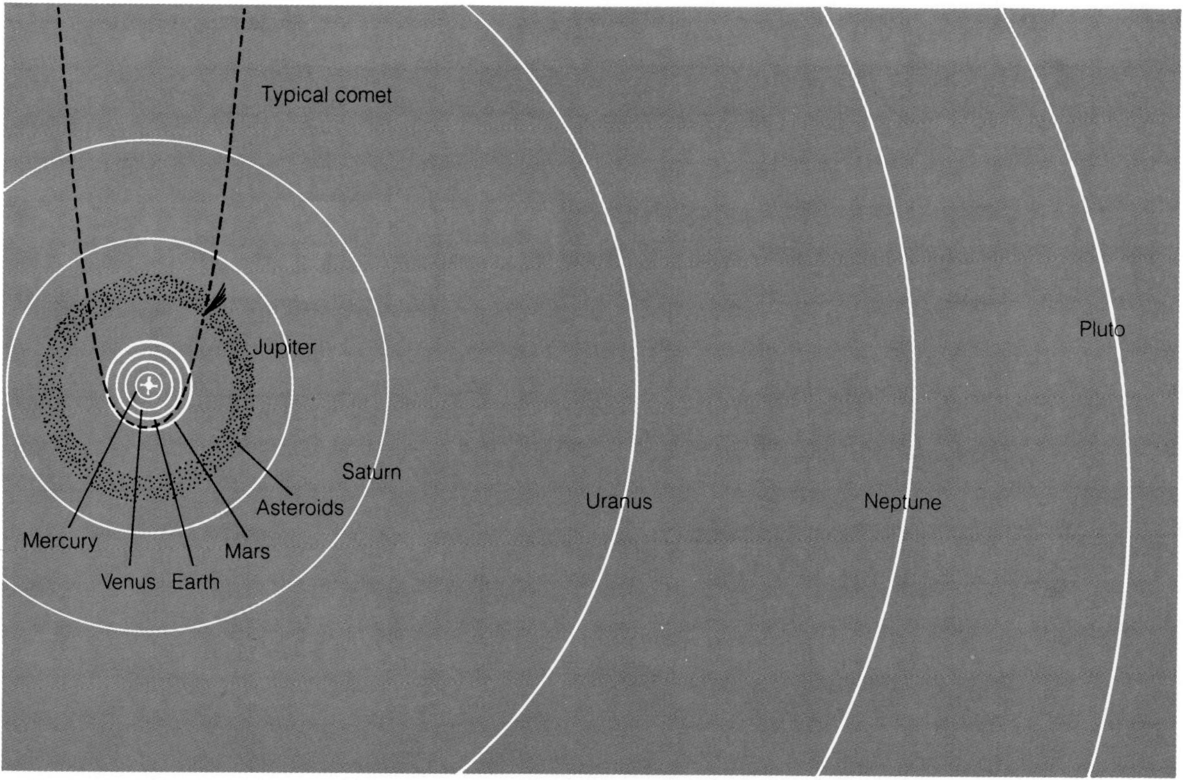

FIGURE 19–3 The solar system. The radii of the orbits are drawn to scale here. Note the eccentricity of Pluto's orbit. At this scale only the sun would be visible. The planets are too small to be seen.

less, would be more difficult to detect, but these observations suggest that extra-solar planets do exist and may be common.

If our theory is right, the planets formed from the nebula around the young sun. Can we find any supporting evidence in the present solar system? To answer that question, we must survey the solar system and identify its most significant characteristics, for if our theory is correct, these characteristics were determined by the nature of the solar nebula.

19.2 A SURVEY OF THE SOLAR SYSTEM

The solar system is almost entirely empty space (Figure 19–3). Imagine that we reduce the solar system until the earth is the size of a grain of table salt, about 0.3 mm (0.01 inch) in diameter. The moon is a speck of pepper about 1 cm (.4 inch) away, and the sun is the size of a small plum 4 m (13 ft) from the earth. Mercury, Venus, and Mars are grains of salt. Jupiter is an apple seed 20 m (66 ft) from the sun, and Saturn is a smaller seed over 36 m (120 ft) away. Uranus and Neptune are slightly larger than average salt grains, and Pluto, the farthest planet, is a speck of pepper over 150 m (500 ft) from the central plum. These tiny specks of matter scattered around the sun are the remains of the solar nebula.

REVOLUTION AND ROTATION

The planets revolve around the sun in orbits that lie close to a common plane. The orbit of Mercury, the closest planet to the sun, is tipped 7° to the earth's, and Pluto's orbit is tipped 17.2°. The rest of the planets' orbital planes are inclined by no more than 3.4°. Thus, the solar system is basically disk-shaped.

BOX 19–1
The Titius–Bode Rule

The Titius–Bode rule specifies a simple series of steps that produce a list of numbers matching the sizes of the planetary orbits. To construct this list, we write down the number 0 and below that 3. Continuing downward, we make each number twice the preceding number—0, 3, 6, 12, 24, 48, and so forth. After adding 4 to each number, we divide by 10. The resulting numbers approximate the radii of the planetary orbits in astronomical units (see Table 19–1).

The rule describes the inner planets well, even including the asteroid belt, but the match for the outer solar system is not as good. If we leave out Neptune, Pluto fits well, but the omission of Neptune is not justified by any physical evidence.

Some astronomers contend that the agreement is mere chance. There is no physical reason for any of the steps in the Titius–Bode rule. In fact, no matter what the sizes of the orbits, a mathematician could find a sequence of steps that would generate matching numbers. However, the simplicity of the Titius–Bode rule leads many astronomers to view it as significant. It may be telling us that there was nothing random about the formation of the planets.

Table 19–1 Average planetary distance from the sun.

Planet	Titius–Bode Prediction (AU)	Observed (AU)
Mercury	(0 + 4)/10 = 0.4	0.387
Venus	(3 + 4)/10 = 0.7	0.723
Earth	(6 + 4)/10 = 1.0	1
Mars	(12 + 4)/10 = 1.6	1.524
Asteroids	(24 + 4)/10 = 2.8	2.77 ave
Jupiter	(48 + 4)/10 = 5.2	5.203
Saturn	(96 + 4)/10 = 10.0	9.539
Uranus	(192 + 4)/10 = 19.6	19.18
Neptune		30.06
Pluto	(384 + 4)/10 = 38.8	39.44

The rotation of the sun and planets on their axes also seems related to this disk shape. The sun rotates with its equator inclined only 7.25° to the earth's orbit, and most of the other planets' equators are tipped less than 30°. The rotation of Venus and Uranus are peculiar, however. Venus rotates backward compared with the other planets, and Uranus rotates on its side (with its equator almost perpendicular to its orbit). We will discuss the peculiar motions of these planets in detail in Chapters 21 and 23.

Apparently, the preferred direction of motion in the solar system—counterclockwise as seen from the north—is also related to its disk shape. All the planets revolve counterclockwise around the sun, and, with the exception of Venus and Uranus, they rotate* counterclockwise on their axes.

Thus, with only a few exceptions, most of which are understood, revolution and rotation in the solar system follow a disk theme. But sharp eyes might detect another significant pattern in the planetary orbits as shown in Figure 19–3. Each planet is a little less than twice as far from the sun as its inward neighbor. In 1766 Johann Titius found a simple sequence of numbers that reproduces

*Recall from Chapter 2 that astronomers distinguish between the words *revolve* and *rotate*. A planet revolves around the sun but rotates on its axis.

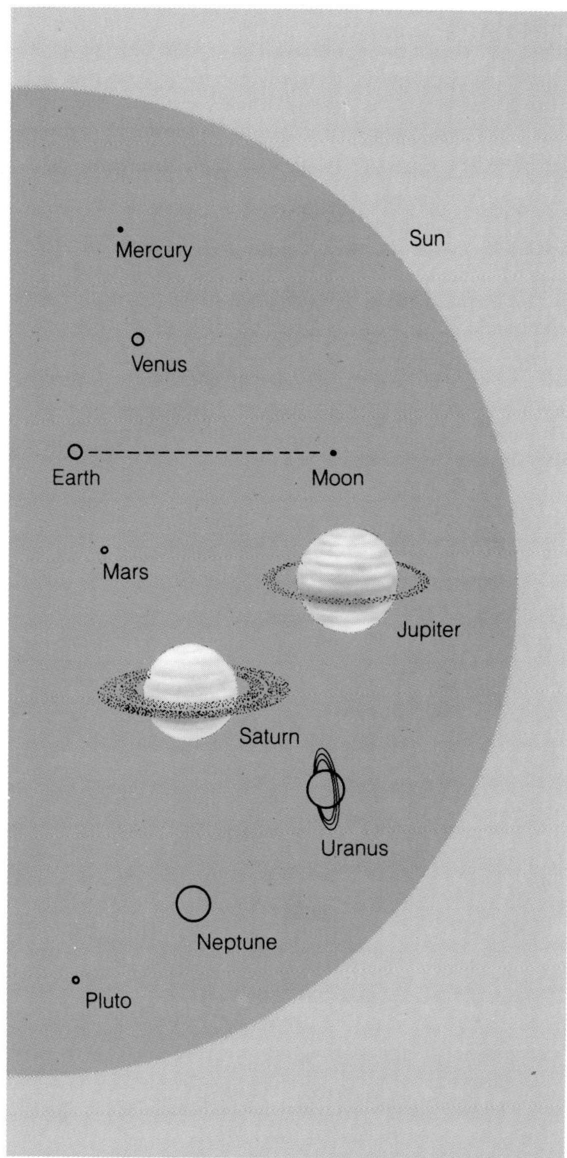

FIGURE 19–4 The relative sizes of the planets compared to the solar disk.

these distances, and, because it was first reported by Johann Bode in 1772, the sequence is now known as the **Titius–Bode rule** (Box 19–1).

No one knows why the Titius–Bode rule generates the distances of the planets, but it is often interpreted to mean that the planets were not formed by some random process. They could not have been captured at random over billions of years, for instance.

TWO KINDS OF PLANETS

Perhaps the most important clue we have to the origin of the solar system is the division of the planets into two categories: terrestrial or Earth-like and Jovian or Jupiter-like (Figure 19–4). The **terrestrial planets** are small, dense, rocky worlds with less atmosphere than the Jovian planets. The terrestrial planets—Mercury, Venus, Earth, and Mars—lie in the inner solar system. By contrast, the **Jovian planets** are large, gaseous, low-density worlds. The Jovian planets—Jupiter, Saturn, Uranus, and Neptune—lie in the outer solar system beyond the asteroids. Note that Pluto does not fit either category very well. It is small like the terrestrial planets but lies far from the sun and has a low density like the Jovian planets.

All the terrestrial planets are scarred by craters. Mercury's surface looks much like the moon's, with thousands of overlapping craters (Figure 19–5). Because Mercury is very hot and only 40 percent larger than the moon, it has held no atmosphere. Venus is nearly as large as Earth and has a thick atmosphere. In fact, its surface is perpetually hidden below a dense layer of clouds, so it is only by mapping its surface with radar that astronomers have found craters on Venus. Earth, too, has some craters, although erosion rapidly wears away such features. Mars is about half Earth's diameter and has a much thinner atmosphere, which permits us to see its surface easily. Space probe photos show that some of the markings visible from Earth are, in fact, craters (Figure 19–6). In addition, planetary probes have found craters on the satellites of Mars, Jupiter, Saturn, and Uranus. This suggests that craters are a characteristic of every object in the solar system with a surface capable of retaining such features.

The terrestrial planets have densities ranging from about 3–5 g/cm^3. Typically, they contain cores of iron and nickel surrounded by a mantle of dense rock. In contrast, the Jovian planets are rich in hydrogen and helium, so their average density is low—less than 1.75 g/cm^3. Saturn's very low density, 0.7 g/cm^3, is less than the density of water, so the planet would float if we could find a bathtub big enough. Nevertheless, all the Jovian planets are massive. Jupiter is over 318 times as massive as Earth, and Saturn is about 95 Earth masses. Uranus and Neptune are smaller—15 and 17 Earth masses, respectively.

Although photographs of Jovian planets reveal swirling cloud patterns (Figure 19–7), their atmospheres are not very deep compared to their radii. In fact, if these so-called gas giants were shrunk to a few centimeters in diameter, their gaseous hydrogen and helium atmo-

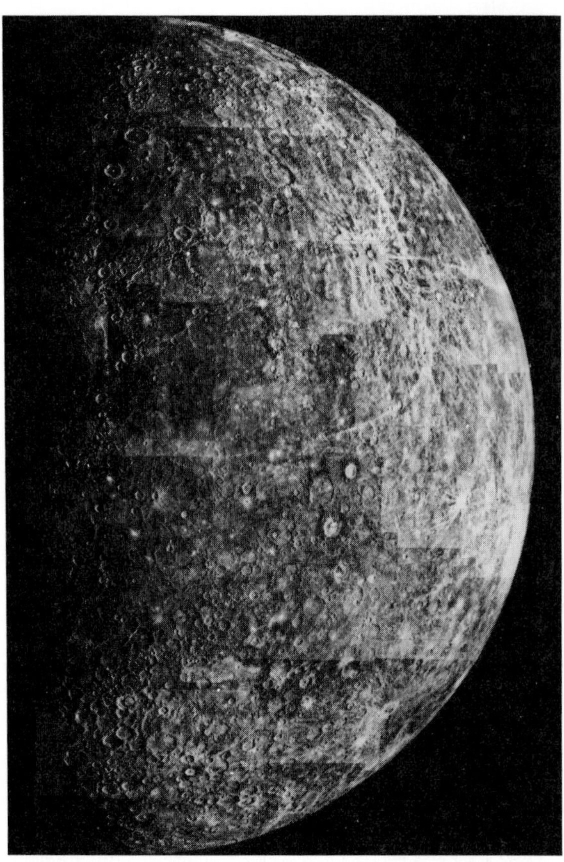

FIGURE 19–5 Mercury, like earth's moon, is heavily cratered. The youngest craters are bright with rays of ejecta. Although it is not easily apparent from this photo, some regions are old, heavily cratered terrain, and other regions are younger and have fewer craters. (NASA/JPL.)

FIGURE 19–6 (a) Venus, seen here at crescent phase, is covered by a thick, cloudy atmosphere. Thus surface features are never visible. (b) Mars has a thin atmosphere, so surface features are visible. This CCD image recorded October 3, 1988 is one of the best ever taken from earth. Spacecraft photos show the features are craters, volcanos, and valleys. (a, Lowell Observatory Photograph; b, Courtesy Stephen M. Larson, Lunar and Planetary Lab.)

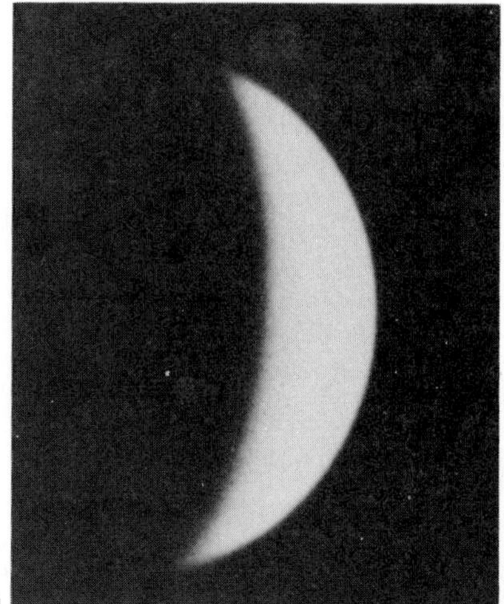

a

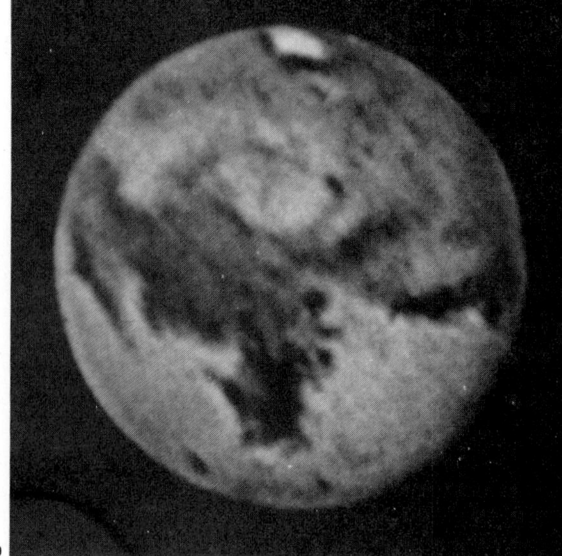

b

spheres would be no deeper than the fuzz on a badly worn tennis ball.

Mathematical models predict that the interiors of the Jovian planets are mostly hydrogen in two different states (Figure 19–8). Beneath the cloud belts of these planets are layers where pressure forces the hydrogen into a liquid state. In the larger Jovian planets, Jupiter and Saturn, the pressure at deeper layers is so high that the hydrogen atoms can no longer hold their electrons. With the electrons free to move about, the material (called **liquid metallic hydrogen**) is an excellent electrical conductor. The centers of the Jovian planets are believed to be hot, earth-sized cores of heavy elements often called "rocky cores," although the heat and pressure would not allow rock to exist as it does on Earth.

Another characteristic of the Jovian planets is their large satellite systems. Jupiter has at least 16 known moons; 4 of them, the **Galilean satellites**—named after Galileo, who discovered them—are visible through a small telescope or even binoculars (Figure 19–9). Saturn and Uranus also have systems of small satellites. In August 1989, Voyager 2 flew past Neptune and discovered six small moons.

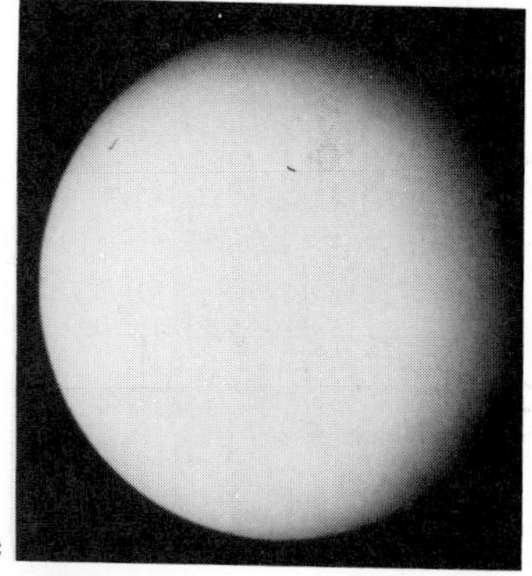

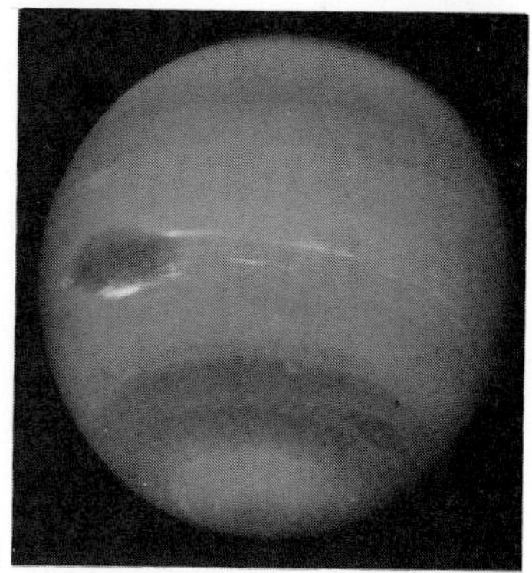

FIGURE 19–7 (a) Jupiter, its cloud belts, and Great Red Spot photographed by Voyager 1. (b) Saturn and its belts and rings photographed by the 100-inch telescope. (c) Uranus photographed by the Voyager 2 spacecraft. Cloud belts are not visible because of methane ice-crystal haze high in the planet's atmosphere. (d) Neptune imaged by Voyager 2. Note the great dark spot at left. (a, NASA/JPL; b, Mount Wilson and Las Campanas Observatories, Carnegie Institution of Washington; c, NASA/JPL; d, NASA.)

Thus Neptune, with two moons previously discovered from Earth, has at least eight satellites.

All of the Jovian planets have ring systems. Saturn has bright rings made of ice particles (Figure 19–10), and Jupiter has rings composed of dark, rocky dust. The rings of Uranus are also dark and rocky, but the particles are bigger than dust. Voyager 2 found Neptune surrounded by a ring system of dark, presumably rocky, dust particles.

We discuss the satellites and rings of the Jovian planets in later chapters. Here it is sufficient to note that the Jovian planets have large numbers of satellites and all four have ring systems. In contrast, the terrestrial planets have few satellites and no rings at all. These facts will help us understand how planets were formed in the solar nebula.

SPACE DEBRIS

The sun and planets are not the only remains of the solar nebula. The solar system is littered with three kinds of space debris: asteroids, comets, and meteoroids. Although this material represents a tiny fraction of the mass of the system, it is a rich source of information about the origin of the planets.

The **asteroids*** are small, rocky worlds, most of which orbit the sun between the orbits of Mars and Jupiter. A few asteroids follow orbits that bring them into the inner solar system, and several occasionally pass within a few tens of millions of miles of Earth. Some are located in Jupiter's orbit, and some have been found as far away as the orbit of Saturn.

About 200 of these objects are more than 100 km (60 miles) in diameter, and more than 2000 are more than 10 km (6 miles). There are probably 500,000 larger than 1 km (0.6 mile) and billions that are smaller. Because even the largest are only a few hundred kilometers in diameter,

*Also called minor planets.

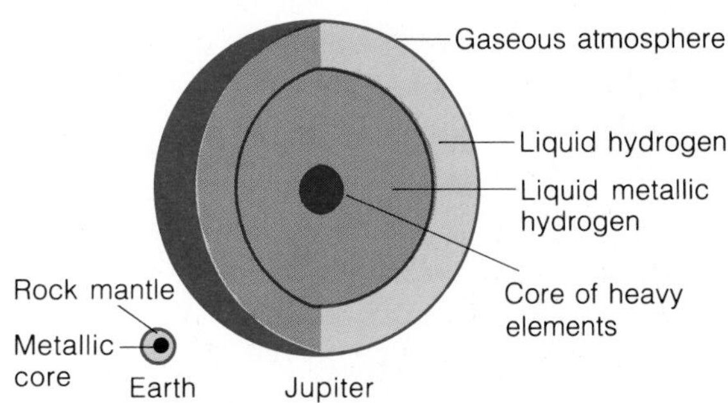

FIGURE 19–8 Cross sections of Earth and Jupiter illustrate the differences between the interiors of terrestrial and Jovian planets. Earth contains a metallic core with a rock mantle. Jupiter, 11.18 times larger in diameter, is composed mostly of liquid hydrogen. A core composed of heavy elements such as iron, nickel, and silicon lies at the center.

FIGURE 19–9 Jupiter has at least 16 satellites, but only the 4 Galilean satellites are visible in small telescopes. (Lick Observatory photograph.)

telescopes reveal no surface features. However, slow variations in the amount of reflected sunlight suggest they are irregular in shape and rotating as they orbit the sun. Only the largest are roughly spherical.

Astronomers may have gotten a close look at a pair of asteroids when the Viking and Mariner space probes to Mars photographed the two small Martian satellites, Deimos and Phobos. These objects are only 12 and 28 km (7.5 and 18 miles) in diameter, and the photos show that they are irregular in shape and heavily cratered (Figure 19–11). Whether these are asteroids that Mars has captured is still open to debate, but many asteroids of the belt must look like Deimos and Phobos.

FIGURE 19–10 The icy rings of Saturn are so reflective that they are easily visible through a small telescope. This photo shows what we might expect to see through a 6-inch telescope. For more detailed photos of the rings, see Chapter 22. (Grundy Observatory photograph.)

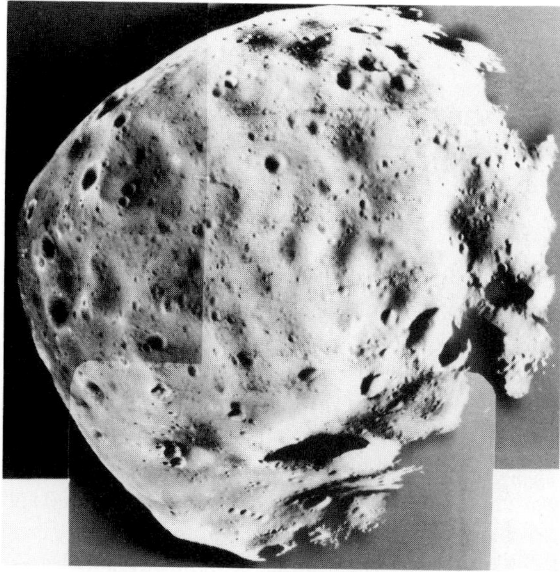

FIGURE 19–11 The asteroids must look much like Phobos, one of the tiny satellites of Mars. (NASA/JPL.)

Apparently, the asteroids are composed of material that failed to build a planet at a distance of 2.8 AU from the sun. When we discuss the formation of planets in the solar nebula, we must be prepared to explain why material at 2.8 AU failed to form a planet.

In contrast to asteroids, the brightest **comets** are impressively beautiful objects (Figure 19–12a). Most comets are faint, however, and are difficult to locate even at their brightest. A comet may take months to sweep through the solar system, during which time it appears as a glowing head with an extended tail of gas and dust.

According to the **dirty snowball theory** of comets, the tail, which can be over an astronomical unit in length for a bright comet, is produced by the nucleus, a ball of dirty ices (mainly water and carbon dioxide) only a few dozen kilometers in diameter. When the nucleus enters the inner solar system, the sun's radiation begins to vaporize the ices, releasing gas and dust. The pressure of sunlight and the solar wind pushes the gas and dust away, forming a long tail that always points away from the sun (Figure 19–12b).

The nuclei of comets are icy bodies left over from the origin of the planets. Thus, we must conclude that at least some parts of the solar nebula were rich in ices. We will see later in this chapter how important ices were in the formation of the Jovian planets, and we will discuss comets in more detail in Chapter 24.

Unlike the stately comets, **meteors** flash across the sky in momentary streaks of light (Figure 19–13). They are commonly called "shooting stars." Of course, they are not stars but small bits of rock and metal falling into the earth's atmosphere and bursting into incandescent vapor about 80 km (50 miles) above the ground because of friction with the air. This vapor condenses to form dust, which settles slowly to the earth adding about 10,000 tons per year to the earth's mass.

Technically the word *meteor* refers to the streak of light in the sky. In space, before its fiery plunge, the object is called a **meteoroid**, and, if any part of it survives its fiery passage to the earth's surface, it is called a **meteorite**. Most meteoroids are specks of dust, grains of sand, or tiny pebbles. Almost all the meteors we see in the sky are produced by meteoroids that weigh less than 1 g. Only rarely is one massive enough and strong enough to survive its plunge and reach the earth's surface.

FIGURE 19–13 A meteor is the streak of glowing gases produced by a bit of material falling into the earth's atmosphere. Friction with the air vaporizes the material about 80 km (50 miles) above the earth's surface. (Lick Observatory photograph.)

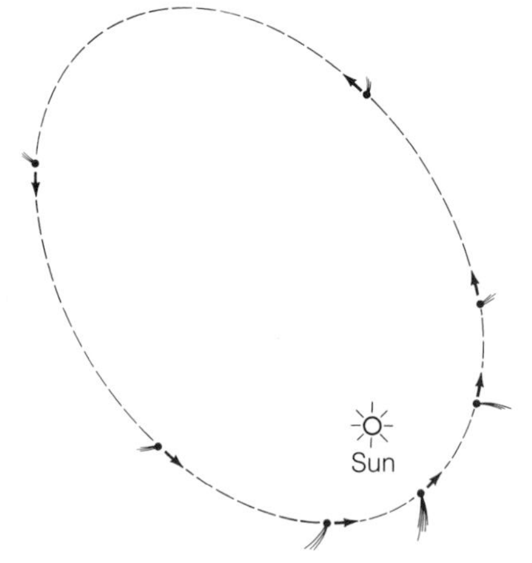

FIGURE 19–12 (a) A comet may remain visible in the evening or morning sky for weeks as it moves through the inner solar system. Comet West was in the sky during March 1976. (b) A comet in a long elliptical orbit becomes visible when the sun's heat vaporizes its ices and pushes the gas and dust away in a tail. (Lick Observatory photograph.)

A few thousand meteorites have been found, and we will discuss their particular forms in Chapter 24. We mention meteorites here for one specific clue they can give us concerning the solar nebula. Meteorites can tell us the age of the solar system.

THE AGE OF THE SOLAR SYSTEM

If the solar nebula theory is correct, the planets should be about the same age as the sun. The most accurate way to find the age of a celestial body is to bring a sample into a laboratory and determine its age by radioactive dating (Box 19–2). So far, the only celestial bodies that can be dated in this way are the earth, the moon, and meteorites.

The oldest earth rocks so far discovered and dated are about 3.9 billion years old. That does not mean that the earth formed 3.9 billion years ago. The surface of the earth is active, and the crust is continually destroyed and reformed from material welling up from beneath the crust (see Chapter 20). Thus, the age of these oldest rocks tells us only that the earth is *at least* 3.9 billion years old.

BOX 19–2
Radioactive Dating

The rate at which radioactive nuclei decay is so predictable we can use it to find the ages of rocks. When a nucleus decays, it emits a particle (such as an electron or positron) and changes into a different nucleus. For example, the carbon isotope ^{14}C decays into the nitrogen isotope ^{14}N. Because this decay occurs at a known rate, measuring the amount of these isotopes in a sample can tell us its age.

Radioactive decay is based on probability and is similar to the flipping of a coin. We cannot predict whether the flip of a coin will result in a head or a tail, but we can predict that a thousand flips will result in about half heads and half tails. Similarly, we cannot predict when a nucleus will decay, but if we are given a large sample of radioactive nuclei, we can find their **half-life**—the time it takes for half of the nuclei in the sample to decay. The half-life of ^{14}C is 5730 years. After one half-life, 50 percent of the carbon nuclei in a sample will have turned into nitrogen, and 50 percent will remain carbon. After another half-life, half of the remaining carbon will become nitrogen, and only 25 percent of the original carbon will remain. Thus, if we know how much carbon was present to start and how much now remains, we can find the number of half-lives that have elapsed.

However, ^{14}C dating is not useful in astronomy because of the isotope's short half-life. It is difficult to date objects older than about 50,000 years using carbon dating. The ages of astronomical samples—moon rocks, for example—are almost always greater than a billion years. Instead of ^{14}C, astronomers must use longer-lived isotopes. Potassium (^{40}K), for example, decays with a half-life of 1.3 billion years. Of the decaying nuclei, 89 percent become calcium (^{40}Ca), and 11 percent become argon (^{40}Ar). The most difficult part of the analysis is discovering how much potassium, calcium, and argon were present in the sample to start with, but that can be found from studies of the abundances of the elements. Once that problem is solved, the potassium–argon clock can tell us the age of the sample.

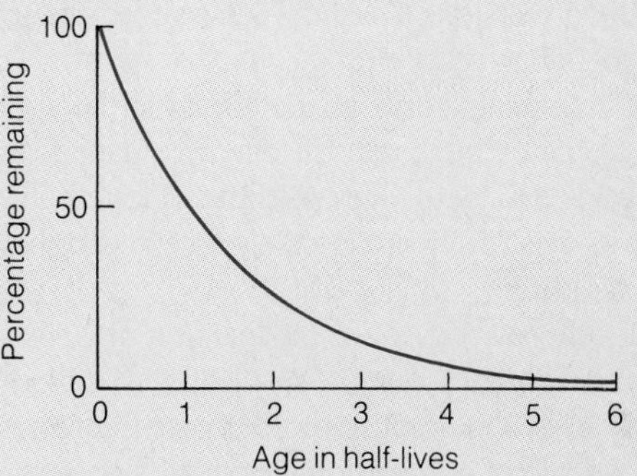

FIGURE 19–14 The number of nuclei remaining in a radioactive sample decreases such that 50 percent are left after one half-life, 25 percent after two half-lives, 12.5 percent after three half-lives, and so on.

For example, suppose that we have corrected for the argon and calcium that were originally in a meteorite, and we find that it now contains 77 ^{40}Ar nuclei and 623 ^{40}Ca nuclei for every 100 nuclei of ^{40}K. Adding 77 plus 623 plus 100, we find that 800 nuclei of ^{40}K were originally present and only 100 now remain. That is, only 12.5 percent of the potassium has not decayed. Figure 19–14 shows that 12.5 percent of the decaying nuclei remain when the age of the sample is three times the half-life. Thus, the meteorite must be 3.9 billion years old.

To be more precise, this tells us how long ago the meteorite last melted. Argon is a gas, and if the material were ever heated, the gas would escape. Once the matter cools, argon begins to accumulate as the potassium decays, and the clock ticks off the years. Thus, melting the sample resets the radioactive clock to zero.

Potassium–argon is one of many radioactive clocks. Uranium (^{238}U) decays to an isotope of lead (^{206}Pb) with a half-life of 4.5 billion years, and rubidium (^{87}Rb) decays to strontium (^{87}Sr) with a half-life of 47 billion years. Any such decay with a long half-life can be used as a radioactive clock to date astronomical samples.

One of the most exciting goals of the Apollo lunar landings was bringing lunar rocks back to earth's laboratories, where they could be dated. Because the moon's surface is not being recycled like earth's, some parts of it might have survived unaltered since early in the history of the solar system. Dating the rocks showed the oldest to be 4.48 billion years old. Thus, the solar system must be *at least* 4.48 billion years old.

Another important source for determining the age of the solar system is meteorites. Radioactive dating of meteorites yields a range of ages with the oldest about 4.6 billion years old. This figure is widely accepted as the age of the solar system.

One last celestial body deserves mention: the sun. Astronomers estimate the age of the sun as about 5 billion years, but this is not a radioactive date because they cannot obtain a sample of solar material. Instead, they estimate the sun's age from the radioactive ages of the earth, the moon, and meteorites. Computer models of the sun give only approximate ages, but they generally agree with the age of the earth.

Apparently, all the bodies of the solar system formed at about the same time some 4.6 billion years ago. This is the last item we add to the list of significant properties in Table 19–2.

Table 19–2 Characteristic properties of the solar system.

1. Disk shape of the solar system
   - Orbits in nearly the same plane
   - Common direction of rotation and revolution
2. Titius–Bode rule
3. Two planetary types:
   - Terrestrial—inner planets: high-density
   - Jovian—outer planets: low-density
4. Planetary ring systems for Jupiter, Saturn, and Uranus
5. Space debris—asteroids, comets, and meteors
   - Composition
   - Orbits
6. Common ages of about 4.6 billion years for the earth, moon, meteorites, and sun

19.3 THE STORY OF PLANET BUILDING

The challenge for modern planetary astronomers is to compare the characteristics of the solar system to the solar nebula theory and tell the story of how the planets formed.

THE CHEMICAL COMPOSITION OF THE SOLAR NEBULA

Everything we know about the nature of the solar system and the nature of star formation suggests that the solar nebula was a fragment of an interstellar gas cloud. Such a cloud would have been mostly hydrogen with some helium and tiny traces of the heavier elements.

This is precisely what we see in the composition of the sun (see Table 7–2). Analysis of the solar spectrum shows that the sun is mostly hydrogen, with 25 percent of its mass being helium and only about 2 percent being heavier elements. Of course, nuclear reactions have fused some hydrogen into helium, but this happens in the sun's core and has not affected its surface composition. Thus, the composition reflected in its spectrum is essentially the composition of the gases from which it formed.

This must have been the composition of the solar nebula. We can see that in the Jovian planets where hydrogen and helium are abundant and heavier elements are rare. Yet the Jovian planets do not have as large a percentage of hydrogen and helium as does the sun. These gases are light-weight and escape into space more easily than the heavier elements. Thus, we can conclude that the Jovian planets have lost a small amount of the lighter elements originally present in the solar nebula.

The terrestrial planets, in contrast, contain very little hydrogen and helium. These small planets have such low masses they were unable to keep the hydrogen and helium from leaking into space. Indeed, because of their small masses they were probably unable to capture very much of these light-weight gases from the solar nebula. The terrestrial planets are dense worlds because they are composed of the heavier elements from the solar nebula.

Although we can see how the chemical composition of the solar nebula is reflected in the present composition of the sun and planets, a very important question remains: How did gas and dust in the solar nebula come together to form the solid matter of the planets? We must answer that question in two stages. First, we must understand how gas and dust formed billions of small solid particles, and then we must explain how these particles built the planets.

Table 19–3 Observed and uncompressed densities.

Planet	Observed Density (g/cm^3)	Uncompressed Density (g/cm^3)
Mercury	5.44	5.4
Venus	5.24	4.2
Earth	5.52	4.2
Mars	3.93	3.3
(The moon)	3.36	3.35

Table 19–4 The condensation sequence.

Temperature (K)	Condensate	Planet (Estimated Temperature of Formation) (K)
1500	Metal oxides	Mercury (1400)
1300	Metallic iron and nickel	
1200	Silicates	
1000	Feldspars	Venus (900)
680	Troilite (FeS)	Earth (600) Mars (450)
175	H_2O ice	Jovian (175)
150	Ammonia-water ice	
120	Methane-water ice	
65	Argon–neon ice	Pluto (65)

THE CONDENSATION OF SOLIDS

The key to understanding the process that converted the nebular gas into solid matter is the variation in density among solar system objects. We have already noted in Table 19–2 that the four inner planets are high-density terrestrial bodies, whereas the outermost planets are low-density giant planets. This division is due to the different ways gases condensed into solids in the inner and outer regions of the solar nebula.

Even among the four terrestrial planets, we find a pattern of subtle differences in density. Merely listing the observed densities of the terrestrial planets does not reveal the pattern because Earth and Venus, being more massive, have stronger gravity and have squeezed their interiors to higher densities. We must look at the **uncompressed densities**—the densities the planets would have if their gravity did not compress them. These densities (Table 19–3) show that the closer a planet is to the sun, the higher its uncompressed density.

This density variation probably originated when the solar system first formed solid grains. The kind of matter that condensed in a particular region would depend on the temperature of the gas there. In the inner regions, the temperature may have been 1500 K or so. The only materials that could form grains at this temperature are compounds with high-melting points such as metal oxides and pure metals, which are very dense. Farther out in the nebula it was cooler, and silicates (rocky material) could condense. These are less dense than metal oxides and metals. In the cold outer regions, ices of water, methane, and ammonia could condense. These are low-density materials.

The sequence in which the different materials condense from the gas as we move away from the sun is called the **condensation sequence** (Table 19–4). It suggests that the planets, forming at different distances from the sun, accumulated from different kinds of materials. Thus, the inner planets formed from high-density metal oxides and metals, and the outer planets formed from low-density ices.

We must also remember that the solar nebula did not remain the same temperature throughout the formation of the planets, but may have grown progressively cooler. Thus, a particular region of the nebula may have begun by producing solid particles of metals and metal oxides but, after cooling, began producing particles of silicates. Allowing for the cooling of the nebula makes our theory much more complex, but it also makes the processes by which the planets formed much more understandable.

THE FORMATION OF PLANETESIMALS

In the development of a planet, three groups of processes operate. First, grains of solid matter grow larger, eventually reaching diameters ranging from a few centimeters to kilometers. These objects, called **planetesimals**, are believed to be the bodies that the second group of processes collect into planets. Finally, a third set of processes clears away the solar nebula. The study of planet building is the study of these three groups of processes.

According to the solar nebula theory, planetary development in the solar nebula began with the growth of dust grains. These specks of matter, whatever their composition, grew from microscopic size by two processes: condensation and accretion.

A particle grows by **condensation** when it adds matter one atom at a time from a surrounding gas. Thus, snowflakes grow by condensation in the earth's atmosphere. In the solar nebula, dust grains were continuously bombarded by atoms of gas, and some of these stuck to the grains. A microscopic grain capturing a gas atom increases its mass by a much larger fraction than a gigantic boulder capturing a single atom. Thus, condensation can increase the mass of a small grain rapidly, but as the grain grows larger, condensation becomes less effective.

The second process is **accretion**, the sticking together of solid particles. In building a snowman, we roll a ball of snow across the snowy ground so that it grows by accretion. In the solar nebula, the dust grains were, on the average, no more than a few centimeters apart, so they collided with each other frequently. Their mutual gravitation was too small to hold them to each other, but other effects may have helped. Static electricity generated by their passage through the gas could have held them together, as could compounds of carbon that might have formed a sticky surface on the grains. Ice grains might have stuck together better than some other types. Of course, some collisions might break up clumps of grains; on the whole, however, accretion must have increased grain size. If it had not, the planets would not have formed.

There is no clear distinction between a very large grain and a very small planetesimal, but we can consider an object a planetesimal when its diameter becomes a centimeter or so. Objects this size and larger were subject to new processes that tended to concentrate them. One important effect may have been that the growing planetesimals collapsed into the plane of the solar nebula. Dust grains could not fall into the plane because the turbulent motions of the gas kept them stirred up, but the larger objects had more mass and the gas motions could not have prevented them from settling into the plane of the spinning nebula. This would have concentrated the solid particles into a thin plane about 0.01 AU thick and would have made further planetary growth more rapid.

This collapse of the planetesimals into the plane is analogous to the flattening of a forming galaxy. However, an entirely new process may have become important once the plane of planetesimals formed. Computer models show that the rotating disk of particles should have been gravitationally unstable and would have broken up into small clouds (Figure 19–15). This would further concentrate the planetesimals and help them coalesce into objects up to 100 km (60 miles) in diameter. Thus, the theory predicts that the nebula became filled with trillions of planetesimals ranging in size from pebbles to tiny planets. As the largest began to exceed 100 km in diameter, new processes began to alter them, and a new stage in planet building began, the growth of protoplanets.

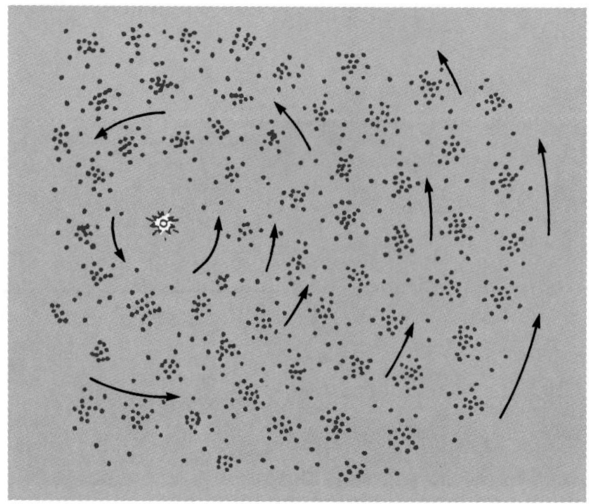

FIGURE 19–15 Gravitational instabilities in the rotating disks of planetesimals may have forced them to collect in clumps, accelerating their growth.

THE GROWTH OF PROTOPLANETS

The coalescing of planetesimals eventually formed **protoplanets**, massive objects destined to become planets. As these larger bodies grew, new processes began making them grow faster and altered their physical structure.

If planetesimals collided with each other at orbital velocities, it is unlikely that they would have stuck together. The average orbital velocity in the solar system is about 30 km/sec (67,000 mph). Head-on collisions at this velocity would have pulverized the material. However, the planetesimals moved the same direction in the nebular plane and thus avoided head-on collisions. Instead, they merely rubbed shoulders at low relative velocities. Such collisions would more likely fuse them together than shatter them.

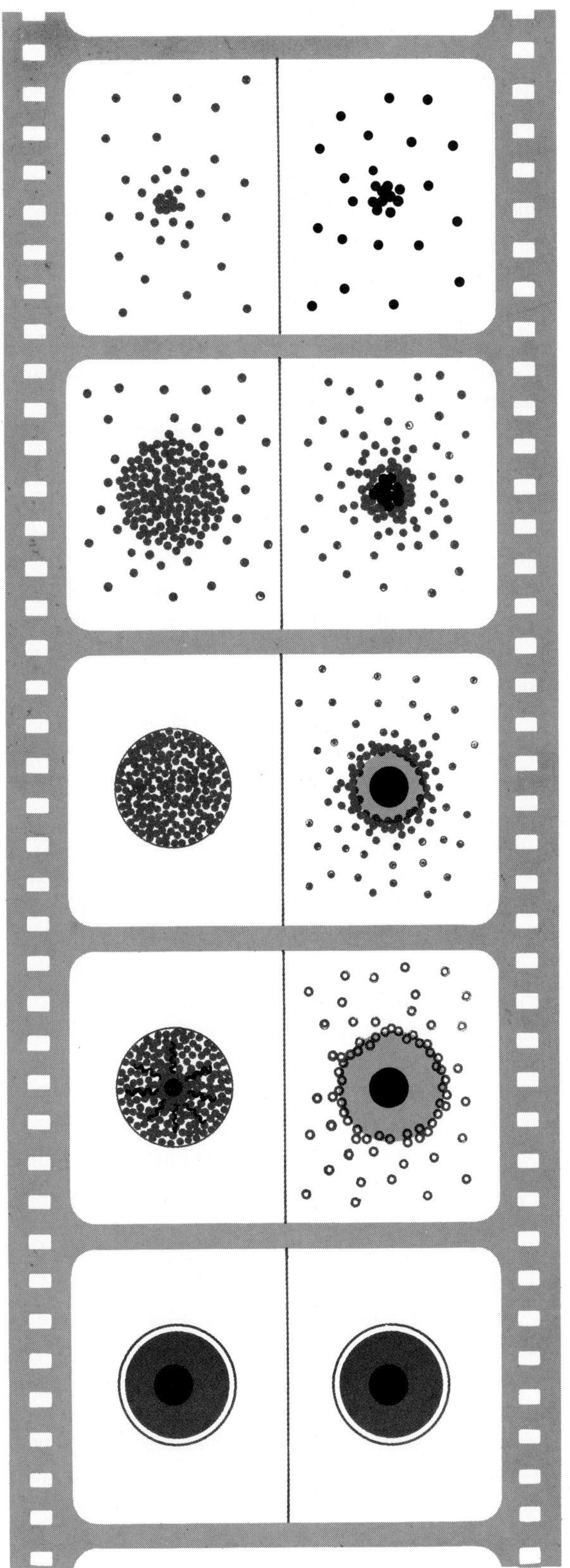

FIGURE 19–16 At left, a planet grows by homogeneous accretion, is heated by radioactive decay, and differentiates. At right, a planet grows by heterogeneous accretion, accumulating first a metallic core, then a mantle, and then a crust.

In addition, some adhesive effects probably helped. Sticky coatings and electrostatic charges on the surfaces of the smaller planetesimals probably aided formation of larger bodies. Collisions would have fragmented some of the surface rock, but, if the planetesimals were large enough, their gravity would have held onto some fragments, forming a layer of soil composed entirely of crushed rock. Such a layer on the larger planetesimals may have been effective in trapping smaller bodies.

The largest planetesimals would grow the fastest because they had the strongest gravitational field. Not only could they hold on to a cushioning layer to trap fragments, but their stronger gravity could attract additional material. These planetesimals probably grew quickly to protoplanetary dimensions, sweeping up more and more material. When massive enough, they trapped some of the original nebular gas to form primitive atmospheres. At some point, they crossed the boundary between planetesimals and protoplanets.

No one knows exactly how the protoplanets grew to planetary dimensions. Nevertheless, we can examine an older and simpler theory of how the earth might have grown and then see how that theory has been modified by newer ideas.

According to the **homogeneous accretion** theory, the protoplanets, including the earth, grew by the accretion of planetesimals that had the same general composition. These accumulated to form a planet-size ball of material that was of homogeneous composition throughout (Figure 19–16). Had the planet accumulated rapidly, the energy released by the in-falling planetesimals, the **heat of formation**, would have melted the planet; its denser material would have sunk to the center, and the lighter material would have floated to the surface. Because the homogeneous accretion theory holds that the planet remained homogeneous during formation, it assumes that the planet formed slowly to give the heat time to radiate into space.

Once the planet formed, heat began to accumulate in its interior from the decay of short-lived radioactive elements such as ^{26}Al and others. This heat eventually melted the planet and allowed it to differentiate. **Differentiation** is the separation of material according to density. When the homogeneous planet melted, the heavy metals such as iron and nickel settled to the core, while the lighter silicates floated to the surface to form a low-density crust.

According to the homogeneous accretion theory, the earth's present atmosphere was not its first. The first atmosphere consisted of gases trapped from the solar nebula—mostly hydrogen and helium. They were driven off by the heat, aided perhaps by outbursts from the infant sun, and new gases baked from the rocks to form a secondary atmosphere. This creation of a planetary atmosphere from the planet's interior is called **outgassing.**

The homogeneous accretion theory is being challenged by a modified version called the **heterogeneous accretion** theory. This theory assumes that the solar nebula cooled during the formation of the planets so that they did not accumulate from planetesimals of common composition. As planet building began, the first particles to condense in the inner solar system were metals and metal oxides, so the protoplanets began by accreting metallic cores. Later, as the nebula cooled, more silicates could form, and the protoplanets added silicate mantles and crusts. In addition, the planets probably grew rapidly enough so that the heat released by the in-falling material melted the planets and they further differentiated as they formed.

The heterogeneous accretion theory suggests that our present atmosphere was not produced entirely by outgassing. Much of our atmosphere and our rich supply of water may have accumulated late in the formation of the planets as the earth swept up volatile-rich planetesimals formed in the cooling solar nebula. Such icy planetesimals may have formed in the outer parts of the solar nebula and were then scattered by encounters with the Jovian planets.

Clearly, the heterogeneous accretion theory is much more sophisticated than the homogeneous accretion theory. Probably neither model is entirely correct, but they illustrate the processes that created the planets in our solar system. So far we have applied these processes primarily to the earth, but we can now ask how well they explain the entries in Table 19–2.

EXPLAINING THE CHARACTERISTICS OF THE SOLAR SYSTEM

Table 19–2 contains the list of distinguishing characteristics of the solar system. Any theory of the origin of the solar system should explain these characteristics.

The disk shape of the solar system is inherited from the solar nebula. The sun and planets revolve and rotate in the same direction because they formed from the same rotating gas cloud. The orbits of the planets lie in the same plane because the rotating solar nebula collapsed into a disk, and the planets formed in that disk.

The Titius–Bode rule (Box 19–1) is more difficult to explain. We don't understand why the planets are spaced in the way they are, but we can suspect that the spacing is related to the way the growing protoplanets dominated regions of the solar nebula.

The condensation sequence explains how the terrestrial planets formed in the inner part of the solar system. Most of the solar nebula consisted of gases that could not condense at the higher temperatures near the sun, so the earthlike planets grew from what little metals and silicates were able to condense there.

In contrast, the Jovian planets formed in the outer solar nebula where the lower temperature allowed the gas to form large amounts of ices, perhaps three times more ices than silicates. Thus, the Jovian planets grew rapidly and became massive. Some theoretical models indicate that Jupiter formed from this abundance of material in less than a thousand years, although most theories predict many millions of years for the formation of the Jovian planets.

The heat of formation (the energy released by in-falling matter) was tremendous for these massive planets. Jupiter must have grown hot enough to glow with a luminosity about 1 percent that of the present sun. However, because it never got hot enough to generate nuclear energy as a star would, it cooled. Jupiter is still hot inside. In fact, both Jupiter and Saturn radiate more heat than they absorb from the sun, so they are evidently still cooling.

When the Jovian planets grew sufficiently large, they attracted vast amounts of nebular gas and thus grew rich in light gases such as hydrogen and helium. The terrestrial planets could not do this because they never became massive enough and because the gas in the inner nebula was hotter and more difficult to trap.

The large satellite systems of the Jovian planets may be made up, in part, of captured planetesimals. The inner planets have only three satellites (Earth has one and Mars two) apparently because the terrestrial planets are less massive and less able to capture satellites.

Saturn's rings may have survived from the formation of the planet, but the other ring systems are more delicate (Chapters 22 and 23). They may be recent accumulations of trapped debris. In any case, we can understand why Jovian planets have ring systems while terrestrial planets do not. Jovian planets are far from the sun where sunlight and the solar wind are too weak to blow away ring particles quickly.

Many astronomers see the formation of the Jovian planets and their satellites as miniature versions of the

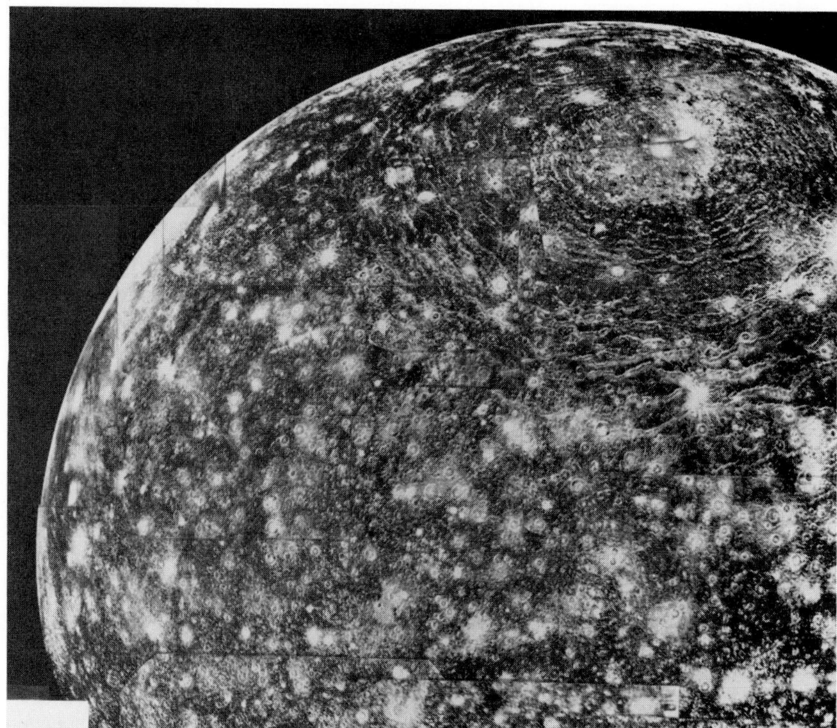

FIGURE 19–17 The heavily cratered surface of Jupiter's satellite Callisto is shown in this composite of images made by the Voyager spacecraft. Nearly all solid surfaces in the solar system are blasted with impact scars, showing that the solar system once contained much more debris than at present. The smallest features visible in this photo are about 7 km in diameter. (NASA)

origin of the solar system from the solar nebula. A rotating gas cloud flattens into a disk with a massive body growing in the center. Some ring systems may be fossils of these flattened gas clouds, and the condensation of the Jovian satellites may have been similar to the condensation of the planets in the solar nebula.

Our general understanding of the origin of the solar system also explains the origin of asteroids, meteors, and comets. They appear to be the last of the debris left behind by the solar nebula. These objects are such important sources of information about the history of our solar system that we will discuss them in detail in Chapter 24. But for now, what happened to the solar nebula?

CLEARING THE NEBULA

The planets appear to have grown in the solar nebula at the same time that the sun was forming about 4.6 billion years ago. Indeed, much of the planetary growth may have taken place in darkness because the sun may not have become luminous yet. But once the sun became hot and luminous, it began to clear the nebula of gas and dust and brought planet building to a halt.

Four effects helped to clear the nebula. The most important was **radiation pressure.** When the sun became a luminous object, light streaming from its surface pushed against the particles of the solar nebula. Large bits of matter like planetesimals and planets were not affected, but low-mass specks of dust and individual gas atoms were pushed outward and eventually driven from the system. This is not a sudden process, and it may not have occurred at the same time everywhere in the nebula. Before sunlight could begin clearing the outer nebula, it first had to push its way through the inner nebula.

The second effect that helped clear the nebula was the solar wind, the flow of ionized hydrogen and other atoms away from the sun's upper atmosphere. This flow is a steady breeze that rushes past the earth at about 400 km/sec (250 miles/sec). When the sun was young, it may have had an even stronger solar wind, and irregular fluctuations in its luminosity, like those observed in T Tauri stars, may have produced surges in the wind that helped push dust and gas out of the nebula.

The third effect for clearing the nebula was the sweeping up of space debris by the planets. Some astronomers believe that this was very efficient; others suspect that it was less important than the first two processes. Certainly it helped. The extensive cratering of the terrestrial planets and the satellites of Earth, Mars, Jupiter, and Saturn is evidence that the last planetesimals were gobbled up by the planets about 4 billion years ago (Figure 19–17). Since then the solar system has been relatively clear of solid debris.

The fourth effect was the ejection of material from the solar system by close encounters with planets. If a small object such as a planetesimal passes close to a planet, it can gain energy from the planet's gravitational field and be thrown out of the solar system. Ejection is most probable for encounters with massive planets, so the Jovian planets were probably very efficient at ejecting the icy planetesimals that formed in their region of the nebula.

Together these four effects cleared away the solar nebula. This may have taken considerable time, but eventually the solar system was relatively clear, the planets could no longer gain mass, and planet building ended.

SUMMARY

We can reconstruct the process by which the solar system formed by studying the characteristic properties of the system. These properties suggest that the sun and planets formed at about the same time from the same cloud of gas and dust—the solar nebula.

One of the most striking properties of the solar system is its disk shape. The orbits of the planets lie in nearly the same plane, and they all revolve around the sun in the same direction, counterclockwise as seen from the north. This is also true of many of the satellites of the planets. With only two exceptions the planets rotate counterclockwise around axes roughly perpendicular to the plane of the solar system. This disk shape and the motion of the planets appear to have originated in the solar nebula.

Another striking feature of the solar system is the division of the planets into two families. The terrestrial planets, which are small and dense, lie in the inner part of the system. The Jovian planets are large, low-density worlds in the outer part of the system. In general, the closer a planet lies to the sun, the higher its uncompressed density.

The solar system is now filled with smaller bodies such as asteroids, comets, and meteors. The asteroids are small, rocky worlds, most of which orbit the sun between Jupiter and Mars. They appear to be material left over from the formation of the solar system.

Another important characteristic of the solar system bodies is their similar ages. Radioactive dating tells us that the earth, moon, and meteorites are no older than about 4.6 billion years. Thus, it seems our solar system took shape about 4.6 billion years ago.

The solar nebula theory proposes that the solar system began as a contracting cloud of gas and dust that flattened into a rotating disk. The center of this cloud became the sun, and the planets eventually formed in the disk of the nebula. Because the nebula was disk-shaped, the planetary orbits lie in nearly the same plane, and most rotation in the solar system has the same direction—the direction in which the nebula was originally rotating.

According to the condensation sequence, the inner part of the nebula was so hot only high-density minerals could form solid grains. The outer regions, being cooler, condensed to form icy material of lower density. The planets grew from these solid materials with the denser planets forming in the inner part of the nebula and the lower-density Jovian planets forming farther from the sun.

Planet building probably began as dust grains, which grew by condensation and accretion into planetesimals ranging from a centimeter to a few kilometers in diameter. These planetesimals settled into a thin plane around the sun and accumulated into larger bodies, the largest of which grew the fastest and eventually became protoplanets.

According to the homogeneous accretion theory, the terrestrial planets grew slowly at low temperature from the accumulation of planetesimals of homogeneous composition. Later the planets grew hot from the decay of radioactive elements, and their interiors melted and differentiated into core, mantle, and crust. The heterogeneous accretion theory supposes that the planets formed rapidly, grew hot from their heat of formation, and differentiated as they formed. The first planetesimals were metals and metal oxides, but, as the nebula cooled, more silicates condensed and built mantles and crusts.

The Jovian planets probably grew rapidly from icy material and became massive enough to attract and hold vast amounts of nebular gas. Heat of formation raised their temperatures very high when they were young, and Jupiter and Saturn still radiate more heat than they absorb from the sun.

Once the sun became a luminous object, it cleared the nebula, as its light and solar wind pushed material out of the system. The planets helped by absorbing some planetesimals and ejecting others from the system. Once the solar system was clear of debris, planet building ended.

NEW TERMS

solar nebula theory
Titius–Bode rule
terrestrial planets
Jovian planets
liquid metallic hydrogen
Galilean satellites
asteroids
comet
dirty snowball theory
meteor, meteoroid, meteorite
half-life
uncompressed density
condensation sequence
planetesimal
condensation
accretion
protoplanet
homogeneous accretion
heat of formation
differentiation
outgassing
heterogeneous accretion
radiation pressure

QUESTIONS

1. What produced the helium now present in the sun's atmosphere? In Jupiter's atmosphere? In the sun's core?
2. In your opinion, if a planetary system originated with one of the first stars to form in our galaxy, how would the process of planetary development differ from the formation of our solar system? (HINT: To what population would the star belong?)
3. What observational evidence do we have that stars commonly form with disks of gas and dust surrounding them?
4. If the Hubble Space Telescope searches among all the nearest stars and finds no planetary systems, the solar nebula theory will be put in doubt. Why?
5. According to the solar nebula theory, why is the sun's equator nearly in the same plane as that of the earth's orbit?
6. If you visited another planetary system circling a G star like our sun, would you expect the outer planets to be Jovian or might this be peculiar to our solar system?
7. Why is almost every solid surface in the solar system scarred by craters?
8. If you visited the planetary system described in Question 6 and determined its age by radioactive dating, would you expect to get 4.6 billion years or some other number? Why?
9. What is the difference between condensation and accretion? Between homogeneous accretion and heterogeneous accretion?
10. Why are the terrestrial planets so much less massive than the Jovian planets?
11. How does the infrared radiation from β Pictoris suggest that that star may be in the process of forming planets?

PROBLEMS

1. If you observed the solar system from the nearest star (1.3 pc), what would the maximum angular separation be between Jupiter and the sun? (HINT: See Box 3–1.)
2. The brightest planet in our sky is Venus, which is sometimes as bright as apparent magnitude -4 when it is at a distance of about 1 AU. How many times fainter would it look from a distance of 1 parsec (206,265 AU)? What would its apparent magnitude be? (HINTS: Remember the inverse square law; see Box 2–1.)
3. If a planet existed beyond Pluto, where would it lie according to the Titius–Bode rule? (Searches for such a planet have found nothing.)
4. What is the smallest diameter crater you can identify in Figure 19–5a? (HINT: See Appendix C to find the diameter of Mercury in kilometers.)
5. A sample of a meteorite has been analyzed and the result shows that out of every 1000 nuclei of ^{40}K originally in the meteorite, only 100 have not decayed. How old is the meteorite? (HINT: See Figure 19–14.)
6. In Table 19–3 which object's observed density differs least from its uncompressed density? Why?
7. What composition might we expect for a planet that formed in a region of the solar nebula where the temperature was about 100 K?
8. Suppose that the earth grew to its present size in 1 million years through the accretion of planetesimals averaging 100 g each. On the average, how many planetesimals did the earth capture per second? (HINT: See Appendix C to find the earth's mass.)
9. If you stood on the earth during its formation as described in Problem 8 and watched a region covering 100 m^2, how many impacts would you expect to see in an hour? (HINTS: Assume that the earth had its present radius. The surface area of a sphere is $4\pi R^2$.)
10. The velocity of the solar wind is roughly 400 km/sec. How long does it take to travel from the sun to Pluto?

RECOMMENDED READING

BEATTY, J. KELLY, BRIAN O'LEARY, and ANDREW CHAIKIN *The New Solar System.* New York: Cambridge University Press, 1981.

CAMERON, A. G. W. "The Origin and Evolution of the Solar System." *Scientific American 233* (Sept. 1975), p. 32.

CHAIKIN, ANDREW "Pieces of the Sky." *Sky and Telescope 63* (April 1982), p. 344.

COUPER, HEATHER "In Search of Solar Systems." *New Scientist 106* (13 Nov. 1986), p. 34.

FALK, S. W., and D. N. SCHRAMM "Did the Solar System Start with a Bang?" *Sky and Telescope 58* (July 1979), p. 18.

GILLETT, STEPHEN "The Rise and Fall of the Early Reducing Atmosphere." *Astronomy 13* (July 1985), p. 66.

HARTMANN, W. K. "In the Beginning." *Astronomy 4* (June 1976). p. 6.

———. "Cratering in the Solar System." *Scientific American 236* (Jan. 1977), p. 84.

———. *Moons and Planets,* 2nd ed. Belmont, Calif.: Wadsworth, 1983.

HOYLE, FRED *The Cosmogony of the Solar System.* Short Hills, N.J.: Enslow, 1979.

JAKI, S. L. "The Titius–Bode Law: A Strange Bicentenary." *Sky and Telescope 43* (May 1972), p. 280.

MARAN, STEPHEN P. "Where Do Comets Come From?" *Natural History 91* (May 1982), p. 80.

MARVIN, URSULA B. "Search for Antarctic Meteorites." *Sky and Telescope 62* (Nov. 1981). p. 423.

McBRIDE, KEN "Looking for Extra Solar Planets." *Astronomy 12* (Oct. 1984). p. 6.

MURRAY, BRUCE, MICHAEL C. MALIN, and RONALD GREELEY *Earthlike Planets.* San Francisco: W. H. Freeman, 1981.

PONNAMPERUMA, CYRIL, ed. *Comparative Planetology.* New York: Academic Press, 1978.

REEVES, H. "The Origin of the Solar System." *Mercury 6* (March/April 1977), p. 80.

RINGWOOD, A. E. *Origin of the Earth and Moon.* New York: Springer-Verlag, 1979.

WETHERILL, G. W. "The Allende Meteorite." *Natural History 87* (Nov. 1978), p. 102.

———. "Apollo Objects." *Scientific American 240* (March 1979), p. 54.

———. "The Formation of the Earth from Planetesimals." *Scientific American 244* (June 1981), p. 163.

———. "Dating Very Old Objects." *Natural History 91* (Sept. 1982), p. 14.

WOOD, J. A. *The Solar System.* Englewood Cliffs, N.J.: Prentice-Hall, 1979.

CHAPTER 20

THE EARTH AND MOON

We don't usually think of both the earth and moon as planets, but that is what they are. The earth is one of the nine planets that circle the sun in our solar system, and the moon, although technically a satellite of the earth, is big enough to have passed through the developmental stages of a planet. Thus, we can think of it as a planet.

Our goal is to examine the history the earth and moon using the principles of **comparative planetology**, the study of planets by the comparison of one with another. Before spacecraft visited the moon and other planets, it was almost impossible to compare the planets in any detail. But now that we have photos and measurements sent back to the earth by distant planetary probes and now that our explorers have visited the moon, we can begin to see the similarities and differences between our planet and other planets.

We begin our study of comparative planetology with the familiar Earth. One reason is that the earth is the planet we know best. We can study it carefully and then draw comparisons to other worlds. But another reason is that our home is a planet of extremes. Its interior is molten and generates a magnetic field. Its crust is active, with moving sections that push against each other and trigger earthquakes, volcanoes, and mountain building. Even the earth's atmosphere is extreme. Processes have altered the earth's air from its original composition to a highly unusual, oxygen-rich sea of gas. Once we understand the earth's complex properties, the remaining planets in our system should be easier to comprehend.

In our study of the earth, we will find a four-stage history of planetary development. The moon and all the terrestrial planets have passed through these stages, although differences in the way the planets were altered by these stages have produced dramatically different worlds. The moon, for

That's one small step for man . . . one giant leap for mankind.

Neil Armstrong, on the moon

Beautiful, beautiful. Magnificent desolation.

Edwin E. Aldrin, on the moon

example, is much like the earth, but its evolution has been dramatically altered by its smaller size. Thus, as we explore the solar system, we will discover not entirely new processes, but rather familiar effects working in slightly different ways.

20.1 PLANET EARTH

Like all the terrestrial planets, the earth formed from the inner solar nebula about 4.6 billion years ago. As it took form, it began to change, passing through four developmental stages (Figure 20–1). Processes that occurred during these four stages determined the present condition of the earth's interior, crust, and atmosphere.

FOUR STAGES OF PLANETARY DEVELOPMENT

If modern astronomers are correct and the planets were formed by heterogeneous accretion (Chapter 19), then a core of metal and metal oxides may have accumulated before the solar nebula cooled enough to allow silicates to add a mantle around the core. Thus, the earth may have begun its existence with an iron core, although a later process may have added to that core and built a low-density crust. If the earth formed by homogeneous accretion, then all of the particles that went into its formation had about the same composition, and the dense core and low-density crust developed later as the materials separated.

In either case, differentiation became important as the interior of the planet grew hot from the heat released by in-falling material and from the decay of radioactive elements. The hot material was at least partially fluid, and the denser materials such as iron could sink toward the center while the lower-density materials such as silicates (rocky minerals) could float to the surface to form a thin, brittle crust.

The second stage, cratering, began as the crust formed. Because the solar nebula was filled with rocky debris, the young Earth was heavily battered by meteorites that pulverized the newly formed crust and created a moonlike landscape. The last of the large meteorites blasted out crater basins hundreds of kilometers in diameter. As the solar nebula cleared, the amount of debris decreased, and the level of cratering fell rapidly to its present slow rate.

The third stage, flooding of the basins, began as the decay of radioactive elements heated the earth's interior.

FIGURE 20–1 The four stages of planetary development. First: Differentiation into core and crust. Second: Cratering. Third: Flooding of the lowlands by lava or water or both. Fourth: Slow surface evolution.

Lava welled up through fissures in the crust and flooded the deeper basins. Later, as the atmosphere cooled, water condensed and fell as rain, filling the basins and forming the first oceans.

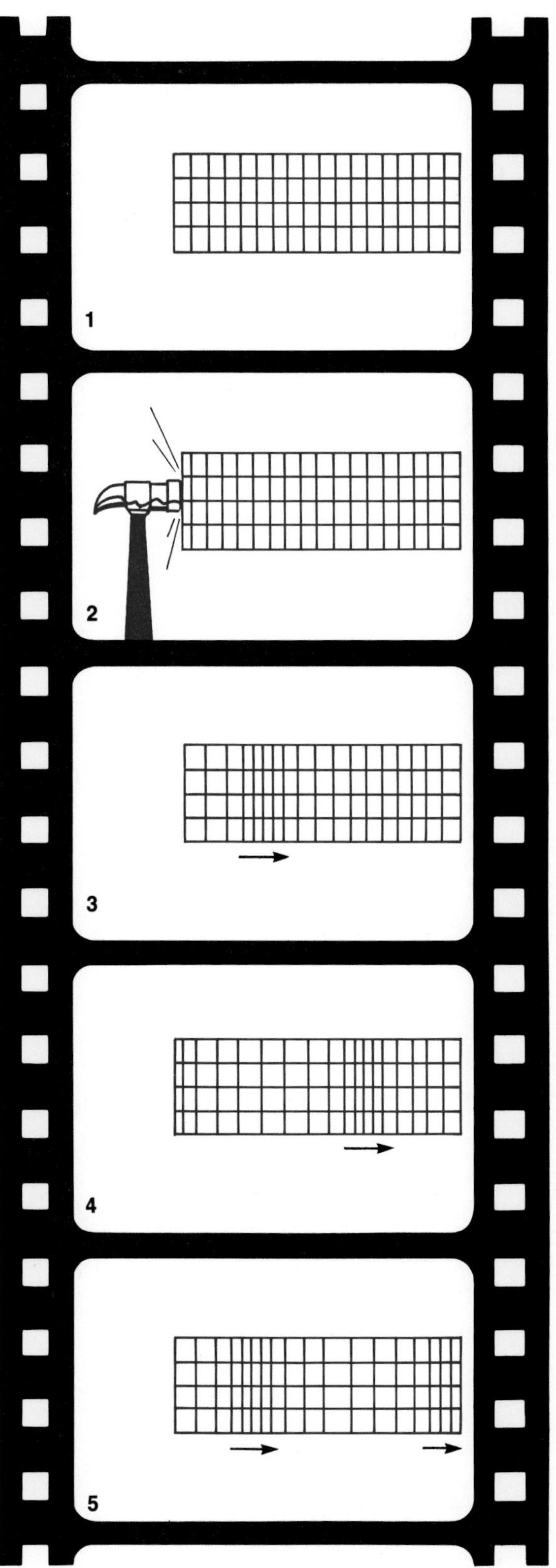

The fourth stage, slow surface evolution, has continued for the last 3.5 billion years or more. The earth's surface is in constant motion as sections of crust slide over each other, build mountains, and shift continents. In addition, moving water and air erode the surface and wear away geological features.

All terrestrial planets pass through these four stages, but some planets, because of their mass or temperature, have emphasized some stages over others. Our goal in this section is to study the earth's interior, crust, and atmosphere to establish a base for comparison with other planets. Only by understanding the earth in detail can we understand planets in general.

THE EARTH'S INTERIOR

The theory of the origin of planets from the solar nebula predicts that the earth should have melted and differentiated into a dense metallic core with a low-density silicate crust. But did it differentiate? Clearly the earth's surface is made of silicates, but what of the interior?

High temperature and tremendous pressure in the earth's interior make any direct exploration impossible. Even the deepest oil wells extend only a few kilometers down and don't reach through the crust. It is quite impossible to drill far enough to sample the earth's core. Yet earth scientists have studied the interior and found clear proof that the earth did differentiate.

This analysis of the earth's interior is possible because earthquakes produce two kinds of seismic waves that travel through the interior and eventually register on seismographs all over the world. The ***P*** or **pressure waves** are much like sound waves in that they travel as a region of compression. As a *P* wave passes, particles of matter vibrate back and forth parallel to the direction of wave travel (Figure 20–2). In contrast, the ***S*** or **shear waves** move as displacements of particles perpendicular to the waves' direction of travel (Figure 20–3). Thus *S* waves distort the material but do not compress it. Normal sound waves are pressure waves, whereas the vibrations in a bowl of jelly are shear waves. Because *P* waves are compression waves, they can move through a liquid, but *S* waves cannot. A glass of water can't shimmy like jelly because a liquid does not have the rigidity required to transmit *S* waves.

FIGURE 20–2 *P* or pressure waves, like sound waves in air, travel as regions of compression.

The *P* and *S* waves caused by an earthquake do not travel in straight lines or at constant speed within the earth. The waves may reflect off boundaries between layers of different density, or they may be refracted as they pass through such a boundary. In addition, the gradual increase in density toward the earth's center forces the paths of waves to curve (Figure 20–4). This complex behavior enables scientists to construct a picture of the earth's interior by measuring the arrival times of the different types of waves from distant earthquakes.

Such studies show that the interior consists of three parts, a central core, a thick mantle, and a thin crust. A critical fact about the core is that *S* waves cannot penetrate it. When an earthquake occurs, no direct *S* waves register on seismographs on the opposite side of the earth, as if the core were casting a shadow (Figure 20–4). This shows that the core is liquid, and the size of the shadow fixes the size of the core at about 55 percent of the earth's radius. Theoretical calculations predict that the core is hot (about 4000 K), dense (about 14 g/cm^3), and composed of iron and nickel.

Although the core is hot, it is not entirely molten. Nearer the center the material is under higher pressure, which in turn raises the melting point so high the material cannot melt at the existing temperature. Thus, there is an inner core of solid iron and nickel. Estimates suggest the inner core's radius is about 22 percent that of the earth.

The **mantle** is the layer of dense rock and metal oxides that lies between the molten core and the crust. The paths of seismic waves in the mantle show that it is not molten, but it is not precisely solid either. Mantle material behaves as a **plastic**, a material with the properties of a solid but capable of flowing under pressure. The asphalt used in paving roads is a common example of a plastic. It shatters if struck with a sledgehammer, but it bends under the weight of a heavy truck. Just below the earth's crust, where the pressure is less than at greater depths, the mantle is most plastic.

The earth's rocky crust is quite thin, about 70 km (44 miles) under the continents and only 10 km (6 miles) under the oceans. Owing to its lower density (2.5–3.5 g/cm^3), this material floats on the denser mantle (3.5–5.8 g/cm^3). Unlike the mantle, the crust is brittle and breaks much more easily than the plastic mantle. We will see that

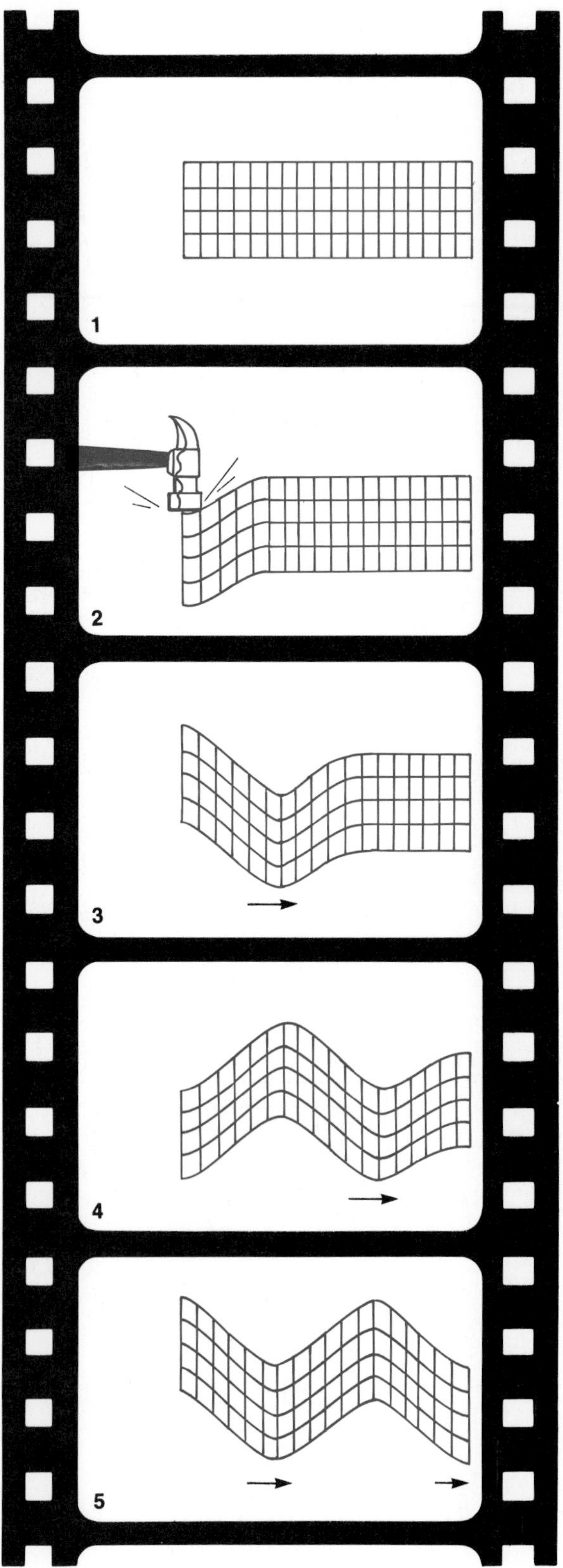

FIGURE 20–3 *S* or shear waves, like vibrations in a bowl of jelly, travel as displacements perpendicular to the direction of travel.

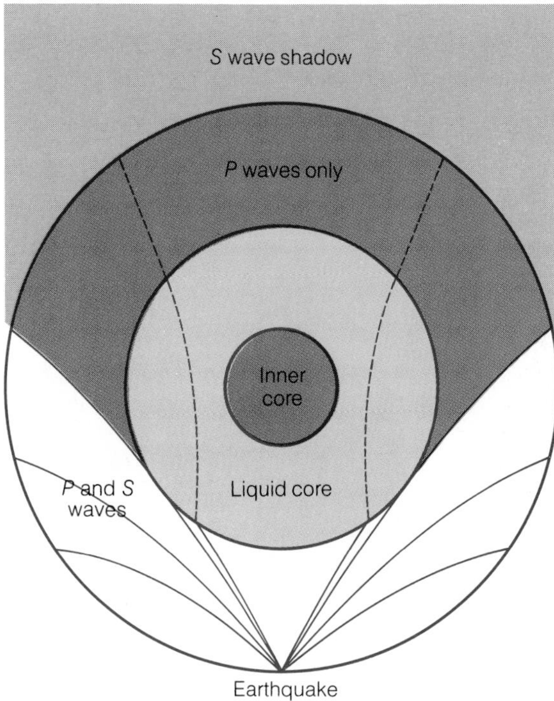

FIGURE 20–4 *P* and *S* waves give us clues to the earth's interior. That no direct *S* waves reach the far side shows that the earth's core is liquid. The size of this *S* wave shadow tells us the size of the core.

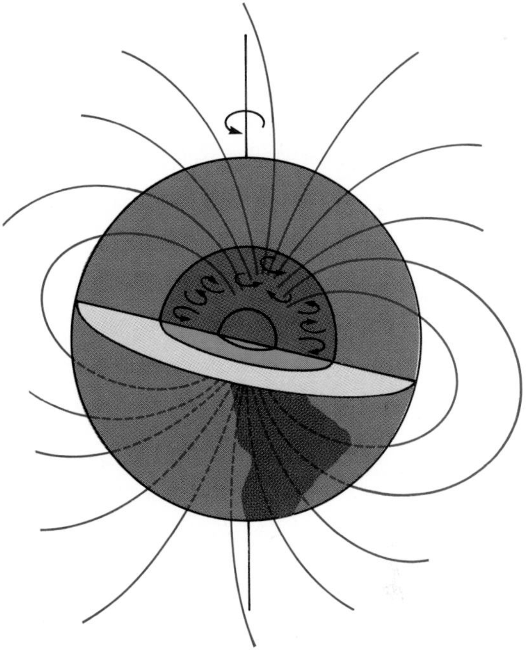

FIGURE 20–5 The dynamo effect couples convection in the liquid core with the earth's rotation and produces electric currents that are believed responsible for the earth's magnetic field.

this is a characteristic property of many planetary crusts.

Apparently the earth's magnetic field is a direct result of its rapid rotation and its molten core. The origin of planetary magnetic fields is not yet well understood, but the best current theory is the **dynamo effect.** It supposes that the liquid core is stirred by convection. The rotation of the earth couples this motion into a circulation that generates electric currents throughout the core. Because the highly dense, molten, iron–nickel alloy is a better electrical conductor than copper, the material commonly used for electrical wiring, the currents can flow freely and generate a magnetic field (Figure 20–5).

The earth's magnetic field protects it from the solar wind. Blowing outward from the sun at about 400 km/sec, the solar wind consists of ionized gases carrying a small part of the sun's magnetic field. When the solar wind encounters the earth's magnetic field, it is deflected like water flowing around a boulder in a stream. The surface where the solar wind is first deflected is called the **bow shock**, and the cavity dominated by the earth's magnetic field is called the **magnetosphere** (Data File 2b). High-energy particles from the solar wind leak into the magnetosphere and become trapped within the earth's magnetic field to produce the **Van Allen belts** of radiation. We will see in later chapters that all planets that have magnetic fields have bow shocks, magnetospheres, and radiation belts.

Although we are confident that the earth's magnetic field is generated within its molten core, many mysteries remain. For example, rocks retain traces of the magnetic field in which they solidify, and some contain fields that point backward. That is, they imply that the earth's magnetic field was reversed at the time they solidified. Careful analysis of such rocks indicates that the earth's field has reversed itself every million years or so, with the north magnetic pole becoming the south magnetic pole and vice versa. These reversals are poorly understood, but may be related to changes in the core convection.

Convection in the earth's core is important because it generates the magnetic field. As we will see in the next section, convection in the mantle constantly remakes the earth's surface.

DATA FILE 2

Earth

Average distance from the sun	1.00 AU (1.495979 × 10^8 km)
Eccentricity of orbit	0.0167
Maximum distance from the sun	1.0167 AU (1.5210 × 10^8 km)
Minimum distance from the sun	0.9833 AU (1.4710 × 10^8 km)
Inclination of orbit to ecliptic	0°
Average orbital velocity	29.79 km/sec
Orbital period	1.00 y (365.26 days)
Period of rotation (with respect to the sun)	$24^h00^m00^s$
Inclination of equator to orbit	23°27′
Equatorial diameter	12,756 km
Mass	5.976 × 10^{24} kg
Average density	5.497 g/cm^3
Surface gravity	1.0 earth gravities
Escape velocity	11.2 km/sec
Surface temperature	−50° to 50°C (−60° to 120°F)
Average albedo	0.39
Oblateness	0.0034

a. A typical view of earth. Liquid water oceans cover 75 percent of the surface.

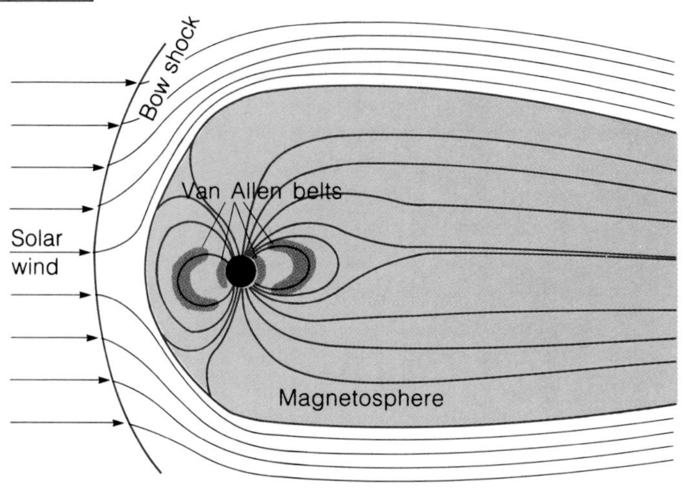

b. Earth's magnetic field dominates space around it, deflecting the solar wind and forming radiation belts.

c. The earth photographed from lunar orbit. Weather patterns and oceans dominate the planet. (NASA.)

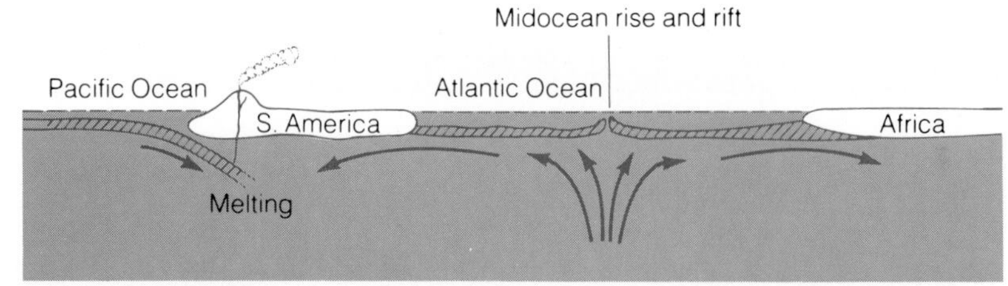

FIGURE 20–6 Currents of moving mantle material rise at midocean and spread the seafloor, moving the continents apart.

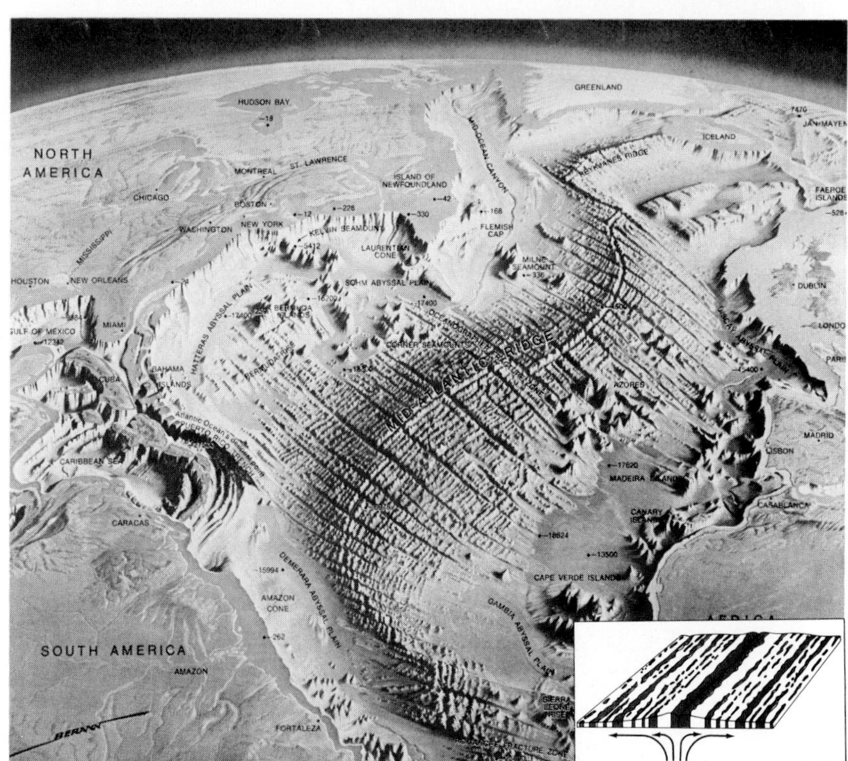

FIGURE 20–7 If we could drain the Atlantic Ocean, we would see the midocean rift snaking along the crest of the midocean rise from Iceland into the South Atlantic. The ocean floor spreads apart along the rift, pushing the continents apart. As magma rises into the rift and solidifies, it records the alternation of Earth's magnetic field (inset). These alternating bands of normal (dark) and reversed (white) magnetism in the ocean floor are clear evidence of seafloor spreading. (Courtesy Alcoa.)

PLATE TECTONICS

The earth's surface is active. It is continually destroyed and renewed as large sections of crust move about. Geologists refer to this phenomenon as **plate tectonics.** *Plate* refers to the sections of moving crust, and *tectonics* comes from the Greek word for "builder." Plate tectonics is the builder of the earth's surface, because interactions between plates destroy old terrain, push up mountains, and create new crust.

The energy that moves the plates comes from the hot interior. Convection currents of hot mantle material rise from deep layers and spread out under the crust, while other cooler regions sink. This circulation in the mantle apparently drags along large crustal sections at speeds of a few centimeters per year (Figure 20–6).

The key to understanding plate tectonics lies hidden in the ocean floor. The crust there is thin and composed primarily of dense **basalt**, rock characteristic of solidified lava. These seafloors are not at all flat. Running down the center of the ocean beds are undersea mountain ranges called **midocean rises**, and splitting these rises are chasms called **midocean rifts** (Figure 20–6). The rocks of the midocean rises are young, but that only hints at the expla-

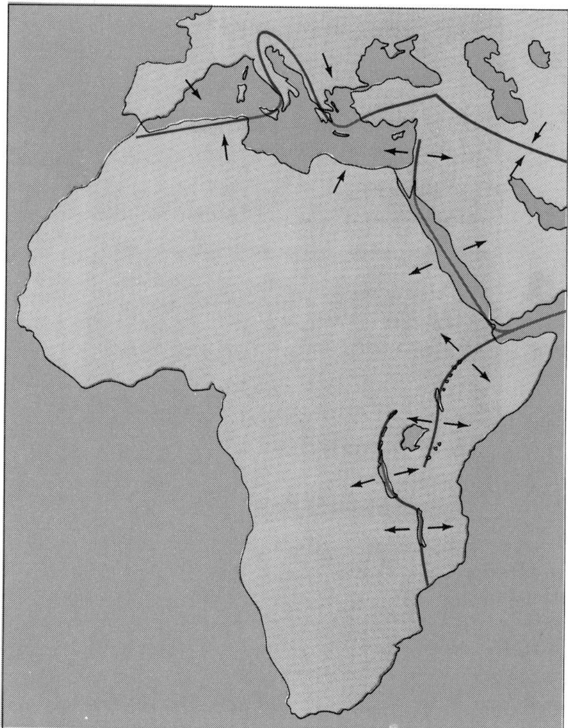

FIGURE 20–8 The Red Sea and valleys to the south are rift valleys produced by the separation of crustal plates. The compression of the crust in the area of the Mediterranean accounts for its mountain ranges and volcanism.

nation. The real evidence appeared when scientists measured the residual magnetism of the seafloor and found that some sections retained fields that were the reverse of the earth's present field. Evidently these regions solidified from molten rock during the earth's periodic magnetic reversals. That these regions show alternating, symmetric bands of magnetism running parallel to the midocean rifts indicates that the earth is creating new crust along the rifts while the seafloors spread outward.

The Atlantic Ocean serves as a good example (Figure 20–7). Molten material wells up along the midocean rift and solidifies to form young basalt. As the crustal plates spread apart at about 2–4 cm/y, the midocean rift opens and new material rises to form more crust. Thus, midocean rises are composed of young rock. Parallel magnetic bands form because the earth's field reverses every million years or so, and the solidifying rock records these changes in alternating strips (Figure 20–7) just as the moving tape in a tape recorder makes a record of the changing magnetic field in the recording head.

If the earth is adding crust in one region, it must be destroying it in another. Crust vanishes in trenches along the coasts of some continents where the spreading seafloor slides downward (Figures 20–6 and 20–7). The collision of moving plates often crumples the crust and pushes up mountains. In addition, the descending seafloor melts along with its accumulated sediment and releases low-density molten rock that rises to the surface and produces volcanism. The Andes Mountains along the west coast of South America and the associated volcanism are due to the descent of the Pacific Ocean floor beneath the continent. The same process produces the volcanoes (including Mt. St. Helens) and earthquakes of Washington state, western Canada, Alaska, and Japan.

The seafloor does not always slip below a continent. The floor of the Atlantic Ocean is locked to North and South America and is pushing the continents westward. Tracing this continental drift backward in time, earth scientists find that North and South America were in contact with Europe and Africa only 200 million years ago. Even a quick glance at a map shows how well these continents fit together (Figure 20–7).

This motion continues today. The plates drift at the urging of the currents in the mantle. They split apart to form new seas, come together to crumple the crust into mountains, and slip against each other to generate earthquakes. The first sign of a continent's splitting is a long, straight, deep depression called a **rift valley.** Africa has recently split from Arabia, opening a rift valley now filled by the Red Sea. Other rift valleys extend southward from the Red Sea across eastern Africa (Figure 20–8). In contrast, the collision of Africa with the east coast of North America roughly 250 million years ago folded the crust into the Appalachian Mountains, just as the collision of India with southern Asia is now building the Himalaya Mountains. In addition, sections of crust in direct contact may slip past each other, as in the case of the Pacific plate carrying part of southern California northward along the San Andreas Fault and generating frequent earthquakes.

Plotting the location of earthquakes and volcanism on a world map reveals the edges of the moving plates (Figure 20–9). The Pacific plate, bounded by a ring of earthquake zones, descends under the Eurasian plate along the North Pacific and Japanese islands. This active zone of circum-Pacific earthquakes and volcanoes is often called the "ring of fire." Earthquakes and volcanoes also occur along the midocean rises where new crust forms and the seafloor spreads.

Because the earth's crust is active, all geological features gradually change. The oldest existing portions of the crust, the Canadian shield and portions of South Africa

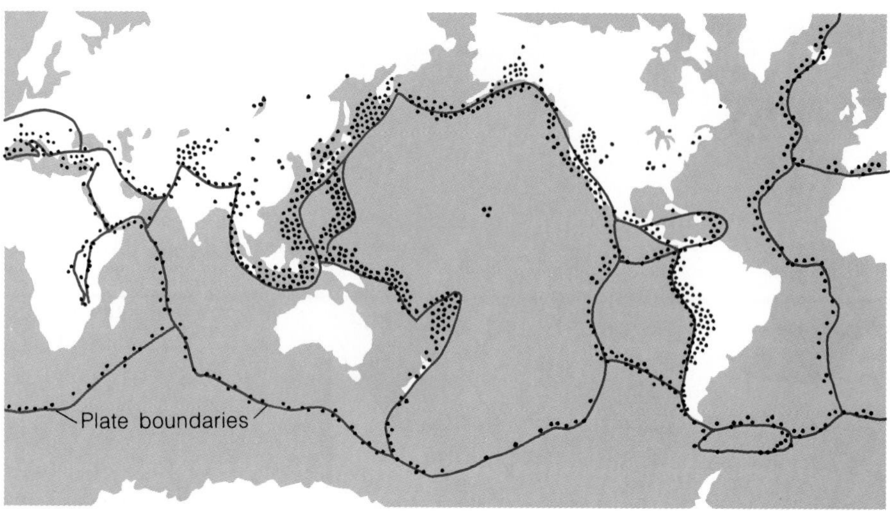

FIGURE 20–9 Earthquakes (shown as dots here) tend to outline the crustal plates. Volcanism follows a similar pattern, and much of the Pacific is ringed by volcanic peaks that extend from Chile through Central America, Alaska, and south through Japan. This pattern of volcanism is termed the "ring of fire." The volcanism and associated earthquakes in Hawaii are produced where a hot spot in the mantle penetrates upward through a plate. A few earthquakes, such as those near St. Louis and the American east coast, are caused by secondary faults produced by the tension and compression of plate motions.

and Australia, are only about 3.9 billion years old. The constant churning of the earth's surface has wiped away all record of older crust.

THE ATMOSPHERE

We cannot complete our four-stage history of the earth without mentioning the atmosphere. Not only is it necessary for life, but also it is intimately related to the crust. It affects the surface by eroding geological features through wind and water, and in turn the chemistry of the earth's surface affects the composition of the atmosphere.

Until recent years, earth scientists believed that the earth's atmosphere had developed slowly by outgassing, the release of gases from a planet's interior. According to this picture, the earth formed by homogeneous accretion (Chapter 19) and slowly grew into a large, cold ball of matter whose composition did not vary from surface to center. This early earth was supposed to have attracted gases such as hydrogen, helium, methane, and ammonia from the solar nebula to form a **primeval atmosphere**. The slow decay of uranium, thorium, and potassium heated the earth's interior, melted it, caused it to differentiate into core, mantle, and crust, and drove out gases that diluted the primeval atmosphere. These products of outgassing became a **secondary atmosphere** rich in CO_2 and H_2O.

A better understanding of how the solar system formed suggests that the earth formed by heterogeneous accretion. The first materials to accumulate would have been metals and silicates that were mostly free of volatiles. Even so, in the heterogeneous accretion theory, the earth formed hot; any volatiles present would have been driven out of the rock immediately, and the atmosphere would have grown from the very beginning.

More importantly, the heterogeneous accretion theory holds that the cooling solar nebula condensed volatile-rich materials late in the development of the planets. Thus, the earth may have later been bombarded by lower-density, volatile-rich planetesimals that built the atmosphere and oceans. It is fascinating to speculate that these icy planetesimals may have formed in the outer solar system and have been scattered into the inner solar system by the growing Jovian planets. Such icy planetesimals moving through the inner solar system are now known as comets, so we might wonder if the earth's oceans and atmosphere were brought to the earth by a bombardment

Table 20–1 The earth's atmosphere.

Gas	Percent by Weight
N_2	75.5
O_2	23.1
Ar	1.29
CO_2	0.05
Ne	0.0013
He	0.00007
CH_4	0.0001
Kr	0.0003
H_2O (vapor)	1.7–0.06

of comets. If this speculation is true, the night sky must have been illuminated with comets falling through the inner solar system.

The earth may have existed for half a billion years before the condensation of the volatile-rich planetesimals, but once the earth had an atmosphere, the mix of gases began to change. The young atmosphere must have been rich in water vapor, carbon dioxide, and other volatiles. As the atmosphere cooled, the water condensed to form the first oceans, and, as the oceans grew bigger, carbon dioxide began to dissolve in the water. Carbon dioxide is soluble in water—which is why carbonated beverages are so easy to manufacture—but the first oceans could not have absorbed all of the atmospheric carbon dioxide if the gas had not reacted with dissolved substances in the seawater to form silicon dioxide, limestone, and other mineral sediments. Thus, chemical reactions in the oceans transferred the carbon dioxide from the atmosphere to the seafloor.

Any methane or ammonia in the atmosphere would have been broken up by ultraviolet radiation from the sun. The earth's lower atmosphere is now protected from ultraviolet radiation by a layer of ozone (O_3) 15–30 km above the surface. However, the atmosphere of the young earth did not contain free oxygen, so an ozone layer could not form, and the sun's ultraviolet radiation penetrated deep into the atmosphere. The energetic ultraviolet photons broke up water, methane, and ammonia molecules, and the hydrogen escaped to space. Carbon and oxygen formed carbon dioxide and dissolved in the oceans, leaving nitrogen to accumulate in the atmosphere (Table 20–1).

The origin of the atmospheric oxygen is linked to the origin of life, the subject of Chapter 25, but it is sufficient here to note that life must have originated within a billion years of the earth's formation. This life did not significantly alter the atmosphere, however, until nature invented photosynthesis about 3.3 billion years ago. Photosynthesis absorbs carbon dioxide from the air or water and utilizes it for plant growth, releasing oxygen. Because oxygen is a very reactive element, it combines easily with other compounds, and thus the oxygen abundance grew slowly at first. Apparently the development of large, shallow seas along the continental margins half a billion years ago allowed ocean plants to manufacture oxygen faster than chemical reactions could consume it. Atmospheric oxygen then increased rapidly, and it is still increasing at the rate of about 1 percent every 36 million years.

In this section we have studied the processes that are responsible for the four-step history of the earth. The differentiation of the crust and interior, plate tectonics, and the development of the atmosphere have shaped the surface of our planet and made it a comfortable home for life. But what of other planets? In the next section we will explore the surface of another world—the moon.

20.2 THE MOON

On other planets the balance between processes differs from that on the earth. For instance, plate motion on the earth moves large sections of crust. Some other planets may have begun plate formation but never reached the stage of plate motion; others, like the moon, have solid crusts that appear never to have broken into plates.

The moon is an especially valuable world to compare with the earth. The moon is the only celestial body that humans have visited, and it is the only body beyond the earth from which scientists have samples of known origin that can be analyzed and dated in terrestrial laboratories. These samples and other data will show us how the moon passed through the four stages of planetary evolution. Also, as we will see, the moon's surface has not been as severely altered as the earth's, so we can hope to learn about the early history of the solar system by studying the moon's ancient crust.

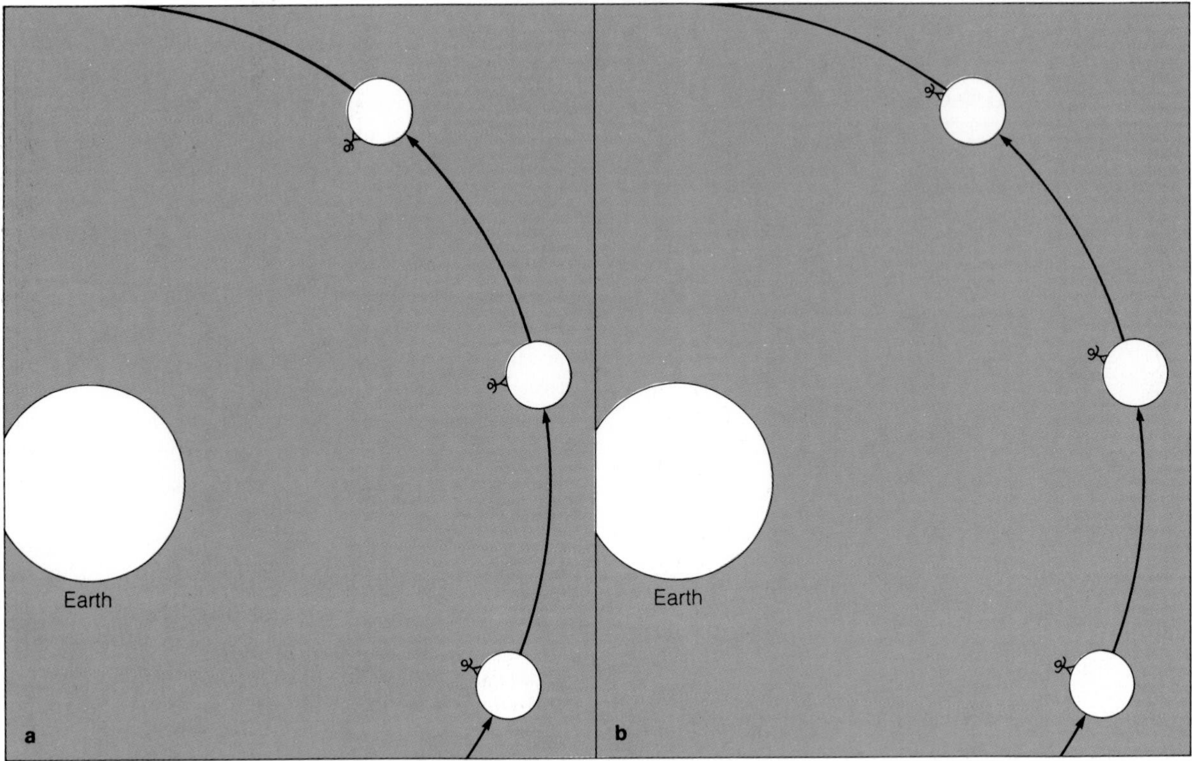

FIGURE 20–10 Tidal forces have slightly distorted the moon from a spherical shape, and its rotation is now tidally locked to the earth. It rotates on its axis and revolves around its orbit with the same period keeping the same side toward the earth (a). An astronaut on the near side of the moon would always be able to see the earth overhead. If the moon did not rotate on its axis (b) but maintained a fixed orientation in space as it moved along its orbit, it would not keep the same side toward earth.

THE VIEW FROM EARTH

From the earth we always see the same side of the moon. It probably rotated faster long ago, but the earth's tidal forces have slowed its rotation so that it now rotates on its axis with the same period that it revolves around its orbit—27.322 days. An astronaut standing on the near side of the moon would find the earth perpetually overhead (Figure 20–10). Thus, the near side of the moon is always visible from the earth.

Details on the moon's face are easily visible in a small telescope because the moon has no atmosphere. It is only 1738 km in radius, about 27 percent the radius of the earth. The escape velocity from its surface is only about 21 percent of the escape velocity from the surface of the earth, so it has been unable to hold any gases. Near the **terminator**, the dividing line between daylight and darkness, the shadows are stark and black because there is no air to scatter the light. Also, stars disappear behind the **limb** of the moon, the edge of its disk, without dimming. Clearly, the moon is an airless (and soundless) world.

Yet the face of that world is marked by regions, each called a **mare**, Latin for "sea" (Data File 3a). The dark maria (plural of *mare*) that make up the face of the man in the moon were so named because early observers thought the great smooth expanses were seas. Even a small telescope will show that the maria are marked by ridges, faults, smudges, and craters and therefore can't be water. The maria are actually great lowland plains covering 17 percent of the moon's surface.

Not many decades ago astronomers argued about the nature of the maria. Some said that they were great solidified lava flows, but others argued that they were vast oceans of dust blasted from the lunar mountains by billions of years of meteorites and drawn to the lowlands by the lunar gravity. Science fiction writers wrote fasci-

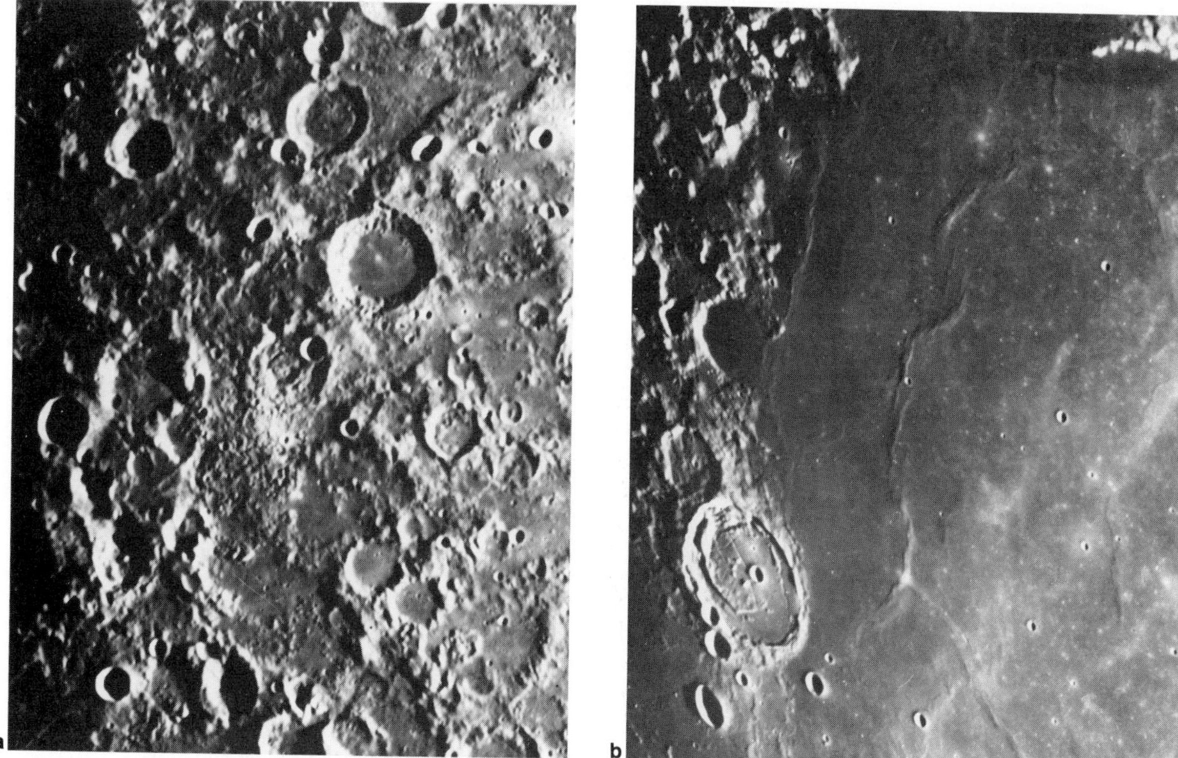

FIGURE 20–11 (a) The lunar highlands are the oldest parts of the moon and are heavily battered by craters. (b) The maria are younger and contain few craters. They were formed by lava flows that occurred after the end of most cratering. (Lick Observatory photographs.)

nating stories about the dust oceans. For example, *A Fall of Moon Dust* by Arthur C. Clarke tells of tourists in a lunar sightseeing boat trapped when their boat sinks in the dust to the bottom of a mare. Although it is a wonderful rescue story, it can never happen. We know now that the maria are great lava flows with only a few centimeters of loose dust on their surfaces.

Besides the maria, the most striking feature of the moon is the millions of impact craters. The lighter parts of the moon's face are heavily cratered highlands. Although they look mountainous at first glance, there are no folded mountains on the moon. Its crust has never wrinkled into folded mountain ranges like the earth's crust. The mountainous regions of the highlands are actually millions of meteorite craters, one on top of the other (Figure 20–11a).

The crater density in different regions of the moon gives us a way of estimating relative ages of different regions of the lunar surface. The highlands must be old to have accumulated so many craters, but the maria can't be that old. They have only a few craters (Figure 20–11b). These are called **relative ages** because we cannot attach a true age to any feature without a sample for laboratory analysis. However, crater counts can tell us that one part of the moon's crust is older or younger than another part.

Most of the craters on the moon were formed long ago, perhaps during the last stages of the formation of the moon as the last debris fell into the surface. But some craters are quite young. Such craters are usually very bright and marked by bright **rays** extending across the surface. The crater Copernicus is the center of a vast ray system, and rays from Tycho extend halfway around the moon (Data File 3a). These rays were apparently formed by falling **ejecta**, material ejected by the impact that formed the crater. Older craters had rays, but exposure to sunlight and micrometeorites gradually obliterated the rays. Thus, the presence of bright rayed craters shows that some

DATA FILE 3
The Moon

Average distance from the earth	384,400 km (center to center)
Eccentricity of orbit	0.055
Maximum distance from the earth	405,500 km
Minimum distance from the earth	363,300 km
Inclination of orbit to ecliptic	5°9′
Average orbital velocity	1.022 km/sec
Orbital period (sidereal)	27.321661 days
Orbital period (synodic)	29.5305882 days
Inclination of equator to orbit	6°41′
Equatorial diameter	3476 km
Mass	7.35×10^{22} kg (0.0123 $M_\oplus$)
Average density	3.36 g/cm^3 (3.35 g/cm^3 uncompressed)
Surface gravity	0.167 earth gravities
Escape velocity	2.38 km/sec (0.21 $V_\oplus$)
Surface temperature	−170° to 130°C (−274° to 266°F)
Average albedo	0.07

a. Photo composite of the moon (inverted as in a telescope) shows major maria and craters. (Lick Observatory photograph.)

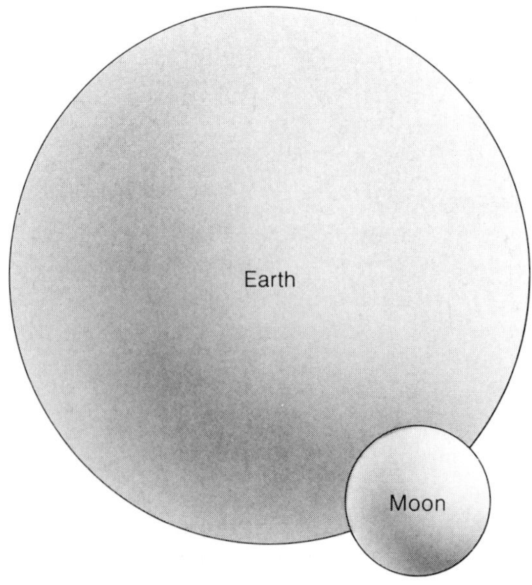

b. The moon is roughly 27 percent the diameter of the earth.

▲ **c.** A typical lunar scene. The airless sky is black, and the surface has been pulverized by meteorite impacts. (NASA.)

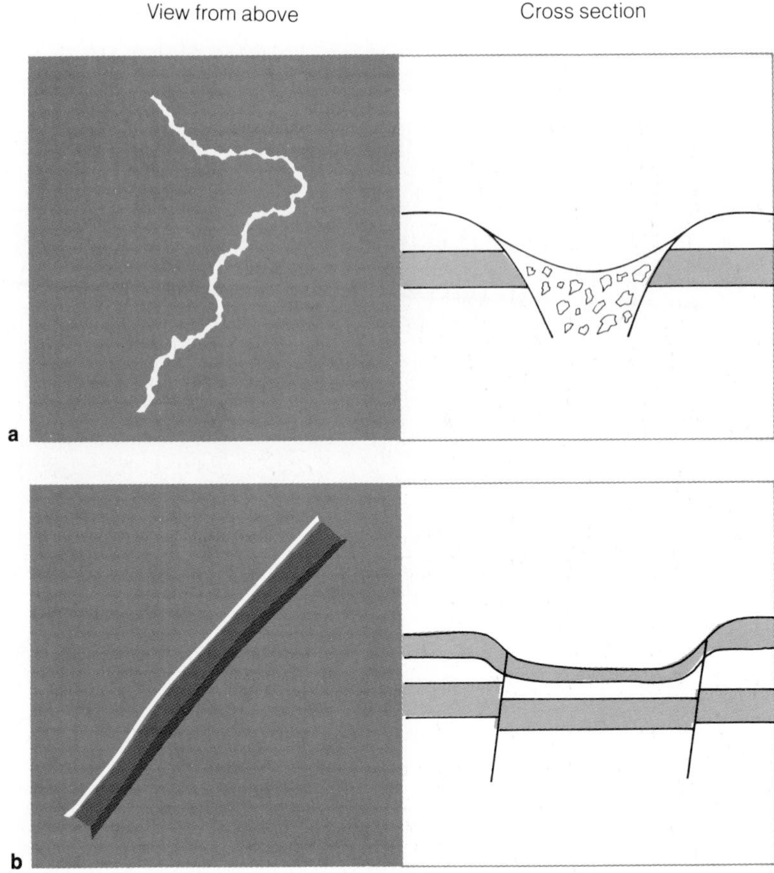

FIGURE 20–12 (a) A sinuous rille is the winding, rubble-filled channel cut by flowing lava. (b) A graben rille is straight and caused by faults in the lunar crust.

cratering is still occurring on the moon, but most of the craters formed long ago.

A few decades ago, some astronomers argued that the craters were volcanic, but that has not proven to be true. There are traces of volcanism on the moon, however. Small domes appear to be volcanic vents, and narrow, winding valleys called **sinuous rilles** appear to be channels cut by flowing lava (Figure 20–12a). The moon is also cut by numerous faults that show that stresses have developed and broken the crust. Some of these are long, narrow valleys called **graben rilles** (Figure 20–12b), where sections of crust have dropped. The presence of these rilles and faults shows that the lunar surface has been volcanically active, but we see nothing to suggest that the lunar surface was ever divided into moving plates as is the earth's. Whatever volcanic activity occurred was not produced by the collision of plates.

The view from the earth is stark and fascinating, but we can learn only so much by looking. The only way to really understand the moon is to go there.

THE APOLLO MISSIONS

On May 25, 1961, President John Kennedy addressed Congress and committed the United States to landing a human being on the moon by 1970. The reasons for that decision are complicated and related more nearly to economics and the stimulation of technology than to science, but the Apollo program became a fantastic adventure in science, a flight to the moon that changed how we think about the earth.

Flying to the moon is not particularly difficult; with powerful enough rockets and enough food, water, and air, it is a simple trip. Landing on the moon is more difficult but not impossible. The moon's gravity is only one-sixth that of the earth, and there is no atmosphere to disturb the trajectory of the spaceship. Getting to the moon is simple and landing is possible; the difficulty is doing both on one trip. The spaceship must carry food, water, and air for a number of days in space plus fuel and rockets for midcourse corrections and for a return to the earth.

FIGURE 20–13 The Apollo astronauts took two spaceships to the moon. While one astronaut stayed in orbit in the command module, the other two rode the lunar lander down to the surface. (NASA.)

All of this adds up to a ship that is too massive to make a safe landing on the lunar surface. The solution was to take two spaceships to the moon, one to ride in and one to land in (Figure 20–13).

The command module was the long-term home and command center for the trip. Three astronauts had to live in it for a week, and it had to carry all the life-support equipment, navigation instruments, computers, power packs, and so on for a week's jaunt in space. The lunar landing module (LM for short) was tacked to the front of the command module like a bicycle strapped to the front of the family camper. It carried only enough fuel and supplies for the short trip to the lunar surface, and it was built to minimize weight and maximize maneuverability.

The weaker gravity of the moon made the design of the LM simpler. Landing on the earth requires reclining couches for the astronauts, but the trip to the lunar surface involved smaller accelerations. In an early version of the LM, the astronauts sat on what looked like bicycle seats, but these were later scrapped to save weight. The astronauts had no seats at all in the LM, and once they began their descent and acquired weight, they stood at the controls held by straps, riding the LM like two daredevils riding a rocket surfboard.

Lifting off of the lunar surface, the LM saved weight by leaving the larger descent rocket and support stage behind. Only the astronauts, their instruments, and the cargo of rocks returned to the command module orbiting above. Again the astronauts in the LM blasted up from the lunar surface standing at the controls. The rocket engine that lifted them back into lunar orbit was not much bigger than a dishwasher.

The most complicated parts of the trip were the rendezvous and docking between the tiny remains of the LM and the command module. That was aided by radar systems and computers, and once completed, the astronauts could rejoin their colleague in the command module, transship their moon rocks, and jettison the LM. Only the command module returned to the earth.

The first manned lunar landing was made on July 20,

Table 20–2 Manned Apollo missions.

Apollo Mission	Astronauts: Commander LM Pilot CM Pilot	Date	Mission Goals	Sample Weight (kg)	Typical Samples	Ages (10^9 y)
7	Schirra Eisele Cunningham	Oct. 1968	First test of command module in orbit			
8	Borman Lovell Anders	Dec. 1968	First manned orbit of moon in command module			
9	McDivitt Scott Schweickart	March 1969	First manned tests of lunar module in earth orbit			
10	Stafford Young Cernan	May 1969	Manned lunar module tests in lunar orbit Descent to 17-km altitude			
11	Armstrong Aldrin Collins	July 1969	First manned landing Mare Tranquillitatis	21.7	Mare basalts	3.48–3.72
12	Conrad Bean Gordon	Nov. 1969	Visit Surveyor 3 Sample Oceanus Procellarum (mare)	34.4	Mare basalts	3.15–3.37
13	Lovell Swigert Haise	April 1970	Destination Fra Mauro Explosion aborts landing			
14	Shepard Mitchell Roosa	Feb. 1971	Fra Mauro, Imbrium ejecta sheet	42.9	Breccias	3.85–3.96
15	Scott Irwin Worden	July 1971	Edge of Mare Imbrium and Apennine Mts. Hadley Rille	76.8	Mare basalts Highland anorthosite	3.28–3.44 4.09
16	Young Duke Mattingley	April 1972	Sample highland crust Cayley formation (ejecta) Descartes	94.7	Highland basalt Breccia	3.84 3.92
17	Cernan Schmitt Evans	Dec. 1972	Sample highland crust Dark halo craters Taurus–Littrow	110.5	Mare basalt Highland breccia Fractured dunite	3.77 3.86 4.48

FIGURE 20–14 (a) Astronaut Aldrin stands on the dusty surface of Mare Tranquillitatis in the lunar lowlands, site of the first lunar landing. The flat lava plain is almost featureless, and the horizon is straight. (b) In the lunar highlands at Taurus–Littrow the surface is mountainous and irregular. Note the horizon. Apollo 17 scientist-astronaut Harrison Schmitt is shown passing huge boulders. (NASA.)

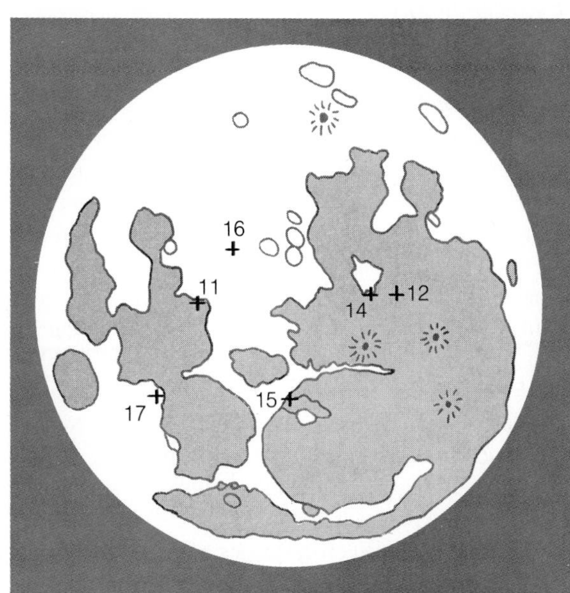

FIGURE 20–15 The landing sites of the six Apollo missions that reached the moon. For the names of lunar features see Data File 3a.

1969. While astronaut Michael Collins waited in orbit, Neil Armstrong and Edwin Aldrin, Jr., took the LM down to the surface. Although much of the descent was controlled by computers, the astronauts had to override a number of computer alarms and take control of the LM to avoid a boulder-strewn crater bigger than a football field. Climbing down the ladder, Armstrong and then Aldrin stepped onto the lunar surface.

Between July 1969 and December 1972, a total of 12 people reached the lunar surface and collected a total of 380 kg (840 lb) of rocks and soil (Table 20–2). The flights were carefully planned to visit different regions and thus develop a comprehensive history of the lunar surface.

The first flights went into relatively safe landing sites (Figure 20–14)—Mare Tranquillitatis for Apollo 11 and Oceanus Procellarum for Apollo 12. Apollo 13 was aimed at a more complicated site, but an explosion in an oxygen tank on the way to the moon ended all chances of a landing and nearly cost the astronauts their lives. They succeeded in using the life support in the LM to survive till they returned to the earth.

The last four Apollo missions, 14 through 17, sampled geologically important places on the moon. Apollo 14 visited the Fra Mauro region, which is covered by ejecta from the impact that dug the basin now filled by Mare Imbrium. Apollo 15 visited the edge of Mare Imbrium at the foot of the Apennine Mountains and examined Hadley Rille. Apollos 16 and 17 visited highland regions to sample the older parts of the lunar crust (Figures 20–14 and 15). Almost all of the lunar samples from these six landings are now held at the Planetary Materials Laboratory at the Johnson Space Center in Houston. They are a national treasure containing clues to the beginnings of our solar system.

LUNAR GEOLOGY

Technically, the word *geology* means "earth-study" or "earth-knowledge," but we can now apply it to the study of the moon because we have been there, and we have samples to examine just as we examine earth rocks. Those samples can tell us about the age, composition, and processes that have created the lunar terrain. Ultimately, they may make it possible for us to tell the story of the lunar surface.

We can find the age of a moon rock—the length of time since it solidified from a molten state—by radioactive dating (Box 19–2). The oldest rocks come from the lunar highlands and have ages of 4–4.3 billion years. The oldest rock of all, a small green fragment called 72417, was found by Apollo 17 astronauts and has an age of 4.6 billion years. This is quite unusual. Most of the old rocks are no older than 4.3 billion years, and these are mostly fragments cemented together. Samples from the maria have ages of 3.8–3.1 billion years. No rock was found younger than this.

Although moon rocks look like earth rocks, their chemical composition is clearly different. The moon is richer in refractory elements such as minerals of calcium, aluminum, and titanium, but poorer in volatile materials like water, sodium, and potassium. This suggests that the material in the moon has been exposed to a high temperature perhaps before the moon formed.

Moon rocks are so distinctive that mineralogists now believe they have identified at least three meteorites, found on the earth, that are fragments of the lunar surface (Figure 20–16). These samples were evidently blasted into space by impacts on the lunar surface and then fell to the earth's surface where they were eventually recognized.

The physical characteristics of moon rocks tell us a great deal. There are no sedimentary rocks, of course, because these form in water and there has never been water on the moon. The lunar samples are all igneous rocks, minerals that have solidified from a molten state. Samples from the highlands are often a low-density, light-colored kind of rock called **anorthosite**. This is the kind of rock that would float to the top of a molten mass. Samples from the maria are basalts, that is, solidified lavas. Some of these are called **vesicular basalts**, meaning that they contain holes that formed when the lava flowed up to the lunar surface and gas bubbles formed (Figure 20–17a). However, most of the samples are **breccias**, cemented masses of fragments (Figure 20–17b). These clearly show that the lunar surface has been battered to fragments by meteorite impact.

FIGURE 20–16 The meteorite ALHA81005 was found in Antarctica in January 1982 and later identified as a fragment of the lunar surface. (NASA, courtesy of Ursula B. Marvin.)

To develop a history of the lunar surface, we must combine what we know about the lunar rocks with what we can learn about the moon's interior. The average density of the moon is only 3.36 g/cm^3, which means it cannot contain a large iron core. In addition, the lunar magnetic field is less than 0.0001 that of the earth, so it cannot have a large molten core. If it did, the dynamo effect, even with the moon's slow rotation, would produce a measurable magnetic field. However, some lunar samples have remnant magnetic fields as strong as 0.04 times the earth's. This might be left over from an earlier magnetic field on the moon generated by meteorite impacts or produced by the magnetic fields of the sun or the earth. No one knows for sure, so no one can be sure about the moon's core. However, it seems likely that if the moon has a core, it is small and relatively cool.

Seismographs left on the lunar surface recorded about 3000 moonquakes per year. These were small, about as large as a big firecracker, but by triangulating from three seismographs, geologists located the sources of the quakes at depths of 700–1200 km, down in a region near the suspected core (Figure 20–18). This region may be semi-plastic. Because the quakes are more common when the moon is closest to earth, it seems that tidal stresses trigger at least some quakes.

The interior of the moon is now almost dead, which is what we would expect for a small body that loses its

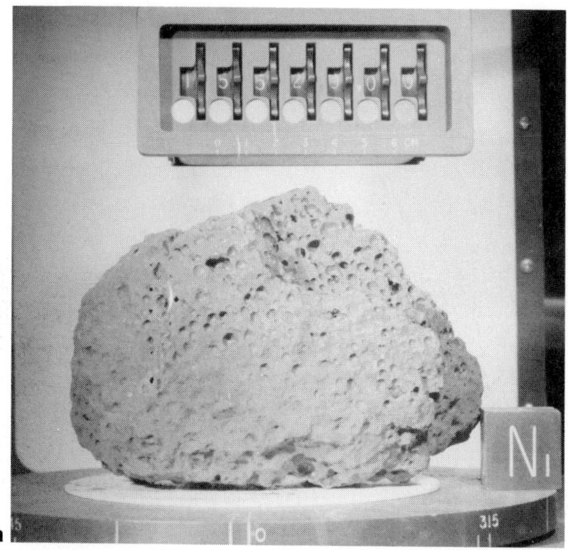

FIGURE 20–17 Lunar rocks. (a) A vesicular basalt contains numerous bubbles frozen into the rock when it solidified from a surface lava flow. (b) A breccia is a rock composed of cemented fragments. In this sample the light areas are pieces of anorthosite heavily pitted by tiny micrometeorite craters. The dark areas are vesicular glass. (NASA.)

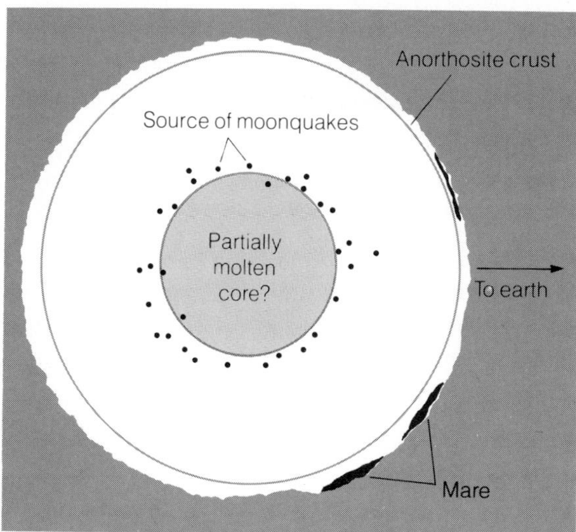

FIGURE 20–18 Cross section of moon shows deep location of moonquakes near what may be a partially molten core. Note that the lunar crust is thinner on the side toward Earth.

heat easily. Thus, the history of the moon is dominated by one important fact: It is a small world.

Almost no rocks have been found older than 4.3 billion years, and that suggests the lunar surface must have been molten when it formed. The surface was a great sea of magma melted perhaps by radioactive elements like aluminum-26, by the heat generated by in-falling debris, or perhaps by magnetic fields in the young solar system. Whatever the cause, the lunar surface was molten from 4.6 to about 4.4 billion years ago.

In the first of the four stages of planetary evolution, the moon cooled and differentiated as the lower-density minerals floated to the top like the slag in an open hearth furnace. This material solidified to form a low-density anorthosite crust. Because the moon is small, it cooled rapidly and its crust grew thick.

The second stage of planetary evolution is cratering. This process was intense in the early solar system. Most rocks older than 4 billion years are smashed to fragments, and the entire crust is fractured to a depth of about 2 km (Color Plate 32). Until recently this intense cratering was thought to have been caused by the last of the debris left over from planet formation, but that is no longer clear. The moon rocks tell us that the cratering lasted until about 4 billion years ago and then tapered off to its present low level by about 3.8 billion years ago. This is very late for planet-building debris to remain free in the solar system. An alternate theory is that this bombardment was cataclysmic with long periods of inactivity interspersed with sudden injections of debris into the inner solar system. Some theorists now speculate that occasional collisions between comets and satellites of the Jovian planets could have scattered debris throughout the young solar system.

The cratering not only battered the crust, it excavated vast crater basins. The basin now filled by Mare Imbrium, for instance (Data File 3a), was excavated by the impact of a meteorite about the size of Rhode Island. The resulting crater was 1300 km in diameter, and the impact smashed the crust into mountain ranges up to 5 km high. Ejecta from this impact covers 16 percent of the lunar surface. Apollo 14 landed at Fra Mauro to sample it.

The Imbrium Basin is difficult to study because it has been flooded with lava, but a similar basin lies at the western limb of the moon and has not been flooded. Called Mare Orientale, it is a great crater consisting of three circular mountain ranges that were smashed into the lunar crust by the impact (Figure 20–19). The Imbrium Basin has similar rings, although they are mostly buried under successive lava flows.

The moon would have been a dangerous place to live during this cratering phase. Not only might you have been hit by a falling meteorite, you might have been buried by scattered ejecta or jolted by shocks from distant impacts. You could have felt the Imbrium or Orientale impacts anywhere on the moon, but had you been standing on the side of the moon opposite one of these impacts, you would have been at the focus of seismic waves traveling around the moon from different directions. As a result, the surface under your feet would have jerked up and down by as much as 10 m. Both places on the moon opposite the Imbrium Basin and the Orientale Basin are strangely disturbed landscapes called **jumbled terrain.** We will find similar effects on other planets and moons.

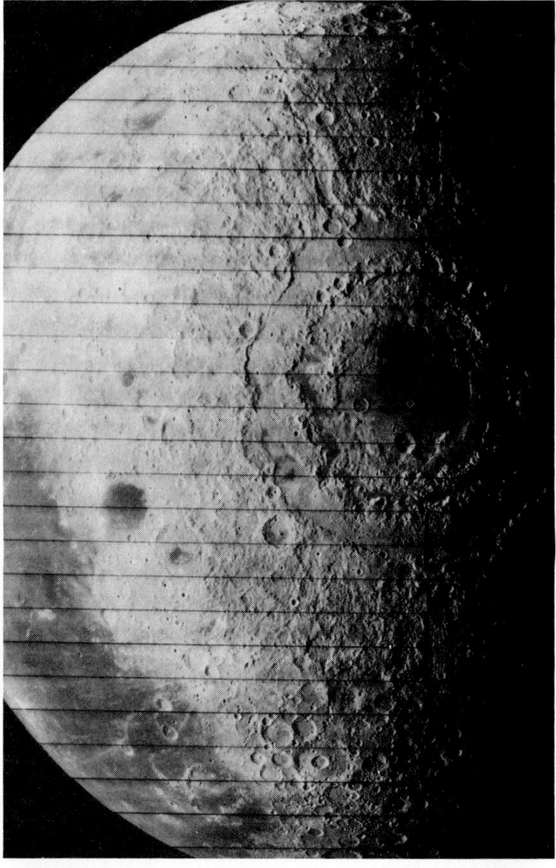

FIGURE 20–19 Mare Orientale, the large bullseye feature just right of center, is an impact basin nearly 1000 km (630 mi) in diameter. It lies just beyond the western edge of the moon as seen from the earth. Note that lava flows have flooded small areas of the basin. (NASA/JPL.)

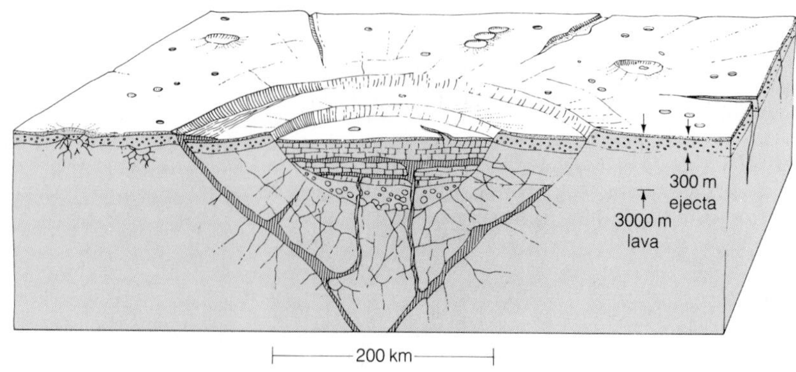

FIGURE 20–20 Major impacts on the moon broke the crust to great depth and produced large basins. Lava flowed up through the broken crust and flooded the basins to produce maria.

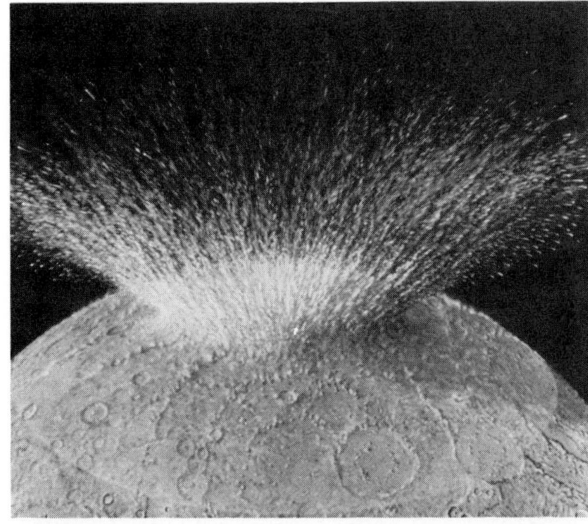

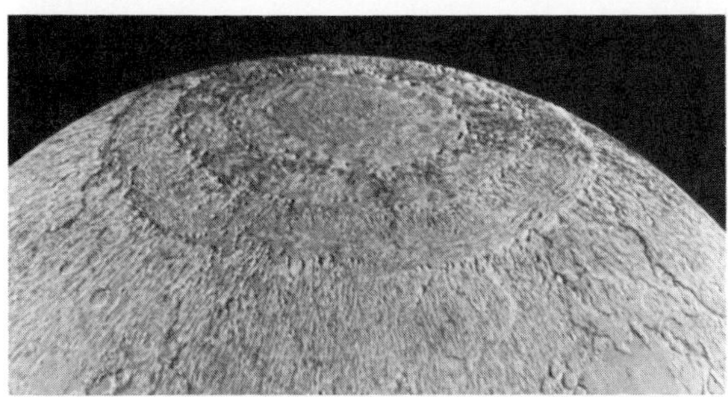

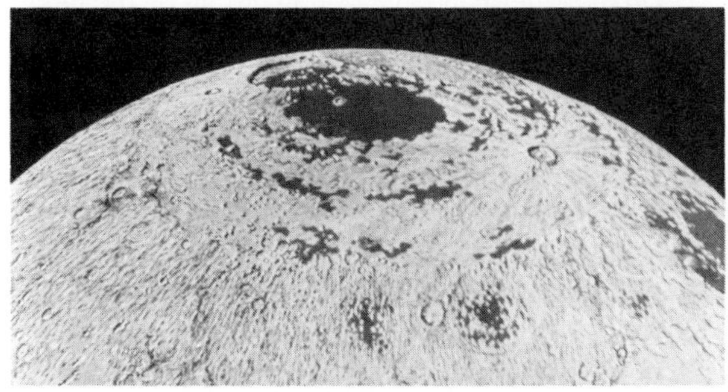

The Imbrium event occurred about 4 billion years ago, and the Orientale event sometime later. We know that because ejecta from the Orientale event lies on top of ejecta from the Imbrium event. Other impacts had dug large basins and fractured the crust. Gradually, the decay of uranium, thorium, and potassium began to melt the moon's mantle, and the molten lava followed the fractures up from depths of 200–300 km into the basins forming the maria (Figure 20–20). Some of the maria are round, following the outline of the basin they fill, but others are irregular in shape, showing that the lava overflowed the basin boundaries. This third stage, flooding, lasted from about 3.8 to about 3.1 billion years ago (Figure 20–21).

This flooding did not occur all at once. Studies of the maria show that they consist of successive lava flows one on top of another. In some cases, the lava flows pushed against obstacles or pressed the crust downward and developed wrinkle ridges. In other cases, the lava poured out and melted tunnels through the existing crust. These lava tubes have been seen in volcanic eruptions in Hawaii, but on the moon they are larger. After the lava drained away, the roof of the tube collapsed and formed a long, winding sinuous rille. Apollo 15 visited the margin of Mare Imbrium, and astronauts Scott and Irwin studied the collapsed lava tube called Hadley Rille (Figure 20–22).

Almost all of this flooding occurred on the side of the moon facing the earth. The back side of the moon is heavily cratered, but only a few of the largest craters con-

FIGURE 20–21 The evolution of the lunar surface. Mare Imbrium began when a meteorite as large as Rhode Island excavated a great crater basin. Lava welled up to flood the basin, starting about 3.8 billion years ago and lasting until no less than 2 billion years ago. Since then only scattered meteorites have modified Mare Imbrium. (Courtesy Don Davis.)

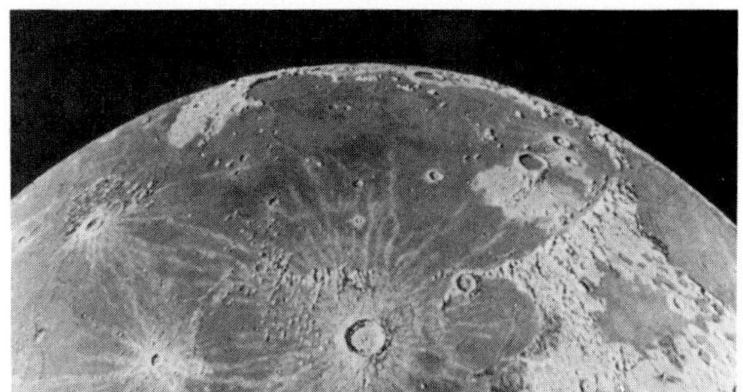

FIGURE 20–22 Apollo 17 visited Hadley Rille at the edge of Mare Imbrium. (NASA.)

tain any lava flows. Even Mare Orientale near the west limb is only slightly flooded. The difference is caused by the thickness of the crust. On the side toward the earth, the lunar crust is only about 60 km thick, but on the far side it is more than 100 km thick (Figure 20–18). Apparently, this difference is related to the tidal forces the earth exerts on the moon, but the details are not understood.

The evolution of the moon might have continued had it been larger, but it is too small to remain hot and too small to keep an atmosphere. It cooled, and its crust grew thick. There are no rift valleys on the moon and no folded mountains because the moon's crust has never divided into moving plates. In addition, the lowlands of the moon were never flooded by water because the moon never had an atmosphere. Thus, the moon is a dead world preserved halfway through stage three.

Although the moon is geologically dead, it is still evolving slowly. The constant blast of micrometeorites grinds the surface of exposed rock to dust, and the dust collects in a deep **regolith**, a soil composed of rock fragments. The surface lavas of the maria are battered to the consistency of talcum powder. The footprints left by the 12 astronauts will last for millions of years, but they will eventually be erased by the constant flux of meteorites.

On the other hand, the overall terrain on the moon is almost fixed. A billion years from now the North American continent may be gone, recycled into the earth's crust; the descent stage of the Apollo 17 lunar lander will be nothing but a peculiar contamination in the regolith at Taurus–Littrow; but the lunar mountains will look much the same.

THE ORIGIN OF THE MOON

The Apollo missions to the moon told us a great deal, but they failed to answer the most basic question: "How did the moon form?" The three traditional theories do not work, but a new hypothesis may solve the problem.

The first of the three traditional theories, the **fission hypothesis**, supposes that the moon formed by the fission of the earth. If the young earth spun fast enough, tides raised by the sun might make it break into two parts (Figure 20–23a). If this separation occurred after the earth differentiated, the moon would have formed from crust material. Thus, the theory explains the moon's low density.

But the fission theory has problems. No one knows why the young earth should have spun so fast, nearly ten

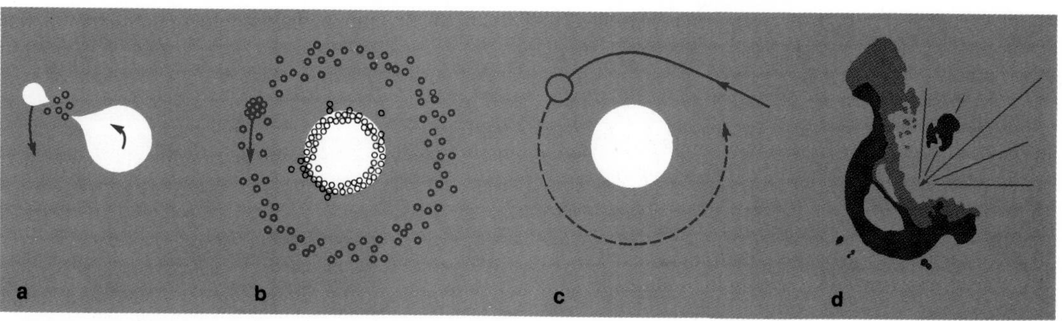

FIGURE 20–23 Four theories of the moon's origin. (a) Fission theories suppose that the earth and moon were once one body and broke apart. (b) Co-accretion theories suppose that the moon formed from material near the earth. (c) Capture theories suggest that the moon formed elsewhere and was captured by the earth. (d) The large-impact hypothesis suggests a Mars-size planetesimal struck the young earth, ejecting mantle material into orbit where it formed the moon. (d, Adapted from computations by Marlin Kipp and Jay Melosh.)

times faster than today, nor where all that angular momentum went after the fission. In addition, the moon's orbit is not in the plane of the earth's equator, as it would be if it had formed by fission.

The second traditional theory is the **co-accretion** (or double-planet) **hypothesis.** It supposes that the earth and moon condensed as a double planet from the same cloud of material (Figure 20–23b). However, if they formed from the same material they should have the same chemical composition and density, which they don't. The moon is very poor in iron and related elements such as titanium, but it also contains almost exactly the same ratios of oxygen isotopes as does the earth's mantle. Co-accretion cannot explain these compositional differences.

The third theory is the **capture hypothesis.** It supposes that the moon formed somewhere else and was later captured by the earth (Figure 20–23c). We might imagine the moon forming inside the orbit of Mercury, where the heat would prevent the condensation of solid metallic grains and only high-melting-point metal oxides could have solidified. A later encounter with Mercury could "kick" the moon out to the earth.

The capture theory is not popular because it requires highly unlikely events involving interactions with Mercury and the earth to move the moon from place to place. Scientists are always suspicious of explanations that require a chain of unlikely coincidences. Also, upon encountering the earth, the moon would be moving so rapidly that the earth's gravity would be unable to capture it without ripping the moon to fragments through tidal forces.

Until recently, this left astronomers with no acceptable theory to explain the origin of the moon, and they occasionally joked that the moon could not exist. But astronomers attending a conference in late 1984 reached some agreement on a new theory that explains some of the moon's most puzzling properties.

The **large-impact hypothesis** (Figure 20–23d) supposes that the moon formed from debris ejected into a disk around the earth by the impact of a large body. The impacting body may have been as large as Mars. Instead of saying that the earth was hit by a large body, we may be more nearly correct to say that the earth and moon resulted from the collision and merger of two very large planetesimals. The resulting large body became the earth, and the ejected debris formed the moon.

This would explain a number of things. The collision had to occur at a steep angle to eject enough matter to make the moon. That is, the objects did not collide head-on. Such a glancing collision would have spun the resulting material rapidly and would explain the high angular momentum in the earth–moon system. If the colliding planetesimals had already differentiated, the ejected material would be mostly iron-poor mantle and crust. Calculations show that the iron core of the impacting body would have fallen into the earth within 4 hours. This would explain why the moon is so poor in iron and why the abundances of other elements are so similar to rocks from the earth's mantle. Also, the material that eventually became the moon would have remained in a disk long enough for volatile elements, which the moon lacks, to be lost to space.

Many astronomers are climbing on the large-impact bandwagon. It is exciting and satisfying to finally have a hypothesis that survives first examination (Table 20–3).

Table 20–3 Theories of lunar origin: a report card.*

Criteria	Fission	Co-accretion	Capture	Large-Impact
Mass of moon	D	B	B	I
Earth–moon angular momentum	F	F	C	B
Lack of lunar volatiles	B	C	C	B
Lack of lunar iron	A	D	F	I
Oxygen isotopes	A	A	B	B
Allows lunar magma ocean	A	C	D	A
Physical plausibility	F	C	D−	I

*Grades assigned A through F with I representing an incomplete grade yet to be determined. (Based on an analysis by John Wood.)

Nevertheless, not all astronomers agree that the problem is solved, and careful analysis of lunar samples and simulations of the collision may yet rule out the theory. But, for the moment at least, most astronomers can agree that the moon exists.

SUMMARY

All terrestrial planets pass through a four-stage history: (1) differentiation; (2) cratering; (3) flooding of the crater basins by lava, water, or both; (4) slow surface evolution. The importance of each of these processes in the evolution of a planet depends on the planet's mass and temperature.

The earth is the largest terrestrial planet in our solar system, and it passed through all four stages. Seismic studies show that the earth has differentiated into a metallic core and a silicate crust. Currents in the molten portion of the core produce earth's magnetic field.

Because the earth is still partially molten, its surface is active. Plate tectonics refers to the motion of large crustal sections. As new crust appears in the midocean rifts, it pushes the plates apart and destroys old crust where plates slide over each other.

Current studies suggest that the earth formed hot through heterogeneous accretion. In that case, the earth did not capture a primeval atmosphere from the solar nebula but rather acquired a water-rich atmosphere later when the nebula cooled. This atmosphere gradually changed as ultraviolet light from the sun broke up molecules, light gases escaped, CO_2 dissolved in the oceans, and life began producing oxygen.

The moon is, at first glance, dramatically different from the earth. We see two distinct kinds of terrain on the moon. The maria, named after seas, are lowland plains with few craters. They are great lava flows and are younger than the highlands. These highlands are lighter in color and heavily cratered.

Samples returned by the Apollo mission show that the maria are dark basalts, whereas the highlands are light anorthosite. Many samples are breccias, which show how severely the moon's crust has been fractured.

The ages of the mare samples range from 3.1–3.8 billion years, but the highland samples are 4.0–4.3 billion years old. Few samples are older than 4.3 billion years, although at least one was found to be 4.6 billion years old. This suggests that the moon's surface was molten until about 4.4 billion years ago.

As the anorthosite crust solidified, cratering battered it and excavated great crater basins. Later, as radioactivity melted the mantle, molten rock rose through fractures in the crust and flooded the basins to produce the maria. By the time the maria began to form, cratering had declined so the maria are marked by few craters.

The lunar surface now evolves very slowly. Because the moon is small, it has cooled rapidly, its crust has grown thick, and it has never divided into moving plates. Now only meteorites alter the surface.

A recent theory suggests that the earth–moon system formed by the collision of two large planetesimals. The differentiated bodies formed a single large object that became the higher-density, iron-rich earth. Ejected mantle material, poor in iron, formed a disk and eventually condensed to form the lower-density, iron-poor moon.

NEW TERMS

comparative planetology	limb
pressure *(P)* waves	mare
shear *(S)* waves	relative age
mantle	ray
plastic	ejecta
dynamo effect	sinuous rille
bow shock	graben rille
magnetosphere	anorthosite
Van Allen Belts	vesicular basalt
plate tectonics	breccia
basalt	jumbled terrain
midocean rise	regolith
midocean rift	fission hypothesis
rift valley	co-accretion hypothesis
primeval atmosphere	capture hypothesis
secondary atmosphere	large-impact hypothesis
terminator	

QUESTIONS

1. How did the earth and moon resemble each other during the first, second, and third stages of planetary development? How did they differ?
2. What evidence do we have that the earth contains a molten, metallic core?
3. Why are there so few craters on the earth compared with the surface of the moon?
4. From what you know of the evolution of the earth, what would our atmosphere be like today if there had never been any oceans? If there had never been any life?
5. How do we find the relative ages of the maria and highlands?
6. How do we know that Copernicus is a young crater?
7. Why did the first Apollo missions land on the maria? Why were the other areas of more scientific interest?
8. How does the range of ages of lunar samples suggest that the moon formed with a molten surface?
9. Why are so many lunar samples breccias?
10. What do the vesicular basalts tell us about the evolution of the lunar surface?
12. What evidence would we expect to find on the moon if it had been subjected to plate tectonics? Do we find such evidence?
13. Cite objections to the fission, co-accretion, and capture hypotheses.
14. How does the large-impact hypothesis explain the moon's lack of iron?

PROBLEMS

1. If the Atlantic seafloor is spreading at 30 mm/year and is now 6400 km wide, how long ago were the continents in contact?
2. Why do small planets cool faster than large planets? Compare surface area to volume.
3. The smallest detail visible through earth-based telescopes is about 1 second of arc in diameter. What size is this on the moon? (HINT: See Box 3–1)
4. Midocean rifts and the trenches where the seafloor slips downward are 1 km or less wide. Could earth-based telescopes resolve such features on the moon? Why are we sure such features are not present on the moon?
5. The Apollo command module orbited the moon about 200 km above the surface. What was its orbital period? (HINT: See Box 5–1.)
6. From a distance of 200 km above the surface of the moon, what is the angular diameter of an astronaut in a space suit? Could we have seen the astronauts from the command module? (HINT: See Box 3–1.)
7. What is the escape velocity from the surface of the moon? (HINT: See Box 5–2.)

RECOMMENDED READING

BEATTY, J. KELLY "The Making of a Better Moon." *Sky and Telescope 72* (Dec. 1986), p. 558.

———. "Lunar Meteorites: Three and Counting." *Sky and Telescope 68* (Sept. 1984), p. 224.

BEATTY, J. KELLY, BRIAN O'LEARY, and ANDREW CHAIKIN *The New Solar System,* 2nd ed. Cambridge, Mass.: Sky Publishing, 1982. BEN-AVRAHAM, ZVI

BEN-AVRAHAM ZVI "The Movement of Continents." *American Scientist 69* (May/June 1981), p. 291.

BRUSH, STEPHEN G. "Nickel for Your Thoughts: Urey and the Origin of the Moon." *Science 217* (3 Sept. 1982), p. 891.

CADOGAN, PETER H. "The Moon's Origin." *Mercury 12* (March/April 1983), p. 34.

———. *The Moon: Our Sister Planet.* New York: Cambridge University Press, 1981.

CARRIGAN, CHARLES R., and DAVID GUBBINS "The Source of the Earth's Magnetic Field." *Scientific American 240* (Feb. 1979), p. 198.

CHAIKIN, ANDREW "Pieces of the Sky." *Sky and Telescope 63* (April 1982), p. 344.

CLOUD, PRESTON *Cosmos, Earth and Man.* New Haven, Conn.: Yale University Press, 1977.

———. "Beyond Plate Tectonics." *American Scientist 68* (July/Aug. 1980), p. 381.

CORTRIGHT, EDGAR M. *Apollo Expeditions to the Moon.* Washington, D.C.: NASA, 1975.

CROWE, MICHAEL J. "New Light on the Moon Hoax." *Sky and Telescope 62* (Nov. 1981), p. 428.

EVANS, DAVID S. "The Great Moon Hoax." *Sky and Telescope 62* (Sept. 1981), p. 196, and *62* (Oct. 1981), p. 308. (See also *62* [Nov. 1981], p. 501.)

GILLETT, STEPHEN "The Rise and Fall of the Early Reducing Atmosphere." *Astronomy 13* (July 1985), p. 66.

HARTMANN, W. K. "The Moon's Early History." *Astronomy 4* (Sept. 1976). p. 6.

———. "Cratering in the Solar System." *Scientific American 236* (Jan. 1977), p. 84.

———. "The Early History of Planet Earth." *Astronomy 6* (Aug. 1978), p. 6. Reprinted in *Astronomy: Selected Readings.* ed. M. A. Seeds. Menlo Park, Calif.: Benjamin/Cummings, 1980, pp. 3–15.

———. *Moons and Planets*, 2nd ed. Belmont, Calif.: Wadsworth, 1983.

MARAN, STEPHEN P. "Quakes on the Moon." *Natural History 91* (Feb. 1982), p. 82.

MAXWELL, JOHN C. "What is the Lithosphere?" *Physics Today 38* (Sept. 1985), p. 32.

MURRY, BRUCE, MICHAEL C. MALIN, and RONALD GREELEY *Earthlike Planets.* San Francisco: W. H. Freeman, 1981.

REGISTER, BRIDGET MINTZ "The Fate of the Moon Rocks." *Astronomy 13*(Dec. 1985), p. 14.

RINGWOOD, A. E. "Terrestrial Origin of the Moon." *Nature 322* (24 July 1986), p. 323.

RUBIN, ALAN E. "Whence Came the Moon." *Sky and Telescope 68* (Nov. 1984), p. 389.

STEVENSON, P. J. "Models of the Earth's Core." *Science 214* (6 Nov. 1981), p. 611.

VAN ANDEL, T. *New Views on an Old Planet.* New York: Cambridge University Press, 1985.

WASHBURN, MARK "The Moon—A Second Time Around." *Sky and Telescope 69* (March 1985), p. 209. (See also *69* [May 1985], p. 395.)

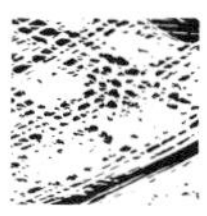

CHAPTER 21

MERCURY, VENUS, AND MARS

There wasn't a breath in that land of death . . .

Robert Service
THE CREMATION OF SAM MCGEE

Without a spacesuit you could live only about 30 seconds on the surface of Mars. The temperature might not be too bad. On a hot summer day at noon the temperature might reach 20°C (68°F), but the air is 95 percent carbon dioxide, deadly dry, and contains almost no oxygen. In addition, the atmospheric pressure at the surface of Mars is only 0.01 that at the surface of the earth. Stepping out onto the surface of Mars without a spacesuit would be about the same as opening the door of an airplane flying at 30,000 m (100,000 ft) above the earth. (Commercial jets do not generally fly above 12,000 m, or 38,000 ft.)

A hike on Mercury would be even less fun. Mercury has no atmosphere at all, and the temperature in direct sunlight can be as high as 430°C (806°F). At night the temperature can drop to − 180°C (− 292°F). You would definitely want a spacesuit to visit Mercury.

No spacesuit would save you on Venus. The atmospheric pressure is 90 times Earth's, and the air consists almost entirely of carbon dioxide with traces of various acids. Worse yet, the surface is hot enough to melt lead.

Why are the planets such rotten places for picnics? The terrestrial planets are all made of silicates and metals, and they all lie in the inner solar system. We might expect them to be rather similar. How did they get to be so different from each other? As we will see in this chapter, the answer to that question is very complex.

21.1 MERCURY

The planet Mercury is named after the fleet-footed messenger of the gods, presumably because the planet circles the sun with a period of only 87.969 days, faster than any other planet in the solar system.

Because Mercury is so close to the sun, it is difficult to observe (Figure 2–16). For centuries astronomers have drawn maps of what they see through their telescopes, but even the best of these maps, drawn in the early 1970s, shows little more than dark smudges (Figure 21–1).

Yet Mercury is an important planet in our study of our solar system. Its rotation and revolution illustrate the importance of tides, and its geology—as revealed by photographs taken by the Mariner 10 spacecraft—provides an interesting contrast with the geology of Earth and moon.

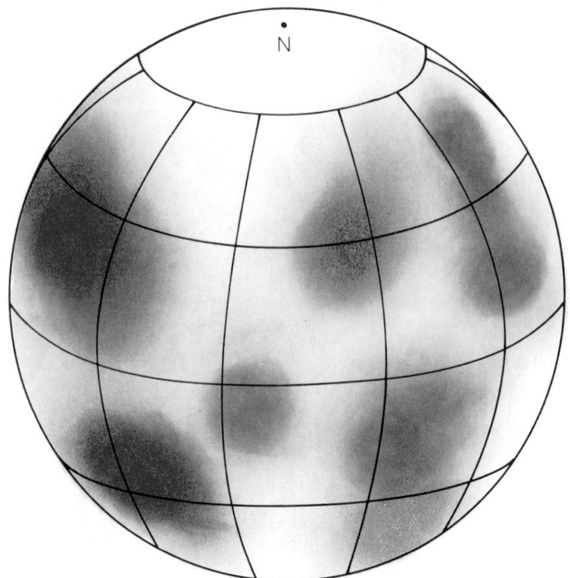

FIGURE 21–1 Mercury as mapped from Earth. Because of the difficulty in observing Mercury, even the best visual observations reveal little. The orientation of this map is the same as that in Figure 21–5. (Adapted from 1972 data by John Murray.)

ROTATION AND REVOLUTION

During the 1880s the Italian astronomer Giovanni Schiaparelli sketched the faint features he thought he saw on the disk of Mercury and concluded that the planet was tidally coupled to the sun just as Earth's moon is tidally coupled to Earth. That is, he concluded that Mercury rotated on its axis to keep the same side facing the sun throughout its orbit. This is actually a very good guess because, as we will see, **tidal coupling** between rotation and revolution is common in the solar system. But the rotation of Mercury is more complex than Schiaparelli thought.

In 1962 radio astronomers detected radio emission from the planet and concluded that the dark side was not as cold as it should have been if the planet kept one side perpetually in darkness. In 1965 radio astronomers made radar contact with Mercury. They used the 305-m Arecibo dish (Figure 6–21) to transmit a pulse of radio energy at Mercury and then waited for the reflected signal to return. Doppler shifts in the reflected radio pulse showed that the planet was rotating with a period of only about 59 days, much shorter than the orbital period of 87.969 days.

Apparently Mercury is tidally coupled to the sun such that it rotates, not once per orbit, but 1.5 times per orbit (Figure 21–2). Thus, its period of rotation is two-thirds of its orbital period, or 58.65 days. This means that a mountain on Mercury directly below the sun at one place in its orbit will point away from the sun one orbit later and toward the sun after the next orbit.

The complex tidal coupling between the rotation and revolution of Mercury is an important illustration of the power of tides. Just as the tides in the earth–moon system have slowed the moon's rotation and locked it to Earth, so have the sun–Mercury tides slowed the rotation of Mercury and coupled its rotation to its revolution. We can refer to such a relationship as a **resonance**. We will see many such resonances as we explore the solar system.

In addition to its rotation, Mercury's orbital motion is also peculiar. In Chapter 5 we discussed the orbital motion of Mercury and saw how its elliptical orbit precesses faster than can be explained by Newton's laws but at just the rate predicted by Einstein's theory of general relativity. Thus the orbital motion of Mercury is taken as a confirmation of the curvature of space-time as predicted by general relativity.

THE EXPLORATION OF MERCURY

Although Mercury's rotation and orbit were understood, modern astronomers knew little about Mercury as a planet until it was visited by the Mariner 10 spacecraft, launched from Earth on November 3, 1973. After calibrating its cameras on Earth and moon, it coasted away toward the sun, passing and photographing Venus en route to Mercury. It flew by Mercury in March 1974, passing only 700 km above

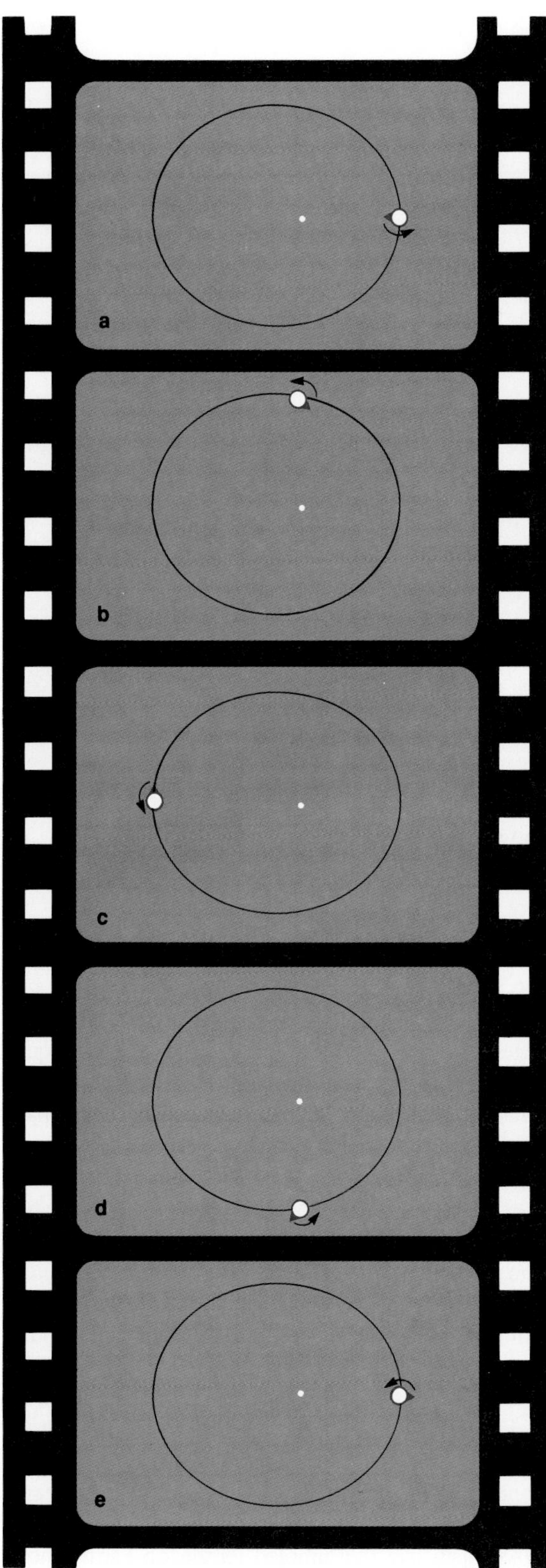

FIGURE 21–2 Mercury's rotation is tidally coupled to the sun, but unlike the moon, which rotates to keep the same face toward Earth, Mercury rotates 1.5 times during each orbit. A mountain facing the sun (a) will face away from the sun after one orbit (e), and will face the sun again after the next orbit.

the dark side of the planet. Mariner 10 then looped around the sun and met Mercury again in September and again in March 1975.

During these flybys, instruments aboard the spacecraft searched for traces of a magnetic field and an atmosphere. Planetary astronomers did not expect Mercury to have a magnetic field because it is small and rotates slowly. As we saw in Chapter 20, Earth's magnetic field is generated by the dynamo effect operating in the rotating, molten core. Mercury is smaller than Earth and should have lost its internal heat so rapidly that, like the moon, it should not have a molten core. Also, it rotates much more slowly than does Earth, so the dynamo effect was not expected to generate any magnetic field.

Contrary to expectations, Mariner 10 detected a magnetic field around Mercury, which deflected the solar wind to form a magnetosphere similar to Earth's (Chapter 20). The magnetic field was weak—too weak, in fact, to trap high-energy particles to form radiation belts like Earth's Van Allen belts—but the existence of any magnetic field was unexpected.

Some theorists suggest that Mercury's magnetic field is only a remnant frozen into the planet when it solidified, whereas others argue that a weak dynamo effect is working in the planet's core. If the field is a remnant, then the rocky crust must be at least 100 times richer in iron than Earth's, which seems unreasonable. However, if there is a dynamo in the planet's core, then some energy source must be providing the heat to cause the circulation needed to generate the field. No such source of energy is presently known.

Mercury's magnetic field serves to remind us that we do not understand how planetary magnetic fields are generated. We will find other magnetic oddities as we explore the other planets.

Not only did Mariner 10 find a magnetic field where theory predicted there should have been none, but also it found an atmosphere where none was expected. Mercury is a small world with an escape velocity of only 4.3 km/sec compared with Earth's 11.2 km/sec. At such a small distance from the sun, Mercury's temperature can range from −173°C (−279°F) in shadow to 430°C (806°F) in

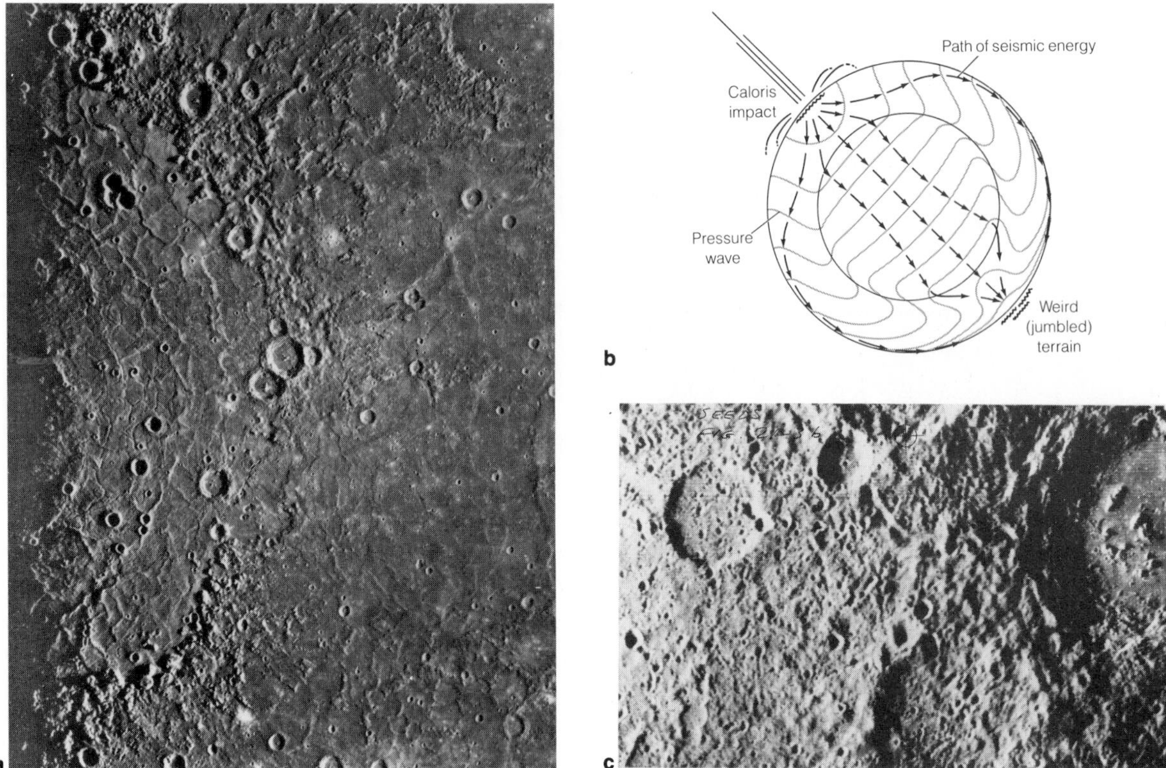

FIGURE 21–3 The Caloris Basin on Mercury, shown partially in darkness in this Mariner 10 photomosaic (a), was produced by a large impact and subsequent lava flows. The focusing of seismic waves (b) produced by the impact created jumbled terrain on the opposite side of the planet (c). (NASA.)

direct sunlight. At such high temperatures the atoms and molecules of any atmosphere should travel faster than escape velocity, and thus the atmosphere should have been lost to space long ago.

The atmosphere that Mariner 10 detected consists of hydrogen and helium gas with a density about 10^{15} times less than Earth's atmosphere at sea level. These gases are apparently trapped from the solar wind, and, as they leak away to space, are replenished by more hydrogen and helium. Thus, Mercury is able to maintain a tenuous atmosphere by "borrowing" hydrogen and helium from the solar wind.

THE SURFACE OF MERCURY

In addition to measurements of magnetic field and atmosphere, Mariner 10 photographed Mercury's surface. As the NASA planners had expected, Mercury looks generally like the moon (Data File 4). It is heavily battered, with craters of all sizes, including some large basins. Some craters are obviously old and degraded; others seem quite young and have bright rays of ejecta. However, a quick glance at photos of Mercury shows no large, dark maria like the moon's flooded basins.

The largest basin on Mercury is called Caloris Basin after the Latin word for "heat," recognition of its location at one of the two "hot poles," which face the sun at alternate perihelions. At the times of the Mariner encounters, the Caloris Basin was half in shadow (Figure 21–3). Although half cannot be seen, the low illumination is ideal for the study of the lighted half.

The Caloris Basin is a gigantic impact basin 1300 km in diameter with concentric mountain rings up to 3 km high. The impact threw ejecta 600–800 km across the planet, and the focusing of seismic waves on the far side produced peculiar terrain (called "weird" terrain) that looks much like the jumbled terrain on Earth's moon (21–3b). Like Mare Orientale on the moon, the Caloris

FIGURE 21–4 A lobate scarp (arrow) crosses craters, indicating that Mercury cooled and shrank, wrinkling its crust after many of its craters had formed. (NASA.)

Basin is partially filled with lava flows. Some of this lava may be material melted by the energy of the impact, but some may be lava from inside the planet that leaked up through cracks in the crust. The weight of this lava and the response of the crust have produced deep cracks in the central lava plains. The geophysics of such large, multiringed crater basins is not well-understood at present, but Caloris Basin seems to be the same kind of structure as Mare Orientale and the Imbrium Basin on the moon.

When planetary scientists began looking at the Mariner photographs in detail, they discovered something not seen on the moon. Mercury is marked by great curved cliffs called **lobate scarps** (Figure 21–4). These seem to have formed when the planet cooled and shrank by a few kilometers, wrinkling its crust as a drying apple wrinkles its skin. Some of these scarps are as high as 3 km and reach hundreds of kilometers across the surface. Other faults in Mercury's crust are straight and may have been produced by tidal stresses generated when the sun slowed Mercury's rotation.

THE PLAINS OF MERCURY

The most striking difference between Mercury and the moon is that Mercury lacks the great dark lava plains so obvious on the moon. Under careful examination, the Mariner 10 photographs show that Mercury has plains, two different kinds in fact, but they are different from the moon's. Understanding the differences is the key to understanding the history of Mercury.

Much of Mercury's surface is old, cratered plains (Figure 21–5), but other areas called **intercrater plains** are less heavily cratered. These plains are marked by meteorite craters less than 15 km in diameter and secondary craters produced by chunks of ejecta from larger impacts. Unlike the heavily cratered regions, the intercrater plains are not totally saturated with craters. This suggests that the intercrater plains were produced by lava flows that buried older terrain.

In contrast, smaller regions called **smooth plains** appear to be even younger than the intercrater plains.

DATA FILE 4
Mercury

Average distance from the sun	0.387 AU (5.79×10^7 km)
Eccentricity of Orbit	0.2056
Maximum distance from the sun	0.467 AU (6.97×10^7 km)
Minimum distance from the sun	0.306 AU (4.59×10^7 km)
Inclination of orbit to ecliptic	7°00′16″
Average orbital velocity	47.9 km/sec
Orbital period	87.969 days (0.24085 y)
Period of rotation	58.646 days (direct)
Inclination of equator to orbit	0°
Equatorial diameter	4878 km (0.382 $D_\oplus$)
Mass	3.31×10^{23} kg (0.0558 $M_\oplus$)
Average density	5.44 g/cm^3 (5.4 g/cm^3 uncompressed)
Surface gravity	0.38 earth gravities
Escape velocity	4.3 km/sec (0.38 $V_\oplus$)
Surface temperature	(−279° to 806°F) −173° to 430°C
Average albedo	0.1
Oblateness	0

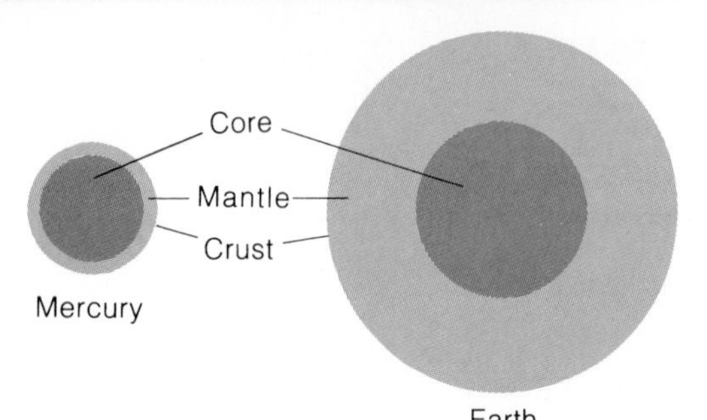

a. In proportion to its size, Mercury has a very large metallic core.

b. The smooth plains near Caloris Basin. (NASA.)

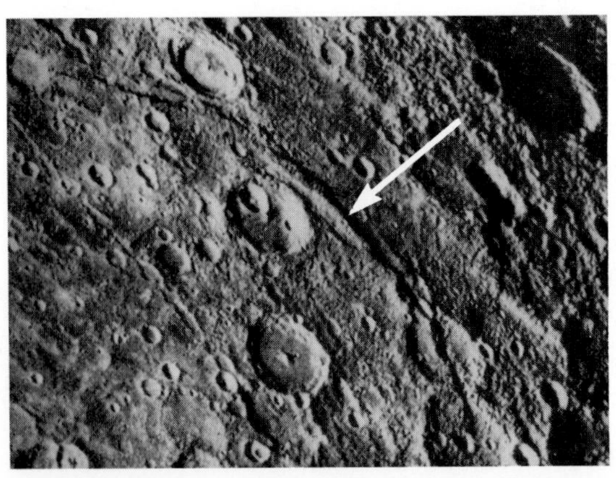

c. Hero Rupes, a lobate scarp, cuts through craters in smooth plains. (NASA.)

d. Photomosaic of Mercury. Caloris Basin is in shadow at left. (NASA.)

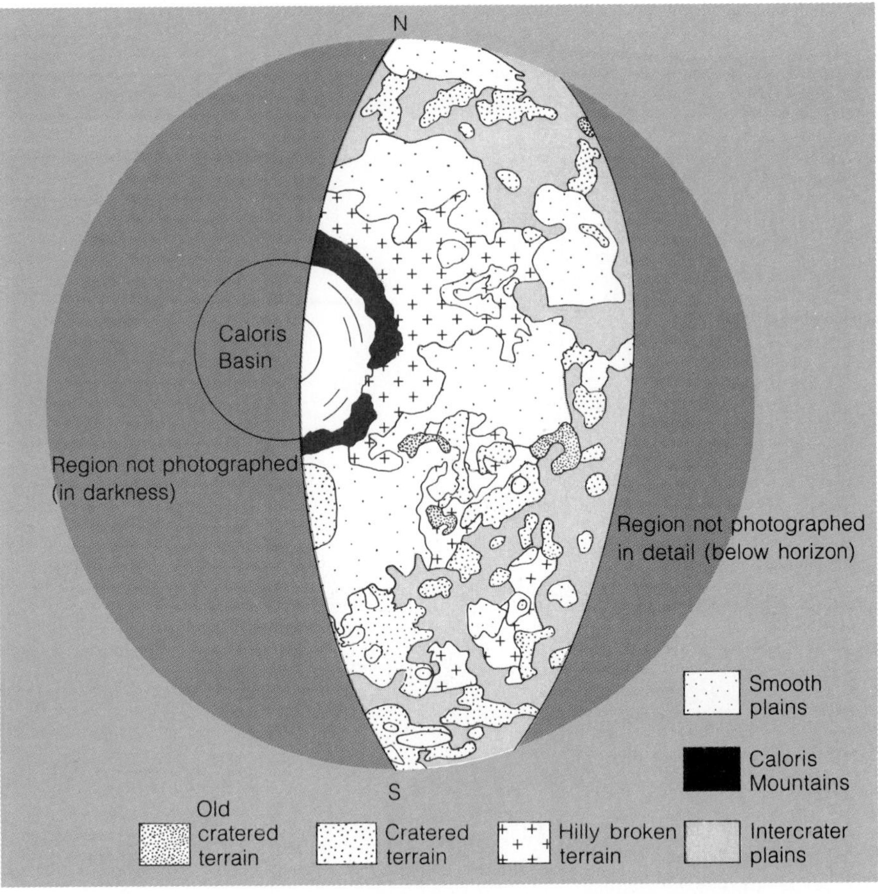

FIGURE 21–5 A geological map of part of Mercury shows the Caloris Basin at left and the hilly, broken terrain associated with that impact. The intercrater plains are interrupted by smooth plains and more heavily cratered terrain.

They have fewer craters and appear to be ancient lava flows. Much of the region around the Caloris Basin is composed of these smooth plains (Figure 21–5), and they appear to have formed soon after the Caloris impact.

Given the available evidence, most planetary astronomers believe that the plains of Mercury are solidified lava flows, in which case they are much like the maria on Earth's moon. Unlike the maria, Mercury's lava plains are not significantly darker than the rest of the crust. Except for a few bright crater rays, Mercury's surface is a uniform gray with an albedo of only about 0.1. Thus, Mercury's lava plains are not as dramatically obvious on photographs as the much darker maria on our own moon.

A HISTORY OF MERCURY

Mercury's plains hint at the history of the planet. In the first stage of planetary formation, the planet developed a dense core. We know the planet now has a large metallic core because its mass divided by its volume yields an average density of 5.44 g/cm^3, which implies a large amount of metal. Spectroscopic observations from Earth show that the crust is composed of silicates, probably basalts, so the metal must be in a large core (Data File 4b).

Like Earth's moon, Mercury was subjected to intense bombardment by debris in the young solar system. We don't know the ages of features on Mercury, but we can assume that cratering, the second stage of planetary evolution, occurred on Mercury over the same period as did the cratering on the moon—from formation to about 4 billion years ago. Presumably, this cratering ended as the planets swept up the last of the debris left over from planet building.

Mercury's crust still carries the scars of this cratering, but it is subtly different from the terrain on the moon. Because of Mercury's stronger gravity, the ejecta from the impacts did not cover as large of an area. Ejecta from a crater on Mercury would be thrown only 65 percent as far as ejecta from a crater of the same size on the moon. Lava flows during the cratering period may have covered older crust and then been partially cratered by the declining bombardment, thus producing the intercrater plains. Sometime near the end of cratering a planetesimal over 100 km in diameter (perhaps the size of Rhode Island) smashed into the planet and blasted out Caloris Basin. At about the same time, the cooling core contracted, and the crust broke to form the lobate scarps.

The third stage, flooding by lava, may have begun with the formation of the intercrater plains, but lava flows became more common soon after the Caloris impact. In fact, that impact may have triggered the flows. These upwellings of molten lava occurring after the end of cratering, produced the regions known as the smooth plains. This stage of flooding by lava ended quickly, perhaps because the shrinkage of the planet squeezed off the crustal fractures through which the lava reached the surface. Because it has no atmosphere, Mercury has never been subjected to flooding by water, and its surface has evolved very slowly since the end of the lava flows.

The fourth stage in Mercury's evolution, slow surface evolution, is now limited to micrometeorites, which grind the surface to dust; rare larger meteorites, which leave bright rayed craters; and the slow but intense cycle of heat and cold, which weakens the rock at the surface. The planet's crust is now thick, and, although its core may be partially molten, the surface is not actively erasing craters and modifying the landscape. To find such an active planetary crust, we must look at a larger planet such as Venus.

21.2 VENUS

An astronomer once became annoyed when someone referred to Venus as a planet gone wrong. "No," she argued, "Venus is probably a fairly normal planet. It is the earth that is peculiar. The universe probably contains more planets like Venus than like the earth." That may be true, but many astronomers still think of Venus as a once earth-like planet that has developed into a very unpleasant place.

In many ways, Venus is a twin of Earth, and we might expect the two planets to be quite similar (Data File 5). Venus is nearly 95 percent the diameter of Earth and has a similar average density. It formed in the same general part of the solar nebula, so we might expect it to have a composition similar to Earth's. Planets of Earth's size cool slowly, so we should expect that Venus has a molten metallic interior and an active crust.

Unfortunately, we cannot see the surface of Venus because of a perpetual layer of thick white clouds that completely envelops the planet (Data File 5a). From the time of Galileo to the early 1960s, astronomers could only speculate about Earth's twin, and science fiction writers supposed that Venus was a steamy swamp planet inhabited by strange creatures.

During the last few decades astronomers have peered below the clouds using radar, 18 spacecraft have flown past or orbited Venus, and 17 probes have landed on its surface. The resulting picture of Venus is dramatically different from the murky swamps of fiction. In fact, the surface of Venus is dryer than any desert on Earth and hot enough to melt lead.

If Venus is not a planet gone wrong, it is certainly a planet gone down a different evolutionary path. How did Earth's twin become so different?

THE ROTATION OF VENUS

One way Venus differs from Earth is that it rotates backward. Except for Uranus, which rotates on its side, all other planets in the solar system rotate from west to east (counterclockwise as seen from the north). In 1962 radar observations showed that Venus rotates from east to west (retrograde) with a period of 243.01 days.

This rotation may be linked to Earth. As Venus revolves around the sun and rotates on its axis, it turns nearly the same face toward Earth each time the two planets come near each other. This might arise if Venus were tidally locked to Earth, but a perfect resonance would require a period of rotation of 243.16 days. Venus rotates slightly too slowly to be tidally coupled with Earth.

An alternate explanation is that the rotation of Venus was altered during the final accretion of the planet. A very large planetesimal striking the planet off center could alter its rotation. This arbitrary explanation seems inadequate to many planetary astronomers.

THE ATMOSPHERE OF VENUS

Although Venus is Earth's twin in size, the composition, temperature, and density of its atmosphere make it the most inhospitable of planets. About 96 percent of its atmosphere is carbon dioxide, and 3.5 percent is nitrogen. The remaining ½ percent is water, sulfuric acid (H_2SO_4),

DATA FILE 5

Venus

Average distance from the sun	0.7233 AU (1.082×10^8 km)
Eccentricity of orbit	0.0068
Maximum distance from the sun	0.7282 AU (1.089×10^8 km)
Minimum distance from the sun	0.7184 AU (1.075×10^8 km)
Inclination of orbit to ecliptic	2°23′40″
Average orbital velocity	35.03 km/sec
Orbital period	0.61515 y (224.68 days)
Period of rotation	243.01 days (retrograde)
Inclination of equator to orbit	177°
Equatorial diameter	12,104 km (0.95 $D_\oplus$)
Mass	4.870×10^{24} kg (0.815 $M_\oplus$)
Average density	5.24 g/cm^3 (4.2 g/cm^3 uncompressed)
Surface gravity	0.903 earth gravities
Escape velocity	10.3 km/sec (0.92 $V_\oplus$)
Surface temperature	745 K (472°C or 882°F)
Albedo (cloud tops)	0.76
Oblateness	0

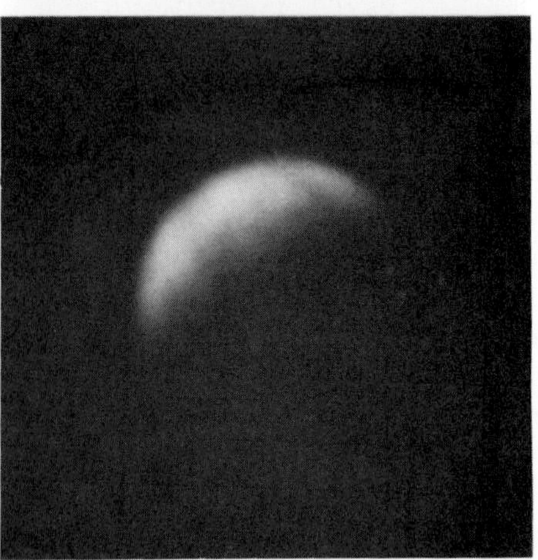

a. Ultraviolet photos of Venus taken from Earth reveal changing weather patterns on three separate days. (Lick Observatory photograph.)

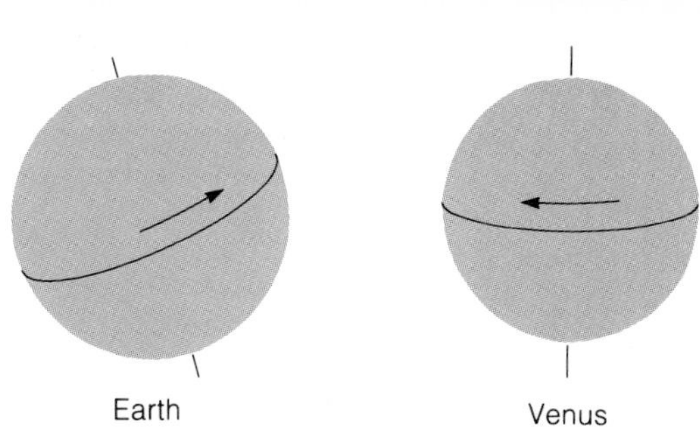

b. Earth and Venus are nearly the same size, but Venus rotates in the retrograde direction.

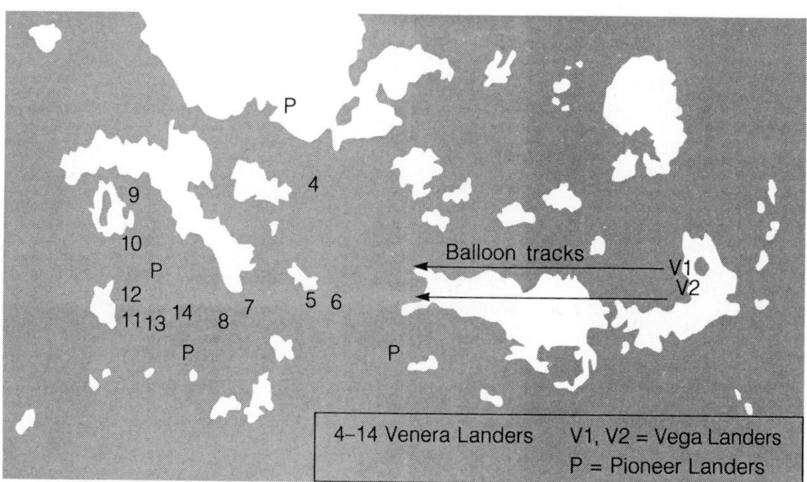

c. Landing sites of spacecraft on Venus. Arrows show the paths of balloons dropped into the atmosphere by the Vega spacecraft.

hydrochloric acid (HCL), and hydrofluoric acid (HF) (Table 21–1). In fact, the thick clouds that hide the surface are believed to be composed of sulfuric acid droplets and microscopic sulfur crystals.

Soviet and American spacecraft have dropped probes into the atmosphere of Venus, and those probes have radioed data back to Earth as they fell toward the surface. These studies show that the cloud layers are much higher and much more stable than those on Earth. The highest layer of clouds, the layer we see from Earth, extends from 68–58 km (Figure 21–6). For comparison, the highest clouds on Earth do not extend higher than about 16 km. Another cloud layer lies between 56–52 km and a third, the densest, between 52–48 km. Thin haze extends down to about 33 km, below which the air is surprisingly clear.

These cloud layers are highly stable because the atmospheric circulation on Venus is much more regular than that on Earth. The heated atmosphere at the **subsolar point**, the point on the planet where the sun is directly overhead, rises and spreads out in the upper atmosphere. Convection circulates this gas toward the dark side of the planet and the poles where it cools and sinks. This circulation produces 300 km/h jet streams in the upper atmosphere, which carry the atmosphere from east to west (the same direction the planet rotates) so rapidly that the entire atmosphere seems to rotate with a period of only 4 days. As evidence of these circulations we can see a distinctive Y pattern in ultraviolet images of the cloud pattern (Figure 21–7).

The high-speed winds in the upper atmosphere were studied in June 1985 when the Vega 1 and Vega 2 spacecraft, on their way to rendezvous with Comet Halley, flew plast Venus and dropped instrumented balloons into the upper atmosphere. Floating at an altitude of 55 km, the 3.5-m diameter balloons found winds as high as 240 km/h. In the 46 hours the balloons survived, they were carried one-third of the way around Venus by the high-altitude winds (Data File 5c).

Table 21–1 The atmosphere of Venus.

Gas	Percent by Volume
Carbon dioxide (CO_2)	96
Nitrogen (N_2)	3.5
Water vapor (H_2O)	0.1 ?
Carbon monoxide (CO)	0.002
Sulfur dioxide (SO_2)	0.0002
Argon (Ar)	0.00007
Hydrochloric acid (HCl)	0.00002
Neon (Ne)	0.000005
Hydrofluoric acid (HF)	0.0000001

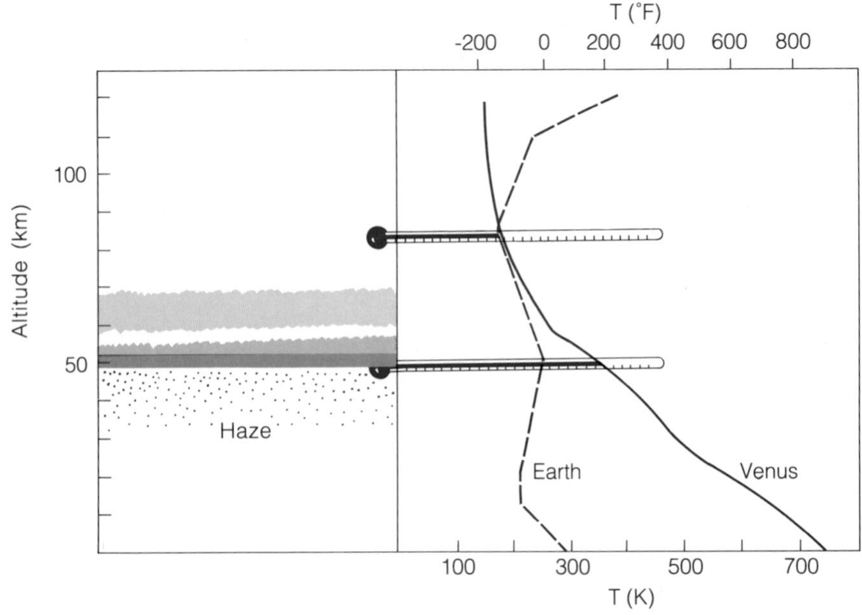

FIGURE 21–6 The four principal cloud layers in the atmosphere of Venus are shown at the left. Thermometers inserted into the atmosphere would register temperature as indicated in the graph at the right. The dashed line shows the temperature in Earth's atmosphere for comparison.

The details of this atmospheric circulation are not well-understood, but it seems that the slow rotation of the planet is an important factor. On Earth, large-scale circulation patterns are broken into smaller cyclonic disturbances by Earth's rapid rotation. Because Venus rotates more slowly, its atmospheric circulation is much less disturbed.

Although the upper atmosphere is cool, the lower atmosphere is quite hot (Figure 21–6). Instrumented probes that have reached the surface report that the temperature is 745 K (880°F) and the atmospheric pressure is 90 times that of Earth's. Earth's atmosphere is 1000 times less dense than water, but on Venus the air is only 10 times less dense than water. If we could survive the unpleasant composition, intense heat, and high pressure, we could strap wings to our arms and fly.

The present atmosphere of Venus is the result of the planet's proximity to the sun. It is 30 percent closer than Earth and receives twice as much solar energy per square centimeter. Thus when it formed, it was warmer than the Earth, and this prevented liquid water from condensing and forming oceans, or it evaporated the oceans soon after they formed. In either case, Venus lacked oceans to absorb carbon dioxide and convert it into mineral deposits. While Earth was removing carbon dioxide from its atmosphere via its oceans, Venus was adding more and more to its atmosphere via volcanism and outgassing. When its atmosphere grew rich in carbon dioxide, its fate was sealed.

THE VENUSIAN GREENHOUSE

A carbon dioxide atmosphere can trap heat by a process called the **greenhouse effect.** When sunlight shines through the glass roof of a greenhouse, it heats the benches and plants inside (Figure 21–8). The warmed interior radiates heat in the form of infrared radiation, but the infrared photons cannot get out through the glass. Heat is trapped within the greenhouse, and the temperature climbs until the glass itself grows warm enough to radiate heat away as fast as the sunlight enters.* In the case of a planet,

*A greenhouse also grows warm because the walls prevent the warm air from mixing with the cooler air outside.

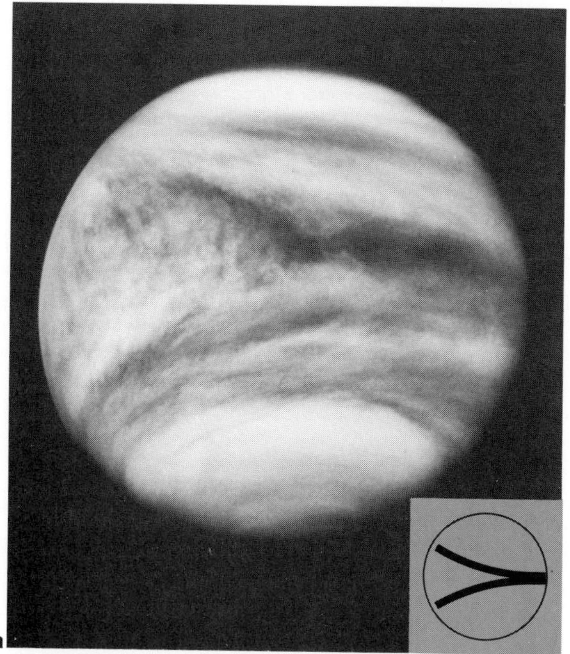

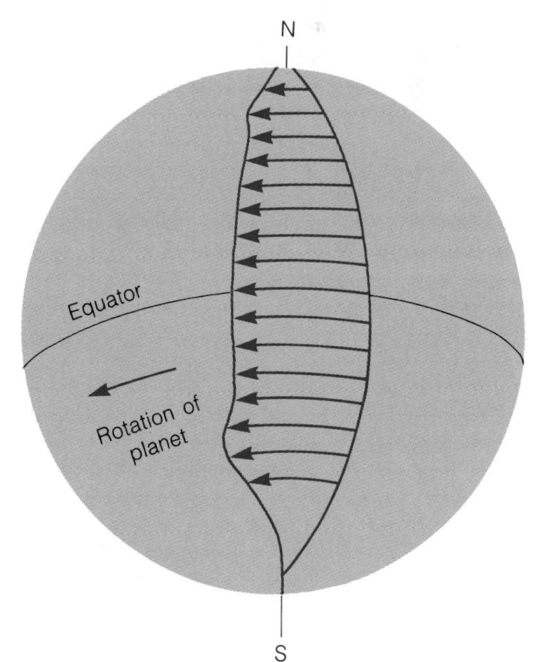

FIGURE 21–7 Photographed in the ultraviolet by Mariner 10, the Venusian clouds show variations in brightness due to differences in composition (a). The cloud circulation patterns are apparent as a Y shape (inset) caused by the high-altitude jet streams blowing from east to west around the planet (b.) (a, NASA.)

carbon dioxide in the atmosphere admits sunlight and the surface grows warm. However, carbon dioxide is not transparent to infrared radiation, so heat is trapped, and the planet's surface temperature rises.

Venus was caught by a runaway greenhouse effect. The rising temperature baked more carbon dioxide out of the surface, and the atmosphere became even less transparent to infrared, which forced the temperature even higher. The surface is now so hot even chlorine and fluorine have baked out of the rock and formed hydrochloric acid and hydrofluoric acid vapor.

Earth avoided this runaway greenhouse effect because it was farther from the sun and cooler. Thus, it could form and preserve liquid-water oceans to absorb the carbon dioxide, which left a nitrogen atmosphere that was relatively transparent in some parts of the infrared. If all the carbon dioxide in Earth's sediments were put back into the atmosphere, our air would be as dense as that of Venus.

This points out the risk we run by burning fossil fuels and adding carbon dioxide to Earth's air. Since the late 1950s, the Scripps Institution of Oceanography has monitored the abundance of carbon dioxide in Earth's atmosphere. The abundance of carbon dioxide has been increasing year by year (Figure 21–8c). Our burning coal, oil, and natural gas has added at least 40 gigatons (40×10^9 tons) of carbon to the atmosphere in 20 years.

This rise in carbon dioxide is increasing the greenhouse effect on Earth and is believed to be causing a slight warming of our climate. One study has found a 0.7°C increase in the average temperature since 1861. In addition to the danger of reducing agriculture's ability to feed the 5 billion people on Earth, we should be concerned about melting part of the ice caps. That would raise the sea level and inundate low-lying regions. For example, a rise of 8 m would flood the White House, and a rise of only 1 m would be disasterous to many ocean-side cities. Currently, earth's oceans appear to be rising at the rate of about 2.4 mm per year. Although it is unlikely that we will soon trigger a runaway greenhouse effect on Earth, we may have already added enough carbon dioxide to the atmosphere to change our climate for the next century.

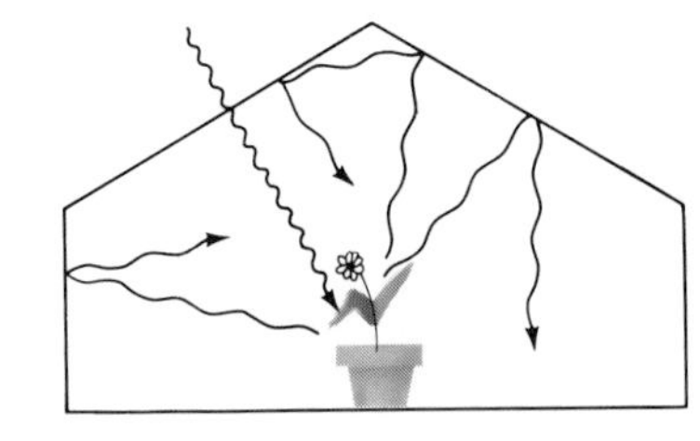

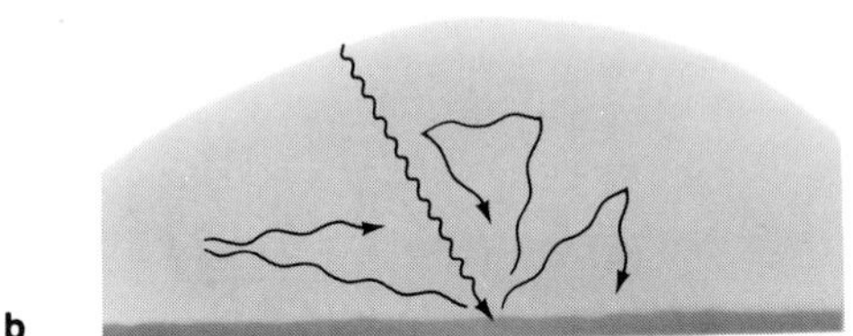

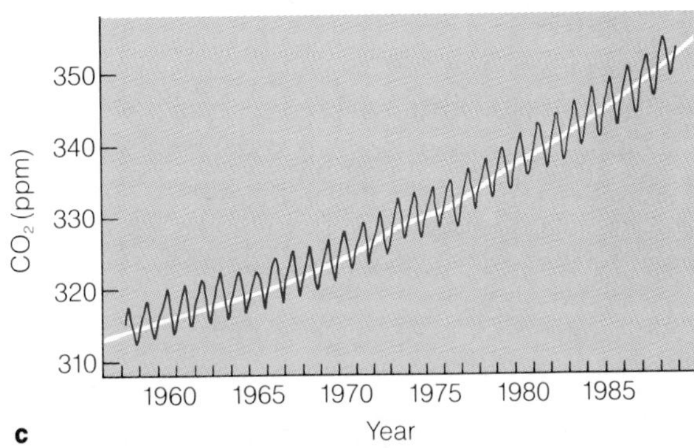

FIGURE 21–8 The greenhouse effect. (a) Short-wavelength light can enter a greenhouse and heat its contents, but the longer-wavelength infrared cannot get out. (b) The same process heats Venus because its atmosphere of CO_2 is not transparent to infrared. Because we are burning fossil fuels, we are increasing the abundance of CO_2 in Earth's atmosphere (c). The wavy line shows the seasonal variation caused by photosynthesis; the smoother line is the averaged trend.

THE SURFACE OF VENUS

Although we cannot see below the clouds of Venus, we can discuss the surface of Venus with some assurance. In this section, we will discuss the features we find there, and in the next section we will trace the history of the geology of Venus.

Most of what we know about Venus comes from radar maps. Since 1965 radio astronomers have been able to map portions of Venus by bouncing radio signals off the surface. Such studies are done only when the planets are nearest each other, and the travel time for the radio signals is then about 5 minutes. The intensity, time delay, and frequency spread of the reflected signal contain enough information to allow a computer to draw a radar map of the reflecting region, showing details as small as 2 km in

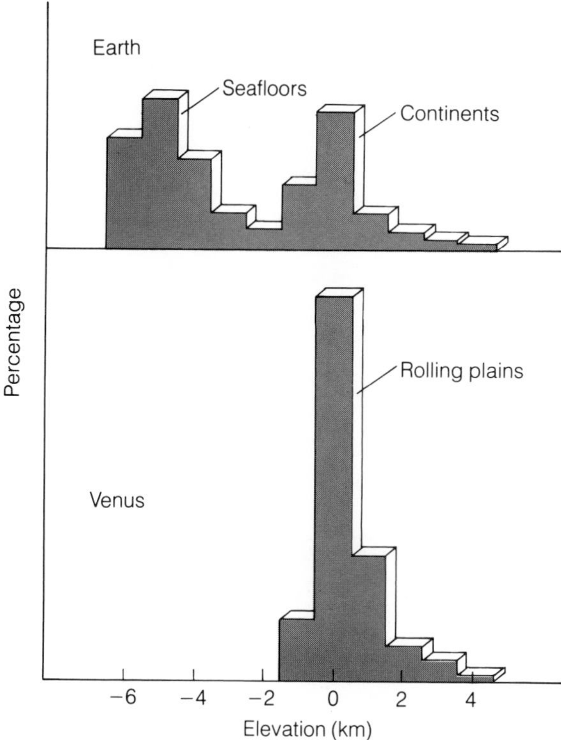

FIGURE 21–9 Earth's surface contains two major types of terrain, but Venus's surface has only one major type—rolling plains.

diameter (Color Plate 35). The maps reveal impact craters, volcanoes, mountains, and valleys.

In December 1978 two Pioneer spacecraft arrived at Venus and began mapping the surface by radar. The resulting radar map is shown in Color Plate 33. It has a resolution of 25 km and covers most of the planet, including many areas that could not be mapped from Earth. The Soviet Venera 15 and 16 spacecraft took up orbits around Venus in October 1983 and mapped about one-fourth of the surface centered on the North Pole at resolutions as good as 1.3 km.

The terrain of Venus differs from that of Earth in a fundamental way. Seafloors cover 65 percent of Earth, and continents, which rise 4 km higher, cover 35 percent. High mountains and deep trenches in the seafloor are quite rare. Thus, Earth's surface is divided into two major types of terrain—seafloors and continents.

Venus is different in that it has only one major type of terrain. Roughly 70 percent of the planet is covered by gently **rolling plains** (Figure 21–9). Only 20 percent is lowlands, and 10 percent is highlands.

We should not think of the rolling plains as ocean beds. Color radar maps such as Color Plate 33 often depict the rolling plains blue, but remember that there is no liquid water on Venus at all. The rolling plains are a vast, dry desert, which we should think of as the average surface of the planet. Fully 80 percent of the planet's terrain lies within 2 km of this average radius.

The highlands resemble continents only in that they rise above the average surface (Figure 21–9 and Color Plate 33). One of these areas, Ishtar Terra,* is larger than the United States and contains some of the most dramatic terrain in the solar system. At its eastern edge a great mountain called Maxwell Montes thrusts up 12 km. (Mt. Everest, at 29,028 feet, is only 8.8 km high.) Radar images show a circular feature 80 km in diameter at the summit (Color Plate 35), which many experts believe is evidence that Maxwell is a volcano. Bounded by lower mountain ranges on the north and west, the western portion of Ishtar Terra contains a great plateau called Lakshmi Planum. Its southern edge is marked by steep flanks rising 4 km above the rolling plains.

Beta Regio is the smallest of the three major highland areas, but it may be the most interesting. It can be studied in some detail by radar imaging from Earth, and those studies suggest it is a volcanic region containing two volcanic peaks named Theia Mons and Rhea Mons. The newest studies show that the two peaks are associated with a system of faults and valleys. It seems possible that the volcanism was triggered when the crust broke under tension and opened a path for magma from the interior.

The highlands seem related to volcanism, but the lowlands may be related to cratering. Impact craters have been found in many places over the rolling plains, but the generally circular outline of the lowlands, their smoothness, and their depression about 1.5 km below the rolling plains suggest old crater basins flooded by lava. The present radar maps do not have sufficient resolution to clear up the issue.

Any discussion of the history of the highlands, rolling plains, and lowlands of Venus must be speculative because we have very little firsthand evidence. Of course, no one has ever visited the surface of Venus, but a few robot spacecraft have landed. The heat and atmospheric pressure destroyed the spacecraft in about an hour, but in that hour they transmitted data and pictures back to Earth (Figure 21–10). Only four of the spacecraft carried cameras, but those reveal a rocky, desert world.

*Ishtar was the Babylonian goddess of love. Except for Maxwell and Beta Regio, place names on Venus are female.

a

b

FIGURE 21–10 The surface of Venus as photographed by Venera spacecraft. (a) Venera 13 photographed a site marked by loose soil; (b) Venera 14 landed on an older region containing cracked slabs (see Color Plate 34). (Sovfoto.)

If you ever visit Venus and go walking just east of the volcanoes of Beta Regio, keep a sharp lookout for the hulk of Venera 9. It landed there in October 1975. The photos it sent back in the hour that it survived (Figure 21–10) show a surface of jumbled angular stone blocks as large as 40 cm (16 inches) in diameter. The angular stones and the lack of dust and soil suggest that the region is young. That is, the rocks were broken up not long ago. A few of the blocks contain holes characteristic of vesicular basalts—typical of volcanism.

The remains of Venera 10 sit 2200 km away, where it landed at about the same time as Venera 9. The terrain around Venera 10, however, looks older, and it has more rounded rocks and more loose soil and dust.

In March 1982 Venera 13 and 14 landed on Venus and sent back more photos. Venera 13 landed in the rolling plains near Beta Regio and sent back photos of an old surface of loose soil and dust. Venera 14 landed 1000 km away in a lowland region and photographed a surface that looks like cracked slabs of lava or cemented volcanic ash. The lack of dust and loose debris suggests that the area is young.

In addition to photographing the surface, seven Soviet landers have carried instruments to analyze the soil. Although the results from these probes are difficult to match to known kinds of rock on Earth, it is clear that all six probes with functioning instruments landed on basalt surfaces. Most of the basalts seem similar to those found in earth's seabeds.

Because basalts are produced by solidified magma, these results suggest that the surface of Venus has been dominated by volcanism and that most of the surface is covered by basaltic lava flows of various ages. The history of the surface of Venus has evidently been dominated by volcanism.

A HISTORY OF VENUS

The four-stage history of Venus is not well-understood. We are not sure how it formed and differentiated, was cratered, flooded, or evolved a surface.

Venus formed near Earth in the solar nebula, so we might expect it to be a similar planet differentiated into a silicate mantle and a molten iron core. Such a molten core should have a magnetic field, but no spacecraft has ever detected such a field. The field must be at least 25,000 times weaker than Earth's. Some theorists believe that even the slow rotation of Venus would be sufficient to

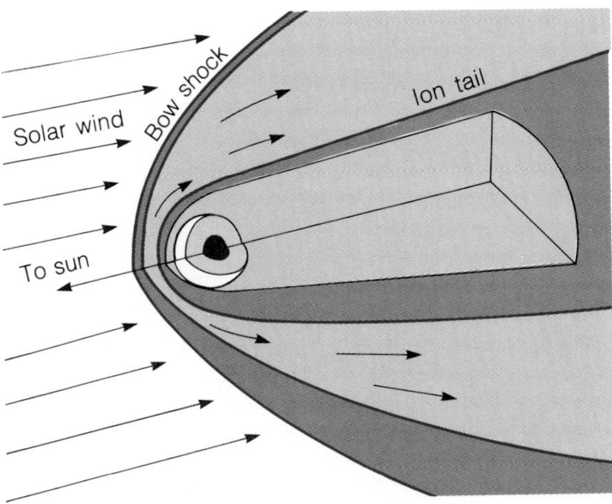

FIGURE 21–11 By analogy with Earth, the interior of Venus should contain a molten core (estimated here), but no spacecraft has detected a planetary magnetic field. Thus Venus is unprotected from the solar wind, which strikes the planet's upper atmosphere and is deflected into an ion tail.

generate a magnetic field if the planet had a molten, conducting core, so the lack of a field remains unexplained.

Because the planet lacks a magnetic field, the solar wind is deflected by the uppermost layers of the planet's atmosphere, thus forming a bow shock where the solar wind is slowed and deflected (Figure 21–11). The magnetic field carried by the solar wind drapes over Venus like seaweed over a fish hook, forming a long tail within which ions flow away from the planet. We will see in Chapter 24 that comets, which also lack magnetic fields, interact with the solar wind in the same way.

The interaction of Venus with the solar wind is clear evidence that the planet has little or no magnetic field. Thus, we know little about the differentiation of the planet into core and mantle. The size of the core shown in Figure 21–11 is estimated by analogy with Earth.

We know a little bit more about the cratering history of Venus. Radar maps have found many craters ranging from hundreds of kilometers down to the resolution limit. Recent results from the Soviet radar-mapping satellites suggest that craters are less common than they should be on an old crust. Soviet scientists estimate that the average age of the crust they have mapped may be about 1 billion years. This suggests a planet whose surface is more active than our moon but less active than Earth.

The third stage of planetary evolution is flooding by lava and water. We have only slight evidence that lava flooding has occurred. The round lowlands may resemble crater basins, and the composition of surface rocks resembles the products of lava flooding, although this may have originated in more recent volcanism.

Flooding by water is a possible episode in the history of Venus. We do have evidence that Venus was once rich in water, but it may never have been a liquid on the surface. As a Pioneer spacecraft descended through the atmosphere in 1978, it detected about 100 times more deuterium per hydrogen atom than on Earth. This abundance of heavy hydrogen suggests that Venus was once a very wet planet. But the water was broken up by ultraviolet light, and the hydrogen was lost to space. Because the deuterium atoms are twice as heavy as hydrogen, they are lost more slowly and are now more abundant.

The oxygen from this water has apparently formed compounds in the surface rocks. At room temperature these oxides would make the rocks rust red, but at the higher temperatures on Venus the minerals are dark gray. These conclusions are based on the analysis of rock color as photographed by the Venera spacecraft and on laboratory experiments, which subjected appropriate oxides to high temperatures (Color Plate 34).

If it was cool enough when it was young, Venus may have had oceans. The minimum amount of water needed to explain the present abundance of deuterium could have covered the entire surface of Venus in an ocean 10 m deep. For contrast, the present water vapor in the Venusian atmosphere would make a planet-wide ocean only 30 cm thick, whereas water on Earth would cover Earth with a layer 3 km deep. If Venus was not initially too hot to prevent the water from forming oceans, it gradually grew hotter because it is closer to the sun. Any oceans evaporated, and the water was lost as the greenhouse effect made the planet even hotter. Had Earth been just 5 percent closer to the sun, it might have suffered a similar fate.

The last stage of planetary evolution is slow surface evolution. Weathering and erosion may not be very important on Venus because it is so dry. Water is the principal erosive agent on Earth. Also, the thick atmosphere of Venus has very low winds at the surface.

The evolution of Earth's surface is dominated by plate tectonics, but the radar maps of Venus show only slight evidence of such activity (Figure 21–12). No systems or rift valleys outline moving plates, and no linear midocean rises suggest plate spreading. The current maps, however, may not have sufficient resolution to detect such features. A radar-mapping satellite called Magellan will map the planet with much better resolution of 0.3 to 0.1 km. Launched by the Space Shuttle in April 1989, Magellan is scheduled to reach Venus in August 1990.

Fully 70 percent of the heat from Earth's interior flows out through the midocean rifts. Lacking these, Venus seems to get rid of its interior heating through volcanism. Maxwell Montes and the two peaks of Beta Regio appear to be volcanic. They are not, however, the steep-sided peaks that are characteristic of the volcanoes that erupt along the edges of colliding plates on Earth. Mt. St. Helens and Mt. Fuji are such steep-sided cones. Instead, the volcanoes on Venus are low-profile cones, termed shield volcanoes (Figure 21–13). On Earth, such volcanoes occur over hot spots above rising currents of molten rock in the mantle. Hawaii is an example. Thus, the volcanoes of Venus suggest hot spot activity, not plate tectonics.

It is possible that the volcanoes of Venus are erupting now. Pioneer Orbiter detected a 90 percent decrease in atmospheric SO_2 since 1978. This compound breaks down over time, so it must have been produced recently, probably in a major volcanic outburst. Venus may be one of the most volcanically active planets in our solar system.

With all of this heat flowing outward, why doesn't the crust break into moving plates? Again the answer may be the heat at the surface. The crust is so hot and dry that it is 12 percent less dense than Earth's crust. It may be too buoyant to sink, and if one plate can't slip under another plate, plate tectonics cannot get started. Also, the crust contains very little water. On Earth, water weakens rock structures and lubricates faults.

To astronomers trying to understand the laws of planetary evolution, Venus is a disappointment. Although it is Earth's twin in size, it is dramatically divergent in its atmosphere and geology. Ironically, the smaller Mars has a more earthlike climate.

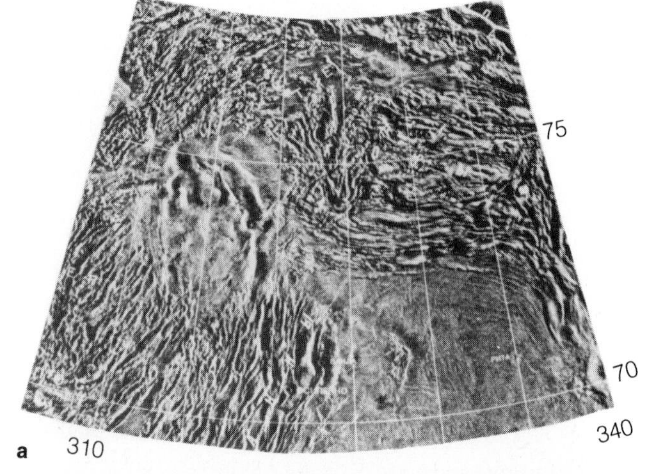

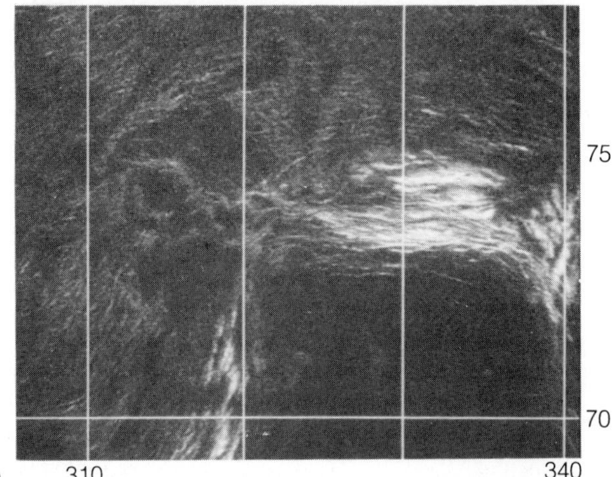

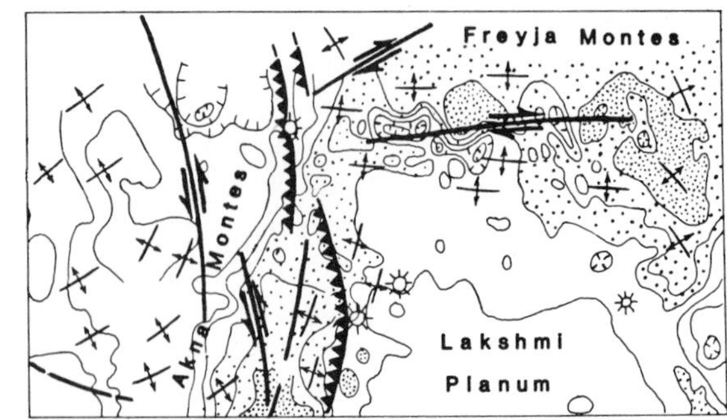

FIGURE 21–12 Two radar images of the mountains north and west of the high plateau Lakshmi Planum show evidence of crust wrinkling like a rug pushed up on a slippery floor. The radar image in (a) was made by the Venera orbiter and shows the folded mountain ranges, whereas the radar image in (b) was made using the Arecibo radio telescope. A topographical map (c) shows the location of faults and the direction of motion in the crust. (a, USSR Academy of Sciences; b and c, Arecibo Observatory, Puerto Rico.)

21.3 MARS

Mercury and the moon are small. Venus and Earth are, for terrestrial planets, large. But Mars occupies an intermediate position. It is twice the diameter of the moon but only 53 percent Earth's diameter. Its small size has allowed it to cool faster than Earth, and much of its atmosphere has leaked away. Its present carbon dioxide atmosphere is only 1 percent as dense as Earth's.

In some ways Mars is much like Earth. A day on Mars is nearly the same length as an Earth day—24 hours and 40 minutes—and its year lasts 1.88 Earth years. Also, just as Earth's axis is tipped 23.5°, Mars's is tipped 24°. As the northern and southern hemispheres turn alternately toward the sun, seasonal changes are visible even through a small telescope. As spring comes to the southern hemisphere,

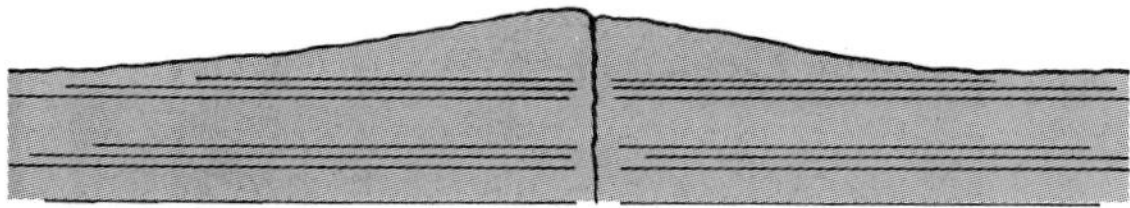

FIGURE 21–13 Shield volcanoes may be very large, but because they form from highly fluid lava, their slopes are not steep.

the white polar cap shrinks and the grayish surface markings grow darker, and, according to some observers, greener. At one time these seasonal changes led some to believe that plant life flourished on Mars when spring thawed the polar cap. We will see later that this is not so.

In other ways, Mars is very different from Earth. Its surface has never broken into moving plates like those that create the topography on Earth. Instead, Mars is a one-plate planet, its crust frozen into a solid layer but bearing signs of geological activity that include some of the largest volcanoes in the solar system.

THE CANALS OF MARS

Even before the space age brought us photographs of the Martian surface, the planet Mars was a mysterious landscape in the public mind. In the century following Galileo's first astronomical use of the telescope, astronomers discovered dark markings on Mars and bright polar caps (Data File 6a). Timing the motions of the markings, they concluded that a Martian day is about 24 hours long, and the similarity with the Earth's day further supported the belief that Mars is another Earth.

Apparent proof that Mars is peopled by intelligent beings appeared late in the summer of 1877, when the Italian astronomer Giovanni Virginio Schiaparelli, using a telescope only 8.75 inches in diameter, thought he glimpsed fine straight lines visible when Earth's atmosphere was unusually still. When he reported his discovery, he used the Italian world *canali,* which means "channel." But the word was translated into English as "canal," an artifically dug channel, and the "canals of Mars" were born. Many astronomers could not see the canals, whereas others drew maps showing hundreds.

In the decades that followed Schiaparelli's discovery, many people assumed that the canals were real water courses built by an intelligent race on Mars to carry water from the polar caps to the lower latitudes. Much of this excitement was generated by Percival Lowell, a wealthy Bostonian who founded Lowell Observatory in 1894, principally for the study of Mars. He not only mapped hundreds of canals, but publicized his results (Data File 6b). By 1907 the general public was so sure that life existed on Mars that the *Wall Street Journal* could suggest that the most extraordinary event of the past year was "the proof by astronomical observations . . . that conscious, intelligent human life exists upon the planet Mars." Further sightings of bright clouds and flashes of light on Mars confirmed this belief, and some urged that gigantic geometrical diagrams be traced in the Sahara Desert to signal to the Martians that earth too is inhabited. All seemed to agree that the Martians are older and wiser than humans.

This fascination with men from Mars was not a passing fancy. Beginning in 1912 Edgar Rice Burroughs wrote a series of 11 novels about the adventures of the earthman John Carter lost on Mars. Burroughs made the geography of Mars, named by Schiaparelli after mythical lands, into household words. He also gave his Martians green skin.

By Halloween night of 1938, people were so familiar with life on Mars that they were ready to believe Earth had been invaded. When a radio announcer repeatedly interrupted dance music to report the landing of a spaceship in New Jersey, the emergence of monstrous beings, and their destruction of whole cities, thousands of otherwise sensible people rushed to their cars and fled in panic, not knowing that Orson Welles and other actors were dramatizing H. G. Wells' book *The War of the Worlds.*

The fascination with Mars, its canals, and its little green men lasted right up to July 15, 1965, when Mariner 4, the first spacecraft to fly past Mars, radioed back photos of the surface and proved that there are no canals and no Martians. Apparently, the canals are optical illusions produced by the human brain's astounding ability to assemble a field of disconnected marks into a coherent image. If your brain could not do this, the photos on these pages would be nothing but swarms of black dots, and the screen of a television set would never make sense. The brain of an astronomer looking for something at the edge of visibility is capable of connecting random markings on Mars into the straight lines of canals.

THE EXPLORATION OF MARS

The Martian surface has been photographed by a number of spacecraft ranging from Mariner 4, which flew by in 1965, to the Viking spacecraft, which orbited Mars in 1976. These photos show that Mars is a cratered world (Figure 21–14) like the moon, but that it also has gigantic volcanoes, deep canyons, dry channels, and vast dust storms.

DATE FILE 6

Mars

Average distance from the sun	1.5237 AU (2.279×10^8 km)
Eccentricity of orbit	0.0934
Maximum distance from the sun	1.6660 AU (2.492×10^8 km)
Minimum distance from the sun	1.3814 AU (2.066×10^8 km)
Inclination of orbit to ecliptic	1°51′09″
Average orbital velocity	24.13 km/sec
Orbital period	1.8808 (686.95 days)
Period of rotation	$24^h\ 37^m\ 22.6^s$
Inclination of equator to orbit	23°59′
Equator diameter	6796 km (0.53 $D_\odot$)
Mass	0.6424×10^{24} kg (0.1075 $M_\odot$)
Average density	3.94 g/cm^3 (3.3 g/cm^3 uncompressed)
Surface gravity	0.379 earth gravities
Escape velocity	5.0 km/sec (0.45 $V_\odot$)
Surface temperature	−140°–20°C (−220°–68°F)
Average albedo	0.16
Oblateness	0.009

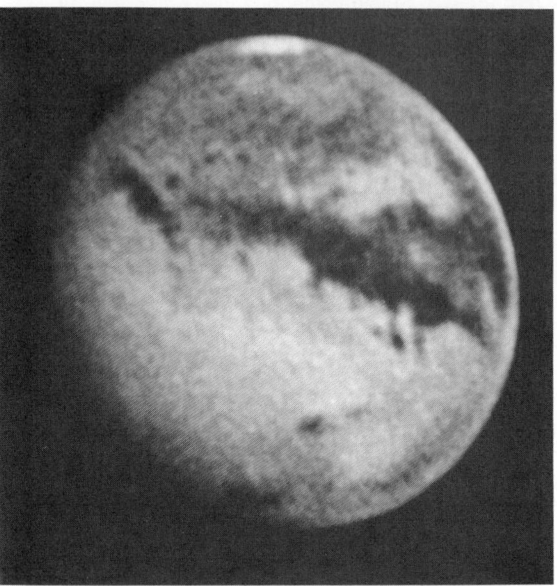

▲ **a.** This computer-enhanced CCD image of Mars reveals about as much detail as is visible to visual observers using the finest telescopes. Compare with globe below. (Courtesy Stephen M. Larson, Lunar and Planetary Lab.)

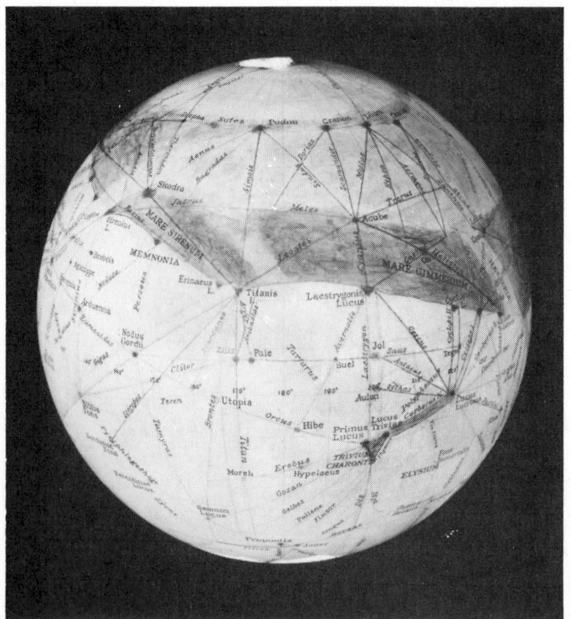

▶ **b.** A globe of Mars showing the canals mapped by Percival Lowell. (Lowell Observatory photograph.)

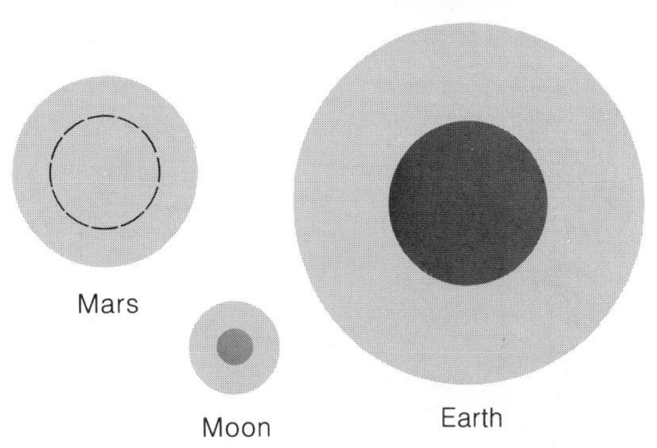

c. Mars is intermediate in size, and the nature of its core is unknown.

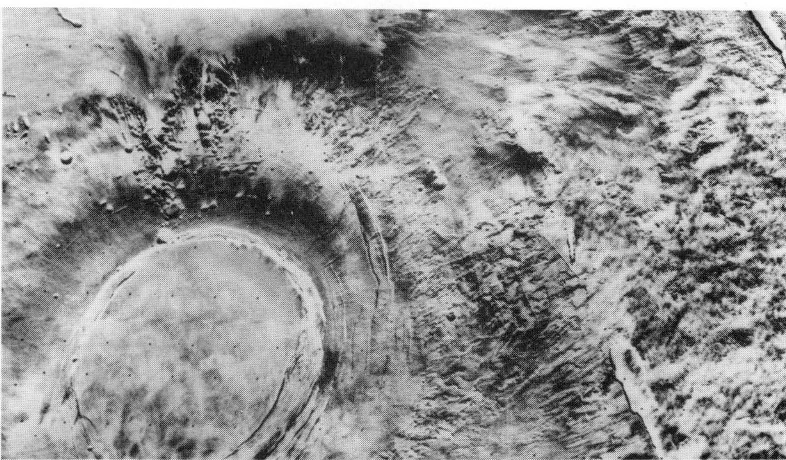

d. Seen from orbit, a volcanic caldera is marked by fractures, collapsed lava tubes (upper left), and smaller impact craters. (NASA.)

e. The Martian surface. Meteorite impacts have broken the surface rock to produce a rugged desert. See Color Plate 37. (NASA.)

The seasonal dust storms are intrinsically interesting because they explain the apparent seasonal changes in the dark markings. Apparently, yearly dust storms cover outcrops of dark rock, and, as spring begins, the winds gradually strip away the dust and the markings grow darker. Because the entire planet has a red color, probably due to iron oxides, the dark markings look greenish in contrast.

Although the canals and seasonal changes caused by vegetation proved illusory,* Mariner 9 and Vikings 1 and 2, which arrived in orbit in June 1976, have shown that Mars is geologically fascinating. The southern hemisphere is old and heavily cratered, giving it a lunar appearance (Figure 21–15). The northern hemisphere, however, has few craters, and those few are much sharper and less eroded and thus younger. The region has been smoothed by repeated lava flows that have buried the original surface, and two areas of volcanic cones have been found. In addition, the northern hemisphere is the site of deformed crustal sections in the form of uplifted blocks and collapsed depressions that suggest geological activity.

Martian volcanoes are of the shield type, showing that the lava flowed easily. Nevertheless, the largest volcano, Olympus Mons, is a vast structure (Figure 21–16). Its base is 600 km (370 miles) in diameter, and it towers 25 km (16 miles) above the surface. In contrast, the largest volcano on Earth is Mauna Loa in Hawaii. It rises only 10 km (6 miles) above its base on the Pacific Ocean floor, and its base is only 225 km (140 miles) in diameter. The weight of Mauna Loa is too great to be supported by the seafloor. It has pushed the seafloor down just as a bowling ball resting on a bed depresses the surface. This depression is visible as an undersea moat ringing Mauna Loa, but no depression is visible ringing Olympus Mons, even though it is 2.5 times higher than Mauna Loa. Evidently the crust of Mars is thicker than Earth's.

Other evidence shows that the Martian crust is thinner than the moon's. The Mariner spacecraft discovered a vast valley at least 4000 km (2500 miles) long and, at its widest, 600 km (400 miles) wide. At its deepest point it is over 6 km (4 miles) deep, making it over four times deeper than the Grand Canyon. On Earth it would stretch from New York to Los Angeles. The valley, now named Valles

*We will discuss the search for life on Mars (and elsewhere) in Chapter 25.

a

b

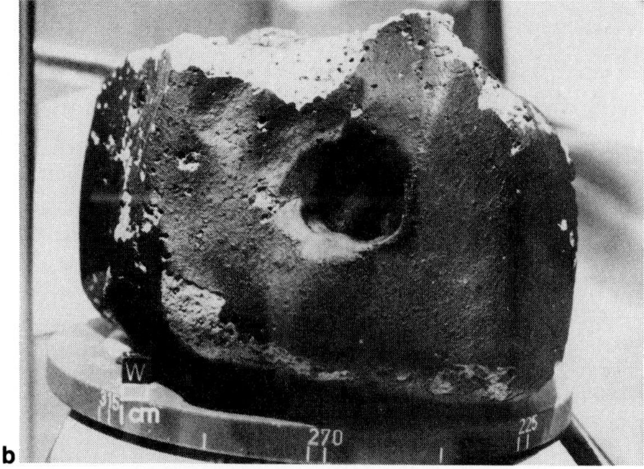

FIGURE 21–14 Impact cratering on Mars. (a) The impact basin Argyre is 740 km in diameter and has a rim composed of uplifted crustal blocks. The large crater at the upper left is Galle with a diameter of 210 km. Note the haze above the horizon. Impacts on Mars may occasionally eject fragments of the crust into space, and a few such fragments could have fallen to Earth as meteorites. (b) This stony meteorite collected in Antarctica has chemical and physical characteristics that suggest it originated on Mars. (NASA.)

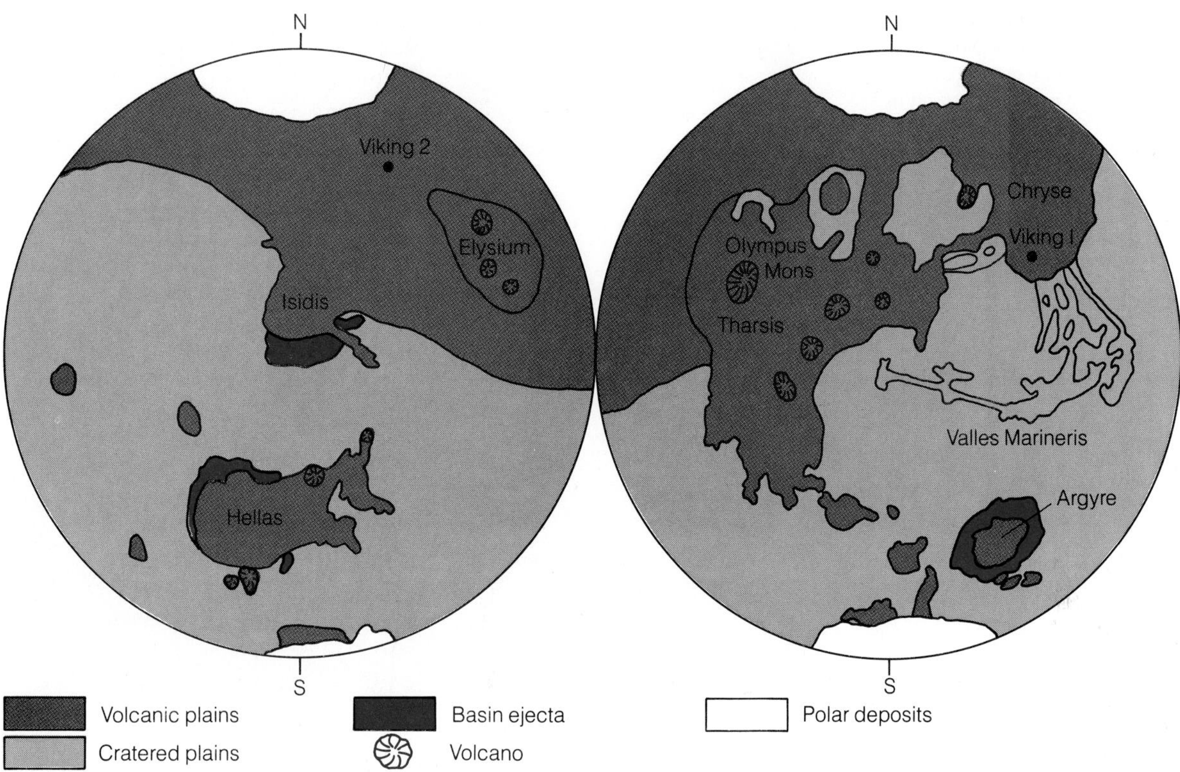

FIGURE 21–15 Geological map of Mars. Most of the southern hemisphere is old, cratered terrain, whereas most of the northern hemisphere is younger, volcanic plains. Note the volcanic areas, Tharsis and Elysium, and the large basins, Hellas and Argyre.

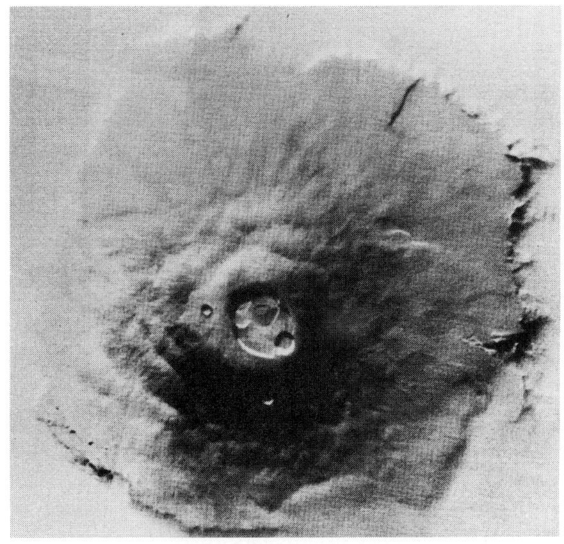

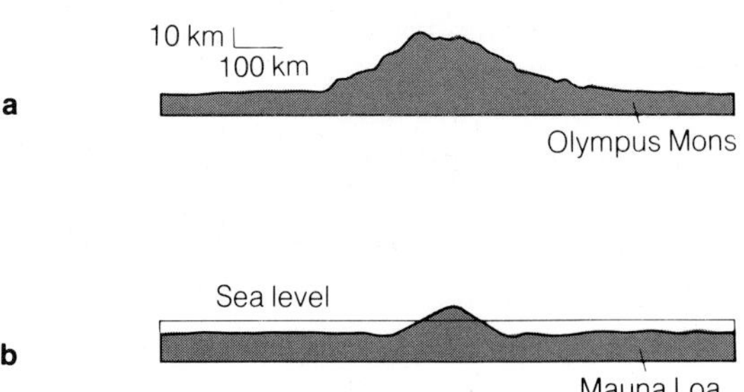

FIGURE 21–16 Olympus Mons, the tremendous shield volcano on Mars, is much larger than Mauna Loa, largest volcano on Earth. Mauna Loa has sunk into the crust, producing a moat around its base. Olympus Mons has no moat, suggesting the Martian crust is stronger than Earth's. (NASA.)

FIGURE 21–17 Valles Marineris, the huge canyon of Mars, would stretch entirely across the United States. It is 4000 km (2500 miles) long and up to 6 km (4 miles) deep. The three circular features at the left are volcanic peaks of the Tharsis region. (Artwork by Stephen P. Meszaros, NASA/Goddard.)

Marineris (Figure 21–17), is apparently a block of crust that has dropped downward along parallel faults. We see no such features on the moon, so we can conclude that the crust of Mars is thinner than the moon's crust. This is what we would expect for a planet that is larger than the moon but smaller than Earth.

Valles Marineris begins in an area of volcanism called the Tharsis bulge and may be related to the stresses created in the crust by this volcanic activity. Olympus Mons is only one of a number of volcanoes in the Tharsis region (Color Plate 36). The area bulges about 10 km (6 miles) higher than the average surface of Mars and has few craters, showing that it is geologically young. Some of the volcanoes in this region, such as Olympus Mons, appear to be less than a billion years old, but others may be very old volcanoes whose flanks have been buried by lava from younger volcanoes. The Tharsis bulge itself may be almost as old as the surface of the planet, but recent volcanism has wiped out surface features from more than about 1 billion years ago.

A similar uplifted volcanic plain, the Elysium region, lies almost halfway around the planet. It is more heavily cratered and appears to be slightly older than the Tharsis region.

We have no evidence that Martian volcanoes are still active, although that is a possibility. Craters in the youngest lava flows on Olympus Mons suggest that the volcano may have been active as recently as a few hundred million years ago. The Elysium volcanoes appear to be inactive.

Although planetary geologists were fascinated to see signs of geological activity on Mars, they were especially excited to see dry streambeds (Figure 21–18) much like the dry arroyos common in America's Southwest. Some of them look like the runoff channels produced by streams and rivers on Earth, but others look like outflow channels produced by sudden flooding. From the number of superimposed craters, experts estimate that the runoff channels formed almost 4 billion years ago, while some of the outflow channels are a billion years younger.

The outflow channels could have been formed by sudden flooding produced when subsurface ice was melted by volcanism or meteorite impact. We can see evidence of this permafrost where meteorite impacts or geological activity has melted it, causing floods and flow formations (Figure 21–19).

The runoff channels, however, are harder to explain. Liquid water cannot exist on Mars today. If water cut the runoff channels, then Mars must have had a thicker atmo-

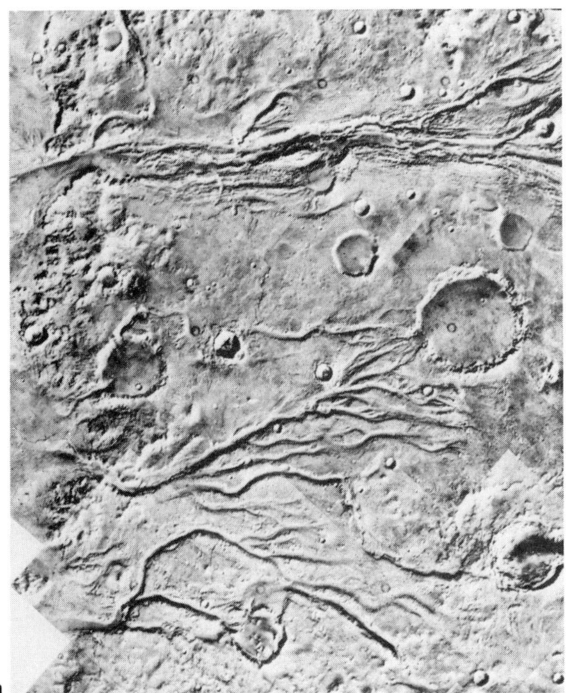

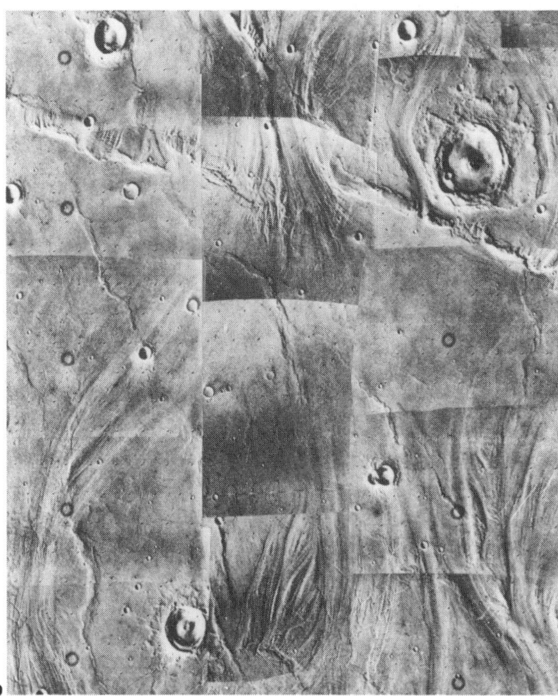

FIGURE 21–18 (a) Dry river channels flow through a cratered Martian plain. (b) Signs of erosion due to a flooding liquid (probably water) mark this cratered region of Mars. (NASA/JPL.)

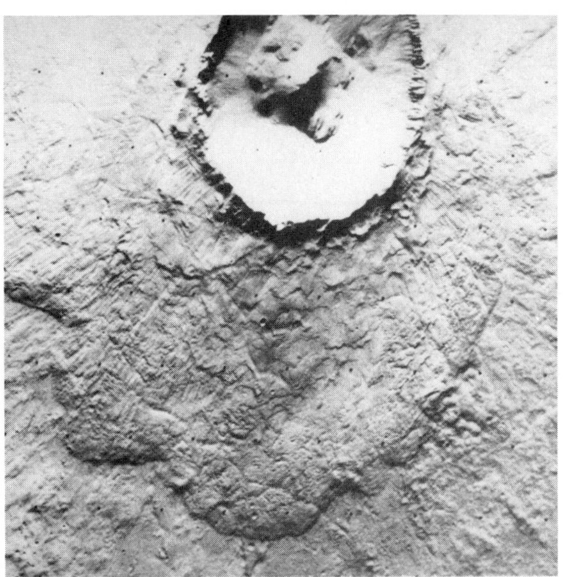

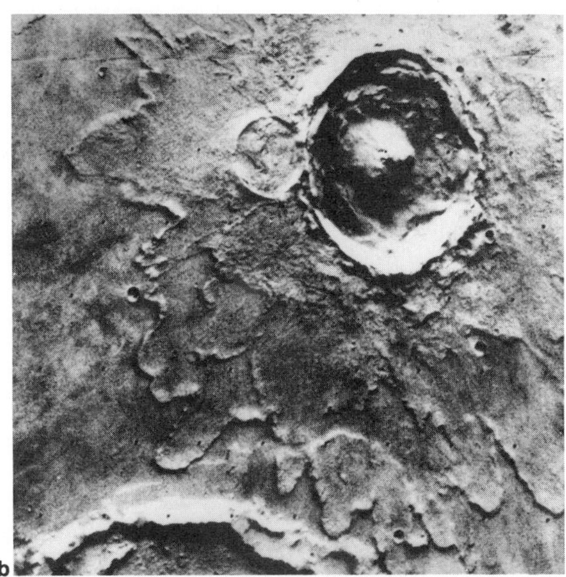

FIGURE 21–19 Some Martian craters such as Arandas (a) and Yuty (b) appear to have formed flow patterns rather than ejecta blankets. This suggests that the impacts melted water frozen in the crust. (NASA.)

sphere to keep the water from boiling away. Although some studies suggest that the runoff channels were not produced by moving water, most experts agree that at least some were. These streambeds tell us that Mars once had a thicker atmosphere, and we can find further evidence near the polar caps.

The Martian polar caps consist of frozen carbon dioxide (dry ice) with frozen water beneath. When spring begins in a hemisphere, the corresponding polar cap begins to shrink as the carbon dioxide turns directly into gas. In the case of the northern cap, the water never melts or vaporizes but stays behind in a smaller permanent ice cap. The region around this permanent cap is marked by layered terrain (Figure 21–20), evidently deposited by wind-borne dust that accumulates on the ice cap. When the cap shrinks, the material accumulates in layers 10–50 m (13–30 ft) thick. The variation in the layering suggests that some periodic change in climate, perhaps due to orbital changes, may affect the frequency and intensity of planetwide dust storms and thus alter the rate at which material is deposited. Note that Earth's climate is suspected of following similar cycles due to changes in its motion. (See pages 36–38.)

THE ATMOSPHERE OF MARS

If you visited Mars you would find that the sky is pink because of fine red dust carried into the atmosphere by winds. The dust is believed to be red because of iron oxides in the soil (Color Plate 37).

The air itself is mostly carbon dioxide with small amounts of nitrogen and argon (Table 21–2). It contains almost no water vapor or oxygen, and its density at the surface of the planet—only 1 percent that of Earth's atmosphere—does not provide enough pressure to prevent liquid water from boiling into vapor. If you visited Mars without a spacesuit, your own body heat would make your blood boil.

To understand Mars, we must ask why its atmosphere is so thin and dry and why the surface is rich in oxides while the atmosphere contains almost no oxygen at all. The answers lie in the origin and evolution of the Martian atmosphere.

The gases in the Martian atmosphere were outgassed from its interior. Because Mars formed farther from the sun, it may have incorporated more volatiles. But, being smaller, it may not have outgassed as much as Earth. In any case, this outgassing occurred early in the planet's history. Mars has cooled and now releases little gas.

a

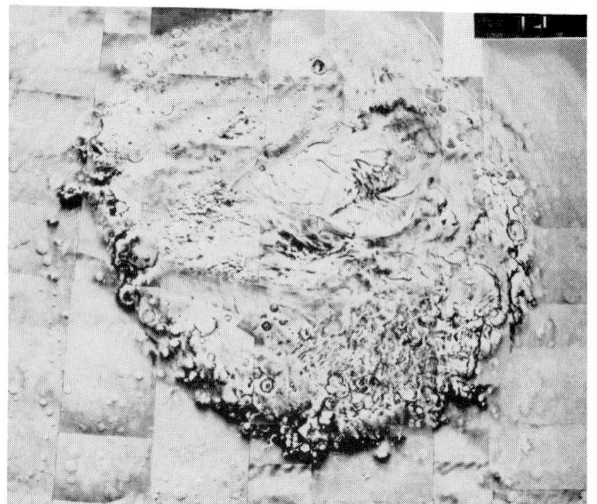

b

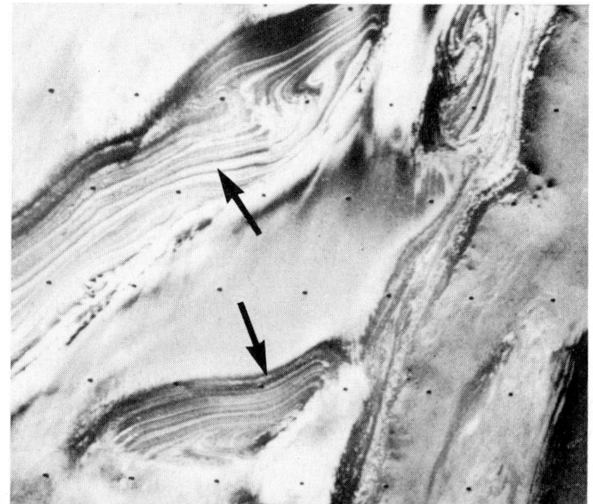

c

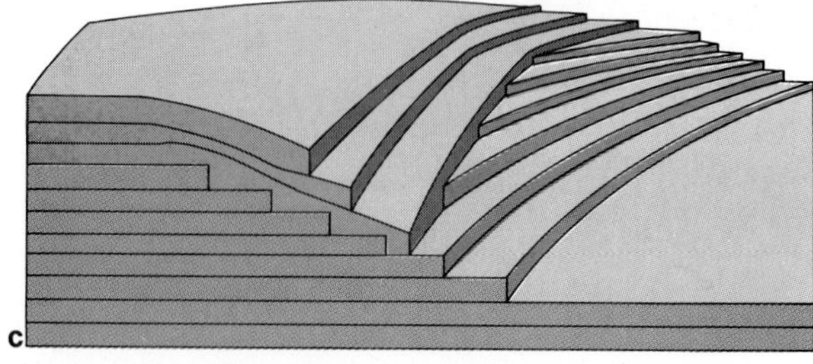

FIGURE 21–20 (a) A computer-enhanced photomosaic of the south polar cap of Mars shows the complexity of the frozen layers. (b) A summer photo taken near the North Pole shows frozen water left behind when the dry ice vaporized. The varied extent of the layered terrain (arrows) is apparently caused by periodic changes in the Martian climate. (c) In places different sets of layers are superimposed. Layers are 10–50 m thick. (a and b, NASA; c, adapted from a diagram by J. A. Cutts, K. R. Blasius, G. A. Briggs, M. H. Carr, R. Greeley, and H. Masursky.)

Table 21–2 The atmosphere of Mars.

Gas	Percent by Volume
Carbon dioxide (CO_2)	95
Nitrogen (N_2)	2.7
Argon (Ar)	1.6
Carbon monoxide (CO)	0.6
Oxygen (O_2)	0.15
Water vapor (H_2O)	0.03 variable

How much atmosphere a planet has depends on how rapidly it releases internal gas and how rapidly it loses gas from its atmosphere. Mars has lost significant amounts of carbon dioxide to its polar caps and water to permafrost in its soil, but it has also lost gases to space.

How rapidly a planet loses gas depends on its mass and temperature. The more massive the planet, the higher its escape velocity (Box 5–2), and the more difficult it is for gas atoms to leak into space. Mars has a mass less than 11 percent of Earth's, and its escape velocity is only 5 km/sec, less than half Earth's.

The temperature of a planet's atmosphere is also important. If the gas is hot, its molecules have a higher average velocity and are more likely to exceed escape velocity. Thus a planet that is near the sun and very hot is less likely to retain an atmosphere than a more distant, cooler planet. However, the velocity of a gas molecule also depends on the mass of that molecule. On average a low-mass molecule travels faster than a massive molecule. Thus, a planet loses its lightest gases more easily because they travel the fastest.

Applying these factors to individual planets, we find that all of the terrestrial planets are too small to retain hydrogen and helium. Those molecules are light and escape into space. Although Venus and Earth can hold onto their remaining gases, Mercury and the moon are too small and have lost all of their air. Mars is barely able to retain water vapor, methane, and ammonia, but it has a better gravitational grip on carbon dioxide.

The argon in the Martian atmosphere is further evidence of a denser blanket of air in the past. Argon atoms are heavy and would not be lost easily. In addition, argon is inert and would not form compounds in the soil. The 1.6 percent argon in the present atmosphere suggests that Mars once had an atmosphere 10 to 20 times denser than it does now.

Mars should have been able to retain some gases such as water vapor, but we would not expect to find much of this gas in its atmosphere now. Such molecules can be broken up by ultraviolet radiation. On Earth the ozone layer protects the lower layers of the atmosphere from ultraviolet photons, but Mars never had an oxygen-rich atmosphere and so never had an ozone layer. Ultraviolet photons from the sun can penetrate deep into the atmosphere and break up molecules such as water. The hydrogen can escape to space, and the oxygen can form oxides. Thus, molecules too heavy to leak into space can be lost if they break into lighter fragments.

Much of the water on Mars may have survived by being hidden from the ultraviolet radiation. The Viking landers found that the soil contains about 1 percent water chemically combined with minerals. This water would not be broken up and lost.

Additional water may be protected by being frozen in the soil or in polar caps. The temperature on Mars only rarely rises above freezing. A hot summer day might be rather pleasant at 20°C (68°F), but at night the thin atmosphere allows the heat to radiate to space, and the temperature falls dramatically. Just before dawn the temperature is typically −88°C (−126°F). These low temperatures would freeze the water out of the atmosphere, and thus Mars may have much of the water it originally outgassed. Future generations may find a way to release that water and make Mars into a new planet for colonization.

A HISTORY OF MARS

The four-stage history of Mars is a case of arrested development. The planet began by differentiating into a crust and core. However, the crust of Mars is rich in iron oxides, and it has a magentic field no stronger than 0.004 times Earth's. It is large enough to be molten at its center, and it spins fast enough for the dynamo effect to generate a strong magnetic field. The weak field suggests that Mars did not form a core of conducting iron compounds.

The Martian crust was thinner when it was young, and impacts broke the crust, triggering lava flows that flooded some of the largest basins such as Hellas and Argyre. One team of astronomers has suggested that a single major impact caused the volcanism so common in the northern hemisphere. Others suggest that the formation of the core caused the formation of the Tharsis bulge and its associated volcanism.

Stresses in the crust caused faulting and the formation of valleys such as Valles Marineris, but the planet cooled too fast to form moving plates. Valles Marineris might be the beginning of a rift valley, but it does not outline a plate. Nor do we see folded mountains that mark colliding plates.

The large size of the Martian volcanoes also tells us that the planet has no moving plates. On Earth, volcanoes like those that formed the Hawaiian Islands occur over rising currents of hot material beneath the crustal plate. Because the plate moves, the hot material breaks through the crust repeatedly and forms a chain of volcanoes instead of a single large feature. The Hawaiian Islands are merely the most recent of a series of volcanic islands called the Hawaiian-Emperor island chain stretching nearly 3800 km (2300 miles) across the Pacific Ocean. The lack of plate motion on Mars could have allowed Martian volcanoes to grow to gigantic proportions.

The fourth stage in the history of Mars has been one of slow decline. As the small planet cooled rapidly, its crust grew thicker than Earth's and volcanism declined.

In addition, the climate seems to have changed. Recently calculated models suggest that Mars may have once rotated at a much steeper angle to its orbit, about 45°, but the rise of the Tharsis bulge tipped its axis to its present 25°. If this is true, then Mars may once have begun with a thicker atmosphere. But the rise of the Tharsis bulge changed the axis of rotation and made the climate colder. Much of the carbon dioxide and water froze out in the polar caps, and at least some of the water became trapped in the soil as permafrost below the surface.

Our history of Mars must include a phenomenon that is not necessary in our histories of Mercury and Venus. Unlike those worlds, Mars has somehow acquired moons. These moons, and the other moons we will study in the outer solar system, contain important clues to the origin of the planets.

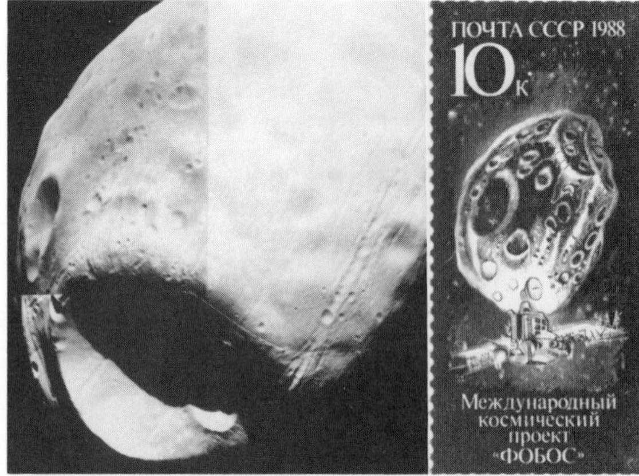

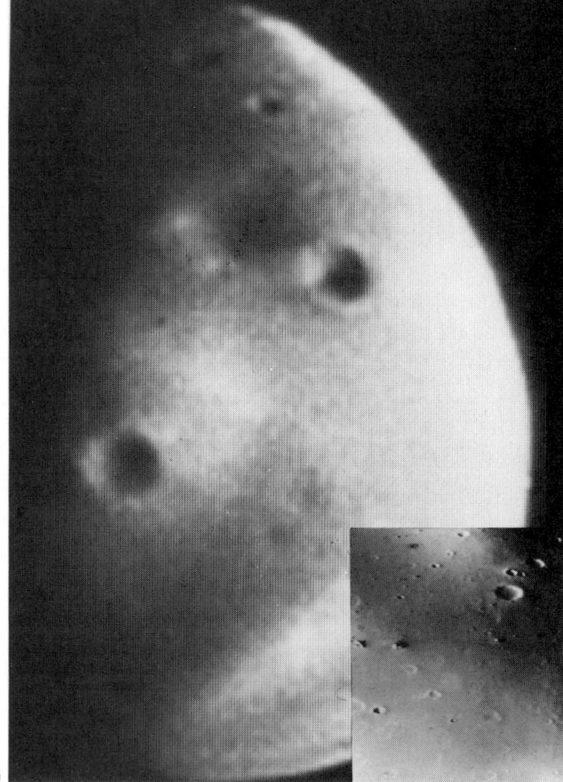

FIGURE 21–21 (a) Phobos, the larger satellite of Mars, is irregular in shape, heavily cratered, and marked by grooves. The inset shows the Soviet stamp honoring the ill fated Phobos probes. (b) Deimos, smaller than Phobos, looks more uniform because of dust covering smaller features (inset). (NASA.)

21.4 THE MOONS OF MARS

If we could camp overnight on Mars, we might notice its two small moons, Phobos and Deimos. Phobos, shaped like a flattened loaf of bread 20 km × 23 km × 28 km (Figure 21–21a), would appear less than half as large as the earth's full moon. Deimos, only 12 km in diameter and three times farther from Mars, would look only one-fifteenth the diameter of earth's moon (Figure 21–21b).

Both moons are tidally locked to Mars, keeping the same side facing the planet as they orbit. Also, both moons revolve around Mars in the same direction that Mars rotates, but Phobos follows such a small orbit that it revolves faster than Mars rotates. Thus Phobos rises in the west and sets in the east.

ORIGIN AND EVOLUTION

Deimos and Phobos are typical of the small, rocky moons in our solar system. They are very dark gray, with albedos of only about 0.06, and they have low densities, about 2 g/cm^3.

Many of the properties of these moons hint that they are captured asteroids. In the outer parts of the asteroid belt almost all asteroids are dark, low-density objects. Massive Jupiter, orbiting just outside the asteroid belt, can scatter such bodies throughout the solar system, so we should not be surprised if Mars, next planet inside the asteroid belt, has captured a few of these as satellites.

However, capturing a passing asteroid into a closed orbit is not so easy that it happens often. The asteroid approaches the planet along a hyperbolic orbit and, if it is unimpeded, swings around the planet and disappears back into space. To convert the hyperbolic orbit into a closed orbit, the planet must slow the asteroid as it passes. Tidal forces might do this, but they would be rather weak. Interactions with other moons or grazing collisions with a thick atmosphere might also slow the asteroid.

Both satellites have been photographed by the Mariner and Viking spacecraft, and those photos show that the satellites are heavily cratered. Such cratering could have occurred while the moons were still in the asteroid belt or in orbit around Mars. In any case, the heavy battering has broken the satellites into iregular chunks of rock, and they cannot pull themselves into smooth spheres because their gravity is too weak to overcome the structural strength of the rock. We will discover that low-mass moons are typically irregular in shape whereas more massive moons are more spherical.

Photos of Phobos reveal a unique set of narrow, parallel grooves (Figure 21–21a). Averaging 150 m wide and 25 m deep, the grooves run from Stickney, the largest crater, to an oddly featureless region on the opposite side of the satellite. One theory suggests that the grooves are deep fractures produced by the impact that formed the crater. The featureless region opposite Stickney may be similar to the jumbled terrain on Earth's moon and the weird terrain on Mercury. All were produced by the focusing of seismic waves from a major impact on the far side of the body. High-resolution photographs show that the grooves are lines of pits, suggesting that the pulverized rock material on the surface has drained into the fractures, or that gas, liberated by the heat of the impact, escaped through the fractures and blew away the dusty soil.

Deimos not only has no grooves, it also looks smoother because of a thicker layer of dust on its surface (Figure 21–21b). This material partially fills craters and covers minor surface irregularities. It seems certain that Deimos experienced collisions in its past, so fractures may be hidden below the debris.

The debris on the surfaces of the moons raises an interesting question. How can the weak gravity of small bodies hold any fragments from meteorite impacts? Escape velocity on Phobos is only 12 m/sec. An athletic astronaut could almost jump into space. Certainly most fragments from impacts should escape, but some could fall back and accumulate on the surface.

Deimos, smaller than Phobos, has a smaller escape velocity. But it has more debris on its surface because it is farther from Mars. Phobos is close enough to Mars that most ejecta from impacts on Phobos will be drawn into Mars. Deimos, being farther from Mars, is able to keep a larger fraction of its ejecta. In fact, Phobos is so close to Mars that tides are making its orbit shrink, and it will fall into Mars or be ripped apart by tidal forces within about 100 million years.

Launched in 1988, two Soviet spacecraft attempted to rendezvous with Phobos, analyze its surface by vaporizing samples with a laser beam, and explore its surface with a lander that could hop from place to place. Phobos 1, the first spacecraft, was accidentally disabled by an error in transmitted instructions soon after launch. Phobos 2 reached Mars in March 1988, rendezvoused with Phobos, and photographed it from a distance before radio contact with earth was lost. The laser and landers were never used.

Deimos and Phobos illustrate three principles. First, some satellites are probably captured asteroids. Second, small satellites tend to be irregular in shape and heavily cratered. And third, tidal forces can affect small moons and gradually change their orbits. We will find even stronger tidal effects in Jupiter's satellite system (Chapter 22).

SUMMARY

All terrestrial planets pass through a four-stage history: (1) differentiation; (2) cratering; (3) flooding of the crater basins by lava, water, or both; and (4) slow surface evolution. The importance of each of these processes in the evolution of a planet depends on the planet's mass and temperature.

Mercury is a small world, only about 40 percent larger than Earth's moon, and, like the moon, it has lost any atmosphere it might have had and has cooled rapidly. Debris from the formation of the solar system cratered Mercury's surface as it did the moon, and lava flows buried parts of that terrain under the intercrater plains. The Caloris Basin, formed near the end of cratering, is the largest crater basin. Soon after the Caloris impact, lava flows created the smooth plains. These lava plains have about the same color as the intercrater plains, so the contrast between plains and highlands is not as obvious on Mercury as on the moon.

The interior of Mercury must contain a large metallic core to account for the planet's high density. A slight shrinkage in the diameter of the planet at the time of the cooling of the core may have led to wrinkling of the crust. This would account for the lobate scarps that mark all of the photographed parts of Mercury.

Venus is Earth's twin in size and density, but the planet has evolved along divergent lines because it is slightly closer to the sun. The higher temperature evaporated any early oceans and prevented the absorption of carbon dioxide from the atmosphere. The accumulating carbon dioxide created a greenhouse effect that produced very high surface temperatures.

The crust of Venus is now dry and of lower density than Earth's crust, and this seems to prevent it from developing plate tectonics. Instead, the planet is losing its heat through hot spot volcanism, which has produced very high volcanoes.

A number of spacecraft have landed on the surface of Venus, but none has survived for more than 2 hours or so. The surface temperature is about 745°C (880°F), and the atmosphere contains traces of sulfuric, hydrochloric, and hydrofluoric acids. Four Russian spacecraft have taken photographs on the surface and transmitted them back to Earth. These photos show rocky terrain of broken and weathered rock. Analysis of rock samples at the landing sites suggests that the rocks are basalts.

The interior of Venus should resemble that of Earth, but no spacecraft has ever detected a magnetic field around Venus. It should have a field if it has a molten metallic core like Earth's. Another mystery associated with Venus is its retrograde rotation. This may have been produced by a major impact during accretion.

Mars is smaller than Earth but larger than Mercury. Although early astronomers saw canals on Mars and thought it to be inhabited, we now know that it is a dry desert world. Its southern hemisphere is a heavily cratered, old region; the northern hemisphere has been resurfaced by lava flows. Much of Mars has been altered by volcanism, and some of its volcanoes are very large. Some may still be active.

Mars may once have had liquid water flowing over its surface. Photographs returned by Mariner 9 and the Viking Orbiters show extensive dry streambeds. Most of these seem to have been cut by sudden flooding rather than by gradual runoff. Whatever water Mars still has is now frozen in the soil or in the polar caps.

The Martian atmosphere is now very thin because the low gravity has been unable to retain most gases. But features such as the streambeds and the layered terrain near the polar caps suggest that the Martian climate may change cyclically, perhaps due to orbital changes.

Deimos and Phobos, the two small moons of Mars, may be captured asteroids. Like most small moons, they are irregular in shape because their gravities are too weak to deform their rock into spheres.

NEW TERMS

tidal coupling
resonance
lobate scarp
intercrater plains
smooth plains
subsolar point
greenhouse effect
rolling plains

QUESTIONS

1. How does the tidal coupling between Mercury and the sun differ from that between Phobos and Mars?
2. What is the difference between the intercrater plains and the smooth plains in terms of time of formation?
3. What evidence do we have that Mercury has a partially molten, metallic core?
4. From your knowledge of comparative planetology, propose explanations for the absence of lobate scarps on Earth, the moon, and Phobos.
5. From what you know of Earth, Venus, and Mars, do you expect the volcanoes on Venus and Mars to be active or extinct? Why?
6. What evidence do we have that Venus and Mars once had more water than at present? Where did that water come from? Where did it go?
7. What features will you look for in future radar maps of Venus to search for plate tectonics?
8. Why doesn't Mars have mountain ranges like Earth's? Why doesn't Earth have large volcanoes like Mars?
9. What were the canals on Mars? How do they differ from the dry streambeds on Mars?
10. Propose two different explanations for the nearly pure carbon dioxide atmospheres of Venus and Mars. How did Earth avoid such a fate?
11. What evidence do we have that the climate on Mars has changed?
12. Phobos rises in the west, and Deimos sets in the west. Which, in its motions, is most like our own moon? Why?

PROBLEMS

1. How long would it take radio signals to travel from Earth to Venus and back if Venus were at its farthest point from Earth? Why are such observations impractical?

2. Repeat Problem 1 for Mercury.
3. We have sent a spacecraft to land on Mercury, and it has transmitted radio signals to us at a wavelength of 10 cm. If we see Mercury at its greatest angular distance west of the sun, to what wavelength must we tune our radio telescope to detect the signals? (HINTS: See Data File 4 and Box 7–3.)
4. The Pioneer Venus Orbiter circled Venus with a period of 24 hours. What was its average altitude above the surface? (HINT: See Box 5–1.)
5. The smallest feature visible through an earth-based telescope has an angular diameter of about 1 second of arc. If a canal on Mars was just visible when Mars was at its closest to Earth, what is its linear diameter? (HINT: See Box 3–1.)
6. What is the maximum angular diameter of Phobos as seen from Earth? What surface features should we be able to see using earth-based telescopes? (HINT: See Box 3–1.)
7. Tides are making the orbit of Earth's moon larger, but tides are making the orbit of Phobos smaller. Explain this by redrawing Figure 3–6 to show Phobos orbiting Mars faster than Mars rotates. Do the tidal bulges lead or lag the satellite?
8. Assume that Deimos is 12 km in diameter and has a density of 2 g/cm^3. What is its mass and escape velocity? (HINTS: The volume of a sphere is $4/3\ \pi r^3$. See Box 5–2.)

RECOMMENDED READING

BAKER, VICTOR R. *The Channels of Mars.* Austin: University of Texas Press, 1982.

BEATTY, J. KELLY "Venus: The Mystery Continues." *Sky and Telescope 63* (Feb. 1982), p. 134.

———. "Report from a Torrid Planet." *Sky and Telescope 63* (May 1982), p. 452.

———. "The Amazing Olympus Mons." *Sky and Telescope 64* (Nov. 1982), p. 420.

———. "Radar Views of Venus." *Sky and Telescope 67* (Feb. 1984), p. 110.

———. "A Radar Tour of Venus. *Sky and Telescope 69* (June 1985), p. 507.

BEATTY, J. KELLY, BRIAN O'LEARY, and ANDREW CHAIKIN *The New Solar System,* 2nd ed. Cambridge, Mass.: Sky Publishing, 1982.

BURGESS, ERIC *Venus: An Errant Twin.* New York: Columbia University Press, 1985.

CARR, MICHAEL H. "The Geology of Mars." *American Scientist 68* (Nov./Dec. 1980), p. 626.

———. *The Surface of Mars.* New Haven, Conn.: Yale University Press, 1981.

———. "The Surface of Mars: A Post-Viking View." *Mercury 12* (Jan./Feb. 1983), p. 2.

CORDELL, BRUCE M. "Mercury: The World Closest to the Sun." *Mercury 13* (Sept./Oct. 1984), p. 136.

———. "Mars, Earth, and Ice." *Sky and Telescope 72* (July 1986), p. 17.

CROSS, CHARLES A., and PATRICK MOORE *The Atlas of Mercury.* New York: Crown, 1977.

GREELEY, RONALD *Planetary Landscapes.* Boston: Allen and Unwin, 1985.

GUEST, JOHN, PAUL BUTTERWORTH, JOHN MURRAY, and WILLIAM O'DONNELL *Planetary Geology.* New York: Wiley, 1979.

HARTMANN, W. K. "Cratering in the Solar System." *Scientific American 236* (Jan. 1977), p. 84.

———. *Moons and Planets,* 2nd ed. Belmont, Calif.: Wadsworth, 1983.

HEAD, JAMES W., SANDRA E. YUTER, and SEAN C. SOLOMON "Topography of Venus and Earth: A Test for the Presence of Plate Tectonics." *American Scientist 69* (Nov./Dec. 1981), p. 614.

HOYT, WILLIAM GRAVES *Lowell and Mars.* Tucson: University of Arizona Press, 1976.

LANZEROTTI, LOUIS J., and STAMATIOS M. KRIMIGIS "Comparative Magnetospheres." *Physics Today 38* (Nov. 1985), p. 24.

MARAN, STEPHEN P. "Very Volcanic Venus." *Natural History 94* (Nov. 1985), p. 34.

MILLMAN, PETER M. "Names on Other Worlds." *Sky and Telescope 67* (Jan. 1984), p. 23.

MURRAY, BRUCE, MICHAEL C. MALIN, and RONALD GREELEY *Earthlike Planets.* San Francisco: W. H. Freeman, 1981.

PARKER, DONALD C., CHARLES F. CAPEN, and JEFF D. BEISH "Exploring the Martian Arctic." *Sky and Telescope 65* (March 1986), p. 218.

PRINN, RONALD G. "The Volcanoes and Clouds of Venus." *Scientific American 252* (March 1985), p. 46.

SCHUBERT, GERALD, and CURT COVEY "The Atmosphere of Venus." *Scientific American 245* (July 1981), p. 66.

SMITH, DAVID H. "Testing Relativity with DI Herculis." *Sky and Telescope 71* (March 1986), p. 236.

TAYLOR, HARRY A., and PAUL A. CLOUTIER "Venus: Dead or Alive." *Science 234* (28 Nov. 1986), p. 1087.

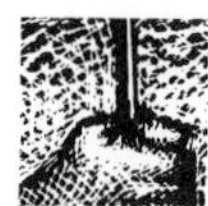

CHAPTER 22

JUPITER AND SATURN

There is something fascinating about science. One gets such wholesale returns of conjecture out of such a trifling investment of fact.

Mark Twain
LIFE ON THE MISSISSIPPI

The sulfuric acid fogs of Venus seem totally alien, but compared to the planets of the outer solar system, Venus is a tropical oasis. Of the five planets beyond the asteroid belt, four have no solid surface, and one is so far from the sun that it is gripped by deadly cold.

The four Jovian planets—Jupiter, Saturn, Uranus, and Neptune—are strikingly different from the terrestrial planets. Their thick atmospheres of hydrogen, helium, methane, and ammonia are filled with clouds of ammonia and methane crystals. Explorers descending into the atmosphere of a Jovian planet would find the atmosphere and its clouds blending gradually with the planet's liquid hydrogen interior. Thus, the Jovian planets have no solid surface and are not subject to four-stage development through which terrestrial planets pass.

Of the Jovian planets, Jupiter and Saturn are the largest (Color Plate 38). In some ways, such as composition and size, they are very similar. We will discover, however, that they differ dramatically in ways that will help us understand the origin, structure, and evolution of such giant planets.

22.1 JUPITER

Jupiter is the most massive of the Jovian planets, containing 71 percent of all planetary matter in our solar system. This high mass accentuates some processes that are less obvious or nearly absent on the other Jovian planets. We will examine Jupiter in detail so we can use it as a standard in our comparative study of other Jovian planets.

The most striking feature of Jupiter is its varicolored cloud belts, which make it look like a child's rubber ball (Data File 7a). We will examine these belts later and see that they are related to weather patterns on Earth. However, first we must descend through the 1000-km deep atmosphere to examine the interior of the planet.

THE INTERIOR OF JUPITER

Jupiter is a giant planet. It is over 11 times the diameter of Earth and contains 317.83 times more mass. Most of this mass is hydrogen and helium, and this has led astronomers to refer to Jupiter as a gas giant planet. But the interior of Jupiter is not gaseous. The pressure in the interior is so great that the hydrogen becomes a liquid. Thus, Jupiter is a liquid planet (Data File 7b).

Two phenomena easily visible from Earth show that Jupiter is not a rigid sphere. The planet rotates quite rapidly, in only $9^h50^m30^s$, and this rapid rotation has flattened the liquid planet from a perfect sphere. Its equatorial diameter exceeds its polar diameter by 6.37 percent. Astronomers refer to this by saying the **oblateness** of Jupiter is 0.0637. Clearly, we should expect a rapidly rotating, liquid planet to be oblate, and we will discover that the other Jovian planets are also oblate.

Another characteristic of a liquid planet is that it need not rotate as a rigid body. The equator rotates with a period of $9^h50^m30^s$, but latitudes farther north and south rotate with slightly different periods. Thus, Jupiter rotates differentially. Recall that the sun also rotates differentially. Studies of Jupiter's magnetic field show that the deep interior rotates with a period of $9^h55^m30^s$.

The chemical composition of Jupiter resembles the sun in that hydrogen and helium atoms are very common and carbon, nitrogen, oxygen, and the other heavy elements have similar relative abundances. Because Jupiter is cooler than the sun, the atoms can form stable molecules (Table 22–1) such as methane (CH_4), ammonia (NH_3), water, and other compounds. During the last decade infrared observations have identified a great many molecules in Jupiter including ethane (C_2H_6), acetylene (C_2H_2), phosphine (PH_3), carbon monoxide (CO), and hydrogen cyanide (HCN).

Why does Jupiter have so much hydrogen? Its sunlike composition is a clue that it has retained most of its light elements, unlike the smaller planets whose gravities are too weak. Thus, Jupiter is a lump of the original solar nebula, compressed by its own gravity, but more or less unaltered by the loss of the lighter gases.

Table 22–1 Composition of the Jovian planets.

Molecule	Jupiter (%)	Saturn (%)
H_2	90	94
He	10	6
H_2O	0.0001	—
CH_4	0.07	0.05
NH_3	0.02	0.02

Careful measurements of the amount of energy leaving Jupiter reveal that it emits about twice as much energy as it absorbs from the sun. This is evidently heat left over from the formation of the planet. In Chapter 19 we concluded that Jupiter should have grown very hot when it formed, and it might still be as hot as 30,000 K at its center. As this heat leaks away, the planet will contract slightly, producing more heat. In a sense, Jupiter's formation is still in progress—it is still contracting.

Theoretical models of Jupiter predict that it contains a core of dense material whose composition resembles earth rocks and metals. Although this is commonly referred to as a "rocky" core, the high temperature and pressure make the material unlike any rock on Earth's surface. The behavior of materials under such high pressures is not well-understood, so we can't be sure it is a liquid. The core, however, must be a hot mixture of iron, nickel, silicon, and the other heavy elements, which settled out of the liquid and gaseous layers above. Although this core is larger than the earth, it occupies less than 1 percent of Jupiter's total mass and thus has little effect on the planet.

Jupiter's internal pressure is as high as 300 million times the pressure of Earth's atmosphere, and this pressure alters the structure of the hydrogen atoms in the deep interior. Under very high pressure, electrons are free to move from atom to atom, and this makes the material a very good conductor of electricity. Although the material remains a liquid, it conducts electricity as well as a metal does on Earth, so physicists refer to such material as **liquid metallic hydrogen**.

Jupiter's strong magnetic field is generated in this vast globe of conducting liquid. It spins in less than 10 hours and is stirred by convection currents. The resulting

DATA FILE 7
Jupiter

Average distance from the sun	5.2028 AU (7.783 × 10^8 km)
Eccentricity of orbit	0.0484
Maximum distance from the sun	5.455 AU (8.160 × 10^8 km)
Minimum distance from the sun	4.951 AU (7.406 × 10^8 km)
Inclination of orbit to ecliptic	1°18′29″
Average orbital velocity	13.06 km/sec
Orbital period	11.867 y (4334.3 days)
Period of rotation	$9^h50^m30^s$
Inclination of equator to orbit	3°5′
Equatorial diameter	143,800 km (11.18 $D_\oplus$)
Mass	1.899 × 10^{27}kg (317.83 $M_\oplus$)
Average density	1.34 g/cm^3
Gravity at base of clouds	2.54 earth gravities
Escape velocity	61 km/sec (5.4 $V_\oplus$)
Temperature at cloud tops	− 110°C (− 166°F)
Albedo	0.51
Oblateness	0.0637

a. Jupiter as seen from Earth. The Great Red Spot is at the lower right near the shadow of a satellite. (Palomar Observatory photograph.)

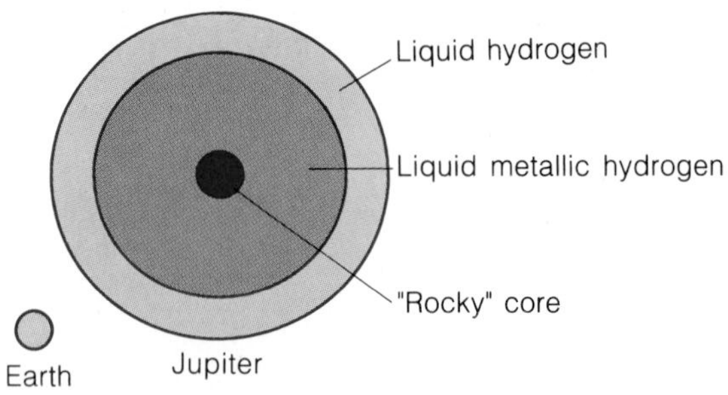

b. A theoretical model of the interior of Jupiter. Note the oblateness of the planet.

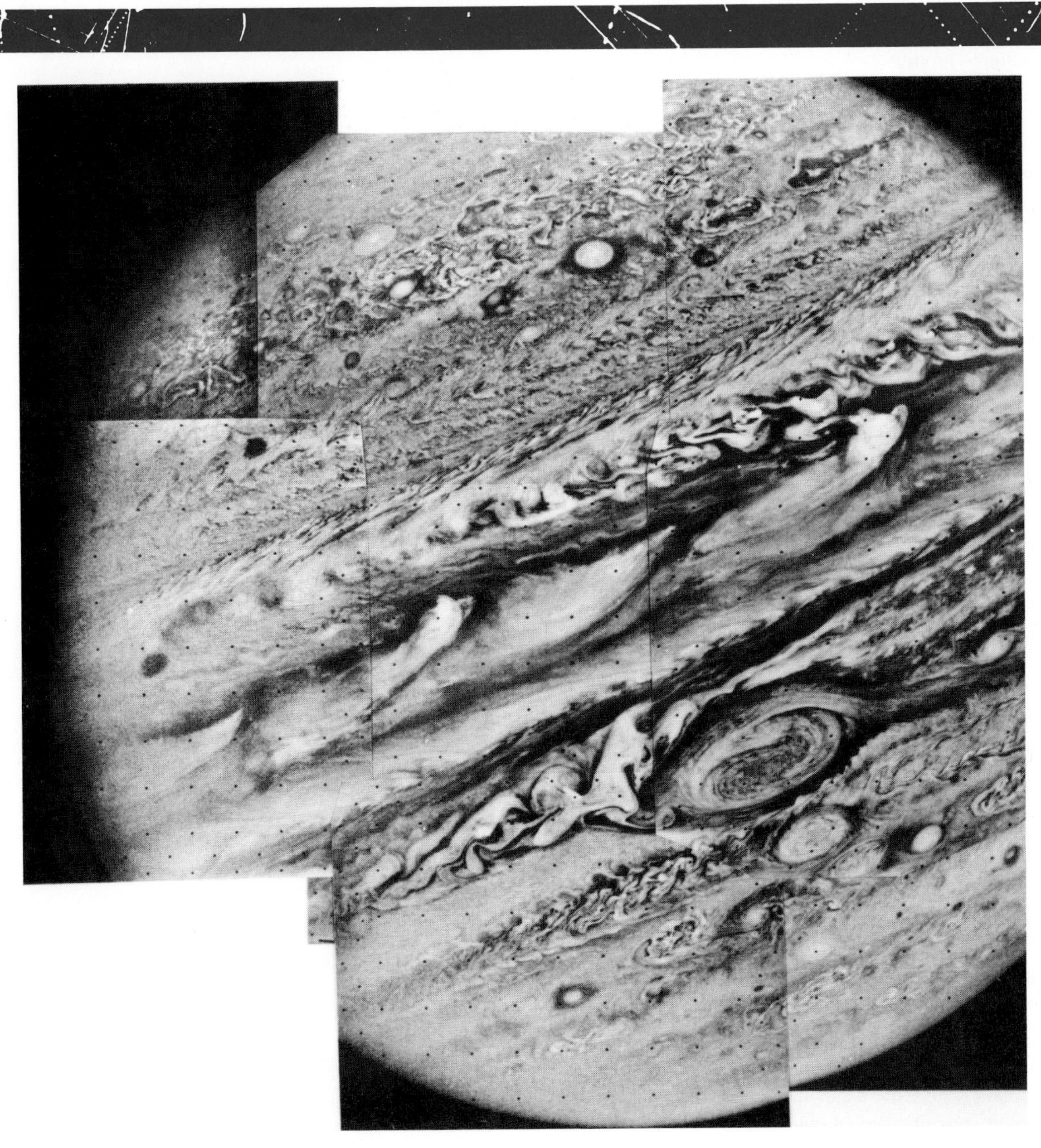

c. Computer enhanced image by the Voyager 1 spacecraft shows complex circulation. (NASA.)

dynamo effect produces the strongest planetary magnetic field in the solar system. The interaction of that field with the solar wind produces some dramatic effects that are unique to Jupiter.

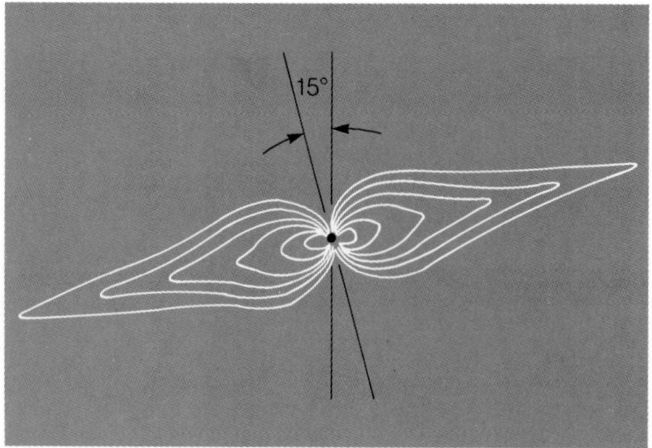

FIGURE 22–1 Jupiter's intense magnetic field traps high-energy particles into radiation belts. The belts are flattened by Jupiter's rapid rotation, and the magnetic field is tipped 15° from the axis of rotation.

RADIATION BELTS

Pioneer and Voyager spacecraft flew past Jupiter in the 1970s and detected a magnetic field 10 to 20 times stronger than Earth's and extending through a vast region around the planet. Outside this magnetosphere, the motions of particles are dominated by the solar wind. Jupiter's magnetosphere is 100 times larger than Earth's because its magnetic field is stronger than Earth's and because the solar wind is weaker at the distance of Jupiter.

Unlike Earth's magnetosphere, Jupiter's is highly flattened because of the rapid rotation of the planet. Jupiter's interior rotates in only 9 hours and 55.5 minutes and drags the magnetic field around with it.

Charged particles from the solar wind leak into planetary magnetospheres and become trapped in belts of radiation (Figure 22–1). This happens on Earth and produces the Van Allen belts. Jupiter's trapped radiation is much more intense because the field is much stronger and can trap higher-energy particles. But the Pioneer and Voyager spacecraft, which passed through these belts, found the radiation much stronger than expected—strong enough to damage electronic circuits and kill humans. These particles may be coming from Jupiter's peculiar satellite Io, which follows an orbit inside the strongest part of the radiation belt. The particles gain their energy from the rapid rotation of the magnetosphere.

We know that the magnetic field rotates in 9 hours and 55.5 minutes because it emits radio signals that fluctuate with that period. This is called the **decimeter radiation** because it is strongest at a wavelength of 0.1 m. It is produced by synchrotron emission from relativistic electrons trapped in the magnetic field. (See Box 13–1.)

Jupiter's **decameter radiation** is strongest at wavelengths of tens of meters and consists of bursts of noise lasting about 1 second. They may be produced by lightning bolts a million times stronger than a large bolt on Earth. Such lightning has been seen on the dark side of Jupiter (Figure 22–2).

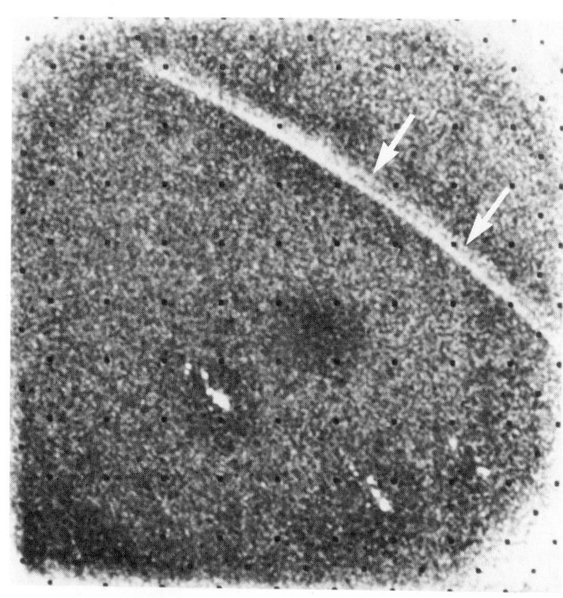

FIGURE 22–2 After passing close to Jupiter, the Voyager 1 spacecraft looked back at its dark side and detected auroras (arrows) along the edge of the disk near the planet's north magnetic pole and lightning bolts among the cloud belts (lower center). (NASA.)

We would expect lightning to be erratic, but the decameter bursts are strongest at times when the satellite Io is in certain positions with respect to Earth. The lightning discharges occur continuously, but the decameter radiation seems to be guided into beams by an interaction between Io and the magnetosphere. We hear the bursts best when one of those beams points our way.

Earth and Jupiter share another magnetic phenomenon, the aurora—a faint glow produced in the upper atmosphere when charged particles in the magneto-

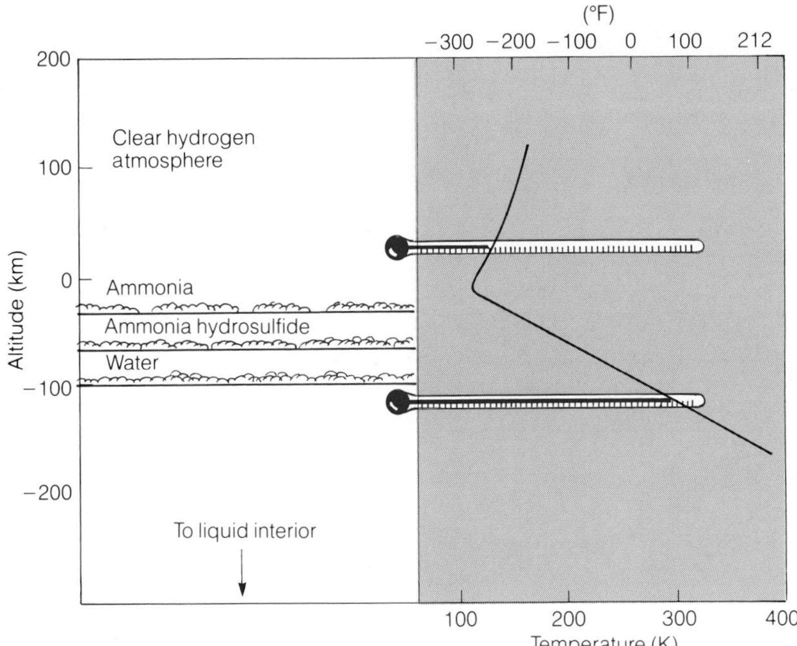

FIGURE 22–3 The three principal cloud layers on Jupiter are shown at left. Thermometers inserted into the Jovian atmosphere would register temperature as in the graph at right.

sphere leak down into the atmosphere under the magnetic poles (Figure 22–2). The auroral display on Jupiter is apparently continuous and intense. One reason it is intense is that Io acts as a source of charged particles within the magnetosphere. We will see later in this chapter that Io is a satellite with a number of active volcanoes spewing gases and particles into space. In fact, Io and its orbit are surrounded by a cloud of sodium, potassium, and sulfur vapor that is either vented from its volcanoes or blasted from its surface by radiation (Color Plate 40).

ATMOSPHERE

The apparent surface we see on Jupiter is only the highest layer of clouds composed of ammonia crystals (Figure 22–3). The temperature there is about 150 K (− 190°F), and the atmosphere, mostly hydrogen, is only about 70 percent as dense as Earth's is at sea level. Through occasional holes in this cloud layer, infrared astronomers have seen deeper, warmer layers. Models suggest that the next cloud layer is composed of ammonia hydrosulfide crystals with a temperature of about 200 K (− 100°F). Ammonia hydrosulfide decomposes into ammonia and hydrogen sulfide, the gas that gives rotten eggs their smell. Deeper still lies a layer of liquid ammonia droplets and water crystals, and below that the atmosphere gradually merges with the liquid hydrogen interior.

The clouds we see are organized into dark **belts** and light **zones**. Even a small telescope can reveal the belt–zone pattern on Jupiter. The belts are generally brown or red, although they may shade to blue-green, and the zones are yellow-white (Color Plate 39). The highest clouds on Jupiter are composed of ammonia crystals, and deeper cloud layers may be made up of ammonium hydrosulfide (NH_4SH) and water. All these compounds are white, so the colors must come from impurities in the clouds. Some theorists believe that ultraviolet light from the sun and energy from lightning bolts in Jupiter's atmosphere can break down methane, ammonia, and other molecules, especially those containing sulfur and phosphorus. Fragments of these molecules could join to form short-lived, complex compounds, which would have colors ranging from light yellow through red to deep brown. The exact nature of these compounds is not known.

Observations made by the Pioneer and Voyager spacecraft show that the zones are cooler and higher than the belts. The zones are apparently high-pressure regions of rising gas, cooling as it rises and expands. The belts are low-pressure areas in which the gas sinks and grows warmer. At the boundaries between belts and zones, jet streams blow at high velocity (Figure 22–4).

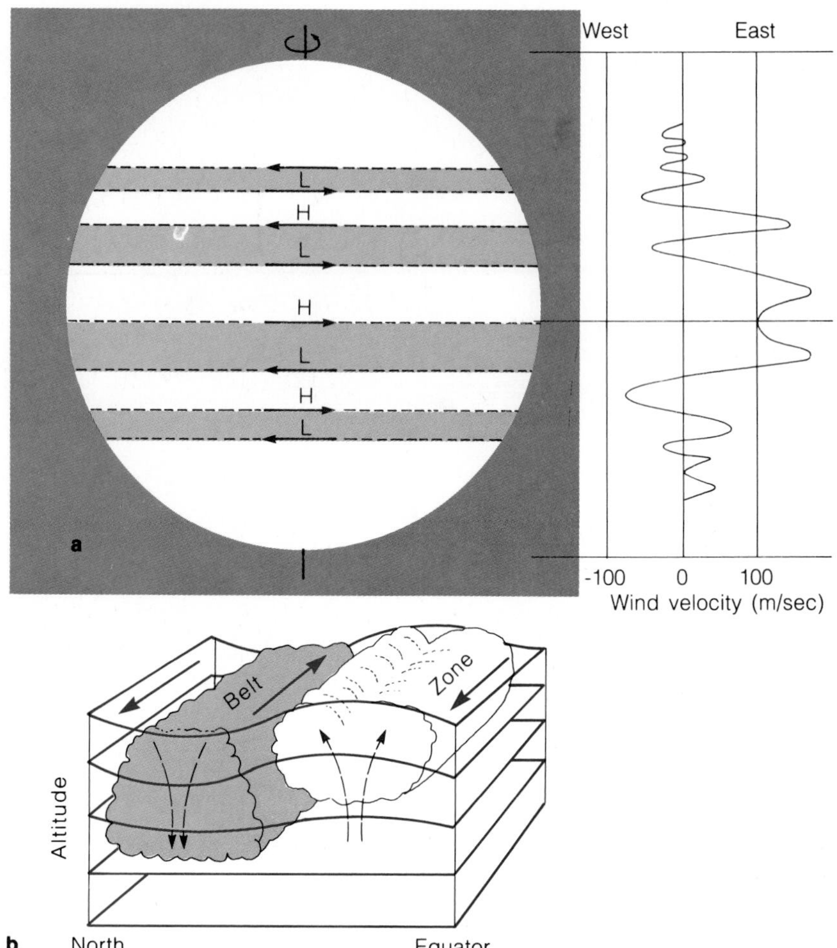

FIGURE 22–4 The dark belts and light zones on Jupiter (a) are low- and high-pressure regions bounded by high-speed winds. Rising gas in the zones condenses to form clouds at high attitudes (b) where they are bright. Sinking gas in the belts does not form high-level clouds, thus allowing us to see to deeper, darker cloud layers. This circulation, coupled with the rapid rotation of the planet, produces high-speed winds between the belts and zones. The dark blue lines in part (b) are lines of constant pressure.

On Earth the temperature difference between the equator and the poles drives a wavelike wind pattern that organizes such high- and low-pressure regions into cyclonic circulations. On Jupiter the equator and poles have about the same temperature, perhaps due to heat rising from the interior. Thus, there is no wave circulation, and Jupiter's rapid rotation draws the high- and low-pressure regions into bands that circle the planet. On Earth the high- and low-pressure regions are bounded by winds induced by the wave circulation. On Jupiter the same circulation appears as high-speed winds that blow around the planet at the boundaries of the belts and zones.

We don't know how these belts and zones can be maintained. Although they change color and brightness occasionally, they have not changed position for at least 80 years. One theory is that jet streams at the belt–zone boundaries are linked to circulation patterns deep in the liquid interior (Figure 22–5a). Recall from Chapter 19 that the atmosphere of Jupiter is in proportion no deeper than the fuzz on a badly worn tennis ball. An alternative theory is that the circulation pattern is maintained from outside by the heat input from the sun. Models of Jupiter's atmosphere that ignore any influence from the interior can mimic belt-zone circulation (Figure 22–5b).

Due to be launched by the Space Shuttle in late 1989, the Galileo spacecraft may help us choose between these theories. On its 6-year journey, it will loop past Venus once and Earth twice, picking up energy for the trip through the asteroid belt, where it will photograph the asteroids Gaspra and Ida before reaching Jupiter. Once at Jupiter it will send a probe on an hour long plunge into the atmosphere where it will radio back to Earth its measurements of density, composition, and temperature before being crushed by the pressure. The main section of the spacecraft will spend 20 months studying the planet and its satellites.

Long before the Galileo probe reaches Jupiter in 1995, astronomers on Earth will be using the Hubble Space

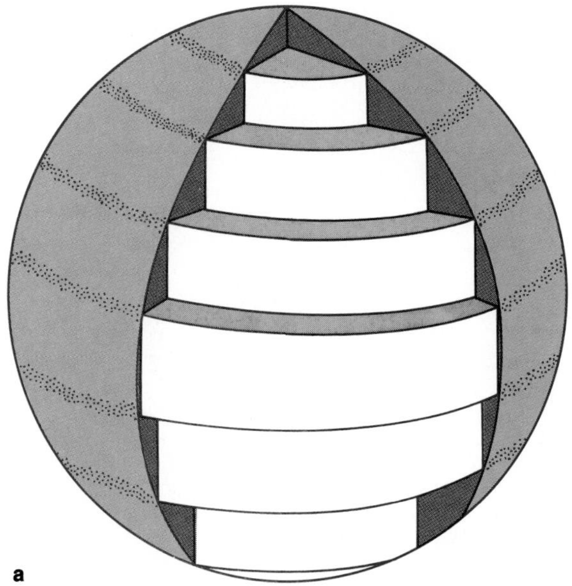

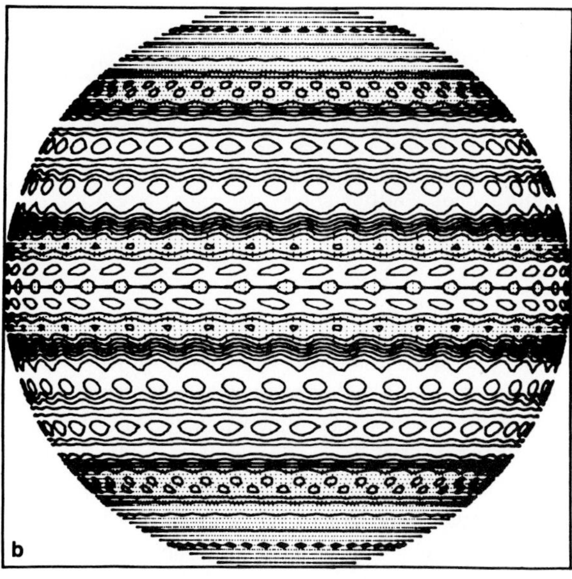

FIGURE 22–5 Two theories for the origin of belts and zones: (a) A cutaway drawing of Jupiter shows the cylindrical circulation patterns believed to occur in its liquid interior. The sizes of these cylinders may fix the positions of the belts and zones. (b) A computer model of the Jovian atmosphere, which assumes no connection with circulation in the liquid interior. The lines represent motions in the atmosphere and reproduce the observed belt-zone circulation. (b. Courtesy Gareth Williams.)

FIGURE 22–6 The Galileo Probe, shown here being prepared for tests in a thermal-vacuum chamber, will plunge into Jupiter's atmosphere. Before being destroyed by the pressure and temperature of Jupiter's lower atmosphere, it will report measurements to an orbiting satellite, which will relay the data to Earth. (Hughes Aircraft Co.)

Telescope due for a 1990 launch. It will be able to monitor the weather patterns on Jupiter, and such long-term studies may help resolve the mystery of the origin of the belts and zones.

Mixed in among the belts and zones are light and dark spots, a few larger than Earth. Some appear and disappear in a few days; others last for years. The largest is the Great Red Spot, which has been in one of the southern zones for more than 300 years. It is 14,000 km wide and up to 40,000 km long. Although it sometimes grows lighter or darker and drifts in longitude, it remains one of the dominant features of Jupiter (Figure 22–7).

Pioneer and Voyager photographs and measurements show that the Great Red Spot is a few degrees cooler than the zone in which it lies and about 8 km higher. The photographs also show that it rotates counterclockwise like a ball rolling between two boards. This rotation suggests that it is a high-pressure area in which the gases are rising and cooling. Smaller spots come and go in a few days to a few years, but the Great Red Spot is apparently such a large cyclonic disturbance that it can remain stable over very long periods.

The swirling cloud layers of Jupiter are visible through a small telescope, but another feature of Jupiter—Jupiter's ring—is almost invisible from Earth.

JUPITER'S RING SYSTEM

When Voyager 1 flew through the system, it was programmed to search for a possible ring as it passed through Jupiter's equatorial plane. The photograph it sent back to Earth showed that Jupiter does indeed have a ring with a thickness of less than 30 km and an outer edge at 1.81 planetary radii (Figure 22–8). The ring is very tenuous, at least 100 times less opaque than Saturn's rings.

The Voyager 2 photo in Figure 22–8b proves that many of the particles must be very small—no larger than 10 μm (10^{-5} m) in diameter. The photo was taken looking back at Jupiter with the sun behind the planet, and the rings look bright because they are scattering sunlight forward toward the spacecraft. This efficient **forward scattering** of light occurs because the particles have diameters that are roughly the same as the wavelength of the light. Particles with diameters much larger than the wavelength of the light are not efficient at forward scattering. Thus, if Jupiter's ring were composed of basketballs, the ring would not look bright in Figure 22–8b. Rather, the ring particles must be about the size of the particles in cigarette smoke.

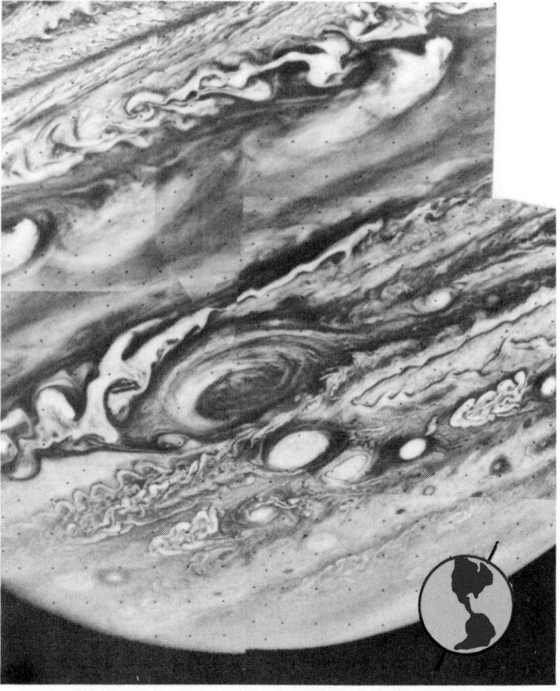

FIGURE 22–7 Jupiter's swirling cloud belts and zones. The Great Red Spot circulates counterclockwise. Smaller spots are also cyclonic disturbances. Earth is shown for scale. (NASA.)

Computer-enhanced images show that the ring is made up of four parts (Figure 22–9). A broad, main ring extends from 1.81–1.72 Jupiter radii, about 6000 km. Embedded in this ring is a brighter component at 1.79 radii, and a thin sheet of ring particles extends down to the cloud tops. The fourth part of the ring is a halo of dust particles. The halo, up to 20,000 km thick near the planet, is probably formed by dust grains that build up electrostatic charges and are pulled out of the ring by Jupiter's powerful magnetic field.

The material is dark and reddish. Thus, it cannot be ice but must be silicates, perhaps from a moon that came too close to Jupiter and was pulled apart by tidal forces.

The **Roche limit** is the distance from a planet within which a moon cannot hold itself together by its own gravity. Imagine a moon constructed of loose rocks held together by mutual gravity. If that moon orbits far from its parent planet, the moon's gravity will be adequate to hold itself together. But if it comes within the Roche limit, tidal forces will be greater than the moon's own gravitational force, and the moon will disintegrate. Artificial earth satellites can orbit inside Earth's Roche limit and not be pulled apart because the satellites are held together by bolts and welded seams, not by their own gravity. Natural moons are held together by their own gravity, and tides will destroy them if they come within the Roche limit.

If a moon and planet have similar densities, then the Roche limit is 2.44 planetary radii. Jupiter's rings, and the rings of Saturn and Uranus, lie inside this Roche limit.

It is possible that such material is left over from the formation of the solar system, but the small dust specks cannot be that old because a number of processes remove such small specks from the ring. The pressure of sunlight and the magnetic forces in the magnetosphere alter the orbits of small particles, and, if they don't fall into Jupiter, they will be blown out of the solar system. In addition, the deadly radiation grinds the dust specks down to nothing in a century or so. Thus, the tiny particles we see in the ring must be continuously supplied from some source.

One set of sources are moonlets as small as a centimeter or as large as 10 km in diameter. They can't be much larger or they would have been seen. If such moonlets exist in the ring, they must be structurally strong

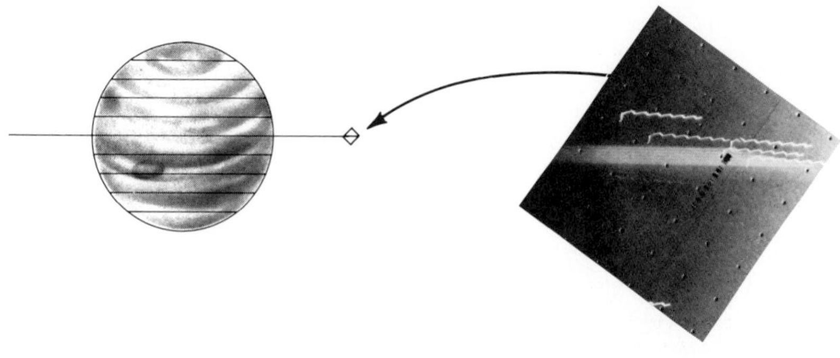
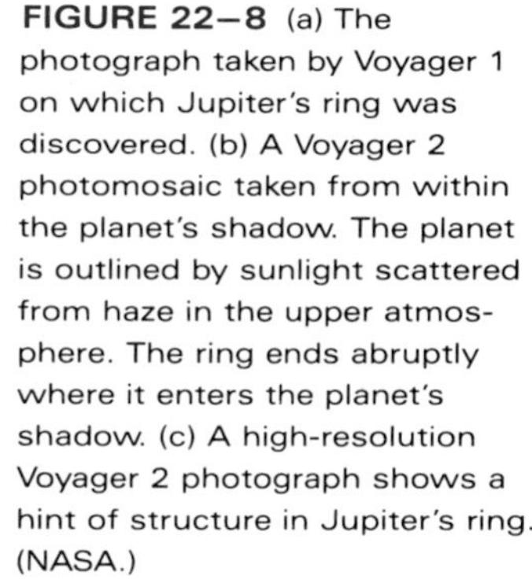
a

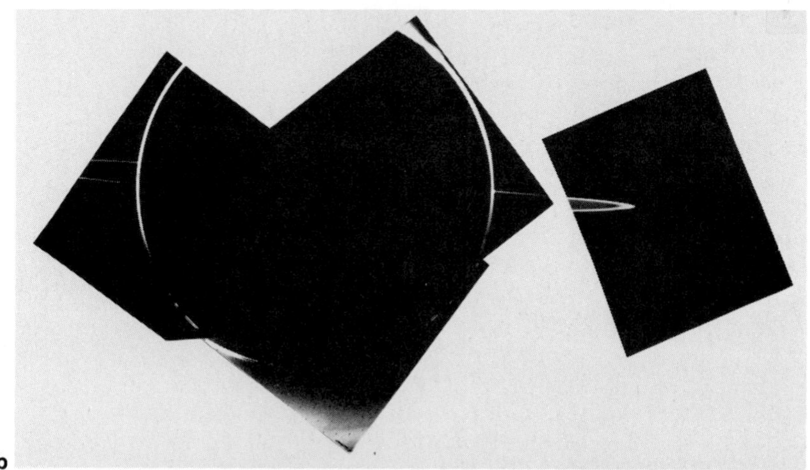
b

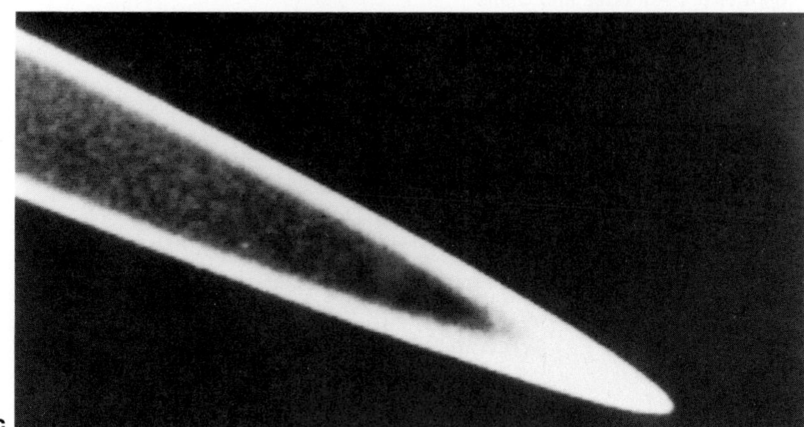
c

FIGURE 22–8 (a) The photograph taken by Voyager 1 on which Jupiter's ring was discovered. (b) A Voyager 2 photomosaic taken from within the planet's shadow. The planet is outlined by sunlight scattered from haze in the upper atmosphere. The ring ends abruptly where it enters the planet's shadow. (c) A high-resolution Voyager 2 photograph shows a hint of structure in Jupiter's ring. (NASA.)

enough to withstand the tidal forces. Nevertheless, bombardment by micrometeorites or by dust from the volcanoes on Io may be chipping off a constant supply of dust specks to keep the rings populated.

Two such moonlets may have been found. Adrastea is a moon 25 km in diameter that orbits Jupiter just at the edge of the ring. In fact, the gravitational effects of Adrastea may keep shepherding stray particles back into the ring and thus give the ring a sharp edge. Metis is another small satellite that lies at 1.79 Jupiter radii, just where the bright ring is located. Perhaps particles blasted from Metis keep Jupiter's ring bright at this radius.

Jupiter is important as a planet not only because it is the standard of comparison in our study of the Jovian

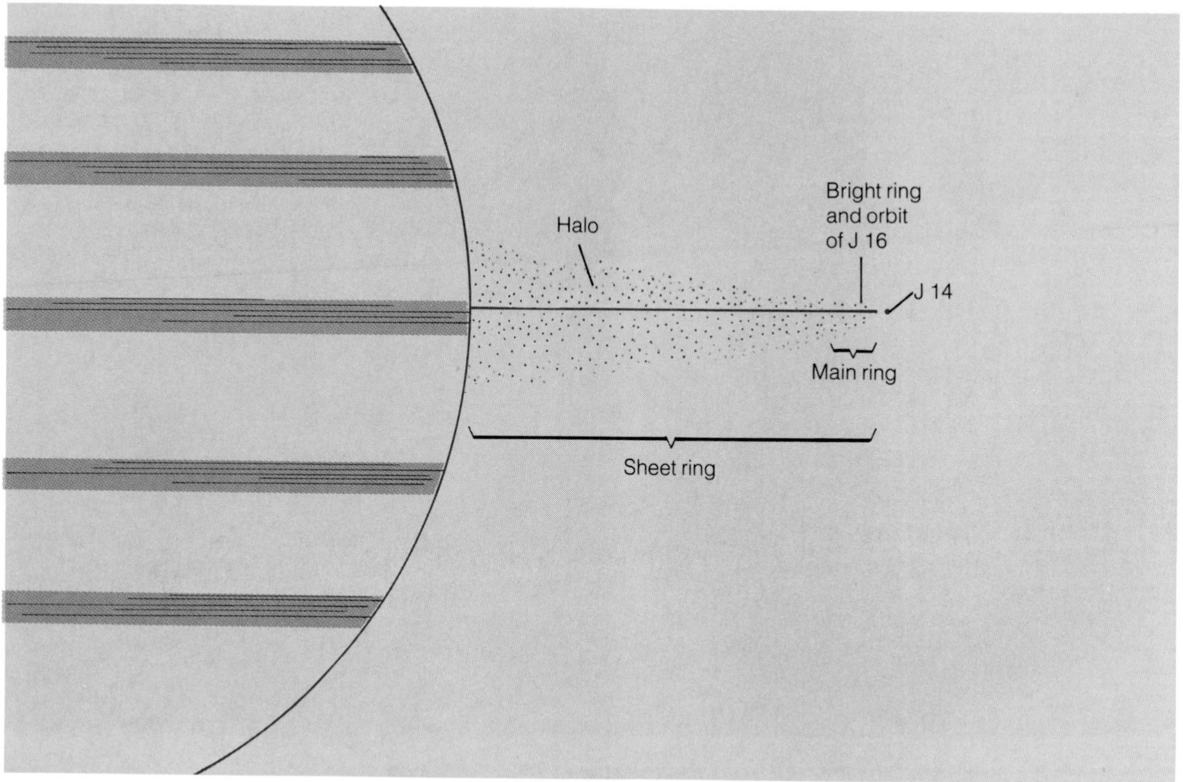

FIGURE 22–9 An edge-on view of the rings of Jupiter shows the four ring components and the position of the two satellites Adrastea and Metis.

planets, but also it is host to a family of at least 16 moons. Like a miniature solar system, the Jupiter family includes a diverse mix of characteristics (Color Plate 41).

22.2 JUPITER'S FAMILY OF MOONS

We can divide Jupiter's satellites into two groups. The smaller moons resemble the moons of Mars and may be captured asteroids. The four larger moons, the moons seen by Galileo, are clearly not captured asteroids but complicated worlds in their own right.

Our study of the moons of Jupiter will illustrate three important principles in comparative planetology. First, a body's composition depends on the temperature of the material from which it formed. This is illustrated by the prevalence of ice as a building material in the outer solar system. The second principle is that cratering can tell us the age of a surface. Finally, we will see that tides are a powerful influence on the evolution of moons.

CALLISTO

The outermost of Jupiter's large moons, Callisto is superficially similar to Earth's moon. It is 44 percent larger, heavily cratered, and, like all of Jupiter's large satellites, tidally locked to its planet. Unlike our moon, it is not solid rock.

The average density of Callisto is only 1.79 g/cm^3, which means it cannot contain all rock. It probably has a rocky core extending about 75 percent of the way to the surface, but that is topped by a mantle of water at least partially frozen (Figure 22–10). The surface crust is rich in frozen water—at least 90 percent— and contains dust and rock fragments mixed in.

Photographs of the crust show that Callisto is a dead world. Its surface is heavily cratered, and there are no crater-free regions like the moon's maria (Figure 22–11). Since Callisto's crust first hardened, the only activity has been the continuous bombardment of meteorites, including some very large impacts. Valhalla is a crater 350 km in diameter and surrounded by a bull's-eye of ridges

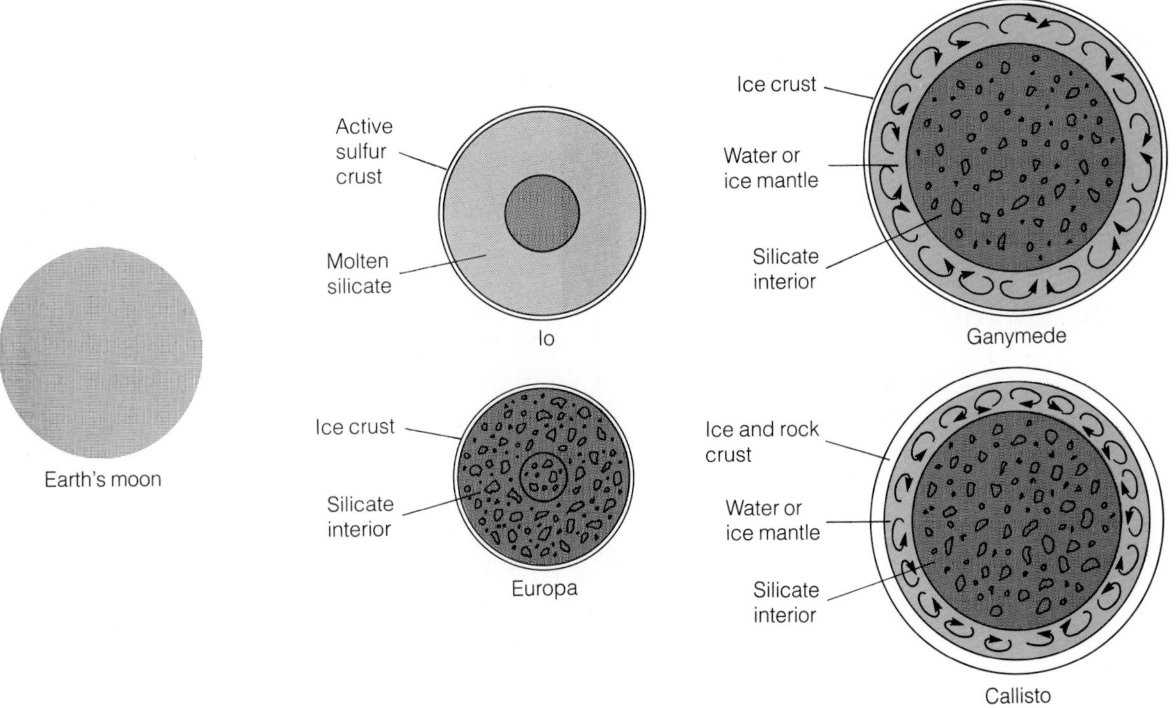

FIGURE 22–10 The relative sizes and interiors of the Galilean moons of Jupiter.

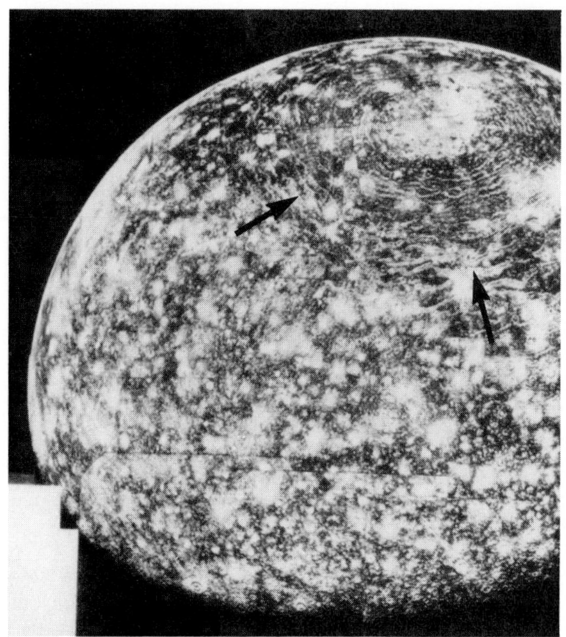

FIGURE 22–11 Callisto's dark surface is dirty ice broken by heavy cratering. The ringed plane, Valhalla (arrows), is the largest impact feature known in the solar system. (NASA.)

extending 2000 km from the center. In some ways it resembles the ringed basins on Earth's moon and on Mercury.

Although craters are common and some are very large, Callisto has little relief—that is, there are few high mountains or ridges. This is probably due to a combination of three effects. Impacts in ice may not dig craters with the same depth proportions that we see on rocky worlds. Also, once excavated, a crater may slump due to the slow, plastic flow of ice. The largest features may sink back into the icy crust for lack of support from below, especially if the mantle is partially fluid.

Most of Callisto's craters are quite old, but a few are young, bright scars. Evidently, an impact ejects from below ice that is much cleaner than the surface. This raises a question: How can an icy world have such a dark surface? It reflects only 17 percent of the light that strikes it (albedo = 0.17). For one thing, the in-fall of meteoric material over billions of years has added dark material to the crust. However, when a meteorite strikes, even if it is microscopic, it excavates many times its mass from the crust. This should stir the crust and mix the dark material with the brighter ice the same way stirring pancake batter can mix in blueberries. Something else must be going on.

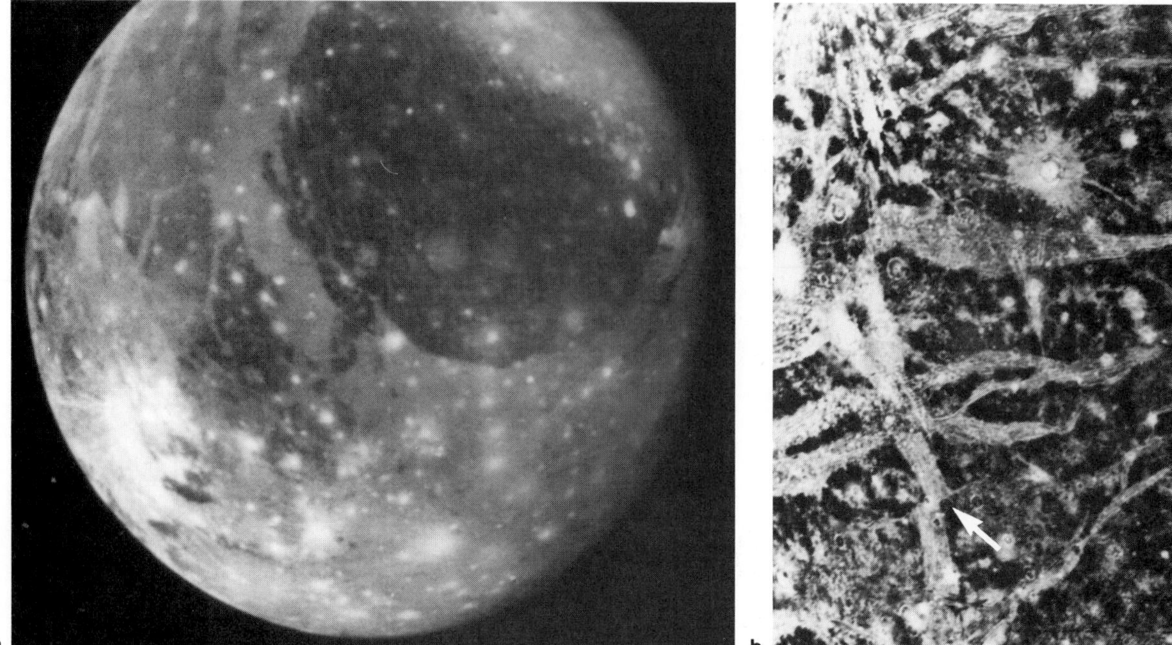

FIGURE 22–12 (a) Ganymede's surface is divided into dark cratered terrain and lighter grooved regions. (b) Note the fault cutting across the grooves (arrow). The grooves are ancient breaks in the icy crust. (NASA.)

The second effect that makes an icy moon dark is the energy of the impacts. Each time a meteorite strikes the surface, the energy of motion becomes heat. In an icy surface, this heat can vaporize ice and leave dust, rock, and meteoric debris behind in a dark film of grime on the surface. Thus, the older an icy region is, the darker and thicker is the layer of grime on the surface of the ice. Fresh impacts break through that dirty surface to cleaner ice below.

GANYMEDE

The next satellite after Callisto is Ganymede, an ice world very similar to Callisto in some ways and dramatically different in other ways. Ganymede is about 50 percent bigger than Earth's moon and slightly larger than Callisto, and it has a density of 1.9 g/cm^3, only a bit more than Callisto's density. Models of Ganymede's interior suggest that it is similar to Callisto; it has a rock core, a water-ice mantle, and an icy crust.

Despite these similarities, one glance at a photograph of Ganymede shows that it is different from Callisto (Figure 22–12). About 66 percent of Ganymede's surface is marked by light, **grooved terrain.** These grooves are typically over 100 km long, 3–10 km apart, and about 300 m high. The slopes are so shallow, only about 5 percent, that we would not find it difficult to walk from the top of one down through the valley and up to the top of the next. The only parts of the crust that resemble Callisto are the old, dark, cratered regions between the bands of grooves.

The explanation of the light, grooved terrain on Ganymede seems to lie below the surface. As the moon cooled, the interior expanded, and the tension fractured the icy crust time after time, producing patterns of grooves. At the same time, some parts of the interior were warm enough for water to melt, and a water-rich slurry leaked out onto the surface to freeze into ice-rich plains of lighter coloration than the older crust. Spacecraft photographs of the surface reveal a few faults indicative of ancient stresses in the crust (Figure 22–12b, arrow).

Ganymede is no longer active. The number of craters in the grooved terrain shows that the grooves formed at least a few billion years ago. We can see bright ray patterns around young craters but no bright, young grooves. The

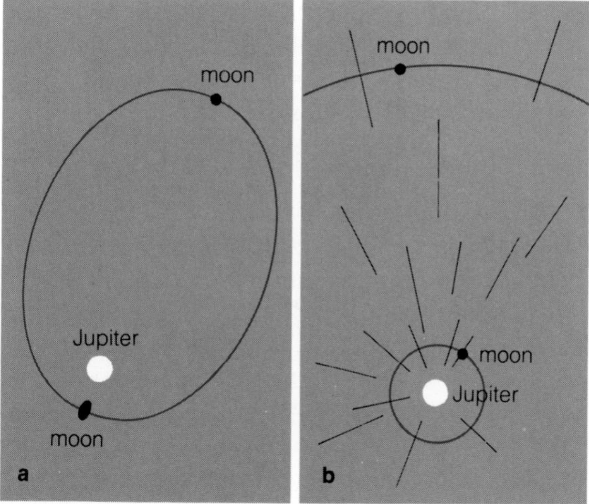

FIGURE 22–13 Two effects on planetary satellites. (a) Tidal heating occurs when changing tides cause friction within a moon. (b) The focusing of meteoroids exposes satellites in small orbits to more impacts than satellites in larger orbits.

FIGURE 22–14 Europa's surface is marked by long, narrow lines that may be the scars of past fractures in a planetwide ice sheet. No craters are evident, suggesting an active surface. (NASA.)

icy crust has grown thick enough to withstand stresses without breaking.

What could have heated Ganymede that didn't heat Callisto? Both should have incorporated about the same proportion of radioactive elements, but Ganymede is slightly larger and thus loses its internal heat more slowly. Also, Ganymede is closer to Jupiter and feels tides squeezing it out of spherical shape. As it moves along its elliptical orbit, the tides flex it and friction generates heat (Figure 22–13a). This is called **tidal heating.**

Another source of heat for Ganymede is the focusing of meteorites by Jupiter's gravity—an effect that will be more important for the satellites closer to the planet. Jupiter's gravitational field is large and strong, and it can pull in a great many meteorites (Figure 22–13b). Ganymede's surface must have been bombarded by a larger number of meteorites than the surface of Callisto. Also, Jupiter's gravity boosts the velocity of these meteorites so they hit with greater energy and release more heat in the crust.

Whatever generated the heat in Ganymede, that heat slowed the cooling and kept the mantle liquid longer than on Callisto. Whereas the crust of Callisto grew thick and rigid, the crust of Ganymede remained thin enough to crack repeatedly and generate the grooved terrain. This effect of internal heat is even more dramatic for the satellites closer to Jupiter—Europa and Io.

EUROPA

Unlike Callisto and Ganymede, Europa is a rocky world. Its density of 3.03 g/cm^3 shows that it contains a large proportion of rock. On the other hand, earth-based infrared observations have detected water ice, and the Voyager photos show that the surface is a bright icy plain (Figure 22–14). Thus, Europa must be a rocky satellite with a thin mantle of ice or water topped by an icy crust.

Europa is obviously active. Although the best Voyager photographs could have detected features as small as 100 m in diameter, it found almost no craters at all. Obviously, Europa has been hit by meteorites, especially because it is deep within Jupiter's focusing gravitational field. Somehow, Europa's crust is erasing craters as fast as they form. In addition, Europa's surface is quite bright, with an albedo of 0.69. Old icy crusts are dark, so Europa's surface must be young ice.

The youth of the ice crust is evident in the fine lines that mark the surface of Europa (Figure 22–14). These lines are typically 10–50 km wide and 1000 km long. Some wind more than halfway around the planet. These appear to be cracks in the icy crust, filled by water from below that froze to form fresh crust.

Thus, Europa is an active world. Its source of heat may be tides caused by nearby Jupiter, which melt the water mantle and crack the icy crust. Such tidal heating would be more effective on Europa than on Ganymede because Europa is much closer to Jupiter.

IO

The innermost of Jupiter's large moons is also the most peculiar. Earth-based observations show that Io and its orbit are surrounded by a cloud of sulfur, oxygen, and sodium (Figure 22–15a). Also, Io occasionally grows brighter in **infrared outbursts.** Despite these oddities, astronomers were not prepared for Io's strange appearance.

The Voyager 1 images, computer-enhanced to bring out color, showed Io mottled with red, white, orange, and black like a badly made pizza. Its true colors more nearly resemble the pale yellow of sulfur, an important clue to the cause of its odd appearance. (See Figure 22–15b and Color Plate 41). Careful analysis of the photographs revealed volcanoes spewing out gas and solid debris up to 300 km above the surface (Figure 22–15c). Io is almost airless, so much of the debris falls back to the surface unimpeded by atmospheric friction. For example, Voyager 1 photos showed a heart-shaped ring of debris surrounding the volcano now known as Pele (upper right in Color Plates 42 and 44). Eight volcanoes were active on Io at the time of the Voyager encounters.

Although Io is rich in volcanoes, it is poor in impact craters. In fact, the Voyager photos show no impact craters at all. Because Jupiter focuses meteorites toward the center of its gravitational field, Io should suffer five times the impacts per square meter that Callisto suffers. The lack of craters on Io is further proof that the surface is active. Apparently, the craters are covered by debris from the volcanoes. The escape velocity on Io is only 2.5 km/sec, so some of the material from the volcanic plumes escapes into the Jovian magnetosphere. But some falls back, blanketing the surface with at least 1 mm of material per year.

In addition to explaining the lack of craters, Io's volcanoes also explain the sulfur and sodium cloud around its orbit and its infrared outbursts. Much of the material in the volcanic plumes is sulfur vapor or the gas sulfur dioxide. These gases escape at the rate of about 1 ton per second and supply the sulfur cloud with fresh atoms. Sodium and other elements are probably present in Io as impurities and are vented through its volcanoes or blasted from its surface by the radiation in Jupiter's powerful radiation belts. (See Color Plate 40.) Io's part of the universe is thus quite dirty and smelly with sulfur. Amalthea, a small rocky moon orbiting just inside the orbit of Io, has a reddish orange surface, which is apparently due to sulfur contamination from Io (Color Plate 43.).

The infrared outbursts visible on Io can also be explained by volcanism. A sudden outpouring of molten sulfur or sulfur vapor at a temperature of 600 K (1100°F) could produce a sudden increase in infrared brightness. Another possibility suggests that great lakes of molten sulfur grow cool near their tops but remain warm near their bottoms. A sudden release of infrared radiation would occur if the cool surface of the lake sank and exposed the hotter sulfur below. This overturning of a molten sulfur lake has not been observed, but at least one infrared hot spot has been identified as a lake of molten sulfur.

In addition to explaining certain of Io's properties, the volcanism is interesting in itself. Earth's volcanoes eject lava and ash because of the pressure of the water that is dissolved in the lava and that flashes into steam when the pressure is suddenly released. But observations show that Io is very dry. Its volcanoes draw their power from other gases. Some of the volcanoes seem to be marked by white deposits that suggest a frost of sulfur dioxide. This material as a gas could power the volcanoes. But other volcanoes are surrounded by darker deposits and may draw their power from deeper, hotter sources of molten sulfur. Certainly the varied colors on the surface of Io suggest the various forms of sulfur and sulfur compounds.

Volcanoes require an energy source, so we must ask where Io gets its energy. It is much too small to have remained hot since it formed, or to have retained heat from internal radioactivity. The energy it vents through its volcanoes is about three times the energy it could generate by radioactive decay.

One suggested energy source involves electrostatic discharges on Io's surface. Because Io moves rapidly through Jupiter's magnetic field, the little moon develops a very large electrostatic charge, which flows back to Jupiter along the **Io flux tube**, a conducting channel of ionized gas between Io and Jupiter. The theory suggests that the volcanoes on Io were produced by the 10 million amperes of electric current flowing out of Io. Detailed calculations and analysis of the volcanoes shows that although the flux tube is real and explains some Jovian phenomena such as the decameter radiation, it cannot generate the volcanoes. Io must have some other energy source.

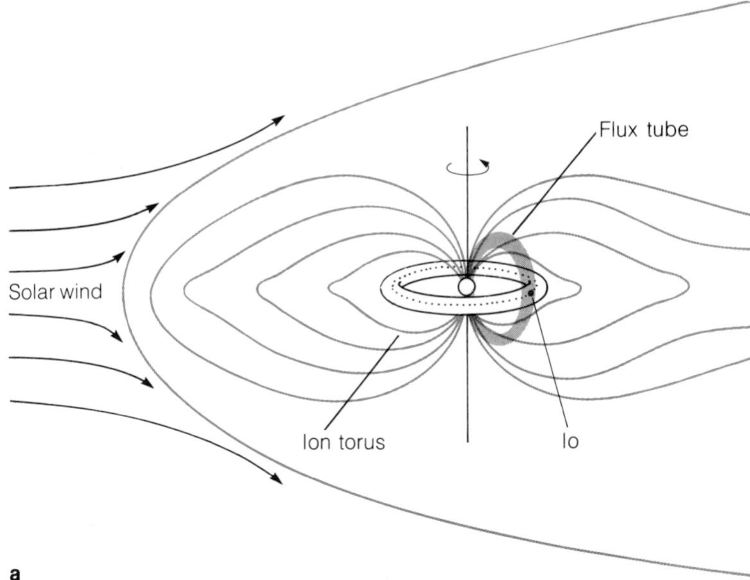

FIGURE 22–15 (a) The orbit of the moon Io is surrounded by ions of sulfur, oxygen, and sodium. The Io flux tube is a channel of ionized gases through which powerful electric currents flow from Jupiter through Io. (b) Io's surface is free of impact craters because new features are rapidly covered by material from at least eight active volcanoes. (c) Seen at the limb of the moon, one of its volcanoes blasts material as high as 300 km above the surface. (NASA.)

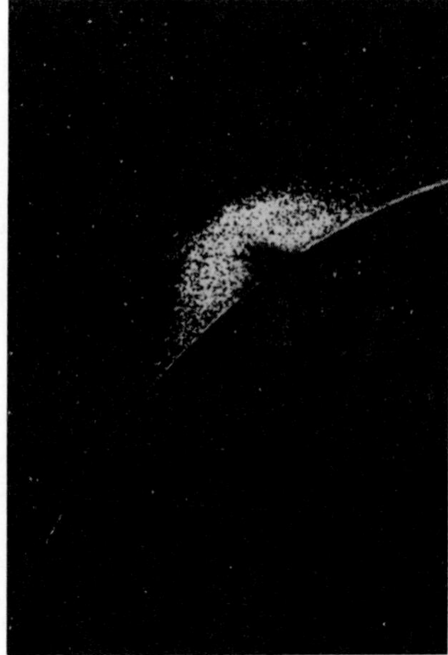

Io seems to be hot because of changing tidal distortion caused by Jupiter. Europa and Ganymede force Io's orbit into an ellipse, and as its distance from Jupiter changes, the size of the tidal bulge on Io changes. Its surface rises and falls by a hundred meters or so. The friction is sufficient to melt its interior. Models of Io's volcanism suggest that the little moon has a molten silicate interior with a partially molten sulfur-rich mantle (Figure 22–16). Where molten sulfur bursts through the solid crust we see a volcano. Over a geologically short period, the debris from the volcanoes buries the original crust so deep that it melts and merges with the molten mantle. Thus, Io recycles its crust through its volcanoes again and again.

Clearly, Io is a very active world. Voyager 1 found eight active volcanoes, but when Voyager 2 arrived 4 months later, six were still active, one had ceased, and one could not be observed (Table 22–2). One of the active volcanoes, Loki, had increased the height of its erupting plume from 100 to 210 km and developed a more complex pattern. In addition to active volcanoes, Io is marked by about

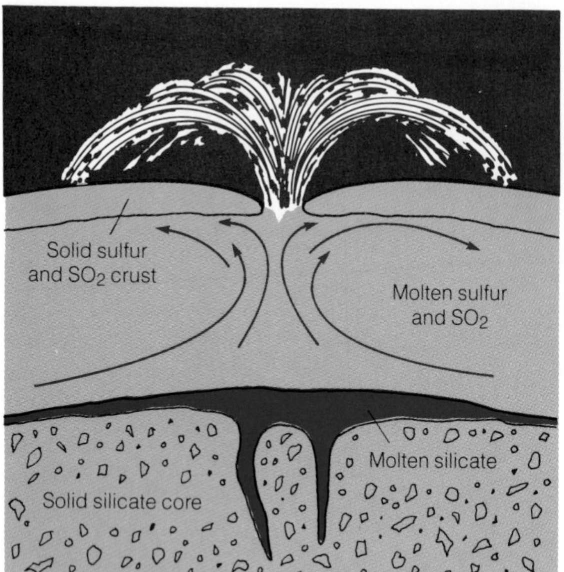

FIGURE 22–16 The volcanoes on Io appear to be caused by heat from the interior causing hot spots in the outer layers of sulfur and sulfur dioxide (SO_2).

Table 22–2 Volcanoes on Io.

Name	Height During Voyage 1 Flyby (km)	Activity During Voyager 2 Flyby
Pele	280	ceased
Loki	100	increased
Prometheus	70	increased
Volund	95	no data
Amirani	80	unchanged
Maui	80	unchanged
Marduk	120	unchanged
Masubi	70	unchanged

(From Morrison and Samz, *Voyage to Jupiter,* NASA, 1980.)

170 spots believed to be currently inactive volcanoes. Such an active, thin crust should not be able to support large mountains, but a few mountains, one as high as 9 km, do exist. These may mark irregularities in the silicate core, or they may be evidence of stresses in the active crust.

Io is the most active world in our solar system, and the other Galilean satellites are similarly peculiar. They illustrate common geological processes twisted and exaggerated by the conditions in the Jupiter system. But the Galilean satellites are not the only moons in Jupiter's family.

THE MINOR MOONS OF JUPITER

The 12 remaining satellites of Jupiter are smaller than the Galilean satellites, and we do not have good data concerning their surfaces. Nevertheless, their orbits and sizes can tell us a few things about their origin.

Four of the minor moons lie inside the orbit of Io. They are small moons, which follow nearly circular orbits in the plane of Jupiter's equator. We can guess that their orbits have been altered by tidal forces, and their small size suggests that they have cooled long ago and are no longer active. By orbiting so close to Jupiter, they probably have lost most ices and are mostly rock.

A second set of four small moons follows similar elliptical orbits, between 156–165 Jupiter radii outside the Galilean satellites. They are probably related objects. Some theorists believe that they are the four largest fragments of a moon that broke up or of an asteroid that collided with a moon.

Farther from Jupiter is a third set of four moons with orbits between 291–333 Jupiter radii. Again, the four orbits are very similar in that they have about the same eccentricity and inclination, and all four moons orbit Jupiter in the retrograde direction. These objects may be fragments of a larger body.

Just as we suspected that the two moons of Mars might be captured asteroids, we must suspect that at least some of Jupiter's moons are captured. We must also consider the possibility that a moon could be struck and broken up by a passing asteroid or comet. We will see further evidence of this when we study the moons of Uranus.

Table 22–3 The Galilean satellites.*

Name	Radius (km)	Density (g/cm^3)
Io	1816	3.55
Europa	1563	3.04
Ganymede	2638	1.93
Callisto	2410	1.81

*For comparison, the radius of Earth's moon is 1738 km, and its density is 3.36 g/cm^3.

THE ORIGIN OF THE GALILEAN MOONS

Whatever the origin of the minor moons, it seems clear that the four Galilean moons are primordial. That is, they formed with Jupiter. They have nearly circular, direct orbits that lie in Jupiter's equatorial plane. In addition, the average density of these satellites decreases as we go outward in the system, just as the density of the planets in the solar system decreases as we go away from the sun (Table 22–3). When Jupiter formed, it was very hot, and the gas cloud around it became a miniature of the solar nebula. The gas cloud grew flat and gave birth to satellites having nearly circular, coplanar orbits. Because the inner part of the gas cloud was hot, the satellites that formed there have higher densities than the satellites that formed in the outer part of the cloud. This explains why Callisto and Ganymede are ice and Io and Europa contain less water.

However, we can also explain the densities of the Galilean satellites by using tides. We have seen how flexing tides can heat Io, and a similar process may keep Europa warm. Such tidal heating could have kept Ganymede warm long enough for its icy crust to break and refreeze to produce the grooved terrain. Thus, the degree of tidal heating depends on the radius of the satellite's orbit. The satellites near Jupiter might have experienced sufficient tidal heating to drive off water and other volatiles, whereas the more distant satellites experienced less tidal heating and kept their volatiles.

Which theory explains the observed densities of the Galilean satellites? We can't be sure, but many astronomers think of the formation of the Galilean satellites as a miniature of the formation of the solar system, and this approach neglects tidal heating. Nevertheless, such heating has been very important in making Io a dry world and may have had some smaller effects on the other satellites.

Jupiter's satellites are fascinating worlds with geologies unlike anything on Earth. Even while astronomers struggled to understand the processes on Jupiter's moons, they wondered what the Voyager spacecraft would find at Saturn.

22.3 SATURN

Saturn has played second fiddle to its own rings since Galileo first viewed it through a telescope in 1609. The rings are dramatic, strikingly beautiful, and easily seen through even a small telescope, but Saturn itself is a fascinating planet. Only slightly smaller than Jupiter, it is similar in many ways. Earth-based telescopes show that it has faint cloud belts in its atmosphere and is slightly oblate. But as is usually the case in comparative planetology, it is the differences between Saturn and Jupiter that are most informative.

SATURN THE PLANET

A quick glance at a photo of Saturn suggests that it is a bland Jupiter (Data File 8a and Figure 22–17). It has belts and zones, but they are not as prominent nor as colorful as those on Jupiter (Color Plate 45). Also, the belts seem less turbulent and contain only rare spots. Only about ten have been seen from Earth.

The photos taken by the Voyager spacecraft in 1979 show that Saturn's belts and zones are complex and that small spots are common. The reason Saturn looks so bland is related to its distance from the sun. Because it is almost twice as far away as Jupiter, it is colder and its cloud layers lie deeper in the atmosphere beneath a methane haze (Figure 22–18). This haze hides much of the detail.

Measurements made by the Voyager spacecraft assure us that the atmospheric circulation on Saturn is similar to that on Jupiter. Belts and zones are low- and high-

DATA FILE 8
Saturn

Average distance from the sun	9.5388 AU (14.27×10^8 km)
Eccentricity of orbit	0.0560
Maximum distance from the sun	10.07 AU (15.07×10^8 km)
Minimum distance from the sun	9.005 AU (13.47×10^8 km)
Inclination of orbit to ecliptic	2°29′17″
Average orbital velocity	9.64 km/sec
Orbital period	29.461 y (10,760 days)
Period of rotation	$10^h13^m59^s$
Inclination of equator to orbit	26°24′
Equatorial diameter	120,660 km (9.42 $D_\oplus$)
Mass	5.69×10^{26} kg (95.147 $M_\oplus$)
Average density	0.69 g/cm^3
Gravity at base of clouds	1.16 earth gravities
Escape velocity	35.6 km/sec (3.2 $V_\oplus$)
Temperature at cloud tops	− 180°C (− 292°F)
Albedo	0.61
Oblateness	0.102

▲ **a.** Saturn as seen from Earth. (Mount Wilson and Las Campanas Observatories, Carnegie Institution of Washington.)

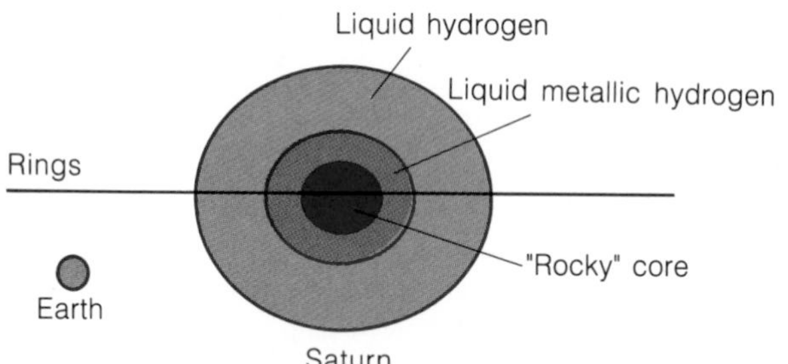

▲ **b.** A theoretical model of the interior of Saturn. Note the oblateness of the planet.

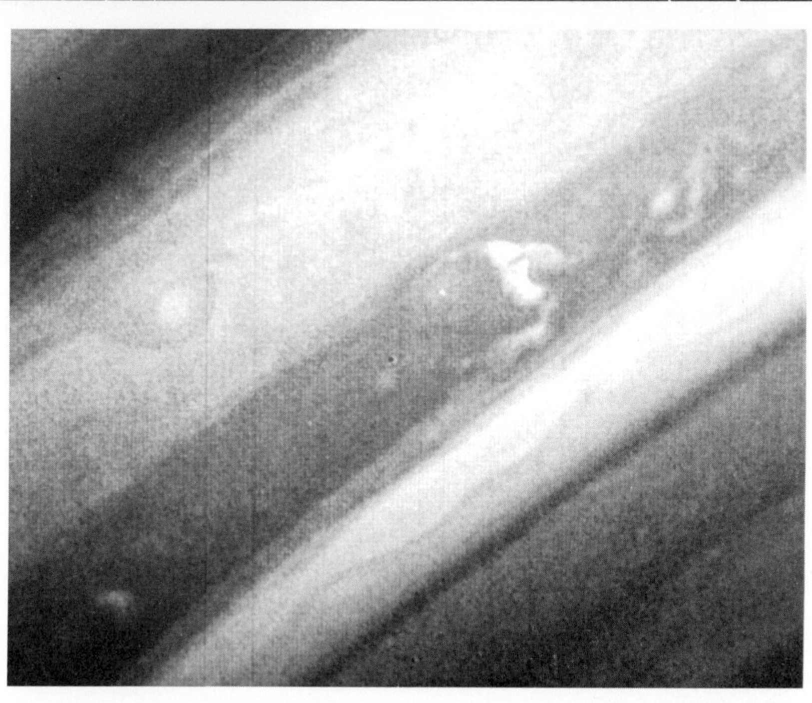

c. Voyager photograph of cloud detail in Saturn's atmosphere. (NASA.)

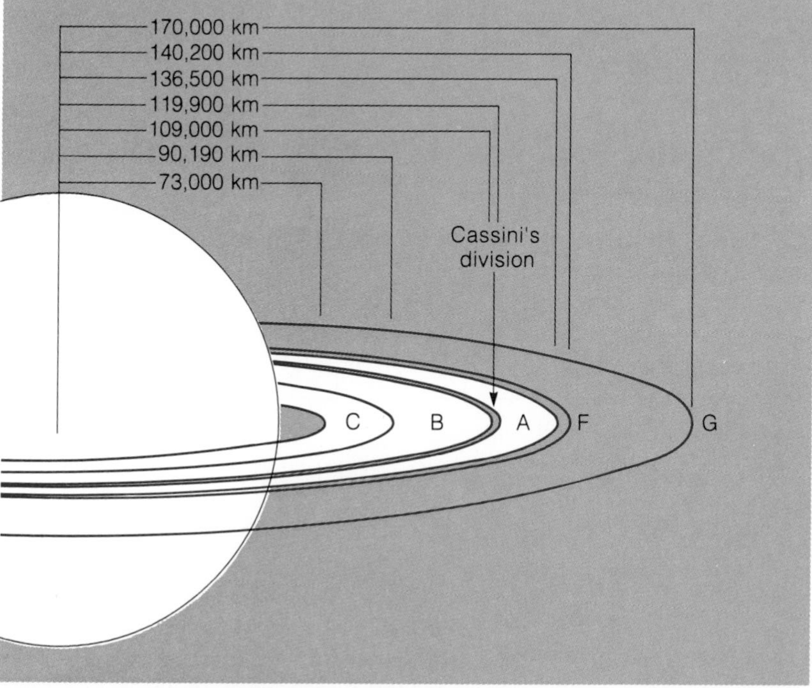

d. Ring dimensions and identifications.

FIGURE 22–17 Although Saturn has belts and zones, they are partially hidden below a layer of methane haze. (NASA/JPL.)

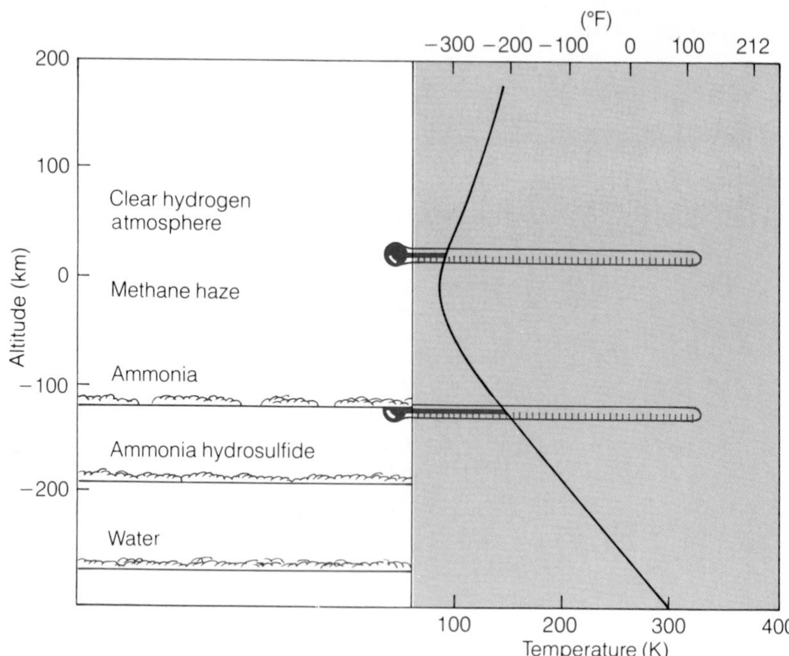

FIGURE 22–18 The three principal cloud layers on Saturn (left) lie deep in the atmosphere due to the lower temperature (right). Compare this diagram with Figure 22–3.

pressure areas, and high-speed winds bound the belts and zones. On Saturn, however, the winds can reach much higher velocities. The equatorial jet stream, for example, races eastward around the planet at almost 500 m/sec (1000 mph), over three times faster than the winds on Jupiter.

The lack of color in Saturn's belts and zones may also be related to its lower temperature. It may be too cold for the color compounds to form on Saturn as they do on Jupiter.

We know that the composition of Saturn is similar to Jupiter's—that is, it is the solar composition (Table 22–1). Ammonia is underabundant in Saturn's upper atmosphere, but this is probably due to the low temperature. Much of the ammonia has frozen out.

The solar composition, coupled with the observed

FIGURE 22–19 A computer-enhanced photograph taken by Voyager 2 shows that Saturn's rings are divided into hundreds of ringlets. (NASA/JPL.)

density of 0.68 g/cm^3—less than the density of water—permits us to model the interior of Saturn (Data File 8b).

Like Jupiter, Saturn should have a liquid metallic region, and it may have a silicate core. If its composition is truly solar, it should contain heavy elements that would settle to the center.

Infrared observations show that Saturn, like Jupiter, emits 2.5–3 times as much energy as it gets from the sun. Thus, its interior must be hot. Some of this energy may be left over from its formation, but Saturn is less massive than Jupiter and should have produced less heat. It should also have lost its internal heat faster. Another process may be producing heat inside Saturn. The interior may be cool enough to allow helium to condense and fall through liquid hydrogen like raindrops. This would release enough gravitational energy to account for Saturn's excess heat. Voyager measurements of the density of the Saturnian atmosphere show that helium has an abundance of only 11 percent as compared with 19 percent for Jupiter. Evidently some of the helium has sunk deeper into the interior.

Of course, liquid metallic hydrogen in the interior—forced into convection by the heat flowing outward and spun by the rotation of the planet—should produce a magnetic field. Voyager did, in fact, detect such a field, rotating with a period of $10^h39^m25^s$, apparently the period of rotation of the core. However, the field is about 20 times weaker than that around Jupiter, apparently because the liquid metallic region is smaller. This means that the radiation belts around Saturn are smaller and less intense than those around Jupiter.

In addition, Voyager spacecraft found that Saturn's magnetic field is more uniform and is tipped only 0.7° with respect to the axis of rotation. From the dynamo theory and practical experience, planetary astronomers expect that the magnetic fields of planets will be tipped at some angle to the axes of rotation. Jupiter's field is tipped 11°, for example. Saturn's field may be more regular because of the great depth of the liquid metallic layer.

THE RINGS OF SATURN

Galileo first saw the rings of Saturn in 1609, but perhaps because of the poor quality of telescopes then, he did not recognize the rings as a disk. He thought Saturn to be three objects—a large central object and two smaller ones on either side. In 1659 Christiaan Huygens recognized the rings as a disk surrounding but not touching the planet, and in 1675 Giovanni Cassini discovered a dark gap in the rings, now known as Cassini's division.

The rings of Saturn are designated by letters of the alphabet (Data File 8d). The A ring lies outside Cassini's division and the bright B ring inside. The C ring, or Crepe ring*, is a faint ring lying inside the B ring. The outermost edge of the A ring has a radius of 136,500 km, or about 21 times the radius of Earth. A journey in a commercial airliner from the outer edge of the A ring to the inner edge of the C ring would take about 5 days (Color Plate 46).

These rings are made up of billions of particles orbiting Saturn in its equatorial plane. James Clerk Maxwell (now honored with a mountain on Venus) proved in 1859 that the rings would be ripped apart if they were solid because the outer parts must orbit slower than the inner parts in accord with Kepler's third law. The thickness of the rings was estimated to be 2 km because they disappear every 15 years when Earth passes through Saturn's equatorial plane and we observe the rings edge-on. Infrared observations show that the orbiting particles are frozen water with temperatures of 70 K (− 330°F).

Thus, as the Voyager spacecraft approached Saturn in 1979, scientists expected to see a uniform plane of icy particles with only a few gaps. The two Voyager spacecraft revealed rings containing roughly 1000 ringlets, some no wider than 2 km (Figure 22–19). A few were slightly elliptical, and a few others contained kinks and condensations.

*The name is derived from the black cloth traditionally associated with mourning.

FIGURE 22–20 The radial spokes in the bright B ring are visible in this photo sequence as dark streaks oriented radially. The black dots are reference points in the camera. (NASA/JPL.)

Since 1867 astronomers had thought they knew why Saturn's rings had a gap between the A ring and the B ring. In that year Daniel Kirkwood demonstrated that resonances between the orbital motion of asteroids and Jupiter could produce gaps in the asteroid belt (Chapter 24). That same theory predicted that a gap should occur in the rings of Saturn where a ring particle would orbit Saturn twice while the moon Mimas orbits once. On alternate orbits the ring particle would feel the gravitational tug of Mimas pulling in the same direction, the total effect would accumulate, and the orbit of the ring particle would be changed. Thus, no particle could remain in orbit at that resonance. When astronomers calculated the position of this gap (using Kepler's third law), they found that it coincided with Cassini's division between the A and B rings. Other gaps visible from Earth seemed to match resonances with other satellites.

But the Voyager spacecraft found a thousand gaps (Figure 22–19), and they did not all fall at the location of resonances. Such resonances do occur, but clearly the rings are much more complex than had been thought.

In addition, Voyager 1 discovered **spokes** oriented radially in the rings (Figure 22–20). The strong differential rotation of the rings should erase such features very quickly, but they somehow regenerate. The forward scattering of light showed that they are made of microscopic particles, and many astronomers now believe that they are linked to electromagnetic effects involving Saturn's magnetic field, which pulls the charged particles up out of the ring plane.

Photographs and measurements revealed the existence of other rings around Saturn. The E ring, detected from Earth, was traced from 3–8 Saturn radii and shown to be most dense near the orbit of the satellite Enceladus. This moon may somehow be ejecting ice particles to maintain the E ring. Voyager images also revealed a D ring inside the C ring and two narrow rings labeled F and G (Data File 8d).

The F ring posed a difficult problem. Voyager 1 photos showed it was narrow and irregularly twisted (Figure 22–21a). Planetary rings should gradually spread out due to collisions between particles and should not develop irregular twists if the particles move in simple orbits. The mystery was eventually resolved when photos showed two small satellites orbiting, one inside and one outside the F ring (Figure 22–21b). These **shepherd satellites** interact gravitationally with wandering particles and return them to the ring, thus keeping it confined. The distortions in the F ring appear to be caused by gravitational interactions when the inner shepherd moon overtakes and passes the outer moon. Another satellite at the edge of the A ring ushers straying particles back and thus keeps the edge of the A ring sharp.

The discovery of the shepherd satellites gave astronomers hope of understanding the hundreds of ringlets within the main rings by supposing that small moons are embedded within the rings. Although these moons would be inside the Roche limit of Saturn, they could resist being pulled apart by tides if they were small and made of strong material. The Voyager spacecraft did not record any images of moons within the rings. However, a recently developed technique for analyzing the radio transparency of the rings has allowed a team of astronomers to detect irregularities in the edges of some rings. They conclude that two moons

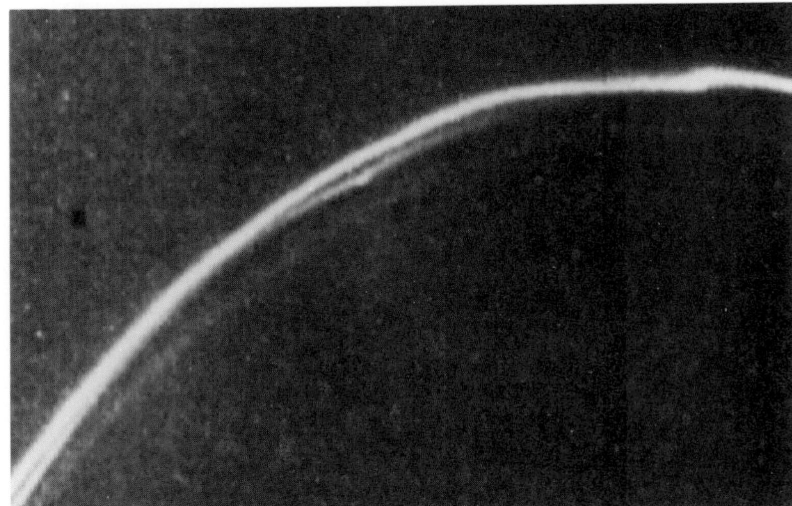

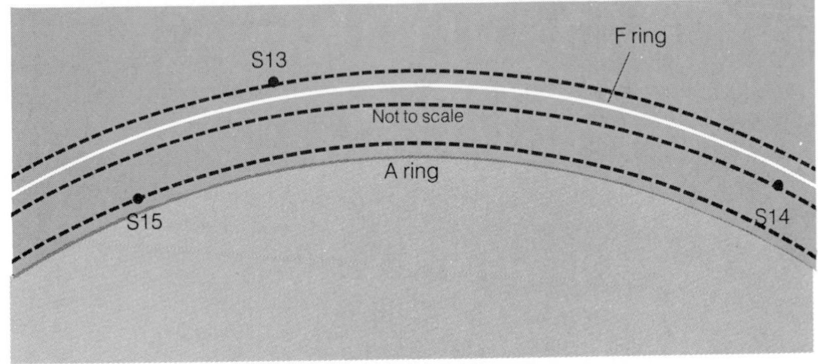

FIGURE 22–21 (a) A Voyager 1 photograph of the F ring shows a number of strands with kinks and twists. (b) Another Voyager 1 image revealed shepherd satellites (arrows) orbiting just outside and inside the F ring keep the ring narrow. (c) These satellites control the F ring and another satellite just outside the A ring keeps its edge sharp. (NASA.)

10 km and 12 km in diameter must exist within Cassini's division. Other moons of similar size probably also lurk within the rings.

Some of the grooves seen in the rings have their origin in instabilities that can occur in material rotating differentially in a disk. The spacing of some ringlets shows that spiral density waves (the same phenomena seen in spiral galaxies) can be set up by satellite resonances. These waves can clear gaps and create entire families of ringlets.

Spiral density waves move slowly outward, but other waves, called bending waves, move inward from resonances. These bending waves cause the surface of the rings to be corrugated (Figure 22–22). This observation helps resolve a question concerning the thickness of the rings. Some photographs by the spacecraft show that the rings are as thin as 10 m, yet observations from Earth yield a thickness of 1–2 km. Apparently, the amplitude of the bending waves makes the rings look much thicker than they really are, just as corrugated cardboard is much thicker than the paper in the corrugations.

The Voyager observations were astounding in many ways, but there was one thing the spacecraft could not detect—individual ring particles. Only a few experiments involving the forward scattering of light and the transmission of radio signals through the rings can help us understand what the particles are like. Dust particles are

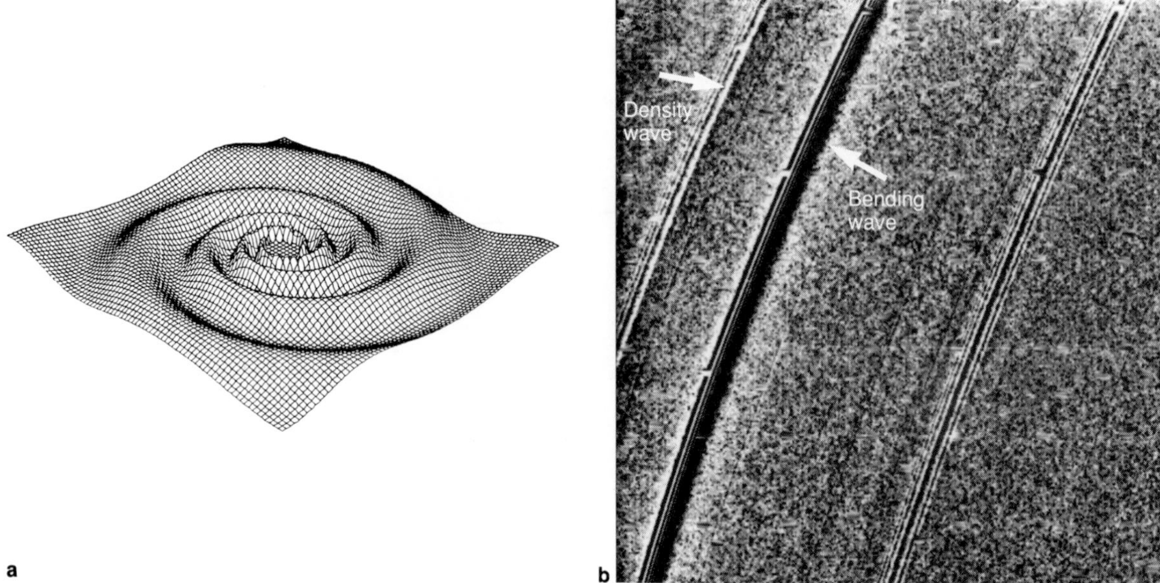

FIGURE 22–22 (a) A computer-generated diagram of a spiral bending wave. Such waves in Saturn's rings are much more tightly wound and much lower in amplitude than the example shown here. (b) Waves in Saturn's rings. A spiral density wave moves outward while a darkly shadowed spiral bending wave moves inward to the right. Another set of waves lies to right of center. (a, Courtesy Frank Shu and Academic Press; b, NASA.)

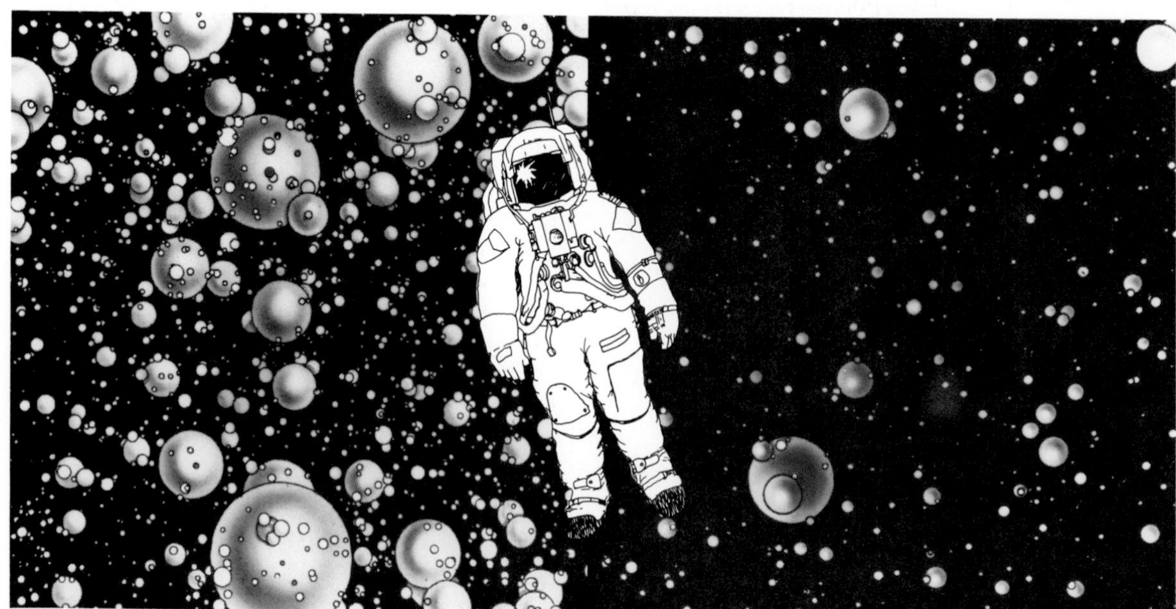

FIGURE 22–23 Swimming through Saturn's rings might be possible. These computer-generated views of particles in the A ring (left) and the C ring (right) show how particle size and number differ through the rings. Particles smaller than 2 cm in diameter are not shown. The actual particles are believed to be irregular ice chunks and not the perfect spheres shown. Astronaut included for scale. (Courtesy NASA/JPL.)

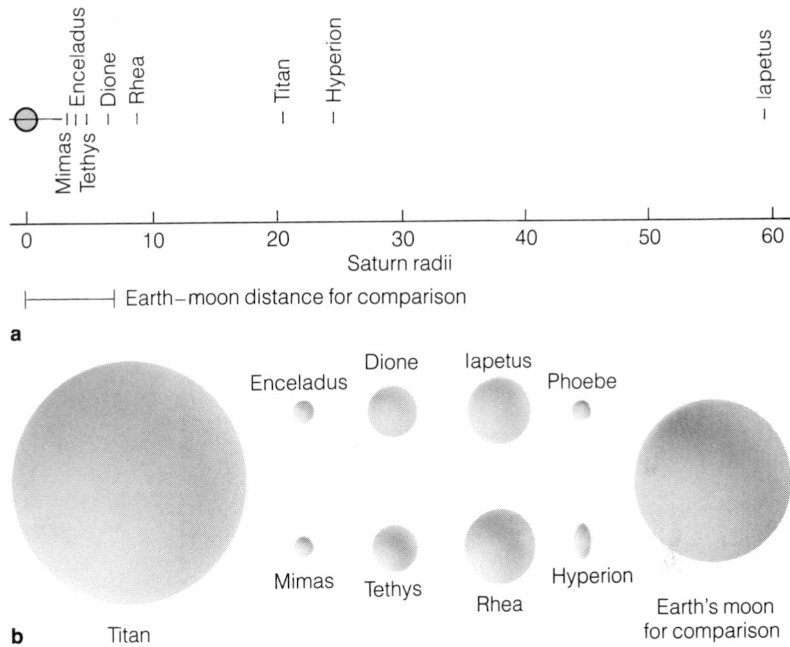

FIGURE 22–24 (a) Most of Saturn's satellites follow orbits within 10 Saturn radii. Phoebe lies at a distance of 215 Saturn radii. (b) In size, the moons are all smaller than Earth's moon except for Titan.

very common, especially in the B ring and Cassini's division, making these features bright in forward scattered light. Particles a few centimeters in diameter are common in some regions, such as the C ring, and almost absent from others, such as Cassini's division. Chunks the size of a person are much rarer, and boulders the size of a house are separated by many meters.

An astronaut in a spacesuit could probably swim through the densest parts of the ring by pushing against the larger particles and drifting from particle to particle (Figure 22–23).This would not be dangerous because the orbiting particles in any given region travel at nearly the same speed. When particles do collide, they merely nudge each other. Occasionally, more violent collisions may produce a few bits of dust that slowly drift inward through the rings and finally fall into the planet. In the thinner parts of the rings, the particles are so widely spaced that ring swimming would be difficult.

The different distributions of particle sizes through the rings are responsible for differences in the appearance of the A, B, and C rings. But no one presently understands what forces sort the particles by size in the different regions. Nor do we understand the processes that must break up particles and reform them.

Saturn's rings are both elegant and mysterious. We do not entirely understand the processes that govern their structure, nor do we understand where they came from. They could be primordial, or they could be temporary.

If the rings are primordial, then they are composed of material left over from the formation of Saturn. But we suspect that the giant planets formed hot, and material near them would have been heated with volatiles driven away. Saturn, being smaller than Jupiter and farther from the sun, might have been cooler. Another objection to the primordial hypothesis is that the distended atmosphere of the forming planet would have slowed the orbital motion of ring material and it would have fallen into the planet.

If the rings are not primordial, then we must suppose they were formed by material added to the system later. An icy moon that came within Saturn's Roche limit could have broken up. One interesting suggestion is that Saturn's rings are composed of debris from comets that have collided with its satellites. Such collisions must occur occasionally, and they would produce plenty of debris to keep Saturn's elegant rings richly supplied with fresh ice.

22.4 THE ICY MOONS OF SATURN

Seventeen known satellites orbit Saturn—far too many for us to discuss individually—but those satellites share characteristics that hint at their origin and evolution. Most are small, dead, icy worlds, but one is large enough to keep an atmosphere and may have oceans (Figure 22–24).

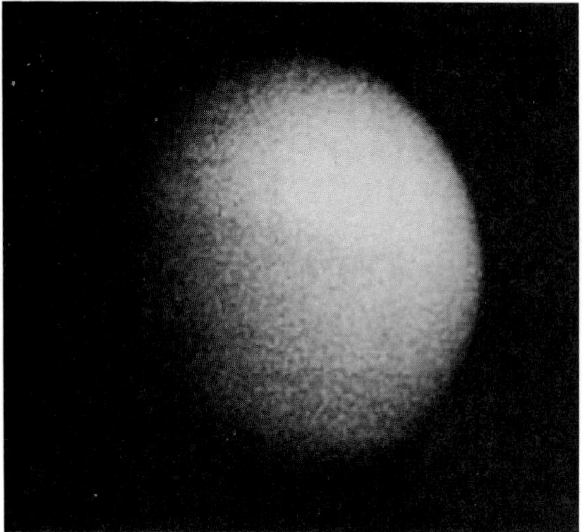

FIGURE 22–25 Titan's atmosphere is a nearly featureless haze. (NASA.)

Table 22–4	Organic compounds detected on Titan by the Voyager infrared spectrometer.
C_2H_6	ethane
C_2H_2	acetylene
C_2H_4	ethylene
C_3H_4	methylacetylene
C_3H_8	propane
C_4H_2	diacetylene
HCN	hydrogen cyanide
HC_3N	cyanocetylene
C_2N_2	cyanogen

TITAN

With a diameter of 5150 km, Saturn's giant satellite Titan is about 6 percent larger than the planet Mercury. Through even a large telescope, however, it is nothing more than a dot of light, so we knew little about it until the Voyager spacecraft visited Saturn in 1980 and 1981. The photographs and measurements radioed back to Earth revealed that Titan is a complex world with a thick atmosphere (Figure 22–25).

Spectra taken in 1944 revealed the presence of methane, and measurements made in the 1970s detected hydrocarbon molecules and suggested that the atmosphere is very hazy. The Voyager photographs show the moon wrapped in a deep haze layer that completely hides the surface. Detailed analysis of Voyager data shows that the atmosphere contains about 3 percent methane, 85 percent nitrogen, and as much as 12 percent argon. The atmosphere is about 1.6 times denser than Earth's.

This abundance of nitrogen on Titan probably came from ammonia (NH_3). Ammonia is a very common building material in the outer solar system, and it was probably quite abundant in the icy planetesimals from which Titan formed. As Titan outgassed this ammonia, sunlight broke it up and the hydrogen gas leaked into space, leaving nitrogen behind.

We might wonder how a world as small as Titan manages to retain any atmosphere at all. It is only slightly large than Mercury and not as large as Jupiter's moon Ganymede. Neither of those bodies has an atmosphere. But Titan lies almost twice as far from the sun as does Ganymede, so Titan receives only one-fourth as much sunlight and is much colder. Although Titan is small, it is so cold that the molecules in its atmosphere do not travel fast enough to escape easily.

The temperature on the surface of Titan is about 95 K (– 290°F), and the atmospheric pressure is high enough for methane to exist as a liquid on the surface. Thus, methane on Titan may behave as does water on Earth—that is, it may exist as a gas in the atmosphere and as a liquid on the surface. We don't know how much liquid methane might exist on Titan because of the thick clouds, but it could be anything from small ponds to vast oceans a kilometer deep.

If humans ever sail methane oceans on Titan, they will find it a messy, smelly experience. Liquid methane is a clear, odorless fluid, but sunlight stimulates methane molecules in the upper atmosphere to form other organic molecules* (Table 22–4). These condense into microscopic smog particles and settle down through the atmosphere, taking about a year to reach the surface. This continuous rain of organic compounds probably covers the surface with a tarry, smelly goo anywhere from 10 m to 1 km thick.

*Organic molecules are common in living things on earth but do not have to be derived from living things. One chemist defines an organic molecule as "any molecule with a carbon backbone."

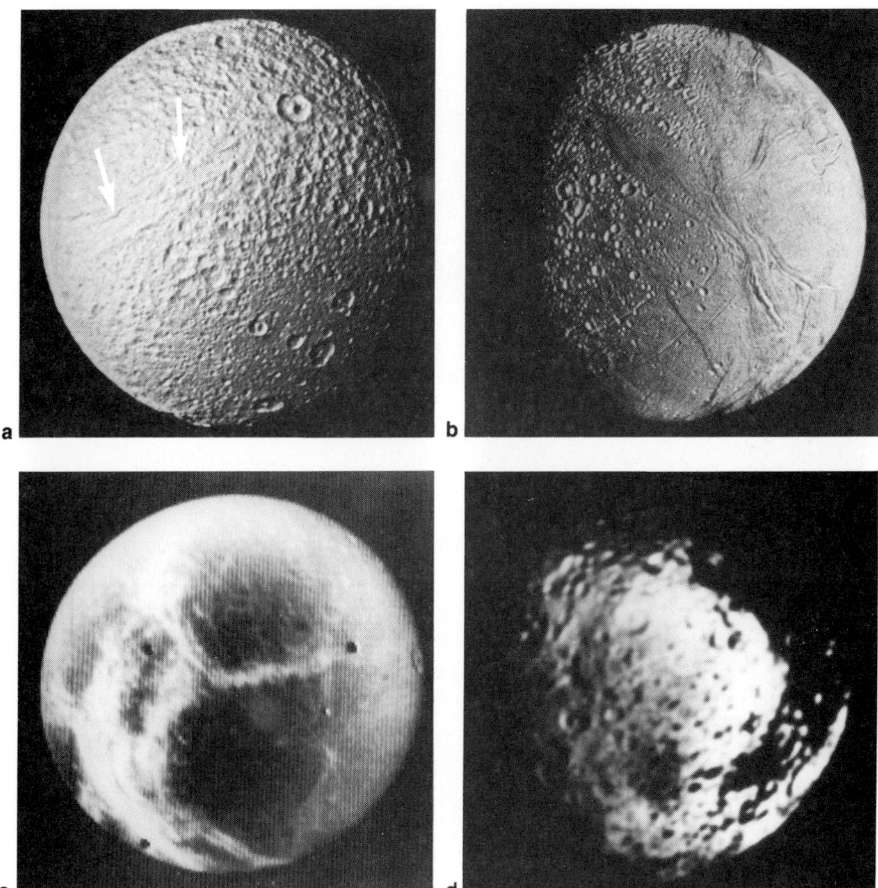

FIGURE 22–26 (a) Heavily cratered Tethys appears to be a dead world, although it is marked by a deep valley (arrows). (b) Some portions of Enceladus have fewer craters. These areas apparently have been resurfaced recently, which suggests that Enceladus is an active moon. (c) Although Dione is cratered much like Tethys, the trailing side is marked by wisps. (d) The leading side of Iapetus (on the right) is much darker than the trailing side. (NASA.)

The goo on Titan is exciting for two reasons. First, such organic molecules may have been the precursors of life on Earth, and similar chemical reactions on Titan may have created life there. Although some scientists argue that the temperature on Titan is too low, future explorers will want to visit Titan to look for life. We will consider this further in Chapter 25.

Another reason the goo is exciting is that it tells us something about Titan as a world. The processes that convert methane to organic molecules could exhaust the methane in Titan's atmosphere in only 10^6 years, so we must suppose the methane is being supplied from Titan itself. Titan's density is 1.90 g/cm^3. If it were not compressed by its own gravity, its density would be slightly more than 1.2 g/cm^3—the density of a 50:50 mix of rock and ice. The methane and ammonia so abundant on Titan are probably trapped in an ice mantle many hundred kilometers deep.

Titan's interior may be hot enough to make its icy surface active. It is large enough to have differentiated into a rocky core and an icy mantle, and the heat from the interior may drive unearthly volcanoes of ice, methane, and ammonia. We won't know until we send a spacecraft down through Titan's dirty atmosphere.

THE SMALLER MOONS

In addition to Titan, the Voyager spacecraft imaged 8 smaller Saturnian moons. These moons have low densities and are evidently formed of mixtures of rock and ice.

Almost all of these moons are heavily cratered and appear to be geologically dead (Figure 22–26a). But a few show signs of past activity. Tethys is marked by a crack 3 km deep trailing 1500 km across its surface—three-quarters of the way around the moon. Similarly, Mimas, the innermost of the larger moons, is marred by a crack 2 km deep. These cracks may have formed when the interiors of the icy moons froze and expanded.

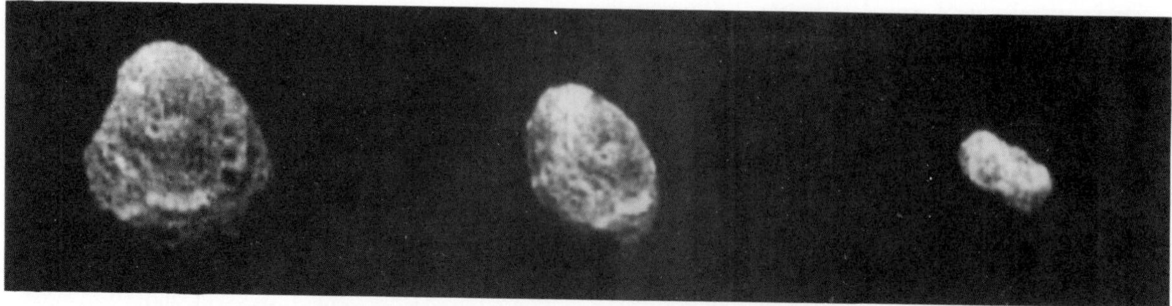

FIGURE 22–27 Hyperion is a small satellite that has not been able to pull itself into a spherical shape. This photo shows three views from slightly different distances. (NASA.)

Because most of the moons are heavily cratered we can conclude that they are geologically dead, but Enceladus may be an active moon (Figure 22–26b). Its surface contains areas of fewer craters, and some regions contain parallel grooves similar to those on Ganymede. These may be cracks filled by frozen water from the interior. The light yellow color of Enceladus also suggests that its surface is not old. It seems possible that eruptions of liquid water occasionally resurface parts of Enceladus with fresh ice. This would explain why the ice crystals of Saturn's extended E ring are more common near the orbit of Enceladus.

Whenever we see an active moon, we must wonder about the source of its internal energy. Enceladus is too small to have been heated significantly by radioactive decay, so we can suspect that tidal heating is responsible.

Another oddity found among the Saturnian moons is an asymmetry between the leading and trailing sides of some of the moons. Like the moons of Jupiter, all of Saturn's moons are tidally locked to Saturn and orbit with one side leading and one side trailing. Rhea's trailing side is marked by **wispy terrain**, which may be surface ice deposited by venting gas. Dione, too, has wisps on its trailing side (Figure 22–26c). If such wisps were produced when the moons were young and active, the heavier cratering could have erased wisps on the leading sides of the moons.

Iapetus is another asymmetric moon. Its trailing side is about as bright as dingy snow, but its leading side has an albedo of only 5 percent—about as dark as fresh asphalt on a highway (Figure 22–26d and Color Plate 47). The origin of this dark material is unknown, but theorists suggest that exposure to sunlight could convert methane trapped in the icy crust into dark organic compounds. The heavier cratering on the leading side could have driven off water and exposed a deeper layer rich in methane.

Saturn's moon Hyperion is odd because it is not spherical. It is shaped like a battered hockey puck with diameters of 250 km, 230 km, and 110 km (Figure 22–27). Mimas and Enceladus are only slightly larger and are quite spherical, so we must suppose that the gravitational field of a larger moon is sufficient to overcome the structural strength of ice and squeeze the moon into spherical shape. A small moon like Hyperion, located far from its parent planet where tidal heating is unimportant, may remain cold, rigid, irregular, and undifferentiated.

THE MINOR SATELLITES

Understanding the origin of the satellite systems is more difficult because the moons interact gravitationally in complicated ways. Planetologists did not realize how common these interactions are until the Voyager spacecraft visited Saturn and revealed small moons sharing orbits in peculiar ways.

We have previously discussed the small satellite that shepherds the outer edge of the A ring and the pair of moons that shepherds the F ring. An even more peculiar pair of moons is now known as the coorbital satellites. These two irregularly shaped moonlets have orbits separated by only 100 km. Because one moon is about 200 km in diameter and the other is about 100 km in diameter, they cannot pass in their orbits. Instead, the innermost moon gradually catches up with the outer moon. As the moons draw closer together, the gravity of the trailing moon slows the leading moon and makes it fall into a lower orbit. Simultaneously, the gravity of the leading

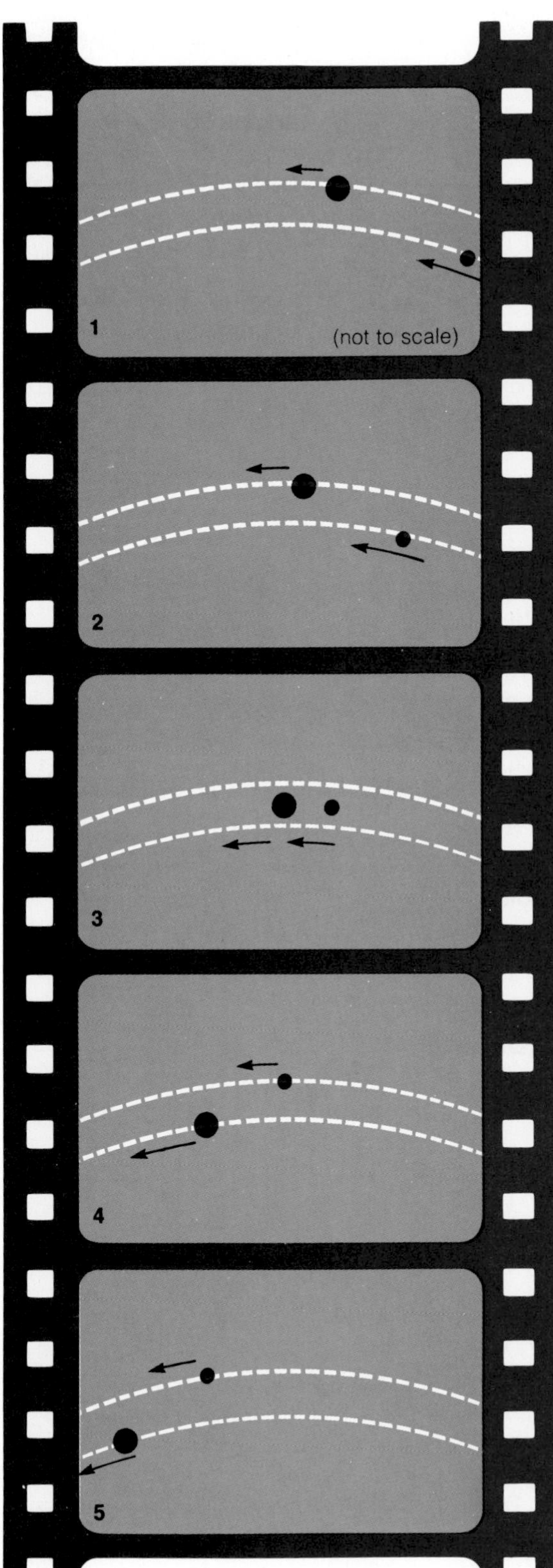

FIGURE 22–28 As the trailing coorbital satellite overtakes the leading satellite, gravitational forces make the moons trade orbits. Because the inner orbit is slightly faster, the innermost moon always gains on the outermost.

moon pulls the trailing moon ahead, and it rises into a higher orbit (Figure 22–28). The higher orbit has a longer period, so the trailing moon begins to fall behind the leading moon, which is now in a smaller, faster orbit. Thus, the moons exchange orbits and draw apart only to meet again and again. It seems very likely that these two moons are fragments of a larger moon destroyed by a major impact.

In addition to the waltzing coorbital moons, the Voyager spacecraft discovered small moons trapped at specific points in the orbits of Dione and Tethys. These points, known as the **Lagrangian points**, lie 60° ahead of and 60° behind a moon as it orbits its parent planet and represent points of orbital stability within which small moonlets can become trapped. (We will see in Chapter 24 that asteroids are trapped in the Lagrangian points of Jupiter's orbit around the sun.) The satellite Helene orbits Saturn 60° ahead of Dione. Voyager 2 searched for a satellite at the trailing Lagrangian point but did not find anything. Tethys has two Lagrangian satellites, one ahead and one behind. Unlike Helene, which wanders as much as 14° from its Lagrangian point, Calypso and Telesto do not wander far.

SUMMARY

The Jovian planets are so massive that they have been able to retain much of their internal hydrogen and helium. This excess of light elements gives the Jovian planets densities much lower than those of the terrestrial planets. In their interiors, the tremendous pressure forces the hydrogen into a liquid state, and in Jupiter and Saturn the pressure is so high that hydrogen becomes a liquid metal, which is a good conductor of electricity. Electric currents within this material produce magnetic fields that trap particles from the solar wind to form radiation belts around the planets. Both Jupiter and Saturn are known to have radiation belts similar to Earth's Van Allen belts, and Jupiter's belts are very intense.

The atmosphere of Jupiter is marked by high- and low-pressure regions in the form of belts and zones, which encircle the planet. Chemical reactions in the clouds are believed to produce compounds that color the belts. Saturn's belts and zones are not as well defined as those on Jupiter, and those of Uranus and Neptune are barely visible.

The Great Red Spot on Jupiter is believed to be a cyclonic disturbance much like an earthly hurricane. It has lasted for at least 300 years. The Voyager spacecraft revealed that Jupiter's atmosphere contains a number of similar spots that are smaller and shorter-lived.

Jupiter's ring is made up of dark, silicate dust specks, probably blasted from small satellites near the outer edge of the ring. The dust falls into Jupiter or is destroyed by radiation in less than 100 years, so it must be constantly resupplied.

Saturn's rings are icy and highly complex. Much of the structure within the rings appears to be created by the gravitational influence of Saturn's moons, although the planet's magnetic field may produce the spokes in the ring. A few small moons have been located shepherding the narrow F ring and the outer edge of the A ring. The icy material in the rings could be left over from the formation of Saturn, but it could also be debris left behind by the break up of one or more satellites.

Jupiter has a number of small moonlets—some in retrograde orbits—that are probably captured asteroids. But the four Galilean satellites seem to have formed at about their present locations. The density of these moons decreases with increasing distance from Jupiter, suggesting that they formed in a small nebula around the condensing planet. The heat from Jupiter would have driven off volatiles in the inner nebula and made the inner moons denser than the outer moons. This is a miniature of the formation of the planets from the solar nebula. This process does not seem to have occurred in the Saturn system, perhaps because Saturn is less massive and did not become as hot as Jupiter.

Jupiter's satellites have clearly been affected by tidal heating. Io is violently active with sulfur volcanoes. Europa is less active, but the lack of craters on its surface suggests that its icy crust is still active enough to erase craters. Ganymede and Callisto are both heavily cratered, inactive worlds, but the grooved terrain on Ganymede seems to date from a period when it was more active. How much tidal heating contributed to the activity of the outer satellites is not known.

If Titan, the largest satellite of Saturn, is active, we cannot tell because of its thick atmosphere of nitrogen and methane. Titan's low temperature allows it to keep gases that would escape from a warmer world. In fact, liquid methane may exist at the surface. Also, the surface may be dirtied by a thick layer of organic goo produced in the upper atmosphere by the interaction of sunlight with methane.

Some of Saturn's satellites are clearly inactive. Mimas, Tethys, and Dione are heavily cratered, icy bodies. Nevertheless, the presence of long, deep valleys formed by cracks in the icy crust shows that the worlds once had more internal heat.

Enceladus seems to be a very active moon. It is cratered, but some regions are obviously young. Water-driven volcanoes may periodically resurface portions of Enceladus and spew water vapor into space, where it freezes to form particles in Saturn's E ring. The E ring is most dense at the orbit of Enceladus.

The debris in space craters the surfaces of moons most heavily on the side facing forward in their orbits. The trailing side is less cratered and may preserve features from the early days of the moon, such as the wisps on the trailing side of Rhea. Another moon with a hemispheric asymmetry is Iapetus, bright on the trailing side and very dark on the leading side. Whether the black deposits are debris from some outside source or material from the interior is not known.

Many of Saturn's moons are small, irregular moons that are either captured bodies or the fragments of collisions. Hyperion is very irregular in shape.

The orbital dynamics of the smaller moons is complex. The coorbital moons share nearly identical orbits, whereas other moons occupy the Lagrangian points in the orbits of larger moons.

NEW TERMS

oblateness	grooved terrain
liquid metallic hydrogen	tidal heating
decimeter radiation	infrared outburst
decameter radiation	Io flux tube
belts	spoke
zones	shepherd satellite
forward scattering	wispy terrain
Roche limit	Lagrangian point

QUESTIONS

1. Why is Jupiter oblate? Do you expect all Jovian planets to be oblate? Why or why not?
2. How do the interiors of Jupiter and Saturn differ? How does this affect their magnetic fields?
3. Describe two theories to explain the persistence of the Jovian belts and zones.
4. How can we be certain that Jupiter's ring does not date from the formation of the planet?
5. If Jupiter had a satellite the size of our own moon orbiting outside the orbit of Callisto, what would you predict for its density and surface features?
6. Why are there no craters on Io or on Europa? Why should we expect Io to suffer more impacts per square kilometer than Callisto?
7. Why are the belts and zones on Saturn less distinct than those on Jupiter?

8. If Saturn had no moons, what do you suppose its rings would look like?
9. Suggest two theories for the origin of Saturn's rings.
10. How can Titan keep an atmosphere when it is smaller than airless Ganymede?
11. If you piloted a spacecraft to visit Saturn's moons and wished to land on a geologically old surface, what features would you look for? What features would you avoid?
12. Why does the leading side of some satellites differ from the trailing side?
13. Why do we expect small satellites to be irregular in shape? Give examples and exceptions.
14. How might the origin of Saturn's coorbital satellites be similar to the origin of the outer two groups of Jupiter's satellites? Why?

PROBLEMS

1. What is the maximum angular diameter of Jupiter as seen from Earth? What is the minimum? Repeat this calculation for Saturn and Titan. (HINT: See Box 3–1.)
2. The highest-speed winds on Jupiter are in the equatorial jet stream, which has a velocity of 150 m/sec. How long does it take for these winds to circle Jupiter?
3. Measure the photograph in Data File 7a and compute the oblateness of Jupiter.
4. What is the orbital velocity and period of a ring particle at the outer edge of Jupiter's ring? At the outer edge of Saturn's A ring? (HINT: See Box 5–1.)
5. What is the angular diameter of Jupiter as seen from the surface of Callisto? (HINT: See Box 3–1.)
6. What is the escape velocity from the surface of Ganymede if its mass is 1.5×10^{26} g and its radius is 2628 km? (HINT: See Box 5–2.)
7. If we photograph the spectrum of Saturn and its rings, we should find light from one edge of the rings red-shifted and light from the other edge blue-shifted. If we observe at a wavelength of 500 nm, what difference in wavelength should we expect between the two edges of the rings? (HINT: See Problem 4 and Box 7–3.)
8. What is the difference in orbital velocity between particles at the outer edge of Saturn's B ring compared with particles at the inner edge of the B ring? (HINT: See Data File 8 and Box 5–1.)
9. Draw a circle of the proper size to represent the sun on Figure 22–24a.
10. What is the difference in orbital velocity between the two coorbital satellites if the semimajor axes of their orbits are 151,400 km and 151,500 km? The mass of Saturn is 5.7×10^{26} kg. (HINT: See Box 5–1.)

OBSERVATIONAL ACTIVITY: OBSERVING THE RINGS OF SATURN

The observational activities in this book have taken you outside to look at the sky. However, most astronomical observations are made from photographs or other records. This activity invites you to analyze a photograph taken from Earth and another radioed back to Earth from Saturn by the Voyager 2 spacecraft.

Dimensions of Saturn's Rings Examine Figure 19–7b. This photograph of Saturn was taken from earth when the planet was at a distance of 8.5 AU. The scale of the photograph is 0.60 seconds of arc per millimeter. Use a millimeter rule to measure the photograph to the nearest tenth of a millimeter. Use the scale of the photograph to convert to seconds of arc, and then use the small-angle formula (Box 3–1) to calculate the following dimensions:

Radius of planet = ________km
Radius of outer edge of A ring = ________km
Radius of Cassini's division = ________km
Radius of inner edge of B ring = ________km

Compare your results with the values given in Data File 8.

Ringlets Examine the photograph shown in Figure 22–19. This Voyager 2 photograph was taken from a distance of 2.7 million km, and the scale is 10.5 seconds of arc per millimeter.

Develop your own methods for the analysis of this photo and answer the following questions:

1. Saturn's rings contain hundreds of ringlets. From your measurements, what is the narrowest observed ringlet width?
2. The television camera that recorded this photograph produced an image with 1000 dots across the width of the picture. No object in the rings smaller than this could be recorded. What is the diameter of the smallest object that could have been detected in this photograph?

Describe how you analyzed the photograph to answer the two questions above and explain why you can or cannot be sure you have detected the narrowest ringlet features in Saturn's rings.

RECOMMENDED READING

BEATTY, J. KELLY "Rendezvous with a Ringed Giant." *Sky and Telescope 61* (Jan. 1981), p. 7.

———. "Voyager at Saturn, Act II." *Sky and Telescope 62* (Nov. 1981), p. 430.

BEATTY, J. KELLY, BRIAN O'LEARY, and ANDREW CHAIKIN *The New Solar System,* 2nd ed. Cambridge, Mass.: Sky Publishing, 1982.

BERRY, RICHARD "A Closer Look at Saturn's Rings." *Astronomy 10* (Feb. 1982), p. 74.

BURGESS, ERIC *By Jupiter: Odysseys to a Giant.* New York: Columbia University Press, 1982.

BURNHAM, ROBERT "The Saturnian Moons." *Astronomy 9* (Dec. 1981), p. 6.

CHAIKIN, ANDREW "Voyager Among the Ice Worlds." *Sky and Telescope 71* (April 1986), p. 338.

CHAPMAN, CLARK R. *Planets of Rock and Ice.* New York: Scribner, 1982.

COOPER, HENRY S. F. *Imaging Saturn: The Voyager Flights to Saturn.* New York: Holt, Rinehart & Winston, 1985.

CUZZI, JEFFREY N. "Ringed Planets: Still Mysterious." *Sky and Telescope 68* (Dec. 1984), p. 511, and *69* (Jan. 1985), p. 19.

ELLIOT, J., and R. KERR *Rings: Discoveries from Galileo to Voyager.* Cambridge, Mass.: MIT Press, 1984.

———. "How Jupiter's Ring was Discovered." *Mercury 14* (Nov./Dec. 1985), p. 162.

GLASS, BILLY P. *Introduction to Planetary Geology.* Cambridge, Mass.: Cambridge University Press, 1982.

HARTMANN, WILLIAM K. "The View from Io." *Astronomy 9* (May 1981), p. 17.

INGERSOLL, ANDREW P. "Jupiter and Saturn." *Scientific American 245* (Dec. 1981), p. 90.

JOHNSON, TORRENCE V., and CLAYNE M. YEATES "Return to Jupiter: Project Galileo." *Sky and Telescope 66* (Aug. 1984), p. 99.

LANZEROTTI, LOUIS J., and STAMATIOS M. KRIMIGIS "Comparative Magnetospheres." *Physics Today 38* (Nov. 1985), p. 24.

MARAN, STEPHEN P. "Inside Jupiter's Rings." *Natural History 91* (Aug. 1982), p. 64.

MORRISON, DAVID "The New Saturn System." *Mercury 10* (Nov./Dec. 1981), p. 162.

———. *Voyages to Saturn.* Washington, D.C.: NASA, 1982.

———. "The Enigma Called Io." *Sky and Telescope 69* (March 1985), p. 198, and *69* (May 1985), p. 395.

MORRISON, DAVID, and JANE SAMZ *Voyage to Jupiter.* Washington, D.C.: NASA, 1980.

MORRISON, N., and S. GREGORY "The Exotic Atmosphere of Titan." *Mercury 14* (Sept./Oct. 1985), p. 154.

MULHOLLAND, DERRAL "Ice Planet." *Science '82 3* (Dec. 1982), p. 64.

OSTERBROCK, D. "The Nature of Saturn's Rings: James Keeler and the Doppler Principle." *Mercury 14* (March/April 1985), p. 46.

OWEN, TOBIAS "Titan." *Scientific American 246* (Feb. 1986), p. 98.

POLLACK, JAMES B., and JEFFREY N. CUZZI "Rings in the Solar System." *Scientific American 245* (Nov. 1981), p. 105.

SIMON, SHERIDAN "The View from Europa." *Astronomy 14* (Nov. 1986), p. 98.

SODERBLOM, LAWRENCE A., and TORRENCE V. JOHNSON "The Moons of Saturn." *Scientific American 246* (Jan. 1982), p. 101.

SQUYRES, STEVEN W. "Ganymede and Callisto." *American Scientist 71* (Jan./Feb. 1983), p. 56.

WALDROP, M. MITCHELL "Saturn Redux: The Voyager 2 Mission." *Science 213* (11 Sept. 1981), p. 1236.

———. "The Puzzle That Is Saturn." *Science 213* (18 Sept. 1981), p. 1347.

YEATES, CLAYNE M., and THEODORE C. CLARKE "Galileo Mission to Jupiter." *Astronomy 10* (Feb. 1982), p. 6.

See also *Science 188* (1 June 1975), *203* (23 Feb. 1979), *204* (1 June 1979), *205* (6 July 1979), *206* (23 Nov. 1979), *207* (25 Jan. 1980), *212* (10 April 1981), and *215* (29 Jan. 1982).

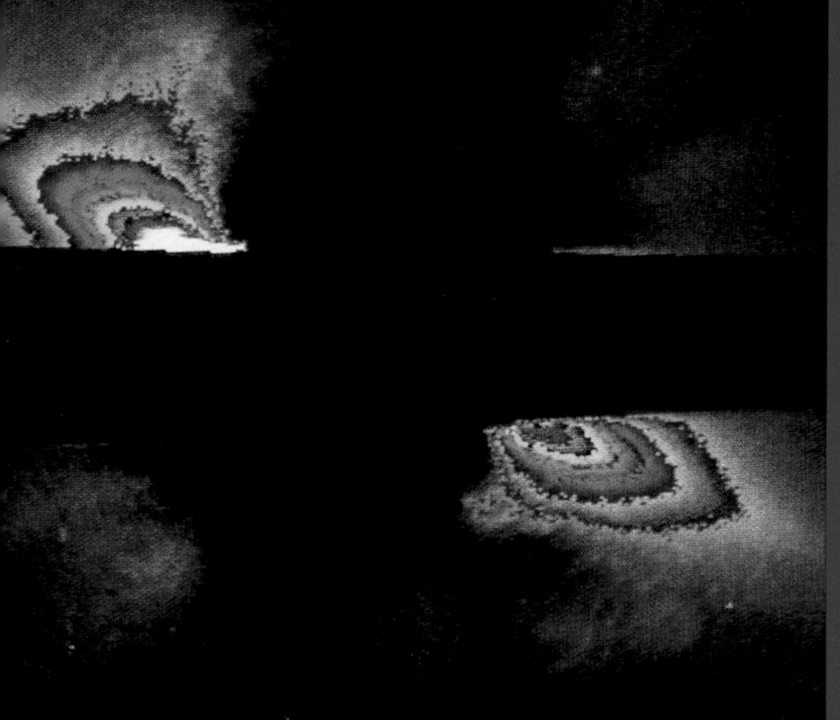

▲30

▼31

30. This false-color image shows a nearly edge-on disk of dusty material orbiting the nearby star Beta Pictoris. The star was hidden behind the horizontal bar across the center, and scattered light from the star was subtracted from the image. Such dusty disks are thought to resemble the solar nebula from which the sun and planets formed (pp. 230, 400). (Courtesy Francesco Paresce and Christopher Burrows, Space Telescope Science Institute and European Space Agency.)

31. A false-color radio map shows the distribution of the molecule CO near the star HL Tauri. The location of the gas in what appears to be a disk suggests that HL Tauri is a young system still surrounded by a disk of gas much like the sun's solar nebula (pp. 230, 400). (Courtesy Anneila I. Sargent and Steven Beckwith.)

32. The moon, like all of the planets in the solar system, formed from the heavier elements in the solar nebula. The surface of the moon has been modified only by volcanism and by meteorite impacts that have fractured the crust to great depth. The large rock visible here was ejected from a meteorite impact hundreds of kilometers away (p. 439). (NASA.)

▼32

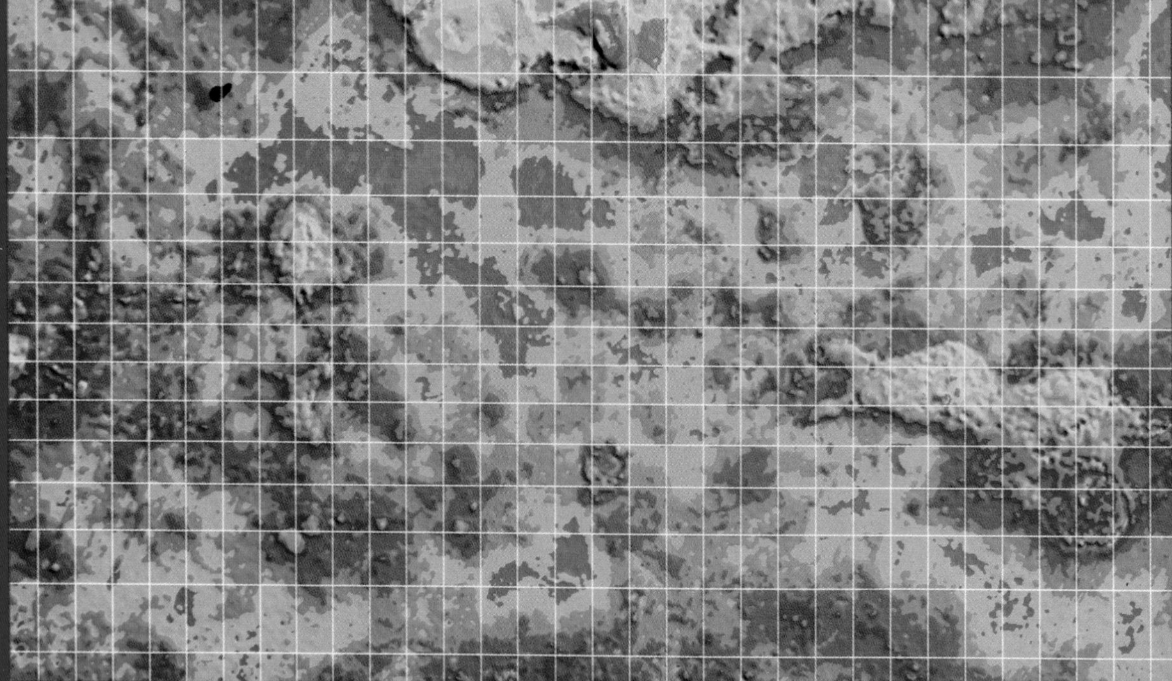

▲33

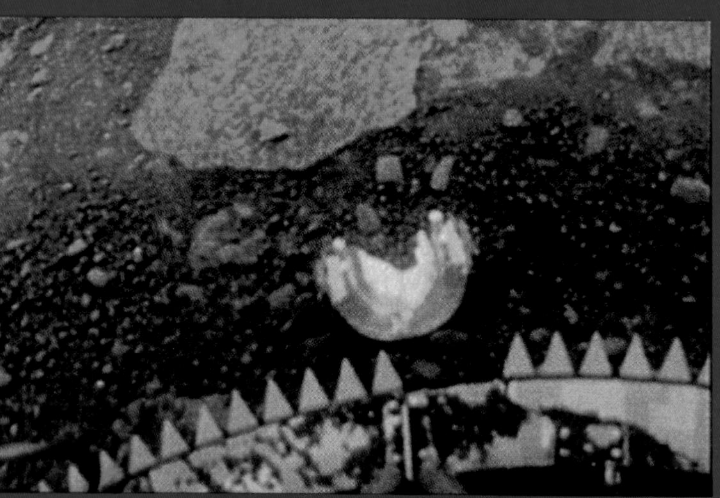

▲34a

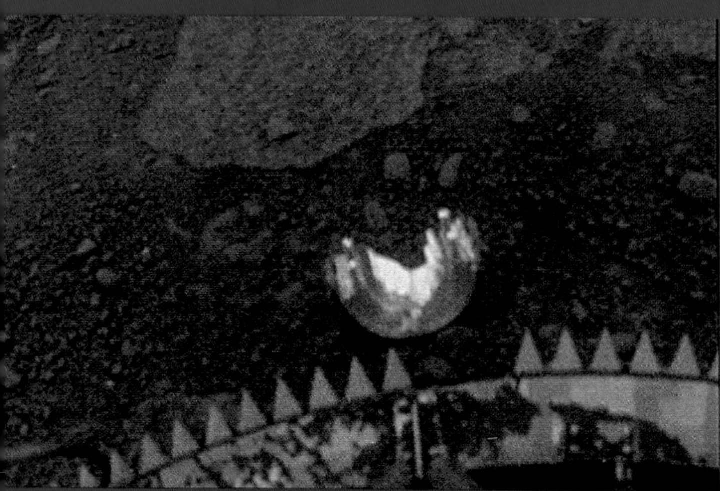

▲34b

33. The topography of Venus is revealed in a radar map made by the Pioneer Venus Orbiter. The highest regions are red; the lowest violet. The high volcanic peak Maxwell is located at top center with Aphrodite Terra at right center (p. 461). (NASA image courtesy Gordon Pettengill.)

34. The color of the surface of Venus appears orange in the original Venera 13 photograph (a), but analysis shows the color is produced by the thick atmosphere. The corrected image (b) reveals that the surface minerals are dark gray, which is consistent with iron oxides at high temperatures (p. 462). (Courtesy of C. M. Pieters through the Brown/ Vernadsky Institute to Institute Agreement and the U.S.S.R. Academy of Sciences.)

35. This radar image of Maxwell on Venus shows a system of fracture patterns in the surrounding terrain and a circular feature, possibly a volcanic caldera, near the 12-km-high summit. Made from the earth, this image has a resolution of about 2 km (p. 461). (Courtesy D. B. Campbell.)

36. Made of 102 separate images recorded by the Viking 1 orbiter, this mosaic of the face of Mars shows three volcanic features in the Tharsis rise at the left. The great canyon system Valles Marineris stretches across the lower third of the mosaic with erosion features at and above center (p. 470). (NASA/USGS courtesy Alfred S. McEwen.)

37. Mars, known since ancient times as the red planet because of its appearance to the naked eye, is indeed red. This Viking lander image shows the red soil and the sky colored red by suspended dust. Iron oxides are believed responsible for the red color (p. 472). (NASA.)

▼35

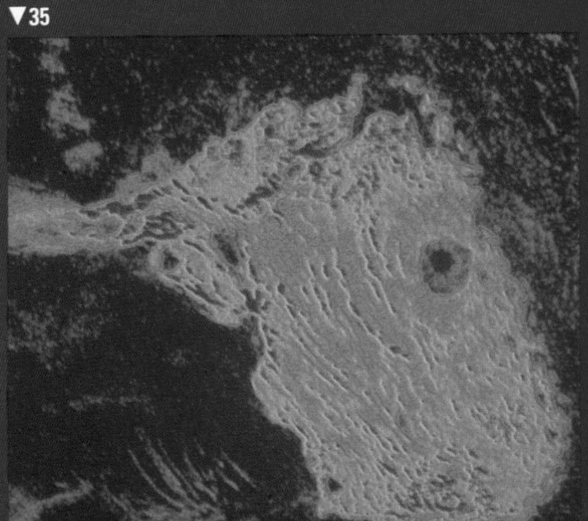

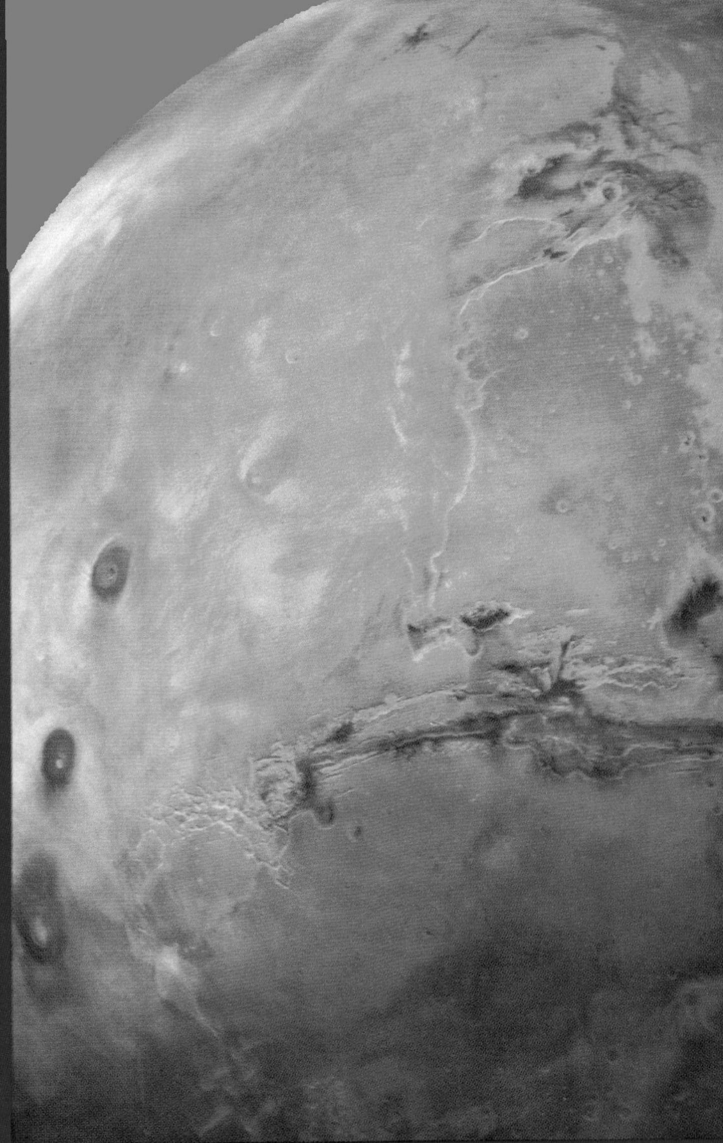

▲36

▼37

38. Photographs of Jupiter, Saturn, and Earth have been reduced to the same scale for comparison. Note the contrast between the well-defined cloud belts of Jupiter and the indistinct belts of Saturn (p. 479). (Stephen P. Meszaros, NASA/ Goddard Space Flight Center.)

39. The Great Red Spot in Jupiter's atmosphere (extending beyond right edge) is, like the many smaller spots, a cyclonic circulation. The image of the earth is reproduced to the same scale for comparison (p. 483). (Stephen P. Meszaros, NASA/ Goddard Space Flight Center.)

40. Io is surrounded by a cloud of glowing sodium vapor as shown in this photograph taken from the earth. The sodium and other atoms are believed emitted by the volcanic activity on Io (pp. 483, 492). (JPL/NASA, Bruce A. Goldberg.)

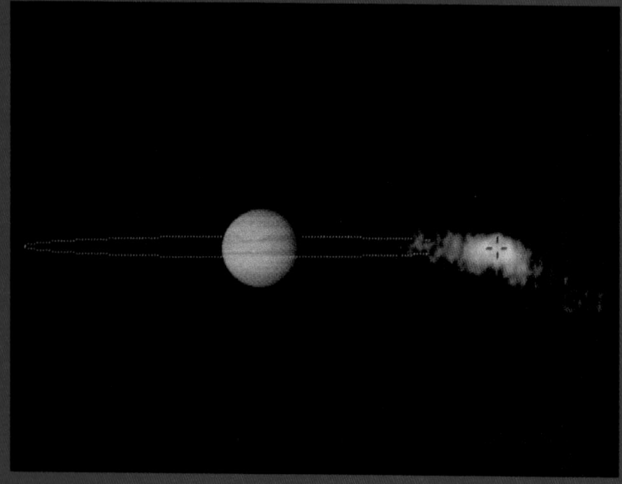

41. The Galilean satellites of Jupiter. Clockwise from upper left: Io, Europa, Callisto, and Ganymede. All but Io are rich in ice (pp. 488, 492). (NASA.)

42. This computer-enhanced image of Jupiter's satellite Io shows an erupting volcano, Loki, at the edge of the disk. Such plumes of sulfur ash fall back to blanket the satellite's surface in concentric rings around volcanic vents, such as Pele at lower left. The light areas appear to be sulfur dioxide frost (p. 492). (NASA image processed by Alfred McEwen.)

▼39

▲41

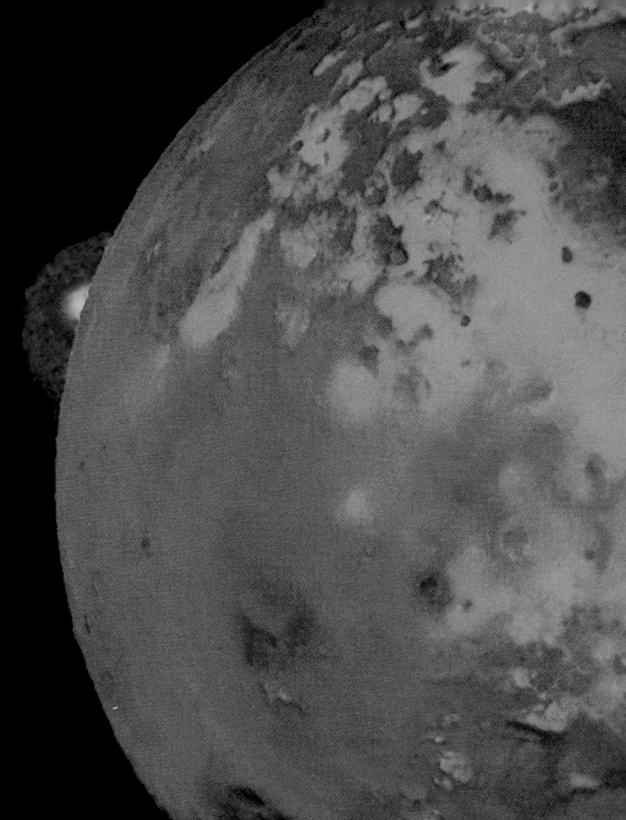

▲42

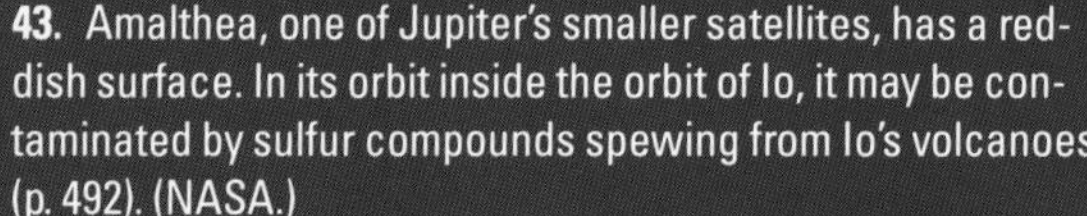

43. Amalthea, one of Jupiter's smaller satellites, has a reddish surface. In its orbit inside the orbit of Io, it may be contaminated by sulfur compounds spewing from Io's volcanoes (p. 492). (NASA.)

44. The red surface of Io is marked by active volcanic calderas venting sulfur ash and sulfur dioxide gas. The dark channels appear to be the paths of lava flows (p. 492). (NASA.)

▼43

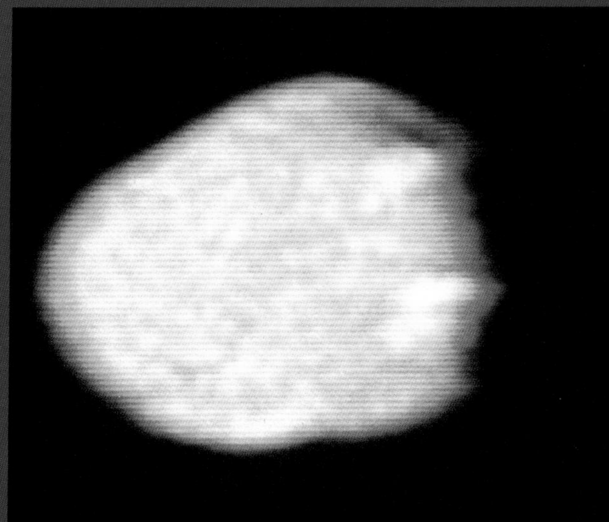

44►

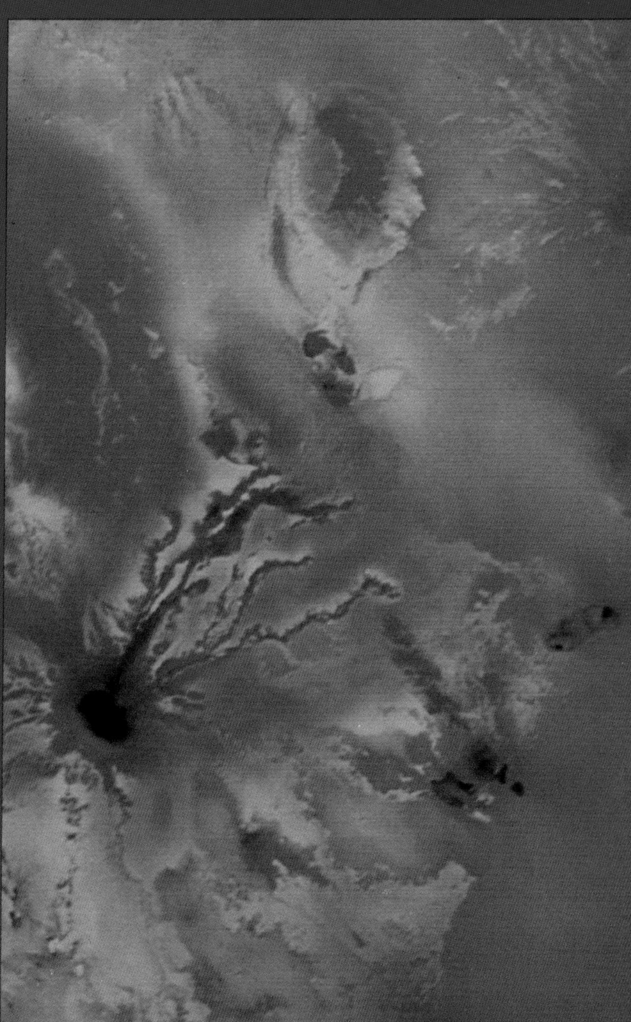

▲45

▼46

45. The rings of Saturn consist of nearly 1000 concentric ringlets. This computer-enhanced photograph shows the narrow F ring beyond the A ring and spokes in the B ring. Note the transparency of the rings at left (p. 495). (NASA.)

46. A computer-enhanced image of Saturn's rings shows the complex structure. Note the ringlets within Cassini's division. Earth reproduced for scale. A 2-cm wavelength radio image (inset) shows the contrast between the radio bright planet and the rings (p. 499). (NASA; inset courtesy NRAO/AUI, observers I. de Pater and J. R. Dickel.)

47. Mysterious Iapetus has a very bright surface except on its leading side, where it is a very dark red-brown. This asymmetry between leading and trailing sides is common among Saturn's moons, but its exaggeration on Iapetus is not understood (p. 506). (NASA.)

▼47

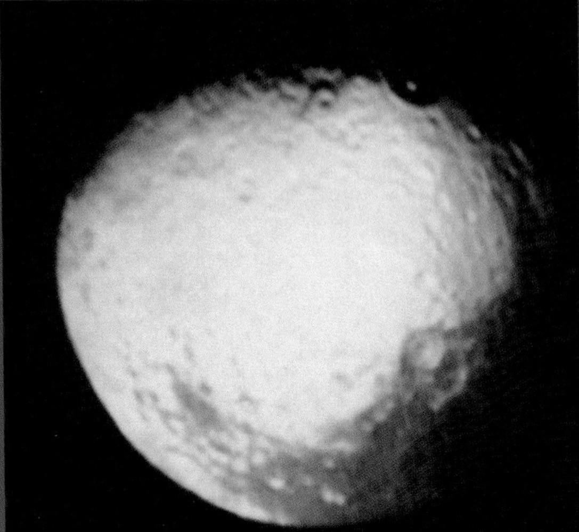

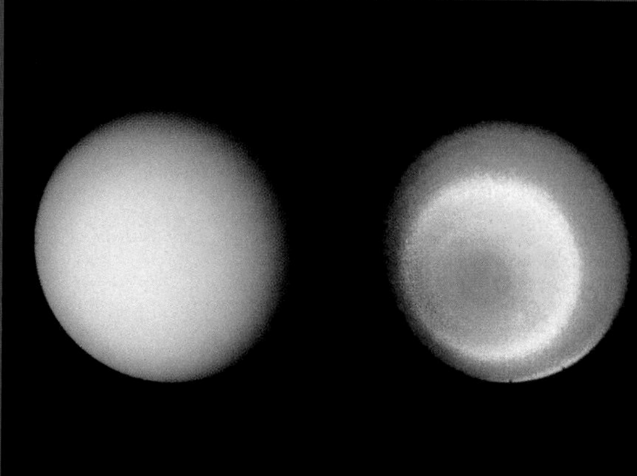

▲48

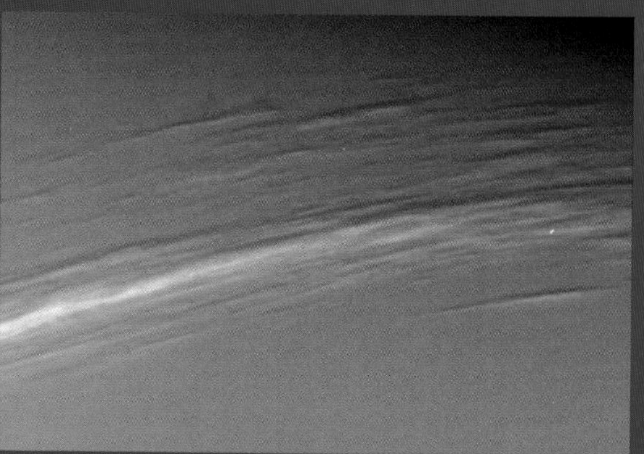

▲49

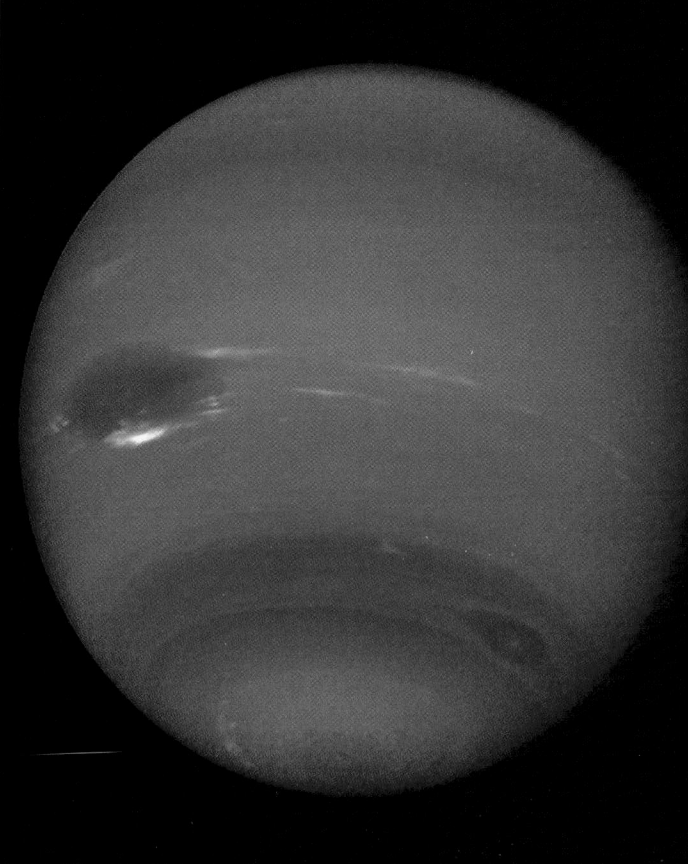

▲50

48. The planet Uranus as photographed by the Voyager 2 spacecraft (left). The blue-green color arises from methane in the planet's atmosphere. No belt-zone circulation is visible, but extreme image enhancement (right) reveals atmospheric banding concentric with the planet's axis of rotation (p. 517). (NASA.)

49. Two hours before closest approach, Voyager 2 snapped this photo of methane ice crystal clouds on Neptune. High speed winds stretch the clouds into streaks parallel to the equator. The clouds are bright because they extend up to 50 km (31 miles) above the average cloud deck. Such high level clouds are brightly lit by sunlight (p. 527). (NASA/JPL.)

50. Neptune's blue-green color is caused by methane in its atmosphere which absorbs red light. The Great Dark Spot at the left is larger than the earth, and one of the smaller spots is visible at lower left. Such spots are believed to be cyclonic storms. The planet's south pole is at the bottom (p. 527). (NASA/JPL.)

51. Triton, Neptune's largest satellite, has its south pole (bottom) exposed to sunlight. The bright terrain around the pole appears to be frozen nitrogen. Note the linear faults north of the pole and the dark smudges within the polar cap. These smudges have been interpreted as surface debris from liquid nitrogen volcanoes (p. 530). (NASA/JPL.)

▼51

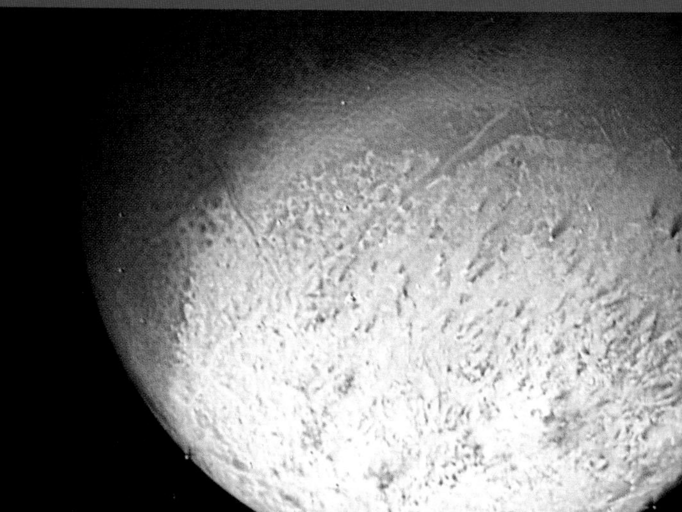

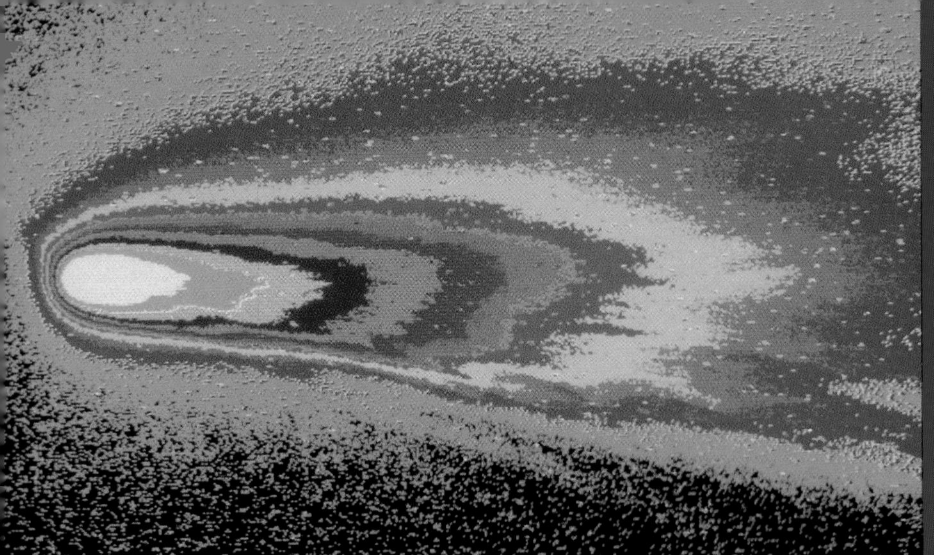

▲52

▼53

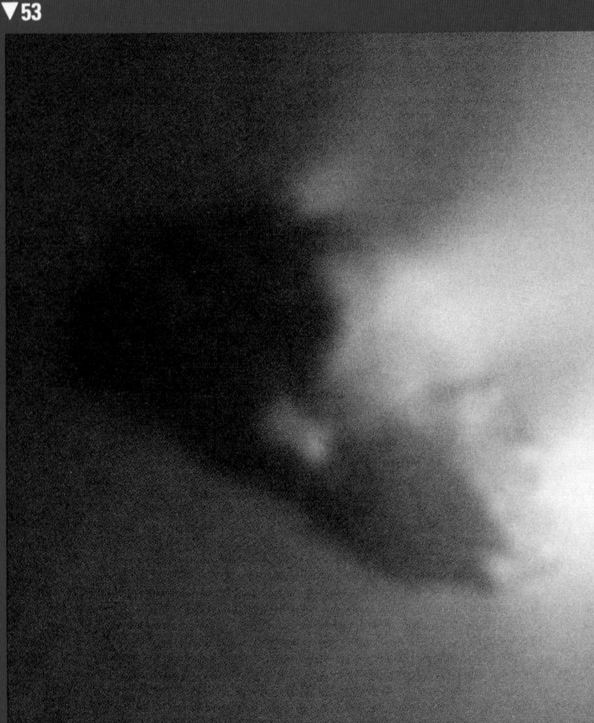

52. In reality, comets are pale, white objects, but digital enhancement of this photo of Comet Halley produces a false-color map of intensity. Such images reveal subtle detail such as streamers in the tail apparently produced by jets issuing from the nucleus (p. 550). (Copyright ©1986 Royal Observatory, Edinburgh.)

53. Sixty separate images from the Giotto spacecraft have been computer enhanced and merged to form this image of the icy nucleus of Comet Halley. Jets of gas and dust spew from active regions on the irregular surface. A crust of carbon-rich compounds is believed responsible for the dark color (pp. 555, 557). (Photo courtesy Harold Reitsema, Ball Aerospace. Copyright ©1986 Max Planck Institut fur Astrophysik.)

54. In this visible light photograph of Comet West, the tail is a transparent haze of gas and dust through which distant stars are visible (p. 549). (Celestron International.)

55. Looking much as it did to the naked eye, Comet Halley passes in front of the Milky Way. The clouds of gas, dust, and stars in the Milky Way are a thousand or more light-years distant. The comet is only a few light-minutes from the earth (p. 549). (National Optical Astronomy Observatories.)

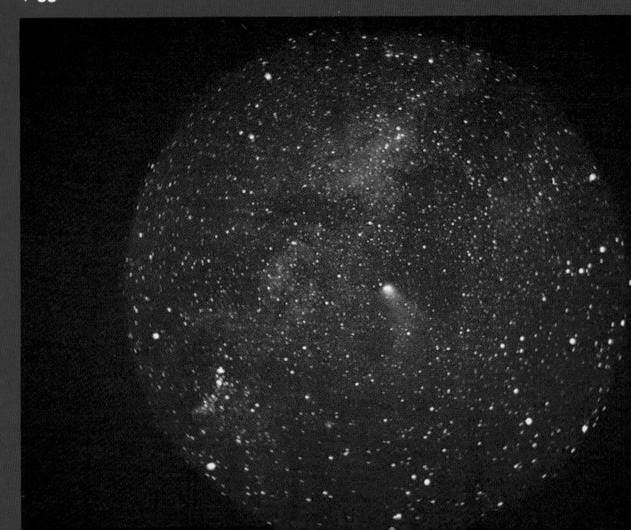

▼54

▼55

CHAPTER 23

URANUS, NEPTUNE, AND PLUTO

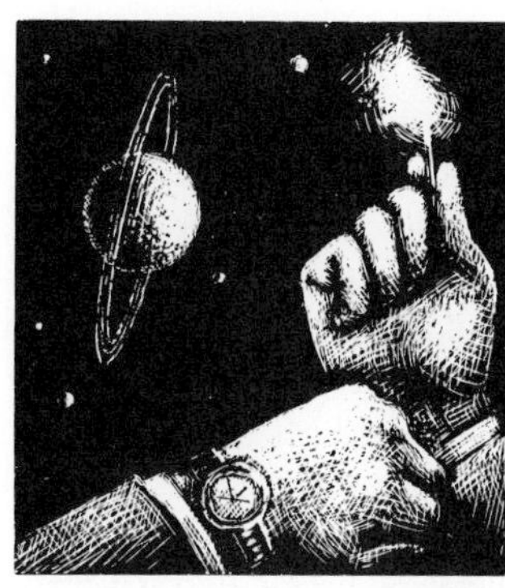

A good many things go around in the dark besides Santa Claus.

Herbert Hoover

Out in the darkness beyond Saturn, out where sunlight is 1000 times fainter, there are things going around that Aristotle, Galileo, and Newton never imagined. They knew about Mercury, Venus, Mars, Jupiter, and Saturn, but our solar system includes three more planets, planets that were not discovered until the invention of the telescope.

Uranus and Neptune are Jovian planets, nearly the same size, and about four times Earth's diameter. Thus, they are about half the size of Saturn. Through earth-based telescopes we see nothing but tiny disks, and until recently what little we knew of these planets was deduced from a few observations compared with Jupiter and Saturn. But in January 1986, the Voyager 2 spacecraft flew past Uranus and returned a wealth of photographs and measurements. Voyager 2 is now on its way to an August 1989 rendezvous with Neptune. Thus, we now can discuss Uranus and Neptune in some detail.

Pluto, however, is a more difficult subject. Through an earth-based telescope it looks like nothing more than a faint point of light, and the best estimates of its size suggest that it is smaller than Earth's moon. Voyager 2 will leave the solar system without passing near Pluto, so we may have to wait many decades before we can obtain photographs of its surface. Nevertheless, we may be able to guess at its nature by applying the principles of comparative planetology.

Uranus, Neptune, and Pluto differ from the other planets because they were not known from antiquity. They had to be discovered. The stories of the discoveries of these planets highlight the different ways that scientific discoveries are made.

23.1 URANUS

In March 1781 Benjamin Franklin was in France raising money, troops, and arms for the American Revolution. George Washington and his colonial army were only 6 months away from the defeat of Cornwallis at Yorktown and the end of the war. In England King George III was beginning to show signs of madness. And a German-speaking music teacher in the English resort city of Bath was about to discover Uranus.

THE DISCOVERY OF URANUS

William Herschel (Figure 15–3) came from a musical family in Hanover, Germany, but immigrated to England as a young man and eventually obtained a prestigious job as the organist at the Octagon Chapel in Bath.

To compose exercises for his students and choral and organ works for the chapel, Herschel studied the mathematical principles of musical harmony from a book by Professor Robert Smith of Cambridge. The mathematics was so interesting that Herschel searched out other books by Smith, including a book on optics. Of course, an eighteenth-century book on optics written by a professor in England relied heavily on Isaac Newton's optics and described the principles of the reflecting telescope. (See Chapter 6.) Herschel and his brother Alexander began building telescopes, and William went on to study astronomy in his spare time.

Herschel's telescopes were similar to Newton's in that they had metal mirrors, but they were much larger. Newton's telescope had a mirror about 1 inch in diameter, but Herschel developed ways of making much larger mirrors. He soon had telescopes as long as 20 feet, although one of his favorite telescopes was 7 feet long with a mirror 6.2 inches in diameter. Using it he began a research project that led to the discovery of Uranus.

Herschel did not set out to search for a planet; he was trying to detect stellar parallax produced by the motion of the earth around the sun. No one had yet detected this effect, although by the 1700s nearly all astronomers accepted the motion of the earth. Galileo had pointed out that parallax might be detected if a nearby star and a very distant star lay so nearly in the same direction that they looked like a very close double star through a telescope. In such a case, the orbital motion of the earth would produce a parallactic shift in the position of the nearby star with respect to the more distant star (Figure 23–1). Herschel began to examine all stars brighter than eighth magnitude to search for double stars, which might show parallax. That project alone took over 2 years.

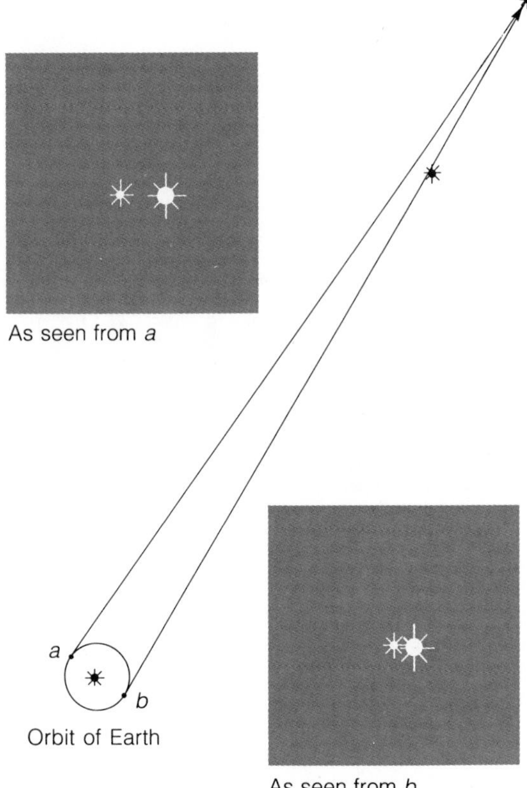

FIGURE 23–1 A double star consisting of a nearby star and a more distant star could be used to detect stellar parallax. The separation between the stars would be different as seen from different places in Earth's orbit. The effect is actually much too small to detect by eye.

On the night of March 13, 1781, Herschel set up his 7-foot telescope and continued his work. He later wrote, "In examining the small stars in the neighbourhood of H Geminorum, I perceived one that appeared visibly larger than the rest." As seen from Earth, Uranus is never larger in angular diameter than 3.7 seconds of arc, so Herschel's detection of the disk illustrates the quality of this telescope and his eye (Figure 23–2). At first he suspected that the object was a comet, but other astronomers quickly showed that it was a planet orbiting the sun beyond Saturn.

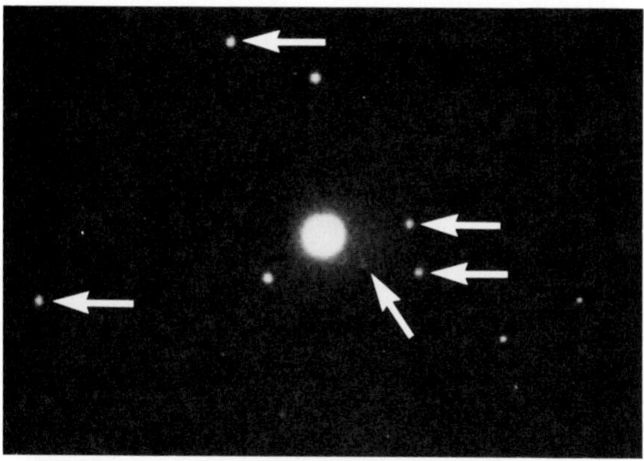

FIGURE 23–2 Through a telescope, Uranus is a featureless disk never more than 3.7 seconds of arc in diameter. This CCD image made with the 4-m telescope at Cerro Tololo Inter-American Observatory in Chile shows the disk of Uranus and its five major satellites (arrows). Other images are faint stars. (Courtesy Cerro Tololo Inter-American Observatory.)

The discovery of Uranus made Herschel world famous. Since antiquity astronomers had known the five planets—Mercury, Venus, Mars, Jupiter, and Saturn—and they supposed that the universe was complete. Herschel's discovery extended the classical universe. The English public accepted Herschel as their astronomer–hero, and, having named the new planet *Georgium Sidus* (George's Star) after King George III, Herschel received a royal pension. The former music teacher was welcomed into court society where he later met and married a wealthy widow and took his place as one of the great English astronomers in history. We saw in Chapter 15 how he later built large telescopes on his estate and, with his sister Caroline, a talented astronomer herself, attempted to map the extent of the universe.

European astronomers were less than thrilled that an Englishman had made the great discovery, and even some professional English astronomers called Herschel's discovery a lucky accident. Herschel responded in his own defense by making three points. First, he had built some of the finest-quality telescopes then in existence. Second, he had been conducting a systematic research project and would have found Uranus eventually because he was inspecting all of the brighter stars visible with his telescope. And third, he had great experience seeing fine detail with his telescopes. As a musician he knew the value of practice and applied it to the business of astronomical observing. In fact, astronomers had seen Uranus more than 17 times before Herschel, but each time they plotted it on their charts as a star.

This illustrates one of the ways scientific discoveries are made. Often discoveries seem accidental, but on closer examination we find that the scientist has earned the right to the discovery through many years of study and preparation. To quote a common saying, "Luck is what happens to people who work hard."

European astronomers, especially the French, insisted that the new planet not be named after an English king. The French stubbornly called the planet Herschel, as did many other non-English astronomers. Some years later Bode (of the Titius–Bode rule) suggested the name Uranus, the oldest of the Greek gods.

Over the half century following the discovery of Uranus, astronomers noted that Newton's laws did not quite predict the observed position of the planet. Tiny variations in the orbital motion of Uranus eventually led to the discovery of Neptune, a story we will tell later in this chapter.

THE MOTION OF URANUS

Uranus follows a slightly elliptical orbit with an average distance from the sun of 19.18 AU and an orbital period of 84.013 years (Date File 9). Compare this with the orbital period of Saturn, 29.461 years. Unlike Saturn, Uranus rotates on its side.

Earth rotates approximately upright in its orbit. That is, its axis of rotation is inclined only 23.5° from the perpendicular to its orbit. Uranus rotates on an axis that is inclined 97.9° to the perpendicular to its orbit (Figure 23–3). This peculiar axis of rotation produces strange seasons on the planet and may hint at an important process in the formation of the planets.

Uranus is now positioned in its orbit such that its South Pole faces the inner solar system. Thus, its southern hemisphere is bathed in sunlight, and a creature living on Uranus (an unlikely possibility as we will discover later) would see the sun near the planet's south celestial pole (Figure 23–4a). In 21 earth-years Uranus will have moved one-quarter of the way around its orbit, and the sun will be on its celestial equator (Figure 23–4b). Our Uranian creature would see the sun rise and set with each rotation of the planet. In another 21 earth-years, the sun

DATA FILE 9
Uranus

Average distance from the sun	19.18 AU (28.69×10^8 km)
Eccentricity of orbit	0.0461
Maximum distance from the sun	20.1 AU (30.0×10^8 km)
Minimum distance from the sun	18.3 AU (27.4×10^8 km)
Inclination of orbit to ecliptic	0°46′23″
Average orbital velocity	6.81 km/sec
Orbital period	84.013 y (30,685 days)
Period of rotation	17^h14^m
Inclination of equator to orbit	97°55′
Equatorial diameter	51,118 km (4.01 $D_\oplus$)
Mass	8.69×10^{25} kg (14.54 $M_\oplus$)
Average density	1.19 g/cm^3
Gravity	0.919 earth gravities
Escape velocity	22 km/sec (1.96 $V_\oplus$)
Temperature above cloud tops	−221°C (−366°F)
Albedo	0.35
Oblateness	0.024

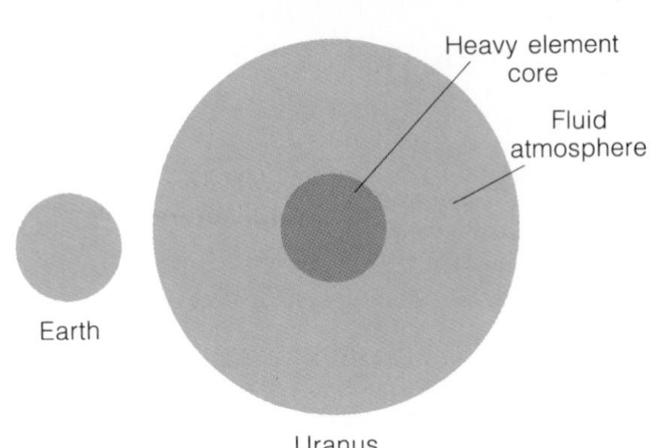

a. A theoretical model of the interior of Uranus.

b. Crescent Uranus seen by Voyager 2 as it entered the planet's shadow. (NASA.)

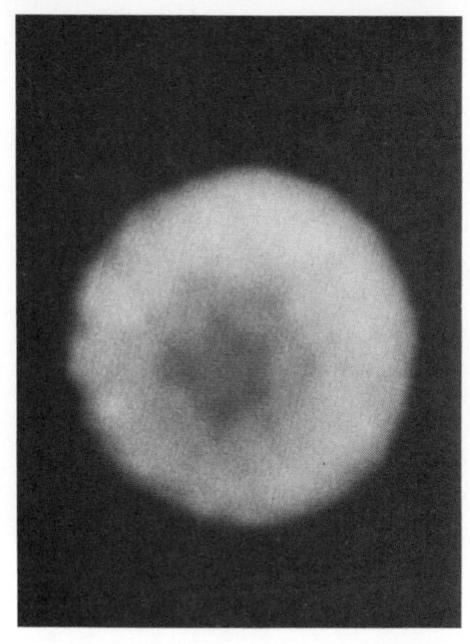

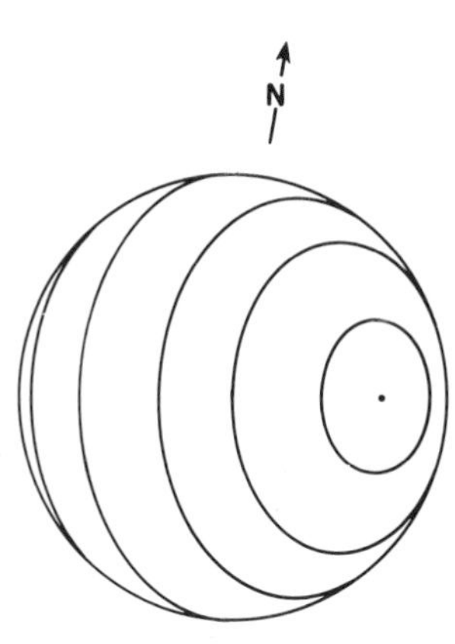

c. Uranus as seen from Earth. The diagram shows location of its rotation axis. (Lunar and Planetary Laboratories.)

d. Voyager 2 at Uranus. Note the cloud belts on the planet in this preencounter artist's rendition. (NASA.)

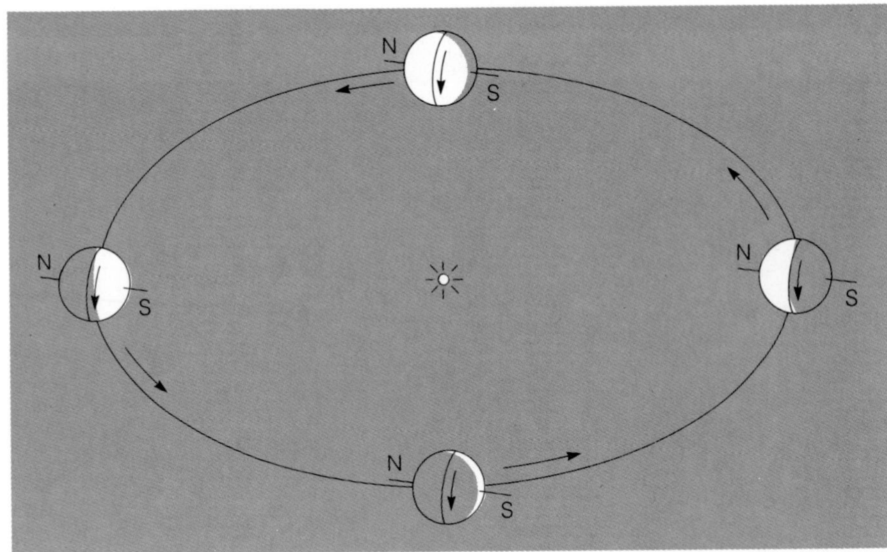

FIGURE 23–3 Uranus rotates on an axis that is tipped 97.9° from the perpendicular to its orbit. It is presently located such that its South Pole faces toward the sun, producing summer at southern latitudes. As it orbits the sun, it maintains the direction of its axis in space thus producing seasons. (Compare with Figure 23–4.)

would have reached a point near the planet's north celestial pole, and the southern hemisphere would be experiencing a lightless winter, which would last 21 earth-years. Thus, the ecliptic as seen from Uranus passes very near the planet's north and south celestial poles.

The high inclination of the axis of rotation has never been fully explained. Planetary astronomers generally propose that as Uranus formed from the icy planetesimals of the outer solar nebula, it had grown to nearly full size when it was struck at a sharp angle by a very large planetesimal, perhaps as large as Earth. That impact could have altered the axis of rotation. Such impacts may have been much more common in the outer solar system where large amounts of ices could condense. We will see more evidence for large-impact events when we consider the satellites of Uranus. But first we must consider the interior and atmosphere of the planet.

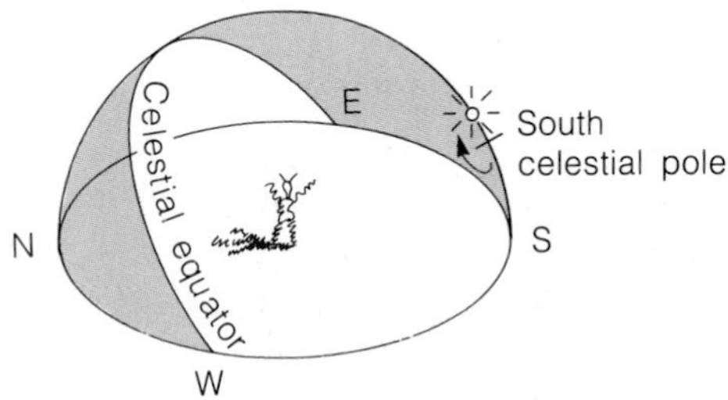

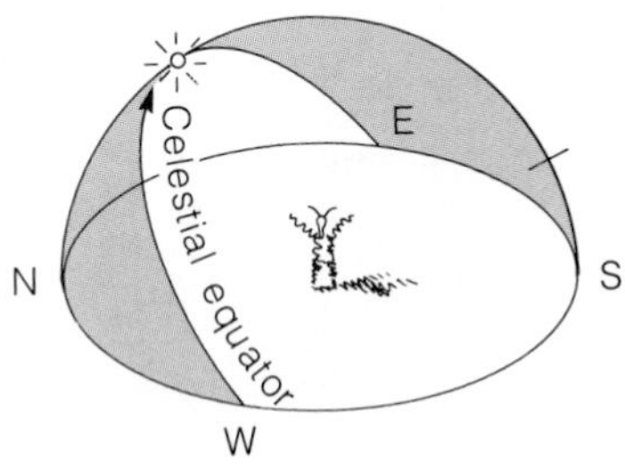

FIGURE 23–4 Because Uranus is inclined farther than 90° from the perpendicular to its orbit, it rotates from east to west. An alien creature living in the atmosphere (there is no surface) of Uranus' southern hemisphere would currently find the sun very near the south celestial pole and thus would have continuous sunlight. But in 21 earth-years the sun will move to the celestial equator of Uranus, and the alien would see the sun rise in the west and set in the east. In another 21 earth-years, the sun will be near the north celestial pole and will not rise as seen from the southern hemisphere. (Compare with Figures 23–3 and 2–15.)

THE ATMOSPHERE OF URANUS

Like Jupiter and Saturn, Uranus has no surface. The gases of its atmosphere—mostly hydrogen, 15 percent helium, and a few percent methane, ammonia, and water vapor—blend gradually with a fluid interior. But unlike Jupiter and Saturn, Uranus has almost no visible cloud features in its atmosphere. Nevertheless, Voyager photographs and measurements suggest that the atmospheric circulation pattern is similar to that found in the two larger Jovian planets.

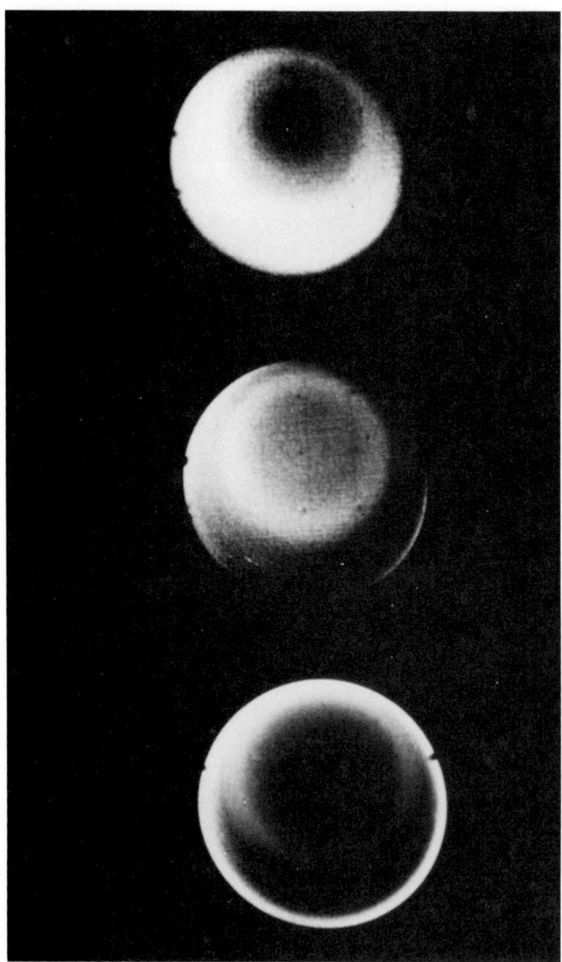

FIGURE 23–5 Although nearly featureless in visible light, the atmosphere of Uranus shows bands in these computer-enhanced images recorded through filters. The top image was made through a violet filter, the middle image through an orange filter, and the bottom image through a filter that admits light near a wavelength of 0.619 μm, a wavelength that is strongly absorbed by gaseous methane. The bright ring circling the south pole (middle and bottom images) is probably produced by rising gas currents that form high clouds. (NASA.)

Seen from Earth, Uranus is a small, featureless disk. As Voyager 2 drew closer to the planet in 1985, astronomers examined the photos to see if Uranus had belt–zone circulation like Jupiter and Saturn. Early photos showed no clouds at all. Uranus is far from the sun, and its atmosphere is so cold that rising currents of gas condense and form clouds at lower levels in its atmosphere. Water clouds form at the lowest layers, with ammonia and then methane clouds at higher and colder levels. But even the highest and coldest methane clouds lie below a deep atmosphere of hydrogen. Small amounts of methane absorb red light, giving the planet its characteristic blue color (Color Plate 48).

Although no cloud belts were obvious, Voyager photographs made through filters can be computer enhanced to reveal evidence of circulation patterns. Figure 23–5 shows three images of Uranus recorded through a violet filter, an orange filter, and a filter that admits light near a wavelength of 0.619 μm, a wavelength that is strongly absorbed by gaseous methane. Regions that appear bright through the methane filter must be cloud layers that lie higher in the atmosphere and are less dimmed by methane absorption. The bright ring circling the South Pole is clearly visible in the orange and methane filters and is probably produced by rising gas currents that form high clouds.

A few cloud features, no more than 5 percent brighter than the disk of the planet, have been detected in highly computer-enhanced images. A bright cloud is visible in the methane-filter image in Figure 23–5, for example, and a series of images reveal that these clouds circulate from west to east around the planet (Figure 23–6). The winds near the equator blow slower than the planet's interior rotates, but at higher latitudes the winds blow faster.

This appears to be a form of belt–zone circulation, which is a bit surprising. Because Uranus is so highly inclined to its orbit, it experiences extreme seasons, and its poles receive, on the average, more solar energy than does the equator. Thus, we might expect its atmospheric circulation to be different from that of Jupiter and Saturn. That its atmospheric circulation is concentric with its axis of rotation suggests that solar energy, although it supplies the energy for the circulation, does not determine its geometry. Rather, the nature of the planet's weather patterns are determined by its rotation.

Apparently, solar heating is not very important to the Uranian seasons. Sunlight is almost 400 times weaker at Uranus than at Earth. Also, atmospheric circulation distributes thermal energy around the planet and prevents seasonal temperature changes of more than a few degrees Kelvin. Infrared observations reveal that the atmosphere above the planet's poles is nearly the same temperature, although the North Pole has been in darkness for some years.

If we could descend into the Uranian atmosphere in a spaceship, we would find the outer layers rich in hydrogen and helium. Just above the highest methane clouds the temperature is −220°C (−364°F), and the pressure is about 0.1 times Earth's atmosphere at sea level. Temperature and pressure increase downward. Below the methane clouds our spaceship would pass through ammonia clouds and deeper still water clouds.

Below the water clouds the temperature passes 370°C (700°F) and the pressure exceeds 217 earth-atmospheres. This is known as the **critical point** for water, and beyond this point it is not possible to distinguish between liquid water and water vapor. Outside the porthole of our spaceship/submarine, we would see a fluid of H_2O with hydrogen, helium, methane, and ammonia mixed in. As we descend farther, the pressure and temperature continue to increase, and we are in the fluid interior of the planet.

THE INTERIOR OF URANUS

Before Voyager 2 flew past Uranus, theorists expected the interior of the planet to consist of a heavy element ("rocky") core surrounded by a mantle of water, ammonia, and methane ices, topped by a deep atmosphere of hydrogen and helium. However, studies of the Uranium magnetic field suggest that the interior is fluid rather than ice.

The average density of Uranus, 1.19 g/cm^3, shows that it is rich in hydrogen and helium. Theoretical models of the interior reveal that the pressure is not high enough to convert hydrogen into liquid metallic hydrogen, the highly conductive fluid present in Jupiter and Saturn. Convection currents in the rotating mass of liquid metallic hydrogen appear to be the source of Jupiter and Saturn's magnetic field. Because Uranus lacks liquid metallic hydrogen, we might expect it to have a weak magnetic field.

As Voyager 2 passed Uranus, it detected a magnetic field roughly as strong as Earth's, but that field was startlingly different from any other magnetic field in our solar system. Our experience has been that planetary magnetic fields, presumably generated by the dynamo effect, are inclined slightly to the axis of rotation of the planet. Earth's is inclined 11.7°. But the Uranian magnetic field is inclined by a whopping 60°. In addition, the field is offset from the center by about 30 percent of a planetary radius. Some planetary astromers have suggested that we are living at a time when Uranus' field is reversing, as Earth's does occasionally. A few theorists suggest that the dynamo effect is operating nearer the surface, which would require a large fluid mantle. A fluid composed of water, ammonia, and methane under high pressure would be a very good conductor of electricity and would be capable of supporting the dynamo effect.

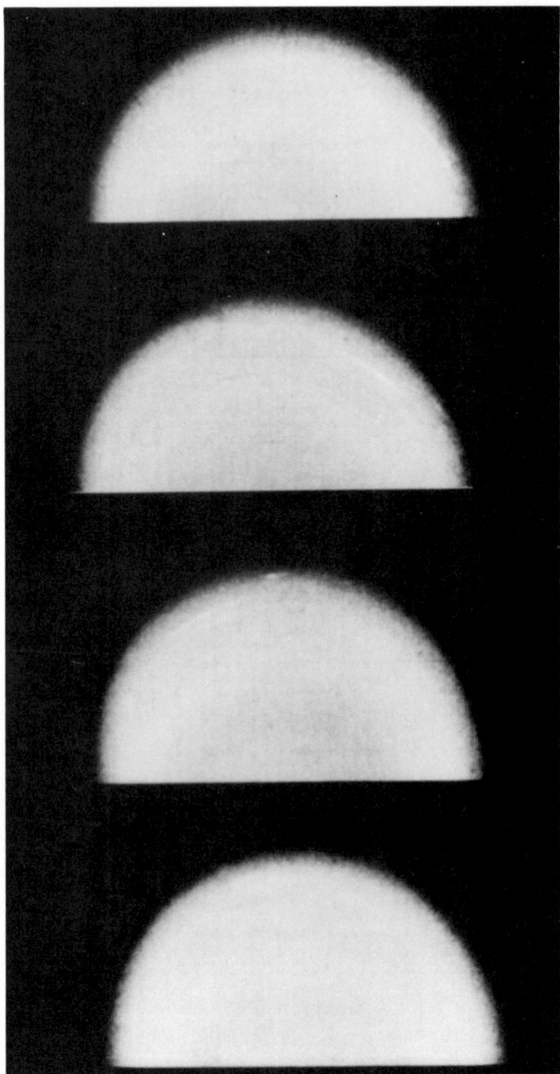

FIGURE 23–6 Extreme computer enhancement of these Voyager 2 images of Uranus reveal the presence of bright cloud features in its atmosphere. This time-lapse series shows the rotation of the planet. (NASA.)

The magnetic field of Uranus creates a magnetosphere as it deflects the solar wind (Figure 23–7), and high-energy particles are trapped to form radiation belts. Energy from the radiation belts is believed to leak into the upper atmosphere where it breaks up hydrogen molecules and produces ultraviolet photons in the process.

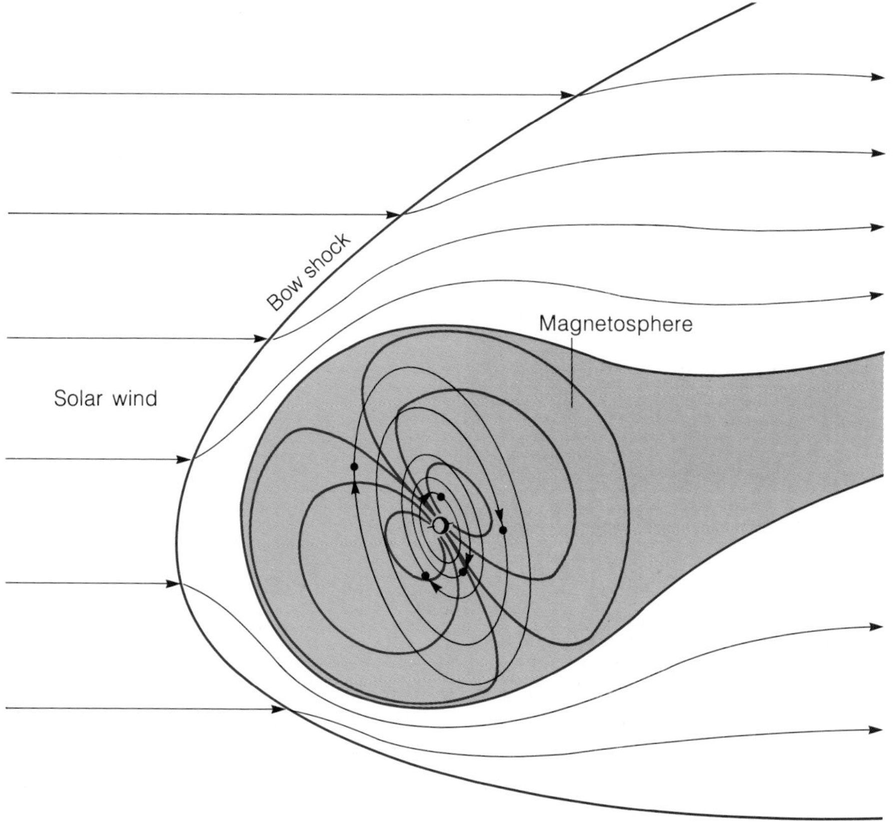

FIGURE 23–7 The magnetosphere of Uranus is produced by the highly inclined magnetic field of the planet. The orbits of the five moons as well as the rings of Uranus (inside the smallest orbit) all lie within the radiation belts trapped in the magnetosphere.

This **electroglow** was seen by the IUE satellite orbiting Earth and has now been detected in the atmosphere of Jupiter and Saturn. We will see later how hydrogen atoms thus released leak out of the upper atmosphere and affect the rings of Uranus.

High-speed electrons spiraling along the magnetic field produce synchrotron radio emission just as on Jupiter. Voyager 2 detected the rotation of this magnetic field with a period of 17.24 hours, thus revealing the rotation period of the planet's interior.

The rotation period combined with the planet's oblateness (0.024) hints that the interior must be fluid. The older model with an icy interior, given a rotation period of 17.24 hours, would not produce the observed oblateness. A less centrally concentrated interior with a fluid mantle is needed.

Unlike the interiors of Jupiter and Saturn, Uranus does not appear to have a great deal of heat stored in its interior. It emits only slightly more energy than it receives from the sun. This suggests that it has lost most of its internal heat. Yet it must have some internal heat to cause convection in the fluid mantle and thus drive the dynamo effect. Radioactive decay in the core is a possibility, although some astronomers have suggested that the slow settling of heavy elements through the fluid mantle could release sufficient energy to heat the interior.

The Uranian magnetic field is both a puzzle and a key. Modern astronomy does not understand how planetary magnetic fields are generated because the region involved is deep within the planet where it is difficult to study. The dynamo theory seems barely adequate to explain the known planetary fields, and some characteristics of the Uranian field are yet to be understood. On the other hand, if we can solve the puzzle of the highly inclined, offset field of Uranus, we may better understand the dynamo effect in general.

Notice from Figure 23–7 that the rings and satellites of Uranus orbit within the magnetosphere. The radiation there is severe enough to influence the characteristics of these smaller bodies.

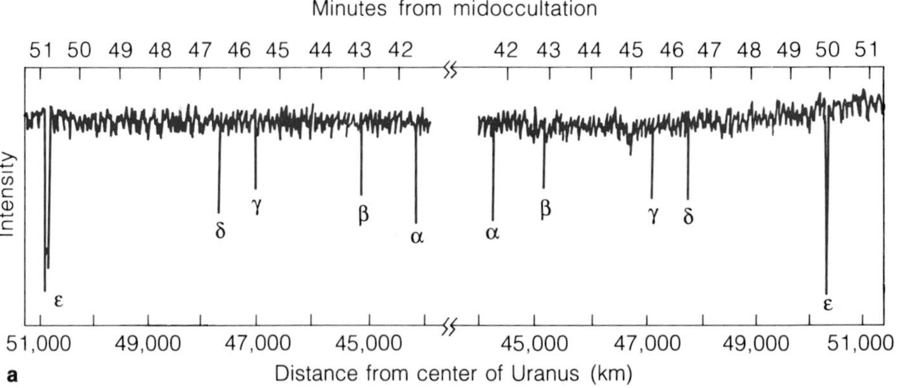

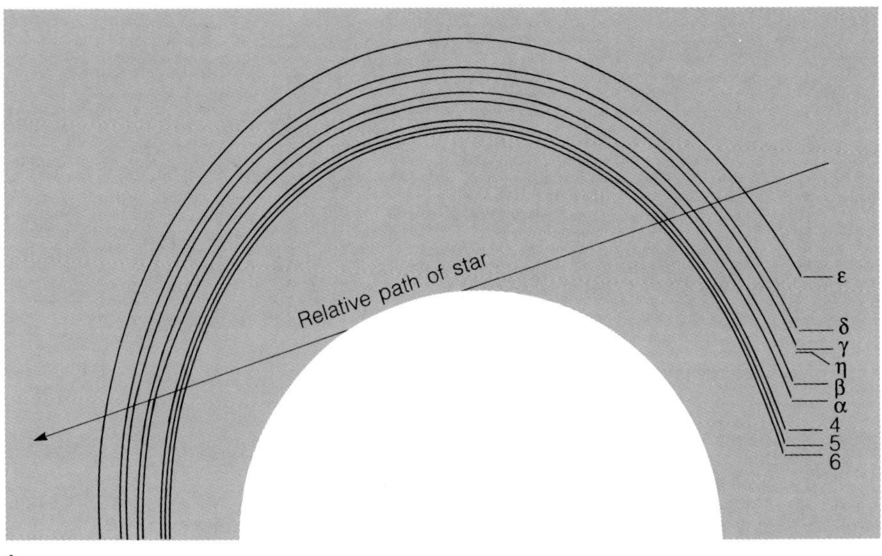

FIGURE 23–8 The rings of Uranus were discovered when Uranus crossed in front of a star, and the dimming of the starlight revealed the rings (a). The original charts show the intensity of the star's light interrupted by the five principal rings as Uranus approached the star (left) and after it had occulted the star (right). The calculated radius of the rings is given at the bottom of the chart. Note that the ε ring is eccentric. (b) As Uranus moved in front of the star, the relative path of the star crossed through the rings. Although only five rings were discovered in the first occultation, a total of nine were detected in subsequent occultations, and a tenth was found by Voyager 2. The η ring and the γ ring are drawn here as a single line. (Adapted from diagrams by James Elliot.)

THE RINGS OF URANUS

Until March 10, 1977, no one knew that Uranus had rings. Late that night James Elliot and his associates began preparations that would lead to the discovery of rings circling Uranus—a discovery that, like Herschel's discovery of the planet, is more than accidental.

Elliot's team intended to study the Uranian atmosphere by observing an **occultation**, the passage of the planet in front of a star. As the edge of the planet crosses in front of the star, various layers in the planet's atmosphere absorb starlight. Therefore, the way in which the starlight fades away can reveal details of the structure of the planet's atmosphere. To reduce the contribution of light from the planet, the team of astronomers chose to observe in the infrared where Uranus emits little radiation. On this late night in March, they were aboard the Kuiper Airborne Observatory (Figure 6–24a) flying at 41,000 feet above the Indian Ocean.

As they calibrated their equipment about 50 minutes before the occultation was to occur, they saw momentary dips in the brightness of the star. At first they suspected a malfunction in the telescope's automatic tracker, but roughly 50 minutes after the occultation they saw the same pattern of dips in reverse order (Figure 23–8). They soon concluded that they had seen the star occulted by a series of narrow rings around Uranus.

Since 1977 the rings have been detected from Earth and have been studied in detail by the Voyager 2 spacecraft (see Figure 23–9). A total of ten narrow rings

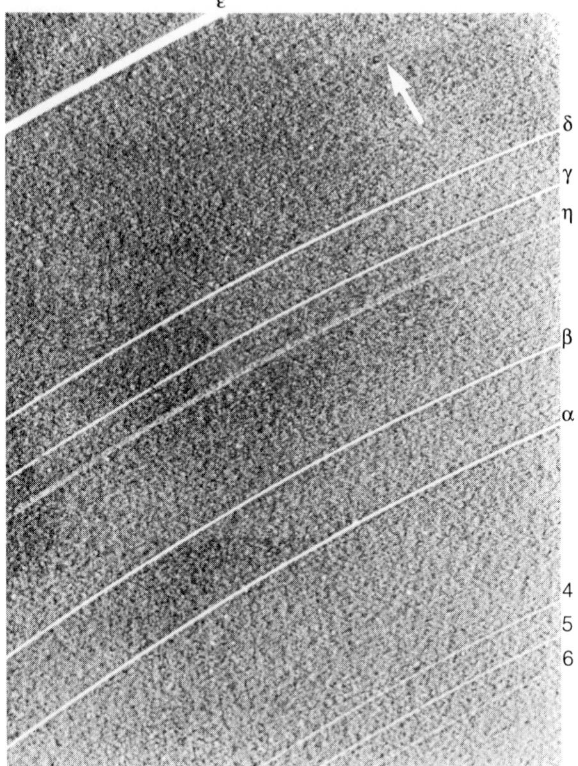

FIGURE 23–9 The nine rings of Uranus detected from Earth are shown in this Voyager 2 image. Voyager found another ring (arrow) halfway from the ϵ ring to the next innermost ring (δ). Tip this page and view the newly discovered ring along its length. Compare this image with Figure 23–8a and try to locate the η ring in the occultation data. (NASA.)

are now known (Date File 9c) orbiting in the plane of the planet's equator. With widths ranging from 10–100 km, most of the rings are round, but the outermost and brightest, the ϵ ring, is elliptical.

Unlike the rings of Saturn, which are reflective and broad, the rings of Uranus are dark and narrow. The particles in the rings have an albedo of about 4 percent, about that of coal. This black color may be caused by the methane-rich composition of the ices at the distance of Uranus. All the rings lie between 1.60–1.95 planetary radii, within the radiation belts in Uranus' magnetosphere. If the ring particles are composed of water and methane ices, the radiation could decompose some of the methane and release carbon to produce a very dark crust. We will see that this may also darken the surfaces of the Uranian moons.

The narrowness of the rings was a puzzle at first. Planetary rings tend to spread because collisions between ring particles, drag from the solar wind and the planet's upper atmosphere, and the pressure of sunlight disturb the orbits of the ring particles. Unrestrained, a planetary ring should gradually spread outward into a uniform, thin disk of particles too tenuous to detect.

The narrow rings of Uranus must be confined, and Voyager 2 detected two shepherd moons (no more than 25 km in diameter)—one inside and one outside the ϵ ring (Figure 23–10a). Particles wandering away from the ring are forced back into the ring by gravitational interaction with the satellites. The eccentricity and variation in width of the ϵ ring is apparently caused by the eccentricity of the shepherd satellite orbits. (Figure 23–10b). Traces of structure in the rings may be density waves generated by the satellites. Presumably, shepherd satellites confine the other rings, but the diameters of these moons must be 10 km or less to escape detection. Note that the rings and shepherd satellites lie inside the Roche limit of the planet, so the satellites must be structurally strong enough to withstand the tidal forces that would otherwise pull them apart.

The rings of Uranus illustrate the importance of satellites in confining planetary rings. The narrow rings are anchored to the small shepherd satellites, and the natural tendency of the rings to spread pushes the satellites outward. Of course, the satellites are much more massive than the tenuous rings and respond slowly. The entire ring system of Uranus, for example, would make an icy satellite only 15 km in radius. Nevertheless, over long periods the small satellites would be pushed outward by the spreading of the rings were it not for the presence of the larger satellites. Many of the small satellites are believed to be locked into resonances with larger satellites, and this anchors the small moons. The spreading pressure of the rings is transferred through resonances to the larger and more massive satellites, and thus the rings can be confined for long periods of time. Without satellites, no planet could keep a ring system for long.

It may seem fortuitous that small moons happen to take up orbits at the edges of the rings, but this puts the cosmic cart before the horse. Ring particles tend to become trapped in stable locations within the system of small satellites. Thus, rings exist between closely orbiting satellites where the orbits of ring particles are most stable.

The origin of the rings is unknown, but it is clear that the rings are eroding. When Voyager 2 looked back

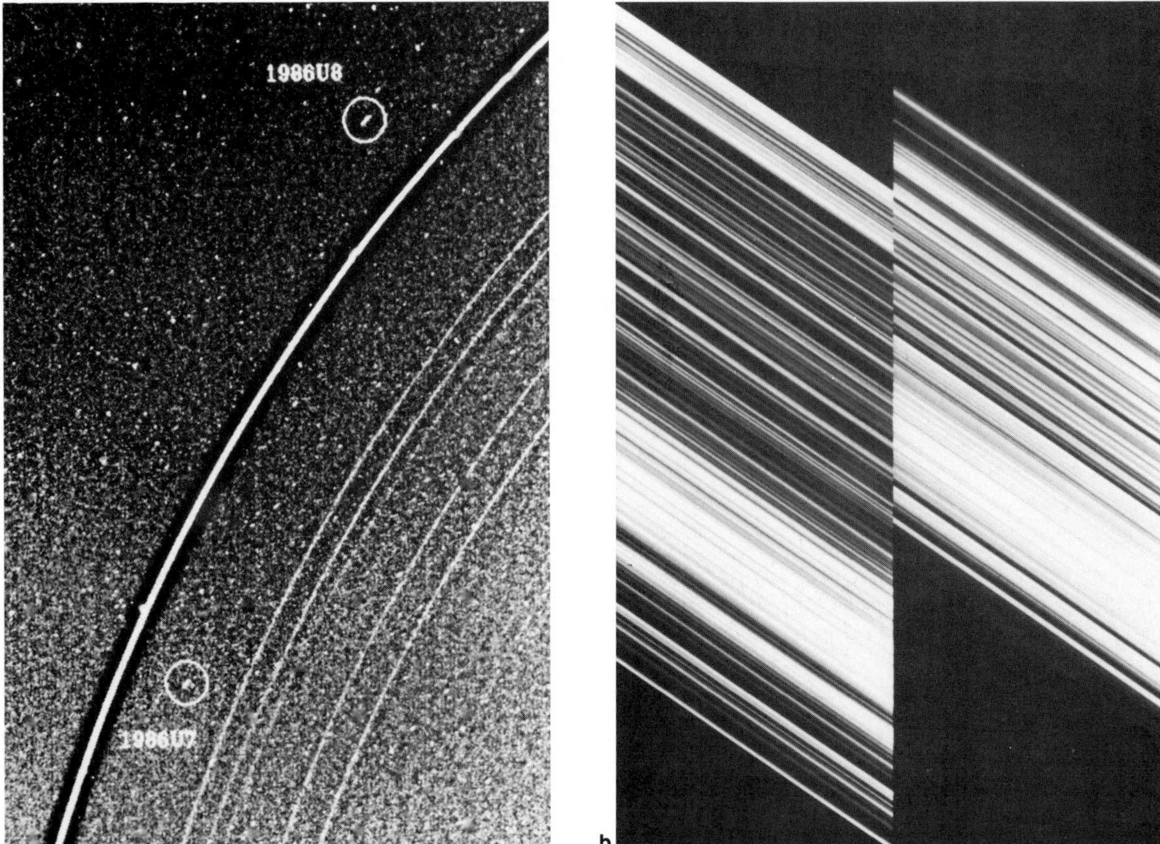

FIGURE 23–10 (a) Voyager 2 discovered two shepherd satellites (circled), now known as Ophelia and Cordelia, that usher the particles in the ϵ ring into a narrow band. Because the satellites have slightly elliptical orbits, the ϵ ring is slightly elliptical and is wider where it is farther from the planet. (b) Two slices through the ϵ ring were made by Voyager 2 by observing the light of a star as it passed behind opposite sides of the ring. The eccentricity, variation in width, and structure within the ring are evident. (NASA.)

at the rings, it found them almost undetectable in forward scattered light (Figure 23–11a). This means the rings are nearly free of small particles. The ϵ ring, for example, contains very few particles smaller than a beach ball, and the entire ring system contains no more than 0.1 percent dust. (Saturn's rings contain several percent dust in the thickest places and Jupiter's ring 50 percent.) Some tenuous dust bands were detected, including one broad band 2500 km wide inside the innermost of the narrow rings Figure 23–11b). Apparently, this dust is meeting resistance from low-density hydrogen gas escaping from the planet's atmosphere, and the dust particles are spiraling into the planet. This process is presumably removing dust from the rings as fast as it can be produced by collisions between ring particles.

It seems possible that the more massive ϵ ring could have survived 4.5 billion years since the formation of the solar system, but the less massive rings must have short lifetimes. It may be that these rings formed more recently by the disruption of a small icy moon. A moon that wandered within the planet's Roche limit could be destroyed by tides, but many astronomers now suspect that collisions between the icy heads of comets and the satellites of the outer planets could be common occurrences. Such an impact any time in the last few billion years could have generated enough icy debris to produce the rings of Uranus.

If such impacts occur, might we see evidence of it among the existing Uranian moons? The answer seems to be a resounding yes.

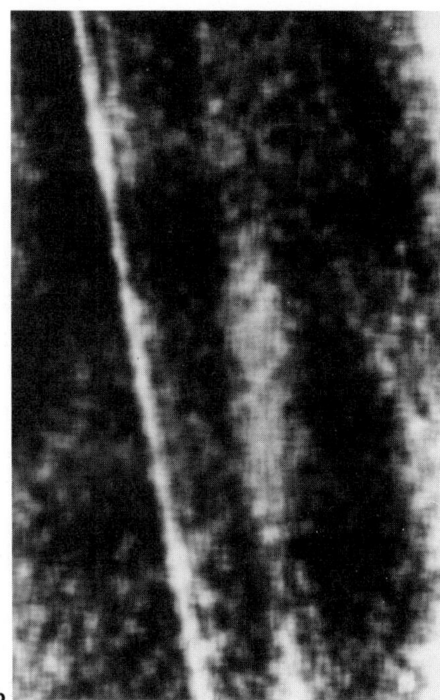

a b

FIGURE 23–11 (a) Viewed from their dark side, the rings of Uranus glow very faintly in forward scattered light. This lack of strong forward scattering shows that the rings contain little dust. The ε ring crosses the top of the frame. Short diagonal lines are trailed images of stars. (b) This highly enhanced image shows the dust band just inside the brighter, previously known rings. (NASA.)

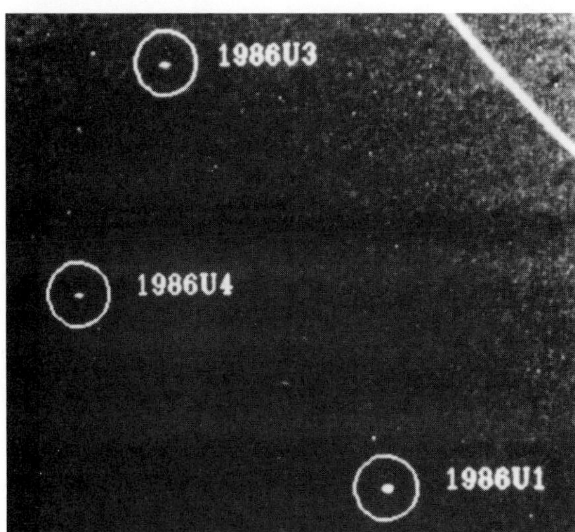

FIGURE 23–12 Three of the ten newly discovered, small moons of Uranus orbit just outside the ε ring. 1986 UI, known officially as Puck, is the largest with a diameter of 170 km and is roughly spherical. The other new satellites are smaller. The images here are elongated slightly, due to motion during the time exposure needed to record the black moons. (NASA.)

23.2 THE MOONS OF URANUS

The five largest moons of Uranus, from outermost inward, are Oberon, Titania, Umbriel, Ariel, and Miranda. Umbriel and Ariel are names from Alexander Pope's *The Rape of the Lock*, and the rest are from Shakespeare's *A Midsummer Night's Dream* and *The Tempest* (see Figure 23–7 to locate the orbits). Spectra show they contain frozen water, and their surfaces are very dark, so astronomers assumed they were made of dirty ices. Little more was known of the moons before Voyager 2 flew through the system.

Before the Voyager 2 cameras could begin photographing the surfaces of the five known moons, Voyager discovered ten additional satellites orbiting in the planet's equatorial plane inside the orbit of Miranda (Figure 23–12). The first, identified as 1986 U1 but now known as Puck (after the character from *A Midsummer Night's Dream*), is about 170 km in diameter and spherical. The remaining nine new satellites, including the ε ring shepherd satellites, are smaller, 40–80 km in diameter. Like Puck, they are as dark as coal. Apparently they, like the ring particles, have been darkened by the radiation in the radiation belts.

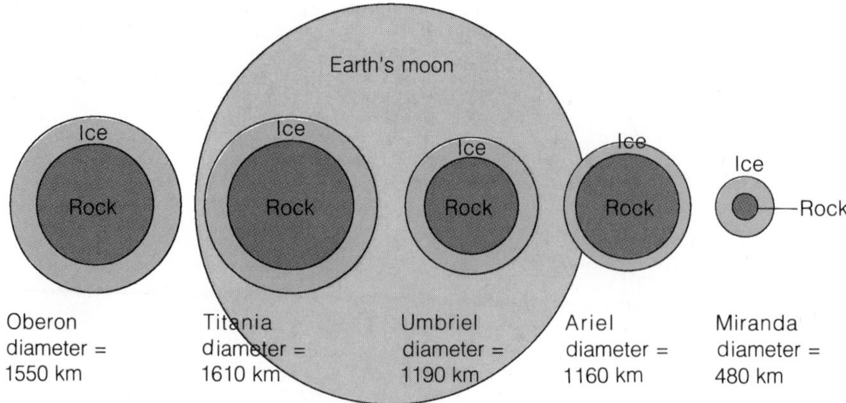

FIGURE 23–13 The density of the five major satellites of Uranus suggest that they contain relatively large rocky cores with mantles of dirty ice. The size of Earth's moon is shown for comparison.

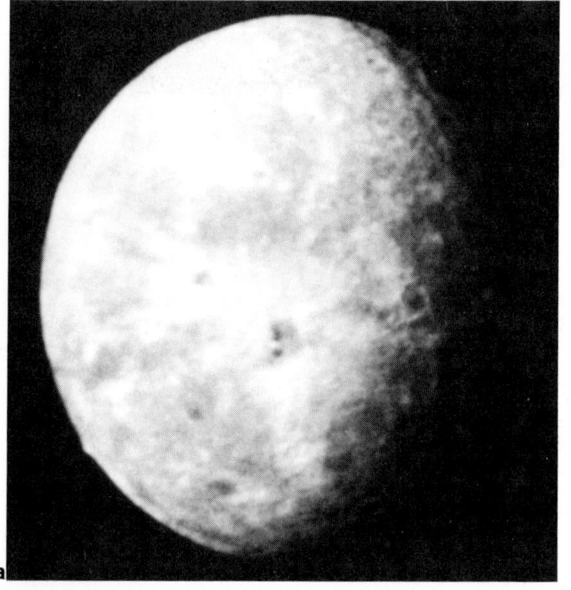

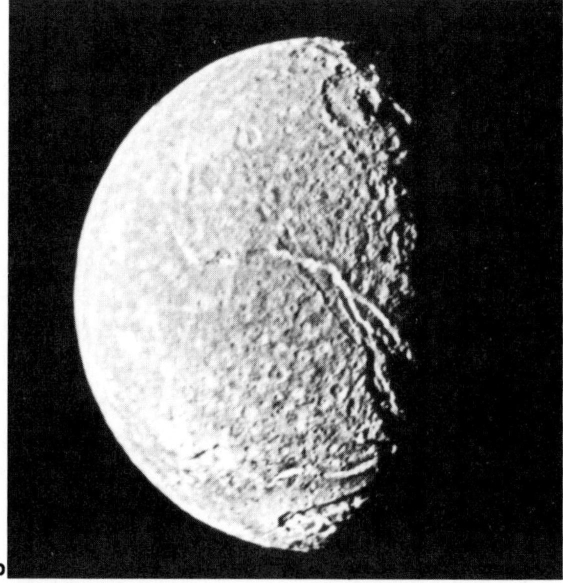

FIGURE 23–14 (a) Oberon imaged by Voyager 2 appears to have an old, inactive, heavily cratered surface. (b) Titania, 1600 km in diameter, has a cratered surface crossed by deep grooves and valleys up to 100 km across—signs of past activity. (NASA.)

The five large satellites are all tidally locked to Uranus, and thus their South Poles are currently pointed toward the sun; it was these hemispheres that were photographed by Voyager 2. Their densities suggest that they contain relatively large rock cores surrounded by icy mantles (Figure 23–13).

Oberon, the outermost of the large moons, has a dark, cratered surface, but it was once an active moon (Figure 23–14a). A large fault crosses the sunlit hemisphere, and dark material, perhaps dirty water "lava," appears to have flooded the floors of some craters.

Titania is the largest of the five moons and has a heavily cratered surface, but it lacks the largest craters (Figure 23–14b). This suggests that soon after it formed, after most of the large debris had been swept up, Titania underwent an active phase in which internal melting flooded the surface with water and erased the early craters. Since then more craters have formed but few as large as those that were erased. Another clue to Titania's past activity is the network of faults, which cross its surface.

Umbriel, third moon inward, is a very dark, cratered world with no sign of faults or surface activity (Figure

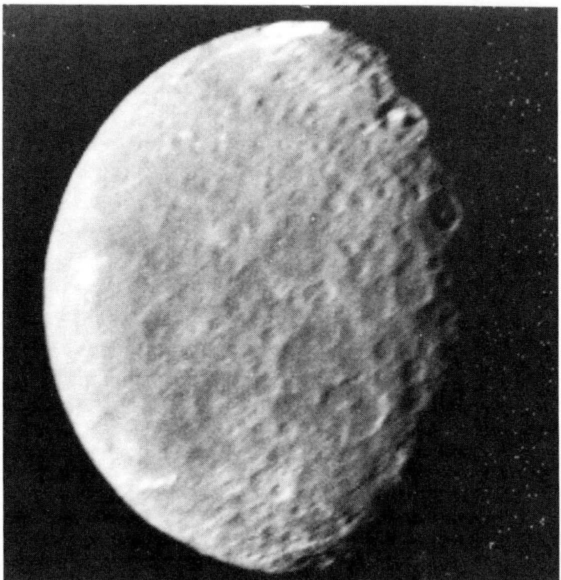

FIGURE 23–15 Umbriel is a dark, cratered moon, which shows no sign of the faults and valleys seen on the other moons. The bright ring at the top of the disk is a large crater with a bright floor. Note the lack of faults and valleys. (NASA.)

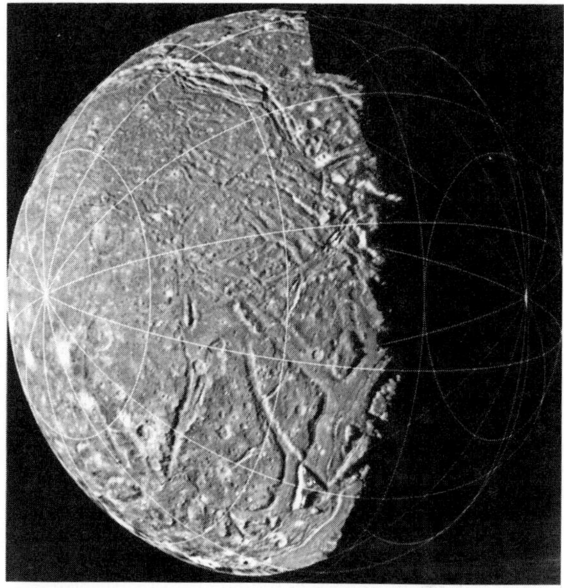

FIGURE 23–16 A number of Voyager images were computer-processed and combined to produce this map of Ariel. Less than half the size of Titania, Ariel has a bright surface crossed by faults over 10 km deep and broad valleys, which appear to have been smoothed of earlier craters. (U.S. Geological Survey, Flagstaff, Arizona.)

23–15). It is the darkest of the moons with an albedo of only 0.16 compared with 0.25–0.45 for the other moons. Planetary astronomers suspect that Umbriel's crust is a dark mixture of rock and ice. In one region a bright crater floor suggests that some clean ice may lie at shallow depths in some regions.

Ariel has the brightest surface of the five major satellites and shows clear signs of geological activity. It is crossed by faults over 10 km deep, and some regions appear to have been smoothed by resurfacing (Figure 23–16). Crater counts show that the smoothed regions are younger than the other regions. Apparently, past melting in the interior has caused geological activity on the moon's surface. Ariel may have been subject to tidal heating caused by an orbital resonance with Miranda and Umbriel.

The smallest of the five satellites, Miranda, discovered from Earth is the most unusual. In fact, its surface is so peculiar one astronomer referred to it as a "moon designed by a committee." Miranda is marked by oval patterns of grooves, which have been dubbed *circi maximi* after the ancient Roman chariot-racing tracks (Figure 23–17). These features also are called ovoids. The origin of the ovoids is unknown, but the most popular theory is that Miranda was struck by a large body, broke into pieces, and then was pulled back together gravitationally. If Miranda had differentiated into a rocky core and icy mantle before the impact, the fragments would have been a mixture, and the reassembled moon might have contained large blocks of rock embedded in its icy crust. As these heavy blocks sank and lighter ice rose, the ovoids could have formed.

Certainly, Miranda has had a violent past. Near the equator a huge cliff rises 20 km. An astronaut who slipped from the top of the cliff would fall for 10 minutes before hitting the surface.

The discovery of the peculiar surface on Miranda has forced planetary astronomers to consider the probability of major impacts destroying the icy moons in the outer solar system. Comets sweep through the satellite systems occasionally, and a moon such as Miranda should have experienced more than one shattering impact since the formation of the solar system. Major impacts that shatter icy moons and allow them to re-accrete may be an impor-

a

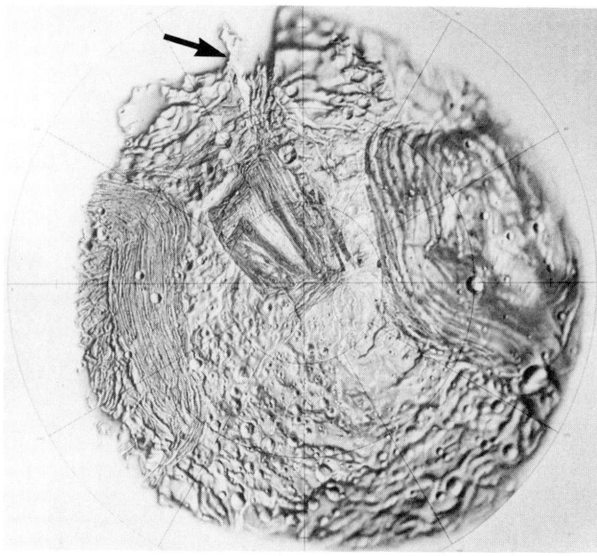

b

FIGURE 23–17 Miranda, the smallest of the five major moons of Uranus, is only 484 km in diameter, but its surface shows signs of activity. This photomosaic of Voyager 2 images reveals that it is marked by great oval systems of grooves. In the map (b), the smallest features detected are about 1.5 km diameter. The high cliff marked by an arrow near the top of the map is 20 km high. (U.S. Geological Survey, Flagstaff, Arizona. Map by P. Bridges.)

tant geological process in the outer solar system. Some planetary astronomers now believe that many of the smaller, icy moons of Saturn and Uranus have had such experiences. We will see more hints of catastrophic collisions when we discuss the moons of Neptune.

23.3 NEPTUNE

Uranus and Neptune are often discussed together. They are about the same size and density and are probably very similar planets in some ways. They do differ, however, in certain respects. Neptune has a peculiar ring and a peculiar satellite system. Even the discovery of Neptune was different than the discovery of Uranus.

THE DISCOVERY OF NEPTUNE

In 1843 the young English astronomer John Couch Adams (1819–1892) completed his degree in astronomy and immediately began the analysis of one of the great problems of nineteenth-century astronomy. William Herschel had discovered Uranus in 1781, but earlier astronomers had seen the planet as early as 1690 and plotted it on charts as a star. When nineteenth-century astronomers tried to combine this data, they found that no planet obeying Newton's laws of motion and gravity could follow such an orbit.

Some astronomers had suggested that the gravitation attraction of an undiscovered planet was causing the discrepancies. Adams began with the variation in the positions of Uranus, never more than 2 minutes of arc, and by October 1845 had calculated the orbit and position of the undiscovered planet. He sent his prediction to the Astronomer Royal, Sir George Airy, who passed it on to an observer. Neither Airy nor his observer took the prediction too seriously, and they were very dilatory in making the search.

Meanwhile the French astronomer Urbain Jean Leverrier (1811–1877) made the same calculations and sent his prediction to Johann Galle at the Berlin Observatory. Galle received Leverrier's prediction on the afternoon of September 23, 1846, and, after searching for 30 minutes that evening, found Neptune. It was only 2° from Adam's predicted position.

When the English announced Adam's work, the French suspected that he had plagiarized the calculations, and the controversy was bitter. In fact, both men had performed the calculations correctly. But they could have been beaten to the discovery had only Galileo paid less attention to Jupiter and more attention to what he saw in the background.

Recent studies of Galileo's notebooks show that he saw Neptune on December 27, 1612, and again on January 28, 1613, but he plotted it as a star in the background of drawings of Jupiter. It is interesting to speculate about the response of the Inquisition had Galileo proposed that a planet existed beyond Saturn. Unfortunately for history (perhaps fortunately for Galileo), he did not recognize Neptune as a planet, and its discovery had to wait 234 years.

The discovery of Neptune was a triumph for Newtonian physics: The three laws of motion and the law of gravity had proven sufficient to predict the position and orbit of an unseen planet. Thus, the discovery of Neptune was fundamentally different than the discovery of Uranus—the existence of Neptune was predicted from basic laws. Discovery by prediction is important in science because confirmation of the prediction serves as confirmation of the basic laws.

This story of planetary discovery by accident and by prediction can take strange twists. Both Adams and Leverrier assumed that the new planet obeyed the Titius–Bode rule (Box 19–1), which it does not. But by good luck, the two astronomers happened to live at a time when that assumption did not lead to serious errors in their calculations. We will meet yet another twist in the story when we discuss the discovery of Pluto.

THE ATMOSPHERE AND INTERIOR OF NEPTUNE

Little was known about Neptune before the Voyager 2 spacecraft swept past it in August 1989. Seen from earth, Neptune is a tiny blue-green dot never more than 2.3 seconds of arc in diameter. Astronomers knew it was almost four times the diameter of the earth and about 4 percent smaller in diameter than Uranus. Spectra revealed that its blue-green color was caused by methane in its hydrogen-rich atmosphere. Methane absorbs red light making Neptune look blue-green. Neptune's density showed that it was a Jovian planet rich in hydrogen, but almost no detail is visible from earth, so even its period of rotation was poorly known.

Voyager 2 passed only 4905 km above Neptune's cloud tops, closer than any spacecraft had ever come to one of the Jovian planets. The Voyager photos revealed that Neptune is marked by dramatic belt-zone circulation parallel to the planet's equator (Data File 10b). At least four cyclonic disturbances mark the belts, and the largest, dubbed the Great Dark Spot, looks similar to the Great Red Spot on Jupiter (Color Plate 50). As on the other Jovian planets, high winds circle the planet parallel to its equator (Color Plate 49). Now that we have seen belt-zone circulation on all four Jovian planets, it seems clear that the atmospheric circulation is dominated by the rotation of the planet and not by the inclination of the planet in its orbit. Even Uranus, inclined over 90°, has faintly detectable belt-zone circulation parallel to its equator.

The interior of Neptune is presumed to contain a rocky core with an icy mantle topped with a deep layer of liquid hydrogen (Data File 10a). Though Voyager 2's cameras could not probe below the clouds, other instruments detected a complex magnetic field weaker than that of Uranus. Neptune's magnetic field is tipped 50° from the axis of rotation and offset from the center of the planet much like the field of Uranus. This suggests that the high inclination of Uranus is not responsible for its peculiar field. Astronomers assume that planetary magnetic fields are generated by the dynamo effect in the conducting cores of planets. However, the highly inclined, offset fields of Uranus and Neptune may originate in a dynamo effect in a conducting shell not far below the cloud belts.

Periodic changes in the magnetic field show that Neptune's interior rotates with a period of 16 hours 3 minutes. Cloud patterns rotate with slightly different periods depending on atmospheric winds.

NEPTUNE'S RINGS

No rings orbiting Neptune are directly visible from earth, but on a number of occasions in the last decade, Neptune crossed in front of (occulted) distant stars. Sometimes astronomers noticed the light of the star dim for a few seconds before or after the main occultation. This suggested that the planet had rings, but the stars did not always dim as they should. Astronomers proposed that Neptune had segments of rings rather than complete rings.

Voyager 2's photographs revealed that the ring segments were slightly denser arcs in rings that completely circled the planet (Figure 23–18). Two narrow rings have radii of 63,000 km and 53,000 km with a fainter, broader, ring halfway between. The most detailed photos revealed two more broad, faint bands. The narrow rings are brightest when illuminated from behind, so we can conclude they contain large amounts of dust but the broader rings seem to contain less dust. Because the rings are not very reflective, we can guess they are rocky rather than icy.

DATA FILE 10
Neptune

Average distance from the sun	30.0611 AU (44.971 × 10^8 km)
Eccentricity of orbit	0.0100
Maximum distance from the sun	30.4 AU (45.4 × 10^8 km)
Minimum distance from the sun	29.8 AU (44.52 × 10^8 km)
Inclination of orbit to ecliptic	1°46′27″
Average orbital velocity	5.43 km/sec
Orbital period	164.793 y (60,189 days)
Period of rotation	16^h3^m
Inclination of equator to orbit	28°48′
Equatorial diameter	49,500 km (3.93 $D_⊕$)
Mass	1.030 × 10^{26} kg (17.23 $M_⊕$)
Average density	1.66 g/cm^3
Gravity	1.19 earth gravities
Escape velocity	25 km/sec (2.2 $V_⊕$)
Temperature at cloud tops	−216°C (−357°F)
Albedo	0.35
Oblateness	0.027

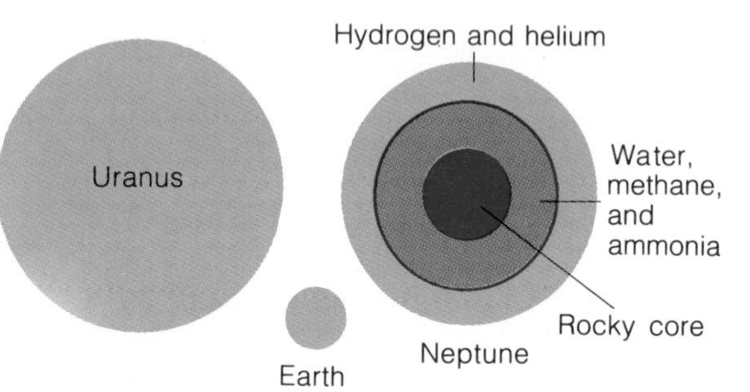

a. A model of the interior of Neptune, containing water and ammonia ices.

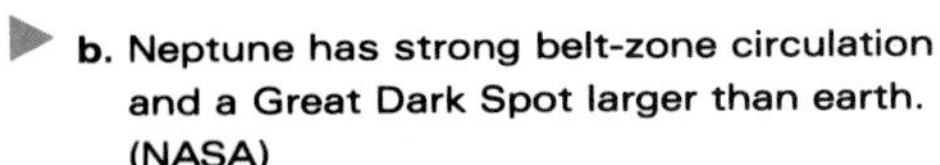

b. Neptune has strong belt-zone circulation and a Great Dark Spot larger than earth. (NASA)

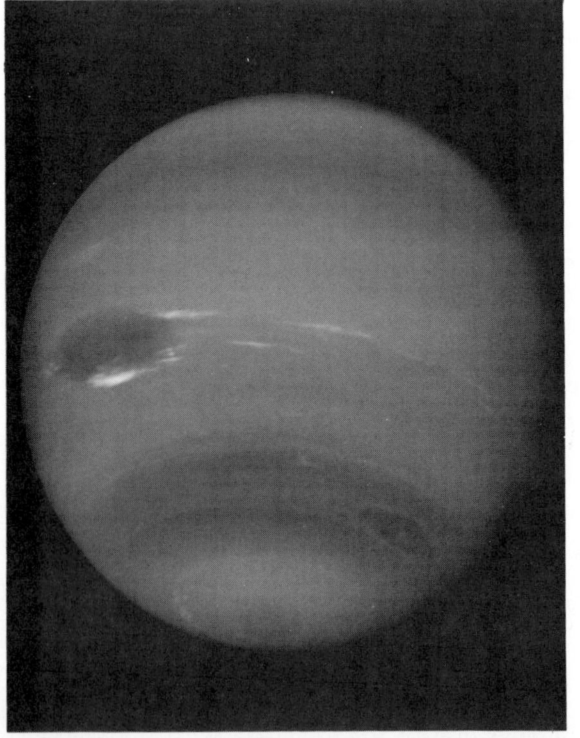

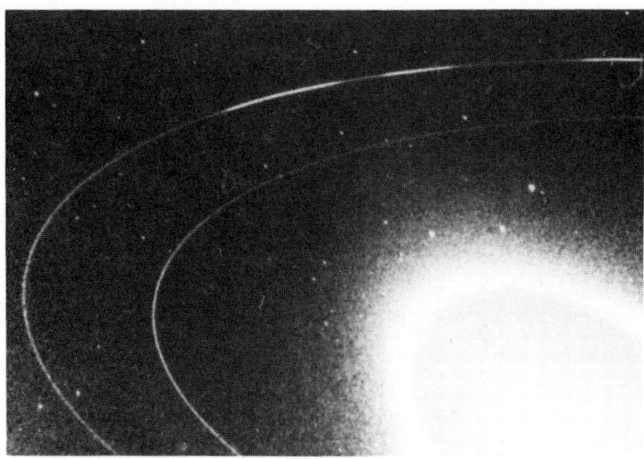

FIGURE 23–18 The two brightest and narrowest of Neptune's rings are bright in this backlit photograph. Note the denser arcs in the outer ring. These are believed to be caused by the gravitational influence of satellites, but the mechanism is not well understood. The overexposed image of the planet is at lower right. (NASA/JPL.)

The odd structure of the rings with the denser ring arcs and narrow rings is partially explained by small moons imbedded among the rings. Some of these moons appear to shepherd the rings and keep them narrow. Some models suggest a moon about 200 km in diameter in an inclined orbit could produce the ring arcs, but such a satellite was not found. Thus the cause of the ring arcs is still being debated.

23.4 THE MOONS OF NEPTUNE

Neptune has only two moons visible from earth, but Voyager 2 discovered six more and gave us a close look at the largest moon Triton.

THE ORBITS OF THE MOONS

Even before Voyager 2 visited Neptune, earth's astronomers knew there was something odd about Neptune's moons. Triton has a nearly circular orbit but it travels backward—clockwise as seen from the north. This makes Triton the only large satellite in the solar system with a retrograde (backward) orbit. The other satellite visible from earth, Nereid, goes the right direction, but its orbit is highly elliptical and very large. Nereid takes 359.4 days to orbit Neptune once. Many astronomers speculated that

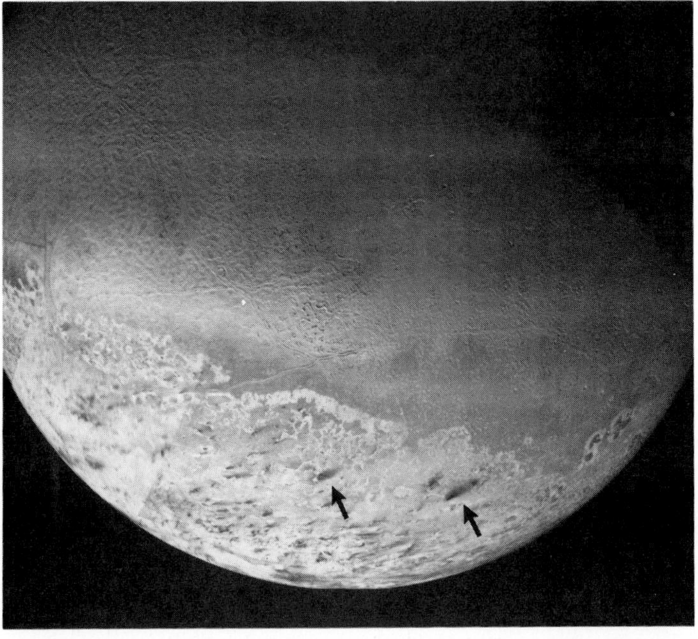

FIGURE 23–19 Triton's south pole (bottom) has been exposed to sunlight for the last 30 years, and deposits of frozen nitrogen in the polar cap appear to be vaporizing, perhaps to refreeze at the north pole. Note the linear faults and lack of craters. The dark smudges (see arrows) are produced when liquid nitrogen below the crust vaporizes and drives nitrogen volcanism. (NASA/JPL.)

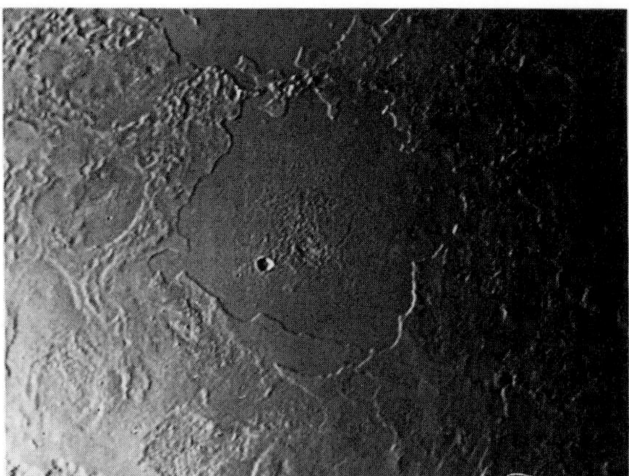

FIGURE 23–20 These two roughly round basins on Triton may be old impact basins each about 190 km (120 miles) in diameter. They have been flooded repeatedly by a liquid, probably a slurry of water ice, methane, and ammonia. Such flooding demonstrates the importance of surface activity in erasing craters. Note the small number of impact craters on the surface. (NASA/JPL.)

the orbital motion of Triton and Nereid were disturbed long ago by an encounter with a massive planetesimal.

Voyager 2 found six more moons orbiting Neptune, but they all lay close to the planet among the rings. No new moons were found beyond the orbit of Triton. Some astronomers suggest that Triton in its retrograde orbit would have consumed any moons near it.

TRITON

The Voyager 2 photographs revealed that Triton is a highly complex moon (Color Plate 51). Although it is only 1360 km in radius (78% the radius of earth's moon) it is so cold (37 K or −393°F) it can hold an atmosphere of nitrogen with some methane. Its air is 10^5 times less dense than earth's atmosphere. Although a few wisps of haze were visible, the atmosphere is transparent, and the surface was easily visible.

The surface is evidently composed of ices. Frozen water is a major constituent, but methane and ammonia are also present. Some regions look dark and may be older terrain that has grown dark as sunlight has converted methane into dark organic compounds. The south pole of Triton (Figure 23–19) has been turned toward the sun for the last 30 years, and deposits of nitrogen frost appear to be vaporizing there and refreezing in the darkness of the north pole.

Many features on Triton suggest it has had an active past. Few craters mark the surface, suggesting that whatever craters formed in the early history of the moon have been erased by later activity. Also, long, linear features appear to be fractures in the icy crust where ices have welled up and frozen. Some roughly round basins appear to have been flooded time after time by liquids from the interior (Figure 23–20). Most exciting of all are the dark smudges visible in the bright nitrogen ices near the south pole. The smudges are caused by deposits of liquid nitrogen below the surface. When warmed by the sun, the nitrogen vaporizes and breaks through the surface to form nitrogen volcanos. Two such nitrogen plumes were seen extending nearly 8 km above the surface. Sunlight could convert some of the methane carried along with the nitrogen into dark organic materials that would account for the smudges.

If Triton's orbit was disturbed early in its history, it may have been thrown into an elliptical orbit at first and tidal heating could have made it so active it erased any early craters. Tidal forces would have made the orbit grow more circular until today when the orbit is too circular to generate tidal heat. The main form of activity now may be the cycle of vaporizing and condensing nitrogen in its crust.

23.5 PLUTO

Pluto, the farthest outpost of our solar system, is a small, icy world. Discovered through a combination of ingenuity, perseverance, and luck, it is now known to have a satellite.

THE DISCOVERY OF PLUTO

Percival Lowell was fascinated with the idea that an intelligent race built the canals he thought he could see on Mars (Chapter 21). Lowell founded Lowell Observatory primarily for the study of Mars. Later, to improve the reputation of his observatory, he began to search for a planet beyond Neptune.

Lowell used the same method that Adams and Leverrier had used to predict the position of Neptune. Working from the observed irregularities in the motion of Neptune, Lowell predicted the location of an undiscovered planet of about 7 earth masses, which, he concluded, would look like a 13th magnitude stellar image in eastern Taurus. Lowell searched for the planet photographically until his death in 1916.

In the late 1920s, 22-year-old amateur astronomer Clyde Tombaugh began using a homemade 9-inch telescope to sketch Jupiter and Mars from his family's wheat farm in western Kansas. He sent his drawings to Lowell Observatory, and the observatory director hired him (without an interview) to resume Lowell's search for the undiscovered planet.

Tombaugh began photographing the sky along the ecliptic, obtaining two 14 × 17-inch glass plates exposed 2–3 days apart. To compare the plates, he mounted them in a blink comparator, which allowed him to examine star images, first on one plate and then on the other while looking through a microscope eyepiece. Star images do not move, but a planet would reveal itself by its motion.

For almost a year Tombaugh photographed the sky and blinked the plates, which contained as many as 400,000 stars per plate. Then on February 18, 1930, a quarter of the way through a pair of plates, he found a 15th magnitude image that moved. He later remembered, "Oh, I thought I had better look at my watch, this could be an historic moment. It was within about two minutes of four PM [MST]." The discovery was announced on March 13, the 149th anniversary of the discovery of Uranus and the 75th anniversary of the birth of Percival Lowell. The planet was named Pluto after the god of the underworld and, in a way, after Lowell, because the first two letters in Pluto are the initials of Percival Lowell.

The discovery of Pluto seemed a triumph of discovery by prediction, but Tombaugh sensed something was wrong from the first moment he saw the image. It was moving in the right direction by the right amount, but it was 2.5 magnitudes too faint. Clearly, Pluto was not the 7 earth-mass planet that Lowell had predicted. The faint image implied that Pluto was a small planet with a mass too low to seriously alter the motion of Neptune.

Later analysis has shown that the variations in the motion of Neptune, which Lowell used to predict the location of Pluto, were random errors of observation and could not have led to a trustworthy prediction. The discovery of the new planet only 6° from Lowell's predicted position was apparently an accident, which proves that if you search long enough, you are likely to find something.

PLUTO AND CHARON

Most of the planetary orbits in our solar system are nearly circular, but Pluto's is quite elliptical (Date File 11b). Pluto's average distance from the sun is 39.44 AU, but it can come as close as 29.64 AU. In fact, from January 21, 1979, to March 14, 1999, Pluto will be closer to the sun than Neptune. The planets will never collide, however, because Pluto's orbit is inclined 17° to the plane of the solar system. Most of the planetary orbits are inclined by only a few degrees, so the high inclination of Pluto's orbit, combined with its eccentricity, seems peculiar.

Because Pluto is so distant, we know very little about it. Seen from Earth, Pluto's angular diameter is slightly larger than 0.1 second of arc, and even the best photographs do not show more than a tiny dot whose diameter is not apparent. Not until its moon was discovered did we begin to understand what Pluto is really like.

In 1978 James W. Christy of the United States Naval Observatory, while examining a photographic plate, discovered a faint image beside Pluto's (Date File 11c). The image turned out to be a moon orbiting the planet once every 6.387 days at an average distance of about 19,700 km (12,000 miles). The new moon was named Charon after the mythological ferryman who transports souls across the river Styx into the underworld.

The discovery of a new object in the solar system is always exciting, but the discovery of Charon was especially important because it made it possible to determine Pluto's mass. The same analysis that reveals the total mass of binary star systems reveals that the total mass of Pluto and its moon is about 7×10^{-9} solar masses or about 0.002 earth mass.

Also, beginning in 1985 astronomers began seeing eclipses and occultations as Pluto and its moon crossed in front of each other. These events occur for only a few years out of every 124 as Earth passes through the plane of Charon's orbit.

The eclipses that have been detected to date allow for the calculation of the diameters and densities of Pluto and Charon. Pluto has a radius of 1145 km, and Charon has a radius of 642 km. Thus, Charon is half as large in diameter as its planet. Combining the radii of the two bodies with the mass of the system gives a density of

DATA FILE 11
Pluto

Average distance from the sun	39.44 AU (59.00 × 10^8 km)
Eccentricity of orbit	0.2484
Maximum distance from the sun	49.24 AU (73.66 × 10^8 km)
Minimum distance from the sun	29.64 AU (44.34 × 10^8 km)
Inclination of orbit to ecliptic	17°9′3″
Average orbital velocity	4.73 km/sec
Orbital period	247.7 y (90,465 days)
Period of rotation	$6^d9^h21^m$
Inclination of equator to orbit	118° (estimated)
Equatorial diameter	2294 km (0.18 $D_\oplus$)
Mass	1.2 × 10^{22} kg (0.002 $M_\oplus$)
Average density	1.84 g/cm^3
Surface gravity	0.06 earth gravities
Escape velocity	1.2 km/sec (0.11 $V_\oplus$)
Surface temperature	−230°C (−382°F)
Albedo	0.4

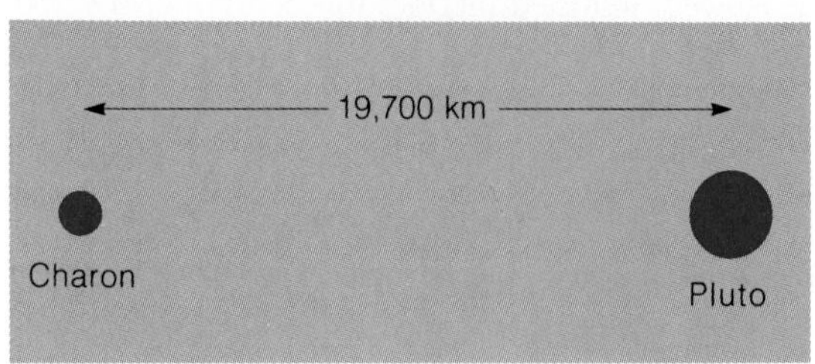

a. Pluto and its moon Charon.

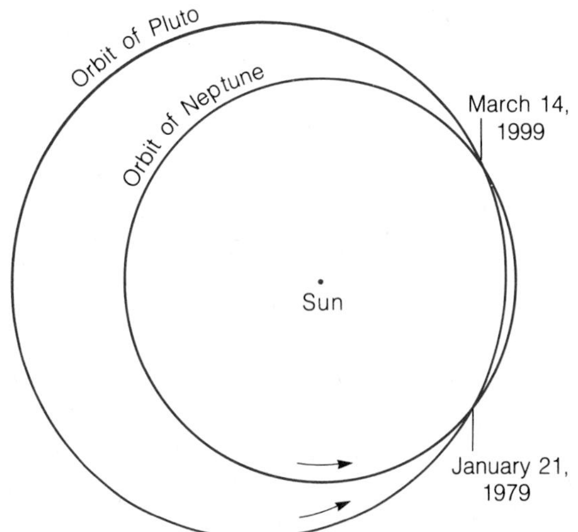

b. Currently, Pluto is closer to the sun than Neptune. This will not occur again until September 2226.

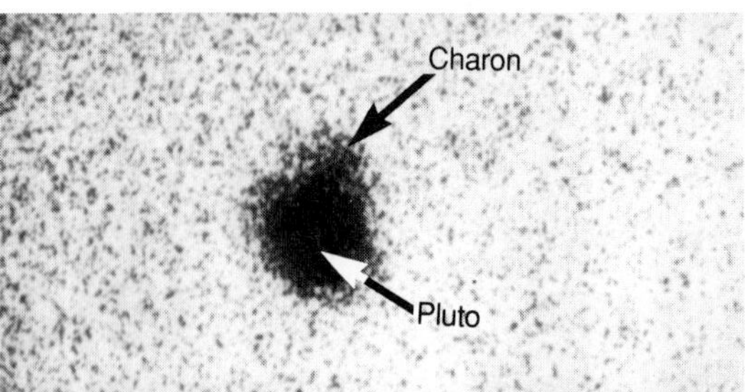

c. Photographic plate (negative print) on which Charon was discovered. (U.S. Naval Observatory photograph.)

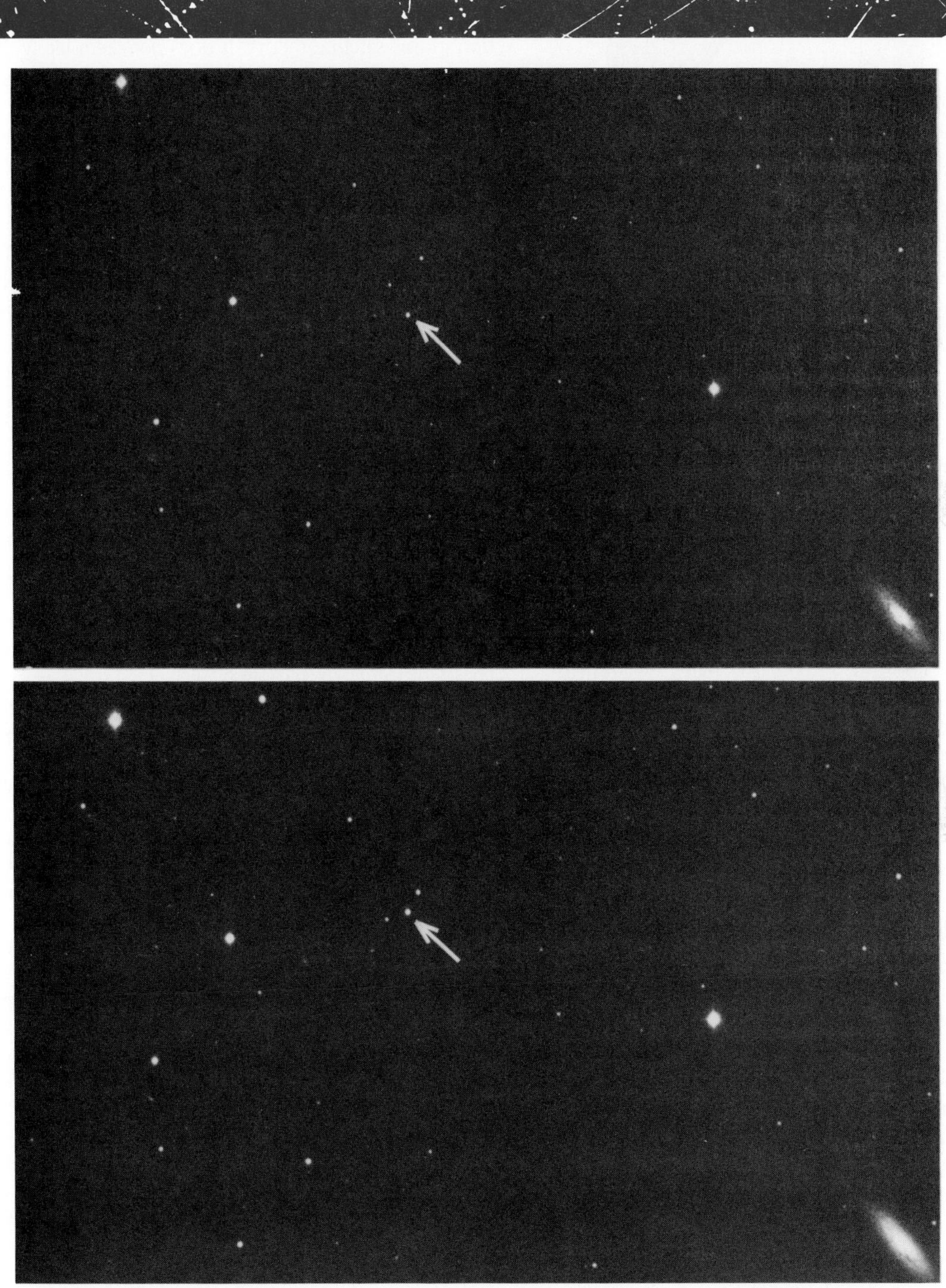

d. The motion of Pluto is apparent on photos taken 24 hours apart. (Lick Observatory photographs.)

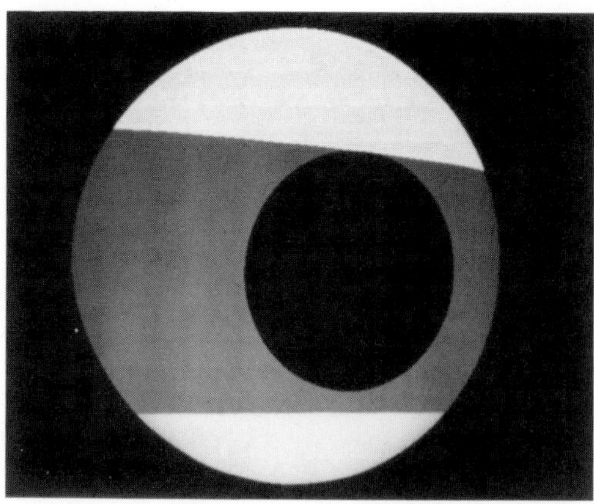

FIGURE 23–21 A map of features on Pluto produced by the analysis of the brightness changes that occur as Charon crosses in front of the planet. (Courtesy Marc W. Buie, Institute for Astronomy, University of Hawaii.)

about 1.84 g/cm^3. Such a density suggests that Pluto is more than half rock with a mixture of ices.

Charon circles Pluto every 6.39 days, and the two bodies are locked together, each rotating to keep the same side facing the other body. When Charon crosses in front of Pluto, it "scans" across the disk, blocking our view of a small region of the planet. Although we cannot see features on the planet, the variation in brightness as Charon crosses Pluto's disk contains information about the brightness of different regions of the disk. Using a supercomputer and the eclipse events observed so far, astronomers have been able to construct a map of Pluto (Figure 23–21), which shows bright polar caps and a darker equatorial band. A large dark spot lies near the equator. Such features are presumably related to surface features such as crater basins.

Spectroscopic observations of Pluto show traces of frozen methane, and laboratory studies indicate that most gases at Pluto's low-surface temperature of −230°C (−382°F) would freeze. The planet probably has an atmosphere of low-density methane. Some changes in brightness in recent years suggest that seasons do occur on Pluto and may be related to the growth of methane-frost polar caps.

Pluto's eccentric, highly inclined orbit and its small size have led some astronomers to suggest that Pluto is not really a planet, but an escaped satellite of Neptune. Indeed, studies of the satellites of the Jovian planets reveal that some of them are low-density, icy bodies like Pluto. Yet the presence of Charon orbiting Pluto raises an objection to this satellite theory. It does not seem likely that Pluto could have held its moon while it was itself a satellite of Neptune; so we must suppose that it captured its moon after it escaped from Neptune or that Charon is a fragment ripped from Pluto by some cosmic collision that knocked Pluto from its orbit around Neptune.

Clyde Tombaugh searched for more planets beyond Pluto and found none although he should have been able to detect any planet as large as Neptune out to a distance of 270 AU. There may be no planets beyond Pluto, but our exploration of the solar system is not yet over. The asteroids, meteorites, and comets discussed in the next chapter may hold the most valuable clues to the origin of the solar system.

SUMMARY

Although Uranus was discovered unexpectedly in 1781, in 1846 the position of Neptune was predicted from the orbital motion of Uranus, and Pluto was discovered in 1930 based on predictions derived from the motions of Neptune.

Uranus and Neptune are commonly classified as Jovian planets. Their low densities imply that they are rich in hydrogen and helium with earth-sized cores of heavy elements.

The axis of rotation of Uranus is tipped 97.9° from the perpendicular to its orbit, so its South Pole now faces the sun. When Voyager 2 flew past Uranus, it discovered that methane-cloud belts lie deep in the hazy atmosphere. Thus, the rotation of the planet, rather than solar energy, dominates the atmospheric circulation.

The magnetic field of Uranus is about as strong as Earth's, but it is inclined by 60° to the axis of rotation. This and the planet's oblateness suggest that the interior of the planet is made up of a rocky core surrounded by a mantle of liquid water and ammonia. The dynamo effect operating in this fluid mantle could generate the magnetic field. Another possibility is that the field is in the process of reversing.

The rings of Uranus are very narrow and dark. Two shepherd satellites were discovered orbiting on either side of the ϵ ring, and other shepherd satellites are believed to bound the other rings. These satellites and the larger satellites appear to anchor the rings and prevent them from spreading and dissipating.

The dark color appears to come from the methane-rich ices of the belt particles darkened by the radiation belts around the planet. The surfaces of the moons are also very dark.

Seen in forward scattering, the rings of Uranus are very dark and thus must be very poor in small particles. Some dust was detected, but it seems to be falling into the planet due to the resistance of low-density hydrogen gas leaking out of the upper atmosphere.

Voyager 2 discovered ten new satellites orbiting Uranus, all of them small and dark. The five largest moons show signs of impact cratering and some surface activity in the past. The innermost moon, Miranda, is marked by large ovals called *circi maximi,* or ovoids. These may be scars left when the little moon was shattered by a major impact and re-accreted. As heavy blocks of rock sank, rising ice might have formed the ovoids. A number of the icy moons in the outer solar system may have been shattered and re-formed in the past.

Little is known about Neptune. It is about the size and mass of Uranus, so its atmosphere and interior may be similar. Occultation observations suggest that Neptune may have a ring made up of short arcs. Neptune's two moons follow peculiar orbits. Triton, large enough to have an atmosphere and seasons, orbits in the retrograde direction; Nereid follows a highly elliptical orbit. We will learn more when Voyager 2 flys by Neptune in August 1989.

Thanks to its elliptical orbit, Pluto is now closer to the sun than Neptune. Pluto is a small world about the size of Earth's moon, and it has a satellite of its own called Charon. As that satellite crosses in front of the planet, the total light of the system varies, thus allowing us to find the diameter of the planet and its moon and to map the surface of Pluto with very low resolution. The density of Pluto suggests that it is made up of more than 50 percent rock mixed with ices.

NEW TERMS

critical point

electroglow

occultation

QUESTIONS

1. Why might it be unfair to describe William Herschel's discovery of Uranus as accidental? Why might it be unfair to describe the discovery of the rings of Uranus as accidental?
2. Describe the location of the equinoxes and solstices in the Uranian sky.
3. Suggest a phenomenon that could explain the inclination of the rotation axis of Uranus, the ovoids on Miranda, the orbits of Neptune's satellites, and the existence of Pluto.
4. Why is belt–zone circulation so difficult to detect on Uranus? Do you think Voyager 2 will detect it in the atmosphere of Neptune? Why or why not?
5. What evidence do we have that the mantle inside Uranus is fluid?
6. If Voyager 2 finds that Neptune has no small satellites at all, will you expect it to have rings like those orbiting Uranus? Why or why not?
7. If Voyager 2 does find small moons orbiting Neptune, what composition would you expect them to have? Why might their surface brightness depend on whether Neptune has a magnetic field?
8. Both Uranus and Neptune have a blue-green tint when observed through a telescope. What does that suggest about the compositions of their atmospheres?
9. From your knowledge of Saturn's satellite Titan, what do you expect Voyager 2 photos of Triton will reveal? Compare Io and Nereid in the same way.
10. If you visited Pluto and found Charon a full moon directly overhead, where would Charon be in the sky when it was at new moon? At first quarter?

PROBLEMS

1. What is the maximum angular diameter of Uranus as seen from Earth? Of Neptune? Of Pluto? (HINT: See Box 3–1.)
2. One way to recognize a distant planet is by its motion along its orbit. If Uranus circles the sun in 84 years, how many seconds of arc will it move in 24 hours? (This does not include the motion of Earth.)
3. What is the orbital velocity of Miranda around Uranus?
4. What is the escape velocity from the surface of Miranda? (HINTS: See Appendix C for satellite data. Assume the density is 2 g/cm^3. See Box 5–2.)
5. The magnetosphere of Uranus rotates with the planetary interior in 17.24 hours. What is the velocity of the outer portion of the magnetic field just beyond the orbit of Oberon? (HINT: The circumference of a circle is $2\pi r$.)
6. If the ϵ ring is 50 km wide and the orbital velocity of Uranus is 6.81 km/sec, how long a blink should we expect to see when the ring crosses in front of a star? Is this consistent with Figure 23–8a?
7. What is the escape velocity from the surface of an icy moon with a diameter of 20 km? (HINTS: The density of ice is 1 g/cm^3. The volume of a sphere is $(4/3)\pi r^3$. See Box 5–2.)
8. What is the difference in the orbital velocities of the two shepherd satellites 1986U7 and 1986U8? (HINTS: See Appendix C for satellite data. See Box 5–1.)

9. Repeat Problem 2 for Pluto. That is, ignoring the motion of Earth, how far across the sky would Pluto move in 24 hours?
10. Given the size of Triton's orbit and its orbital period, calculate the mass of Neptune. (HINT: See Box 5–1.)

RECOMMENDED READING

BEATTY, J. KELLY "Voyager 2's Triumph." *Sky and Telescope 72* (Oct. 1986), p. 336.

———. "A Place Called Uranus." *Sky and Telescope 71* (April 1986), p. 333.

———. "Uranus' Last Stand." *Sky and Telescope 71* (Jan. 1986), p.11.

———. "Pluto and Charon: The Dance Begins." *Sky and Telescope 69* (June 1985), p. 501.

BERRY, RICHARD "Voyager: Discovery at Uranus." *Astronomy 14* (May 1986), p. 6.

———. "Mysterious Pluto." *Astronomy 8* (July 1980), p. 14.

BROWN, ROBERT HAMILTON "Exploring the Uranian Satellites." *The Planetary Report 6* (Nov./Dec. 1986), p. 4.

CHAIKIN, ANDREW "Voyager Among the Ice Worlds." *Sky and Telescope 71* (April 1986), p. 338.

CHAPMAN, CLARK R. "Encounter! Voyager 2 Explores the Uranian System." *The Planetary Report 6* (March/April 1986), p. 8.

CROSSWELL, KEN "Pluto: Enigma on the Edge of the Solar System." *Astronomy 14* (July 1986), p. 6.

ELLIOT, J. *The Ring Tape: A Voice Recording.* Cambridge, Mass.: MIT Press, 1984.

———. "Uranus: The View from Earth." *Sky and Telescope 70* (Nov. 1985), p. 415.

ELLIOT, JAMES, and RICHARD KERR *Rings: Discoveries from Galileo to Voyager.* Cambridge, Mass.: MIT Press, 1985.

GROSSER, MORTON *The Discovery of Neptune.* Cambridge, Mass.: Harvard University Press, 1962.

HARRINGTON, R., and B. HARRINGTON "The Discovery of Pluto's Moon." *Mercury 8* (Jan./Feb. 1979), p. 1.

———. "Pluto: Still an Enigma After 50 Years." *Sky and Telescope 59* (June 1980), p. 452.

HENBEST, NIGEL "Uranus After Voyager." *New Scientist 106* (31 July 1986), p. 42.

HUNT, GARRY "Voyager 2 Investigates the Atmosphere of Uranus." *The Planetary Report 6* (Nov./Dec. 1986), p. 14.

INGERSOLL, ANDREW P. "Uranus." *Scientific American 256* (Jan. 1987), p. 38.

MINER, ELLIS D. "Voyager 2 and Uranus." *Sky and Telescope 70* (Nov. 1985), p. 420.

MOORE, PATRICK "The Naming of Pluto." *Sky and Telescope 68* (Nov. 1984), p. 400.

NESS, NORMAN F. "The Magnetosphere of Uranus." *The Planetary Report 6* (Nov./Dec. 1986), p. 8.

PORCO, CAROLYN "Voyager 2 and the Uranian Rings." *The Planetary Report 6* (Nov./Dec. 1986), p. 11.

SIDGWICK, J. B. *William Herschel: Explorer of the Heavens.* London: Faber and Faber, 1954.

STEVENSON, DAVID "An Ocean in Uranus." *The Planetary Report 6* (Nov./Dec. 1986), p. 11.

TOMBAUGH, CLYDE W. "The Discovery of Pluto: Some Generally Unknown Aspects of the Story." *Mercury 15* (May/June 1986), p. 66.

———. "The Discovery of Pluto: Part II." *Mercury 15* (July/Aug. 1986), p. 98.

———. *Out of the Darkness: The Planet Pluto.* Harrisburg, Pa.: Stackpole Books, 1980.

See *Science 233* (4 July 1986).

CHAPTER 24

METEORITES, ASTEROIDS, AND COMETS

When they shall cry "PEACE, PEACE" then cometh sudden destruction! COMET'S CHAOS?—What Terrible events will the Comet bring?

From a religious pamphlet predicting the end of the world because of the appearance of Comet Kohoutek, 1973

In the first months of 1910, Comet Halley passed through the inner solar system and was spectacular in the sky. On the night of May 19, the earth actually passed through the tail of the comet, and millions of people panicked. The spectrographic discovery, only a few years earlier, of cyanide gas in the tails of comets led many to believe that all life on the earth would end. Householders in Chicago stuffed rags around doors and windows to keep out the gas, and many bought supplies of bottled oxygen. Con artists in Texas sold comet pills and inhalers to ward off the noxious fumes. An Oklahoma newspaper reported (in what was apparently a hoax) that a religious sect called the Select Followers tried to sacrifice a virgin to the comet and were prevented by the last-minute arrival of the sheriff.

Throughout history comets have been seen as portents of doom. Even in 1973–74 Comet Kohoutek (Figure 24–1) was hailed by one mystic as "Star-Seed," an interstellar spaceship bringing light and love to earth. One group tried to contact the comet by ESP, while others claimed the comet announced the second coming of Christ.

In 1986 Comet Halley returned and, being fainter and farther south than in 1910, did not stimulate much public concern. Astronomers, on the other hand, studied the comet closely and confirmed the theory that a comet is a lump of dirty ices vaporizing in the glare of the inner solar system (Chapter 19).

To an astronomer, a comet is a messenger from the age of planet building. By studying comets we can learn about the conditions in the solar nebula from which the planets formed. In addition to the planets, the solar nebula left behind lumps of ice, which we see as comets, and lumps of

FIGURE 24–1 Comet Kohoutek was predicted to be a very impressive comet but did not develop as expected. Nevertheless, its appearance in 1973 caused a number of predictions of the end of the world. (Palomar Observatory photograph.)

rock, which we see as asteroids. Because we cannot easily visit comets and asteroids, we begin our discussion with the fragments of those bodies which fall into our atmosphere—the meteorites.

24.1 METEORITES

On the afternoon of November 30, 1954, Mrs. E. Hulitt Hodges of Sylacauga, Alabama, lay napping on her living room couch.* An explosion and a sharp pain jolted her awake, and she found that a meteorite had smashed through the ceiling and struck her left leg. Mrs. Hodges is the only person known to be injured by a meteorite.

Meteorite impacts on homes are not common, but statistical calculations show that a meteorite should damage a building somewhere in the world about once every 16 months. The earth gains about 10,000 tons of matter per year from meteorites. Most are small specks of dust, but some are quite large.

We have two questions to consider about meteorites: Where in the solar system does this material originate? What can meteorites tell us about the origin of the solar system? To answer those questions, we must consider the orbits of meteoroids and their mineral content.

METEOROID ORBITS

Meteoroids are much too small to be visible through even the largest telescopes, but once they enter the earth's atmosphere at speeds of 10–30 km/sec* and are heated to incandescence by friction, we can determine their speed and direction of travel. Thus, we can backtrack the flaming trail of a meteor to discover the orbit of the meteoroid. These orbits hint at the origins of the meteorites.

One way to backtrack meteor trails is to observe meteor showers. On any clear night, you can see 3 to 15 meteors an hour, but on some nights there are meteor showers during which you can see as many as 60 meteors per hour.

The meteors in a meteor shower are related to each other. To confirm this, try observing a meteor shower.

*Across the street from the Comet Drive-in Theater.

*A rifle bullet travels at about 0.7 km/sec.

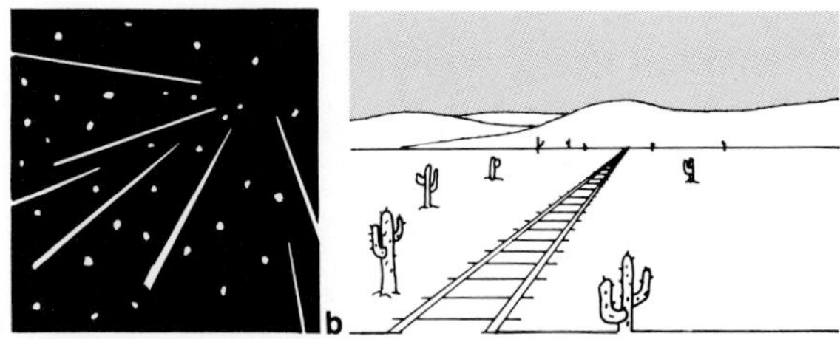

FIGURE 24–2 (a) Meteors in a meteor shower enter the earth's atmosphere along parallel paths, but perspective makes them appear to diverge from a radiant point. (b) Similarly parallel railroad tracks appear to diverge from a point on the horizon.

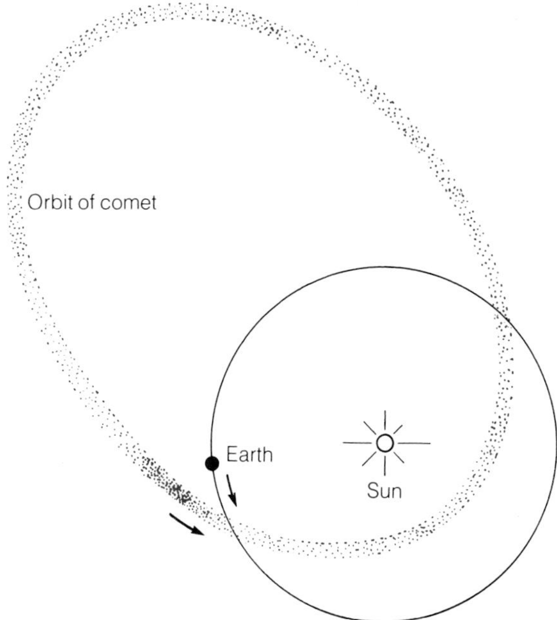

FIGURE 24–3 As a comet's ices evaporate, it releases rocky and metallic bits of material that spread along its orbit. If the earth passes through such material, it experiences a meteor shower.

Select a shower from the table in Appendix C, and on the appropriate night stretch out in a lawn chair and watch the sky. When you see a meteor, plot its path on a star chart. You will discover that most of the meteors you see seem to come from a single area of the sky, the **radiant** of the shower (Figure 24–2a). In fact, meteor showers are named after the constellation from which they seem to radiate. Thus, the Perseids are a meteor shower in August that radiates from the constellation Perseus.

Meteors in a shower appear to come from a radiant because they are all traveling through space along parallel paths. When they encounter the earth and are vaporized at altitudes of 80 km or so, we see their fiery tracks in perspective; they appear to come from a single radiant point, just as railroad tracks seem to come from a single point on the horizon (Figure 24–2b).

Studies of meteor shower radiants reveal that these meteors are produced by bits of matter orbiting the sun along the paths of comets. The vaporizing head of the comet releases bits of debris that become spread along the entire orbit (Figure 24–3). When the earth passes through this stream of material, we see a meteor shower. In some cases the comet has wasted away and is no longer visible, but in other cases the comet is still prominent although somewhere else along its orbit. For example, in May the earth comes near the orbit of Comet Halley, and we see the Eta Aquarids shower. Around October 20 the earth passes near the other side of the orbit of Comet Halley, and we see the Orionids shower.

Even when there is no shower, we still see meteors, and these too may be related to comets. Such meteors are called **sporadic meteors** and are caused by random bits of matter entering the earth's atmosphere. Many of these specks of debris come from comets.

If a meteor is photographed from two locations on the earth, it is possible to determine its altitude, speed, and direction. This means it is possible to find out what its orbit looked like before it entered the earth's atmosphere. In many cases the orbits appear to be related to cometary orbits, but in some cases the meteoroids appear to have originated in the asteroid belt. Thus, we can be confident that the meteors in showers and some sporadic meteors originate in comets, but it also seems likely that some sporadic meteors come from the asteroid belt.

METEORITES ON THE GROUND

When a meteorite is massive enough and mechanically strong enough to survive its passage through the earth's atmosphere, it finally reaches the earth's surface. A very large meteorite might hit with the energy of a very large nuclear weapon and dig a crater many kilometers in diameter. However, most are small, and, because their energy of motion is spent in the atmosphere, they fall to the earth with no more energy than a falling hailstone.

For an example of the impact of a large meteorite, we can study the Barringer Meteorite Crater in Arizona. The crater is 1.2 km in diameter and 200 m deep. It was produced 25,000 years ago by the impact of an iron mete-

FIGURE 24–4 The Barringer Meteorite Crater in Arizona is nearly a mile in diameter. It was caused by the impact of a massive iron meteorite with a blast equivalent to a 3-megaton hydrogen bomb. It is approximately 25,000 years old. (U.S. Geological Survey.)

a b

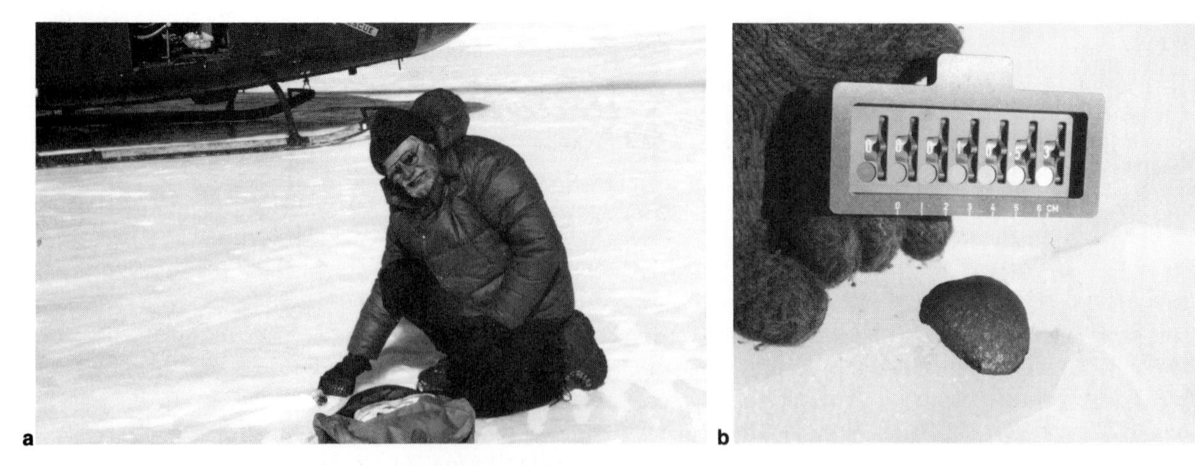

FIGURE 24–5 (a) W. A. Cassidy, leader of several expeditions to Antarctica, holds a numbering device near an achondritic meteorite exposed on the icy surface. (b) An iron meteorite shown as it was found. (NASA photographs, courtesy Ursula B. Marvin.)

Table 24–1 Proportions of meteorites.

Type	Falls (%)	Finds (%)
Stony	92	26
Iron	6	66
Stony–iron	2	8

FIGURE 24–6 The Widmanstätten pattern is evident in this iron meteorite, which has been sliced open, polished, and etched with acid. The pattern shows that the iron solidified from a molten state very slowly. (Griffith Observatory.)

orite that must have been as large as a big house (Figure 24–4).

The best place to look for meteorites turns out to be certain areas of Antarctica—not because more meteorites fall there, but because the meteorites are easy to recognize on the icy terrain. Most meteorites look like earth rocks, but on the ice of Antarctica there are few natural rocks. Also, the slow creep of the ice toward the sea concentrates the meteorites in certain areas where the ice runs into mountain ranges, slows down, and evaporates. Thus, teams of scientists can recover meteorites from the ice (Figure 24–5).

Meteorites that are seen to fall are called **falls**. A fall is known to have occurred at a given time and place, and thus the meteorite is well-documented. A meteorite that is discovered but was not seen to fall is called a **find**. Such a meteorite could have fallen thousands of years ago.

Studies of meteorites reveal three major classes: irons, stones, and stony-irons. **Iron meteorites** make up 66 percent of finds (Table 24–1) but only 6 percent of falls. Why? Because an iron meteorite does not look like a rock. If you trip over it on a hike, you are more likely to carry it home and show it to your local museum. Also, some stone meteorites deteriorate rapidly when exposed to weather; irons survive longer. Thus, there is a **selection effect** that makes it more likely that iron meteorites will be found. That only 6 percent of falls are irons shows that iron meteorites are fairly rare.

When iron meteorites are sliced open, polished, and etched with nitric acid, they reveal regular bands called **Widmanstätten patterns** (Figure 24–6). These patterns are caused by alloys of iron and nickel called kamacite and taenite. These two alloys formed as the molten iron cooled and solidified, and the size and shape of the bands indicate that the molten metal cooled very slowly, no faster than 20 K per million years.

A lump of molten metal floating in space would cool very quickly, so the presence of the Widmanstätten pattern tells us that the molten metal must have been well-insulated to have cooled so slowly. Such slow cooling is typical of the interiors of bodies 30–50 km in diameter. On the other hand, the iron meteorites show no evidence of the very high pressures that would develop in larger bodies. Thus, the iron meteorites appear to have formed from the interiors of molten objects of planetesimal size.

In contrast to irons, **stony meteorites** are common. Among falls, 92 percent are stones. Although there are many different types of stony meteorites, we will classify them into three subgroups depending on their physical and chemical content.

Roughly 80 percent of all meteorite falls are stony meteorites called **chondrites**. These are dark gray granular rocks continuing **chondrules**, round bits of glassy rock no larger than 5 mm across (Figure 24–7a). Chondrules must have cooled very quickly (in minutes) but their origin is not clear. They may be bits of molten rock splashed into space by impacts between planetesimals. Another possibility is that they are bits of rock that condensed from the solar nebula and were later melted by a sudden flare-up of the young sun or by electrical discharges (lightning). Whatever their origin, they are very old, and, because melting the meteoroid would have destroyed them, their presence in chondrites indicates that these meteorites have never been melted. Chondrites can never have been hotter than 1000–1300 K.

On the other hand, the chondrites have been heated to moderate temperatures for long periods sometime in their past. This is shown by studies of the granulation and

FIGURE 24–7 (a) A stony meteorite sliced open and polished to show the small, spherical inclusions called chondrules. (b) A sliced portion of the Allende meteorite showing round chondrules and irregularly shaped white inclusions called CAIs. (a, Smithsonian Institution; b, NASA.)

chemical composition of the rock. Also, the chondrites contain no volatiles. These easily vaporized compounds appear to have been baked out of the chondrites or were never present because the chondrites formed at too high a temperature.

In contrast, meteorites called **carbonaceous chondrites** are rare, only about 5.7 percent of falls, but they are important because they contain volatiles. They are rich in water and compounds of carbon that would have been driven out if they had ever been heated above about 500 K. Thus, if the chondrites have never been hot, the carbonaceous chondrites have never even been warm. They are the least modified of all the meteorites.

One of the most important meteorites ever recovered was a carbonaceous chondrite that was seen falling on the night of February 8, 1969, near the Mexican village of Pueblito de Allende. The brilliant fireball was accompanied by tremendous sonic booms and showered an area about 50 km by 10 km with over 4 tons of fragments. About 2 tons were recovered.

Studies of the Allende meteorite disclosed that it contained—besides volatiles and chondrites—small, irregular inclusions rich in calcium, aluminum, and titanium. Now called **CAIs** for calcium–aluminum-rich inclusions, these bits of matter are highly refractory (Figure 24–7b). That is, they can survive very high temperatures. If we could scoop out a ton of the sun's surface matter and cool it, the CAIs would be the first particles to form. As the temperature fell, other materials would condense in accord with the condensation sequence described in Chapter 19. When the material finally reached room temperature, we would find that all of the hydrogen, helium, and a few other gases like argon and neon had escaped and that the remaining lump, weighing about 18 kg (40 pounds), had almost the same composition as the Allende meteorite, including CAIs. Thus, the Allende meteorite seems to be a very old sample of the solar nebula.

Unlike the carbonaceous chondrites, stony meteorites called **achondrites** are highly modified. They contain no chondrules and no volatiles. This suggests that they have been hot enough to melt chondrules and drive off volatiles, leaving behind rock with compositions similar to the earth's lavas.

Iron meteorites and stony meteorites make up most falls, but 2 percent of falls are meteorites that are made up of mixed iron and stone. These **stony–iron meteorites** appear to have solidified from a region of molten iron and rock—the kind of environment we might expect deep inside a planetesimal with a molten iron core and a rock mantle.

THE ORIGINS OF METEORITES

The properties of meteorites suggest that they formed in the solar nebula. Their radioactive ages are about 4.6 billion years (Chapter 19), and some appear not to have been modified since they condensed from the solar nebula. Others have been heated or melted sometime after formation.

Meteorites almost certainly do not come from comets. Most cometary particles are very small specks of low-density, almost fluffy, material. When these enter the atmosphere, we see them incinerated as meteors, and most meteors are produced by this cometary debris. But such meteoroids are small and weak and do not survive to reach the ground. Although most meteors come from comets, most meteorites are stronger chunks of matter—more like fragments of asteroids.

Thus, when we talk of the origin of the meteorites, we are concerned with the history of asteroidlike planetesimals rich in metals and silicates, not the cometlike planetesimals that were rich in ices. Indeed, many of the properties of the meteorites suggest that they originated in stony–iron planetesimals, which evolved in complicated ways and were eventually broken up by collisions. Here we will concentrate on the answers to two questions: How did these planetesimals evolve? When did these planetesimals break up?

The iron and achondritic meteorites show that at least some of the planetesimals melted, but what produced this heat? Planets the size of the earth can accumulate a great deal of heat from the decay of radioactive atoms such as uranium, thorium, and the radioactive isotope of potassium. The heat is trapped deep underground by many kilometers of insulating rock. But in a small planetesimal, the insulating layers are not as thick, and the heat leaks out into space as fast as the slowly decaying atoms can produce it. If a small planetesimal is to melt, it must have a more rapidly decaying heat source.

Recent studies have shown that some meteorites may have contained the radioactive element aluminum-26. These atoms are now gone, decayed to magnesium-26, but if they were once present, their rapid decay could have melted the center of a planetesimal as small as 20 km in diameter. This is a sufficient heat source to melt the parent bodies of the meteorites.

Once the planetesimals were melted and the aluminum-26 had decayed, they cooled and slowly solidified. In the meantime, however, the planetesimals would have differentiated. The heavy metals would have sunk to the center to form a molten metal core, and the less dense silicates would have floated upward to form a stony mantle. In some cases, the surface may not have been melted. If this planetesimal were broken up (Figure 24–8), fragments from the center would look much like iron meteorites with their Widmanstätten patterns, and fragments from the outer portions might look like stony–irons and achondrites.

The chondrites and carbonaceous chondrites might have formed as smaller bodies farther from the sun. Planetesimals in the outer asteroid belt would have been cooler and could have retained volatiles more easily. They could not have been large, or heating and differentiation would have altered them.

The evidence seems strong that the meteorites are the result of the breaking up of larger bodies. For example, some meteorites are breccias—collections of fragments cemented together. Breccias are found on the earth and are very common on the moon, but studies of the meteoric breccias show that they were produced by impacts. A collision between planetesimals produces fragments, and the slower moving particles fall back to the surface of the planetesimal to form a regolith, a soil of broken rock fragments. Later impacts may add to this regolith and stir it. Still later, an impact may be so violent that the fragments are pressed together and momentarily melt where they touch. Almost instantly, the material cools and the fragments weld themselves together to form a breccia. Most of the breccias were formed as the planetesimals formed. Much later, as the planetesimals were broken up by collisions, the layers of breccia were exposed, shattered, and became brecciated meteoroids.

The collisions that broke up the planetesimals could not have happened long ago when the solar system was young. Small meteoroids would have been swept up by the planets in only a billion years or less. Also, cosmic rays striking meteoroids in space produce isotopes such as helium-3, neon-20, and argon-38. Studies of these atoms in meteorites show that most meteorites have not been exposed to cosmic rays for more than a few tens of millions of years. The meteorites must have been buried

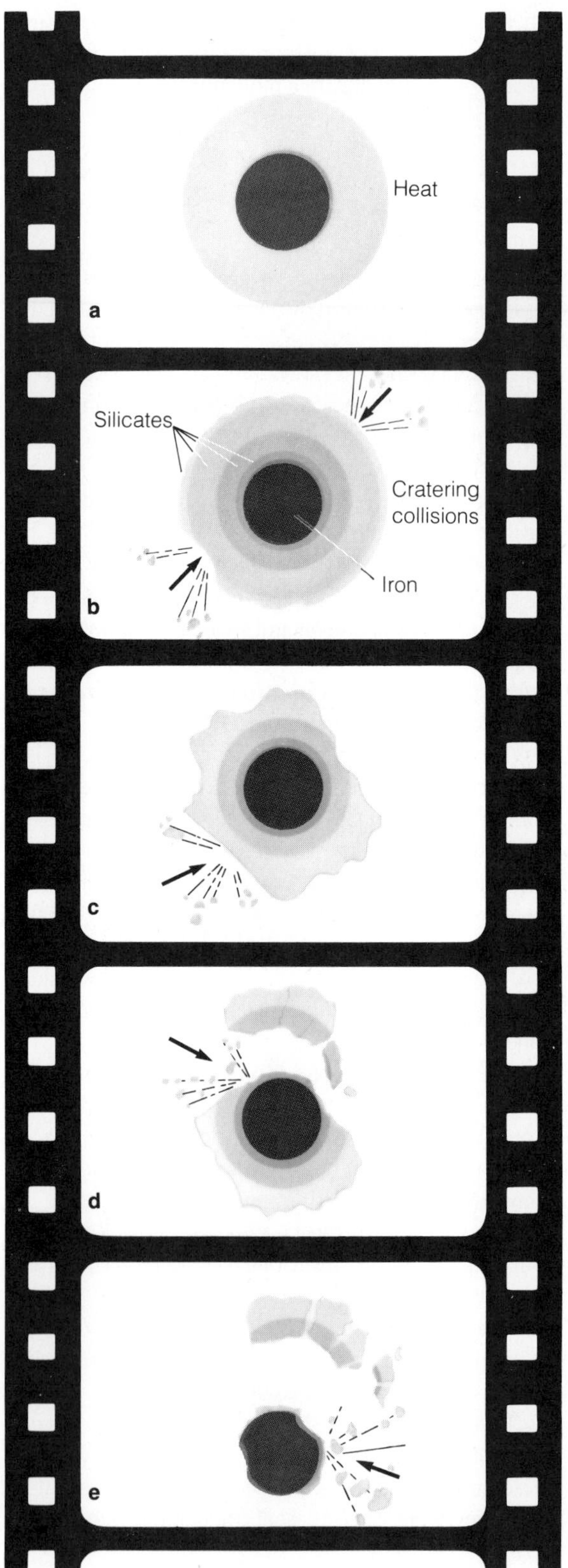

FIGURE 24–8 Production of various kinds of meteoroids from asteroids. (a) If it melted, an asteroid could differentiate into (b) an iron core surrounded by layers of different silicate compositions. (c, d, e) Cratering, collisions, and fragmentation could break up these layers and produce various kinds of meteorites. (Based on "The Nature of Asteroids" by C. R. Chapman. Copyright © 1975 by *Scientific American, Inc.* All rights reserved.)

under protective layers until recently. Thus, the 3000 or so meteorites now in museums all over the world must have broken off of planetesimals somewhere in our solar system within the last billion years. Where are these planetesimals?

24.2 ASTEROIDS

Until recently few astronomers knew or cared much about asteroids. They were small chunks of rock drifting between the orbits of Mars and Jupiter and occasionally marred long-exposure photographs by drifting past more interesting objects. Asteroids were more irritation than fascination.

Now we know different. The asteroids seem to be the last remains of the planetesimals that built the planets 4.6 billion years ago. Almost all of the original planetesimals were incorporated into planets or ejected from the solar system, but out in the asteroid belt, trapped in an orbital dustbin, are a few small fragments. The study of the asteroids is now an exciting exploration of the ancient past of our planetary system.

ASTEROID ORBITS

The first asteroid was discovered on January 1, 1801 (the first night of the nineteenth century), by the Sicilian monk Giuseppe Piazzi. It was later named Ceres after the patron saint of Sicily, and we know now that it is a spherical ball of rock about 1000 km in diameter.

Astronomers were excited by Piazzi's discovery. The Titius–Bode rule had predicted that a planet existed between Mars and Jupiter at an average distance from the sun of 2.77 AU. Ceres fit. Its average distance from the sun is 2.766 AU. It was a bit small to be a planet, and because

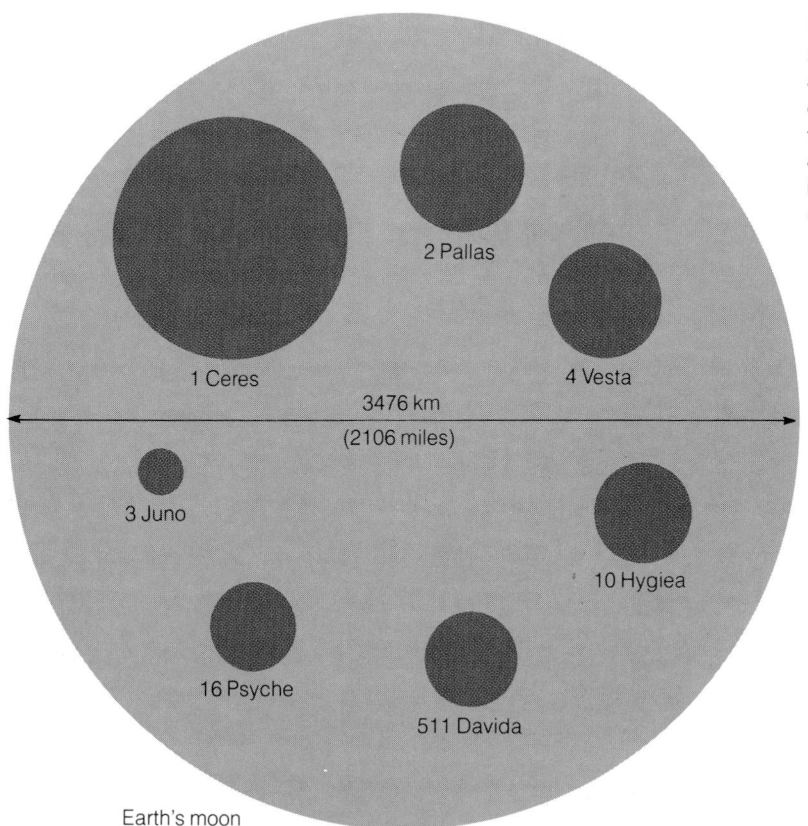

FIGURE 24–9 The relative sizes of some of the larger asteroids are shown here as circles compared with the size of the earth's moon. Smaller asteroids may be highly irregular in shape, and even Pallas is slightly ellipsoidal.

three more objects—Pallas, Juno, and Vesta—were discovered in the following years, astronomers realized that Ceres and the other asteroids were not true planets.

Today over 3500 asteroids have accurately known orbits, and about 6000 more have poorly known orbits. In 1983 the Infrared Astronomy Satellite (Chapter 6) mapped the sky and detected 35,618 unidentified objects. Astronomers reducing the data expect that many of these images are multiple sightings of asteroids, both known and newly discovered.

When an asteroid is discovered, the discoverer is allowed to choose a name for it, and asteroids have been named for spouses, lovers, dogs, Greek gods, politicians, and more.* Once an orbit has been calculated, the asteroid is assigned a number listing its position in the Soviet catalog known as the *Ephemerides of Minor Planets.* Thus, Ceres is known as 1 Ceres, Pallas as 2 Pallas, and so on. Only three of these are larger than 400 km in diameter, and most are much smaller (Figure 24–9).

Most of the these objects orbit the sun in the asteroid belt between Mars and Jupiter, and we might suspect that massive Jupiter was responsible for their origin. Certainly the distribution of asteroids in the belt shows the importance of Jupiter's gravitation. Certain regions of the belt, called **Kirkwood's gaps** (Chapter 22) after their discoverer, are almost free of asteroids (Figure 24–10). These gaps lie at certain distances from the sun where an asteroid would find itself in resonance with Jupiter. For example, if an asteroid lay 3.28 AU from the sun, it would revolve twice around its orbit in the time it took Jupiter to revolve once around the sun. Thus, on alternate orbits the asteroid would find Jupiter at the same place in space tugging outward. The cumulative perturbations would rapidly change the asteroid's orbit until it was no longer in resonance with Jupiter. Our example is in a 2:1 resonance, but gaps occur in the asteroid belt at many resonances, such as 3:1, 5:2, and 7:3.

*Some sample asteroid names: Olga, Chicago, Vaticana, Noel, Ohio, Tea, Gaby, Fidelio, Hagar, Geisha, Dudu, Tata, Mimi, Dulu, Tito, Zulu, and Beer.

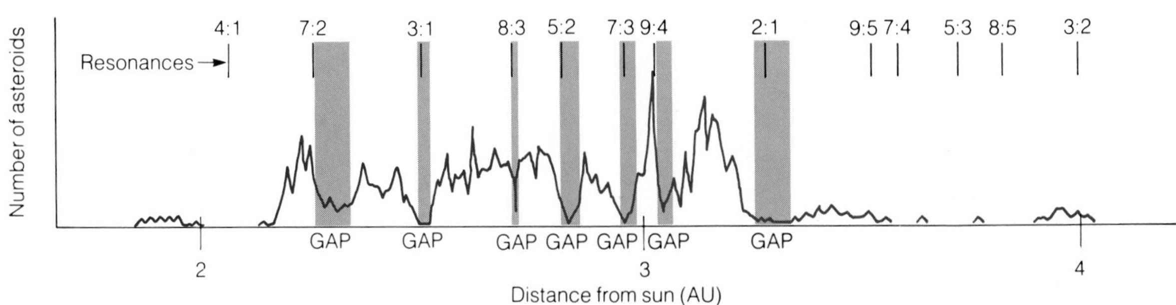

FIGURE 24–10 Kirkwood's gaps are regions in the asteroid belt where there are few asteroids. These gaps are caused by orbital resonances with Jupiter.

Recent research shows that the motions of asteroids in Kirkwood's gaps is described by a newly developed theory in mathematics that deals with chaotic behavior. The smooth motion of water sliding over the edge of a waterfall decays rapidly into a chaotic jumble. The same mathematical theory of chaos, which describes the motion of the water, shows how the slowly changing orbit of an asteroid within one of Kirkwood's gaps can suddenly change into a long elliptical orbit that carries the asteroid into the inner solar system where it is likely to be removed by a collision with Mars, Earth, or Venus. This ejection process may also explain how Jupiter's gravity can throw so many stony meteoroids from the asteroid belt into the inner solar system.

Jupiter is also a dominant influence on two groups of asteroids that do not lie in the asteroid belt but have become trapped in the Lagrangian points (Chapter 22) along Jupiter's orbit. These points lie 60° ahead and 60° behind the planet, and are regions where the gravitation of the sun and Jupiter combine to trap small bodies (Figure 24–11). Like cosmic sinkholes, the Lagrangian points have trapped chunks of debris now called **Trojan asteroids** and named after the heros of the Trojan war (588 Achilles, 624 Hektor, 659 Nestor, and 1143 Odysseus, for example). Although about 15 Trojans have been found and named, astronomers estimate that 700 may be trapped in the Lagrangian points.

Another peculiar asteroid is Chiron, a 350-km diameter object found in 1977. It follows an orbit that carries it from near the orbit of Uranus to just inside the orbit of Saturn. Recent observations suggest that Chiron may be enclosed in a gaseous cloud, which some astronomers interpret to mean that Chiron is rich in volatiles such as frozen water. If so, Chiron may be more comet than asteroid.

Like Chiron and the Trojans, the **Apollo** and **Amor objects** do not lie in the asteroid belt. The Amor objects follow orbits that cross the orbit of Mars, and the Apollo objects cross the orbit of the earth. These Apollo–Amor objects are dangerous. Jupiter's influence makes their orbits precess, so they are bound to collide with a planet eventually. Earth is hit by an Apollo object once every 250,000 years on average. With a diameter up to about 2 km, they hit with the power of a 100,000-megaton bomb and can dig craters 20 km in diameter. About 30 Apollo objects are known, and none will hit Earth in the foreseeable future. The bad news is that an estimated 1000 of these objects are as yet undiscovered.

On March 23, 1989 an undiscovered asteroid several hundred meters in diameter rushed past Earth at a distance of 300,000 km (500,000 mi)—only twice the distance to the moon. It was not seen until April 6 because it approached from the daylight side of the Earth. Had it struck the Earth on land, it would have devastated an area many kilometers in radius, but an impact at sea would have been even worse because of the tidal waves it would have created. The object, now known as 1989FC, has an orbital period that is only about 15 days longer than 1 earth year, so it will pass close to the earth every 25 to 30 years.

Apollo–Amor objects are also important because they may be one of the sources of meteorites. Collisions between asteroids in the main belt may not have enough energy to eject fragments into earth-crossing orbits. But the Apollo–Amor objects rush across the asteroid belt like runaway trucks crossing lanes on a highway. The inevitable collisions are energetic and could eject fragments into earth-crossing orbits.

Collisions are surely common among the asteroids, and we can see evidence of catastrophic collisions, impacts powerful enough to break up an asteroid. Early in this century the Japanese astronomer Kiyotsugu Hirayama discovered that some groups of asteroids share similar orbits. They have the same average distance from the sun,

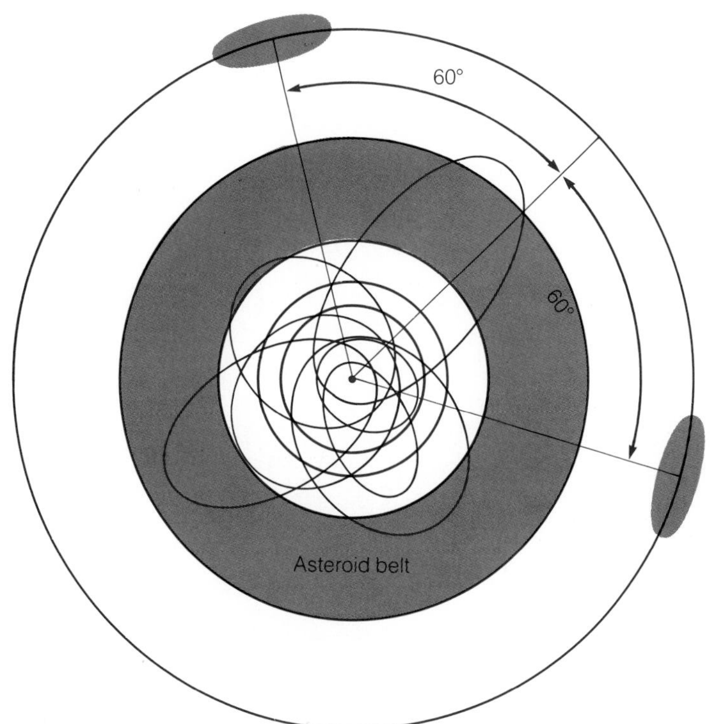

FIGURE 24–11 The Trojan asteroids are trapped in the Lagrangian points 60° ahead of and 60° behind Jupiter. The Apollo asteroids move along orbits that cross Earth's orbit. The orbits of only 7 of the 31 known Apollo asteroids are shown here.

the same eccentricity, and the same inclination. These **Hirayama families**, as they are now called, appear to be the fragments from catastrophic collisions that broke a single asteroid into fragments. The fragments still travel along similar orbits around the sun.

In 1983 the Infrared Astronomy Satellite (Chapter 6) detected the infrared glow of sun-warmed dust scattered in bands throughout the asteroid belt. These dust bands appear to be the products of past collisions. The dust will eventually be blown away, but collisions occur constantly in the asteroid belt, so new dust bands will presumably be produced as the present bands dissipate.

COMPOSITION AND ORIGIN

We now suspect that the asteroids are fragments of planetesimals left over from the formation of the planets. The most popular theory to account for the asteroids supposes that Jupiter's gravity stirred the planetesimals in the asteroid belt and caused collisions at unusually high velocities. These impacts tended to break up the planetesimals rather than assemble them into a planet. Much of the material has been lost—swept up by planets, captured as satellites, or ejected from the solar system.

In our mind's eye we can see the asteroids, great irregular fragments of rock with numerous craters, tumbling ponderously through space. In fact, the asteroids are not just broken rocks. They are complicated little worlds, devoid of atmosphere, of course, but heirs to a complicated geological history that astronomers are only now beginning to suspect.

In the 1930s astronomers discovered that some asteroids are redder than others, but only in the last decade has infrared photometry made it possible to study the color of the light reflected from asteroids and compare it with the colors reflected from various kinds of minerals and meteorites (Figure 24–12a). This has led to the discovery of at least three different kinds of asteroidal worlds.

The S type asteroids are bright, with albedos of about 0.1–0.2, and tend to be reddish. In a diagram of albedo versus ultraviolet minus visual color, the S type asteroids lie toward the upper right (Figure 24–13). At present the data cannot tell us whether they are composed of mixed silicates and metals or of chondritic minerals.

The C type asteroids are very dark, with albedos of less than 0.06. That is, they reflect less than 6 percent of the light that hits them and thus are darker than a lump of sooty coal. Their properties match those of carbonaceous chondrites.

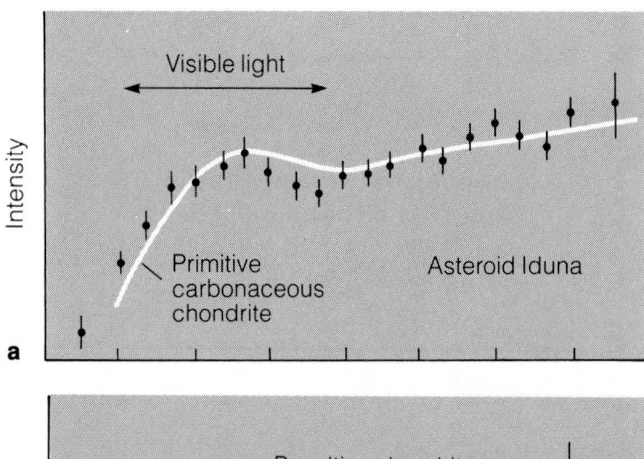

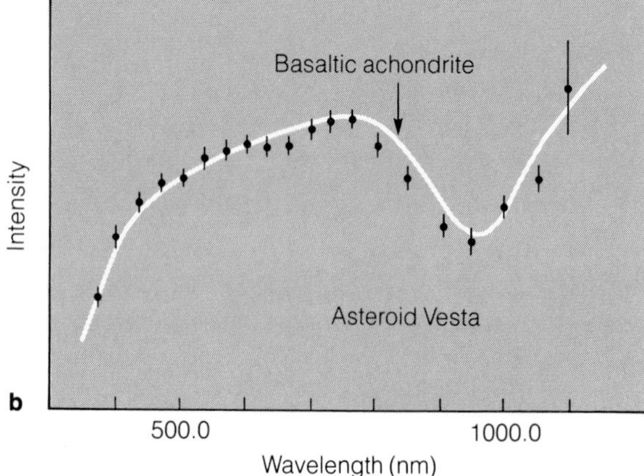

FIGURE 24–12 Measurements of the reflected light in the visible and infrared parts of the spectrum can be compared with light reflected from meteorites in the laboratory. (a) The asteroid Iduna matches a carbonaceous chondrite, but the surface of Vesta (b) reflects light as does a basaltic achondrite. This hints that Vesta is an evolved body whose interior has melted and flooded its surface with lava flows.

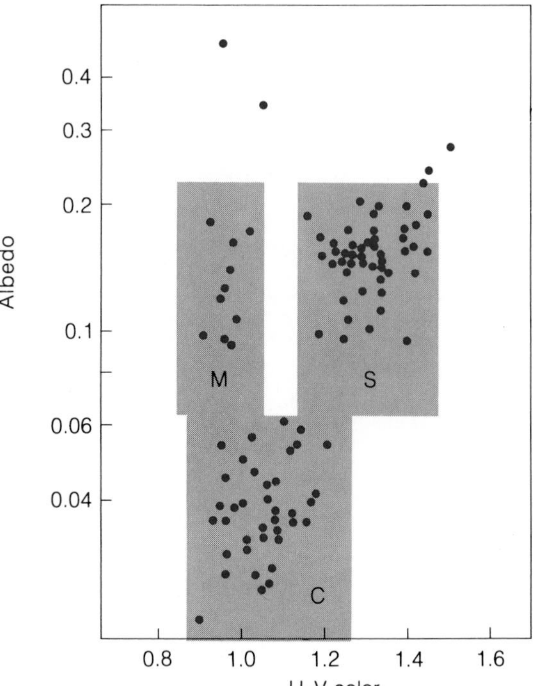

FIGURE 24–13 The three principal types of asteroids. In a diagram of albedo versus ultraviolet minus visual color index, the most reflective asteroids lie near the top and the reddest at the right. The S types, which are mixed silicates and metals or perhaps chondritic, are clearly redder than the M types, which seem to be metallic. The darkest asteroids, the C types, are believed to be similar to carbonaceous chondrites. (Diagram adapted from a figure by B. Zellner.)

The M type asteroids are moderately bright, with albedos about the same as those of the S types, but the M types are not as red. Thus, we can separate the two types in the albedo–color diagram (Figure 24–13). Their optical properties suggest that they are composed of iron and nickel alloys and contain very little silicates.

A few other types of asteroids are known, and a number of individual asteroids have been found that are unique, but these three classes contain a majority of the known asteroids.

How did these three types originate? A clue lies in their distribution in the asteroid belt. The S type asteroids are much more common in the inner belt. In fact, there are almost none beyond a distance of about 3.45 AU. The C type are rather rare in the inner belt but are very common in the outer belt. Many astronomers believe that this reflects the differences in temperature during the formation of the planetesimals. It was cooler in the outer belt, and the planetesimals there tended toward the composition of carbonaceous chondrites. In the warmer regions of the inner belt, the composition of the growing planetesimals was more like that of the chondrites.

However, other astronomers argue that the influence of Jupiter and the fragmentation of the original planetesimals have erased all traces of the conditions of formation. Instead, they argue, the different types of asteroids are the result of different geological histories. The M type, for instance, might be the exposed iron–nickel cores of fragmented planetesimals. The truth may be a blend of these two interpretations, but we can see clear evidence of geological development in some of the larger asteroids.

Vesta, for example, is a small world about 550 km in diameter. The light that it reflects is very similar to the light reflected from certain achondritic meteorites that resemble basalts (Figure 24–12b). Basalts are rocks that solidified from certain kinds of lava flows, so this suggests that Vesta has a surface that has been modified by flowing lava. Until recently, astronomers could not account for the heat to melt such a small planet, but now they suspect that the decay of aluminum-26 could be the energy source. Thus, it seems likely that Vesta has melted, differentiated, and that molten rock has oozed up to its surface to form basaltic lava flows. Vesta isn't just a chunk of rock.

If these processes can alter Vesta, they can alter some of the other asteroids. Ceres is almost twice as large as Vesta, and Pallas is about the same size. More than 200 asteroids are larger than 100 km in diameter. Studies show that aluminum-26 decays rapidly enough to melt the interiors of any planetesimal larger than about 20 km in diameter, so a great many of the asteroids should show signs of geological activity like that on Vesta. However, not many of the larger asteroids appear to be basaltic. Ceres, for instance, reflects light like carbonaceous chondrites.

Clearly, the history of asteroids is very complex and may not be unraveled until we finally visit the asteroid belt. The Galileo spacecraft will fly past two asteroids in the early 1990s on its 6-year journey to Jupiter. If all goes well, in 1991 Galileo will photograph Gaspra, an asteroid 16 km (10 mi) in diameter from a distance of 620 miles. In late 1992, it will photograph Ida, an asteroid about 32 km (20 mi) in diameter. By analogy with the two moons of Mars, which may be captured asteroids, we can guess that we will see dark, irregular shapes, heavily cratered by impacts, and blanketed with a regolith of pulverized rock.

24.3 COMETS

If you listen to an old-timer talk about Comet Halley in 1910, you may think that modern comets are cheap imitations. But many of the people who remember Comet Halley from 1910 remember it brighter than it really was, and some people remember the wrong comet. In January 1910, while the world was looking forward to the approach of Comet Halley, an unexpected bright comet appeared in the sky. Known as the Great January Comet of 1910, it was brighter than Comet Halley was a few months later. Many people misremember the January Comet as Comet Halley.

When Comet Halley returned in 1986, public expectations ran high, and public disappointment ran deep. Because it remained farther from the earth and in the southern sky, the comet was not as bright as it was in 1910. If you missed seeing the comet or if you saw it and weren't impressed, take heart. Bright comets appear unexpectedly once or twice a decade. You will probably see a bright comet sometime during your life.

Nothing in astronomy is quite so beautiful as a major comet sweeping through the inner solar system (Color Plates 54 and 55). Whereas meteors shoot across the sky like demented fireflies, a comet moves with the stately grace of a great ship at sea, its motion hardly apparent.

OBSERVATIONAL PROPERTIES OF COMETS

We saw in Chapter 19 how the overall properties of comets could be explained by the dirty snowball hypothesis. This proposes that a comet is a lump of dirty ices, mostly water ice, orbiting the sun in a long elliptical orbit. Only when that nucleus of ice comes within a few astronomical units of the sun, does the ice vaporize to form a tail of gas and dust pointing away from the sun.

Now we are ready to compare this hypothesis in detail to the observed properties of comets and especially with Comet Halley (Data File 12), which was studied exhaustively during its return in 1986. Two Soviet spacecraft, Vega 1 and Vega 2, flew past the comet at distances of 8900 km and 7900 km. Two Japanese spacecraft, Sakigake and Suisei, studied the comet from 20 times farther away, but the European Space Agency's Giotto spacecraft flew within about 600 km of the comet's heart. The data from those missions paints a newly detailed picture of what a comet is like.

We can think of a comet as comprising a head and a tail. The visible head, or **coma**, of a comet is a vast gas and dust cloud, which can be up to 15 times the diameter of the earth (Figure 24–14). The Vega spacecraft approaching Comet Halley encountered the first dust particles when they were still 47 earth radii from the nucleus. Spectra made from the earth (Figure 24–15) and measurements by spacecraft show that the gas is made of the atoms hydrogen, carbon, nitrogen, and oxygen linked to form various ions and molecules (Table 24–2). These atoms are what we would expect to find if ultraviolet radiation from the sun broke down vaporized ices consisting of water (H_2O) and carbon dioxide (CO_2). The ices may also contain ammonia (NH_3) and methane (CH_4).

In 1970 observations made from earth orbit revealed a vast cloud of hydrogen surrounding the nucleus of bright comets. This sphere is larger than the sun and must be continually resupplied with fresh hydrogen coming from the head of the comet at the prodigious rate of 10^{29} atoms

DATA FILE 12

Comet Halley

Average distance from the sun	17.8 AU (2.66×10^9 km)
Eccentricity of orbit	0.97
Maximum distance from the sun	35.1 AU (5.24×10^9 km)
Minimum distance from the sun	0.53 AU (0.08×10^9 km)
Inclination of orbit to ecliptic	162° (retrograde)
Orbital velocity:	
Maximum	56 km/sec
Minimum	0.85 km/sec
Orbital period	76.1 y (27,800 days)
Period of rotation	52.5 hours (uncertain)
Size	16 km × 8 km × 7 km
Mass	$5 \times 10^{13} - 2 \times 10^{14}$ kg (estimated)
Average density	0.1–0.25 g/cm^3 (estimated)
Surface gravity	0.0004 times earth's
Escape velocity	6.3 m/sec (0.0006 $V_{\oplus}$)
Surface temperature	300 to 400 K (27° to 127°C or 80° to 160°F)
Albedo	0.04

a. A photo enhanced to reveal streamers of gas twisted by the solar wind. (Color Plate 52.) (Courtesy David Malin. Copyright © 1986 Royal Observatory, Edinburgh.)

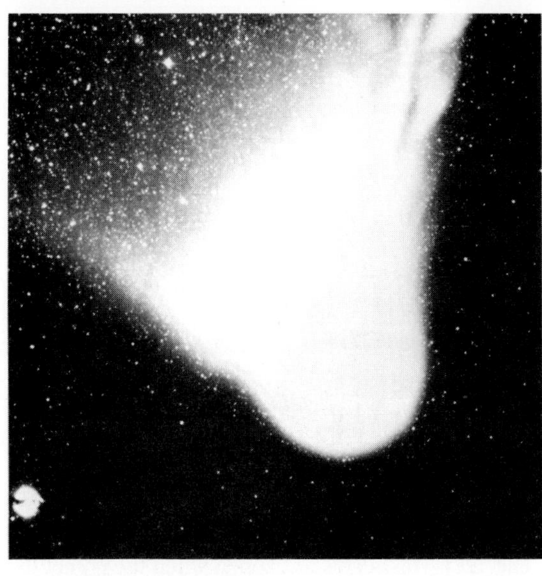

b. Five days after the closest approach to Earth in April 1986, Comet Halley passed nearly in front of the peculiar radio galaxy Centaurus A at lower left. (See Figures 17–3 and 17–4.) (National Optical Astronomy Observatory.)

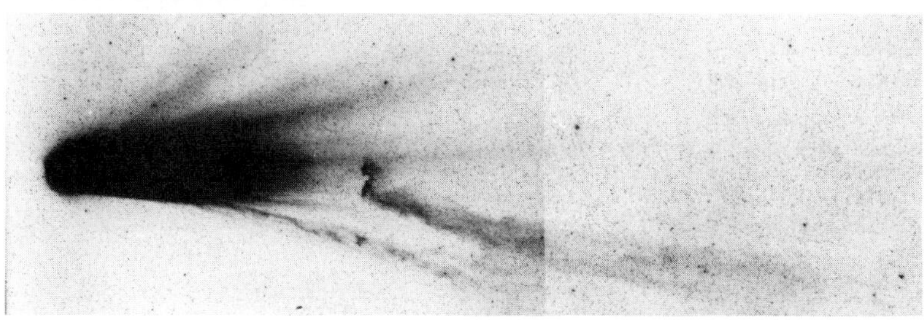

c. A negative image shows curving dust tails and a disconnection in the gas tail. (Courtesy David Malin. Copyright © 1986 Royal Observatory, Edinburgh.)

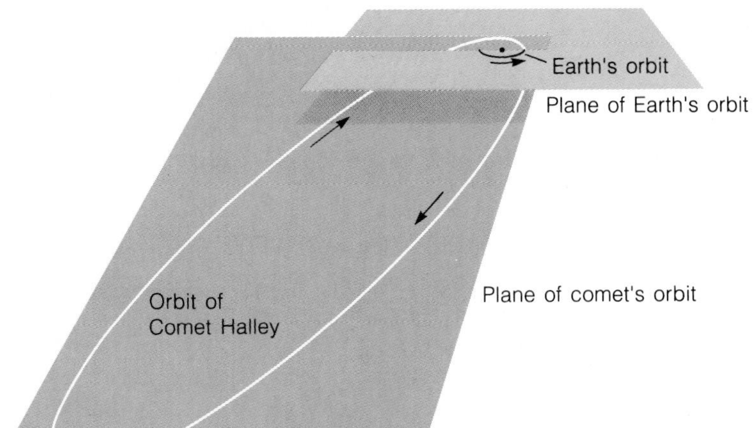

d. The orbits of Earth and of Comet Halley. Note that the comet's orbit is retrograde.

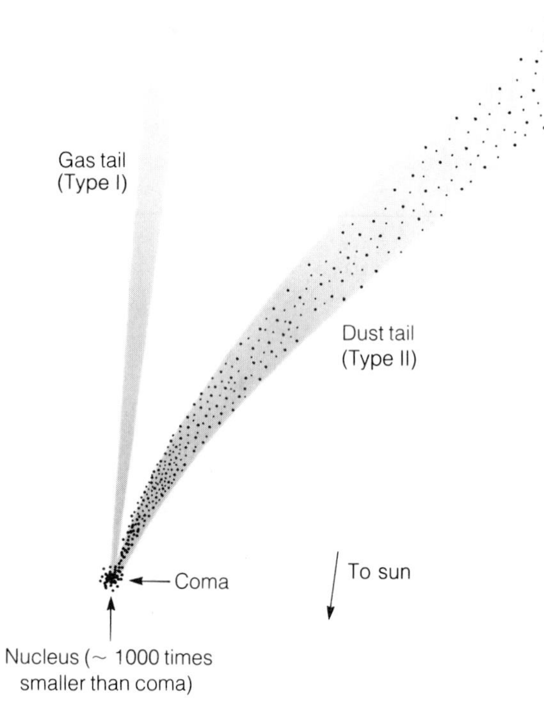

FIGURE 24–14 The principal parts of a comet. Not shown is the extended hydrogen cloud. At this scale it would be about 3 m in diameter.

Table 24–2 Elements and ions detected in comets.

Coma	Gas tail
H, OH, O	CO^+, $CO_2{}^+$
C, C_2, C_3, CH	H_2O^+, OH^+
CN, CO, CS, S	CH^+, CN^+, $N_2{}^+$
NH, NH_2, HCN, CH_3CN	C^+, Ca^+
Na, Fe, K, Ca, V	
Cr, Mn, Co, Ni, Cu	

per second. Apparently these atoms come from disrupted water molecules.

The tail of a comet springs from the coma and typically extends 10--100 million km through space. The longest tails reach a length of 1 AU, about 150 million km. Seen from the earth, the tail of a major comet can span 30° in the sky. This immense structure contains nothing more than very tenuous gas and scattered bits of microscopic dust. Comets have been described as "as close to nothing as something can be and still be something."

Comets have two kinds of tails. A **type I tail**, also known as a **gas tail**, is straight and streaked with wisps and irregularities, which make it look like a streamer of smoke caught in a smooth, fast wind (Data File 12a). The spectra of type I tails show emission lines of ionized atoms excited by the ultraviolet radiation from the sun (Table 24–2). This ionized gas is carried away from the head by the solar wind and its embedded magnetic field pushing the gas away from the sun. In fact, type I tails provide a way to study the solar wind.

Occasionally, a comet will disconnect its gas tail, allow it to blow away, and begin developing a new tail (Data File 12c). This seems to happen when the comet moves across a boundary in the solar wind into a region where the magnetic field is reversed. This illustrates the importance of the magnetic field in the solar wind in controlling the structure of gas tails. In September 1985 a spacecraft called ICE (International Cometary Explorer) flew through the tail of the comet Giacobini–Zinner and confirmed the magnetic nature of the gas tail. The spacecraft found the magnetic field of the solar wind draped over the head of the comet like seaweed draped over a fishhook. The gas tail was composed of ionized gas trapped in this magnetic tail streaming outward with the solar wind (Figure 24–16).

A **type II tail**, also known as a **dust tail**, is smooth and featureless. Its spectrum is merely the spectrum of the sun. This tells us that the type II tail is made up of dust particles that reflect sunlight and that the tail does not contain atoms that would produce emission lines. The dust particles are typically about 1 μm in diameter (Figure 24–17). (One micrometer or micron, abbreviated 1 μm, is a millionth of a meter.) The dust is pushed outward by the pressure of sunlight and forms the long, featureless type II tails, also known as dust tails (Data File 12c). These tails are curved because the comet moves along its orbit with a velocity comparable to the velocity of the dust. Like the stream of water from the moving nozzle of a lawn sprinkler, the dust tail appears to curve.

The dust in a comet's coma poses a serious threat to spacecraft. The Vega probes were partially damaged by

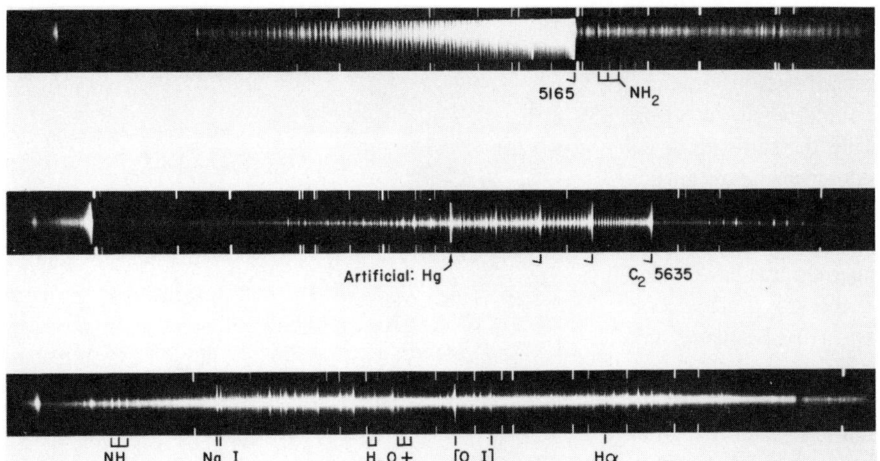

FIGURE 24–15 The spectrum of Comet Kohoutek is shown here in three sections. Comparison lines mark the top and bottom of each section with the comet's spectrum in between. Note the bands of emission lines from molecular carbon (C_2) and the evidence of water ions (H_2O^+). (The line labeled Hg is caused by light pollution from mercury-vapor streetlights.) (Lick Observatory photograph.)

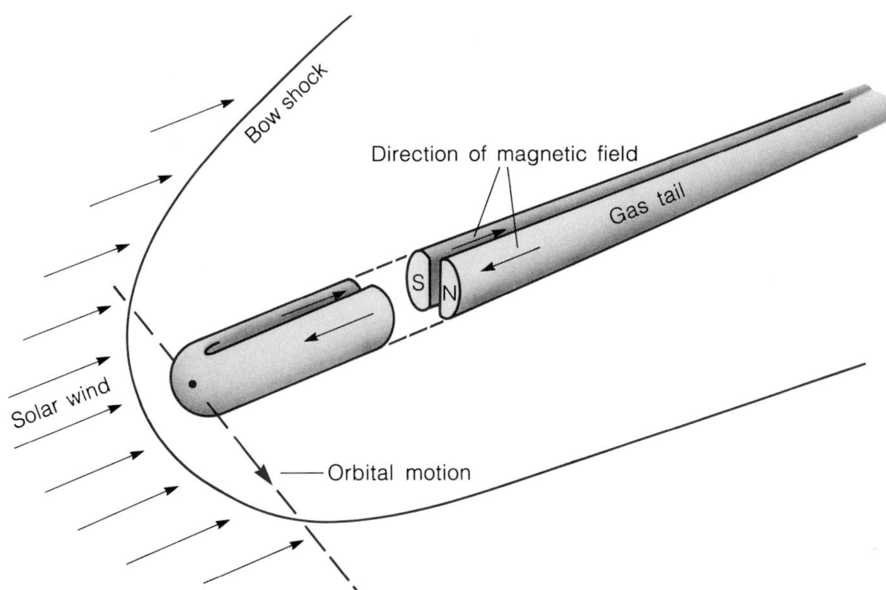

FIGURE 24–16 Cross section of the magnetic field in a cometary gas tail. When the ICE spacecraft flew through the gas tail of Comet Giacobini–Zinner, it found the magnetic field from the outward flowing solar wind draped over the nucleus of the comet like seaweed over a fishhook. On one side of the tail, the magnetic field is oriented with magnetic north toward the nucleus, whereas on the other side of the tail the field is oriented with magnetic south toward the nucleus.

the dust they encountered some 8000 km from the nucleus of Comet Halley, but the Giotto probe, traveling at 70 km/sec (155,000 mph), passed only 600 km from the nucleus. It was struck by 12,000 dust specks and was finally knocked tumbling and partially damaged when it plunged through a denser dust jet only seconds before its closest approach. Astronauts may someday visit the nucleus of a comet, but they will have to approach slowly because of the dust.

The dust released by comets becomes scattered throughout the solar system and can be collected by spacecraft. The Infrared Astronomy Satellite found that this dust was much more abundant than expected. IRAS observed a number of comets and found that the comets themselves and their orbits were rich in dust. Spacecraft measurements of the dust grains around Comet Halley revealed that the dust is partially silicate material, but it has a high content of carbon and organic compounds and thus resembles the meteorites known as carbonaceous chondrites. Comet dust falls into the earth's atmosphere continuously. During the summer of 1986 a team of geologists reported vacuuming up about 10 kg of comet dust from the bottom of an ice lake in Greenland. Such dust must have collected on the ice over thousands of years.

Both dust tails and gas tails point generally away from the sun. Thus, a comet approaches the sun with its tail streaming out behind, but it recedes from the sun tail first (Figure 19–12). Also, a single comet can have both a dust tail and a gas tail at the same time (Figure 24–18).

Many comets (IRAS data suggest most) have no visible tail. This may arise when a comet exhausts its supply of ices. A comet that passes close to a massive planet such as Jupiter can be deflected into a smaller orbit that keeps it in the inner solar system (Figure 24–19). Nearer the sun, its ices vaporize rapidly until little is left but a ball of dust and debris. A number of these "burnt-out" comets are known, and IRAS found that although such comets are poor in ice and have no visible tails, they are, indeed, rich in dust.

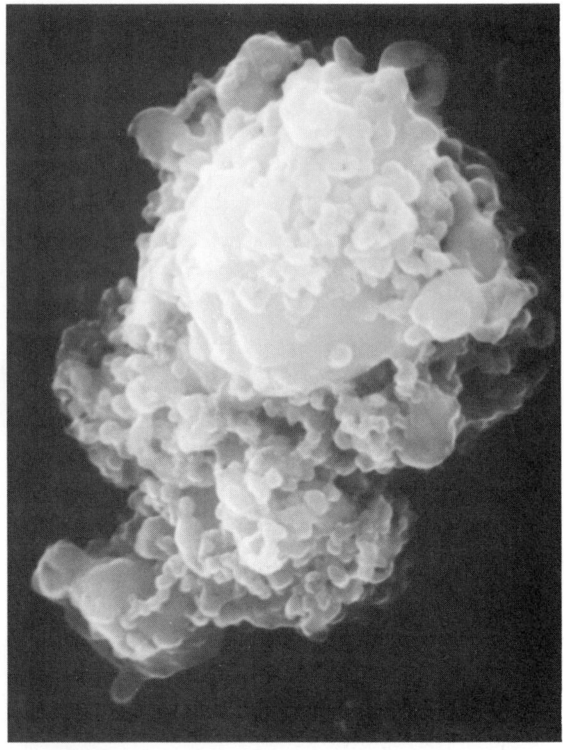

FIGURE 24–17 A cosmic dust speck (5 × 6 μm) collected by a NASA aircraft flying at an altitude of 18 km (59,000 ft). Comets release vast numbers of such dust particles. (NASA.)

THE ORBITS OF COMETS

If we look at the orbits of comets instead of their heads and tails, we discover some family groups. Of about 600 comets with known orbits, 100 or so have been trapped into small orbits with periods of less than 100 years. These short-period comets do not have randomly oriented orbits

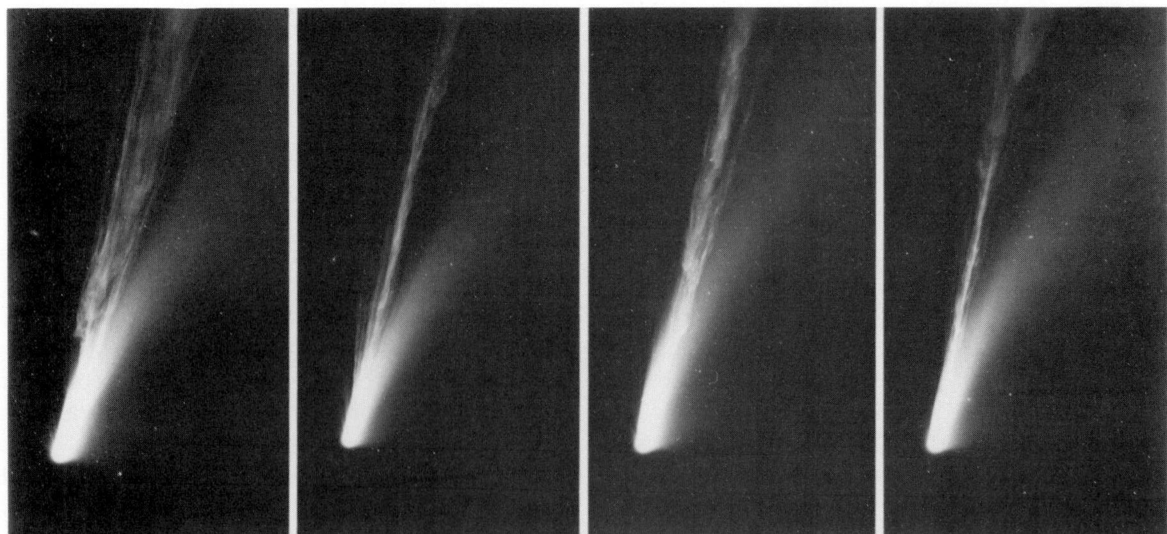

FIGURE 24–18 Comet Mrkos displayed both a gas tail and a dust tail. Changes are apparent in these photos taken on August 22, 24, 26, and 27, 1957. (Palomar Observatory photograph.)

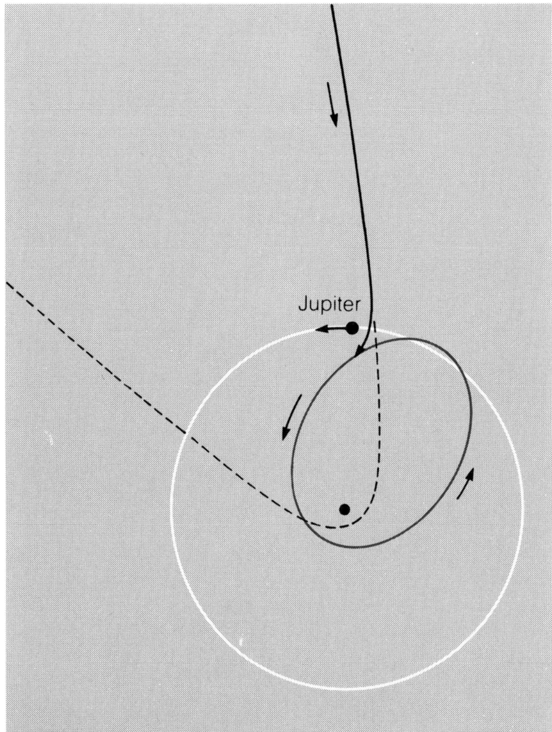

FIGURE 24–19 A comet that normally passes the sun in a long elliptical orbit (the dashed line) can be deflected by the gravitational attraction of Jupiter and trapped in a smaller orbit (the dark ellipse).

like other comets. Because of their past interactions with planets, their orbits tend to lie within 30° of the plane of the solar system and carry them around the sun in the same direction as the planets.

Comet Halley, with a period of 76 years, is a short-period comet, but its orbit is retrograde. It never travels farther from the sun than about 35 AU, just beyond the distance of Neptune. It is presently falling outward along its orbit, and will not return to the inner solar system until 2062.

Long-period comets have periods longer than 200 years and move in random orbits. As many circle the sun clockwise as counterclockwise, and their orbits are tipped at all angles. Comet Kohoutek, which was visible in 1973–74, is such a comet; it has a period of 80,000 years and a maximum distance from the sun of 4000 AU–about 100 times the distance from the sun to Pluto.

There is almost always a comet in the sky, but most are too faint to see without a good telescope. About half a dozen new ones are discovered each year, many by amateur astronomers, and another half dozen known comets reappear. IRAS data and other infrared observations of known comets suggest that many comets pass the sun and we never see them. They are too poor in ices to have bright comas and tails.

Sungrazers are comets that come very close to the sun, and some have slipped into the solar system unnoticed only to appear during total solar eclipses. Some have vaporized entirely in the sun's atmosphere (Figure 24–20). The Solwind satellite detected six sungrazers during 5 years in orbit, and the Solar Max satellite has detected seven in only 13 months.

The orbits of sungrazing comets seem to make up a family, and some astronomers believe they were all formed by the breakup of a larger comet. As we will see in the next section, the icy heart of a comet can break.

THE NUCLEI OF COMETS

The nucleus of a comet is so small that it cannot be resolved by any earth-based telescope, and thus its size and composition must be inferred from the gases and dust that it releases. Until recently, astronomers thought that cometary nuclei were only a few kilometers in diameter. But when the Vega and Giotto spacecraft sent back pictures of Comet Halley's nucleus (Color Plate 53), it proved to be shaped like a peanut 16 km long, 8 km wide, and 7 km thick with a very black crust.

The nucleus of Comet Halley has a density of only 0.1–0.25 g/cm^3, which is surprisingly low. The density of water ice is about 1 g/cm^3. Thus the icy nucleus must be less like a hard-packed snowball of ice and more like a fluffy snowbank containing large amounts of empty volume.

Unlike freshly formed snowbanks on earth, the nucleus of Comet Halley is covered by a black crust that insulates the ices below. Vega spacecraft measured the temperature of the crust as 300–400 K (27°–127°C). This is far above the temperature of vaporizing ice and shows that the crust insulates the interior from the sun's heat. Judging by the albedo of the crust (0.04) and the composition of the dust found near the nucleus, the crust is apparently made up of carbon-rich material mixed with silicates and may vary from a few centimeters to a kilometer thick. Vapor and dust spew primarily from faults in the crust, which make up no more than 15 percent of the surface area.

It now seems clear that the vaporization of the ice is only the simplest part of the process that releases gases into the coma. The composition of the ice is complex, and a number of chemical reactions, some aided by ultraviolet light from the sun, break down parent molecules into the simpler forms we detect in the coma and gas tail.

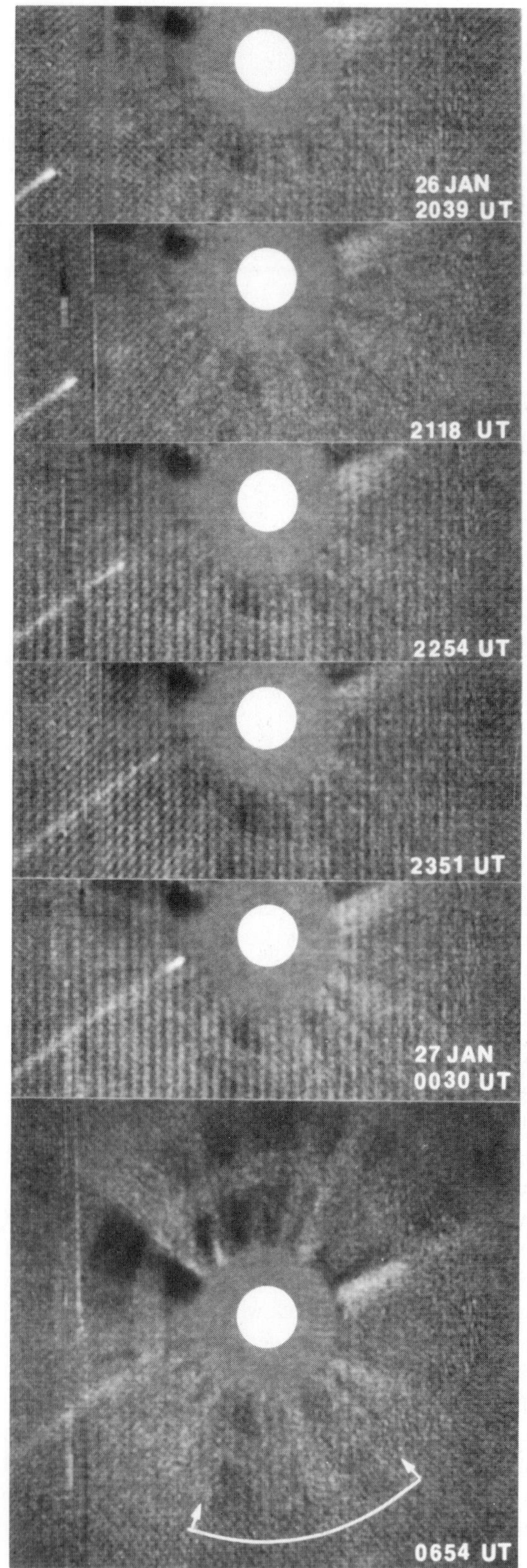

FIGURE 24–20 A comet strikes the sun. This sequence of photographs was taken by the Naval Research Laboratory's satellite Solwind, which carried telescopes to study the sun's corona and solar wind. The comet approaches the sun from the left, but fails to reappear on the other side (shown by arrows). The comet must have collided with the sun or vaporized in the corona. (Naval Research Laboratory.)

For the first time in Comet Halley, water vapor, carbon dioxide, and carbon monoxide have been positively identified as parent molecules. Also detected was the CN molecule, a possible fragment of hydrogen cyanide (HCN). Methane (CH_4) was not detected in Comet Halley, although many theorists expected methane molecules to be trapped in the spaces between the frozen water molecules. Apparently, methane is not very abundant in cometary ice or is transformed into other molecules as it is released from the icy nucleus.

As the ices vaporize, gases vent through faults in the crust and produce jets that are visible in the coma before the solar wind blows them back to form the tail (Figure 24–21a). These jets can act like small rocket motors and slightly modify the orbit of the comet. This effect was used to estimate the mass of Comet Halley's nucleus (Data File 12).

Studies of the motions of these jets have also made it possible to find the period of rotation of some cometary nuclei. The nucleus of Comet Halley, for example, rotates with a period of 52.5 h. This rotation was detected in photographic data from 1910 and spacecraft data from 1986. But earth-based observations suggest a period of 7.4 days. Apparently the peanut-shaped nucleus can rotate with one period and precess with the other like a badly thrown football.

From an analysis of 34 photographs of Comet Halley in 1910, astronomers Zdenek Sekanina and Stephen Larson were able to construct a map showing the location of the active regions on the nucleus. These active features are shown in Figure 24–21b, plotted on an idealized spherical nucleus. Vega and Giotto photographs show that some of these same regions were active again in 1986 (Figure 24–21c). An active site on the nucleus turns on rapidly when the rotation of the nucleus carries it into sunlight, and it turns off as soon as it is carried into darkness.

The violence of gases bursting through the crust plus tidal forces can break up a cometary nucleus. A number of comets have been observed to break into two or more parts while near the sun (Figure 24–22). At least one comet has been seen to break up on passing near Jupiter, presumably because of tidal forces.

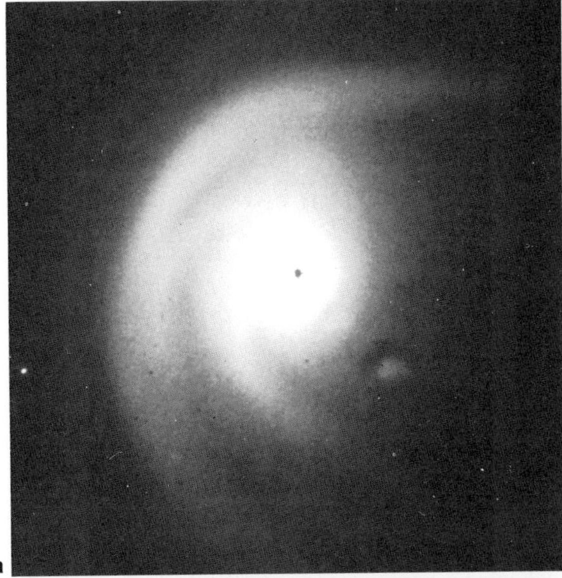
a

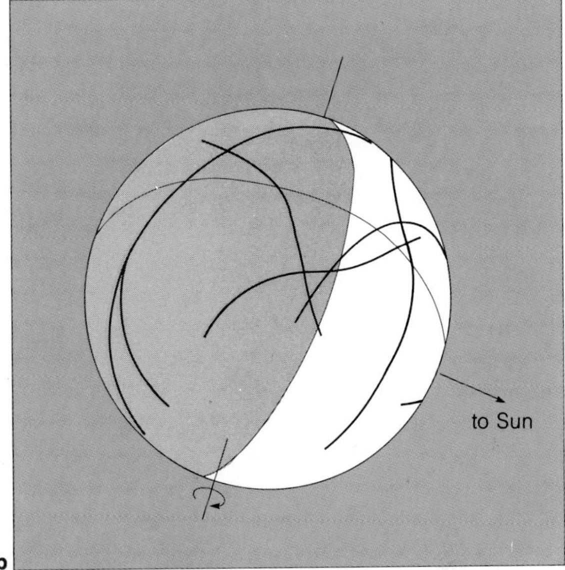

b

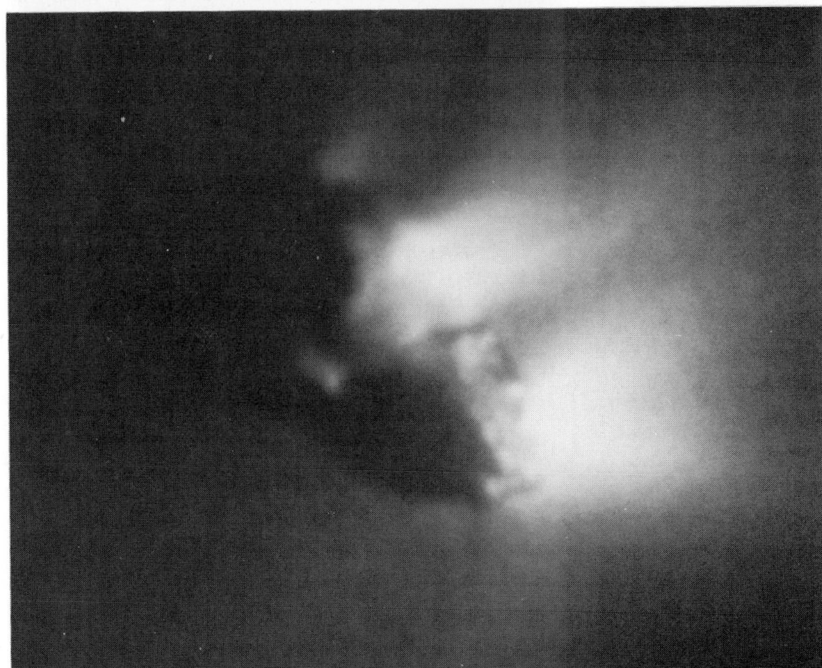
c

FIGURE 24–21 (a) Curved gas and dust jets are often visible in the comas of comets, as shown in this image of Comet Halley from March 14, 1986. (b) Analysis of 1910 photographs suggests that jets spewed from linear active regions, here shown on an idealized spherical nucleus. (c) Sixty separate images from the Giotto spacecraft have been merged to form this image of the icy nucleus of Comet Halley and the dusty jets venting from active regions on its sunward side. (a, Courtesy Steven Larson; c, photo courtesy Harold Reitsema, Ball Aerospace. Copyright © 1986 Max Planck Institute für Astrophysik.)

A comet near the sun is literally a snowball in Hades. Each time it swings past the sun, the nucleus can lose ices equivalent to a surface layer 1 meter thick. Even if its does not break up, it must eventually lose most of its ices and become a burnt-out comet after 1000 orbits or so. A number of such comets are known, and many resemble asteroids. Comet Schwassman–Wachmann-1 is 40 km in diameter, follows a nearly circular orbit beyond Jupiter, and usually looks like an asteroid. Occasionally, however, it has outbursts and develops a coma. Similarly, asteroids are known with a faintly cometary appearance. This suggests that comets and asteroids are not fundamentally different, but are similar objects that differ in their content of ices. Dead comets may look much like asteroids, and some astronomers believe that the Apollo asteroids that cross the earth's orbit are exhausted comets.

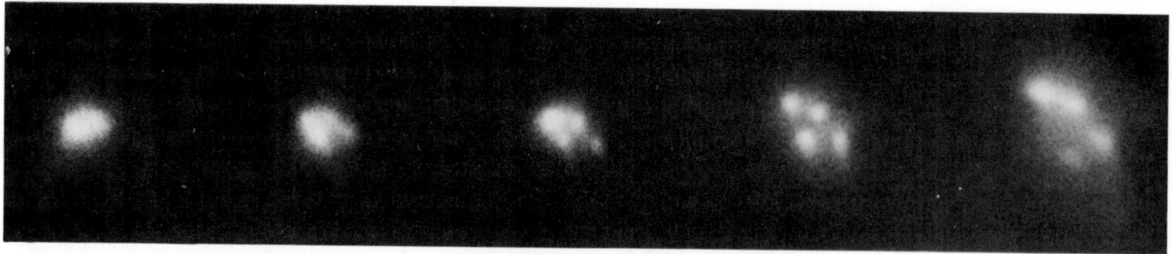

FIGURE 24–22 The breakup of the nucleus of Comet West. These photographs were taken (left to right) on March 8, 12, 14, 18, and 24, 1976. They show the nucleus separating into at least four fragments. (Courtesy Dennis Dawson and New Mexico State University.)

THE ORIGIN OF COMETS

The most interesting question we can ask about comets concerns their origin. Since they gradually vaporize in the sun's heat, the comets we see now cannot be very old. There must be a continuous supply of new comets to make up for those that exhaust their ices.

The Dutch astronomer Jan Oort proposed the theory now known as the **Oort cloud theory**. Oort suggested that icy bodies orbit the sun in a hollow, spherical cloud extending from 20,000 to 100,000 AU (Figure 24–23). As many as 2 trillion icy snowballs may orbit in this cloud. So far from the sun, they would lack comas and tails and would be invisible. They could last for billions of years with rare close encounters with passing stars deflecting a few icy bodies into the inner solar system where we see them as comets.

If the Oort cloud does exist, how did the icy bodies form? They could not have formed at their present location because the solar nebula would not have been dense enough so far from the sun. Also, had they formed from the solar nebula, we would expect them to be distributed in a disk and not in a sphere. Rather, astronomers think the future comets formed as icy planetesimals in the outer solar system near the present orbits of Saturn and Uranus. As the Jovian planets grew more massive, they swept up some of these planetesimals and ejected others to form the Oort cloud. If this is true, then comets are icy planetesimals left over from the solar nebula.

This could account for the long-period comets with their randomly inclined orbits, but the short-period comets must have had a different history. Because the orbits of short-period comets lie in the plane of the solar system, astronomers thought they were long-period comets whose orbits had been altered by the Jovian planets. But mathematical models show this is not possible. Some astronomers now think the short-period comets originate from an unseen belt of icy bodies beyond the orbit of Neptune. Confined to the plane of the solar system, these could be the innermost core of the Oort cloud.

Since the time of Edmund Halley, astronomers have wondered if comets could come from distant stars. It is possible but unlikely that many of the comets we see are interstellar travelers. But a new suggestion is related. If the Oort cloud does exist, then the motion of the sun must carry it along around the galaxy. As the sun passes through giant molecular clouds, the Oort cloud might lose most of its contents because of the gravity of the giant cloud. But the Oort cloud might gain new icy bodies by picking up new snowballs that had formed in the cold, dense gas of the giant cloud. This is a highly speculative conjecture, but it illustrates the most interesting thing about comets—we don't know quite what they are.

COMETS AND DINOSAURS

Comets must hit the earth occasionally. Although a comet is "close to nothing," its nucleus is a ball of ice a dozen kilometers in diameter that would strike the earth with tremendous impact. Many scientists suspect that such impacts throw vast quantities of dust into the atmosphere and cause widespread changes in climate. The extinction of entire species, such as the dinosaurs, may have been caused by comets.

In 1980 Luis and Walter Alvarez announced the discovery of an excess of the element iridium in sediments laid down at the end of the Cretaceous period—just when the dinosaurs and many other species became extinct. Iridium is rare in the earth's crust but common in meteorites, which led the Alvarezes to suggest that a major meteorite impact at the end of the Cretaceous threw vast amounts of dust, rich in iridium, into the atmosphere.

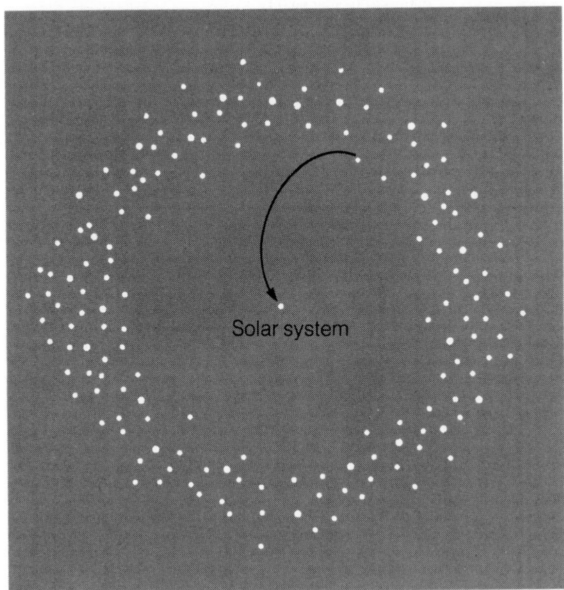

FIGURE 24–23 The Oort cloud of icy bodies is believed to extend out to about 100,000 AU from the sun. The effects of the motions of nearby stars could throw bodies into the inner solar system, where they become comets.

This dust plunged the earth into a winter that lasted many years, killing off many species of plants and animals, including the dinosaurs.

This theory was rejected at first, but soon scientists found the iridium anomaly in sediments of the same age all over the world. Others found related elements and mineral forms typical of meteorite impacts. At the same time, theorists studying nuclear weapons predicted that a nuclear war would throw so much dust into the atmosphere that our planet would be plunged into a "nuclear winter" that could last a number of years. Within a few years, scientists generally agreed that the Cretaceous extinctions were caused by one or more major impacts.

At first the impacts were blamed on large meteorites, objects big enough for us to call them asteroids such as the Apollo objects. But work published in 1984 hinted that comets are the projectiles involved. The new work found that extinctions seem to occur about every 26 million years and that large-impact craters on the earth seem to have been made in episodes every 27–30 million years. The dependability and interpretation of these statistical results are much more controversial than the iridium anomaly.

Some scientists were quick to suggest that periodic storms of comets could cause the periodic impacts, but two different theories were suggested to explain how comet storms could occur periodically.

One hypothesis suggests that the sun is a member of a binary star system in which the companion star has such low mass and is so distant that we have not found it yet. This star travels around the sun in a very large, elliptical orbit that carries it up to 3 light-years from the sun. But at its closest approach, which would occur every 26–30 million years, the star would pass through the Oort cloud, and its gravity would perturb as many as a billion comets into the inner solar system, all within a few million years. The night sky would be glorious with comets, but roughly ten would strike the earth with dramatic consequences.

This deadly companion star, dubbed Nemesis after the goddess of divine retribution, has not been found and may not exist. Some astronomers believe that a star in such a large orbit would not remain stable over even a few orbital periods. Nevertheless, a number of teams of astronomers are now searching for it. It should be near its farthest distance from the sun now and is thus a very faint object. However, being a low-mass star, it should be very cool, and in that case its image may exist unrecognized in the data from the Infrared Astronomy Satellite.

The second hypothesis to explain the periodic trigger of comet showers involves the motion of the sun around the center of the galaxy. The sun's orbit is not flat. Because of the mass of the galactic disk, the sun follows an orbit that carries it up and down through the disk with a period of 27–30 million years—about the right period given the accuracy with which the numbers are known.

How could the passage of the sun through the galactic plane affect comets in the Oort cloud? The plane of the galaxy is littered with giant molecular clouds. These cool clouds contain great mass, and their gravity could disturb the motions of the dirty snowballs in the Oort cloud, thus triggering a comet storm.

One objection to this hypothesis is that we are presently passing through the disk of the galaxy but are about halfway between mass extinctions, assuming that such extinctions are indeed periodic. The proposed trigger seems out of phase with the effect.

The controversy continues, but it is an interesting example of how scientists work with new ideas. The iridium anomaly and its meaning for the Cretaceous extinctions are now widely accepted. The periodic nature of mass extinctions and impacts is viewed with caution and, in some cases, skepticism. Suggestions that showers of comets caused these extinctions are highly speculative, and most scientists are waiting for better evidence before they take such theories seriously.

PERSPECTIVE

COLLISION WITH A COMET

The solar system is a dirty place, filled with meteoroids, specks of dust, asteroids, and comets. It is also a dangerous place. These projectiles travel at high speeds and hit with titanic impact. The earth has been hit before and will be hit again, so it is especially interesting to consider a recent event that seems related. This particular event also illustrates an important aspect of the scientific process, and, of course, it makes a good story.

SIBERIA 1908

On the morning of June 30, 1908, scattered reindeer herders and homesteaders in central Siberia were startled to see a brilliant blue-white fireball brighter than the sun streak across the sky. Still descending, it exploded with a blinding flash and an intense pulse of heat.

> The whole northern part of the sky appeared to be covered with fire. . . . I felt great heat as if my shirt had caught fire . . . there was a . . . mighty crash. . . . I was thrown on the ground about [7 m] from the porch. . . . A hot wind, as from a cannon, blew past the huts from the north.

The blast was heard up to 1000 km away, and the resulting pulse of air pressure circled the earth twice. For a number of nights following the blast, European astronomers, who knew nothing of the explosion, observed a glowing reddish haze high in the atmosphere.

Travel in the wilderness of Siberia was difficult early in this century; moreover, World War I, the Bolshevik Revolution, and the Russian Civil War prevented expeditions from reaching the site before 1927. When at last an expedition arrived, they found that the blast had occurred above the Stony Tunguska River valley and had flattened trees in an irregular pattern extending out 30 km (Figure 24–24). The trees were knocked down radially away from the center of the blast, and limbs and leaves had been stripped away. The trunks of trees at the very center of the area were still standing, although they had lost all their limbs. No expe-

FIGURE 24–24 The aftermath of the Tunguska blast. Trees were knocked down, pointing radially away from the center of the explosion. (Sovfoto.)

dition has ever found a crater, so it seems the explosion, equaling a 12-megaton nuclear weapon, occurred at least a few kilometers above the ground.

Later studies suggested that the area of the explosion had been contaminated by radiation. Some explorers reported the direct detection of radiation; others found that the residents of the area recalled mysterious diseases among their reindeer herds. Trees in the area seemed to grow with unusual vigor, a characteristic of plants exposed to radiation.

FLYING SAUCERS AND BLACK HOLES

The Tunguska event is so peculiar it has spawned a number of peculiar explanations. One of the most bizarre is the suggestion that the explosion was caused by an alien spaceship attempting to land with a malfunctioning nuclear reactor. The reactor exploded, producing the Tunguska blast and destroying all evidence of the spaceship. This theory has been reported as fact by flying saucer experts and has been the subject of sensational books, articles, and newspaper stories.

In fact, there is no evidence that the explosion involved nuclear materials. Reports of radiation in the area are exaggerated, misunderstood, fabricated, or mistaken. There is no evidence of unusual radioactivity in the soil or in the trees that have grown there since the explosion. There is nothing to suggest that the reindeer herds were harmed by radioactivity, and there are no traces of radiation-induced sickness or mutations in the population. The vigorous growth of the trees in the area is the result of the clearing of undergrowth and the fertilization of the soil by the forest fire that followed the explosion.

Other sensational explanations have been proposed. One team of physicists suggested that the earth had been hit by a piece of antimatter, which had exploded when it encountered the normal matter in the earth's atmosphere. Another team suggested that a miniature black hole, about 10^{15} g and the diameter of an electron, had penetrated the earth, entering at Tunguska and exiting unnoticed in the north Pacific.

These more responsible suggestions have not gained much support because, like the spaceship hypothesis, they violate an important principle of science. They ask us to assume more than we need to. Because the explosion was not a nuclear detonation, we do not need to assume that spaceships exist and visit the earth, that large chunks of antimatter float through space, or that miniature black holes are common. There is a simpler explanation, and science always prefers the simplest adequate explanation for any given phenomenon.

DANGER: FALLING COMETS

Soon after the world heard of the Tunguska event, astronomers began to suspect that the earth had been hit by the head of a comet. Opponents have raised objections, some sincere and some contrived, but it is now widely accepted that the Tunguska event was the explosion of the head of a small comet in the earth's atmosphere. That hypothesis explains a great deal and requires no unreasonable assumptions.

The approaching comet was not seen because it was coming from the direction of the sun. Witnesses reported the fireball in the atmosphere came from the east, which in the morning means from the direction of the sun. If the object were a comet, perhaps a sungrazer, whatever tail it had would have been projected straight at the earth and would not have been visible. More likely, the object was the exhausted core of a comet and had no appreciable tail. In fact, some astronomers now believe the object was a fragment of the short-period Comet Encke.

A fragment of a partially exhausted comet should consist of silicate material resembling carbonaceous chondrites mixed with ices. Modern studies of the area show that thousands of tons of powdered material resembling carbonaceous chondrites are scattered in the soil. A chunk of dirty ice 50 m in diameter would contain about the right mass and would be totally invisible approaching from the sun. Also, a comet fragment might be so fragile it would explode in the atmosphere, not dig a crater the way a meteorite would.

Perspective—Collision with a Comet *(continued)*

A smaller explosion may have occurred in 1965 near the Canadian town of Revelstoke. The mysterious blast was equal to about 10 kilotons of TNT, equivalent to the Hiroshima bomb but about 1000 times less violent than the Tunguska blast. In the following days, searchers found the snow dusted with soot, which proved to be powdered carbonaceous chondrite.

The falling comet hypothesis does not ask us to make any fantastic assumptions, and it explains the observed phenomena, even the reddish glow over Europe. Some astronomers suspect that the glow was the faint traces of the comet's tail and scattered dust trapped high in the atmosphere and illuminated by the summer sun shining over the North Pole.

Compelling as this evidence may seem, not all astronomers agree that the object was a spent comet. In a recent paper, Z. Sekanina published an intensive analysis of the Tunguska event and argues that the object was not cometary.* A fragile comet head could not have survived the forces in the atmosphere, and its speed and direction argue against a cometlike orbit. Instead, Sekanina suggests that the object was a small Apollo asteroid 90–190 m in diameter.

Any astronomical explanation may seem disappointing at first. We naturally like exotic stories and fantastic theories, but if we are to do good science, we must prefer the simplest explanation that covers the facts. And after all, a falling comet or asteroid is fairly fantastic. An alien spaceship might explode on the earth only once, but meteorites fall all the time—perhaps one like the Tunguska event every 2000 years or so. We can only wonder when and where the next one will hit.

**The Astronomical Journal 88* (Sept. 1983), p. 1382.

SUMMARY

Most of the meteors we see, and certainly those in showers, are caused by small particles released by comets. When the earth crosses the orbit of a comet, we see a meteor shower. Most meteorites, however, are more massive, stronger fragments that probably originated in the asteroid belt.

The different kinds of meteorites could have originated in a planetesimal-size body that was heated by the decay of short-lived radioactive nuclei like aluminum-26. Differentiation would form an iron–nickel core that would cool slowly. Fragments from this core would show the Widmanstätten pattern common in iron meteorites. The outer parts of such a body might resemble the stony–iron and achondritic meteorites.

Chondrites with their beadlike chondrules may have formed as smaller bodies farther from the sun. The carbonaceous chondrites, being rich in volatile compounds, probably formed farther from the sun. The presence of volatiles, chondrules, and CAIs shows that the carbonaceous chondrites are the least altered of the meteorites.

Collisions must have been common among the planetesimals, just as they are still common in the asteroid belt. We find breccias in some meteorites that must have been formed in impacts long ago. In the asteroid belt today we can find evidence that the asteroids are fragmented. The Hirayama families appear to be fragments with similar orbits.

Of the three kinds of asteroids, the S type are more common in the inner belt, whereas the C type are more common in the outer belt. This may represent a difference in temperature at the time of formation or different evolution. The M type appear to be the exposed iron–nickel cores of broken planetesimals.

Some of the asteroids show clear evidence of geological processes. The reflected light from Vesta, for instance, looks much like the light reflected from achondritic meteorites that resemble basalts. Thus it seems that Vesta has melted and has had lava flows on its surface.

Jupiter clearly dominates the asteroid belt and may have prevented the formation of a planet there long ago. It now causes Kirkwood's gaps and traps the Trojan asteroids at the Lagrangian points along its orbit.

Jupiter is also good at trapping comets in the inner solar system. The short-period comets rapidly waste away as their ices become vaporized, and their exposed cores may become Apollo–Amor asteroids with Earth- and Mars-crossing orbits.

According to the dirty snowball theory, a comet is a ball of dirty ices, which produces a coma as it approaches within 3 AU of the sun. The pressure of the solar wind pushes ionized gas back into a type I, or gas tail; the pressure of sunlight pushes dust particles back into a type II, or dust tail.

When Comet Halley passed through the inner solar system in 1986, five spacecraft flew past it and discovered that its nucleus was larger and blacker than expected. The nucleus was about 16 km × 8 km × 7 km and covered with a black crust having an albedo of only 0.04. The crust is probably made of material resembling carbonaceous chondrites. Gas and dust jets vent from active regions, which cover only 15 percent of the surface.

According to the Oort cloud theory, comets occupy a hollow spherical cloud centered on the sun and extending out about 100,000 AU. Perturbations from the motions of nearby stars may cause a few of these icy bodies to fall into the inner solar system where they become comets. These dirty snowballs may have originated at their present distance or as icy planetesimals among the Jovian planets. In either case, comets are almost unaltered samples of the original solar nebula.

The earth runs into space debris quite often. The Apollo asteroids must hit the earth with great regularity, and the heads of short-period comets must also intersect the earth now and then. One such collision, the Tunguska event, occurred as recently as 1908.

NEW TERMS

radiant	achondrite
sporadic meteor	stony–iron meteorite
fall	Kirkwood's gaps
find	Trojan asteroids
iron meteorite	Apollo/Amor objects
selection effect	Hirayama families
Widmanstätten pattern	coma
stony meteorite	gas tail (type I)
chondrite	dust tail (type II)
chondrule	sungrazer
carbonaceous chondrite	Oort cloud theory
CAI	

QUESTIONS

1. If most falls are stony meteorites, why are most finds iron?
2. How do observations of meteor showers reveal one of the sources of meteoroids?
3. How can most meteors be cometary if most, perhaps all, meteorites are asteroidal?
4. Why are meteorites easier to find in Antarctica? Why are stony meteorites better represented among these finds than among finds in the United States?
5. What evidence do we have that some meteorites have originated inside large bodies?
6. Why couldn't uranium, thorium, and radioactive potassium have melted the planetesimals? What evidence do we have that some asteroids have had molten interiors?
7. How do Kirkwood's gaps resemble Cassini's division? How do the Trojans resemble some of the satellites of Saturn?
8. The first asteroids discovered were much larger than the typical asteroid known today. How does this illustrate a selection effect?
9. Describe three types of asteroids (S, M, and C) and explain a theory to account for their differences.
10. How might the Apollo and Amor objects have originated?
11. Why doesn't a type II comet tail ever suffer a disconnection?
12. If comets are icy planetesimals left over from the formation of the solar system, why haven't they all vaporized by now?
13. How could Valhalla on Callisto, the ovoids on Miranda, and the extinction of dinosaurs on Earth be related to a common phenomenon?

PROBLEMS

1. Large meteorites are hardly slowed by the earth's atmosphere. Assuming that the atmosphere is 100 km thick and that a meteorite falls vertically toward the ground, how long does it spend in the atmosphere? (HINT: How fast do meteoroids travel?)
2. If a single asteroid 1 km in diameter were fragmented, how many meteoroids 1 m in diameter could it yield? (HINT: The volume of a sphere is $\frac{4}{3}\pi r^3$.)
3. What is the orbital velocity of a meteoroid orbiting the sun at the distance of the earth? (HINT: See Box 5–1.)
4. If a half-million asteroids each 1 km in diameter were assembled into one body, how large would it be? (HINT: The volume of a sphere is $\frac{4}{3}\pi r^3$.)
5. The asteroid Vesta has a mass of 2×10^{20} kg and a radius of about 250 km. What is its escape velocity? Could you jump off the asteroid? (HINT: See Box 5–2.)

6. The asteroid Pallas has a mass of 2.5×10^{20} kg. What is the orbital velocity of a small satellite orbiting 500 km from the center of Pallas? (HINT: See Box 5–1.)
7. What is the maximum angular diameter of Ceres as seen from the earth? Could earth-based telescopes detect surface features? Could the Hubble Space Telescope (Chapter 6)? (HINT: See Box 3–1.)
8. At what distances from the sun would you expect to find Kirkwood's gaps where the orbital period of asteroids is one-half of and one-third of the orbital period of Jupiter? Compare your results with Figure 24–10. (HINT: Use Kepler's third law.)
9. If the velocity of the solar wind is about 400 km/sec and the visible tail of a comet is 100 million km long, how long does it take an atom to travel from the nucleus to the end of the visible tail?
10. If you saw Comet Halley when it was 0.7 AU from the earth and it had a visible tail 5° long, how long was the tail in kilometers? Suppose that the tail was not perpendicular to the line of sight. Is your answer too large or too small? (HINT: See Box 3–1.)
11. Calculate the orbital velocity of a comet while it is in the Oort cloud. (HINT: See Box 5–1.)
12. The mass of a comet's nucleus is about 10^{14} kg. If the Oort cloud contains 200 billion cometary nuclei, what is the mass of the cloud in earth masses? (HINT: Mass of the earth = 6×10^{24} kg.)

OBSERVATIONAL ACTIVITY: OBSERVING METEORS

Meteors or shooting stars are very common, and you may see one by accident any night. However, with a little planning, you can make systematic observations of meteors that will reveal something about their origins.

Observations Select a cloudless night when the moon will not be out and try to find a location away from bright lights. You can see meteors from the center of a city, but you will see more out in the country away from city lights. Equip yourself with a reclining lawn chair, flashlight, clipboard, star charts, pencil, and refreshments.

Allow your eyes to adjust to the dark and avoid looking at bright lights. A moonless night is best. Tape a few layers of brown grocery-bag paper over your flashlight to make it dimmer and redder; it will not spoil your night vision while you study star charts and make notes. If you must observe from a city, choose a site where buildings or trees block your view of street lights.

Relax and watch the sky. You don't need a telescope or binoculars because you want to watch a large area of the sky continuously. When you see a meteor, note its position among the constellations and then sketch its path on your star charts.

In an hour or two you may begin to see a pattern appearing on your star chart. Most of the meteors may seem to be radiating from a specific region of the sky. These meteors are part of a meteor shower, and the region from which they seem to come is the radiant of the shower. Meteors that do not seem to radiate from the radiant are not members of the shower.

Planning Ahead On any night you can see from 5 to 15 meteors an hour from a dark site, but these will not necessarily be part of a meteor shower. If you want to see more meteors and you want to be sure to find them coming from a radiant point, use the table of meteor showers in Appendix C to select a night when a shower is in progress.

If you can arrange it, observe after midnight. Before midnight you are on the trailing side of the earth, and meteoroids must catch up with the earth. After midnight the rotation of the earth has carried you to the leading side, and you are overtaking meteoroids.

Observing meteors is one of the most enjoyable activities in astronomy. Don't delay observing just because there is not a shower tonight. There are always meteors to be seen.

RECOMMENDED READING

ADAMS, MARK "Observing Fallen Stars." *Mercury 9* (March/April 1980), p. 31.

BEATTY, J. KELLY "Crater Hunting in Brazil." *Sky and Telescope 59* (June 1980), p. 464.

———. "Comet G–Z: The Inside Story." *Sky and Telescope 70* (Nov. 1985), p. 426.

———. "The High Road to Halley." *Sky and Telescope 71* (March 1986), p. 244.

———. "An Inside Look at Halley's Comet." *Sky and Telescope 71* (May 1986), p. 438.

BERRY, RICHARD, and RICHARD TALCOT "What Have We Learned from Comet Halley?" *Astronomy 14* (Sept. 1986), p. 6.

———. "Search for the Primitive." *Astronomy 15* (June 1987), p. 6.

BORTLE, JOHN E. "Comets and How to Hunt Them." *Sky and Telescope 61* (Feb. 1981), p. 123.

———. "How to Observe Comets." *Sky and Telescope 61* (March 1981), p. 210.

———. "Comet Halley: A Newsmaker for 2000 Years." *Sky and Telescope 70* (Aug. 1985), p. 126.

BRACHER, K. "1910 Comets." *Mercury 14* (Jan./Feb. 1985), p. 18.

BRANDT, JOHN C., and MALCOLM B. NEIDNER, JR. "The Structure of Comet Tails." *Scientific American 254* (Jan. 1986), p. 49.

CASSIDY, WILLIAM A., and LOUIS A. RANCITELLI "Antarctic Meteorites." *American Scientist 70* (March/April 1982), p. 156.

CHAIKIN, ANDREW "Pieces of the Sky." *Sky and Telescope 63* (April 1982), p. 344.

———. "Target: Tunguska." *Sky and Telescope 67* (Jan. 1984), p. 18.

DAVIES, JOHN "Can Comets Become Asteroids?" *Astronomy 13* (Jan. 1985), p. 66.

———. "Is 3200 Phaethon a Dead Comet?" *Sky and Telescope 70* (Oct. 1985), p. 317.

DAVIS, JOEL "Asteroids: The Flying Mountains." *Astronomy 8* (May 1980), p. 66.

DUNHAM, EDWARD "Measuring the Diameter of Juno." *Sky and Telescope 59* (April 1980), p. 276.

FERRIN, IGNACIO R., and EDGAR GUZMAN "How a Cometary Nucleus Turns On." *Sky and Telescope 62* (Aug. 1981), p. 103.

GEHRELS, TOM "Asteroids and Comets." *Physics Today 38* (Feb. 1985), p. 11.

HARTMANN, WILLIAM K. "Vesta: A World of Its Own." *Astronomy 11* (Feb. 1983), p. 6.

KNACKE, ROGER "Comet Dust and the Comet Connection." *Sky and Telescope 68* (Sept. 1984), p. 206.

———. "Sampling the Stuff of a Comet." *Sky and Telescope 73* (March 1987), p. 246.

LEMAIRE, T. R. *Stones from the Stars.* Englewood Cliffs, N.J.: Prentice-Hall, 1980.

MARAN, STEPHEN P. "A Near Miss." *Natural History 90* (March 1981), p. 98.

———. "Where Do Comets Come From?" *Natural History 91* (May 1982), p. 80.

———. "Sungrazing Comets." *Natural History 92* (March 1983), p. 68.

———. "The Comet That Wouldn't Die." *Natural History 94* (Jan. 1985), p. 84.

———. "On the Trail of Comet G–Z." *Sky and Telescope 70* (Sept. 1985), p. 198.

———. "Off the Main Drag." *Natural History 95* (June 1986), p. 70.

———. "Gaps in the Asteroid Belt." *Natural History 95* (Aug. 1986), p. 62.

MARVIN, URSULA B. "Search for Antarctic Meteorites." *Sky and Telescope 62* (Nov. 1981), p. 423.

McSWEEN, HARRY Y., and EDWARD M. STOLPER "Basaltic Meteorites." *Scientific American 242* (June 1980), p. 54.

MEADOWS, JACK *Space Garbage.* London: George Philip, 1985.

MENDIS, D. ASOKA "The Science of Comets: A Post Encounter Assessment." *The Planetary Report 7* (March/April 1987), p. 5.

MICHELS, D. J., N. R. SHEELEY, JR., R. A. HOWARD, and M. J. KOOMEN "Observations of a Comet on a Collision Course with the Sun." *Science 215* (26 Feb. 1982), p. 1097.

OTTEWELL, GUY *Mankind's Comet.* Greenville, S.C.: Astronomical Workshop, 1985.

PELLEGRINO, CHARLES "The Fallen Sky." *Astronomy 9* (April 1981), p. 66.

RAYMO, CHET "Comet Halley: An Appreciation." *Sky and Telescope 72* (July 1986), p. 6.

REDDY, RANCIS "Backtracking the Comets." *Astronomy 10* (Aug. 1982), p. 6.

RICHIE, DAVID *Comets: The Swords of Heaven.* New York: Plume, 1985.

ROMER, ALFRED "Halley's Comet." *The Physics Teacher 22* (Nov. 1984), p. 488.

RUBIN, ALAN E. "Chondrites and the Early Solar System." *Astronomy 12* (Feb. 1984), p. 17.

RUSSELL, JOHN A. "Did the Perseids Peak in 1980?" *Sky and Telescope 63* (July 1982), p. 10.

SAGAN, CARL, and ANN DRUYAN *Comet.* New York: Random House, 1985.

TISKE, RICHARD "Asteroid Collisions with Earth." *Astronomy 10* (Jan. 1982), p. 18.

WAGNER, JEFFREY K. "The Sources of Meteorites." *Astronomy 12* (Feb. 1984), p. 6.

WASSON, JOHN T. *Meteorites: Their Record of Early Solar-System History.* New York: Freeman, 1985.

WEISSMAN, P. R. "Realm of the Comets." *Sky and Telescope 73* (March 1987), p. 238.

WETHERILL, GEORGE W. "Apollo Objects." *Scientific American 240* (March 1979), p. 54.

WHIPPLE, FRED L. "Flying Sandbanks Versus Dirty Snowballs: Discovery of the Nature of Comets." *Mercury 15* (Jan./Feb. 1986), p. 2.

———. "The Black Heart of Comet Halley." *Sky and Telescope 73* (March 1987), p. 242.

See also *Natural History 90* (April 1981).

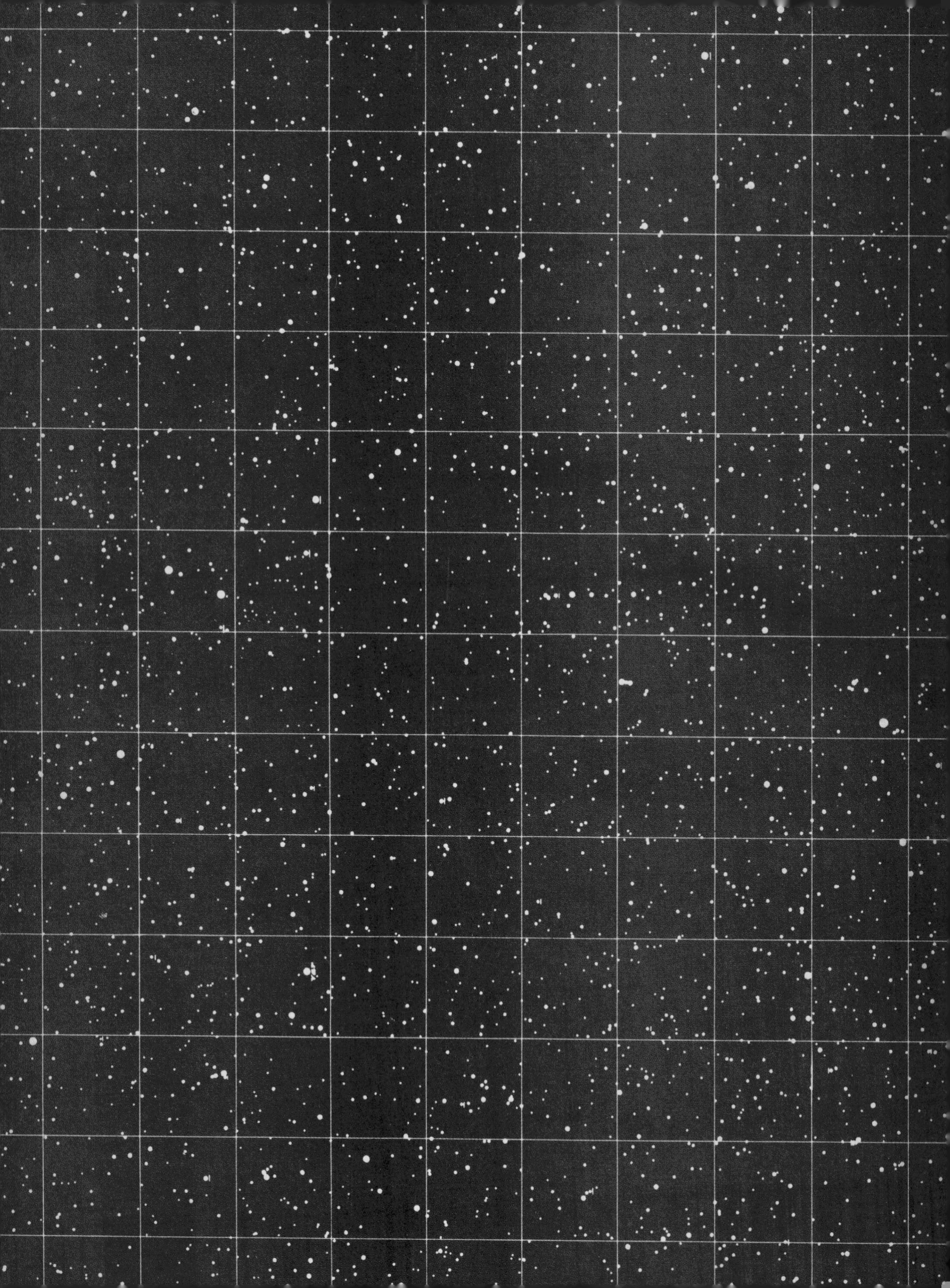

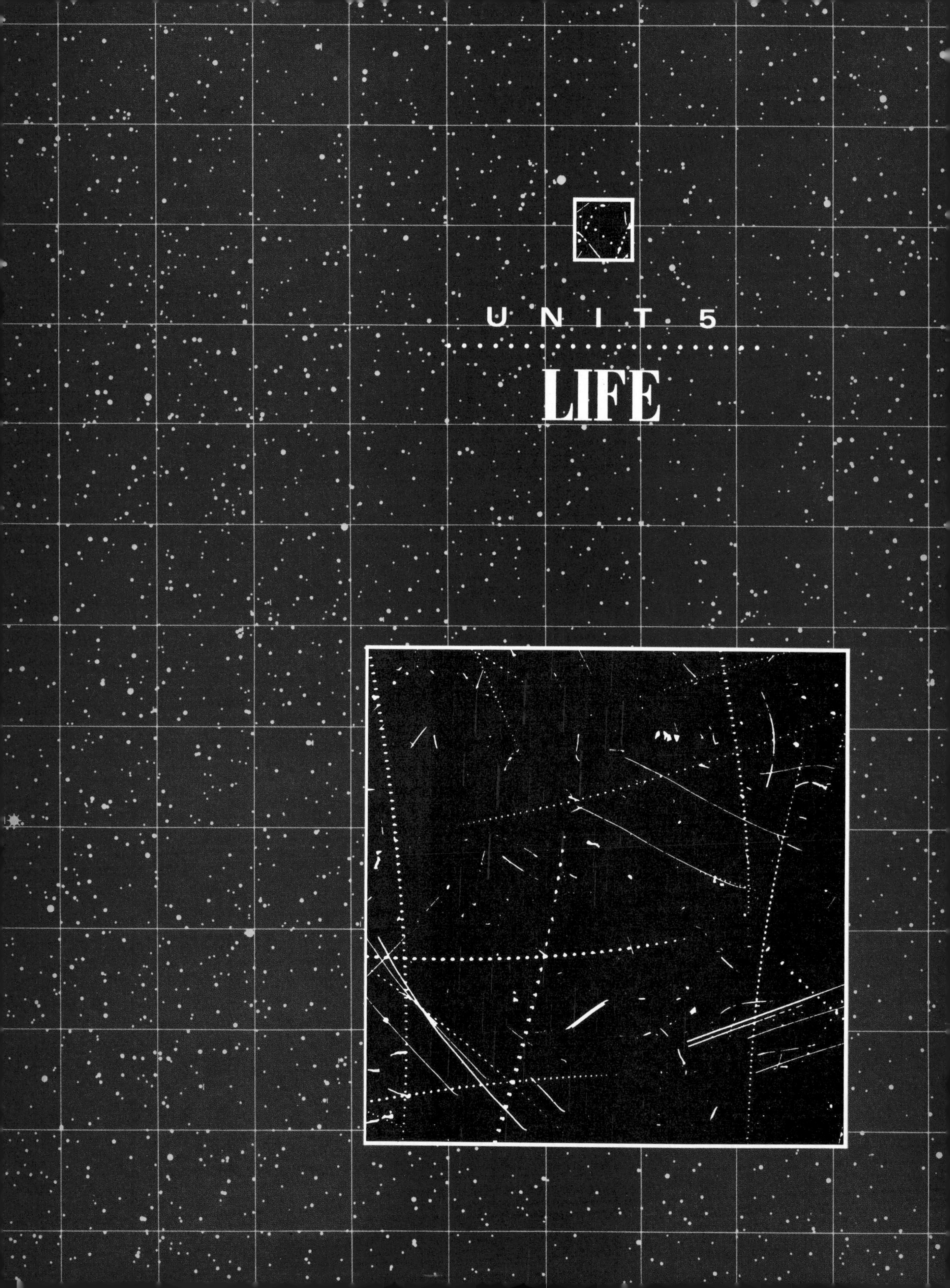

UNIT 5

LIFE

CHAPTER 25

LIFE ON OTHER WORLDS

Did I solicit thee from darkness to promote me?

John Milton
PARADISE LOST

As living things, we have been promoted from darkness. We are made of heavy atoms that could not have formed at the beginning of the universe. Successive generations of stars fusing light elements into heavier elements have built the atoms so important to our existence. When a dark cloud of interstellar gas enriched in these heavy atoms fell together to form our sun, a small part of the cloud gave birth to the planet we inhabit.

Are there intelligent beings living on other planets? That is the last and perhaps the most challenging question in our study of astronomy. We will try to answer it in three steps, each dealing with a different aspect of life.

First, we must decide what we mean by life. A living thing is not so much a physical object as a process. We are not simply the matter that forms our bodies, but rather a tremendously complex system that has the ability to duplicate and protect itself. Thus, life is based on information that contains the directions for the processes of duplication and preservation.

Our second step is to study the origin of life. Direct investigation is limited to Earth, but if we can understand how life began here, we may better estimate the chances that it occurred elsewhere. We will find that life on Earth probably began with simple chemical reactions that happened naturally. If these gave rise to life on Earth, similar reactions may have provided the spark on other worlds.

Our third step is to study evolution, the process by which life improves its ability to survive. The survival of stable species has transformed the simple organisms that began in Earth's oceans into a wide variety of creatures with special adaptations. The rose's thorn, the rabbit's long ears, and the human's intelligence are protective adaptations. Evolution is so natural a process that it surely works on any planet where life begins, and if we

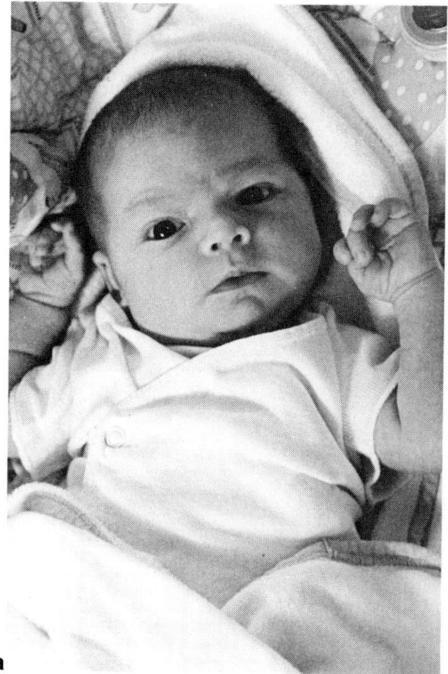

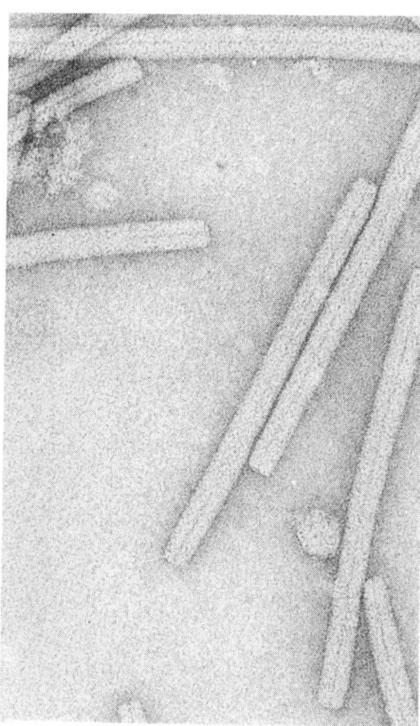

FIGURE 25–1 All living things on earth are based on carbon chemistry. (a) Katie, a complex mammal containing about 30 astronomical units of DNA. (b) Tobacco mosaic virus. Each rod is a single spiral strand of RNA about 0.01 mm long surrounded by a protein coat. (a, Michael Seeds; b, L. D. Simon.)

assume that intelligence is a valuable trait, then intelligent beings may eventually emerge.

If life is common in the universe, where might we look for it, and how might we communicate with other intelligent beings? Certainly the prospects of finding life, intelligent or otherwise, on any of the other planets in our solar system are bleak. If we are to find extraterrestrial life, we must go beyond our solar system and search among any planets that may orbit other stars.

Communication with intelligent races on other worlds may be possible, but we cannot expect to travel between solar systems. Interstellar distances are so great that only in science fiction do spaceships flit from star to star. However, it may be possible to communicate via radio. If civilizations can survive for long periods of time at a technological level at which they can build large radio telescopes, then we may be able to send and receive messages. Such messages would mark a turning point in the history of humanity. If life is common in the universe, such signals may be detected during our lifetime.

25.1 THE NATURE OF LIFE

What is life? Philosophers have struggled with that question for thousands of years, so it is unlikely that we will answer it here. But we must agree on a working model of life before we can speculate on its occurrence on other worlds. To that end we will identify in living things three properties—a process, a physical base, and a unit of controlling information.

The life process is aimed at survival. Living things extract energy from their environment and use that energy to modify their surroundings to make their own preservation more likely. For example, human beings obtain energy by eating and breathing, and they use that energy to build houses, cities, and stable societies to protect themselves. The same can be said of a bacterium absorbing food and reinforcing its cell structure.

This apparently selfish protection of the individual is aimed at the preservation of the race through safe reproduction. The ability to reproduce is one of the distinguishing characteristics of living organisms, and any organism that does not, in some way, ensure safe reproduction will not survive many generations. The entire life process is aimed at safe reproduction because any other target is self-destructive.

The physical basis of life on Earth is the chemistry of the carbon atom (Figure 25–1). Because of the way this atom bonds to other atoms, it can form long, complex, stable chains that are capable of extracting, storing, and utilizing energy. Other chemical bases may exist. Science fiction stories and movies abound with silicon creatures, living things whose body chemistry is based on silicon rather than carbon. However, silicon forms weaker bonds than carbon does, and it cannot form double bonds as easily. Consequently, it cannot form the long, complex, stable chains that carbon can. Silicon is 135 times more common on Earth than carbon, yet there are no silicon people among us. All earth life is carbon-based. Thus, the likelihood that distant planets are inhabited by silicon people seems small, but we cannot rule out life based on noncarbon chemistry.

In fact, nonchemical life might be possible. All that nature requires is some mechanism capable of supporting the extraction and utilization of energy that we have identified as life. One could at least imagine life based on electromagnetic fields and ionized gas. No one has ever met such a creature, but science fiction writers conjure up all sorts.

Clearly we could range far in space and time theorizing about alien life, but to make progress we must discuss what we know best, carbon-based life on Earth. We must try to understand how it works and how it came to exist. Only then can we consider life on other worlds.

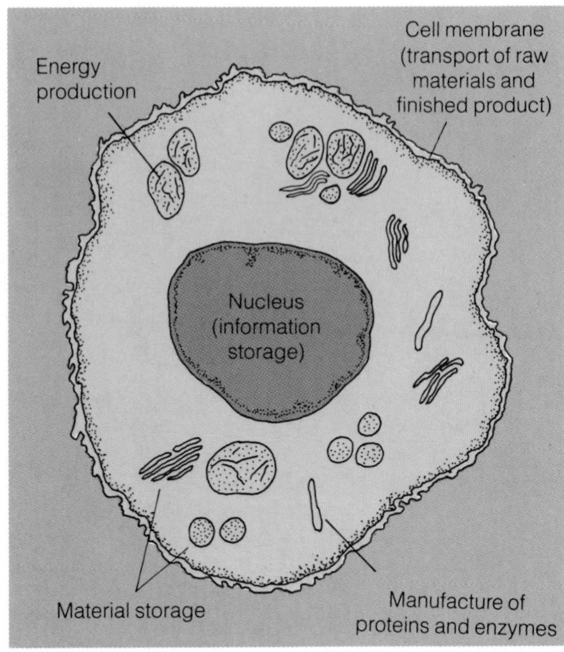

FIGURE 25–2 A living cell is a self-contained factory that absorbs raw materials from its surroundings and uses them to maintain itself and manufacture finished products for the use of the organism as a whole.

THE DNA CODE

The key to understanding life is information—the information the organism uses to control its utilization of energy. We must discover how life stores and uses that information and how the information changes and thus preserves the species.

The unit of life on Earth is the cell (Figure 25–2), the self-contained factory capable of absorbing nourishment from its surroundings, maintaining its own existence, and performing its task within the larger organism. The foundation of the cell's activity is a set of patterns that describe how it is to function. This information must be stored in the cell in some safe location, yet it must be passed on easily to new cells and be used readily to guide the cell's activity. To understand how matter can be alive, we must understand how the cell stores, reproduces, and uses this information.

The information is stored in long carbon-chain molecules called **DNA (deoxyribonucleic acid)**, most of which reside in the cell nucleus. The structure of DNA resembles a long twisted ladder. The rails of the ladder are made of alternating phosphates and sugars; the rungs are made of pairs of molecules called bases (Figure 25–3). Only four kinds of bases are present in DNA, and the order in which they appear on the DNA ladder represents the information the cell needs to function. One human cell stores about 1.5 m of DNA, containing about 4.5 billion pairs of bases. Thus, 4.5 billion pieces of information are available to run a human cell. That is enough to record all of the works of Shakespeare over 200 times. Because the human body contains about 60×10^{12} cells, the total DNA in a single human would stretch 9×10^{13} m, about 600 AU.

Storing all this data in each cell does the organism no good unless the data can be reproduced and passed on to new cells. The DNA molecule is especially adapted for duplicating itself by splitting its ladder down the center of the rungs, producing two rails with protruding bases (Figure 25–4). These quickly bond with the proper bases, phosphates, and sugars to reconstruct the missing part of the molecule, and, presto, the cell has two complete copies of the critical information. One set goes to each of the newly forming cells. Thus, DNA is the genetic information passed from parent to offspring.

Segments of DNA molecules are patterns for the production of **proteins.** Many proteins are structural molecules—the cell might make protein to repair its cell wall, for example. **Enzymes** are special proteins that control

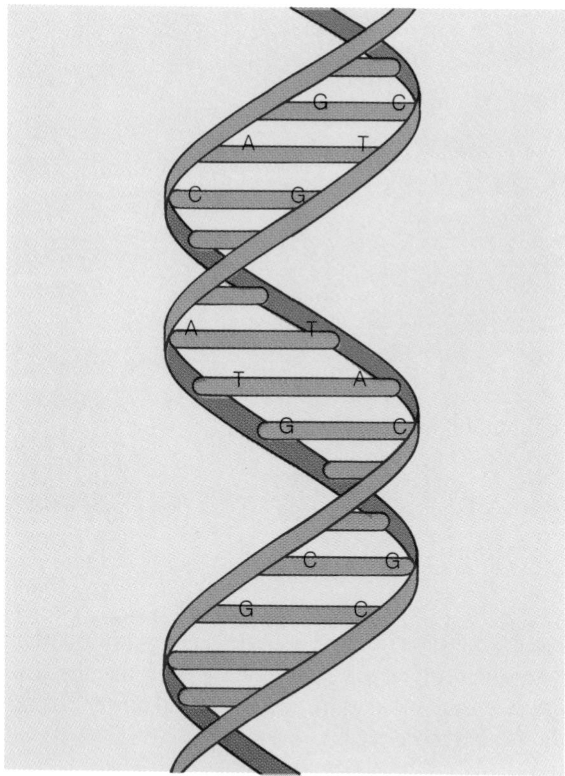

FIGURE 25–3 The DNA molecule consists of two rails of sugars and phosphates (dark) and rungs made of bases [adenine (A), cytosine (C), guanine (G), thymine (T)].

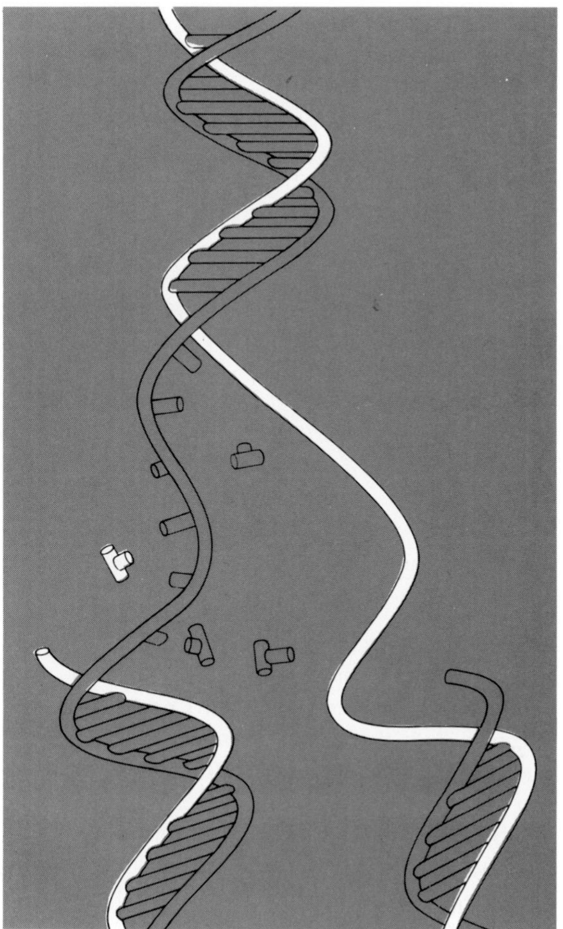

FIGURE 25–4 The DNA molecule can duplicate itself by splitting in half (top), assembling matching bases, sugars, and phosphates (center), thus producing two DNA molecules (bottom). The actual duplication process is significantly more complex than in this schematic diagram.

other processes, growth for example. Thus, the DNA molecule contains the recipes to make all of the different molecules required in an organism.

Actually, the cell does not risk its precious DNA patterns by involving them directly in the manufacture of protein. DNA stays safely in the cell nucleus where it produces a copy of the patterns by assembling a long carbon-chain molecule called **RNA (ribonucleic acid)**. RNA carries the information out of the nucleus and then assembles the proteins from simple molecules called **amino acids**, the basic building blocks of protein. Thus, RNA acts as a messenger, carrying copies of the necessary plans from the central office to the construction site.

Although the information coded on DNA must be preserved for the survival of the organism, it must be changeable or the species will become extinct. To see why, we must study evolution, the process that rewrites the data in DNA.

THE EVOLUTION OF LIFE-FORMS

Every living thing on Earth is part of a web of interdependence. Not only now, but ever since life began, life-forms have depended on each other for food and shelter. This exposes life to a serious danger in that gradual changes in climate may destroy one life-form and endanger hundreds of others. A slight warming of the climate, for example, might kill a species of plant, starving the rabbits, deer, and other herbivores and leaving the hawks, wolves, and mountain lions with no prey. If a species is to survive in such a world, it must be able to adapt to changing

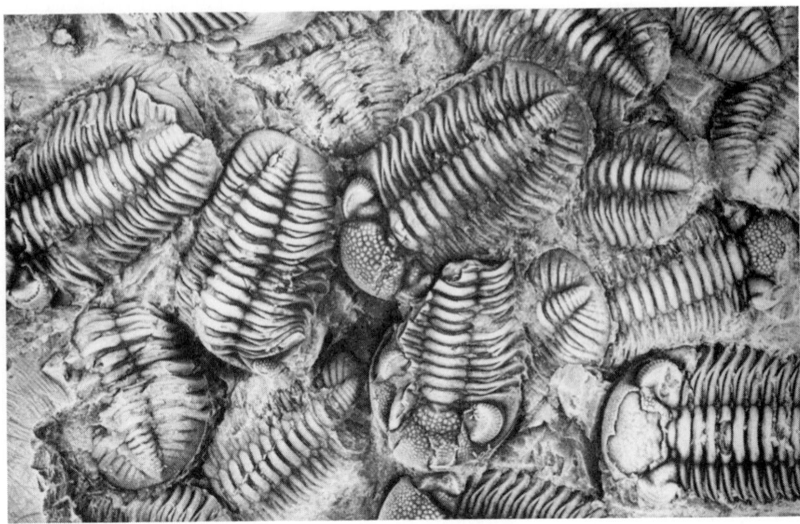

FIGURE 25–5 Trilobites made their first appearance in the Cambrian oceans about 600 million years ago. These fossils, about 400 million years old, came from Ontario, Canada. (Smithsonian Institution Photo No. 76-17821.)

conditions, and that means the data coded in DNA must change. The species must evolve.

Species evolve because of a process called **natural selection**. Each time an organism reproduces, its offspring receive the data stored in DNA, but some variation is possible. For example, most of the rabbits in a litter may be normal, but it is possible for one to get a DNA recipe that gives it stronger teeth. If it has stronger teeth, it may be able to eat something other than the plant the others depend on, and if that plant is becoming scarce, the rabbit with stronger teeth has a survival advantage. It can eat other plants and will be healthier than its litter mates and have more offspring. Some of these offspring may also have stronger teeth as the altered DNA data are handed down to the new generation. Thus, nature selects and preserves those attributes that contribute to the survival of the species. Those that are unfit die. Natural selection is merciless to the individual, but it gives the species the best possible chance to survive.

The only way nature can obtain new DNA patterns from which to select the best is to alter actual DNA molecules. This can happen through chance mismatching of base pairs—errors—in the reproduction of the DNA molecule. Another way this could occur is through damage to reproductive cells from exposure to radioactivity. Cosmic rays or natural radioactivity in the soil might play this role. In any case, an offspring born with altered DNA is called a **mutant**. Most mutations are fatal, and the individual dies long before it can have offspring of its own. But rarely a mutation may give a species a new survival advantage. Then natural selection makes it likely that the new DNA message will survive and be handed down, making the species more capable of surviving.

25.2 THE ORIGIN OF LIFE

Clearly the carbon chemistry of life on Earth is extremely complex. How could it have ever gotten started? Obviously, 4.5 billion chemical bases didn't just happen to drift together to form the DNA formula for a human. The key is evolution. Once a life-form begins to reproduce itself, natural selection preserves the most advantageous traits. Over long periods of time spanning thousands, perhaps millions, of generations, the life-form becomes more fit to survive. This nearly always means the life-form becomes more complex. Thus, life could have begun as a very simple process that gradually became more sophisticated as it was modified by evolution.

We begin our search on Earth, where fossils and an intimate familiarity with carbon-based life gives us a glimpse of the first living matter. Once we discover how earthly life could have begun, we can look for signs that life began on other planets in our solar system. Finally, we can speculate on the chances that other planets, orbiting other stars, have conditions that give rise to life.

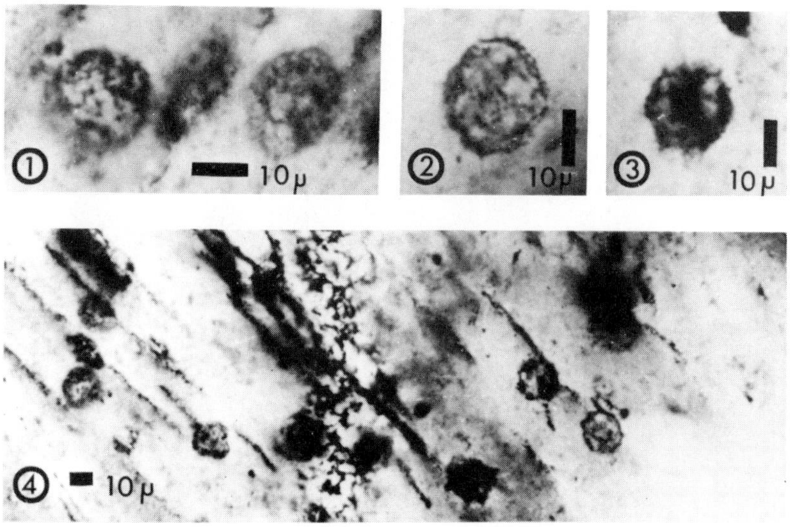

FIGURE 25–6 Among the oldest fossils known, these microscopic spheres lie in the Precambrian fig-tree chert near Barberton, South Africa. They are at least 3.1 billion years old. The black bars are 10 μm long. (E. S. Barghoorn.)

THE ORIGIN OF LIFE ON EARTH

The oldest fossils hint that life began in the oceans. The oldest easily identified fossils appear in sedimentary rocks that formed between 0.6–0.5 billion years ago—the **Cambrian period.** Such Cambrian fossils were simple ocean creatures, the most complex of which were trilobites (Figure 25–5), but there are no Cambrian fossils of land plants or animals. Evidently, land surfaces were totally devoid of life until only 400 million years ago.

Precambrian deposits contain no obvious fossils, but microscopes reveal microfossils that were the ancestors of the Cambrian creatures. Fig-tree chert* in South Africa is 3.0–3.3 billion years old, and the Onverwacht shale, also found in South Africa, may be as old as 3.6 billion years. Both contain structures that appear to be microfossils of bacteria or simple algae such as those that live in water (Figure 25–6). Apparently, life was already active in Earth's oceans a billion years after Earth formed.

The key to the origin of this life may lie in an experiment performed by Stanley Miller and Harold Urey in 1952. This **Miller experiment** sought to reproduce the conditions on Earth under which life began (Figure 25–7). In a closed glass container, the experimenters placed water (to represent the oceans), the gases hydrogen, ammonia, and methane (to represent the primitive atmosphere), and an electric arc (to represent lightning bolts). The apparatus was sterilized, sealed, and set in operation.

After a week, Miller and Urey stopped the experiment and analyzed the material in the flask. Among the many compounds the experiment produced, they found four amino acids that are common building blocks in protein, various fatty acids, and urea, a molecule common to many life processes. Evidently, the energy from the electric arc had molded the atmospheric gases into some of the basic components of living matter. Other energy sources such as hot silica (to simulate hot lava spilling into the sea) and ultraviolet radiation (to simulate sunlight) give similar results.

Recent studies of the composition of meteorites and models of planet formation suggest that Earth's first atmosphere did not resemble the gases used in the Miller experiment. Earth's first atmosphere was probably composed of carbon dioxide, nitrogen, and water vapor. This change, however, does not invalidate the Miller experiment. When such gases are processed in a Miller apparatus, a tarry gunk rich in organic molecules coats the inside of the chamber.

The Miller experiment did not create life, nor did it necessarily imitate the exact conditions on the young Earth. Rather, it is important because it shows that complex organic molecules form naturally in a wide variety of circumstances. The chemical deck is stacked to deal nature a hand of complex molecules. If we could travel back in time, we would probably find Earth's oceans filled with a

*Chert is a rock form, which resembles flint.

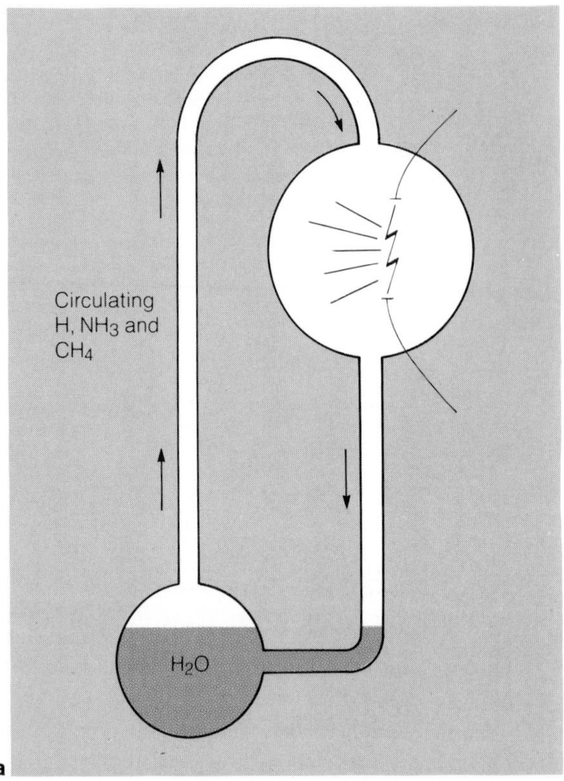

FIGURE 25–7 The Miller experiment (a) circulated gases through water in the presence of an electric arc. This simulation of primitive conditions on Earth produced amino acids, the building blocks of protein. (b) Stanley Miller with a Miller apparatus (b, Courtesy of Stanley Miller.)

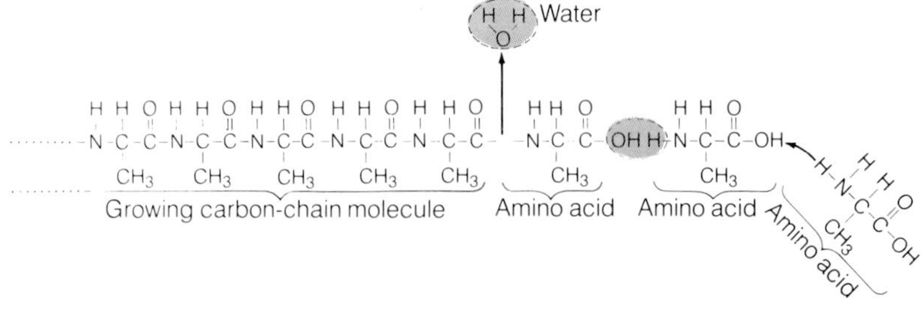

FIGURE 25–8 Amino acids can link together through the release of a water molecule to form long carbon-chain molecules. The amino acid in this hypothetical example is alanine, one of the simplest.

rich mixture of organic compounds in what some have called the **primordial soup**.

The next step on the journey toward life is for the compounds dissolved in the oceans to link up and form larger molecules. Amino acids, for example, can link together to form proteins. This linkage occurs when amino acids join together end to end and release a water molecule (Figure 25–8). Among the ways this could have happened, retreating ocean tides might have left behind pools of water to evaporate in the sun. As the water evaporated, it would leave a concentrated broth of amino acids, facilitating linkups between molecules. Sunlight would drive off water molecules from between the amino acids, and long protein chains would grow until the returning tide swept them all back into the sea.

Although these proteins might have contained hundreds of amino acids, they were not alive. Not yet. Such molecules did not reproduce, but merely linked

FIGURE 25–9 A sample of the Murchison meteorite, a carbonaceous chondrite, which fell in 1969 near Murchison, Australia. Analysis of the meteorite's interior revealed evidence of amino acids. That the first building blocks of life originated in space is unknown, but amino acids found in meteorites illustrate how commonly these and other complex molecules occur through nonorganic means. (Courtesy of Chip Clark, National Museum of Natural History.)

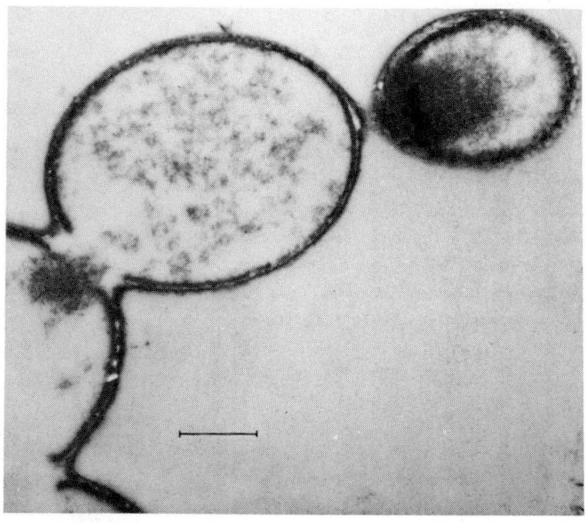

FIGURE 25–10 Proteinlike material added to water forms microspheres. Although these bodies do not contain DNA or related genetic information, they have many of the properties of cells such as a double-layered boundary similar to a cell membrane. Thus, the first cell structures may have originated through the self-ordering properties of the first complex molecules. The horizontal bar is 1 μm long. (Courtesy of S. W. Fox from S. W. Fox and K. Dose, *Molecular Evolution and the Origin of Life*, rev. ed. New York: Marcel Dekker, 1977.)

together and broke apart at random. However, because some molecules are more stable than others and because some molecules bond together more readily than others, this blind **chemical evolution** led to the concentration of the varied smaller molecules into the most stable larger forms. Eventually, somewhere in the oceans, a molecule took shape that could reproduce a copy of itself. At that point the chemical evolution of molecules became the biological evolution of living things.

An alternative theory supposes that primitive living things such as reproducing molecules did not originate on Earth but came here in meteorites or comets. Radio astronomers have found a wide variety of organic molecules in the interstellar medium, and some studies have found similar compounds inside meteorites (Figure 25–9). Such molecules form so readily that we would be surprised if they were not present in space. However, a few investigators have speculated that living, reproducing molecules originated in space and came to Earth as a cosmic contamination. If this is true, every planet in the universe is contaminated with the seeds of life. However entertaining this theory may be, it is presently untestable, and an untestable theory is of little use in science. Experts studying the origin of life proceed on the assumption that life began as reproducing molecules in Earth's oceans.

Which came first, reproducing molecules or the cell? Because we think of the cell as the basic unit of life, this question seems to make no sense, but in fact the cell may have originated during chemical evolution. If a dry mixture of amino acids is heated, the acids form long, proteinlike molecules that, when poured into water, collect to form microscopic spheres that function in ways similar to cells (Figure 25–10). They have a thin membrane surface, absorb material from their surroundings, grow in size, and divide and bud just as cells do. However, they contain no large molecule that copies itself. Thus, the structure of the cell may have originated first and the reproducing molecules later.

An alternative theory supposes that the replicating molecule developed first. Such a molecule would be exposed to damage if it were bare, so the first to manufacture or attract a protective coating of protein would

FIGURE 25–11 (a) A 3.5 billion-year-old fossil stromatolite from western Australia is one of the oldest known fossils. Stromatolites were formed, layer by layer, by mats of blue-green algae or bacteria living in shallow water. (b) Such algae may have been common in shallow seas when Earth was young. Such stromatolites are still being formed today in similar environments. (Mural by Peter Sawyer; photos courtesy of Chip Clark, National Museum of Natural History.)

have a significant survival advantage. If this is the case, the protective cell membrane is a later development of biological evolution.

The first living things must have been single-celled organisms much like modern bacteria and simple algae. Some of the oldest fossils known are **stromatolites**, structures produced by communities of blue-green algae or bacteria, which grew in mats and, year by year, deposited layers of minerals, which were later fossilized. One of the oldest such fossils known is 3.5 billion years old (Figure 25–11). If such algae were common when Earth was young, it may have been able to produce a small amount of oxygen in the early atmosphere. Recent studies suggest that only 0.1 percent would have been sufficient to provide an ozone screen, which would protect organisms from the sun's ultraviolet radiation.

How evolution shaped creatures to live in the ancient oceans, molded them into multicellular organisms, and developed sexual reproduction, photosynthesis, and respiration is a fascinating story (Box 25–1), but we cannot explore it in detail here. We can see that life could begin through simple chemical reactions building complex molecules, and that once some DNA-like molecule formed, it protected its own survival with selfish determination.

BOX 25–1

Geologic Time

Humanity is a very new experiment on planet Earth. All of the evolution that leads from the primitive lifeforms in the oceans of the Cambrian period 0.6 billion years ago, to fishes, amphibians, reptiles, and mammals can fit on a single chart like Figure 25–12. But the 3-million-year history of humanity is so short that it appears as a single, thin line at the top of the diagram. If we tried to represent the entire geologic history of Earth in such a chart (Figure 25–13), the portion describing the rise of life on the land would be an unreadably small segment.

One way to represent the evolution of life is to compress the 4.6-billion-year history of Earth into a 1-year-long film. In such a film, Earth forms as the film

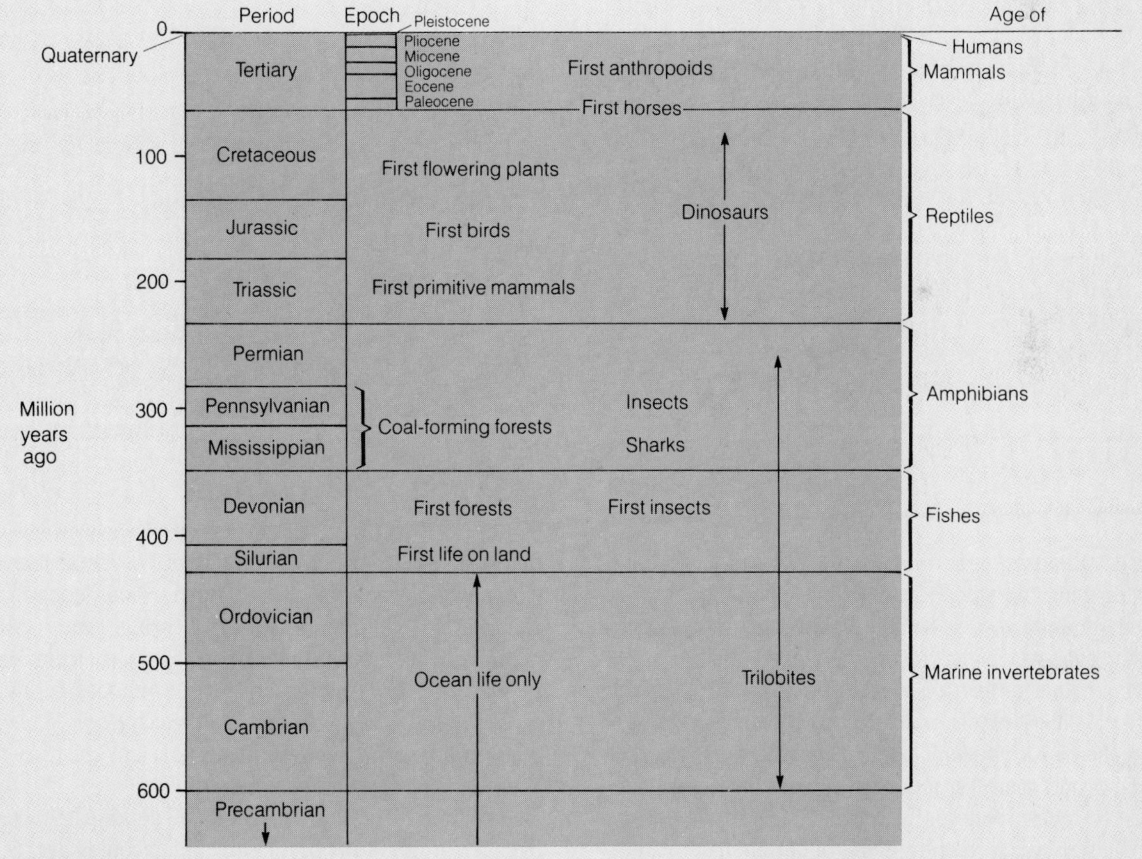

FIGURE 25–12 When the ages of life on Earth are plotted on a single chart, the age of humanity becomes no more than a thin line at the top. (Compare with Figure 25–13.)

Geologic Time (continued)
begins on January 1, and through all of January and February it cools and is cratered and the first oceans form. But those oceans remain lifeless until sometime in March or early April, when the first living things develop. The 4-billion-year history of Precambrian evolution lasts until the film reaches mid-November, when primitive ocean life begins to evolve into complex organisms such as trilobites.

If we examine the land instead of the oceans, we find a lifeless waste. But once our film shows plant and animal life on the land, about November 28, evolution proceeds rapidly. Dinosaurs, for example, appear about December 12 and vanish by Christmas evening, as mammals and birds flourish.

Throughout the 1-year-run of our film there are no humans, and even during the last days of the year as the mammals rise and dominate the landscape, there are no people. In the early evening of December 31, vaguely human forms move through the grasslands, and by late evening they begin making stone tools. The Stone Ages lasts till about 11:45 PM, and the first signs of civilization, towns and cities, do not appear until 11:54 PM. The Christian era begins only 14 seconds before the New Year, and the Declaration of Independence is signed with but 1 second to spare.

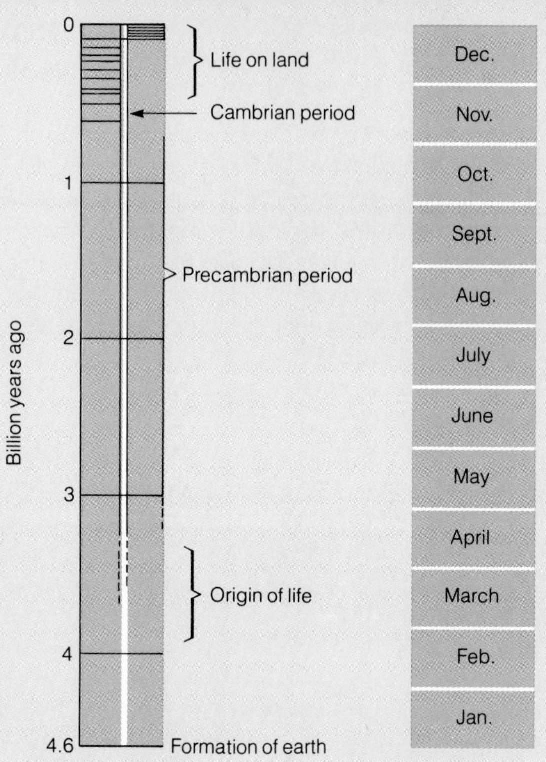

FIGURE 25–13 If the entire history of Earth were compressed into 1 year, life would originate in March, but would not emerge from the sea until late November.

Over billions of years the genetic information stored in living things kept those qualities that favored survival and discarded the rest. As Samuel Butler said, "The chicken is the egg's way of making another egg." In that sense, all living matter on Earth is merely the physical expression of DNA's mindless determination to continue its existence.

Perhaps this seems harsh. Human experience goes far beyond mere reproduction. *Homo sapiens* has art, poetry, music, philosophy, religion, science—all of the great, sensitive accomplishments of our intelligence. Perhaps that is more than mere reproduction of DNA, but intelligence, the ability to analyze complex situations and respond with appropriate action, must have begun as a survival mechanism. For example, a fixed escape strategy stored in DNA is a disadvantage for a creature that frequently moves from one environment to another. A rodent that always escapes from predators by automatically climbing the nearest tree would be in serious jeopardy if it met a hungry fox in a treeless clearing. Even a faint glimmer of intelligence might allow the rodent to analyze the situation and, finding no trees, to choose running over climbing. Thus intelligence, of which *Homo sapiens* is so proud, may have developed in ancient creatures as a way of making them more versatile.

If life could originate on Earth and develop into intelligent creatures, perhaps the same thing could have happened on other planets. This raises three questions. First, could life originate if conditions were suitable? Second, if life begins on a planet, will it evolve toward intelligence? The answer to both questions seems to be yes. The direction of chemical and biological evolution is directed toward survival, which should lead to versatility

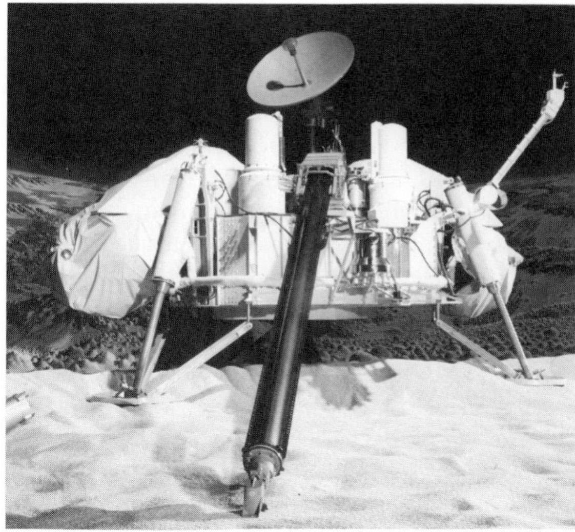

FIGURE 25–14 Viking 1 and Viking 2 landers reached the Martian surface in 1976. Among other experiments, they searched for signs of life in soil samples collected by a remote-controlled arm. (NASA.)

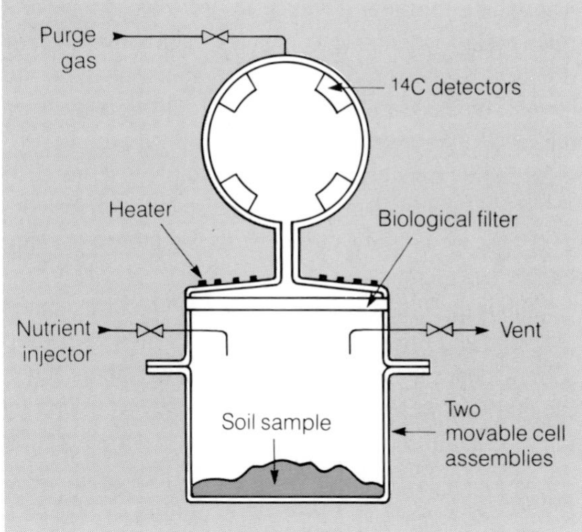

FIGURE 25–15 The labeled-release experiment aboard the Viking landers exposed soil samples to a nutrient containing radioactive carbon-14. Radioactivity in the gas given off by the sample indicates the level of biological activity. (Martin Marietta Aerospace.)

and intelligence. But what of the third question: Are suitable conditions so rare that life almost never gets started? The only way to answer that is to search for life on other planets. We begin, in the next section, with the other planets in our solar system.

LIFE IN OUR SOLAR SYSTEM

Although life based on other than carbon chemistry may be possible, we must limit our discussion to life as we know it and the conditions it requires. The most important condition is the presence of liquid water, not only as part of chemical reactions, but also as a medium to transport nutrients and wastes within the organism. This means the temperature must be moderate. Thus, our search for life in the solar system must look for a planet where liquid water could exist.

The water requirement automatically eliminates a number of worlds. The moon and Mercury are airless, and thus liquid water cannot exist on their surfaces. Venus has some water vapor but it is much too hot for liquid water, and in the outer regions of the solar system the temperature is much too low.

Mars gives us good reason to hope for life. It is a temperate planet with summer temperatures at high noon that are not unlike a pleasant autumn day (17°C, or 62°F). But the thin atmosphere provides little insulation so that temperatures at night drop to about − 88°C (− 126°F). This is unpleasantly cold, but life-forms on Earth have evolved to survive under harsh circumstances, so life-forms on Mars might have evolved ways of coping with the midnight cold. Even if there is no life there now, Mars may once have had a thicker atmosphere, liquid water, and a more moderate temperature range. Life might have developed under such circumstances, but withered as the atmosphere leaked away. Such life may have left seeds or spores that would germinate with the return of acceptable conditions.

Searching for life on Mars is difficult as long as humans cannot visit the planet. The solution is robots that land on the surface and perform automatic experiments under remote guidance from Earth. The Viking 1 and Viking 2 spacecraft (Figure 25–14) landed safely on the Martian surface in the summer of 1976, each carrying three experiments designed to search for life in the Martian soil by exposing soil samples to water, light, and nutrients.

The labeled-release experiment (Figure 25–15) placed a thimbleful of soil in a container and dampened it with a rich broth of nutrients that any earth organism would have loved. Some of the carbon atoms in this broth were radioactive carbon-14, and the assumption was that any living things in the soil would absorb the nutrients and release carbon dioxide containing radioactive carbon. Counters measuring the amount of radioactivity in the gases escaping from the container could thus measure the level of biological activity. As a control, the experiment could be repeated with a soil sample first heated to 160°C (320°F) to kill any life. The radioactivity in the control phase would show the scientists what to expect if no life were present.

Both the Viking 1 and Viking 2 landers functioned properly, and their remote-controlled arms scooped up soil and deposited it in the experimental chambers. When the nutrients were added, the radioactivity went up rapidly, and when the sample was heated, the radioactivity declined. This sounds encouraging but the radioactivity release came much too soon. Living things should have taken some time to absorb and process the molecules in the nutrient solution. The experiment was repeated a number of times with new samples, but the results were always the same. Most experts believe that the released gases came from simple chemical reactions in the soil.

Two other automatic experiments on the landers searched for signs of photosynthesis and respiration (the absorption of carbon dioxide gas and the release of oxygen). All three experiments were repeated a number of times, but in every case the results were negative or borderline. There was no indication of anything more than simple chemical reactions.

Although the Viking lander experiments were highly ingenious, they could detect no life on Mars. There are, however, two hopes. Life on Mars may be so unlike us that it did not respond to the experimental conditions. It may have found the nutrient broth unacceptable, perhaps even poisonous. Alternatively, Mars may have had life in the distant past, but that life became extinct when the atmosphere leaked away and the water vanished. An astronaut visiting Mars may someday find fossils in the dry streambeds, proving that life can begin on other planets.

The disappointing news from Mars leaves us with only a few remaining candidates in our solar system. We might hope for life on Jupiter because its present atmosphere is composed of gases quite similar to the primeval atmosphere of Earth. Energy sources such as solar ultraviolet radiation and tremendous lightning bolts could make complicated molecules in the atmosphere. In some of the deeper layers, water exists as vapor and as droplets, and the temperature 50 km (30 miles) below the cloud tops is a pleasant 27°C (80°F). Under these circumstances, life may have begun and evolved into forms like the plankton that lives in Earth's oceans. Some have even speculated about Jovian equivalents to fish and sharks—floating creatures that feed on the plankton and on each other. This is obviously speculation, and we will have no evidence until the Galileo Probe descends into the Jovian clouds.

A few of the satellites of the Jovian planets might have conditions that could support life. Saturn's moon Titan has an atmosphere rich in methane, which can become complex organic molecules under the stimulation of sunlight. In Chapter 22, we considered the possibility that organic smog settles to Titan's surface to form a layer of tarry goo. Certainly any life that might originate in that goo would be different from life as we know it. The surface of Titan has a temperature of − 178°C (− 288°F) and may be covered by liquid methane.

Neptune's icy moon Triton also has an atmosphere containing methane, but Voyager 2 flew past in August 1989 and revealed that it is as cold as 37 K (−393°F). This seems too cold for life based on chemical reactions.

Slightly nearer to Earth, Jupiter's moon Europa has been mentioned as a possible abode of life. Its icy crust may conceal a liquid water mantle, and if that water has never been frozen, living things may have developed beneath the ice. Again, the only way to be sure is to drill through the ice to reach any water that may lie there. The chance of life on Europa is probably slim, but it will be many years before that chance can be tested conclusively.

So far as we know now, the solar system is bare of life except for our planet. Consequently, our search for life in the universe takes us to other planetary systems.

LIFE IN OTHER PLANETARY SYSTEMS

Might life exist in other solar systems? To consider this question let us try to decide how common planets are and what conditions a planet must fulfill for life to originate and evolve to intelligence. The first question is astronomical; the second is biological. Our ability to discuss the problem of life outside our solar system is severely limited by our lack of experience.

In Chapter 19 we concluded that planets form as a natural result of star formation. In addition, the process that gives rise to planets is probably related to the process that forms binary star systems. Had Jupiter been 100 times more massive, it would have been a star instead of a planet. Because about half of all stars are members of binary or multiple star systems, it seems that this process is very common, implying that planetary systems are also common.

If a planet is to become a suitable home for life, it must have a stable orbit around its sun. This is simple in a solar system like our own, but in a binary system most planetary orbits are unstable. Most planets in such systems would not last long before they were swallowed up by one of the stars or ejected from the system.

Thus, it seems that single stars are the most likely to have planets suitable for life. Because our galaxy contains about 10^{11} stars, half of which are single, there should be roughly 5×10^{10} planetary systems in which we might look for life.

A few million years of suitable conditions does not seem to be enough time to originate life. Our planet required at least 0.5–1 billion years to create the first cells and 4.6 billion to create intelligence. Clearly, conditions on a planet must remain acceptable over a long time. This eliminates giant stars that change their luminosity rapidly as they evolve. It also eliminates massive stars that remain stable on the main sequence for only a few million years. If life requires a few billion years to originate and evolve to intelligence, no star hotter than about F5 will do. This is not really a serious restriction because upper main-sequence stars are rare anyway.

In previous sections we decided that life, as we know it at least, requires liquid water. That requirement defines a **life zone** (or ecosphere) around each star, a region within which a planet has temperatures that permit the existence of liquid water.

The size of the life zone depends on the temperature of the star (Figure 25–16). Hot stars have larger life zones because the planets must be more distant to remain cool. But the short main-sequence lives of these stars make them unacceptable. M stars have small life zones because they are extremely cool—only planets very near the star receive sufficient warmth. However, planets that are close to a star would probably become tidally coupled, keeping the same side toward the star. This might allow the water and atmosphere to freeze in the perpetual darkness of the planet's night side and end all chance of life. Also, M stars are subject to sudden flares that might destroy life on a planet close to the star. Thus, the life zone restricts our search for life to main-sequence G and K stars. Some of the cooler F stars and warmer M stars might also be good candidates.

Even a star on the main sequence is not perfectly stable. Main-sequence stars gradually grow more luminous as they convert their hydrogen to helium, and thus the life zone around a star gradually moves outward. A planet might form in the life zone and life might begin and evolve for billions of years only to be destroyed as the slowly increasing luminosity of its star moved the life zone outward, evaporated the planet's oceans, and drove off its atmosphere. If a planet is to remain in the life zone for 4–5 billion years, it must form on the outer edge of the zone. This may be the most serious restriction we have yet discussed.

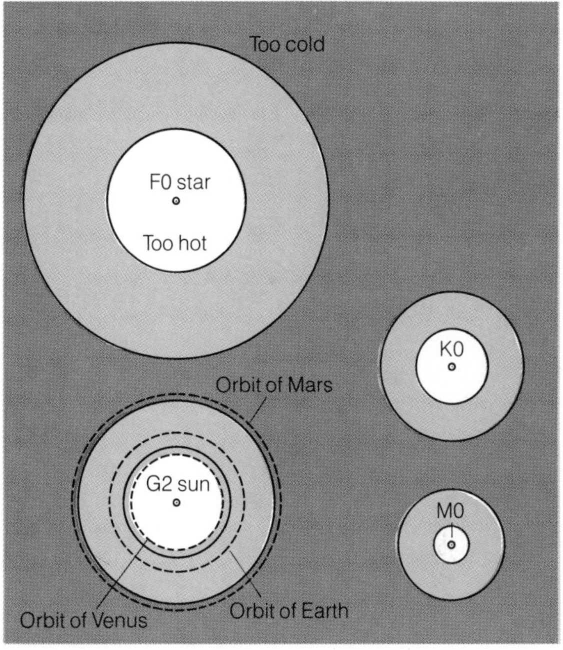

FIGURE 25–16 The size of the life zone around a star depends on the temperature of the star. Stars hotter than about F5 do not remain stable long enough for life to develop.Stars cooler than about M0 may not support life because their life zone is too near the star and the planet's rotation becomes locked to the star.

If all of these requirements are met, will life begin? Early in this chapter we decided that life could begin through simple chemical reactions, so perhaps we should change our question and ask, What could prevent life from beginning? Given what we know about life, it should arise whenever conditions permit, and our galaxy should be filled with planets that are inhabited with living creatures. Then why haven't we heard from them?

25.3 COMMUNICATION WITH DISTANT CIVILIZATIONS

If other civilizations exist, perhaps we can communicate with them in some way. Sadly, travel between the stars appears more difficult in real life than in science fiction.

It may in fact be almost impossible. If we can't physically visit, perhaps we can communicate by radio. Again, nature places restrictions on such conversations, but the restrictions are not too severe. The real problem lies with the nature of civilizations.

TRAVEL BETWEEN THE STARS

Roaming among the stars is in practice tremendously difficult because of three limitations: distance, speed, and fuel. The distances between stars are almost beyond comprehension. It does little good to explain that if we represent the sun by a golf ball in New York City, the nearest star would be another golf ball in Chicago. It is only slightly better to note that the fastest commercial jet would take about 4 million years to reach the nearest star.

The second limitation is a speed limit—we cannot travel faster than the speed of light. Although science fiction writers invent hyperspace drives so their heroes can zip from star to star, the speed of light is a natural and unavoidable limit that we cannot exceed. This, combined with the large distances between stars, makes interstellar travel very time consuming.

The third limitation says that we can't even approach the speed of light without using a fantastic amount of fuel. Even if we ignore the problem of escaping from Earth's gravity, we must still use energy stored in fuel to accelerate to high speed and to decelerate to a stop when we reach our destination. To return to Earth, assuming we wish to, we have to repeat the process. These changes in velocity require a tremendous amount of fuel. If we flew a spaceship as big as a large yacht to a star 5 light-years (1.5 pc) away and wanted to get there in only 10 years, we would use 40,000 times as much energy as the United States consumes in a year.

Travel for a few individuals might be possible if we accept very long travel times. That would require some form of suspended animation (currently unknown) or colony ships that carry a complete, though small, society in which people are born, live, and die generation after generation. Whether the occupants of such a ship would retain the social characteristics of humans over a long voyage is questionable.

These three limitations not only make it difficult for us to leave our solar system, but they would also make it difficult for aliens to visit Earth. Reputable scientists have studied UFOs and related phenomena and have never found any evidence that Earth is being visited or has ever been visited by aliens from other worlds. Thus, it seems unlikely that humans will ever meet an alien face to face. The only way we can communicate with other civilizations is via radio.

RADIO COMMUNICATION

Nature places two restrictions on our ability to communicate with distant societies by radio. One has to do with simple physics, is well-understood, and merely makes the communication difficult. The second has to do with the fate of technological civilizations, is still unresolved, and may severely limit the number of societies we can detect by radio.

Radio signals are electromagnetic waves that travel at the speed of light. Because even the nearest civilizations must be a few light-years away, this limits our ability to carry on a conversation with distant beings. If we ask a question of a creature 4 light-years away, we will have to wait 8 years for a reply. Clearly, the give-and-take of normal conversation will be impossible.

Instead, we could simply broadcast a radio beacon of friendship to announce our presence. Such a beacon would have to consist of a pattern of pulses obviously designed by intelligent beings to distinguish it from natural radio signals emitted by nebulae, pulsars, and so on. For example, pulses counting off the first dozen prime numbers would do. In fact, we are already broadcasting a recognizable beacon. Short-wavelength radio signals, such as TV and FM, have been leaking into space for the last 20 years or so. Any civilization within 20 light-years might already have detected us.

If we intentionally broadcast such a signal, we could give listening aliens a good idea of what humanity is like by including coded data in the signal. For example, in 1974 at the dedication of the new reflecting surface of the 1000-ft radio telescope at Arecibo, Puerto Rico, radio astronomers transmitted a series of pulses toward the globular cluster M 13 in Hercules (Figure 25–17). The number of data points in the message was 1679, a number selected because it can be factored only into 23 and 73. When the signal arrives at the globular cluster 26,000 years from now, any aliens who detect it will be able to arrange the data in only two ways—23 rows of 73 data points each, or 73 rows of 23 points each. The first way yields nonsense, but the second produces a picture that

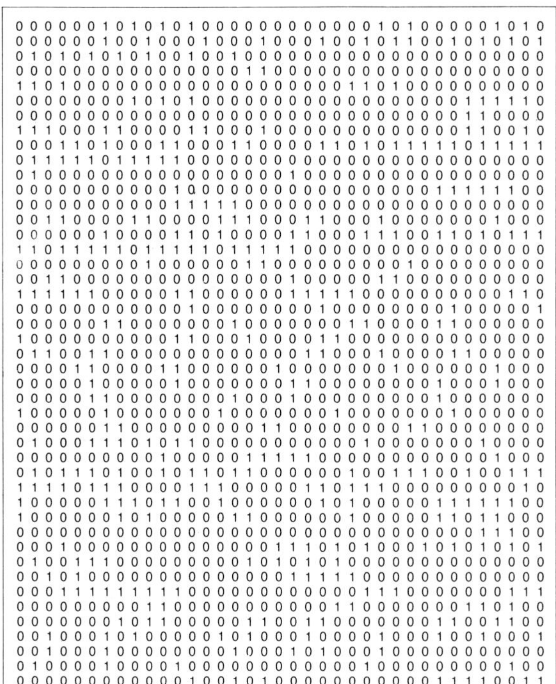

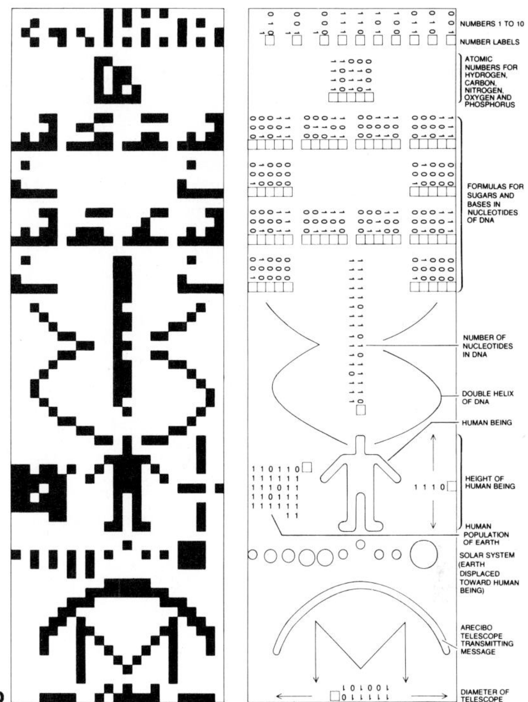

FIGURE 25–17 (a) The Arecibo message of pulses transmitted toward the globular cluster M 13 is shown as a series of 0s and 1s. (b) Arranged in 73 rows of 23 pulses each and represented as light and dark squares, the message would tell aliens about human life. (From Carl Sagan and Frank Drake, "The Search for Extraterrestrial Intelligence." Copyright © 1975 by *Scientific American, Inc.* All rights reserved.)

describes our solar system, the chemical basis of our life-form, the general shape and size of the human body, and the number of humans on Earth. Whether there will still be humans on earth in 52,000 years when any reply to our message returns cannot be predicted.

It took about 10 minutes to transmit the Arecibo message. If more time were taken, a more detailed picture could be sent, and if we were sure our radio telescope were pointed at a listening civilization, we could send a long series of pictures. With pictures we could teach them our language and tell them all about our life, our difficulties, and accomplishments.

If we can think of sending such signals, aliens can think of it too. If we point our radio telescopes in the right direction and listen at the right wavelength, we might hear other intelligent races calling out to each other. This raises two questions. Which stars are the best candidates, and what wavelengths are most likely? We have already answered the first question. Main-sequence G and K stars have the most favorable characteristics. But the second question is more complex.

Only certain wavelengths would be useful for communication. We cannot use wavelengths longer than about 30 cm because the signal would be lost in the background radio noise from our galaxy. Nor can we go to wavelengths much shorter than 1 cm because of absorption within our atmosphere. Thus, only a certain range of wavelengths, a radio window, is open for communication (Figure 25–18).

This communications window is very wide, so a radio telescope would take a long time to tune over all the wavelengths searching for intelligent signals. However, nature may have given us a way to narrow the search. Within the communications window lie the 21-cm line of neutral hydrogen and the 18-cm line of OH. The interval between these two lines has been dubbed the **water hole**

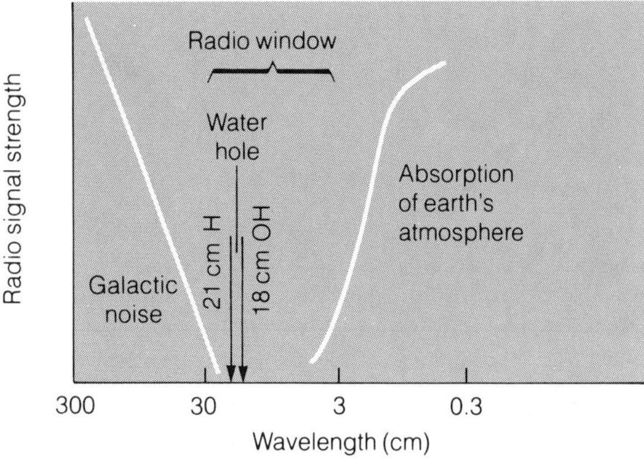

FIGURE 25–18 The wavelength range between 30 cm and 3 cm is a window for possible communication between civilizations. The water hole between the radio emissions of H and OH is an especially likely wavelength range.

because the combination of H and OH yields water (H_2O). Water is the fundamental solvent in our life-form, so it might seem natural for similar water creatures to call out to each other at wavelengths in the water hole. But even a silicon creature would be familiar with the 21-cm line of hydrogen. Thus, they too might select wavelengths near the water hole.

This discussion is not just speculation. A number of searches for extra-terrestrial radio signals are now under way. META, Megachannel Extra-Terrestrial Assay, is now searching the entire sky using a 26-m radio antenna at Harvard University. Funded by the Planetary Society, the project uses a special receiver to simultaneously search 8.4 million adjacent radio frequencies in the water hole. META has been in operation since September 1985.

One ingenious search is called SERENDIP, Search for Extraterrestrial Radio Emission from Nearby Developed Intelligent Populations. This search uses a 65,000-channel receiver attached to the 92-m radio dish at Green Bank, West Virginia, but rather than dominating the telescope, SERENDIP rides piggyback. Wherever the radio astronomers point their telescope, SERENDIP samples the signal received looking for traces of intelligent signals.

The most sophisticated search is currently being developed by NASA, a search dubbed SETI, Search for Extra-Terrestrial Intelligence. It will use a 10-million-channel receiver and the 34-m radio dishes that are part of the Deep Space Network to survey the entire sky in 5 to 7 years. In addition it will use the 305-m Arecibo telescope to listen carefully to near-solar-type stars within 80 ly.

The Soviet Union is also conducting searches and is building a new radio telescope in south-central Asia specifically to search for extra-terrestrial signals. That telescope, scheduled to begin operation in the early 1990s, will be isolated from much of the radio noise that might be mistaken for true signals from another race.

The technology exists, but the most serious restriction on the search is the unanswered social question. How long does a civilization remain at a high enough technological level to engage in galactic communication? If other planets are like Earth, life takes 4.6 billion years to reach a technological level. If a society destroys itself within 100 years of the invention of radio (by nuclear war, nuclear pollution, chemical pollution, overpopulation, and so on), the chances of our communicating with them are very small. Most life-forms in the galaxy would be on the long road up to civilization or on the path leading down from the collapse of their technology. But if a technological society can solve its internal problems and remain stable for a million years, at least some would be at that stage at the proper time to communicate with us. Estimates of the number of communicative civilizations in our galaxy range from 10 million to 1 (Box 25–2).

Are we the only thinking race? If we are, we bear the sole responsibility to understand and admire the universe. Then we are the sole representatives of that state of matter called intelligence. The mere detection of signals from another civilization would demonstrate that we share the universe with others. Although we might never leave our solar system, such communication would end the self-centered isolation of humanity and stimulate a reevaluation of the meaning of our existence. We may never realize our full potential as humans until we communicate with nonhumans.

SUMMARY

To discuss life on other worlds, we must first understand something about life in general, life on Earth, and the origin of life. In general, we can identify three properties in living things—a process, a physical basis, and a controlling unit of information. The process must extract energy from the surroundings, maintain the organism, and modify the surroundings to promote the organism's survival. The physical basis is the arrangement of matter and energy that implements the life process. On Earth all life is based on carbon chemistry. The controlling information is the data necessary to maintain the organism's function. Data for earth life are stored in long carbon-chain molecules called DNA.

BOX 25–2

Communicative Civilizations in Our Galaxy

Simple arithmetic can give us an estimate of the number of technological civilizations with which we might communicate. The formula for the number of communicative civilizations in a galaxy, N_c, is

$$N_c = N^* \cdot f_P \cdot n_{LZ} \cdot f_L \cdot f_I \cdot F_S$$

N^* is the number of stars in a galaxy, and f_P represents the probability that a star has planets. If all single stars have planets, f_P is about 0.5. The factor n_{LZ} is the average number of planets in a solar system suitably placed in the life zone, f_L is the probability that life will originate if conditions are suitable, and f_I is the probability that the life-form will evolve to intelligence. These factors can be roughly estimated, but the remaining factor is much more uncertain.

F_S is the fraction of a star's life during which the life-form is communicative. Here we assume a star lives about 10 billion years. As explained in the text, if a society survives at a technological level for only 100 years, our chances of communicating with it are small. But a society that stabilizes and remains technological for a long time is much more likely to be in the communicative phase at the proper time to signal to us. If we assume technological societies destroy themselves in about 100 years, F_S is 100 divided by 10 billion, or 10^{-8}. But if societies can remain technological for a million years, then F_S is 10^{-4}. The influence of this factor is shown in Table 25–1.

If the optimistic estimates are true, there may be a communicative civilization within a few dozen light-years of us, and we could locate it by searching through only a few thousand stars. On the other hand, if the pessimistic view is correct, we may be the only planet in our galaxy capable of communication. We may never know until we understand how technological societies function.

Table 25–1 The number of technological civilizations per galaxy.

		Estimates	
		Pessimistic	Optimistic
N^*	Number of stars per galaxy	2×10^{11}	2×10^{11}
f_P	Fraction of stars with planets	0.01	0.5
n_{LZ}	Number of planets per star that lie in life zone for longer than 4 billion years	0.01	1
f_L	Fraction of suitable planets on which life begins	0.01	1
f_I	Fraction of life-forms that evolve to intelligence	0.01	1
F_S	Fraction of star's life during which a technological society survives	10^{-8}	10^{-4}
N_c	Number of communicative civilizations per galaxy	2×10^{-5}	10×10^{6}

The DNA molecule stores information in the form of chemical bases linked together like the rungs of a ladder. When these patterns are copied by RNA molecules, they can direct the manufacture of proteins and enzymes. Thus, DNA information is the chemical formulae the cell needs to function. When a cell divides, the DNA molecule splits lengthwise and duplicates itself so that each of the new cells has a copy of the information. Errors in the duplication or damage to the DNA molecule can produce mutants, organisms that contain new DNA information and have new properties. Natural selection determines which of these new organisms are most suited to survive, and the species evolves to fit its environment.

The Miller experiment duplicated conditions in Earth's primitive environment and suggests that energy sources such as lightning could have formed amino acids and other complex molecules. Chemical evolution would have connected these together in larger and more complex, but not yet living, molecules. When a molecule acquired the ability to produce copies of itself, natural selection perfected the organism through biological evolution. Although this may have happened in the first billion years, life did not become diverse and complex until the Cambrian period about 0.6 billion years ago. Life emerged from the oceans about 0.4 billion years ago, and humanity developed only a few million years ago.

It seems unlikely that there is life on other planets in our solar system. Most of the planets are too hot or too cold. Mars may have had life long ago if its atmosphere was thicker and liquid water existed on its surface, but the Viking landers performed three kinds of experiments to look for life and found none. We can imagine how lightning bolts in Jupiter's atmosphere might have spawned life, but there are no data available on complex molecules in the Jovian atmosphere.

To find life we must look beyond our solar system. Because we suspect that planets form from the leftover debris of star formation, we suspect that most stars have planets. The rise of intelligence may take billions of years, however, so short-lived massive stars and binary stars with unstable planetary orbits must be discarded. The best candidates are G and K main-sequence stars.

The distances between stars are too large to permit travel, but communication by radio could be possible. A certain wavelength range called a radio window is suitable, and a small range between the radio signals of H and OH, the so-called water hole, is especially likely.

NEW TERMS

DNA (deoxyribonucleic acid)
protein
enzyme
Cambrian period
Miller experiment
primordial soup
RNA (ribonucleic acid)
amino acid
natural selection
mutant
chemical evolution
stromatolite
life zone
water hole

QUESTIONS

1. If life is based on information, what is that information?
2. What would happen to a life-form if the information handed down to offspring was always the same? How would that endanger the future of the life-form?
3. How does the DNA molecule produce a copy of itself?
4. Why do we believe that life on Earth began in the sea?
5. What is the difference between chemical evolution and biological evolution?
6. How does intelligence make a creature more likely to survive?
7. In your opinion, where in our solar system is the most likely place to find life beyond Earth?
8. What role did the control play in the Viking labeled-release experiment?
9. Why are upper main-sequence stars unlikely sites for intelligent civilizations?
10. What would you change in the Arecibo message if humanity lived on Mars instead of Earth?
11. How does the stability of technological civilizations affect the probability that we can detect them?
12. Make as strong an argument as you can that we are alone in our galaxy.

PROBLEMS

1. A single human cell encloses about 1.5 m of DNA containing 4.5 billion base pairs. What is the spacing between these base pairs in nanometers? That is, how far apart are the rungs on the DNA ladder?
2. If we represent the history of Earth by a line 1 m long, how long a segment would represent the 400 million years since life moved onto the land? How long a segment would represent the 3-million-year history of humanity?
3. If a human generation (the time from birth to childbearing) is 20 years, how many generations have passed in the last 1 million years?

4. If a star must remain on the main sequence for at least 5 billion years for life to evolve to intelligence, how massive could a star be and still harbor intelligent life on one of its planets? (HINT: See Box 11–2.)
5. If there are about 1.4×10^{-4} stars like the sun per cubic light-year, how any lie within 100 ly of Earth? (HINT: The volume of a sphere is $\frac{4}{3}\pi r^3$.)
6. Mathematician Karl Gauss suggested planting forests and fields in a gigantic geometric proof to signal to possible Martians that intelligent life existed on Earth. If Martians had telescopes, which, like ours, could resolve details no smaller than 1 second of arc, how large would the smallest element of Gauss' proof have to be? (HINT: See Box 3–1.)
7. If we detected radio signals with an average wavelength of 20 cm and suspected that they came from a civilization on a distant planet, roughly how much of a change in wavelength should we expect to see because of the orbital motion of the distant planet? (HINT: See Box 7–3.)
8. Calculate the number of communicative civilizations per galaxy from your own estimates of the factors in Box 25–2.

RECOMMENDED READING

BAUGHER, JOSEPH F. *On Civilized Stars.* Englewood Cliffs, N.J.: Prentice-Hall, 1985.

BEATTY, J. KELLY "The New, Improved SETI." *Sky and Telescope 65* (May 1983), p. 411.

CRICK, FRANCIS *Life Itself.* New York: Simon & Schuster, 1981.

DAWKINS, R. *The Selfish Gene.* New York: Oxford University Press, 1976.

DICK, STEVEN J. *Plurality of Worlds: The Origins of the Extra-terrestrial Life Debate from Democritus to Kant.* Cambridge, Mass.: Cambridge University Press, 1982.

FEINBERG, G., and R. SHAPIRO, eds. *Life Beyond Earth: The Intelligent Earthling's Guide to Life in the Universe.* New York: Morrow, 1980.

FINLEY, DAVID "The Search for Extra Solar Planets." *Astronomy 9* (Dec. 1981), p. 90.

FINNEY, BEN R., and ERICH M. JONES, eds. *Interstellar Migration and the Human Experience.* Berkeley: University of California Press, 1985.

FOX, SIDNEY "From Inanimate Matter to Living Systems." *The American Biology Teacher 43* (March 1981), p. 127.

GILLETT, STEPHEN "The Rise and Fall of the Early Reducing Atmosphere." *Astronomy 13* (July 1985), p. 66.

GOLDSMITH, D., ed. *The Quest for Extra Terrestrial Life: A Book of Readings.* Mill Valley, Calif.: University Science Books, 1980.

GOLDSMITH, D., and T. OWEN *The Search for Life in the Universe,* 2nd ed. Menlo Park, Calif.: Benjamin/Cummings, 1987.

McDONOUGH, THOMAS R. *The Search for Extraterrestrial Intelligence.* New York: Wiley, 1987.

MOOD, STEPHANIE "Life on Europa?" *Astronomy 11* (Dec. 1983), p. 16.

OLSON, EDWARD C. "Intelligent Life in Space." *Astronomy 13* (July 1985), p. 6.

PAPAGIANNIS, M. D. "The Search for Extra Terrestrial Civilizations—A New Approach." *Mercury 11* (Jan./Feb. 1982), p. 12.

———. "Bioastronomy: The Search for Extra Terrestrial Life." *Sky and Telescope 67* (June 1984), p. 508.

———, ed. *The Search for Extraterrestrial Life: Recent Developments.* Boston: D. Reidel, 1985.

PARKER, B. "Are We the Only Intelligent Life in Our Galaxy?" *Astronomy 7* (Jan. 1979), p. 6. Reprinted in *Astronomy: Selected Readings.* ed. M. A. Seeds. Menlo Park, Calif.: Benjamin/Cummings, 1980, p. 155.

REGIS, E., ed. *Extraterrestrials: Science and Alien Intelligence.* Cambridge, Mass.: Cambridge University Press, 1985.

SAGAN, C., and T. PAGE, eds. *UFO's: A Scientific Debate.* New York: Norton, 1972.

SCHECHTER, MURRAY "Planets in Binary Star Systems." *Sky and Telescope 68* (Nov. 1984), p. 394.

SCHORN, R. A. "Extraterrestrial Beings Don't Exist." *Sky and Telescope 62* (Sept. 1981), p. 207.

STRAND, LINDA JOAN "The Search for Life on Mars: Shots in the Dark." *Astronomy 11* (Dec. 1983), p. 66.

———. "The Star Tar in the Jupiter Jars." *Astronomy 12* (June 1984), p. 66.

TIPPLER, FRANK "Extraterrestrial Beings Do Not Exist." *Physics Today 34* (April 1982), p. 9. But see *35* (March 1982), p. 26.

WOESE, CARL R. "Archaeobacteria." *Scientific American 244* (June 1981), p. 98.

CHAPTER 26

AFTERWORD

Supernatural is a null word.

Robert A. Heinlein
THE NOTEBOOKS OF LAZARUS LONG

Our journey is over, but before we part company, there is one last thing to discuss—the place of humanity in the universe. Astronomy gives us some comprehension of the workings of stars, galaxies, and planets, but its greatest value lies in what it teaches us about ourselves. Now that we have surveyed astronomical knowledge, we can better understand our own position in nature.

To some, the word *nature* conjures up visions of furry rabbits hopping about in a forest glade dotted with pastel wildflowers. To others, nature is the blue-green ocean depths filled with creatures swirling in a mad struggle for survival. Still others think of nature as windswept mountaintops of gray stone and glittering ice. As diverse as these images are, they are all earthbound. Having studied astronomy, we can view nature as a beautiful mechanism composed of matter and energy interacting according to simple rules to form galaxies, stars, planets, mountaintops, ocean depths, and forest glades.

Perhaps the most important astronomical lesson is that we are a small but important part of the universe. Most of the universe is lifeless. The vast reaches between the galaxies appear to be empty of all but the thinnest gas, and the stars, which contain most of the mass, are much too hot to preserve the chemical bonds that seem necessary for life to survive and develop. Only on the surfaces of a few planets, where temperatures are moderate, could atoms link together to form living matter.

If life is special, then intelligence is precious. The universe must contain many planets devoid of life, planets where the wind has blown unfelt for billions of years. There may also exist planets where life has developed but has not become complex, planets on which the wind stirs wide plains of grass and rustles dark forests. On some planets, insects, fish, birds, and animals may watch the passing days unaware of their own existence. It is intelligence, human or alien, that gives meaning to the landscape.

Science is the process by which intelligence tries to understand the universe. Science is not the invention of new devices or processes. It does not create home computers, cure the mumps, or manufacture plastic spoons—that is engineering and technology, the adaptation of scientific understanding for practical purposes. Science is understanding nature, and astronomy is understanding on the grandest scale. Astronomy is the science by which the universe, through its intelligent lumps of matter, tries to understand its own existence.

As the primary intelligent species on this planet, we are the custodians of a priceless gift—a planet filled with living things. This is especially true if life is rare in the universe. In fact, if Earth is the only inhabited planet, our responsibility is overwhelming. In any case, we are the only creatures who can take action to preserve the existence of life on Earth, and ironically, it is our own actions that are the most serious hazards.

The future of humanity is not secure. We are trapped on a tiny planet with limited resources and a population growing faster than our ability to produce food. In our efforts to survive, we have already driven some creatures to extinction and now threaten others. If our civilization collapses because of starvation, or if our race destroys itself somehow, the only bright spot is that the rest of the creatures on Earth will be better off for our absence.

But even if we control our population and conserve and recycle our resources, life on Earth is doomed. In 5 billion years the sun will leave the main sequence and swell into a red giant, incinerating Earth. However, Earth will be lifeless long before that. Within the next few billion years the growing luminosity of the sun will first alter Earth's climate and then boil its atmosphere and oceans. Our Earth is, like everything else in the universe, only temporary.

To survive, humanity must leave Earth and search for other planets. Colonizing the moon and other planets of our solar system will not save us, since they will face the same fate as Earth when the sun dies. But travel to other stars is tremendously difficult and may be impossible with the limited resources we have in our small solar system. We and all of the living things that depend on us for survival may be trapped.

This is a depressing prospect, but a few factors are comforting. First, everything in the universe is temporary. Stars die, galaxies die, perhaps the entire universe will fall back in a "big crunch" and die. That our distant future is limited only assures us that we are a part of a much larger whole. Second, we have a few billion years to prepare, and a billion years is a very long time. Only a few million years ago our ancestors were learning to walk erect and communicate with each other. A billion years ago our ancestors were microscopic organisms living in the primeval oceans. To suppose that a billion years hence we humans will still be human, or that we will still be the dominant species on Earth, or that we will still be the sole intelligence on Earth is the ultimate conceit.

Our responsibility is not to save our race for all eternity but to behave as dependable custodians of our planet, preserving it, admiring it, and trying to understand it. That will call for drastic changes in our behavior toward other living things and a revolution in our attitude toward our planet's resources. Whether we can change our ways is debatable—humanity is far from perfect in its understanding, abilities, or intentions. However, we must not imagine that we and our civilization are less than precious. We have the gift of intelligence, and that is the finest thing this planet has ever produced.

APPENDIX A

ASTRONOMICAL COORDINATES AND TIMEKEEPING

ASTRONOMICAL COORDINATES

To describe the location of an object in the sky, we must specify its distance from known reference points and lines. The main-reference line is the celestial equator, and the distances refer not to distance through space but to angular distance across the sky. Thus, the location of celestial objects on the sky by their celestial coordinates is analogous to the location of cities on the earth by their latitude and longitude.

LATITUDE AND LONGITUDE

We imagine that Earth is girded by a mesh of lines that define the latitude and longitude of every spot on the surface. An observer's latitude is measured north or south from the earth's equator. People living on the earth's equator, in Quito, Ecuador, for instance, have latitude 0°, whereas an encampment of explorers at the North Pole are at latitude 90°N. The people of New York City live at latitude 40°45′N and those in Los Angeles at 34°05′N. Because latitude is measured north from the equator, people living south of the equator have negative latitudes. New Zealand, for example, has a latitude of 40°S or −40°. Thus, we can draw lines around the earth parallel to the equator and refer to them as circles of latitude (Figure A–1a).

Longitude is measured east or west of Greenwich, England. New York is 74° west of Greenwich. A north–south line through New York City extending from the North Pole to the South Pole is a line of longitude, and anyone on that line has the same longitude as New York City. Consequently, we can cover the earth with north–south lines of longitude that are parallel near the equator and converging near the poles (Figure A–1b).

It is common to express longitude as a time instead of an angle because the earth rotates through 360° in 24 hours. Thus an angle of 15° is equivalent to an angle of 1 hour. Instead of saying New York City is 74° west of Greenwich, we could say it is 4^h56^m west where the superscripts refer to hours and minutes.

CELESTIAL COORDINATES

A similar system of lines on the sky defines the celestial coordinate system; however, instead of latitude we refer to a star's **declination**, and instead of its longitude we refer to its **right ascension**. These are often abbreviated

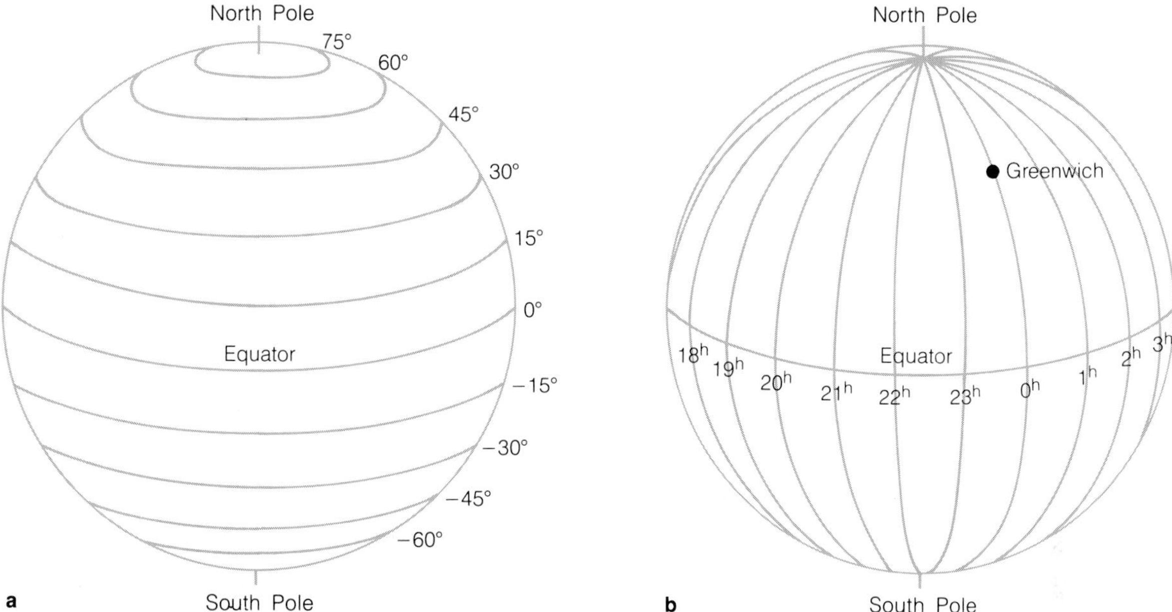

FIGURE A–1 (a) Lines on the earth parallel to earth's equator are lines of constant latitude. (b) North-south lines extending from the earth's North Pole to its South Pole are lines of constant longitude. Greenwich, England, is the reference point from which longitude is measured.

Dec. and R. A. In some books and tables, you may find declination abbreviated δ and right ascension abbreviated α.

Declination is measured north or south from the celestial equator. A star on the celestial equator has declination 0°, and a hypothetical star at the north celestial pole has declination 90°. Polaris, the North Star, is not precisely at the pole, so its declination, 89°11′, is slightly less than 90°. Stars located south of the celestial equator have negative declinations, such as Rigel at −8°13′. Thus, lines drawn around the celestial sphere parallel to the celestial equator are lines of constant declination (Figure A–2a).

Before we can measure right ascension, we must have some reference mark on the sky from which to measure. Longitude is measured east or west from Greenwich, but there are no cities on the sky, and the stars are all moving slowly as they orbit the center of our galaxy. Faced with this dilemma, astronomers choose as their reference point the vernal equinox, the point on the celestial equator where the sun crosses into the northern sky in the spring. Even the vernal equinox is moving slowly because of precession (Chapter 2).

Unlike longitude, right ascension is always measured eastward from the reference mark and is nearly always expressed in units of time. We can say the star Sirius has a right ascension of 6^h44^m, meaning it lies 6 hours and 44 minutes east of the vernal equinox. We divide the celestial equator in 24 equal divisions called hours of right ascension, and we draw 24 equally spaced north–south lines from the north celestial pole to the south celestial pole to serve as lines of constant right ascension (Figure A–2b).

Using this mesh of declination and right ascension lines, we can define the declination and right ascension of any object. For instance, the nearest star to the sun, α Centauri, is located at R. A. 14^h38^m, Dec. −60°46′. Notice that celestial coordinates tell us not the distance to the object, but its direction in space.

We could also give the right ascension and declination of the sun, moon, and planets, but because these objects move relatively rapidly, their coordinates change from day to day. We would have to list them in an almanac giving the location of the object on specific dates.

Because the earth rotates eastward on its axis each day, the sky appears to rotate westward around us once

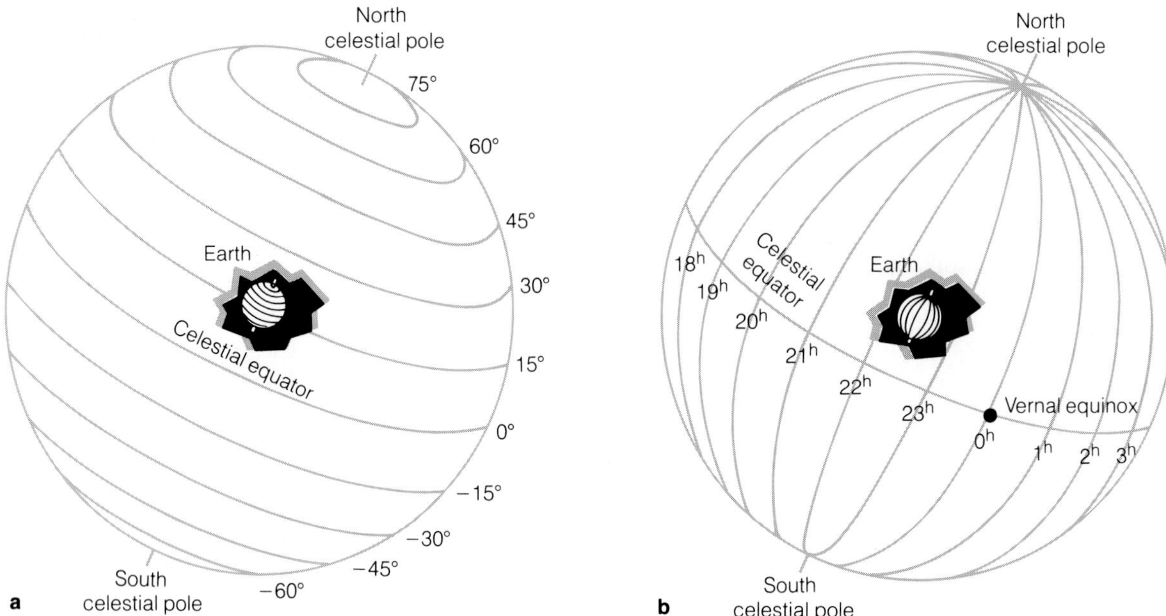

FIGURE A–2 The celestial coordinate system is the projection on the sky of latitude and longitude. (a) On the celestial sphere, lines of constant declination are parallel to the celestial equator. (b) Lines of constant right ascension run north-south from celestial pole to celestial pole.

a day. The rotation of the celestial coordinate system is the basis of our systems of timekeeping.

TIMEKEEPING

Have you ever noticed that sundials are almost always wrong? They almost never agree with our wristwatches. To account for this disparity, we must consider one of the oldest branches of astronomy—timekeeping. The shadow on a sundial, the time on our wristwatches, and even the dates on our calendars are astronomical phenomena. They are directly related to the rotation of the earth on its axis and its revolution around the sun.

THE DAY

We commonly define a day to be the interval between successive midnights, but we must be careful. There is more than one kind of day, depending on how time is measured.

The kind of time we usually deal with is related to the sun and its position in the sky relative to a line called the **local celestial meridian.** This line runs from the south point on the horizon to the zenith directly overhead, to the north point on the horizon, to the **nadir** directly underfoot, and back to the south point. We use this reference line every day when we refer to the time. If the sun is east of the local celestial meridian, we say the time is ante meridiem (before the meridian) or AM. If the sun is west of the meridian, we say the times are post meridiem (past the meridian) or PM. When the sun is on the meridian, the time is noon; when the sun is on the meridian below the horizon, it is midnight.

Our clocks keep track of the location of the sun with respect to the local celestial meridian. If we ignore for the moment certain irregularities, we can say that a 24-hour clock displays a time equal to 12 hours plus the **hour angle** of the sun. This hour angle is measured westward from the upper half of the local celestial meridian to the sun. Like right ascension the sun's hour angle is usually measured in units of time. Thus, at 3 PM by our wristwatches, a 24-hour clock reads 15 hours, telling us that the sun is 3 hours past the meridian (Figure A–3).

Using this convention for the measurement of time, we can define a **mean solar day** to be the average interval between successive passages of the sun across the lower half of the local celestial meridian (that is, midnight). If

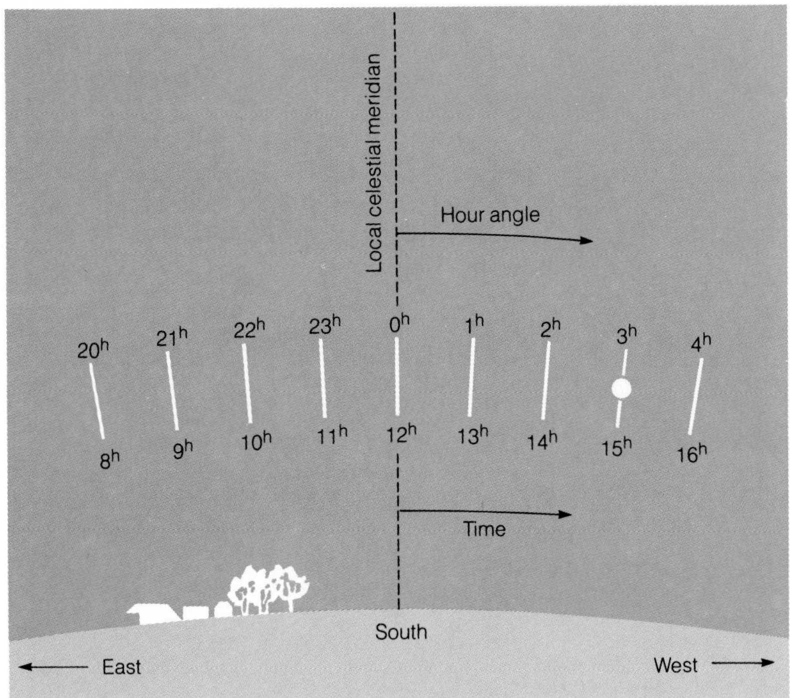

FIGURE A–3 We measure time by the hour angle of the sun, the angular distance from the local celestial meridian west to the sun. When the sun has an hour angle of 3 hours, we say the time is 15:00 hours (on a 24-hour clock).

we divide this interval into hours, minutes, and seconds, we have the basis for our system of timekeeping.

However, the solar day is not the only kind of day. We could define a day to be the interval between the passages of a certain star across the local celestial meridian. This would be called a **sidereal day** (pronounced to rhyme with *ethereal*), and a clock keeping the sidereal day would display the hour angle of that particular star. Rather than use a star, it is easier to use the vernal equinox as the basis for the sidereal day.* The sidereal day is the interval between successive passages of the vernal equinox across the upper part of our local celestial meridian, and **sidereal time** equals the hour angle of the vernal equinox. This is equal to the right ascension of stars on the upper part of the local celestial meridian.

The sidereal day is shorter than the solar day because of the eastward motion of the sun. To see how this comes about, imagine that we start our clocks at noon on the day of the vernal equinox. Both the sun and the vernal equinox would be on the local celestial meridian, so we would set a normal clock to read 12:00 noon and we would set our sidereal clock to read 0 hours. The next day when our sidereal clock reads 0 hours again, the vernal equinox would be back on the local celestial meridian. However, during the interval of one day, the earth would have moved along its orbit and, consequently, the sun would have moved eastward about 1° along the ecliptic. It would be not at the vernal equinox, but about 1° east. Because the earth requires 4 minutes to rotate 1°, our normal clock would read 11:56 AM (Figure A–4).

Thus, a sidereal clock that keeps track of the stars must run 4 minutes a day faster than a normal clock that keeps track of the sun. If we tried to run our lives according to a sidereal clock, it would gain 4 minutes a day, and in 6 months, it would be 12 hours fast. We would be eating supper at sunrise and breakfast at sunset.

MEAN SOLAR TIME

Sidereal time is useful in an observatory where astronomers must keep track of the positions of the stars, but we must regulate our lives by solar time. Both our wristwatches and our sundials keep solar time, but they usually

*Precession causes the vernal equinox to move slowly westward with respect to the stars, but this causes an error of only 0.008 second in the length of a sidereal day.

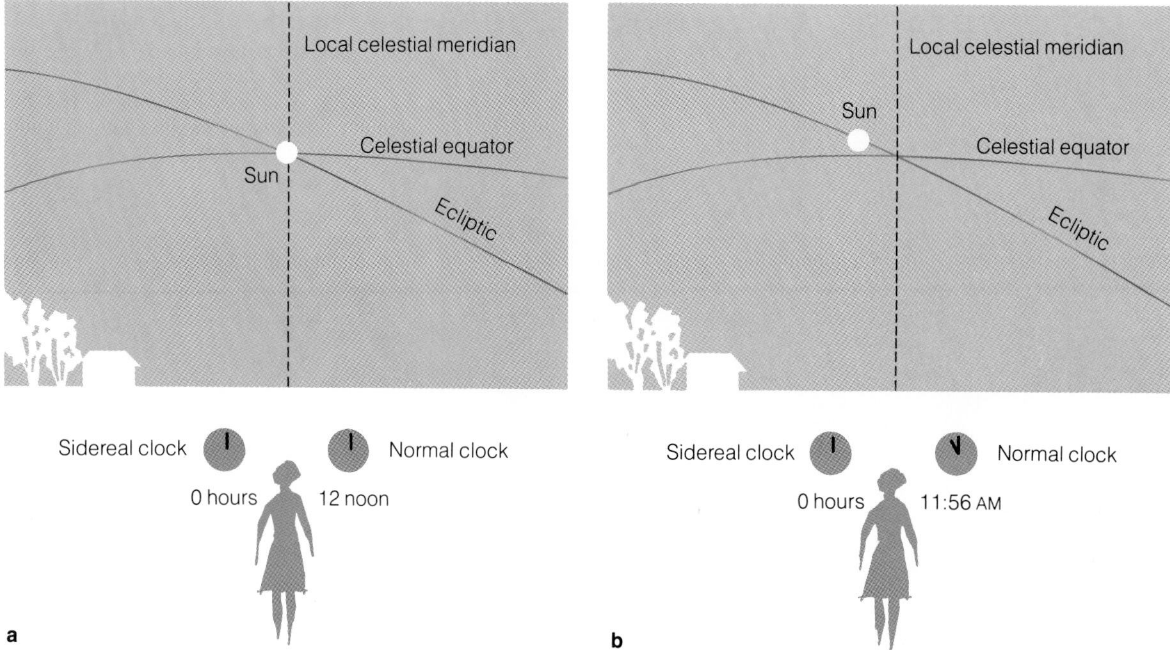

FIGURE A–4 (a) About noon on a day in late March when the vernal equinox and the sun are on the local celestial meridian, we set a normal clock to 12:00 noon and a sidereal clock to 0:00 hours. (b) One day later the earth has rotated once on its axis, and the vernal equinox is back on the local celestial meridian. Our sidereal clock reads 0:00 hours. However, the sun has moved eastward along the ecliptic, and our normal clock reads 11:56 AM.

differ slightly because one keeps zone time and one keeps apparent time.

The time on a sundial is indicated by a shadow cast by the sun, so sundial time is **local apparent time.** We call it *local time* because sundials at different locations show different times. A sundial in Pittsburgh is 20 minutes behind a sundial in Philadelphia. We call sundial time *apparent* because it depends on the actual position of the sun with no corrections for irregularities.

In fact, the eastward motion of the sun along the ecliptic is not perfectly uniform. The earth's orbit is slightly elliptical, and the earth is closer to the sun in January and farther away in July. When it is closer to the sun, the earth moves faster along its orbit, and when it is farther away, it moves slower. The sun, therefore, moves faster along the ecliptic in January and slower in July.

Another irregularity in the sun's motion is related to the inclination of the ecliptic to the celestial equator. When the sun is moving parallel to the celestial equator, around the times of the solstices, it moves about 1° per day eastward. However, around the times of the equinoxes, its motion is not directed parallel to the celestial equator. In the spring, the sun's motion is directed partially toward the east and partially toward the north as it returns to the northern hemisphere of the sky. In the fall, part of the sun's motion is toward the south. At these times, its eastward progress can be as small as 0.9° per day (Figure A–5a).

Still another irregularity in the motion of the sun arises because the sun does not follow the celestial equator. If it moved 1° eastward along the celestial equator, its right ascension would increase by 4 minutes. However, around the times of the solstices, the sun lies 23.5° north or south of the celestial equator. There the lines of right ascension are closer together, and when the sun moves 1° eastward along the ecliptic, its right ascension can increase by as much as 4.4 minutes (Figure A–5b).

Sundial time refers to the sun's hour angle, and that depends on the sun's right ascension, so these irregularities in the eastward motion of the sun can affect the

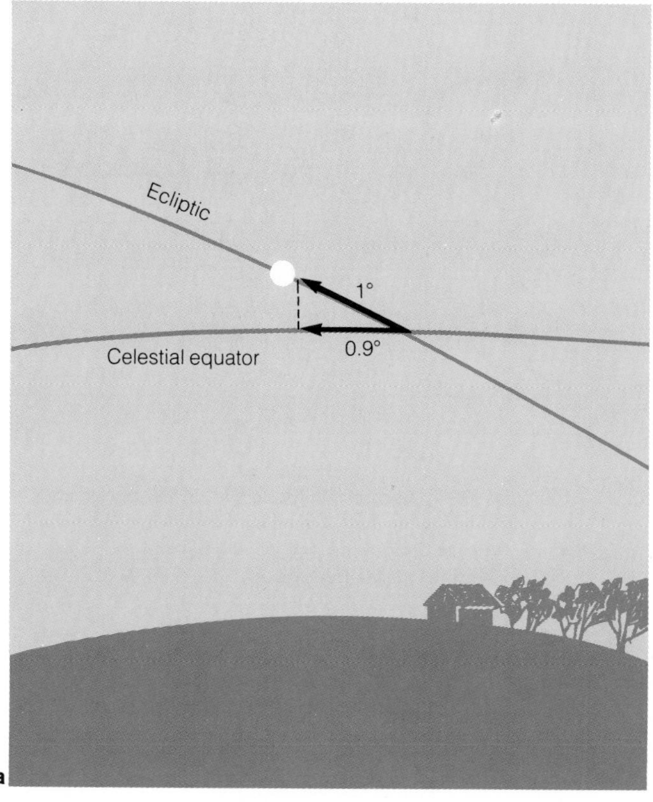

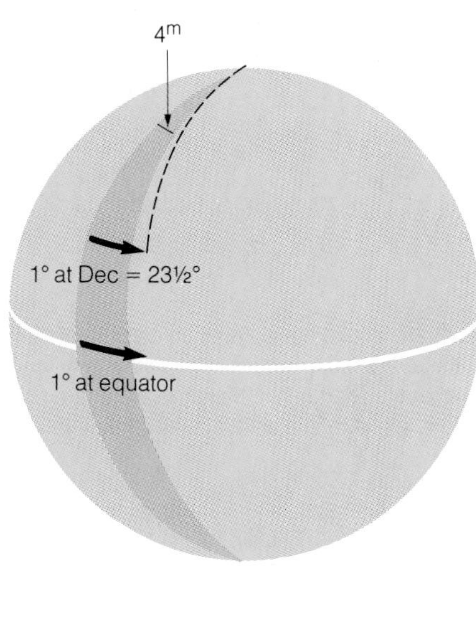

FIGURE A–5 Two things affect the eastward motion of the sun. (a) In the spring and fall, its motion is not directed exactly east, and its net progress eastward is less than 1° per day. (b) Also, lines of constant right ascension are closer together at higher declinations. Thus 1° of travel covers 4 minutes of right ascension at the equator, but 1° at a declination of 23.5° covers more than 4 minutes. Both of these effects are part of the equation of time. (See Figure A–6.)

sundial. The sun sometimes gets ahead of its average place and sometimes falls behind. Consequently, a sundial is sometimes fast and sometimes slow.

Modern civilization requires clocks that run at a constant rate, so we have replaced sundials and apparent time with mechanical or electronic clocks and mean time. **Local mean time** is the time that a sundial would keep if the sun moved at a constant rate eastward around the sky. The difference between local apparent time and local mean time is called the **equation of time** (Figure A–6). It never amounts to more than 17 minutes. To correct the time on a sundial, subtract the equation of time for the current date. This corrected time is local mean time.

Even with this correction, sundial time does not agree with our wristwatches. Both local apparent time and local mean time are *local.* They depend on the location of the sundial. To compare local mean time to our wristwatches, we must discuss time zones.

ZONE TIME

According to a sundial, noon occurs when the sun is on the local celestial meridian. If one family lives 80 km west of another family, their local celestial meridians differ. The sun would first cross the local celestial meridian of the eastern family, and their sundial would read noon. However, the western family would see the sun slightly east of the local meridian, and their sundial would say it

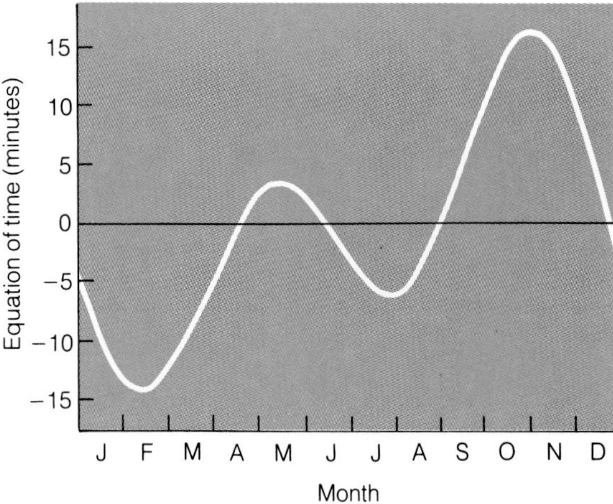

FIGURE A–6 The equation of time is the difference between local apparent time as shown on a sundial and local mean time. To correct sundial time, subtract the equation of time for the appropriate date.

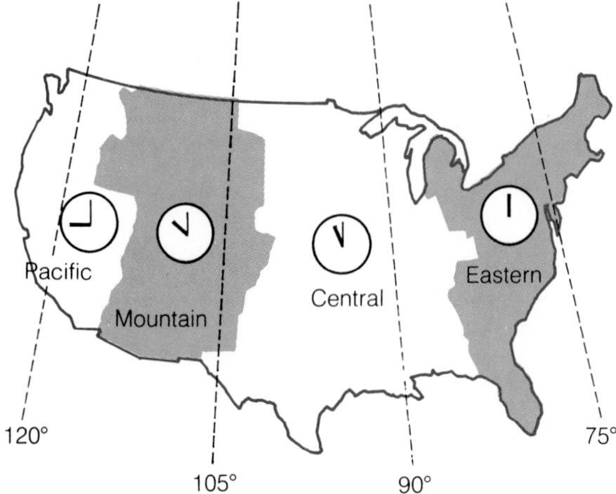

FIGURE A–7 For convenience we divide the earth into time zones and all clocks within a zone keep the same time as a clock on the zone's central meridian. Adjacent zones differ by 1 hour.

is a few minutes before noon. Thus, every person with a different longitude has a different time called local time.

This made no difference in the distant past, but with the advent of rapid communication and transportation it became very inconvenient if the clocks in one town were a few minutes different from the clocks in the next town west. To solve this problem, time zones were established in 1883, and every clock in a given time zone was set to the same time. Philadelphia and Pittsburgh have local times that differ by about 20 minutes, but they both lie in the Eastern Time Zone so the clocks in both cities read the same time. The **standard time** adopted in each time zone is the time kept by clocks on its central meridian (Figure A–7). Thus, when it is 12 noon in New York, it is 11:00 AM in Kansas, 10:00 AM in Colorado, and 9:00 AM in California.

When we compare our wristwatches with a sundial, we compare zone time to local apparent time. Once we subtract the equation of time for that day from the sundial time, we are comparing zone time with local mean time. Local mean time is slow (early) by 4 minutes for every degree of longitude west of the central meridian of the time zone. For instance, if you live 3° of longitude west of the central meridian in your time zone, local mean time is 12 minutes slower (earlier) than your zone time. Of course, if you live east of the central meridian of your time zone, your local mean time is faster (later) than your zone time.

Let us review the three kinds of time we have discussed. Zone time, the time kept by our wristwatches, is the time a sundial would keep if it were on the central meridian of the time zone and if the sun moved at a constant rate eastward around the sky. Local mean time is the time a sundial would keep anywhere in the time zone assuming the sun moved at a constant rate eastward around the sky. Obviously, sundials at different longitudes would have different local mean times. Local apparent time is the time that sundials really keep.

One kind of time is especially important in astronomy. Universal Time (UT) is the same as the local mean time on the prime meridian passing through Greenwich, England—that is, it is the same as Greenwich Mean Time (GMT). Most astronomers use UT in their work no matter what time zone they live in, and most astronomical almanacs list the times of celestial events in UT. Universal Time is expressed on a 24-hour clock with 0 hours occurring at 7 PM Eastern Standard Time.

In a sense, our wristwatches are astronomical instruments that indicate the position of the sun in the sky. In the same sense, our calendar is an astronomical instrument that indicates the position of the earth in its orbit.

THE CALENDAR

Each year at the time of the vernal equinox, the earth returns to the same place in its orbit, and the cycle of the seasons begins again. To early cultures, this cycle had life-and-death importance. On it depended the migration of animals, the flooding of rivers, and the planting and harvesting of crops. An error in the prediction of the seasons could cause starvation. Consequently, even the most primitive cultures developed calendars to try to keep track of the seasons.

In addition to charting the seasons, a calendar is useful in establishing the times of religious holidays, determining dates for civil matters, and documenting historical accounts. The Mayan calendar, for instance, contained a 365-day cycle for the seasons, a sacred 260-day cycle for religious and astrological purposes, and a long count of days mainly for historical purposes. The Jewish calendar, like many others, uses the 29½-day cycle of the lunar phases to divide the year into months. Jewish holy days are thus related to the phases of the moon, and Passover, for instance, always occurs near full moon.

Our modern calendar has its roots in a lunar calendar used by the Romans. It originally contained 10 months, but it was later modified to contain 12. Because the period of lunar phases is not an integral number of days, the months had to alternate in length between 29 and 30 days to produce an average length of 29½ days. This ensured that each month would begin with a new moon.

Unfortunately, 12 months of 29½ days average length make a year of only 354 days, whereas the seasons repeat with a period of about 365¼ days. Each year, the calendar ended about 11 days too soon, and in 3 years it was a full month fast. To remedy this, the priests in charge inserted a thirteenth month at 3-year intervals, but they used this power arbitrarily for political purposes, and soon the calendar was hopelessly confused.

In 46 BC Julius Caesar demanded a calendar reform that led to the adoption of a system now known as the **Julian calendar.** Caesar ordered that the calendar be 365 days in length and that it contain 12 months. A few days had to be added to various months to raise the total from 354 to 365, and this forced the abandonment of the lunar cycle. Although it retained months, the Julian calendar was not a lunar calendar.

Unhappily for Caesar and his calendar makers, the seasons do not repeat every 365 days, but every 365.242199 days.* Thus, every year the calendar ended about one-quarter day early, and in 4 years, the calendar was a full day in error. To account for this, February was given an extra day every fourth year (those evenly divisible by four)—now called leap years. According to the Julian calendar, any year evenly divisible by four was a leap year, and thus the average length of the calendar was 365.25 days. Actually, this was 11 minutes and 14 seconds too long; after 128 years, the calendar was ending a full day too late.

By 1582 the error had accumulated to 10 days, and Pope Gregory XIII ordered another reform that produced the **Gregorian calendar,** which is in use today. The extra 10 days were dropped from October of that year; October 4 was followed by October 15. Also, century years that were not evenly divisible by 400 were not considered leap years. Thus, 1600 was a leap year but 1700 was not. This made the average length of the calendar 365.2425 days and reduced the error to only 1 day in 3322 years.

Catholic countries adopted the new calendar immediately, but other countries continued to use the Julian calendar. It was possible to leave a Catholic country, cross the border into a Protestant country, and, according to the calendars, arrive 10 days before you left. England did not adopt the Gregorian calendar until 1752, and Russia, influenced by the Eastern Church, held out until after the revolution of 1917.

Our modern calendar is based on the Gregorian calendar. A slight modification makes years evenly divisible by 4000 regular years rather than leap years. This reduces the error to less than 1 day in 20,000 years.

*This is called the **tropical year.** It differs from the **sidereal year** (the true orbital period of the earth) 365.256366 days, due to precession.

NEW TERMS

declination
right ascension
local celestial meridian
nadir
hour angle
mean solar day
sidereal day
sidereal time
local apparent time
local mean time
equation of time
standard time
Julian calendar
tropical year
sidereal year
Gregorian calendar

PROBLEMS

1. What are the celestial coordinates of an object that lies directly north of the vernal equinox and 15° from the north celestial pole?
2. What are the celestial coordinates of an object that lies 25° directly south of the summer solstice?
3. If a sundial reads 3:15 on January 1, what is the local mean time? What is the hour angle of the sun? (HINT: See Figure A–6).

Fold at latitude and glue to a cardboard base.

Glue to base.

Punch hole

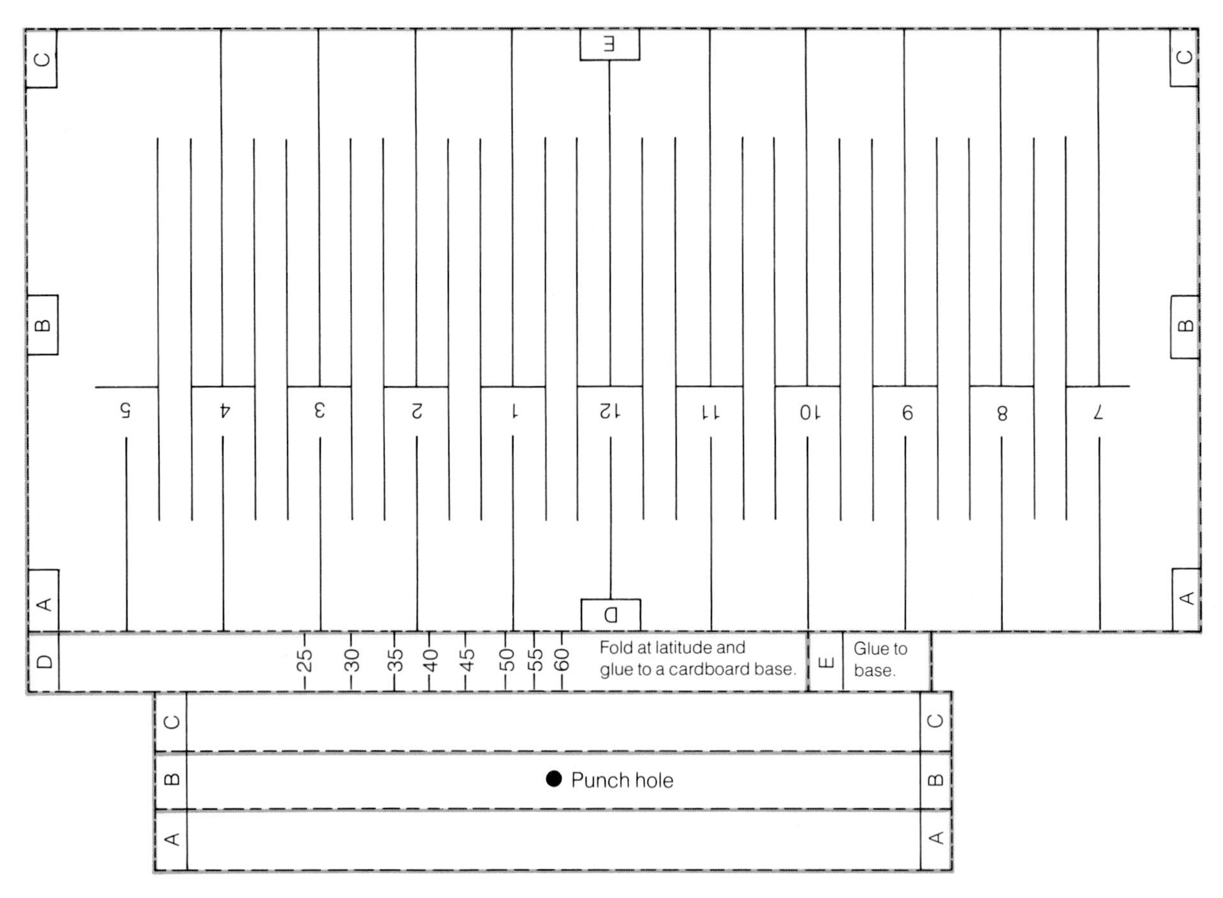

FIGURE A–8 Parts for a sundial.

4. On November 2, a sundial reads 12:37. If it is located 5° east of the central meridian of its time zone, what is the zone time? (HINT: See Figure A–6).
5. At exactly 9:00 PM last night you looked due east and saw a star just above the horizon. If you look tonight at exactly 9:00 PM will the star be higher, lower, or in the same place? Why?

ACTIVITY: YOUR OWN SUNDIAL

Most sundials are wrong, not because they keep local apparent time, but because they are made or installed incorrectly. Some cast-metal sundials sold by garden supply companies do not have the numbers correctly spaced around the dial. Other sundials may be made correctly, but the owners may not have installed them level and properly oriented north–south. Often these garden sundials are made for the wrong latitude!

In fact, installing and using a sundial is not difficult, and this activity will help you build your own sundial, correctly designed for your latitude. The simple steps describe how to install your sundial and use it to tell the time on any sunny day.

Assembly and Installation The sundial shown in Figure A–8 is especially designed for use at any latitude. Only one simple adjustment is needed.

Cut out the six parts along the blue lines and use glue or tape to attach the tabs. Begin by bending the rectangular dial to form a half cylinder with the printed scale inside. Attach tabs A, B, and C to hold the cylinder in shape and carefully punch a small hole in the center brace. Sunlight shining through this hole will indicate the time on the dial. Attach tab E to form a hinge at the bottom of the dial and attach tab D at the top. Fold this long brace at the number equal to your latitude and then attach the

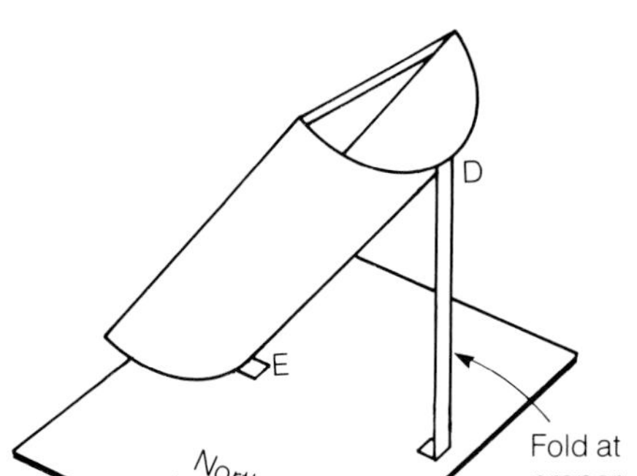

FIGURE A–9 The assembled sundial.

dial to a cardboard base. You may need to reinforce the long brace with cardboard. Your sundial should look like Figure A–9.

If you live south of the earth's equator, you will need to exchange tabs E and D. This will adjust your sundial to the southern latitude.

To install your sundial, you need a level surface exposed to sunlight, a windowsill perhaps. Place your sundial in the sun and turn it until the long axis of the cylinder points north.

But how do you find true north? This is actually a simple thing to discover. The sun is exactly south of you at 12 noon local apparent time, and you can find out when this occurs by a simple two-step calculation. Use today's date to find the equation of time from Figure A–6 and subtract this from 12:00. Now add 4 minutes for every degree of longitude west of the prime meridian in your time zone. (The prime meridians are shown in Figure A–7). The result is the zone time of local apparent noon.

Try this example. I live at longitude 76° and the date is October 1. From Figure A–6 I find that the equation of time is +10 minutes, and subtracting from 12:00 tells me that the sun is directly south of the prime meridian at 11:50 AM local mean time. Because I live 1° west of the prime meridian of the eastern time zone, I add 4 minutes to get 11:54 AM Eastern Standard Time. At that time the sun will be directly south, and I can orient my sundial north–south by turning it until it reads 12:00 noon.

Finding the Time Reading the time from your sundial is very simple. Note the time indicated by the dot of light on the dial and subtract the equation of time for the date. Then add 4 minutes for each degree of longitude west of the prime meridian in your time zone. The result is Standard Zone Time.

RECOMMENDED READING

BARTKY, I. R., and I. HARRISON "Standard and Daylight-Saving Time." *Scientific American 240* (May 1979), p. 46.

BERRY, R. "Celestial Software, Part 2: Telling Time." *Astronomy 6* (July 1979), p. 47.

BURNHAM, R. "Building a Sundial." *Astronomy 8* (March 1980), p. 47.

GOULDSMIT, SAMUEL A., and ROBERT CLAIBORNE *Time.* New York: Time, 1966.

HAZELRIGG, H. C. "Atkinson Sundial at Indiana University." *Sky and Telescope 57* (Feb. 1979), p. 137.

KLUEPFEL, CHARLES "How Accurate *Is* the Gregorian Calendar?" *Sky and Telescope 64* (Nov. 1984), p. 417.

MAYALL, R. NEWTON "A Bit of Porcelain." *Sky and Telescope 63* (Jan. 1982), p. 16.

MOYER, GORDON "Luigi Lilio and the Gregorian Reform of the Calendar." *Sky and Telescope 64* (Nov. 1984), p. 418.

ROHR, RENE R. J. *Sundials: History, Theory, and Practice.* Toronto: University of Toronto Press, 1965.

"Telling Time with Dolphins." *Sky and Telescope 56* (Oct. 1978), p. 304.

APPENDIX B

SYSTEMS OF UNITS

INTRODUCTION

The metric system is used worldwide as the system of units not only in science but also in engineering, business, sports, and daily life. Developed in eighteenth century France, the metric system has gained acceptance in almost every country in the world because it simplifies computations.

A system of units is based on the three fundamental units for length, mass, and time. Other quantities such as density and force are derived from these fundamental units. In the English (or British) system of units (commonly used in the United States) the fundamental unit of length is the foot composed of 12 inches. The metric system is based on the decimal system of numbers, and the fundamental unit of length is the meter composed of 100 centimeters.

To see the advantage of having a decimal-based system, try computing the volume of a bathtub that is 5′9″ long, 1′2″ deep, and 2′10″ wide. In the metric system the length of the tub is 1 m and 75 cm or 1.75 m. The other dimensions are 0.35 m and 0.86 m, and the volume is just $1.75 \times 0.35 \times 0.86$ cm^3. To make the computation in English units, we must first convert inches to feet by dividing by 12. We can convert centimeters to meters by the simpler process of moving the decimal point. Thus, the computation is much easier if we measure the tub in meters and centimeters instead of feet and inches.

Because the metric system is a decimal system, it is easy to express quantities in larger or smaller units as is convenient. We can express distances in centimeters, meters, kilometers, and so on. The prefixes specify the relation of the unit to the meter. Just as a cent is 1⁄100 of a dollar, a centimeter is 1⁄100 of a meter. A kilometer is 1000 m, and a kilogram is 1000 g. The meanings of the commonly used prefixes are given in Table B–1.

THE SI UNITS

Any system of units based on the decimal system would be easy to use, but by international agreement, the preferred set of units, known as the *Système International d'Unités* (SI units) is based on the meter, kilogram, and second. These three fundamental units define the rest of the units as given in Table B-2.

The SI unit of force is the newton (N) named after Isaac Newton. It is the force needed to accelerate a 1 kg mass by 1 m/sec^2, or the force roughly equivalent to the weight of an apple at the earth's surface. The SI unit of energy is the joule (J), the energy produced by a force of

Table B–1 Metric prefixes.

Prefix	Symbol	Factor
mega	M	10^6
kilo	k	10^3
centi	c	10^{-2}
milli	m	10^{-3}
micro	μ	10^{-6}
nano	n	10^{-9}

Table B–2 SI metric units.

Quantity	SI Unit	English Unit
length	meter (m)	foot
mass	kilogram (kg)	slug (sl)
time	second (sec)	second (sec)
force	newton (N)	pound (lb)
energy	joule (J)	foot-pound (fp)

Table B–3 Conversion factors.

1 inch = 2.54 centimeters	1 centimeter = 0.394 inch
1 foot = 0.3048 meter	1 meter = 39.36 inches = 3.28 feet
1 mile = 1.6093 kilometers	1 kilometer = 0.6214 mile
1 slug = 14.594 kilograms	1 kilogram = 0.0685 slug
1 pound = 4.4482 newtons	1 newton = 0.2248 pound
1 foot-pound = 1.35582 joules	1 joule = 0.7376 foot-pound
1 horsepower = 745.7 joules/sec	1 joule/sec = 1 watt

1 N acting through a distance of 1 m. A joule is roughly the energy in the impact of an apple falling off a table.

EXCEPTIONS

Units help us in two ways. They make it possible to make calculations, and they help us to conceive of certain quantities. For calculations the metric system is far superior, and we will use it for our calculations throughout this book.

But Americans commonly use the English system of units, so for conceptual purposes we can express quantities in English units. Instead of saying the average person would weigh 133 N on the moon, we could express their weight as 30 lb. Thus, we will commonly give quantities in metric form followed by the English form in parenthesis. Thus, we might say the radius of the moon is 1738 km (1080 miles).

In SI units, density should be expressed as kilograms per cubic meter, but no human can enclose a cubic meter in their hand, so this unit does not help us grasp the significance of a given density. In this book we will refer to density in grams per cubic centimeter. A gram is roughly the mass of a paperclip, and a cubic centimeter is the size of a small sugar cube, so we can conceive of a density of 1 g/cm^3, roughly the density of water. This is not a bothersome departure from SI units because we will not make complex calculations using density.

CONVERSIONS

To convert from one metric unit to another (from meters to kilometers, for example), we have only to look at the prefix. However, converting from metric to English or English to metric is more complicated. The conversion factors are given in Table B–3.

Example: The radius of the moon is 1738 km. What is this in miles? Table B-3 indicates that 1 mile equals 1.609 km, so

$$1738 \text{ km} \times \frac{1 \text{ mile}}{1.609 \text{ km}} = 1080 \text{ miles}$$

Table B–4 Temperature scales.

	Kelvin (K)	Celsius (°C)	Fahrenheit (°F)
Absolute zero	0 K	−273°C	−459°F
Freezing point of water	273 K	0°C	32°F
Boiling point of water	373 K	100°C	212°F
Conversions:	K = °C + 273	°C = $\frac{5}{9}$(°F − 32)	°F = $\frac{9}{5}$°C + 32

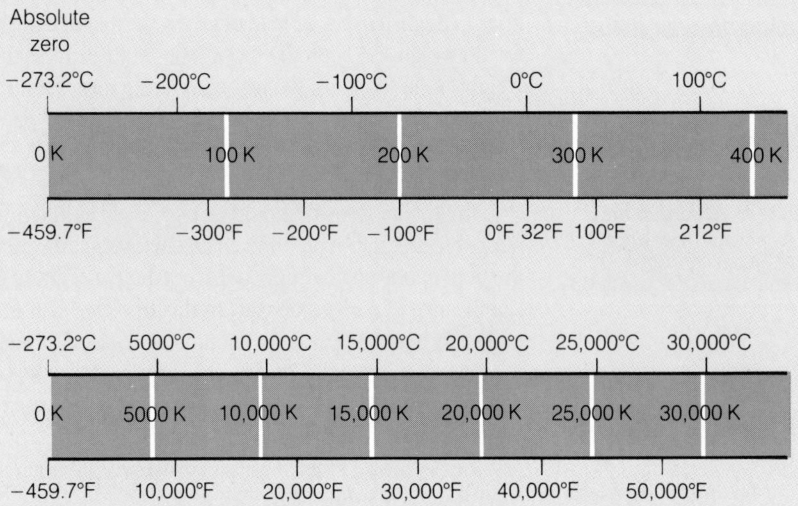

Absolute zero

−273.2°C −200°C −100°C 0°C 100°C

0 K 100 K 200 K 300 K 400 K

−459.7°F −300°F −200°F −100°F 0°F 32°F 100°F 212°F

−273.2°C 5000°C 10,000°C 15,000°C 20,000°C 25,000°C 30,000°C

0 K 5000 K 10,000 K 15,000 K 20,000 K 25,000 K 30,000 K

−459.7°F 10,000°F 20,000°F 30,000°F 40,000°F 50,000°F

TEMPERATURE SCALES

In astronomy, as in most sciences, temperatures are expressed on the Kelvin scale, although the centigrade (or Celsius) scale is also used. The Fahrenheit scale commonly used in the United States is not used in scientific work.

Temperatures on the Kelvin scale are measured from absolute zero, the temperature of an object that contains no extractable heat. In practice, no object can be as cold as absolute zero, although laboratory apparatus have reached temperatures less than 10^{-6} K. The scale is named after the Scottish mathematical physicist William Thomson, Lord Kelvin (1824–1907).

The centigrade scale refers temperatures to the freezing point of water (0°C) and to the boiling point of water (100°C). One degree centigrade is 1/100th the temperature difference between the freezing and boiling points of water. Thus the prefix "centi." The centigrade scale is also called the Celsius scale after its inventor, the Swedish astronomer Anders Celsius (1701–1744).

The Fahrenheit scale fixes the freezing point of water at 32°F and the boiling point at 212°F. Named after the German physicist Gabriel Daniel Fahrenheit (1686–1736), who made the first successful mercury thermometer in 1720, the Fahrenheit scale is used only in the United States.

It is easy to convert temperatures from one scale to another using the information given in Table B-4.

POWERS OF 10 NOTATION

Powers of 10 make writing very large numbers much simpler. For example, the nearest star is about 43,000,000,000,000 km from the sun. Writing this number as 4.3×10^{13} km is much easier.

Very small numbers can also be written with powers of 10. For example, the wavelength of visible light is about 0.0000005 m. In powers of 10 this becomes 5×10^{-7} m.

The powers of 10 used in this notation appear in the table at right. The exponent tells us how to move the decimal point. If the exponent is positive, we move the decimal point to the right. If the exponent is negative, we move the decimal point to the left. Thus, 2×10^3 equals 2000.0, and 2×10^{-3} equals 0.002.

	⋮	
10^5	=	100,000
10^4	=	10,000
10^3	=	1,000
10^2	=	100
10^1	=	10
10^0	=	1
10^{-1}	=	0.1
10^{-2}	=	0.01
10^{-3}	=	0.001
10^{-4}	=	0.0001
	⋮	

APPENDIX C

ASTRONOMICAL DATA

UNITS USED IN ASTRONOMY

1 Ångstrom (Å)	$= 10^{-8}$ cm
	$= 10^{-10}$ m
1 astronomical unit (AU)	$= 1.495979 \times 10^{11}$ m
	$= 92.95582 \times 10^{6}$ miles
1 light-year (ly)	$= 6.3240 \times 10^{4}$ AU
	$= 9.46053 \times 10^{15}$ m
	$= 5.9 \times 10^{12}$ miles
1 parsec (pc)	= 206265 AU
	$= 3.085678 \times 10^{16}$ m
	= 3.261633 ly
1 kiloparsec (kpc)	= 1000 pc
1 megaparsec (Mpc)	= 1,000,000 pc

CONSTANTS

astronomical unit (AU)	$= 1.495979 \times 10^{11}$ m
parsec (pc)	= 206265 AU
	$= 3.085678 \times 10^{16}$ m
	= 3.261633 ly
light-year (ly)	$= 9.46053 \times 10^{15}$ m
velocity of light (c)	$= 2.997925 \times 10^{8}$ m/sec
gravitational constant (G)	$= 6.67 \times 10^{-11}$ N · m²/kg²
mass of earth	$= 5.976 \times 10^{24}$ kg
equatorial radius of earth	= 6378.164 km
mass of sun ($M_\odot$)	$= 1.989 \times 10^{30}$ kg
radius of sun ($R_\odot$)	$= 6.9599 \times 10^{8}$ m
solar luminosity ($L_\odot$)	$= 3.826 \times 10^{26}$ J/sec
mass of moon	$= 7.350 \times 10^{22}$ kg
radius of moon	= 1738 km
mass of H atom	$= 1.67352 \times 10^{-27}$ kg

THE NEAREST STARS

Name	Absolute Magnitude (M_v)	Distance (ly)	Spectral Type	Apparent Visual Magnitude (m_v)
Sun	4.83		G2	−26.8
α Cen A	4.38	4.3	G2	0.1
B	5.76	4.3	K5	1.5
Barnard's Star	13.21	5.9	M5	9.5
Wolf 359	16.80	7.6	M6	13.5
Lalande 21185	10.42	8.1	M2	7.5
Sirius A	1.41	8.6	A1	− 1.5
B	11.54	8.6	white dwarf	7.2
Luyten 726–8A	15.27	8.9	M5	12.5
B (UV Cet)	15.8	8.9	M6	13.0
Ross 154	13.3	9.4	M5	10.6
Ross 248	14.8	10.3	M6	12.2
ϵ Eri	6.13	10.7	K2	3.7
Luyten 789-6	14.6	10.8	M7	12.2
Ross 128	13.5	10.8	M5	11.1
61 CYG A	7.58	11.2	K5	5.2
B	8.39	11.2	K7	6.0
ϵ Ind	7.0	11.2	K5	4.7
Procyon A	2.64	11.4	F5	0.3
B	13.1	11.4	white dwarf	10.8
Σ 2398 A	11.15	11.5	M4	8.9
B	11.94	11.5	M5	9.7
Groombridge 34 A	10.32	11.6	M1	8.1
B	13.29	11.6	M6	11.0
Lacaille 9352	9.59	11.7	M2	7.4
τ Ceti	5.72	11.9	G8	3.5
BD + 5° 1668	11.98	12.2	M5	9.8
L 725–32	15.27	12.4	M5	11.5
Lacaille 8760	8.75	12.5	M0	6.7
Kapteyn's Star	10.85	12.7	M0	8.8
Kruger 60 A	11.87	12.8	M3	9.7
B	13.3	12.8	M4	11.2

THE CONSTELLATIONS

		Approximate Position	
		Right Ascension (h)	Declination (°)
Andromeda (And)	The Princess	1	+40
Antlia (Ant)	The Air Pump	10	−35
Apus (Aps)	The Bird of Paradise	16	−75
Aquarius (Aqr)	The Water Bearer	23	−15
Aquila (Aql)	The Eagle	20	+5
Ara (Ara)	The Altar	17	−55
Aries (Ari)	The Ram	3	+20
Auriga (Aur)	The Charioteer	6	+40
Boötes (Boo)	The Bear Driver	15	+30
Caelum (Cae)	The Sculptor's Chisel	5	−40
Camelopardus (Cam)	The Giraffe	6	+70
Cancer (Cnc)	The Crab	9	+20
Canes Venatici (CVn)	The Hunting Dogs	13	+40
Canis Major (CMa)	The Greater Dog	7	−20
Canis Minor (CMi)	The Smaller Dog	8	+5
Capricornus (Cap)	The Sea Goat	21	−20
Carina (Car)	The Keel	9	−60
Cassiopeia (Cas)	The Seated Queen	1	+60
Centaurus (Cen)	The Centaur	13	−50
Cepheus (Cep)	The King	22	+70
Cetus (Cet)	The Whale	2	−10
Chamaeleon (Cha)	The Chameleon	11	−80
Circinus (Cir)	The Compasses	15	−60
Columbia (Col)	The Dove	6	−35
Coma Berenices (Com)	Berenice's Hair	13	+20
Corona Australis (CrA)	The Southern Crown	19	−40
Corona Borealis (CrB)	The Northern Crown	16	+30
Corvus (Crv)	The Crow	12	−20
Crater (Crt)	The Cup	11	−15
Crux (Cru)	The Southern Cross	12	−60
Cygnus (Cyg)	The Swan	21	+40
Delphinus (Del)	The Dolphin	21	+10
Dorado (Dor)	The Swordfish	5	−65
Draco (Dra)	The Dragon	17	+65
Equuleus (Equ)	The Foal	21	+10
Eridanus (Eri)	The River	3	−20
Fornax (For)	The Laboratory Furnace	3	−30
Gemini (Gem)	The Twins	7	+20
Grus (Gru)	The Crane	22	−45
Hercules (Her)	Hercules	17	+30
Horologium (Hor)	The Clock	3	−60
Hydra (Hya)	The Water Serpent	10	−20
Hydrus (Hyi)	The Water Snake	2	−75

		Approximate Position	
		Right Ascension (h)	Declination (°)
Indus (Ind)	The American Indian	21	−55
Lacerta (Lac)	The Lizard	22	+45
Leo (Leo)	The Lion	11	+15
Leo Minor (LMi)	The Lion Cub	10	+35
Lepus (Lep)	The Hare	6	−20
Libra (Lib)	The Scales	15	−15
Lupus (Lup)	The Wolf	15	−45
Lynx (Lyn)	The Lynx	8	+45
Lyra (Lyr)	The Lyre	19	+40
Mensa (Men)	The Table Mountain	5	−80
Microscopium (Mic)	The Microscope	21	−35
Monoceros (Mon)	The Unicorn	7	−5
Musca (Mus)	The Fly	12	−70
Norma (Nor)	The Carpenter's Square	16	−50
Octans (Oct)	The Octant	22	−85
Ophiuchus (Oph)	The Serpent Holder	17	0
Orion (Ori)	The Great Hunter	5	0
Pavo (Pav)	The Peacock	20	−65
Pegasus (Peg)	The Winged Horse	22	+20
Perseus (Per)	The Hero	3	+45
Phoenix (Phe)	The Phoenix	1	−50
Pictor (Pic)	The Painter's Easel	6	−55
Pisces (Psc)	The Fishes	1	+15
Piscis Austrinus (PsA)	The Southern Fish	22	−30
Puppis (Pup)	The Stern	8	−40
Pyxis (Pyx)	The Compass Box	9	−30
Reticulum (Ret)	The Net	4	−60
Sagitta (Sge)	The Arrow	20	+10
Sagittarius (Sgr)	The Archer	19	−25
Scorpius (Sco)	The Scorpion	17	−40
Sculptor (Scl)	The Sculptor's Workshop	0	−30
Scutum (Sct)	The Shield	19	−10
Serpens (Ser)	The Serpent	17	0
Sextans (Sex)	The Sextant	10	0
Taurus (Tau)	The Bull	4	+15
Telescopium (Tel)	The Telescope	19	−50
Triangulum (Tri)	The Triangle	2	+30
Triangulum Australe (TrA)	The Southern Triangle	16	−65
Tucana (Tuc)	The Toucan	0	−65
Ursa Major (UMa)	The Greater Bear	11	+50
Ursa Minor (UMi)	The Smaller Bear	15	+70
Vela (Vel)	The Sail	9	−50
Virgo (Vir)	The Maiden	13	0
Volcans (Vol)	The Flying Fish	8	−70
Vulpecula (Vul)	The Fox	20	+25

PROPERTIES OF MAIN-SEQUENCE STARS

Spectral Type	Absolute Visual Magnitude (M_v)	Luminosity*	Temp. (K)	λ max (nm)	Mass*	Radius*	Average Density (g/cm^3)
O5	−5.8	501,000	40,000	72.4	40	17.8	0.01
B0	−4.1	20,000	28,000	100	18	7.4	0.1
B5	−1.1	790	15,000	190	6.4	3.8	0.2
A0	+0.7	79	9900	290	3.2	2.5	0.3
A5	+2.0	20	8500	340	2.1	1.7	0.6
F0	+2.6	6.3	7400	390	1.7	1.4	1.0
F5	+3.4	2.5	6600	440	1.3	1.2	1.1
G0	+4.4	1.3	6000	480	1.1	1.0	1.4
G5	+5.1	0.8	5500	520	0.9	0.9	1.6
K0	+5.9	0.4	4900	590	0.8	0.8	1.8
K5	+7.3	0.2	4100	700	0.7	0.7	2.4
M0	+9.0	0.1	3500	830	0.5	0.6	2.5
M5	+11.8	0.01	2800	1000	0.2	0.3	10.0
M8	+16	0.001	2400	1200	0.1	0.1	63

*Luminosity, mass, and radius are given in terms of the sun's luminosity, mass, and radius.

THE BRIGHTEST STARS

Star	Name	Apparent Visual Magnitude (m_v)	Spectral Type	Absolute Visual Magnitude (M_v)	Distance (ly)
α CMa A	Sirius	−1.47	A1	1.4	8.7
α Car	Canopus	−0.72	F0	−3.1	98
α Cen	Rigil Kentaurus	−0.01	G2	4.4	4.3
α Boo	Arcturus	−0.06	K2	−0.3	36
α Lyr	Vega	0.04	A0	0.5	26.5
α Aur	Capella	0.05	G8	−0.6	45
β Ori A	Rigel	0.14	B8	−7.1	900
α CMi A	Procyon	0.37	F5	2.7	11.3
α Ori	Betelgeuse	0.41	M2	−5.6	520
α Eri	Achernar	0.51	B3	−2.3	118
β Cen AB	Hadar	0.63	B1	−5.2	490
α Aql	Altair	0.77	A7	2.2	16.5
α Tau A	Aldebaran	0.86	K5	−0.7	68
α Cru	Acrux	0.90	B2	−3.5	260
α Vir	Spica	0.91	B1	−3.3	220
α Sco A	Antares	0.92	M1	−5.1	520
α PsA	Fomalhaut	1.15	A3	2.0	22.6
β Gem	Pollux	1.16	K0	1.0	35
α Cyg	Deneb	1.26	A2	−7.1	1600
β Cru	Beta Crucis	1.28	B0.5	−4.6	490

THE GREEK ALPHABET

Α, α	alpha	Η, η	eta	Ν, ν	nu	Τ, τ	tau
Β, β	beta	Θ, θ	theta	Ξ, ξ	xi	Υ, υ	upsilon
Γ, γ	gamma	Ι, ι	iota	Ο, ο	omicron	Φ, φ	phi
Δ, δ	delta	Κ, κ	kappa	Π, π	pi	Χ, χ	chi
Ε, ε	epsilon	Λ, λ	lambda	Ρ, ρ	rho	Ψ, ψ	psi
Ζ, ζ	zeta	Μ, μ	mu	Σ, σ	sigma	Ω, ω	omega

PROPERTIES OF THE PLANETS AND SATELLITES

PLANETS: PHYSICAL PROPERTIES (EARTH = ⊕)

Planet	Equatorial Radius (km)	(⊕ = 1)	Mass (⊕ = 1)	Average Density (g/cm^3)	Surface Gravity (⊕ = 1)	Escape Velocity (km/sec)	Sidereal Period of Rotation	Inclination of Equator to Orbit
Mercury	2439	0.382	0.0558	5.44	0.378	4.3	58.646^d	0°
Venus	6052	0.95	0.815	5.24	0.903	10.3	244.3^d	177°
Earth	6378	1.00	1.00	5.497	1.00	11.2	$23^h56^m04.1^s$	23°27′
Mars	3398	0.53	0.1075	3.94	0.379	5.0	$24^h37^m22.6^s$	23°59′
Jupiter	71,900	11.18	317.83	1.34	2.54	61	$9^h50^m30^s$	3°5′
Saturn	60,330	9.42	95.147	0.69	1.16	35.6	$10^h13^m59^s$	26°24′
Uranus	25,559	4.01	14.54	1.19	0.919	22	17^h14^m	97°55′
Neptune	24,750	3.93	17.23	1.66	1.19	25	16^h3^m	28°48′
Pluto	1147	0.18	0.0022	1.84	0.06	1.2	$6^d9^h21^m$	118°?

PLANETS: ORBITAL PROPERTIES

Planet	Semimajor Axis (a) (AU)	(10^6 km)	Orbital Period (P) (y)	(days)	Average Orbital Velocity (km/sec)	Orbital Eccentricity	Inclination to Ecliptic
Mercury	0.3871	57.9	0.24084	87.969	47.89	0.2056	7°0′16″
Venus	0.7233	108.2	0.61515	224.68	35.03	0.0068	3°23′40″
Earth	1	149.6	1	365.26	29.79	0.0167	0°
Mars	1.5237	227.9	1.8808	686.95	24.13	0.0934	1°51′09″
Jupiter	5.2028	778.3	11.867	4334.3	13.06	0.0484	1°18′29″
Saturn	9.5388	1427.0	29.461	10,760	9.64	0.0560	2°29′17″
Uranus	19.18	2869.0	84.013	30,685	6.81	0.0461	0°46′23″
Neptune	30.0611	4497.1	164.793	60,189	5.43	0.0100	1°46′27″
Pluto	39.44	5900	247.7	90,465	4.74	0.2484	17°9′3″

SATELLITES OF THE SOLAR SYSTEM

Planet	Satellite	Radius (km)	Distance from Planet (10^3 km)	Orbital Period (days)	Orbital Eccentricity	Orbital Inclination
Earth	Moon	1738	384.4	27.322	0.055	5°8′43″
Mars	Phobos	14 × 10	9.38	0.3189	0.018	1°.0
	Deimos	8 × 6	23.5	1.262	0.002	2°.8
Jupiter J16	Metis	20	126	0.29	0.0	0°.0
J15	Adrastea	18	128	0.294	0.0	0°.0
J5	Amalthea	135 × 78	182	0.4982	0.003	0°.45
J14	Thebe	38	223	0.674	0.0	1°.3
J1	Io	1820	422	1.769	0.000	0°.3
J2	Europa	1565	671	3.551	0.000	0°.46
J3	Ganymede	2640	1071	7.155	0.002	0°.18
J4	Callisto	2420	1884	16.689	0.008	0°.25
J13	Leda	~4	11,110	240	0.146	26°.7
J6	Himalia	~85	11,470	250.6	0.158	27°.6
J10	Lysithea	~10	11,710	260	0.12	29°
J7	Elara	~30	11,740	260.1	0.207	24°.8
J12	Ananke	8?	20,700	617	0.169	147°
J11	Carme	12?	22,350	692	0.207	163°
J8	Pasiphae	14?	23,300	735	0.40	147°
J9	Sinope	10?	23,700	758	0.275	156°
Saturn S15	Atlas	25 × 10	137.7	0.601	0.002	0°.3
S14	(1980 S27)	70 × 40	139.4	0.613	0.003	0°.0
S13	(1980 S26)	55 × 35	141.7	0.629	0.004	0°.05
S11	Epimetheus	70 × 50	151.4	0.694	0.009	0°.34
S10	Janus	110 × 80	151.4	0.695	0.007	0°.14
S1	Mimas	196	185.54	0.942	0.020	1°.5
S2	Enceladus	250	238.04	1.370	0.004	0°.0
S3	Tethys	530	294.67	1.888	0.000	1°.1
S16	Calypso	17 × 11	294.67	1.888	0.0	~1° ?
S17	Telesto	17 × 13	294.67	1.888	0.0	~1° ?
S4	Dione	560	377	2.737	0.002	0°.0
S12	Helene	18 × 15	377	2.74	0.005	0°.15
S5	Rhea	765	527	4.518	0.001	0°.4
S6	Titan	2575	1222	15.94	0.029	0°.3
S7	Hyperion	205 × 110	1484	21.28	0.104	~0°.5
S8	Iapetus	730	3562	79.33	0.028	14°.72
S9	Phoebe	110	12,930	550.4	0.163	150°

Planet	Satellite	Radius (km)	Distance from Planet (10^3 km)	Orbital Period (days)	Orbital Eccentricity	Orbital Inclination
Uranus	Cordelia	25	49.8	0.3333	~0	~0°
	Ophelia	25	53.8	0.375	~0	~0°
	Bianca	25	59.1	0.433	~0	~0°
	Cressida	30	61.8	0.462	~0	~0°
	Desdemona	30	62.7	0.475	~0	~0°
	Juliet	40	64.4	0.492	~0	~0°
	Portia	40	66.1	0.512	~0	~0°
	Rosalind	30	69.9	0.558	~0	~0°
	Belinda	30	75.2	0.621	~0	~0°
	Puck	85 ± 5	85.9	0.762	~0	~0°
U5	Miranda	242 ± 5	129.9	1.414	0.017	3°.4
U1	Ariel	580 ± 5	190.9	2.520	0.003	0°
U2	Umbrial	595 ± 10	266.0	4.144	0.003	0°
U3	Titania	805 ± 5	436.3	8.706	0.002	0°
U4	Oberon	775 ± 10	583.4	13.463	0.001	0°
Neptune						
	1989N6	25	48.2	0.296	~0	~0°
	1989N5	45	50.0	0.312	~0	~0°
	1989N3	70	52.5	0.333	~0	~0°
	1989N4	80	62.0	0.396	~0	~0°
	1989N2	100	73.6	0.554	~0	~0°
	1989N1	200	117.6	1.121	~0	~0°
N1	Triton	1360	354.59	5.877	0.00	160°
N2	Nereid	170	5510.66	359.4	~0.76	27.7°
Pluto						
P1	Charon	642	17	6.3861	~0	120°

METEOR SHOWERS

Shower	Dates	Hourly Rate	Radiant R.A.	Radiant Dec.	Associated Comet
Quadrantids	Jan. 2–4	30	15^h24^m	50°	
Lyrids	April 20–22	8	$18^h\ 4^m$	33°	1861 I
η Aquarids	May 2–7	10	22^h24^m	0°	Halley?
δ Aquarids	July 26–31	15	22^h36^m	−10°	
Perseids	Aug. 10–14	40	$3^h\ 4^m$	58°	1982 III
Orionids	Oct. 18–23	15	6^h20^m	15°	Halley?
Taurids	Nov. 1–7	8	3^h40^m	17°	Encke
Leonids	Nov. 14–19	6	10^h12^m	22°	1866 I Temp
Geminids	Dec. 10–13	50	7^h28^m	32°	

THE MESSIER OBJECTS

M	NGC*	Right Ascension (1950)	Declination (1950)	Apparent Visual Magnitude (m_v)	Description
1	1952	$5^h31.5^m$	+21°59′	8.4	Crab Nebula in Taurus; remains of supernova
2	7089	$21^h30.9^m$	−1°02′	6.4	Globular cluster in Aquarius
3	5272	$13^h39.8^m$	+28°38′	6.3	Globular cluster in Canes Venatici
4	6121	$16^h20.6^m$	−26°24′	6.5	Globular cluster in Scorpius
5	5904	$15^h16.0^m$	+2°16′	6.1	Globular cluster in Serpens
6	6405	$17^h36.8^m$	−32°10′	5.3	Open cluster in Scorpius
7	6475	$17^h50.7^m$	−34°48′	4.1	Open cluster in Scorpius
8	6523	$18^h00.1^m$	−24°23′	6.0	Lagoon Nebula in Sagittarius
9	6333	$17^h16.3^m$	−18°28′	7.3	Globular cluster in Ophiuchus
10	6254	$16^h54.5^m$	−4°02′	6.7	Globular cluster in Ophiuchus
11	6705	$18^h48.4^m$	−6°20′	6.3	Open cluster in Scutum
12	6218	$16^h44.7^m$	−1°52′	6.6	Globular cluster in Ophiuchus
13	6205	$16^h39.9^m$	+36°33′	5.9	Globular cluster in Hercules
14	6402	$17^h35.0^m$	−3°13′	7.7	Globular cluster in Ophiuchus
15	7078	$21^h27.5^m$	+11°57′	6.4	Globular cluster in Pegasus
16	6611	$18^h16.1^m$	−13°48′	6.4	Open cluster with nebulosity in Serpens
17	6618	$18^h17.9^m$	−16°12′	7.0	Swan or Omega Nebula in Sagittarius
18	6613	$18^h17.0^m$	−17°09′	7.5	Open cluster in Sagittarius
19	6273	$16^h59.5^m$	−26°11′	6.6	Globular cluster in Ophiuchus
20	6514	$17^h59.4^m$	−23°02′	9.0	Trifid Nebula in Sagittarius
21	6531	$18^h01.6^m$	−22°30′	6.5	Open cluster in Sagittarius
22	6656	$18^h33.4^m$	−23°57′	5.6	Globular cluster in Sagittarius
23	6494	$17^h54.0^m$	−19°00′	6.9	Open cluster in Sagittarius
24	6603	$18^h15.5^m$	−18°27′	11.4	Open cluster in Sagittarius
25	IC4725†	$18^h28.7^m$	−19°17′	6.5	Open cluster in Sagittarius
26	6694	$18^h42.5^m$	−9°27′	9.3	Open cluster in Scutum
27	6853	$19^h57.5^m$	+22°35′	7.6	Dumb-bell Planetary Nebula in Vulpecula
28	6626	$18^h21.4^m$	−24°53′	7.6	Globular cluster in Sagittarius
29	6913	$20^h22.2^m$	+38°21′	7.1	Open cluster in Cygnus
30	7099	$21^h37.5^m$	−23°24′	8.4	Globular cluster in Capricornus
31	224	$0^h40.0^m$	+41°00′	4.8	Andromeda galaxy
32	221	$0^h40.0^m$	+40°36′	8.7	Elliptical galaxy; companion to M31
33	598	$1^h31.0^m$	+30°24′	6.7	Spiral galaxy in Triangulum
34	1039	$2^h38.8^m$	+42°35′	5.5	Open cluster in Perseus
35	2168	$6^h05.7^m$	+24°21′	5.3	Open cluster in Gemini
36	1960	$5^h33.0^m$	+34°04′	6.3	Open cluster in Auriga
37	2099	$5^h49.1^m$	+32°33′	6.2	Open cluster in Auriga
38	1912	$5^h25.3^m$	+35°47′	7.4	Open cluster in Auriga
39	7092	$21^h30.4^m$	+48°13′	5.2	Open cluster in Cygnus

*New General Catalogue number.

†Index Catalogue (IC) number.

M	NGC*	Right Ascension (1950)	Declination (1950)	Apparent Visual Magnitude (m_v)	Description
42	1976	$5^h32.9^m$	−5°25′	4.0	Orion Nebula
43	1982	$5^h33.1^m$	−5°19′	9.0	Northeast portion of Orion Nebula
44	2632	$8^h37.0^m$	+20°10′	3.7	Praesepe; open cluster in Cancer
45	—	$3^h44.5^m$	+23°57′	1.6	The Pleiades; open cluster in Taurus
46	2437	$7^h39.5^m$	−14°42′	6.0	Open cluster in Puppis
47	2422	$7^h34.3^m$	−14°22′	5.2	Loose group of stars in Puppis
48	2458	$8^h11.0^m$	−5°38′	5.5	Open cluster in Hydra
49	4472	$12^h27.3^m$	+8°16′	8.5	Elliptical galaxy in Virgo
50	2323	$7^h00.6^m$	−8°16′	6.3	Loose open cluster in Monoceros
51	5194	$13^h27.8^m$	+47°27′	8.4	Whirlpool spiral galaxy in Canes Venatici
52	7654	$23^h22.0^m$	+61°20′	7.3	Loose open cluster in Cassiopeia
53	5024	$13^h10.5^m$	+18°26′	7.8	Globular cluster in Coma Berenices
54	6715	$18^h51.9^m$	−30°32′	7.3	Globular cluster in Sagittarius
55	6809	$19^h36.8^m$	−31°03′	7.6	Globular cluster in Sagittarius
56	6779	$19^h14.6^m$	+30°05′	8.2	Globular cluster in Lyra
57	6720	$18^h51.7^m$	+32°58′	9.0	Ring Nebula; planetary nebula in Lyra
58	4579	$12^h35.2^m$	+12°05′	8.2	Barred spiral galaxy in Virgo
59	4621	$12^h39.5^m$	+11°56′	9.3	Elliptical spiral galaxy in Virgo
60	4649	$12^h41.1^m$	+11°50′	9.0	Elliptical galaxy in Virgo
61	4303	$12^h19.3^m$	+4°45′	9.6	Spiral galaxy in Virgo
62	6266	$16^h58.0^m$	−30°02′	6.6	Globular cluster in Ophiuchus
63	5055	$13^h13.5^m$	+42°17′	10.1	Spiral galaxy in Canes Venatici
64	4826	$12^h54.2^m$	+21°57′	6.6	Spiral galaxy in Coma Berenices
65	3623	$11^h16.3^m$	+13°22′	9.4	Spiral galaxy in Leo
66	3627	$11^h17.6^m$	+13°16′	9.0	Spiral galaxy in Leo; companion to M65
67	2682	$8^h48.4^m$	+12°00′	6.1	Open cluster in Cancer
68	4590	$12^h36.8^m$	−26°29′	8.2	Globular cluster in Hydra
69	6637	$18^h28.1^m$	−32°24′	8.9	Globular cluster in Sagittarius
70	6681	$18^h40.0^m$	−32°20′	9.6	Globular cluster in Sagittarius
71	6838	$19^h51.5^m$	+18°39′	9.0	Globular cluster in Sagitta
72	6981	$20^h50.7^m$	−12°45′	9.8	Globular cluster in Aquarius
73	6994	$20^h56.2^m$	−12°50′	9.0	Open cluster in Aquarius
74	628	$1^h34.0^m$	+15°32′	10.2	Spiral galaxy in Pisces
75	6864	$20^h03.1^m$	−22°04′	8.0	Globular cluster in Sagittarius
76	650	$1^h38.8^m$	+51°19′	11.4	Planetary nebula in Perseus
77	1068	$2^h40.1^m$	−0°12′	8.9	Spiral galaxy in Cetus
78	2068	$5^h44.2^m$	+0°02′	8.3	Small reflection nebula in Orion
79	1904	$5^h22.1^m$	−24°34′	7.5	Globular cluster in Lepus
80	6093	$16^h14.0^m$	−22°52′	7.5	Globular cluster in Scorpius
81	3031	$9^h51.7^m$	+69°18′	7.9	Spiral galaxy in Ursa Major
82	3034	$9^h51.9^m$	+69°56′	8.4	Irregular galaxy in Ursa Major
83	5236	$13^h34.2^m$	−29°37′	10.1	Spiral galaxy in Hydra
84	4374	$12^h22.6^m$	+13°10′	9.4	S0 type galaxy in Virgo
85	4382	$12^h22.8^m$	+18°28′	9.3	S0 type galaxy in Coma Berenices

M	NGC	Right Ascension (1950)	Declination (1950)	Apparent Visual Magnitude (m_v)	Description
86	4406	$12^h23.6^m$	+13°13′	9.2	Elliptical galaxy in Virgo
87	4486	$12^h28.2^m$	+12°40′	8.7	Elliptical galaxy in Virgo
88	4501	$12^h29.4^m$	+14°42′	10.2	Spiral galaxy in Coma Berenices
89	4552	$12^h33.1^m$	+12°50′	9.5	Elliptical galaxy in Virgo
90	4569	$12^h34.3^m$	+13°26′	9.6	Spiral galaxy in Virgo
91*	4571(?)	—	—	—	
92	6341	$17^h15.6^m$	+43°12′	6.4	Globular cluster in Hercules
93	2447	$7^h42.4^m$	−23°45′	6.0	Open cluster in Puppis
94	4736	$12^h48.6^m$	+41°24′	8.3	Spiral galaxy in Canes Venatici
95	3351	$10^h41.3^m$	+11°58′	9.8	Barred spiral galaxy in Leo
96	3368	$10^h44.1^m$	+12°05′	9.3	Spiral galaxy in Leo
97	3587	$11^h12.0^m$	+55°17′	12.0	Owl Nebula; planetary nebula in Ursa Major
98	4192	$12^h11.2^m$	+15°11′	10.2	Spiral galaxy in Coma Berenices
99	4254	$12^h16.3^m$	+14°42′	9.9	Spiral galaxy in Coma Berenices
100	4321	$12^h20.4^m$	+16°06′	10.6	Spiral galaxy in Coma Berenices
101	5457	$14^h01.4^m$	+54°36′	9.6	Spiral galaxy in Ursa Major
102*	5866(?)	—	—	—	
103	581	$1^h29.9^m$	+60°26′	7.4	Open cluster in Cassiopeia
104	4594	$12^h37.4^m$	−11°21′	8.3	Spiral galaxy in Virgo
105	3379	$10^h45.2^m$	+13°01′	9.7	Elliptical galaxy in Leo
106	4258	$12^h16.5^m$	+47°35′	8.4	Spiral galaxy in Canes Venatici
107	6171	$16^h29.7^m$	−12°57′	9.2	Globular cluster in Ophiuchus

*Items of doubtful identification.

GLOSSARY

absolute bolometric magnitude The absolute magnitude we would observe if we would detect all wavelengths.

absolute visual magnitude (M_v) Intrinsic brightness of a star. The apparent visual magnitude the star would have if it were 10 pc away.

absorption line A dark line in a spectrum. Produced by the absence of photons absorbed by atoms or molecules.

absorption spectrum (dark line spectrum) A spectrum that contains absorption lines.

acceleration A change in a velocity. Either a change in speed or direction. (See **velocity**.)

acceleration of gravity A measure of the strength of gravity at a planet's surface.

accretion The sticking together of solid particles to produce a larger particle.

accretion disk The whirling disk of gas that forms around a compact object such as a white dwarf, neutron star, or black hole as matter is drawn in.

achondrites Stony meteorites containing no chondrules or volatiles.

achromatic lens A telescope lens composed of two lenses ground from different kinds of glass and designed to bring two selected colors to the same focus and correct for chromatic aberration.

active galacitic nucleus The central energy source of an active galaxy.

active galaxy A galaxy that is a source of excess radiation, usually radio, X rays, gamma rays, or some combination.

active optics Optical elements whose position or shape is continuously controlled by computers.

alt–azimuth mounting A telescope mounting capable of motion parallel to and perpendicular to the horizon.

amino acids Carbon-chain molecules that are the building blocks of protein.

Angstrom (Å) A unit of distance. 1 Å $= 10^{-10}$ m. Often used to measure the wavelength of light.

angular momentum The tendency of a rotating body to continue rotating. Mathematically it is the product of mass, velocity, and radius.

annular eclipse A solar eclipse in which the solar photosphere appears around the edge of the moon in a bright ring or annulus. The corona, chromosphere, and prominences cannot be seen.

anorthosite Rock of aluminum and calcium silicates found in the lunar highlands.

aphelion The orbital point of greatest distance from the sun.

apogee The orbital point of greatest distance from Earth.

Apollo/Amor objects Asteroids whose orbits cross that of Earth/Mars.

apparent relative orbit The orbit of one star in a visual binary with respect to the other star as seen from Earth. (See **true relative orbit.**)

apparent visual magnitude (m_v) The brightness of a star as seen by human eyes on Earth.

archaeoastronomy The study of the astronomy of ancient cultures.

associations Groups of widely scattered stars (10–1000) moving together through space. Not gravitationally bound into clusters.

asterism A named group of stars not identified as a constellation, e.g., the Big Dipper.

asteroids Small rocky worlds most of which lie between Mars and Jupiter in the asteroid belt.

astrometric binary A binary star identified by its irregular proper motion.

astronomical unit (AU) Average distance from the earth to the sun; 1.5×10^8 km, or 93×10^6 miles.

atmospheric window Wavelength regions in which our atmosphere is transparent—at visual wavelengths, infrared and at radio wavelengths.

aurora The glowing light display that results when a planet's magnetic field guides charged particles toward the north and south magnetic poles, where they strike the upper atmosphere and excite atoms to emit photons.

autumnal equinox The point on the celestial sphere where the sun crosses the celestial equator going southward. Also, the time when the sun reaches this point and autumn begins in the northern hemisphere—about September 22.

Bailey's beads Bright spots of the solar surface visible at the edge of the moon during a total solar eclipse.

Balmer series Spectral lines in the visible and near-ultraviolet spectrum of hydrogen produced by transitions whose lowest orbit is the second.

barred spiral galaxy A spiral galaxy with an elongated nucleus resembling a bar from which the arms originate.

barycenter The center of motion around which a pair of binary stars move.

basalt Dark, igneous rock characteristic of solidified lava.

belts Dark bands of clouds that circle Jupiter parallel to its equator. Generally red, brown, or blue-green. Believed to be regions of descending gas.

β Canis Majoris variables Short-period variable stars that do not lie in the instability strip.

bipolar flows Oppositely directed jets of gas ejected by some protostellar objects.

big bang theory The theory that the universe began with a violent explosion from which the expanding universe of galaxies eventually formed.

binary stars Pairs of stars that orbit around their common center of mass.

binding energy The energy needed to pull an electron away from its atom.

black body radiation Radiation emitted by a hypothetical perfect radiator. The spectrum is continuous, and the wavelength of maximum emission depends only on the body's temperature.

black dwarf The end state of a white dwarf that has cooled to low temperature.

black hole A mass that has collapsed to such a small volume that its gravity prevents the escape of all radiation. Also, the volume of space from which radiation may not escape.

BL Lac object Objects that resemble quasars. Thought to be highly luminous cores of distant galaxies.

blue shift The shortening of the wavelengths of light observed when the source and observer are approaching each other.

Bok globules Small, dark clouds only about 1 light-year in diameter that contain 10–1000 $M_\odot$ of gas and dust. Believed related to star formation.

bolometric magnitude The magnitude we would measure if we could detect electromagnetic radiation of all wavelengths.

bow shock The boundary between the undisturbed solar wind and the region being deflected around a planet or comet.

breccia A rock composed of fragments of earlier rocks bonded together.

bright line spectrum See **emission spectrum.**

brown dwarf A very cool, low-luminosity star whose mass is not sufficient to ignite nuclear fusion.

burster A source of bursts of X rays or, in some cases, gamma rays. Believed associated with neutron stars.

butterfly diagram See **Maunder butterfly diagram.**

CAI Calcium-aluminum–rich inclusions found in some meteorites. Believed to be very old.

Cambrian period A geological period 0.6–0.5 billion years ago during which life on Earth became diverse and complex. Cambrian rocks contain the oldest easily identifiable fossils.

capture hypothesis The theory that the moon formed elsewhere in the solar system and was later captured by the earth.

carbonaceous chondrites Stony meteorites that contain both chondrules and volatiles. They may be the least altered remains of the solar nebula still present in the solar system.

carbon detonation The explosive ignition of carbon burning in some giant stars. A possible cause of some supernova explosions.

carbon–nitrogen–oxygen (CNO) cycle A series of nuclear reactions that use carbon as a catalyst to combine four hydrogen atoms to make one helium atom plus energy. Effective in stars more massive than the sun.

Cassegrain telescope A reflecting telescope in which the secondary mirror reflects light back down the tube through a hole in the center of the objective mirror.

celestial equator The imaginary line around the sky directly above the earth's equator.

celestial sphere An imaginary sphere of very large radius surrounding the earth and to which the planets, stars, sun, and moon seem to be attached.

center of mass The balance point of a body or system of bodies.

Cepheid variable stars Variable stars with a period of 1–60 days. Period of variation related to luminosity.

Chandrasekhar limit The maximum mass of a white dwarf, about 1.4 $M_\odot$. A white dwarf of greater mass cannot support itself and will collapse.

charge-coupled device (CCD) An electronic device consisting of a large array of light-sensitive elements used to record very faint images.

chemical evolution The chemical process that led to the growth of complex molecules on the primitive earth. This did not involve the reproduction of molecules.

chondrite A stony meteorite that contains chondrules.

chondrules Round glassy bodies in some stony meteorites. Believed to have solidified very quickly from molten drops of silicate material.

chromatic aberration A distortion found in refracting telescopes because lenses focus different colors at slightly different distances. Images are consequently surrounded by color fringes.

chromosphere Bright gases just above the photosphere of the sun. Responsible for the emission lines in the flash spectrum.

circular velocity The velocity required to remain in a circular orbit about a body.

circumpolar constellation Any of the constellations so close to the celestial pole that they never set (or never rise) as seen from a given latitude.

closed orbit An orbit that returns to its starting point. A circular or elliptical orbit. (See **open orbit.**)

closed universe A model universe in which the average density is great enough to stop the expansion and make the universe contract.

cluster method The method of determining the masses of galaxies based on the motions of galaxies in a cluster.

CNO cycle See **carbon–nitrogen–oxygen cycle.**

co-accretion hypothesis The theory that the earth and moon formed together.

cocoon The cloud of gas and dust around a contracting protostar that conceals it at visible wavelengths.

collisional broadening The smearing out of a spectral line because of collisions among the atoms of the gas.

color index A numerical measure of the color of a star.

coma The glowing head of a comet.

comet One of the small icy bodies that orbit the sun and produce tails of gas and dust when they near the sun.

compact object A star that has collapsed to form a white dwarf, neutron star, or black hole.

comparative planetology The study of planets by comparing the characteristics of different examples.

comparison spectrum A spectrum of known spectral lines used to identify unknown wavelengths in an object's spectrum.

condensation The growth of a particle by addition of material from surrounding gas, atom by atom.

condensation sequence The sequence in which different materials condense from the solar nebula as we move outward from the sun.

conic sections The family of curves generated by slicing a cone–the circle, ellipse, hyperbola, and parabola.

constellation One of the stellar patterns identified by name, usually of mythological gods, people, animals, or objects. Also, the region of the sky containing that star pattern.

continuity of energy law One of the basic laws of stellar structure. The amount of energy flowing out of the top of a shell must equal the amount coming in at the bottom plus whatever energy is generated within the shell.

continuity of mass law One of the basic laws of stellar structure. The total mass of the star must equal the sum of the masses of the shells, and the mass must be distributed smoothly throughout the star.

continuous spectrum A spectrum in which there are no absorption or emission lines.

Copernican principle The belief that the earth is not in a special place in the universe.

corona The faint outer atmosphere of the sun. Composed of low-density, very hot, ionized gas.

coronagraph A telescope designed to photograph the inner corona of the sun.

coronal hole An area of the solar surface that is dark at X-ray wavelengths. Thought to be associated with divergent magnetic fields and the source of the solar wind.

cosmic rays Atomic nuclei that enter Earth's atmosphere at nearly the speed of light. Some originate in solar flares and some may come from supernova explosions, but their true nature is not well-understood.

cosmological principle The assumption that any observer in any galaxy sees the same general features of the universe.

cosmological test A measurement or observation whose result can help us choose between different cosmological theories.

cosmology The study of the nature, origin, and evolution of the universe.

coudé focus The focal arrangement of a reflecting telescope in which mirrors direct the light to a fixed focus beyond the bounds of the telescope's movement, typically in a separate room. Usually used for spectroscopy.

Coulomb barrier The electrostatic force of repulsion between bodies of like charge. Commonly applied to atomic nuclei.

critical density The average density of the universe needed to make its curvature flat.

critical point The temperature and pressure at which the vapor and liquid phases of a material have the same density.

cultural shock The bewildering impact of an advanced society upon a less sophisticated one.

dark line spectrum See **absorption spectrum.**

dark matter The matter believed to exist to make up the missing mass in the universe.

dark nebula A nebula consisting of dust and gas blocking our view of more distant stars.

decameter radiation Radio signals from Jupiter with wavelengths of about 10 m.

decimeter radiation Radio signals from Jupiter with wavelengths of about 0.1 m.

declination A coordinate used on the celestial sphere just as latitude is used on earth. An object's declination is measured from the celestial equator—positive to the north and negative to the south.

declination axis The pivot in a telescope mounting that allows the telescope to move north–south.

deferent In the Ptolemaic theory, the large circle around the earth along which the center of the epicycle moved.

degenerate matter Extremely high-density matter in which pressure no longer depends on temperature due to quantum mechanical effects.

density wave theory Theory proposed to account for spiral arms as compressions of the interstellar medium in the disk of the galaxy.

diamond ring A momentary phenomenon seen during some total solar eclipses when the ring of the corona and a bright spot of photosphere resemble a large diamond set in a silvery ring.

differential rotation The rotation of a body in which different parts of the body have different periods of rotation. This is true of the sun, the Jovian planets, and the disk of the galaxy.

differentiation The separation of planetary material according to density.

diffraction fringe Blurred fringe surrounding any image caused by the wave properties of light. Because of this, no image detail smaller than the fringe can be seen.

dilation of time The slowing of time as recorded by moving clocks.

direct orbit An orbit that carries a moon or planet in the same direction as most other motion in the solar system—counterclockwise as seen from the north.

dirty snowball theory The hypothesis that comets are kilometer-size balls of ices with embedded impurities.

disk component All material confined to the plane of the galaxy.

distance indicators Objects whose luminosities or diameters are known. Used to find the distance to a star cluster or galaxy.

distance modulus ($m - M_v$) The difference between the apparent and absolute magnitude of a star. A measure of how far away the star is.

diurnal motion Daily motion as in the rising and setting of the sun.

DNA (deoxyribonucleic acid) The long carbon-chain molecule that records information to govern the biological activity of the organism. DNA carries the genetic data passed to offspring.

Doppler broadening The smearing of spectral lines because of the motion of the atoms in the gas.

Doppler effect The change in the wavelength of radiation due to relative radial motion of source and observer.

double exhaust model The theory that double radio lobes are produced by pairs of jets emitted in opposite directions from the centers of active galaxies.

double galaxy method A method of finding the masses of galaxies from orbiting pairs of galaxies.

double-line spectroscopic binary A spectroscopic binary star in which spectral lines from both stars are visible in the spectrum.

double-lobed radio source A galaxy that emits radio energy from two regions (lobes) located on opposite sides of the galaxy.

double stars A pair of stars close together in the sky. Not all double stars are necessarily in orbit around each other.

dust tail (type II) The tail of a comet formed of dust blown outward by the pressure of sunlight. (See **gas tail.**)

dwarf nova A star that undergoes novalike explosions every few days or weeks. Believed associated with mass transfer onto a white dwarf in a binary system.

dynamo effect The theory that the earth's magnetic field is generated in the conducting material of its molten core.

eclipse season That period when the sun is near a node of the moon's orbit and eclipses are possible.

eclipse year The time the sun takes to circle the sky and return to a node of the moon's orbit. 346.62 days.

eclipsing binary A binary star system in which the stars eclipse each other.

ecliptic The apparent path of the sun around the sky.

ejecta Pulverized rock scattered by meteorite impacts on a planetary surface.

electroglow The ultraviolet radiation produced in the upper atmospheres of Jupiter, Saturn, and Uranus by high-energy particles in the planets' magnetospheres.

electromagnetic radiation Changing electric and magnetic fields that travel through space and transfer energy from one place to another. For example, light, radio waves, etc.

electrons Low-mass atomic particles carrying negative charges.

electron volt (eV) A unit of energy equal to the energy produced by an electron accelerated through a voltage difference of 1 volt.

ellipse A closed curve enclosing two points (foci) such that the total distance from one focus to any point on the curve back to the other focus equals a constant.

elliptical galaxy A galaxy that is round or elliptical in outline. It contains little gas and dust, no disk or spiral arms, and few hot, bright stars.

emission line A bright line in a spectrum caused by the emission of photons from atoms.

emission nebula A cloud of glowing gas excited by ultraviolet radiation from hot stars.

emission spectrum (bright line spectrum) A spectrum containing emission lines.

energy level One of a number of states an electron may occupy in an atom, depending on its binding energy.

energy machine An object that releases energy. Commonly used to refer to the source of energy in active galactic nuclei.

energy transport Energy must flow from hot regions to cooler regions by conduction, convection, or radiation.

enzymes Special proteins that control processes in an organism.

epicycle The small circle followed by a planet in the Ptolemaic theory. The center of the epicycle follows a larger circle (deferent) around Earth.

equant The point off center in the deferent from which the center of the epicycle appears to move uniformly.

equation of time The difference between apparent solar time and mean solar time.

equatorial mounting A telescope mounting that allows motion parallel to and perpendicular to the celestial equator.

ergosphere The region surrounding a rotating black hole within which one could not resist being dragged around the black hole. It is possible for a particle to escape from the ergosphere and extract energy from the black hole.

escape orbit An orbit that does not return to its starting point. (See **open orbit.**)

escape velocity The initial velocity an object needs to escape from the surface of a celestial body.

ether The medium through which light traveled according to late-nineteenth-century physics. Accepted as nonexistent by modern physics.

evening star Any planet visible in the sky just after sunset.

event horizon The boundary of the region of a black hole from which no radiation may escape. No event that occurs within the event horizon is visible to a distant observer.

excited atom An atom in which an electron has moved from a lower to a higher orbit.

extinction The dimming of light by intervening material. Commonly used to refer to the dimming of light by the interstellar medium.

eyepiece A short-focal-length lens used to enlarge the image in a telescope. The lens nearest the eye.

faculae Bright areas of the solar surface associated with sunspots and prominences.

fall A meteorite seen to fall. (See **find.**)

false color A representation of graphical data in which the colors are altered or added to reveal details.

field (gravitational, electric, or magnetic) A way of explaining action at a distance. A particle produces a field of influence to which another particle in the field responds.

filar micrometer An instrument that permits precise measurements at the telescope of the position of visual binary stars and similar objects.

filtergram A photograph (usually of the sun) taken in the light of a specific region of the spectrum—e.g., an H-alpha filtergram.

find A meteorite which is found but was not seen to fall. (See **fall.**)

fission hypothesis The theory that the moon formed by breaking away from the earth.

flare A violent eruption on the sun's surface.

flash spectrum The emission spectrum of the chromosphere that is visible for the few seconds during a total solar eclipse when the moon has covered the photosphere but has not yet covered the chromosphere.

flatness problem In cosmology the circumstance that the early universe must have contained almost exactly the right amount of matter to close space-time (to make space-time flat).

flat universe A model of the universe in which space-time is not curved.

flocculent Woolly, fluffy. Used to refer to certain galaxies that have a woolly appearance.

focal length The focal length of a lens is the distance from the lens to the point where it focuses parallel rays of light.

focus (of an ellipse) One of two points inside an ellipse that satisfy the condition that the distance from one focus to any point on the ellipse to the other focus is a constant.

forward scattering The optical property of finely divided particles to preferentially direct light in the original direction of the light's travel.

galactic cannibalism The theory that large galaxies absorb smaller galaxies.

galactic corona The low-density extension of the halo of a galaxy. Now suspected to extend many times the visible diameter of the galaxy.

Galilean satellites The four largest satellites of Jupiter, named after their discoverer, Galileo.

gas tail (type I) The tail of a comet produced by gas blown outward by the solar wind. (See **dust tail.**)

Gauss (G) A unit used to measure the strength of a magnetic field.

general relativity Einstein's second theory of relativity, which proposes that a gravitational field is a curvature of space-time caused by the presence of a mass.

geocentric universe A model universe with the earth at the center, such as the Ptolemaic universe.

geosynchronous orbit An eastward orbit whose period is 24 hours. A satellite in such an orbit remains above the same spot on the earth's surface.

giant molecular clouds Very large, cool clouds of dense gas in which star formation occurs.

giant stars Large, cool, highly luminous stars in the upper right of the H–R diagram. Typically 10 to 100 times the diameter of the sun.

glacial period An interval when ice sheets cover large areas of the land.

glitch A sudden change in the period of a pulsar.

globular cluster A star cluster containing 50,000 to 1 million stars in a sphere about 75 ly in diameter. Generally old, metal-poor, and found in the spherical component of the galaxy.

graben rille A linear feature on a planetary surface caused by faulting and sinking of portions of the crust.

grand unified theories (GUTs) Theories that attempt to unify (describe in a similar way) the electromagnetic, weak, and strong forces of nature.

granulation The fine structure visible on the solar surface caused by convective cells below.

grating A piece of material in which numerous microscopic parallel lines are scribed. Light encountering a grating is dispersed to form a spectrum.

gravitational lens effect The focusing of light from a distant galaxy or quasar by an intervening galaxy to produce multiple images of the distant body.

gravitational red shift The lengthening of the wavelength of a photon due to its escape from a gravitational field.

gravitational wave The transport of energy by the motion of waves in a gravitational field. Predicted by general relativity.

grazing incidence optics Reflecting mirrors using very shallow angles of incidence to focus X rays.

greatest elongation The maximum angular separation between an object and the sun. Typically said of Mercury or Venus.

greenhouse effect The process by which a carbon dioxide atmosphere traps heat and raises the temperature of a planetary surface.

Gregorian calendar The calendar now in use, instituted by Pope Gregory XIII in 1582.

grooved terrain Regions of the surface of Ganymede consisting of parallel grooves. Believed to have formed by repeated fracture and refreezing of the icy crust.

ground state The lowest permitted electron orbit in an atom.

half-life The time required for half of the atoms in a radioactive sample to decay.

halo The spherical region of a spiral galaxy containing a thin scattering of stars, star clusters, and small amounts of gas.

head–tail radio galaxy A radio galaxy with a contour consisting of a head and a tail. Believed caused by the motion of an active galaxy through the intergalactic medium.

heat of formation In planetology, the heat released by the infall of matter during the formation of a planetary body.

heliocentric universe A model of the universe with the sun at the center, such as the Copernican universe.

helioseismology The study of the interior of the sun by the analysis of its modes of vibration.

helium flash The explosive ignition of helium burning that takes place in some giant stars.

Herbig–Haro objects Small nebulae that vary irregularly in brightness. Believed associated with star formation.

Hertzsprung–Russell diagram A plot of the intrinsic brightness versus the surface temperature of stars. It separates the effects of temperature and surface area on stellar luminosity. Commonly absolute magnitude versus spectral type, but also luminosity versus surface temperature or color.

heterogeneous accretion The formation of a planet by the accumulation of planetesimals of different composition—e.g., first iron particles, then silicates. (See **homogeneous accretion.**)

high-velocity star A star with a large space velocity. Such stars are halo stars passing through the disk of the galaxy at steep angles.

Hirayama families Families of asteroids with orbits of similar size, shape, and orientation. Believed to be fragments of larger bodies.

homogeneity The assumption that, on the large scale, matter is uniformly spread through the universe.

homogeneous accretion The formation of a planet by the accumulation of planetesimals of the same composition. (See **heterogeneous accretion.**)

horizon problem In cosmology the circumstance that the primordial background radiation seems much more isotropic than could be explained by the standard big bang theory.

horizontal branch In the H–R diagram of a globular cluster, the sequence of stars extending from the red giants toward the blue side of the diagram. Includes RR Lyrae stars.

horoscope A chart showing the positions of the sun, moon, planets, and constellations at the time of a person's birth. Used in astrology to attempt to read character or foretell the future.

horseshoe orbit The path followed by a small moon when it occupies the same orbit as a larger moon. The small moon moves along a horseshoe-shaped orbit with respect to the larger moon.

hot spot In geology a place on the earth's crust where volcanism is caused by a rising convection cell in the mantle below. In radio astronomy, a bright spot in a radio lobe.

hour angle The angle from the local celestial meridian westward to the right ascension of a star or body.

H–R diagram See **Hertzsprung–Russell diagram.**

H II region A region of ionized hydrogen around a hot star.

Hubble constant (*H*) A measure of the rate of expansion of the universe. The average value of velocity of recession divided by distance. Presently believed to be about 50 km/sec/megaparsec.

Hubble law The linear relation between the distance to a galaxy and its radial velocity.

hydrostatic equilibrium The balance between the weight of the material pressing downward on a layer in a star, and the pressure in that layer.

image intensifier An electronic device that increases the brightness of telescope images.

inflationary universe A version of the big bang theory that includes a rapid expansion when the universe was very young.

infrared cirrus A fine network of filaments covering the sky detected in the far infrared by the IRAS satellite. Believed associated with dust in the interstellar medium.

infrared outburst A sudden brightening of an object at infrared wavelengths.

infrared radiation Electromagnetic radiation with wavelengths intermediate between visible light and radio waves.

instability strip The region of the H–R diagram in which stars are unstable to pulsation. A star passing through this strip becomes a variable star.

intercrater plains The relatively smooth terrain on Mercury.

interglacial period A period when ice sheets melt back and the climate is warmer.

interstellar absorption lines Dark lines in some stellar spectra that are formed by interstellar gas.

interstellar medium The gas and dust distributed between the stars.

interstellar reddening The process in which dust scatters blue light out of starlight and makes the stars look redder.

inverse square law The rule that the strength of an effect (such as gravity) decreases in proportion as the distance squared increases.

Io flux tube A tube of magnetic lines and electric currents connecting Io and Jupiter.

ion An atom that has lost or gained one or more electrons.

ionization The process in which atoms lose or gain electrons.

iron meteorite A meteorite composed mainly of iron-nickel alloy.

irregular galaxy A galaxy with a chaotic appearance, large clouds of gas and dust, both population I and population II stars, but without spiral arms.

isotopes Atoms that have the same number of protons but a different number of neutrons.

isotropy The assumption that in its general properties the universe looks the same in every direction.

joule (J) A unit of energy equivalent to a force of 1 newton acting over a distance of 1 meter. One joule per second equals one watt of power.

Jovian planets Jupiter-like planets with large diameters and low densities.

Julian calendar The calendar established in 46 BC by Julius Caesar. It included a leap day every 4 years.

jumbled terrain Strangely disturbed regions of the moon opposite the locations of the Imbrium Basin and Mare Oriental.

Keplerian motion Orbital motion in accord with Kepler's laws of planetary motion.

Kerr black hole A solution to the equations of general relativity that describes the properties of a rotating black hole.

kiloparsec (kpc) A unit of distance equal to 1000 pc or 3260 light-years.

Kirchhoff's laws A set of laws that describe the absorption and emission of light by matter.

Kirkwood's gaps Certain distances in the asteroid belt at which few asteroids orbit.

Lagrangian points Points of stability in the orbital plane of a planet or moon. One is located 60° ahead and one 60° behind the orbiting bodies.

large-impact hypothesis The theory that the moon formed from debris ejected during a collision between the earth and a large planetesimal.

life zone A region around a star within which a planet can have temperatures that permit the existence of liquid water.

light curve A graph of brightness versus time commonly used in analyzing variable stars and eclipsing binaries.

light-gathering power The ability of a telescope to collect light. Proportional to the area of the telescope objective lens or mirror.

lighthouse theory The theory that a neutron star produces pulses of radiation by sweeping radio beams around the sky as it rotates.

light-year A unit of distance. The distance light travels in 1 year.

limb The edge of the apparent disk of a body, as in "the limb of the moon."

line of nodes The line across an orbit connecting the nodes. Commonly applied to the orbit of the moon.

line profile A graph of light intensity versus wavelength showing the shape of an absorption line.

liquid metallic hydrogen A form of hydrogen under high pressure that is a good electrical conductor.

lobate scarp A curved cliff such as those found on Mercury.

local apparent time The time defined by the true location of the sun in the sky. The time kept by a sundial.

local celestial meridian A north–south line around the sky passing through the zenith and nadir.

local mean time The time defined by the location the sun would have if it moved at a constant rate along the ecliptic.

long-period variable Variable stars with periods ranging from 100 days to over 400 days.

look-back time The amount by which we look into the past when we look at a distant galaxy. A time equal to the distance to the galaxy in light-years.

luminosity The total amount of energy a star radiates in 1 second.

luminosity class A category of stars of similar luminosity. Determined by the widths of lines in their spectra.

lunar eclipse The darkening of the moon when it moves through the earth's shadow.

Lyman series Spectral lines in the ultraviolet spectrum of hydrogen produced by transitions whose lowest orbit is the ground state.

Magellanic Clouds Small, irregular galaxies that are companions to the Milky Way. Visible in the southern sky.

magnetosphere The volume of space around a planet within which the motion of charged particles is dominated by the planetary magnetic field rather than the solar wind.

magnifying power The ability of a telescope to make an image larger.

magnitude scale The astronomical brightness scale. The larger the number, the fainter the star.

main sequence The region of the H–R diagram running from upper left to lower right, which includes roughly 90 percent of all stars.

mantle The layer of dense rock and metal oxides that lies between the molten core and the surface of the earth. Also, similar layers in other planets.

mare (sea) One of the lunar lowlands filled by successive flows of dark lava.

mass A measure of the amount of matter making up an object.

mass function A measure of the ratio of the masses in a single-line spectroscopic binary. Also includes the inclination, which is unknown for such systems.

mass–luminosity relation The more massive a star is, the more luminous it is.

Maunder butterfly diagram A graph showing the latitude of sunspots versus time. First plotted by W. W. Maunder in 1904.

Maunder minimum A period of less numerous sunspots and other solar activity between 1645–1715.

mean solar day The average time between two passages of the sun across the local celestial meridian.

megalith A very large stone used in a prehistoric structure such as Stonehenge.

megaparsec (Mpc) A unit of distance equal to 1 million pc.

metals In astronomical usage, all atoms heavier than helium.

meteor A small bit of matter heated by friction to incandescent vapor as it falls into Earth's atmosphere.

meteorite A meteor that has survived its passage through the atmosphere and strikes the ground.

meteoroid A meteor in space before it enters the earth's atmosphere.

midocean rift Chasms that split the midocean rises where crustal plates move apart.

midocean rise One of the undersea mountain ranges that push up from the seafloor in the center of the oceans.

Miller experiment An experiment that reproduced the conditions under which life began on earth and manufactured amino acids and other organic compounds.

minute of arc An angular measure. Each degree is divided into 60 minutes of arc.

missing mass Unobserved mass in clusters of galaxies believed to provide sufficient gravity to bind the cluster together.

model In science a mental conception of how a specific aspect of nature works. Could be expressed mathematically.

molecular cloud An interstellar gas cloud that is dense enough for the formation of molecules. Discovered and studied by the radio emissions of such molecules.

molecule Two or more atoms bonded together.

momentum The tendency of a moving object to continue moving. Mathematically, the product of mass and velocity.

morning star Any planet visible in the sky just before sunrise.

mutant Offspring born with altered DNA.

nadir The point on the celestial sphere directly opposite the zenith.

nanometer A unit of length equal to 10^{-9}m.

natural motion In Aristotelian physics, the motion of objects toward their natural places—fire and air upward and earth and water downward.

natural selection The process by which the best traits are passed on, allowing the most able to survive.

nebula A cloud of gas and dust in space.

neap tides Ocean tides of low amplitude occurring at first- and third-quarter moon.

neutrino A neutral, massless atomic particle that travels at the speed of light.

neutron An atomic particle with no charge and about the same mass as a proton.

neutron star A small, highly dense star composed almost entirely of tightly packed neutrons. Radius about 10 km.

Newton (N) A unit of force. One newton is the force needed to accelerate a mass of 1 kilogram by 1 meter per second in 1 second.

Newtonian focus The focal arrangement of a reflecting telescope in which a diagonal mirror reflects light out the side of the telescope tube for easier access.

node The points where an object's orbit passes through the plane of the earth's orbit.

noncosmological red shift A galaxy red shift caused by something other than the expansion of the universe. No conclusive examples are known.

north celestial pole The point on the celestial sphere directly above the earth's North Pole.

north circumpolar constellation One of the constellations near the north celestial pole. Such constellations never set as seen from moderate northern latitudes.

nova From the Latin "new," a sudden brightening of a star, making it appear as a "new" star in the sky. Believed associated with eruptions on white dwarfs in binary systems.

nuclear bulge The spherical cloud of stars that lies at the center of spiral galaxies.

nucleosynthesis The production of elements heavier than helium by the fusion of atomic nuclei in stars and during supernovae explosions.

nucleus (of an atom) The central core of an atom containing protons and neutrons. Carries a net positive charge.

objective lens In a refracting telescope, the long-focal-length lens that forms an image of the object viewed. The lens closest to the object.

objective mirror In a reflecting telescope, the principal mirror (reflecting surface) that forms an image of the object viewed.

oblateness The flattening of a spherical body. Usually caused by rotation.

oblate spheroid A sphere flattened such that its polar diameter is smaller than its equatorial diameter.

occultation The passage of a larger body in front of a smaller body.

Olbers' paradox The conflict between observation and theory as to why the night sky should or should not be dark.

135 km/sec arm A receding cloud of neutral hydrogen lying on the far side of the galactic center.

Oort cloud theory The hypothesis that the source of comets is a swarm of icy bodies believed to lie in a spherical shell 50,000 AU from the sun.

opacity The resistance of a gas to the passage of radiation.

open (escape) orbit An orbit that does not return to its starting point. An escape orbit. (See **closed orbit.**)

open star cluster A cluster of 10 to 10,000 stars with an open, transparent appearance. The stars are not tightly grouped. Usually relatively young and located in the disk of the galaxy.

open universe A model universe in which the average density is less than the critical density needed to halt the expansion.

optical binary A binary star in which the stars are only apparently associated. One star is nearby and one is more distant.

oscillating universe theory The theory that the universe begins with a big bang, expands, is slowed by its own gravity, and then falls back to create another big bang.

outgassing The release of gases from a planet's interior.

parallax (*p*) The apparent change in the position of an object due to a change in the location of the observer. Astronomical parallax is measured in seconds of arc.

parsec (pc) The distance to a hypothetical star whose parallax is one second of arc. 1 pc = 206,265 AU = 3.26 ly.

partial eclipse A lunar eclipse in which the moon does not completely enter the earth's shadow. A solar eclipse in which the moon does not completely cover the sun.

Paschen series Spectral lines in the infrared spectrum of hydrogen produced by transitions whose lowest orbit is the third.

path of totality The track of the moon's umbral shadow over the earth's surface. The sun is totally eclipsed as seen from within this path.

penumbra The portion of a shadow that is only partially shaded.

penumbral eclipse A lunar eclipse in which the moon enters the penumbra of the earth's shadow but does not reach the umbra.

perfect cosmological principle The belief that, in general properties, the universe looks the same from every location in space at any time.

perigee The orbital point of closest approach to the earth.

perihelion The orbital point of closest approach to the sun.

period–luminosity diagram A graph showing the relation between period of pulsation and intrinsic brightness among Cepheid variable stars.

permitted orbit One of the energy levels in an atom that an electron may occupy.

photometer An instrument used to measure the intensity and color of starlight.

photon A quantum of electromagnetic energy. Carries an amount of energy that depends inversely on its wavelength.

photosphere The bright visible surface of the sun.

plage The bright area surrounding sunspots.

planetary nebula An expanding shell of gas ejected from a star during the latter stages of its evolution.

planetesimal One of the small bodies that formed from the solar nebula and eventually grew into protoplanets.

plastic A material with the properties of a solid but capable of flowing under pressure.

plate tectonics The constant destruction and renewal of earth's surface by the motion of sections of crust.

polar axis The axis around which a celestial body rotates.

poor galaxy cluster An irregularly shaped cluster that contains fewer than 1000 galaxies, many spiral, and no giant ellipticals.

population I Stars rich in atoms heavier than helium. Nearly always relatively young stars found in the disk of the galaxy.

population II Stars poor in atoms heavier than helium. Nearly always relatively old stars found in the halo, globular clusters, or the nuclear bulge.

position angle The angular direction of one body with respect to another. Measured from north toward the east. Typically used in the study of visual binaries.

precession The slow change in the direction of the earth's axis of rotation. One cycle takes nearly 26,000 years.

pressure broadening The blurring of spectral lines due to the gas pressure in a star's atmosphere.

pressure (*P*) waves In geophysics, mechanical waves of compression and rarefaction that travel through the earth's interior.

primary minimum In the light curve of an eclipsing binary, the deeper eclipse.

prime focus The point at which the objective mirror forms an image in a reflecting telescope.

primeval atmosphere Earth's first air, composed of gases from the solar nebula.

primordial background radiation Radiation from the hot clouds of the big bang explosion. Because of its large red shift it appears to come from a body whose temperature is only 2.7 K.

primordial black holes Low-mass black holes that may have formed during the big bang explosion.

primordial soup The rich solution of organic molecules in the earth's first oceans.

prolate spheroid A sphere stretched along its polar axis so its polar diameter is greater than its equatorial diameter.

prominences Eruptions on the solar surface. Visible during total solar eclipses.

proper motion The rate at which a star moves across the sky. Measured in seconds of arc per year.

proteins Complex molecules composed of amino acid units.

proton A positively charged atomic particle contained in the nucleus of atoms. The nucleus of a hydrogen atom.

proton–proton chain A series of three nuclear reactions that build a helium atom by adding together protons. The main energy source in the sun.

protoplanet Massive objective resulting from the coalescence of planetesimals in the solar nebula and destined to become a planet.

protostar A collapsing cloud of gas and dust destined to become a star.

pulsar A source of short, precisely timed radio bursts. Believed to be spinning neutron stars.

quantum mechanics The study of the behavior of atoms and atomic particles.

quasar (quasi-stellar object or QSO) Small, powerful sources of energy believed to be the active cores of very distant galaxies.

quasi-periodic object Certain x-ray sources that "flicker" rapidly for short intervals.

radial velocity (V_r) That component of an object's velocity directed away from or toward the earth.

radial velocity curve A graph of the velocity of recession or approach of the stars in a spectroscopic binary.

radiant The point in the sky from which meteors in a shower seem to come.

radiation pressure The force exerted on the surface of a body by its absorption of light. Small particles floating in the solar system can be blown outward by the pressure of the sunlight.

radio galaxy A galaxy that is a strong source of radio signals.

radio interferometer Two or more radio telescopes that combine their signals to achieve the resolving power of a larger telescope.

rays Ejecta from meteorite impacts forming white streamers radiating from some lunar craters.

recombination The stage within 10^6 years of the big bang when the gas became transparent to radiation.

recurrent novae Stars that erupt as novae every few dozen years.

red dwarf Cool, low-mass stars on the lower main sequence.

red shift The lengthening of the wavelengths of light seen when the source and observer are receding from each other.

reflecting telescope A telescope that uses a concave mirror to focus light into an image.

reflection nebula A nebula produced by starlight reflecting off of dust particles in the interstellar medium.

refracting telescope A telescope that forms images by bending (refracting) light with a lens.

regolith A soil made up of crushed rock fragments.

relative age The age of a geological feature referred to other features. For example, relative ages tell us the lunar maria are younger than the highlands.

relativistic jet model An explanation of superluminal expansion based on a high-velocity jet from a quasar directed approximately toward the earth.

relativistic red shift The red shift due to the Doppler effect for objects traveling at speeds near the speed of light.

resolving power The ability of a telescope to reveal fine detail. Depends on the diameter of the telescope objective.

resonance The coincidental agreement between two periodic phenomena. Commonly applied to agreements between orbital periods, which can make orbits more or less stable.

rest mass The mass of a particle as measured by an observer not moving with respect to the particle.

retrograde loop The apparent backward (westward) motion of planets as seen against the background of stars.

retrograde motion Backward motion. In the sky, westward motion.

retrograde orbit In the solar system, an orbit that carries a moon or planet in the opposite direction from most other motion in the solar system—that is, retrograde orbits are clockwise as seen from the north.

rich galaxy cluster A cluster containing over 1000 galaxies, mostly elliptical, scattered over a volume about 3 Mpc in diameter.

rift valley A long, straight, deep valley produced by the separation of crustal plates.

right ascension (R.A.) A coordinate used on the celestial sphere just as longitude is used on earth. An object's right ascension is measured eastward from the vernal equinox.

ring galaxy A galaxy that resembles a ring around a bright nucleus. Believed to be the result of a head-on collision of two galaxies.

RNA (ribonucleic acid) Long carbon-chain molecules that use the information stored in DNA to manufacture complex molecules necessary to the organism.

Roche limit The minimum distance between a planet and a satellite that holds itself together by its own gravity. If a satellite's orbit brings it within its planet's Roche limit, tidal forces will pull the satellite apart.

Roche surface The outer boundary of the volume of space that a star's gravity can control within a binary system.

rolling plains The most common type of terrain on Venus.

rotation curve A graph of orbital velocity versus radius in the disk of a galaxy.

rotation curve method A method of determining a galaxy's mass by observing the orbital velocity and orbital radius of stars in the galaxy.

RR Lyrae variable stars Variable stars with periods of from 12–24 hours. Common in some globular clusters.

Sagittarius A The powerful radio source located at the core of the Milky Way galaxy.

Saros cycle An 18-year 11-day period after which the pattern of lunar and solar eclipses repeats.

Schmidt camera A photographic telescope that takes wide-angle photographs.

Schwarzschild radius (R_s) The radius of the event horizon around a black hole.

scientific notation The system of recording very large or very small numbers by using powers of 10.

secondary atmosphere The gases outgassed from a planet's interior. Rich in carbon dioxide.

secondary minimum In the light curve of an eclipsing binary, the shallower eclipse.

secondary mirror In a reflecting telescope, the mirror that reflects the light to a point of easy observation.

second of arc An angular measure. Each minute of arc is divided into 60 seconds of arc.

seeing Atmospheric conditions on a given night. When the atmosphere is unsteady, producing blurred images, the seeing is said to be poor.

selection effect An influence on the probability that certain phenomena will be detected or selected, which can alter the outcome of a survey.

self-sustaining star formation The process by which the birth of stars compresses the surrounding gas clouds and triggers the formation of more stars. Proposed to explain spiral arms.

semimajor axis Half of the longest axis of an ellipse.

setting circle One of two circular scales on a telescope used for setting right ascension and declination.

Seyfert galaxy An otherwise normal spiral galaxy with an unusually bright, small core that fluctuates in brightness. Believed to indicate the core is erupting.

shear (S) waves Mechanical waves that travel through earth's interior by the vibration of particles perpendicular to the direction of wave travel.

shepherd satellite A satellite that, by its gravitational field, confines particles to a planetary ring.

shield volcanoes Wide, low-profile volcanic cones produced by highly liquid lava.

shock wave A sudden change in pressure that travels as an intense sound wave.

sidereal day The period of rotation of the earth with respect to the stars.

sidereal drive The motor and gears on a telescope that turn it westward to keep it pointed at a star.

sidereal period The period of rotation or revolution of an astronomical body referred to the stars.

sidereal time Time based on the rotation of the earth with respect to the stars. The sidereal time at any moment equals the right ascension of objects on the upper half of the local celestial meridian.

sidereal year The period of the earth's revolution around the sun with respect to the stars.

single-line spectroscopic binary A spectroscopic binary in which lines of one star are visible in the spectrum.

singularity The object of zero radius into which the matter in a black hole is believed to fall.

sinuous rille A narrow, winding valley on the moon caused by ancient lava flows along narrow channels.

smooth plains Apparently young plains on Mercury formed by lava flows at or soon after the formation of the Caloris Basin.

solar constant A measure of the energy output of the sun. The total solar energy striking 1 cm^2 just above earth's atmosphere in one minute.

solar eclipse The event that occurs when the moon passes directly between the earth and sun, blocking our view of the sun.

solar granulation The patchwork pattern of bright areas with dark borders observed on the sun. The tops of rising currents of hot gas in the convective zone.

solar nebula theory The theory that the planets formed from the same cloud of gas and dust that formed the sun.

solar wind Rapidly moving atoms and ions that escape from the solar corona and blow outward through the solar system.

south celestial pole The point on the celestial sphere directly above the earth's South Pole.

special relativity The first of Einstein's theories of relativity, which dealt with uniform motion.

spectral class or type A star's position in the temperature classification system O, B, A, F, G, K, and M. Based on the appearance of the star's spectrum.

spectral sequence The arrangement of spectral classes (O, B, A, F, G, K, M) ranging from hot to cool.

spectrograph A device that separates light by wavelength to produce a spectrum.

spectroscopic binary A star system in which the stars are too close together to be visible separately. We see a single point of light and only by taking a spectrum can we determine that there are two stars.

spectroscopic parallax The method of determining a star's distance by comparing its apparent magnitude with its absolute magnitude as estimated from its spectrum.

spherical component The part of the galaxy including all matter in a spherical distribution around the center (the halo and nuclear bulge).

spicules Small, flamelike projections in the chromosphere of the sun.

spiral arms Long spiral patterns of bright stars, star clusters, gas, and dust, that extend from the center to the edge of the disk of spiral galaxies.

spiral galaxy A galaxy with an obvious disk component containing gas; dust; hot, bright stars; and spiral arms.

spiral tracers Objects used to map the spiral arms (e.g., O and B associations, open clusters, clouds of ionized hydrogen, and some types of variable stars).

spoke A radial feature in the rings of Saturn.

sporadic meteor A meteor not part of a meteor shower.

spring tides Ocean tides of high amplitude that occur at full and new moon.

standard time The local mean time on the central meridian of the time zone.

starburst galaxy A bright blue galaxy in which many new stars are forming. Believed caused by collisions between galaxies.

steady state theory The theory (now generally abandoned) that the universe does not evolve.

stellar density function A description of the abundance of stars of different types in space.

stellar model A table of numbers representing the conditions in various layers within a star.

stellar parallax (p) A measure of stellar distance. (See **parallax.**)

stony–iron meteorite A meteorite that is a mixture of stone and iron.

stony meteorite A meteorite composed of silicate (rocky) material.

stromatolite A layered fossil formation caused by ancient mats of algae or bacteria, which build up mineral deposits season after season.

subsolar point The point on a planet that is directly below the sun.

summer solstice The point on the celestial sphere where the sun is at its most northerly point. Also, the time when the sun passes this point about June 22. Summer begins in the northern hemisphere.

sungrazer A comet that comes very close to the sun.

sunspots Relatively dark spots on the sun that contain intense magnetic fields.

supercluster A cluster of galaxy clusters.

superconductor A material that can conduct electricity with essentially zero resistance.

supergiant stars Exceptionally luminous stars 10 to 1000 times the sun's diameter.

supergranule A large granule including many smaller granules on the sun's surface.

superluminal expansion The apparent expansion of parts of a quasar at speeds greater than the speed of light.

supernova remnant The expanding shell of gas marking the site of a supernova explosion.

supernova (type I) The explosion of a star, believed caused by the transfer of matter to a white dwarf.

supernova (type II) The explosion of a star, believed caused by the collapse of a massive star.

synchrotron radiation Radiation emitted when high-speed electrons move through a magnetic field.

synodic period The period of rotation or revolution of a celestial body with respect to the sun.

terminator The dividing line between daylight and darkness on a planet or moon.

terrestrial planets Earthlike planets—small, dense, rocky.

third contact During an eclipse, the moment when the edge of the sun reappears from behind the moon, or when the leading edge of the moon reaches the edge of the umbra.

3-kpc arm A cloud of neutral hydrogen moving outward from the nucleus of our galaxy at about 53 km/sec. It lies 3 kpc from the center of the galaxy.

tidal coupling The locking of the rotation of a body to its revolution around another body.

tidal heating The heating of a planet or satellite because of friction caused by tides.

time dilation The slowing of moving clocks or clocks in strong gravitational fields.

Titius–Bode rule A simple series of steps that produces numbers approximately matching the sizes of the planetary orbits.

total eclipse A solar eclipse in which the moon completely covers the bright surface of the sun, or a lunar eclipse in which the moon completely enters the dark shadow of the earth.

transition The movement of an electron from one atomic orbit to another.

transverse velocity The velocity of a star perpendicular to the line of sight.

triaxial ellipsoid A geometrical solid whose three axes are unequal.

trilithon Literally, "three stones." Any of the five large arches at Stonehenge composed of two uprights and a horizontal top piece.

triple alpha process The nuclear fusion process that combines three helium nuclei (alpha particles) to make one carbon nucleus.

Trojan asteroids Small rocky bodies caught in Jupiter's orbit, the Lagrangian points 60° ahead of and behind the planet.

tropical year The time from one vernal equinox to the next.

true relative orbit The orbit of one star in a visual binary with respect to the other star after correction for orbital inclination. (See **apparent relative orbit.**)

T Tauri stars Young stars surrounded by gas and dust. Believed to be contracting toward the main sequence.

tuning fork diagram A system of classification for elliptical, spiral, and irregular galaxies.

turn-off point The point in an H–R diagram where a cluster's stars turn off of the main sequence and move toward the red giant region, revealing the approximate age of the cluster.

21 cm radiation Radio emission produced by cold, low density hydrogen in interstellar space.

twin paradox The seeming contradiction when one twin travels near the speed of light and returns younger than the twin who stayed behind. (See **dilation of time.**)

ultraviolet radiation Electromagnetic radiation with wavelengths shorter than visible light but longer than X rays.

umbra The region of a shadow that is totally shaded.

uncompressed density The density a planet would have if its gravity did not compress it.

uniform circular motion The classical belief that the perfect heavens could only move by the combination of uniform motion along circular orbits.

universality The assumption that the physical laws observed on earth apply everywhere in the universe.

Van Allen belts Radiation belts of high-energy particles trapped in the earth's magnetosphere.

variable star A star whose brightness changes periodically.

velocity A rate of travel that specifies both speed and direction.

velocity dispersion method A method of finding a galaxy's mass by observing the range of velocities within the galaxy.

vernal equinox The place on the celestial sphere where the sun crosses the celestial equator moving northward. Also, the time of year when the sun crosses this point, about March 21, and spring begins in the northern hemisphere.

Very Long Baseline Interferometry (VLBI) The use of radio telescopes located thousands of miles apart to resolve detail in radio sources.

vesicular basalt A porous rock formed by solidified lava with trapped bubbles.

vidicon A type of vacuum tube used to record television images.

violent motion In Aristotelian physics, motion other than natural motion. (See **natural motion.**)

visual binary A binary star system in which the two stars are separately visible in the telescope.

VLBI See **Very Long Baseline Interferometry.**

water hole The interval of the radio spectrum between the 21-cm hydrogen radiation and the 18-cm OH radiation. Likely wavelengths to use in the search for extraterrestrial life.

wavelength The distance between successive peaks or troughs of a wave. Usually represented by λ.

wavelength of maximum (λ_{max}) The wavelength at which a perfect radiator emits the maximum amount of energy. Depends only on the object's temperature.

white dwarf stars Dying stars that have collapsed to the size of the earth and are slowly cooling off. At the lower left of the H–R diagram.

Widmanstätten patterns Bands in iron meteorites due to large crystals of nickel–iron alloys.

winter solstice The point on the celestial sphere where the sun is farthest south. Also the time of year when the sun passes this point, about December 22. Winter begins in northern hemisphere.

wispy terrain Light coloration found on the trailing sides of some planetary satellites.

Zeeman effect The splitting of spectral lines into multiple components when the atoms are in a magnetic field.

zero-age main sequence (ZAMS) The locus in the H–R diagram where stars first reach stability as hydrogen burning stars.

zodiac The band around the sky centered on the ecliptic within which the planets move.

zone of avoidance The region around the Milky Way where almost no galaxies are visible because our view is blocked by dust in our galaxy.

zones Yellow-white regions that circle Jupiter parallel to its equator. Believed to be areas of rising gas.

ANSWERS TO EVEN-NUMBERED PROBLEMS

Chapter 1 (page 16)
2. 3475 km
4. about 1.3 sec
6. about 28 million
8. about 20

Chapter 2 (page 40)
2. 2755
4. 14 mag or 400,000

Chapter 3 (pages 60–61)
2. a. full; b. first quarter; c. gibbous waxing; d. crescent waxing
4. 12 hours, about 12.4 hours, because the moon moves eastward about 13° per day.
6. about 31 minutes of arc
10. about August 10, 2026

Chapter 4 (pages 84–85)
2. Jupiter, Saturn, Uranus, Neptune, and Pluto. None of these could reach crescent phase.
4. Mars, about 18 seconds of arc; Saturn's rings, about 44 seconds of arc
6. about 5.2 years

Chapter 5 (pages 106–107)
2. 19.6 m/sec, 39.2 m/sec
4. about 3 km/sec, about 24 hours (Don't forget to calculate the orbital radius from earth's center.)
6. Escape velocity only 0.09 km/sec less

Chapter 6 (pages 130–131)
2. short radio waves
4. about 560,000 times
6. No. His resolving power was about 5.8 sec of arc.
8. 1.3 cm
10. about 50 cm. (From 400 km above, a human is about 0.25 sec of arc in diameter.)

Chapter 7 (page 153)
2. 150 nm
4. by a factor of 16
6. 250 nm
8. a. B; b. F; c. M; d. K
10. about 0.58 nm

Chapter 8 (page 181)
2. 9×10^{16} joules
4. 0.22 kg
6. 36 km
8. about 3.6 times
10. 400,000 years

Chapter 9 (page 201)
2. 63 pc, 2nd mag absolute
4. about B7
6. about 1580 $L_\odot$
8. 160 pc
10. a, c, c, d

Chapter 10 (page 218)
2. 1.28 $M_\odot$
4. 6.48 $M_\odot$, mass ratio 1.85, masses are 4.21 $M_\odot$ and 2.27 $M_\odot$
6. a = 24.8″ or 84.5 AU so the total mass is about 1.4 $M_\odot$
8. 3.69 days (0.0101 yr), 1.19 $M_\odot$, 0.57 $M_\odot$, 0.62 $M_\odot$
10. 1.38×10^6 km (about 0.002 $R_\odot$)

Chapter 11 (pages 244–245)
2. 1/5.75th or about 17 percent
4. about 1.5 million years
6. about 40 billion $L_\odot$ (Find surface area in square meters, then multiply by σT^4.)
8. about 1000 sec
10. 128 $L_\odot$, 2187 $L_\odot$
12. 100 nm

Chapter 12 (pages 262–263)
2. about −4.5
4. 1000 pc
6. about 300 million years
8. about 0.125 pc

Chapter 13 (page 279)
2. about 1.75 ly
4. about 16,000 years
6. about 940 years ago (about 1050 AD)
8. 220 km/sec
10. about 300 years ago

Chapter 14 (page 301)
2. 7.1×10^{25} joules/sec or about 0.19 $L_\odot$ (Find surface area and multiply by σT^4.)
4. 814 km/sec (assuming mass = 1 $M_\odot$)
6. about 12 sec of arc
8. about 490 sec (orbital velocity about 2,600 km/sec)

Chapter 15 (page 328)
2. less than 11 percent
4. about 4 million years
6. 6300 pc
8. 25 pc
10. about 20 times
12. 30 K
14. about 0.001 sec of arc

Chapter 16 (page 350)
2. 2.6 Mpc
4. 2×10^{11} $M_\odot$
6. 2.2×10^{11} $M_\odot$ (Assume escape velocity is 420 km/sec at 11 kpc.)

Chapter 17 (page 372)
2. 7.8 million years
4. about 200 Mpc (Angular diameter is about 50 sec of arc.)
6. 52,000 km/sec (about 17 percent of the speed of light)
8. −13.5 (slightly brighter than the full moon assuming no obscuration due to interstellar dust)
10. 0.16
12. 1.5 sec of arc, 4100 Mpc

Chapter 18 (pages 394–395)
2. 17.5 billion years, 11.7 billion years
4. 2×10^{-31} g/cm^3
6. 75 km/sec/Mpc
8. 53 Mpc (The remnant is expanding at about 3×10^{11} km per year.)

Chapter 19 (page 417)
2. about 22.6 magnitude (It will look $206{,}265^2$ times fainter, which is 26.6 magnitudes.)
4. about 20 km
6. The moon. Its gravity is least.
8. about 2×10^{12} per second
10. about 0.46 years (5.6 months)

Chapter 20 (page 445)

4. No. Their angular diameter would be only 0.5 sec of arc. They are not visible in photos from orbit around the moon.
6. 0.5 sec of arc (assuming an astronaut seen from above is 0.5 meters in diameter)

Chapter 21 (pages 476–477)

2. 11.5 minutes
4. 33,400 km (39,400 from center of Venus)
6. 0.04 sec of arc, none
8. 1.4×10^{16} kg (as much mass as 10 mountains combined); 13 m/sec (about 28 mph)

Chapter 22 (page 509)

2. 35 earth days
4. 31 km/sec, 7.2 hours; 16.7 km/sec, 14.3 hours
6. 2.8 km/sec (6100 mph)
8. 1.85 km/sec
10. 5.2 m/sec

Chapter 23 (page 535)

2. 42 seconds of arc
4. 0.25 km/sec (570 mph)
6. about 7 seconds, yes
8. 0.42 km/sec
10. 1.04×10^{26} kg (17.3 $M_{\oplus}$)

Chapter 24 (pages 563–564)

2. 1 billion
4. diameter 80 km
6. 0.18 km/sec
8. 3.28 years, 2.50 years
10. 0.06 AU (9 million km)
12. 3.3 $M_{\oplus}$

Chapter 25 (pages 586–587)

2. 9.6 cm, 0.65 mm
4. 1.3 $M_{\odot}$
6. 380 km

INDEX

Numbers in **boldface** refer to pages where the item appears in boldface and is defined. Page numbers S–1 through S–12 refer to the supplement on observing the sky at the back of the book. Because most concepts are illustrated, no distinction is made between references to text and references to illustrations.

B

C

D

H

N

X

Y

Z

SUPPLEMENT

OBSERVING THE SKY

Observing the sky with the naked eye is of no more importance to modern astronomy than picking up pretty pebbles is to modern geology. But the sky is a natural wonder unimaginably bigger than the Grand Canyon, the Rocky Mountains, or any other natural wonder that tourists visit every year. To neglect the beauty of the sky is equivalent to geologists neglecting the beauty of the minerals they study. This supplement is meant to act as a tourist's guide to the sky. We analyzed the universe in the regular chapters, but here we will admire it.

The brighter stars in the sky are visible even from the centers of cities with their air and light pollution. But in the countryside only a few miles beyond the cities, the night sky is a velvety blackness strewn with thousands of glittering stars. From a wilderness location, far from the city's glare, and especially from high mountains, the night sky is spectacular.

USING STAR CHARTS

The constellations are a fascinating cultural heritage of our planet, but they are sometimes a bit difficult to learn because of the earth's motion. The constellations above the horizon change with the time of night and the seasons.

Because the earth rotates eastward, the sky appears to rotate around us westward. A constellation visible in the southern sky soon after sunset will appear to move westward, and in a few hours, it will disappear below the horizon. Other constellations will rise in the east, so the sky changes gradually through the night.

In addition, the orbital motion of the earth makes the sun appear to move eastward among the stars. Each day the sun moves twice its own diameter eastward along the ecliptic, and each night at sunset, the constellations are about 1° farther toward the west.

Orion, for instance, is visible in the southern sky in January, but as the days pass, the sun moves closer to Orion. By March, Orion is difficult to see in the southwest sky soon after sunset. By June, the sun is so close to Orion it sets with the sun and is invisible. Not until late July is the sun far enough past Orion for the constellation to become visible rising in the eastern sky just before dawn.

Clearly the rotation and orbital motion of the earth requires that we use more than one star chart to map the sky. Which chart we select depends on the time of night and the time of year. The charts given in this book show the evening sky for each month.

To use the charts, select the chart for the appropriate month and hold it overhead as shown in Figure

FIGURE S–1 To use the star charts in this book, select the appropriate chart for the date and time. Hold it overhead and turn it until the direction at the bottom of the chart is the same as the direction you face.

S–1. If you face south, turn the chart until the words *Southern Horizon* are at the bottom. If you face other directions, turn the chart appropriately.

THE CONSTELLATIONS THROUGHOUT THE YEAR

The following paragraphs describe the evening sky during the fall, winter, spring, and summer as seen by an observer at the average latitude of the United States. If your latitude is significantly different, you will need to make some adjustment to locate constellations.

These charts show more than just constellations. They show the Milky Way, the band of light produced by the hundred billion stars in our wheel-shaped galaxy. They also show bright star clusters and nebulae. In addition, the maps contain asterisms. An **asterism** is a named pattern of stars that is not a recognized constellation. The Big Dipper, for instance, is an asterism. It is part of the constellation Ursa Major, the Great Bear.

FALL

In early October the evening sky is dominated by a brilliant star, Vega, high in the west. This star is part of a small constellation called Lyra, the Lyre. To the east is Cygnus, the Swan, also known as the Northern Cross. The swan flies southwest along the Milky Way with its long neck extended and the bright star Deneb marking its tail. In fact, *Deneb* is derived from the Arabic for "the hen's tail" or "the hindmost." Vega, Deneb, and the bright star Altair mark the vertices of the Summer Triangle shown on the summer and fall charts.

Later in the fall, the Great Square of Pegasus is high in the east. As we saw in Figure 2–3, only three of these four stars belong to Pegasus. The other, Alpheratz, is part of Andromeda. Late in the autumn, when Andromeda is higher in the evening sky, and when there is no moon, the naked eye can pick out the Great Galaxy in Andromeda as a faint hazy patch. (See Figure 15–1). This galaxy is similar to our own Milky Way galaxy, and the light we see left the Andromeda galaxy over 2 million years ago.

WINTER

The winter sky is dominated by Orion, the Hunter, outlined by brilliant stars high in the southern sky. The upper left corner of the constellation is marked by a brilliant red star, Betelgeuse, and its lower right corner by a brilliant blue star, Rigel. Below the diagonal line of stars marking Orion's belt hangs the sword, and among the stars of the sword lies the Orion Nebula, a great swirl of gas and dust surrounding hot, blue stars. The nebula is visible to the naked eye and is impressive through binoculars or a small telescope.

Southeast of Orion is Canis Major, the Great Dog, containing Sirius, the brightest star in the sky. Canis Minor, the Small Dog, is located northeast of Sirius, and the three stars Betelgeuse, Sirius, and Procyon mark the corners of the asterism called the Winter Triangle. Just north of the Winter Triangle lies the elongated box shape of Gemini. The sun is located at the extreme western edge of Gemini at the time of the summer solstice.

Orion and his hunting dogs face westward, meeting the charge of Taurus, the Bull. (See Figure 2–1.) the V-shaped cluster of stars that outlines the forehead of the bull are known as the Hyades, and slightly farther west, on the bull's broad back, is the smaller cluster called the

Pleiades, the seven sisters of Greek mythology. Binoculars or a small telescope will reveal hundreds of stars in the Pleiades.

SPRING

Leo, the Lion, is the central figure of the spring sky. The sickle shape of the lion's mane with its bright star Regulus stands almost overhead at sunset in late April. To the west of Leo are the faint stars of Cancer, the Crab. Although Cancer is not impressive to the naked eye, binoculars or a small telescope reveal a large cluster of stars called the Praesepe, the Beehive. Beyond Cancer is Gemini, the Twins, with its two bright stars Castor and Pollux, named after the famous twins of Greek mythology.

East of Leo lies the kite shape of Bootes, the Bear Driver, with its bright star Arcturus. This star is the fourth brightest star in the sky and the brightest in the northern hemisphere. In the past it had such names as "Watcher," "Guardian," and "Keeper of Heaven." Bootes itself derives its designation as the Bear Driver from its position following Ursa Major around the sky.

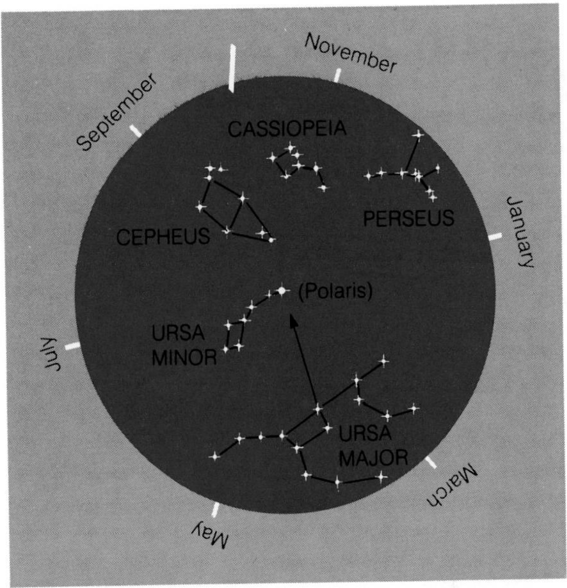

FIGURE S–2 The northern constellations. To use the chart, face north soon after sunset and hold the chart in front of you with the current date at the top.

SUMMER

The evening sky in summer is rich with beautiful stars and constellations. Bootes, the Bear Driver, with its bright star Arcturus are in the west, and east of them is Corona Borealis, the Northern Crown. To the Pawnee, Corona Borealis was the "Circle of Chiefs," but to the Blackfeet, it was the home of the Spider God.

High in the summer sky shines Vega in the constellation Lyra, and just east is Cygnus, the Swan, also known as the Northern Cross. South of Cygnus is Aquila, the Eagle, with its bright star Altair. Vega, Deneb, and Altair mark the corners of the asterism known as the Summer Triangle.

The summer sky glows with the light of the Milky Way stretching from the north through Cygnus to Sagittarius and Scorpius in the south. On a dark night the naked eye can make out the great rift, an obscuring band of interstellar dust that blocks the light of the Milky Way from Serpens to Cygnus. The center of our galaxy lies in Sagittarius, and that area is filled with star clusters that are impressive through binoculars or a small telescope. At the winter solstice the sun is located just northwest of the teapot-shaped Sagittarius.

THE NORTHERN SKY

A few constellations of the northern sky appear in Figure S–2. These are called **north circumpolar constellations** because, as seen from the average latitude of the United States, they never disappear below the horizon. Thus, on any night of the year, at any time of night, these constellations are visible. To use the chart in Figure S–2, face north soon after sunset and hold the chart directly in front of you. Turn it until the current date is on top, and the chart will represent the sky before you.

Locate the Big Dipper and note that it is not a constellation. It is an asterism composed of some of the stars in Ursa Major, the Great Bear. In the same way, the Little Dipper is part of Ursa Minor, the Small Bear. The star at the bend in the handle of the Big Dipper is Mizar, and its companion, just visible to the naked eye, is Alcor. The Arabic name for Alcor was Saidak, meaning "proof," referring to the use of this star as proof of good eyesight. To the Iroquois, the three stars in the Big Dipper's handle were warriors hunting the great bear, and little Alcor was their cooking pot.

The Big Dipper is so easy to find it can be used to locate other objects. The two stars at the end of the cup

Table S–1 Greatest elongations of Mercury.

Evening Sky	Morning Sky
Jan. 26, 1988	March 8, 1988
May 19, 1988*	July 6, 1988
Sept. 15, 1988	Oct. 26, 1988*
Jan. 9, 1989	Feb. 18, 1989
May 1, 1989*	June 18, 1989
Aug. 29, 1989	Oct. 10, 1989*
Dec. 23, 1989	Feb. 1, 1990
April 13, 1990*	May 31, 1990
Aug. 11, 1990	Sept. 24, 1990*
Dec. 6, 1990	Jan. 14, 1991
March 27, 1991*	May 12, 1991
July 25, 1991	Sept. 7, 1991*
Nov. 19, 1991	Dec. 27, 1991
March 9, 1992*	April 23, 1992
July 6, 1992	Aug. 21, 1992*
Oct. 31, 1992	Dec. 9, 1992
Feb. 21, 1993*	April 5, 1993
June 17, 1993	Aug. 4, 1993*
Oct. 14, 1993	Nov. 22, 1993
Feb. 4, 1994	March 19, 1994
May 30, 1994	July 17, 1994
Sept. 26, 1994	Nov. 6, 1994*
Jan. 19, 1995	March 1, 1995
May 12, 1995*	June 29, 1995
Sept. 9, 1995	Oct. 20, 1995*

*Most favorable elongations.

Table S–2 Greatest elongations of Venus.

Evening Sky	Morning Sky
April 3, 1988	Aug. 22, 1988
Nov. 8, 1989	March 30, 1990
June 13, 1991	Nov. 2, 1991
Jan. 19, 1993	June 10, 1993
Aug. 24, 1994	Jan. 13, 1995
April 1, 1996	Aug. 20, 1996
Nov. 6, 1997	March 27, 1998
June 11, 1999	Oct. 30, 1999
Jan. 17, 2001	June 8, 2001
Aug. 22, 2002	Jan. 11, 2003
March 29, 2004	Aug. 17, 2004
Nov. 3, 2005	

are known as the Pointers because a line through them points to Polaris, the North Star, located at the end of the handle of the Little Dipper. Continuing past Polaris, the line comes to Cepheus, the King, and Cassiopeia, the Queen. Following the curve of the Big Dipper's handle leads to Arcturus in Bootes, and on to Spica. (See, for example, the star chart for May.)

FINDING PLANETS

The planets are both easy and difficult to find. They are easy to find because they are nearly all bright and easy to see with the naked eye. Only Uranus, Neptune, and Pluto are too faint to be seen without a telescope. The planets are difficult to find because they are moving along their orbits and thus move gradually eastward along the ecliptic. This section contains tables to help you find the brighter planets.

MERCURY AND VENUS

The inner planets, Mercury and Venus, follow orbits inside that of Earth and are sometimes visible in the western sky just after sunset or in the eastern sky just before sunrise (Figure 2–16). Tables S–1 and S–2 will help you locate them. Mercury is especially difficult to spot because it never gets far from the sun. You will have to look for it within a few days of its **greatest elongation** (greatest angular distance from the sun) as shown in Table S–1. Venus is not so difficult to see; look for it within a month or two of the times in Table S–2.

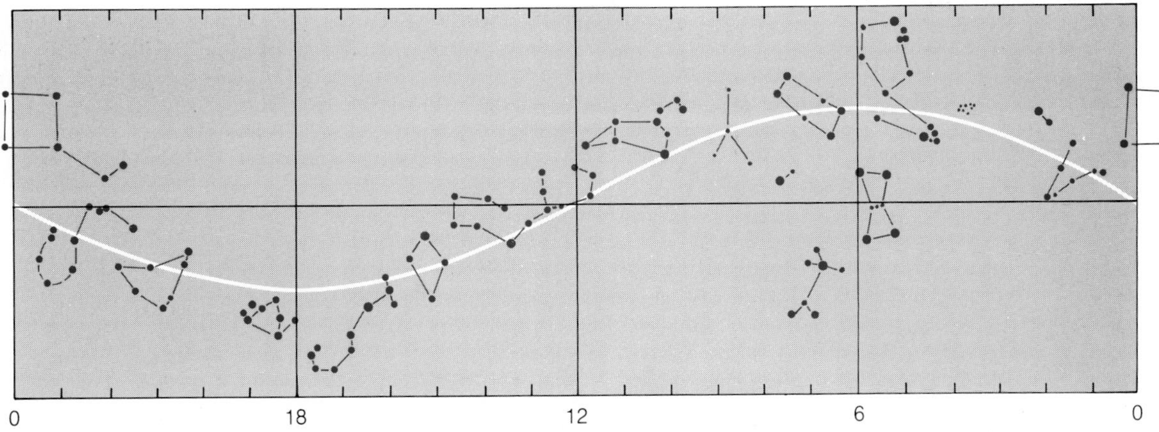

FIGURE S–3 Plot the planets on the ecliptic using the data in Table S–3 to discover what constellations they currently occupy.

MARS, JUPITER, AND SATURN

The outer planets—Mars, Jupiter, and Saturn—move rather slowly along the ecliptic. To find them, look up their current right ascensions in Table S–3 and plot them and the sun on the ecliptic in Figure S–3. Their orbits may carry them a few degrees from the ecliptic, but the diagram will tell you which constellation they occupy. Look for bright, starlike objects that don't fit into the star pattern of the constellation. Mars stands out because of its red color. The position of the sun in the diagram will help you decide what parts of the ecliptic are visible at any given time.

Once you find the planets, look at them through a small telescope or a good pair of binoculars firmly braced. The four brightest moons of Jupiter and the rings of Saturn are visible through binoculars if they can be held steady, and you can see much more through a small telescope. Do not, however, look at the sun through any optical instrument; its brilliance can burn your eyes.

LOOKING FURTHER

Whatever your interest in astronomy as a science, do not neglect its cultural value. Some of the following references will introduce you to the lore of the constellations and star names. These represent a blend of cultures and mythologies spanning thousands of years.

As you learn the constellations, note the changing shape of the moon and the slow wandering of the planets. The cyclical patterns made by the sun, moon, and planets as they move along the ecliptic have been a part of human existence since the first glimmer of intelligence on our planet. *The Observer's Handbook* for the current year will help you locate the planets and the moon.

NEW TERMS

asterism

north circumpolar constellation

greatest elongation

RECOMMENDED READING

ALLEN, RICHARD H. *Star Names: Their Lore and Meanings.* New York: Dover, 1963.

CHAMBERLAIN, VON DEL "Prehistoric American Astronomy (c. 1054 AD)." *Astronomy 4* (July 1976), p. 10.

MURIDEN, JAMES *Astronomy with Binoculars.* New York: Crowell, 1979.

PERCY, JOHN R., ed. *The Observer's Handbook.* Toronto: Royal Astronomical Society of Canada, current year.

PHILLIPS, HENRY J. "Sky Lore of Indigenous Americans," *Astronomy 4* (July 1976), p. 12.

WHITNEY, CHARLES A. *Whitney's Star Finder.* New York: Knopf, 1980.

Table S–3 Midmonth right ascensions of the Sun, Mars, Jupiter, and Saturn.

Date	Sun	Mars	Jupiter	Saturn	Date	Sun	Mars	Jupiter	Saturn
Jan. 1988	19:45	16:20	1:20	17:50	Jan. 1992	19:45	18:20	10:50	20:30
Feb.	21:55	17:40	1:40	18:00	Feb.	21:55	20:00	10:40	20:40
March	23:40	19:00	2:00	18:00	March	23:40	21:30	10:30	21:00
April	1:30	20:20	2:40	18:10	April	1:30	23:00	10:20	21:10
May	3:25	21:40	3:10	18:10	May	3:25	0:40	10:20	21:10
June	5:30	23:10	3:40	18:00	June	5:30	2:00	10:30	21:10
July	7:35	0:10	4:00	17:50	July	7:35	3:30	10:50	21:10
Aug.	9:35	0:50	4:20	17:40	Aug.	9:35	4:50	11:10	21:00
Sept.	11:30	0:40	4:20	17:40	Sept.	11:30	6:00	11:40	20:50
Oct.	13:20	0:00	4:20	17:50	Oct.	13:20	7:00	12:00	20:40
Nov.	15:20	0:00	4:10	18:00	Nov.	15:20	7:40	12:20	20:50
Dec.	17:30	0:40	3:50	18:10	Dec.	17:30	7:40	12:40	21:00
Jan. 1989	19:45	1:50	3:40	18:30	Jan. 1993	19:45	7:10	13:00	21:20
Feb.	21:55	3:00	3:50	18:40	Feb.	21:55	6:40	13:00	21:40
March	23:40	4:10	4:00	18:50	March	23:40	7:00	12:50	21:50
April	1:30	5:20	4:20	18:50	April	1:30	7:40	12:30	22:00
May	3:25	6:40	4:50	19:00	May	3:25	8:40	12:20	22:10
June	5:30	8:00	5:20	18:50	June	5:30	9:50	12:20	22:10
July	7:35	9:10	5:40	18:40	July	7:35	10:50	12:30	22:10
Aug.	9:35	10:30	6:10	18:30	Aug.	9:35	12:00	12:40	22:00
Sept.	11:30	11:50	6:40	18:30	Sept.	11:30	13:20	13:10	21:50
Oct.	13:20	13:10	6:40	18:30	Oct.	13:20	14:40	13:30	21:50
Nov.	15:20	14:30	6:40	18:40	Nov.	15:20	16:20	14:00	21:50
Dec.	17:30	15:50	6:30	19:00	Dec.	17:30	18:00	14:20	21:50
Jan. 1990	19:45	17:30	6:10	19:10	Jan. 1994	19:45	19:30	14:40	22:10
Feb.	21:55	19:00	6:00	19:20	Feb.	21:55	21:10	14:50	22:20
March	23:40	20:10	6:00	19:40	March	23:40	22:40	14:50	22:30
April	1:30	21:50	6:20	19:40	April	1:30	0:10	14:40	22:40
May	3:25	23:20	6:40	19:40	May	3:25	1:40	14:30	22:50
June	5:30	0:50	7:00	19:40	June	5:30	3:00	14:10	23:00
July	7:35	2:10	7:30	19:30	July	7:35	4:30	14:10	23:00
Aug.	9:35	3:30	8:00	19:20	Aug.	9:35	6:00	14:20	22:50
Sept.	11:30	4:30	8:20	19:20	Sept.	11:30	7:20	14:40	22:40
Oct.	13:20	5:00	8:40	19:10	Oct.	13:20	8:30	15:00	22:40
Nov.	15:20	4:40	8:50	19:20	Nov.	15:20	9:40	15:30	22:30
Dec.	17:30	4:00	8:50	19:30	Dec.	17:30	10:10	16:00	22:40
Jan. 1991	19:45	3:50	8:40	19:50	Jan. 1995	19:45	10:20	16:20	22:50
Feb.	21:55	4:30	8:20	20:00	Feb.	21:55	9:50	16:40	23:00
March	23:40	5:20	8:20	20:20	March	23:40	9:10	16:50	23:10
April	1:30	6:20	8:20	20:20	April	1:30	9:10	17:00	23:20
May	3:25	7:30	8:20	20:20	May	3:25	9:50	16:50	23:40
June	5:30	8:40	8:40	20:20	June	5:30	10:40	16:30	23:40
July	7:35	10:00	9:00	20:20	July	7:35	11:40	16:20	23:40
Aug.	9:35	11:20	9:40	20:10	Aug.	9:35	12:50	16:20	23:40
Sept.	11:30	12:40	10:00	20:00	Sept.	11:30	14:10	16:30	23:30
Oct.	13:20	13:50	10:20	20:00	Oct.	13:20	15:30	16:50	23:20
Nov.	15:20	15:20	10:40	20:00	Nov.	15:20	17:20	17:10	23:20
Dec.	17:30	16:40	11:00	20:10	Dec.	17:30	19:00	17:40	23:20

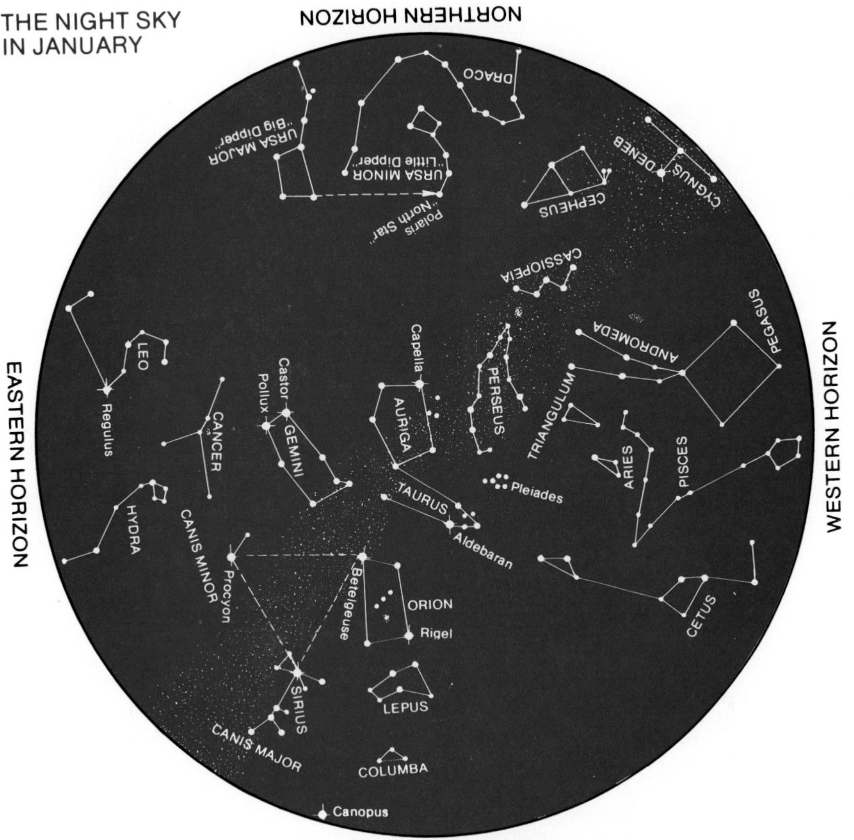

Chart time (local Standard):
10:00 PM First of month
9:00 PM Middle of month
8:00 PM End of month

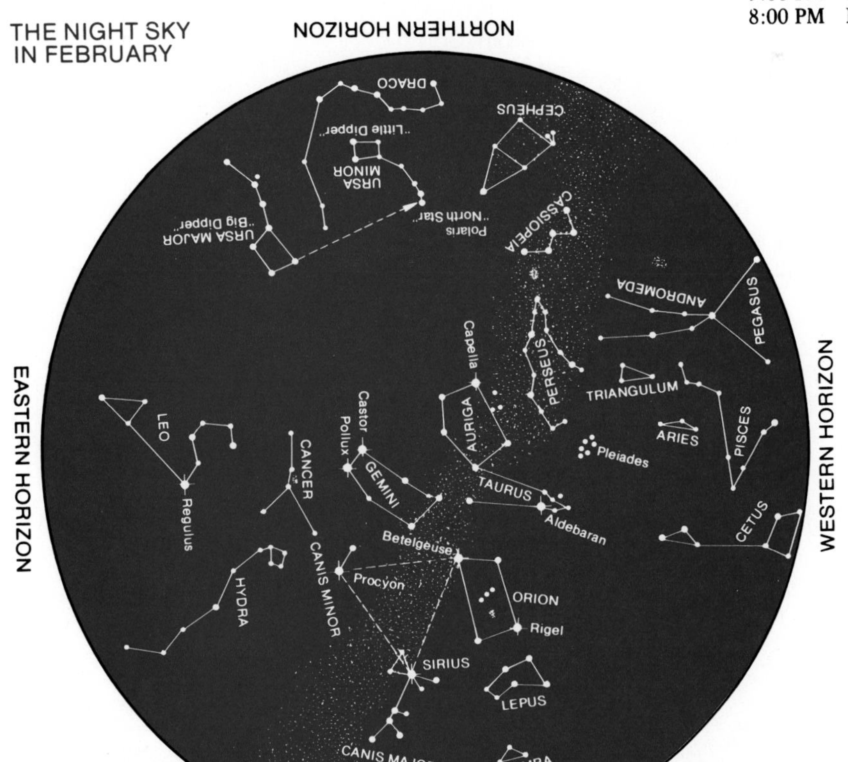

THE NIGHT SKY IN MARCH

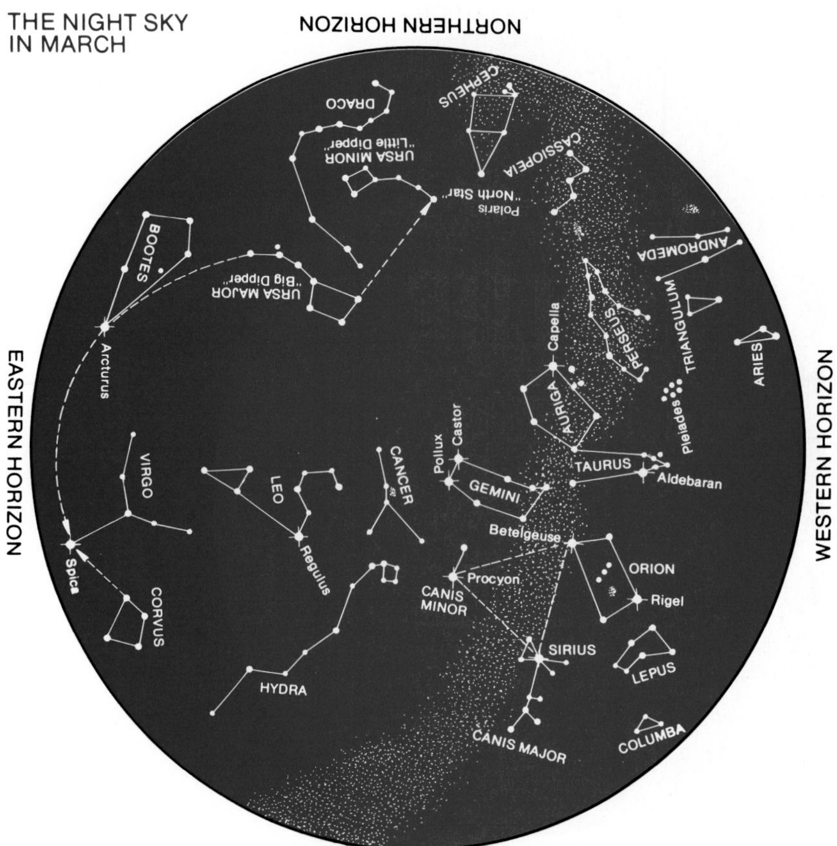

Chart time (local Standard):
10:00 PM First of month
9:00 PM Middle of month
8:00 PM End of month

THE NIGHT SKY IN APRIL

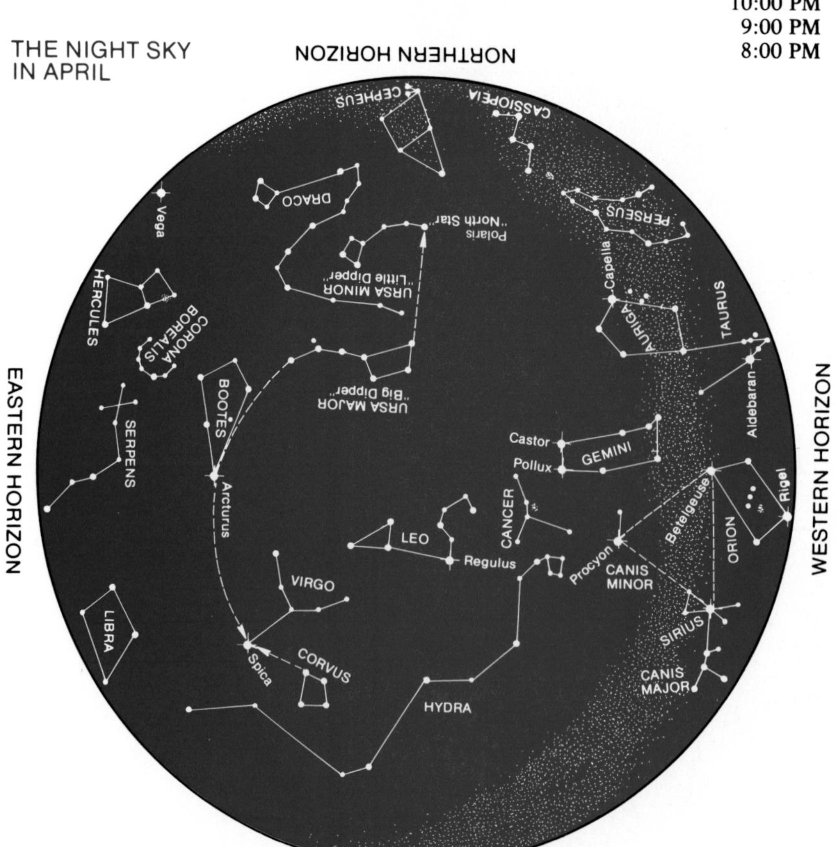

THE NIGHT SKY IN MAY

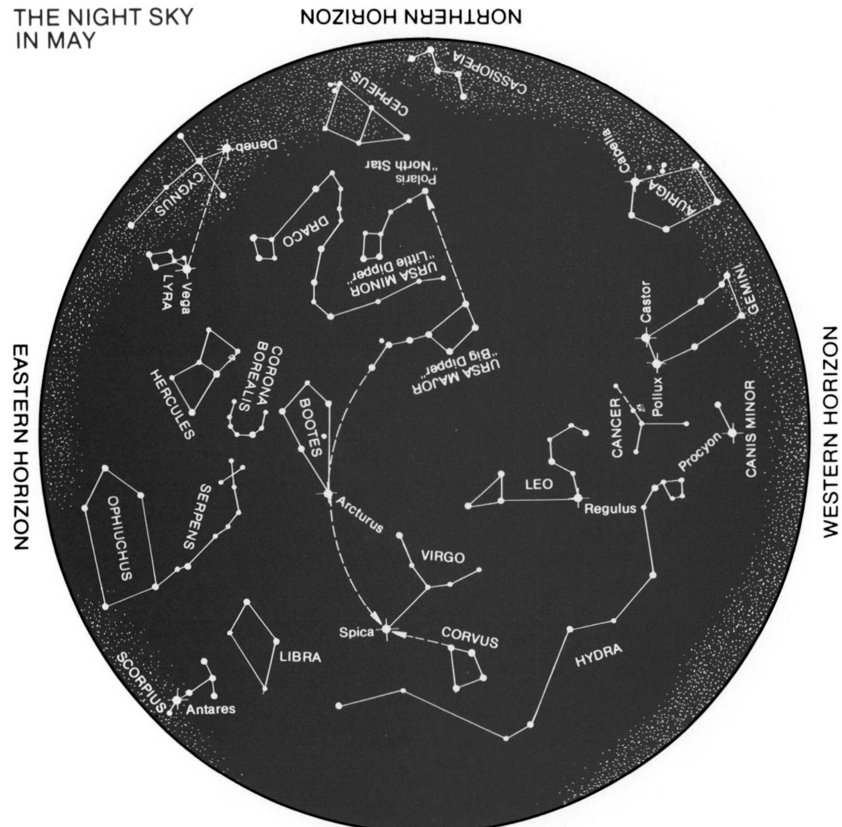

Chart time (local Standard):
10:00 PM First of month
9:00 PM Middle of month
8:00 PM End of month

THE NIGHT SKY IN JUNE

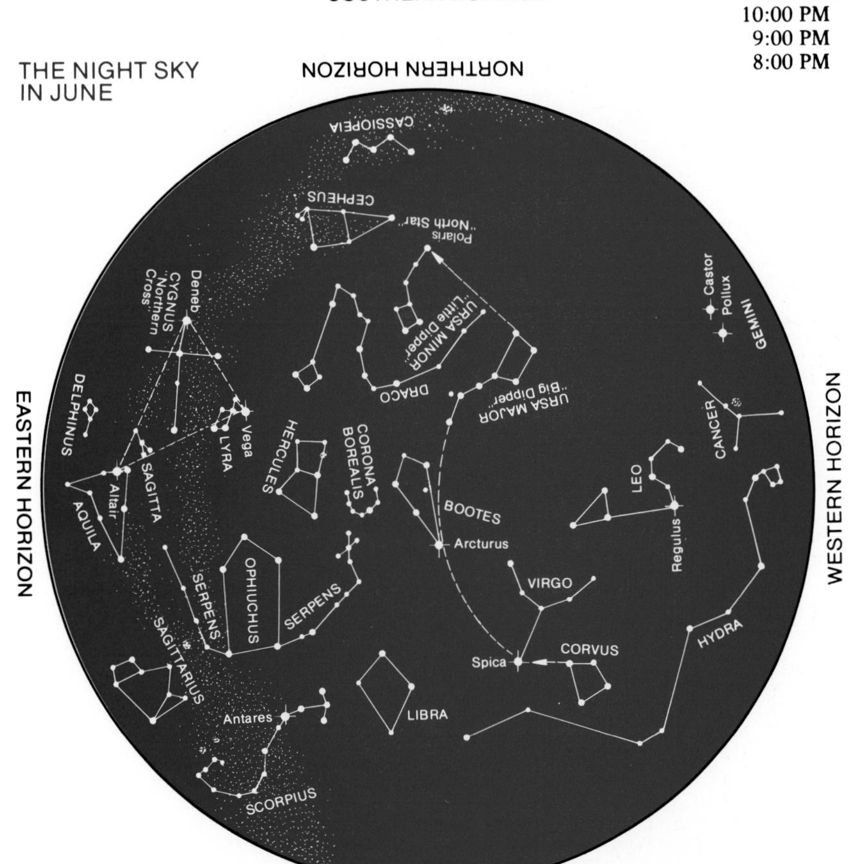

S-10 THE NIGHT SKY IN JULY

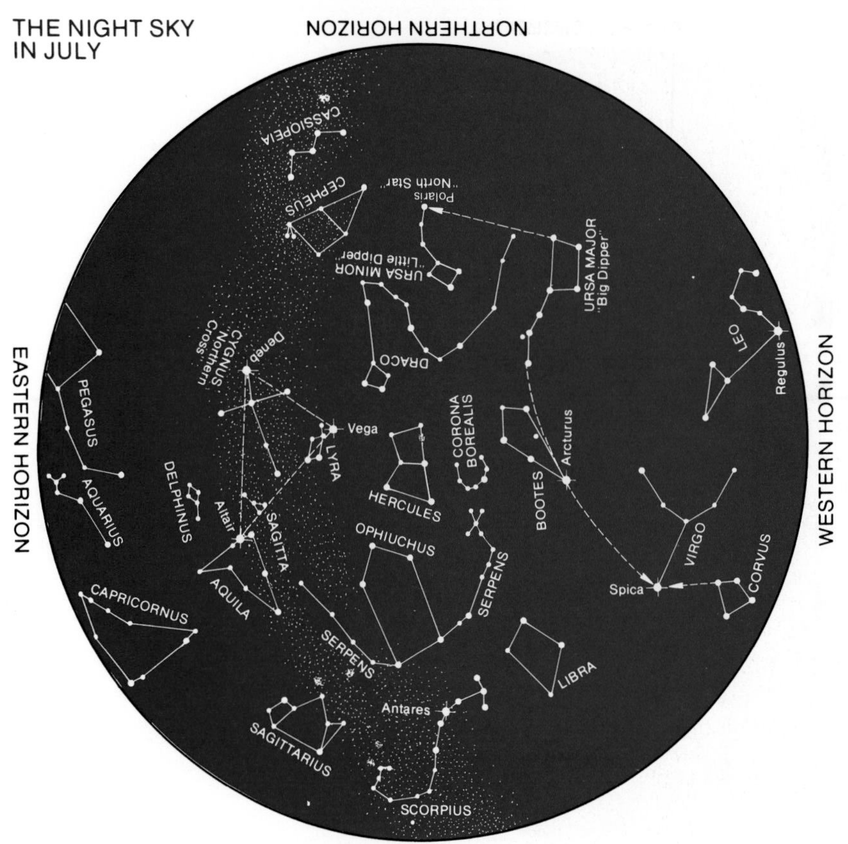

Chart time (local Standard):
10:00 PM First of month
9:00 PM Middle of month
8:00 PM End of month

THE NIGHT SKY IN AUGUST

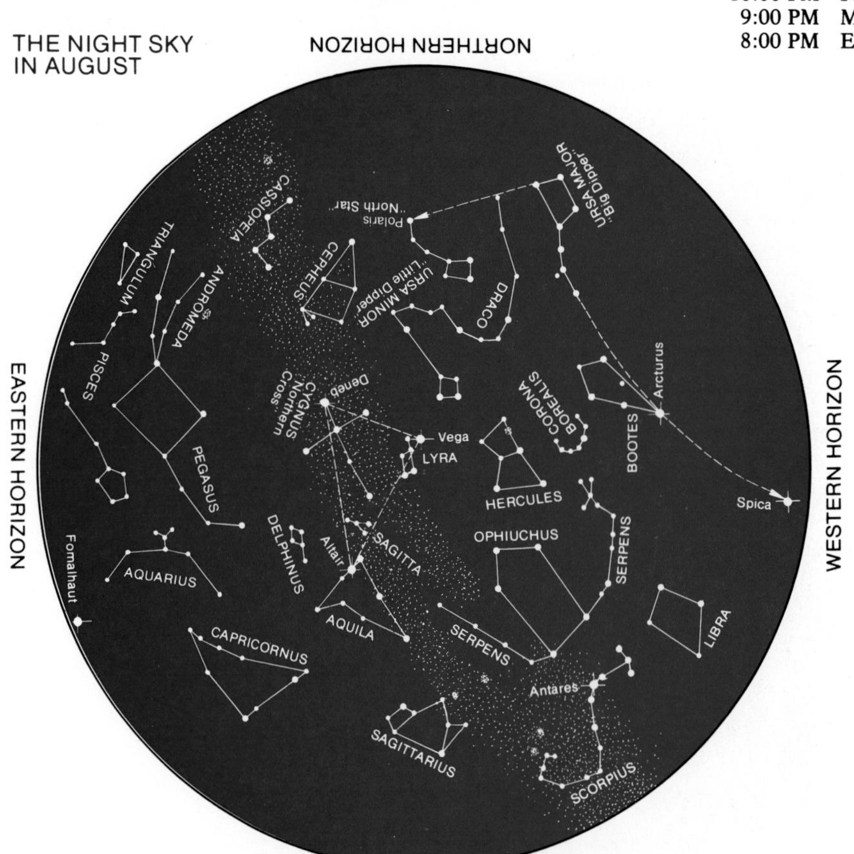

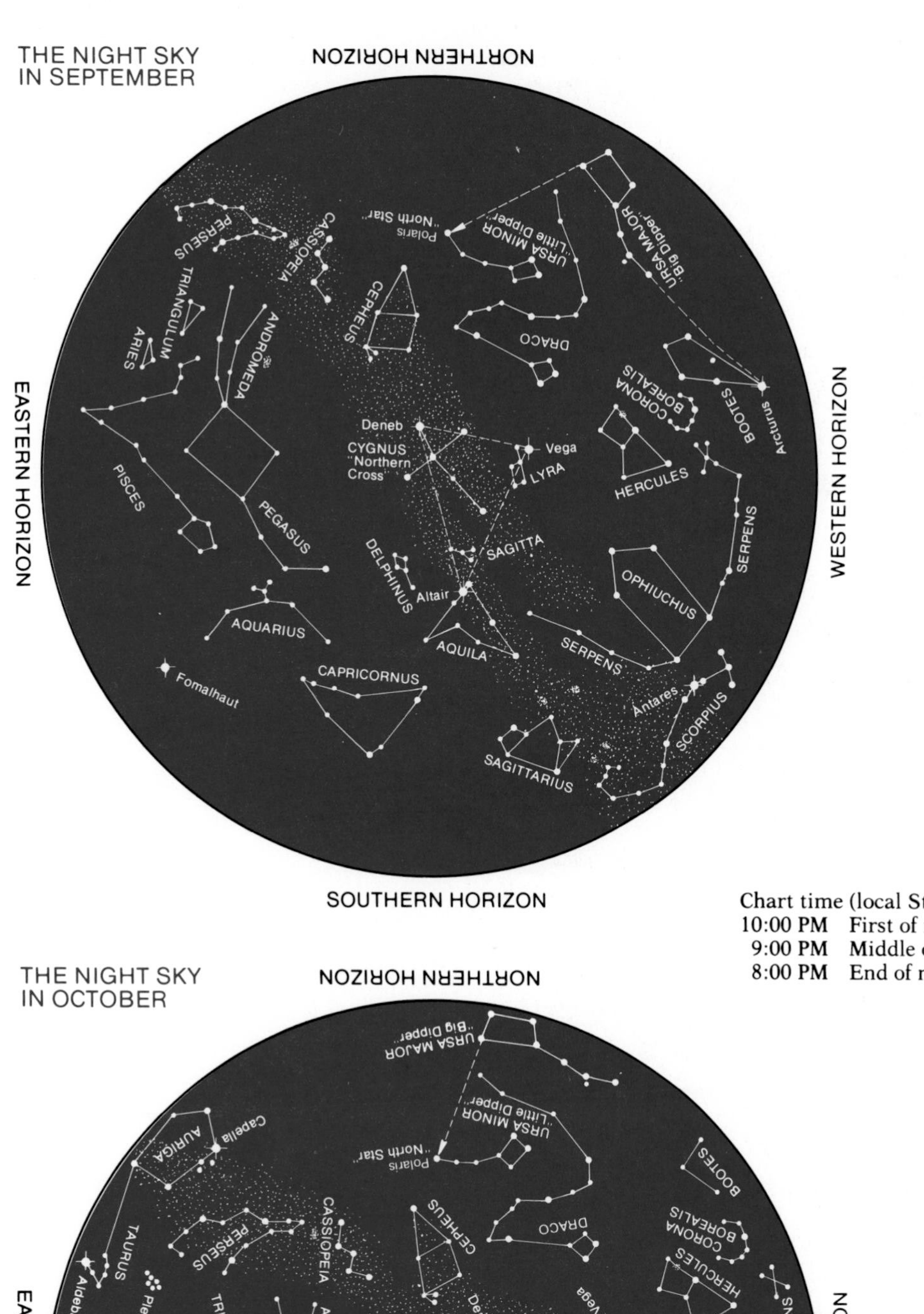

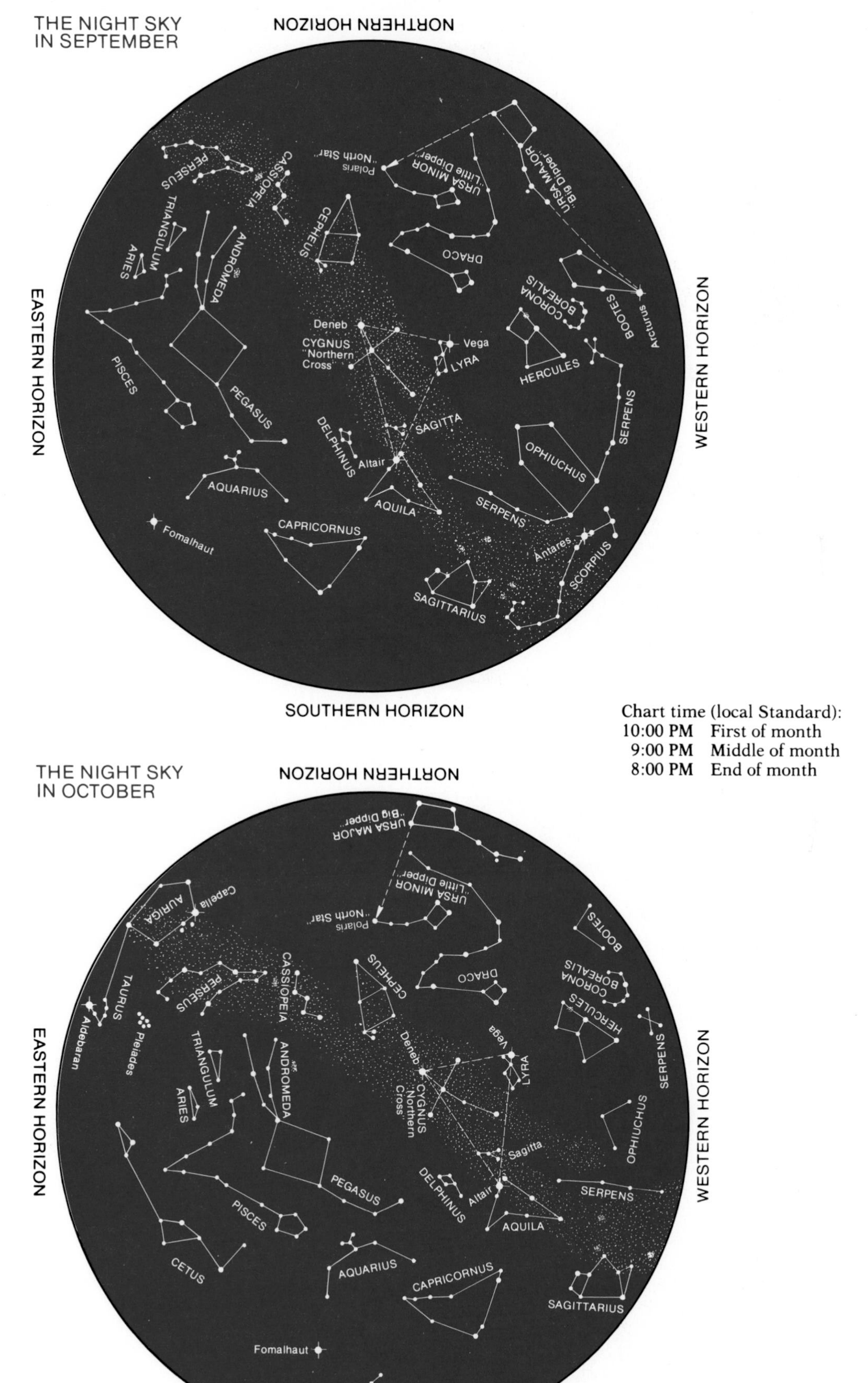

Chart time (local Standard):
10:00 PM First of month
9:00 PM Middle of month
8:00 PM End of month

THE NIGHT SKY
IN OCTOBER

NORTHERN HORIZON

EASTERN HORIZON

WESTERN HORIZON

SOUTHERN HORIZON

THE NIGHT SKY IN NOVEMBER

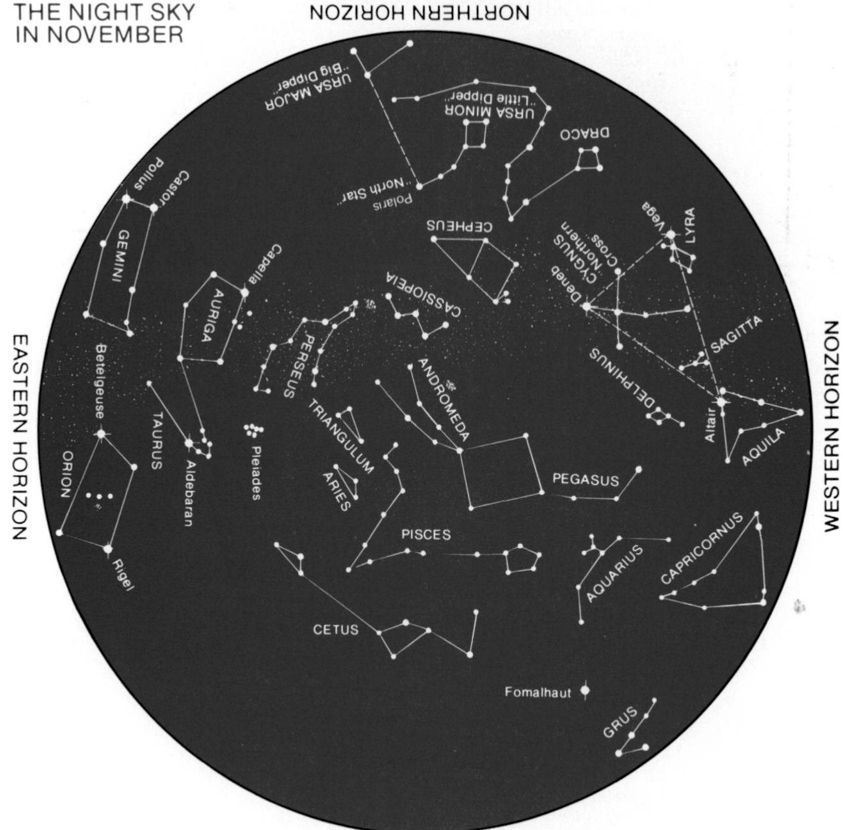

Chart time (local Standard):
10:00 PM First of month
9:00 PM Middle of month
8:00 PM End of month

THE NIGHT SKY IN DECEMBER